Physics of
Plasma–Wall Interactions
in Controlled Fusion

NATO ASI Series

Advanced Science Institutes Series

A series presenting the results of activities sponsored by the NATO Science Committee, which aims at the dissemination of advanced scientific and technological knowledge, with a view to strengthening links between scientific communities.

The series is published by an international board of publishers in conjunction with the NATO Scientific Affairs Division

A	**Life Sciences**	Plenum Publishing Corporation
B	**Physics**	New York and London
C	**Mathematical and Physical Sciences**	D. Reidel Publishing Company Dordrecht, Boston, and Lancaster
D	**Behavioral and Social Sciences**	Martinus Nijhoff Publishers
E	**Engineering and Materials Sciences**	The Hague, Boston, and Lancaster
F	**Computer and Systems Sciences**	Springer-Verlag
G	**Ecological Sciences**	Berlin, Heidelberg, New York, and Tokyo

Recent Volumes in this Series

Series B: Physics

Physics of Plasma–Wall Interactions in Controlled Fusion

Edited by

D. E. Post
Princeton University Plasma Physics Laboratory
Princeton, New Jersey, USA

and

R. Behrisch
Max-Planck-Institut für Plasmaphysik
Garching/Munich, Federal Republic of Germany

Springer Science+Business Media, LLC

Proceedings of a NATO Advanced Study Institute entitled
Physics of Plasma–Wall Interactions in Controlled Fusion,
held July 30–August 10, 1984,
in Val-Morin, Quebec, Canada

Library of Congress Cataloging in Publication Data

NATO Advanced Study Institute (1984: Val-Morin, Québec)
  Physics of plasma–wall interactions in controlled fusion.

  (NATO ASI series. Series B, Physics; v. 131)
  "Published in cooperation with NATO Scientific Affairs Division."
  "Proceedings of a NATO Advanced Study Institute entitled Physics of
Plasma-wall Interactions in Controlled Fusion, held July 30–August 10, 1984,
Val-Morin, Quebec, Canada"—Verso t.p.
  Includes bibliographical references and index.
  1. Plasma–wall interactions—congresses. 2. Controlled fu-
sion—Congresses. I. Post, D. E. (Douglas Edmund), 1945-   . II. Behrisch,
Rainer. III. North Atlantic Treaty Organization. Scientific Affairs Division. IV.
Title. V. Series.
QC718.5.P5N38   1984   621.48′4   85-24417

ISBN 978-1-4757-0069-5   ISBN 978-1-4757-0067-1 (eBook)
DOI 10.1007/978-1-4757-0067-1

© 1986 Springer Science+Business Media New York
Originally published by Plenum Press, New York in 1986
Softcover reprint of the hardcover 1st edition 1986

PREFACE

Controlled thermonuclear fusion is one of the possible candidates
for long term energy sources which will be indispensable for our
highly technological society. However, the physics and technology
of controlled fusion are extremely complex and still require a
great deal of research and development before fusion can be a
practical energy source.

For producing energy via controlled fusion a deuterium-tritium gas
has to be heated to temperatures of a few 100 Million °C corresponding to about 10 keV. For net energy gain, this hot plasma has
to be confined at a certain density for a certain time. One promising scheme to confine such a plasma is the use of intense magnetic fields. However, the plasma diffuses out of the confining
magnetic surfaces and impinges on the surrounding vessel walls
which isolate the plasma from the surrounding air. Because of this
plasma wall interaction, particles from the plasma are lost to the
walls by implantation and are partially reemitted into the plasma.
In addition, wall atoms are released and can enter the plasma.
These wall atoms or impurities can deteriorate the plasma
performance due to enhanced energy losses through radiation and
an increase of the required magnetic pressure or a dilution of the
fuel in the plasma. Finally, the impact of the plasma and energy on
the wall can modify and deteriorate the thermal and mechanical properties of the vessel walls.

The subject of this NATO advanced Study Institute is the physics
and technology important for control of the plasma-wall interaction.
Research in this field has been important since the beginning of
fusion research. Wall released impurities have always been a major
concern in each experiment and considerable effort has been devoted
to this problem. In the 1950's and 1960's the fusion community was
one of the driving forces in the advance of ultra high vacuum technology. Metallic vacuum vessel walls have replaced glass and wall
cleaning techniques have been developed. For reducing and control-
ing the plasma wall interaction, limiters and divertors have been
introduced.

Further, fusion research has stimulated a high level of fundamental research in the interaction of energetic particles with solid surfaces. Significant progress has been made in understanding and characterizing the basic processes such as sputtering, ion reflection, trapping and desorption. Complementary theoretical models and computer codes have been developed to understand and predict the properties of the boundary plasma. Only a detailed understanding of all these processes will allow practical systems for particle and impurity control to be designed, tested and built for the forthcoming machines. Over the last 5 years, the field has matured considerably. There have been substantial developments in the basic science, in theory and modelling of plasma wall interaction processes, and in the experimental application of impurity control techniques. Particularly noteworthy has been the development of the "high recycling" divertor and the pumped limiter, which offers promising solutions to the particle and impurity control problem.

The aim of this NATO Advanced Study Institute was to present a comprehensive set of lectures on all the major aspects of the plasma-wall interaction problem. The Institute begins with lectures covering the basic processes and science, then covers theory and modelling, applications in current experiments, and summarizes with the design aspects for future experiments. In addition to the lectures printed in this volume, approximately 35 participants presented their work in two poster sessions.

The Institute was held in the Auberge Far Hills Inn in Val Morin, Quebec, Canada, near Montreal, from July 30 to August 10, 1984. We would like to express our gratitude to the NATO Scientific Affairs Division for its support, which allowed the meeting to take place, and to Dr. Craig Sinclair of NATO for his advice and encouragement. The Institute was further supported by the Canadian Material Research Council (NCR), the Natural Sciences and Engineering Research Council of Canada (NSERC), the U.S. Department of Energy, INRS-Energy, University of Quebec, and the Princeton University Plasma Physics Laboratory. In particular, Dr. Charles Daughney of the Canadian NRC and Dr. Erol Oktay of the US Department of Energy were very helpful.

The members of the Scientific Organizing Committee, Dr. R. Conn, Dr. F. Engelmann, Dr. E. Hintz, Dr. G.M. McCracken, and Dr. B. Stansfield assisted in preparing and organizing the program of the lectures. We wish to thank all lecturers who not only gave excellent lectures but also prepared the manuscripts which comprise this book. Dr. D. Manos served as treasurer, handling the complicated funding arrangements from five different sources, two different banks and currencies, and widely diverse disbursements.

The conference secretary, Mrs. E. Carey, handled the applications, prepared much of the conference material, and helped to manage the conference. Mrs. Carol Phillips was the program secretary and organized the announcements, handled the preparation of the lecture notes for the Institute and assisted in running the meeting. She had a major role in editing the materials for this book. All of these contributions are greatfully acknowledged.

The local arrangements at the conference site and the local activities were very successfully organized by Dr. B. Stansfield, who headed the local committee of Dr. C. Boucher, Dr. K. Dimoff, Dr. B. Gregory, Dr. G. Ross, Dr. B. Terreault, Dr. J. Vitali, Dr. C. Neufield, and Dr. Tarasick, from the INRS-Energy, University of Quebec at Varennes. They contributed immensely to the success of the Advanced Institute. The hospitality and fine service of Mrs. L. Pemberton-Smith and her staff at the Auberge Far Hills Inn made the lectures go smoothly and made the conference a delightful experience for the conference organizers and the conference participants.

We are also grateful to Prof. Charles Joachain of the University of Bruxelles, for the example he provided as Director in organizing another Advanced Study Institute on Atomic Processes in Fusion Plasmas for which D. Post served a co-director.

D.E. Post

R. Behrisch

Directors of the
Advanced Study Institute

CONTENTS

INTRODUCTION TO THE PHYSICS OF PLASMA WALL INTERACTIONS IN

CONTROLLED FUSION

D.E. Post,[1] R. Behrisch,[2] and B. Stansfield[3]

[1] Plasma Physics Laboratory, Princeton University
    Princeton, New Jersey 08544 USA

[2] Max-Planck-Institut für Plasmaphysik
    EURATOM-Association, D-8046 Garching, FRG

[3] INRS Energie, Varennes, Québec, Canada

In the last fifteen years, great strides have been made in controlled fusion research. The plasma temperatures, beta, and confinement needed for a reactor have been approached in recent experiments (Table 1). In the next few years, experiments on JET (EC), TFTR (USA), JT-60 (Japan), and T-15 (USSR) are expected to answer many of the remaining physics questions as to whether it is possible to obtain reactor grade plasmas in large tokamaks. The major physics questions are:

(1) confinement,
(2) beta,
(3) heating and current drive, and
(4) impurity and particle control.

For an ignition experiment energy confinement times, τ_E , of 1 to 2 seconds at plasma densities of $n_e \simeq 10^{14}$ cm^{-3} will be needed. Present experiments on TFTR /1,2/ and JET /1,2/ (Fig. 1) have reached confinement times up to 0.8 seconds with ohmic heating. Record values of $n\tau_E = 8 \times 10^{13}$ cm^{-3} sec have been obtained on Alcator-C /2/ (Fig. 2). On smaller experiments, the energy confinement time achieved during ohmic heating is usually reduced by a factor of two or three when the high power auxiliary heating necessary to obtain high temperatures and a high beta is applied. The best confinement time obtained in tokamak experiments with high power auxiliary heating

1

Table I: Current Tokamak Data Base*

CONFINEMENT	$\tau_E \sim 0.2$ sec with auxiliary heating (D-III, ASDEX, TFTR)
	$\tau_E \sim 0.4$ to 0.8 sec with ohmic heating (TFTR, JET)
BETA	$\beta \sim 4.5$ to 5 % (D-III, PBX)
T_E	4 - 5 keV (D-III, T-10)
T_i	7 - 8 keV (PLT, PDX, TFTR)
$n\tau_E$	$\sim 8 \times 10^{13}$ cm^{-3} sec (ALCATOR-C)
HEATING	8 MW Beams (PDX, D-III)
	5 MW ICRF (PLT)
CURRENT RAMP UP	100 kA (PLT)
TRANSFORMER RECHARGE	400 kA (ASDEX)
IMPURITY CONTROL	
Limiters	$Z_{eff} \leq 2$ almost all tokamaks
Divertors	$Z_{eff} \leq 2$ all divertors (D-III, ASDEX, PDX)
Pumped Limiters	ISX, PLT, PDX, TEXTOR

---

* Results described in /1, 2/.

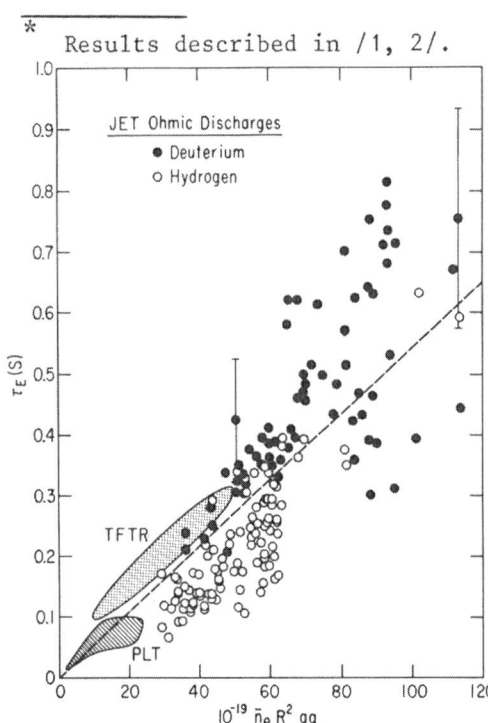

Fig. 1:

Measured confinement times for ohmic heating in JET, TFTR /1,2/ and PLT as a function of $n_e R^2$ aq, where n_e is the plasma density, R is the major radius, a is the minor radius, and q is the safety factor.

is 0.07 - 0.09 seconds on ASDEX /1, 2/ and D-III /1, 2/. During the next two or three years, high power heating will become available on TFTR, JET, JT-60, and T-15, and many of the questions about confinement scaling should be answered (Table 2). Up to now, the basic physics of magnetic plasma confinement is still not

Table II: Tokamak Data Base Additions (1984-1987)

CONFINEMENT	TFTR, JET, high power heating JT-60, T-15 (1987), D-III-D
BETA	JET, Big-D, ASDEX, PBX
T_E, T_i	TFTR (27 MW beams), JET (25 MW) JT-60 (30 MW), D-III-D (8-15 MW)
$n\tau_E$	All the big machines
CURRENT RAMP UP	PLT, ASDEX, ALCATOR-C
HEATING	TFTR (27 MW beams) JET (ICRF 15 MW + 10 MW beams) JT-60 (20 MW beams, 10 MW LH, 20 MW ICRF, not simultaneously) D-III-D (14 MW beams, 2 MW ECRH) PLT (5 MW ICRF) ASDEX (3 MW ICRF)
IMPURITY CONTROL	Limiters: All large tokamaks Pumped Limiters: PLT, PBX, TEXTOR, Big-D Divertors: D-III-D, ASDEX, PBX
Plus	ASDEX/Upgrade, Tore-Supra (1987 - on)

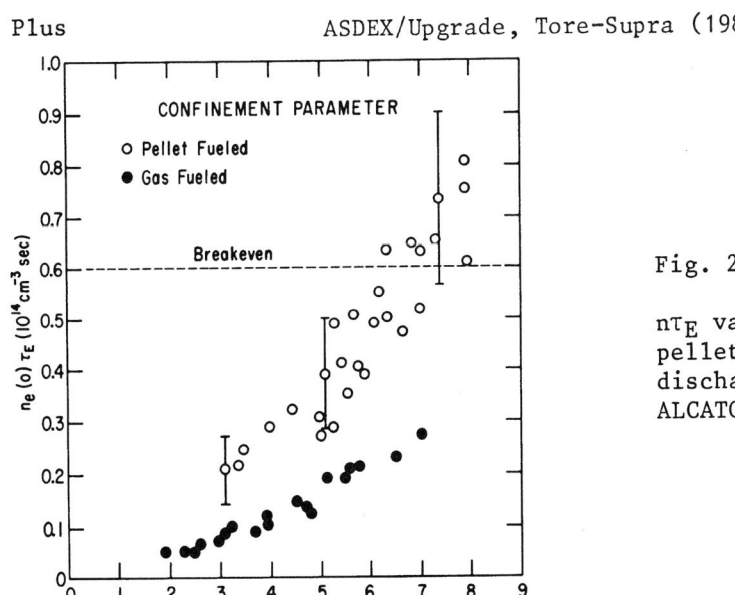

Fig. 2:

$n\tau_E$ values for pellet fuelled discharges in ALCATOR-C /2/.

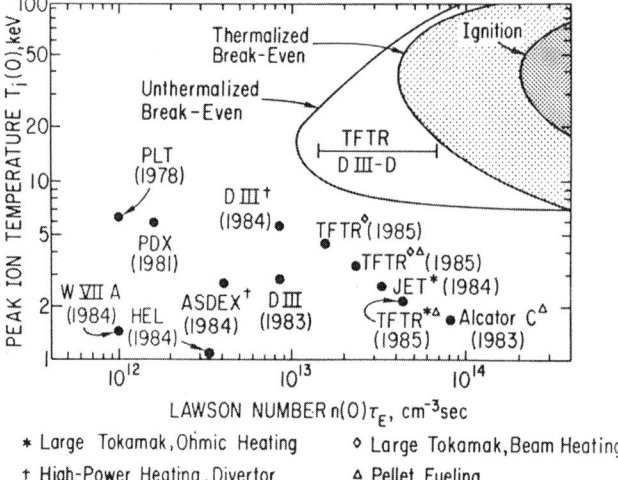

Fig. 3:   The achieved ion temperature as a function of the "Lawson"
          number, $n(o)\tau_E$ where $n(o)$ is the central electron density
          and τ_E is the energy confinement time. Also shown are tar-
          get parameters for the fusion program.

understood and is one of the major subjects of intense research.
However, the progress over the last few years has been substantial
(Fig. 3).

For a reactor experiment, the plasma beta (the ratio of the plasma
pressure to the magnetic field pressure) needs to be of the order
of 6 % or higher. Beta's of about 4.5 - 5 % have been achieved on
D-III /1, 2/ and PBX /4/ using a specially shaped plasma. By
optimizing the shaping, beta's of $\sim$ 6 % should be obtainable. With
its large size, high current, substantial enlongation, and high
heating power, the JET experiment should be able to test the useful beta limits for tokamaks. In addition, experiments with extreme shaping, such as PBX /2/ (Fig. 4), are underway, and may
allow the attainment of beta's of 10 % or more.

Neutral beam heating is the most commonly used heating method /5/.
Electron and ion temperatures of > 5 keV have been obtained on
D-III recently with 7 MW of injection /1, 2/. On earlier experiments, ion temperatures of 6-7 keV have been obtained on PLT /4/
and PDX /5/. Reactor experiments require temperatures in the
10-15 keV range. Because of their greater simplicity, radio frequency heating methods are preferred for future experiments. RF experiments in the 1-5 MW level have been carried out on PLT, TFR,
ASDEX and ALCATOR-C (Fig. 5). With additional heating the

4

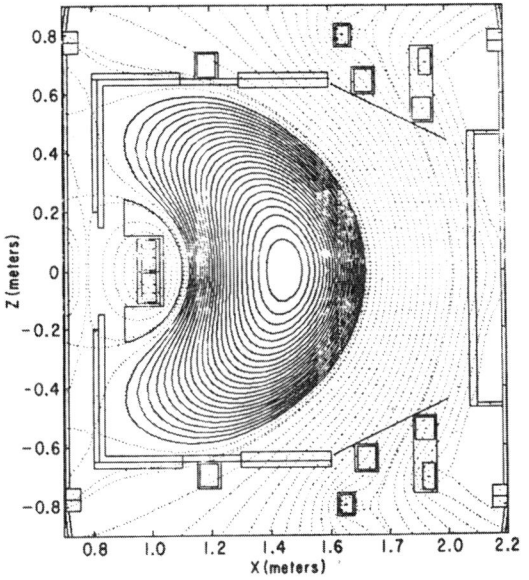

Fig. 4:

Magnetic flux plot of "bean" shaped plasma in PBX experiment /4/.

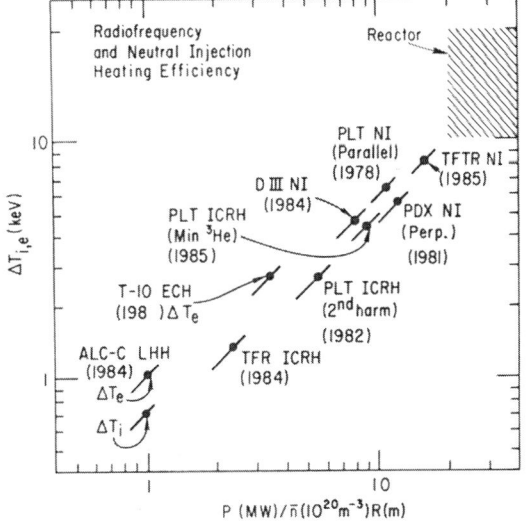

Fig. 5:

Temperature increases for neutral beam and radio frequency heating as a function of the power normalized by density and major radius /2, 6/.

plasma confinement time in tokamaks generally decreases while wall
released impurities increase. However, this may be avoided in divertor tokamaks, where no decrease in the plasma confinement with
neutral beam heating is often observed ("H-Mode" /2/). Higher power
heating experiments are planned for all large tokamaks as TFTR, JET,
JT-60 and T-15 in the next few years (Table II).

Recent results obtained with current drive experiments using RF,
particularly lower hybrid ($\sim$ 2 GHz), have been quite promising.
Current drive can be used to supplement or replace the ohmic heating
transformer normally used on tokamaks. Thus, the pulse length can
be increased and/or the machine made smaller. Unfortunately, beams
are not very efficient for current drive ($\sim$ 0.1 amp/watt), and RF
is not much better. The efficiency scales as T_e/n_e, so if n_e becomes
very low, the efficiency can be increased. This is the basis of a
scheme to reduce the volt seconds required from the ohmic transformer by first using RF to initially establish the current at low
density, then to increase the density and use the transformer to
sustain the current. This has been demonstrated in a preliminary
way on a small scale (100 kA) on PLT (Fig. 6). Lower hybrid waves
have also been used on ASDEX /8/, to maintain the current while
recharging the ohmic transformer.

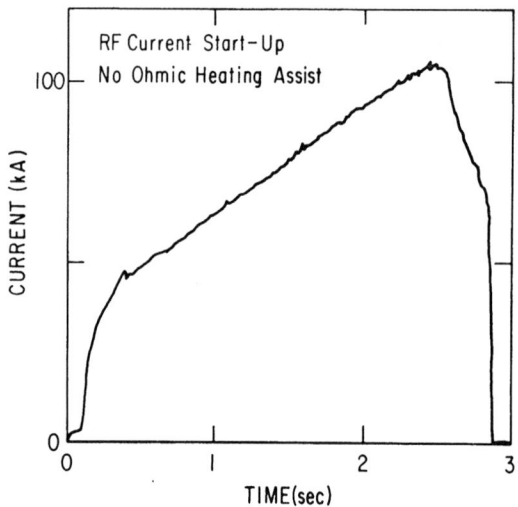

Fig. 6:

Plasma current on
PLT established with
100 kW of lower
hybrid waves with
no inductively
driven current /7/.

There are three major impurity and particle control issues:
control of the hydrogen ions, control of impurities, and materials
for first wall components.

The first major issue is the control of the DT-particle density
and composition. In current plasma experiments the discharge times
are still short, the duty factors are small (10^{-3} - 10^{-4}), the
plasma densities are mostly low, especially during the start up
phase, and the walls are not saturated with hydrogen and are
often partially covered with getters. A fraction of the plasma
particles hitting the walls are trapped, resulting in a recycling factor below one and significant pumping by the wall. Thus
the particle density, which is largely determined by the particles
recycling from the walls, can be adjusted by hydrogen fuelling
alone, either by gas puffing or injection of solid hydrogen pellets.
In forthcoming fusion experiments the plasma density will be
greater, and the discharge times will be longer. The duty factor
will be greater (1 - 0.1) and the wall will become saturated with
implanted hydrogen. As a consequence, detrapping and hydrogen release will result in a recycling factor above one. This effect
must be controlled by appropriate wall materials and operation
temperatures. In addition effective pumping at the plasma boundary
using pumped limiters or divertors will be necessary. Pumping at
the plasma boundary will also be needed to remove the He ash formed
in the D,T reactions and replace it by D and T.

A very critical issue to achieve ignition in coming fusion experiments is the control of non-hydrogenic atoms, i.e. their release
at the solid walls and their diffusion into and out of the central
plasma. The total level of these impurity atoms in the central
plasma must be low enough so that the electrons from the impurities do not significantly dilute the D-T fuel for a fixed beta,
and that the energy lost by radiation is much below the energy
produced by fusion reactions. The impurity levels in current experiments are mostly still acceptable. Z_{eff}'s of 2-3 can generally
be obtained with the use of graphite limiters or divertors. However, the impurity concentration usually increases with high power
additional heating. The release of the impurities at the surfaces
of the vessel walls takes place by different processes, such as
desorption, sputtering, evaporation, electrical arcs or gross
melting resulting in droplet emission. Once released, a portion of
the impurities will diffuse through the boundary plasma into the
central plasma. The impurities are also redeposited at the walls
due to their limited confinement in the plasma. In order to control the impurity sources the boundary plasma must be controlled
so that the wall bombardment takes place only at very low energies
and so that the heat deposition is very uniform in time and space.
Control of these processes is a major issue in running and
planning future plasma machines. The generation of experiments
coming on line now (TFTR, JET, JT-60, and T-15) will tell us much
more about the feasibility of impurity control with limiters.
Experiments on ASDEX, ASDEX-Upgrade, and D-III-D will continue to
test divertors.

The question of material selection, especially for the areas of high plasma particle and heat loads, is a major issue for the experiments under construction and for coming fusion devices. This has involved the selection of material with low Z atoms and with low erosion coefficients. The wall material is further modified by the erosion, redeposition and implantation from the plasma. Finally the vessel walls must also be able to sustain high heat loads for longer discharge times. In today's experiments most of the heat load bearing structures are inertially cooled during the pulse. As the plasma pulses become longer than the thermal time constants of such structures (5-10 seconds), the cooling must also become steady state.

PARAMETERS NEEDED FOR A TOKAMAK REACTOR CORE

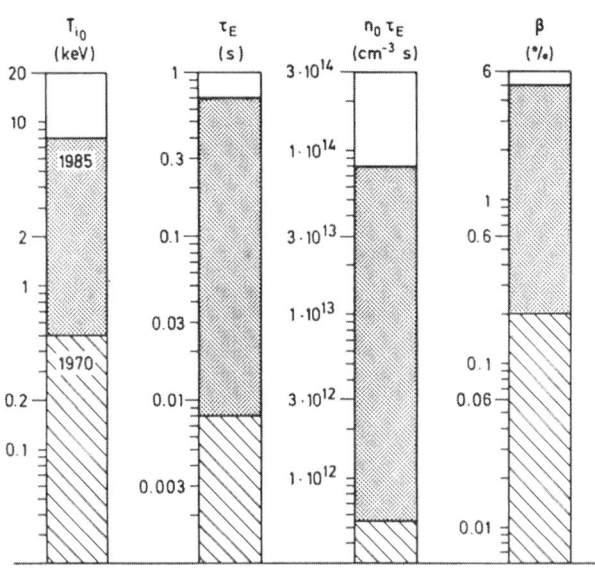

Fig. 7: Relative progress in the key plasma parameters for tokamaks from 1970 through 1985 compared to the requirements for an ignition experiment /9/.

The progress in tokamak plasma parameters over the last 14 years has been dramatic (Fig. 7). Ion temperatures have progressed from < 300 eV to 8 keV, confinement times from 8 msec to 600-800 msec, $\overline{n}\tau_E$'s from 10^{11} cm^{-3} sec to 8×10^{13} cm^{-3} sec, and beta's from $0.1 - 0.3$ % to 5 %. Further progress is to be expected from experiments on JET, TFTR, JET-60, and T-15. Design work has started on the next generation of machines: NET (EC), FER (Japan), OTR (USSR).

Impurity and particle control as well as minimising the wall erosion and damage is a key issue in each of these design efforts and is the major subject of this NATO ASI. In the course of 24 lectures, we will examine the different aspects of these topics in fusion experiments. It is appropriate to do so now because the impurity control field has matured to the point where a relatively coherent overview is possible. The list of physics and engineering topics involved in the design and performance of impurity and particle control systems is large (Fig. 8). The lectures are organized in such a way as to:

HIGH BETA PLASMA

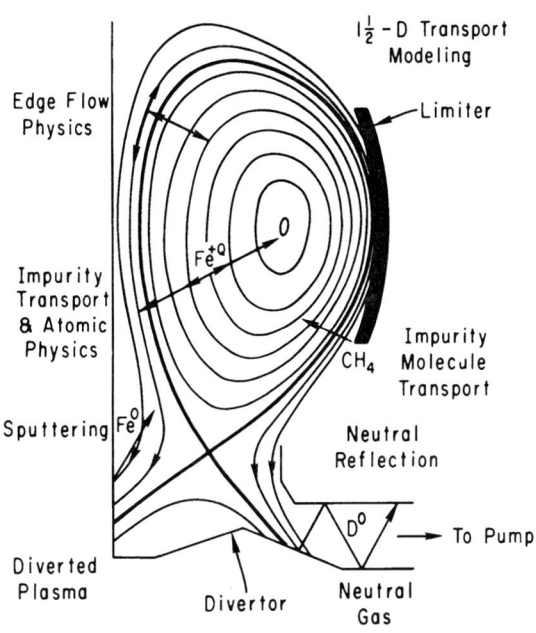

Fig. 8:

Schematic illustration of the diverse physics ingredients to a model for the performance of an impurity and particle control system.

(1) provide an introduction to fusion research and the associated impurity and particle control problems,

(2) discuss the basic plasma physics, atomic physics, surface physics, and material properties,

(3) discuss how these separate topics apply to fusion experiments, and

(4) tie everything together in large global models for the performance of impurity and particle control systems.

The first lecture provides an introduction to fusion research. The general concepts of tokamaks, mirrors, etc. and the experimental results will be described. The lectures also introduces the nature

of the impurity and particle control problems associated with both
current experiments and the design of future experiments.

The next four lectures review the plasma physics aspects of plasma
wall interactions. The first lecture covers simple sheath theory,
particularly Langmuir sheaths and probe theory. These lectures are
followed by a discussion of presheath and sheath physics as they
pertain to particle and energy transport and edge fluxes. In particular, boundary conditions for computational models are discussed.
Probe measurement techniques in the far scrape-off layer plasma
outside the separatrix (Fig. 9) and results are presented in the
third lecture of this group. The theory and operation of electro-

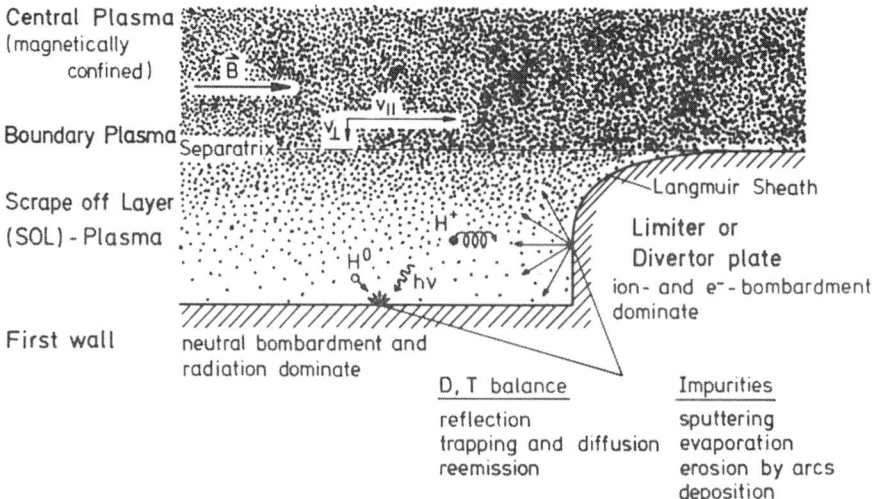

Fig. 9:   Schematic of the surface processes at different areas
          of the vessel walls.

static, bolometric and surface probes are covered and probe measurements from various experiments are reviewed. The last lecture in
this series covers nonprobe plasma edge measurements. Techniques
such as spectroscopy, interferometry, and laser scattering are described and assessed, and characteristic results are presented.

Atomic processes important at the plasma edge are then outlined and
discussed. These can be broadly divided into (1) atomic and molecular reactions of hydrogen and helium, and (2) impurity processes.
The details of hydrogen recycling and impurity transport depend
crucially on the dynamics of how molecules and atoms are converted
to ions.

10

The next series of lectures cover surface effects. The importance of these problems at the different wall areas is shown schematically in Fig. 9. With respect to the plasma particle balance, backscattering of ions at surfaces is covered, followed by a lecture on implantation and trapping, diffusion, and surface recombination. Three lectures deal with the major impurity release processes such as physical sputtering, desorption, chemical sputtering, and arcing. Secondary electron emission, which influences the sheath potentials and the ignition and burning of electrical arcs, are also described. Bulk material properties are covered. These issues are crucial for both mechanical strength properties and for heat transfer. The lecture covers melting, sublimation, thermal conductivity, thermal stress, thermal expansion, and general strength of materials issues.

All of the above topics are brought together in the last series of lectures and applied to fusion experiments. Both experimental results and modeling are covered. The series opens with a discussion of classical transport theory for the plasma edge. Next follows a discussion of models for neutral gas transport at the plasma edge. This lecture covers both the physics of the transport with some typical results and an outline of the numerical techniques for solving the problem. One and two-dimensional models for plasma transport particularly coupled to the neutral transport, are then described. Some of the numerical techniques are also covered.

The lectures on the experiments begin with a discussion of particle confinement and control in existing tokamaks. The role of wall conditioning for recycling and refuelling is covered. Then the theory and operation of divertors is presented. The design issues including the magnetic field requirements are covered for both poloidal and bundle divertors. Then the practical aspects of designing and fabricating limiters and divertor plates are described. Particular emphasis is placed on mechanical, electrical, and thermal problems. Then advanced limiters, including pumped limiters, are described. Both the design consideration and the experience of current tokamaks are covered.

The impurity and particle control problems specific to high power auxiliary heating both with neutral beams and with RF are introduced. Then the impurity and particle control problems with mirrors are covered. The final lecture is on the impurity and particle control problems associated with ignition experiments such as INTOR will be detailed.

The purpose of the institute is to serve as a general introduction to the field for new research workers and to provide a description of each separate research topic to workers expert in one area but not in other areas.

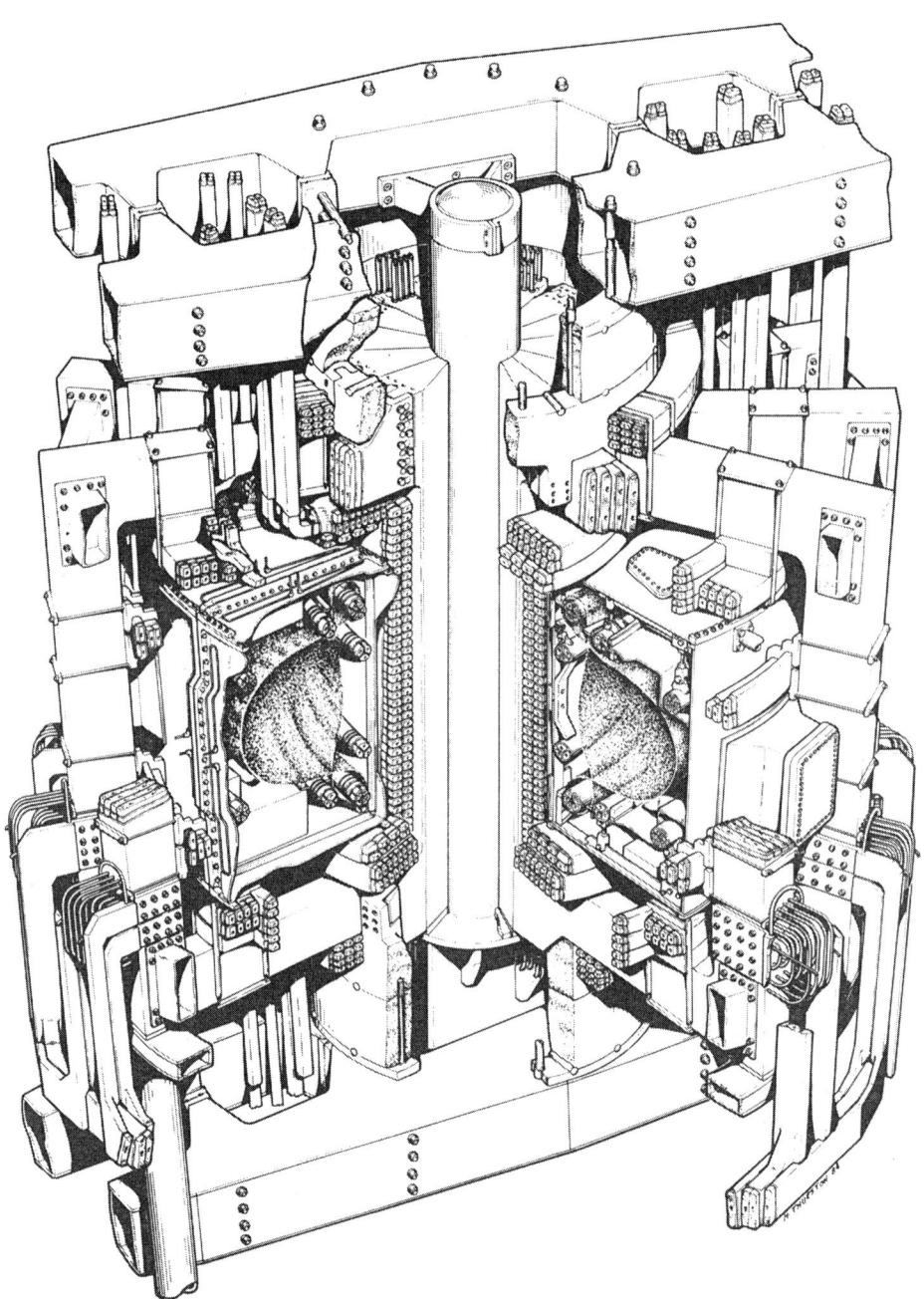

Fig. 10: Diagram of the Varennes Tokamak

It is appropriate that the course of lectures be held in Canada near Montreal. The plasma physics group at Montreal is constructing an experiment, the Tokamak de Varennes, which has long pulse and impurity control issues as a major focus of the experimental program (Table III).

Table III: Main Parameters of the Tokamak de Varennes

Major radius	0.85 m
Minor radius	0.27 m
Toroidal field	1.5 T
Maximum Pulse length	30 sec
Plasma current	300 kA

Additional features (double null poloidal divertor, replaceable liner, large pumping system).

The flat top of the toroidal field coils is 30 seconds, making current drive and long pulse issues very interesting for the machine program. With adequate current drive, the high duty factor and long pulse issues associated with recycling and impurity control can be studied. The combination of the divertor and a large pumping system will aid in the current drive studies. Extensive diagnostics for plasma wall interaction studies are being developed, including a laser fluorescence system for neutral hydrogen and impurity measurements, infrared scattering for measuring the turbulence level of the edge plasma, and a surface analysis station for studying the deposition of impurities onto the first wall.

ACKNOWLEDGEMENT

Part of this work was supported by U.S. Department of Energy Contract No. DE-AC02-76-CHO-3073.

REFERENCES

1. Proceedings of the Sixth International Conference on Plasma Surface Interactions, Nagoya, Japan, 1984,
   J.Nucl.Mater. 128&129 (1984).

2. Plasma Physics and Controlled Nuclear Fusion Research,
   Vol. 1 to 3, Proc. of the Tenth Int. Conf. London, UK, Sept. 1984, Nuclear Fusion, Suppl. 1985, IAEA, Vieanna, Austria

13

3. D. Post and R. Pyle, "Neutral Beam Heating", Atomic and Molecular Physics of Controlled Fusion, eds. C. Joachin and D. Post, Plenum, 1983

4. Eubank et al., Phys. Rev. Lett. 43:270 (1979)

5. R. Fonck et al., J.Nucl.Mater. 111&112:343 (1982)

6. P. Colestock, "Review of ICRF Experiments", IEEE Trans. on Plas. Sci. PS-12:64 (1984)

7. Jobes et al., Phys. Rev. Lett. 52:1005 (1984)

8. F. Leuterer et al., "Lower Hybrid Current Drive in the ASDEX Tokamak, Proceedings of the 12th European Conference on Controlled Fusion and Plasma Physics, Budapest (1985)

9. Courtesy of H. Furth.

INTRODUCTION: APPROACHES TO CONTROLLED FUSION

AND ROLE OF PLASMA-WALL INTERACTIONS

Folker Engelmann

FOM-Instituut voor Plasmafysica "Rijnhuizen"
Association EURATOM-FOM, Nieuwegein, The Netherlands; and
The NET Team, Garching, Federal Republic of Germany

ABSTRACT

The various approaches to thermonuclear fusion and the underlying physics concepts are introduced, and the role and nature of
plasma-wall interaction is discussed with emphasis on tokamaks and
the reactor regime.

OVERVIEW

Plasma-wall interaction is a critical issue on the way to the
realization of thermonuclear fusion. Its impact on plasma conditions
is considerable already in present-day plasma confinement devices.

It is the aim of this Introduction to discuss the role and the
nature of plasma-wall interaction in fusion devices after having
shortly described the physics requirements for fusion and the various
approaches used to generate fusion plasmas. Tokamaks are taken as an
example. The points to be touched upon include the ways to control
plasma-wall interaction, the possible edge plasma regimes, particle
recycling at the walls and its consequences for the wall (erosion,
migration and redeposition of wall material) and for the plasma (impurity generation, transport and radiation) as well as the issue of
particle exhaust. Thereby, fusion reactor conditions are emphasized.
References will be limited to papers of general character, while for
those referring to more specific topics the references of the specialized lectures should be consulted.

PHYSICS REQUIREMENTS FOR FUSION

The basis of energy production by nuclear fusion is the fact that the binding energy per nucleon varies from nucleus to nucleus, and is particularly high for moderately light nuclei, like e.g. He^4. Examples for energy producing fusion reactions of practical interest are (see, e.g. Thompson, 1965)

$$D + T \rightarrow (He^4 + 3.5 \text{ MeV}) + (n + 14.1 \text{ MeV}) \tag{1}$$

and

$$D + D \rightarrow (He^3 + 1 \text{ MeV}) + (n + 3 \text{ MeV})$$
$$\rightarrow (T + 0.8 \text{ MeV}) + (p + 2.5 \text{ MeV}). \tag{2}$$

There are two conditions to be fulfilled for being able to generate fusion reactions at a useful rate. The first is that the typical energy of the nuclei to be fused is high enough to overcome the Coulomb barrier between them. This, for singly charged nuclei, is of the order of

$$\mathcal{E}_{Coul} = e^2/r_{nucl} \tag{3}$$

with e the elementary charge and r_{nucl} the size of the nuclei, that is for heavy hydrogen isotopes

$$r_{nucl} \approx 10^{-12} \text{ cm} . \tag{4}$$

Equations (3) and (4) yield for the kinetic energies required for the reactions (1) and (2)

$$\mathcal{E}_{kin} \approx \mathcal{E}_{Coul} \approx 150 \text{ keV} , \tag{5}$$

corresponding to temperatures of

$$T \approx 100 \text{ keV} \approx 10^9 \text{ }^\circ K \tag{6}$$

if thermal states are considered. This crude evaluation, to a good approximation, yields the temperature for which the fusion cross-sections of the DT and DD reactions are maximum (cf. Fig. 1). This maximum is close to $\sigma = \pi r^2_{nucl}$. In fact, for T = 100 keV, one has

$$<\sigma v_r> \approx \pi r^2_{nucl} \bar{v} \approx 6 \cdot 10^{-16} \text{ cm}^3/s \tag{7}$$

with $\bar{v} \approx 2 \cdot 10^8$ cm/s the thermal velocity of the deuterons. Effectively, for the DT reaction the cross-section is somewhat larger, while for DD it is about an order of magnitude smaller. This, and the larger energy produced per reaction, constitutes an important advantage of the DT reaction. In practice, due to the tunnel effect and the presence of suprathermal particles, somewhat lower temperatures than that given by condition (6) are sufficient for the fusion reaction rates to be close to maximum (see Fig. 1). Specifically for

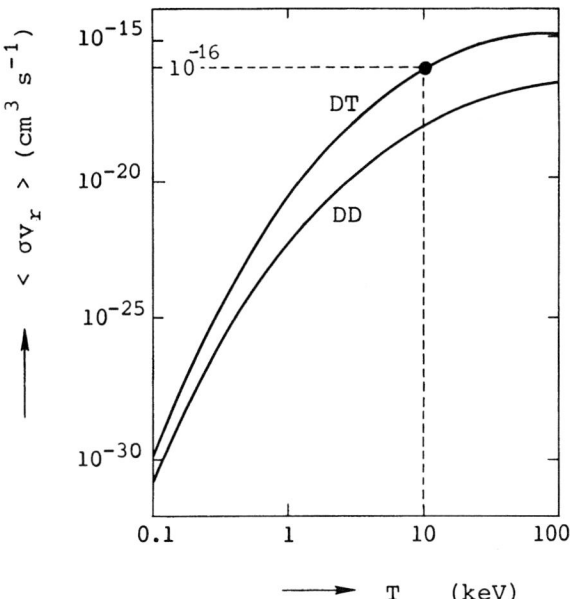

Fig. 1.   Average of the product σv_r of the fusion cross-section (σ) and the relative velocity (v_r) of the interacting particles for Maxwellian particle distributions as a function of temperature (T), for the DT and DD reactions (see, e.g., Artsimovich, 1964).

the DT reaction, the <u>temperature required</u> is only

$$\| \; T \approx 10 \text{ keV} \approx 10^8 \; {}^\circ K. \; \| \tag{8}$$

The <u>second condition</u> expresses the fact that fusion energy production is of interest only if it is, at least, larger than the energy needed to make the reactions possible. This implies

$$P_{fus} \; \tau_E > 3 \, nT \; , \tag{9}$$

where n is the plasma density, τ_E the energy confinement time of the plasma, and P_{fus} the fusion power density which for a 1 : 1 mixture of D and T is

$$P_{fus} = (n^2/4) \; \langle \sigma v_r \rangle_{DT} \; \mathcal{E}_{DT} \tag{10}$$

with $\mathcal{E}_{DT}$ = 17.6 MeV the energy produced per reaction. For T = 10 keV where $\langle \sigma v_r \rangle$ = 10^{-16} cm^3/s, condition (9) yields the so-called <u>break-even criterion</u>

$$\| \; n\tau_E > 0.7 \cdot 10^{14} \text{ cm}^{-3} \text{ s.} \; \| \tag{11}$$

It implies a requirement on the quality of plasma confinement. The ignition condition which ensures that a confined plasma is kept at thermonuclear temperatures by self-heating through the charged particles generated in fusion reactions, has an analogous structure, but is more stringent. For a DT plasma, it is obtained when $\mathcal{E}_{DT}$ in Eq. (10) is replaced by the α-particle energy $\mathcal{E}_{\alpha}$ = 3.5 MeV, yielding

$$\| \ n\tau_E > 3 \cdot 10^{14} \ cm^{-3} \ s. \ \|$$ (12)

From the preceding and the consideration that Coulomb interactions between charged particles are very efficient in creating thermal states (the rate coefficient $<\sigma v_r>$ for elastic scattering of light ions is, for T = 10 keV, of the order of 10^{-12} cm^3 s^{-1}), it follows that for generating fusion power one has to work with a hot fully ionized gas or plasma which must be sufficiently confined. It is obviously impossible to do this by just keeping the plasma within material walls.

APPROACHES TO PLASMA CONFINEMENT

There are two essentially different schemes to confine hot plasmas, namely just by inertial effects far away from walls or by using magnetic fields.

Inertial confinement

If one relies on particle inertia, the confinement is necessarily limited in time. This implies that a fusion reactor scheme based on this principle requires pulsed operation, the pulse length being equal to the plasma life time τ. The latter is also a measure for the energy confinement time, i.e., $\tau_E \approx \tau$, and is just given by

$$\tau \approx \Delta / \bar{v} ,$$ (13)

with Δ the plasma diameter and $\bar{v}$ the typical particle velocity, i.e., $\bar{v} \approx 10^8$ cm/s for T = 10 keV. Hence, the confinement condition $n\tau \gtrsim 10^{14}$ s/cm^3 takes the form

$$n\Delta \gtrsim 10^{22} \ cm^2 ,$$ (14)

showing that very high densities are required to keep Δ reasonably small. This is also necessary to avoid too large input energies $\mathcal{E}_{in}$ for generating the hot plasma. In fact, $\mathcal{E}_{in}$ is given by

$$\mathcal{E}_{in} = 3nT \cdot \frac{4\pi}{3} \left(\frac{\Delta}{2}\right)^3$$ (15)

and observing that for the fusion gain

$$Q = \mathcal{E}_{fus} / \mathcal{E}_{in}$$ (16)

18

one has $Q = \Delta / \Delta_{min}$, with, according to Eq. (14),

$$n\Delta_{min} \approx 10^{22} \text{ cm}^2 , \tag{17}$$

one finds

$$\mathcal{E}_{in} \approx 2 \cdot 10^{51} Q^3 / n^2 \quad \text{[Joule]} \tag{18}$$

where n is in cm^{-3}. E.g., for reaching $Q = 10$ with an input energy $\mathcal{E}_{in} = 10^6$ Joule, hence, plasma densities above 10^{24} cm^{-3}, that is almost 100 times the density of solid DT, are required and the plasma size Δ has to be of the order of 0.1 cm. The life time in this case is $\tau \approx 10^{-9}$ s.

Practical ways of creating plasmas of this kind are compressing small pellets of DT to the high densities and temperatures required by focussing beams of laser light or of fast electrons or ions onto the pellet (for more details, see Miley, 1979). Using laser beams has the disadvantage of a low efficiency of laser light production and of a comparatively bad coupling of the power to the pellet. Particle beams, and in particular ion beams, have a better potential in both respects, but less work has been done so far. The plasma parameters achieved in pellet compression experiments are quite close to the target values required, namely $n \approx 3 \cdot 10^{24}$ cm^{-3} and $T \approx 1$ keV, but the plasma size has been very small, i.e. of the order of 10^{-3} cm (see, e.g., Hauer, 1981).

In inertial confinement fusion, plasma-wall interaction does not affect the plasma behaviour, as the vessel walls are far away from the pellet and the pulse duration is very short. However, the vessel wall has to take high pulsed heat and particle loads appearing as a consequence of pellet disintegration. Typical values are 10^3 Joule cm^{-2} in 10^{-9} to 10^{-8} s. This implies that there is a life time problem for the vessel walls. Gas (Conn et al., 1978) and liquid (Maniscalo et al., 1978) mantle concepts have been proposed to solve this issue.

Magnetic confinement

Plasmas confined by a magnetic field necessarily are in a completely different regime. In fact, at least in principle, magnetic confinement, although imperfect and thus to be supplemented by confining material walls, can ensure a steady plasma state so that continuous reactor operation is a possibility. Furthermore, the densities of magnetically confined plasmas have to be comparatively low. In practice the plasma pressure always has to be small compared to the pressure of the confining magnetic field, i.e.,

$$\beta \equiv \frac{2nT}{B^2 / 2\mu_o} \ll 1 , \tag{19}$$

19

with μ_o the permeability in vacuum. Since the confining magnetic
field B, for technical reasons, is limited to about 5 T, this implies for a plasma temperature of 10 keV

$$n \ll 3 \cdot 10^{15} \ cm^{-3} \ . \tag{20}$$

In practice, one expects the density of a magnetically confined thermonuclear plasma to be somewhat above $10^{14} \ cm^{-3}$.

The main limitation of magnetic plasma confinement is the fact
that it is effective only in two dimensions, namely perpendicular to
the direction of the magnetic field B. While a particle is forced to
spiral around the magnetic field on a Larmor orbit, cf. Fig. 2, its
motion along the magnetic field remains free. Therefore, all confinement geometries in which the magnetic field lines are not within a
finite volume ("open confinement systems") suffer from end losses.
Linear θ-pinches in which the plasma is compressed by currents circulating in poloidal (θ) direction around the plasma axis are the
simplest example of devices of this kind. Another open confinement
geometry, advantageous because of its stability properties, is the
magnetic cusp. Also the Z-pinch, a linear system with axial currents
which very efficiently is produced in plasma focus devices, may be
mentioned here although in this scheme the magnetic field lines tend
to close in poloidal direction.

The most important configuration of this category is the magnetic mirror (cf. Fig. 3). Here, a limited confinement parallel to B is
achieved by a field increase towards the ends of the device, due to
the fact that the magnetic moment

Fig. 2.   Motion of a particle in a magnetic field B; $r_L = v_\perp m/eB$ is
the Larmor radius with $v_\perp$, e and m the particle velocity
perpendicular to B, charge and mass, respectively.

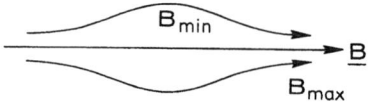

Fig. 3.   Basic magnetic mirror geometry.

$$\mu = v_\perp^2 / B \tag{21}$$

of a charged particle ($v_\perp$ = particle velocity perpendicular to $\underline{B}$) is
adiabatically conserved in a strong magnetic field. Together with the
conservation of the particle kinetic energy, i.e., $v_\perp^2 + v_\parallel^2$ = const,
this leads to a reduction of the velocity $v_\parallel$ parallel to $\underline{B}$ for a particle moving in the direction of increasing B that confines particles
for which the ratio $v_\parallel / v_\perp$ is below a critical limit, namely

$$v_\parallel / v_\perp < [(B_{max} - B_{min}) / B_{min}]^{1/2} . \tag{22}$$

On the other hand, particle scattering induced by binary collisions
and/or fluctuating fields, generated by instabilities, leads to a
continuous loss of particles from the confined area of velocity space
into the unconfined one, the so-called loss cone, making simple
mirror confinement rather imperfect.

Present-day mirror devices have a much more complex geometry
than the simple basic concept of Fig. 3 (see, e.g., Cohen, 1980;
TASKA Team, 1982). In fact, in order to ensure gross plasma stability,
the configuration must be modified such that, on the average, the magnetic field is minimum in the centre of the device, and to reduce end
losses an electric field can be used which is created through closing
the main mirror on both ends by a mirror configuration whose strong
fields are able to confine a high pressure plasma ("tandem mirror").
Adding an additional mirror between the main one and the end plugs
(see Fig. 4) is anticipated to allow to enhance the effectiveness of
the electrostatic confinement by the creation of a thermal barrier
for electrons which makes it possible to keep the electrons in the
plug mirrors hotter than in the central cell. The electrostatic confinement is based on the fact that in the presence of electron density and temperature gradients along the field lines, an equilibrium
potential ϕ develops which is given by

$$e(\phi - \phi_e) = T_e \ln \frac{n}{n_c} , \tag{23}$$

with ϕ_e and n_c the potential and the plasma density in the central
cell. Hence, ϕ is high where n and/or T_e is large, which permits to
improve ion confinement in the central cell by generating a dense
and hot plasma in the plug mirrors, e.g., by intense neutral injection. The intermediate thermal barrier allows to reduce the end plug
density n_p below that in the central cell, if the electron temperature in the plug is kept higher than the electron temperature in the
central cell. A disadvantage of the tandem mirror configuration is
that particle drift motions tend to enhance the transport across the
magnetic field induced by collisions (Ryutov and Stupakov, 1978) in
a way similar to what occurs in toroidal geometry (see below). The
presence of a loss cone makes the velocity distribution of mirror
plasmas anisotropic, which can cause plasma instability and enhanced

21

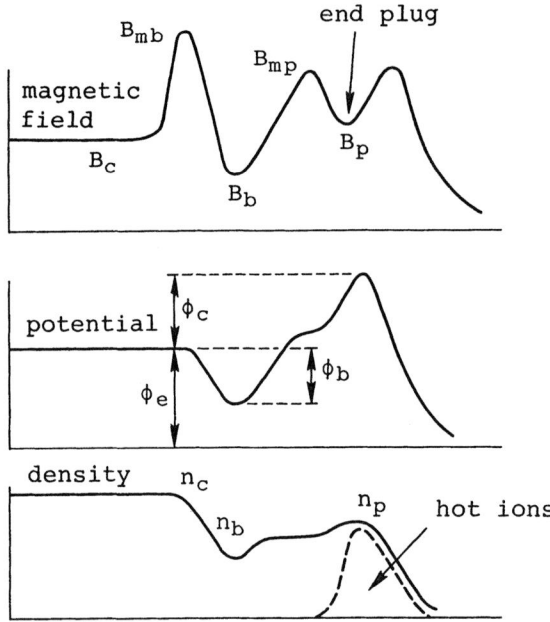

Fig. 4.   Magnetic field B, electrostatic potential ϕ and plasma density n, along the axis of the device, in a tandem mirror with thermal barrier.
The subscripts refer to "central cell" (c), "barrier" (b), and "end plug" (p), with m indicating mirror fields; ϕ_e is the reference potential in the central cell.
(Source: Baldwin and Logan, 1979).

losses (Baldwin, 1977). The most important unstable wave mode is the drift cyclotron loss cone mode. It was shown that this instability is avoided if there is a small quantity of warm plasma that affects the ion distribution in the loss cone in an appropriate way. As a consequence, mirror losses were reduced to collisional values. Plasma heating in mirrors is mainly done by beam injection, but also RF methods are being applied.

The best performance has been obtained in the tandem mirror TMX at Livermore (Simonen, 1981; Simonen, 1983). Typical ion energies, in the central plasma, are 0.25 keV, the electron temperature being about equal to this value. In the plug mirrors, the typical ion energy reaches 13 keV, and the plasma density $0.4 \cdot 10^{14}$ cm^{-3}. In the central plasma a plasma density of $0.3 \cdot 10^{14}$ cm^{-3} was attained. A maximum potential difference $\phi_c \approx 0.3$ kV was created. The overall confinement parameter is $n\tau_E \approx 10^{11}$ cm^{-3} s. The improvement of confinement, by adding the electrostatic potential barrier, is about an order of magnitude. In the central plasma the plasma β has reached values as high as 40%. Note, however, that these values were not achieved simultaneously on the same shot. In TMX-U, recently potential differences between the end plug and the central cell as high as $\phi_c \approx 1$ kV were obtained and a thermal barrier of $\phi_b \approx 0.5$ kV was generated (Turner et al., 1984).

End losses can be avoided when the magnetic field lines are confined within a finite volume ("closed confinement systems"). The simplest and most important example of this kind of confinement geometry is the <u>toroidal</u> one (cf. Fig. 5). The price to be paid for obtaining a closed geometry is that the magnetic field becomes space-dependent which leads to the appearance of particle drift motions that tend to make the plasma escape perpendicularly to the confining field (see Figs. 5 and 6). In this, the appearance of a "vertical" electric field E_v is one essential ingredient. Therefore, the situation is strongly improved when a poloidal magnetic field B_p is added to the toroidal field B_t. In this case the magnetic field lines become screw-like and lie on closed and nested magnetic surfaces (cf. Fig. 7). This provides a connection between the top and bottom half of the toroidal plasma column along magnetic field lines which short-circuits the field E_v. However, a residual electric field, inducing enhanced "toroidal" trans-

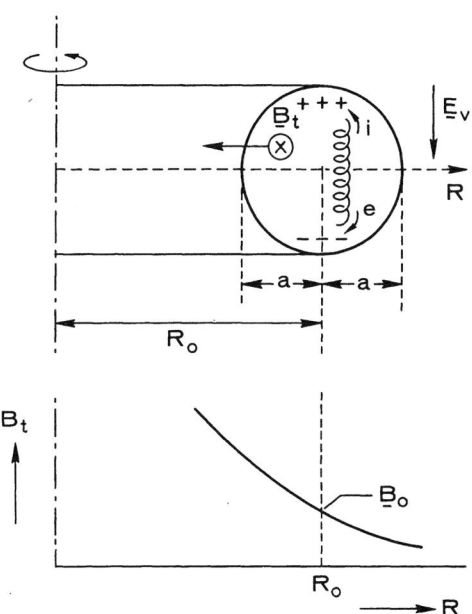

Fig. 5. Vertical particle (e = electrons, i = ions) drift motion and electric field E_v in a purely toroidal magnetic field; a and R_o are the minor and major radius of the torus, respectively, B_t is the toroidal magnetic field which has the value B_o on the axis of the plasma column.

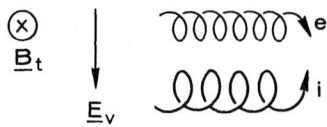

Fig. 6. Horizontal electron (e) and ion (i) drift motion in a purely toroidal magnetic field as a consequence of the presence
of the vertical electric field E_v.

port across the magnetic surfaces, remains also in this case, as dis-
sipative mechanisms like Coulomb collisions and possibly plasma turbulence lead to a finite plasma resistivity. Actually, toroidal
transport is also influenced by gradients of the plasma density and
temperature within magnetic surfaces (for more details, see Hinton
and Hazeltine, 1976; Tang, 1978; Engelmann, 1981).

A poloidal magnetic field can be generated by inducing a toroidal current in the plasma column and/or using suitable coils that
allow appropriate external currents to flow in toroidal direction.
The first case corresponds to the tokamak concept. However, to ensure
magnetohydrodynamic plasma stability, the poloidal field B_p must not
exceed the limit imposed by the Kruskal-Shafranov condition

$$q(r) \equiv r B_t / R_o B_p > 1 , \tag{24}$$

with r the minor radius of any magnetic surface and R_o the plasma
major radius. Relation (24), of course, also limits the plasma current in a tokamak-like confinement scheme. Devices of this kind are
the tokamak proper and the toroidal screw-pinch. Both devices have
a limited pulse length, the plasma current being driven inductively

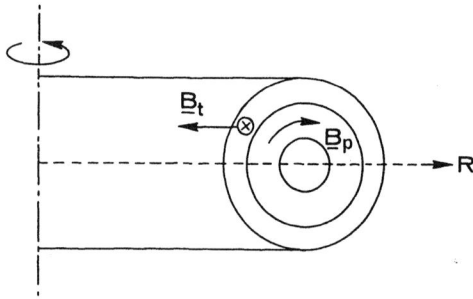

Fig. 7. Toroidal geometry in the presence of toroidal (B_t) and poloidal (B_p) magnetic field components: closed and nested
magnetic surfaces are formed.

so that the flux swing of the transformer sets an upper bound to the operation time. Only by applying non-inductive current drive methods, such as injection of particle beams or RF waves, is steady-state operation possible, but under reactor-grade conditions this is anticipated to require a very large power limiting the fusion gain Q to about 10. Tokamak plasmas are primarily heated ohmically by the plasma current induced, but since the efficiency of ohmic heating decreases with increasing electron temperature T_e like $T_e^{-3/2}$, thermonuclear temperatures can be obtained only by applying intense additional heating by neutral beams or RF waves. The screw-pinch mainly relies on heating by fast compression which limits the life time of the hot plasma to about the energy confinement time τ_E. Somewhat similar is the reversed-field pinch for which, however, condition (24) is violated, B_p and B_t being of the same order, and gross stability is obtained upon reversal of the toroidal field in the outside part of the plasma column. If the aspect ratio R_o/a, with a the minor plasma radius, is about 1, one talks about compact tori. As the geometry of the plasma edge in this case is essentially spherical, these schemes combine the advantage of a closed confinement geometry with a simple topology of the device. On the other hand, plasmas of this type are rather difficult to produce and to sustain. Specifically, a low aspect ratio tokamak is called a spheromak (see Fig. 8), while field reversed mirror configurations have $B_t = 0$.

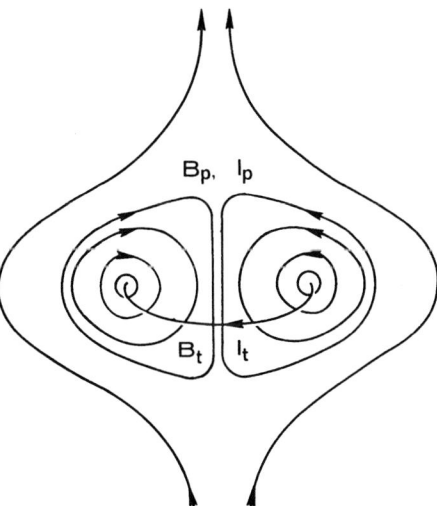

Fig. 8.   Spheromak geometry; B_t and B_p are the toroidal and poloidal magnetic field, I_t and I_p the toroidal and poloidal plasma currents; for $B_t = I_p = 0$, the field-reversed mirror configuration is obtained.

The most prominent example of a device where the poloidal field can be totally created by external currents is the stellarator, of which different variants exist. At present, mostly helical windings around the plasma are used to create the poloidal field, but applying twisted toroidal field coils is an attractive alternative. A stellarator configuration necessarily is non-axisymmetric. It offers the advantage of continuous operation. The same is true for bumpy torus devices where plasma equilibrium is ensured by a strong corrugation of the toroidal magnetic field, thus incorporating features of mirror confinement into a large aspect ratio toroidal geometry. Also multipoles with internal ring-shaped conductors generate closed confinement geometries. Also stellarator plasmas are often heated ohmically, in which case effectively a stellarator/tokamak hybrid configuration is obtained. Recently, neutral beam and RF heating was applied to generate stellarator plasmas.

The most advanced magnetic confinement scheme is the tokamak (for more details, see Furth, 1975; Rawls, 1979; INTOR Workshop, 1980, 1982, 1983). Its performance, measured in terms of the plasma temperature and plasma confinement obtained, has come close to thermonuclear requirements. The highest plasma ion temperature, achieved in the PLT device at Princeton using neutral injection for plasma heating, is 7 keV (Eubank et al., 1979; Stodiek et al., 1981). As this experiment was done at relatively low plasma density, namely $n \approx 0.4 \cdot 10^{14}$ cm^{-3}, the collisionality $\nu_* \equiv qR_0/\lambda$ (λ = particle mean free path) of these plasmas was very low, about the same as it will be in a fusion reactor plasma. On the other hand, plasma confinement was modest in this regime. For this to be good, one has to work at higher density for which tokamaks with high confining magnetic field are particularly suited. The largest values of the confinement parameter $n\tau_E$ thus have been obtained in ohmically heated discharges in ALCATOR C where the product of the peak density $\hat{n}$ in the discharge and the energy confinement time τ_E has reached $\hat{n}\tau_E = 0.8 \cdot 10^{14}$ cm^{-3} s (Greenwald et al., 1984), that is a value which approximately corresponds to the requirements of the break-even criterion, Eq. (11). In this case, the plasma temperature was around 1 keV.

The performance of stellarators has recently made considerable progress (see, e.g., Rau, 1981). An important fact is that pure stellarator operation, with effectively vanishing plasma current, was reached with plasma parameters similar to those in comparable tokamaks, the plasma heating being achieved by injection of neutral particle beams and electron cyclotron heating. In these discharges, the plasma temperature reaches the 1 keV range and the energy confinement parameter is $n\tau_E \approx 0.02 \cdot 10^{14}$ cm^{-3} s.

Plasma-wall interaction in magnetic confinement devices is an inevitable consequence of the imperfection of this confinement scheme. Therefore, the volume containing the plasma must be delimited by a vessel and/or plates onto which particles escaping from the plasma interior impinge. These particles carry an appreciable part of the energy which is exhausted from the plasma. The walls, hence, are subject to heat and particle loads as long as the discharge lasts. This wall load again is a limiting factor for the life time of the vessel wall and the plates collecting the particle flux. On the other hand, the plasma-wall interaction causes a release of non-hydrogenic atoms from the walls which, upon ionization, contaminate the hydrogen plasma and enhance radiative power losses. The positive side of imperfect confinement is that in a fusion reactor the "ash products" (in a DT reactor, helium) generated in the plasma interior can migrate to the plasma edge from which they can be exhausted. The main difference between open and closed confinement systems is that in the former a large part of the particle fluxes leaves the plasma along field lines and can be dumped onto plates quite far away from the main plasma region (see, e.g., TASKA Team, 1982), while in the latter all fluxes are necessarily radial and guiding particles onto plates (limiters, divertors) is possible only in the edge region. Therefore, all the plasma-wall interaction here takes place relatively close to the hot plasma, favouring plasma contamination. Only by very complex magnetic configurations ("local" divertors, especially the bundle divertor and its improvements) can the edge plasma be guided outside the discharge chamber (see, e.g., Harbour, 1981). It is, however, improbable that such a configuration could be used in a fusion reactor.

In the following Chapter an overview of the features and implications of plasma-wall interaction in magnetic confinement devices will be given, taking the tokamak as an example.

PLASMA-WALL INTERACTION IN TOKAMAKS

Due to the presence of particle and energy transport across the magnetic field, the wall of a tokamak device is subject to a continuous flow of charged hydrogenic (and other) particles which also have to carry that fraction of the energy lost from the plasma which is not converted into radiation in the discharge. This fraction usually is between 30 and 70%. Of course, also the radiated power has to be absorbed by the vessel wall, leading to an additional heat load on the vessel.

The particles impinging on the wall transfer to this (part of) the kinetic energy they carry. Charged particles are neutralized during their interaction with the wall. Hence, a cloud of neutral particles forms in front of the wall. This cloud interacts with the plasma: charge exchange leads to the formation of neutrals of higher energy which in part go back to the wall contributing to the wall load, while ionization processes finally cause the neutrals to reconvert into charged particles. This phenomenon is called particle recycling. In a steady state, the flux Γ of charged particles towards the wall must be compensated by the neutral flux Γ_o from the wall towards the plasma. The radial profiles of the plasma and neutral densities as appearing due to the particle recycling are schematically shown in Fig. 9.

In order to keep the consequences of plasma-wall interaction (plasma contamination and wall erosion due to the release of wall material) within acceptable limits, the plasma-wall contact must occur in a controlled way. This can be achieved (see Fig. 10) either by concentrating the plasma-wall contact on a limiter plate that defines the outer boundary of the area of the discharge where the magnetic surfaces are closed (i.e., where particles cannot reach a material surface by flowing along field lines), or by diverting the plasma from the edge region onto a remote material surface ("divertor plate"), the discharge being limited magnetically, in the simplest case by a separatrix. Examples of limiter and divertor configurations, integrated into a blanket/shield structure of a tokamak reactor are shown in Figs. 11a and 11b.

Let us now consider in some more detail the consequences of plasma-wall interaction (see also Fig. 12). The recycling of charged hydrogenic particles on the wall and limiter/divertor plates, implying a continuous bombardment of these material surfaces by energetic (charged and neutral) particles, leads to the desorption of light

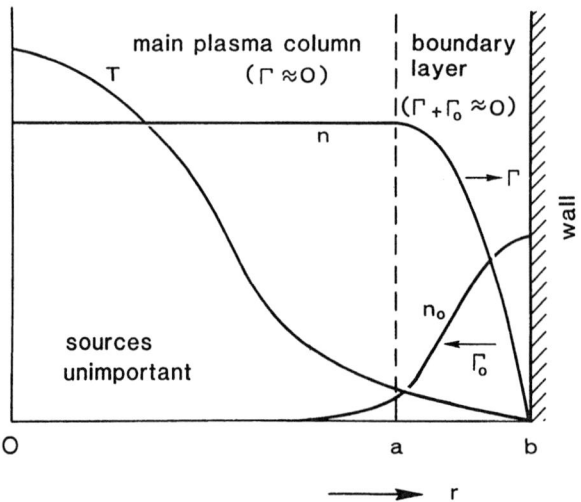

Fig. 9.   Schematical plot of radial profiles of a steady-state discharge (n = plasma density, n_o = neutral density, T = temperature, Γ = plasma flux, Γ_o = neutral flux).

1. "simple" wall contact

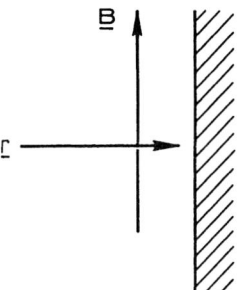

2. wall contact concentrated at limiter

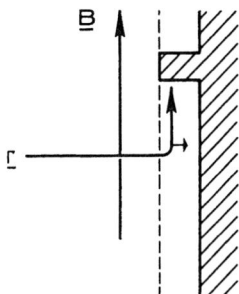

3. wall contact concentrated in divertor

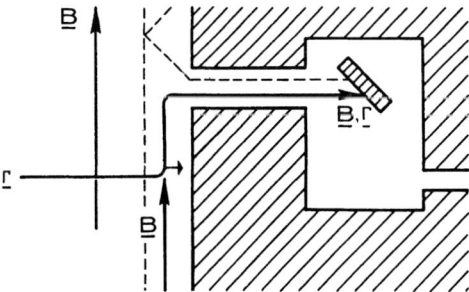

Fig. 10. Different configurations for plasma-wall contact (schematic; $\underline{B}$ = magnetic field, Γ = plasma flux): simple wall contact (1) does not allow for controlled plasma-wall interaction, therefore a limiter (2) or a divertor (3) is needed.

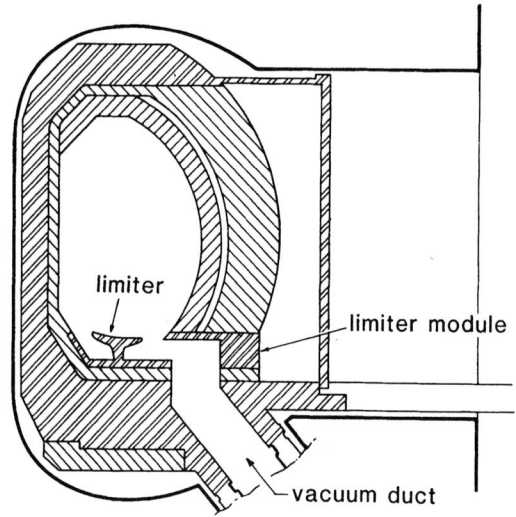

Fig. 11a. INTOR concept for blanket/shield structure with plasma-wall
interaction controlled by a pumped limiter (from INTOR
Workshop, 1983).

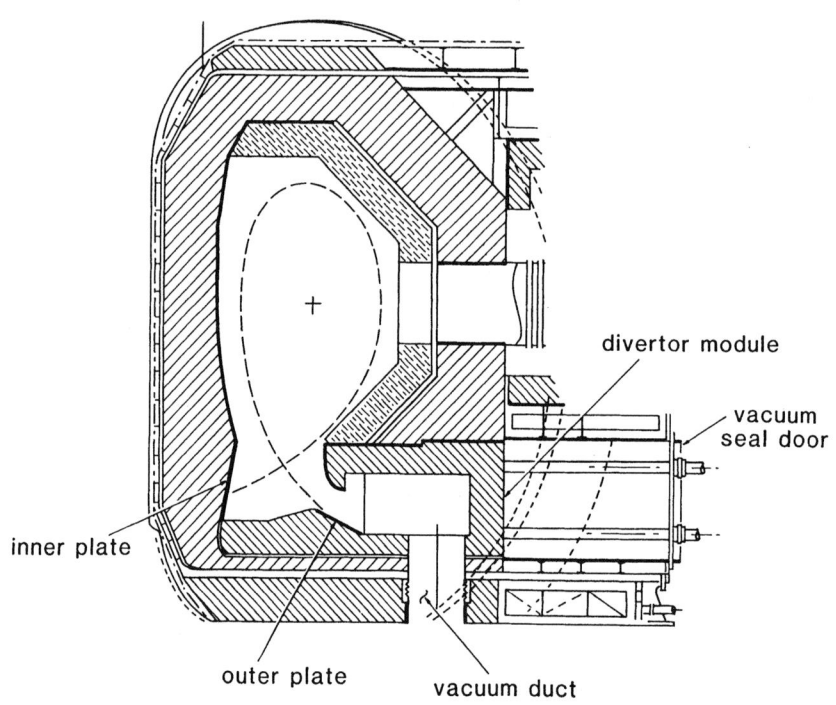

Fig. 11b. INTOR concept for blanket/shield structure with plasma-wall
interaction controlled by a single-null poloidal divertor
(from INTOR Workshop, 1983).

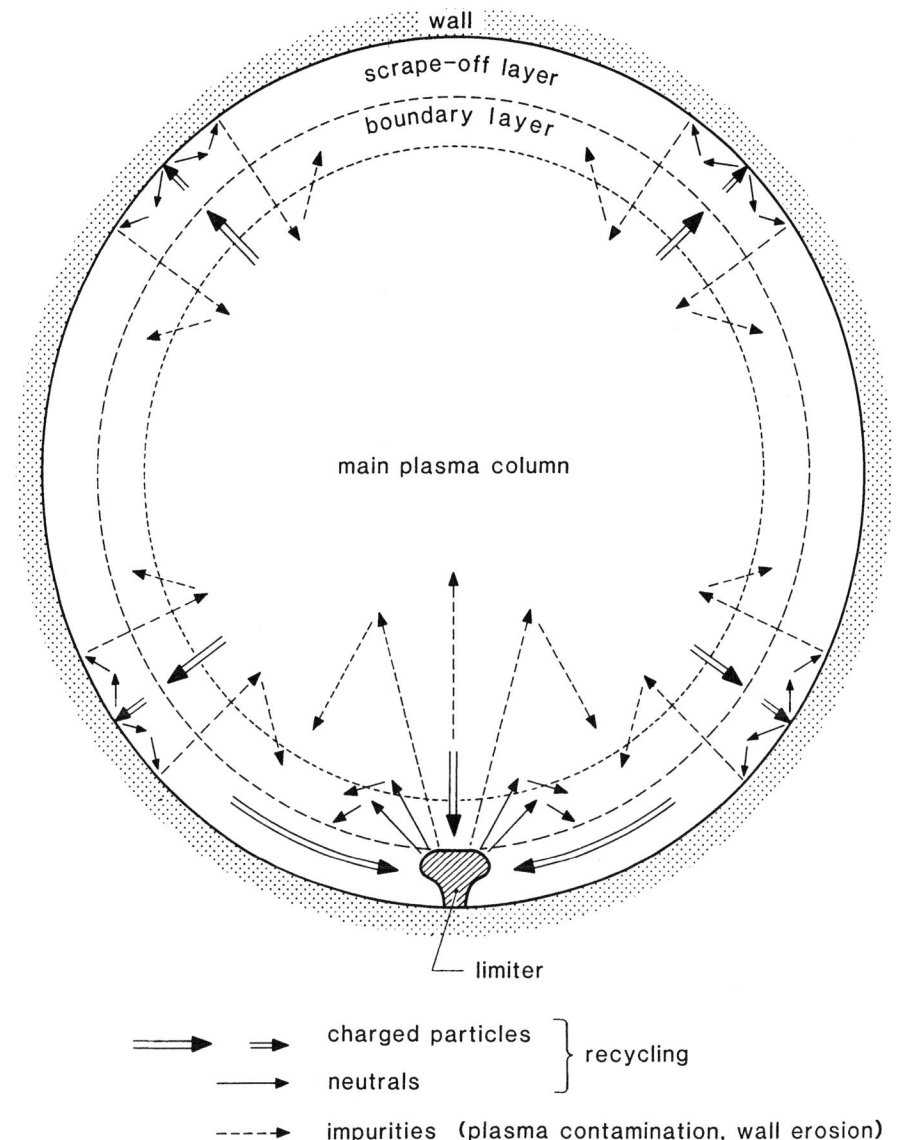

wall

scrape-off layer

boundary layer

main plasma column

limiter

charged particles
neutrals } recycling

- - - → impurities (plasma contamination, wall erosion)

Fig. 12. Plasma-wall interaction and its consequences (schematic).

31

impurities, like oxygen, present on the wall as well as to sputtering of wall material. Part of these non-hydrogenic atoms, upon ionization, diffuse into the plasma and reach higher and higher ionization states as they come in contact with hotter plasma. As long as the ionization is not complete, these impurity ions emit line radiation which even for a low impurity concentration may enhance the power loss of the plasma appreciably. To a lesser extent this also applies to fully stripped ions which enhance losses by bremsstrahlung. It must be noted that the cooling of the plasma and, in particular, of the edge plasma by enhanced radiative losses has a beneficial consequence if it is not excessive: in fact, it reduces the power to be exhausted via plasma-wall interaction, thus mitigating the latter. There is, hence, a kind of self-regulating mechanism that prevents the plasma-wall interaction from becoming very strong (Gibson and Watkins, 1977; Lackner, 1983). However, it must be recognized that this mechanism also may just lead to cooling and quenching the discharge if radiative losses become too strong.

The enhancement of the power loss is particularly large for ions of high nuclear charge Z. This is exemplified in Fig. 13, following Jensen, et al., 1977 (see also INTOR Workshop, 1980). As a consequence, the maximum permitted impurity concentration in a tokamak reactor is extremely low for high-Z impurities (typically 10^{-4} for tungsten), while the concentration of medium-Z impurities like iron may reach about 10^{-3} and that of low-Z impurities like oxygen and carbon may attain several percent (see Fig. 14).

The large majority of the atoms sputtered from the wall, due to the fact that their mean free path in the plasma is short, does not penetrate far into the plasma, but after ionization immediately turns back to the wall. This implies that the erosion of the wall and of the limiter/divertor plates, on the average, is much weaker than estimated from the strength of the primary sputtering process. However, since the impurity ions nevertheless can move somewhat along the magnetic field in the plasma in front of the wall and some particles penetrate deep into the plasma, there will be a migration of wall material all over the wall surface, and the profile of redeposition of wall material may be different from the sputtering profile. It is thus the combined effect of sputtering, migration and redeposition which will effectively determine the life time of the wall and of the limiter/divertor plates in a tokamak reactor. Also the answer to the question whether it is at all possible to have certain parts of the wall structure, like the plates, made out of materials differing from that of the main wall, depends on it. On the other hand, since the migration of wall material is a very slow process, quantitatively little is known at present on this effect.

As far as the strength of the sputtering process is concerned, it is important to note that the ambipolarity of the charged particle flux onto the wall, dominated by flows along the magnetic field lines

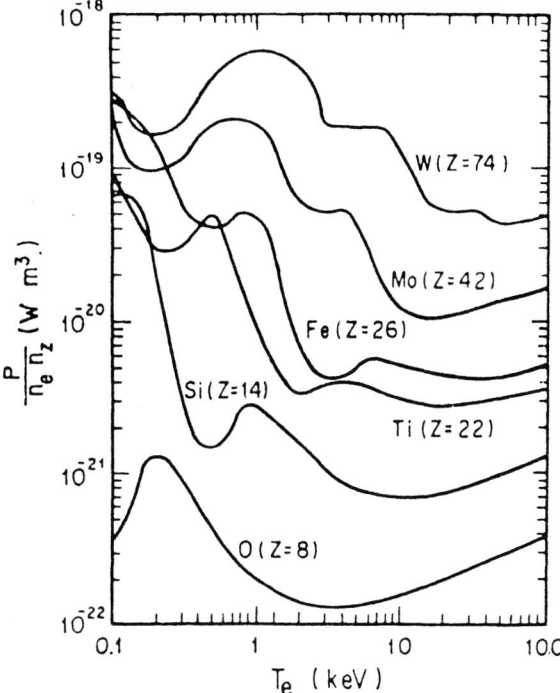

Fig. 13. Radiated power P, divided by the product of electron density n_e and impurity density n_z, as a function of electron
temperature T_e for various impurities in coronal equilibrium; Z = nuclear charge of the ion (source: Jensen et al.,
1977).

in the scrape-off layer, requires the presence of an electrostatic
sheath in front of material walls that decelerates electrons and accelerates ions. The potential drop in this sheath is about 3 times
the electron temperature (in eV) in front of the wall. This leads to
an appreciable increase in the energy of hydrogenic ions impinging
on the wall with respect to their kinetic energy at the edge of the
sheath. This energy increase is still stronger for multiply charged
impurity ions, which therefore can contribute appreciably to sputtering even when the impurity concentration is low.

The primary sputtering rate, under tokamak reactor conditions,
is quite high. This is shown in Fig. 15 where the erosion rate (not
corrected for redeposition, duty cycle and availability) is plotted
as a function of the energy of hydrogenic particles impinging on three
different material surfaces (stainless steel, aluminium, molybdenum)
when the total power flux exhausted by these particles is 10 MW and
30 MW, respectively. Therefore, acceptable conditions are anticipated

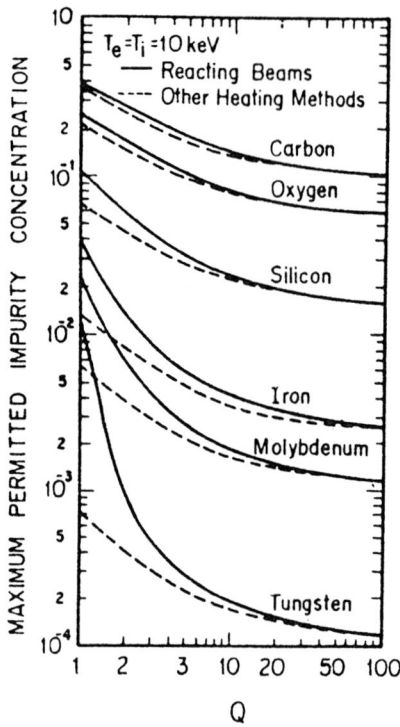

Fig. 14. Maximum permitted impurity concentration for various impurities as a function of the fusion gain Q (INTOR Workshop, 1980); note that the values shown for Q = 100 are representative for the ignited regime.

to be attained only if the plasma temperature in front of the material surfaces is low, typically below 10 to 30 eV according to the material, and/or the power exhausted by particle convection is not large. The important role of redeposition of wall material for the life time of the wall and limiter/divertor plates also becomes transparant from this quantification of wall erosion by sputtering. The conditions for the plasma temperature at the wall are the least stringent for high Z materials. This tendency also implies that the "factor of merit" defined as the ratio of the maximum permitted impurity concentration and the sputtering yield and measuring the "compatibility" of a wall material with the plasma conditions, is largest for high-Z materials, like tungsten, in this low temperature regime (while the situation is opposite for high temperatures in front of the wall); see Fig. 16 and for more details Bohdansky, 1981. At present it is an open question which choice of the wall and plate materials is best for a tokamak reactor, because the plasma edge regime to be anticipated is not sufficiently known.

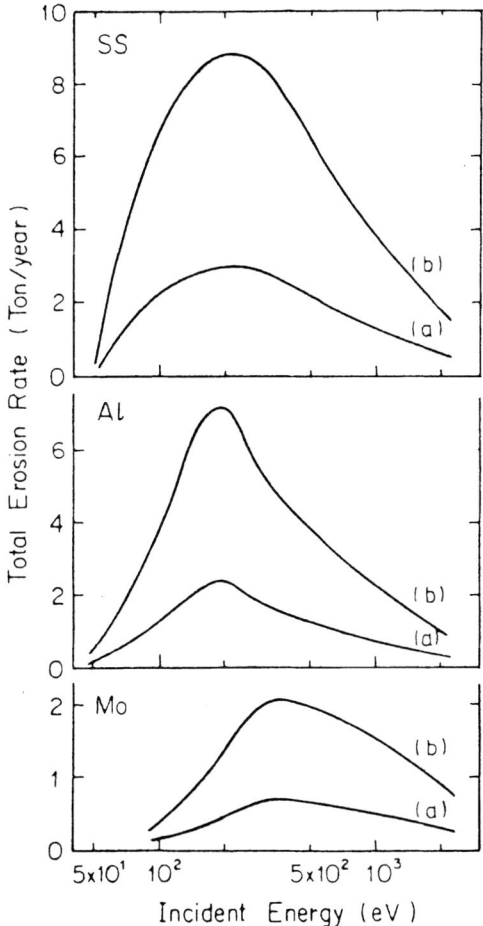

Fig. 15. Erosion rate due to sputtering by hydrogenic particle
fluxes carrying 10 MW (a) and 30 MW (b), respectively,
to a material wall consisting of stainless steel (SS)
aluminium (Al) and molybdcnum (Mo), as a function of
the energy of the incident particles (INTOR Workshop,
1980).

Essentially two regimes appear possible (see, e.g., Lackner,
1983), distinguished by the relative importance of radiative power
exhaust. A low edge temperature can be obtained when the edge density
is high, ensuring a strong particle recycling and, hence, a low energy exhausted per particle. If, in addition, there is a strong radiation cooling of the edge, the edge temperature might even be lower
still. The first regime is probably possible only in a divertor configuration and might allow using tungsten divertor plates (INTOR
Workshop, 1982; 1983), while the second could correspond to limiter

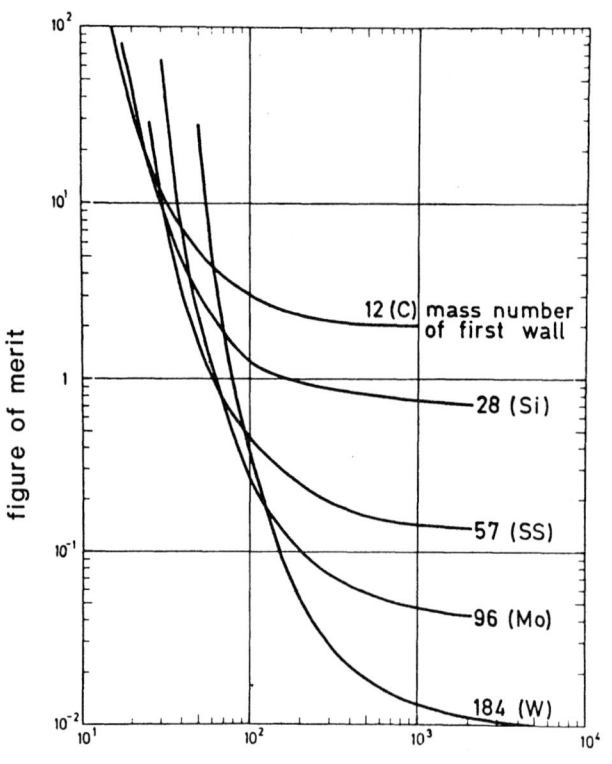

Fig. 16. Figure of merit for various surface materials as a function of the plasma temperature in front of the surface (source: Bohdansky, 1981).

or divertor bounded discharges using stainless steel or the like for the plates. Whether the latter situation is consistently possible, is however uncertain. In fact, a plasma with a strongly radiating edge tends to be thermally unstable; moreover, the probably existing limitations on the edge plasma density in tokamaks are certainly the more stringent the more power is radiated from the plasma. For low edge temperature the wall material can be stainless steel. If, on the other hand, sufficiently low edge temperatures could not be ensured, the limiter/divertor plates as well as the wall would have to be made of low-Z material (carbides, beryllium). In this case most of the power exhaust from the plasma would be through particles impinging on the plates, radiation losses remaining comparatively small. However, the contribution of the low-Z impurity pressure to the plasma kinetic pressure might become too high for being acceptable under these conditions. From this discussion, it is apparent that finding the optimum way for controlling the plasma-wall interaction is a critical issue on the way to a tokamak reactor.

A problem related to plasma-wall interaction is the exhaust of the helium ash from a reactor. This requires pumping the neutral gas generated during particle recycling out of the reactor at a rate that avoids helium accumulation in the reactor core beyond a concentration of 5 to 10%. Although the limited knowledge of the transport properties of helium and of the physics of the edge plasma make a final quantification of the required conditions impossible, it is anticipated that at least in a divertor configuration with high particle recycling a modest pumping speed of the order of 10^5 1/s could be sufficient. This is mainly due to the fact that the recycling fluxes are much larger (typically by two to three orders of magnitude) than the radial particle fluxes from the plasma core to the edge required for exhaust, but also the larger mean free path of neutral helium compared to neutral hydrogenic particles, allowing neutral helium to reach the pumping ducts more easily, is helpful (for more details, see INTOR Workshop, 1982; 1983). Concepts of pumped limiter and divertor configurations in a reactor are shown in Figs. 11a and b.

FINAL REMARK

In order to solve the many physics problems still open with respect to the interplay of the edge plasma with material walls and to optimize the control of plasma-wall interaction in magnetic confinement devices and particularly in tokamaks, still a great experimental and theoretical effort is necessary. This must concentrate on reactor-grade plasma edge conditions, that is discharges where the boundary layer is well separated from the main plasma column having a power flux approaching 30 W/cm^2 into the boundary layer and with a density close to 10^{14} particles/cm^3.

ACKNOWLEDGEMENT

This work was performed as part of the research programme of the association agreement of Euratom and the "Stichting voor Fundamenteel Onderzoek der Materie" (FOM) with financial support from the "Nederlandse Organisatie voor Zuiver-Wetenschappelijk Onderzoek" (ZWO) and EURATOM.

REFERENCES

Artsimovich, L.A., 1964, "Controlled Thermonuclear Reactions", Oliver
   and Boyd, Edinburgh and London.
Baldwin, D.E., 1977, Rev. Mod. Phys., 49:317.
Baldwin, D.E., and Logan, B.G., 1979, Phys. Rev. Lett., 43:1318.
Bohdansky, J., 1981, Plasma Wall Interaction in Tokamaks, in "Plasma
   Physics for Thermonuclear Fusion Reactors", Harwood Acad. Publ.
   for the Commission of the European Communities, Brussels and
   Luxembourg.
Cohen, B.I., Ed., 1980, "Status of Mirror Fusion", Lawrence Liver-
   more National Laboratory, UCAR 10 049-80, Livermore.
Conn, R.W., et al., 1978, The SOLASE Laser Fusion Reactor Design and
   its Technological Implications, in "Fusion Reactor Design Con-
   cepts", IAEA, Vienna.
Engelmann, F., 1981, Transport in Toroidal Magnetoplasmas, in "Modern
   Plasma Physics", IAEA, Vienna.
Eubank, H., et al., 1979, PLT Neutral Beam Heating Results, in "Proc.
   of the 7th Int. Conf. on Plasma Physics and Contr. Nuclear Fu-
   sion Res.", Vol. I, IAEA, Vienna.
Furth, H.P., 1975, Nucl. Fusion, 15:487.
Gibson, A., and Watkins, M.L., 1977, in "Proc. of the 9th European
   Conf. on Contr. Fusion and Plasma Physics", Prague, Vol. I, p.31.
Greenwald, M. et al., 1984, Plasma Fusion Center, Massachusetts In-
   stitute of Technology, PFC/JA-84-3, Cambridge.
Harbour, P.J., 1981, Divertor Problems, in "Plasma Physics for Ther-
   monuclear Fusion Reactors", Harwood Acad. Publ. for the Commis-
   sion of the European Communities, Brussels and Luxembourg.
Hauer, A., "Survey of Atomic Physics Issues in Experimental Inertial
   Confinement Fusion Research", XII ICPEAG, Gatlinburg.
Hinton, F.L., and Hazeltine, R.D., 1976, Rev. Mod. Phys., 48:239.
INTOR Workshop, 1980, Report on Zero Phase, IAEA, Vienna;
                   1982, Report on Phase One, IAEA, Vienna;
                   1983, Report on Phase II A-Part I, IAEA, Vienna.
Jensen, R.V., et al., 1977, Nucl. Fusion, 17:1187.
Lackner, K., et al., 1983, Plasma Phys. and Contr. Fusion, 26:105.
Maniscalco, J.A., et al., 1978, Design Studies of a Laser Fusion
   Power Plant, in "Fusion Reactor Design Concepts", IAEA, Vienna.
Miley, G., Ed., 1979, "Proc. of the IEEE Minicourse on Inertial Con-
   finement Fusion", Montreal.
Rau, F., Ed.,1981, "Stellarators: Status and Future Directions",
   Max-Planck-Institut für Plasmaphysik, IPP 2/254, Garching bei
   München.
Rawls, J., 1979, "Status of Tokamak Research", DoE/ER - 0034, US De-
   partment of Energy, Washington DC.
Ryutov, D.D., and Stupakov, G.V., 1978, Sov. Phys.-Dokl., 23:412.
Simonen, T.C., Ed., 1981, "Summary of TMX Results", Lawrence Liver-
   more Laboratory, UCRL 53120, Livermore.
Simonen, T.C., et al., 1983, TMX Tandem Mirror Experiments and Ther-
   mal-Barrier Theoretical Studies, in "Proc. of the 9th Int.

Conf. on Plasma Physics and Contr. Nuclear Fusion Res.", Vol. I, IAEA, Vienna,

Stodiek, W., et al., 1981, Transport Studies in the Princeton Large Torus, in "Proc. of the 8th Int. Conf. on Plasma Physics and Contr. Nuclear Fusion Res.", Vol. I, IAEA, Vienna.

Tang, W.M., 1978, Nucl. Fusion, 18:1089.

TASKA Team, 1982, "TASKA – A Tandem Mirror Fusion Engineering Facility", Institut für Technische Physik, Kernforschungszentrum Karlsruhe, KfK 3311, UWFDM – 500, Karlsruhe.

Thompson, W.B., 1965, Introduction to Controlled Thermonuclear Research, in "Plasma Physics", IAEA, Vienna.

Turner, W.C., and TMX-U Group, 1984, TMX-U Thermal Barrier Tandem Mirror Experiment, invited paper at the "International Conference on Plasma Physics", Lausanne.

THE PLASMA SHEATH

Peter C. Stangeby

University of Toronto
Institute for Aerospace Studies
Toronto, Canada

and

Princeton Plasma Physics Laboratory
Princeton, NJ 08544 USA

1.   INTRODUCTION AND OUTLINE

Outline: Section 2 simply describes, without attempt at explanation, <u>what</u> happens electrically when a plasma is in contact with a solid surface. The practical implications of the interaction are briefly described.   Section 3 provides a physical explanation of <u>why</u> these effects occur and deduces initial estimates of the plasma-solid voltage difference and the spatial extent of this voltage drop. Section 4 deduces the Bohm Criterion (ion drift velocity out of the plasma must equal the ion acoustic speed) using ion fluid models.   The Criterion is obtained from both the plasma and the 'sheath equations separately.   Section 5 deduces simple formulae for the particle and energy flux which is transmitted by a sheath, both for electrically floating and biased objects.   Section 6 gives a brief indication of how the sheath particle and energy transmission characteristics influence the modeling of the edge plasma in magnetically confined plasma devices, while Section 7 gives a similar brief introduction to their use in the interpretation of plasma probe data.   Section 8 indicates various refinements on sheath theory and concludes with recommended formulations for floating potential, particle and energy transmission coefficients for the sheath.

Some caveats: The plasma/sheath analysis presented in this Chapter does not explicitly include the presence of a magnetic field. It is therefore suitable for describing the flow of plasma along magnetic field lines to a surface, since such flow is not impeded by the magnetic field. Such parallel-field plasma flow is generally a good first approximation to the interaction of plasma with probes and other solid objects, such as limiters, inserted into a plasma. When the plasma flow must cross magnetic field lines in order to reach the object, then some corrections can be made to the simple theory; see Chapter by Chodura. Also, if the ion or electron Larmor radii are large compared with the object size, then the simple theory presented in this Chapter is not applicable.

Some omissions: Because of space limitations, only the steady-state, wall sheath will be considered here, i.e., the case of a solid on one side of the sheath, with plasma on the other side. There is thus no treatment of other important types of sheath: (a) dynamic sheaths which occur when, for example, a very rapidly varying voltage is applied to an object immersed in a plasma (monotonic time variation or oscillatory), or when a plasma expands into a vacuum, etc., (b) double sheaths, or free-standing sheaths, which are bounded on each side by plasmas and which may occur, for example, at magnetic mirrors in fusion devices.

2. THE BASIC FACTS

A plasma is generally highly conducting, and we may often consider it to be an equipotential at the plasma potential Φ_p. An object placed in the plasma will generally not assume this potential. Rather, when a plasma is in contact with any solid object, such as a limiter, divertor plate or diagnostic probe, a voltage difference spontaneously develops between the plasma and the object called the floating potential, Φ_f (often, for convenience we take plasma potential as reference potential thus making $\Phi_p = 0$). The object will generally float electrically at a potential which is negative relative to the plasma with $|\Phi_f| =$ a few kT_e/e where T_e is the electron temperature of the plasma actually contacting the object. Of course, one may bias the object relative to the plasma at any desired potential using an external circuit provided that one has a means of applying this extra potential difference. This latter situation is generally only achievable for small objects, typically probes, where one can use other (large) objects contacting the plasma, e.g., limiters or divertor plates, as the bias reference. Taken in aggregate, the solid surfaces contacting the plasma will float negatively relative to the plasma, or alternatively, one may say that the plasma will adjust to assume a positive potential relative to the aggregate of surfaces.

The potential drop, Φ_f, occurs in a thin <u>sheath</u> which is established between the plasma and the solid. The sheath thickness is of order the <u>Debye length</u>

$$\lambda_D = (\varepsilon_o kT_e/ne^2)^{1/2} \simeq 743\ T^{1/2}\ (eV)\ n^{-1/2}\ (cm^{-3})\ cm. \quad (1)$$

Example: Typical tokamak edge plasma with $n = 10^{12}\ cm^{-3}$, $T_e = 15$ eV, then $\lambda_D = 0.003$ cm.

In the plasma itself, $n_e = n_i$ to a high order, i.e., the plasma is <u>quasi-neutral</u>. The sheath, by contrast, has a net positive charge per unit volume since the plasma electrons are repelled by the negative potential on the solid. The areal negative charge density on the solid surface approximately equals the areal positive charge density in the sheath. The sheath thus acts to <u>shield</u> the plasma from the potential on the solid surface. This shielding effect also occurs if the object is biased more negatively than Φ_f; the sheath thickness increases in this case, but is still usually very small compared with plasma dimensions. If the object is biased positively relative to the plasma, then the sheath disappears and the random, Maxwellian flux of electrons strikes the surface, unattenuated (in the simplest cases).

The shielding effect of the sheath is imperfect and a small residual field, the <u>pre-sheath</u>, penetrates deep into the plasma. In the simplest case this pre-sheath field extends all the way to the symmetry point between two opposite-facing solid surfaces. The potential drop in the pre-sheath is small, ~1/2 kT_e/e and acts to draw ions from the plasma into the sheath. This accelerating field is just such as to cause the ion drift velocity, at the sheath/plasma interface, to equal the ion acoustic speed

$$C_s = [k(T_e + T_i)/m_i]^{1/2}. \quad (2)$$

This is the (generalized) <u>Bohm Criterion</u>. A (collisionless) plasma itself cannot support potential differences greater than ~1/2 kT_e/e without breaking down into a space-charge zone and thus ions cannot be accelerated to supersonic velocities in terms of C_s) by <u>plasma</u> fields.

The potential variations in sheath and pre-sheath, ion drift speed and n_e, n_i variations are shown in Fig. 1.

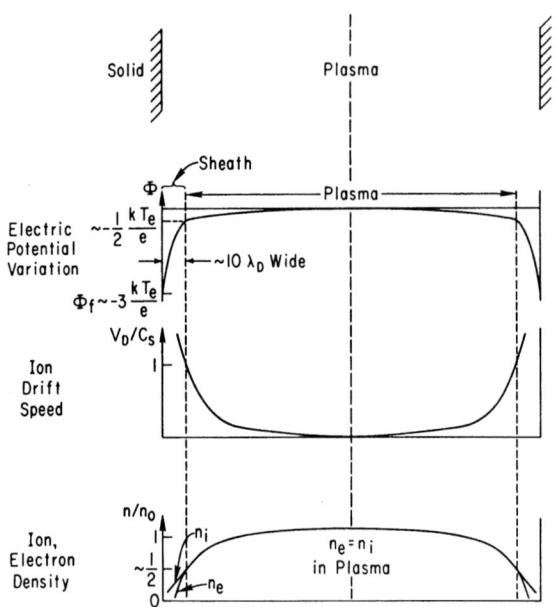

Fig. 1.   Schematic of the variation of electric potential, ion
drift speed and ion/electron densities in the plasma
between two semi-infinite planes.   So long as (i) a
magnetic field is perpendicular to the plane surfaces
and (ii) it is strong enough that Larmor radii are
small compared with object dimension then the
magnetic field will not effect the analysis.   The
thickness of the sheath is exaggerated for clarity.

The existence of the potential drop Φ_f has a number of
consequences: (a)   Ions are accelerated through the sheath and
thus impact the solid surface with an energy which is greater
than that associated with T_i.   This generally increases
sputtering.   It also influences backscattering/retention/release
and thus the ability of the solid and plasma to come into
equilibrium with regard to the recycle of particles (hydrogen in
the case of fusion devices).   (b)   The sheath controls the rates
at which particles and energy are removed from the plasma by the
solid surface.   One wishes then to know the sheath transmission
factors as a boundary conditions for modeling of the edge
plasma.   (c)   For probe analysis one also requires knowledge of
these transmission factors for all potentials, in addition to
Φ_f.

The fact that the potential drop generally occurs in a very thin layer simplifies sheath analysis; for example, one can usually neglect ionization and many other effects within the sheath itself.

3. A PRELIMINARY ANALYSIS OF THE SHEATH

Why can't the plasma simply contact the solid without generating a potential difference?

Consider the solid suddenly introduced into the plasma at t = 0, with the solid initially biased to be at the plasma potential (thereafter the solid potential is allowed to adjust itself). The random flux of electrons strikes the solid at flux density $1/4\ n_e\bar{c}_e$ while the ions strike it at flux density $1/4\ n_i\bar{c}_i$ where $\bar{c}_{e,i} = (8\ kT_{e,i}/\pi m_{e,i})^{1/2}$.[1] We can easily show that in the plasma $n_e = n_i$, to a high order. Consider a plasma of density 10^{20} ions/m^3 between two flat metal plates across which is suddenly applied a voltage which pulls all of the electrons out of the plasma before the heavier ions have a chance to move. What voltage is required? We use Maxwell's equation,[2] $dE/dx = en/\varepsilon_o$ to obtain a required field of about 10^{12}V/m! Clearly it is not practical to generate such voltages; the potential energy represented by such charge separation is also unphysically large. Hence, unless $T_i \gg T_e$, the electron flux to the solid will greatly exceed the ion flux and within a very short time the solid will gain a negative charge. This is the origin of the potential difference which is generated between plasma and solid.

How large is the potential difference? If the solid is floating relative to the plasma it must, in the steady state, receive equal currents of positive and negative charge, i.e., equal ion and electron particle fluxes if z = 1, where z = charge on each ion. The negative potential on the solid will then adjust itself so as to reduce the electron flux to the surface until it is equal to the ion flux.

We may now obtain a first estimate of the magnitude of Φ_f. The electrons from the plasma find themselves in a retarding electrostatic field. From elementary statistical mechanics we know that in a repelling, conservative force-field, a Maxwellian distribution remains Maxwellian but the number density is simply reduced.[3] Example: the vertical variation of air density in the earth's atmosphere. In the present case the flux of electrons reaching the solid is thus reduced to $1/4\ n\bar{c}_e$ exp $(e\Phi_f/kT_e)$; note $\Phi_f < 0$. (Strictly, this simple relation for the electrons only holds if all electrons are reflected and the electron distribution is therefore fully Maxwellian. The effect of the distortion to the distribution caused by the escaping electrons on the calculation of floating potential has been shown by Self[4] to be negligible. For objects which are biased more positively the effect can be more substantial; Andrews and Varey[5] have provided the corrections).

Considering now the ion flux to the surface: the situation is more complex since the ions are attracted and their distribution is distorted from Maxwellian. As indicated in Section 2, the ions actually enter the principal potential drop region with a net drift velocity C_s and are then accelerated freely to potential Φ_f. The ion flux to the surface is therefore simply nC_s. We mustn't jump ahead, however, and we want to actually <u>prove</u> this result for the ion flux. A crude estimate would be to take the ion flux into the accelerating region (hence into the solid surface) as the free Maxwellian value, $1/4\ n\bar{c}_i$, neglecting pre-sheath accelerations. (This approximation will be seriously invalid when $T_i < T_e$, but otherwise will turn out not to be too gross.)

Thus in steady state we have

$$1/4\ n\bar{c}_e \exp\ (e\Phi_f/kT_e) \simeq 1/4\ n\bar{c}_i \tag{3}$$

thus

$$\Phi_f \simeq \frac{kT_e}{e}\ \ell n\ \left(\frac{T_i m_e}{T_e m_i}\right)\ . \tag{4}$$

Example: H^+ ions, $T_e = T_i$ then $\Phi_f \simeq -3.8\ kT_e/e$.

For this example, one can see that the impact energy of the ion on the solid is substantially increased over the average ion energy in the plasma.

From where does the ion obtain the increased impact energy? This energy comes from the electrons as one may now see. Consider those electrons at the plasma/sheath interface which have enough energy to escape over the potential barrier Φ_f to reach the solid surface. For a floating surface, the wall flux of such electrons equals the wall flux of ions. In traversing the sheath, these electrons will lose an amount of energy $e\Phi_f$ which is, of course, precisely the energy gained by the ions in the same transversal. The sheath thus acts as a mechanism for <u>transferring energy</u> from the energetic tail of the electron distribution, to the ions.

Although the ion impact energy exceeds the electron impact energy, one should note that the power removal rate from the population of electrons in the plasma exceeds that from the ions. Thus a surface in contact with a plasma exerts a preferential cooling effect on the electrons.[6]

We turn next to estimating the sheath thickness. We will approximate the sheath as a planar diode in which only one charge species is present (the ions) emitted at one surface (the plasma/sheath interface) and collected at the other (the solid surface). How can we justify neglecting the electrons in the sheath? Because the electron density falls off exponentially as $\Phi(x)$ decreases through the sheath $\{n_e(x) \propto \exp [e\Phi(x)/kT_e]\}$ while the ions in being accelerated by $\Phi(x)$ fall off more slowly in density, namely $n_i(x) \propto [-\Phi(x)]^{1/2}$.[7] The latter result is seen as follows: Neglecting local ionization the ion flux, $n_i(x)V_i(x)$, is constant through the sheath, where $V_i(x)$ is the local ion velocity. Neglecting any initial ion velocity we then have $V_i(x) = [-2e\Phi(x)/m_i]^{1/2}$. Thus, within a short distance of the sheath/plasma interface, the electron density becomes negligible compared with the ion density.

Next we employ Poisson's equation[8] which relates potential variation to space charge density

$$\frac{d^2\Phi}{dx^2} = -\frac{e}{\varepsilon_o} (n_i - n_e) \tag{5}$$

$$\approx -\frac{e}{\varepsilon_o} n_i(x) \tag{6}$$

The ion current density to the wall

$$j^+ = e n_i(x) V_i(x) \tag{7}$$

is constant throughout the sheath hence giving

$$\frac{d^2\Phi}{dx^2} = \frac{j^+}{\varepsilon_o} (2e/m_i)^{-1/2} (-\Phi)^{-1/2} . \tag{8}$$

The solution to this equation is the famous Child-Langmuir[9] relation for current in a space-charge limited diode

$$j^+ = 4\varepsilon_o (2e/m_i)^{1/2} (-\Phi_d)^{3/2}/(9d^2) , \tag{9}$$

where d = thickness of acceleration zone, Φ_d is the total voltage drop and the boundary conditions $\Phi = d\Phi/dx = 0$ at the emitting surface are assumed. This relation thus gives the current which can be drawn between two plates, when the space charge between the plates (rather than the current emitted by one of them) limits the achievable current density. Only one charged particle species was assumed to be present in the original Child-Langmuir derivation, and since we neglected the presence of electrons in the sheath we have simply regained this classical result. It is however, only a first approximation for the wall sheath; see next Section.

We now insert our estimates of j^+ and Φ_d for the sheath to calculate the sheath thickness. We use for the present our rough estimate

$$j^+ = 1/4 \ n\bar{c}_i e \qquad\qquad (10)$$

where n represents the (constant) plasma density and $\Phi_d = -\eta kT_e/e$, where for floating conditions in a hydrogen plasma with $T_e = T_i$ we found $\eta \simeq 4$. Inserting these values into Eq. (9) gives

$$d = 1.3 \ \eta^{3/4} \ \lambda_D$$

$$\simeq 3.6 \ \lambda_D \ \text{for} \ \eta = 4 \ . \qquad\qquad (11)$$

We have thus deduced that the sheath is of order a few Debye lengths[10] in thickness. Even for very large applied (negative) potentials, $\eta \gg 1$, the sheath is still generally very thin compared to typical plasma scale lengths.

The foregoing derivation is clearly an approximate one since (a) electrons have been neglected in the sheath, (b) the initial velocity of the ions entering the sheath has been taken as zero [which would require that $n_i(0) = \infty$ if flux continuity is to hold] instead of $V_i(0) = C_s$. A complete treatment, however, only changes the numerical factor in Eq. (11) somewhat. Since for most practical purposes the precise thickness of the sheath is not important we will not attempt any further refinement of this first estimate of sheath thickness.

4.   THE BOHM CRITERION

In this section we deduce two results: (a)   that the voltage drop between the plasma and object cannot be accommodated within the plasma itself and thus that a charge layer or sheath must occur between the two to accommodate the drop, (b) that a small pre-sheath electric field extends into the plasma and is sufficient to accelerate the ions such that by their point of entering the sheath they have attained a drift velocity equal to the ion acoustic speed.   The latter result is known as the Bohm Criterion.[11]

We deduced in Section 3 that a substantial voltage drop must exist between a plasma and an electrically floating object with which it is in contact.   Can this potential drop occur in the plasma itself?   That is, in a region where $n_e = n_i$? The answer is no.   We will show, in fact, that a collisionless plasma cannot contain a potential drop greater than ~1/2 kT_e/e. (This also assumes either no magnetic field or at least we restrict our attention to motion along the B-field.)

We will apply the fluid equations to the plasma, e.g., we will assume that all of the ions at point x in the plasma have the same drift velocity V(x).   We assume one dimensional, steady-state flow.   This analysis could therefore be applied to the case of plasma flow along magnetic field lines to solid surfaces such as limiters, divertor plates, probes, etc.   We must allow for a source of ion pairs since particles are removed by the solid surface and steady-state is assumed.   The equation of continuity[12] is thus

$$\frac{d}{dx} (nV) = S ,   \tag{12}$$

where S = volume source rate of ion pairs (ion pairs/m^3/s)

$$n = n_e = n_i .$$

S might, for example, represent local ionization of neutral atoms by impact with plasma electrons--the situation in low pressure, partially-ionized discharges such as neon lights, etc.[13]   S can also represent the appearance of ion pairs in a magnetic flux tube via cross-field diffusion.   This latter case would characterize the scrape-off region of magnetically-confined plasma devices where ion pairs may enter the edge flux tube primarily from the main or core plasma,[6,14-22] rather than being created locally by neutral ionization.   Note also that S may vary, S(x).

The momentum equation[12] for the ions is

$$m_i V \frac{dV}{dx} = -\frac{dp_i}{dx} + enE - m_i VS \tag{13}$$

where $p_i = nkT_i$, ion pressure $E = -d\Phi/dx$ electric field in the plasma.

The first term in Eq. (13) gives the convective rate of change of momentum. The last term allows for the drag on the ion flow when a new ion is created at point x with zero velocity and has to be brought up to velocity V(x). For a derivation of Eqs. (12) and (13) from basic principles, see Ref. 23.

We can write similar equations for the electrons; however, since they are in a retarding field, it is convenient to employ the result mentioned earlier, i.e., one has the Boltzmann Relation[3] for the electrons in the plasma:

$$n_e(x) = n_o \exp [e\Phi(x)/kT_e] , \tag{14}$$

where n_o is the reference density, conveniently taken to be that at the symmetry point (Fig. 1), i.e., at a great distance from the object. This relation permits a replacement of the third term in Eq. (13)

$$enE = -en \frac{d\Phi}{dx} = -kT_e \frac{dn}{dx} . \tag{15}$$

It is convenient to consider the ion flow to be either isothermal (T_i = constant) or adiabatic (no energy addition) thus one has that

$$\frac{dp_i}{dx} = \gamma_s kT_i \frac{dn}{dx} , \tag{16}$$

where $\gamma_s = 1$, 5/3 for isothermal, adiabatic flow. One may therefore rewrite the ion momentum conservation equation as

$$V \frac{dV}{dx} = -\frac{c_s^2}{n} \frac{dn}{dx} - \frac{SV}{n} \tag{17}$$

where

$$C_s = [(\gamma_s kT_i + kT_e)/m_i]^{1/2}, \tag{18}$$

50

the ion acoustic speed at the sheath/plasma interface.[24] For isothermal flow C_s = C = constant.

One thus sees that in the plasma the natural ion velocity for normalization is the ion acoustic speed and one may define a local ion _Mach number_ M(x) ≡ $V(x)/C_s$. We will see that something rather spectacular occurs when the ions "break the sound barrier", namely the plasma is terminated and forms a sheath (Ref. 25). We combine these equations to give

$$\frac{dM}{dx} = \frac{S}{nC_s} \frac{1+M^2}{1-M^2} .$$ (19)

We assign V(0) = M(0) at the symmetry point. This last equation is a combination of the fundamental expressions of the conservation of mass and momentum and it provides us with the basic picture of what happens to the ion flow in the plasma. Thus from Eq. (19) we have the obvious result that dM/dx > 0, i.e., the ion velocity increases as one approaches the surface. (Note that we used the fact that S/nC_s is inherently positive and that the flow starts subsonically at the stagnation point.) Most importantly we have from Eq. (19) that as M → 1, dM/dx → ∞, i.e., the ions are abruptly accelerated to very high velocities and the plasma solution fails, "blows up". This corresponds physically to the termination of the plasma and the start of the sheath with its large electric fields and high ion acceleration.[4,11,25-30]

The foregoing equations have simple solutions[31] for relating the plasma density and potential to the local Mach number

$$\frac{n(M)}{n_o} = \frac{1}{(1+M^2)}$$ (20)

$$\Phi(M) = -\frac{kT_e}{e} \ln (1+M^2) .$$ (21)

[To calculate n(x), Φ(x), V(x), etc. requires knowledge about the spatial variation of S(x)]. We thus have the useful fluid model pre-sheath results that at the plasma/sheath edge where M = 1, then n ≡ n_{se} = 1/2 n_o and Φ = -0.69 kT_e/e.[32] That is, the plasma density at the sheath edge is half that far from the surface and the voltage drop in the plasma cannot exceed ~0.7 kT_e/e. We thus have obtained our result that the potential drop that we know must exist between the plasma and the solid cannot be accommodated in the plasma itself.

Instead of employing the fluid equations to describe the ion flow to the surface one can use equations which allow for individual particle motion (as is most appropriate when the ions are collisionless). This type of analysis was first carried out (for the case of $T_i = 0$) by Tonks and Langmuir[26] in their classic 1929 paper on plasma flow to a surface. They showed that the plasma solution "blows up" when the potential reaches a value $(0.854-1.418)kT_e/e$ (depending on geometry and other details) giving an average ion velocity at the plasma/sheath interface of $(1.144-1.213)C_s$.

The result that $V_i = C_s$ at the plasma/sheath interface is the all important <u>Bohm Criterion</u> and was first derived <u>explicitly</u> by David Bohm (Ref. 11) in the 1940's by considering the <u>sheath</u> equations rather than the <u>plasma</u> equations as we have done above. Because of the historical importance of this result, we now deduce the Bohm Criterion from an analysis of the <u>sheath</u>. Our analysis will in fact be a more refined version of the one employed in Section 3 to estimate the sheath thickness. We now allow for (a) electrons in the sheath, (b) a finite ion velocity for the ions entering the sheath. We follow Bohm[11] in assessing the case of $T_i = 0$ and all ions have the same (drift) velocity at any given point in the sheath. The ion momentum and particle conservation equations therefore give the particularly simple result that

$$n_i(x) = n_{se} \ [\Phi_o/\Phi(x)]^{1/2}, \qquad\qquad (22)$$

where n_{se} = ion (and electron) number density at the plasma/sheath interface, $|\Phi_o| = 1/2 \ m_i v_o^2/e$, the pre-sheath potential drop (assumed to be experienced by all ions) and which we are attempting to calculate, V_o = the monoenergetic ion velocity at the plasma/sheath interface. As noted earlier we can neglect the creation of new ions in the sheath since it is so thin. The electrons obey, to a good approximation,[3,4]

$$n_e(x) = n_{se} \ \exp \ [e(\Phi-\Phi_o)/kT_e] \ . \qquad\qquad (23)$$

We now insert Eqs. (22) and (23) into Poisson's Eq. (5) and integrate once to obtain the electric field in the sheath:

$$E^2(x) = \frac{2}{\varepsilon_o} n_{se} e\left\{2(\Phi_o\Phi)^{1/2} + \frac{kT_e}{e} \exp\left[e\left(\frac{\Phi-\Phi_o}{kT_e}\right)\right]\right\} + C' ,$$

$$(24)$$

where C' is the constant of integration. To find C', Bohm reasoned that $E(0) \simeq 0$ by noting that in the plasma itself we know that Φ only changes over scale lengths which are (generally) extremely large compared with the sheath thickness, i.e., in the plasma $E \simeq 0$ at least compared with typical values in the sheath where $E \sim kT_e/e\lambda_D$. Thus to obtain the physically satisfactory situation of a continuous electric field at the plasma/sheath interface we take in our sheath analysis that $E(0) = 0$. This gives

$$E^2(x) = \frac{2}{\varepsilon_o} n_{se} e\left\{-2\Phi_o[(\Phi/\Phi_o)^{1/2}-1] + \frac{kT_e}{e}\left[\exp\left(\frac{e(\Phi-\Phi_o)}{kT_e}\right)-1\right]\right\}.$$

$$(25)$$

We now consider $E(x)$ for small $\Delta\Phi \equiv \Phi-\Phi_o$ and expanding Eq. (25) find

$$E^2(x) \simeq \frac{n_{se}}{\varepsilon_o} e \left(\frac{e}{kT_e} - \frac{1}{2\Phi_o}\right)(\Delta\Phi)^2 .$$

$$(26)$$

Thus the electric field is only real provided

$$|\Phi_o| \geq \frac{kT_e}{2e} ,$$

$$(27)$$

that is the ion velocity at the plasma/sheath interface, V_o, must satisfy

$$V_o \geq (kT_e/m_i)^{1/2}$$

$$= C_s \text{ for } T_i = 0 .$$

$$(28)$$

(Aside: The small difference between the pre-sheath potential drop calculated using the plasma fluid equations, $-0.69 \, kT_e/e$ and using the sheath equations, $-0.5 \, kT_e/e$, is due to the simplifying assumption used in the latter case that the ions originated from a point source in the plasma.)

We thus find that a physically realistic, steady-state sheath solution requires that the ions enter the sheath with a speed of <u>at least</u> C_s, while from the plasma analysis, we found that no steady-state solution is possible for ion drift velocity <u>exceeding</u> C_s. Thus we may conclude that the plasma terminates and the sheath commences when the ion velocity equals the ion acoustic speed, precisely. The <u>sheath</u> is thus seen to be analogous to the <u>shock</u> which can form at supersonic velocities in conventional fluid flow.[25]

5. SIMPLE EXPRESSIONS FOR THE FLOATING POTENTIAL, PARTICLE AND HEAT FLUX DENSITIES THROUGH THE SHEATH[33]

In this section we seek to establish expressions--if possible, simple and convenient ones--for <u>nine</u> key sheath quantities, namely: Φ_f and the electron (and ion) particle (and energy) flux densities to floating (and biased) surfaces.

We seek to relate these nine quantities to the plasma density and temperatures (ion, electron), and in the case of the biased surface to the bias voltage.

The Bohm Criterion is the starting point in this undertaking since it provides a value for the ion particle flux density from a plasma to a surface (thus also the electron particle flux density if the surface is floating). We can now replace our first estimate of j^+, $1/4 \, n\bar{c}_i e$, with

$$j^+ = en_{se}C_s \, . \qquad (29)$$

If in flowing through the plasma the ions do not suffer momentum loss due to collisions with other particles, and if T_e and T_i are spatially constant, then we can also relate j^+ to conditions far from the surface,[31,32] namely,

$$j^+ \simeq 1/2 \, en_o C_s(T_e, \, T_i) \, . \qquad (30)$$

In more complex plasmas the relation between n_o and n_{se} can be quite different and temperature variations (Refs. 19, 20, 22, 34, 35) in the flow direction can also occur. In such cases detailed <u>modeling of the plasma</u>[19,20,35-37] (pre-sheath) is required in order to relate n, T_e, and T_i at the sheath edge to values far away. Such modeling is the subject of other chapters and will not be dealt with further here. Equation (29) is generally true, however, that is one can express the particle flux to the surface in terms of the local plasma density and temperature.

We are now in a position to deduce a more accurate value for the floating voltage Φ_f. We wish to allow for secondary electron emission from the surface, arising from electron impact, since this is significant even at rather low energies, $T_e \gtrsim 30$ eV (by contrast, ion-induced secondary electron emission is usually only important for ion impact energies of $\gtrsim 1$ keV). We thus have that the secondary electron current density is

$$j_{SEC}^- = \gamma_e j_{TOT}^- = \gamma_e (j_{NET}^- + j_{SEC}^-) \tag{31}$$

where

γ_e = secondary electron emission coefficient (includes both true s.e.e. and electron backscatter),

j_{TOT}^- = total electron current density striking the surface,

$j_{NET}^- \equiv j_{TOT}^- - j_{SEC}^- = (1-\gamma_e) j_{TOT}^-$.

We thus have that

$$j_{NET}^- = 1/4 \; n_{se} \; \bar{c}_e (1-\gamma_e) \; \exp \; (e\Phi_f/kT_e) \; . \tag{32}$$

For floating conditions we have that

$$j_{NET}^- = j^+ \; , \tag{33}$$

hence equating Eqs. (29) and (32) one has

$$\frac{e\Phi_f}{kT_e} = 0.5 \; \ln \; \left[\left(2\pi \; \frac{m_e}{m_i}\right)\left(1 + \frac{T_i}{T_e}\right)(1-\gamma_e)^{-2} \right] \; . \tag{34}$$

One may note that, if we put $\gamma_e = 0$, our first estimate of Φ_f, Eq. (4), is not greatly different from our more refined value, provided also $T_i \gtrsim T_e$. Equation (34) does not, of course, include the pre-sheath voltage drop and one may add ~1/2 kT_e/e to find the potential difference between the surface and the plasma far from the surface. The value of Φ_f, Eq. (34), is shown in Fig. 2. As can be seen, Φ_f is reduced by increasing

T_i/T_e or by increasing secondary electron electron emission. The often-quoted statement that "the floating voltage is about $3kT_e/e$" can thus be in error.

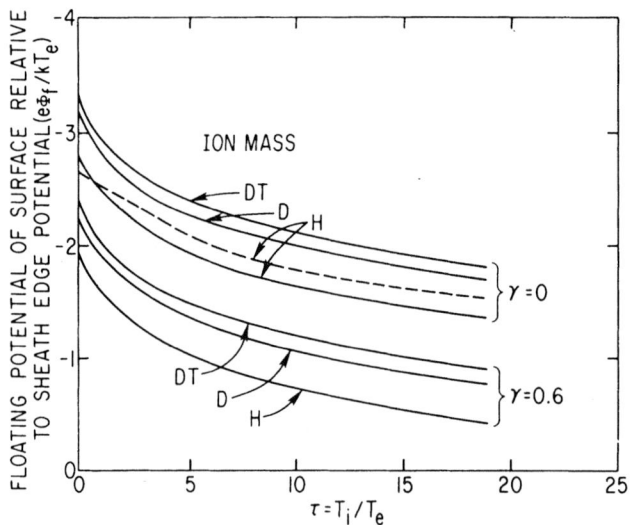

Fig. 2.   The voltage difference between a floating surface and the potential at the plasma/sheath interface (normalized), i.e., excluding the pre-sheath voltage, Eq. (34). Dotted line from Emmert et al.[86]

We consider next the energy transmission of the sheath. As indicated earlier, the electron distribution at the sheath solid interface is still Maxwellian, at least in the forward direction. We may therefore use the fact that for a Maxwellian distribution the energy flux in the x-direction is just 2 kT times the particle flux in the x-direction:

$$\iiint v_x 1/2m(v_x^2+v_y^2+v_z^2)f_{max}\,dv_x\,dv_y\,dv_z = 2kT \iiint v_x f_{max}\,dv_x\,dv_y\,dv_z$$

$$= 2kT\ (1/4\ n\bar{c}) \qquad (35)$$

If we wish to calculate the power removal rate from the plasma electron distribution, then we must note that these escaping electrons actually possessed a higher kinetic energy as they were removed from the plasma, namely, one higher by the amount $e\Phi_f$. Thus the electron power flux density <u>removed from the plasma</u> is

$$Q_e = (2kT_e - e\Phi_f) \frac{\bar{j}_{TOT}}{e} + e\Phi_f \frac{\bar{j}_{SEC}}{e} \ . \tag{36}$$

(Note that since Φ_f is -ve, the $-e\Phi_f$ term is +ve.)

The last term in Eq. (36) represents the energy reinjected <u>into</u> the plasma by secondary electrons accelerated to energy $-e\Phi_f$. We neglect the thermal energy of the secondary electrons since it is only a few electron volts. Thus we can write

$$Q_e = (\frac{2kT_e}{1-\gamma_e} - e\Phi_f) \frac{j^+}{e} \ . \tag{37}$$

Consider now the ion energy flow. To calculate this we need to know the actual ion velocity distribution at the sheath edge--for more on this see Section 8. As a first approximation we can neglect the pre-sheath in which case the relation between particle and energy flow is given by the Maxwellian distribution result:

$$Q_i = 2kT_i \frac{j^+}{e} \ . \tag{38}$$

One may add pre-sheath contributions to Q_e and Q_i, however, these contributions are comparatively small and their values depend on assumptions about pre-sheath conditions[33]; see Section 8 for further discussion on this point.

It is generally found to be useful to define the <u>sheath energy transmission factor,</u> which is the ratio of the power flux to (kT_e) × (particle flux).

Thus the electron energy transmission factor

$$\delta_e \equiv \frac{Q_e}{kT_e(j^+/e)} = \frac{2}{1-\gamma_e} - \frac{e\Phi_f}{kT_e} \tag{39}$$

and so

$$\delta_e = \frac{2}{1-\gamma_e} - 0.5 \, \ell n \left[\left(2\pi \frac{m_e}{m_i} \right) \left(1 + \frac{T_i}{T_e} \right) \left(1 - \gamma_e \right)^{-2} \right] \qquad (40)$$

while for the ions

$$\frac{Q_i}{kT_e(j^+/e)} \equiv \frac{2T_i}{T_e} . \qquad (41)$$

The total energy transmission factor $\delta = \delta_e + \delta_i$ is thus

$$\delta = \frac{2T_i}{T_e} + \frac{2}{1-\gamma_e} - 0.5 \, \ell n \left[\left(2\pi \frac{m_e}{m_i} \right) \left(1 + \frac{T_i}{T_e} \right) \left(1 - \gamma_e \right)^{-2} \right] . \qquad (42)$$

Relation (42) is shown in Fig. 3.

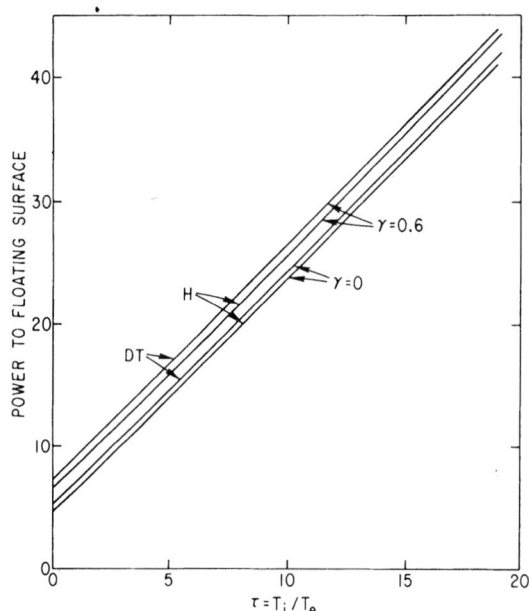

Fig. 3. Power flux density to a floating surface (normalized), Eq. (42). Note that the refinements of Eq. (50) are not included here.

Note: When $\gamma_e = 1 - [(2\pi m_e/m_i)(1 + T_i/T_e)]^{1/2}$ then $\Phi_f = 0$; for example for $T_i = 10 T_e$ and H^+ ions this occurs for $\gamma_e = 0.8$. Then the electrons reach the surface unimpeded by any sheath, so $Q_e = 1/4 \, n \bar{c}_e \, 2kT_e$ [equivalent to Eq. (39)] and the surface heating rate is extremely high.

The inclusion of the effect of secondary electron emission in the calculation of $e\Phi_f$ (thus also of δ_e and δ) warrants further discussion. In their pioneering work on this problem, Hobbs and Wesson[34] fully accounted for the existence of secondary electrons by allowing for the presence of injected secondary electrons in the plasma far from the sheath (thus reducing the density of primary electrons there below the ion density). It turns out that this has a negligible influence on Φ_f, provided $\gamma_e \neq 1$, and the Hobbs and Wesson value for Φ_f is the same as that given in Eq. (34), (taking $T_i = 0$, their assumed value). These authors also found the (electron) heat flux density through the sheath to be

$$Q_e = 1/4 \ n_o \bar{c}_e \ 2kT_e \ F(\gamma_e) \tag{43}$$

where they define

$$F(\gamma_e) \equiv \left(\frac{\pi m_e}{8m_i}\right)^{1/2} \left[\ell n \ \left(\frac{(1-\gamma_e^2)}{2\pi m_e/m_i}\right) + \frac{5-\gamma_e}{1-\gamma_e} \right] . \tag{44}$$

For $T_i = 0$ Eq. (37) gives the same value for Q_e as obtained by Hobbs and Wesson except that:

(a) These authors (incorrectly) take $j^+ = e n_o c_s$, neglecting the pre-sheath density drop; thus the n_o in Eq. (43) should read n_{se}.

(b) Hobbs and Wesson include a pre-sheath energy contribution, $1/2 \ kT_e$ in Eq. (43).

Clearly neither Eqs. (42) nor (43) can be applied for strong secondary electron emission, $\gamma_e \gtrsim 1$, since a singularity occurs at $\gamma_e = 1$. Hobbs and Wesson, in fact, showed that for $\gamma_e \gtrsim 0.8$, an electron space charge layer will occur at the surface inhibiting any further secondary emission. Thus these equations do not apply for strong secondary electron emission which can set in at $T_e \gtrsim 100$ eV.

We should, in principal, also include secondary electron emission due to ion, photon, metastable atom impact etc.[34] Usually ions will not create significant amounts of secondary electrons for fusion edge conditions since the required energy, $\gtrsim 1$ keV, normally implies intolerable levels of sputtering and surface heating. Photon fluxes are only ~1 W/cm^2 on average in the edge region (contrasted with the 10's – 1000's W/cm^2 of charged particle power flux along magnetic field lines to surfaces) and so usually one can neglect this process compared with electron impact s.e.e.; however, near probes, limiters etc. high gas levels are often present (due to the release of hydrogen initially deposited in the solid as ions), and so strong _local_ sources of radiation can exist arising from electron impact with this gas. These photons may increase s.e.e. locally to substantial levels.

It is left as an exercise to show that Eqs. (31)-(34) can be generalized to allow for these additional s.e.e. processes by simply replacing γ_e with γ where[34]

$$\gamma = (\gamma_e + \gamma_i + j)/(1 + \gamma_i + j)$$

and

$$\gamma_i \equiv \text{ion s.e.e. coefficient}$$

$$j \equiv j_{ph}/n_{se} c_s$$

$$j_{ph} \equiv \text{s.e.e. flux density due to photons, metastable atoms, etc.}$$

Now that we have dealt with the particle and energy flow to an electrically _floating_ surface, let us examine the case of an electrically _biased_ surface.[33,38,39] We will restrict our attention to the case of surfaces which are still biased negatively with respect to the plasma. As we shall see, as the surface potential approaches the plasma potential, the heat flux can reach extremely high values.

So long as the surface is negative with respect to the plasma, a sheath still exists, although its thickness varies. We thus have that the ion current is still given by $j^+ = en_{se}c_s$. The net electron current is

$$j_{NET}^- = 1/4\, n_{se}\bar{c}_e\, e\, (1-\gamma)\, \exp\, (e\Phi/kT_e), \tag{45}$$

where Φ is the (negative) potential applied to the surface relative to plasma potential. The net total current density can therefore be written

$$\frac{j_{TOT}(\Phi)}{1/4\, en_{se}\bar{c}_e} = \left[\left(1+\frac{T_i}{T_e}\right)\left(\frac{2\pi m_e}{m_i}\right)\right]^{1/2} - (1-\gamma)\, \exp\, \left(\frac{e\Phi}{kT_e}\right). \tag{46}$$

Relation (46) is shown in Fig. 4. The net total current reaches the "saturation ion current" j_{SAT}^+, which is just the j^+ of Eq. (29), at sufficiently negative potentials. Increasing the potential above the floating potential causes an exponential increase in the electron current. In the simplest model, the electron current attains the saturation value $j_{SAT}^- = 1/4 \, n_{se} \bar{c}_e e$ at $\Phi = 0$ and this then remains constant for all $\Phi > 0$. Often a true saturation electron current is not observed but the current continues to rise with Φ; however, the increase is slower than exponential. The solid-plasma interaction is difficult to model for $\Phi > 0$ since no sheath is present and the applied electric field penetrates far into the plasma causing a significant perturbation. In some circumstances, for example, operation in a strong magnetic field parallel to the direction of plasma collection, there can be a reduction[40] of the maximum electron current from the value of $1/4 \, n_{se} \bar{c}_e e$. (See Section 7.4.)

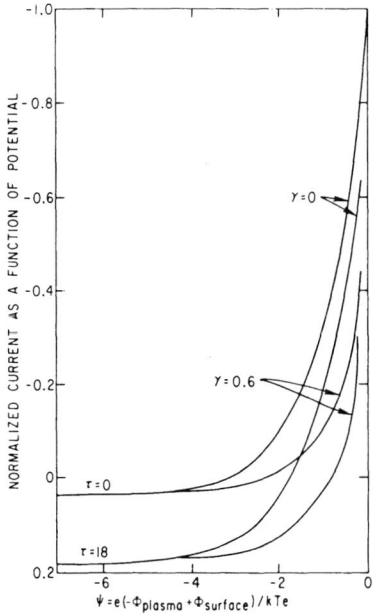

Fig. 4. The normalized current as a function of the applied potential to a surface, i.e., the Langmuir J-Φ characteristic, Eq. (46).

Next consider the power flux density to an electrically biased surface. The ion power is

$$Q_i(\Phi) = j^+(2kT_i - e\Phi) \tag{47}$$

while for the electrons

$$Q_e(\Phi) = 2kT_e \, j^-_{TOT} \,. \tag{48}$$

Note that we use j^-_{TOT}, in the last expression rather than j^-_{NET}, since every electron striking the surface deposits, on average, $2kT_e$ there. Again, we neglect the thermal energy of the emitted secondaries. Thus we may write the total power flux density in the form of a transmission coefficient

$$\frac{Q(\Phi)}{kT_e(j^+/e)} \equiv \delta(\Phi) = -\frac{e\Phi}{kT_e} + \frac{2T_i}{T_e} + 2\left[\left(1 + \frac{T_i}{T_e}\right)\left(\frac{2\pi m_e}{m_i}\right)\right]^{-1/2} \exp\left(\frac{e\Phi}{kT_e}\right). \tag{49}$$

Note that γ does not appear in this last expression. Relation (49) is shown in Fig. 5.

Note that the heat flux is nearly a minimum at the floating potential.[38] At potentials below floating, the power flux increases slightly due to the increasing impact energy of the ions. For potentials above floating, the electron particle, and thus energy, flux increases exponentially fast. Operation at potentials near or above the plasma potential is often hazardous due to this strong heating.

The foregoing analysis applies to the case of no magnetic field. Provided ion motion is along B-field lines, this theory should also apply when B is finite. For ion motion oblique to magnetic field lines, as occurs when the plane of the collecting surface is not parallel to the B-field, further analysis[41-43] is required. This has been carried by Chodura[42,43] and will be discussed in the next lecture.

Further refinements that can be added to the δ-values already calculated include:

(a)   Ion backscatter.   The impacting ions tend to be

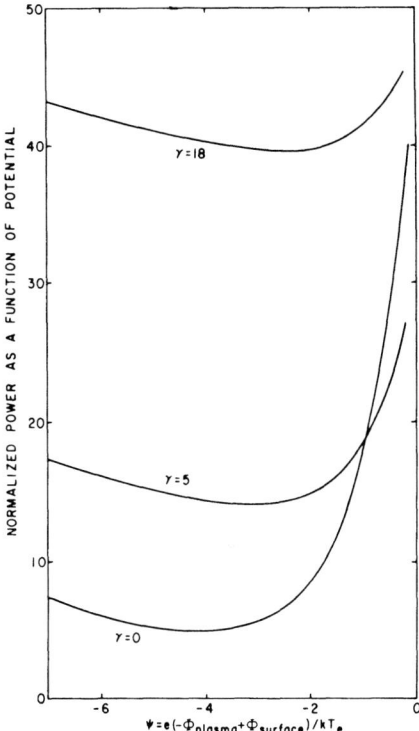

Fig. 5. The normalized power flux density as a function of the potential applied to a surface, Eq. (49). $\tau = T_i/T_e$. Note that refinements are neglected here: $R_{iE} = R_{iN} = R_{eN} = \chi_i = \chi_r = 0$.

reflected from the surface (usually as neutrals), depositing only a fraction R_{iE} of their impact energy there. Values of R_{iE} are dependent on the energy, mass, and incidence angle of the impacting ion and the material of the solid; values of R_{iE} are given in various references.[44-46]

(b) Ionization. Recombination energy χ_i should be added.

(c) Atom. Atom recombination energy χ_r should be added (if molecular formation ensues). Here also an ion <u>particle</u> reflection coefficient R_{iN}, should be allowed for.

63

(d)   Electron backscatter.[47]   For purposes of computing *particle* fluxes (hence floating potential) it is not necessary to distinguish between electron-induced secondary electron emission and electron backscatter.   For purposes of calculating heat flux and δ_e, however, the distinction can be worth making since a secondary electron only removes a few eV from the solid, while the backscattered electron can return a sizeable fraction of $2kT_e$ to the plasma.

(e)   Pre-sheath   contributions.   As   indicated   above, pre-sheath contributions can also be added.   In the simplest case this introduces $1/2\ kT_e$ to δ_e, but for more   complex   pre-sheaths,   other   values   may   be appropriate, see Section 8.

Including these latter refinements one may rewrite Eq. (42) as

$$\delta kT_e = \left[2kT_i - e\Phi_f\left(\frac{m_e}{m_i}, \frac{T_i}{T_e}, \gamma\right)\right](1-R_{iE})$$

$$+ \frac{2kT_e}{1-\gamma}\ (1-R_{eE})$$

$$+ \ \varepsilon_{pre.sh.}$$

$$+ \ \chi_i$$

$$+ \ \chi_r(1-R_{iN})\ . \tag{50}$$

Example: H^+ on W, T_i = 20 eV, T_e = 10 eV.   Then χ_i = 13.6 eV,

χ_r = 2.2 eV, $\gamma \cong$ 0.3, $R_{iN} \cong$ 0.5, $R_{iE} \cong$ 0.3,

$R_{eE} \cong$ 0.15, $e\Phi_f$ = -2.1 kT_e, and thus

δkT_e = 42.7 + 24.3 + (~5) + 13.5 + 1.1,

hence $\delta \simeq$ 8.7.

Ideally one should employ values of R_{iE}, R_{eE}, R_{iN}, and γ which are appropriate for the actual incident angle of the particles. In practice, surface roughness implies that one should probably use average values. The effect on heat flux and δ of terms which are independent of T_e and T_i, such as χ_i, is only significant when T_e and T_i are rather small, i.e., generally smaller than χ_i. For many edge plasma conditions, e.g., the foregoing example, the T-independent contributions to δ are not great and can be ignored, to first order. In some divertor devices, however, very low temperature (and high density) plasmas have been achieved near the surface, with $T < 5$ eV. In such cases one should include further T-independent contributions to δ, for example, the energy removed from the solid for each secondary electron emitted, a few eV. For a non-floating surface the solid will also again (lose) the electron work function energy corresponding to the net gain (loss) of an electron from the plasma. For the case of high density edge plasmas, neutrals emitted from the solid into the plasma will be dissociated (into Frank-Condon atoms of a few eV energy), electronically and vibrationally excited and ionized all quite close to the surface. One should then include in the calculation of heat flux to the surface contributions due to Frank-Condon atoms, photons, excited neutrals etc., impacting on the surface. Clearly the calculation of heat flux to a surface for high density, low temperature edge plasmas requires special treatment, and this will not be dealt with further here.

6. APPLICATIONS: EDGE MODELING

For many magnetic confinement configurations one can consider the plasma as consisting of two regions the core plasma and the edge or scrape-off plasma, see Fig. 6. The behavior of the scrape-off layer, SOL, is strongly influenced by the presence of solid surfaces inserted into this plasma, since plasma motion along magnetic field lines to these surfaces is scarcely impeded. In the simplest magnetic topology these inserted solids are called limiters, while if the magnetic field lines near the edge are distorted and drawn away from the main vessel in the divertor arrangement, then the solid surfaces are called divertor target plates. In either case we can model the SOL as unimpeded, one-dimensional flow to a solid surface and thus use the results we have obtained in Section 5. This topic of plasma edge modeling[14-22,35-38] is dealt with more extensively in other lectures and here we merely wish to gain some basic insight into the influence of the sheath on SOL properties. We will consider a few basic results.

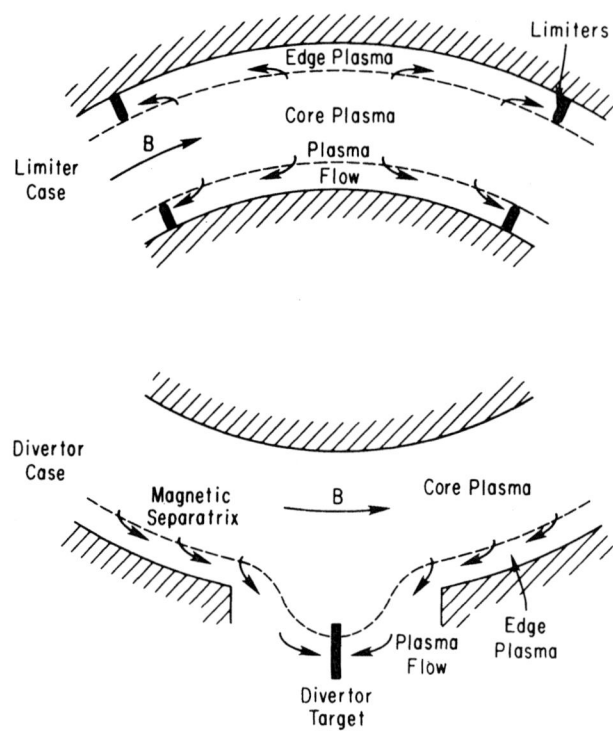

Fig. 6.   Schematic indicating the difference between limiter- and divertor-edge plasmas.

6.1   The Particle Flux Scrape-Off Width

If there is no local ionization of neutral particles in the SOL, then plasma is entirely supplied to the SOL by cross-field diffusion from the core plasma. An important question is: How far does the plasma extend out beyond the leading edge of the limiter? That is, over what radial extent of the limiter surface are the plasma particles deposited?

Consider Fig. 7 where we define the distance between two facing limiters to be the connection length, 2L. We may write the cross-field particle flux into the SOL as

$$D_\perp \frac{dn}{dr} \, LW, \tag{51}$$

66

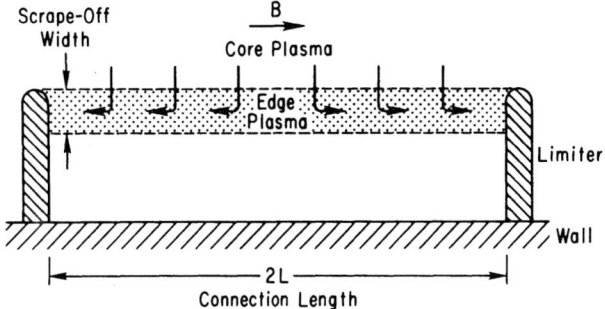

Fig. 7. Schematic of the plasma flow from the core plasma into the scrape-off edge plasma by cross-field diffusion and the rapid parallel-field flow of the plasma to the scrape-off surfaces (limiters in this case). In a fusion device, the magnetic flux tube between limiting surfaces is rarely straight as shown here but twists, e.g., around a torus. The connection length is the distance between surfaces measured <u>along</u> the curved flux tube.

which must equal the plasma flux reaching the limiter surface

$$1/2 \; nC_s \; \lambda_\Gamma \; W \; , \tag{52}$$

where

dn/dr = radially outward density gradient driving the cross-field diffusion,

$D_\perp$ = cross-field diffusion coefficient,

W = (poloidal) width of limiter,

λ_Γ = (radial) particle flux <u>scrape-off length</u> of SOL.

We now approximate $dn/dr \simeq n/\lambda_\Gamma$ and thus obtain the simple estimate that

$$\lambda_\Gamma = (2D_\perp \; L/C_s)^{1/2} \; . \tag{53}$$

Typically[20] $D_\perp \simeq 1$ m^2/s, L $\simeq$ 1-10 m, $C_s = 10^4$-10^5 m/s hence $\lambda_\Gamma = 0.5$-5 cm. Thus the plasma is removed in a surprisingly <u>thin layer</u> resulting in rather high particle flux densities to the leading edge of the limiters. The SOL is thin because of the very high ion velocities along the magnetic field ($\sim C_s = 10^4$-10^5 m/s) compared with the effective cross-field drift velocity ($\sim D_\perp/\lambda_\Gamma \simeq 10^2$ m/s). Example of particle flux densities at the leading edge of the limiter: Suppose in the core plasma of V = volume = 10^2 m^3, $\bar{n}_e = 10^{20}$ m^{-3} and particle confinement time τ_p = 1 sec, thus plasma outflow to limiter(s) = $\bar{n}_e V/\tau_p = 10^{22}$ ion pairs/sec. If there is a single, poloidal limiter of circumference 10 m (both sides) and if $\lambda_\Gamma = 10^{-2}$ m, then the plasma flux at the leading edge of the limiter is 10^{23} ion pairs/m^2/s $\simeq$ 1 amp/cm^2. If the particle flux falls off radially as exp $(-r/\lambda_\Gamma)$ then provided the wall is located just a few scrape-off lengths behind the leading edge of the limiter, it will receive negligible plasma flux. Thus the limiter acts to <u>limit</u> the effective radius of the plasma to approximately its own radius, <u>protecting</u> the wall from plasma exposure.

6.2 The Energy Scrape-Off Width

We consider the simple case where cross-field heat convection can be neglected compared with the cross-field heat conduction into the SOL which is at the rate

$$n\chi_\perp \frac{d(kT)}{dr} \ LW \ . \tag{54}$$

in the absence of energy losses to the side walls, this must equal the energy flow through the sheath to the limiter at rate

$$1/2 \ nC_s \ \delta \ kTW \ \lambda_Q \ , \tag{55}$$

where $n\chi_\perp$ = cross-field heat conductivity, λ_Q = energy scrape-off width, and we take the simple case of $T_e = T_i = T$. If we now approximate the radial temperature gradient as d(kT)/dr $\simeq$ kT/λ_Q then we obtain the estimate

$$\lambda_Q = (2\chi_\perp L/\delta C_s)^{1/2} \ . \tag{56}$$

Generally, it is found that $\chi_\perp$ is about equal to or somewhat larger than $D_\perp$,[20] while as we have seen δ is of order 10 or larger. [One should note that in magnetic confinement plasma physics it has not as yet been possible to calculate $D_\perp$ or $\chi_\perp$ from basic principles, although a large portion of the plasma theoretical effort expended over the past four decades has focused on this difficult problem. Instead measured values are used for these quantities. In the edge region, the empirical finding of Bohm[40] is often applicable $D_\perp = D_B = 6.25 \times 10^{-2}$ T (ev) B^{-1} (Tesla) $m^2/s \cong 1 \ m^2/s$, while $\chi_\perp \cong (1-10) \ D_\perp$]. Thus we see the result that $\lambda_Q \leq \lambda_\Gamma$, a fact which is generally observed in SOL's.[38,48-50] This same inequality can be obtained by a second approach: Suppose the plasma density in the SOL varies radially as exp $(-r/\lambda_n)$ and the plasma temperature as exp $(-r/\lambda_T)$. Then since the particle flux varies as $n(r)C_s[T(r)] \propto n(r) \ T^{1/2}(r)$ we find that $\lambda_\Gamma = (1/\lambda_n + 0.5/\lambda_T)^{-1}$. The energy flux varies as $n(r)C_s(r)kT(r)$. Hence $\lambda_Q = (1/\lambda_n + 1.5/\lambda_T)^{-1}$, again implying $\lambda_Q \leq \lambda_\Gamma$. Note the general ordering: $\lambda_Q \leq \lambda_\Gamma \leq \lambda_n, \ \lambda_T$.

Since $\lambda_Q \leq \lambda_\Gamma$, the concentration of heat deposition at the leading edge of the limiter is perhaps even more troublesome than the concentration of particle flux, and observed heat fluxes[49,51] can reach the level of kilowatts per cm^2--near or exceeding engineering limits.[52,53]

6.3 Ionization-Sustained Plasma

We now consider just the opposite case to that of Section 6.1 and hypothesize that there is no cross-field diffusion of particles to supply the SOL with plasma, but instead plasma is entirely supplied within the SOL by the ionization of neutrals resulting from their impact with SOL plasma electrons. There now need be no radial gradient of n to "draw in" plasma from the core plasma, i.e., the SOL becomes thicker [its width is now limited by the radial decay of $T(r)$: when T_e is too low to ionize neutrals, the plasma terminates]. This tendency is a general one, namely, any ionization within the SOL makes it thicker, i.e., more extended radially.[18,21]

Let us simplify the radial variation of T for this case to be a constant out to $r = \lambda_T$, then zero. In this case particle balance gives

$$n_e n_n \overline{\sigma v_i} LW\lambda_T = 1/2 \ n_e C_s W\lambda_T , \qquad (57)$$

where n_n = neutral density and $\overline{\sigma v_i (T_e)}$ is the electron ionization rate, a known function of T_e for any neutral species.[53] Rewriting we find

69

$$\frac{C_s}{2\sigma v_i} = n_n L \ . \tag{58}$$

Note that the left-hand side of Eq. (58) is simply a function of T (we assume $T_e = T_i$ for simplicity). Hence the important result for an ionization-sustained plasma: $\underline{T_e}$ is not an independent variable, but is a function of the $\overline{\text{product } n_n L}$, i.e., $T_e(n_n L)$ (Ref. 54). The example for hydrogen is shown in Fig. 8. This situation is one encountered in low pressure, partially ionized electric discharges (B = 0) such as fluorescent lights, etc.: as one lowers the gas density one drives up the electron temperature.[54] Such ionization-sustained plasmas can also occur in magnetic confinement device SOL's.[55]

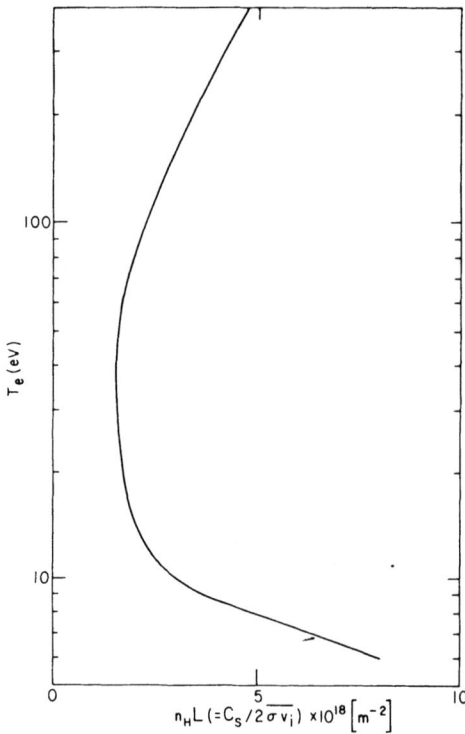

Fig. 8.   The electron temperature required to sustain a plasma dominated by local ionization and containing hydrogen atoms of uniform density n_H. The plasma volume can be considered to be a cylinder of length L, its axis aligned parallel to a magnetic field with particle loss out one end of the cylinder only, Eq. (58).

70

For example, the ASDEX divertor tokamak can be operated with little net particle flow from the core into the SOL and then particle loss through the divertor sheath is balanced by SOL ionization [although in this case ionization occurs only over length L_D, the length of the divertor chamber, where the neutrals are semi-trapped and n_n is high enough to satisfy Eq. (58)]. The core plasma must, of course, still supply (by conduction) the energy lost through the sheath, plus ionization and other inelastic losses in the SOL.

In Fig. 9 we have applied this Eq. (58) to ASDEX divertor conditions and compared the results from this simple model to the observed results. In the case of H_2 gas entering a dense plasma the neutrals are not wall-temperature molecules, but are Frank-Condon atoms of average energy $\varepsilon \approx 3$ eV (as the molecules enter the plasma they are impact-dissociated by the plasma electrons, giving them this dissociation energy). Thus n_{H_2} and n_H [$= n_n$ for Eq. (58)] are related by $n_H/n_{H_2} \approx (kT_{H_2}/\varepsilon)^{1/2}$.

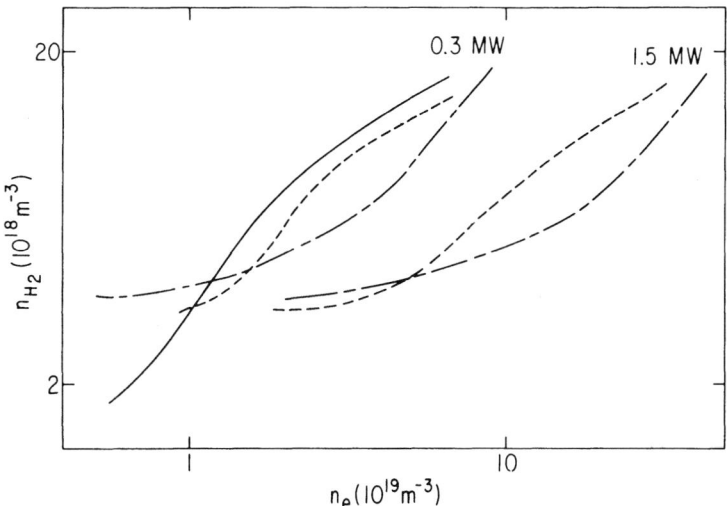

Fig. 9. Solid line is experimental result from ASDEX divertor experiment[55] relating the H_2 density in the divertor chamber to the plasma density in the plasma flux tube in the divertor chamber. Experimental data for ohmic heating only (0.3 MW). Power refers to heat flow along scrape-off tube into divertor chamber plasma. - - - - - - Theoretical predictions from a computer code (SOLID)[55]. —— - —— - Predictions from simple model of section 6.3 (using L = 5 m and volume of divertor plasma = 0.125 m^3).

6.4 The Average Density and Temperature of the SOL

Assuming no inelastic processes in the SOL then the sheath transmission of particles and energy establishes the average plasma density and temperatures of the SOL. Suppose that the particle loss from the core plasma into the SOL is known $\Gamma_c = (\bar{n}_{core} \, Vol/\tau_p)$ and <u>also the</u> energy flow from the core plasma into the core $P_c = (3nkT_{core} \, Vol/\tau_E)$. Then we have

$$\Gamma_c = 1/2 \; nC_s \, \lambda_\Gamma \, W \; , \qquad\qquad (59)$$

$$P_c = 1/2 \; n \, C_s \, \delta \, kTW\lambda_Q \; . \qquad\qquad (60)$$

Since we already have values for λ_Γ and λ_Q, these two equations specify the average n and T in the SOL.

We may also observe the tendency of the sheath to cool the SOL electrons more than the ions[6,22]: Ignoring heat conduction into the SOL, the ion heat convection is $2kT_i^{in} \, \Gamma$ where Γ is the flux of ions and electrons from the core plasma, through the SOL to the limiter and T_i^{in} is the temperature of ions as they enter the SOL from the core. A similar relation applies to the electrons. As noted in Section 5 the ion energy removal rate at the sheath is $2k\overline{T}_i^{out} \, \Gamma$ where T_i^{out} is the average ion temperature at the limiter. The electron removal rate is $\delta_e k\overline{T}_e^{out} \, \Gamma$ with $\delta_e \gtrsim 5$. Thus, provided there are no electron-ion collisions in the SOL and no other energy loss/gain processes there, we find that the ions are not cooled in passing through the SOL, $T_i^{in} \simeq T_i^{out}$ while the electrons are cooled, $\overline{T}_e^{out} \simeq (2/\delta_e) \, T_e^{in}$. [In order to achieve this low <u>average</u> $\overline{T}_e^{out}$, $T_e(r)$ must decay radially more quickly that $T_i(r)$ and in fact such differences in radial temperature profiles have been observed.[48,56]] Note, however, that electron heat <u>conduction</u> reduces this effect, and of course collisional processes will also strongly modify this effect.

Ions backscattering as fast neutral atoms from the solid also have a tendency to keep the SOL ions hot: these neutrals have energies of $R_{iE}(2kT_i-e\Phi_f)$ and so if they are ionized or charge-exchange in the SOL, they create ions of energy greater than kT_i, typically, thus heating the SOL ions. Of course, the SOL plasma is often opaque to such fast neutrals, i.e., they pass through too quickly to be ionized.

6.5 Tendency of Sheath to Dominate Energy Balance in the SOL

Turning then to the matter of inelastic collisional energy losses/gains in the SOL, we will note that the sheath energy loss rate is so great that it can often dominate energy balance in the SOL even when inelastic events are frequent.[22]

Consider the extreme case examined in Section 6.3 where <u>all</u> ions are supplied to the SOL by (inelastic) electron impact ionization of neutral gas. For each ion pair produced the sheath will transmit the following energy to the limiter: $\sim 2kT_i + \delta_e kT_e + \chi_i + \chi_r$, neglecting reflection. The only other energy sink will be the side walls which will receive the radiation associated with the electron excitation of neutrals plus the kinetic energy of fast neutrals exiting from the SOL.

Considering these latter two quantities:

(a) For SOL temperatures $\gtrsim 10$ eV, the radiation energy generated per ion pair created is ~ 30 eV.[20,57]

(b) For each ion pair produced from neutral atoms there tends, on average, to be of order one charge exchange event[53] where a plasma ion at temperature T_i is converted into a neutral at T_i and this energy is lost to the side wall.

Thus, for example, if $T_e = T_i = 20$ eV then for each ion pair produced the sheath transmits ~ 150 eV to the limiter while the walls receive only ~ 50 eV. Thus we see that even in a SOL where <u>particle</u> balance is <u>totally</u> dominated by inelastic processes, the sheath still dominates <u>energy</u> balance. For the more general case where particles are also received from the core plasma the sheath domination of SOL energy balance will, of course, be still greater. In modeling the SOL one can thus usually neglect the effect of inelastic processes on energy balance provided $T \geq 10$ eV, a regime very commonly encountered.

When the SOL electron temperature drops below 10 eV, the ratio of electronic excitation rates to ionization rates increases dramatically, and the SOL energy balance can then be <u>dominated by radiation</u> loss to the walls. Such SOL's have recently been observed[58,59] and their attainment is clearly most desirable since this arrangement allows the power from the core into the SOL to be deposited over the large wall area, rather than channeled into the small area at the limiter tip.

We thus note that there is a <u>temperature threshold</u> effect for changing the limiter-vs-wall energy balance in the SOL, an important factor effecting the problem of <u>limiter heat removal</u>. The limiter <u>sputtering</u> problem is also subject to a similar and linked threshold effect. If one adds in extra energy-carrying ion pairs to the SOL by increasing ionization there, then the average particle temperature is reduced proportionately as the number of carriers is increased, (assuming fixed power from the core plasma into the SOL). Unfortunately, sputter-yields tend to fall off approximately linearly with decreasing ion impact energy, until a sputtering threshold energy is reached.[53] Thus unless one achieves a sufficient increase in the number of carriers to drop T to this sputtering threshold and/or the radiation threshold, one still ends up with the same number of sputtered particles from the limiter (for a given power from the core plasma).

6.6 Unipolar Arcs

One of the sources of contamination in plasmas arises from the formation of arcs on metallic surfaces. Although arcs have been studied for over a century and their gross characteristics are well known,[60] the basic explanation for this unusual phenomenon is still not entirely in hand. Experimentally, it is known that when a dc voltage is applied between two metal plates, ~1 cm apart, and with some gas present, a glow discharge is established characterized by anode-cathode voltages of 100's to 1000's of volts and currents of $\sim 10^{-12}$ to $\sim 10^{-1}$ amps. The electron emission process is simply secondary electron emission caused by ion, photon and metastable atom impact on the cathode. The entire cathode tends to emit and so the current density is low. The plates are typically cold. The plasma ions are primarily from the fill gas plus a small amount of sputtered plate material.

Now if the external circuit permits a current increase to $\gtrsim 1$ amp (total current) then a surprising thing happens: suddenly the voltage required to sustain the discharge drops drastically, to ~10 V. The cathode emitting area collapses to virtually a point, with current densities of 10^4-10^6 A/cm^2. While this effect has been known experimentally for over a century and has been widely studied, the basic processes are still not well understood. Evidently the arc spot involves all four states of matter in close proximity--solid, liquid, gas, plasma. Large quantities of the cathode material are eroded, typically 0.1-1 atoms per emitted electron; thus, for example, one amp-hour of arcing ejects gram quantities of solid (indeed the arc can operate with plasma created from the ejected material alone, no gas fill being required).

Since dc voltages are not generally applied between metallic components of fusion devices one might hope that such an undesirable process would not occur here. Unfortunately, this proves not to be the case[53] and arc tracks are widely observed in fusion devices and are associated with plasma contamination, particularly at the start of discharges. How can this happen? The answer is that unipolar arcs are involved.

The explanation for unipolar arcs was first given by Robson and Thonemann[61] in 1959 and is quite simple. For convenience we insert the values $T_i = \gamma = 0$ into Eq. (34) for the floating potential and assume D^+ ions; thus $\Phi_f = -3.2\ kT_e/e$. Now, since an arc only requires $\Phi_c \approx 10$ volts to operate, even a quite cool plasma will cause a solid surface to float sufficiently negatively to sustain an arc. Of course, the current circuit must be completed, but this is readily achieved if $|\Phi_f| > |\Phi_c|$ and the surface is electrically conducting, e.g., a metal. The arc circuit can be completed if the metal potential, Φ_m, decreases to $\Phi_m = \Phi_c$, thus the electron flow from the plasma to the surface is increased, Eq. (45), (i.e., the surface repels electrons less effectively), thereby allowing a larger current to be emitted from the arc spot into the plasma. Provided Φ_m remains negative, the ion current density is constant at j_{SAT}^+. When $\Phi_m = \Phi_f$, then the total currents from the plasma to the metal are (no arc spot yet):

$$J^- = J_{SAT}^+ = j_{SAT}^+ A = 1/4\ n\bar{c}_e\ e\ A\ \exp\ (e\Phi_f/kT_e)$$

where A = total area of metal. When $\Phi_m = \Phi_c$ then J_{SAT}^+ is unchanged but now

$$J^- = 1/4\ n\bar{c}_e\ eA\ \exp\ (e\Phi_c/kT_e),$$

($\Phi_c < 0$) hence the total, net current to the metal is

$$J_{NET} = 1/4\ n\bar{c}_e\ eA\ [\exp\ (e\Phi_f/kT_e) - \exp\ (e\Phi_c/kT_e)]$$

This, of course, is also the total electron current emitted from the arc spot. Now if J_{NET} exceeds the minimum current required to sustain an arc, ~ 1 A, the unipolar arc can occur. Example: $T_e = 10$ eV, $\Phi_c = 10$ V, $A = 1$ cm^2, $n = 10^{12}$ cm^{-3} then $J_{NET} \sim 2.5$ A. Clearly, most edge plasmas are capable of supporting unipolar arcs. The ability to sustain an arc can be reduced by making all metal components of small area and keeping them mutually isolated, electrically. Some trigger mechanism appears to be necessary to precipitate the collapse of a normal sheath into a unipolar arc sheath, and contaminants, surface protrusions etc. appear to play a role. The theory of the unipolar arc continues to be developed.[62]

7. APPLICATIONS: PROBE THEORY

Experimentally one of the simplest plasma diagnostic techniques consists of inserting a small object--a probe--into the plasma and measuring the current and heat flux to the probe as a function of the voltage applied between it and the plasma. It has often been observed, however, that the price paid for the experimental attractiveness of probes is that the data interpretation can be difficult. It is also evident that the probe disturbs the plasma, and this must be taken into account. One of the most obvious examples of plasma disturbance is that the ions are caused to flow toward the probe (by the pre-sheath field) whereas in the probe's absence this flow would not occur. We take this particular disturbance into account, of course, when we write the ion flux to the probe as $j^+ = 1/2 \, n_0 C_s e$ (the velocity C_s results from the pre-sheath field caused by the probe and the factor of 1/2 indicates the local depression in plasma density caused by insertion of the probe). We will consider other disturbance effects, Section 7.4.

As with plasma flow to limiters, we will assume that when a probe is inserted into a magnetized plasma, that unimpeded plasma motion to the probe occurs along the field lines, see Fig. 10. We assume that plasma is supplied to the probe magnetic flux tube by cross-field flow from the rest of the (relatively) undisturbed plasma.[63] We assume that the direct cross-field flow of plasma onto the probe collector can be neglected compared with the parallel flow to the collector. Ideally, the probe face should be perpendicular to the B-field, in which case we can directly use our results of Section 5. If the probe face is oblique to the B-field or multifaceted then the results of Section 5 will still be valid to first order, since it has been shown that the Bohm Criterion is still valid in such cases,[41-43] and so, for example, Φ_f is not a strong function of the angle between the surface and the B-field, see Chapter by Chodura. One must, however, take the probe's effective collection area to be the projected area, $A_\perp$, of the surface on the plane perpendicular to $\vec{B}$. Also, if the angle between the surface and the magnetic field is shallow, secondary electrons cannot escape, owing to their small Larmor radius; hence $\gamma \simeq 0$.

In order to apply the results of Section 5 to the analysis of probes in a magnetic field one further condition must be met, namely that $A_\perp^{1/2} \gg r_{i,e}$, the ion (electron) Larmor radius. Generally $r_i \gg r_e$ of course, and for hydrogen we have

$$r_i \simeq 10^{-2} \, T_i^{1/2} \, (\text{eV})/B \, (\text{T}) \, \text{cm} . \tag{61}$$

Example: T_i = 100 eV, B = 1 T, r_i = 1 mm. It is worth noting that the probe detection element must, in some circumstances, be rather large to avoid finite ion Larmor radius effects.[64] When the latter occur, i.e., $r_i \gtrsim A_\perp^{1/2}$, ion collection by the probe is fairly difficult to model and depends on the actual collector

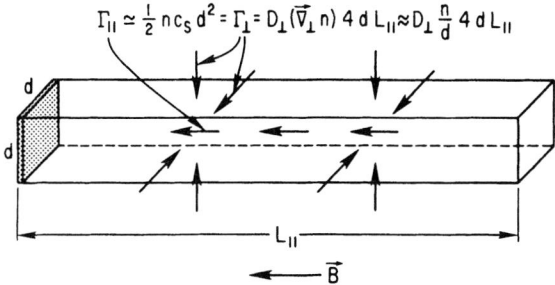

Fig. 10.   Schematic of particle flows induced by insertion of a solid surface into a plasma with a magnetic field. Since parallel-field motion is rapid, initially a long flux tube is almost evacuated of plasma, setting up in the steady-state a cross-field density gradient of order n/d which supplies plasma to the flux tube. A natural "collection" or "disturbance" length $L_\parallel$ is thus defined.

shape. Paradoxically our simple analysis of Section 5 applies to the extremes of B = 0 or B very large. For tokamak probe applications we will then be assuming the applicability of the strong magnetic field approximation.

7.1 Measuring T_e

A probe which measures collected current as a function of its voltage is called a Langmuir Probe.[65-67] From Section 5 we

77

have that the total current density to the probe as a function of its voltage is

$$j_{TOT} = en_{se}C_s - 1/4\ n_{se}\bar{c}_e\ (1-\gamma)\ \exp\ (e\Phi/kT_e).\qquad(62)$$

See Fig. 4. The first term is called the <u>ion saturation current</u> j_{SAT}^+, and is readily identified from examining the measured <u>J-Φ</u> <u>characteristic</u> since $j_{TOT} \rightarrow j_{SAT}^+$ for large negative Φ.

We may therefore plot experimental values for $\ln$ $(j_{TOT} - j_{SAT}^+)$ vs Φ and from the slope of this plot deduce T_e.

Note that the Φ in Eq. (62) is the voltage of the probe relative to the local plasma. This latter potential is not, however, directly accessible to us and the voltage applied to the probe is in fact between the probe and some reference such as the limiters. In Section 7.5 this question of the reference point for a single probe is examined in more detail and it is concluded that usually--but not always--the $\ln\ (j_{TOT} - j_{SAT}^+)$ vs Φ_{app} plot gives T_e correctly. One can also avoid reference problems by using <u>double</u> or <u>triple probes,</u> see Chapter on edge probes.

7.2 Measuring Plasma Density

As noted in Section 7.1, the ion saturation, $j_{SAT}^+ = en_{se}C_s$ $(= 1/2\ en_oC_s$ for the collisionless pre-sheath) is almost always well defined and is easily measured from probe J-Φ characteristics. If T_e has already been measured, see Section 7.1, and <u>if</u> $T_e = T_i$, or one otherwise knows T_i,[38,39,56,68-74] then one has a value for C_s. Hence, from j_{SAT}^+ the local plasma density, n_{se}, is inferred and then depending on the pre-sheath conditions assumed, the undisturbed density n_o can be found. We will show in Section 7.4 that for a probe the ion flow is often collisionless, and so one can usually take $n_{se} \simeq 1/2\ n_o$.

The principal difficulty to be noted, however, is that we have assumed a knowledge of T_i. Unfortunately, there is no easy way to infer T_i from the J-Φ characteristic. One might try to measure the electron saturation current, $j_{SAT}^- = 1/4\ n_o\bar{c}_e e$, and then (since T_e is available) one could deduce n_o without requiring a knowledge of T_i. In practice, however, j_{SAT}^- is often either (a) not clearly discernible from the J-Φ characteristic,[64] (b) impractically large from the viewpoint of probe over-heating or the available probe power supply, (c) not in fact equal to $1/4\ n_o\bar{c}_e e$. The latter effect can occur for probes used in magnetic fields,[39,40] as we are concerned with, and is discussed further in Section 7.4.

In the absence of any information on T_i, it is common practice to simply assume $T_i = T_e$ and use j_{SAT}^+ to infer n_o. Fortunately T_i only enters the calculation via a square root and the error in n_o is unlikely to be more than a factor of 3 at worst.

7.3 Measuring T_i

Various probe techniques have been employed to measure T_i,[38,39,56,68-74] see Chapter on edge probes. Here we briefly consider the bolometer[38,39,45,49,51,56] (which can be operated simultaneously as a Langmuir probe). From Section 5 we have that the heat flux density to the bolometer when floating is given by $Q = \delta k T_e \, j_{SAT}^+/e$. If we know j_{SAT}^+ and T_e from the Langmuir probe characteristic then the bolometer gives δ, hence from Eq. (42), T_i is deduced. (One can then use this value of T_i to deduce n_o from j_{SAT}^+.[39])

This analysis is based on the assumption of Maxwellian ions and electrons in the plasma. When fast ion or electron components are present, e.g., from neutral beam injectors, electron runaway processes etc., then some of the heat flux registered by the bolometer must be assigned to these components.[38,75] The use of bolometer-Langmuir probes in such situations is discussed in the Chapter on edge probes.

Note also that if the bolometer is not floating but is, for example, swept in voltage (as a Langmuir probe) then the heat flux is increased, see Fig. 5, and an appropriate correction is required in deducing T_i.[39]

7.4 Probe Collection/Disturbance Lengths[76,77]

An absorbing probe depresses the plasma density relative to that far away. In a non-magnetic plasma this density depression is more or less spherically symmetrical,[67] while in a magnetized plasma the depression tends to be localized along the magnetic flux tube subtended by the probe.[39,40,63] Thus a transverse density gradient is established between this flux tube and adjacent, relatively undisturbed flux tubes which causes plasma to cross-field diffuse into the probe's flux tube. This is indeed the source of ions for the probe, see Fig. 10. How far does this disturbance extend from the probe? It extends to infinity but, obviously at decreasing magnitude. We may approximate this situation by defining $L_{\parallel i}$ as the length over which the disturbance is "significant", i.e., over this length the cross-field density gradient is of order n/d, where d is the size of an assumed square probe of area $A_\perp = d^2$. Thus particle balance, see Fig. 10 gives

$$D_\perp \frac{n}{d} L_{\|i} \, 4d = 1/2 \, nC_s d^2 \tag{63}$$

thus

$$L_{\|i} = \frac{C_s d^2}{8 D_\perp} . \tag{64}$$

$L_{\|i}$ is thus also the <u>ion collection length</u> of the probe. This length can be surprisingly great; for example, if $d = 5$ cm, $C_s = 10^5$ m/s, $D_\perp = 1$ m^2/s then $L_{\|i} = 30$ m. Note that in this calculation one should use the size of the probe head itself--not just the collecting element--since plasma flows to all solid surfaces.

A number of implications may be considered:

(a) The probe does not provide a point measurement of particle or heat flux but one which is an <u>average</u> over length $L_{\|i}$. Thus if there are any gradients of T or n in the flow direction, then analysis of the probe data will provide <u>average</u> values of T and n over this region.

(b) If $L_{\|i}$ is sufficiently great then ion flow along the flux tube will be collisional. As indicated, Section 8, data interpretation is easier when flow to the probe is collisionless. The ion-ion mean free path[78] is

$$\lambda_{ii} \simeq 10^{16} \, T_i^2 \, (eV)/n_i \, (m^{-3}) \, m \tag{65}$$

Example: $T_i = 50$ eV, $n_i = 10^{18}$ m^{-3} then $\lambda_{ii} = 25$ m.

(c) If $L_{\|i}$ is larger than the distance from the probe to the next solid surface (typically a limiter) then the simple analysis of Section 5 will require significant corrections. This problem is examined in Refs. (76,77) and will not be dealt with further here. With regard to the drawing of net electron currents to the probe: If $j_{SAT}^- = 1/4 \, n\bar{c}_e e$, the disturbance length for electron collection is given by

$$L_{\|e} \simeq \bar{c}_e d^2/16 \, D_\perp . \tag{66}$$

Example: $d = 5$ cm, $D_\perp = 1$ m^2/s, $\bar{c}_e = 4\times10^6$ m/s then $L_{\|e} \simeq 600$ m.

As intuitively expected, the drawing of j^-_{SAT} is much more disturbing to the plasma than the drawing of j^+_{SAT}. (It would, however, be more realistic to consider an example of $d \simeq 2$ mm, say, since one would not bias the entire probe head into electron collection but only the smaller collecting element itself.) We may draw some conclusions from the great magnitude of $L_{\parallel e}$:

(a) As with ions the probe measures plasma quantities such as T_e which are in fact averages over length $L_{\parallel e}$. One cannot detect fine-grain, parallel-field variations in T_e or n_e unless the probe is very small.

(b) Net electron flow to the probe is much more likely to be collisional than is the case for ion flow. The electrons lose momentum in collisions with the quasi-stationary ions. It can be shown[39,40] that this results in a reduction in j^-_{SAT} to a value

$$ j^-_{SAT} = (\frac{r}{1+r})\ 1/4\ n_o \bar{c}_e e \ , \tag{67} $$

where

$$ r = \frac{16}{\pi}\ (1 + \frac{T_i}{T_e})(\frac{D_\perp}{D_\parallel})^{1/2}\ (\frac{\lambda_{ei}}{d}) \ , \tag{68} $$

where $D_\parallel$ = parallel-field electron diffusion coefficient $\approx 1/4\ \lambda_{ei}\ \bar{c}_e$ (if only classical e-i collisions are important) λ_{ei} = electron-ion mean free path. Further details are given in Ref. (39).

It is in fact often observed to be the case for probes used in strong magnetic fields that the ratio of j^-_{SAT}/j^+_{SAT} is substantially depressed from the normal value of ~50 for a hydrogen plasma with $T_e \approx T_i$.[40,64,79] Not only is j^-_{SAT} reduced from its normal value but to a varying degree so are the electron currents drawn at all probe voltages.[40] Since these electron currents are used to measure T_e, erroneous values of T_e can be indicated by a large probe[64] [large d gives small r, thus depressed j^-, Eq. (67)]. This provides an additional reason for employing small collectors for measuring electron properties--although one should avoid reducing d to the point where $d \simeq r_i$.

7.5 Sheath Across an Orifice or Aperture

Various types of probe employ an aperture or orfice separating the plasma being sampled from the detection element, see Chapter on edge probes. The full analysis of such probes is left to that Chapter; we note here, however, that the size of the aperture, d_a, is quite important. If the sheath thickness d > d_a, see Eq. (11), then the sheath "bridges" over the aperture and the plasma does not "know" that the opening is there, i.e., the ions are accelerated through the usual sheath voltage drop and then pass through the orifice toward the detector. On the other hand, if d_a > d them plasma penetrates through the orifice and reaches the detection element, i.e., the sheath appears at the detection element rather than at the orifice. Depending on what diagnostic technique is desired one arrangement or the other will be appropriate, see Chapter on edge probes.

7.6 The Reference Point for Single Probe Operation[76]

Probe theory employs the quantity Φ, the voltage between the probe and the local plasma potential. In practice the applied (measured) voltage, Φ_{app}, is usually applied between the probe and some object contacting the plasma, say the limiter(s). Thus we have that

$$\Phi_{app} = \Phi + \Delta\Phi_\ell + \Delta\Phi_{p\ell} \tag{69}$$

where $\Delta\Phi_\ell$ = potential difference between the limiter (or other reference object) and the plasma in its vicinity. $\Delta\Phi_{p\ell}$ = potential drop <u>in</u> the plasma between the plasma adjacent to the probe and that adjacent to the reference object.

In plotting the "J-Φ characteristic" of a probe one is, of course, plotting J vs Φ_{app}. Fortunately $\Delta\Phi_\ell$ and $\Delta\Phi_\ell$ are usually constant as Φ_{app} (hence j_{TOT}) is varied and so the plot of ℓn $(j_{TOT} - j_{SAT}^+)$ vs Φ_{app} simply has an offset which does not effect the deduction of T_e from the slope. Note two points, however:

(a) Usually $\Delta\Phi_{p\ell} \simeq 0$, however, $|\Delta\Phi_\ell|$ is about the same magnitude as the difference between the probe (when it is floating) and its adjacent plasma. Hence the probe will usually be found to float (as indicated by $j_{TOT} = 0$) when $\Phi_{app} \simeq 0$. Obviously this should not be taken to imply that the probe floats at zero potential difference between itself and its local plasma, although superficially it looks that way.

(b)   It is not the case that $\Delta\Phi_\ell$ and $\Delta\Phi_{p\ell}$ will always be
        constant as Φ_{app}, and thus j_{TOT}, are varied. The
        object used as reference for the probe must supply,
        of course, the current drawn by the probe. How are
        we sure that we are not seeing the J-Φ characteristic
        of this object? We will be in just this predicament
        unless the reference object is quite large compared
        with the probe. When it is large--such as a limiter
        --then it can supply all the current needed for the
        probe by only slight changes in $\Delta\Phi_\ell$.

$\Delta\Phi_{p\ell}$ will also depend on j_{TOT} if the plasma resistance, R_p,
is high. In such a case the J-Φ characteristic will simply be
$\Phi_{app} \simeq j_{TOT}R_p$ and one will not measure T_e from it. Fortunately,
plasma resistance is usually extremely low compared to the
effective sheath resistance R_s. R_s is not a constant but is of
order

$$R_s \simeq \frac{kT_e/e}{nC_s eA_\perp} \gtrsim 10\text{'s of ohms,}$$

typically. If the electrical path from probe to reference
object crosses magnetic field lines then R_p may become large
compared with R_s. This situation does not arise if the probe is
deployed outside the limiter radius when the limiters are used
as reference; however, if the probe is inserted inboard of the
limiters it may require its own reference object, for example,
the probe housing surrounding the actual collector.

The foregoing provides only a first look at the use of
probes in magnetic fields. This matter will be expanded in
detail in the Chapter on edge probes.

8.   SOME REFINEMENTS TO SHEATH THEORY AND IMPROVED EXPRESSIONS
        FOR THE FLOATING POTENTIAL, PARTICLE AND HEAT FLUX
        DENSITIES THROUGH THE SHEATH

8.1   Generalizing the Bohm Criterion

In Section 4 we examined the simplest formulation of the
Bohm Criterion, i.e., the condition imposed on the ion velocity
at the plasma/sheath interface. Since Bohm's work in 1949[11]
this Criterion has been generalized in a number of ways. These
results can be only briefly reviewed here:

(a)   It has been shown using the fluid equations that the
        "blow-up" of the plasma solution at the plasma/sheath
        interface, which we first noted in Section 4, can be

geometrically generalized, and that this interface is mathematically equivalent to the Mach or sonic surface in conventional fluid flow.[80] This type of result appears to be a general property of plasma fluids which satisfy relatively simple conservation equations, e.g., Eq. (13). It has also been shown that a number of additional physical processes, e.g., collisions, can be included without altering the "blow-up" result[81]; however, special spatial variations of physical processes, e.g., ionization and electron heat conductivity, imply the possibility of achieving a smooth transition to supersonic flow within the plasma itself, see Chapter by Chodura. Whether such smooth transitions are physically realizable is not yet established and if they do exist they would appear to be special cases.

(b)   Without using the fluid equations, but using kinetic theory formulations (where the ion velocity distribution is allowed to be general), the plasma "blow-up" result has also been obtained.[82] Also using kinetic theory formulations, the plasma/sheath interface has been identified as a Mach or sonic surface by a quite different method[30]: Namely, by identifying it as the first point in the accelerating flow where waves can no longer propagate back upstream (a fundamental property of the sonic point). Both these last two analysis[30,82] give the result that at the plasma/surface interface, the ion velocity distribution must satisfy

$$\int_0^{+\infty} \frac{f_i(v_x)dv_x}{v_x^2} = \frac{m_i}{kT_e} \,. \tag{70}$$

where v_x is in the surface-direction. Note: this result is obtained by considering the _plasma_ equations and gives an equality (rather than a "$\leq$") result--the same situation we encountered in our simpler approach of Section 4. Note: Equation (70) requires $f(0) = 0$ for $v_x < 0$, i.e., no backward-going ions--a physically intuitive result of course. Note for monoenergetic ions Eq. (70) gives our earlier result, $v_i = C_s$.

(c)   The first authors to obtain a Generalized Bohm Criterion similar to Eq. (70) were Harrison and Thompson in 1959.[29] These authors examined the _sheath_ equations (following Bohm's original approach), but allowed for a general ion velocity distribution (rather than Bohm's monoenergetic ion assumption). They then obtained the same result as Eq. (70) but with "$\leq$" in place of the equality (approaching the problem from the _sheath_ side, evidently always results in this type of ambiguity).

(d) The first authors to obtain the equality of Eq. (70) appear to have been Cavaliere, Engelmann and Onori in 1965,[83] who arrived at the result by requiring that the plasma and sheath equations match at the interface. Boozer[84] (1975) and Riemann[85] (1977) similarly arrived at the equality result, Eq. (70).

We next examine whether this Generalized Bohm Criterion is actually obeyed. We will find that indeed it is, at least for all cases so far analyzed, the sole exception being the case where the ion distribution is assumed to be a drifting Maxwellian at the sheath edge; see Section 8.3. We will consider two cases: ions collisionless everywhere (Section 8.2) and ions collisional before they reach the sheath (Section 8.3). The sheath itself is usually much thinner than any collisional mean free path, and so it is generally taken to be collisionless.

8.2 Fully Collisionless Ions

In 1929 Tonks and Langmuir[26] analyzed the case for fully collisionless, (pre-sheath and sheath), $T_i = 0$, ions; see Fig. 11. The resulting ion distribution at the plasma/sheath interface was explicitly worked out by Harrison and Thompson,[29] and was found to satisfy[30] Eq. (70).

Bohm,[40] in another of his many basic contributions to plasma boundary analysis, worked out a collisionless spherical probe theory. This ion distribution at the sheath edge also satisfies[30] Eq. (70).

Recently, Emmert et al.,[86] extended the fully collisionless analysis of Tonks and Langmuir to cover the case of $T_i \neq 0$, (plane geometry only). This analysis provides the ion velocity distribution at each point, and thus it can be checked against the Generalized Bohm Criterion, Eq. (70). Also particle and energy fluxes to the wall can be found, hence Φ_f also.

Not surprisingly, the results of Emmert et al. involve rather complex formula, (see Eqs. 72-77) and it has not as yet been proven in general that their ion distribution at the sheath edge satisfies Eq. (70). (Emmert et al., themselves, make no reference to this Generalized Bohm Criterion.) However, the present author has tested the case of $T_e = T_i$, $z = 1$, and found that the ion distribution of Emmert et al. at the sheath edge

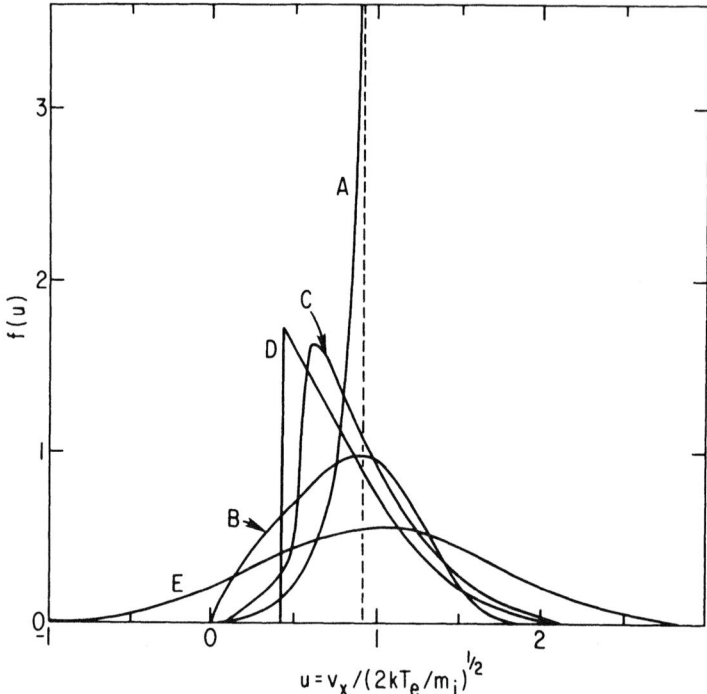

Fig. 11.   Various ion velocity distributions at the plasma/sheath interface. Only the distribution in the direction toward the surface is shown. All distributions normalized, $\int_{\infty}^{\infty} f(u)du = 1$. (A) Tonks and Langmuir[26]; collisionless ions, $T_i = 0$. (B) Reimann[89]; collisional ions, $T_i = 0$. (C) Emmert et al.[86]; collisionless ions, only $T_e = T_i$ case shown. (D) Chekmarev et al.[90]; collisional ions, only $T_e = T_i$ case shown. (E) A drifting Maxwellian, drift velocity $v_x = C_s$, i.e., $f(u) \propto \exp[-(u-1)^2]$.

satisfies Eq. (70) to within a few percent; this particular ion distribution is also shown in Fig. 11.

Because Emmert et al. have provided a complete solution for the fully collisionless case (plain geometry), this work is quite important for plasma edge analysis. It is useful therefore to summarize their results here and to compare them with our simple results of Section 5:

(a)   the pre-sheath voltage drop $\Phi_{p.s.}$ normalized as

$$\psi_1 \equiv e\Phi_{p.s.}/kT_e \tag{71}$$

is given by the solution of the transcendental equation

$$1 = [2/(\pi z T_e/T_i)]^{1/2} \exp [-(1+zT_e/T_i)\psi_1] \, D(\psi_1^{1/2})$$

$$+ \operatorname{erf} [(z\psi_1 T_e/T_i)^{1/2}] \, , \tag{72}$$

where erf (x) is the error function, $D(x) = \int_0^x \exp (t^2) dt$ is the Dawson function, and z is the ion charge. ψ_1 is found to be a monotonically decreasing function of T_i/zT_e falling from a value of 0.854 at $T_i/zT_e = 0$ to zero as $T_i/zT_e = \infty$.

(b) The floating <u>sheath</u> voltage drop (leaving out the pre-sheath) is given by

$$\frac{e\Phi_f}{kT_e} = -\ln \left[\left(\frac{m_i}{m_e} \frac{z}{4\pi}\right)^{1/2} \left[\frac{1}{z+T_i/T_e} \frac{\pi}{2 \exp (-\psi_1) D(\psi_1^{1/2})}\right] + \psi_1 \right. \, ,$$

$$\tag{73}$$

(c) the ion and electron particle flux density to a floating surface is given by

$$\frac{j^+}{e} = \left(\frac{n_o}{z}\right)\left(\frac{kT_i}{2\pi m_i}\right)^{1/2} \beta \tag{74}$$

where

$$\beta = 4(zT_e/\pi T_i)^{1/2}(1+T_i/zT_e) \exp (-\psi_1) D(\psi_1^{1/2}) \, , \tag{75}$$

(d) The ion power flux <u>entering</u> the sheath (hence including the pre-sheath gain) is

$$Q_i = 2kT_i \, \mu \, \frac{j^+}{e} \, , \tag{76}$$

where

$$\mu = 1 - \frac{zT_e}{2T_i} \left(1 - \frac{(\psi_1)^{1/2}}{\exp (-\psi_1) D(\psi_1^{1/2})}\right) \, . \tag{77}$$

87

(e)   the electron power flux density entering the sheath
       is

$$Q_e = (2kT_e + e\Phi_{ps} - e\Phi_s) \frac{j^+}{e} .$$ (78)

We may now compare our relatively simple results from
Section 5 with these exact results for the collisionless pre-
sheath case:

(a)   The simple theory gives $\Phi_{p.s.}$ as either -0.69 kT_e/e
       (plasma equations) or -0.5 kT_e/e (sheath equations)
       compared with -0.854 $kT_e/e \leq \Phi_{p.s.}(T_i/zT_e) \leq 0$ from
       the exact theory. Since the pre-sheath contribution
       to δ is comparatively small anyway, these differences
       are of second order.

(b)   The floating sheath voltage drop itself is compared
       with the simple result in Fig. 2 and as can be seen,
       the agreement is to within <10%.

(c)   The ion particle flux given in Eq. (74) can be
       rewritten in terms of C_s using a factor $f(T_i/T_e)$:

$$\frac{j^+}{e} = f(T_i/T_e)n_oC_s,$$ (79)

       where it has been shown[39] that $f(T_i/T_e)$ increases
       monotonically from a value of 0.487 at $T_i/T_e = 0$ to
       0.798 at $T_i/T_e \rightarrow \infty$, (case of $z = 1$). This compares
       with the simple formulation $j^+ = 0.5$ en_oC_s.

(d)   The ion power flux: Since $\mu(T_i/T_e)$ decreases
       monotonically from a value of ~1.15 at $T_i/T_e = 1$ to
       unity as $T_i/T_e \rightarrow \infty$, this result is essentially the
       same as that used in the simple theory Eq. (41)
       [a small pre-sheath contribution is included in
       Eq. (76)].

(e)   The terms in the expression for the electron power
       flux are the same in both cases.

We may therefore conclude that, at least as far as the
collisionless case is concerned, the convenient formula obtained
in Section 5 for the sheath are first-order accurate; they also
have the benefit of including secondary electron emission. The
pre-sheath voltage drop is less well reproduced by the simple
theory, but this quantity has little effect on the transmission
factors.

88

It is rather surprising that a fluid model, which is strictly only valid in a collisional situation, should reproduce collisionless results so closely. This appears to be a particular manifestation of a more general finding in plasma edge modeling, namely that plasma fluid models appear to be able to describe even collisionless cases rather well; see, e.g., the invited paper[87] by D. E. Post to the Sixth International Conference on Plasma Surface Interactions, 1984.

8.3 Collisional Ions in the Plasma

Here the fluid model should do particularly well, one would think, but in fact a problem occurs if one proceeds in the most direct and straightforward way! The fluid model implies that at the plasma/sheath interface the ion velocity distribution is a Maxwellian, of temperature T_i, drifting at velocity C_s, i.e.,

$$f(v_x, v_y, v_z) = (2\pi kT_i/m_i)^{-3/2} \exp \left\{-(2kT_i/m_i)[(v_x - C_s^2) + v_y^2 + v_z^2]\right\} . \tag{80}$$

Fig. 11 shows such a distribution. This distribution is not, however, acceptable: (a) it has backward going ions, which is unphysical since the surface absorbs all ions (b) it does not satisfy the Generalized Bohm Criterion, Eq. (70). If we were to assume a drifting Maxwellian anyway, we obtain the ion energy flux density through the plasma/sheath interface to be[43]

$$Q_i = \left(\frac{5}{2} kT_i + \frac{1}{2} m_i C_s^2\right) j^+/e$$

$$= \left(\frac{5}{2} kT_i + \frac{1}{2} \gamma_s kT_i + \frac{1}{2} kT_e\right) j^+/e \tag{81}$$

where γ_s = the ratio of specific heats for the ions. One should note that the T_i and T_e used in Eq. (81) should be the values just $\underline{at}$ the plasma/sheath interface, and not the values at the distant reference (stagnation) point, T_{io}, T_{eo}. As with all adiabatic flow the ions cool as they are accelerated. For conventional fluid flow, with γ_s = 5/3 one finds[88] $T_i(M=1)/T_{io}$ = 3/4; thus Eq. (81) would give

$$Q_i = \left(\frac{5}{2} kT_{io} + \frac{1}{2} kT_e\right) j^+/e . \tag{82}$$

This last relation doesn't in fact differ much from our simple value in Section 5: $5/2 \ kT_{io}$ instead of $2 \ kT_{io}$ (the $1/2 \ kT_e$ factor representing the ion energy gained from the <u>electrons</u> in the pre-sheath). This result, however, still leaves a somewhat larger uncertainty unresolved: generally, computer code models of the SOL "follow" the ions and electrons along the B-field flux tube toward the sheath, accounting for e-i energy transfer, inelastic atomic energy losses, etc. and calculating T_e and T_i at each point along the flux tube; at the plasma/sheath interface these codes end up with final temperatures (T_{ef}, T_{if}) and boundary conditions are required. If one were to simply take the drifting Maxwellian result, then one would use Eq. (81) inserting values of T_{ef}, T_{if}. Just such a boundary condition has in fact been employed (Refs. 19,35), but it appears that this would overestimate Q_i: As can be seen from Fig. 11, the drifting Maxwellian has a relatively large number of fast, forward-going ions and this substantially increases Q_i/j^+.

What is clearly needed is an analysis which correctly "tracks" the ion distribution as the ions move down the flux tube toward the surface, moving from a far-field collisional region, through the last collisional mfp in front of the plasma/sheath interface, then finally into the collisionless sheath itself. Recently just such studies have started to appear in the literature (although, unfortunately, not describing precisely the situation of interest for plasma edge modeling).

Riemann[89] has considered such a collisional situation (plain geometry, $T_i = 0$) where the ions experience charge-exchange collisions with cold neutrals (thus completely halting the ions at each collision--not a good model for the ion-ion collisional case in which we are primarily interested). Riemann's ion velocity distribution at the sheath edge is shown in Fig. 11; he proved that this distribution satisfies the Generalized Bohm Criterion, Eq. (70), precisely. One may note the substantial difference collisions make for the $T_i = 0$ case by comparing the Tonks and Langmuir distribution with Reimann's. The Tonks and Langmuir distribution has a sharp upper cut-off velocity corresponding to ions which have fallen all the way from the midplane between the two surfaces. Riemann assumes an infinite plasma (the collisional mfp provides his scale length instead of the spacing between two planes) and since the ions suffer collisional drag, his electric potential goes to infinity in the far plasma. Since a few ions can reach the sheath edge collisionlessly from the far plasma, his distribution has no upper cut-off in velocity. Riemann's ions thus experience more acceleration from the electric field than do collisionless ions, and since the associated energy gained is taken from the electrons, one finds a larger Q_i/j^+ here. One can extract from Riemann's results that

$$Q_i \simeq 0.9\ kT_e\ j^+/e\ .\qquad\qquad(83)$$

A somewhat larger pre-sheath value than we found earlier (there is no T_i-term here, since $T_i \sim 0$).

Going a step further, Chekmarev, Sklyaroba and Kolesnikova[90] have allowed for finite T_i and followed the ion distribution through three zones: (a) the far-field, collisional, fluid zone (b) the transitional "Knudsen" zone extending one mfp upstream from the plasma/sheath interface (c) the collisionless sheath. Their ion distribution at the plasma/sheath interface is a truncated Maxwellian, rather than a drifted one, i.e., no backward-going ions at all, and no foward-going ones either below a certain cut-off velocity. See Fig. 11. These authors provide two cut-off criteria, one of which can be shown to satisfy the Generalized Bohm Criterion, Eq. (70); it is the latter distribution which is shown in Fig. 11. (Chekmarev et al. do not themselves refer to the Generalized Bohm Criterion.) It is also possible to extract from the results of Chekmarev et al. that

$$Q_i = (2kT_i + fkT_e)j^+/e\ ,\qquad\qquad(84)$$

where for $T_e = T_i$, f = 0.19-0.61 (depending on their truncation cut-off criterion), a result close to that of the simple analysis of Section 5.

One must note, however, that the work of Chekmarev et al. is also only a first step toward our desired goal since:

(a)   In the far zone the ions are assumed to collide with a third species, rather than being self-collisional.

(b)   The solutions obtained have not been shown to be unique.

(c)   It is not clear that energy is conserved through the three zones in their analysis.

8.4   Conclusions

The case of the collisionless pre-sheath or plasma seems to be well in hand: Emmert et al.[86] have provided a full kinetic solution (although neglecting secondary electrons); in addition the simple fluid model of Section 5 gives essentially the same results as Emmert et al., while also including secondary electron emission. Hence, for simplicity, one can use for the particle flux to a floating surface:

$$j^+_{SAT} = 1/2 \, en_o c_s \qquad \text{[Eq. (30)]}$$

for the floating potential (sheath only, excluding pre-sheath):

$$e\Phi_f/kT_e = 1/2 \, \ln\left[\left(2\pi \frac{m_e}{m_i}\right)\left(1 + \frac{T_i}{T_e}\right)\left(1-\gamma\right)^{-2}\right] , \qquad \text{[Eq. (34)]}$$

for the total heat flux density received by a floating surface:

$$Q/(j^+_{SAT}/e) \equiv \delta kT_e = [2kT_i - e\Phi_f + (\sim)1/2 \, kT_e](1-R_{iE})$$

$$+ \frac{2kT_e}{1-\gamma}\,(1-R_{eE}) + \chi_i + \chi_r\,(1-R_{iN}) \qquad \text{[Eq. (50)]}$$

Note that this expression for Q assumes that the plasma ions and electrons are Maxwellian; this is not always the case and there may exist high energy tails to the distributions which can be responsible for a large part of Q.[38,75]

For the heat flux density energy loss rate <u>from</u> the plasma electrons (case of a floating surface):

$$Q_e/(j^+_{SAT}/e) \equiv \delta_e kT_e \simeq -e\Phi_f + \frac{2kT_e}{1-\gamma}\,(1-R_{eE}) + (\sim)1/2kT_e$$

$$(85)$$

and for ions:

$$Q_i/(j^+_{SAT}/e) \equiv \delta_i kT_e \simeq 2\,kT_i . \qquad (86)$$

For the case of biased surfaces, one can incorporate the reflection refinements into the expressions, Eqs. (46)-(49).

For convenient reference the sheath expressions for the floating surface are collected in Table I.

Table I. Sheath Expressions for a Floating Surface in a Collisionless Plasma.

---

1. PARTICLE FLUX DENSITY:

$$j^+_{SAT} = \frac{1}{2} e n_o C_s$$

2. FLOATING POTENTIAL:

$$\frac{e\Phi_f}{kT_e} = \frac{1}{2} \ln \left[\left(2\pi \frac{m_e}{m_i} \right) \left(1 + \frac{T_i}{T_e} \right) (1-\gamma)^{-2} \right]$$

3. ENERGY FLUX DENSITY (LOSS) FROM PLASMA ELECTRONS:

$$\frac{Q_e}{j^+_{SAT}/e} \equiv \delta_e kT_e \simeq -e\Phi_f + \frac{2kT_e}{1-\gamma} (1-R_{eE}) + (\sim) \frac{1}{2} kT_e$$

4. ENERGY FLUX DENSITY (LOSS) FROM PLASMA IONS:

$$\frac{Q_i}{j^+_{SAT}/e} \equiv \delta_i kT_e \simeq 2 kT_i$$

5. ENERGY FLUX DENSITY RECEIVED BY FLOATING SURFACE:

$$\frac{Q}{j^+_{SAT}/e} \equiv \delta kT_e \simeq [2 kT_i - e\Phi_f + (\sim) \frac{1}{2} kT_e](1-R_{iE})$$

$$+ \frac{2kT_e}{1-\gamma} (1-R_{eE}) + \chi_i + \chi_r (1-R_{iN})$$

One may note that the energy loss from the plasma ions and electrons does not equal the energy to a floating surface. For example, the potential energy terms χ_i and χ_r are usually not included as energy content of the plasma particles, and so should not be included as an energy loss either. A second example: Ions backscatter primarily as neutral atoms carrying energy $[2kT_i - e\Phi_f + (\sim)1/2 \ kT_e] \ R_{iE}$ (these atoms may, of course, be subsequently ionized thus returning this energy to the plasma, however, this energy input to the plasma should be calculated separately to establish if and where it occurs).

Turning to the case of the collisional pre-sheath or plasma: The analysis of this case is not complete at this time. It awaits a solution which properly matches the far-field collisional flow to the collisionless sheath. If for the present we may assume that the collisional case is represented to first order by the collisionless results, then the following simple adaptations are indicated:

(a) The sound speed at the sheath edge should be calculated using T_{ef} and T_{if}, the final temperatures computed from the collisional fluid model of the plasma, i.e., $C_s (T_{ef}, T_{if})$.

(b) $j^+_{SAT} = n_{se} C_s$ where $n_{se} \equiv n_f$, the final plasma density computed from the collisional fluid model of the plasma.

(c) The expressions for Φ_f, Q_e, Q_i, Q, would be given by Eq. (34), (50), (85), (88) calculated using T_{ef} and T_{if}.

ACKNOWLEDGMENTS

The author wishes to thank J.W. Davis and I.S. Youle for helpful discussions. The work was supported by the Canadian Fusion Fuels Technology Project and by the U.S. DOE Contract No. DE-AC02-76-CHO-3073.

REFERENCES

1. F.F. Chen "Introduction to Plasma Physics and Controlled Fusion, Vol I. Plasma Physics", Second Edition, Plenum Press, New York, p. 228 (1983).
2. Ref. 1, pg. 54
3. E.H. Holt and R.E. Haskell "Foundations of Plasma Dynamics", MacMillan, New York, p. 134 (1965).
4. S.A. Self, Phys. Fluids 6, 1762 (1963).
5. J.G. Andrews and R.H. Varey, J. Phys. A 3, 413 (1970).
6. A.T. Mense and G.A. Emmert, Nucl. Fusion 19, 361 (1979).
7. Ref. 1, p. 293.
8. Ref. 1, p. 9; Ref. 3, p. 59.
9. Ref. 1, p. 294.
10. Ref. 1, p. 10; Ref. 3, p. 248.
11. D. Bohm, "The Characteristics of Electrical Discharges in Magnetic Fields" Eds A. Guthrie and R.K. Wakerling, McGraw Hill, New York, Chapt. 3 (1949).

12. Ref. 1, Chapt. 3; Ref. 3, Chapt. 6.
13. A. von Engel, "Ionized Gases", Oxford University Press (1965).
14. K. Uehara, Y. Gomay, T. Yamamoto, N. Suzuki, M. Maeno, T. Hirayama, M. Shimada, S. Konoshima, N. Fujisawa, Plasma Phys. 21, 89 (1979).
15. G. Haas, M. Keilhacker, and K. Lackner, J. Nucl. Mater. 76 & 77, 279 (1978).
16. M.R. Gordinier and R.W. Conn, J. Nucl. Mater. 93 & 94, 420 (1980).
17. G. Fuchs and A. Nicolai, Nucl. Fusion 20, 1247 (1980).
18. J.M. Ogden, C.E. Singer, D.E. Post, R.V. Jensen, F.G.P. Seidl, IEEE Trans. on Plasma Science PS9, 274 (1981).
19. P.J. Harbour and J.G. Morgan, "Models and Codes for the Plasma Edge Region", Culham Laboratory Report CLM-R234 (1982).
20. M.F.A. Harrison, P.J. Harbour, and E.S. Hotston, Nucl. Technol./Fusion 3, 432 (1983).
21. P.C. Stangeby, J. Nucl. Mater. 121, 55 (1984).
22. P.C. Stangeby, Phys. Fluids 28, 644 (1985).
23. W.M. Stacey, "Fusion Plasma Analysis", Wiley, New York, p. 78 (1981).
24. Ref. 1, p. 96.
25. P.C. Stangeby and J.E. Allen, J. Phys. A. 3, 304 (1970).
26. L. Tonks and I. Langmuir, Phys. Rev. 34, 876 (1929).
27. J.E. Allen and P.C. Thonemann, Proc. Phys. Soc. B67, 768 (1954).
28. L. C. Woods, J. Fluid Mech. 23, 315 (1965).
29. E.R. Harrison and W.B. Thompson, Proc. Phys. Soc. 72, 2145 (1959).
30. J.E. Allen, J. Phys. D. 9, 2331 (1976).
31. P.C. Stangeby, Phys. Fluids 27, 2699 (1984).
32. E.R. Harrison, "Mean Kinetic Energy of Ions in Low Pressure Plane Symmetric Plasmas", AERE GP/M 203, Harwell, U.K. (1957).
33. P.C. Stangeby, Phys. Fluids 27, 682 (1984).
34. G.D. Hobbs and J.A. Wesson, "Heat Transmission through a Langmuir Sheath in the Presence of Electron Emission", Culham Laboratory Report CLM-R61 (1966); also Plasma Phys. 9, 85 (1967).
35. R. Chodura, K. Lackner, J. Neuhauser, W. Schneider, R. Wunderlich, in Proc. 9th International Conference on Plasma Physics and Controlled Nuclear Fusion Research (1982), (IAEA, Vienna, 1983) Vol. I, 313.
36. M.A. Mahdavi et al., J. Nucl. Mater. 111 & 112, 355 (1982).
37. M. Petravic, D. Post, D. Heifetz, and J. Schmidt, Phys. Rev. Lett. 48, 326 (1982).
38. H. Kimura et al., Nucl. Fusion 18, 1195 (1978).
39. P.C. Stangeby, J. Phys. D. 15, 1007 (1982) and also J. Nucl. Mater. 111 & 112, 84 (1982).

40. D. Bohm, E.H.S. Burhop, and H.S.W. Massey, in "Characteristics of Electric Discharges in Magnetic Fields", Eds, A. Guthrie and R.K. Wakerling, McGraw-Hill, New York, Chapter 2 (1949).

41. U. Daybelge and B. Be in, Phys. Fluids 24, 1190 (1981).

42. R. Chodura, Phys. Fluids 25, 1628 (1982).

43. R. Chodura, J. Nucl. Mater. 111 & 112, 420 (1982).

44. J.P. Biersack and L.G. Haggmark, Nucl. Instrum. Methods 174, 257 (1980).

45. D.M. Manos, R.V. Budny, and S.A. Cohen, J. Vac. Sci. Technol. A1, 845 (1983) (data available from the authors).

46. W. Eckstein and H. Verbeek, IPP Garching Report IPP9/32 (1979).

47. C.F. Barnett, J.A. Ray, E. Ricci, M.I. Wilker, E.W. McDaniel, E.W. Thomas, H.B. Gilbody, "Atomic Data for Controlled Fusion Research", Oak Ridge National Laboratory Report ORNL-5207, Oak Ridge, TN (1977), Vol. II, Section D.

48. P. Staib, J. Nucl. Mater. 111 & 112, 109 (1982).

49. D.M. Manos, R. Budny, T. Satake, and S.A. Cohen, J. Nucl. Mater. 111 & 112, 123 (1982).

50. G. Proudfoot and P.J. Harbour, J. Nucl. Mater. 111 & 112, 87 (1982).

51. P.C. Stangeby, G.M. McCracken, and J.E. Vince, J. Nucl. Mater. 111 & 112, 81 (1982).

52. H. Vernickel, N. Nucl. Mater. 111 & 112, 531 (1982).

53. G.M. McCracken and P.E. Stott, Nucl. Fusion 19, 889 (1979).

54. Ref. 13, p. 243

55. W. Schneider, D. Heifetz, K. Lackner, J. Neuhauser, D. Post, and K.G. Raub, J. Nucl. Mater. 121, 178 (1984).

56. P.C. Stangeby, G.M. McCracken, S.K. Erents, J.E. Vince, and R. Wilden, J. Vac. Sci. Technol. A1, 1302 (1983).

57. R.K. Janev, D.E. Post, W.D. Langer, K. Evans, D.B. Heifetz, and J.C. Weisheit, J. Nucl. Mater. 121, 10 (1984).

58. D.R. Baker, R.T. Snider, and M. Nagami, Nucl. Fusion 22, 807 (1982).

59. Y. Shimomura, M. Keilhacker, K. Lackner, and H. Murmann, Nucl. Fusion 23, 869 (1983).

60. Ref. 13, Chapters 8 and 9.

61. E.A. Robson and P.C. Thonemann, Proc. Phys. Soc. 73, 508 (1959).

62. E. Hantzsche, Beitr. aus der Plasmaphysik 24, 329 (1980).

63. S.A. Cohen, J. Nucl. Mater. 76 & 77, 68 (1978).

64. P.C. Stangeby, G.M. McCracken, S.K. Erents, and G. Mathews, J. Vac. Sci. Technol. A2; 702 (1984).

65. F.F. Chen, in "Plasma Diagnostic Techniques," Eds. R.H. Huddlestone, and S.L. Leonard, Academic Press, New York, Chapt. 3 (1965).

66. J.D. Swift and M.J.R. Schwar, "Electric Probes for Plasma Diagnostics", Iliffe Books, New York (1969).

96

67. J.G. Laframboise, UTIAS Report No. 100, Institute for Aerospace Studies, University of Toronto (1966).

68. S.K. Erents, G.M. McCracken, and J. Vince, J. Phys. D. 11, 227 (1978).

69. G. Staudenmaier, J. Roth, R. Behrisch, J. Bohdansky, W. Eckstein, P. Staib, S. Matteson, S.K. Erents, J. Nucl. Mater. 84, 149 (1979).

70. W.R. Wampler, Appl. Phys. Lett. 41, 335 (1982).

71. W.R. Wampler and D.M. Manos, J. Vac. Sci. Technol. A1, 827 (1983).

72. S.K. Erents and P.C. Stangeby, J. Nucl. Mater. 111 & 112, 165 (1982).

73. P. Staib, J. Nucl. Mater. 93 & 94, 351 (1980).

74. G. Mathews, to be published in J. Phys. D.

75. P.C. Stangeby, J. Nucl. Mater. 128 & 129, 969 (1984).

76. P.C. Stangeby, "Large Probes in Tokamak Scrape-Off Plasmas. The Collisionless Scrape-Off Layer: Operation in the Shadow of Limiters or Divertor Plates", to be published in J. Phys. D.

77. P.C. Stangeby, J. Nucl. Mater 121, 36 (1984).

78. Ref. 1, p. 176; Ref. 3, p. 257.

79. R. Budny and D. Manos, J. Nucl. Mater. 121, 41 (1984).

80. P.C. Stangeby and J.E. Allen, J. Phys. A 3, 304 (1970).

81. J.G. Andrews and P.C. Stangeby, J. Phys. A 3, L39 (1970).

82. K.-U. Riemann, in Proceeding International Conference on Plasma Physics (Nagoya, Japan, 1980) Vol. I, p. 66.

83. A. Cavaliere, F. Engelmann, G. Onori, Internal Report L.G.I. 65/21, Frascati (1965).

84. A. H. Boozer, Princeton University Plasma Physics Laboratory Report MATT 1148 (1975).

85. K.-U. Riemann, Ph.D. Thesis, Ruhr-Universitat, Bochum (1977).

86. G.A. Emmert, R.M. Wieland, T. Mense, and J.N. Davidson, Phys. Fluids 23, 803 (1980).

87. D.E. Post, J. Nucl. Mater. 128 & 129, 78 (1984).

88. A.H. Shapiro, "The Dynamics and Thermodynamics of Compressible Fluid Flow", Ronald Press, New York, Vol. I, p. 84 (1953).

89. K.-U. Riemann, Phys. Fluids 24, 2163 (1981).

90. I.B. Chekmarev, E.M. Sklyaroba, and E.N. Kolesnikova, Beitr. Plasma Phys. 23, 411 (1983).

PLASMA FLOW IN THE SHEATH AND THE PRESHEATH OF A SCRAPE-OFF LAYER

Roland Chodura

Max-Planck-Institut für Plasmaphysik, EURATOM

Association, D-8046 Garching, Germany

ABSTRACT

The theory of plasma-wall transition is reviewed including the effect of a magnetic field oblique to the wall leading to a double structure of the sheath. The implications of the conditions at the sheath edge on the flow in the presheath are discussed.

1. INTRODUCTION

The maintenance of a hot plasma in a fusion device is only possible if the plasma is to a large extend separated from material walls. In closed devices like tokamaks or stellarators this separation is achieved by creating the plasma within a volume of nested magnetic surfaces which are closed in themselves, so that motion along magnetic field lines will not lead to plasma loss. Nevertheless, by collisional diffusion plasma slowly leaks out of the confinement region into a loss region where magnetic field lines and at material walls, e.g. limiters or divertor plates (Fig. 1). In this loss- or scrape-off region the plasma absorbing wall establishes a pressure gradient which accelerates the plasma along the magnetic field beyond sound speed C. Thus, plasma motion in the scrape-off layer is a superposition of slow diffusion across and fast flow along magnetic field.

The scrape-off layer divides into two markedly different regions (Fig. 2):
1. A narrow region ahead of the wall with large gradients of the state variables and supersonic flow velocity, called "sheath";
2. The by far larger region upstream of the sheath with relatively weak gradients and in general subsonic flow, named "pre-sheath".

99

The sheath, due to its small extension, is nearly collisionfree.
It is determined by the electron and ion dynamics in their electric
and magnetic fields. The presheath, in general, is collision dominated. It contains the interactions among plasma particles themselves, i.e. plasma transport and relaxation processes as well as
interactions between plasma and neutrals (e.g. ionization or charge

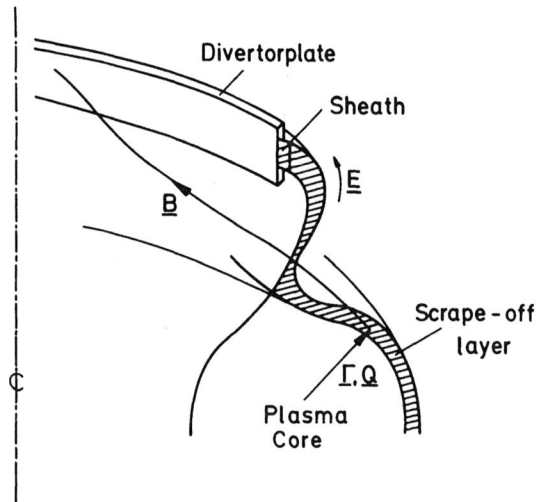

Fig. 1.   Scrape-off layer in a tokamak with divertor.

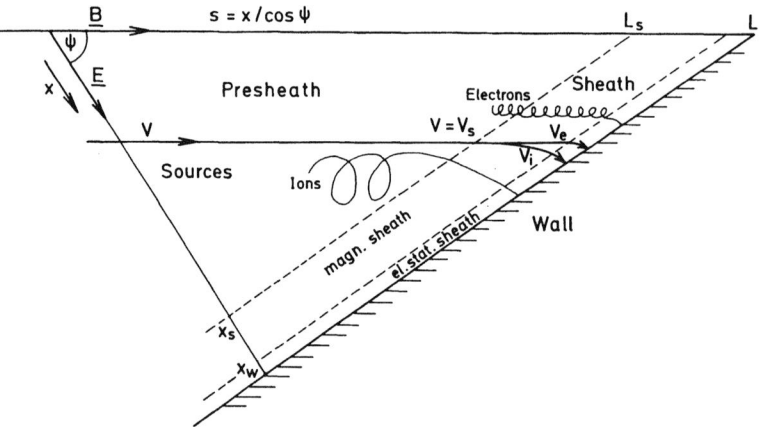

Fig. 2.   Schematic view of presheath and sheath in a scrape-
           off layer.

exchange) which represent volume sources and sinks of plasma parti-
cle-, momentum- and energy flow.

Due to their different masses and pressure gradients ions and
electrons experience different accelerations. This leads to charge
separation. In order to preserve $\nabla \cdot \underline{j}=0$ of current density $\underline{j}$, an electric space charge field $\underline{E}$ is built up in the presheath and sheath
region. Since gradients in the presheath are small, the electric
field is relatively small as well, while it becomes large in the
sheath. Accordingly, the presheath is quasineutral, i.e. ion and
electron charge densities Zn_i and en_e are nearly equal there,
$Zn_i \cong n_e$, where Z is the ion charge number. On the contrary, the plasma in the sheath may become non-neutral, i.e. $Zn_i > n_e$.

In an axisymmetric device such as a tokamak with toroidal limiter or divertor the electric potential field is purely poloidal
while the magnetic field is mainly toroidal. So the angle between
electric and magnetic field may be nearly 90° in practical cases.
If the wall is a good conductor the electric field in the sheath is
normal to the wall.

In the presheath the plasma flow is essentially directed along
the magnetic field. The electric field, in general oblique to the
magnetic field, tends to pull the flow towards its direction. Thus,
in the sheath, where the electric field is fairly strong, the plasma
flow is bent from magnetic field direction to the direction of the
wall normal   (Fig. 2).

In the following section the description of the sheath and the
presheath plasma, as given by P.C. Stangeby in the previous chapter,
is extended in some respects. Sections 2 and 3 present numerical
models of the sheath and the presheath on the basis of the relevant
length scales of the problem. Section 4 investigates the role of
an oblique magnetic field on the properties of the sheath, and
Section 5 discusses consequences of boundary conditions imposed
by the sheath on the 1d plasma flow in the presheath.
More detailed treatments of the presheath (i.e. 2d flow, interaction
with neutrals) will be given in Chapters 10, 11 and 12.

2. LENGTH SCALES OF THE SCRAPE OFF-LAYER

In order to derive a consistent model of a stationary scrape-
off layer one has to consider the relevant length scales of the
system. Some of them are externally given by the device, others are
internally determined by the physical processes in the layer.

External   length scales are:
the length L of the scrape-off layer measured along the magnetic
field lines from a plane of vanishing particle and energy fluxes
up to the wall or probe,

the characteristic dimension d   along the magnetic field of the
particle and energy fluxes fed into the scrape-off layer from the
main plasma, the characteristic dimension of the wall or probe.

Internal length scales are given by transport (i.e. diffusion,
heat conduction, viscosity), relaxation (i.e. Maxwellization, temperature exchange, resistivity) of plasma particles and by their
interaction with neutrals.

The extension of the scrape-off layer perpendicular to the
magnetic field $l_\perp$ is determined by an "anomalous" diffusion coefficient $D_\perp$ which cannot be derived from classical transport theory,

$$D_\perp = 0.1 - 0.4 \text{ m}^2/\text{s}$$

for tokamaks. From particle continuity one gets for $l_\perp$

$$l_\perp^2 \cong \frac{D_\perp}{v_{\parallel}} \, l_{\parallel}$$

where $V_{\parallel}$ and $l_{\parallel}$ are the flow velocity and characteristic length
parallel $\underline{B}$ in the scrape-off layer.

Plasma transport and relaxation along the magnetic field are
governed by Coulomb collisions. They are characterized by the length
of mean free path for 90° deflection λ. For collisions between a
test particle of velocity v and Maxwellian target particles of the
same sort (i.e. electron-electron or ion-ion collisions) with mean
energy $m\langle v_f^2\rangle/2 = 3\,T/2$, λ is given by

$$\lambda = \left(\frac{v}{\sqrt{3}\cdot v_t}\right)^4 \lambda_t \qquad\qquad (2.1)$$

where λ_t is the mean free path length for a test particle with
energy $m\,v^2/2 = 3\,T/2$,

$$\lambda_t = \sqrt{3}\ v_t\ t_c \cong 1.5\cdot 10^{16}\ T^2/n\,. \qquad\qquad (2.2)$$

T and n are temperature and density of the target particles respectively, $v_t = (T/m)^{1/2}$ is the thermal speed and t_c the self-colli-
sion time /1/. T stands for $k_B T$ throughout the text, where k_B is
the Boltzmann constant. In equ.(2.2) T is measured in eV, all other
quantities in mks-units. Electron-ion collisions have a mean free
path only by a factor $1/\sqrt{2}$ shorter than that for electron-electron
colllisions. λ_t is plotted in Fig. 3.

From λ_t one can derive several transport coefficients. Of
special importance for the energy transport in the scrape-off layer
is the electron heat conduction along the magnetic field

$$q_e = - \chi_{e\parallel} \, \nabla_\parallel \, T_e \qquad (2.3)$$

where q_e and T_e are electron heat flux and temperature respectively. Heat conduction $\chi_{e\parallel}$ is given by /2/

$$\chi_{e\parallel} = 2.9 \cdot 10^{-19} \, n_e \, v_{te} \, \lambda_t \cong 1.8 \cdot 10^3 \, T_e^{5/2} . \qquad (2.4)$$

T_e and ∇T_e are measured in eV and eV/m respectively, heat flux q_e in W/m².

Ion parallel heat conductivity $\chi_{i\parallel}$ is smaller by a factor of order square root of the mass ratio $(m_e/m_i)^{1/2}$ than $\chi_{e\parallel}$,

$$\chi_{i\parallel} = 5.1 \cdot 10^{-19} \, n_i \, v_{ti} \, \lambda_t \cong 75 \cdot T_i^{5/2} \qquad (2.5)$$

and therefore less important.

If a certain amount of heat flux $q_{e\parallel}$ has to be transported along B to the wall an electron temperature gradient with characteristic length $\Lambda_{e\parallel}$ will arise where

$$\Lambda_{e\parallel} = \chi_{e\parallel} \, T_e/q_{e\parallel} = 2.9 \cdot 10^{-19} \, (n_e \, v_{te} \, T_e/q_{e\parallel}) \, \lambda_t$$

$$\cong 1.8 \cdot 10^3 \, T_e^{7/2}/q_{e\parallel}. \qquad (2.6)$$

$\Lambda_{e\parallel}$ is plotted in Fig. 3 for $q_{e\parallel} = 10^8$ W/m².

It has to be pointed out, however, that in order equ. (2.3) to be valid electrons which contribute most to the heat flux, i.e. those with $v_q \cong 3.7 \, v_{te}$, have to have a mean free path λ_q smaller than the temperature gradient length $\Lambda_{e\parallel}$. According to (2.1) the mean free path is proportional to v^4. Therefore the applicability of equ.(2.3) is restricted to rather small gradients /3/

$$\Lambda_{e\parallel} > \lambda_q = (v_q/\sqrt{3} \, v_{te})^4 \, \lambda_t = 21 \, \lambda_t$$

or, by equ. (2.6),

$$q_{e\parallel} < 1.0 \cdot 10^{-20} \, n_e \, v_{te} \, T_e . \qquad (2.7)$$

For steeper gradients the heat flux at a point becomes non-local, i.e. it is no more determined by the local temperature gradient at this point but by a weighted mean of the temperature profile over a distance of about $\pm \, 5 \, \lambda_t$ around this point /4/.

Another important length scale is defined by the relaxation of the electron velocity distribution to a nearly Maxwellian one. As will be shown in Section 4 the electron distribution ahead of the sheath is depleted at velocities $|v| > |v_c|$ where $m_e v_c^2/2 \cong 3 \, T_e$ due to losses to the wall. With increasing distance from the wall this loss region is filled up by collisions, i.e. the electron distribution is isotropized by electron-ion and electron-electron-collisions and Maxwellized by electron-electron collisions. Both processes have about the same length scale of

$$\Lambda_{re} = (\frac{v_c}{\sqrt{3}\, v_{te}})^4 \lambda_t \cong 4\, \lambda_t . \tag{2.8}$$

The relaxation of different electron and ion temperatures to a common temperature during the flow to the wall occurs on a much longer length scale of about mass ratio times λ_t,

$$\Lambda_{t\,e,i} \cong \frac{m_i}{m_e} \lambda_t . \tag{2.9}$$

Out of the huge amount of interactions of plasma with neutrals only two important processes are presented in Fig. 3, i.e. ionization and charge exchange collisions: A neutral H atom at Frank-Condon energy of about $m_i\, v_0^2/2 = 4$ eV transverses a mean free path of λ_{ion} before being ionized by collisions with electrons of velocity v_e,

$$\lambda_{ion} = \frac{v_o}{n_e <\sigma_{ion}\, v_e>} , \tag{2.10}$$

and a mean free path λ_{cx} before exchanging its charge with an H+-ion of velocity v_i,

$$\lambda_{cx} = \frac{v_o}{n_i <\sigma_{cx}\, v_i>} . \tag{2.11}$$

Both path lengths are shown in Fig. 3 for thermal electrons and ions of temperatures T.

The shortest scale lengths in the scrape-off layer are those of the sheath. As will be shown in Section 4 the part of the sheath which is determined by the magnetic field B has a characteristic thickness perpendicular to the wall of the order of the ion gyroradius at sound speed $C = (2T/m_i)^{1/2}$. For oblique angles

$$\lambda_m \cong 4C/\omega_{ci} , \tag{2.12}$$

where ω_{ci} is the ion gyrofrequency for H^+ ions, $\omega_{ci} = eB/m_i$. The projection of d_m along $\underline{B}$, $d_m/\cos\psi$, where ψ is the angle between $\underline{B}$ and the wall normal is plotted in Fig. 3 for B = 2 T and $\psi = 85°$.

Immediately ahead of the wall extends the electrostatic Debye sheath over about 10 Debye lengths λ_D perpendicular to the wall,

$$\lambda_{es} \cong 10\, \lambda_D. \tag{2.13}$$

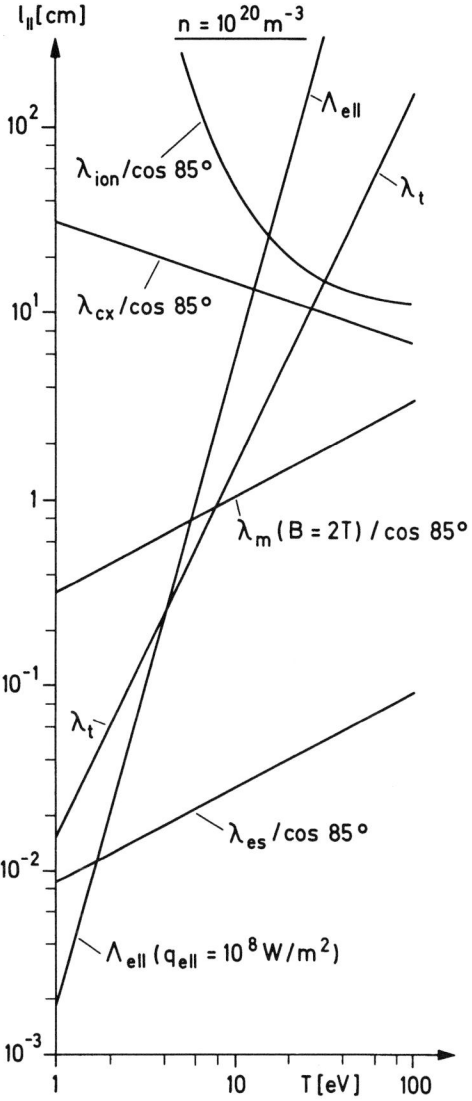

Fig. 3.   Length scales $l_{\parallel}$ (n,T) along the magnetic field in the
scrape-off layer (for hydrogen).

The projection of λ_{es} along B, $\lambda_{es}/\cos \psi$, is plotted in Fig. 3 again for $\psi = 85°$.

From inspection of Fig. 3 one can see that the presheath is more or less collision-dominated and may be described by a fluid model but that parameter regions may exist where the Braginskii transport coefficients may not be applicable and an improved transport theory or even kinetic theory must be applied. The sheath under nearly all conditions is collisionless and must be described by a kinetic model.

All regions of the scrape-off layer with length scales larger than the Debye length λ_D may be treated as being quasineutral, i.e.

$$Z \, n_i \cong n_e \, .$$

Thus only the electrostatic part of the sheath deviates appreciably from electric neutrality.

3. MATHEMATICAL MODELS

In order to describe the plasma flow in the sheath and presheath region of a scrape-off layer, the most general approach is a kinetic model. If collisions can be neglected, as it is the case in the sheath, this kinetic model consists simply of the equation of motion of the plasma particles in the externally imposed and self-created electric and magnetic fields $\underline{E}$ and $\underline{B}$. In general the self-created magnetic field is small as compared to the externally applied one and will be neglected here ("low-ß approach"). On the other hand, the self-created electric field by charge separation in general far exceeds the externally induced electric field. Thus only the self-consistent electric space charge field and a prescribed external magnetic field will be retained.

The equation of motion of plasma particles p (i.e. ions i and electrons e) with mass m_p, charge e_p ($e_e=e$, $e_i=Ze$) and velocity $\underline{v}_p$ is then given by

$$m_p \, \underline{\dot{v}}_p = e_p \, (\underline{E} + \underline{v}_p \times \underline{B}) \tag{3.1}$$

where

$$\underline{E} = - \nabla \phi \, , \quad \Delta \phi = - \frac{e}{\varepsilon_o} \, (Zn_i - n_e) \tag{3.2}$$

$$n_p = \int d^3 v \, f_p. \tag{3.3}$$

$f_p(\underline{x}, \underline{v}, t)d^3x \, d^3v$ is the number of particles p counted in a phase volume $d^3x \, d^3v$. Equations (3.1) to (3.3) represent the collisionless "Vlasov model".

If one intends to apply the kinetic model to the larger length scales of the presheath, one has to include collisions. In the equation of motion (3.1) a term $m_p \, \dot{\underline{v}}_{p \, coll}$ must be added, which

106

accounts for the dynamical friction and stochastic diffusion in velocity space of particle p by collisions with other plasma particles,

$$\dot{v}_{coll} = \dot{v}_{dyn.\ fr.} + \dot{v}_{diff} \ . \tag{3.4}$$

Equ. (3.4) represents the Fokker-Planck ansatz for Coulomb collisions / 5 /.

For a solution of the kinetic eqs. (3.1) to (3.3) one has to impose boundary conditions: For the particles one has to prescribe the velocity distribution $f_p^+ (x_b , v , t)$ at the boundary x_b for instreaming particles. At the plasma side, one has to infer this boundary condition from the knowledge of plasma properties, for instance, by fluid calculations of the main plasma and parts of the scrape-off layer or from symmetry conditions. At the wall boundary conditions are determined by particle and energy reflexion properties of the wall. As boundary condition for the electric field it is assumed that the time-integrated current flow into and out of the system gives rise to a surface charge at the boundary which determines the normal component of the elctric field.

Within the computation area there will exist sources or sinks of particles, momentum and energy (e.g. by ionization or charge exchange).

The particle system is followed in time starting from some initial condition and will in general run into a steady state for stationary boundary conditions and sources.

The results of the computational model will be presented in the following two sections by plotting the final steady-state profiles of moments of the distribution function. The lowest order moments are

density	n	$=$	$\int f\ d^3v$
flux	$\underline{\Gamma}$	$=$	$\int f\ \underline{v}\ d^3v$
flow velocity	$\underline{V}$	$=$	$\underline{\Gamma}/n$
pressure	$\underline{\underline{p}}$	$=$	$m\int f\ (\underline{v}-\underline{V})\ (\underline{v}-V)\ d^3v, \ p = \frac{m}{3}\int f\ (\underline{v}-\underline{V})^2\ d^3v$
"temperature"	$\underline{\underline{T}}$	$=$	$\underline{\underline{p}}/n, \ T = p/n$
momentum flux	$\underline{\underline{P}}$	$=$	$m\int f\ \underline{v}\ \underline{v}\ d^3v = m\ \underline{V}\ \underline{\Gamma} + \underline{\underline{p}}$
heat flux	$\underline{q}$	$=$	$m/2\ \int f\ (\underline{v}-\underline{V})^2\ (\underline{v}-\underline{V})\ d^3v$
energy flux	$\underline{Q}$	$=$	$m/2\ \int f\ v^2\ \underline{v}\ d^3v =$
		$=$	$[(m/2\ V^2 + T/(\gamma-1)]\ \underline{\Gamma} + \underline{\underline{p}} \cdot \underline{V} + \underline{q} \ , \ \gamma = 5/3.$

($\underline{\underline{T}}$ is not a temperature in the thermodynamic sense but indicates the width of the velocity distribution.)

These moments satisfy certain conservation relations, i.e. for particle density $\partial n/\partial t + \nabla \cdot \underline{\Gamma} = S_n$

momentum $\qquad\qquad \partial(m\underline{\Gamma})/\partial t + \nabla \cdot \underline{\underline{P}} = ne\ (\underline{E} + \underline{V} \times \underline{B}) + \underline{S}_m$

energy $\qquad\qquad \partial(nmV^2/2 + p/(\gamma-1))/\partial t + \nabla \cdot \underline{Q} = e\underline{\Gamma} \cdot \underline{E} + S_q$. $\qquad$ (3.6)

S_n, $\underline{S}_m$ and S_q are sources of particle number density, momentum and energy.

The set of conservation equations (3.6) for ions and electrons together with equation (3.2) for the electric field, may serve as a base for a fluid description of the plasma /2/ in the scrape-off layer. To this end the third order moment must be express-able by lower order moments, as for instance in equ. (2.3). This is only possible if the mean free path of plasma particles is suffi-ciently smaller than the macroscopic gradient lengths of the plasma (see Section 2).

If the plasma is quasineutral, equ. (3.2) may be omitted and the electric field is implicitly determined by the quasineutrality condition

$$Zn_i = n_e .\qquad\qquad (3.7)$$

The set of conservation equations (3.6) may serve as a base for a qualitative understanding of the plasma flow in the scrape-off layer. Assume that the flow is stationary ($\partial/\partial t = 0$) and one-dimensional ($\nabla \equiv \partial/\partial x$). If there were

no sources, $\quad S = 0$,

no net space charge, $Zn_i e = n_e e = ne$,

no current, $Ze\ \underline{\Gamma}_i = e\ \underline{\Gamma}_e = \underline{\Gamma}$,

it would follow from (3.6) that

$$\underline{\Gamma}_x = \text{const}$$
$$\underline{\underline{P}}_{xi} + \underline{\underline{P}}_{xe} = \text{const}$$
$$\underline{Q}_{xi} + \underline{Q}_{xe} = \text{const}. \qquad\qquad (3.8)$$

If furthermore there were no transport, i.e. no heat flux, $\underline{q} = 0$, and no viscosity, p isotropic, $p = nT$, then from (3.8)

$$n = \text{const}$$
$$\underline{V} = \text{const}$$
$$T_i + T_e = \text{const}. \qquad\qquad (3.9)$$

Thus, under the assumptions made, the conservation relations would prohibit any spatial change of the fluid variables.

The actual changes of flow variables in the 1d stationary flow of the scrape-off layer thus originates from

1. sources and transport in the presheath
2. the Lorentz force $\underline{j}$ x $\underline{B}$, $\underline{j}$ = e ($Z\underline{\Gamma}_i - \underline{\Gamma}_e$), in the magnetic sheath
3. the electrostatic force e ($Zn_i - n_e$) $\underline{E}$ in the electrostatic sheath.

A two- or more dimensional flow is less restricted by conservation laws. The current component to a floating wall does not necessarily vanish as in 1d, the flow can perform circulating motions without sources etc. The features of 2d flow in the scrape-off layer will be presented in Chapter 10.

In the following two Sections the kinetic and fluid models, as described before, will be applied to investigate the 1d flow in the sheath with an oblique magnetic field and in the presheath.

4. THE SHEATH

Plasma flowing through the scrape-off layer to an absorbing wall passes two regions. In the first, plasma and energy is fed into the layer by diffusion and heat conduction from the main plasma and by ionization within the layer itself. The plasma flow becomes accelerated along the magnetic field to sonic or supersonic velocities at length scales of the sources, the transport processes and the interactions with neutrals. This region is sometimes called "presheath".

The second region, called "sheath" is directly attached to the wall. It is to a large extent determined by the electron reflexion conditions of the wall. Within the sheath plasma flow is accelerated further but in general at much smaller scale lengths than in the presheath, i.e. at the ion gyro radius and the Debye length. Therefore, gradients of flow variables and electric field in the sheath are much larger than in the presheath. Collisions within the sheath in general can be ignored (see Fig. 3).

The structure of the sheath depends on the magnitude and orientation of the magnetic field relative to the wall. For an oblique field the sheath exhibits a double structure of quasineutral flow at the ion gyro scale under the combined action of electric and magnetic forces followed by a region of non-neutral flow at the Debye length scale where the electric force dominates. Without or with wall-perpendicular magnetic field the sheath consists only of the electric Debye part. This case has been discussed in the preceding chapter by P.C. Stangeby. It will be used as a starting point and reference case for the more general case of an oblique magnetic field.

a)   Sheath without or with perpendicular magnetic field (Debye sheath)

We assume an infinitely extended, plane wall at floating potential ϕ_w. The plasma flow through the sheath is then one- dimensional and directed along the wall-normal x ($\psi = 0$ in Fig.2). The wall is assumed to absorb ions totally and electrons to a fraction $1-\gamma_{re}$. The reemitted electron fraction γ_{re} starts with zero energy from the wall. Let us further assume that the flow in the sheath is collisionless. The stationary distribution function f_p of ions and electrons at any point x in the sheath is then determined by the distribution function of instreaming particles at the upstream sheath edge x_S, by the reflexion condition at the wall at $x = x_w$ and by the potential $\phi(x)$.

Electrons entering the sheath at x_s with potential $\phi_s = \phi(x_s)$ may have a Maxwellian distribution

$$f_e^+(x_s, \underline{v}) = f_{es}^M \qquad \text{for } v_x > 0$$

with $\quad f_{es}^M = n_{es}^M [m_e/(2\pi T_{es})]^{3/2} \exp(-\tfrac{1}{2} m_e v^2/T_{es}). \qquad (4.1)$

The distribution for $v_x < 0$ is determined by the reflexion of electrons by the negative sheath potential and by the wall : All electrons entering the sheath with energy $m_e v_x^2/2 < e(\phi_s-\phi_w)$ are reflected in the sheath, electrons with $m_e v_x^2/2 \geq e(\phi_s-\phi_w)$ are either absorbed or inelastically reflected at the wall. Thus, the absorbing wall leads to an electron velocity distribution with a loss region:

$$f_e(x,\underline{v}) = \begin{cases} f_{es}^M \exp[e(\phi-\phi_s)/T_{es}] + f_{re} , & v_x \geq v_c \\ 0 & , \quad v_x < v_c \end{cases} \qquad (4.2)$$

where $v_c = - [2e(\phi-\phi_w)/m_e]^{1/2}$ is the velocity of an electron starting at the wall with $v = 0$, f_{re} is the number of electrons at x_c having been reemitted from the wall.

The distribution function of ions at the sheath edge x_S depends on the conditions in the presheath. Due to the acceleration in the presheath, their distribution is shifted to a positive mean velocity V_S and since no ions are reflected in the sheath or at the wall, f_{is} vanishes for $v_x < 0$.

Figure 4 shows a schematic picture of the changes of ion and electron distributions within the sheath. The negative potential accelerates ions towards the wall, whereas it reflects electrons to a great extent. Therefore, approaching the wall f_e decreases with the Boltzmann factor $\exp[e(\phi- \phi_S)/T_{es}]$ of equ. (4.2) and the loss region at negative velocities grows until at the wall $x = x_w$ only electrons with $v_x \geq 0$ are present, since the wall was assumed to be totally energy absorbing.

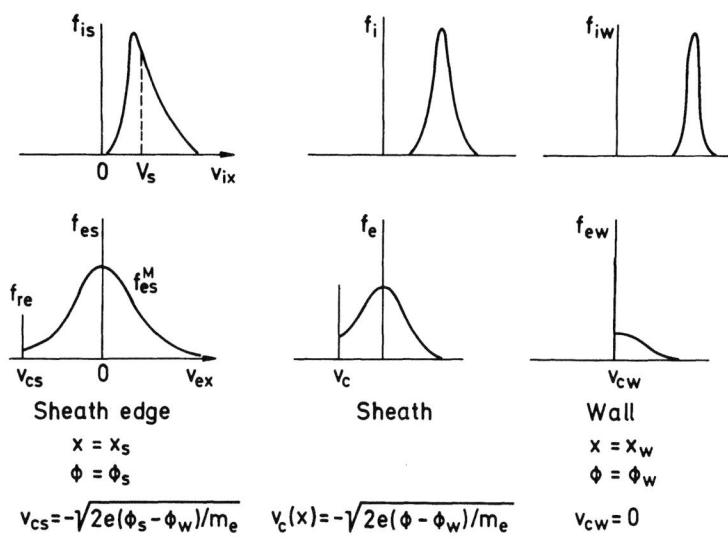

Fig. 4. Schematic velocity distributions of ions and electrons in a sheath without or with perpendicular magnetic field. Electrons are partly reemitted from the wall without initial energy.

Table 1: Moments of ions and electrons in a sheath without or with perpendicular magnetic field. At $x = x_s$: $v_{ix}=V_s$, $T_i=0$, $T_e=T_{es}$, $v_{te}=v_{tes}=(T_{es}/m_e)^{1/2}$. A fraction γ_{re} of wall-incident electrons is re-emitted with zero energy.

Moment	Ions	Electrons
n	Γ_i / v_{ix}	$\frac{1}{2}n^M_{es}\,[1-\mathrm{erf}(\frac{v_c}{\sqrt{2}\,v_{tes}})]\,\exp[e(\phi-\phi_s)/T_{es}]$ $+ \frac{\gamma_{re}}{1-\gamma_{re}}\,\lvert\Gamma_e\rvert/\lvert v_c\rvert$
V_x	$\sqrt{V_s^2 - 2\,Z\,e(\phi-\phi_s)/m_i}$	Γ_e/n_e
Γ_x	Γ_i	$\Gamma_e = \frac{1}{\sqrt{2\pi}}(1-\gamma_{re})n^M_{es}\,v_{tes}\,\exp[(\phi_w-\phi_s)/T_{es}]$
T_{xx}	0	without reflected electrons $T_{es} + m_e\,V_{ex}\,(v_c-V_{ex})$
Q_x	$[m_i\,V_s^2/2-Ze(\phi-\phi_s)]\,\Gamma_i$	$[\frac{2}{1-\gamma_{re}}\,T_{es} + e(\phi-\phi_w)]\Gamma_e$

Table 1 gives the appropriate low-order moments of the velocity distributions of ions and electrons as a function of potential ϕ within the sheath and Fig. 5 shows profiles of these moments over x. Instreaming ions were asssumed to be cold, $T_{is}=0$. The negative sheath potential increases the ion velocity V_{ix}, but, as may be seen from the electron velocity distribution f_e of Fig. 4, even to a larger extent also the mean velocity $V_{ex}=\langle v_{ex}\rangle$, $V_{ex} > V_{ix}$. Correspondingly, since ion and electron flux Γ_x is ambipolar, $n_i > n_e$, i.e. the sheath is charged positively. In the negative sheath potential ion energy flux Q_{ix} is increased at the expense of electron energy flux Q_{ex}, so that $Q_{ix} + Q_{ex} = $ const.

The flow velocity V_s of the instreaming ions at the sheath edge x_s is not completely arbitrary. In order to obtain a monotonic potential drop across the sheath V_s has to exceed a certain limit: This can be shown by inspecting the dispersion property of the flow at x_s. Assume a homogeneous, quasineutral 1d plasma flow with $V_x = V_s$, $T_{is}=0$, $T_e=T_{es}$, and $\gamma_{re}=0$. For a change of the potential ϕ one gets a corresponding change of $n_i(\phi)$ and $n_e(\phi)$ (see Table 1). Inserting n_i and n_e into Poisson's equation (3.2) linearizing for small $\phi-\phi_s$ and assuming $\phi-\phi_s \propto \exp(ikx)$ one gets a dispersion relation

$$k^2 \; \lambda_D^2 = (C_s^2 - V_s^2)/V_s^2 \tag{4.3}$$

as shown in Fig. 6a. $C_s=(ZT_{es}/m_i)^{1/2}$ is the isothermal sound speed at x_s for $T_{is}=0$, λ_D the Debye length. Thus a non-oscillating monotonic change of the potential, i.e. $k^2 < 0$, can only be obtained if

$$V_s \gtreqless C_s \tag{4.4}$$

This inequality is called the Bohm condition /7,8/. It represents a condition for monotonic potential change at the Debye length scale in a 1d plasma flow without sources for cold ions. If ions have a finite thermal spread, the Bohm condition may be generalized to /9/

$$\langle v_{ix}^{-2} \rangle = \frac{1}{n_{is}} \int d^3 v_i \; f_{is}/v_{ix}^2 \leq m_i/(ZT_{es}). \tag{4.4'}$$

The expressions for ion and electron fluxes Γ_{ix} and Γ_{ex} of Table 1 may be used to determine the potential drop across the sheath: From the condition of quasineutrality at x_s, $n_{es} \cong n_{es}^M \cong n_{is}$, and of vanishing net current $Ze\Gamma_i - e\Gamma_e = 0$ one gets the potential drop /6/

$$e(\phi_w - \phi_s) \cong \ln \frac{\sqrt{2\pi} \; V_s}{(1-\gamma_{re})v_{tes}} . \tag{4.5}$$

For $\gamma_{re} = 0$, $T_{is} = 0$, $V_s = C_s$ and H^+ ions $e(\phi_w-\phi_s)/T_{es} = -2.84$.

Inserting n_i and n_e from Table 1 into Poisson's equation (2.3), one gets the potential profile $\phi(x)$ of the Debye sheath. As may be seen from Fig. 5, this profile has an extension of about $\lambda_{es} \cong 10 \, \lambda_D$.

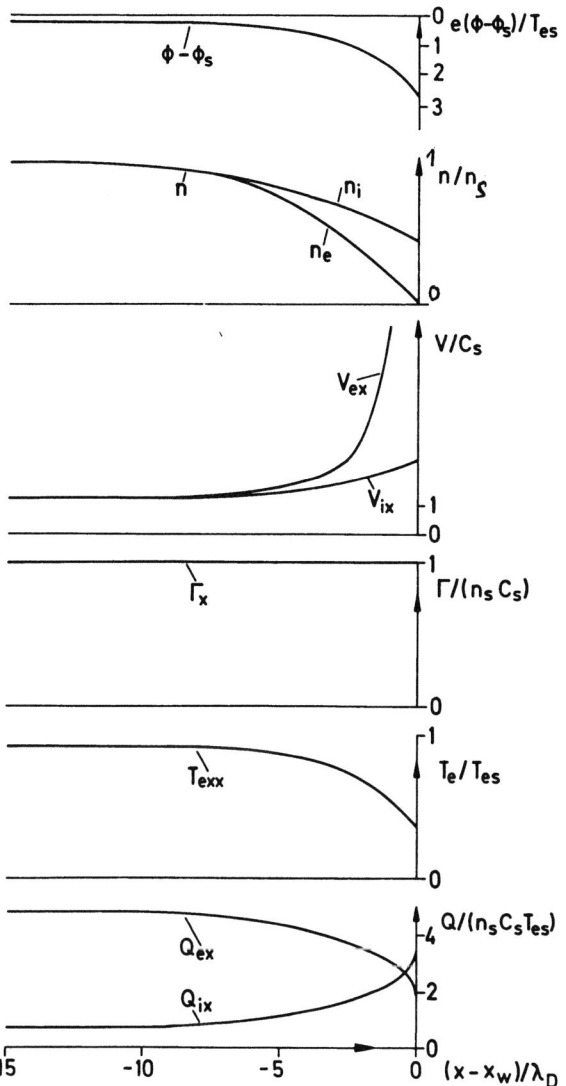

Fig. 5.   Potential and flow variables (density n, mean velocity V, flux Γ, "temperature" T, energy flux Q)
within the sheath for $\psi = 0$, $T_{is} = 0$, $\gamma_{re} = 0$,
$m_i/m_e = 1836$.

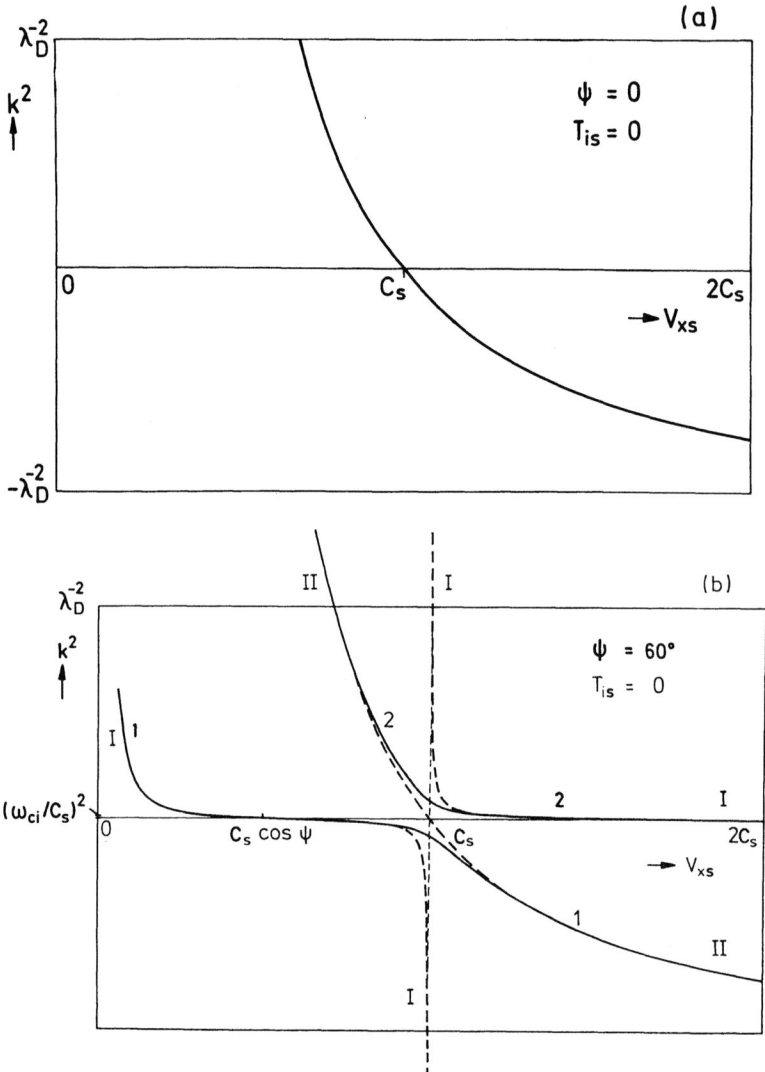

Fig. 6. Dispersion relation k^2 (V_{xs}) for cold ions
a) $B = 0$ or $\psi = 0$, b) $B \neq 0$, $\psi \neq 0$.

b)   Sheath with oblique magnetic field

A magnetic field $\underline{B}$ perpendicular to the wall and thus parallel to the electric field $\underline{E}$ does not change the plasma flow in the sheath. Gyro motion perpendicular to x and flow velocity along x are completely decoupled. On the other hand, a magnetic field intersecting the wall at an oblique angle, i.e. $\psi \neq 0$ in Fig. 2, couples the flow component along the gradients, i.e. along x, to the components perpendicular to this direction and thus changes the structure of the sheath.

In the presheath electric forces are small as compared to magnetic forces. Hence, plasma flows in the pre-sheath nearly parallel to the magnetic field, i.e. obliquely to the electric field along x. In the sheath ions are detracted from B direction by the electric field, they begin to flow obliquely to B. Electrons, due to their small mass, are more strongly coupled to magnetic field lines and at first continue to flow along B until finally they too deviate. Thus, ions and electrons reach the wall at different flow paths. Electrons pulling the ions through the magnetized sheath to the wall via the electric field have to overcome not only the inertial but also the Lorentz force acting on the ions.

The length scale λ_m along x at which cold ions are accelerated under the combined action of the electric and the Lorentz force can be estimated as follows: The component $E_{\perp}$ of $\underline{E}$ normal to $\underline{B}$ forces the ions on gyro orbits with radius

$$l_i \cong E_{\perp}/(B\omega_{ci})$$

where $E_{\perp} = E \sin \psi$ and $\omega_{ci} = ZeB/m_i$ the ion gyro frequency. λ_m is the projection of l_i on x, hence

$$\lambda_m \cong l_i \sin \psi.$$

E may be estimated to be

$$eE \cong T_{es}/\lambda_m.$$

Combining these estimates one gets

$$\lambda_m^2 \cong Z\, T_{es} \sin^2 \psi/(m_i\, \omega_{ci}^2)$$

Thus, the magnetic part of the sheath has a length scale of the ion gyro radius at the ion acoustic speed $C_s = (ZT_{es}/m_i)^{1/2}$.

Further downstream in the sheath again the electrostatic Debye sheath develops. The sheath in an oblique magnetic field thus exhibits a characteristic double structure of scales λ_m and λ_{es} respectively. Since in general

$$\lambda_m \gg \lambda_{es}$$

the magnetic part of the sheath is quasineutral, while the electrostatic part is non-neutral. Figure 7 shows this double structure of

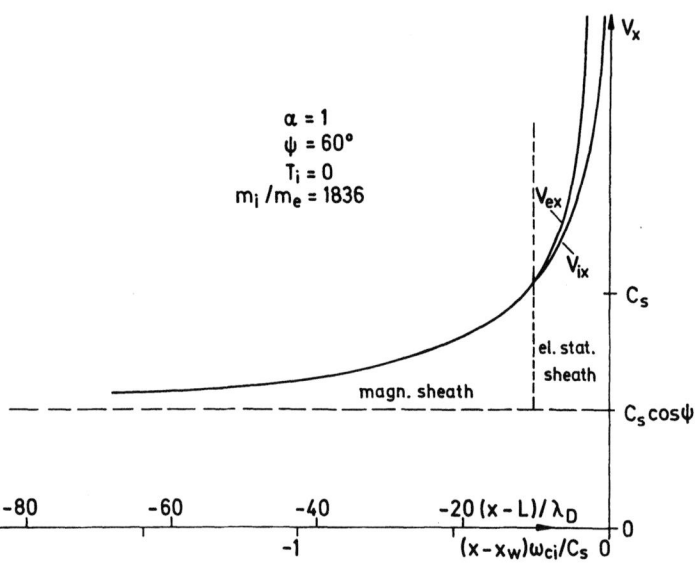

Fig. 7.   Velocity profile $V_x(x)$ in a sheath with oblique
magnetic field.

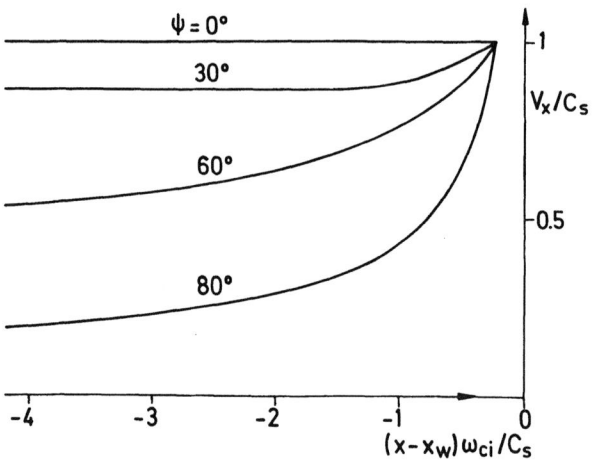

Fig. 8.   Velocity profiles $V_x(x)$ in the magnetic part of the
sheath.

the sheath for the x-component of flow velocity V_x of ions and electrons /10/.

As in the case of perpendicular magnetic field, the ion flow velocity at sheath entrance has to exceed a certain limit to make a monotonic potential drop possible /10/. This can again be shown by the dispersion property of the flow at x_s. A homogeneous plasma flow along the magnetic field with velocity $|V| = V_s$ can be changed by a small disturbance $\phi - \phi_s \propto \exp(ikx)$ only for k-values which obey a dispersion relation $k^2(V_{xs})$ as shown in Fig. 6b. For each value of V_{xs} two values 1 and 2 of k^2 are possible. Curve 2 is always positive corresponding to purely oscillatory modes. Curve 2 can assume negative values $k^2 < 0$ only for

$$V_{xs} \geqq C_s \cos \psi \tag{4.6}$$

or

$$V_s \geqq C_s.$$

Thus for an exponential variation of $\phi - \phi_s$ according to mode 1 (possibly superposed by oscillations of mode 2) the flow velocity V_s along B at the sheath edge must exceed the sound speed C_s irrespective of field angle ψ. Condition (4.6) is the generalization of the Bohm condition (4.4). For $T_{is} \neq 0$ this condition is /11/

$$\langle v_{i\parallel}^{-2}\rangle = \frac{1}{n_{is}} \int d^3v_i \, f_{is}/v_{i\parallel}^2 \leqq m_i/(ZT_{es}) \text{with } v_\parallel = (\underline{v} \cdot \underline{B})/B. \tag{4.6'}$$

For $V_{xs} > C_s \cos \psi$ the decay length k^{-1} of the potential is of the order C_s/ω_{ci} and decreases with increasing V_{xs}. At

$$V_{xs} \cong C_s \tag{4.7}$$

a qualitative change in the dispersion property occurs: Curve 1 departs from the asymptotic mode I (which is a quasineutral electrostatic ion cyclotron mode) and approaches the asymptotic mode II (which is a non-neutral ion acoustic mode). The decay length k^{-1} decreases to a value of the order of the Debye length. Thus condition (4.7) represents the Bohm condition for the onset of the electrostatic part of the sheath in case of an oblique magnetic field.

The dispersion properties of Fig. 6b may be recognized in the profiles of the flow velocity V_x in Fig. 7: In the magnetic sheath the flow is quasineutral, i.e. $V_{ix} = V_{ex} = V_x$. V_x increases from $C_s \cos \psi$ to C_s on a length scale

$$\lambda_m \cong C_s/\omega_{ci}. \tag{4.8}$$

The further growth of V_x at $V_x > C_s$ takes place on the Debye length scale λ_D and the flow becomes non-neutral, i.e. $V_{ix} \neq V_{ex}$.

The profiles of V_x in the magnetic sheath for different angles ψ are shown in Fig. 8. As may be seen, the characteristic thickness of the magnetic sheath is $0 \ldots 4 \ C_s/\omega_{ci}$.

The double structure of the sheath may also be recognized in the potential profiles $\phi(x)$, Fig. 9, for different angles ψ /10/. It may be

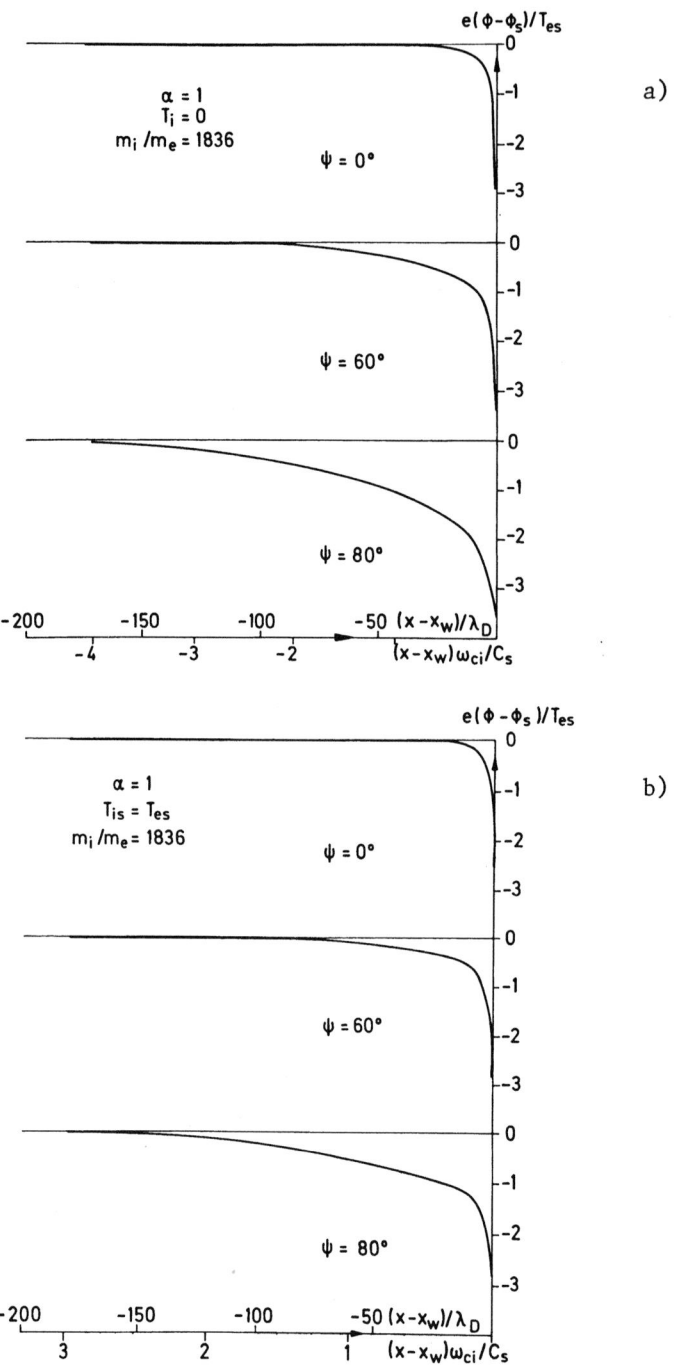

Fig. 9. Potential profiles in the sheath for different angles ψ of the magnetic field, $\gamma_{re} = 0$; a) $T_{is} = 0$, b) $T_{is} = T_{es}$.

118

noticed that the relative portion of potential drop in the magnetic
and electrostatic part of the sheath strongly depends on ψ : The potential drop in the magnetic part of the sheath increases and the
potential drop in the electrostatic sheath decreases with ψ until
for $\psi \to 90°$ the whole potential drop occurs in the magnetic part
and the electrostatic part of the sheath completely disappears. The
reason for this behaviour is, that for increasing field angles
it becomes more difficult for the ions to reach the necessary velocity V_x of equ. (4.7) for the onset of the electrostatic sheath.

Figure 10 shows the dependence of total potential drop $\phi_s - \phi_w$
across the sheath (magnetic and electrostatic) on field angle
and electron reflexion coefficient γ_{re}. The curves $\gamma_{re} = 0$ correspond to the potential profiles of Fig. 9.

As may be noticed the dependence of the sheath potential $\phi_s - \phi_w$
on the angle ψ is amazingly small. This can be explained as follows:
The function of the sheath potential is to adapt the electron flux
component perpendicular to the wall Γ_{ex} to the ion flux component
$Z \Gamma_{ix}$ by electron reflexion. For an ion flow velocity V_s at the
sheath edge the ion flux component is $\Gamma_{ix} = n_s V_s \cos \psi$. Electrons in
the magnetic sheath are magnetized i.e. their velocity distribution
is rotationally symmetric around the direction of the magnetic field.
If the magnetic field would be so strong that their gyroradius
$l_e \ll \lambda_D$, then they would be magnetized in the electrostatic sheath
either and their velocity distribution in the whole sheath would be
the same as that of Fig. 4 provided that the x-direction in Fig. 4
is replaced by the direction of the magnetic field, i.e. v_{ex} changed
to $v_{e\parallel}$. In particular, the electron flux would be

$$\Gamma_{ex} \cong \frac{1-\gamma_{re}}{\sqrt{2\pi}} \; n_s \; v_{tes} \; \exp[e(\phi_w - \phi_s)] \; \cos \psi,$$

Equating Γ_{ex} to $Z \Gamma_{ix}$ results in the same expression (4.5) for the
sheath potential with oblique magnetic field as in the perpendicular
case $\psi = 0$. For finite electron gyroradius l_e the velocity distribution of electrons in the sheath is changed by the more complicated
absorbtion conditions at the wall. The boundary of the loss region
is no longer a straight line $v_{e\parallel} = v_c$ but lies somewhere in the
region $v_e^2 > v_c^2$ (for the special case $\psi = 90°$ this loss region was
determined analytically in /12/, in the particle simulation model
described in Section 3 it is incorporated by the reflexion condition
at the wall).

For decreasing magnetic field strength, i.e. increasing electron
gyroradius l_e electrons would reach the wall more easily, so the
electron repelling sheath potential increases. This is shown in
Fig. 11. As may be seen, for decreasing values of l_e / λ_D the sheath
potential approaches the value of $\psi = 0$, i.e. $\phi_w (\alpha \to \infty) = \phi_w (\psi = 0)$.

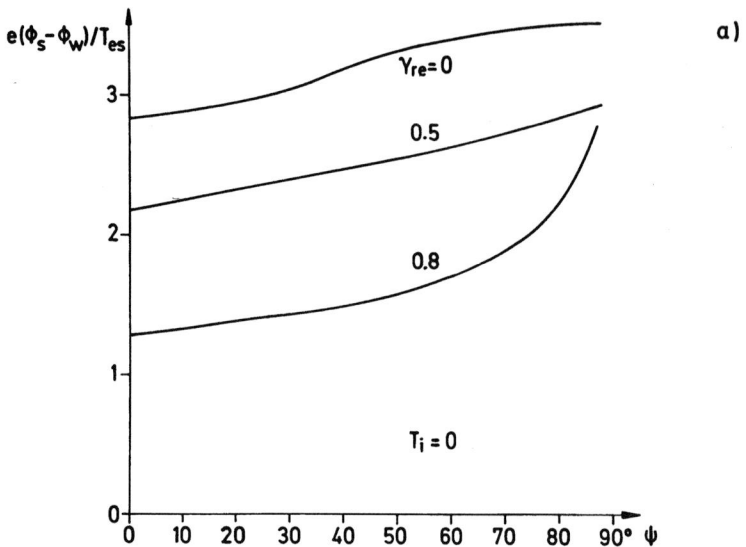

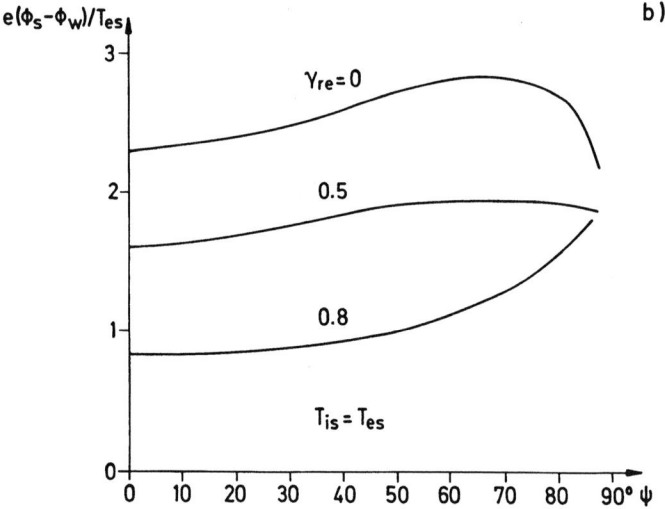

Fig. 10. Potential drop across the sheath for different
angles ψ and secondary electron emission coefficients γ_{re}; a) $T_{is} = 0$, b) $T_{is} = T_{es}$.

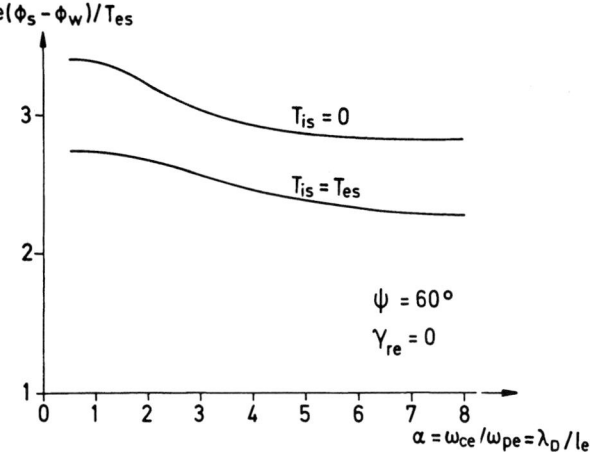

Fig. 11.   Potential drop across the sheath for different
magnetic field strengths. $\alpha = \omega_{ce}/\omega_{pe}$, $\psi = 60°$,
$\gamma_{re} = 0$.

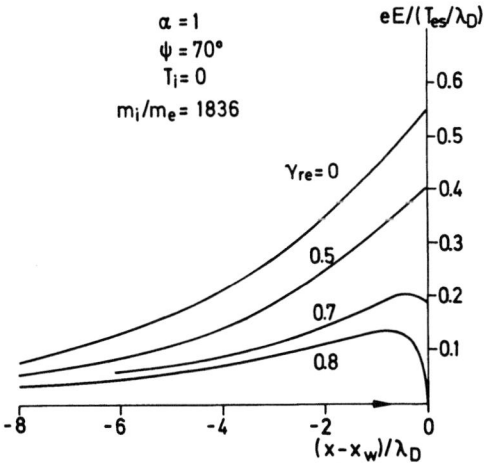

Fig. 12.   Electric field near the wall for different values
of secondary emission coefficient γ_{re}.

If electrons are partly reflected by the wall, e.g. by secondary emission, the potential drop across the sheath is smaller than in the case of a totally absorbing wall since part of the electron reflexion in the sheath is now taken over by the wall (see equ. (4.5) and /13/). In Fig. 10 the effect of reflexion rate γ_{re} on the potential drop $\phi_s - \phi_w$ is shown. It is assumed that reemitted electrons start at the wall with negligible energy. For very oblique angles $\psi > \psi_r$ of the magnetic field the reemitted electrons do not get free of the wall, because their departure Δx from the wall by acceleration along the magnetic field within half a gyro period is less than their gyro-radius l_e, i.e.

$$\Delta x = \frac{1}{2} \frac{e}{m_e} E_w \ (\pi/\omega_{ce})^2 \ \cos \psi \ < \ l_e = E_w \ \sin \psi \ / \ (B\omega_{ce})$$

$$\psi > \psi_r = \text{arc tg } \pi^2/2 = 79° \tag{4.9}$$

where $E_w = E(x_w)$. These electrons therefore hit the wall again and again and are finally absorbed. Therefore, for angles $\psi > \psi_r$ the sheath potential should become independent on γ_{re}. Another limit is reached if the reemission coefficient γ_{re} is increased above a critical value /13/ which is about 0.8 nearly independent on ψ. This is shown in Fig. 12. For increasing values of γ_{re} the electric field at the wall E_w decreases until, at the critical value of γ_{re}, $E_w = 0$. For even larger values of γ_{re}, the electric field E_w becomes instationary and temporarily negative. During the negative field periods electrons are driven back to the wall, so that in the mean no more electrons can escape the wall than at the critical value of γ_{re}.

Figure 13 shows profiles of density n, flow velocity $\underline{V}$, particle flux $\underline{\Gamma}$, temperature T_e, heat flux $\underline{Q}$, and electric potential ϕ across the sheath for a magnetic field angle $\psi = 85°$. For comparison with the case of perpendicular field incidence (Fig. 5), we have assumed equal particle and energy influx in both cases, i.e. the normal components of particle and energy flux Γ_x and Q_x at the sheath edge x_s are the same (the total fluxes at x_s along B then are larger for the oblique case by a factor $1/\cos\psi$). Since the dimension of the sheath in general is small compared to the mean free path for ionization (Fig. 3), we neglect particle and energy production in the sheath. Thus in a stationary 1d sheath

$$Z \ \Gamma_{ix} = \Gamma_{ex} = \Gamma_x = \text{const} \tag{4.10}$$

and

$$S_{qi} = S_{qe} = 0.$$

Within the sheath preferentially the ion flux is turned from the B parallel direction towards the wall normal along x and twisted to some extent into the y direction. So a wall parallel current mainly along the z-direction arises in the sheath.

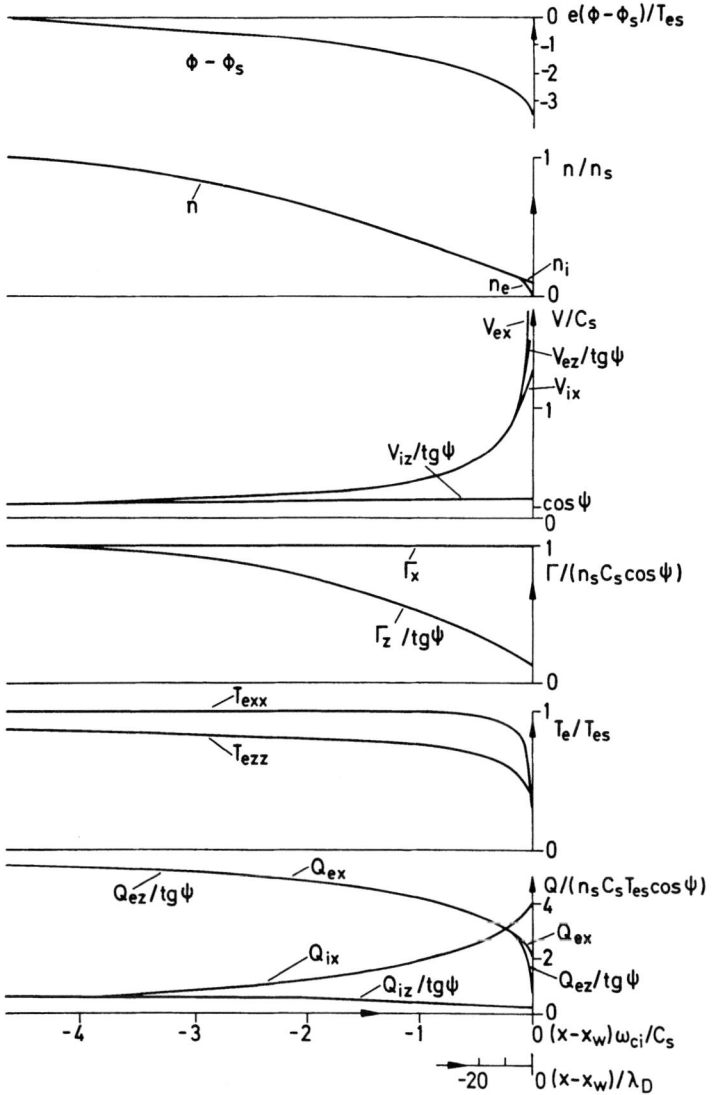

Fig. 13.   Potential and flow variables in a sheath with oblique magnetic field. $\alpha = 1$, $\psi = 85°$, $T_{is} = 0$,
$\gamma_{re} = 0$.

Also the energy flux of the ions is bent more strongly to the wall normal than that of the electrons. Ions gain energy flux in x direction (see eq. (3.6), ' = d/dx),

$$Q'_{ix} = Z e \Gamma_{ix} E \qquad (4.11)$$

from the electrons

$$Q'_{ex} = - e \Gamma_{ex} E. \qquad (4.12)$$

So

$$Q_{ix} = Ze \Gamma_{ix} (\phi_s - \phi) + Q_{ix} (x_s) \qquad (4.13)$$

$$Q_{ex} = -e \Gamma_{ex} (\phi_s - \phi) + Q_{ex} (x_s) \qquad (4.14)$$

and for condition (4.10)

$$Q_{ix} + Q_{ex} = Q_{ix} (x_s) + Q_{ex} (x_s) = \text{const.} \qquad (4.15)$$

Energy flux of ions at the sheath edge $Q_{ix} (x_s)$ /14/ depends on their acceleration and thermalization in the presheath. Bohm's condition eq. (4.6) gives a lower bound of their kinetic energy at x_s. Together with an estimate of the convective energy flux $\gamma_i/(\gamma_i-1) T_{is} \Gamma_{ix}$ for a shifted Maxwellian and a heat conduction flux q_{ix} one gets

$$Q_{ix}(x_s) = \delta_i T_{es} \Gamma_{ex} = (\frac{m_i}{2} V_s^2 + \frac{\gamma_i}{\gamma_i - 1} T_{is}) \Gamma_{ix} + q_{ix}(x_s) \quad . \qquad (4.16)$$

Electron energy flux at the sheath edge

$$Q_{ex}(x_s) = \delta_e T_{es} \Gamma_{ex} \qquad (4.17)$$

is mainly determined by the electron reflexion properties of the sheath and of the wall, i.e. by sheath potential $\phi_s - \phi_w$ and electron reflexion coefficient γ_{re} of the wall. For $\psi = 0$ and electrons reflected from the wall as cold δ_e becomes (see /13/ and preceding Chapter)

$$\delta_e = 2/(1 - \gamma_{re}) + e(\phi_s - \phi_w)/T_{es} \quad . \qquad (4.18)$$

In Fig. 14 δ_e is plotted for arbitrary angles ψ and different electron wall reemission coefficients γ_{re}. For $\gamma_{re} = 0$, δ_e like the sheath potential is insensitive to ψ. With increasing γ_{re} also δ_e is increased because by secondary emission more energy is transported to the wall before an electron is absorbed at the wall. For large angles $\psi > \psi_r$ the secondary electrons are reabsorbed and δ_e approaches the value for $\gamma_{re} = 0$.

From eqs. (4.13) and (4.14) together with (4.16) to (4.17), (4.10) and the results of Fig. 10 and Fig. 14 one gets the energy fluxes of

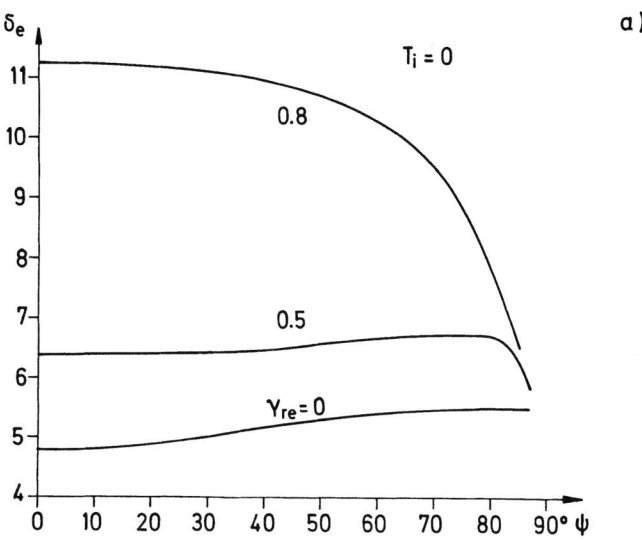

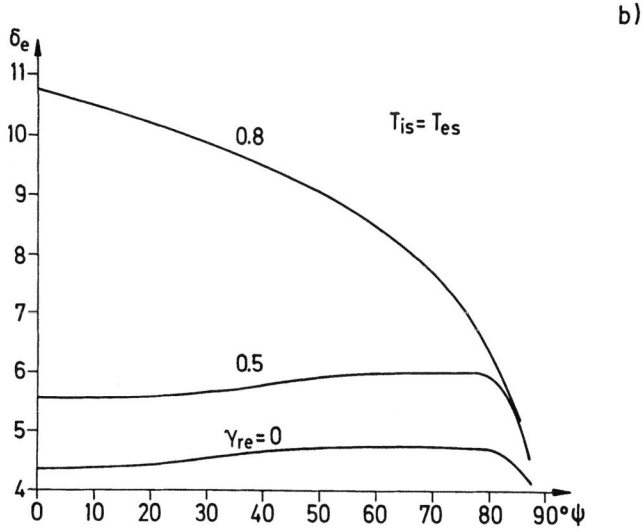

Fig. 14. Electron energy flux at the sheath edge per electron absorbed at the wall for different field angles ψ and secondary electron emission coefficient γ_{re}; a) $T_{is} = 0$, b) $T_{is} = T_{es}$.

ions and electrons to the wall:

$$Q_{ix}(x_w) = [\delta_i + e(\phi_s - \phi_w)/T_{es}] \, T_{es} \, \Gamma_x \qquad (4.19)$$

$$Q_{ex}(x_w) = [\delta_e - e(\phi_s - \phi_w)/T_{es}] \, T_{es} \, \Gamma_x. \qquad (4.20)$$

Ion energy may be partly reflected as energy of recycled neutral atoms (see Chapter 7), while electrons in general deposit all their energy at the wall.

The total amount of energy flux extracted from the scrape-off plasmas and transported through the sheath to the wall by plasma particles, is given by

$$Q_{ix}(x_s) + Q_{ex}(x_s) = (\delta_i + \delta_e) \, T_{es} \, \Gamma_x \quad . \qquad (4.21)$$

For $\gamma_{re} = 0$, $T_{is} = T_{es}$, $\gamma_i = 5/3$ and $q_i = 0$

$$\delta_i + \delta_e \cong 3.5 + 4.3 = 7.8 \quad .$$

Interactions of ions with the wall such as reflexion (as neutrals), sputtering etc., depend on the incidence angle ϑ_i' with the wall normal x. This angle together with the twisting angle φ_i out of the (E, B)-plane is plotted in Fig. 15 for incident cold H^+-ions as function of magnetic field inclination ψ. As may be seen, the ion incidence angle ϑ_i' increases with magnetic field inclination ψ but the ions always hit the wall under a steeper angle than the magnetic field /15/.

What happens to the sheath potential if the magnetic field becomes nearly parallel to the wall? As long as particle flow along B is the dominant transport process to the wall, the sheath potential continuously stays negative because ambipolar flow to the wall demands electron reflexion in the sheath. The case $\psi = 90°$ is singular since no field aligned particle flow reaches the wall and flow ambipolarity to the wall is fulfilled trivially. In this case the sheath potential is determined by diffusion flux across B or, in the absence of diffusion, by the time history of the sheath evolution.

5. THE PRESHEATH

As was pointed out in Section 4, the sheath demands sonic or supersonic plasma flow velocities and a certain amount of electron heat flux at its upstream edge. The accommodation of the flow to these conditions occurs in the so-called presheath region upstream of the sheath. Particle flow in the presheath originates from particles diffusing into the presheath from the main plasma and from ionization processes within the presheath. Energy is fed into the presheath from the core plasma by convection of the diffusing hot particles, and by heat conduction. In contrast to the sheath

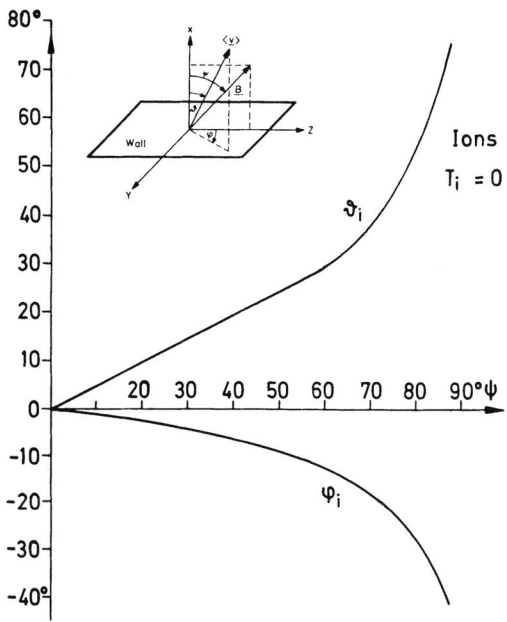

Fig. 15.   Incidence angles (ϑ_i, φ_i) of cold ions at the wall for different field angles ψ ($\alpha = 1$, $\gamma_{re} = 0$).

the presheath in general is collisonal. Therefore collisional transport and relaxation within the flow has to be taken into account.
In addition, the presheath comprises a large amount of interaction
between plasma and neutral gas or impurities. These processes will
be discussed in detail in Chapters 6 and 10 to 14. In this Section
only some general features of the plasma flow in the presheath will
be presented.

a) Collisional Presheath

Let us assume a very simplified presheath: a straight magnetic field
extends along the coordinate s (Fig. 2). Plasma flow along s is one-
dimensional, stationary and quasineutral. The mean free path of
plasma particles is so short that fluid equations (3.6) suffice to
describe the plasma flow. (A kinetic treatment of the collisional
presheath for a special kind of ion collisions and for isothermal
electrons was carried out in / 16 /.) The only relevant transport
process is assumed to be electron heat conduction

$$q_e = - \chi_e \, T_e'$$

(5.1)

where $' = d/ds$. Plasma flow is fed by sources of particles S_{ni} and
S_{ne} ($ZS_{ni} = S_{ne}$), momentum S_{mi} and S_{me} and energy S_{qi} and S_{qe} for
ions and electrons respectively. The sources may be due to diffusion
perpendicular to s or from interaction with neutral gas.

Elimination of the potential ϕ from the conservation eqs.
(3.6) yields the following system of equations for the 1d plasma
flow:

$$\frac{V'}{V} = Z \, \frac{(\frac{\gamma_i+1}{2 \, Z} \, m_i \, V^2 + T_e) \, S_n - V S_m + (\gamma_i - 1) \, S_{qi} + \Gamma \, T_e'}{m_i \, \Gamma \, (c^2 - V^2)}$$

(5.2)

$$\frac{1}{\gamma_i - 1} \, \Gamma \, T_i' = - \Gamma \, T_i \, V'/V + (\frac{m_i}{2} \, V^2 - \frac{1}{\gamma_i - 1} \, T_i) \, S_n + Z \, S_{qi}$$

(5.3)

$$\chi_e \, T_e' = \Gamma \, (\frac{m_i}{2} \, V^2 + \frac{\gamma_i}{\gamma_i - 1} \, T_i + \frac{\gamma_e \, Z}{\gamma_e - 1} \, T_e)/Z - Q.$$

(5.4)

$S_n = ZS_{ni} = S_{ne}$, $S_m = S_{mi} + S_{me}$, S_{qi} and S_{qe} are the amounts of
supplied particles, total momentum and energies to ions and electrons
per length ds. Accordingly, the particle flux

$$\Gamma = \int_0^s S_n \, ds$$

(5.5)

and the total energy flux

$$Q = Q_i + Q_e = \int_0^s (S_{qi} + S_{qe}) \, ds.$$

128

are the accumulated particle and energy supplies. V is the common flow velocity of ions and electrons, $n_i = \Gamma/(ZV)$ and $n_e = \Gamma/V$ their densities, γ the ratio of specific heats, and

$$C^2 = (\gamma_i T_i + Z T_e)/m_i \qquad (5.7)$$

the square of the sound speed C. (This specific form of the sound speed results from the assumptions of adiabatic ions and non-adiabatic electrons).

Inspection of equ. (5.2) shows that particle sources S_n and ion energy sources S_{qi} as well as momentum sinks $S_m < 0$ (for instance by friction with neutral gas) drives the plasma flow velocity towards the sound speed C, i.e. these terms tend to increase V in the subsonic regime V < C and tend to decrease it for supersonic velocities V > C. The electron temperature gradient T_e', which in general is negative, has the opposite effect, i.e. it drives V away from C.

As a boundary condition for eqs. (5.2) to (5.4) at s = 0, we demand symmetry, i.e.

$$T_i'(0) = 0 \qquad (5.8)$$

from which follows

$$T_i(0) = Z \frac{\gamma_i - 1}{\gamma_i} \frac{S_{qi}(0)}{S_n(0)} .$$

$T_e'(0) = 0$ and $V(0) = \Gamma(0)/n_e(0) = 0$ are already guaranteed by eqs. (5.4), (5.5), and (5.6). Thus the quantities

$$n_o = n_e(0)$$

and

$$T_{eo} = T_e(0)$$

have to be determined otherwise.

The boundary conditions at the sheath edge $s = L_s$ are determined by the properties of the sheath. The Bohm condition (4.4) and (4.6) demands for $T_i \ll T_e$ a flow velocity

$$V(L_s) = V_s \geq C(L_s) = C_s \qquad (5.9)$$

(for $T_i \equiv T_e$ conditions (4.4') and (4.6') for perpendicular or oblique magnetic field cannot be easily expressed in terms of low moments). Equation (4.17) together with the values for δ_e from Fig. 14 determines the electron energy flux

$$Q_e(L_s) = \frac{\gamma_e}{\gamma_e - 1} T_{es} \Gamma - \chi_{es} T_{es}' = \delta_e T_{es} \Gamma$$

The latter equation gives

$$T_{es}'/T_{es} = -[\delta_e - \gamma_e/(\gamma_e - 1)] \Gamma/\chi_{es} \qquad (5.10)$$

The two free parameters n_o and T_{eo} must be chosen in such a way that conditions (5.9) and (5.10) are satisfied. Two cases are possible:

1) V is subsonic, V < C, in the whole presheath $0 \leqslant s \leqslant L_s$. Then V approaches C at the sheath edge, $V_s = C_s$. This boundary condition together with (5.10) determines n_o and T_{eo}.

2) V passes the sound speed C and becomes supersonic at the sheath edge, $V_s > C_s$. In this case, at the transition of V through C the numerator together with the denominator of the r.h.s. expression of eq. (5.2) must become zero, in order to keep V' finite. This condition together with the boundary condition (5.10) again determine n_o and T_{eo}.

Which one of the two possibilities, 1) or 2), will be realized, depends on the distribution of the sources S and on the heat conductivity χ_e.

To illustrate the two different flow patterns 1) and 2), we assume two different particle sources: in case 1), particle Source S increases toward the wall, as it would be the case for ionization of neutrals recycled back from the wall,

$$S_n = \Gamma_1/d \ \exp \ [(s-L_s)/d].$$ (5.11)

In the other case, the particle source is located near the point x = 0, i.e. the particle source is remote from the wall,

$$S_n = \Gamma_1/d \ \exp \ (-s/d).$$ (5.12)

It is assumed further that $S_m = S_{qi} = 0$ and $\chi_e = $ const for simplicity.

Figure 16a,b shows the solutions of system (5.2) to (5.4) under the boundary conditions (5.9) and (5.10) for these two cases. It depends on the size of the ratio $\sigma = \Gamma^2/(S_n \chi_e)$ whether or not V passes through the sound speed C. In case 1) σ is bounded, $\sigma = \Gamma d/\chi_e < 2$. The flow velocity rises at scale length d of the extension of the source to sound speed C and does not exceed it. In case 2) σ grows unlimitedly. The flow velocity first grows to sound speed at scale length d and proceeds to grow further at a scale length Γ/χ_e determined by the heat conductivity.

b) Collisionless Presheath

Let us now consider the other limiting case, where ions and electrons can run through the presheath without collisions after they entered the presheath by ionization or perpendicular diffusion. This kinetic problem has been treated analytically by many authors for different starting conditions of the ions and for isothermal electrons, /6, 9, 17/. By using the particle model described in Sect. 3, one can simulate the presheath without making assumptions on the electron distribution. Ions are assumed to start without energy,

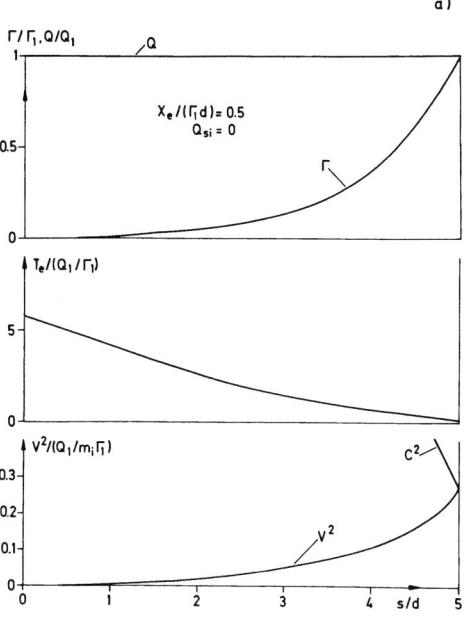

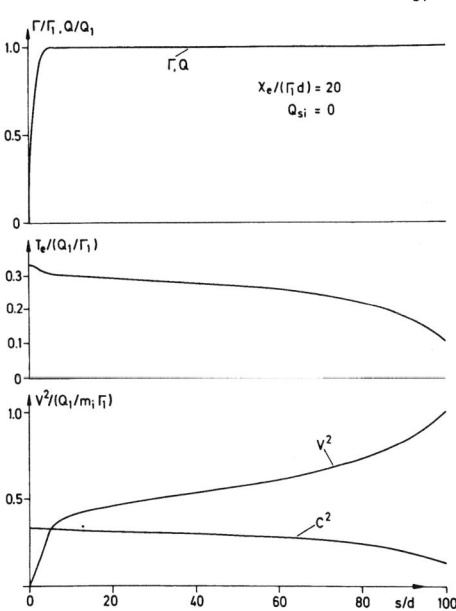

Fig. 16. Flow profiles in the presheath (from fluid equations) for
different locations of the source;
a) $S_n = \Gamma_1/d \exp\left[(s - L_s)/d\right]$,
b) $S_n = \Gamma_1/d \exp(-s/d)$.

electrons with temperature T_{eo} . At the plasma side s = 0 ions
are specularily reflected, lost electrons are replaced by the same
number entering with Maxwellian distribution at temperature T_{eo}.
The calculation area reaches up to the wall, covering the presheath
as well as the sheath.

Figure 17 shows flow profiles for the case of a remote source
given by eq. (5.12).As one can see, the flow velocity V grows up to
isothermal sound speed $C = \sqrt{(T_i + Z\ T_e)/m_i}$ within the source region
and continues to grow at the same length scale d after the source
has ceased. V saturates at about 1.2 C_s. This behaviour of V can be
better understood by inspecting the conservation of total momentum
flow P

$$P = (m_i\ V^2 + T_i + Z\ T_e)\ \tilde{\Gamma}/(ZV) = const = n_o\ T_{eo}\quad.$$

Solving for V yields

$$V = \frac{Z}{2m_i\tilde{\Gamma}}\ (P \mp \sqrt{P^2 - 4m_i(T_i + Z\ T_e)\ \tilde{\Gamma}^2/Z^2}\ )\quad.\qquad (5.13)$$

At the start of the flow the sign of the square root is negative.
Γ rises as well as T_i, the latter because ions start at different
values of electric potential giving rise to a thermal spread of
velocities. $T_e \cong T_{eo}$ in the presheath. At the end of the source
region, T_i has a maximum, the square root in (5.13) becomes zero and
$V = C_s$. For larger s the square root changes sign to + and V continues to rise with decreasing T_i. Nevertheless, the increase of
ion kinetic energy originates mainly from electron thermal energy.

Figure 18 shows the flow velocity V_s at the sheath edge for
different orientation of the magnetic field /18/. For $\psi = 0$ and
$S_n \propto n^\nu$ Harrison and Thompson /9/ obtained a value $V_s/C_s = 1.144$.
Thus, also in the collisionless presheath, the sound velocity is
not necessarily a limit on V.

6. CONCLUSION

In the preceding sections it was tried to derive a coherent description of the 1d plasma flow in the presheath and sheath regions
of the scrape-off layer. This description must be completed in many
respects. One is the inclusion of plasma interaction with neutral
gas in the presheath, another the extension of the models to 2d,
which widens the range of possible flow modes. These topics will be
discussed in the following chapters. A neuralgic point for modelling
the scrape-off layer is the transition between presheath and sheath.
One can avoid difficulties by treating both regions kinetically as
was done in Section 5b. But due to the large length scale and the
large number of physical processes in the presheath it is much more
advantageous to treat this region on a fluid base. This leads to
the problem that kinetic boundary conditions for the sheath must be
derived from the fluid solution of the presheath and vice versa
boundary conditions for the presheath from the kinetic solutions of
the sheath. In order to solve this matching problem satisfactorily,

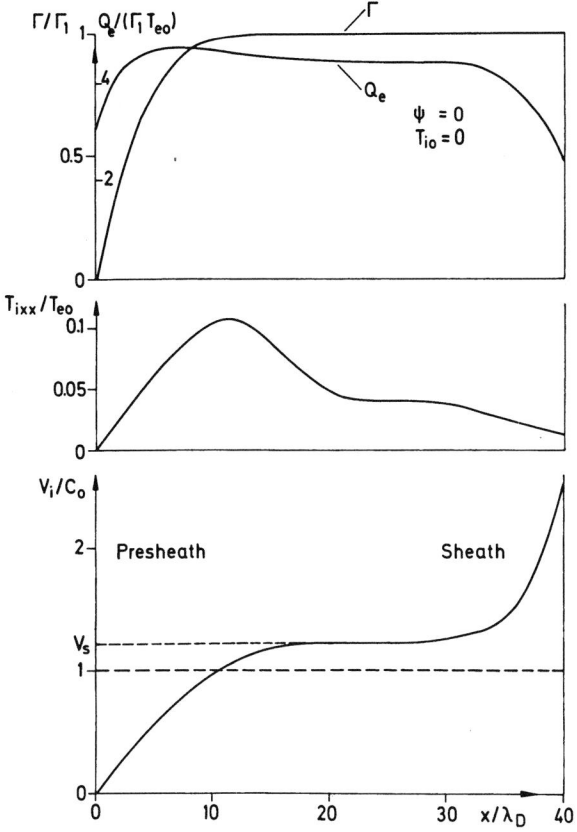

Fig. 17.   Flow profiles (from kinetic model) in a
collisionless presheath and sheath with S_n
as in Fig. 16b ($\psi = 0$).

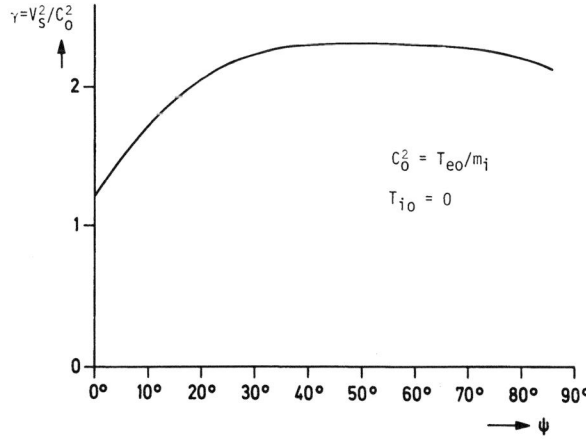

Fig. 18.   Flow velocity at sheath edge V_s for a collisionless presheath as shown in Fig. 17 but
for different field orientations ψ.

one has to put in between the collision-dominated fluid presheath and the collisionless kinetic sheath a transition zone of a collisional kinetic model extending over several mean free paths where the nearly Maxwellian distribution of the presheath adjusts to the truncated distribution within the sheath. This will lead to a more adequate formulation of the presheath boundary conditions.

REFERENCES

/1/ L.Spitzer, Physics of Fully Ionized Gases , Interscience Publishers, New York (1962).
/2/ S.I.Braginskii, Transport Processes in a Plasma in: Review of Plasma Physics , Vol. 1, M.A.Leontovich ed., Consultants Bureau, New York (1965).
/3/ J.P.Matte, and J.Virmont, Electron Heat Transport Down Steep Temperature Gradients, Phys. Rev. Lett. 49: 1936 (1982).
/4/ J.F.Luciani, P.Mora, and J.Virmont, Nonlocal Heat Transport Due to Steep Temperature Gradients, Phys. Rev. Lett. 51: 1664 (1983).
/5/ B.A.Trubnikov, Particle Interactions in a Fully Ionized Plasma, in: Review of Plasma Physics , Vol. 1, M.A.Leontovich ed., Consultants Bureau, New York (1965).
/6/ L.Tonks, and I.Langmuir, A General Theory of the Plasma of an Arc, Phys. Rev. 34: 876 (1929).
/7/ I.Langmuir, The Interaction of Electron and Positive Ion Space Charges in Cathode Sheaths, Phys. Rev. 33: 954 (1929).
/8/ D.Bohm, Minimum Kinetic Energy for a Stable Sheath, in: The Characteristics of Electrical Discharges in Magnetic Fields , A.Guthrie,and R.K.Wakerling eds., McGraw Hill, New York (1949).
/9/ E.R.Harrison, and W.B.Thompson, The Low Pressure Plane Symmetric Discharge, Proc. Phys. Soc., 74, 145 (1959).
/10/ R.Chodura, Plasma-Wall Transition in an Oblique Magnetic Field, Phys. Fluids, 25, 1628 (1982).
/11/ R.Chodura, to be published.
/12/ U.Daybelge, Electric Sheath Between a Metal Surface and a Magnetized Plasma, Phys. Fluids 24, 1190 (1981).
/13/ G.D.Hobbs, and J.A.Wesson, Heat Flow through a Langmuir Sheath in the Presence of Electron Emission, Plasma Phys. 9, 85 (1967).
/14/ P.C.Stangeby, Plasma Sheath Transmission Factors for Tokamak Edge Plasma, Phys. Fluids 27, 682 (1984).
/15/ R.Chodura, Numerical Analysis of Plasma-Wall Interaction for an Oblique Magnetic Field, J.Nucl.Materials 111&112, 420 (1982).
/16/ K.U.Riemann, Kinetic Theory of Plasma Sheath Transition in a Weakly Ionized Plasma, Phys. Fluids 24, 2163 (1981).
/17/ G.A.Emmert, R.M.Wieland, A.T.Mense, and J.N.Davidson, Electric Sheath and Presheath in a Collisionless, Finite Ion Temperature Plasma, Phys. Fluids 23, 803 (1980).
/18/ R.Chodura, Plasma Transport to the Wall Through the Electrostatic Sheath, Proc. 11th Europ. Conf. on Controlled Fusion and Plasma Physics, Vol. 2, 479 (1983).

PROBES FOR PLASMA EDGE DIAGNOSTICS IN MAGNETIC

CONFINEMENT FUSION DEVICES

D.M. Manos and G.M. McCracken*

Plasma Physics Laboratory
Princeton University, Princeton, New Jersey 08544

*UKAEA/Euratom Fusion Association
Culham Laboratory, Abingdon, Oxon OX14 3DB, England

ABSTRACT

Probes are becoming widely used in fusion devices to measure edge plasma parameters. Both electrical probes and surface collector probes can provide useful data and the two techniques are largely complementary. Energy analysers are included among the electrical probes and long term probes placed on the walls for extended periods are included in the surface probes. Methods of measuring erosion and redeposition are also outlined. Some of the problems inherent in probes such as disturbance of the plasma and practical problems of construction are discussed.

1. INTRODUCTION

The use of solid probes to measure plasma parameters has a history going back at least to Langmuir[1]. Their use in tokamaks however has been limited until relatively recently, partly because of the difficulty involved in their proper deployment, and partly due to uncertainty in the interpretation of probe data. There is no doubt that probes affect the plasma and the plasma flux affects the probes. Each of these complex interactions has to be understood if a reliable interpretation of what is happening at the boundary is to be obtained. The presence of surfaces in the plasma boundary as limiters, walls or as divertor neutralizer plates is inevitable. One can make probes large so that they simulate such components or one can make probes small so that they cause the least disturbance possible to the plasma. In either case modelling

135

is necessary in order to interpret the results in terms of the plasma surface interaction.

Probes can be broadly divided into two categories, electrical probes (e.g. Langmuir probe) and surface collector probes. A list of some examples of the two types of probes and the parameters they may potentially measure is given in Table 1. We define electrical probes to include any probe which uses an electrical measurement technique, so the term encompasses gridded energy analysers, ExB probes, heat flux probes, etc. They can, in principle, measure the saturation ion flux, the power deposited, electron and ion temperatures and hence can be used to evaluate the plasma density. At present no electrical probe has been developed which can measure impurities directly, although mass spectrometer techniques are possible. Collector probes, on the other hand are able to detect impurities directly. They rely on exposing a clean, well-characterized surface to the plasma for a known length of time and then removing the sample for analysis using one or more of the wide range of the surface analytical techniques. From the collected number of a given species of atom and the exposure time, the average incident flux can be obtained. The technique is also of value in distinguishing between different hydrogen isotopes. In this case it has been shown that the energy of the ions at the surface can be inferred by measuring the depth distribution of the implanted species. This approach also has considerable potential for determining both the flux and energy distribution of escaping α particles in DT burning plasmas. In order to fully interpret measurements with surface probes it is necessary to have independent measurements of the edge plasma parameters as well. So in most respects the two techniques, electrical measurement and surface analyses, should be looked on as complementary rather than as alternative methods of measuring plasma edge properties.

2. EFFECT OF THE PLASMA ON THE PROBES

One of the first considerations is the type of plasma which may be safely accessed by the probe. It is clearly necessary to avoid excessive heat loads on the surface of a collector in order to avoid evaporation of the collected species or diffusion of it into the bulk of the material. This requirement is most severe when attempting to determine the incident hydrogen flux and energy in collector probes, since the trapped hydrogen can be thermally released. The tolerable heat load on an electrical probe will be higher than for a collector probe, as it is normally only necessary to prevent evaporation or thermionic emission from occurring. The temperature rise of a surface is determined mainly by heat conduction into the bulk since radiation is often negligible for the conditions of interest. The temperature rise can be calculated by solution of the heat conduction equation. The solution for the

Table 1. Type of Probes and Parameters Measured

Electrical Probes	Parameter	Surface Collector Probes	Parameter
Langmuir Probes		Surface analysis	
single	$n_e \ T_e \ V_f$	Impurity mass and flux	Γ_I
double	$n_e \ T_e$	Plasma flux and energy	$\Gamma_H, \Gamma_D, \Gamma_T, T_i$
triple	$n_e \ T_e \ V_f$		
Larmor radius probes	T_i	Thermal desorption probe	$\Gamma_H, \Gamma_D, \Gamma_T, T_i$
Energy analysers Gridded	$N_e(E)$ $N_i(E)$	Alpha particle flux	$\Gamma_\alpha,$
ExB	$N_i(E)$	Metal surface collectors	$N_\alpha(E)$
In situ mass spectrometers		Erosion	
Impurity flux	Γ_I	Thin film or implant marker Thin film activation	
Hydrogen flux	$\Gamma_H, \Gamma_D, \Gamma_T$	Weight loss Collectors	
Heat flux probe		Macroscopic measurement	
Power	P_D		
Partial pressure probes	$\Gamma_H, \Gamma_D, \Gamma_T$		
Carbon resistance probe	$\Gamma_H, \ T_i$		

n_e	electron density
V_f	floating potential
$T_e, \ T_i$	electron and ion temperature
P_D	deposited power
$N_e(E), \ N_i(E)$	electron and ion energy distributions
$\Gamma_I, \ \Gamma_H, \ \Gamma_\alpha$	flux of impurities, hydrogen, α particles.

surface of a semi-infinite solid at a time t after the start of a constant power flux P is[2]

$$\Delta T = 2P \sqrt{t/(\pi KC\rho)} \qquad (1)$$

where K is the thermal conductivity, C the specific heat and ρ the density of the solid. The condition that a solid has a thickness, d, large enough to behave like a semi-infinite solid is

$$t < \frac{d^2 \rho C}{4K}. \qquad (2)$$

(The parameter $D = K/\rho C$ is known as the thermal diffusivity and, for heat diffusion, is the direct analog of the diffusion coefficient in kinetic theory). Materials with good thermal constants are C, Cu, Mo, W. Values of the thermal constants are shown in Table 2. It is seen that carbon, molybdenum and tungsten are comparable. A heat flux of < 500 watts cm^{-2} for 1 sec is tolerable if a surface temperature rise of $300°C$ is not to be exceeded. As discussed later in sections 4.2 and 4.3, other criteria favour carbon as a collector surface. It is seen that pyrolytic carbon has good thermal properties in the direction parallel to the basal plane, although it is very anisotropic. Its thermal conductivity is particularly good at low temperature and it is therefore ideal for collector probes where it is desirable to keep the surface as cool as possible cf. chapter by Smith and Whitely. Copper is an undesirable material close to the plasma because of its high sputter yield and relatively high atomic number.

Table 2. Tolerable Heat Fluxes for Probes

Material	$\sqrt{\pi KC\rho}$ at 300 -1000K	D $cm^2 s^{-1}$ at 300 -1000K	ΔT_1* K	Tolerable Heat Flux for 1 sec		
				$kW\ cm^{-2}$ for ΔT_1	$kW\ cm^{-2}$ ΔT=300K	Heat capacity for 1mm $J\ cm^{-2}$
W	3.3	0.72	2400	4.0	0.5	700
Mo	3.3	0.56	1800	3.0	0.5	560
C(ATJ)	2.9	1.1	1730	2.7	0.43	740
C(Pyrolytic)‖	10.4	12.3	1730	9.7	1.5	740
C(Pyrolytic)⊥	0.7	0.059	1730	0.65	0.1	740
SS	1.0	0.043	1000	0.5	0.15	430

* Temperature rise to that temperature at which the vapour pressure is 10^{-6} torr.

138

For reactor relevant plasmas the pulse length will be much in excess of 1 sec and therefore it is clear that if the critical temperature is not to be exceeded only much lower power levels can be tolerated. There are two approaches to this problem. One is to design a collector which is actively cooled. It can be shown[3] that active cooling, rather than inertial cooling, is appropriate for pulse lengths $\gtrsim$ 10 sec. Clearly if limiters are going to withstand heat loads in reactors then collectors can also be made to do so. However, active cooling of probes will cause significant technical problems. The other approach is to use moving collectors for short times. Either the collector surface can be exposed for a short time and then withdrawn from the plasma or else used as a time-resolved collector rotating behind an aperture. In either case exposure times of 0.1 sec can be easily obtained, increasing the tolerable heat flux by a factor of 3.

We must now consider the plasma conditions accessible to such collectors. The heat flux to a surface is discussed in detail in section 3. It depends strongly on the potential of the surface with respect to the plasma, and is nearly a minimum for the floating potential. The incident power flux P is given in this case by

$$P = \delta \ I_s^+ \ kT_e .\tag{3}$$

where δ is the sheath energy transmission factor, I_s is the ion saturation current given by $I_s \simeq 0.5 \ n_e C_s A$, n_e is the plasma density, A is the probe area and C_s is the ion sound speed given by

$$C_s = \sqrt{k \ (T_e + T_i)/m_i}\tag{4}$$

and T_i and T_e are the ion and electron temperatures respectively. The sheath transmission factor is a complex function which depends on the ratio of T_i/T_e, on secondary electron emission coefficients γ, and on the probe potential with respect to the plasma. For a floating probe, with $T_i/T_e = 1$, and $\gamma = 0.6$, δ is typically ~ 8 for hydrogen. This is made up of 2 kTe each from ions and electrons, $\sim 3 \ kT_e$ due to the sheath potential and ~ 0.5 kTe from the presheath. This subject is discussed in more detail elsewhere (see chapters by Stangeby and Chodura). From (3) using the values for I_s and C_s we obtain

$$P = 1.1 \ \delta \ n_e \ T_e^{3/2} \ (1 + T_i/T_e)^{1/2} \ \text{watts cm}^{-2}\tag{5}$$

for n_e in units of 10^{13} cm^{-3} and T_e and T_i in eV. Normally we have a limitation on the maximum power P_c a material may intercept for a given length of exposure time, e.g. as in Table 2. We thus can rewrite (5) in the form

$$n_c < P_c (1.1 \ \delta \ T_e^{3/2} \ (1 + T_i/T_e)^{1/2})^{-1} .\tag{6}$$

We can then calculate the critical density n_c which should not be exceeded for a given value of T_e. Results are shown in Fig. 1. As the edge temperature increases the tolerable density decreases. Fortunately in practice as the edge density rises the temperature tends to drop thus allowing a reasonably wide range of

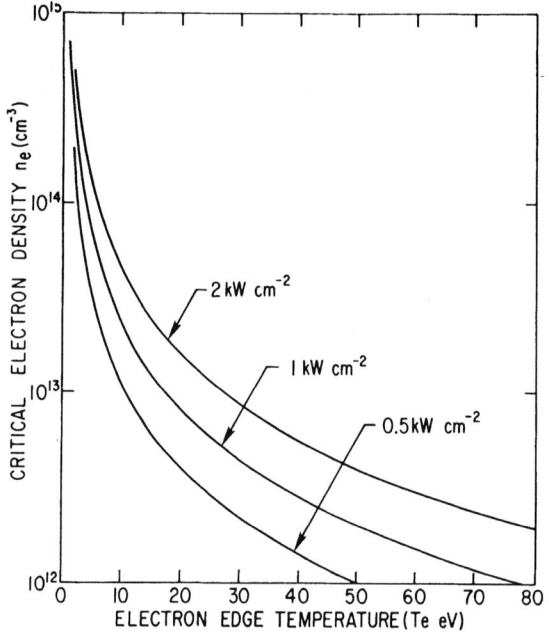

Fig. 1 Maximum plasma density for probe operation as a function of plasma temperature for power fluxes of 0.5, 1.0 or 2.0 kW cm^{-2} ($T_e = T_i$ is assumed).

plasmas to be investigated. A further factor that has to be taken into account is that at low incident ion energies there is a significant fraction of the ion energy backscattered rather than deposited in the solid, see Behrisch (these proceedings). This reduces the heating of the solid. In the case of tungsten at low ion energies the effect can reduce the deposited heat flux by 50%, but in low Z materials such as carbon the correction is much smaller. This effect has to be taken into account when comparing measured heat flux with plasma parameters.

3. ELECTRICAL PROBES

3.1 Single Langmuir Probe

A Langmuir probe is a conductor immersed in a plasma which may be biased relative to a second conductor (counter electrode, wall, limiter, etc.) in contact with the plasma. The theory of the development of the sheath and presheath is developed fully by Stangeby and Chodura (these proceedings). We recall a few essential results in this section. The theory of probes in general is discussed in a number of standard texts[4,5,6].

In the simple theory of a Langmuir sheath the high velocity of electrons compared to ions causes an electrically isolated surface to charge negatively with respect to the plasma. The sheath potential V_f is set up when the probe is floating, that is, when it may draw no net current, forcing the electron and ion fluxes to the surface to be equal. For $T_i = 0$ this result can be written as

$$V_f = \frac{kT_e}{2e} \ln \left(\frac{m_i}{2\pi m_e} \right), \tag{7}$$

where e and m_e are the electron charge and mass and m_i is the positive ion mass. For a hydrogen plasma $V_f \sim 3k\,T_e/e$. The presence of a sheath acts as a thermal barrier by reducing the electron conduction and hence the heat flow, by as much as a factor of 10. It also accelerates the ions into the wall with energies exceeding their thermal energy. In the case of multiply charged ions this can result in a considerable increase in the sputtering rate. As T_e increases the ions and electrons arriving at the surface can produce secondary electrons which will tend to reduce the potential. A model taking into account secondary electron emission but with $T_i = 0$ has been discussed by Hobbs and Wesson[7]. A fuller treatment of the problem has been discussed by Harbour and Harrison[8].

Where the singly charged ions have a finite temperature it is still possible to get a simple form for the sheath potential if it is assumed that the electron energy distribution remains Maxwellian[9]. The electron and ion current densities to the surface are given by

$$j^- = \frac{1}{4} n_e c_e e(1 - \gamma_e) \exp \left(- \frac{eV}{kT_e} \right) \tag{8}$$

$$j^+ = n_e c_s e$$

where

γ_e = secondary electron emission coefficient

$c_e = (8 \ kT_e/\pi \ m_e)^{1/2}$

$c_s = (k \ (T_e + T_i)/m_i))^{1/2}$

V = potential between the plasma and the surface,

and n_e is the density of the electrons (or ions) at the sheath
boundary. It can be shown that n_e at the boundary is approximately equal to 1/2 of its value in the unperturbed plasma at a
large distance (say 100 debye lengths) from the probe surface.

At the floating potential the electron and ion currents must
be equal and we thus obtain

$$1/4 \ n_e c_e e \ (1 - \gamma_e) \exp \left(\frac{-eV_f}{kT_e}\right) = n_e c_s e$$

or

$$V_f = - \frac{kT_e}{2e} \ \ell n \ [\frac{2\pi m_e}{m_i} \ (1 + T_i/T_e)/(1 - \gamma_e)^2] \qquad (9)$$

where we have only included the electron impact secondary electron
emission coefficient. The complete calculation has been made by
Harbour taking into account the correct electron energy distribution[10]. However, the calculation is much more complex and the
result obtained in the simplified treatment is very similar.

Up to now we have been considering the floating potential of
a passive surface. We must remember that the plasma is normally
electrically connected to earth via the limiter (or wall). It
will therefore float at some potential with respect to this
surface. This potential cannot be precisely defined because the
limiter potential is an average of a complicated spatial variation
of sheath potential since it is exposed to the profile variation
of n_e and T_e. We also note that the floating potential of an
object in a strong magnetic field depends upon its size (see
Stangeby these proceedings). This results from the fact that
electrons and ions have greatly different gyroradii. The
different gyro radii affect the ratio of parallel and perpendicular mobility of the species and also change the apparent probe
collection area for ions and electrons. So although a floating
probe will take a value for its potential which is mainly
determined by its local T_e, it will not in general be the same as
the limiter potential. It is frequently found that in scrape-off
plasmas the floating potential of probes relative to the limiter
is less than 10 volts[11].

A probe in the plasma may have potential applied, say with respect to the limiter. It can therefore be driven up or down with respect to the plasma potential from its floating potential. When driven sufficiently negative it will repel all electrons and the current density arriving will just be the ion saturation current density

$$j_{sat} = n_e e\ c_s. \tag{10}$$

The general expression for the net current density to the probe $j = j^+ + j^-$, hence

$$j = j^+ + j^- = n_e e\ C_s - 1/4\ n_e e\ C_e\ (1 - \gamma_e)\ \exp\left(\frac{-eV}{kT}\right), \tag{11}$$

where V is the difference between the potential of the plasma and the probe. In Fig. 2, we show a representative single Langmuir probe characteristic. The zones of behaviour may be described heuristically as follows:

Region C: V << 0:

In this region all of the electrons are repelled and all of the ions are collected. The ions are accelerated in the presheath by the applied voltage and the sheath thickness itself is affected slightly by the applied voltage. Thus there is only a weak dependence on voltage and the formula for "ion saturation" applies.

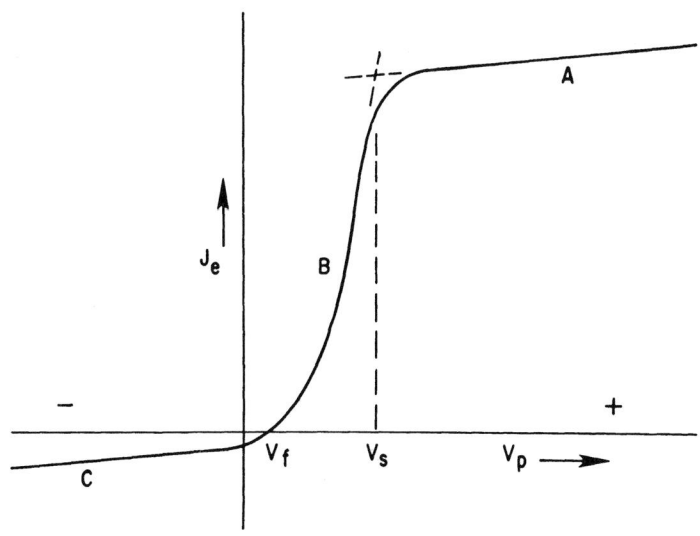

Fig. 2 Schematic of Langmuir probe characteristic illustrating the floating potential V_f, space potential V_s.

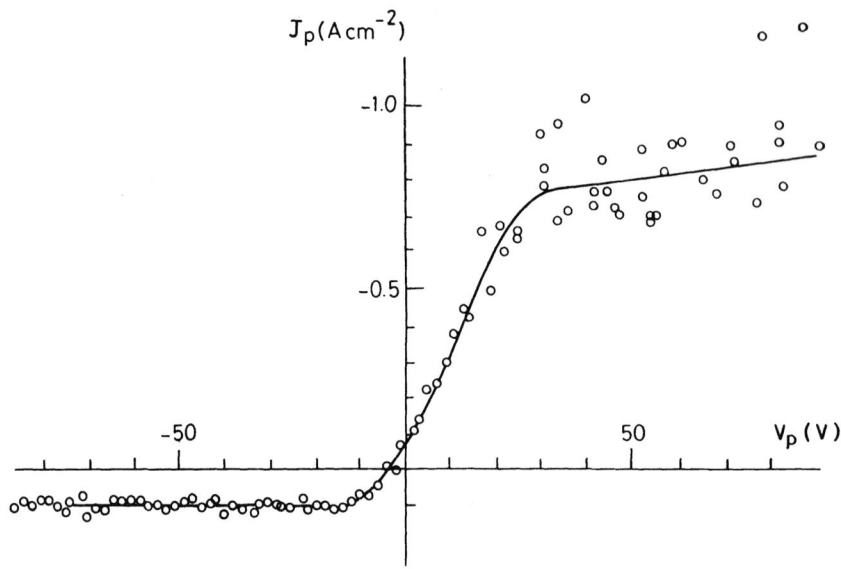

Fig. 3 Langmuir probe characteristic taken in the DITE tokamak
with plasma current = 120 kA, toroidal field = 2T and the
probe 35 mm behind the limiter.

Region B: $V_f \sim V < V_s$:

As the voltage increases some of the electron distribution is
admitted to the probe. The floating potential, V_f, is then
achieved and the net current is zero. As V continues to rise an
exponential increase in the electron current is observed, as
governed by the second term of Eq. (11). It is from this zone
that T_e may be computed from the slope of $\ln(j-j_{sat})$ vs V.

V_s: As V rises further the probe reaches the potential of
the plasma, also called the "space potential" and designated V_s,
the exponential term vanishes and the entire random electron
current is admitted.

Region A: $V > V_s$:

Above V_s one again sees only a slow dependence of current
density on applied voltage as the presheath attraction slowly
reaches further into the bulk plasma. This is the "electron
saturation" regime. In a strong magnetic field the electron
saturation current can be significantly reduced compared to the
zero field case. This is due to depletion of electrons in the
flux tube connected to the probe and the low cross field diffusion

144

rate of electrons. It is discussed in more detail in section 5.1 and by Stangeby and Chodura (these proceedings).

An experimental Langmuir probe characteristic taken on a tokamak with the torus as the reference potential is shown in fig. 3. It is seen that the ion saturation current is very flat to at least - 100 volts. The floating potential is negative with respect to the torus. The noise in the electron saturation current is due to fluctuations; this is typical of behaviour in tokamak boundaries as is the rather low ratio of electron to ion saturation current. It should be noted that a probe can only be driven to electron saturation when it is some distance outside the limiter radius otherwise the current drawn and the power deposited will be excessive.

Langmuir probes have now been used in large numbers of tokamaks[12-18] and increasing use is being made of large arrays of 50 or 60 probes to investigate poloidal variations of plasma edge characteristics[19,20].

3.2 Double Langmuir Probe

There are times when a single probe may be difficult to use. For example, there may be no well-defined counter-electrode (ground) plane as in the case of an electrodeless rf induced plasma in a dielectric container. There may be prohibitively large fluctuations in V_s induced by waves or turbulence in the plasma. These lead to noise in the current which result in large uncertainties in T_e. There are also cases where the plasma is dilute or otherwise easily perturbed and drawing large electron saturation currents is undesirable. In these instances one may employ a double Langmuir probe[21,22] as shown schematically in Fig. 4a.

The double probe uses a small counter electrode as the reference for the return current and the entire system floats at a potential dictated by V_s and T_e. This is shown schematically in Fig. 4b for the case of $V_1 - V_2 > 0$.

The current voltage characteristic equation may be developed by treating each electrode as a single probe, adding the con straint that the system floats, i.e.,

$$j_1 + j_2 = 0 \tag{12}$$

and

$$V_2 = V_1 - V \tag{13}$$

Thus one arrives at the expression relating the applied

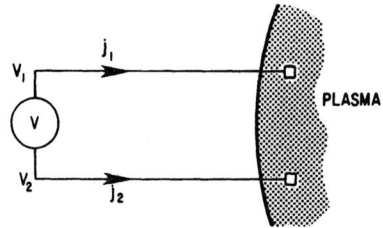

Fig. 4a Schematic of Langmuir double probe.

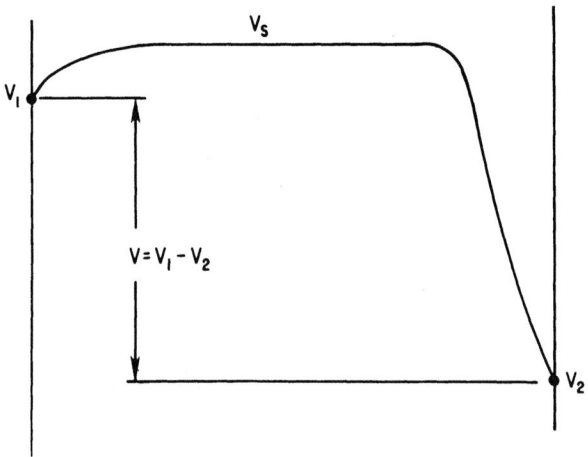

Fig. 4b Potential diagram for the above probe when $V_1 \gg V_2$.

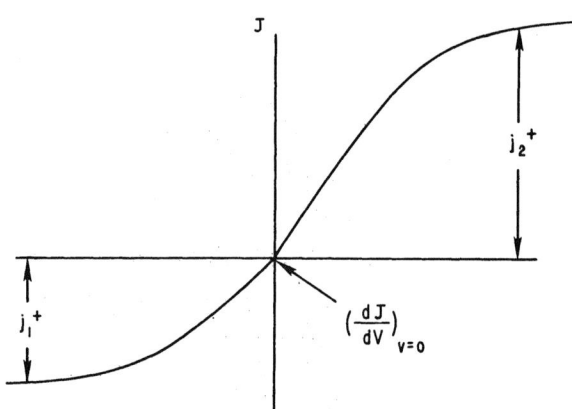

Fig. 4c Representative double probe characteristic.

potential, V, and the actual electrode potential V_1:

$$\exp (eV_1/kT_e) = \frac{2\exp (eV_f/kT_e)}{1 + \exp (-eV/kT_e)} \, . \tag{14}$$

This may be substituted into the expression for the current on electrode 1 and electrode 2 to develop the general characteristic:

$$\frac{J + j_1^+}{j_2^+ - J} = \frac{A_1}{A_2} \exp (eV/kT_e) \tag{15}$$

where J is the current in the external circuit, (j_1 or $-j_2$).
$j_1^+ \, j_2^+$ are the ion saturation currents for electrodes 1 and 2 respectively, i.e., the current observed when $V_1 \ll V_2$ and vice versa.

and A_1 and A_2 are the collection areas of electrode 1 and 2 respectively.

A representative double probe characteristic is shown in Fig. 4c. Differentiating (15) we find

$$\left[\frac{dI}{dV}\right]_{V=0} = \frac{e}{kT_e} \frac{(j_1^+)(j_2^+)}{(j_1^+ + j_2^+)} \, . \tag{16}$$

We note that in the commonly employed special case of $A_1 = A_2$ (therefore $j_1^+ = j_2^+$) the characteristic may be written

$$j_1 = j_+ \tanh (\frac{eV}{2kT_e}) \tag{17}$$

and differentiating yields

$$\frac{kT_e}{e} = \frac{j_+}{2(dJ/dV)_{V=0}} \, . \tag{18}$$

T_e is extracted quite easily from the slope of J vs V at $V = 0$. As for the single probe, n_e is derived from the ion saturation value, the calculated T_e, and the assumed, or independently measured, T_i.

The principal disadvantage of the double probe is that because both electrodes are floating it only samples the tail of the electron energy distribution. For equal area electrodes it samples $\sim$ 14% of the distribution, thus if the distribution is non-Maxwellian erroneous temperatures can be derived. For an asymmetric probe a larger fraction of the distribution is sampled, depending on the ratio of the areas and the ratio of the ion to electron saturation current. Asymmetric double probes have been used routinely on DITE[23]. A further drawback of the double probe is that the floating potential is not measured. However this is easily overcome by using the triple probe.

3.3 Triple Probe

A further modification of Langmuir probes involves adding a third electrode in close proximity to a double probe. This electrode is used to independently measure the floating potential and the entire configuration is called a triple probe[24,25]. Taking the double probe analysis above, we have the result of equation (14)

$$\exp\ (eV_1/kT_e)\ =\ \frac{2\ \exp\ (eV_f/kT_e)}{[1\ +\ \exp\ (-\ eV/kT_e)]}\ , \tag{19}$$

where $V = V_1 - V_2$. Suppose we set the applied potential so that $V \gg kT_e$ then from above we obtain

$$\frac{eV_1}{kT_e}\ =\ \frac{eV_f}{kT_e} + \ell n2 \tag{20}$$

or

$$\frac{kT_e}{e}\ =\ \frac{1}{\ell n2}\ (V_1\ -\ V_f). \tag{21}$$

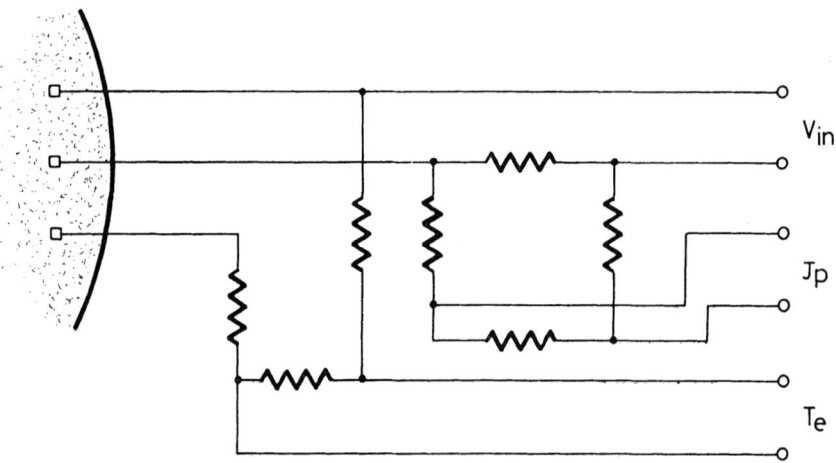

Fig. 5 A simple circuit for measuring the ion probe current and the electron temperature using a combination of a double probe and floating probe[26].

A measurement of the voltage between the positive double probe electrode and the floating electrode yields T_e directly. This can be done by means of a simple electrical circuit[26] as shown in fig. 5. Fig. 6 shows a comparison of T_e derived from the triple probe analysis to that derived from double probe analysis on PDX. The agreement is quite good. A detailed comparison of double and triple probes in magnetic fields has been given by Budny and Manos[16].

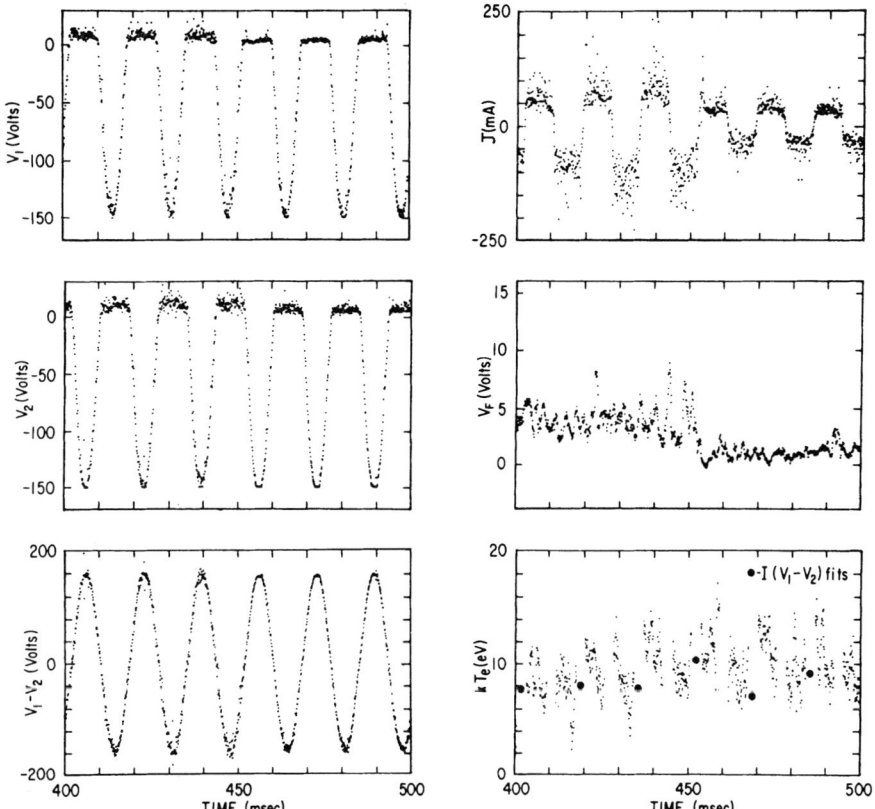

Fig. 6   A comparison of double and triple probe behaviour. The applied voltage (lower left) is measured relatively to the vacuum vessel wall and resolved into two components V_1, V_2. The electron current j, (upper right) and independently measured floating potential on the third wire (middle right) are also shown. At lower right is shown the directly read T_e from the triple probe superimposed on $T_e(0)$ calculated from the double probe characteristics.

3.4   Other Langmuir-like Probes

Another variant on the Langmuir probe is the electron emitting probe[27]. By operating the probe at an elevated temperature, it will emit electrons whenever the probe potential equals or exceeds the plasma potential. The plasma potential can be derived by comparing the current-voltage characteristics of the probe when hot and cold. However operating at positive potential results in high heat flux. Such probes are inherently physically delicate and until recently have been generally of use only in relatively quiescent low density plasmas. They can be constructed to survive in denser plasmas[28], however, and with modifications may be of value even in long pulse machines.

3.5   Heat Flux Probes

With simple electrical measurements it is possible to obtain T_e and the ion saturation current. However without some assumption about T_i it is not possible to derive the density. Frequently it is assumed that $T_i = T_e$ but this is not always justified[29]. One way of obtaining information about T_i in principle is to measure the power (P_D) to a probe. This can be done quite directly by using a thermocouple and measuring the rate of rise of temperature (dT/dt) of a thermally isolated probe[30].

$$P_D = mC \frac{dT}{dt} + \alpha(T - T_B) + \varepsilon\sigma(T^4 - T_B^4) \tag{22}$$

where m is the mass, C the specific heat, α is the thermal conduction drain coefficient, ε the emissivity, σ the Stefan-Boltzman constant, T_B the vessel temperature and T the probe temperature. α can be obtained from the cooling curve after the discharge. For short pulses heat loss by conduction and radiation can often be neglected. By using a material with high thermal conductivity such as tungsten or molybdenum it is quite straightforward to obtain thermal time constants of ~ 1-10 ms. Electrical noise due to inductive effects can be largely eliminated by careful twisting of leads. However noise from the plasma can be a problem, sometimes requiring integration times of ⩾ 100 ms to average out the noise.

A heat flux probe is shown in Fig. 7a. Fig. 7b indicates the maximum operating boundary in the power vs time plane for particular choices of thermocouple types.

The power P_I arriving at a probe at floating potential is given for singly charged ions by

$$P_I = I_s kT_e \left\{ 2(T_i/T_e) + 2(1 - \gamma_e)^{-1} + eV_f/kT_e \right\} \tag{23}$$

where we have neglected the ion neutralization energy.

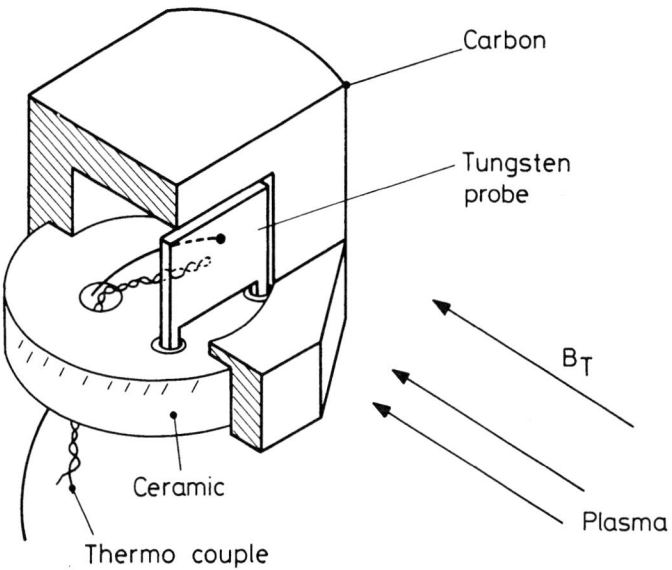

Fig. 7a The combined heat flux probe and single Langmuir probe
construction.

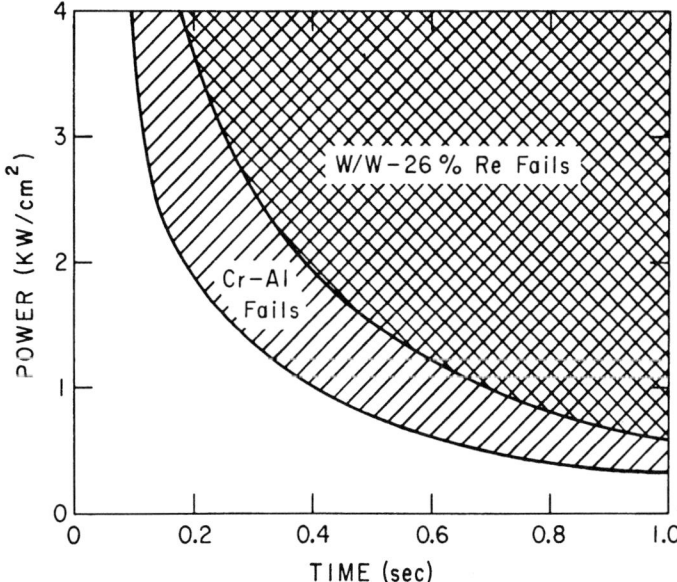

Fig. 7b Calculated operating boundary for a 1 mm thick Ta
collector plate. In the unshaded region the probe
operates normally. In the single shaded region a chromel
alumel thermocouple will melt. In the doubly shaded
region a W/W-26% R_e thermocouple calibration fails but the
device is not destroyed.

Hence

$$\frac{T_i}{T_e} = \frac{e\,P_I}{2kT_e\,I_s} - \frac{e\,V_f}{2kT_e} - \frac{1}{1 - \gamma_e} \qquad (24)$$

We also have from Eq. (9) that

$$\frac{e\,V_f}{kT_e} = -\,1/2\ell n\,\left[2\pi(\frac{m_e}{m_i})(1 + \frac{T_i}{T_e})\,(1 - \gamma_e)^{-2}\right]. \qquad (25)$$

From Eqs. (24) and (25) T_i and V_f can be obtained by successive approximation and hence T_i obtained independently of T_e. However in practice there are a number of uncertainties in this derivation which lead to large errors in T_i and at present it is not a very reliable technique[17]. For example if the electron energy distribution is not Maxwellian but has an unexpected high energy component this will give an unexpectedly high value to P_I and hence to T_i. Uncertainties in the secondary electron emission coefficient or in the ion energy reflection coefficient can also produce substantial errors in the calculated value of T_i. Thus the use of the measured power to the probe to calculate T_i must be used with caution. The technique however, is useful for measuring surface temperature since this parameter is inherently of interest for estimating the effects of evaporation, chemical sputtering, thermal desorption, etc. The measured deposited power is also a useful cross check on the power incident as calculated from independently measured plasma parameters.

3.6   Mach Number Measurements

From a simple analysis of the scrape off layer radially outside the limiter it is easy to show that there will be a stagnation point between the limiters and that the ions will acquire a net drift velocity as they come near to the limiter, (see Stangeby, these proceedings). There will thus be a difference between the flux arriving at a probe placed a distance from the stagnation point, depending on whether it is facing upstream or downstream. The Mach number (M) of the plasma is defined as the ratio of the drift velocity to the sound speed c_s. Under simple conditions the Mach number will increase from zero at the stagnation point to 1 at the sheath edge (according to the Bohm criterion). Taking the fluid model for M < 1 we get a simple form for the ratio of the flux in the upstream to downstream direction[31]

$$\frac{\Gamma_u}{\Gamma_d} = \frac{2 - M}{2 + M}$$

e.g. for M = - 0.5 $\Gamma_u/\Gamma_d = 5/3$

Thus by measuring R the Mach number can be deduced. This technique has been used by Harbour and Proudfoot[32] using 2 double probes facing both directions simultaneously. Significant asymmetries were observed which were interpreted as being due to this flow and Mach numbers deduced. There are however a number of difficulties with this interpretation. Kinetic and fluid models give significantly different results so that the Mach number deduced is model dependent. Both models give unrealistic behaviour for M > 1 so that it is only possible to deduce that flow is supersonic - not what the actual Mach number is. Finally processes other than flow to the limiter can cause plasma drift e.g. plasma rotation due to ion beams or rf heating. Thus at present this technique has to be treated with considerable caution.

3.7 Partial Pressure Probes

A completely different approach to measuring ion flux is shown in fig. 8. An aperture defines the incident flux which enters a small chamber and thermalizes by collisions with the internal surfaces. The partial pressure rise in the chamber is proportional to the incident flux. Although it is not strictly speaking an electrical probe it is a direct reading device and so is included in this section. If we consider the probe to be floating and the chamber to be unpumped except for the aperture then

$$\Delta P = \frac{\Gamma_s}{C} = \frac{\Gamma_s \sqrt{M/2}}{440} \ torr$$

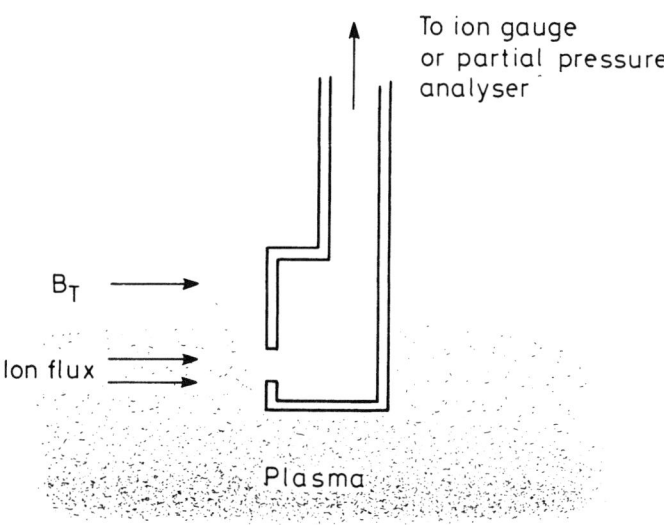

Fig. 8   Schematic of a pressure probe for measuring ion flux.

153

where Γ_s is the ion saturation current in amps cm^{-2}, M is the molecular weight of the incident ion, and we have assumed that the aperture is thin and its conductance C is proportional to its area. Thus a large pressure rise is predicted even for currents at the mA cm^{-2} level. The pressure rise could be enhanced by reducing the conductance using a thick aperture or decreased by adding additional pumping. It is clearly possible to measure this pressure rise using a partial pressure analyser and so to distinguish between hydrogen isotopes and possibly other gaseous species. The technique has been applied to ASDEX[33]. The practical difficulties include getting a partial pressure analyzer sufficiently close to the machine, to reduce the time constant of the vacuum system to make real time measurements during a discharge and at the same time to shield the instrument from electrical and magnetic interference. Another problem is the interpretation of the pressure rise as there is the possibility of either trapping and loss of the incident species or desorption of other gas at the internal surfaces of the chamber. The problems are similar to those in measuring ion beam trapping mass spectrometrically[34]. Conditioning the surface by heating and using repeated discharges until reproducible conditions are obtained can enable a measure of the incident ion flux to be obtained. A further difficulty is that the conductance of the hole for the thermalized species may be reduced by the flux of energetic particles entering the chamber, so-called "plasma plugging". This process is expected to be highly non-linear with little effect at low density and a strong effect at high density. The situation is similar to that for pumped limiters. In pumped limiter experiments on ISX good agreement was obtained between the ion saturation current and the pressure rise[35] with Γ_s in the range 1 to 2 amps cm^{-2}.

3.8 Complications in the Use of Probes

There are numerous complications that arise in the use of probes to diagnose fusion plasmas. An obvious source of trouble is the difficulty involved in the experimental design, construction, and deployment of probes. We defer this topic to the end of this chapter.

Probes are intrinsically perturbing to the distributions they attempt to measure. Probes collect particles in a certain portion of phase space and reflect others, thereby modifying distributions. As discussed by Stangeby[36] the existence of a sheath implies a modification of the plasma density and electric field at its boundary. The electric field in turn seriously modifies the ion distribution at the probe location. Simple estimates of the degree to which a probe intrinsically perturbs the plasma are possible[37]. If the probe collector dimensions, d, become comparable to the collisional mean free path λ_{mfp}, i.e. $d/\lambda_{mfp} \sim 1$, the probe will seriously distort the distribution function of the collected species. If the net probe current drawn is comparable to the volume plasma source rate, the probe is more seriously perturbing. This latter condition rarely applies to modern fusion devices, nevertheless large probes may still seriously affect the density of a strongly magnetized plasma over long parallel distances (see Section 5.1 of this chapter). Stangeby has given an analysis of the various distortions which may occur for large probes[38]. Fortunately it is possible to correct scrapeoff lengths of n_e, T_e, and power extracted from measurements of radial variations rather easily. The analysis also indicates that when a probe is operated in front of a limiter or "back stop", interpretation of the data is facilitated.

As probes are deployed in hotter, denser regions of the plasma, the plasma-surface interactions discussed earlier play a large role in disturbing the plasma. Obviously, evaporated or sputtered probe material may enter the plasma resulting in large changes of n_e or T_e and resulting in global modification of confinement or even major disruption. Sputtering or evaporation may change the probe dimensions, and thus the effective collection areas, in unknown ways. Ion recycling on the probe itself may change the nature of the plasma in the measured region. This is a problem at high densities where re-ionization of ions neutralized at the probe can increase the local density and reduce the temperature. The effect is also likely to increase with probe size. Similar effects occur at the limiter and divertor target plate so the problem is really one of interpretation - relating the results obtained from the probe to the plasma conditions either close to or far from the limiter surface.

On a less dramatic scale, changes in the surface conditions of an electrical probe may lead to varying secondary electron emission, ion particle and energy reflection coefficients, or contact potentials, rendering the analysis difficult. The bombardment of electrical probes by impurities or metastable atoms also complicates the analysis since the secondary electron emission coefficient can be significantly changed. Overheating of the probe can drive it into thermionic emission where the above theory does not apply. Lastly, inherent fluctuations of n_e and/or T_e or T_i can be quite large in fusion plasmas. Such fluctuations are of fundamental physical importance and have been the object of recent studies using probes[39,40]. However, such fluctuations make the simple reduction of data described above quite uncertain. In particular, accurate determination of T_e can become very difficult.

4. MEASUREMENT OF ENERGY DISTRIBUTION

For the techniques described so far it is necessary to assume that the ion and electron energy distributions are Maxwellian. This is by no means certain as high energy components are known to occur under some conditions e.g. runaway electrons at low densities and high energy ions due to neutral injection. As mentioned earlier the comparison of measured power to ion saturation current and electron temperature, that is, a determination of the heat-flux transmission coefficient can indicate the presence of such fast components. In order to determine the identity of such components it is necessary to have some sort of energy analysis. A number of systems have been tried, including retarding field analyzers[41,42,43] and the ExB analyzer[44,45] but none have been widely used in tokamaks.

A problem common to all such analyzers is the proper design of the extraction and retarding field elements[46]. The particle trajectories, determined by the electric fields sampled on passing through the analyzer apertures may be subjected to unwanted divergence or focussing. The measured energy distribution functions can be distorted by defocussing effects such as field penetration[47], by space charge[48], or other geometric effects particular to the analyzer. Accurate description of the behaviour of such systems may require considerable computational modelling.

Complication also arises in using such devices in very strong magnetic fields wherein the alignment is critical. A misalignment of a few degrees to the total field may have serious consequences in some types of energy analyzers. This can lead to problems in mapping radial profiles in certain tokamak discharges if the magnetic field angle changes significantly over the region of interest.

4.1 Retarding Field Analyzers

The well known retarding field analyzer[49] is shown schematically in Fig. 9a. Ions or electrons can be analyzed by retardation in the field between grids. By sweeping the retarding field an integral distribution of ions above a given energy can be obtained. The differential spectrum can be obtained either computationally or electronically; one technique is to use a low amplitude high frequency sweep while simultaneously changing the retarding voltage[50]. While the retarding field analyzer is simple to use for monoenergetic beams problems arise in magnetized plasmas. It is first necessary to separate ions and electrons. If slits or grid meshes of size comparable to the Debye length, λ_D, are used then a sheath is set up and the electron flow into the analyzer is reduced to the value of the ion flow. However such fine grids will withstand only low incident power. A further

156

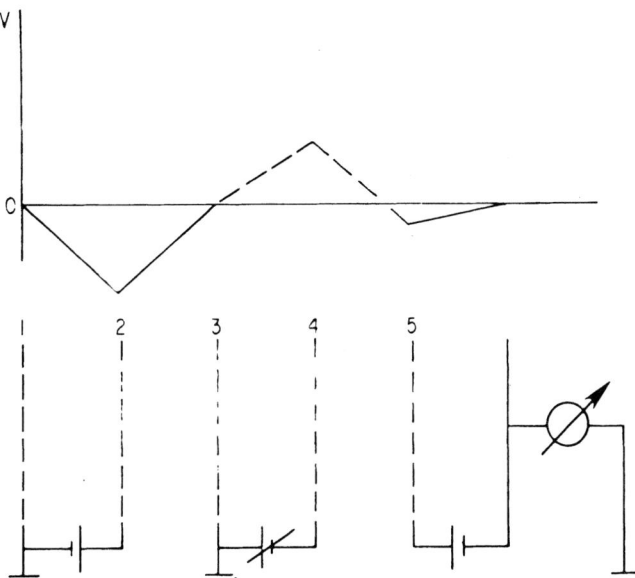

Fig. 9a Schematic of retarding field analyser (R.F.A.) particle
energy analysis. By scanning grid 4 potential an integral
energy distribution is obtained. By reversing polarity
either ions or electrons can be analysed.

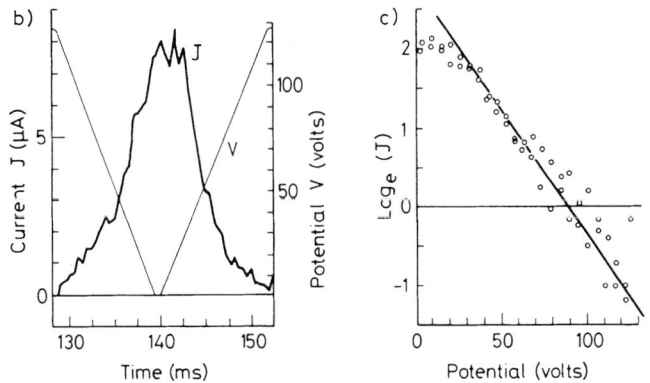

Fig. 9b An integral distribution from a retarding field analyser
resulting from the voltage scan shown. Taken during an
ohmically heated discharge in the DITE tokamak[45].

Fig. 9c A logarithmic plot of the ion current from 9b indicating a
sheath potential of ~ 20 eV and an ion temperature of
32 eV.

157

problem is that when using a small grid mesh or narrow slit the effect of finite ion larmor radius must be taken into consideration. The thickness of the grid must be kept small compared with the hole size in order to prevent the attenuation of the ions with transverse energy. This problem can be alleviated to some extent by calculating the transmission factors of a given geometry as a function of energy and using this to correct experimental data.

Results[45] obtained in a tokamak boundary during ohmic discharges are shown in figs. 9b and 9c. Fig. 9b shows the raw data obtained with a voltage scan from + 120 to - 5 volts. The integral distribution drawn on a logarithmic scale in fig. 9c indicates the acceleration due to the sheath potential and a Maxwellian distribution with an ion temperature of 32 eV.

4.2   The ExB Analyzer

The ExB analyzer has been developed by Staib[44]. The ions enter the analyzer through a small hole or slit. An electric field E is applied at right angles to the existing magnetic confinement field B (Fig. 10a) The ions then have a drift velocity v_D in the ExB direction independent of their charge and mass:

$$V_D = \underset{\sim}{E} \times \underset{\sim}{B}/B^2 \qquad\qquad (26)$$

By having a series of collectors as shown in Fig. 10a the ions in different energy intervals can be detected. For a given geometry the electric field can easily be altered to look at different ranges of energies. Similar problems exist for the ExB analyzers as for the retarding field analyzers with regard to power flux and ion transmission through the defining hole. In addition there can be a greater problem with space charge. When the electrons are removed the space charge of the ions will create an electric field which may be significant compared with the applied field. This additional field may cause spreading of the ions which can make the measured energy distribution appear to be broader than it is. This effect will determine an upper limit to the plasma density which can be measured.

Despite these problems ExB analyzers have been used successfully in tokamak plasmas. By using a slit size of 30 μm, plasmas with density up to 5×10^{18} m^{-3} have been analyzed[45]. Ion temperatures obtained with an ExB analyzer compared with a retarding field analyzer are shown in fig. 10b. Good agreement has been obtained. A characteristic increase in ion temperature when the gas feed is turned off is illustrated. The almost continuous measurement of T_i obtained with the ExB analyzer is a valuable feature of this device.

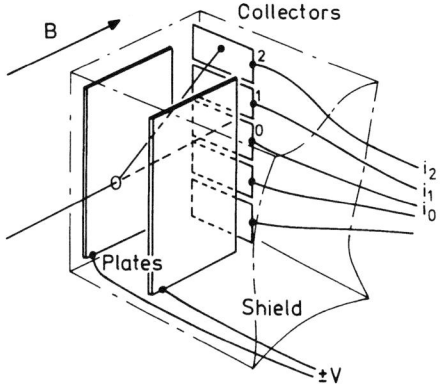

Fig.10a Schematic of an ExB analyser showing the electric field
plates at potential ± V and five collectors.

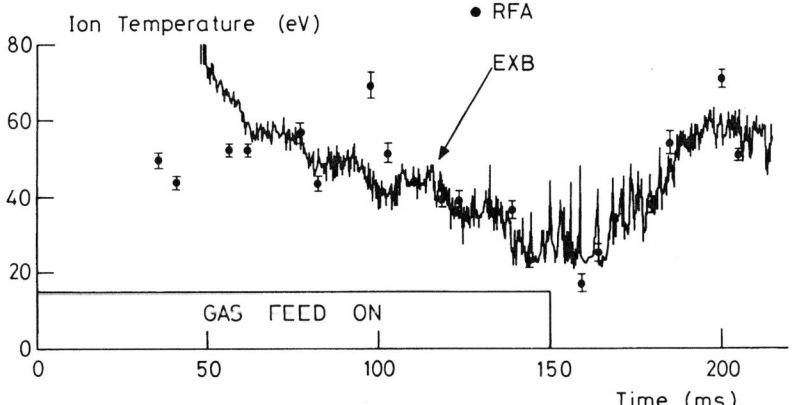

Fig.10b Comparison of results with an ExB analyser and a retarding
field analyser (R.F.A.) taken in the same discharge on the
DITE tokamak[45].

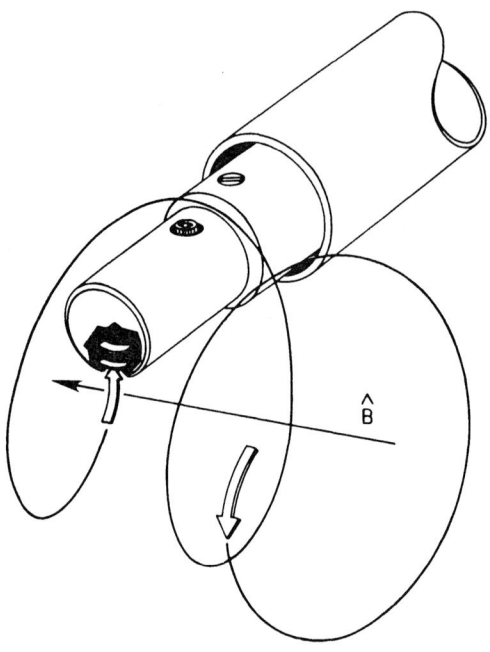

Fig.11a Rotating bolometer (heat flux) probe. The active elements
are hidden behind an aperture in the cap. When the
detectors are at a large angle with respect to the magnetic field only ions with large gyroradii are admitted.

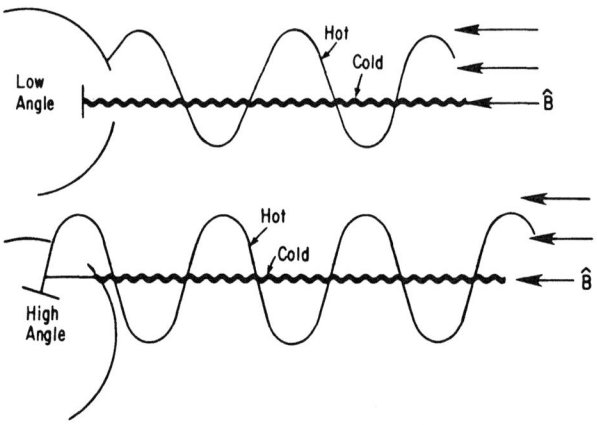

Fig.11b Effect of filtering action on fast and slow ions as the
bolometer is rotated.

4.3 Rotating Calorimeter

Another energy discriminating electrical probe is the rotating calorimeter (or heat flux) probe developed for use on PLT, PDX, and TFTR[51]. It is based on the measurement of heat flux as described in section 3.5. Fig. 11a shows a schematic of the device. It consists of two calorimeter plates with thermocouples attached which are located relatively far behind apertures in a cylindrical carbon housing. The filtering action of the aperture is similar to that described for the collector-aperture probes which are discussed in section 5.2. Fig. 11b shows schematically how the filtering action can be employed to discriminate fast ions from slow ions or electrons.

To extract the actual velocity space distributions associated with a fast ion component is rather complicated. At present this is done via Monte Carlo calculations. Initial distributions of ion velocities are assumed and the anticipated distributions of heat flux vs. probe angle relative to the magnetic field are calculated. The fundamental parameters of the problem are aperture size relative to the ion gyroradius and the ratio of $v_\perp/v_\parallel$. The resulting angular distributions have a rather complicated behaviour in general. However, when the gyroradius

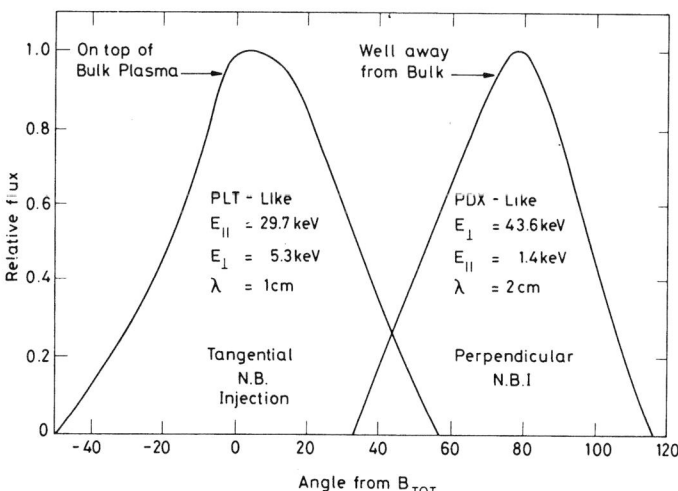

Fig. 12 The calculated distribution of fast ions at the outer midplane of PLT and PDX due to prompt losses of injected neutrals.

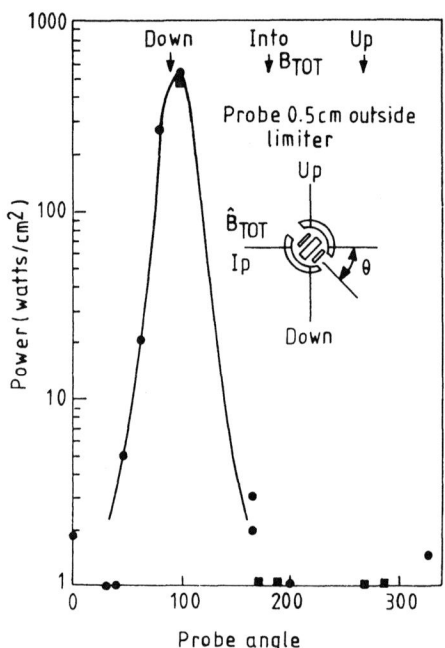

Fig. 13 Measurements of the angular variation of heat flux at the midplane of PDX during 4.2 MW NBI, $B_T = 1.2$ T, $\bar{n}_e \sim 3 \times 10^{18}$ m^{-3}, $I_p = 250$ kA.

is very much larger than the aperture size the resulting angular distribution is simply that associated with a plane parallel flux (zero gyroradius) shifted to a position such that $\Phi_{max} = \tan^{-1} V_\perp/V_\parallel$. Fig. 12 shows results anticipated for distributions of ions which would result at the outer midplane of PLT and PDX in the event of prompt loss during neutral beam injection. The two cases are very easily distinguished. Fig. 13 shows data taken during 4.2 MW neutral beam injection on PDX. The peak at ~ 81° from B_T indicates the occurrence of prompt beam particle loss.

4.4 Larmor Radius Techniques

There are a number of different methods of obtaining information on the energy distribution from the ion larmor radius[52,53,54,55]. One ingenious technique is the Katsumato

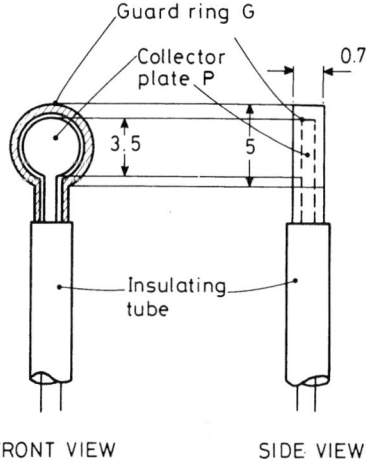

Fig. 14 The Katsumato probe. A collector plate P is mounted with its plane parallel to the magnetic field. A guard ring G shields P from electrons. Dimensions in mm.

probe[55] which is shown schematically in fig. 14. Because of their
small larmor radius electrons are prevented from reaching a centre
disc collector when it is aligned with the magnetic field. Ions
with their larger radius can reach it. A retarding potential
characteristic is taken using the disc electrode and from this ion
energy distributions have been estimated. However, there has been
no treatment of the effect of space charge due to the separation
of ions and electrons and it is unlikely to work satisfactorily
except at low densities. Results have been presented for the
DIVA[56] and TCA[57] tokamaks.

A versatile energy analyzer for electron distributions has
been described by Arion and Ellis[52]. The device is based on the
absorption of ions by small cylindrical channels as plasma
traverses their length, thus the charge separation physics differs
from retarding field in beneficial ways that allow the device to
be made quite small. Stenzel et al[53] have constructed a similar

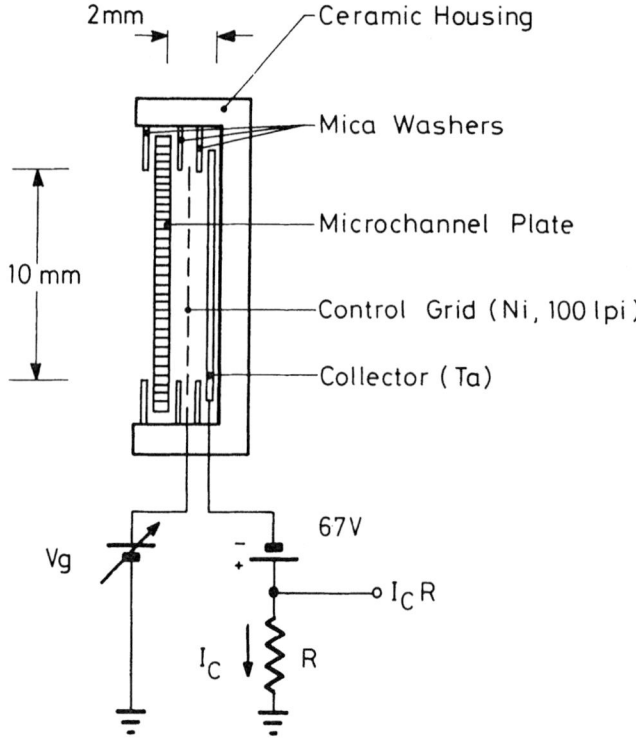

Fig. 15 A microchannel energy analyser. Ions are separated out
because of their large larmor radius and the voltage on a
simple grid is scanned to obtain the electron energy
distribution.

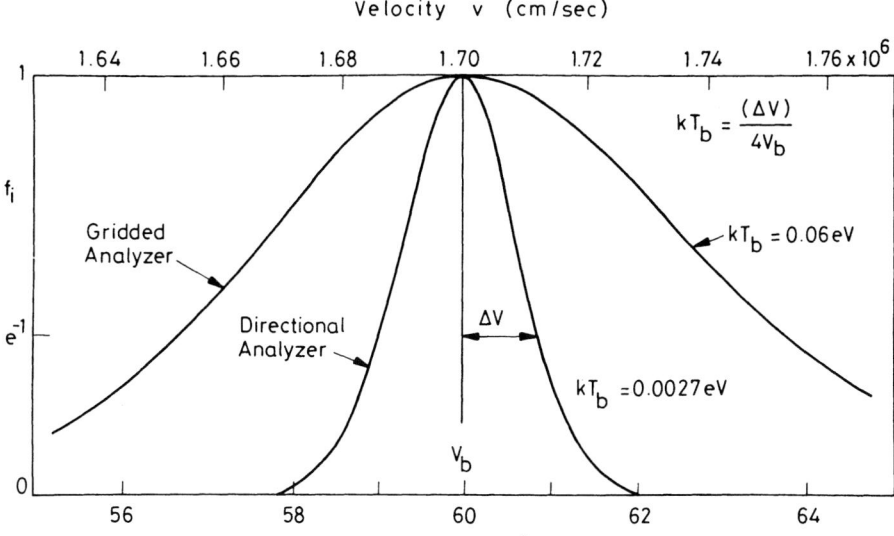

Fig. 16 Comparison of results from a microchannel energy analyser
with a conventional retarding field analyser.

device wherein the channels are provided by a commercial microchannel plate as shown in Fig. 15. Fig. 16 compares the results
of this analyzer to a conventional gridded analyzer immersed in a
60 eV beam generated in a double plasma device. Energetic
electron tails are also easily measured in plasmas of densities up
to 2×10^{12} cm^{-3}.

4.5 In Situ Mass Spectrometers

The main drawback of the electrical probes we have described
so far is that they are insensitive to the ion mass. In principle
the presence of the magnetic field should allow an electrical
configuration which will provide mass analysis. Of the many
configurations possible the presence of the strong magnetic field
throughout the probe appears to restrict us to three possibilities. These are shown in figs. 17, 18 and 19 and have all been
quite widely used in other mass spectrometric applications[58].
Application to plasma analysis introduces many of the same
sampling problems we have discussed for the energy analysers.
However if we assume these problems can be overcome or corrected
we take a plasma sample through a small floating aperture. The
ion beam emerging from the aperture has then an electric field
applied in one of the following ways.

(i) The first approach fig. 17, is the omegatron. This has been
used conventionally both for residual gas analyzers and for
accurate mass determination. An rf field is applied as in a
cyclotron. The frequency of the rf is swept and when it is

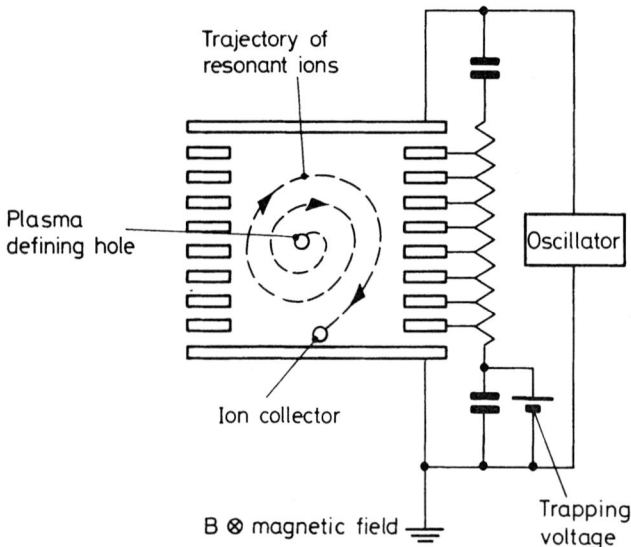

Fig. 17 Schematic of the omegatron mass spectrometer. Resonance
        ions are accelerated by the rf field until they strike the
        collector.

resonant with the ion cyclotron frequency the ion is accelerated
in a spiral. An ion collector near the rf electrodes allows the
resonance peak to be detected. Relatively low potentials are
required and problems occur when used as a residual gas analyzer
with charge build up on electrodes. No application to plasma
analysis has been published.

(ii) The second approach is to use the 180° sector focussing
magnetic field. In order to increase their larmor radius the ions
are accelerated in a strong electric field at right angles to the
magnetic field, fig. 18. The ion then passes through a slit into
a region with zero electric field and describes a circular orbit
until it reaches a detector. The electric field is swept and ions

of increasing mass are successively focussed on the collector
slit. The technique was successfully used for impurity analysis
in the divertor of the C stellerator[59] and more recently has been
used in other relatively low field plasma devices[60]. No success-

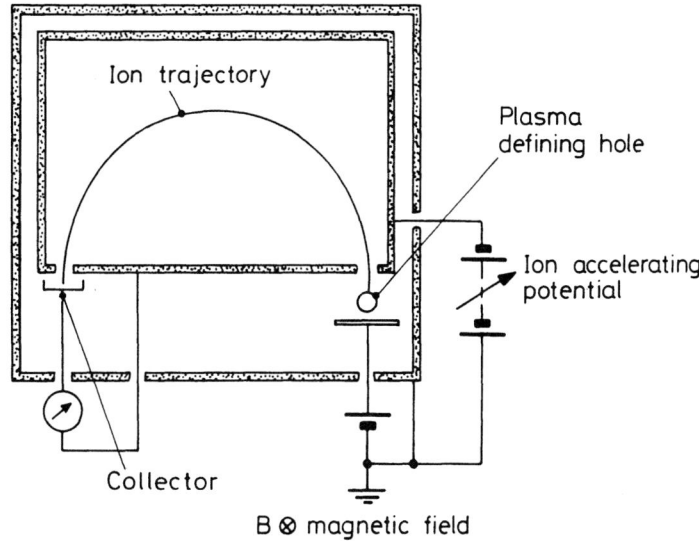

Fig. 18 Schematic of the 180° sector field mass spectrometer.

ful application to a tokamak has been reported. Problems arise
with the focussing and extraction of the ions in the electric
field region. A large parallel velocity results in the ions
continuing to travel along the field while describing the circular
transverse orbit. The time to reach the collector slit in the
plane normal to B must be less than the time taken to travel the
length of the slit parallel to B. In principle this distance
should allow the parallel velocity to be measured. However the
problems which arise from the initial velocity of the ion causing
chromatic aberration are likely to cause serious degradation in
resolution. This arrangement of fields provides first order
focussing in one direction only.

(iii) The cycloidal mass spectrometer has an arrangement of
uniform crossed electric and magnetic fields which in principle
provides both directional and energy focussing, fig. 19. The
problems of ion sampling and acceleration are similar to those
described for the other devices. In addition it is important to

have a uniform electric field. The condition for focussing is

$$\frac{m}{e} = \frac{dB^2}{2\pi E} \tag{27}$$

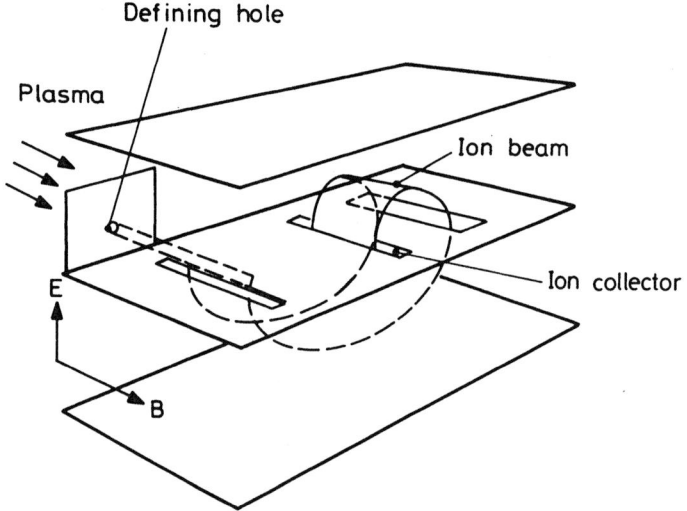

Fig. 19 Schematic of the cycloidal mass spectrometer. Ions are
subjected to a uniform electric and magnetic field and
perform a cycloidal path to the collector.

where d is the distance between the slits, and E and B are the
electric and magnetic fields. There is no known use of this
arrangement for plasma analysis.

5.   FLUX MEASUREMENTS USING SURFACE ANALYSIS

5.1 Theory

     We have already discussed the flux measured by electrical
probes and how this can be interpreted to give the ion density.
However, the analysis assumes that the probe is non-perturbing.
Generally speaking collector probes are larger than electrical
probes and must therefore be considered perturbing. In a magnetic
field flow is predominantly along field lines and the presence of
a probe is therefore perturbing in the sense that it adsorbs the
incident flux. The flow to the probe in this circumstance is

168

replenished by cross field diffusion. It can readily be shown[61]
that for a rectangular flux tube connecting with the probe, in the
absence of source terms within the flux tube,

$$\Gamma_\perp \, A_\perp + \Gamma_\parallel \, A_\parallel = 0,\tag{28}$$

where $\Gamma_\perp$ and $\Gamma_\parallel$ are the fluxes perpendicular and parallel to the
field and $A_\parallel$ and $A_\perp$ are the areas of the cross section and the
side of the flux tube. In what follows it is assumed that
ionization is negligible within the flux tube defined by the
probe. Ionization is negligible whenever the ionization rate
$\langle \sigma_{eii} v_e \rangle \, n_o n_e$. V is small compared to the probe current, where
σ_{eii} is the electron impact ionization cross-section, n_o is the
neutral density, n_e is the electron density, and V is the volume
of the flux tube. When this condition does not apply one must add
the ionization rate as a volume source term in the conservation
equations for particles (equation 28 above) and momentum (see
chapter by Stangeby). The parallel ion and electron flux Γ are
given in general for the collisional case by the equation

$$\Gamma_\parallel = -\, D_\parallel \, [\frac{dn}{dx} + n\mu_\parallel \, E]\tag{29}$$

where $D_\parallel$ and $\mu_\parallel$ are the diffusion coefficients and mobilities
along B. An analysis of the cross field and parallel fluxes leads
to an expression for the electron saturation flux,[62,63]

$$\Gamma_{\parallel,e} = 1/4 \; n_e c_e \left(\frac{r}{1 + r}\right)\tag{30}$$

where

$$r = \frac{16}{\pi} \; \frac{\lambda\sqrt{\alpha}(1 + T_i/T_e)}{d} \, ,\tag{31}$$

$\alpha = D_\perp/D_\parallel$, d is the probe size and λ is the electron mean free
path. Equation (30) is a modified form of equation (8). For
strong magnetic fields $r \ll 1$ the electron saturation current is
reduced by a large factor compared with the value in zero field.
This reduction can affect the voltage range over which
$\ln(I + I_{sat})$ is proportional to V and hence the parameter range
over which T_e should be calculated.

We now consider the ion flux, following the analysis presented by Cohen[61]. The potential variation along the flux tube
towards the probe is shown in Fig. 20. With the probe at large
negative potential the ions get continuously accelerated and the
flux is given by the ion saturation current as discussed in
section 3.1. For a probe at a large positive potential

$$V_p > (T_i/T_e) \, \ln[(1 + r)/r]\tag{32}$$

the ion random flux is reduced by a simple Boltzmann factor

169

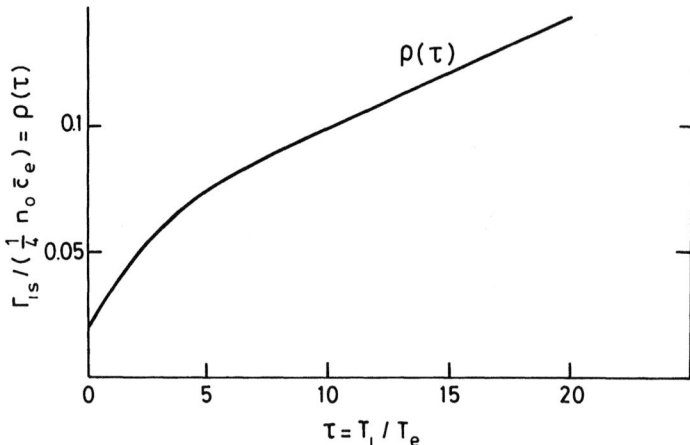

Fig. 21 Ratio of ion saturation current to electron saturation
current as a function of the ratio T_i/T_e for the case of
collisionless flow.

density is reduced to n_1 at the top of the hill

$$n_1 = n_o \exp (- V_H/kT_i). \tag{34}$$

The ion current flowing to the probe becomes

$$\Gamma_i = n_1 \left[\frac{k(T_e + T_i)}{m_i}\right]^{1/2}. \tag{35}$$

The value of n_1 depends on the collisionality of the plasma. For
the case of collisionless ion flow it has been shown that[63]

$$\Gamma_i = \rho(T_i/T_e) \; 1/4 \; n_o c_e \exp (- V_H/kT_i). \tag{36}$$

$\rho(T_i/T_e)$ is a complex function which is shown graphically for D^+
ions in Fig. 21 and c_e is the electron random velocity defined in
section 3.1. In the ion saturation region the ion flux reduces to

$$\Gamma_i = \rho(T_i/T_e) \; 1/4 \; n_o c_e \tag{37}$$

It is seen that the ion flux is quite a strong function of the
T_i/T_e ratio. For D^+ $\rho(T_i/T_e)$ varies from 0.02 at $T_i = 0$ to ~ 0.1
at $T_i = 10 \; T_e$. The general form of $\rho(T_i/T_e)$ can be deduced from
the theory of Emmert et al.[64].

The question of the collisionality of the plasma has been

170

$$\Gamma_i = 1/4 \; n_o \; c_s \; \exp \; (- \; eV_p/kT_i). \tag{33}$$

For intermediate potentials the ions see a potential hill, V_H, between the undisturbed plasma and the probe. This hill can be thought of as due to the resistance of the plasma to the electron flow[61]. The ions are retarded as they approach it and their

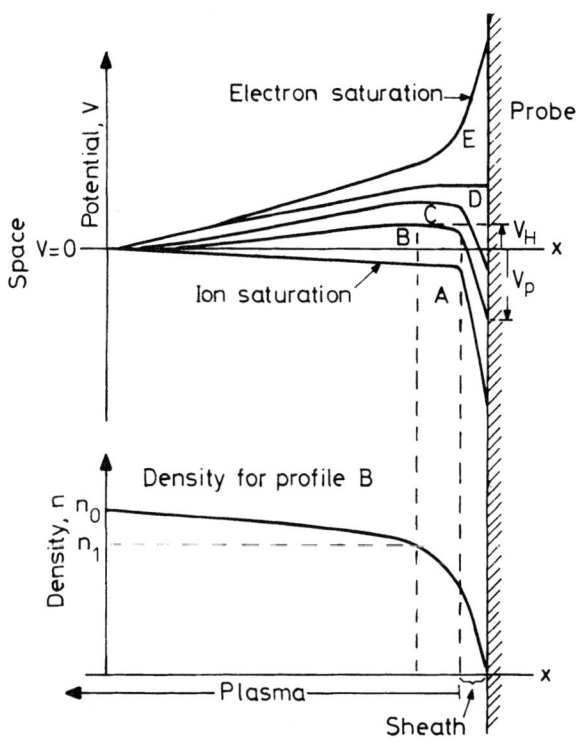

Fig.20 Potential and density variation along a flux tube as a function of distance from a probe surface. Potential distributions are shown for different applied probe potentials A to E.

171

considered by Cohen[61]. The plasma is collisional if the ion-ion mean free path is short compared to the flux collection length, L. The situation is complicated depending on whether impurities are present or not. It has been concluded that in much of the region accessible to probes the plasma is sufficiently collisional to affect the parallel motion of H^+ ions and highly ionized impurities, e.g. 0^{6+}. The flux tube length L is approximately given by

$$L = 0.5 \ d(D_\parallel / D_\perp)^{1/2} \tag{38}$$

and the ion-ion collisional time is given by[65]

$$\tau_{ii} = \frac{3\sqrt{m_i}(kT_i)^{3/2}}{4\sqrt{\pi} \ n \ \ln\Lambda \ e^4} . \tag{39}$$

Where $\ln\Lambda$ is the Coulomb logarithm. If we assume D^+ ions, $\ln\Lambda = 10$ and $D_\perp = 2.5 \times 10^{16}/n$ we obtain

$$nd < 1.1 \times 10^{12} \ T_i^{0.75} \quad (T_i \ \text{in eV}) . \tag{40}$$

Thus for $d = 1$ cm, $T_i = 10$ eV we have $n_e \sim 10^{12} \ cm^{-3}$ as the upper limit that may be regarded as collisionless. The inclusion of other collisions (ion-neutral, ion-impurity, etc.) will further reduce the permitted density.

For operation with collection probes it is most convenient to maintain the probe at floating potential, both to simplify the interpretation and to minimize the heat flux to its surface.

5.2 Surface Collector Probes: Hydrogen Isotopes

The hydrogen isotope flux measurement by surface collection has been used on a number of tokmaks[66]. Like other surface techniques it has the advantage of allowing measurements to be made after the discharge and away from the electrical interference present during the discharge. It also has the advantage of allowing differentiation between various isotopes present. The disadvantage associated with these methods is a lack of real-time feedback to assist in tokamak operations or to guide the experiment in progress. Such a disadvantage is often serious.

The interaction of energetic particles with solid surfaces is reviewed in detail by Behrisch (these proceedings). The use of collector probes for edge plasma analysis requires interpretation of the implantation profiles which result from particle bombardment. We briefly sketch the physical principles involved in the development of the implantation profiles.

172

When an energetic ion or neutral atom strikes a solid it loses energy at an energy dependent rate. At high energies the predominant loss is via interaction with the lattice electrons while at lower energy, loss of energy occurs via collisions with the lattice nuclei, the ions undergoing scattering. The particles of an incident beam come to rest at a distribution of depths. The depth profiles are accurately calculable for amorphous solids by Monte Carlo codes[67] and are fairly well represented by gaussian profiles modified to account for reflection loss at the front surface[68]. The 1st moment (mean) of depth profiles so constructed is known as the "mean projected range" and the 2nd moment (variance squared) is called the "depth straggling". A knowledge of the depth profile can sometimes be related to the energy distribution of the incident particles for a given target if it is assumed that the particles are trapped at their stopping point in the lattice and do not subsequently diffuse. Thus only materials, such as carbon, silicon and beryllium which have very low diffusion coefficients for hydrogen can be used. Carbon and beryllium also have low atomic numbers and thus the fraction of the incident hydrogen flux which is backscattered is low and the sensitivity of the method is enhanced. The most commonly used carbon samples for such probes are carbon ribbon (PAPYEX or POCO), amorphous carbon films, and compression annealed pyrolytic carbon.

As the incident ions are trapped in the solid the concentration increases linearly. Eventually the concentration reaches a saturation level, first at the peak of the range profile and then gradually the saturation region expands, see fig. 22. The mechanism responsible for saturation is not well understood but the saturation level has been measured to be 0.44 hydrogen atoms/ carbon atom[69] at room temperature. As the range of hydrogen ions is roughly proportional to energy the areal density (atoms m^{-2}) at which the surface is saturated is dependant on the incident ion energy. Thus by measuring the trapped fluence as a function of incident fluence the incident flux can be obtained from the initial linear part of the curve and the energy can be obtained from the saturation level.

In order to obtain the incident flux from such measurements it is necessary to know the form of the incident ion energy distribution since this determines the number of incident ions which are backscattered. If the energy distribution form is known, e.g. Maxwellian or monoenergetic, then there are a number of detailed models which can be applied. Wampler et al.[70], have done a careful calibration of these models with experimental data using ion beams to implant samples. They have shown that by taking a series of exposures of samples to different fluences whose ratio is known but whose absolute value is not known, the value of incident flux and the ion energy or temperature (depending on the assumption about the energy distribution) can be derived. In principle this technique is quite easy to apply since samples can be exposed to different numbers of discharges. In practice, there are serious problems with the method. The integration of collected flux over time in the discharge is highly undesirable. Time-resolved

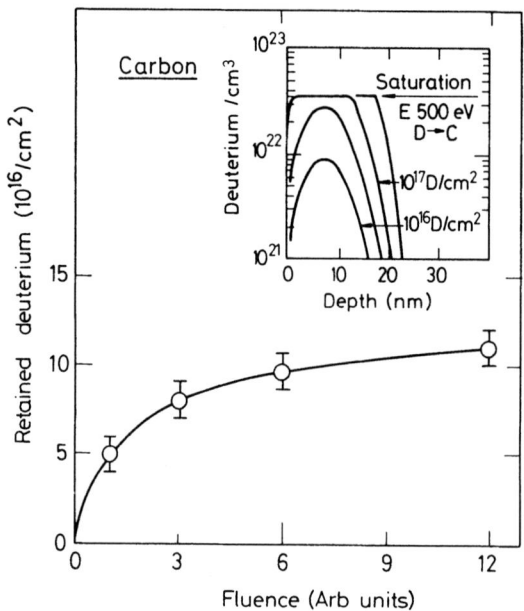

Fig.22 Experimental measurements in HLT of deuterium retained in a carbon probe as a function of incident fluence. The insert shows schematically the depth distribution in the solid illustrating how the retained deuterium increases at first linearly and then saturates.

measurements show that the flux and particle energy change dramatically over the course of a tokamak discharge often showing large start-up and termination flux transients.[71,72,73] These parameters vary a great deal during auxiliary heating as well. Thus a time averaged value may be of little value in interpreting the physics of the edge plasma. There has been a large discrepancy between the flux measured by collector probes and that measured by Langmuir probes when these have been compared[74]. This

could be the result of such time averaging or of the collector
probes losing a large fraction of the low energy component of the
plasma flux.

A variety of methods for measuring the areal density (atoms
m^{-2}) is available. These methods include nuclear reactions[75],
e.g. $^1H(^{15}N,\alpha\gamma)^{12}C$ and $^2H[^3He,p]^4He$, thermal desorption[76], forward
nuclear scattering and Secondary Ion Mass Spectrometry (SIMS).
Thermal desorption is relatively simple, it is sensitive and can
often be carried out "in situ". By using a mass spectrometer all
isotopic species can be detected. The nuclear techniques require
the use of an accelerator and therefore normally require the
collectors to be removed from the tokamak. The hydrogen nuclear
reactions also require high energies ($\sim$ 5 MeV) but the
$^2H[^3He,p]^4He$ reaction is very simple, and has good sensitivity.
Ross and co-workers[77] have recently developed a low energy
(350 keV) helium elastic recoil apparatus for H^+ and/or D^+ depth
profiles with a resolution of $\sim$ 5 nm and an ultimate sensitivity
of 0.1 atomic %. Energy analysis of scattered products or nuclear
resonance methods permit these species to be depth-profiled to an
accuracy of $\sim$ 10 nm, sufficient for many probe applications e.g.
fast neutrals lost during neutral injection heating[78] or minority
ion cyclotron resonance heating[79]. SIMS permits more accurate
depth profile determination ($\leqslant$ 20 Å) and this is more useful for
low energy plasma implants ($T_i \sim$ 50 eV) frequently associated with
the scrape-off layer. The use of two or more of these techniques
on the same samples is often valuable. Fig. 23 shows nuclear
reaction analysis of the retained deuterium during PDX neutral
beam injection, further analyzed by SIMS to obtain depth profiles
at various times indicated on the nuclear reaction analysis, fig.
24. The deep ($E \sim$ 47 keV) beam component is clearly evident. A
similar combination of SIMS and ERD is shown for beam-heated PLT
discharges in fig. 25. In both cases the non-quantitative SIMS
profiles were supplemented by absolute flux measurements of the
other method.

Another method for determining ion energy with surface
collector probes takes advantage of the effect of finite larmor
radius on transmission of ion through an aperture normal to the
total magnetic field[80]. The details of the method depend on the
specific geometry of the aperture as shown in fig. 26. When the
aperture length is large enough to permit more than one full gyro-
rotation of the particle as it traverses the opening the aperture
is said to be thick. Otherwise it is said to be thin.

It is possible to use electrical collectors behind such
apertures by examining the total transmission (collected
current). Krawec[81] has given a detailed analysis of the problem
for both thick and thin apertures and has presented a complete set
of numerical calculations in terms of dimensionless parameters in

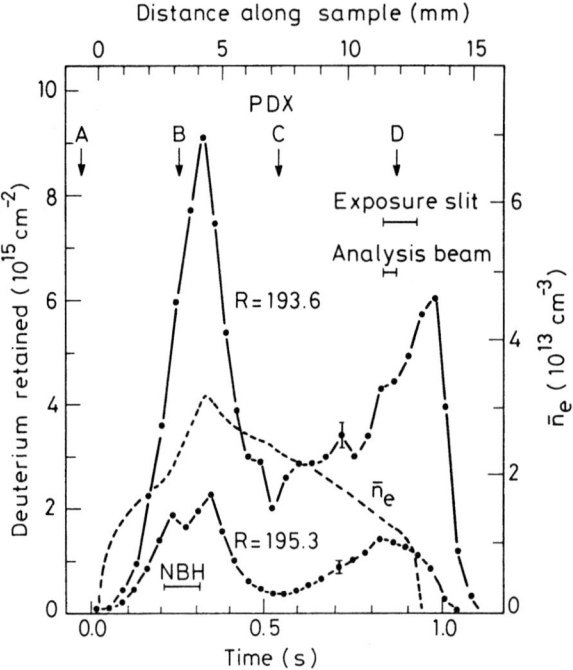

Fig. 23 The deuterium retained in two silicon probes as a function of time during a PDX discharge. The probes are at two different radii. The time when neutral beam heating (NBH) was applied is indicated. Depth profiles were measured at times indicated A,B,C,D.

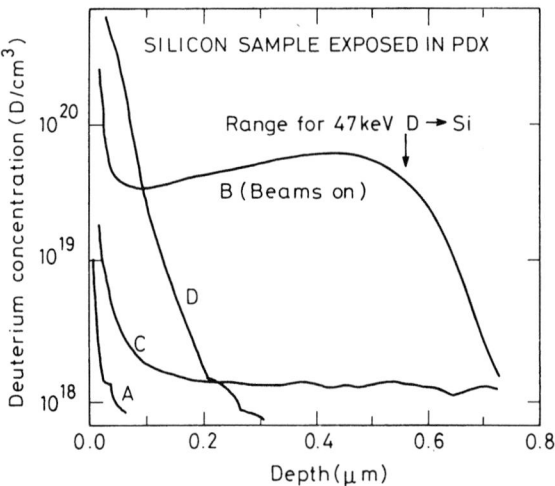

Fig. 24 The depth distribution of deuterium trapped in a silicon sample at various times during exposures in PDX as indicated from fig. 23.

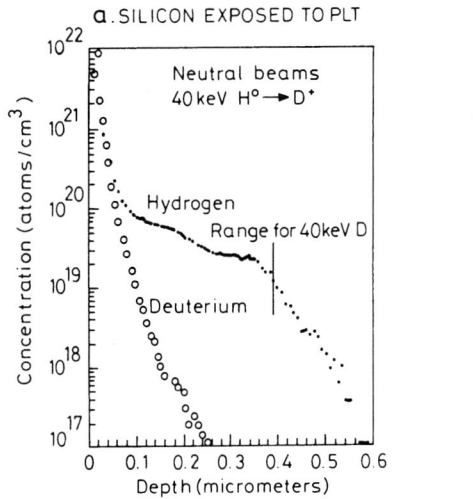

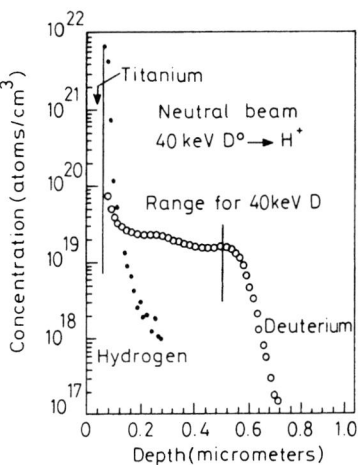

Fig. 25 (a) and (b) The depth distribution of H and D trapped in
silicon during exposure in PLT determined by SIMS. The
calculated range for normally incident 40 keV H(a) and
D(b) ions are indicated.

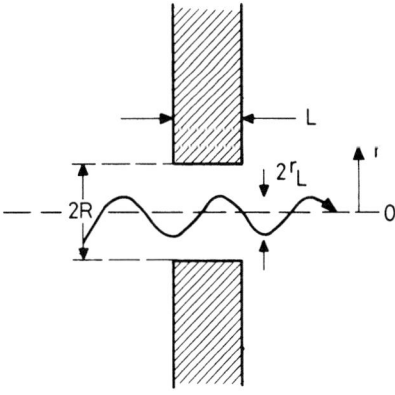

Fig. 26 Schematic of the transmission of ions of larmor radii r_L
going through an aperture of thickness L and radius R.

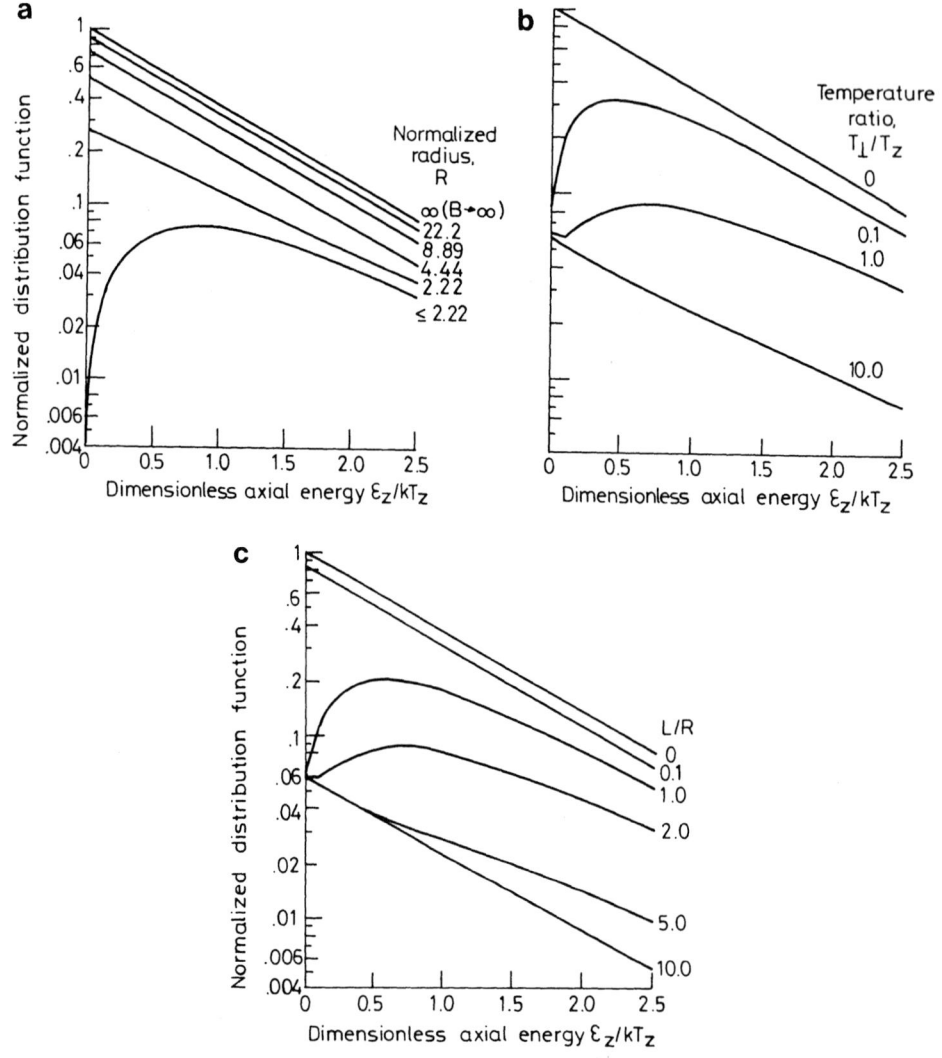

Fig.27   Results of calculations of the transmission of ions of
         larmor radius r through a hole of radius R in a plate of
         thickness L.
(a)   Effect of varying the magnetic field for an L/R ratio 2.0
         and a temperature ratio 1.0.
(b)   Effect of varying the transverse to axial temperature
         ratio L/R and 2.0 normalized aperture radius 0.889.
(c)   Effect of varying the L/R ratio. Temperature ratio 1.0
         normalized aperture radius 0.889.

the problem; examples are shown in fig. 27. On fusion devices
such work has employed only sample collector surfaces for
detection. Staudenmaier[80] and coworkers have considered the thick
aperture case and derived an analytical expression for the transmission as a function of distance from the aperture centreline.
An example of the expected behaviour of a 0.15 cm aperture in a 4T

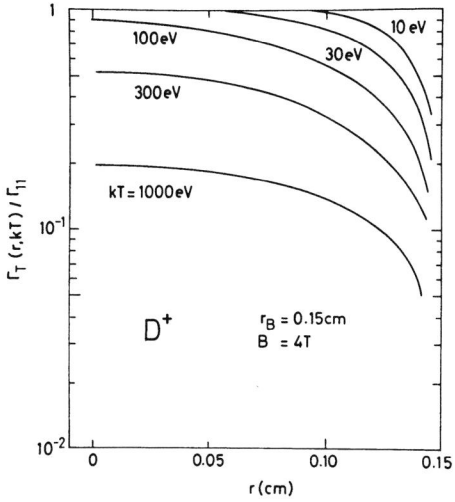

Fig.28 Calculated spatial distribution for deuterium ions of
various temperatures between 10 eV and 1000 eV on a
collector behind a hole of radius 1.5 mm in a toroidal
field of 4 Tesla.

magnetic field is shown for the case of D^+ ions in fig. 28. The
extraction of an ion temperature requires the assumption of an
isotropic Maxwellian ion distribution. Such an assumption is
likely to be untenable in many instances since sheath acceleration, plasma rotation, or auxiliary heating all may

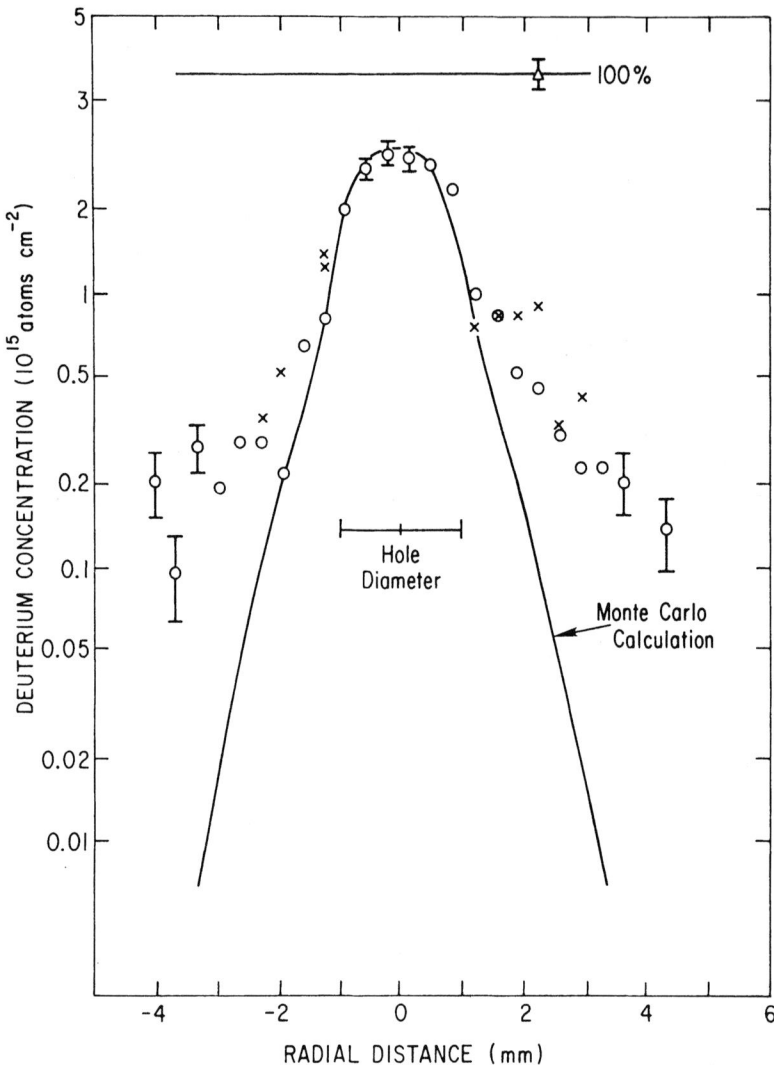

Fig. 29 Experimental measurements of the spatial distribution
behind a hole of 2 mm diameter for deuterium ions in the
DITE tokamak at a field of 2 Tesla. The Monte Carlo
calculation is the distribution expected for ions of
temperature 60 eV.

generate non-Maxwellian, anisotropic ion distributions. Zuhr et
al. have accounted for non-Maxwellian distributions in their
Monte-Carlo orbit calculations[82]. Their calculated results are in
quite good agreement with the data. Less satisfactory agreement
is seen in fig. 29 for D^+ from the DITE tokamak. Multiple ion
reflection[83] may be a complication in this case.

5.3 Carbon Resistance Probe

    A probe technique using a carbon thin film to measure ion
energy has been developed by Wampler[84,85]. Since direct electrical measurements can be made in situ this technique should perhaps
be included in section 3. However because it depends on surface

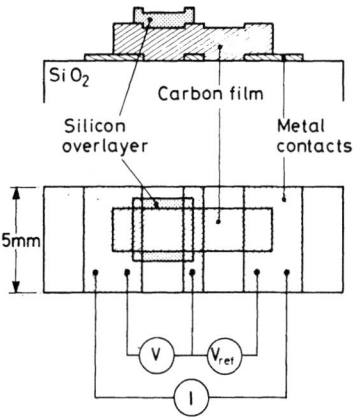

Fig. 30 Schematic of the carbon resistance probe.

modification due to ion bombardment it is more logical to include
it with the other surface techniques. The active element is a
thin film of carbon with a resistance which varies in a well-
characterized manner as it is damaged by an incident particle
flux. A schematic of the probe is shown in fig. 30 and the
technique is similar to the standard 4-Point and Van der Pauw

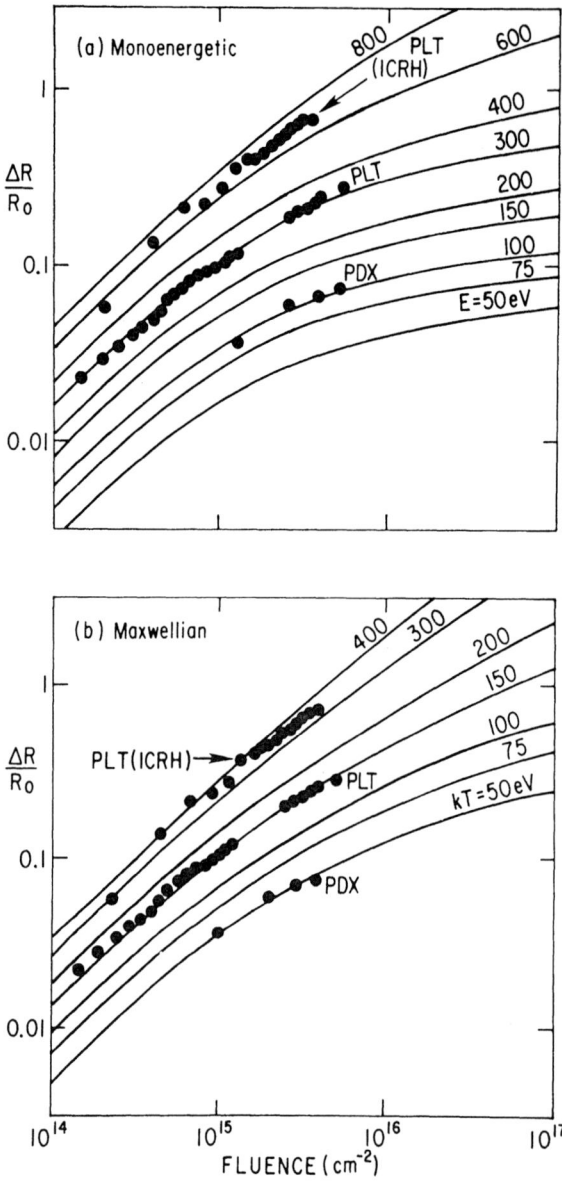

Fig. 31 Experimental results obtained using a carbon resistance
probe in PLT and PDX compared with theoretical results
obtained from ions of various energies. R_o is the
resistance of the undamaged probe.

resistance diagnostic methods of ion-beam dose determination[86] in
the semiconductor manufacturing industry.

The method relies on a change of resistance which occurs when a carbon layer is bombarded. The resistivity is assumed to be given by

$$\rho(x) = \rho_o + A \phi f(x)$$

where ρ_o is the unperturbed resistivity, ϕ is the fluence, A is a proportionality constant, and $f(x)$ is the damage profile induced by the ion or neutral flux. Integrating for total resistance and taking the difference induced by the flux damage one arrives at

$$\Delta R = \frac{A\phi}{d^2} \int_o^d f(x)dx$$

where d is the carbon film thickness. The damage profile $f(x)$, is

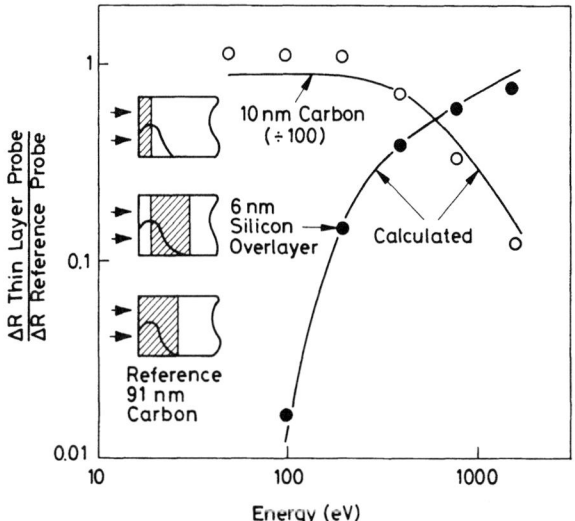

Fig. 32 A comparison of experimental and theoretical estimates of the resistance change in a carbon resistance probe with and without a thin surface layer of silicon.

related to the incident ion energy. The average ion energy may be determined by plotting the resistance as a function of fluence, as shown in fig. 31. We note again that the form of the distribution function must be assumed (Maxwellian or monoenergetic).

A variation of this probe is based upon classical foil methods of particle detection[41]. A thin layer of (high resistance) silicon is coated over the carbon as shown in fig. 32. This admits only particles whose energy exceeds a minimum valueselected by varying the Si layer thickness. By comparing the behaviour of two such detectors with different Si thickness (one of which may be conveniently chosen to be zero) it is possible to determine the average particle energy from a single exposure rather than having to carry out multiple exposures until saturation is reached. The behaviour of a 6 nm overlayer is shown in fig. 32. Such a probe has been used on HLT. We note that the distribution function must still be assumed and further variation is required for bimodal or multi-modal distributions. The detector saturates fairly quickly and must be replaced, or possibly regenerated by annealing. The device has very modest power handling capacity (less than a few tens of watts/cm^2) and thus is of little value for sampling the ion flux near the limiter region. In addition the thin overlayers may erode quickly under such a flux. Therefore its use is mostly restricted to applications where a sensitive charge exchange atom detector is required.

5.4 Surface Collectors: Impurity Fluxes

Impurity concentrations are difficult to measure in the plasma boundary. Although emission spectroscopy can be used in principle, it is often difficult to get optical access and interpretation is difficult without a detailed knowledge of the plasma density and temperature profiles. Another possible technique is laser-resonance-fluorescence which has recently been developed into an exremely sensitive and powerful tool which can measure local impurity densities and also, with more difficulty, velocity distributions[87]. The surface collector method is complementary to these techniques in that it allows measurement to be made of the total population of the parallel flux integrated over all charge and excited states. Time resolution can be obtained by a moving collector. The technique is usually applied with the collector surface normal to the magnetic field so that it collects all the flux travelling parallel to the field on the flux tube connected to the collector[88]. The collector surface must itself be free from impurities and again both carbon and silicon have been used. There is a wide variety of methods by which the implanted plasma and impurity species are analyzed. Table 3 taken from McCracken and Stott[89], lists most of the more common techniques. Complete descriptions of these techniques are available in a number of texts[90]. A recent review by Zuhr et al. also discusses these methods[91]. Therefore we will only summarize their use for probe analysis here.

Both SIMS and Auger spectroscopy are suitable for surface analysis. SIMS has very high sensitivity and good depth

Table 3.   Surface Analytical Techniques

Technique	Primary beam	Primary energy (keV)	Emergent particle	Elements detected	Minimum detectable concentration (monolayers)	Disadvantages	Advantages
XPS X-ray Photo-electron spectroscopy	X-ray	1-2	Electron	$z \geq 6$	10^{-3}	Not quantitative	Sensitive; gives information on chemical structure of surface
AES Auger Electron Spectroscopy	Electron	2-5	Electron	$z \geq 6$	$10^{-3}-10^{-1}$	Detects only surface layer. Needs calibrating	Sensitive; compact; quantitative
ISS Ion Scattering Spectroscopy	Ion	1-50	Ion	$z \geq 6$	10^{-3}	Requires primary ion beam. Has to be calibrated. Detects only surface layer	Sensitive; gives information on surface structure
PIXE Proton Induced X-ray	Ion	1 MeV	X-ray	$z \geq 6$	10^{-2}	Has to be calibrated Requires MeV ion beam No depth resolution	Sensitive; good element resolution
RBS Rutherford Back Scattering Spectroscopy	Ion	0.5-10 MeV	Ion	$z \geq 6$	$10^{-4}-10^{-3}$	Requires MeV ion beam. No light elements. Surface must be reasonably smooth	High sensitivity; quantitative; depth distribution
NRA Nuclear Reaction Analysis	Ion	~ 1 MeV	Ion X-rays	Selected	Varies with reaction	Limited number of elements. Requires MeV beam	Depth profile H,D
SIMS Secondary-Ion Mass Spectroscopy	Ion	1-5	Ion	Selected	$10^{-4}-10^{-2}$	Not quantitative Can be used only for certain elements. Detects only surface layer	High sensitivity; compact Detects H and D
SEM Scanning Electron Microscopy	Electron	5.50	Electron	-	-		Surface topography good resolution; $\leq$ 100 A; good depth of focus
FR Forward Recoil	Ion	> 300	Ion	Light	10^{-2}	Requires flat surfaces	All light elements detected simultaneously

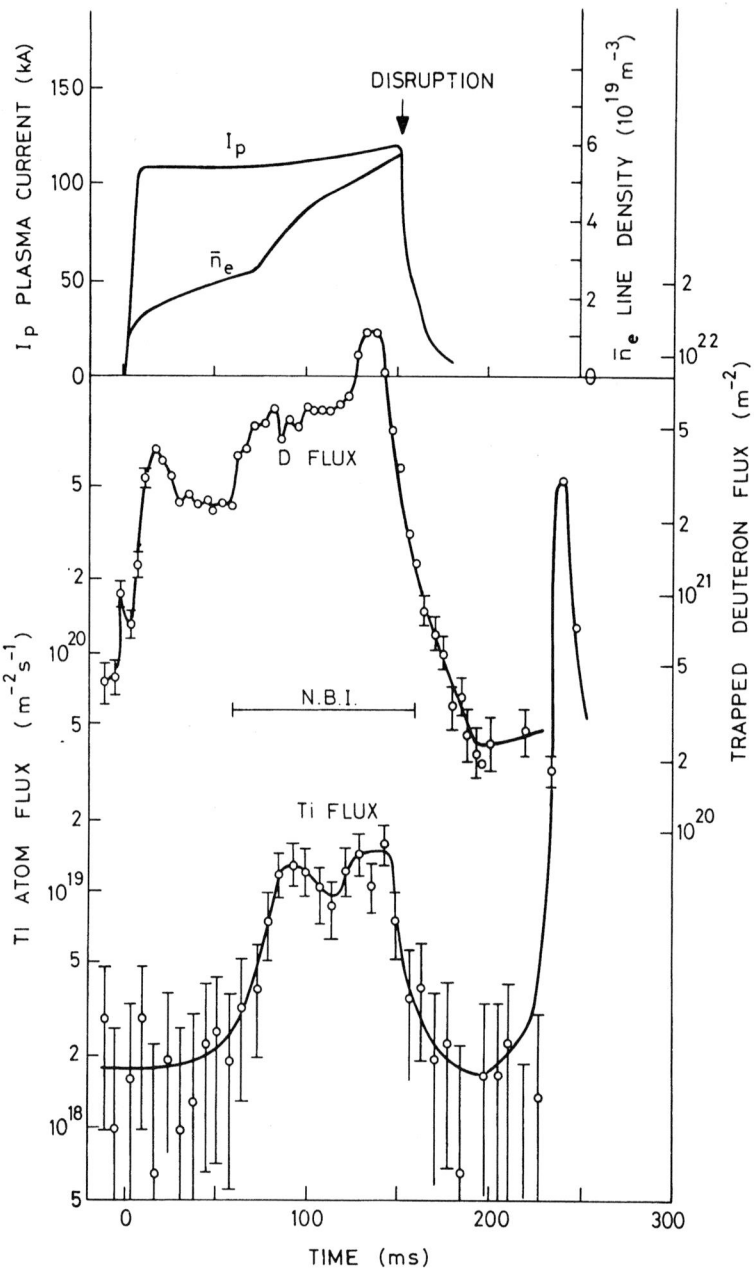

Fig. 33 Measurements of the parallel flux of deuterons and
titanium ions as a function of time in the DITE tokamak
using a carbon surface collector probe. Both the D^+ and
titanium flux increase during <u>hydrogen</u> neutral beam
injection.

resolution but it is difficult to make quantitative. Auger spectroscopy, when combined with sputter erosion, has equally good depth resolution and can be quantitative, but it is less sensitive then SIMS. Auger techniques are most useful for the examination of light impurities. For heavy impurities the most frequently used method is Rutherford Backscattering spectroscopy (RBS). RBS is both sensitive and quantitative. It also has the advantage that its sensitivity is proportional to the square of atomic number (Z) so that the high Z impurities which are most deleterious to the plasma are most easily detected. When performed using heavy ion beams (such as ^{14}N) background scattering via light substrates like C can be eliminated[92].

Transfer of samples in atmosphere is possible, although contamination by lead in the atmosphere can be a problem. Air transfer is not satisfactory for low Z impurities such as carbon and oxygen and contamination can occur even with vacuum transfer. Great care has to be taken with the collector and with the vacuum system to overcome this problem. Time resolved measurements of high Z impurities are shown in fig. 33. The impurity flux is normally large during the starting phase[93]. In this case the flux of both impurities and deuterium ions increases during neutral injection[94]. Results for both time resolved fluxes and radial profiles of impurities have been obtained for a number of machines[95-99]. In general the incident metal fluxes at the limiter radius are in the range 10^{18}-10^{19} ions m^{-2} s^{-1}.

The determination of impurity energy and charge state has not yet been satisfactorily solved. The implantation depth profiles are too small for their measurement to be a satisfactory technique. There have been a few attempts to measure the larmor radius by looking at the spatial distribution behind an orifice in a manner similar to that discussed for hydrogen ions c.f. section 5.2[80,83]. This is quite a simple technique in some cases. However, knowing the larmor radius does not give the energy and charge independently. If the impurities are in energy equilibrium with the plasma the energy can be obtained from the plasma ions (necessarily of charge 1) and hence the mean charge state of the impurities can be obtained. Results for titanium impurities in DITE are shown in fig. 34. This technique is rather time-consuming and has not been widely used. However, since the energy and charge state are important parameters in determining impurity sputtering and there are few alternative techniques available, this method must be considered.

In general the surface collector techniques are rather slow. Nevertheless they have the attraction that they allow measurements to be made away from the radiation from the plasma. This may be a very significant factor in their favour during operation with DT reacting plasmas where large fluxes of neutrons and gamma rays will occur. The techniques are independent of other plasma parameters and give information about the total flux and the energy with which ions arrive at the surface, which are difficult to obtain in any other way.

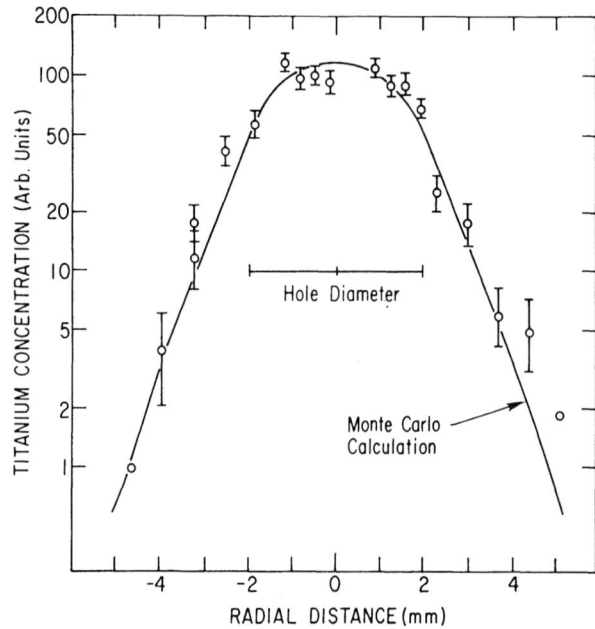

Fig. 34 Experimental measurements of the spatial distribution of titanium atoms on a carbon collector behind a 4 mm diameter hole, compared with a Monte Carlo calculation for 80 eV Ti^{4+} ions.

5.5 Surface Collector Probes: Alpha Particles

The technique for alpha particle collection is very similar to that for hydrogen isotopes and impurities. Helium ions are tightly bound in the lattices of most materials and their trapping behaviour and range-energy relationship have been studied in great detail[100]. Because they are not present as impurities it is possible to detect very low concentrations. The attraction of the surface collection technique is the possibility of determining

188

both flux and energy distribution by measuring the depth distribution in the collector.

The depth distribution will give a measure of the alpha particle confinement. The mean depth of a 3.5 MeV α-particle implanted in nickel is 6 μm as compared to the depth of a 20 keV ion which is 0.085 μm[100]. Thus the fraction of the α flux which

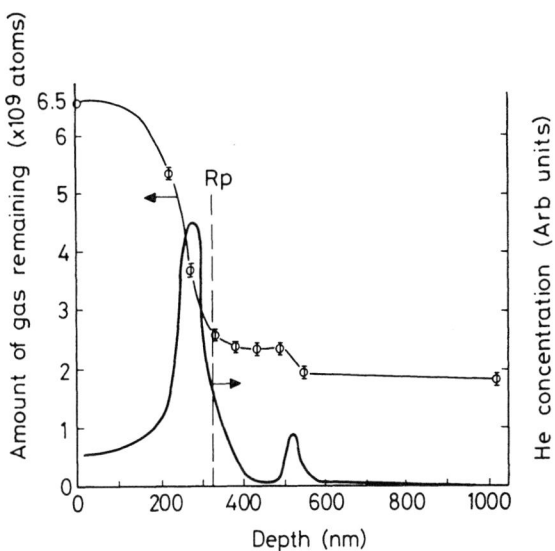

Fig. 35 Depth distribution of helium ions in a nickel foil implanted at 100 keV to a fluence of 2.9×10^{11} atoms cm^{-2}. The measured integral and the derived differential distribution are shown. R_p is the projected range of the He$^+$ ions in nickel.

has not thermalized and contributed to the core heating should be easily distinguished. If the alphas are slowed down in the plasma then it is to be expected that they will ultimately diffuse out with the other plasma ions and hit the limiter. There will only be a very small charge exchange flux to the wall as the cross section for proton charge exchange with helium is typically three

orders lower than for hydrogen[101]. The spatial distribution of
the α's which are lost before slowing down in the plasma has been
discussed by Bauer et al[102]. Their distribution is expected to be
very asymmetric, coming out predominantly along the mid plane.
Thus collectors placed at intervals around the torus poloidally
and toroidally can be used to confirm this asymmetry and to assess
both the energy distribution and the spatial distribution of the
alpha particles arriving at the wall.

A variety of methods can be used for getting the number and
depth distribution of the helium. One of the most sensitive is
thermal desorption which in conjunction with surface lapping has
been used to determine the depth distribution of 10^{11} atoms cm^{-2}
implanted at 100 keV[103]. Results are shown in fig. 35. An
alternative is to use nuclear scattering of protons after catching
the helium in a thin film[104]. There is a nuclear scattering
resonance for helium which significantly enhances the backscattering cross section above the Rutherford cross section. The
use of thin films reduces the backscattering from the substrate
and allows the helium to be clearly observed. Using 2.5 MeV
protons the depth distributions of 50, 100 and 150 keV ^{4}He ion
implanted in thin films of copper have been analyzed. The sensitivity of this technique is low, $\sim 10^{17}$ atoms cm^{-2}, and is
therefore marginally useful for early DT experiments.

The collector technique thus can be used to confirm the
overall flux of alphas (which should be consistent with the
integrated neutron flux), to determine the degree with which they
have been contained by the plasma, and by using a series of
collectors around the torus to determine any poloidal or toroidal
asymmetries.

6.   PROBES FOR EROSION MEASUREMENTS

Erosion of the wall and limiter will take place as a result
of a number of different processes including sputtering, arcing
and evaporation. Arcing and melting can often be recognized by
post exposure analysis but erosion due to sputtering is less
easily recognized. It is clear that the erosion rate will vary
widely depending on the local plasma and energy fluxes. The
removal of wall or limiter material is important not only from the
point of view of plasma contamination but also due to the actual
change in mechanical strength of tokamak components. Very large
erosion rates ~ 10-100 cm yr^{-1} are predicted for the first wall of
reactors due to sputtering if light elements e.g. carbon or
beryllium, are used. Most eroded material it is hoped will be
redeposited[105]. In the steady state in a closed machine on
average the net erosion must be near zero. However because of the
non-uniformities some redistribution of material is inevitable and

this behaviour has been commonly observed in present tokamaks[106]. Let us consider the simple case where we have an impurity flux Γ_I arriving at a surface at the same time as a plasma flux Γ_p. Neglecting impurity-impurity sputtering and assuming plasma sputtering to be proportional to Y.C where Y is the sputtering yield and C is the impurity surface concentration (assumed < monolayer) then the surface concentration will be given by

$$\frac{dC}{dt} = \Gamma_I - C \Gamma_p Y$$

i.e. $C = \dfrac{\Gamma_I}{\Gamma_p Y} [1 - \exp(- \Gamma_p Yt)]$

Thus the surface concentration will build up linearly and then reach a saturation level given by a balance between incident flux and sputter erosion. Obviously much more complex situations can arise but it is clear that in measuring erosion it is necessary to distinguish as far as possible between erosion and deposition. Making measurements as a function of incident fluence is one way in which this can be accomplished. We now consider some techniques which can be used to make sensitive measurements.

6.1 Thin Film Techniques using Surface Analysis

The thickness of a thin film ($\sim$ 20 nm) on a dissimilar substrate can be quite accurately measured using Rutherford backscattering, SIMS or sputter Auger spectroscopy. In successive measurements changes in surface coverage of $\sim 10^{19}$ m^2 can be detected. Using moving probes, time resolution is possible and good spatial resolution using small analysing beams is also straightforward. In addition because RBS detects other elements it is easy to distinguish between erosion of the film and deposition of different atomic species on the surface. Thus if the film is chosen to be an element which is not normally in the tokamak system, it is easy to distinguish between erosion and deposition. This technique has been used with Rutherford backscattering as a sensitive method of measuring sputtering yields[107]. For RBS a heavy film on a light substrate is the simplest to analyse[108]. The major drawback to this technique is that the erosion of the film on a dissimilar substrate may not be representative of bulk material. High heat fluxes may also be perturbing in terms of heating the thin film to high temperatures and possibly causing evaporation.

6.2 Implantation of a Marker Species

This is a development of the thin film approach. Instead of evaporating a thin film a different atomic species is implanted in a standard substrate. This marker can be detected by surface analysis and RBS is again a convenient non-destructive technique. A heavy marker in a light substrate such as carbon is easy to see, fig. 36. In this case the position of the marker in the

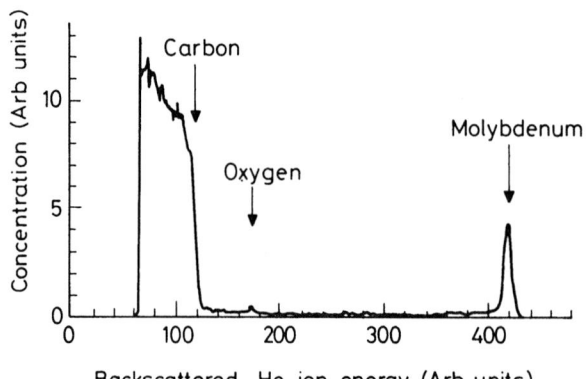

Fig. 36 R.B.S. spectrum of a molybdenum layer on a carbon substrate. A small oxygen impurity peak is also present.

surface is determined from the energy loss of the backscattered atoms in traversing the solid to the marker on the penetrating and backscattered paths. The erosion of the surface is measured by the change in the peak area.

A special case of this technique is the use of an isotopic marker. Implantation of ^{13}C in ^{12}C has been used by Roberto et al[98]. to demonstrate erosion of carbon in the divertor plates of the ASDEX tokamak. The results are shown in fig. 37. The ^{13}C peak was implanted at 35 keV with an incident angle of 45° to a dose of 2×10^{17} ions cm^{-2}. On the control sample the ^{13}C peak is quite clearly separated from the ^{12}C edge. As erosion or deposition occurs the ^{13}C peak moves either away from or towards the ^{12}C edge. A change of about 10 nm in thickness can be detected. The method has again the advantage that any other impurity deposited on the surface can be clearly identified in the RBS spectrum. Measurements of simultaneous erosion and deposition as a function of time are shown in fig. 38. Carbon is clearly eroded during the initial phase of the discharge while iron and titanium are deposited. Later in the discharge there is some indication of carbon deposition.

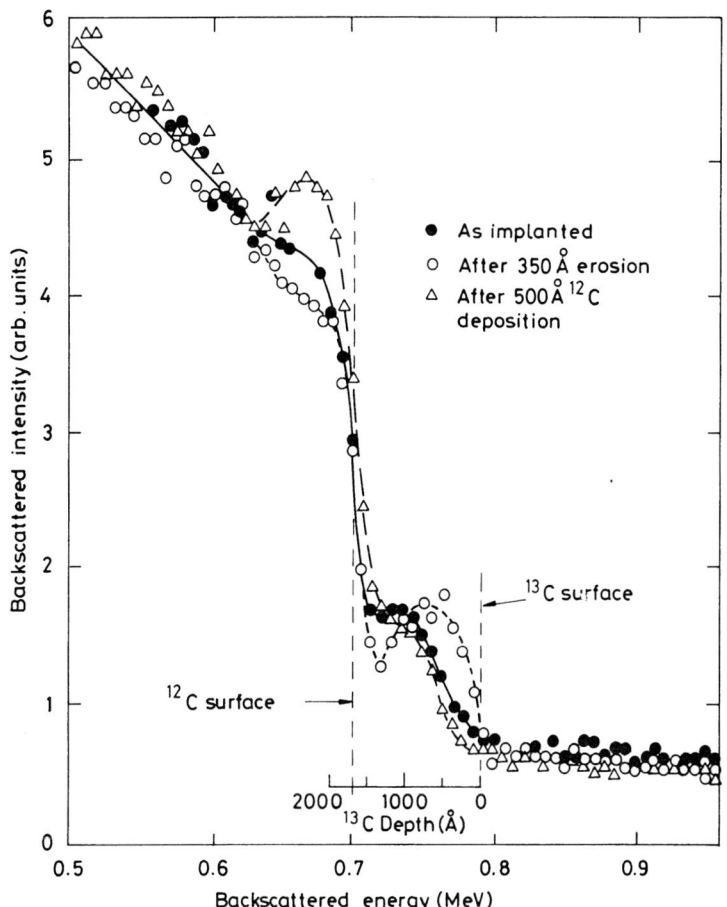

Fig.37   The R.B.S. spectrum of ^{13}C implanted in ^{12}C at 35 keV to a
dose of 3 x 10^{17} atoms cm^{-2}, as implanted and after both
erosion and deposition of carbon.

193

6.3 Thin Film Activation

A third method of measuring erosion is to use a nuclear activation technique. By using a nuclear reaction with a sharp energy threshold and using a beam with an energy just above the threshold

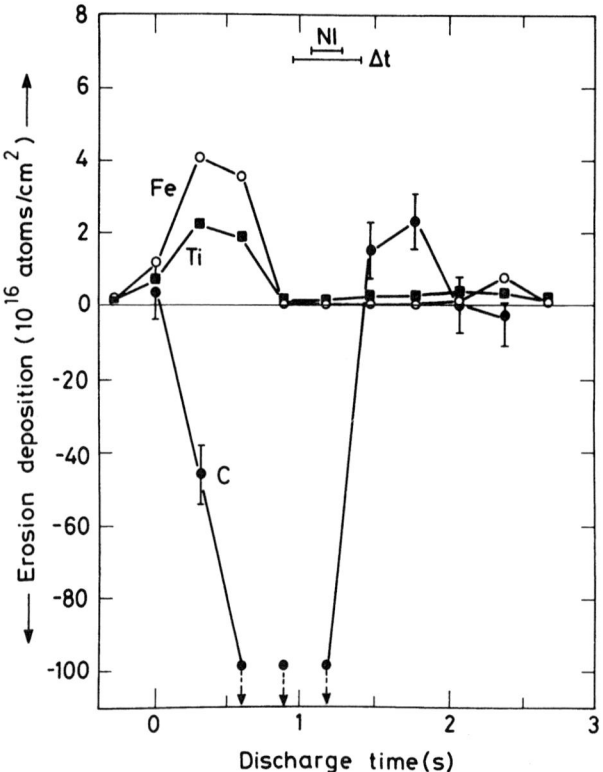

Fig.38 Erosion and deposition as a function of time on a surface collector probe exposed to 10 superimposed discharges in ASDEX with 2.5 MW of neutral beam heating.

a thin layer of the surface can be activated. When the incident beam slows down below the threshold energy activation stops. Many (α,n) and (p,n) reactions are suitable, Table 4, and have threshold energies which can be reached with van de Graaff accelerators[109].

Table 4   Isotopic species suitable for thin film activation

	Activation Half Life Days	Activation Energy, MeV
$^{96}Mo(p,n)^{96}Tc$	4.3	7.5
$^{48}Ti(p,n)^{48}V$	16	6.8
$^{56}Fe(p,n)^{56}Co$	78.8	8.2
$^{65}Cu(p,n)^{65}Zn$	244	5.2
$^{12}C(^{3}He,2\alpha)^{7}Be$	53	16.2

These reactions have been used in a wide range of applications[110] including divertor plate erosion[111]. One attraction of the technique is that real time measurements can be made since the activated species are generally γ emitters with energies of 1 to 2 MeV and can be readily detected with a NaI detector outside the vacuum vessel. Another potential advantage is that the redistribution of the eroded material can be easily detected from the activation of collectors some distance away. The technique is in many ways similar to the use of neutron activation but because only a thin surface layer is activated it is both more sensitive and requires much lower total activation levels, typically 10 μ C. The main drawbacks are that being a radioactive technique safety precautions have to be taken and that sensitivity is somewhat less than the RBS techniques. The activated layer has a thickness of 1 - 100 μm and the erosion level which can be measured is $\gtrsim 0.1$ μm. The technique does not give information directly on whether deposition of other species has occurred. There is no reason however why the two techniques of surface analysis and thin film activation should not be combined.

6.4   Measurement of Eroded Material

Another technique for measuring erosion which has recently been tried[112] is shown in fig. 39. The eroded material is collected and subsequently analysed. In this case electron induced X rays were used. Sputtering from a limiter like probe by ions or from the wall by charge exchange neutrals can be measured. The technique could be used in conjunction with either the thin film activation or the implanted marker techniques.

The above discussion makes it clear that there are many useful experimental approaches to the measurement of erosion. Much wider and more systematic use of these techniques should result in a better understanding of the processes involved and in better estimates of wall lifetimes. In addition the measurement of erosion rates in conjunction with measurements of incident ion and electron fluxes and energies should give an excellent method of cross checking the validity of various boundary layer models.

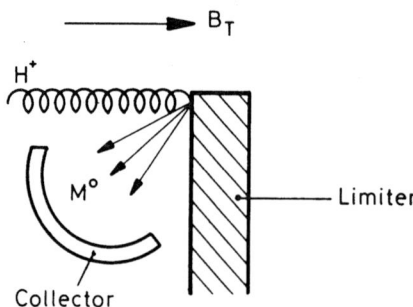

Fig. 39 Schematic of a collector for measuring erosion of the limiter in a tokamak.

7. LONG TERM PROBES

A number of investigations have been made of the condition of tokamak walls after a long period of operation[105]. These show that the material has been widely redistributed by a variety of erosion and redeposition processes. In addition significant physical damage by arcing, melting and evaporation is observed. Careful observations can reveal large non uniformities in the surface interactions. The analyses also indicate that changes in the material properties (particularly surface properties) must have occurred. Such properties including sputtering rates, particle and energy reflection coefficients, secondary electron emission coefficients, etc. are critical inputs to boundary models. Direct measurements of data using samples from the wall would allow an estimate of the error introduced by using the data for pure materials. It is relatively simple to attach samples which can be removed at intervals when access to the torus is possible. Such samples yield a considerable amount of useful data. The sample programme on JET includes:

(1) Examination of samples for physical damage using mainly optical microscope and scanning electron microscope.

Measurement of change of physical properties e.g. surface hardness.

(2) Measurement of impurity deposition and erosion. Methods of measuring erosion have been discussed in section 6.

(3) Measurement of changes in relevant surface properties such as sputtering rate, particle and energy reflection coefficients.

(4) Measurement of the hydrogen and deuterium inventories - to correlate this with the change in surface composition and to try and estimate total incident fluence. This may be obtainable using a diffusion technique proposed by Borgesen et al[114]. The hydrogen or deuterium isotope inventory measurement is particularly important as a cross check for estimates of the tritium inventory during DT operation.

For all these investigations a range of collector materials is desirable - in particular a low Z collector e.g. carbon which can be used to detect the deposition of medium or high Z element deposition; a second clean collector e.g. silicon to measure carbon deposition, and a third probe material which represents the wall material itself. Such probes need only be very simple so that a large number placed in poloidal and toroidal arrays can allow an overall picture of asymmetries. An example of the design for the JET probes is given in fig. 40.

It must be recognized that such probes integrate over a long period of operation and that the resulting surface conditions may possibly depend on the number of malfunctions, e.g. disruptions, or the discharge cleaning technique which is used, rather than on the tokamak discharges themselves. Nevertheless since interpretation of the wall interactions depends in part on knowing the properties of this real wall condition detailed investigations of the type described are highly desirable.

8.   CONSTRUCTION OF AND DEPLOYMENT OF PROBES

Section 2 of this chapter discussed the influence of plasma heat flux on probe head materials. If T_e and T_i are known it is possible to calculate a maximum electron density to which any probe material may be exposed c.f. equation 6. In tokamaks, the deployment of probes may be limited by other considerations. Start-up and termination transients frequently involve very high localized heat flux from either runaway or "epithermal" electrons. Disruptions can deliver large amounts of power at otherwise "safe" radii. Thus even using caution in the deployment of probes, experiments frequently end in the incapacitation of the collector element. This requires probes to be very robust.

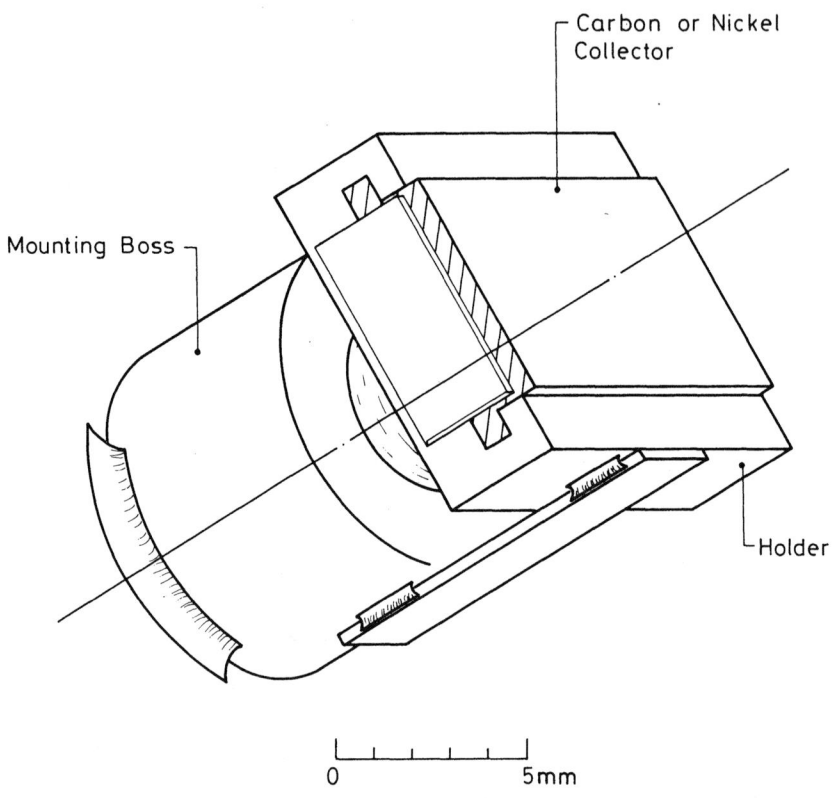

Fig. 40 Schematic of the collector probes used in JET to measure
poloidal and toroidal variation of long term deposition
erosion and damage.

Fig. 41 shows a comparison of a Langmuir probe once used
successfully in the C-stellarator to a triple probe-calorimeter
presently in use on TFTR, where terminal disruptions can carry
large amounts of stored energy to interior wall components.

The construction of probe transport mechanisms requires the
solution of several difficult problems. One must provide a
machine vacuum of high quality; current baseline pressures are
better than 1×10^{-8} torr and must not be degraded by diagnostic
equipment. This usually requires bakeable ($150 < T < 250^\circ C$),
metal-sealed systems. We note that sample exposure analysis for
light impurities places very strict requirements on vacuum. Even
at 10^{-8} torr, monolayer coverage by background gas such as CO can
occur in as little as ~ 100 seconds.

198

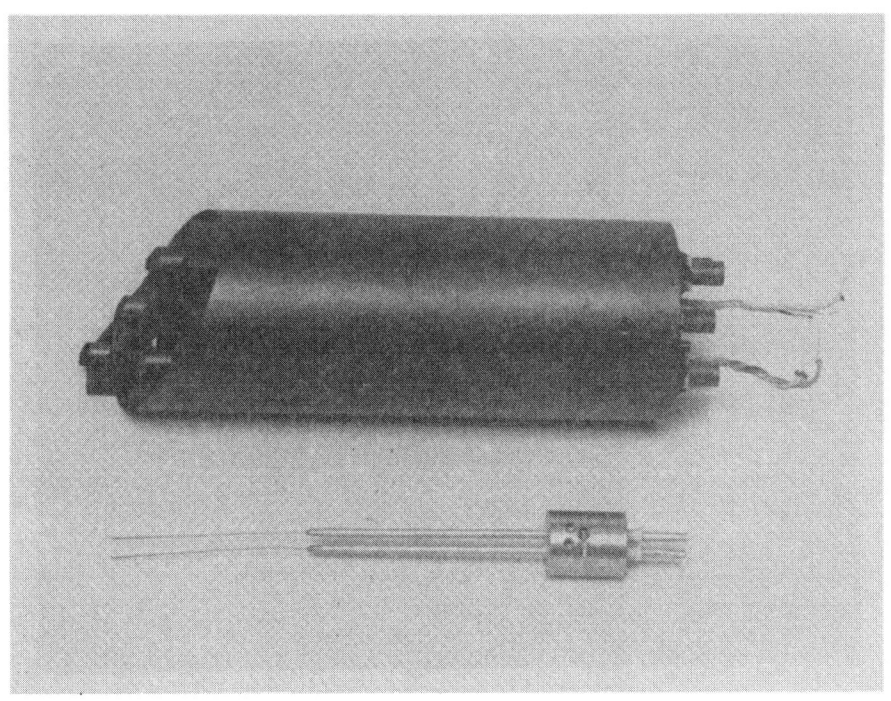

Fig. 41 A comparison of a Langmuir probe used in the C stellerator
(1956) with a recent probe from TFTR. The thin collector
wires of the former are scarcely visible protruding from
their long slender shields.

Probes generally require motion for positioning the head at
various locations and often require rotation of the head or motion
along a second axis. This requires a means to transmit these
motions using drive elements, bearings, and lubricants which are
compatible with the UHV requirements.

Electrical signals must be carried from the probe head to the
outside. In the case of complicated multi-detector heads (e.g.
combined ExB and retarding field analyzer) numerous independent
signal paths are required. These must terminate in a small space
at the probe head where the material temperature can become very
high. Thus special attention to proper insulation material is
required to maintain good vacuum. In some cases signals must be
transmitted many metres in vacuum and conductors must safely
compensate for large motions required by the operation. In future
work the environmental requirements are quite difficult to
fulfill. Fig. 42 shows present day requirements imposed by D-T
operation of TFTR; those for reactor-like operation of JET are
even more stringent. One sees that these small signals (on the
order of millivolts and milliamps) are being acquired less than a

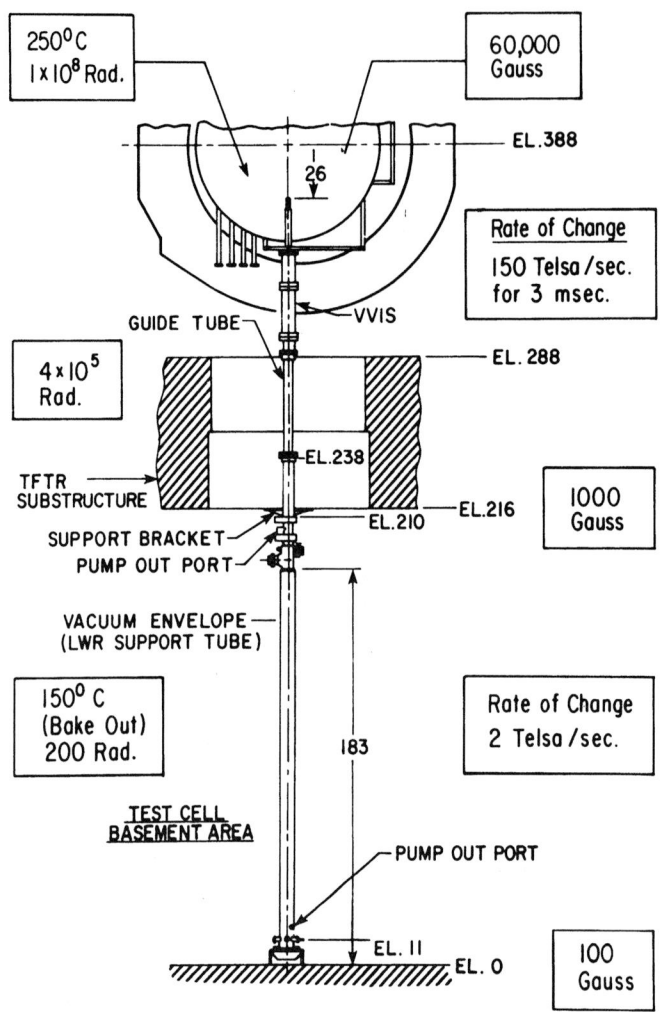

Fig. 42 Schematic diagram showing the requirements for probe operation in TFTR using DT fuel.

metre away from the plasma core where the electro-magnetic interference conditions are severe. Excellent grounding and shielding must be maintained as the probe head moves ~ 5 metres from its rest position.

The solution to these problems cannot be given in general. Each application has its own requirements which are sufficiently

200

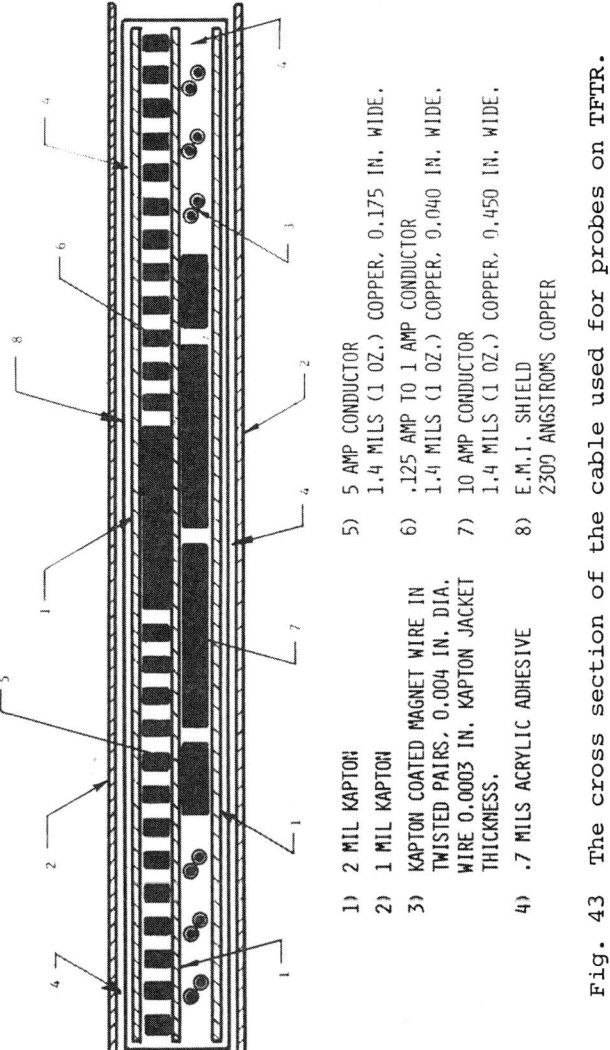

1) 2 MIL KAPTON

2) 1 MIL KAPTON

3) KAPTON COATED MAGNET WIRE IN
   TWISTED PAIRS. 0.004 IN. DIA.
   WIRE 0.0003 IN. KAPTON JACKET
   THICKNESS.

4) .7 MILS ACRYLIC ADHESIVE

5) 5 AMP CONDUCTOR
   1.4 MILS (1 OZ.) COPPER. 0.175 IN. WIDE.

6) .125 AMP TO 1 AMP CONDUCTOR
   1.4 MILS (1 OZ.) COPPER. 0.040 IN. WIDE.

7) 10 AMP CONDUCTOR
   1.4 MILS (1 OZ.) COPPER. 0.450 IN. WIDE.

8) E.M.I. SHIELD
   2300 ANGSTROMS COPPER

Fig. 43   The cross section of the cable used for probes on TFTR.

specific to the confinement device that descriptions of probe drive mechanisms rarely appear in the literature. However, summaries of the solutions for TFTR[115] and JET[113] are available. We treat a few practical problems in the next section.

8.1 High Heat Flux

As shown in fig. 41 one strategy for coping with high heat flux is to make a very large probe nearly entirely from graphite. There are instances where such large probes may be inadmissible, however, and other strategies must be used. One method is to place the probe on a very rapidly moving linear drive. The probe may then be plunged quickly to the desired location and removed quickly. It is also possible to have the probe in place behind a very rugged cover which can be rapidly pulled back a few cm to expose the elements and then replaced. Alternatively, the probe may quickly withdraw a few cm beyond such a stationary shield. One very easy method, already used on PLT, is to rapidly rotate the active elements of the probe (hidden behind apertures) into the flux for a brief period. The same probe allows simple attenuation of the flux by changing the aperture angle between shots.

8.2 Background Noise

There are three major sources of noise or "pick up" in the tokamak environment. The first is rapid fluctuation in the space (or floating) potential of the plasma itself. Such fluctuations are often quite interesting; however in certain measurements they constitute a source of noise. There is little that can be done to reduce this source. The second, magnetic "pick up", can be reduced by careful avoidance of uncompensated loops in the signal path. Tightly twisted pairs of wires in devices such as thermocouples, usually avoids this problem. The third source is tokamak operation electrical systems including auxiliary heating supplies. Avoidance of this source requires attention to proper grounding and shielding techniques. A schematic of the cabling used in the vacuum envelope of the TFTR probes is shown in fig. 43. The wire termination at the probe head requires a continuation of the shielding. Fig. 44 shows a multiconnector coaxial feed-through with which the TFTR probe heads mate. This feed-through is capable of continuous operation up to $400^\circ C$.

8.3 Motion

Linear drives in many probe systems involve long distances, for example, the JET fast transfer system, shown in fig. 45, requires a motion of 22 metres. Such motion in this instance will be provided by a railroad-like system of a carriage propelled along an unlubricated track via linear induction motors.

202

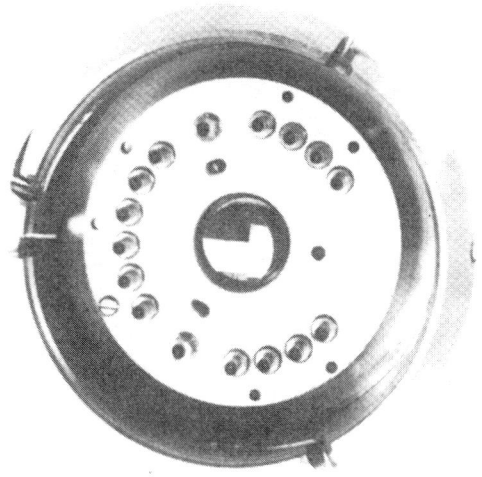

Fig. 44 Multi-connector feed through used for probes on TFTR.

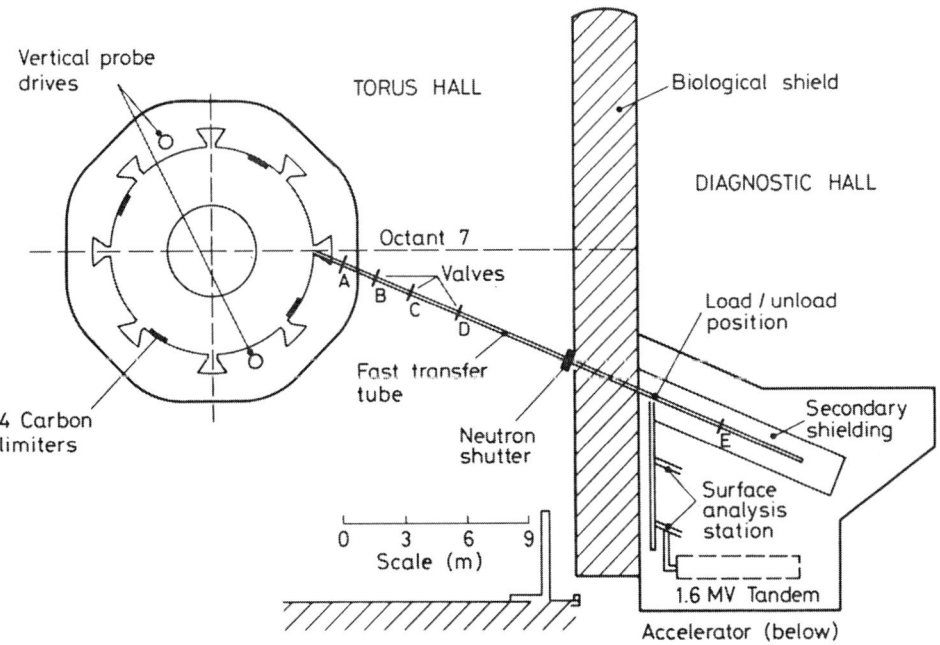

Fig. 45 The layout of the vertical probe drives and the fast
transfer system for taking probes from the plasma edge to
the surface analysis station on JET.

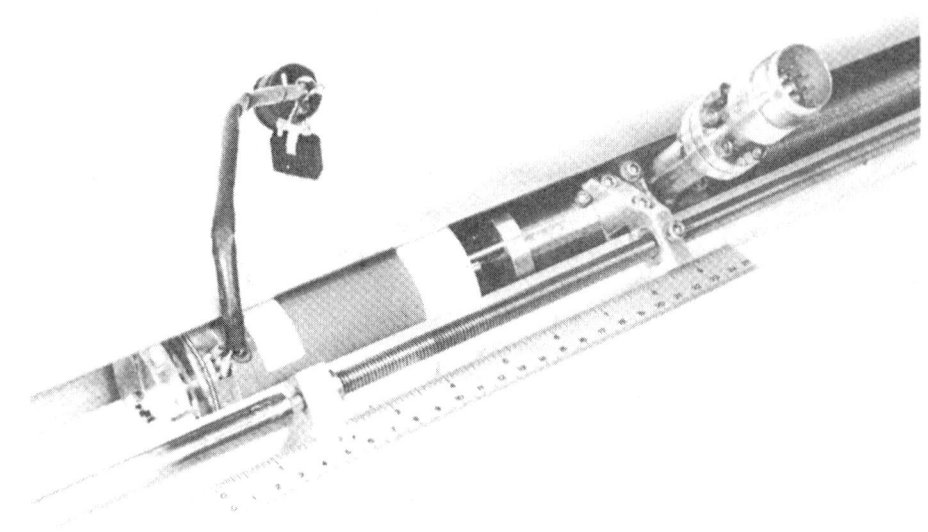

Fig.46   The drive system for moving the probes on PLT.

Shorter distances (5-6 m) may be accommodated by internal rack and pinion drives such as that used on TFTR or external threaded-rod/ball screw drives such as those on PLT and DITE. The latter drives are well-suited to bellows vacuum spaces.   A photograph of such a probe is shown in fig. 46.

Lubrication of interior components is a difficult subject. Most commonly, dry lubricants such as MoS_2, WS_2, Ag or Au are employed on bearing balls and races and gear teeth.   A combination Ni/MoS_2 dry lubricant has been used successfully for high speed probe bearings on PLT.   Excellent results have also been achieved with perfluoropolyethanes such as Bray[116] or Krytox[117].   Whether such lubricants may be safely employed depends on the specific vacuum requirements, operating temperatures and proximity to the plasma.   Often it is necessary to reduce loads and speeds and run dry dissimilar materials for bearings and gears.

Rotary motion is normally achieved via bellows sealed (wobble drives, direct drives, etc.) feed throughs.   Ferrofluidic feed-throughs have been used for very long life and high loads.   These latter require shielding from the magnetic fields of the fusion device.   Unfortunately their application is limited as they cannot be baked due to the high vapour pressure of the ferro-fluid.

204

REFERENCES

1. I. Langmuir and H.M. Mott-Smith, Gen.Elec.Rev. 26, 731 (1923), 27. 449, 583, 616, 726, 810 (1924). Also see I. Langmuir and K.T. Compton, Rev.Mod.Phys. 3, 191 (1931).

2. H. Carslaw and Jaeger, Conduction of Heat in Solids, Oxford Press, Oxford (1959), p 401.

3. H. Vernickel, J.Nucl.Mat. 111 & 112, 531 (1982).

4. L. Schott in Plasma Diagnostics, Ed by W. Lockte-Holtgreven, North Holland, Amsterdam (1968), p 668

5. F.F. Chen in Plasma Diagnostic Techniques, Ed R.H. Huddlestone and S.L. Leonard, Academic Press, New York (1965), p 113.

6. J.D. Swift and M.J.R. Schwar, Electrical probes for plasma diagnostics, Iliffe, London 1970.

7. G.D. Hobbs and J.A. Wesson, Plasma Phys. 9, 85 (1967).

8. P.J. Harbour and M.F.A. Harrison, J.Nucl.Mat. 76 & 77, 513 (1978).

9. P.C. Stangeby, Phys.Fluids 27, 682 (1984).

10. P.J. Harbour, CLM Preprint-535 (1978) unpublished, (available from Culham Lab. on request).

11. P.C. Stangeby, G.M. McCracken, S.K. Erents, J.E. Vince and R. Wilden, J.Vac.Sci. and Technol A1 (1983) 1302.

12. Y. Gomay, N. Fujisawa, M. Maeno, et al., Nuclear Fusion 18 (1978) 849.

13. T. Kobayashi, M. Shimada, S. Sengoku, et al., J.Nucl.Mat. 121 (1984) 17.

14. F. Hoffman, Ch. Hollenstein, B. Joye, et al., J.Nucl.Mat. 121 (1984) 22.

15. D.K. Owens, S.M. Kaye, R.J. Fonck and G.L. Schmidt, J.Nucl.Mat. 121 (1984) 29.

16. R. Budny and D. Manos, J.Nucl.Mat. 121 (1984) 41.

17. G.M. McCracken, S.K. Erents, D.H.J. Goodall, G.F. Matthews, J.W. Partridge, S.J. Fielding and B.A. Powell, J.Nucl.Mat. 128/129 (1984) 150.

18. K. Ertl and the ASDEX team, J.Nucl.Mat. 128/129 (1984) 163.

19. C. Kahn, K.H. Burrell, E. Fairbanks, T. Petrie, M. Shimada, M. Washizu and S. Sengoku, J.Nucl.Mat. 128/129 (1984) 172.

20. B. Lipschultz, et al., to be published.

21. E.O. Johnson and L. Malter, Phys.Rev. 76 (1949) 1411.

22. E.O. Johnson and L. Malter, Phys.Rev. 80 (1950) 58.

23. G. Proudfoot and P.J. Harbour, J.Nucl.Mat. 111 and 112 (1982) 87.

24. M. Kamitsuma, S.L. Chen, J.S. Chang, J.Phys.D. 10, 1065 (1977).

25. S.L. Chen and T. Sekiguchi, J.Appl.Phys. 36 2363 (1965).

26. D.K. Owens and S.M. Kaye, private communication.

27. M.H. Cho, C. Chan, N. Hershkowitz and T. Intrator, Rev.Sci. Instrum. 55, 631 (1984).

28. N. Hershkowitz, B. Nelson, J. Pew and D. Gates, Rev.Sci. Instrum. 53, 29 (1983).
29. S.K. Erents and P.C. Stangeby, J.Nucl.Mat. 111 & 112, 165 (1982).
30. D.M. Manos, R.V. Budny and S.A. Cohen, J.Vac.Sci.Technol., A1, 845 (1983).
31. P.C. Stangeby, Phys. Fluids (1984) in press.
32. P.J. Harbour and G. Proudfoot, J.Nucl.Mat.121 (1984) 222.
33. W. Poschenrieder, private communication.
34. G.M. McCracken and J.H.C. Maple, Brit.J.Appl.Phys. 18 (1967) 919.
35. P. Mioduszewski, L.C. Emerson, J.E. Simpkins, et al., J.Nucl.Mat. 121 (1984) 285.
36. P.C. Stangeby, J.Nucl.Mat. 121, (1984) 36.
37. J.F. Waymouth, Phys.Fluids 7, 1843 (1984)
38. P.C. Stangeby, G.M. McCracken, S.K. Erents and G. Matthews, J.Vac.Sci.Technol. A2 (1984) 702.
39. S.J. Zweben and R.J. Taylor, Nucl.Fusion 21, 193 (1981), 23,513 (1983).
40. P.J. Harbour and G. Proudfoot, IAEA Technical meeting on divertors and impurity control. Editors M. Keilhacker and V. Daybelge, IPP Garching 1981 pg.45.
41. J.E. Osher in Plasma Diagnostic Techniques, eds R.M. Huddlestone and S.L. Leonard, Academic Press, NY (1965).
42. S.S. Medley and D.R.A. Webb, J.Phys.D. 4 (1974) 658.
43. A.W. Molvik UCRL 52981 report
    Lawrence Livermore Laboratory 1981.
44. P. Staib, J.Nucl.Mat. 93 & 94, 351 (1980).
45. G.F. Matthews, J.Phys.D. 17 (1984) 2243, and private communication.
46. M. Caulton, RCA Review 26, 217 (1965).
47. D.W. Mason, Plasma Physics 6, 553 (1964).
48. S. Stephanakis and W.H. Bennett, Rev.Sci.Instrum. 39, 1714 (1968).
49. J.A. Simpson, Rev.Sci.Instrum.32 (1961) 1283.
50. G. Doucas, Int.J. of Mass Spec and Ion Phys.25(1977)71.
51. D. Manos, R. Budny, T. Satake and S.A. Cohen, J.Nucl.Mat. 111 & 112, 130 (1982).
52. D.N. Arion and R.F. Ellis, Rev.Sci.Instrum. 52 1032 (1982).
53. R.L. Stenzel, R. Williams and R. Aguero, Rev.Sci.Instrum. 53 (1982) 1027.
54. R.G. Chambers, Plasma Physics 14 (1972) 747.
55. Katsumato, Japan, J.Appl.Phys.6 (1967) 123.
56. K. Odajima, H. Kimura, H. Maeda and K. Ohasa, Japan, J.Appl.Phys.17 (1978) 1281.
57. A. de Chambrier, G.A. Collins, P.A. Dupecrex et al., J.Nucl.Mat. 128/129 (1984) 310.
58. F.A. White, Mass spectrometry in Science and Technology, John Wiley, New York, 1969, Chapter 2.

59. A. Gibson, A.S. Bishop, E. Hinnov and F.W. Hoffman, MATT report 261, Plasma Physics Laboratory, Princeton University 1964.

60. H. Kojima, H. Sugai, T. Mori, H. Toyada and T. Okuda, J.Nucl.Mat. 128/129 (1984) 965.

61. S.A. Cohen, J.Nucl.Mat. 76 & 77 (1978) 68.

62. D. Bohm, H. Bishop, H. Massey, in Characteristics of Electrical Discharges, eds Gutherlie and Wakerling, (1949).

63. P.C. Stangeby, J.Phys. B 15, 1007 (1982).

64. G.A. Emmert, R.M. Wieland, T. Mense and J.N. Davidson, Phys.Fluids 23, 803 (1980).

65. S.I. Braginskii, Rev. Plasma Physics 1, 205 (1965).

66. P.E. Staib, J.Nucl.Mat. 111 & 112, 109 (1982).

67. G. Staudenmaier, J. Roth, R. Behrisch, J. Bohdansky, W. Eckstein, P. Staib, S. Matteson and S.K. Erents, J.Nucl.Mat.84(1979)149.

68. S.A. Cohen and G.M. McCracken, J.Nucl.Mat. 84, 157 (1979).

69. W.R. Wampler, D. Brice and C. Magee, J.Nucl.Mat. 102, 304 (1981).

70. W.R. Wampler and C. Magee, J.Nucl.Mat. 103 & 104, 509 (1981).

71. S.A. Cohen, H.F. Dylla, W.R. Wampler and C.W. Magee, J.Nucl.Mat. 93 and 94 (1980) 109.

72. C. Sofield, G.M. McCracken, L.B. Bridwell et al., Nucl.Instrum and Meth. 191 (1981) 383.

73. J. Roth, P. Varga, A.P. Martinelli et al., J.Nucl.Mat. 111 and 112 (1982) 123.

74. K.B. Axon, J. Burt, S.K. Erents, et al. The Bundle divertor: A review of experimental results Culham Laboratory Report CLM R235 (1983).

75. J.S. Ziegler et al., Nucl.Instr.Methods 149, 19 (1978).

76. K. Erents, G. McCracken, J. Vince, J.Nucl.Mat. 76 & 77, 623 (1978).

77. G.G. Ross, et al., J.Nucl. Mat. 128/129 (1984) 730.

78. W.R. Wampler, S.T. Picaux, S.A. Cohen, H.F. Dylla, G.M. McCracken and S.Rossnagel, J.Nucl.Mat. 85 & 86, 983 (1979).

79. W.R. Wampler, S. Cohen, D. Manos, C. Magee, J.Vac.Sci. Technol. 20, 1234 (1982).

80. G. Staudenmaier, P. Staib, W. Poschenreider, J.Nucl.Mat. 93 & 94, 121 (1980).

81. R.Krawec, NASA Technical Note - NASA-TN-D5746, April 1970.

82. R. Zuhr, R.E. Clausing, L. Heatherly and R.K. Richards, J.Nucl.Mat. 111 & 112, 177 (1982).

83. C.J. Sofield, G.M. McCracken et al., Nucl.Inst.Methods 191, 1983 (1981).

84. W. Wampler, Appl.Phys.Lett. 41, 335 (1982).

85. W. Wampler and D. Manos, J.Vac.Sci.Technol. A1, 827 (1983).

86. P.L.F. Hemment, in Ion Implantation Techniques, eds.H. Ryssel and H. Glawischnig, Springer-Verlag, NY (1982) p 209.

87. E. Hintz, J.Nucl.Mat. 93 & 94, 86 (1980). See also present volume chapter by Hintz.

88. P. Staib, R. Behrisch, W. Heiland and G. Staudenmaier, Proc.7th Europ.Conf. on Controlled Fusion and Plasma Physics, Lausanne (1975)p.133.

89. G.M. McCracken and P.E. Stott Nuclear Fusion 19(1979) 889.

90. J.C. Riviere, Phil. Trans. Roy. Soc. A305 (1982) 545; D. Briggs and M.P. Seah Editors, "Practical Surface Analysis" John Wiley 1983.

91. R.A. Zuhr, J.B. Roberto and B.R. Appleton, Nuclear Science Applications 1 (1984) 617.

92. G. Dearnaley, G.M. McCracken, J.F. Turner and J. Vince, Nuclear Instrum & Meth. 149(1978) 253.

93. G.M. McCracken, G. Dearnaley, R.D. Gill et al., J. Nucl, Mat 76 & 77 (1978) 431.

94. G.M. McCracken, J.W. Partridge, S.K. Erents et al., J. Nucl.Mat. 111 and 112 (1982) 159.

95. Y. Hori, A. Sagara, Z. Kabeya, J.Nucl.Mat. 111 and 112 (1982) 137.

96. W.R. Wampler, S.T. Picraux, S.A. Cohen et al., J.Nucl.Mat. 93 and 94(1980) 139.

97. H. Wolff, H. Grote, D. Hildebrandt and M. Laux, J.Nucl.Mat. 128/129 (1984) 219.

98. J.B. Roberto, J. Roth, E. Taglauer and O. Holland, J.Nucl.Mat. 128/129 (1984) 244.

99. E. Taglauer, J.Nucl.Mat. 128/129 (1984) 244.

100. J.F. Ziegler,ed., Helium: Stopping Powers and Ranges in all Elemental Matter , Pergamon Press, NY (1978).

101. R.L. Freeman and E.M. Jones, Culham Laboratory report R 137 (1974) Atomic Collision Processes in plasma physics experiments.

102. W. Bauer, K.L. Wilson, C.L.. Bisson, L.G. Haggmark and R.J. Goldston, Nuclear Fusion 19 (1979) 93.

103. S.E. Donnelly, D.S. Whitmell, R.F. Nelson, AERE-Harwell, Report R 7955 (1975).

104. R.S. Blewer, Appl. Phys. Lett. 23, 593 (1973).

105. Ref. 89 page 938.

106. G. Staudenmaier, P. Staib and G. Venus, J. Nucl. Mat. 76 and 77 (1978)

107. R. Weissman and R. Behrisch, Rad.Effects 19 (1973) 69.

108. G. Mezey, J.W. Partridge and G.M. McCracken, Fusion Technology 6 (1984) 459.

109. P.M. Read, J. Asher, T.W. Conlon and C.J. Sofield, J.Nucl.Mat.99 (1981) 235.

110. T.W. Conlon, Wear 29 (1974) 69.

111. D.H.J. Goodall, T.W. Conlon, C. Sofield and G.M. McCracken, J. Nucl. Mat. 76/77 (1978) 492.

112. G. Staudenmaier, J.Vac.Sci.Tech. (1985) to be published.

113. D.H.J. Goodall, W.O. Hoper, G.M. McCracken, R. Behrisch et al., J.Nucl.Mat. 93/94 (1980) 383.

114. P. Borgesen, B.M.V. Scherzer and W. Möller, J.Nucl.Instrum. and Meth. B (1984) in press.

115. R. Mastronardi, R. Cabral and D. Manos, Proc. NASA 16th
     Aerospace Mechanism Conf. Pub. 2221 (1982) p 265.
116. Bray Oil Co., Los Angeles, CA 90032.
117. E.I. Dupont, Wilmington, Delaware.

PLASMA EDGE DIAGNOSTICS USING OPTICAL METHODS

P. Bogen and E. Hintz

Institut für Plasmaphysik
Kernforschungsanlage Jülich GmbH
Association EURATOM-KFA
P.O. BOX 1913, D-5170 Jülich, W.-Germany

Abstract

Based on an analysis of the tasks and the specific needs of plasma edge diagnostics the principles of optical methods, which satisfy these demands, are described. Experimental arrangements and some results are discussed. Main emphasis is given to emission spectroscopy, laser induced fluorescence and atomic beam methods; Thomson scattering of laser radiation and infrared thermography are treated briefly.

I. Introduction

For the plasma core in magnetic confinement devices diagnostic methods have been developed which allow an almost complete description of the plasma state. Plasma edge diagnostics on the other hand are still in an earlier state of development /1/. There are several reasons for this situation. The conditions of the plasma edge differ strongly from those of the core making an extension of most of the existing diagnostics to the plasma edge impossible; furthermore, there are additional plasma parameters requiring new diagnostic methods.

The plasma edge is the transition region between the plasma core and the wall system. Its radial extension can be defined by the ionization length of the atomic hydrogen. A part of the magnetic flux surfaces within the plasma edge intersects the limiter or the divertor plates and, as a consequence, the scrape-off layer is formed. The state of the plasma in this region is far away from thermal equilibrium. A flow of particles and of energy is induced towards the plasma-solid interface, which acts as a sink for the charged particles. Strong gradients are formed in the radial direction and - in contrast to the plasma core - plasma parameters are not constant along a magnetic surface: depending on the actual limiter or divertor configuration, there may be significant variations along the toroidal and poloidal coordinates. Furthermore the limiter, the divertor plates and the remaining parts of the wall system present the main sources of neutral particles, including neutralized plasma particles as well as constituents of the wall material and adsorbed gas particles.

The specific problems of plasma edge diagnostics may be summarized in the following way:
- the values of electron densities and temperature are at least one or two orders of magnitude lower than in the plasma core,
- the neutral and low ionization states of particles have to be detected,
- high spatial resolution is needed in the radial direction,
- plasma parameters have to be measured along the magnetic surface.

As we do not want to introduce possible perturbations on the plasma by the diagnostic methods, e.g. by the introduction of material probes, we would prefer to use optical methods of diagnostics. The diagnostic methods have to be based therefore on ionization, excitation and radiation processes. The discrimination against the radiation from the core plasma may pose serious problems. In order to avoid this background radiation it is desirable to do the observations in the tangential direction only. Appropriate ports have to be available in the vacuum vessel. Certain parts of the plasma edge, e.g. the limiter region, will offer only limited access and may require the application of techniques different from those used for the remaining part of the plasma edge.

The presence of a strong magnetic field, of vessel vibrations and of a high level of x-ray or neutron radiation has to be taken into account during the design of the diagnostic apparatus and in connection with the evaluation of the actual signals - in the same way as it is done in case of the plasma core diagnostics.

In the following we shall first give a survey of the parameters to be measured, including the orders of magnitude to be expected, and on those diagnostic techniques which appear to be especially suited to satisfy the requirements of plasma edge diagnostics. We shall then discuss each of these techniques by describing the principles, the intended applications and the required experimental apparatus, by presenting typical results and by giving an outline of expected future developments.

II. Survey of the plasma parameters to be measured and of the diagnostic technique to be applied

We shall distinguish four groups of particles: electrons (subscript e), hydrogen (H), helium (He) and impurities (I) in the neutral (superscript 0) and in the ionized (+, Z+) state, respectively. The types of impurities to be considered depend on the materials used for the wall system and for the limiter. In all devices the so-called light impurities oxygen and carbon are found in relatively large concentrations (of the order of 1 %) and therefore they require special attention.

For these particle species we want to determine: the density (n), the temperature (T), the drift velocity (V) parallel or perpendicular to the magnetic field and, for the electrons at least, the fluctuation amplitude $\tilde{n}_e$. At the plasma-solid interface it is of special importance to know the fluxes of neutral particles which enter the plasma and the fluxes of charged particles and of energy which are lost by the plasma. The relations between the fluxes going to and leaving the wall are crucial questions of plasma wall interaction.

We expect that the orders of magnitude of these parameters will be within the following ranges:

T_e :   10 - 100 eV;

n_e : $10^{11} - 10^{13}$ cm^{-3}; $n_e = n_{H^+}$

n_I : $10^{-3}n_e$ - a few times $10^{-2}n_e$; $n_{H^0} \approx 10^{-3}n_e$

As a guide line for the diagnostic methods we assume

$T_{H^+} \approx T_{H^0} \approx T_{I^+} \approx T_e$; and

$V_{H^0} \approx V_{H^+} \approx V_{I^+} \approx (T_e/m_H)^{1/2}$

We thus have to pay attention that the respective diagnostic methods satisfy these requirements. The characterstic width of the scrape-off layer is estimated to be 1-2 cm and a space resolution has to be foreseen which is small compared to this length. The needed time resolution and the data acquisition rates, in a similar way, follow from the characteristic times of the processes to be investigated; for instance the energy confinement time τ_E may serve as a characteristic time.

For the measurement of the physical quantities specified above, we want to apply mainly spectroscopic techniques. Commonly emission spectroscopy is used for the measurement of particle densities and temperatures. Its main drawbacks are the following: the radiation processes depend strongly on n_e and T_e; and the emitted radiation is integrated along the line of observation. In case of low n_e and strongly varying n_e and T_e this technique may not be satisfactory. Light scattering techniques e.g. Thomson scattering and fluorescence spectroscopy are more appropriate for our purposes; on the other hand more elaborate apparatus will be needed and may not always be available.

For some of the parameters of interest a local measurement by spectroscopic techniques can also be achieved by the help of collimated particle beams. Beam particles may be excited by electrons and radiate, or they may charge exchange with ions producing excited states which radiate. As the injected beam, its density and its energy can be chosen within certain limits, the excitation process to be utilized can be optimized with respect to its diagnostic purpose. This technique appears to be especially promising for the plasma edge region. By combining the beam method with the method of laser induced fluorescence the measurement of the local density of the beam particles becomes possible and the processes responsible for beam attenuation can also be utilized.

In the course of this lecture we shall discuss in separate sections the principles and the diagnostic possibilities of the above mentioned three techniques, which appear to be especially suited to the measurement of the plasma edge parameters:
   - emission spectroscopy to measure neutral particle fluxes
     and n_e, T_e close to the limiter,
   - laser induced fluorescence, to measure densities and fluxes
     of metal atoms (and ions), of hydrogen, and of light impurities in case n_e is low and in case there exist strong gradients of n_e and T_e,
   - atomic beam methods, to measure n_e, T_e and the density of
     highly ionized impurities and helium.

Other methods like Thomson scattering for the determination of n_e, T_e, and infrared thermography for the measurement of surface temperatures and heat flux densities we shall discuss only briefly.

As pointed out previously plasma edge diagnostics faces two more general problems:
   - the background radiation from the plasma core - to be
     avoided by tangential observation.

214

- the complicated spatial structure of the scrape-off layer, requiring high resolution of diagnostic methods in the radial direction, and the measurement of the plasma parameters along the poloidal and toroidal coordinates.

Possible solutions to these problems shall be outlined using the TEXTOR tokamak /2/ as an example. Fig. 1 shows the porthole arrangement and the limiter configuration in this device. There are several ports permitting tangential observation of the plasma torus - along the poloidal as well as the toroidal direction. Moreover, one sees a few cross-sections where both, radial and tangential observation of the same volume is possible. These are used when laser induced fluorescence or atom beams are applied for plasma edge diagnostics, as will be discussed in more detail later on. Possible experimental arrangements are also illustrated in the figure.

It is normally not possible to employ the same diagnostic method at several positions on a tokamak. For a measurement of the toroidal and poloidal variations of plasma parameters in the

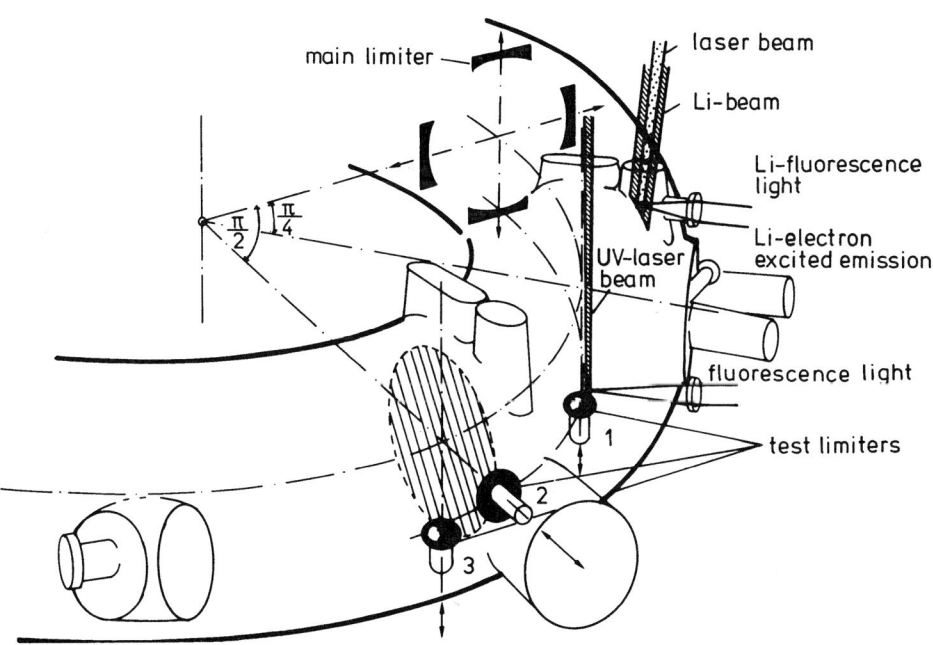

Fig. 1: Limiter configuration and port-hole arrangement for observation of the plasma edge in TEXTOR.

scrape-off layer one may then use a sequence of reproducible
discharges and move the positions of an observation volume and
of the limiter with respect to each other. In TEXTOR a flexible
limiter system permits such variations. All segments of the main
limiter system except the inner one can be moved in the radial
direction. With ohmic heating a single segment can take the full
heat load. By proper positioning of the three segments one can
arrange for each to function as the main limiter. The connection
length of the Li-beam to the upper limiter - serving as main limiter - is very short. If the upper limiter is withdrawn and the
lower limiter acts as main limiter, the Li-beam is almost in the
position of the stagnation point. In this manner, by activating
the outer limiter, the pump limiter (not shown in Fig. 1) or a
test limiter (No. 2) the connection length to the Li-beam can be
varied in a well defined way. An example of this procedure will
be presented in section V.

III. Emission spectroscopy

Emission spectroscopy can give information about the plasma
boundary parameters by a measurement of the atomic line
intensities or of the line profiles. Whereas for the energy
distribution of the free electrons normally a thermal (i.e. a
Maxwellian) distribution is a useful approximation, thermal
equilibrium cannot be expected for the processes determining the
excitation and ionization states. The emission is determined by
electron impact excitation and radiative decay, the ionization
state by electron impact ionization and diffusion. (Radiative or
collisional recombination is usually negligible, since the
recombination times are long compared to the residence times in
the plasma edge, e.g. at $n_e = 10^{12}$ cm^{-3}, $T_e = 20$ eV, they are
about 10 sec for OI, about 100 msec for O VIII /3,4/). In some
cases, also excitation and ionization processes induced by
charge exchange are important. They are considered in section V.
A discussion of molecular line intensities has not been included in this lecture because of the complexity of their interpretation and the lack of observations.

We begin this section with a summary of electron impact excitation and ionization processes which determine the line intensities. There follows a short discussion on the influence of the
Zeeman- and Doppler- effect on the spectral line profile. Then
we describe experimental arrangements for the measurement of
plasma parameters by emission spectroscopy, and finally we
present some applications.

III.1. Excitation and ionization by electron impact

Although the excitation and ionization processes caused by electrons are qualitatively well understood and the general equations to calculate the relevant cross-sections by quantum mechanics are well known /5/, there are only a few cases, for which
the equations have been solved within an accuracy of $\mp$ 20 %, an
accuracy desirable for the diagnostic applications. Unknown
cross-sections σ are either extrapolated from those known by
experiments using semiempirical formulas or in simpler cases
they are calculated by the Born approximation and its extensions, which give normally good results for high collision
energies E, but too high values (typically a factor of 2) at the
maximum of the $\sigma(E)$ curve /3,5/. More details are given in the
chapter of M.F.A. Harrison.

At low electron densities, the spectral line intensities are
determined by the electron impact excitation rate X_{1m} from the
ground state to the excited level m. This rate is given by the
equation

$$X_{1m} = n_e < \sigma_{1m} \vartheta_e > \qquad (1)$$

where ϑ_e is the electron velocity, σ_{1m} is the excitation cross-
section from the groundstate to level m and $<\sigma_{1m} \vartheta_e>$ is the excitation rate coefficient averaged over a Maxwellian distribu-
tion of ϑ_e. The emission coefficient ε , which is the number of
photons emitted per cm^3 s sr by the atoms of density n_A per cm^3,
is obtained, in case that the spontaneous transition to only one
level is possible, by

$$\varepsilon = \frac{1}{4\pi} n_e n_A <\sigma_{1m} \vartheta_e> \qquad (2)$$

The quantity observed in the experiments is the intensity I,
which is ε integrated over the emitting layer from x_1 to x_2.

$$I = \frac{1}{4\pi} \int_{x_1}^{x_2} n_e n_A <\sigma_{1m} \vartheta_e> dx' \quad \left[\text{Photons/cm}^2 \text{ s sr}\right] \quad (3)$$

Excitation rate coefficients of only a few neutral atoms have
been measured (e.g. H, He, Li, Na, Ba /6,7/), otherwise calculated values have to be used. For optically allowed transitions
useful values can often be derived from van Regemorter's semiempirical formula /8/

$$<\sigma_{1m} \vartheta_e> = 32 \cdot 10^{-8} f_{1m} \left(\frac{Ry}{E_{1m}}\right)^{3/2} \beta^{1/2} \exp(-\beta) p(\beta) \qquad (4)$$

217

where f_{lm} is the oscillator strength, R_y the Ryberg constant, E_m the excitation energy, $\beta = E_m/kT$, and $p(\beta)$ is a tabulated function /3,8/, which is $(-\sqrt{3}/2\pi)$ $Ei(-\beta)$ for $\beta \ll 1$. Since the optically allowed transitions for the light elements (e.g. O and C) emit in the vacuum ultraviolet and are therefore difficult to measure, also values of excitation functions for optically forbidden transitions, occuring at longer wavelengths, have to be known. Some values calculated by the normalized Born approximation are given in the book of Sobelman et al. /3/.

In some cases, also cascade transitions from levels with $E_{m\prime} > E_m$ may give appreciable contributions to the intensity. But for the strong resonance lines, the contribution is normally low e.g., the contributions of cascades to the Li resonance line 2P - 2S is calculated to be $\lesssim 7$ % /6/.

As the excitation rate, the ionization rate W at low electron densities is determined by electron impact in the ground state, and the value of W is then given by

$$W = n_e \langle \sigma_i \, v_e \rangle \qquad (5)$$

where σ_i is the ionization cross-section and $\langle \sigma_i \, v_e \rangle$ the ionization rate coefficient averaged over a Maxwellian distribution of v_e at a temperature T_e. Useful experimental values of σ_i and $\langle \sigma_i \, v_e \rangle$ can be found in the literature (summarized up to 1981 in /9/). Values for atoms with low and medium nuclear charge are extrapolated by Lotz /10/, values for some atoms calculated with the Born approximation can be found in /3/.

The ionization rate coefficients are strongly dependent on T_e in case $kT_e < E_i$ (E_i = ionization energy), but they are only slowly varying functions of T_e for $kT_e \gg E_i$, a condition which is usually fulfilled in front of the limiter, but not close to the liner. The ionization rate coefficients increase slightly with density due to stepwise ionization, if $n_e \gtrsim 10^{12}$ cm^{-3}. This effect has been calculated in detail for H /11/ and He /12/, but not for the other elements.

The radial dependence of the ionization of the neutral atoms, injected from the limiter, from the liner, or by an external source, is described by the differential equation

$$d(n_A \, v_A)/dr = W n_A = n_e n_A \langle \sigma_i \, v_e \rangle \qquad (6)$$

or, if v_A is constant, by

$$v_A (d \, \ell n \, n_A/dr) = n_e \langle \sigma_i \, v_e \rangle \qquad (6a)$$

where v_A is the radial velocity component of the neutral atoms. Substituting $n_e \times n_A$ from eq 2 into eq. 6 we obtain

$$d(n_A v_A)/dr = d\Gamma_A/dr = <\sigma_i v_e> 4\pi \varepsilon/<\sigma_{1m} v_e> \qquad (7)$$

and after integration the neutral particle flux density into the plasma

$$n_A v_A = \Gamma_A = \int_0^r 4\pi \varepsilon (<\sigma_i v_e>/<\sigma_{1m} v_e>) \, dr' \qquad (8)$$

$\Gamma_A(r)$ is the flux of atoms through 1 cm^2 of a magnetic surface at radius r per sec. A simple evaluation of this equation is possible, if the ratio $<\sigma_i v_e>/<\sigma_{1m} v_e>$ is independent of radius. Then we obtain

$$\Gamma_A = 4\pi (<\sigma_i v_e>/<\sigma_{1m} v_e>) \int_0^r \varepsilon \, dr' = 4\pi I(r) <\sigma_i v_e>/<\sigma_{1m} v_e> \qquad (9)$$

If the intensity is observed in the radial direction, this formula gives immediately the atom flux density into the plasma. In case the observation is performed tangentially to the toroidal field lines, we obtain the flux/cm, e.g. per cm limiter circumference, as will be discussed below (see applications). The condition, that the ratio $<\sigma_i v_e>/<\sigma_{1m} v_e>$ is independent of radius, may not be fulfilled. Then either a mean value has to be taken for a rough estimate or model calculations have to be performed. Examples of the calculated ratios for an OI and a CrI line are shown in Fig. 2. The strong T_e-dependence of $<\sigma_i v_e>/<\sigma_{1m} v_e>$ of

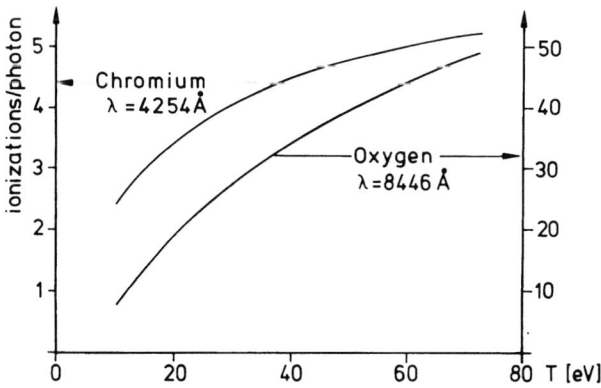

Fig. 2: Number of ionizations per emitted photon for typical oxygen and chromium lines.

219

the OI-line is typical for the excitation of an optically forbidden transition, the weak T_e-dependence in case of the Cr-line is typical for an optically allowed transition.

By eliminating $n_A(r)$ from eq 2 and eq 6, we can also obtain the electron density:

$$n_e(r) = \varepsilon(r) \, v_A / \left\{ \langle \sigma_{1m} v_e \rangle \int_0^r \frac{\langle \sigma_i \, v_e \rangle}{\langle \sigma_{1m} \, v_e \rangle} \varepsilon(r') \, dr' \right\} \qquad (10)$$

Especially simple expressions for $n_e(r)$ can also be obtained for the case where $\langle \sigma_i v_e \rangle / \langle \sigma_{1m} v_e \rangle$ is approximately constant as function of radius

$$n_e(r) = \varepsilon(r) \, v_A / \left\{ \langle \sigma_i v_e \rangle \int_0^r \varepsilon(r') \, dr' \right\} \qquad (11)$$

Approximating the integral $\int_0^r \varepsilon(r') \, dr'$ by $\varepsilon(r) \, \lambda_i$, we arrive at an approximate expression for the "ionization length" λ_i

$$\lambda_i = v_A / n_e \langle \sigma_i v_e \rangle \qquad (12)$$

For v_A, we have to take the mean value of the velocity distribution. In the case of a thermal source this distribution may be a Maxwellian distribution, in the case of sputtered particles it may be approximated by a Thompson distribution /13,14/. The mean velocity may also depend on the source geometry i.e. it may be different for a point source and an infinitely large plane source /15,16/. An exact evaluation of eq. 10 also has to take into account that at high beam attenuation the mean velocity of a beam, e.g. from a thermal source, increases along its path, since the slow atoms are more strongly attenuated than the fast ones.

In order to determine $n_e(r)$ by the use of eqs. 10-12 the emission $\varepsilon(r)$ has to be measured as a function of radius. This can be performed by an observation tangential to the poloidal or toroidal field lines, but not by a radial observation.

III.2. Line profiles

For the line profiles in the plasma edge region, Zeeman effect splitting and the Doppler-effect are important. The measurement of the magnetic field in the boundary by the Zeeman effect is probably not of interest, since these fields can easily be calculated from the fields outside the plasma. For the measurement of the velocity distribution by means of the Doppler effect it is very important to correct for the Zeeman splitting. The splitting $\Delta\lambda$ of a line at a field B (in Gauss) is given by /17/

$$\Delta\lambda = 4.66 \times 10^{-5} \lambda^2 \, B \, (g_{J1} \, m_1 - g_{J2} \, m_2)$$ (13)

where m is the magnetic quantum number, g_J the Lande factor (see e.g. /17/) and λ the wavelength in cm. Transitions with $\Delta m = 0$ give the π-components, polarized parallel to the magnetic field and those with $\Delta m = \mp 1$ the σ-components polarized perpendicular to the magnetic field (or circularly polarized for observation parallel to the field lines).

For a measurement of the broadening or shift of a line due to the Doppler effect, either the π-components have to be selected by using a polarizer or one of the circularly polarized components by using a combination of a $\lambda/4$ plate and a polarizer. If possible, transitions between levels with nearly equal g_J have to be chosen to make the fine structure splitting as small as possible.

All atoms moving with a velocity v_A relative to the observer emit the line shifted by /18/

$$\Delta\lambda = (v_A/c) \, \lambda_o$$ (14)

where λ_o is the wavelength for $v_A = 0$. Therefore the velocity distribution $f(v_A) \, dv_A$ of the atoms results in a wavelength distribution $f(c\Delta\lambda/\lambda_0)$. This way atom velocity distributions can be measured. Unfortunately, the required wavelength resolution is high, e.g. for a velocity of 3×10^5 cm/sec, which is typical for sputtered particles, the resolution $\lambda_0/\Delta\lambda$ has to be considerably higher than 10^5. To obtain such values normally a Fabry-Perot interferometer is used, which needs high photon fluxes for high resolution /19/.

III.3. Experimental arrangements

For the observation of the plasma edge three orthogonal directions are distinguished: the radial direction and the directions tangentially to the poloidal and toroidal magnetic field. Our choice depends on what we want to achieve, e.g. the radial observation offers the best geometry to determine radial fluxes; on the other hand, observation tangential to the toroidal field is the preferred way to look at the plasma-limiter contact and to determine the penetration length of the neutral atoms into the plasma.

For the measurement of hydrogen- and impurity fluxes from the wall, the line radiation in the visible and UV region is observed in the radial direction through a quartz window (e.g. of 6 cm diameter), and the plasma is directly focussed onto the entrance slit of a spectrometer. A disadvantage connected with this set-up is the background radiation due to the emission of the plasma core. Furthermore, the window itself is a perturbation. These disadvantages can be avoided by an observation tangential to the poloidal field. This arrangement is useful e.g. for the observation of a test limiter (see Fig. 1), but it appears to be less useful for the observation of the plasma close to the liner because of the ill-defined optical layer, i.e. an inhomogeneous layer is observed and radial dependences are difficult to take into account.

The TEXTOR limiter can be observed tangential to the toroidal field lines through three different windows of 15 cm diameter. The central one allows the observation of the full poloidal limiter cross-section, whereas the other two windows allow the observation of the upper or lower limiter segments only. In order to take full advantage of the large field of view, a focussing lens has to be placed close to the window to form an intermediate image. The final magnification can then be chosen according to the purpose: a large magnification to see details on a single limiter segment, a low magnification to obtain an overview on the position of the plasma.

For spectral resolution of the light, interference filters (low resolution), a spectrometer, or Fabry-Perot interferometer (high spectral resolution) can be applied. By choice of the right instrument, the spectral resolution can be chosen from $\lambda/\Delta\lambda$ = 100 up to more than 10^6. But the gain in spectral resolution is connected with a loss in received power (for details see e.g. /19/). As light detectors, normally photomultipliers or television cameras are used, but for special purposes, film (to identify the spectral lines) or photodiodes (for the infrared) have been applied.

Presently the light emitted by the plasma in front of the four poloidal limiter segments of TEXTOR is detected (see Fig. 3) either by a CCD-camera (charge coupled device) with an interference filter in front of it (high spatial information, low spectral resolution) or by a photomultiplier attached to a grating spectrometer (low spatial information, high spectral resolution). The CCD-camera is operated in the common TV-mode (20 ms time resolution), and the results are stored on video tape. The solid state design of the camera avoids distortions by magnetic fields, and the signals are proportional to the incident radiation flux. The camera is sensitive in a wavelength range between $0.4\ \mu$ and $1.2\ \mu$ and can be extended to $0.2\ \mu$ by a frequency converter. If neutral atoms or ions have a strong spectral line in this range the whole poloidal cross-section can be observed in the light of this line by using an appropriate interference filter in front of the camera. The video records can be evaluated quantitatively by means of an image processing system /20/.

The grating spectrometer is applied to identify the spectral lines, to measure their spatial dependence in the radial direction for one selected poloidal angle and to measure the hydrogen and impurity fluxes. Using a slotted rotating disc, which scans the entrance slit, radial intensity distributions at high repetition rates can also be obtained with the spectrometer. It allows the measurement of weak lines and those in the UV which cannot be observed with the CCD-camera. The main purpose of the spectrometer, however, is to obtain the line intensities on an absolute scale. The spectrometer therefore is calibrated with the radiation of a carbon arc or of a tungsten ribbon lamp /19/. To obtain a survey on the spectral lines emitted by the edge plasma photographic spectra can also be taken.

The detection limits of the photomultiplier behind the grating spectrometer at a given particle flux depends on a number of parameters: the ratio of ionization to excitation rate coefficients, the solid angle of observation $\Delta\Omega$, the observed area A (corresponding to the slit area in case of radial observation), the transmission of the optical system T, the quantum efficien-

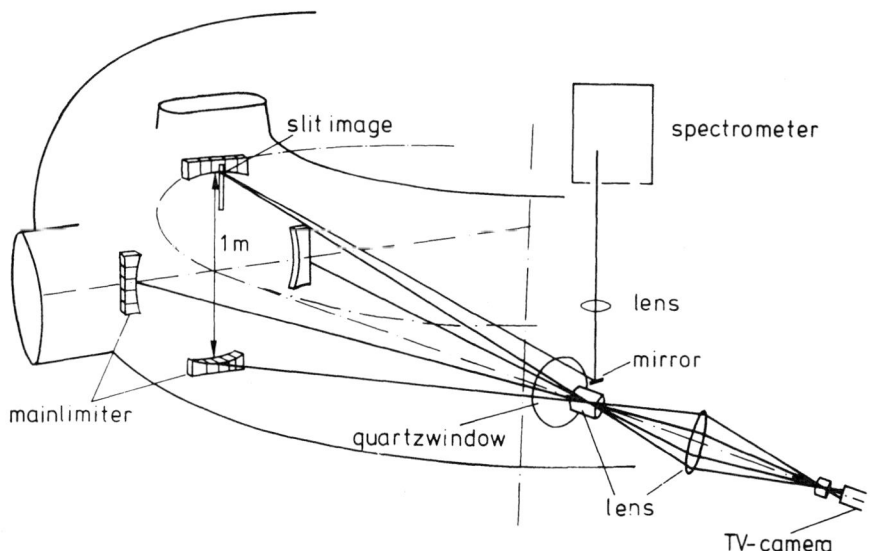

Fig. 3: Experimental arrangement for the observation of the main limiter.

cy η, the required time resolution τ, and the background radiation (which may be the continuum of the plasma volume or a line of a different element). The signal S (number of photoelectrons) can be calculated from the equation

$$S = \frac{1}{4\pi} \Gamma_A \left(\langle \sigma_{1m} v_e \rangle / \langle \sigma_i v_e \rangle \right) \Delta \Omega A T \eta \tau \qquad (15)$$

With $\Delta \Omega = 10^{-2}$ sr, A = 10^{-2} cm^2, T = 0.3 , η = 0.1 and τ = 1 ms we obtain from Eq. 15

$$S = 2.4 \times 10^{-10} \, \Gamma_A \langle \sigma_{1m} v_e \rangle / \langle \sigma_i v_e \rangle \qquad [\text{photoelectrons}] \quad (16)$$

Taking as an example Na-impurities with $\langle \sigma_i v_e \rangle / \langle \sigma_{1m} v_e \rangle$ = 0.4, we see that a flux of 10^{11} atoms/cm^2 gives 60 photoelectrons/msec, which are easily detected. For other elements, this detection limit may be at considerably higher fluxes, depending on the parameter $\langle \sigma_i v_e \rangle / \langle \sigma_{1m} v_e \rangle$ (see also Fig. 2 and Table 1).

The interference by the continuum radiation can be estimated using the emission coefficient for the Bremsstrahlung

$$\varepsilon(\lambda) \approx 7.6 \times 10^{-15} \, Z_{eff} \, n_e^2 \, (\bar{g} / T_e^{1/2}) \Delta\lambda / \lambda \quad [\text{photons/cm}^2 \text{ s sr}] \quad (17)$$

With $Z_{eff} \, n_e^2$ = 2 $\times$ 10^{27}/cm^6, $\bar{g}$ = Gaunt factor = 2, T_e = 400 eV, $\Delta\lambda / \lambda$ = 1/5000, and an emitting layer of 1 m the optical arrangement assumed above gives 90 photoelectrons in 1 ms, i.e. the contributions of the continuum and the line to be measured are in the same order of magnitude.

III.4. Applications

In order to obtain a survey over the emitted spectrum, photographic spectra of the plasma in front of the TEXTOR limiter have been taken with the experimental set-up of Fig. 3. The spectra shown in Fig. 4 are time integrated over an interval from 0.3 s to 1.3 s for discharges of 2 s duration /21/. Spectral lines emitted from the hot 2 m long plasma column between window and limiter can easily be detected by their uniform radial intensity distribution. They are emitted by ions (mainly oxygen) and by hydrogen, whereas the metal atoms (mainly Cr and Fe), which are localized in a well confined zone around the limiter, give rise to short lines indicating their short penetration depth. (This can be well recognized when looking at the weak lines, the strong metal lines are overexposed and give

224

therefore a misleading impression). Neutral oxygen, which is the dominant impurity, has no strong spectral lines in the visible, but only in the infrared and in the vacuum UV.

In order to see the poloidal and radial variation of the line intensities CCD-camera pictures of the four limiter segments in the light of H_α -, and of a CrI-line have been taken (see Fig. 5). The different penetration depths of these atoms can be clearly recognized; they correspond to the different velocities and ionization rate coefficients of the atoms /21/.

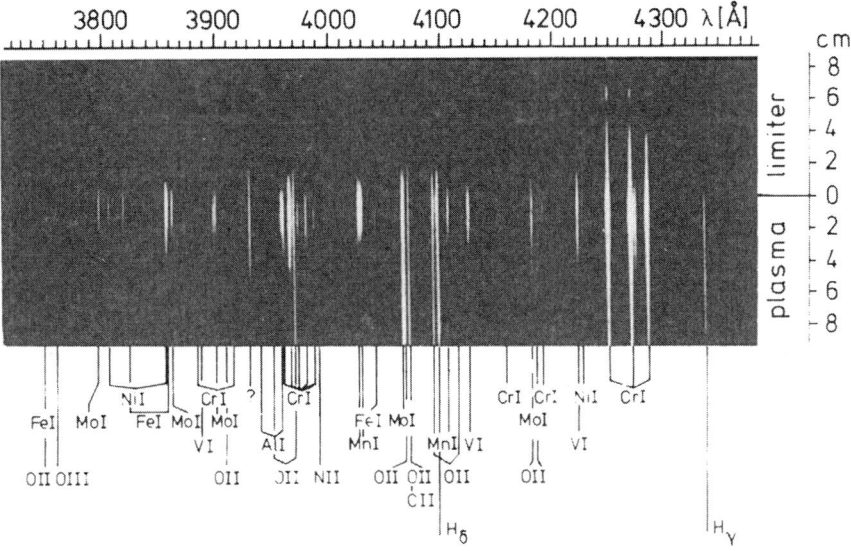

Fig. 4: Spectrum of the plasma in front of the TEXTOR Inconel
limiter /21/.

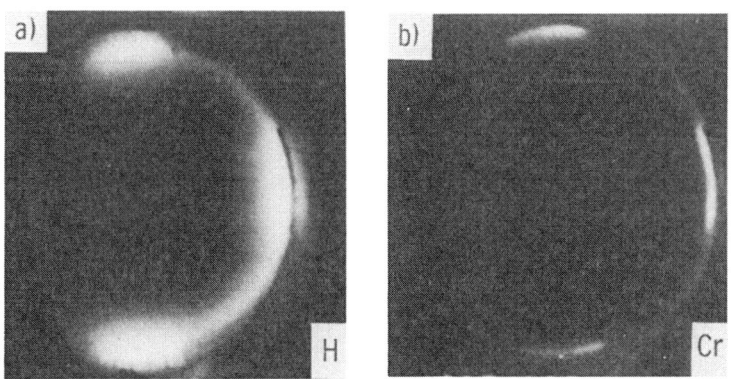

Fig. 5: Photographs of the main limiter in the light of
a) H_α and b) a Cr line.

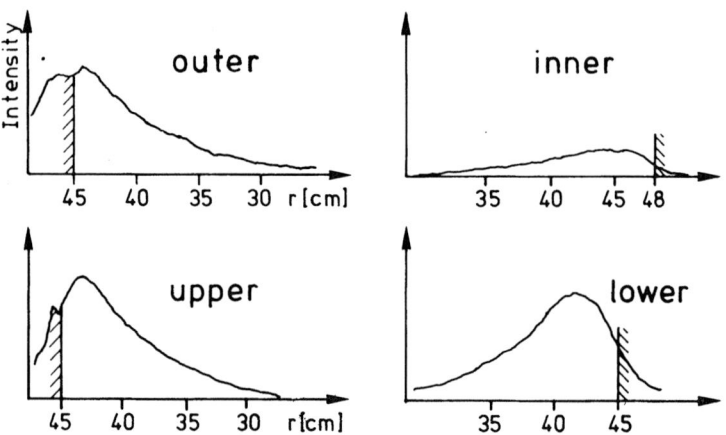

Fig. 6: Radial H_α-intensity profiles in front of the four
limiter segments /21/.

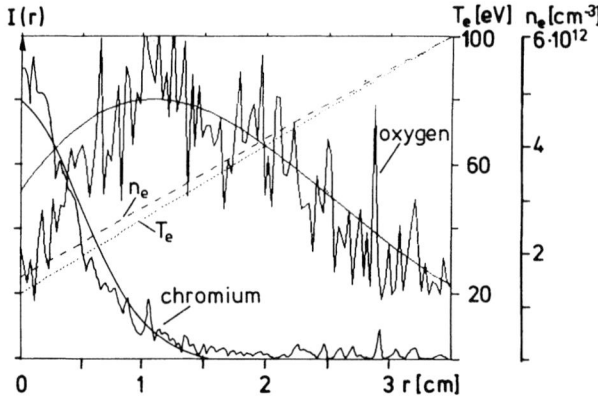

Fig. 7: Intensity profiles of the OI (λ = 8446 Å) and CrI (λ =
4254 Å) lines in front of the upper limiter segment and
calculated with the shown radial n_e- and T_e- dependence
/21/.

In order to obtain quantitative results from these pictures, they are evaluated with an image processing system, which gives plots as shown in Fig. 6, where the H_α-intensity distributions perpendicular to the four limiter segments are recorded /21/. These profiles demonstrate, that the upper, outer and lower limiter segments have nearly equal particle loads, whereas the load on the inner limiter segment is much lower. A typical H_α-intensity decay length of 7 cm is found, from which a mean velocity of 2×10^6 cm/s follows using the mean density of 1×10^{13} cm^{-3} and the ionization rate coefficient of 3×10^{-8} cm^3/s. Charge exchange processes are neglected for this estimate. The same intensity distributions have been obtained by using a spectrometer-photomultiplier arrangement with a rotating slotted disc in front of it. Examples of a CrI and an OI-line are shown in Fig. 7 together with calculated intensity profiles. The mean velocity of sputtered Cr-atoms is known to be approximately 2×10^5 cm/s. With the ionization function of 2.2×10^{-7} cm^3/s for CrI, which is not very sensitive to T_e, and with a decay length of 0.8 cm, a density of 1.5×10^{12} cm^{-3} is found at the limiter edge. With the same T_e and n_e-profile, we have also tried to fit the OI intensity profile, and the result is also given in Fig. 7. The velocity of the OI-atoms is a free parameter, and a value of 5×10^5 cm/s gives the best fit. Obviously the inaccuracy of T_e and n_e measurements originates mainly from the absence of good measurements of the velocity distribution of the sputtered particles.

The hydrogen and impurity emission from the TEXTOR limiter has also been determined with the experimental set-up described above (cf. Fig. 3). The fluxes of H-, Cr-, metastable Cr*-, Na- and O-atoms have been measured. The spectral lines used for this purpose, the calculated number of ionizations per photon, the mean temperature of the emission region, and the fluxes per cm circumference of the limiter are listed in table 1. For Cr, also one of the metastable states has been measured (Cr*); less than 3 % of the total flux is contributed by this state. Summing over all relevant metastable states with E < 2 eV, according to their statistical weight, we obtain an upper limit of the metastable atom contribution of the Cr-flux of 15 %. From the fluxes/cm, the total fluxes are obtained by multiplication with the limiter circumference, taking into account, that the particle load may not be uniform. The result is also given in Table 1.

The measured neutral hydrogen flux per cm of the limiter circumference may be somewhat too low: the emission region of the hydrogen atoms may not be completely within the field of view of our experimental arrangement, since the ionization region extends beyond the field of view over a long distance in

Table 1: Atom species, emission wavelength, number of photons
emitted/cm sec, $\bar{T}$ = mean temperature of emission
region, calculated ionizations per photon, atom fluxes
per cm limiter circumference and total atom fluxes
from the limiter (plasma radius 46 cm).

	$\lambda[\text{Å}]$	number of photons	$\bar{T}$ [eV]	ionizations per photon	atoms/cm sec	atoms/sec
H_α	6563	$5.9 \cdot 10^{17}$	100	14	$8.2 \cdot 10^{18}$	$6.6 \cdot 10^{20}$
OI	8446	$2.7 \cdot 10^{16}$	45	36	$1.0 \cdot 10^{18}$	$8.0 \cdot 10^{19}$
Cr	4254	$8.1 \cdot 10^{15}$	30	4	$3.2 \cdot 10^{16}$	$2.5 \cdot 10^{18}$
Cr^{+++}	5208	$7.6 \cdot 10^{14}$	30	1.4	$1.0 \cdot 10^{15}$	$8.0 \cdot 10^{16}$
Na	5890	$1 \cdot 10^{14}$	30	0.45	$5.0 \cdot 10^{13}$	$4.0 \cdot 10^{15}$

the poloidal and toroidal directions. Furthermore, we disregard
the direct formation of protons from molecules (see chapter of
M.F.A. Harrison), which may lead to an underestimate of the
hydrogen flux.

From the fluxes and the velocities, the neutral impurity
density in front of the limiter can be estimated taking into
account that the flux emitting area extends about 4 cm in
toroidal direction. The impurity concentration calculated this
way i.e. the ratio of impurity density to electron density is
about 2 % for the chromium and 25 % for the oxygen atoms. These
concentrations are considerably higher than those given for the
plasma core.

The metal release at the limiter cannot be explained by
hydrogen sputtering, since the yields are too low. Assuming
that the flux to the limiter has the same composition as that
from the limiter, the metal release could be explained by
sputtering due to the oxygen and metal impurity ions, a result
also found in other experiments /22, 45/. Since the oxygen
fluxes are about ten times larger than the metal fluxes, a
sputtering yield of about 10 % would be needed, which oxygen
ions probably reach at energies of ≈ 200 eV (exact sputtering
yields for oxygen are not available since they depend strongly
on the oxide formation on the limiter surface /103/). This
energy can be gained either by an acceleration of multiply
stripped ions ($Z \approx 3.5$) in the sheath potential or by their
acceleration due to friction with the protons in the scrape-off
layer.

Recently, experimental values of $\langle \sigma_{lm} \, v_e \rangle$ have also been published for Cr-lines /23/, although without error estimates. These values are four times larger than those calculated by the semiempirical eq 4. If these values should be confirmed, the Cr-flux listed in table 1 would have to be reduced by a corresponding factor and as a consequence the O-ion energy needed for sputtering would also be considerably smaller (< 100 eV).

IV. Laser induced fluorescence

As has been shown in the last chapter, the measurement of hydrogen and impurity fluxes, of ionization lengths of neutral particles, and of particle velocities is possible with emission spectroscopy. Several difficulties occur in the interpretation of such measurements: excitation and ionization functions are often not well known, intensities are integrated along the line of sight and depend on n_e and T_e. These difficulties can be avoided by using laser induced fluorescence, although at the expense of a more complicated experimental technique.

In the first part of this section, we will introduce the principles of laser induced fluorescence /17,24,25/ as far as needed for the understanding of the experiments. Simplified rate equations will be used to demonstrate the advantages of laser induced fluorescence for the measurement of densities and velocities: high spatial resolution, high sensitivity, and independence of n_e and T_e. The problems caused by the plasma background radiation, or connected with the signal calibration and with the imperfect technical state of the lasers will also be discussed. In the second part we will describe a few typical applications, especially measurements of small densities of metal atoms and of hydrogen densities and velocities, to elucidate the present state of experimental possibilities.

IV.1 Principles of laser induced fluorescence

In a typical fluorescence experiment (cf. Fig. 8A) a laser is focussed into a plasma volume V, in order to excite the atoms which are in resonance with the laser radiation. They emit fluorescence radiation, part of which is collected by a lens and, after being separated from most of the background radiation by e.g. an interference filter, detected by a photomultiplier. Excitation and fluorescence of the atom may occur either at nearly the same wavelength ("two level atom", Fig. 8B) or the fluorescence wavelength is shifted relative to the laser wavelength ("three level atom", Fig. 8C), when the initial and final state of the atom are different.

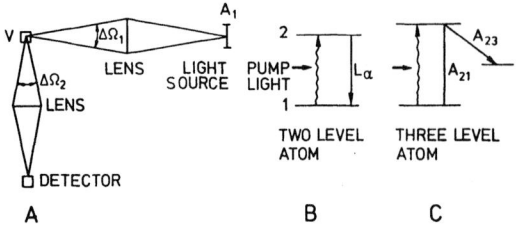

Fig. 8: Principle of fluorescence spectroscopy

IV 1.1) Two level atom

In order to calculate the flux of fluorescence photons, we consider first the simplest case of a two level atom under isotropic illumination by continuum radiation in the vicinity of the resonance transition. Using the Einstein coefficients /18,26/ for absorption B_{12}, for induced emission B_{21}, and for spontaneous emission A_{21} we obtain the occupation densities of level 2 and 1 (n_2 and n_1, respectively), in the presence of the radiation energy density per frequency unit $u(\nu)$, by the rate equation

$$dn_2/dt = u(\nu)(B_{12}n_1 - B_{21}n_2) - A_{21}n_2 \qquad (1)$$

where $n = n_1 + n_2$, $g_1B_{21} = g_2B_{21} = g_2A_{21}c^3 / 8\pi h\nu^3$, and g_1, g_2 denote the statistical weights of level 1 and 2. In case a time independent radiation field is switched on at $t = 0$, the solution of (1) is given by

$$n_2/n = [g_2/(g_1 + g_2)][s/(s+1)]\left\{1 - \exp[-(S+1)A_{21}t]\right\} \qquad (2)$$

where S is the saturation parameter for continuum radiation

$$S = u(\nu)\,(B_{12} + B_{21}) / A_{21} \qquad (3)$$

For practical applications, it is more convenient to use $\phi(\lambda)$ instead of $u(\nu)$, where $\phi(\lambda)$ is the energy flux per unit area and unit wavelength. Then we have

$$S = \phi(\lambda) / \phi_o(\lambda) \qquad (3a)$$

230

where $\phi_o(\lambda)$ is called the saturation power, which is given by

$$\phi_o(\lambda) = [g_1/(g_1 + g_2)]\, 8\pi h c^2/\lambda^5 \qquad (4)$$

The relaxation time for reaching the equilibrium population is shorter than $1/A_{21}$, which is often much shorter than the duration of the light pulse, so that in case of a two level atom we can usually expect time independence of the population and use the steady state equation

$$n_2/n = [g_2/(g_1 + g_2)]\, S/(S+1) \qquad (2a)$$

This means that for high values of S the levels are occupied proportional to their statistical weights. For comparison, the excitation of atoms by electrons is determined by

$$n_2 = n <\sigma_{1m}\, v_e>\, n_e/A_{21}$$

which e.g. for L_α and $n_e = 1 \times 10^{12}/cm^3$ gives only

$$n_2/n = 5 \times 10^{-5}$$

i.e. a factor 10^4 less than laser excitation with $S \gg 1$.

The powers needed for saturation strongly increase with decreasing wave-length according to eq. 4. As an example we give the saturation power for some wavelengths assuming $g_1 = g_2$

λ = 6000 Å	ϕ_o = 100 W/cm^2 Å
λ = 3000 Å	ϕ_o = 3 kW/cm^2 Å
λ = 1200 Å	ϕ_o = 300 kW/cm^2 Å

The intensities to achieve $S \gg 1$ in a two level atom can easily be produced in the visible and near UV, but are difficult to produce in the VUV as e.g. at L_α, as will be shown below. But even with $S < 1$, lasers may produce a much higher ratio of n_2/n than electrons.

The total emitted photon flux /cm^3, $\int \phi\, dt$, is obtained for the two level system by

$$\int \phi\, dt = n_2 A_{21}\, \tau = \frac{g_2\, n}{g_1 + g_2}\, \frac{S A_{21} \tau}{S + 1} \qquad (5)$$

where τ is the duration of the exciting laser pulse, assumed to be long compared to $1/A_{21}$.

If in addition to the density the velocity distribution also has to be determined, the detection limit is shifted to higher densities, since only a small velocity range of the atoms can be excited. The laser spectral width has accordingly to be reduced to a fraction of the Doppler width, depending on the desired accuracy. However, the possible resolution does not only depend on the spectral width, γ_{laser}, and the natural width of the spectral line, γ_{line}, but also on the power of the laser, since a high power laser can saturate the spectral transition even if the tuning is not exactly at resonance, an effect called saturation broadening or power broadening. The upper level of a transition with an energy difference $\hbar\omega_0$ is pumped by a laser of frequency ω to a ratio /24/

$$\frac{n_2}{(n_1+n_2)} = \frac{g_2}{(g_1+g_2)} S_0 \left(\frac{\gamma}{2}\right)^2 / \left\{ (\omega-\omega_0)^2 + (\gamma/2)^2 (1+S_0) \right\} \qquad (6)$$

where $\gamma = \gamma_{laser} + \gamma_{line}$, $S_0 = (2/\pi\gamma)(\phi_L/\phi_0)$ and ϕ_L the total power of the laser/cm^2. This shows that the velocity resolution $\Delta v/c = \Delta\lambda/\lambda = \Delta\omega/\omega$ is restricted to $\Delta\omega = \gamma(1+S_0)^{1/2}$.

IV 1.2) Three level atom

The detection of impurity atoms near the wall using a two level system is often difficult because of the stray light problems. Using a three level system (see Fig. 8c), where the fluorescence wavelength is shifted relative to the exciting wavelength, may for such a situation be a better solution, although the sensitivity is lower and the needed laser powers often are much higher. The rate equations are in this case

$\dot{n}_2 = u(\nu) (B_{12}n_1 - B_{21}n_2) - A_2 n_2$

$\dot{n}_3 = A_{23}n_2$ $\qquad\qquad (7)$

$n = n_1 + n_2 + n_3$

where $A_2 = \sum_i A_{2i}$. In this case, the saturation parameter is decreased by A_{21}/A_2 and given by

$$S = u(\nu) \ (B_{12} + B_{21})/A_2 = \phi(\lambda)/\phi_0(\lambda) \qquad (8)$$

and $\phi_0(\lambda) = \frac{g_1}{g_1 + g_2} (8\pi h c^2/\lambda^5) A_2/A_{21}$ $\qquad (9)$

For $S \gg 1$ and times $t \gg [A_2(S + 1)]^{-1}$, we find the solution for the emitted fluorescence radiation

$$A_{23}n_2(t) = \frac{g_2}{g_1 + g_2} \, n \, A_{23} \, exp\left[-\frac{g_2}{g_1 + g_2} A_{23} t\right] \qquad (10)$$

and using a laser pulse, the duration of which is long compared to $(g_1 + g_2)/g_2 A_{23}$ the total emitted photon number/cm^3 is:

$$\int A_{23}n_2(t) \, dt = n = \int \phi \, dt \qquad (11)$$

If there are other transitions from level 2 than those to level 1 and 3, we have to take into account the branching ratio and obtain

$$\int \phi \, dt = A_{23} \, n / (A_2 - A_{21}). \qquad (12)$$

IV 1.3) Specific problems of laser induced fluorescence

Equations 2-12 do not always provide good estimates for the experiment. Some of our simplifying assumptions may not be justified and the more complicated conditions of the experiment have to be taken into account in the model:
1) The magnetic field is not negligible
2) Illumination is not isotropic and unpolarized
3) Lasers do not emit a true continuum, but modes of a cavity of length L
4) Illumination intensity is not time independent after switching-on the laser

Level splitting by a magnetic field has been discussed in the preceeding section (see eq III, 13). For fluorescence experiments, usually polarized lasers are used with the electric field vector parallel to the magnetic field, so that only the π-components of the line are excited. The distance between the different π-components is small if $g_{J_1} \approx g_{J_2}$. Unfortunately, the different Zeeman components have different absorption coefficients and as a consequence different saturation powers, e.g. for π-components and $J \rightarrow J$ transitions they are proportional to $1/m^2$, m=0 is not pumped at all! On the other hand, for $J \rightarrow (J + 1)$ transitions the levels with m = J + 1 cannot be excited by π-polarized light. Depending on the excitation the fluorescence light can be emitted anisotropically and polarized, and this effect has to be taken into account in the interpretation. Certainly the best way to take these effects into account is to make the calibration and the measurements under similar conditions.

The fact, that the laser emits modes of a cavity and not a true continuum is of no importance as long as the distance of neighbouring lines $\Delta\lambda = \lambda^2/2L$ is smaller than γ, which is the sum of the natural width of the absorption line and the width of the laser line. If γ is smaller than $\Delta\lambda$, the mode structure has to be taken into account. Taking an average energy flux

$$\bar{\phi}(\lambda) = \frac{1}{\Delta\lambda} \int_{\lambda_0}^{\lambda_0 + \Delta\lambda} \phi(\lambda)\, d\lambda \tag{13}$$

and an average saturation parameter $\bar{S} = \bar{\phi}(\lambda)/\phi(\lambda_0)$ we find e.g. for the two level atom, that

$$n_2/n = \left[g_2/(g_1 + g_2) \right] \bar{S} \Big/ \left\{ \bar{S}^2 + 1 + \bar{S}\left(a + \frac{1}{a} \right) \right\}^{1/2} \tag{14}$$

$$a = \coth\left(\pi\gamma/2\Delta\lambda \right). \tag{15}$$

This means, that for saturation

$$\bar{S} \gg \coth\frac{\pi\gamma}{2\Delta\lambda} \approx \frac{2\Delta\lambda}{\pi\gamma} \quad and \quad \bar{S} \gg 1 \tag{16}$$

is necessary, i.e. in case $\Delta\lambda > \gamma$ the power has to be by a factor $\Delta\lambda/\gamma$ higher than for the case $\Delta\lambda < \gamma$.

The laser pulse is generally not time independent and not always sufficiently long to justify the use of the steady state solution. The corrections are of minor importance, if the saturation is high and the laser pulse long compared to the reciprocal transition probabilities.

IV. 1.4) Detection limits

Whereas the detection limits without a plasma background are only determined by the number of electrons produced at the photocathode of the multiplier by the photons of the fluorescence light, the detection limits with plasma are also restricted by the radiation noise of the plasma. As an example we discuss the measurement of atomic hydrogen densities by L_α-fluorescence /27/. The experimental conditions are described by the following set of data: H-density = 10^{10} atoms/cm^3, Doppler profile with T = 30 eV, magnetic field = 20 kGauss, fluorescence volume = 5 cm x 1 cm^2, transmission of the observation optics = 10 %, solid angle of observation = 8 x 10^{-3} sr, quantum yield of the KBr photocathode = 7 %, laser power = 4 x 10^{11} (π polarized) Lα -photons focussed to the fluorescence volume. From these data we calculate an optical depth of 5 x 10^{-4} and (at 90 ° observation) 1300 photoelectrons produced at the photocathode. The background Lα radiation estimated for a tokamak /28/ is 2 x 10^{15} photons/cm^2 s sr. This gives in 10 ns 1100 photoelectrons and a thermal noise of $\sqrt{1100}$ = 33 electrons, from which a signal to noise ratio of 40 follows. This value may be reduced by non-thermal fluctucations, which cannot be predicted reliably. But since the fluorescence signal is higher than the background signal, a reasonable measurement should be possible.

IV. 2) Calibration of the fluorescence signals

In order to obtain the atom density from the fluorescence signal, a calibration is needed. The best calibration will be achieved, if the atoms in question can be provided as a vapour or a gas of known density. The density of a metal vapor can be determined by measuring the flux to a quartz balance at a velocity known from the temperature (see e.g. /29/). Mo-densities and fluxes have been measured by sputtering Mo-atoms under standard conditions where sputtering rates are known /30/.

If conditions with atoms of known densities cannot be established, an absolute measurement of the fluorescence yield is necessary, which then via eq. 5 or 10 yields the particle density. In practice, some of the geometry - and efficiency factors can be eliminated by a comparison with Rayleigh scattering, a technique proven to be useful in Thomson scattering experiments /19/. Especially in the case of low saturation (S ≪ 1), using a laser spectral width small compared to the Doppler width for the excitation of a two level atom, as e.g. for the excitation of the Lα - line, this calibration technique is very useful, since the power dependence cancels out. The Rayleigh scattering cross-section can be calculated from the refractive index n /18/, the resonance cross-section at the wavelength λ_o from the oscillator strength f /18,31/ and the intensity distribution h(λ) of the line /18/:

$$G \text{ (Rayleigh)} = \frac{8\pi^3}{3n_A^2} \frac{(n^2-1)^2}{\lambda^4} \tag{17}$$

$$G \text{ (Resonance)} = (\pi e^2/mc^2) f \lambda_o^2 h(\lambda_o)\Big/\int_L h(\lambda)d\lambda \tag{18}$$

The unknown density of the fluorescing atom n_H can then be calculated from the observed photomultiplier signals ϕ (Rayleigh) and ϕ (Resonance) and from the known density n_A of the Rayleigh scattering gas by the equation

$$\frac{\phi \text{ (Rayleigh)}}{\phi \text{ (Resonance)}} = \frac{\sigma \text{ (Rayleigh)}}{\sigma \text{ (Resonance)}} \frac{n_A}{n_H} \frac{P_A}{P_H} \qquad (19)$$

where P_A and P_H are factors depending on polarization, e.g. $P_A = 1.5$, $P_H = 1.2$ for L_α in case of π-polarized excitation, 90 $^\circ$ observation, and small magnetic fields, but $P_H = 1.5$ for high fields (Paschen-Back effect /17,26/).

For a calibration in the case $S \gg 1$, the number of laser photons $\int Jdt$ and the cross-section A of the laser beam at the fluorescence volume has to be known. For the saturated 3-level system of a metal with a density n_M we obtain /32/:

$$\frac{\int \phi \text{ (Rayleigh) dt}}{\int \phi \text{ (Resonance) dt}} = \frac{\sigma \text{ (Rayleigh)}}{A_{32}/ (A_2 - A_{21})} \frac{n_A}{n_M} \frac{P_A}{P_M} \frac{Q \int Jdt}{A} \qquad (20)$$

where P_M is a polarization factor and Q takes into account the different sensitivity at the laser and fluorescence wavelength. For a determination of the total density, the existence of atoms in metastable states may be important and their contribution to the total number has to be taken into account (see IV. 4.1).

IV 3) Lasers for fluorescence excitation

For the excitation of the fluorescence, a tunable dye laser is at present the best light source. There are mainly three types which are of interest: dye lasers pumped by a YAG-laser, by an excimer laser, or by a flashlamp. Typical specifications are summarized in Table 2. As can be seen from Table 2, the YAG-system is the best one for $\lambda = 6000$ Å and, frequency doubled, for 3000 Å, if the short duration and low repetition rate can be tolerated. The excimer-dye laser system has repetition rates up to 100 Hz and high powers over the whole spectral range from 1 μ - 3200 Å, and frequency doubled down to 2200 Å, but also such a short duration that the time independent eq. 12 often cannot be applied. Only the flashlamp pumped dye laser has longer duration up to the μsec range, but it has considerably less power than the laser pumped dye lasers. Because of their short duration both the YAG and the excimer pumped dye laser have a short cavity ($\approx$25 cm) and therefore the disadvantage of a wide mode separation, whereas the flash lamp pumped dye laser can be built with a long cavity ($\approx$1 m) and has consequently a shorter mode spacing. The choice of the laser system depends on the intended

Table 2: Data for excimer, YAG and Flashlamp pumped dye lasers. The numbers are typical values, not the maximum values. Available energies for 6000 Å, 3000 Å, and 3400 Å are given. (τ = duration of the laser pulse, $\Delta\lambda/\lambda^2$ = mode distance, R = Repetition rate).

	Energy [m Joule]			τ	$\Delta\lambda/\lambda^2$	R
Pump	6000 Å	3000 Å	3400 Å	nsec	cm^{-1}	Hz
Excimer	20	2	20	10	0.02	100
YAG	90	20	10	10	0.02	10
Flashlamp	500	10	10	1000	0.005	10

application, i.e. on the element to be detected, its density, the desired repetition rate and on the background radiation. Continuous wave lasers have only occasionally been applied, because of their low power in the UV.

IV.4. Applications of laser induced fluorescence

Sputtering is the dominant plasma-wall interaction process in magnetically confined plasma according to present knowledge. Using well defined sputtering experiments with low background radiation, laser induced fluorescence has brought considerable progress into the measurement of sputtering yields, of velocity distributions of sputtered atoms, and of the influence of oxidation on sputtering yields /32-38/.

IV. 4.1. Measurements on sputtered iron

Ihe sputtering of Fe (at T = 300 K) has been carefully studied /33,34/ and typical results will be discussed with the aid of Figs. 9-11. By the spin-orbit interaction, the a^5D groundstate is split into five sublevels with excitation energies up to 100 meV. For a quantitative evaluation of the total Fe-density their population has to be known. It has been measured by exciting each level separately. The result shown in Fig. 10 indicates that the occupation of these levels follows a Boltzmann distribution with T=1000 K and that only about 50 % of the atoms are in the groundstate a^5D_4. The metastable states other than a^5D have only a contribution of less than 2 % and can be neglected in the density measurement. Velocity distributions of the sputtered Fe are reproduced in Fig. 11. To avoid saturation broadening, the laser power has to be reduced to less than 500 Watt/cm^2. From these measurements a mean velocity of $\upsilon = 2.5 \times 10^5$ cm/s for sputtered Fe is derived. From υ_{Fe} and n_{Fe} the flux density $\Gamma_{Fe} = n_{Fe} \times \upsilon_{Fe}$ can be calculated. Experiments with the object to measure Fe-fluxes from the tokamak walls have been carried out on ISX-B /39,40/. Measurements on the Fe-fluxes from the TEXTOR limiter are presently being carried out at Jülich.

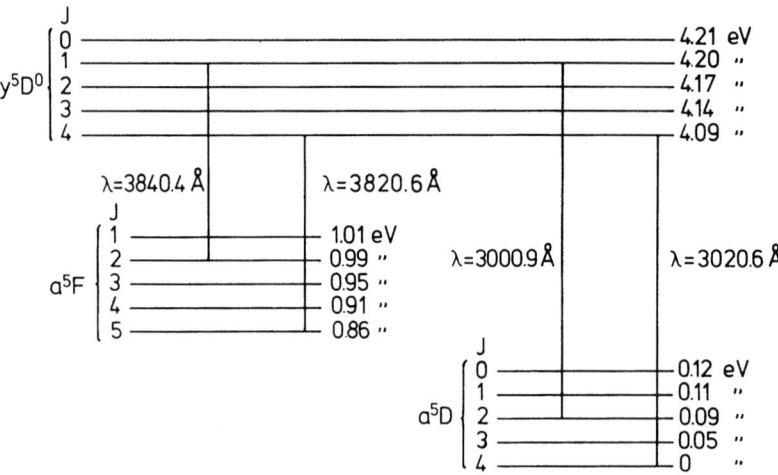

Fig. 9: Fe-energy levels. The wavelength for the excitation from the J = 0 and J = 2 levels and the observed wavelengths are indicated.

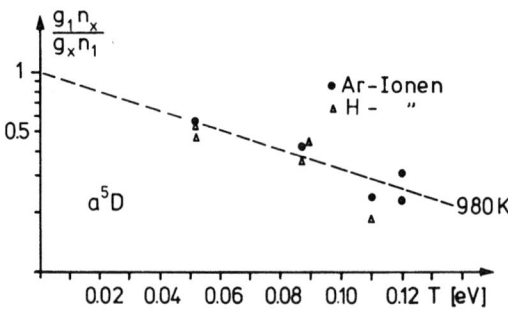

Fig. 10: Population of the J = 4 to J = 0 fine structure levels of the Fe-groundstate as a function of their excitation energy for Fe sputtered by hydrogen or argon ions /15,34/.

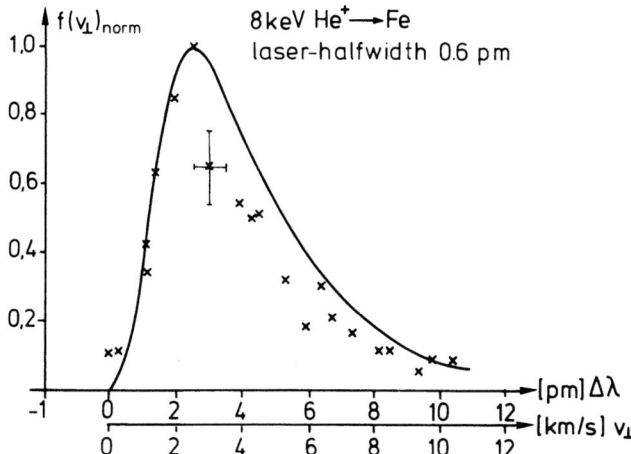

Fig. 11: Velocity distribustion of iron atoms sputtered by H^+
ions. Solid curve: Thompson distribution /13/ with a
surface energy of 4 eV, xxx experimental values /15/.

The utilization of the laser induced fluorescence for plasma
edge diagnostics proved to be more difficult, because of the
background radiation and the complicated obervation geometry.
Nevertheless, a few successful experiments have been performed
/29,30,39-49/. We discuss two examples.

IV. 4.2.) Measurement of Fe-atom fluxes from a test limiter in TEXTOR

The experimental arrangement for the measurement of Fe-atom
fluxes at a TEXTOR test limiter positioned 45° away from the
main limiter, is shown in Fig. 12a /49/. The dye laser, pumped
by a 100 Hz excimer laser, and the data acquisition system were
located outside the concrete shielding about 20 m away from the
fluorescence volume. The laser beam is guided by five mirrors to
the top of the vacuum vessel, where it enters the vacuum through
a quartz window and is guided in radial direction to the test
limiter. The fluorescence light was focussed to a system of five
light pipes connected to photomultiplier tubes.

The observation volume of 0.25 cm^3 was defined in toroidal direction by the laser cross-section and in the radial and poloidal

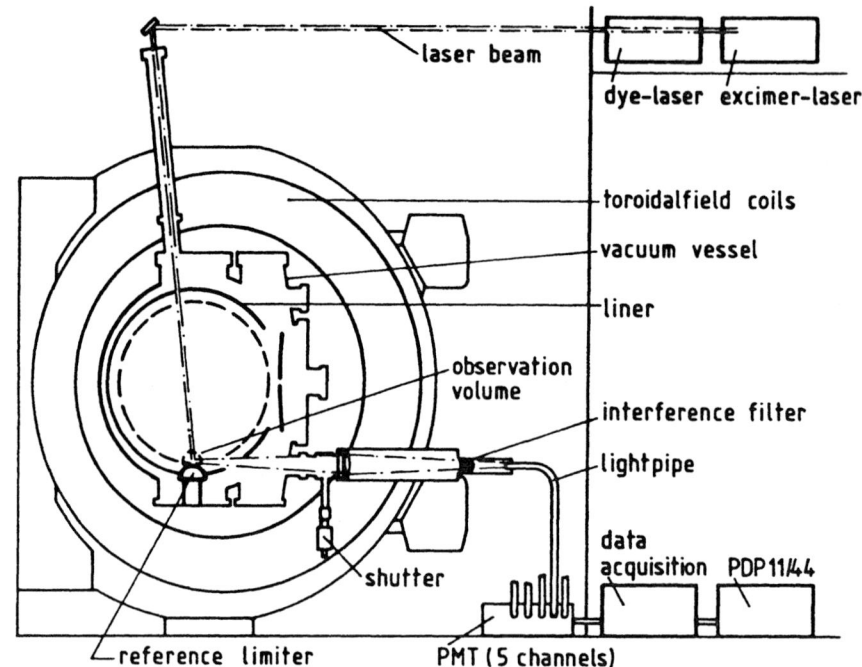

Fig. 12a: Cross-section of the TEXTOR vessel, showing the stainless steel (SS 316) reference limiter and the experimental arrangement with the excimer pumped dye laser
and the 5 channel observation system for the FeI fluorescence.

direction by a diaphragm of 0.5×0.5 cm^2 area. As shown in Fig.
12b the test limiter can be moved in radial direction between r
= 45 cm and r = 55 cm. To scan the impurity densities in the
toroidal and poloidal direction, the laser beam can be shifted
by a mirror. The penetration depth in the radial direction can
be determined by measuring the radiation emitted at five radii,
which have a separation of 1 cm. By means of an interference filter most of the background radiation is rejected. The observation system has been calibrated by a carbon arc.

In Fig. 13, the FeI density in front of the test limiter as a
function of time is shown for the case that the test- and the
main limiter are both at r = 47 cm. The fluorescence volume was
shifted from the apex of the test limiter by 2 cm in the toroidal direction to the point, where the highest densities were
measured. Besides some fast fluctuations, the FeI density is
rather constant during the discharge at about 3×10^{10} atoms
cm^{-3}. The background signal , mainly from Fe-atoms excited by
plasma electrons, is about a factor of 10 smaller than the flu-

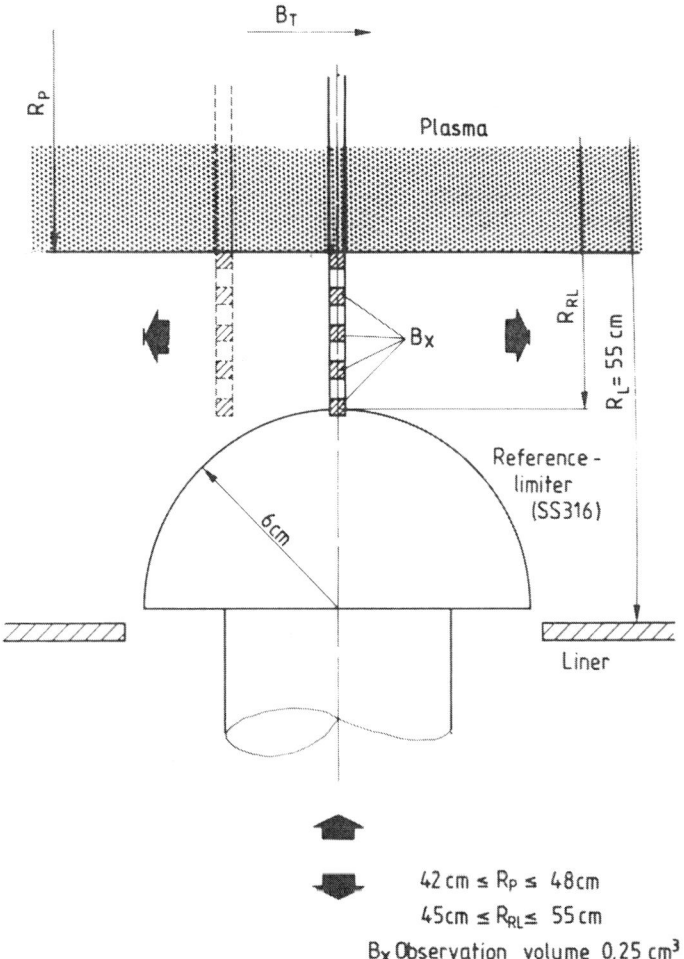

Fig. 12b: Observation geometry at the bottom reference limiter.
The limiter position is given by the radial distance
to the plasma center /49/.

orescence signal. By scanning the toroidal profile of n_{Fe} it has been found that the Fe-atoms are emitted from a rather restricted area of about 1.4 cm "half-width" (see Fig. 14). This flux density distribution of the Fe-atoms is related to that of the primary ions which is determined by the radial profiles of plasma parameters in the scrape-off layer and by the angle

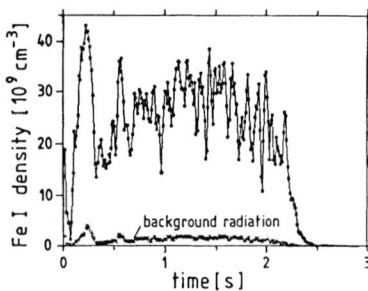

Fig. 13: Typical temporal dependence of the FeI density for an ohmic heated discharge of TEXTOR. Both reference limiter and main limiter are positioned at r = 47 cm. The observation volume was shifted 2 cm from the top to the ion drift side (cf. Fig. 12b) /49/.

between the toroidal magnetic field and the respective surface element of the liner. On the ion drift side, the Fe-density is about a factor of 10 higher than on the electron drift side. This phenomenon can be explained by the fact that the connection length to the main limiter is much shorter on the electron side than on the ion side.

By drawing the test limiter back to the liner, the Fe-atom densities drop by a factor of 10^3 to about 3×10^7 atoms cm^{-3}. The density profile is not peaked any more, but is almost flat. This observation indicates that the Fe-atoms are not released anymore by ions striking the limiter, but only by fast charge exchange neutrals. Taking into account a radial velocity of 2 x

242

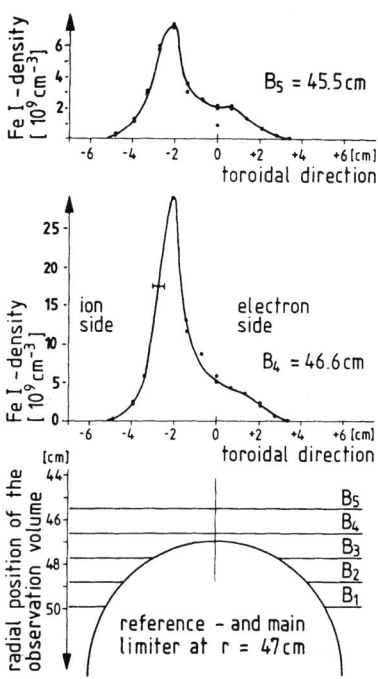

Fig. 14: Spatial distribution of the FeI density in the toroidal
direction during the flat top period. All limiters were
in the same position (r = 47 cm) /49/.

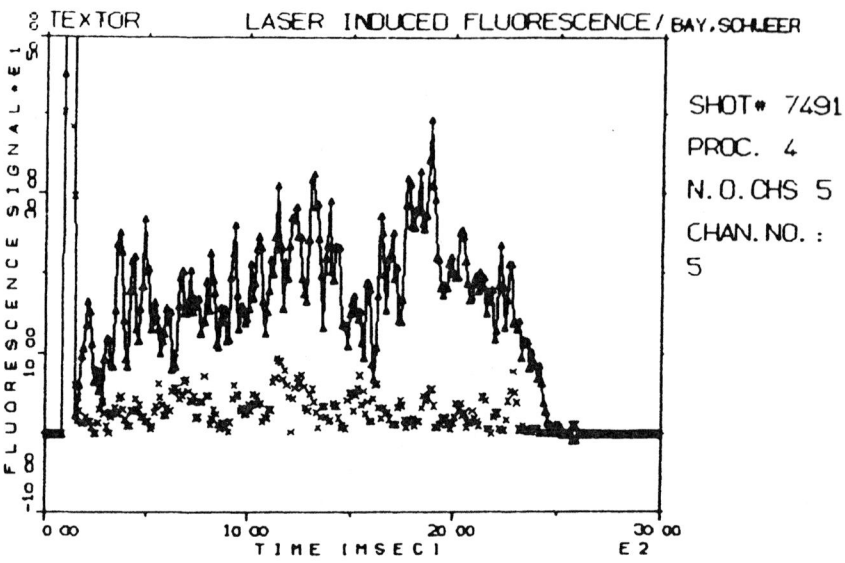

Fig. 15: Ti-fluorescence signal (arbitrary units) observed in front of a SS-test-limiter /49/.

10^5 cm/s and a liner area of 4×10^5 cm^2 a total Fe-flux of 2×10^{18} atoms/s is calculated.

First attempts have been made to use the test limiter as a deposition probe and to study material transport between different parts of the wall system. When the TiC coated head of the pump limiter is used as main limiter, Ti is emitted and a fraction of it is deposited on the test limiter. Fig. 15 shows the Ti detected by fluorescence after a day of operation with the pump limiter withdrawn /49/. The Ti-signal disappears after a few more shots.

IV. 4.3) Laser induced fluorescence in the vacuum ultraviolet

The second example concerns the measurement of atomic hydrogen densities near the wall of a tokamak. Laser induced fluorescence of atomic hydrogen is of special importance, since successful measurements would allow one to determine the neutral hydrogen fluxes into and out of a fusion plasma down to very low energies, where neutral particle spectroscopy cannot be used. In order to achieve this objective first experiments have been started several years ago by building systems for the production

244

of L_α light /50-55/. These suffered, however, from too low an efficiency or from bad reproducibility. Recent developments of excimer pumped dye lasers for the UV /56/ brought considerable progress. By focussing such a laser (cf. Fig. 16) into a phase matched krypton-argon gas mixture (i.e. a mixture, the index of refraction of which is equal at the fundamental frequency and the third harmonic) a considerable L_α-radiation power at the third harmonic of the laser has been detected by a photodiode with Au-cathode, which has about 3 % quantum efficiency at L_α, but less than 10^{-5} % for the fundamental wave ($\lambda = 3647$ Å). Typical pulses with about 4 ns duration and 10^{12} photons are obtained (Fig. 17). The power of 200 Watt after passing the MgF_2 exit window with a spectral width of about 10 mÅ corresponds to

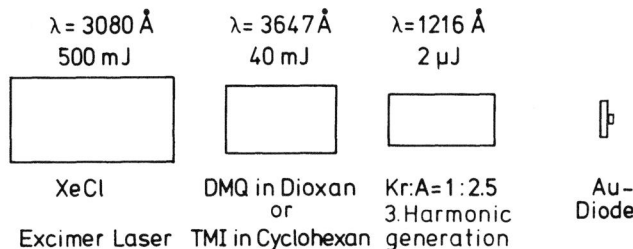

Fig. 16: Experimental arrangement for the production of
L_α - radiation by third harmonic generation /27/.

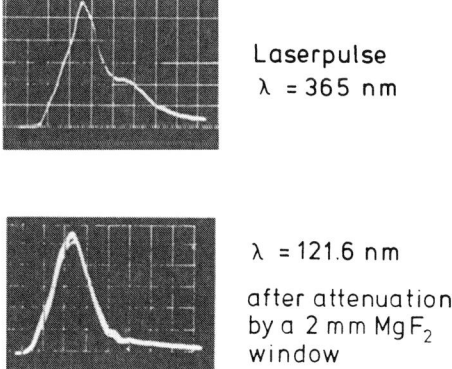

Laserpulse
$\lambda = 365$ nm

$\lambda = 121.6$ nm

after attenuation
by a 2 mm MgF_2
window

Fig. 17: Typical Au-photodiode signals at a time scale of
2 ns/div. Maximum power about 230 Watt.

a saturation parameter $S \approx 0.1$. However, the width is not broad enough to cover the whole velocity distribution of the H-atoms, if the temperature is around 30 eV, as expected for a tokamak boundary. Then the profile has to be scanned shot by shot.

Such a laser system has been used to measure temperatures and densities of atomic hydrogen in the cleaning discharges of the Asdex tokamak /57/. Fig. 18 shows the experimental arrangement /27,58/. The L_α-radiation enters the tokamak vacuum through a 2 mm thick MgF_2 window and illuminates a cross-section of about 1 cm^2 in the plasma boundary. The fluorescence is observed at 5° to the incident beam by a Cassegrain mirror system with 30 cm diameter at a distance of 265 cm from the fluorescence volume. The magnification is 1:1. In front of the image plane, a spectral filter for L_α can be inserted, which is normally an O_2-gascell with MgF_2 window. The fluorescence radiation is detected by a fast solar blind photomultiplier with KBr cathode. KBr is not sensitive above 1700 A, so that further filtering for long wavelengths is not necessary. L_α-profiles of atomic hydrogen produced in the Asdex cleaning discharges (see Fig. 19) have been obtained by scanning the profiles with a boxcar-integrator with a repetition rate of 1.5 Hz. The profile shows a half-width of 22 mA and 68 mA for the glow- and the 50 Hz discharge, respectively. This corresponds to about 400 K and 6000 K, after correcting for the 10 mA width apparatus profile. However, the measured profiles can only poorly be approximated by a Gaussian, as can be seen from Fig. 20. It indicates an excess of hot atoms, which is expected if the atoms are not commpletely thermalized after their production by dissociation of H_2 ("Franck-Condon atoms").

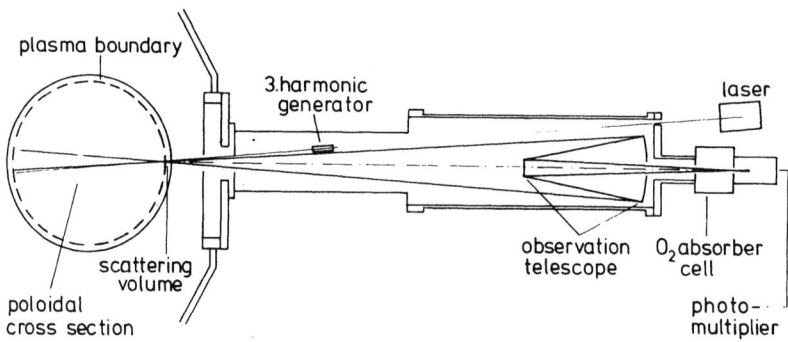

Fig. 18: Experimental arrangement to measure hydrogen densities and velocities in ASDEX /27/.

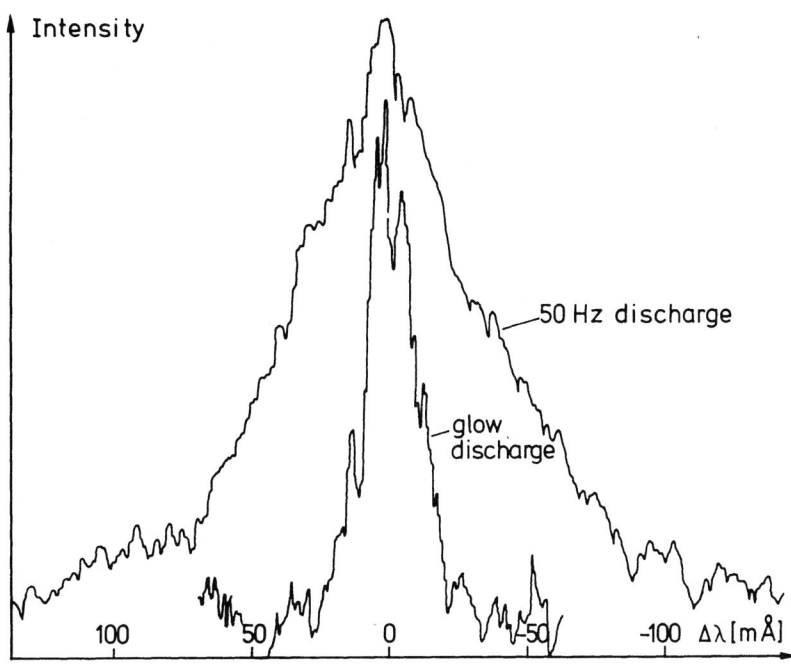

Fig. 19: L_α-profiles as measured in the ASDEX cleaning discharges /27/.

The H-atom density has been inferred from a comparison with experiments with a calibrated atomic hydrogen source /59/. Densities of 5×10^9 atoms cm^{-3} and 2×10^{10} atoms cm^{-3} were found for the glow discharge and the 50 Hz discharge, respectively. Recent experiments have shown that a calibration by Rayleigh scattering in argon at a pressure between 10-100 Torr is also possible. The interference of resonance scattering by impurities (e.g. H_2O) in the argon seems to be negligible.

The excitation by frequency tripled lasers is also possible for light elements as carbon and nitrogen (cf. Fig. 21) and for some metastable states of oxygen. In Fig. 21 also some lines of ionized atoms are indicated, which might be of interest. For the region above 1400 Å, also excitation by Raman shifted lasers has been considered /60/. In the future, the whole spectral region above the LiF limit at 1040 Å might be accessible, if frequency mixing of several dye lasers and two photon resonant pumping will be applied /61,62/.

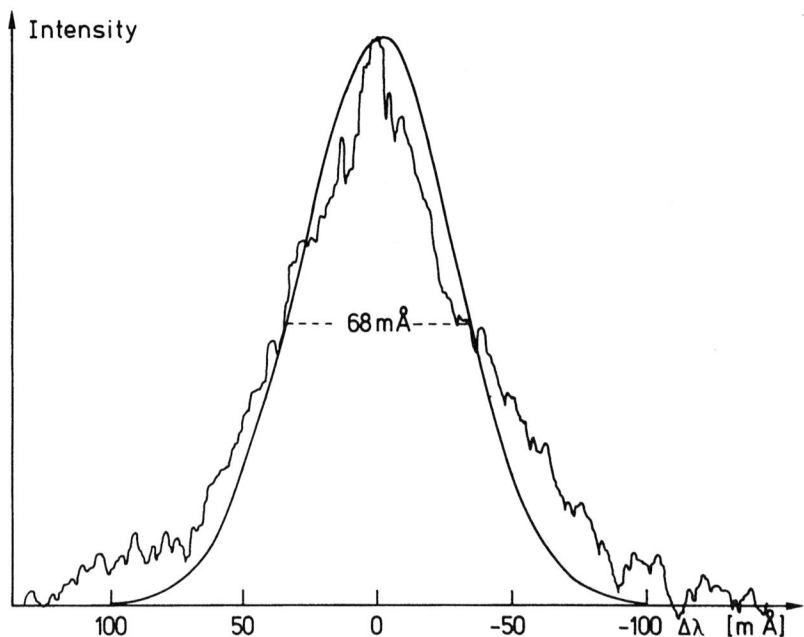

Fig. 20: L_α-profile in the 600 A, 50 Hz cleaning discharge and calculated Gaussian profile for T = 6000 K /27/.

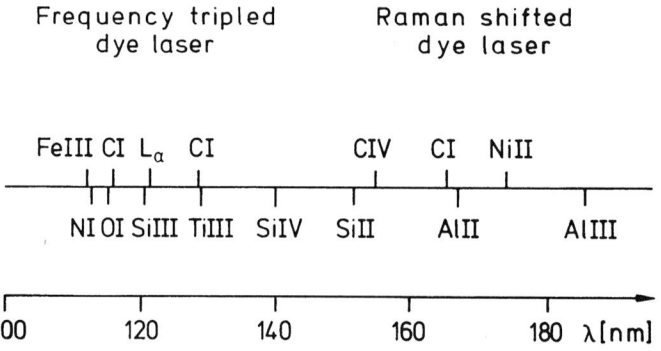

Fig. 21: Possible applications of laser induced fluorescence in the vacuum UV using a frequency tripled or Raman shifted dye laser.

The detection of molecules /63/ near the walls of magnetically confined plasma by laser induced fluorescence might also be of importance in the future, but because of the low oscillator strengths and the splitting over rotational and vibrational levels the sensitivity is certainly much lower.

IV. 5) Conclusions

The densities of most metal atoms in the plasma boundary can be measured by laser induced fluorescence, if the density is higher than about 10^6 atoms cm^{-3}. The population of metastable states with low excitation energies has to be determined, if total atom densities have to be known.

Besides the density, the velocity distributions can also be derived, if the laser spectral width and the saturation broadening is sufficiently smaller than the Doppler width. Since the fluorescence signals are then considerably reduced, densities have to be much higher. The method has been successfully applied to the measurement of velocity distributions of sputtered particles.

Hydrogen and other light elements as C, O, N have to be excited by vacuum UV radiation. Successful experiments have been carried out to excite the hydrogen L_α-line by a frequency tripled dye laser. The detection limit without plasma was 10^7 atoms cm^{-3}, with plasma 10^9 cm^{-3}. The calibration of the fluorescence signals versus density has been performed by Rayleigh scattering in argon or by vapours of known density.

V. Atomic beam methods

The numerous possibilities, which atomic beam methods offer for the determination of plasma parameters cannot be treated in this chapter exhaustively. We shall concentrate on the measurement of a few parameters, the knowledge of which is especially urgent and for which the required experimental techniques are at hand. There exist several survey papers on active particle beam diagnostics (e.g. /64,65/) where the atomic processes which may be utilized for diagnostic purposes are discussed systematically, although the main emphasis is given to properties of the plasma core. It appears quite certain that atomic beam methods are particularly promising for plasma edge diagnostics. This will mainly require improved design of appropriate particle sources and more and better data on the relevant atomic processes.

V.1. Principles

V.1.1. Electron density and electron temperature from excitation and ionization of beam particles by electron impact

As described in section III, $n_e(r)$ near the boundary can be derived from the radial intensity profile of the photons emitted by neutral atoms injected into the plasma, if the corresponding rate coefficient and $n_A(r)$ are known (see eq. III. 2 and

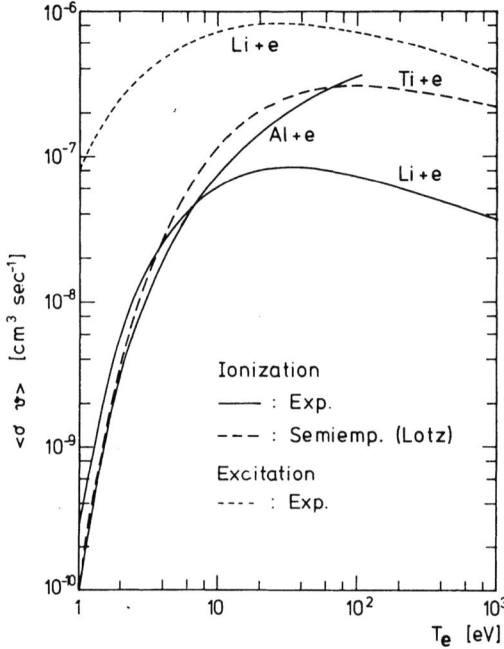

Fig. 22: Rate coefficients for electron impact excitation and ionization /85, 7,9,10,66/.

eqs. III,10-12). The method becomes especially simple in case the rate coefficient $\langle \sigma_{1m} v_e \rangle$ is independent of T_e in the range of practical interest, e.g. for 10 eV $< T_e <$ 100 eV which is typical for the plasma edge of tokamak plasmas. A further simplification occurs, if the species and the velocity of the beam particle have been chosen such that the beam attenuation, while passing through the edge, is negligible. From an inspection of available data on rate coefficients for excitation and ionization by electron impact it becomes apparent that the light alkali metals Li and Na are especially well suited as probing beams. The rate coefficients for Li excitation and ionization are shown in Fig. 22. In case of Li even a thermal beam provided by an oven at about 800 K will penetrate into the edge unattenuated up to $n_{el} \approx 5 \times 10^{11}$ cm^{-2}. At higher n_{el} beam attenuation due to ionization has to be taken into account. This can either be done by measuring the local beam density by means of laser induced fluorescence or by recording the total spatial emission profile. In the latter case n_e is obtained from eq. III, 11

$$ n_e = \frac{\overline{\vartheta}_{Li}}{<\sigma_i \, \vartheta_e>} \quad \frac{\varepsilon(r)}{\int_0^r \varepsilon(r') dr'} \tag{1} $$

In both cases an absolute calibration of the light intensity is not needed. At higher n_e-values corrections in the ionization rate due to the stepwise ionization may be necessary.

Considering also excitation of Li-atoms at considerably higher energy /67/, e.g. 10 - 30 keV, ionization of atoms due to electron impact can be neglected and charge exchange with protons will be the dominant attenuation process /68/. The characteristic attenuation length at $n_e \approx 10^{13}$ cm^{-3} is about 10 cm. The observed intensity is still proportional to n_e. However, due to the high velocity of the Li-atom, the photon emission is spread in space over a length $\vartheta_{Li}/A_{21} = 9.2 \times 10^7/3.7 \times 10^7 = 2.5$ cm.

It is obvious that beams with particle energies in an intermediate range of 1 - 10 eV should be very useful, because then it would be possible to adjust the energy to the n_e-range of interest without losing space resolution.

Assuming now that $n_e(r)$ is known we may use eq. III, 6a for the ionization rate to derive at least approximate values of T_e. With this aim in mind one will choose a particle species which shows a strong dependence of the ionization rate coefficient on T_e and the density of which can be measured using a commercially available tunable laser system, i.e. suitable metal atoms. The ionization rate coefficients for Al and Ti as a function of T_e are plotted in Fig. 22.

For the beam particles to be within the detection limits of the observation system, it is desirable that their density be in the range 10^8 cm^{-3} - 10^9 cm^{-3}. Together with the required particle velocity this will give us the equivalent current density of the beam as an important specification for the particle source.

V.1.2. The measurement of the density and of the temperature of plasma ions using charge transfer processes

Here we are interested in charge transfer processes between an ion species X of charge state Z and a neutral atom Y:

$$X^{Z+} + Y \longrightarrow X^{(Z-1)+} + Y^+$$

i.e. in this process one electron is transferred from Y to X. In general the electron will first go into an excited state (principal quantum number n) and then pass over into the groundstate (or a metastable state) by the emission of photons:

$$X^{Z+} + Y \longrightarrow X^{(Z-1)+*} + Y^+$$

The photon may serve for the detection of X .

The significance of this type of process for fusion plasmas was recognized first by the following reaction:

$$H^+ + H^o \longrightarrow H^o + H^+$$

e.g. an energetic proton is neutralized and a cold atom is ionized. The energetic atom is no longer confined by the magnetic field and may escape from the plasma, or may further penetrate into the plasma until the same process is repeated or ionization due to electron impact takes place. Such processes play an important role at plasma heating by the injection of high energy hydrogen atoms and for the transport of neutral hydrogen atoms and of the energy which they carry. The fluxes of neutral atoms leaving the plasma are of interest for two main reasons:
- They transport energy to the wall and release wall particles;
- their energy distribution is characteristic of that of the protons in the plasma.

The measurement of the H^o-flux in dependence on the particle energy offers one possibility to derive the proton temperature and gives important information on the plasma edge. For the latter purpose one is mainly interested in the energy range E < 200 eV. For E ≲ 10 eV laser induced fluorescence appears to be an appropriate energy analysis- and particle detection method. For the range 10 eV < E < 200 eV time of flight spectroscopy is used for energy analysis and secondary particle production at a target plate for particle detection /69/.

252

The cross-section for the hydrogen proton charge exchange reaction is large (6×10^{-15} cm^2 - 2×10^{-15} cm^2 /70/) in the energy range $0 < E < 1$ keV. The maximum occurs at zero energy due to the fact that the reaction is of the resonance type.

In the course of neutral beam heating experiments it has turned out that charge transfer processes between hydrogen atoms and multiply charged impurities, i.e. of the type

$$X^{z+} + H^o \longrightarrow X^{(z-1)+} + H^+$$

are also of great importance, e.g. because of beam attenuation and because of the production of high energy protons close to the wall. In general the maximum cross-section for this reaction occurs at an energy of the ion which is approximately given by the relation: energy per atomic mass unit $\approx 5 \sqrt{Z}$ keV /71/. For small Z it is of the same order as that for the reactions with protons. Experimental and theoretical investigations indicate that the cross-section increases approximately linearly with Z /71/.

For highly ionized impurities the charge transfer reaction with a hydrogen atom preferentially yields excited states of the impurity ion, which subsequently decay into low lying levels by the emission of photons. If the emission cross-section σ_{em} for the observed transition is known, and the emitted intensity is measured, the density of the impurity in the respective charge state can be calculated.

$$J = \int \varepsilon \, dx = \int n_{I^{z+}} \, n_{H^o} \langle \sigma_{em} \, v_{H^o} \rangle \, \frac{dx}{4\pi} \qquad (2)$$

Such measurements have been performed on the fully ionized ions of C, O by injecting neutral hydrogen beams and observing the resulting H_α - or L_α -like transitions (see eg. /72/). Partial cross-sections predicted by theory (for a review see /71/) were used:

$$\sigma_{em} \approx 3-5 \times 10^{-15} \, cm^2$$

Recently a promising new method for the detection of impurity ions, in particular of non-fully ionized ions has been proposed /73,92/: The excitation of light impurity ions (C,O) by electron capture from 30 keV Li-atoms. The following advantages are expected. Because of the low ionization energy of the Li-atom the excitation of the ion due to collision with this atom will yield relatively high lying excited states which make an efficient discrimination against both, electron impact excitation and H^0-induced excitation possible, cross-sections are large and, due to its low charge number, the Li-ion will cause relatively small detrimental effects to the plasma. Estimates for the most probable excited state and for the associated cross-section are also given in /73/. In particular it is pointed out that the most probable populated quantum number of the impurity ion for the reaction with a 30 keV Li-atom is 1.6 times larger than that for the reaction with a hydrogen atom, the corresponding ratio of partial cross-sections should be 8.

Meanwhile some total and some emission cross-sections for electron capture from Li-atoms have been measured and are presented in Table 3.

Table 3: Cross-sections for electron capture reactions between Li-atoms and light ions.

reaction	wavelength (Å)	$\sigma_{max}(10^{-16}\text{cm}^2)$	reference
H^+ - Li		100	/74/
He^{++} - Li		100	/75/
C^{4+} - Li		195	/76/
He^{++} - Li	1640	100	/77/
C^{3+} - Li	1923	22	/78/
C^{4+} - Li	2530	180	/78/
C^{5+} - Li	2983	157	/78/
C^{6+} - Li	2071	130	/78/
O^{4+} - Li	2450	44	/78/
O^{5+} - Li	2942	91	/78/
O^{6+} - Li	2071	120	/78/
O^{7+} - Li	2524	163	/78/

In order to assess the feasibility of the Li-beam method for impurity ion detection the ratio of signal to background radiation has to be calculated and that Li-atom flux density, which is required to obtain a value of this ratio about equal to one, has to be specified. Such estimates have been performed in /73/ taking into account the various competing excitation and recombination properties at plasma conditions that are typical for the edge of tokamak plasmas. The dominating source for background radiation appears to be H^0 induced charge exchange. An equivalent current density of Li atoms $\Gamma_{Li} \gtrsim 10$ mA/cm^2 is required to obtain a ratio of signal to background radiation larger than one.

We want to point out that the temperature of the ions to be detected by electron capture excitation can be determined - if the signal to noise ratio is sufficiently high - by measuring the spectral profile of the respective line.

We have to discuss the penetration of the 30 keV Li-beam into the plasma. With a total cross-section of $\sigma_{ex} \approx 10^{-14}$ cm^2 for the H$^+$-Li charge exchange process (cf. Table 3) and $n_e \approx 5 \times 10^{12}$ cm^{-3} at the plasma edge we obtain a characteristic attenuation length of the Li-beam of $(n \sigma_{cx})^{-1} \approx 20$ cm; i.e. a 10 % attenuation of the beam will occur after a distance of 2 cm. By using laser induced fluorescence the atom density can be measured locally and from the attenuation rate the proton density can be derived; the accuracy of the measurement will, however, probably not be good enough to obtain by subtraction of n_{H^+} from n_e the density of the electron excess which is due to ionization of impurity atoms.

V.2. Sources of atom particles

Of the experimental apparatus which is used in connection with the application of atom beam methods, the only part, which is not needed for the other techniques too, is the respective atom source. In the following we shall give a brief description of the three types of particle sources which may be especially useful for plasma edge diagnostics and discuss, in particular, those features which are important for the intended application.

V.2.1. Sources for thermal alkali atoms

In principle the source consists of an oven e.g. a cylinder closed at both ends, at temperature T and with a small hole of area A at one of the ends. At a certain value T the alkali metal inside the oven will have a vapour pressure p, which we can take from the vapour pressure curve. The number of particles per second N effusing into a solid angle $d\Omega$ at an angle ϑ with respect to the normal of A is /79/:

$$N(\vartheta)d\Omega = (n\upsilon A /4\pi) \cos\vartheta \, d\Omega \qquad (3)$$

The flux density Γ of particles on the axis of the aperture at a distance d is given by

$$\Gamma = n\upsilon A /4\pi d^2 \qquad (4)$$

Using the relations p = nkT and $\bar{\upsilon} = (8kT/\pi m)^{1/2}$ one obtains

$$\Gamma = (pA/\pi d^2)(2\pi m k T)^{1/2} = 1.115 \times 10^{22} pA/d^2 (MT)^{1/2} cm^{-2} s^{-1} \qquad (5)$$

where p is the pressure in torr, M is the molecular weight and T the temperature in K.

Let us consider a numerical example:
We want to produce a Li beam of density 10^9 cm^{-3} at d = 100 cm
with a source of area 1 cm^2. From the above formulas and from
the vapour pressure curve we then obtain:

$T \approx 8000$ K, $p \approx 10^{-2}$ Torr, $\bar{v} \approx 1.5 \times 10^5$ cm/s.

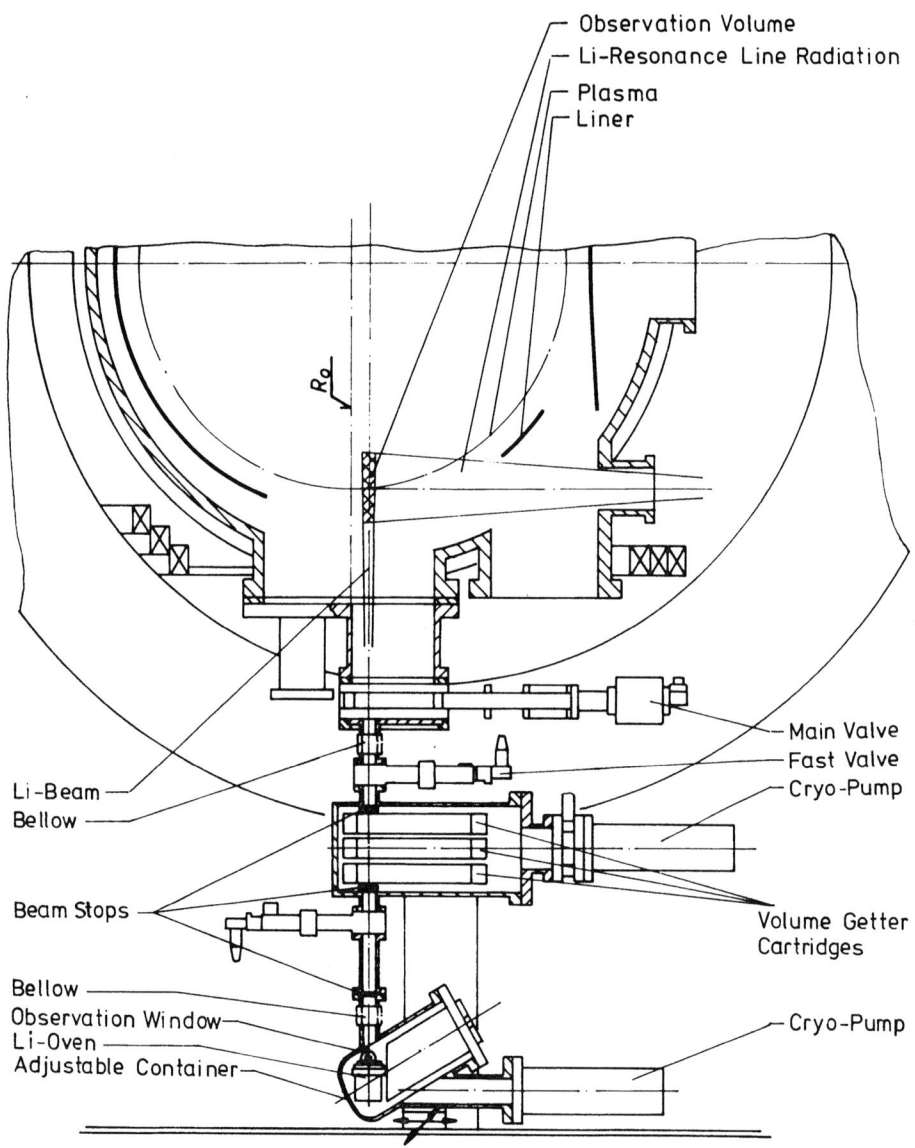

Fig. 23: Thermal Li-beam source and beam line for the measurement of n_e-profiles in the plasma edge of TEXTOR /80/.

Fig. 23 shows the experimental arrangement which has been used /80/ to inject a thermal Li-beam into TEXTOR. The adjustable Li-oven is mounted at a distance of about 140 cm from the observation volume. Special care had to be taken to ensure a good vacuum in the drift space, i.e. $p \lesssim 10^{-6}$ Torr. In particular, one has to avoid the increase of hydrogen density during a tokamak discharge, because this might result in a scattering of beam particles and therefore in a beam attenuation; according to /81/ the inequality $n_{H2} \times d < 4 \times 10^{14}$ cm^{-2} should be satisfied. By activating a fast valve lithium is injected into the tokamak vacuum vessel only during the duration of the discharge. If the lithium metal is consumed, the empty oven can without much delay be exchanged for a filled one.

V.2.2 Sources for atoms in the 10 eV range

Continuous high intensity beams of metal atoms in the energy range $E < 100$ eV are not available. Such beams can however be generated as single pulses of short duration /82/. The energy of the particles can only be varied within narrow limits, but it can be determined in a reliable and simple way by recording the time of flight of the atoms. The beams are generated by focussing the radiation of a high power, short pulse duration laser onto a thin metal film deposited on a glass plate. High energies and good reproducibility are obtained, when the laser strikes the metal film on the side facing the glass. Using e.g. a ruby laser pulse of 10 ns duration and choosing a film thickness between 0.1 μm and 1 μm an energy density of about 10 J/cm² is sufficient to produce metal atoms of the desired energy and density. For the atomic beams which are used for diagnostic purposes it is in general assumed that the neutral beam particles predominantly are atoms in the ground state. This prerequisite has to be checked carefully.

The theoretical description of the ablation process is not possible yet. For the right choice of the appropriate combination of film material and thickness and of the energy density of the laser pulse we have to rely on a data base obtained from experimental investigations. At present only a few such studies oriented towards diagnostic beams have been performed /83/.

For the measurement of n_e and T_e we should like to use beams of Li-, Al- and Ti-atoms, as discussed in V.1. It is desirable to generate beams of Li-atoms (for the n_e-measurement) and of Al- or Ti-atoms (for the T_e-measurement) simultaneously. With the laser ablation technique this can be done by providing film-sandwiches consisting of alternate layers of the two metals of interest or by producing appropriate alloys.

A preliminary study of these possibilities has been performed /84/. Fig. 24 shows the Li-density as a function of time obtained by laser ablation of a 1000 Å thick Li/Al film. The pulse

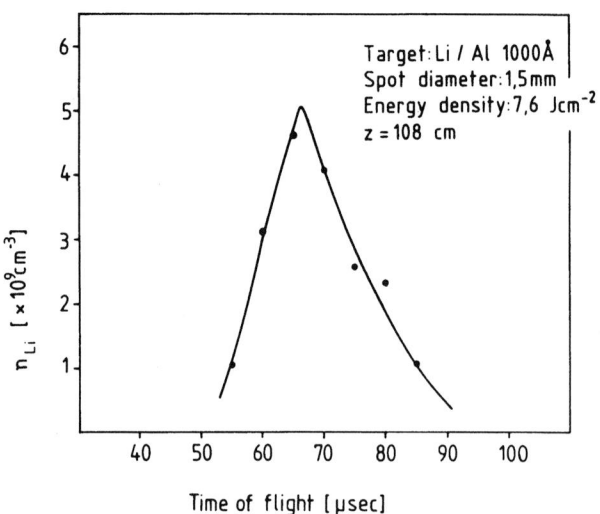

Fig. 24: Density of a Li-beam generated by laser ablation /84/.

is reproducible within 10 % . However, the density of Al-atoms is considerably lower than that of the Li-atoms. More detailed investigations of the dependence of the pulse properties on film composition and thickness are required.

Fig. 25 shows the experimental arrangement for the injection of two component metal beams and for their excitation and observation as it has been prepared for the TEXTOR tokamak /85/. Most of the figure is selfexplanatory. The target can be moved from outside in two directions, which are perpendicular to each other, by means of a manipulator. With a spot size of about 2 mm more than 500 pulses can be produced with one target plate.

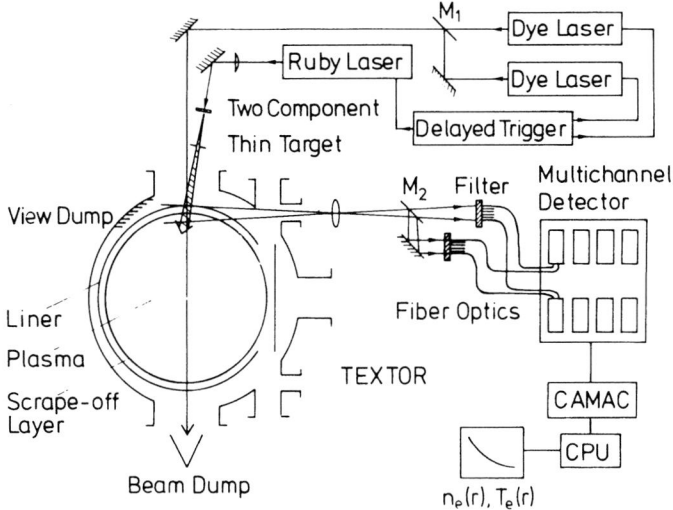

Fig. 25: Combination of atom beam probe and laser induced
fluorescence for the measurement of n_e- and T_e-profiles
in the plasma edge.

V.2.3. Injectors for high energy atoms

In order to utilize charge transfer processes for plasma edge
diagnostics we need beams of H^0- or Li-atoms at energies of the
order 10 keV and with flux densities $\approx 10^{17}/\text{cm}^2\text{s}$ (≈ 10 mA/cm²)
In case of hydrogen such injectors have been developed for plas-
ma heating purposes and are at our disposal.

The principles of a neutral hydrogen beam injector are the same
as those of the Li-beam injector shown in Fig. 26 /86/. By means
of an electrical discharge a hydrogen plasma is generated e.g.
at $n_e \approx 10^{12}$ cm⁻³ and $T_e \approx 10$ eV. The ions flowing to a perfora-
ted metal electrode are accelerated to the desired energy by
means of a potential difference applied to a second electrode. A
large fraction of these ions is neutralized in the charge exchange chamber, which contains a gas at sufficiently high pres-
sure such that $n_{gas} \cdot \ell \cdot \sigma_{cx} \approx 1$ holds, where n_{gas} is the density of
the gas particles, ℓ the length of the chamber, and σ_{cx} the
cross-section for charge exchange between the ions and the gas
particles (in case of protons normally H_2 is used). That fraction of the beam which is not neutralized is deflected by means
of a magnet and its energy is transferred to a beam dump.

259

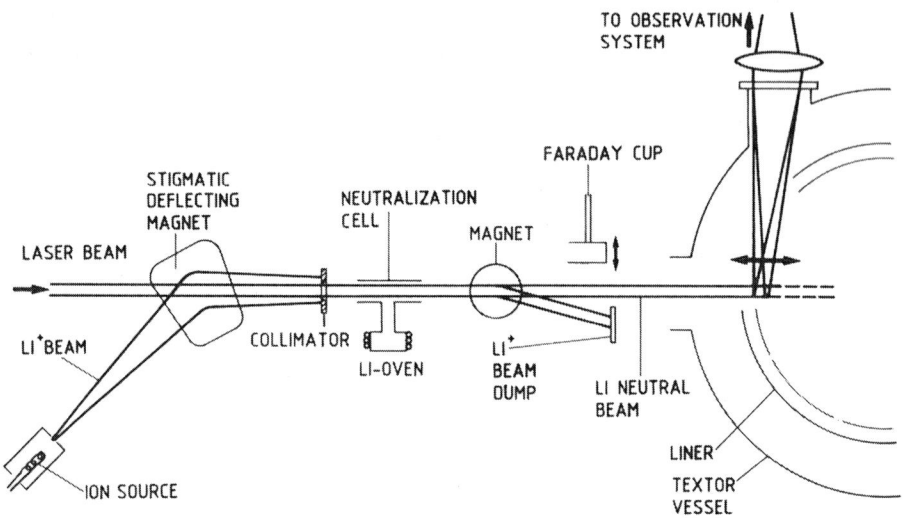

Fig. 26: Experimental set-up for 30 keV Li-neutral beam injector
at TEXTOR /86/.

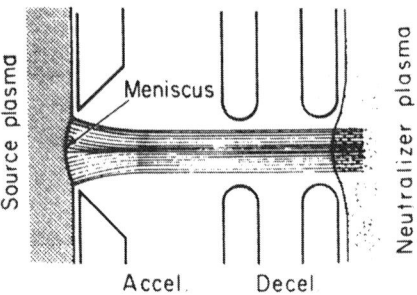

Fig. 27: Extractor system for the H^+-ion source.

The most critical component of the injector is the beam extraction and acceleration system. Fig. 27 shows the geometry of the
most commonly used electrode system, schematically. It is important to note that the ions are extracted from a surface - the
meniscus -, which is free to move under the influence of the

electric field used to extract the ions. The position and shape of the meniscus determines the ion trajectories, which should be parallel to each other and to the axis. The meniscus adjusts itself in such a way that the space charge limited current density is equal to the ion emission from the plasma

$$\dot{f} = \left(\frac{2e}{m_H}\right)^{1/2} \frac{(\Delta \phi_1)^{3/2}}{9 \pi d_1^2}$$ (6)

where $\Delta \phi_1$ is the potential difference and d_1 the distance between the first two electrodes.

For the relation to hold, the electric field must be zero at the plasma boundary and the net space charge density adjusts itself accordingly. d_1 should be small, but large enough to avoid electrical breakdown. The presence of electrons must be avoided in the space charge limited ion flow. To the right of the third electrode the beam has to be neutralized (e.g. by ionizaton of the remaining gas molecules) in order to prevent beam divergence due to space charge. The plasma electrons are prevented entering the acceleration gaps by giving the third electrode a potential slightly positive with respect to the second one. This combination of electrodes is called an "accel-decel" configuration (for a more detailed description and a review of neutral beam injection systems see /87/). For $\Delta \phi_1 \leq 30$ keV a current density up to 500 mA/cm^2 may be achieved. In order to obtain the large currents needed for heating the plasma in nuclear fusion devices electrode structures have been developed which provide a large number of closely spaced holes, each of which generates its own beamlet. For the construction of a high current 30 keV Li-atom injector the same principles can be applied as outlined above for a hydrogen injector. Only the ion source requires major modifications. In order to achieve the necessary vapour pressures of the metal which should be of the order of a mTorr, a high temperature oven containing the metal has to be introduced into the "plasma source". Condensation of the metal on insulating surfaces and on the grids has to be avoided. Such a source has recently been developed /88/. A special characteristic of this source is that the oven temperature (up to 1800 K) is lower than that of the other parts of the source, so that condensation of the vapour is avoided.

The GSI-ion source is part of the Li-beam-injector which will be connected to the TEXTOR tokamak (cf. Fig. 26). A current of 40 mA at a voltage of 30 kV has been achieved with it at a distance of 4 m. The ions will be neutralized in Na-vapour (for Li$^+$-Na, $\sigma_{cx} \approx 6 \times 10^{-15}$ cm^2 /89/, for Li$^+$ - Li, $\sigma_{cx} \approx 7 \times 10^{-15}$ cm^2 /90/). The current can be pulsed with a repetition frequency of 50 Hz. and a duty cycle of 10 %. A magnet provides mass separation and beam focussing. The bending of the beam by the magnet also makes it possible to bring in a laser beam coaxial with the Li-beam. This way the Li-beam intensity can be monitored locally.

V.3. Applications

First results have been reported on the utilization of a thermal Li-beam for the measurement of n_e-profiles /21/. The experimental arrangement shown in Fig. 23 was used.

The intensity emitted by the injected Li-atoms as a function of the radial position is shown in Fig. 28. The strong attenuation of the beam, when it crosses the scrape-off layer, is clearly seen. By application of Eq. III, 11, n_e has been derived as a function of the radius. These n_e-values conform well to those obtained with a HCN-interferometer, which give values up to r = 43 cm with about 30 % accuracy. At the magnetic surface tangent to the limiter, a value of $1.0 \times 10^{12}/cm^3$ is obtained. In the limiter shadow a fast decrease is observed, which follows approximately an exponential with a characteristic length of 1 cm. The simplification made in section III in going from Eq. 10 to Eq. 11 introduces errors only deep in the shadow of the limiter. At r = 48 cm probes indicated T_e = 6 eV; taking into account the proper rate coefficient we concluded that the density at r = 48 cm would have to be increased by 30 %. This error probably is not really relevant; the determination of the position of the observation volume with respect, e.g. to that of the limiter, may be more critical.

Fig. 28 also shows Li-emission signals as a function of time. They demonstrate that n_e at the plasma edge is constant in the time interval from 0.2 sec to 1.5 sec after onset of the discharge and then decreases, as indicated by the decrease of the signal at r = 46,4 cm - because of lower excitation rates - and by its increase at r = 45 cm - because the beam attenuation is reduced.

There is the question to what degree the electron density distribution measured at the position of the Li-beam is representative for the scrape-off layer as it extends along the magnetic field. It would be of interest to compare the two extreme cases; $n_e(r)$ close the the limiter and $n_e(r)$ close to the stagnation point of the plasma flow.

Such measurements have been performed /91/ following the procedure of using the lower or the upper limiter segment as main limiter as described in section II. Fig. 29 shows the results.

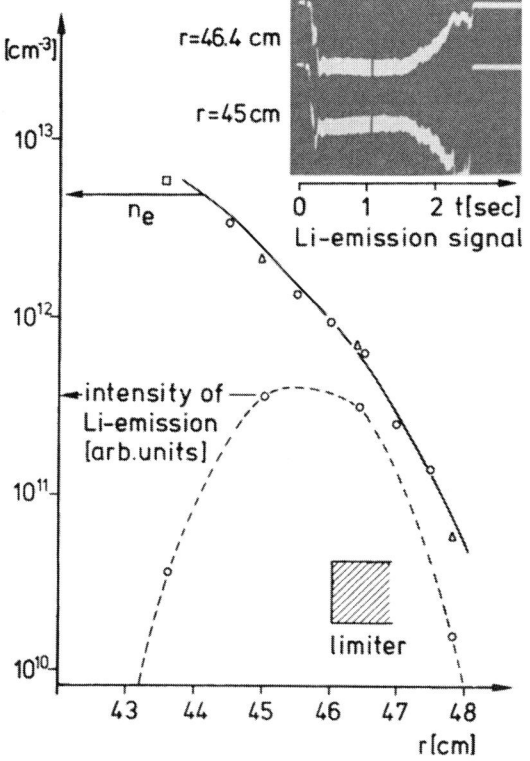

Fig. 28: Intensity profile from the Li-atom beam in the TEXTOR boundary layer together with the n_e-profile derived from it, and the photo-multiplier signals observed at r = 46.4 cm and 45 cm. The n_e-value at 43.5 cm (□) is from HCN interferometer measurements, (Δ) from measured Li-line intensity, (o) interpolated by means of the intensity profile /21/.

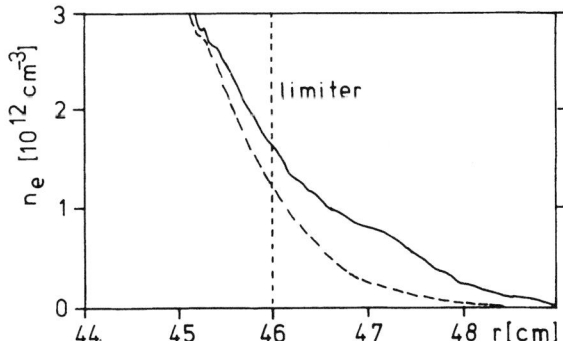

Fig. 29: n_e-profile measured with the thermal Li-atom beam and the CCD-camera. Curve (--) shows the profile for a short connection length to the limiter, curve (——) the profile for a long connection length /91/.

There are clear differences in the electron density profiles. Starting from the stagnation point the electron density in the scrape-off layer decreases towards the limiter and its characteristic decay length gets significantly shorter. According to the estimates reported by /102/ this behaviour is qualitatively what is expected. In particular, the flow velocity $v_{||}$ parallel to B should increase towards the limiter and the decay length, being proportional to $\sqrt{1/v_{||}}$ should decrease.

The spatial intensity profile of the Li-line, which is needed to derive $n_e(r)$, was obtained from the intensity at five spatial points having a separation of 1.4 cm. This space resolution is unsatisfactory. It was considerably improved by recording the radial intensity distribution with the CCD-camera described in section III. With this observation system one gets more than about 10 points/cm which is a number more suited to the problem.

Measurements of the density /92/, the velocity /93/, and the temperature /94/ of fully stripped oxygen, carbon, or helium ions have been reported, based on the excitation of these impurity ions by electron capture from injected hydrogen atoms. Here we are mainly interested in those results, which give information about the plasma edge, and shall discuss therefore the measurements of the radial distribution of the O^{8+} and C^{6+} concentrations which have been performed in the PDX Tokamak /92/.

A highly collimated H^0-beam was injected tangentially into the tokamak with an energy of 25 keV and a total power of 5 - 11 kW. The density of H^0-atoms was of the order of 10^8 cm^{-3}. The beam was scanned across the plasma midplane. The n = 3 to n = 2 transitions were observed by means of a calibrated grazing incidence spectrometer. Fig. 30 shows the observed radial profiles of O^{8+} and C^{6+}. Densities at the radius of the limiter are around 10^{11} cm^{-3} and decay only slowly in the shadow of the limiter, where n_e may be a few times 10^{12} cm^{-3} and T_e a few tens of eV. It is very likely that O^{8+} and C^{6+} ions at such high charge state and large densities after being accelerated by the space charge sheath in front of the limiter will play an important role in impurity generation at the limiter. Further investigations of these important processes are necessary.

By scanning the spectral profiles of suitable Doppler broadened or Doppler shifted lines of He^{++} and O^{8+}, which also were excited by electron capture from hydrogen, measurements of the temperature /94/ and of the drift velocity /93/ of these species have been performed. In case of the temperature measurement the main experimental problem is presented by the background radiation. To measure the drift velocity an accurate wavelength cali-

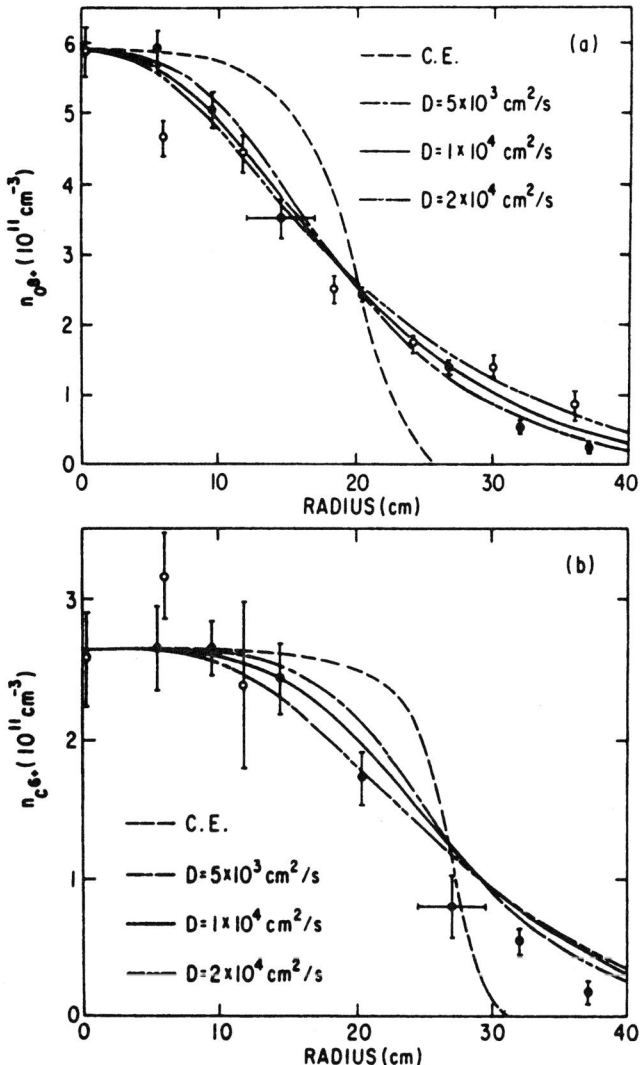

Fig. 30: Radial profiles of fully ionized oxygen and carbon
during the steady-state phase of the discharge. Limiter
radius is at 30 cm. The solid lines are profiles of C^{6+}
and O^{8+} calculated from an impurity transport code.
C.E. = distribution expected from coronal equlibrium
assuming a constant impurity density. D = constant
impurity diffusion coefficient /92/.

bration is required in addition. For this purpose reference lines in the VUV wavelength range are needed which can be provided by the tokamak discharge itself.

VI Other Methods

VI.1. Measurement of $n_e(r)$ and $T_e(r)$ by means of the Thomson scattering of laser light

1.1 Principles

Scattering of electromagnetic waves by electrons provides an excellent non-perturbing method to measure n_e and T_e; the space resolution is good and the interpretation of signals does not depend on the knowledge of plasma parameters. It is for these reasons that Thomson scattering has become the standard method to determine the profiles of n_e and T_e in the plasma core of tokamaks. Since it is difficult to provide the required high power laser radiation in a continuous or quasi-continuous way, such measurements are generally done for a single point in time during the discharge. A vast literature exists on this technique (e.g. /95-97/) and we shall therefore restrict our discussion of this method to a short introduction and to a few comments.

The elementary phenomenon, on which Thomson scattering is based, is the emission of a secondary wave by a free electron at rest, which is set into motion by the oscillating electric field of the primary electromagnetic wave of power P_i. The power P_s of the dipole radiation, emitted at the angle φ relative to the electric field vector E of the primary wave, into the solid angle $d\Omega$ is given in terms of the differential cross-section for Thomson-scattering $d\sigma_T/d\Omega$ by

$$P_s \, d\Omega = (d\sigma_T/d\Omega) \, P_i \, d\Omega \tag{1}$$

$$d\sigma_T/d\Omega = r_e^2 \sin^2\varphi \tag{2}$$

$$r_e^2 = (e^2/mc^2)^2 = 7.94 \times 10^{-26} \, cm^2$$

The total scattering cross-section, the so called Thomson cross section, is obtained by integrating over the entire solid angle

$$\sigma_T = (8\pi/3) \, r_e^2 = 6.65 \times 10^{-25} \, cm^2 \tag{3}$$

266

Considering now the scattering of a light wave by the statistically distributed electrons of a magnetic-field-free plasma in a volume of length l, the intensities of the waves emitted by the single scattering centres are simply added and the total power scattered into the solid angle $d\Omega$ is:

$$P_s \, d\Omega = n_e \, (d\sigma_T/d\Omega) \, P_i \, l \, d\Omega \qquad (4)$$

For order of magnitude evaluations one can use the mean value

$$P_s \, d\Omega = 6.65 \times 10^{-25} \, n_e \, l \, P_i \, d\Omega/4\pi \qquad (5)$$

As the electrons are in motion, the Doppler effect has to be considered. Let $\vec{K_i}$ and $\vec{K_s}$ be the wavevectors of the incident and of the scattered radiation, respectively. The scattering electron, having the velocity $\vec{v_e}$, will see a wave, the frequency of which is shifted by $\vec{K_i}\vec{v_e}$, while the frequency of the emitted wave will be shifted by

$$\Delta\omega = \vec{K_s} \, \vec{v_e} - \vec{K_i} \, \vec{v_e} \qquad (6)$$

With $|K_s| \approx |K_i|$ and with v_{K_e} being the component of $\vec{v_e}$ in the direction of $\vec{K} = \vec{K_i} - \vec{K_s}$ we obtain:

$$\Delta\omega = 2 \, v_{K_e} \, |K_i| \, \sin \Theta/2 \qquad (7)$$

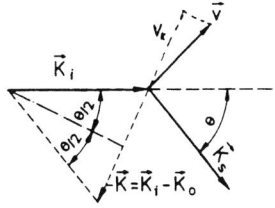

Fig. 31: Scattering geometry with moving electrons.

where Θ is the scattering angle (cf. Fig. 31). The profile of the scattered line will present the velocity distribution of the electrons in the direction of K; the factor $\sin \Theta/2$ causes a "compression" of the profile. In case of a Maxwellilan distribution the FWHM is given by:

$$\Delta\lambda = 4\lambda_i \left[2\ln 2 \ (kT_e/mc^2) \right]^{1/2} \sin \Theta/2 \qquad (8)$$

For the wavelength of the ruby laser and with $\Theta = 90°$, T_e can be calculated from

$$T_e \ [°K] = 11(\Delta\lambda)^2, \quad \Delta\lambda \ in \ Å$$

A magnetic field $\vec{B}$ does not affect the total scattering intensity. It will influence the spectrum of the scattered light significantly only if $\vec{K} \cdot \vec{B} \approx 0$. The spectrum then consists of single lines of frequency-width $k v_e \cos \vartheta$ and spacing $\omega_{ce} = eB/m_e c$ where ϑ is the angle between $\vec{k}$ and $\vec{B}$. If $K v_e \cos \vartheta / \omega_{ce} \gg 1$ the lines overlap and the envelope has a shape corresponding to the spectrum at B = 0.

From the absolute intensity of the scattered light the electron density can be determined. The calibration of the signal is usually accomplished by means of the Rayleigh-scattering from a suitable gas of known density (see section IV).

It is desirable to have an estimate of the scattered light intensity for given plasma conditions. We use eq. 4 and take $\varphi = \pi/2$. With N_i being the number of photons in the incident pulse, T the transmission of the optical system and η the quantum efficiency of the detector, we obtain for the number of detectable scattered photons N_s:

$$N_s \ d\Omega = N_i \ n_e \ r_e^2 \ \ell \ T\eta \ d\Omega$$

Assuming $\eta = 5\ \%$ (at $\lambda = 6943$ Å), T = 10 % and an effective aperture of the light collection system of f/5 one finds that

$$N_s = 1 \times 10^{-29} \ N_i \ n_e \ \ell$$

Assuming further that at least 100 photons are needed to obtain a reasonable signal to noise ratio, a condition for N_i follows:

$$N_i \ n_e \ \ell \approx 1 \times 10^{31}/cm^2$$

In case of a ruby laser (λ = 6943 $\overset{\circ}{A}$) of 10 J energy this leads to

$$n_e \ell \approx 3 \times 10^{11} \, cm^2$$

For a more reliable estimate of the detection limit we would have to consider the noise due to the continuum radiation from the background plasma. Since such estimates, however, are rather unreliable we shall not dwell on it.

VI. 1.2 Applications

Up to now only a few results are reported on the application of Thomson scattering to the plasma edge. For the study of the plasma boundary layer in ASDEX a diagnostic system has been built which was designed specifically for this purpose /98/. Fig. 32 shows the experimental arrangement schematically and gives a summary of the main technical data. To achieve a low detection limit a laser providing a light pulse of high energy

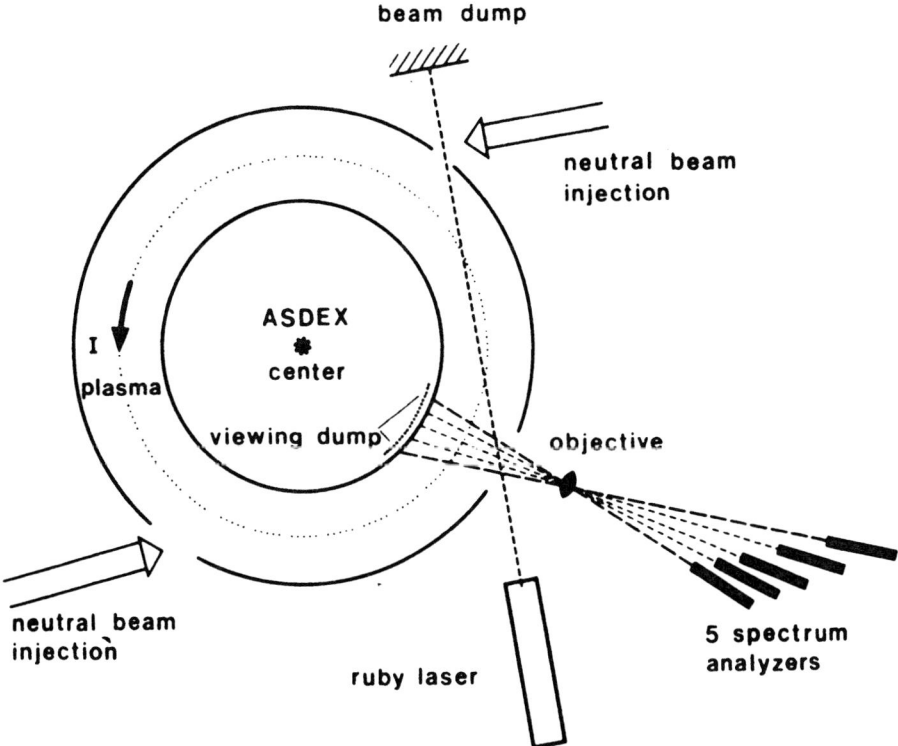

Fig. 32: Schematic set-up of the Thomson scattering system in the plasma boundary of ASDEX. Laser 6943 $\overset{\circ}{A}$, 20 ns, 12 J,$\leq$1 mrad /98/.

content (12 J) and with low beam divergence ($\lesssim 1$ mrad) and a light collection optics of large aperture (f/3.5) have been foreseen. Over the extension of the observation volume (6 cm) a beam diameter of 1 mm is achieved with the aim to keep the background radiation low. The line of observation, on the other hand, crosses the plasma core and the observation system collects the radiation emitted by the 0.8 m plasma column. Fluctuations of the background radiation yield a detection limit for the electron density of about 5×10^{11} cm^{-3}; the electron temperature can be determined in the range 2 eV $\leq T_e \leq$ 1000 eV. Radial profiles are obtained by measuring the scattered light emitted from 5 spatial points separated by about 1 cm.

Fig. 33 shows n_e- and T_e-profiles near the separatrix obtained from several discharges with Ohmic heating only and with slightly different radial positions of the separatrix, to improve the spatial resolution. From the profiles one infers characteristic decay lengths of about 1 cm for T_e and 1.5 cm for n_e, i.e. of the order of the spatial resolution.

The results demonstrate above all that a reliable, space-resolved measurement of the electron temperature is possible, even at the low densities of the plasma boundary. Other methods which have been used or proposed for the measurement of T_e require a number of assumptions to be fulfilled and need confirmation by a more direct method like Thomson scattering.

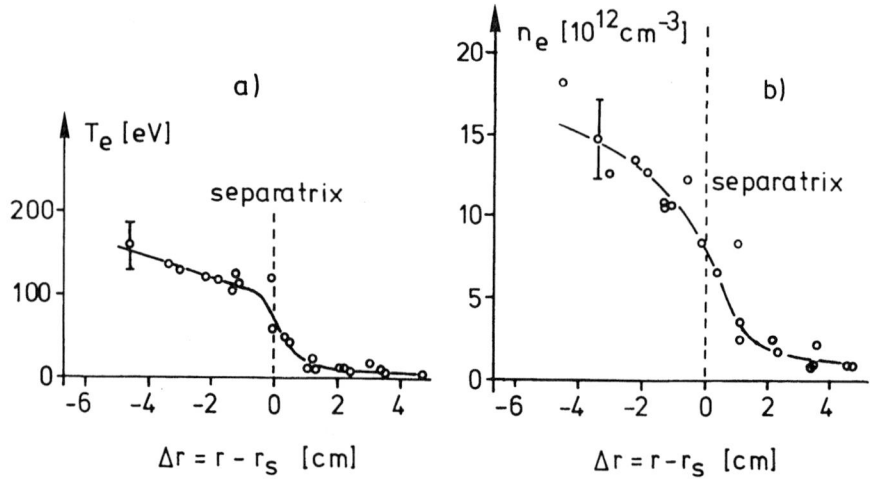

Fig. 33: Electron temperature (a) and density (b) profiles versus distance from the magnetic separatrix in the equatorial plane for an ohmically heated discharge in ASDEX /98/. $n_e = 4 \times 10^{13}$ cm^{-3}, t = 1.05 s, I = 320 kA, B = 2.17 T.

VI.2. Infrared Thermography

The measurement of the temperature distribution $T_S(x,y)$ on the surface of limiters and divertor plates serves several purposes:

- first of all $T_S(x,y)$ characterizes the surface state and determines the rates at which surface processes like evaporation, desorption, and chemical reactions take place.

- from $T_S(t)$ the local heat flux density to the limiter or to the divertor plate $Q_L(x,y)$ and - taking into account the shape of the surface - the heat flux density distribution in the scrape-off layer can be derived.

- Taking the integral of $Q_L(x,y,t)$ over the surface and over a time interval, the energy deposition by the plasma on the limiter is obtained, which is required for an analysis of the energy balance in the plasma.

VI. 2.1 Principles

The basis for thermography is given by Kirchoff's law

$$E(\lambda,T) = K(\lambda,T)\ B(\lambda,T) \tag{11}$$

and Planck's law

$$B(\lambda,T) = (2hc^2/\lambda^5)\ [exp(hc/\lambda kT)-1]^{-1} \tag{12}$$

Here $E(\lambda,T)$ is the emission coefficient which is the power emitted per unit area, wavelength and solid angle and $K(\lambda,T)$ the absorption coefficient. Since $K(\lambda,T)$ depends on the surface structure, it has to be measured in each experiment. It normally depends only weakly on temperature. To do a reliable and accurate T-measurement a good combination of T and λ has to be chosen. At short λ, the plasma background radiation usually dominates; at long λ, the detectors are expensive; furthermore, due to the high reflectivity of metals at these wavelengths, the absorption coefficient is low and very sensitive to surface changes, and $B(\lambda,T) \sim T$, i.e. measurements are not very sensitive to T-changes. Therefore, a range $0.8\,\mu < \lambda < 3\ \mu$ is preferable. Since low temperatures are more difficult to measure, the limiter or the wall are often preheated.

The derivation of the power load from the observed surface temperature rise ΔT_L is in general a non-trivial task. A simple solution is found only in the case of plane geometry, of a constant
power load Q_L switched on at t = 0, and of the thermal properties of the material being constant. The one dimensional thermal
diffusion equation has then the solution /99/

$$\Delta T_L = (Q_L/K)\,(\chi/\pi)^{1/2}\,t^{1/2} \tag{13}$$

which allows the evaluation of Q_L from temperature measurements,
if the heat conductivity χ and the diffusivity k are known. In
case the above mentioned prerequisites are not fulfilled, numerical methods have to be used to obtain $Q_L(t)$ from $T_S(t)$.

VI.2.2 Applications

We want to record a two dimensional light intensity distribution
in the infrared spectral region in order to derive the temperature distribution by which it is generated. Among the commercially available camera systems two have found broader application
for measuring the surface temperature distribution on limiters.
They differ in the way they operate rather fundamentally: One
system uses a single element detector - selected and cooled
according to requirements - and the incoming radiation is
scanned sequentially across the detector by using mechanical
scanners; in the other system a matrix array of detectors
collects the light and transforms the intensity distribution
into an electric charge distribution. The latter is rapidly
shifted into a storage system, from which it is read out
serially, while the detector array is recording the next image.

While in principle the second system offers the advantages of
longer integration time and larger number of picture elements,
it is presently available only with silicon diodes, which are
sensitive in the wavelength region $\lambda < 1$ μ. Its application is
therefore limited to temperatures T > 700 K. Table 4 presents a
survey on those properties of these types of cameras which are
of special interest for the space resolved measurement of
limiter temperatures.

Infrared thermography has been applied successfully to measure
the heat flux density in the scrape-off layer of tokamaks (e.g.
/100/, see also the lecture by Ulrickson, this conference). We
shall discuss the measurements performed on the TEXTOR tokamak
/101, 2/. A test limiter (No 3 of Fig. 1) has been moved into

272

Table 4: Survey on the properties of two commercially available
IR-cameras suitable for the measurement of limiter temperatures.

Observation wavelength	1	5	μ
Detector	Si-diode	InSb-diode	
Detector temperature	25	-196	$^{\circ}$C
Scanning	electronically CCD	mechanically rotating prism	
Manufacturer	Fairchild	AGA	
Repetition rate	30	6.25	Hz
Lines per field/frame	244/488	70/280	
Integration time	33	0.005	ms
Number of resolved points per line	380	100	
Element size	12 x 15 μ^2	100 μ ∅	
Irradiance for signal to noise ratio = 1	1	5	mW/m^2
Aperture of the camera	1 : 0.95	1 : 1.8	
Spectral bandwidth	0.1	1	μ
Spectral radiance for S/N = 1	12	20	$mW/m^2\mu$ sr
Blackbody temperature for S/N = 1	350	-75	$^{\circ}$C
Emissivity of nickel	30	10	%
Nickel temperature for S/N = 1	385	-35	$^{\circ}$C
Plasma bremsstrahlung ($Z=4$, $l=1$, $kT_e=1$ keV, $n_e=2\cdot10^{13}/cm^3$)	320	15	$mW/m^2\mu$ sr
Blackbody temperature of the plasma	450	-80	$^{\circ}$C
Blackbody radiation of 800 $^{\circ}$C corresponds to Ni-radiation at	905	3060	$^{\circ}$C
Sensitivity limit by	plasma radiation	detector noise	

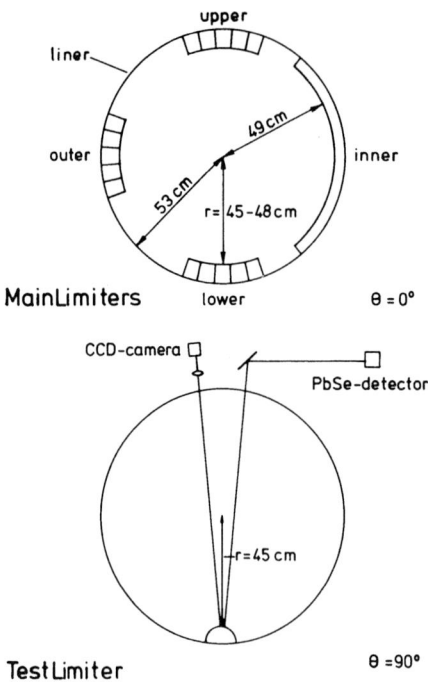

Fig. 34: Poloidal cross-section of the infrared observation
system of the test limiter /101/.

the edge plasma. This limiter, which is preheated to 400 C, can
be observed with a CCD-camera or with a PbSe element, operated
in the photoconductive mode, cf. Fig. 34. The camera serves to
obtain a view of the total surface (at $\lambda \approx 1 \mu$), with high space
resolution (0.5 mm), but moderate time (20 ms) and temperature
resolution (≈ 3 K). The PbSe detector is sensitive in the 3.0 –
3.4 μ range and can resolve temperature changes of 1 K with a
time resolution in the order of 100 μs. For calibration of the
detectors, the test limiter temperature has been measured with a
thermocouple between shots, when a uniform temperature distribution over the test limiter body was reached. The absorption
coefficient was assumed to be temperature independent.

With an exponential decay of the energy flux density $Q_L(r) = F_0 \exp[-(r-a)/d]$ the load on the limiter surface is given by

$$Q_L(r) = F_0 \cos \alpha(r) \exp\left[-(r-a)/d\right] \tag{14}$$

274

Here $\alpha(r)$ is the angle between the magnetic field lines and the normal to the limiter surface. $Q_L(r)$ has two maxima, which are also clearly seen in the experimental results shown in Fig. 35.

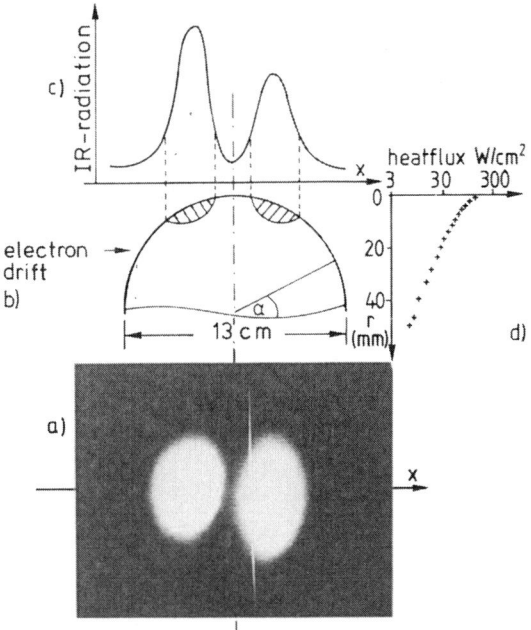

Fig. 35: a) Typical infrared image of the test limiter,
b) side view of the test limiter,
c) line scan of the IR-signal
d) heat flux at the test limiter /101/.

When the main limiters were drawn back and the plasma was shifted against the test limiter, a constant and relatively high power load was obtained, which results in an increase of ΔT_L proportional to $\sqrt{t}$, as expected from eq. 13. Normally the total energy fluxes flowing to the limiters of TEXTOR were only about 10 % of the total power.

References

1. E. Hintz, P. Bogen, J. Nucl. Mater. 128 & 129 (1984) 229
2. H. Soltwisch et al., Plasma Physics and Controlled
    Fusion, 26, 23 (1984).
3. I.I. Sobelman, L.A. Vainshtein, E.A. Yukov, "Excitation
    of Atoms and Broadening of Spectral Lines", Springer
    Series in Chemical Physics 7, 1981.
4. M. Mattioli, Report EUR-CEA-FC-761, Febr. 1975
5. H.S.W. Massey, E.H.S. Burhop and H.B. Gilbody:
    Electronic and Ionic Impact Phenomena,
    Vol. I "Collisions of Electrons with Atoms",
    Oxford, Clarendon Press (1969)
    H.S.W. Massey, E.W. Mc Daniel, B. Bederson:
    Applied Atomic Collision Physics, Vol. 2,"Controlled
    Fusion", Academic Press, New York 1984
6. D. Leep and A. Gallagher, Phys. Rev. A, 10, 1082 (1974)
7. S.T. Chen and A. Gallagher, Phys. Rev. A, 14, 593 (1976)
8. H. van Regemorter, Astrophys. J. 136, 906 (1962)
9. K.L. Bell, H.B. Gilbody, J.G. Hughes, A.E. Kinston, and
    F.J. Smith, J. Phys. Chem. Ref. Data 12, 891 (1983)
    Atomic Data for Controlled Fusion Research, Oak Ridge
    Nat. Lab., ORNL-5206 and 5207 (1977)
10. W. Lotz, Z. Phys. 20, 434 (1969)
11. L.C. Johnson and E. Hinnov, J. Quant. Spec Rad. Transf.
    13, 33, (1973)
12. H.W. Drawin, F. Emard, Z. Naturforschg. 28a, 1422 (1973)
13. M.W. Thompson, Phil. Mag. 18, 383 (1969)
14. R. Behrisch, ed. Sputtering by particle bombardment
    Springer Verlag, Berlin - Heidelberg (1981)
15. B. Schweer, Dissertation, Bochum 1983, Berichte der KFA
    Jülich, Nr. 1876 (1983)
16. E. Hintz, D. Rusbüldt, B. Schweer, J. Bohdansky, J. Roth
    and A.P. Martinelli, J. Nucl. Mater. 93 & 94, 463
    (1980)
17. A.C.G. Mitchell, M.W. Zemansky, "Resonance Radiation and
    Excited Atoms", Cambridge University Press, 1961.
18. A. Unsöld "Physik der Sternatmosphären", Springer
    Verlag, Berlin, 1952
19. W. Lochte-Holtgreven, ed. "Plasma Diagnostics",
    North-Holland Publ. Comp, Amsterdam, 1968.
20. VTE-Digitalvideo, D33 Braunschweig, Germany
    Image processor MBV.
21. A. Pospieszczyk, P. Bogen, U. Samm
    11th Europ. Conf. Controlled Fus. Plasma Phys., Aachen,
    Sept. 1983, Vol. II, 417
    P. Bogen, H. Hartwig, E. Hintz, K. Hoethker, Y.T. Lie,
    A. Pospieszczyk, U. Samm, W. Bieger, J. Nucl. Mater.
    128 & 129 (1984) 157

22. K. Behringer et al. 11th Europ. Conf. Controlled Fusion Plasma Phys., Aachen, 1983, Vol. II, 467.
23. V.V. Melnikov and Yu. M. Smirnov, Opt. Spectrosc. (USSR) 52, 362 (1982)
24. W. Demtröder, "Laser Spectroscopy", Springer Verlag, Berlin (1981)
25. P. Bogen and E. Hintz, Comments Plasma Phys. Controlled Fusion 4, 115 (1978)
26. E.U. Condon and G.H. Shortley, "The Theory of Atomic Spectra "Cambridge, University Press 1957
27. P. Bogen, R.W. Dreyfus, H. Langer, and Y.T. Lie, J. Nucl. Mater. 111 & 112 (1982) 75
28. D. W. Koopmann, T.J. McIlrath, and V.P. Myerscough, J. Quant. Spectrosc. Radiat. Transfer 19, 555 (1978)
29. E. Dullni, P. Bogen, E. Hintz, D. Rusbüldt, B. Schweer, S. Goto, and K.H. Steuer, Phys, Lett. 88A (1982) 40; J. Nucl. Mater. 111 & 112, (1982) 67
31. J. Raeder, C.H. Corliss, W.L. Wiese, and G.A. Martin, Wavelengths and transition probabilities for atoms and atomic ions, Nat. Bur. Stand, (USA), 1980
32. M. Hamamoto et. al., Jap. J. Appl. Phys. 20 (1981) 1709
33. A. Elbern, Appl. Phys. 15 (1978) 111, Dissertation Bochum 1976
34. B. Schweer and H.L. Bay, Appl. Phys. A 29, 53 (1982)
35. H.L. Bay, B. Schweer, P. Bogen, and E. Hintz, J. Nucl. Mater. 111 & 112, (1982) 732
36. M.J. Pellin, C.E. Konsy, M.H. Mendelsohn D.M. Gruen, R.B. Wright and A.B. De Wald, J. Nucl. Mater. 111 & 112 (1982), 738
37. M.J. Pellin, L.E. Young, W.F. Calaway, and D.M. Gruen, Surface Science, to be published
38. Ph. Mertens and P. Bogen, J. Nucl. Mater., 128 & 129 (1984) 551
39. B. Schweer, D. Rusbüldt, E. Hintz, J.B. Roberto, and W.R. Husinsky, J. Nucl. Mater, 93 & 94 (1980) 357
40. C.H. Muller, and K.H. Burrell, Phys. Rev. Lett. 47, 330 (1980)
41. B. Schweer, P. Bogen, E. Hintz, D. Rusbüldt, S. Goto, and K.H. Steuer, J. Nucl. Mater. 111 & 112 (1982) 71
42. E. Dullni, E. Hintz, J.B. Roberto and R.J. Colchin, J. Nucl. Mater., 111 & 112 (1982) 61
43. C.H. Muller, D.R. Eames, K.H. Burell, and S.C. Bates, J. Nucl. Mater., 111 & 112 (1982) 56
44. G.T. Razdobarin et al., Nucl. Fusion 19, 1439 (1979)
45. J. Hackmann, C. Gillet, G. Reinhold, G. Ritter, and J. Uhlenbusch, J. Nucl. Mater., 111 & 112 (1982), 221. G. Reinhold, Dissertation, Universität Düsseldorf, 1984

46. J.B. Roberto, R.E. Clausing, E. Dullni, L.C. Emerson, L. Heatherly, B. Schweer, S.P. Withrow, and R.A. Zuhr, J. Vac. Sci. Technol. A1, 929 (1983)

47. S.A. Moshkalev, G.T. Razdobarin, U.V. Semenov, private communication

48. S. Marlier, H. Ringler, B. Schweer, Verhandl. DPG VI, 19, 1235 (1984)

49. H.L. Bay and B. Schweer, J. Nucl. Mater., 128 & 129 (1984) 257

50. V.S. Burakov et. al., JETP Lett. 26, 403 (1977)

51. R. Mahon, T.J. McIlrath, and D.W. Koopman, Appl. Phys. Lett. 33, 305 (1978)

52. D. Cotter, Opt. comm. 31, 397 (1979)

53. R. Wallenstein, Opt. Comm. 33, 119 (1980)

54. H. Langer, Dissertation München 1980, H. Langer, H. Puell, and H. Röhr, Opt. Comm. 34 (1980) 134

55. M. Maeda et. al. J. Nucl. Mater., 111 & 112 (1982) 95

56. Lambda Physik, D-34 Göttingen, Model EMG 201 and FL 2002

57. M. Keilhacker et.al. in: Proc. 8th Intern. Conf. on Plasma Physics and Controlled Nuclear Fusion Research, Brussels, 1980, Vol. II (IAEA, Vienna, 1981) p. 351

58. R.W. Dreyfus, P. Bogen, H. Langer, AIP Conf. Proc., "Laser Techniques for Extreme Ultraviolet Spectroscopy" (Boulder, 1982)

59. P. Bogen, Y.T. Lie, Appl. Phys. 16, 139 (1978)

60. H. Schomburg, H.F. Döbele and B. Rückle, Appl. Phys. B28, (1982) 201
    H.F. Döbele, and B. Rückle, J. Nucl. Mater., 111 & 112 (1982) 102

61. R. Hilbig, and R. Wallenstein, Appl. Optics 21, 913 (1982)

62. R. Hilbig, and R. Wallenstein, IEEE J. Quantum Electronics 19, 194 (1983)

63. T.A. Miller, Plasma Chemistry and Plasma Processing, 1, 3, (1981)

64. A. Kislyakov and L. Krapnik, Sov. J. Plasma Phys., 7, 478, (1981)

65. D.E. Post, "Particle Diagnostics of Magnetic Fusion Experiments" in: Proc. Nato Advanced Summer Inst. Atomic and Molec. Processes, Contr. Thermonucl. Fus., S. Flavia, 1982

66. L.L. Shimon et al., Sov. Phys. Tech. Phys. 20, 434 (1975)

67. K. McCormick, IPP Rep. III/85 (1983), K. McCormick et al., J. Nucl. Mater., 123 (1984)

68. K. McCormick, IPP Rep. III, 40 (1978)

69. D.E. Voss and S.A. Cohen, J. Nucl. Mater. 93 & 94, 405 (1980)

70. R.L. Freeman and E.M. Jones, Culham Lab. Rep., CLM - R 137 (1974)

71. H.B. Gilbody, Physica Scripta, 23, 143 (1981)

72. A.N. Zino'ev et al., JETP Lett., 32, 539 (1980)
    R.C. Isler et al., Phys. Rev. A, 24, 2701 (1981)
73. H. Winter, Comments At. Mol. Phys., 12, 165, (1982)
74. W. Gruebler et al., Helv. Phys. Acta, 43, 254 (1970)
75. R.W. McCullough et al., J. Phys. B, 15, 111 (1982)
76. D. Dijkhamp et al., J. Phys. B, At. Mol. Phys., 16, 343
    (1983)
77. K. Kadota et al., Phys. Lett., 88A, 135, (1982)
78. A. Brazuk et al., Phys. Lett., 101A, 139, (1984)
79. H. Lew, "Molecular Effusion", in: Atomic Sources and
    Detectors, Methods of Exp. Phys., Vol. 4 A, Acad.
    Press, New York, 1967
80. W. Bieger and K. Höthker, private communication
81. W. Bieger, K. Höthker, Y.T. Lie, private communication
82. J.F. Früchtenicht, Rev. Sci. Instr. 45, 51 (1974)
    E.S. Marmar, J.L. Cecchi, S.A. Cohen, Rev. Sci. Instr.
    46, 1149 (1975)
    D. Manos et al., J. Vac. Sci. Techn. 20, 1230 (1982)
83. Y.T. Lie, A. Pospieszczyk, J.A. Tagle, IX. Intern. Vac.
    Congress, V. Intern. Conf. Solid Surf., Madrid, Sept.
    26 - Oct 1, 1983, to be published in Nucl. Technology/
    Fusion 1984
84. R. Koppmann, A. Pospieszczyk, priv. communication,
85. K. Kadota, A. Pospieszczyk, P. Bogen, E. Hintz, Plasma
    Science 12 (1984) 264
86. H.L. Bay, priv. communication
87. W.B. Kunkel, "Neutral Beam Injection", in: "Fusion"
    Vol. 1B, E. Teller, ed., Academic Press, New York 1981
88. R. Keller, Nucl. Instr. Meth., 189, 97 (1981)
    R. Keller, Symposium on Acceleration Aspects of Heavy
    Ion Fusion, GSI, Darmstadt, (1982)
89. L.J. Pivovar, L.J. Nikolaychuc, A.N. Grigoriev, JETP,
    30, 236 (1970)
90. J. Percel et al., Phys. Rev. Lett., 23, 677 (1969)
91. P. Bogen, H. Hartwig, E. Hintz, K. Höthker, Y.T. Lie,
    A. Pospieszczyk, and U. Samm, private communication
92. R.J. Fonck et al., Phys. Rev. Lett., 49, 737 (1982)
93. R.C. Isler and L.E. Murray, Appl. Phys. Lett., 42, 355
    (1983)
94. R.J. Fonck et al., Appl. Phys. Lett., 42, 239 (1983)
95. H.J. Kunze "The laser as a tool for plasma diagnostics"
    in "Plasma Diagnostics", W. Lochte-Holtgreven, editor,
    North-Holland Publ., Amsterdam 1968
96. S.A. Ramsden, "Light scattering experiments" in "Physics
    of Hot Plasmas", B.J. Ryle and R.C. Taylor, editors,
    Oliver and Boyd, Edinburgh, 1970.
97. A.W. De Silva and G.C. Goldenbaum, "Plasma diagnostics
    by light scattering," chapter 3 in "Methods of
    Experimental Physics", Vol. 9, part A, H.R. Griem,
    editor, Academic Press, New York, 1970

98. H. Murmann, M. Huang, IPP-Report, III/95, 1983
99. H.S. Carslaw, J.C. Jaeger, Conduction of Heat in Solids, Oxford University Press, 1959
100. TFR Group, Report EUR-CEA-FC 1114 (1981)
101. U. Samm, 11th Europ. Conf. Contr. Fusion and Plasma Physics, Aachen 1983, Vol. II, 413
102. M.F.A. Harrison et al., Culham Laboratory, Report, CLM-R211, 1981
103. E. Hechtl, J. Bohdansky, and J. Roth, J. Nucl. Mater., 103 & 104 (1981), 333.

280

ATOMIC AND MOLECULAR COLLISIONS IN

THE PLASMA BOUNDARY

M.F.A. Harrison

UKAEA/Euratom Fusion Association

Culham Laboratory, Abingdon, Oxon OX14 3DB, England

ABSTRACT

The objective of this paper is to provide an introduction to those aspects of atomic collision physics which underly the unavoidably generalised base of cross section data and scaling relationships which is currently employed in plasma modelling. Both experimental and theoretical methods are outlined and, where practicable, general trends in collisional behaviour are illustrated by examples of measured data. Atomic and molecular processes are considered on the basis of their particular relevance to the plasma edge region so that the discussion emphasises the properties of collisions in the regimes of low plasma temperature and low charge state of impurity ions. Nevertheless the basic concepts apply with equal validity throughout the plasma. Particular attention is devoted to recycling of hydrogen atoms and molecules because of its powerful influence upon plasma properties adjacent to boundary surfaces. References are selected with the objective of providing easy access to detailed reviews on topics which perforce cannot be included in this brief account. The method of presentation is firstly to discuss the general roles of atomic and molecular collisions in the plasma edge, then to identify the types of collision involved and subsequently to describe the methods adopted to calculate or measure the relevant cross sections. Cross section data are introduced in increasing order of the complexity of their atomic interactions. Finally the influence of the plasma environment upon atomic collision rates is discussed.

1. INTRODUCTION

Radiative power losses arising from collisions of hot plasma electrons with impurity ions are important consequences of atomic collisions in fusion plasmas. So also are power losses and impurity release subsequent to charge exchange collisions between protons and hydrogen* atoms. However, until recently, emphasis has been placed upon the effects of atomic interactions within the hot core of a magnetically confined plasma and relatively little attention has been devoted to atomic and molecular collisions which occur in the region close to the boundary surface of the confinement vessel. The need to control impurity release in high power, long duration experiments coupled to the interest in divertors and pumped-limiters, for both experiments and reactor concepts, has stimulated studies of the boundary plasma and of the atomic processes which are important in this region.

The residence time for plasma particles within a confinement device must of course be finite so that the surface of the vessel is inevitably bombarded by plasma ions and electrons. Transport of plasma particles within the boundary is predominantly in the direction of the magnetic field so that the flux of escaping charged particles is strongly peaked at the divertor target or limiter plate. In the regime of present interest, incident ions are neutralised at the surface as a consequence of ion-surface interactions and (depending upon the ion energy together with the atomic species of both ion and surface) a substantial fraction of the incident ion flux can return as energetic backscattered atoms. In fusion relevant plasmas the predominant ion is a proton and the predominant backscattered particle a hydrogen atom. In steady state conditions there is conservation of particles so that those protons which do not contribute to backscattering are re-emitted as low energy detrapped neutrals which tend to be hydrogen molecules whose kinetic energy corresponds to the surface temperature. These neutral hydrogen particles traverse the plasma sheath which is collisionless and enter the boundary plasma which, in many envisaged and existing devices, is sufficiently dense and hot for ionisation to occur in the close proximity of the surface. This gives rise to a high degree of localised recycling of hydrogen plasma to the surface.

Localised recycling adjacent to the plasma collection surfaces enhances the fluxes of electrons and ions which are available to convect energy across the plasma sheath and to the surface. Depending upon the degree of plasma collisionality

---

* It is implicit, unless stated otherwise, that the discussion applies equally to all isotopes of hydrogen.

within the recycling region, this enhancement of particle fluxes ensures that a powerful flow of energy can reach the boundary surface without incurring the penalty of a high sheath temperature and sheath potential. The energy of ions incident upon the surface and the consequent yield of impurity atoms sputtered from the surface are thereby reduced. The boundary plasma drifts in the direction of the magnetic field and its maximum velocity (at the plasma sheath edge) is about equal to the ion sound speed. Localisation of sources of ionised hydrogen due to recycling in this downstream region results in a low flow velocity in the plasma upstream of the collection surface and a rapid acceleration of the flow within the recycling region. This spatial distribution of drift velocity impacts substantially upon the ability of the drifting boundary plasma to entrain ions and thereby to sweep them to the plasma collection surface. Impurity ion transport within the boundary is affected, the present understanding being that impurities which are ionised within the recycling region will be swept to the plasma collection surface but that this beneficial action is less likely in the upstream regions of the boundary plasma.

In addition to its effects upon plasma particle transport, hydrogen recycling provides a powerful local sink for plasma electron energy. Not only is electron energy dissipated by ionisation (although this energy is subsequently returned to the plasma collection surface in the form of the potential energy carried by the incident protons) but energy is also lost by excitation of neutral hydrogen. The plasma is transparent to most atomic radiation and hydrogen can radiate powerfully in the low temperature, high density, recycling region. Charge transfer between low temperature protons and hydrogen atoms affects the distribution of plasma ion energy and also the transport properties of the plasma ions both along and across the magnetic field. Indeed the influence of 'atomic and molecular processes is so substantial that in present high recycling divertor experiments (e.g. ASDEX) most of the energy entering the divertor is dissipated by atomic and molecular processes and only a small fraction is carried to the divertor target by charged particles.

The previous discussion has emphasised the role of localised recycling of hydrogen caused by plasma impact upon a boundary surface but similar atomic processes are relevant to issues associated with fuelling by gas puffing and, albeit with somewhat different emphasis, to fuelling by pellet injection.

Impurity atoms present in the boundary are also subjected to collisions with the charged particles of the plasma. In the case of helium the most significant effect is upon the gas exhaust capabilities of a reactor. Ionisation within the boundary plasma causes helium to recycle to the plasma collection surface in a

manner somewhat comparable to hydrogen so that the plasma acts as
a powerful pump for gas. This pumping action opposes that of the
vacuum pumps which must perforce be placed at the wall of the
reactor in order to exhaust the helium.

Heavier impurity elements can be present in the plasma due to
sputtering of the boundary surfaces and interest ranges from low
atomic number elements such as beryllium and carbon through medium
number elements such as silicon and iron to the high atomic number
refractory metals such as tungsten. In addition, gaseous
impurities such as oxygen are frequently encountered in experiments and it is conceivable that noble gases such as argon may be
deliberately injected in order to cool the boundary plasma. The
radiating ability of impurities increases markedly with increasing
atomic number. It is also strongly related to the distribution of
charge states amongst the impurity ions which is itself influenced
by the residence time of these ions within the plasma. Both the
energy with which impurity ions impact upon the plasma collection
surface and the likelihood that they are entrained within the
drifting boundary are sensitive to the charge state of the
impurity. A knowledge of the charge state history of the impurity
species is therefore crucial for the understanding of impurity
control.

The interactive coupling between plasma properties and atomic
processes has been comprehensively reviewed by Drawin[1] in the
context of both a hot and cold plasma environment. In the more
restricted region of the cool edge plasma detailed discussions of
most of the mechanisms which link plasma conditions to atomic
processes can be found in reviews by Harrison[2,3] but these earlier
papers have been directed specifically towards the interests of
the specialist in atomic collision physics. The objective of the
present paper is to reverse the emphasis. It is hoped that the
material selected will provide an informative background to those
aspects of atomic collision physics which underly the cross section data and scaling relationships which are currently employed
in plasma modelling. Both experimental and theoretical methods
are outlined and the general characteristics of collision processes are illustrated by examples of measured data. Emphasis is
placed upon interactions which are of significance in the plasma
edge but the basic concepts apply throughout the whole of the
plasma. The depth of discussion is perforce restricted but the
reader is referred to review articles which provide ready access
to detailed information.

2.   ATOMIC COLLISIONS IN THE BOUNDARY PLASMA

The rate of atomic or molecular collisions with either a
plasma electron or ion can be expressed as

$$\nu_A = n_A \, n \langle \sigma v \rangle \qquad\qquad (2.1)$$

where n_A and n are respectively the densities of the atomic species and of the plasma particles and $\langle \sigma v \rangle$ is the rate coefficient, i.e. the product of the reaction cross section $\sigma(v)$ and collision velocity v averaged over a distribution which can generally be assumed to be Maxwellian. Collisions with plasma electrons (e.g. ionisation, excitation and molecular dissociation) tend to be dominant because $v_e \gg v_i$ but there are cases where ion-atom interactions have large cross sections at low collision velocity. Of particular significance are charge exchange collisions of the type $X + X^+ \rightarrow X^+ + X$ which are both symmetric (in atomic species) and resonant (in potential energy).

3. PROCESSES INVOLVING HYDROGEN ATOMS

The subject of collisions is conveniently introduced by reference to the hydrogen atom because this atom has only one bound electron.

3.1 Atomic Stucture of the Hydrogen Atom

It is desirable to refresh our appreciation of the simpler features of atomic structure in order to understand the significance of various processes and to understand the practical issues which impact upon the accuracy and availability of both the theoretical and experimental data base.

The energy of the single electron moving in an orbit around the positively charged nucleus is determined by four quantum numbers. The "principle" quantum number n describes the scale of the motion and the energy. The ground state of the atom is n = 1 and, at the ionisation threshold, n → ∞. By analogy with classical mechanics, n determines the major axis of eliptic orbits around the Bohr atom. The "azithmutal" quantum number ℓ determines the angular momentum (i.e. $\ell(h/2\pi)$ atomic units) and ℓ has the values (n − 1) (n − 2) ... 0. Electrons with azithmutal quantum numbers equal to 0, 1, 2, 3, 4, etc. are referred to as s, p, and d etc. in compliance with the terminology "sharp", "principle" and "diffuse", etc. which derives from optical spectroscopy. The electron has a "magnetic" quantum number m_ℓ which describes its energy in a magnetic field and m_ℓ takes the values ℓ, $(\ell - 1)$, $(\ell - 2)$ ... $- \ell$ so that the total number of energy states available to an electron with azithmutal quantum number ℓ is, in the presence of a magnetic field, equal to $(2\ell + 1)$.

Spontaneous transitions between levels are governed by selection rules which, whilst not absolutely rigid, nevertheless

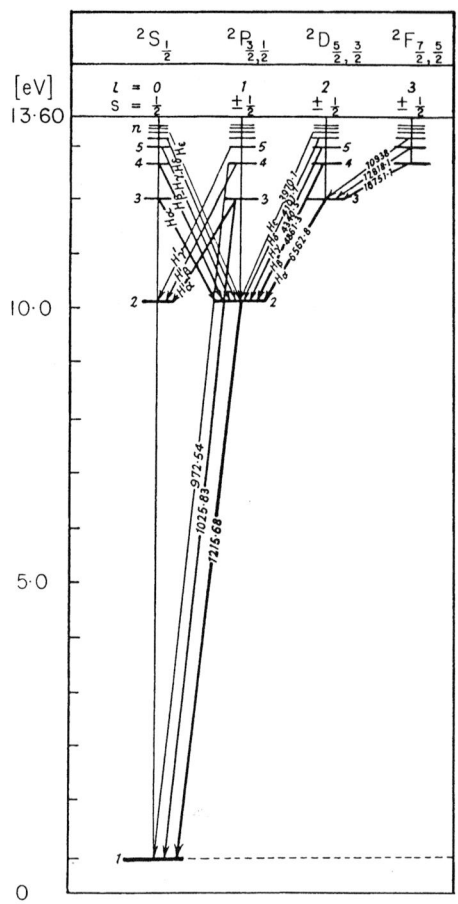

Fig. 1 The energy levels of the hydrogen atom in a field free
environment.
The illustration is based upon Grotian[4]; transitions
between sub-levels are not shown. Wavelengths are in Å.
Note that the Lyman series corresponds to transitions from
$n \geqslant 2$ to $n = 1$ and the Balmer series to $n \geqslant 3$ to $n = 2$.
The Balmer series contains the H_α and H_β lines which arise
respectively from $n = 3 \to 2$ and $n = 4 \to 2$ transitions.

determine the most prominent features of spontaneous radiative
transitions within the atom. There is no restriction on the
principle quantum number so that Δn can range from 0 to ∞. A
transition is unlikely whenever the condition $\Delta \ell = \pm 1$ is
violated. For example, transition between s to p and p to d
levels are allowed whereas those between s to s and s to d are
forbidden. These criteria give rise to the transitions of the
hydrogen atom which (in a field free environment) are shown in
Figure 1. Note that transitions from the level $[n = 2; \ell = 0]$ to

286

the ground state are forbidden. This level (which is more generally designated 2s or $2^2S_{\frac{1}{2}}$) is metastable, its lifetime in a field free environment approaches 0.1 s.

The orbiting electron also spins around its own axis and a fourth quantum number $s = \pm \frac{1}{2}$ must be included to allow for the "mechanical momentum" of the spinning electron. The total mechanical moment of the atom arises therefore from a vectorial combination of the azithmutal and spin moments, namely

$$\vec{j} = \vec{l} + \vec{s}$$

where $\vec{l}$ is the momentum vector corresponding to $l(h/2\pi)$. Quantisation of the total moment is described by the "total" or "inner" quantum number j. Since the direction of electron spin can only be "co" or "counter" to the direction of its motion in orbit it is obvious that $\vec{s} = \pm \frac{1}{2}$ (units of momentum) so that $j = l \pm \frac{1}{2}$. With the exception of the s levels (for which $l = 0$) all of the levels of the hydrogen atom are split into two sublevels (also called terms) which are separated by a small energy difference. This "multiplicity" is not shown in Figure 1.

The selection rule for j is $\Delta j = \pm 1$ or 0. The multiplicity of the level is given by

$(2s + 1)$ when $l > s$ or $(2l + 1)$ when $l < s$

so that the s levels of hydrogen are not split.

3.2 Electron Collisions with Hydrogen Atoms

Those plasma electrons whose energy exceeds the ionisation threshold of the atom (E_i) may impart sufficient energy to the bound electron for it to be removed completely from the influence of the Coulomb field of the proton. Ionisation from the ground state

$$e + H(1s) \rightarrow e + e + H^+ \tag{1}$$

provides a substantial sink for kinetic energy of the plasma electrons because $E_i = 13.6eV$. Moreover the energy of the ejected electron is small (1 to 2 eV is typical for collisions pertinent to the boundary plasma) so that ejected electrons tend to dilute the energy content of the plasma. Recoil of the ion has a negligible effect upon the plasma.

The threshold energy E_{pq} for excitation from a lower atomic level p to upper level q is less than the ionisation threshold energy so that ionisation by plasma electrons is always accompanied by excitation. The excitation process

$$H(p) + e \rightarrow H(q) + e \qquad (2)$$

causes the plasma electron to lose an amount of kinetic energy (equal to the energy difference E_{pq} between the levels p and q) and it is obvious that only those plasma electrons whose energy is greater than E_{pq} can participate in such collisions. The excited state q has a finite lifetime associated with its spontaneous radiative decay,

$$H(q) \rightarrow H(p') + h\nu, \qquad (3)$$

to a lower level p' (for examples see Ref. 5). Here the photon energy $h\nu$ corresponds to the difference in the energy levels $q \rightarrow p'$ and, depending upon the decay characteristics, p' may or may not be the same level as p. The plasma is generally optically transparent to atomic radiation so that the energy associated with reaction (3) is lost to the walls of the vessel where it is absorbed. The spatial distribution of emitted photons is related to the direction of the colliding electron but the photon distribution within the bulk plasma can be assumed to be uniform in space. The collision scatters the plasma electron but the atom motion is unaffected.

The average time for collisions between plasma electrons and an atom in an excited level q is

$$\tau_q = (n \langle \sigma v_e \rangle_q)^{-1}. \qquad (3.1)$$

In plasmas where $n > 10^{14}/cm^3$, this collision time may be appreciably less than the lifetime for spontaneous radiative decay of all but the lower excited states of hydrogen. In such conditions super-elastic collisions

$$e + H(q) \rightarrow e + \Delta E + H(p') \qquad (4)$$

become important. The collision does not yield a photon but the potential energy stored in the excited atom H(q) is returned to the plasma electron. These de-exiting collisions reduce the population of excited hydrogen atoms.

Ionisation of excited H atoms is discussed in Section 7, the cross sections are large because the scaling is of the form $\sigma(v_e)_n \propto n^4$. Moreover the ionisation threshold energy decreases as $(E_i)_n \propto n^{-2}$ so that even electrons in the low energy tail of the plasma thermal distribution are able to ionise excited atoms. Ionisation of excited hydrogen

$$e + H(q) \rightarrow e + e + H^+ \qquad (5)$$

provides a second route by which the population of excited H atoms

288

is reduced. Collisional radiative (or multi-step) processes involving a balance between reactions (2), (3), (4) and (5) powerfully reduce radiative power losses from hydrogen atoms in a low temperature, high density edge plasma (see Refs. 2, 3, 6, 7 and 8). These effects are discussed in Section 10.1. The electron impact ionisation rate coefficient $S_i(g) = \langle \sigma_i v_e \rangle$ for ground state hydrogen and the coefficient S_{CR}, which includes enhancement of ionisation arising due to collisional radiative effects, are shown in Figure 2.

Protons can be destroyed by two-body radiative recombination with an electron

$$e + H^+ \rightarrow H(q) + h\nu. \tag{6}$$

The photon carries away the excess energy of the interaction, i.e. the kinetic energy of the electron plus the energy of ionisation. The recombination rate coefficient for hydrogen $\alpha(T_e)$ is shown in Figure 3. It is very small except at low electron temperature ($k_B T_e < 1$ eV) and high electron density so that the characteristic recombination time

$$\tau_\alpha = [n \, \alpha(T_e)]^{-1} \tag{3.2}$$

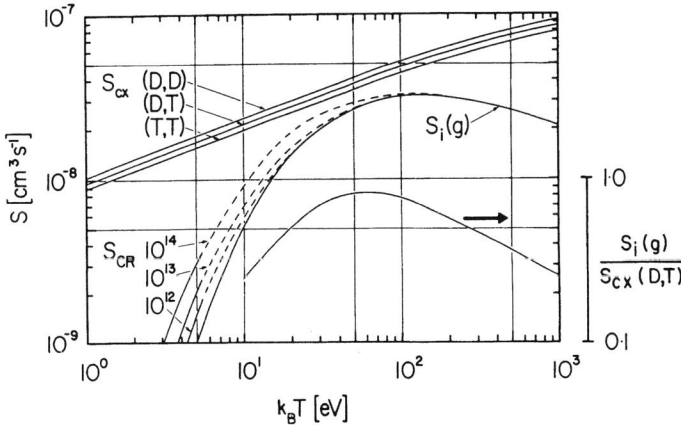

Fig. 2 Rate coefficients for electron ionisation and for proton charge exchange in collisions with hydrogen atoms. Data are taken from Harrison[3].
S$_i$(g) refers to ionisation from the ground state of hydrogen whereas S_{CR} are collisional radiative ionisation coefficients for the electron density range 10^{12} to 10^{14}/cm^3. The charge exchange rate coefficient is $S_{cx} = \langle \sigma_{cx} v_i \rangle$ and G is the ratio $S_i(g)/S_{cx}$.

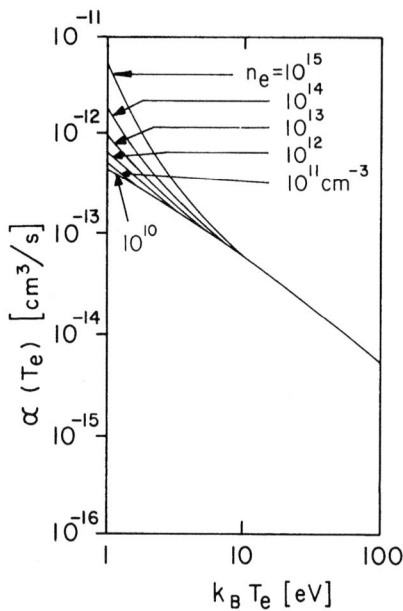

Fig.3   Collisional radiative recombination rate coefficient for
         e + H$^+$ collisions.
         Data are taken from Janev et al[8].

is generally considerably longer than the time that the recycling
proton resides within the plasma. Two-body radiative recombination of hydrogen is therefore not likely to be substantial within
the boundary plasma.

    Electron-proton recombination can in principle arise as a
consequence of three-body collisions,

$$e + e + H^+ \rightarrow H(q) + e, \tag{7}$$

but the electron density within the boundary plasma is insufficient for this process to be significant.

    It is worthwhile noting that radiation due to free-free
collisions

$$e + H^+ \rightarrow \text{bremsstrahlung radiation} \tag{8}$$

is negligible in the boundary because of the relatively low
electron temperature.

3.3 Charge Transfer between Hydrogen Atoms and Protons

Electron loss by a hydrogen atom and the equivalent process of electron capture by a plasma proton has unique significance because the rate coefficient for this species-symmetric and energy resonant reaction

$$H + H^+ \rightleftarrows H^+ + H \tag{9}$$

is so large that it strongly influences the behaviour of the hydrogen recycling within the boundary plasma. By contrast, charge exchange between H^+ and H_2 molecules is neither symmetric nor resonant and such contributions are sufficiently small to be neglected. However electron impact dissociation of molecules (which is discussed in Section 5.2) does produce H atoms which can subsequently charge exchange with plasma protons.

During a charge exchange collision the parent hydrogen atom loses little kinetic energy but it becomes charged and the subsequent motion of the daughter proton is constrained by the magnetic field. The reverse applies to the parent plasma proton which becomes a daughter charge exchange atom whose energy is equal to that of the parent proton but whose trajectory tends to be randomly directed because it is no longer constrained by the magnetic field. When viewed in a specific direction, the depth of penetration into the plasma of many generations of daughter charge exchange atoms prior to their ionisation is reduced by the scattering action of charge exchange. In cases where there are many successive scattering events it is reasonable to determine the effective range, Δ_{cx}, on the basis of a diffusive transport of the daughter atoms. If the plasma is homogeneous it can be argued that

$$\Delta_{cx} = \left(\frac{S_{cx}}{3S_i}\right)^{\frac{1}{2}} \lambda_{cx}. \tag{3.3}$$

Here $S_{cx} = \langle \sigma_{cx} v_i \rangle$ and $S_i = \langle \sigma_i v_e \rangle$ are respectively the rate coefficients for charge exchange and for electron impact ionisation and

$$\lambda_{cx} = \frac{\bar{v}_o}{n \, S_{cx}} \tag{3.4}$$

is the mean free path for charge exchange of hydrogen atoms whose mean velocity is $\bar{v}_o$. The rate coefficients are shown in Figure 2 for a homogeneous Maxwellian plasma wherein $T_e = T_i$. The ratio $G = (S_i/S_{cx})$ is less than unity at all plasma temperatures and so it is evident that scattering of charge exchange daughter atoms appreciably attenuates the effective ionisation range of the neutral hydrogen.

4.   ATOMIC PROCESSES INVOLVING HELIUM AND OTHER IMPURITY SPECIES

4.1   Structure of the Helium Atom and Ion

The helium atom has two electrons and in such simple atoms it is permissible to neglect coupling between the spin and momentum vectors associated with a particular electron and to account for the presence of two electrons by adding the azithmutal momentum and spin moment vectors independently.   Thus $\vec{L} = (\vec{\ell}_1 + \vec{\ell}_2)$ and $\vec{S} = (\vec{s}_1 + \vec{s}_2)$.   The total angular momentum is therefore $\vec{J} = \vec{L} + \vec{S}$.   This is an example of LS (or Russel-Saunders) coupling* and the azithmutal quantum number L has the integral values $(\ell_1 - \ell_2)$, $(\ell_1 - \ell_2 + 1)$ .... $(\ell_1 + \ell_2)$.   The spin quantum numbers are S = 0 or 1 because the electron spins may be aligned either anti-parallel or parallel.

The selection rules are

$$\Delta L = 0, \pm 1; \; \Delta S = 0, \; \Delta J = 0 \pm 1 \; (\text{but } 0 \rightarrow 0 \text{ is forbidden})$$

and the multiplicity is

$$(2S + 1) \text{ when } L \geqslant S \text{ or } (2L + 1) \text{ when } L < S.$$

Levels corresponding to L = 0, 1, 2, 3, etc. are designated S, P, D, F, etc. and the conventional spectroscopic notation is

$$n^{(2S + 1)}(L)_J \quad \text{e.g. } 1\,^1S_0, \; 3\,^3P_2, \text{ etc.}$$

This form of notation is also applied to the hydrogen atom although this atom has but one electron.

In the case of the helium atom there are two multiplicities namely 1 and 3.   The energy levels, which are illustrated in Figure 4, clearly show the singlet and triplet branches.   Both the $2\,^1S_0$ and the $2\,^3S_1$ levels are metastable but the metastability of the $2\,^3S_1$ state is stronger because, not only does a transition to the ground state require that this electron violates the $\Delta\ell = \pm 1$ rule, but it also involves a change in multiplicity, i.e. a change in the spin direction of the bound electron.

The helium ion, He^+, is hydrogenic and its terms differ from the hydrogen atom only to the extent that the energy levels are scaled by a factor $Z^2 = 4$ where Z is the atomic number of

---

* For more complex coupling between jj the reader is referred to textbooks on atomic spectra and structure, for example Herzberg[9], Candler[10].

helium. The same scaling rule can be applied to other hydrogenic ions, e.g. C^{5+}, O^{7+}, etc.

4.2 Structure of Complex Atomic Species

Differences in the structure of the simple species H (or He^+) and He are adequate to provide a framework for most of the following

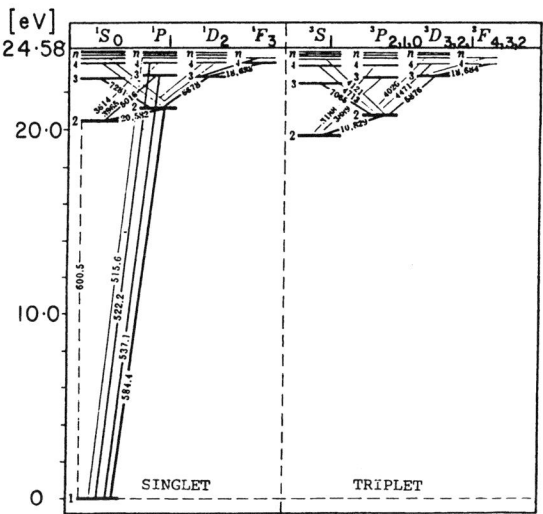

Fig. 4   The energy levels of the helium atom in a field free environment.
Illustration is based upon Grotian[4], transitions between sub-levels are not shown. Wavelengths are in Å.

discussion. It is necessary only to note that the Pauli exclusion principle requires that no two electrons bound to the same atom may have identical quantum numbers. This ensures that two electrons with the same value of n and ℓ must have $s = +\frac{1}{2}$ and $s = -\frac{1}{2}$ and leads to the concept of electron shells which are related to the principle quantum number n, e.g.,

n	1	2	3
Shell designation	K	L	M
Number of electrons	2≡(2s)	8≡(2s+6p)	18≡(2s+6p+10d)

Note that the individual electrons are described in the same manner as in the hydrogen atom. Electrons in the K shell are most strongly bound and the binding energy decreases progressively as n increases.

4.3 Electron Collisions with Impurity Species

Collision processes similar to those in hydrogen atoms occur in the case of impurity elements but the situation is more complex because of the greater number of bound electrons. Electrons in the outermost shell of the atom (or ion) are less tightly bound than those in inner shells and so the outermost electrons participate most readily in excitation and ionisation. However, in many species, there are more inner electrons so that the net contribution from inner shells may well exceed that of the outer (for example see the total electron impact cross ionisation cross section for $Fe^+ \rightarrow Fe^{2+}$ shown in Figure 16).

The presence of many bound electrons increases the number of possible collision processes. One such example is that the plasma electron may, in a single collision, eject more than one of the bound electrons of the impurity species X^{z+}, namely

$$e + X^{z+} \rightarrow e + ae + X^{(z + a)+} \tag{10}$$

However, multiple ionisation is not likely to be particularly significant in the cool boundary plasma where ionisation by electron impact most probably proceeds in a stepwise manner i.e.

$$X^o \rightarrow X^+ \rightarrow X^{2+} ---\rightarrow .$$

A more fundamentally significant process is the excitation of auto-ionising states

$$e + X^{z+} \rightarrow e + X^{z+*} \rightarrow e + e + X^{(z + 1)+} \tag{11}$$

This occurs when an inner bound electron is excited to a level whose bound energy, $E_*(in)$, exceeds the ionisation threshold of the outer electron, $E_i(out)$. When the inner electron level decays, the energy associated with its photon can be coupled to an outer bound electron which is then ejected with an energy $[E_*(in) - E_i(out)]$. The contribution of autoionisation to the total electron impact cross section for ionisation can be seen in Figure 20.

The reverse of autoionisation is dielectronic recombination

$$e + X^{z+} \rightarrow e + X^{z+*} \equiv \left(X^{(z - 1)+**}\right) \rightarrow X^{(z - 1)+*} + h\nu . \tag{12}$$

In this type of collision a plasma electron loses energy by exciting the ion X^{z+} but, after the collision, the incident electron has insufficient energy to escape from the Coulomb field of the ion. Thus, for a short time, there exists a doubly excited species $\left(X^{(z-1)+**} \right)$. If the system can become stable by emitting a photon without suffering auto-ionisation then the electron and ion will have recombined. The effect is really a by-product of excitation and its contributions are significant when the electron energies lie close to the excitation threshold. The rate coefficient, illustrated here in Figure 5 for the case of $Ne^{6+} \rightarrow Ne^{5+}$, follows somewhat the shape of an excitation coefficient and it peaks at a relatively high electron temperature. It thus differs significantly from the radiative recombination coefficient which decreases monotonically with increasing temperature. Even so, dielectronic recombination within the boundary is unlikely to have a substantial effect upon plasma conditions because of the relatively short residence time of the impurity ions.

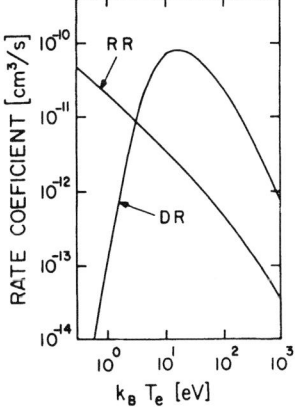

Fig. 5 Comparison of the rate coefficient for collisional dielectronic recombination compared with that for collisional radiative recombination.
Data are for $Ne^{6+} \rightarrow Ne^{5+}$ and are taken from Jacobs et al[11].

Curve RR shows the collisional radiative coefficient but only the most dominant contribution to dielectronic recombination is shown by curve DR, i.e. those due to transitions $n'\ell' \rightarrow n\ell$ where $n' = n = 2$, $\ell' = 2p$ and $\ell = 2s$.

4.4 Collisions between Hydrogen and Impurity Ions

Collisions between impurity ions and their associated atoms can generally be neglected. For example the rate coefficient for the symmetrical, resonant charge exchange reaction in helium,

$$He + He^+ \underset{\leftarrow}{\rightarrow} He^+ + He \tag{13}$$

is large even at low collision velocity but the effect upon He atom transport is slight because of the relatively small concentration of He^+ in the boundary plasma.

Collisions between H atoms and impurity ions cannot be symmetric but in some cases they tend to be energy resonant. A typical example, discussed in Section 9, is

$$C^{6+} + H \rightarrow C^{5+}(n = 4) + H^+ \tag{14}$$

for which the cross section data are shown in Figure 28. At relevant H atom collision energies ($\sim$ 100 eV) the rate coefficient is $\sim 10^{-8}$ cm^3/s which exceeds by many orders that for two body radiative recombination (i.e. e + C^{6+}) which at $k_B T_e$ = 100 eV is about 10^{-11} cm^3/s. This type of collision is often called "charge exchange recombination" but its formal name is "electron capture into excited states". The excited C ion subsequently emits a photon when it decays. The influence of collisions of this type upon the charge state population of impurity ions is considered in Section 10.2.

Cross sections for the reverse type of reaction, e.g.

$$C^{5+} + H^+ \rightarrow C^{6+} + H \tag{15}$$

tend to be small in the proton energy regime of interest because Coulomb repulsion between the colliding ions reduces the interaction probability at low energy. In addition, the tendency to energy resonance can (as in the case of reaction 15) be dominated by a specific excited state of the impurity ion and such excited ions constitute but a small fraction of the impurity population.

Other processes such as proton impact ionisation,

$$H^+ + X^{z+} \rightarrow H^+ + X^{(z + 1)+} + e \tag{16}$$

or hydrogen atom stripping

$$H + X^{z+} \rightarrow H^+ + X^{z+} + e \tag{17}$$

can be neglected in the low ion temperature region of the boundary. It should however be stressed that such neglect is not

valid if there are significant numbers of energetic particles
present in the boundary region, for example, atoms and ions from
injected beams, energetic particles in banana orbits (particularly α-particles). Indeed reactions of the type (14), (16) and
(17) have been invoked in numerous diagnostic studies based upon
injected beams of atoms or ions.

5.   PROCESSES INVOLVING HYDROGEN MOLECULES

5.1   The Structure of the Hydrogen Molecule

At infinite separation the components of the H_2 molecule have
the properties of individual atoms. However at close internuclear
distances there is a complex interplay of forces between the two
nuclei, between the two electrons and between each electron and
the nucleus of the other atom. The two charged nuclei exert a
repulsive force which is always dominant at close internuclear
separation but electrons in certain configurations can exert an
opposing attractive force so that a stable molecule can be
formed. Energy transferred during electron impact can change the
symmetry of the electron configuration so that the repulsive force
is no longer opposed and the molecule dissociates into two atoms.
Dissociation is sometimes accompanied by emission of a photon. A
stable configuration is also possible when only one electron and
two protons are bound so that a stable H_2^+ molecule can be formed
by electron impact upon H_2.

The electronic levels of the hydrogen molecule are too
complex to discuss in detail. Suffice to say here that the desig-
nations Σ, Π, Δ correspond to S, P, D terms in helium but that
they are determined by momentum quantum numbers which refer to the
direction of the molecular axis. The multiplicity relates to the
parallel or anti-parallel directions of the electron spin in the
two atoms which make up the molecule. The subscripts u (ungerade
= uneven) and g (gerade = even) refer to the symmetry effects which
influence transition probabilities. In homonuclear molecules
(such as H_2) the transitions g → u, u → g are allowed but neither
g → g nor u → u are allowed. The reader is referred to Gaydon[12]
for a concise introduction to the subject.

A diagram of the potential energy curves (i.e. the interaction energy as a function of inter-nuclear separtion) is shown
in Figure 6. The molecule rotates about its axis but in the
present context the effect can be neglected. The nuclei also
vibrate in the direction of the molecular axis and this effect is
important. Vibration is quantised and the range of inter-nuclear
separation associated with the lowest vibrational state of H_2 is
indicated in Figure 5 by the vertical lines A and B. The time for
an electronic transition ($\sim 10^{-16}$s) is much smaller than that of

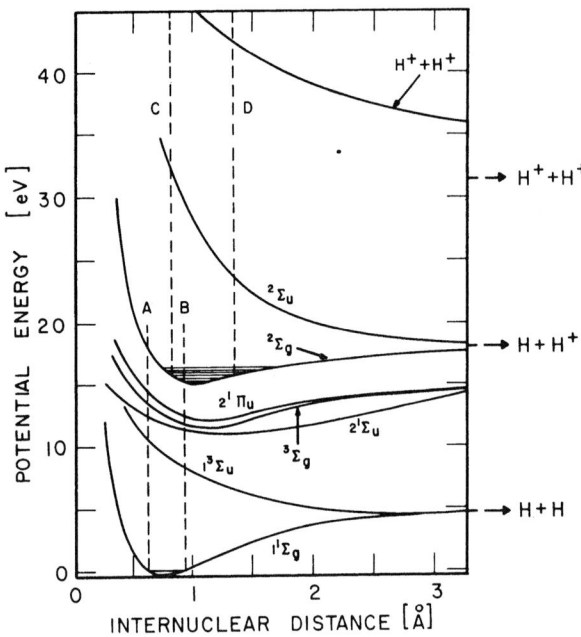

Fig. 6 Potential energy of H_2 and H_2^+ molecules as a function of
internuclear separation.
The Frank-Condon region of H_2 is shown by the vertical
lines A and B and that for H_2^+ by C and D.
Data on energy levels are taken from Oak Ridge National
Laboratory Report ORNL 3113 (Ref. 13).
The arrows on the righthand side show the potential energy
at infinite internuclear separation.

a period of nuclear vibration ($\sim 10^{14}$ s) so that there is negligible motion of the nuclei during an electronic transition. An
incident electron is most likely to encounter the nuclei at the
turning points in their motion which coincide with A or B. A
valuable criterion introduced independently by Frank and Condon is
that transitions between electronic levels are most likely to
occur where the vertical lines A or B intersect the potential
energy curves which identify the levels.

    To describe the effects of electron collisions with molecular
hydrogen let us consider the scenario in which the energy of the
incident electron is progressively increased. The first transition from the $1^1\Sigma_g$ ground state of H_2 occurs at about 8.5 eV where
the vertical B line intersects the $1^3\Sigma_u$ state of H_2. This state
is repulsive and dissociates into two ground state H atoms. At
the Frank-Condon edge (B) the potential energy of the $1^3\Sigma_u$ state
is about 4 eV higher than that of the $1^1\Sigma_g$ curve when this is at

infinite internuclear separation. Thus the transition, which extracts about 8.5 eV from the kinetic energy of the plasma electrons, causes a decrease of about 4 eV in the potential energy of the H + H system which reappears in the form of kinetic energy which is equally shared between the H_2 molecule dissociation products.

When the incident electron energy is increased to 11.75 eV the B line intersects the $2\,^1\Sigma_u$ curve. The molecule can then radiate by a transition to the ground state (i.e. u → g is allowed and there is no change in multiplicity). At a slightly higher energy the $2\,^3\Sigma_g$ state is excited and this can radiate to the $1\,^3\Sigma_u$ repulsive state. Thus this latter transition produces a photon and also results in dissociation. Proceeding to higher energies results in the formation of the $2\,^1\Pi_u$ state which can radiate to the ground state. Eventually, at higher electron energy, an electron is ejected leaving the molecule in the stable $^2\Sigma_g$ ground state of H_2^+.

It is essential to note that internuclear separation of the ground vibrational state of H_2^+ does not coincide with that of H_2 and according to the Frank-Condon criteria, whenever an H_2^+ ion is formed by electron impact upon vibrationally unexcited neutral H_2, the ion is inevitably vibrationally excited. The peak of the vibrational distribution is expected to coincide with the state $\nu \approx 2$ so that a second Frank-Condon region, denoted in Figure 6 by the vertical lines C and D, must be used to describe electronic transitions in H_2^+.

Electron impact upon an H_2^+ ion can excite the lowest lying repulsive $^2\Sigma_u$ state. The intersection point of this curve and the Frank-Condon edge D indicates that the $^2\Sigma_u$ state dissociates into an H(1s) atom and a proton, each particle having about 4.5 eV energy. The incident electron energy required to dissociate H_2^+ is extremely sensitive to the distribution of the vibrational states of H_2^+, a small population of the higher states dominates the transition probability. Transitions to higher electronic levels of H_2^+ occur at higher incident electron energy, some transitions give rise to radiation but all higher states are, in effect, repulsive and yield an excited H* atom and a proton. At the highest relevant energy (~ 28 eV) the molecule breaks up into two protons each having about 6 eV energy.

The preceding scenario is somewhat simplistic. Transitions at the inner edge of the Frank-Condon region (lines A and C) have not been considered. Moreover, a significant issue in the context of the boundary plasma is the lack of knowledge regarding the population of the vibrational states of neutral H_2 molecules involved in recycling.

5.2 Electron Collisions with Hydrogen Molecules

Collisions between plasma electrons and neutral H_2 can, in progressive order of their threshold energies, give rise to the following interactions;

$$e + H_2 \rightarrow H + H^- \qquad \text{[dissociative attachment]} \qquad (18)$$

$$e + H_2 \rightarrow e + H + H \qquad \text{[dissociation]} \qquad (19)$$

$$e + H_2 \rightarrow e + H_2^* \rightarrow e + H_2 + h\nu \qquad \text{[excitation]} \qquad (20)$$

$$e + H_2 \rightarrow e + H_2^* \rightarrow e + H + H + h\nu \qquad \text{[dissociative excitation]} \qquad (21)$$

$$e + H_2 \rightarrow H_2^+ + e + e \qquad \text{[ionisation]} \qquad (22)$$

$$e + H_2 \rightarrow H^+ + H + e + e \qquad \text{[dissociative ionisation]} \qquad (23)$$

$$e + H_2 \rightarrow H^+ + H^+ + 2e + e \qquad \text{[dissociative ionisation]} \qquad (24)$$

Excitation of the electronic levels of the molecule together with ionisation of the molecule act as energy sinks for the plasma electrons but in addition the plasma electrons dissipate energy in collisions which result in dissociation. In the context of plasma transport the molecule can be regarded as a potential source of H atom (or proton) momentum. For example when the H_2 molecule is dissociated into H + H by electron impact then two atoms each with an energy of 2.2 eV are released for the expenditure of 8.8 eV electron energy. The dissociation products can be assumed to have a random spatial distribution within the plasma.

Rate coefficients of these reactions are shown in Figure 7 and the most significant are (a) dissociation S_d^o [this coefficient includes contributions from reactions (19) and (21)], (b) ionisation S_{io}^o [arising from reaction (22)] and (c) dissociative ionisation S_{di}^o [arising from reaction (23)].

In the case of H_2^+ the possible reactions, again in progressive ranking of threshold energy, are

$$e + H_2^+ \rightarrow H + H \qquad \text{[dissociative recombination]} \qquad (25)$$

$$e + H_2^+ \rightarrow e + H^+ + H \qquad \text{[dissociation]} \qquad (26)$$

$$e + H_2^+ \rightarrow e + H_2^{+*} \rightarrow e + H^+ + H + h\nu \qquad \text{[dissociative excitation]} \qquad (27)$$

$$e + H_2^+ \rightarrow e + e + H^+ + H^+ \qquad \text{[dissociative ionisation]} \qquad (28)$$

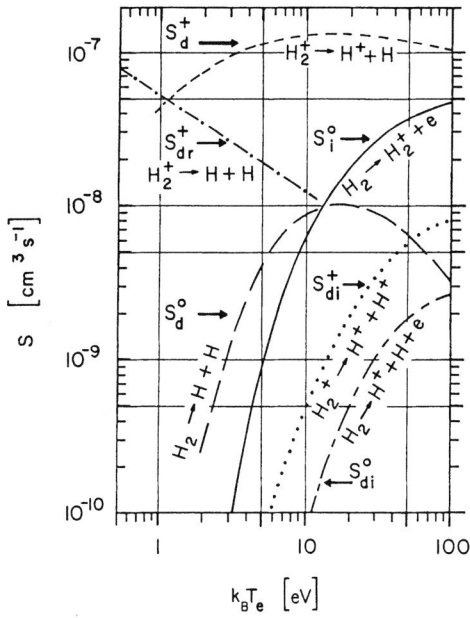

Fig. 7   Dominant rate coefficients for electron collisions with H_2 and H_2^+ plotted as a function of electron temperature. Symbols are defined in the text and data are taken from Harrison[3].

The rate coefficient for dissociation S_d^+ [reaction (26) and (27)] is dominant whereas the rate coefficients for dissociative recombination S_{dr}^+ [reaction (25)] can be neglected except at very low temperature. Dissociative ionisation S_{di}^+ [reaction (28)] has only a minor influence upon the characteristics of the boundary plasma.

It is sometimes convenient to express the rate coefficient $S^o(H^+)$ for the total formation of protons due to collision of electrons with H_2 in the form

$$S^o(H^+) \approx S_i^o \left(\frac{S_d^+ + 2S_{di}^+}{S_d^+ + S_{di}^+} \right) + S_{di}^o \qquad (5.1)$$

and the coefficient for the formation of H atoms, $S^o(H)$ in the form

$$S^o(H) \approx 2S_d^o + S_{di}^o + S_i^o \left(\frac{S_d^+}{S_d^+ + S_{di}^+} \right). \qquad (5.2)$$

Collisions between H_2^+ and H_2 give rise to the molecule H_3^+,

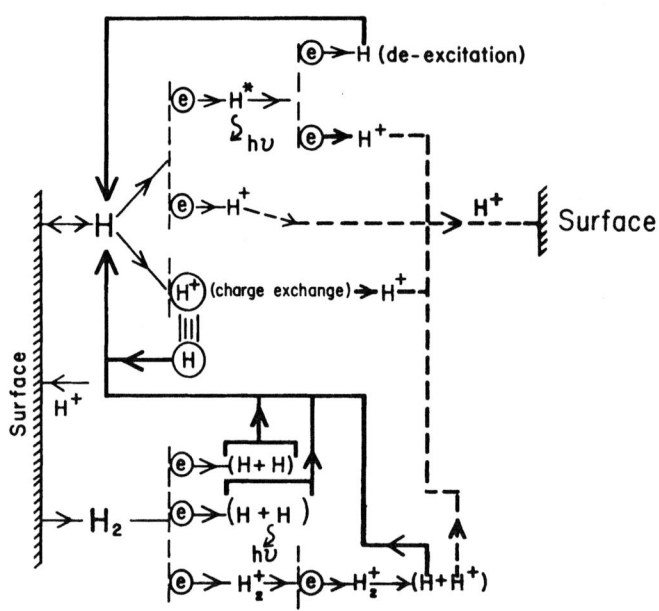

Fig. 8   Schematic representation of the dominant atomic and
         molecular collision processes associated with the
         recycling of hydrogen.

$$H_2^+ + H_2 \rightarrow H_3^+ + H, \qquad\qquad (29)$$

but the rate coefficient is about 10^{-2} less than that for the
destruction of H_2^+ by electron impact and so the reaction will be
significant only when the density of neutral H_2 is much higher
than that of the plasma electrons.

The atomic and molecular processes which are dominant in
hydrogen recycling are indicated schematically in Figure 8.

6.   INTRODUCTION TO ATOMIC COLLISION PHYSICS

Some background knowledge of the physics processes which
influences the magnitude and energy dependence of atomic collision
probabilities helps the plasma modeller to identify the most
significant interactions which pertain in a particular plasma
environment. This paper aims to fulfil this requirement by
providing a highly simplified outline of the nature of the more
common atomic interactions and of some generalised formulae which
are frequently used to predict cross section data. No attempt is
made to provide a comprehensive data base but trends which are
inherent to specific types of collision processes are illustrated

302

by examples of data for species which have particular relevance to the boundary plasma.

It must be stressed that this approach neglects many fundamental details of atomic collision physics but the field of is very well documented elsewhere. The most comprehensive discourse on theory and experiment is provided in 5 volumes by Massey, Burhop and Gilbody[14]. The present author has found McDaniel[15] to be particularly helpful but the choice of literature is wide and personal taste is in some measure invidious. There are several collections of review articles which relate specifically to the atomic and molecular needs of fusion. The proceedings of earlier NATO Advanced Study Institutes[16,17,18] are valuable examples. The multifarious influence of atomic processes on both natural and man made environments have recently been comprehensively reviewed in a series of 5 volumes entitled Applied Atomic Collision Physics edited by Massey, McDaniel and Bederson[19] and volume 2 deals specifically with nuclear fusion.

The complexity of an atomic collision depends upon the type of particles involved. In order of increasing complexity these are firstly photon and secondly electron collisions with either atoms or ions. Then follows collisions with molecules and finally collisions involving two particles each of which has atomic structure (one such example being charge exchange between H atoms and partially stripped impurity ions). The energy at which particles collide is also important; in general the interactions are complicated in the relatively low energy regime which is pertinent to the boundary plasma. For simplicity the subject is introduced by a discussion of inelastic collisions of electrons with atoms or ions but the principles are also applicable to electron-molecule and to ion-atom or ion-ion collisions. The influence of the plasma environment upon the basic collision processes and the methods employed to estimate collision rates are discussed briefly in Section 10.

6.1 Theory of Inelastic Electron Collisions

The inelastic process of excitation or ionisation requires that the incident electron transfers sufficient of its kinetic energy to a bound electron for the latter to be raised to a higher excited state or else to be ejected from the influence of the Coulomb field of the charged nucleus. The first theoretical treatment of ionisation (Thomson[20]) is based upon a classical model in which a stationary electron is approached from infinity by a moving electron with energy E. At a particular impact parameter (indicated by r in Figure 9) the amount of energy exchanged due to a single Coulomb collision is

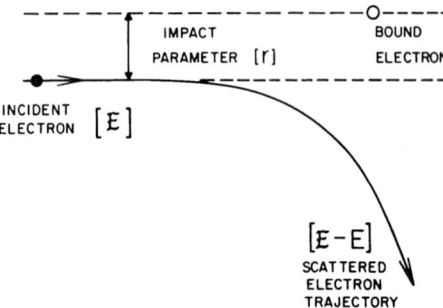

Fig. 9   Trajectory of an inelastically scattered electron.
The impact parameter is r and the threshold energy of the
atomic transition is E.

$$\Delta E = \frac{E}{1 + (rE/e^2)^2} \qquad\qquad (6.1)$$

where e is the electronic charge. Thomson assumed that the
probability of ionisation is unity when ΔE is equal to or greater
than the theshold energy for ionisation E_i. The classical
electron impact cross section can then be expressed as

$$\sigma_{cl} = 4\xi \left(\frac{E_{i,H}}{E_i}\right)^2 \left(\frac{E_i}{E}\right) \left(1 - \frac{E_i}{E}\right) \pi a_o^2, \qquad\qquad (6.2)$$

where $E_{i,H}$ is the ionisation energy of the H atom, $a_o = 0.53$ x
10^{-8} cm is the radius of the Bohr atom and ξ is the effective
number of bound electrons which can contribute to the interaction. The characteristics of this cross section are that its
magnitude increases linearly with excess incident energy$(E - E_i)$
in the regime close to theshold. The energy dependence becomes
progressively weaker at higher energies so that the cross section
peaks at $E = 2E_i$ and when $E \gg E_i$ the cross section decreases as
(E_i/E).

The preceding aproach is clearly over simplistic. In classical terms the collision should be treated as a many body problem
(three body even in the simplest case of a hydrogen atom) but,
more significantly, the scale of the collision system requires
that the problem be treated by quantum theory. Nevertheless the
simple classical cross section given in Eq. (6.2) has formed the
basis of many semi-empirical expressions used in plasma
modelling. For example it shows clearly the scaling relationship

$$\sigma_{scaled} = \sigma \left(\frac{E_i}{E_{i,H}}\right)^2 \xi^{-1} = \text{function } (E /E_i) \qquad\qquad (6.3)$$

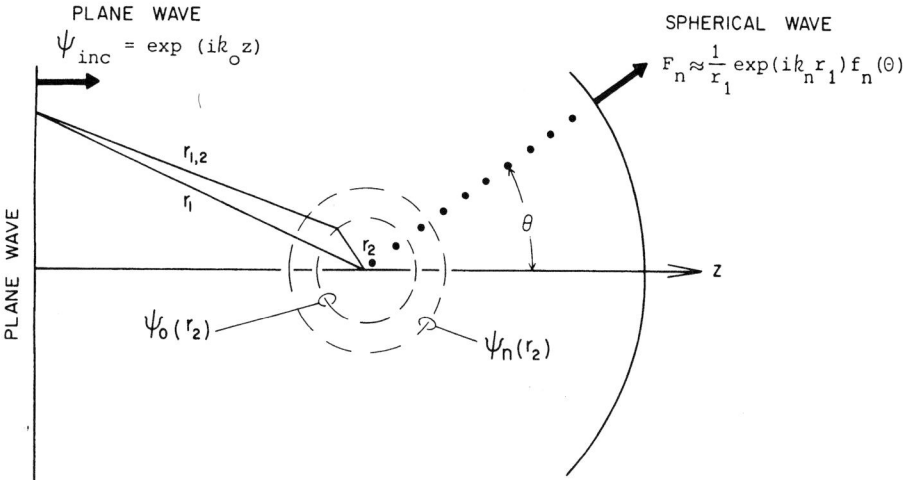

Fig. 10 Illustration of inelastic scattering of electrons based
upon the quantum theory treatment.

Such scaling, which is supported by quantum theory, is used extensively for comparing experimental data and for extrapolation to
species for which no measurement is available.

The quantum theory concept of inelastic electron-atom (or
electron-ion) collisions is illustrated schematically in Figure
10. The method can be briefly outlined as follows. Consider that
the atom (or ion) resides in an infinitely extending, uniform
intensity beam of mono-energetic electrons. The incoming electrons
which move in the z direction have an energy E when they are at
large distances from the atom and here they are represented by a
plane wave whose wave number is

$$k_o = 2\pi/\lambda_o = 2\pi \, m_e \, v_o/h = (2\pi/h) \sqrt{2m_e E} \qquad (6.4)$$

where m_e is the electron mass, v_o the initial velocity and h is
Planck's constant. The wave function of these electrons is
$\psi_{inc}(z) = \exp(ik_o z)$. The time independent wave equation of the
collision system which includes both the incident electron and the
atom (which for simplicity is taken to be an H atom) is

$$\left[-\frac{h^2}{8\pi^2 m_e} \left(\nabla_1^2 + \nabla_2^2 \right) + \left(E_{SYS} - V(r_1, r_2) \right) \right] \Psi(r_1, r_2) = 0 \quad (6.5)$$

Here the suffixes 1 and 2 refer respectively to the incident and
to the atomic electron, the energy of the system is

$$E_{SYS} = E + E_{a,o} \qquad (6.6)$$

where $E_{a,o}$ is the potential energy of the atom in its initial state. The interaction potential energy operator is

$$V(r_1, r_2) = - \frac{e^2}{r_1} - \frac{e^2}{r_2} + \frac{e^2}{r_{1,2}} \tag{6.7}$$

where r_1 and r_2 are respectively the co-ordinates of the incident and atomic electron and $r_{1,2}$ is the distance between electrons. The function $\Psi(r_1, r_2)$ may be expanded over the excited and continuum states of the atom in the form

$$\Psi(r_1, r_2) = \left(\sum_n + \int \right) \phi_n(r_2) \, F_n(r_1) \tag{6.8}$$

where the summation pertains to the bound states and the integration to the continuum states of the atomic electron. The functions of $\phi_n(r_2)$ are the wave functions of the hydrogen atom in state n. It can be shown that at large values of r_1 the function $F_n(r_1)$ is described by a wave number

$$k_n = (2\pi/h) \sqrt{2m_e(E_o - E_n)} \tag{6.9}$$

This wave function therefore corresponds to free electrons which have lost an amount of kinetic energy corresponding to a change in the internal energy of the atom equal to a transition from the initial o state to the final n state. Clearly this represents the inelastically scattered incident electrons. At large values of r_1 the wave function which represents electrons that have been inelastically scattered through an angle θ must have the form of an outgoing spherical wave,

$$F_n \sim r_1^{-1} \exp(ik_n r_1) \, f_n(\theta) \tag{6.10}$$

whereas the elastically scattered electrons are represented by

$$F_n \sim \exp(ik_o z) + r_1^{-1} \exp(ik_o r_1) f_o(\theta). \tag{6.11}$$

The number of inelastically scattered electrons which cross a unit area of the surface of a sphere (of radius large r_1) in unit time is proportional to $k_n r_1^{-2} |f_n|^2$ whereas the associated flux density of electrons incident upon the atom is proportional to k_o. The differential cross section $I_{on}(\theta) d\Omega$ for those transitions o → n which cause scattering into a solid angle $d\Omega$ can be defined as the ratio of these flux densities so that

$$I_{on}(\theta) \, d\Omega = \frac{k_n}{k_o} \, |f_n(\theta)|^2 d\Omega \tag{6.12}$$

Integration over the angles θ and ϕ of the spherical polar co-ordinate system yields the total cross section

$$\sigma_{on} = \int_0^{2\pi} \int_0^{\pi} I_{on}(\theta) \, \sin\theta \, d\theta \, d\phi \tag{6.13}$$

Solution of the problem therefore requires that the asymptotic form (i.e. $r_1 \to \infty$) of the function $f_n(\theta)$ be determined but this is not ammenable to precise calculation because this would involve an infinite number of coupled differential equations associated with the atomic states n. Success in this field has therefore been in large measure due to the informed approximations that have been invoked. The most frequently used approach is the Born approximation[21]. The basic simplifications made by Born are that (a) the incident electrons can be represented by a plane wave which is not distored by the influence of the unscreened charge of nucleus, (b) transition from an initial state (o) to a final (n) state of the atom is direct so that the effects of intermediate states are not significant and finally (c) the outgoing wave of inelastically scattered electrons is not distorted by interactions with the atom (or ion) in its final state. In effect the neglect of wave distortion implies that Born's approximation relates specifically to high energy collisions. Nevertheless the Born approximation and its many variants have been remarkably successful even at modest collision energies. The approach is used here to illustrate the high energy dependence of cross sections for excitation and ionisation.

It is convenient to transform from angular to momentum variables such that the change in momentum $(h/2\pi)$ K of the incident electron which is scattered through an angle θ can be expressed in terms of

$$K = (k_o{}^2 + k_n{}^2 - 2k_o\, k_n \cos \theta)^{1/2}. \tag{6.14}$$

The limits of K correspond to

$$K_{max} = k_o + k_n(\theta=\pi) \text{ and } K_{min} = k_o - k_n(\theta=\pi).$$

However Bethe[22] argued that an upper limit

$$K_o = (2\pi/h)\sqrt{m_e E_{i,o}}$$

can be imposed because at high incident energy the electrons lose only a small fraction of their momentum and are scattered through only a small angle; here $E_{i,o}$ is the ionisation threshold energy of the atom in its initial state. The total cross section for excitation of the o to n state by high energy electrons can be expressed (see Refs. 14 and 15) as

$$\sigma_{on} \sim \frac{128\pi^5 m_e{}^2 e^4}{k_o{}^2 h^4} \int_{K_{min}}^{K_o} \left[K^{-1}|\chi_{on}|^2 + \frac{K}{4}|(\chi^2)_{on}|^2 + \ldots \right] \, dK \tag{6.15}$$

where the matrix elements χ_{on}, χ_{on}^2 are given by

$$(\chi^s)_{on} = \int \chi^s \psi_n^* \psi_o \, dr_2.$$ (6.16)

Here $\psi_o(r_2)$ is the wave function of the atom in its initial state and $\psi_n^*(r_2)$ is the complex conjugate of the wave function of the final state n. For optically allowed transitions ($\Delta \ell = 1$) the first (electric dipole) term of Eq. (6.15) does not vanish and so the remaining terms in the expansion can be neglected. Thus

$$\sigma_{on}^{dipole} \sim \frac{16\pi^3 e^4}{h^2 v_o^2} |\chi_{on}|^2 \log \left(\frac{2m_e v_o^2}{E_{on}} \right)$$ (6.17)

where χ_{on} is the energy difference between levels o and n. Note that $\sigma_{on}^{dipole} \propto v_o^{-2} \log v_o^2$ [i.e. $E^{-1} \log E$].

In the case of optically forbidden transitions ($\Delta \ell = 0$ or 2) then the dipole moment in Eq. (6.15) vanishes and the quadrupole moment becomes dominant so that

$$\sigma_{on}^{quad} \sim \frac{32\pi^5 m_e e^4}{h^4 v_o^2} |(\chi^2)_{on}|^2 E_{i,o}$$ (6.18)

and the high energy dependence is proportional to v_o^{-2}, [i.e. E^{-1}].

The high energy behaviour of the cross section for ionisation is comparable with that for the excitation of allowed transitions

$$\sigma_i \sim \frac{2\pi e^4}{m_e v_o^2} \frac{c}{E_{i,o}} \log \left(\frac{2m_e v_o^2}{C} \right).$$ (6.19)

If k_i is the wave number of the ejected electron then

$$c = \int |\chi_{o,k_i}|^2 \, d\,k_i$$ (6.20)

and the energy C is about one tenth of $E_{i,o}$.

Much effort has been expended in extending the quantum theory approach to lower energies. It is clear from classical arguments that inelastic collisions in the low energy regime (i.e. where E is only slightly greater than E_n) occur via closely coupled inter-actions between the incident and the bound electron. Moreover the trajectory of the incident electron (and also of the slow atomic electron ejected in ionisation) will be significantly influenced by both electron-electron interactions and by the unscreened field of the nucleus. These problems have been studied in detail. The incident electron has been allowed to see the partially screened

Coulomb field of the nucleus (Coulomb-Born), polarization of the atomic charge distribution by the presence of the incident electron has been considered, the influence of the many interacting states of the atom (or ion) has been assessed (Close Coupling), the incident electron has been allowed to change places with the bound electron (Exchange), the final (n) states have been coupled to the scattered electron and the transitory trapping of the incident electron within the partially screened field of the nucleus has been investigated (Dielectronic effects). Recent general surveys of theoretical treatments of electron collisions can be found in Joachain[23].

For present purposes it is sufficient to note that treatments based upon quantum theory demonstrate, in the specific cases of ionisation and of excitation of allowed transitions, that the classical approach over-estimates the coupling of energy when the impact parameter of the incident electron is small but it underestimates the energy coupling at larger impact parameters. As a consequence the peak value of the classical Thomson cross section is too large and too close to the threshold and, in the high energy regime ($E > E$), the classical energy dependence $[\sigma_{cl} \propto E^{-1}]$ is too strong. The high energy behaviour is more accurately described by the Born approximation as $\sigma_B \propto E^{-1} \log E$. Excitation of a forbidden transition which does not also involve a change in multiplicity is predicted to have a different high energy dependence, namely $\sigma_{on} \propto E^{-1}$ and in this respect it is comparable to the classical behaviour. If a disallowed transition also involves a change in multiplicity, then the dependence $\sigma_{on} \propto E^{-3}$ is expected at high energy and the cross section is very strongly peaked in the regime close to the threshold. The marked difference at low collision energy is to be expected because a change in multiplicity involves a change in the spin of the bound electron and this is likely to occur in closely coupled interactions between the bound and incident electron.

6.2 Semi-empirical Cross Sections used in Plasma Modelling

Despite the substantial number of detailed calculations of excitation and ionisation cross sections the plasma modeller is often forced to employ less precise methods. Data for large numbers of atomic processes are required and furthermore data in the low energy regime are of greatest significance (see the discussion of rate coefficients in Section 10). In addition to the complexities encountered at low collision energy, quantum theory calculations become significantly less certain as the number of bound electrons increases. The modeller thererefore tends to use semi-empirical data which are based upon the general energy dependences identified by theory but which are quantified by comparison with experiment. A number of semi-empirical methods have been evolved and these have been reviewed by Kato[24] and by Itikawa and

Kato[25]. Most ionisation data are derived from modifications of the classical cross section σ_{cl} such that (a) the semi-empirical cross sections display an $[E^{-1} \ln (E)]$ dependence at high energy and (b) that the position and magnitude of their peaks comply more closely with the trends identified in measured data. The semi-empirical formulation of Lotz[26] is probably the most widely applied and it takes the form

$$\sigma_i = \sum_{j=1}^{J} a_j \xi_j \frac{\ln(E/E_j)}{E \, E_j} \left\{ 1 - b_j \exp\left[-c_j (E/E_j - 1) \right] \right\} \qquad (6.21)$$

Here E_j is the binding energy of a "j" electron in the j-th sub-shell (j has the values $1 \to J$ where $j = 1$ corresponds to the outermost sub-shell), ξ_j is the number of equivalent electrons in the j-sub-shell and a_j, b_j, c_j are fitting parameters derived by comparison with the limited experimental data base available to Lotz. This approach has served well in the cases where the configuration of atomic electrons do not differ substantially from the data base available to Lotz. Comparison with the more recently expanded base of measured data shows that re-appraisal of the Lotz fitting parameters is required at least for lowly charged ions of dominant impurity species (see the typical case of $Fe^+ \to Fe^{2+}$ shown in Figure 16).

For each ionisation cross section needed for plasma modelling there are many excitation cross sections which must describe the dominant excitation processes experienced by an ion in each of its charge states. The requirement for a very wide base of data has motivated the search for a simple treatment of excitation cross sections and the most frequently employed methods are based on the Bethe approximation. The wave length of high energy electrons is much smaller than atomic dimensions and this causes the dominant interactions to occur outside the range of the atom wave functions. In effect the influence of the charge of the incident electron upon the atomic electron can be neglected and for, these long range collisions, the interaction potential operator [see Eq. (6.8)] becomes equal to the absorption oscillator strength of the atomic electron. The high energy collision can thus be regarded as a radiative process in which (a) the incident electron enters the electric field due to the atomic electron, (b) the incident electron then emits a photon and (c) this photon is absorbed by the atom and gives rise to a transition $o \to n$. The semi-empirical approach which has evolved from this concept[27,28,29] is usually referred to as the Gaunt factor (or $\bar{g}$) formula and it takes the form

$$\sigma_{on} = \frac{8\pi}{\sqrt{3}} \left(\frac{E_{i,H}}{E_{on}} \right)^2 \left(\frac{E_{on}}{E} \right) f_{on} \, \bar{g} \left(E_{on}/E \right) \pi a_o^2 \qquad (6.22)$$

where f_{on} is the absorption oscillator strength and $\bar{g}$ an empirial form of the Kramers-Gaunt $\bar{g}$ factor[30]. There is considerable doubt regarding the selection of $\bar{g}$. The limited amount of experimental data (Refs. 31,32,33,34) indicate that $\bar{g} \sim 1$ for $\Delta n = 0$ transitions in ions of charge state greater than $z \sim 5$ and in addition $\bar{g}$ tends to be invariant with E. (This trend is also supported by theory). For more lowly charged ions $\bar{g} \to 0.2$ at low energies and moreover it displays a dependence upon (E/E_{on}). In cases where $\Delta n \neq 0$, there are indications that $\bar{g}$ is less than 0.2. Dunn[33] cautions that the $\bar{g}$ formula is likely to be accurate only to a factor two or three.

6.3 Experimental Methods

In order to measure absolutely the cross section for a collision between an incident particle (A) and a target particle (B) it is necessary to know the densities n_A and n_B of the particles and to measure the rate ν_{AB} at which collisions occur within a well defined volume V. The cross section at a collision energy E corresponding to a collision velocity v_o is given by

$$\sigma(E) = \frac{\nu_{AB}}{n_A n_B V v_o}. \tag{6.23}$$

Collision rates are determined by observing the absolute rate of formation of collision products (A' and B') and this is most readily achieved if the collision changes the charge state of the particle. For example, the B' product ions which are formed by electron impact ionisation in the reaction

$$e + B^{z+} \to e + e + B^{(z + 1)+} \tag{i}$$

or by charge exchange in the reaction

$$A^+ + B \to A + B^+. \tag{ii}$$

The neutral A' products of the incident A^+ atoms in reaction (ii) are also readily detected if their energies are in excess of a few hundred electron volts. Product photons are much more difficult to observe in a meaningful manner. Firstly, photon emission may be angularly distributed relative to the collision velocity vector $\vec{v}_o$. Secondly, wavelength resolution is required and this severely reduces the fraction of product photons that can be detected.

Many cross section measurements are based upon the principles of the beam-static target technique. In this method a mono-energetic beam of flux I_A (type A particle/s) is directed through a thermal gas target of uniform density n_B. The cross section is determined from the expression

$$\sigma(E) = \frac{\nu_{AB}}{I_A} \frac{1}{l n_B} \tag{6.24}$$

where ν_{AB} is the absolute rate of production of B' products (and less generally A' products) within a well defined length l of the beam. The line density (ln_B) is maintained at a small value in order that multiple collisions can be neglected. This method has yielded data for electrons, ions and atoms incident upon (a) rare gas targets and (b) molecular gas targets. To apply the technique to ions or atoms incident upon atomic hydrogen the target cell (which for such measurements is made from tungsten) is heated to a temperature in excess of 1800 K in order to dissociate virtually all of the H_2 molecular gas.

Studies of collisions which involve atomic oxygen have been performed by replacing the gas cell by a beam of thermal energy oxygen atoms which is formed by allowing atoms to effuse through a small hole in the wall of a radio-frequency discharge tube. Beams of atomic hydrogen are also formed by atom effusion from a hot tungsten furnace which is fed with molecular hydrogen. This technique of thermal atom beams crossed with electron (or ion) beams was pioneered by Fite and his colleagues (see for example Ref. 35) and it has been used for studies of electron impact ionisation and excitation and also proton and O^+ charge exchange. The density of particles in the beams is appreciably less than in the static gas cell, indeed it is frequently smaller than the density of residual gas in the apparatus. Beam modulation techniques are used to distinguish the signals (i.e. amplified currents of ions or photons which arise from v_{AB}) from the larger backgrounds due to collisions of the incident particles with residual gas.

Cross sections for collisions between electrons and ions, ions and ions and in some cases electrons and ground state or excited atoms are measured using fast colliding beams. The term fast is used to distinguish these targets ion beams of 1 to 20 keV energy from the thermal energy atom beams. Fast neutral beams are produced by charge exchange in a gas or vapour cell (see Figure 11). The fast beam technique which was pioneered by Dolder and Harrison and their colleagues (see for example Ref. 36) is the main source of cross section data for ionisation and excitation of impurity ions and also for electron collisions with H_2^+. The target beam density is considerably smaller than in thermal beam experiments (ranging from 10^3 to 10^6 type B particles/cm^3) and quite sophisticated beam pulsing techniques are used to distinguish the count rates of particles which arise due to beam collisions from those which arise due to collisions with residual gas or surfaces within the apparatus. Fast beam techniques offer a number of advantages; the detection efficiency of both charged and neutral particles is high and absolute measurements can be made. Coincidence counting of the A' and B' products can also be used to further distinguish the collision process. Unfortunately the technique offers little scope for studies of molecules because, unlike atoms, fast neutral molecules are not readily formed by charge capture.

312

The angle of intersection of the beams can be varied in order to attain a low or high collision velocity within the centre of mass frame whilst retaining a high velocity within the laboratory frame. The angle of intersection ranges from $\phi \rightarrow 0$ (merged beams), to $\phi = 5^{\circ}$ to 20° (inclined beams) and to $\phi = 90^{\circ}$ (crossed beams). These various configurations are illustrated in Figure 11. The scanning shutters S shown in Figure 11 are used to measure the current density distributions in both of the beams because the assumption, implied in Eq. (6.24) that the target density n_B is uniform is not valid when the target is a fast beam. The fast colliding beams method has, to date, been restricted to target ion charge states z less than 5. Measurements of excitation cross

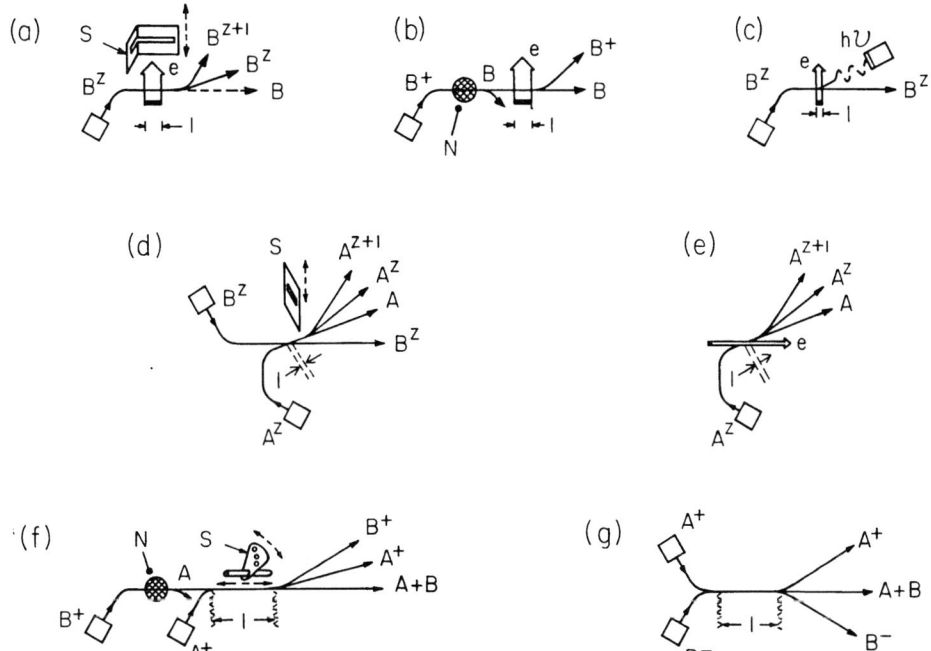

Fig. 11 Configurations used in colliding beam experiments.
(a) Crossed electron-ion beam, (b) crossed electron-fast atom beams, (c) crossed electron-ion beams used for excitation studies, (d) inclined ion-ion beams, (e) inclined electron-ion beams, (f) merged ion-atom beams and (g) merged positive and negative ion beams.
S is a shutter used to measure the profiles of current densities (and hence the profiles of particle densities), N is a neutraliser gas cell in which target atoms are formed by charge capture and l is the length of the collision path. For details see Harrison (Ref. 37).

sections for ions are particularly difficult because of the low
density of the fast target beam and the low overall efficiency of
wave-length selective detectors for photons in the far ultraviolet and soft X-ray region of the spectrum.

There are many reviews of experimental measurements. Kieffer
and Dunn[38] have surveyed beam-static target experiments and early
crossed beams experiments. Recent studies of inelastic collisions
of electrons with ions are reviewed in Refs. 31, 32, 33 and 34 and
these papers also provide bibliographies of earlier but pertinent
review articles. Recent reviews of experimental methods used in
charge exchange measurements have been provided by de Heer[39].

7.   MEASURED DATA FOR ELECTRON IONISATION AND EXCITATION

The measured cross section for ionisation of the ground state
hydrogen atom is shown in Figure 12. It displays the high energy
dependence predicted by the Born approximation but at lower
energies the agreement with quantum theory is less satisfactory.
The cross section recommended by Bell et al.[40] is derived from a

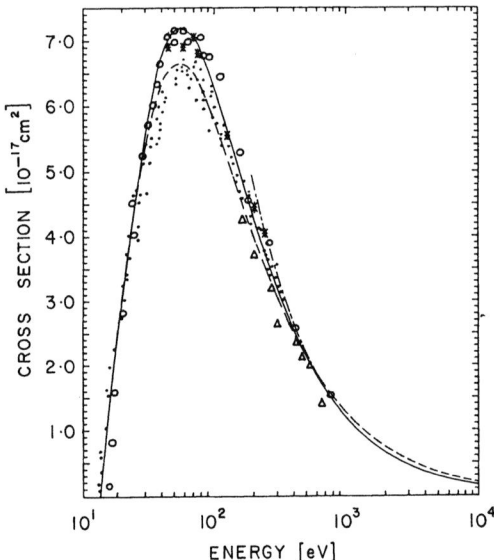

Fig. 12   Cross section for e + H(1s) → 2e + H$^+$.
          Data taken from an assessment by Bell et al. (Ref. 40).
          The solid line is the cross section recommended in Ref.
          40 and the dashed line is the semi-empirical Lotz cross
          section calculated using Eq. (6.21).
          The open circles are the measured data of Fite and
          Brackmann[35]. Other symbols are described in Ref. 40.

314

critical appraisal of both experimental and theoretical data.

The scaling relationship expressed in Eq. (6.3) is well demonstrated by the comparison of the scaled ionisation cross section for atomic hydrogen with the isoelectronic ion He^+ which can be seen in Figure 13. The ionisation threshold energies of these simple one electron atomic species scale as $(E_{iH}/E_{iHe}+) = Z^{-4}$ where Z is the atomic number. Consequently the magnitude of the cross section of He^+ ion is expected to be 16 times smaller in magnitude than that of the H atom. This relationship holds well at incident energies in excess of $(E/E_i) \sim 5$ but not at lower energy. The reason is that during an electron-ion collision the incident electron enters the relatively long range Coulomb field of the nucleus and it is thereby accelerated. Interchange of energy with the bound electron takes place only after the incident electron has experienced some degree of acceleration and, as a consequence, there is a greater probability of imparting energy to the bound electron. This causes the ionisation cross section for ions to be more peaked in the low energy regime. The trend is quite general and has prompted comparison of data along isoelectronic sequences so that differences in the screening of the nucleus by the bound electrons in partially stripped ions are minimised. An example for the beryllium sequence taken from Bell et al.[40] is presented in Figure 14. Such scaling should be applied with caution because indirect contributions to the total ionisation cross section by processes such as excitation of

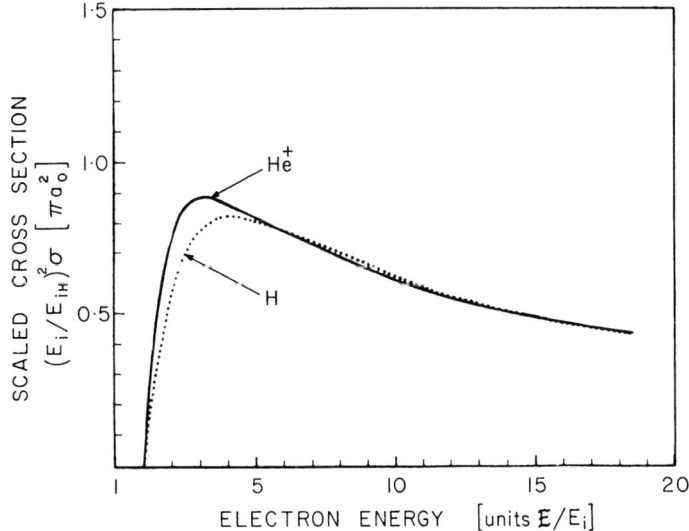

Fig. 13 Comparison of the measured cross sections for $e + H(1s) \rightarrow 2e + H^+$ and $e + He^+(1s) \rightarrow 2e + He^{2+}$. Data for H(1s) are from Fite and Brackman[35] and those for $He^+(1s)$ are from Dolder et al. (Ref. 36).

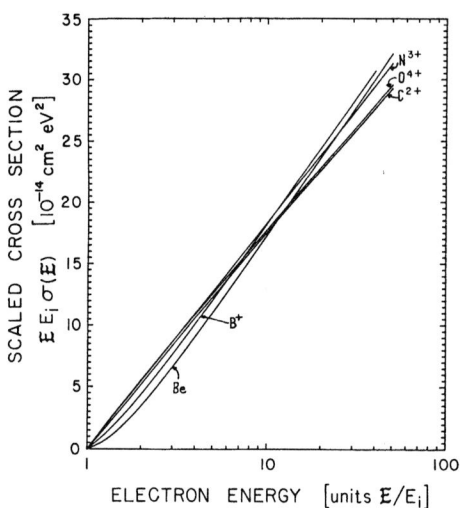

Fig. 14   Scaled cross section for ionisation of beryllium like
          ions.
          Data taken from Bell et al. (Ref. 40).

autoionising transitions do not scale with ionic charge state z in
such a simple manner (see for example Crandall[31]).

Simple scaling based upon the energy levels of the Bohr atom
implies that the ionisation cross section of excited hydrogenic
species will scale as $\sigma_n(E/n^2) \propto n^4 \sigma_o(E)$. Thus the cross section
for H(2s) at the scaled energy (E/n^2) is expected to be about 16
times larger than that of the H(1s) ground state atom. Measured
data are presented in Figure 15 and comparison with the H(1s)
cross section shown in Figure 12 shows that there is reasonable
support for such scaling.

Attention has already been directed in Section 4.3 to the
existance of a multiplicity of interactions which can contribute
to the total ionisation cross sections of multi-electron atoms and
ions. A typical and relevant example of such effects is portrayed
by the ionisation process $e + Fe^+ \rightarrow e + e + Fe^{2+}$ whose cross section appears in Figure 16. Comparison of the measured data with
the scaled plane-wave-Born approximation of McGuire[43] for the
outer and inner shell electrons clearly demonstrates that inner
shell ionisation is the dominant process.

The somewhat similar energy dependence of the cross sections
for ionisation and for excitation of an allowed transition is
demonstrated in Figure 17 by the cross section for excitation of
(1s to 2p) transitions in atomic hydrogen. Also shown is the

316

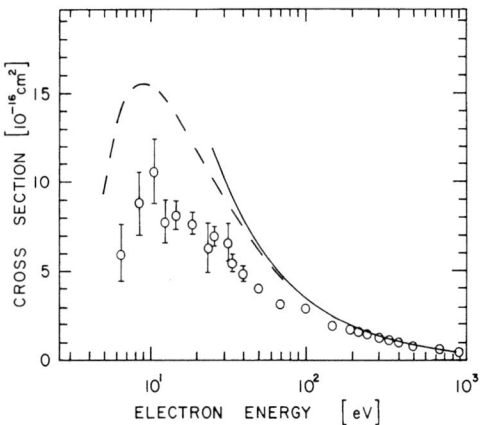

Fig. 15   Cross section for ionisation of metastable H atoms
          $(e + H(2s) \rightarrow e + e + H^+)$.
          Experimental data are taken from Defrance et al. (Ref.
          41). The solid curve is a Bethe approximation
          calculation and the dashed curve a Born approximation
          calculation.

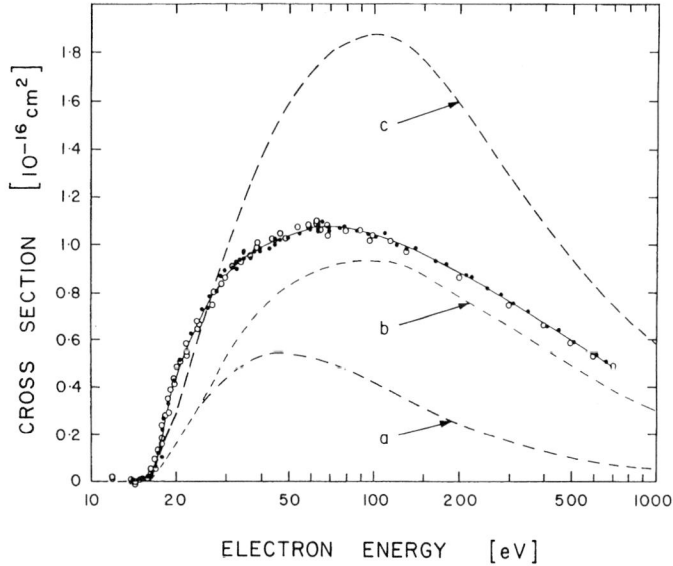

Fig. 16   Cross section for $e + Fe^+ \rightarrow 2e + Fe^{2+}$.
          Data from Montague et al. (Ref. 42). Curves (a) and (b)
          are respectively scaled Born approximations (Ref. 43) for
          outer and for outer plus inner electrons, (c) is the Lotz
          cross section.

cross section for excitation of the disallowed transition (1s to 2s) which has a markedly different energy dependence. The very strong energy dependence of cross sections for excitation of a disallowed transition which also involves a change in multiplicity

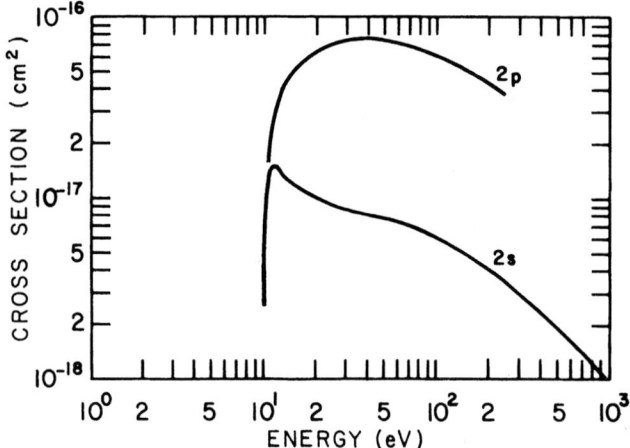

Fig. 17 Excitation cross sections for e + H(1s) → e + H(2p) and e + H(1s) → e + H(2s) [plus cascade contributions from upper levels to the 2s level].
Data are taken from the compilation by Barnett et al. (Ref. 44).

are shown in Figure 18. The data are for excitation of 1^1S ground state helium atoms to the 4^3S and 4^3P states. For comparison, data for allowed transitions to the 3^1P state are also shown.

The low energy dependence of the excitation cross·section of an ion differs significantly from that of an atom. The difference is attributable to the requirement that total angular momentum must be conserved within the collision system. Consider an incident electron whose energy at infinite separation from the atom is exactly equal to the threshold (E_{on}) corresponding to an electron transition which produces a precise change in angular momentum [$\ell_{on} (h/2\pi)$] of the atom. This electron is brought to rest when it excites the atom so that it has zero angular momentum after the collision. In order to conserve the total angular momentum of the system, the incident electron [whose velocity must be $v_e = \sqrt{2E_{on}/m_e}$] is allowed only one specific value of impact parameter. However, a free electron is entitled to have an infinite number of impact parameters; consequently the excitation

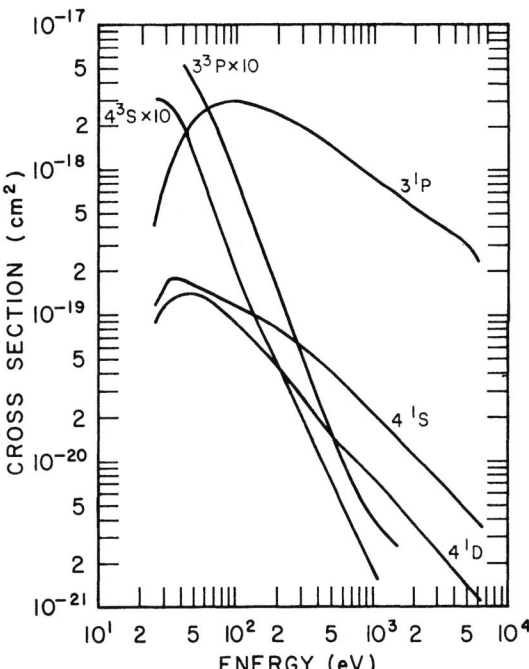

Fig.18   Cross sections for e + He → e + He (4^{1}S; 3^{1}P; 4^{1}D; 4^{3}S, 3^{3}P).

Data are taken from the compilation presented in Ref. 44.

cross section of the atom becomes infinitely small when $E \to E_{on}$. This constraint is removed when the electron collides with an ion because prior to the interaction the incident electron enters the Coulomb field of the ion and is accelerated to energies in excess of E_{on}. As a consequence, excitation of the ion does not bring the electron to rest and so the mobile scattered electron can carry away any angular momentum which is surplus to the collision. Electron impact excitation cross sections of ions are actually finite at energies immediately above the threshold. This can be seen clearly in the measurement of the excitation cross section of the He$^+$(1s → 2s) transition which is shown in Figure 19.

The abrupt onset of excitation is also evident in the excitation of autoionising transitions in ions. Evidence of the effect is apparent in the total ionisation cross sections shown in Figure 20. In certain cases (e.g. Ba$^+$) contributions from autoionisation are the dominant ionisation process. Burgess and Chidichimo[46] have taken account of auto-ionisation in an empirical manner and have proposed an expression for the total ionisation of ions whose charge state exceeds z = 2.

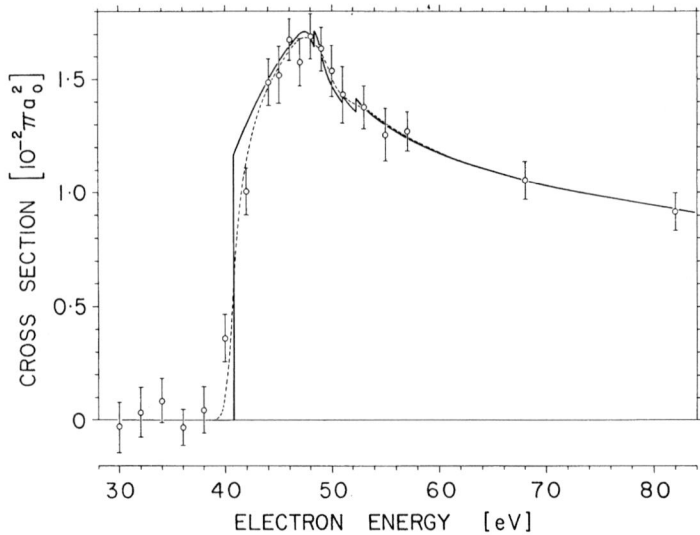

Fig. 19 Cross section for e + He$^+$ (1s) → e + He$^+$ (2s) [plus cascade contributions to the 2s level].
Data from Dance et al. (Ref. 45).

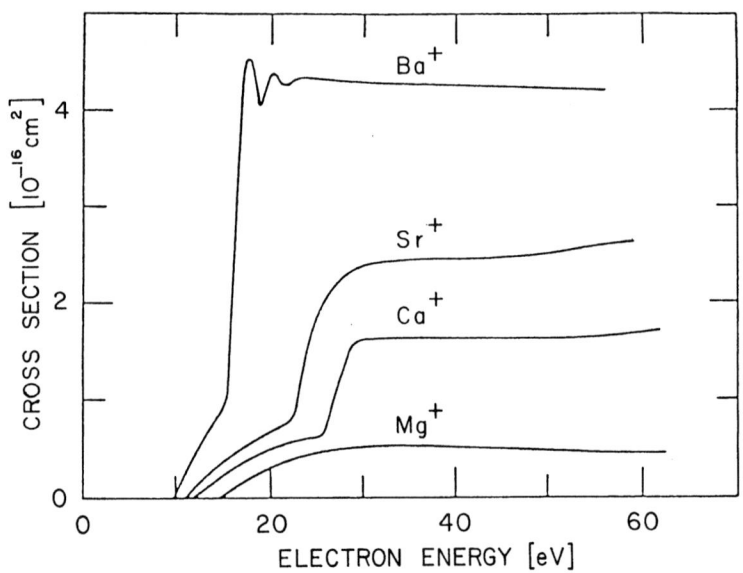

Fig. 20 Measured cross sections for electron impact ionisation of Mg$^+$, Ca$^+$, Sr$^+$ and Ba$^+$.
Data taken from the review by Dolder (Ref. 34).

320

8. ELECTRON COLLISIONS WITH MOLECULES

The principles of collision physics outlined in Section 9 are applicable but the electronic charge distribution of the molecule is more complex. This can readily be appreciated from the wave equation of a molecule in a electronic state n and vibrational state υ, namely

$$\Psi_{n\upsilon JM} = \psi_n(r_e R)\, \chi_{n\upsilon}(R)\, \rho_{JM}(\theta\phi) \qquad (8.1)$$

Here J and M are the rotational quantum numbers (it being assumed that the contribution due to nuclear rotation around the molecular axis is zero), r_e denotes the coordinates of the molecular electrons relative to the nuclei, R is the nuclear separation and θ and ϕ are the polar angles of the nuclear axis relative to a fixed position in space (for example the direction of the incident electron). The term ψ_n gives rise to the electronic energy $E_n(R)$, shown for H_2 by the curves in Figure 6, and the other terms account for vibrational and rotational states.

In the case of neutral molecular hydrogen there exists a substantial amount of measured data for excitation and also for the various collisions (discussed in Section 5.2) which give rise to either H_2^+ or H^+. There are but sparse data for the important process of dissociation into (H + H) atoms because in these measurements it is rather difficult to detect the low energy H atoms. Data for H_2^+ can be obtained from fast colliding beam experiments and, with the exception of photon production, the measured data base is well established.

It must be reiterated that H_2^+ which arises from electron impact upon H_2 will be formed in a distribution of vibrationally excited states (see Section 5.1) and the cross sections for both excitation and ionisation increase strongly with increasing υ. The effect can be seen in Figure 21 which shows the measured cross section for the production of H^+ in electron collisions with H_2^+. It is evident that the cross section increases with decreasing electron energy in the low energy regime below the threshold corresponding to the $\upsilon = 0$ ground vibrational state. However it seems reasonable to assume that the vibrational population in the plasma edge can be determined by the appropriate Frank-Condon factors given by Dunn[48]. Somewhat similar populations are likely to exist in crossed beams experiments so that the existing base of measured data is reasonably applicable to the edge plasma. When applying theoretical data for H_2^+ collisions it is necessary to ensure that the data correspond to the appropriate distribution of vibrational states (see the discussion in Harrison[3]).

A concise but comprehensive survey of electron collisions

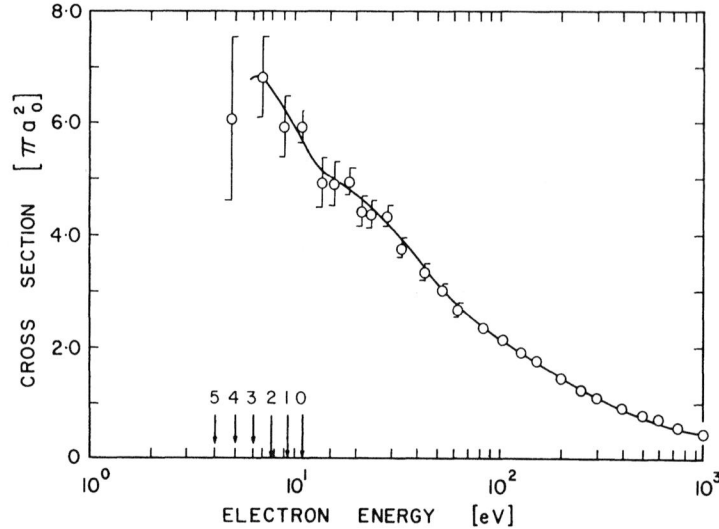

Fig. 21   Cross section for production of protons by electron
          impact on H_2^+.
          Data taken from Dance et al.[47] The cross section
          corresponds to [reactions (26+27) + twice reaction
          (28)]. The threshold energies from the $\mathrm{V} = 0$, 1, 2, 3, 4
          and 5 vibrational levels are shown.

with neutral and ionised H_2, D_2, T_2 together with some simple
hydrocarbons can be found in de Heer[49].

9.   COLLISIONS BETWEEN HEAVY PARTICLES

     This category embraces collisions between two particles each
of which have bound electrons or, in the case of a fully stripped
ion, can capture an electron into a bound state. The consequences
of the collision can be the exchange of one or more electrons
between the particles (i.e. electron capture), the ejection of
bound electrons (ionisation of the target or stripping of an
energetic incident particle), direct excitation of the electronic
states of the particles and indirect excitation which arises due
to electron capture into excited states. When the colliding
particles are widely separated each has the electronic charge
distribution of the individual atomic species but as the particles
approach each other these distributions may overlap and the collision system can be considered as a transitory molecule. After the
collision, the product species move apart and each of them take up
the characteristics of individual atomic species. It is therefore

322

convenient, especially at low collision velocity, to model the
collision on the concept of transitions between the states of the
molecules which describe the collisions partners before and after
their interaction. The internuclear separation is here related to
the distance between the colliding particles and the time during
which the transitory molecule exists i.e. the collision time,
t_{coll}, is related to the collision velocity and the dimension of
the system. Except at very high collision velocities, the wave
length of the incident heavy particle is appreciably larger than
the dimensions of the collision system and the particle trajectories can therefore be described by classical mechanics but the
transitions which involve the bound electrons must be treated by
quantum theory. This is in contrast to electron collisions where
the impact parameter must be quantised because of the short wavelength of the incident electron. At high velocities quantum
theory must also apply to heavy particle collisions. Indeed,
when $E >$ 100 keV, the cross sections for proton impact ionisation
and excitation are similar to those for electron impact collisions
at the same velocity and can be well described by the Born
approximation.

At such high energies the collision time is much shorter than
the time required for transitions between the states of the transitory molecule so that molecular effects tend to be insignificant.
However, when the particles move slowly, the electronic transition
time ($t_e \sim h/\Delta E$ where ΔE is the difference in potential energy
between the molecular states) can be much less than the collision
time $t_{coll} = a/v$. Here a is the range of the collision which can
be related to molecular dimensions (typically 2 to 3 Å) and v is
the collision velocity. At low collision velocity the transfer of
momentum to the bound electrons is insufficient for direct
excitation and ionisation so that collisions in this regime are
predominantly related to charge exchange interactions of the type

$$A^+ + B \rightarrow A + B^+ + \Delta E. \tag{30}$$

Here ΔE is the difference in potential energy between the left
hand and right hand sides of reaction (39) and it can be either
positive (exothermal) or negative (endothermal). In either case
the imbalance is transferred into a change in kinetic energy of
the colliding particles.

Recent reviews of the theories which are appropriate to the
various regimes of heavy particle collisions have been provided by
Brandsden[50] and a critical appraisal of theoretical data for
charge exchange between H atoms and highly charged ions can be
found in Janev and Brandsden[51]. A critical survey of experimental
data for both electron capture and ionisation during collisions
with hydrogen is also available (Gilbody[52]). Both experimental

and theoretical data have been reviewed by de Heer[39] who also discusses the experimental methods employed.

9.1 Electron Capture by Singly Charged Ions

If the presence of non-thermalised energetic particles is neglected then the upper limit of collision energies in the edge plasma is only a few hundred electron volts so that direct ionisation and excitation tend to be insignificant in collisions between heavy particales. Charge capture from recycling hydrogen atoms is important and it is therefore the main topic of the following discussion. For the sake of conciseness interactions involving hydrogen molecules and negative ions are not considered. A brief survey of molecular processes which are likely to be significant in the edge plasma can be found in Janev et al[8].

In the case of low-energy collisions Massey[53] has postulated that any perturbations of the molecular states caused by the collision will have little effect (because the perturbation frequency cannot resonate with the transition frequency) whenever the collision time, t_{coll}, is substantially greater than t_e. The implication of this aidiabatic criterion is that the cross section for the collision is small if

$$a \left| \Delta E \right| / h \nu \gg 1 \tag{9.1}$$

Other workers, notably Hasted[54], have invoked the corollary of the aidiabatic criterion (namely the diabatic condition) which infers that a charge exchange cross section reaches its peak at a velocity $\hat{v}$ given by

$$\frac{a}{\hat{v}} \left(\frac{\Delta E}{h} \right) = 1 \tag{9.2}$$

According to Eq.(9.2) the peak occurs at a collision energy $\hat{E}$ given by

$$\hat{E} = 36 \left| \Delta E \right|^2 m_i a^2 \quad [eV] \tag{9.3}$$

where m_i is the mass (in units of the electron mass m_e) of the incident particle and a is in units of a_o. A typical example of this condition is the cross section for the reaction

$$He^+(1s) + H(1s) \rightarrow He(1s^2) + H^+ + (11 \text{ eV}) \tag{31}$$

which is shown in Figure 22. The value of $\hat{E}$ is about 20 keV/amu which is consistent with Eq.(9.3). At collision energies below $\hat{E}$, the interaction is increasingly aidabatic and the cross section may have the form

$$\sigma_{cx} \approx A \exp \left(- B \Delta E / h \nu \right) \tag{9.4}$$

324

where A and B are constants. At energies $E > \hat{E}$ the collision time becomes small and the cross section decreases with increasing collision energy as can be seen in Figure 22.

Clearly the case of symmetric, resonant charge exchange is a special case and the simplest of such reactions,

$$H^+ + H(1s) \underset{\leftarrow}{\rightarrow} H(1s) + H^+ + (\Delta E = 0) \qquad (32)$$

plays many roles in fusion research. Because the reaction is energy resonant Eq. (9.2) retains physical significance only when $\hat{v} = 0$ and this implies that the cross section is a maximum at zero collision energy. The measured data, which clearly demonstrate this characteristic, are shown in Figure 23. The aidiabatic condition becomes progressively less relevant with increasing collision energy and the theoretical cross section determined by a semiclassical impact parameter treatment, has the form

$$\sqrt{\overline{\sigma}_{cx}} = a - b \ln v \qquad (9.5)$$

where a and b are constants. The curve, which is fitted to the measured data in Figure 23, demonstrates this type of energy dependence.

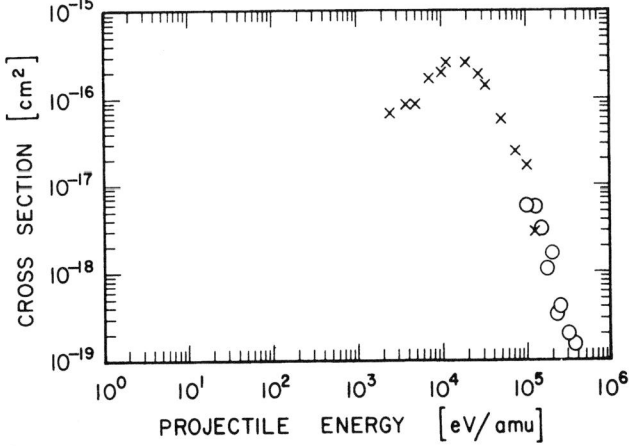

Fig.22   Cross section for $He^+(1s) + H(1s) \rightarrow He(1s^2) + H^+$. The circles show measured data of Olson et al.[55] and the crosses measured data of Hvelplund and Andersen (Ref. 56).

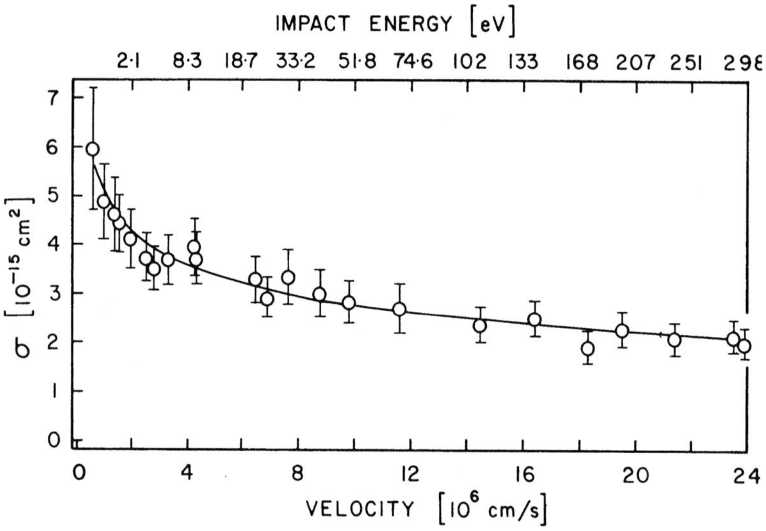

Fig. 23   Cross section for $H^+ + H(1s) \rightleftarrows H(1s) + H^+$
Experimental data are from Newman et al.[57]. The fitted
curve is based upon the expression
$\sigma_{cx} = (7.07 - 1.83 \log_{10} v)^2$ which is given by Greenland
(Ref. 58).

Collisions involving dissimilar particles are not in general
energy resonant but there are many examples of accidental resonance. Some are of relevance to the boundary plasma and one such
is

$$0^+(^4S) + H(1s) \rightarrow 0(^3P_J) + H^+ \quad \begin{bmatrix} -0.01 \text{ eV } (J=0) \\ +0 \text{ eV } \quad (J=1) \\ +0.02 \text{ eV } (J=2). \end{bmatrix} \quad (33)$$

The characteristics of resonant charge capture are clearly evident
in the measured cross section shown here in Figure 24. However
this cross section is smaller than the symmetric resonant case of
$H^+ + H$. The difference can be ascribed to the fact that H(1s) has
only a single level and so each H(1s) interaction with H^+ has a
zero energy defect. By contrast the levels of the atomic species
involved in the capture process $0^+ + H$ are multiple so that
$\Delta E \sim 0$ is not valid for transitions between some sub-levels. Thus
it is necessary to weight the electron capture probability by a
factor which is less than unity and which is related to the
statistical weights of the atomic systems involved in both the
forward and the reverse direction of the collision process. The
reverse reaction ($H^+ + 0$; Ref.60) exhibits the same energy dependence but its magnitude is smaller by a factor of about 8/9.

326

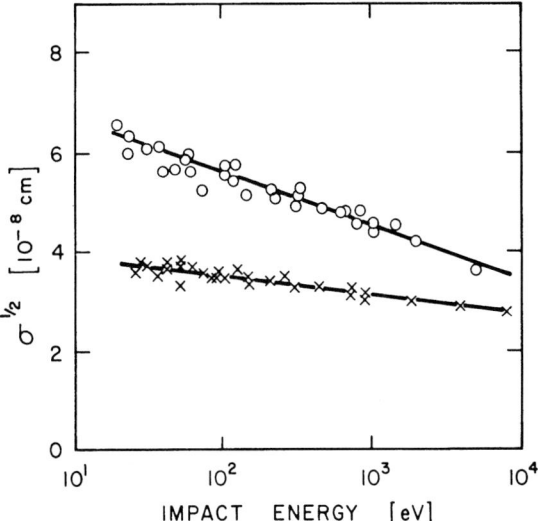

Fig. 24 Measured cross sections for $H^+ + H \rightarrow H + H^+$ and for $O^+ + H \rightarrow O + H^+$.
Data taken from Ref. 59; the circles refer to H^+ and the crosses to O^+.

9.2 Charge Exchange Involving Multiply Charged Ions

It will be noted that for reactions (31), (32) and (33) the value of ΔE has been taken to be equal to the difference between the potential energy of the isolated atomic species. For this assumption to be valid, the value of ΔE must tend to be independent of the internuclear separation so that $d\Delta E/dR \rightarrow 0$. The molecular potential energy curves which describe many collisions involving singly ionised ions and neutral particles tend to display this characteristic and a typical case for $He^+ + H$ is shown in a qualitative manner in Figure 25(a). There are however cases where ΔE varies strongly with R and this is most evident in collisions in which both particles are ionised and therefore experience Coulomb repulsion either before or after the interaction. Typical examples of such collisions are:

$$A^{(z+1)} + B \rightarrow A^{z+} + B^+ + \Delta E(R) \tag{34}$$

or

$$A^{z+} + B^+ \rightarrow A^{(z+1)+} + B + \Delta E(R) \tag{35}$$

where $z \geqslant 1$. The potential energy curves for these reactions are of the form shown qualitatively in Figure 25(b). The intersection of the curves occurs at

$$R_c \approx (z-1)/\Delta E \qquad\qquad (9.6)$$

where ΔE is the value at infinite separation. The parameters in Eq. (9.6) are in atomic units (a u), namely length in a_o, z in units of the electronic charge e and the unit of energy is $(2 \times E_{i,H} = 27.21$ eV). It is clear that charge exchange is most

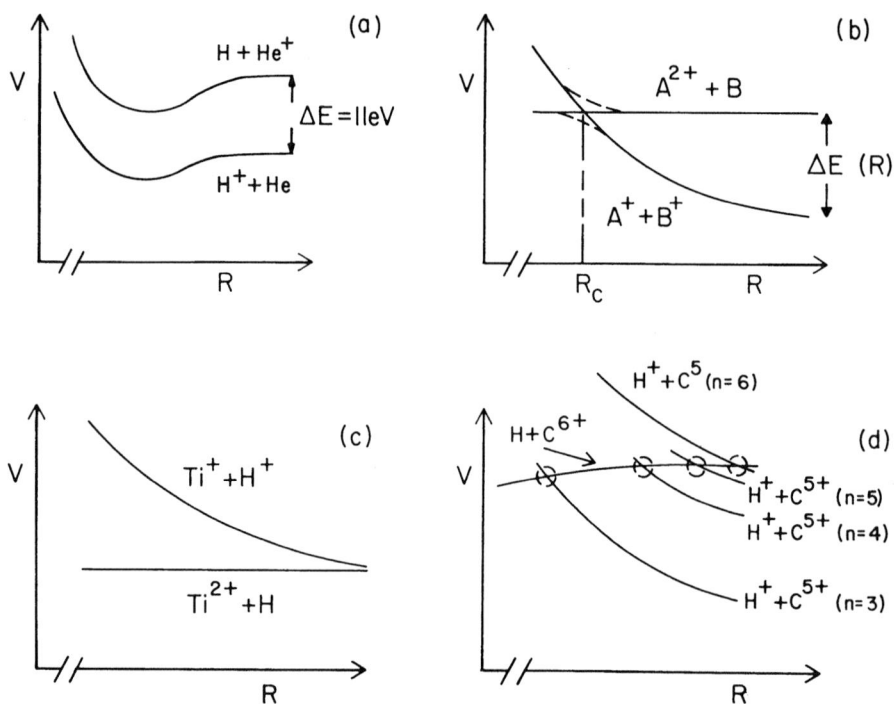

Fig. 25 Qualitative representations of potential diagrams for charge exchange collisions.
(a) $He^+ + H(1s) \rightarrow He(1s^2) + H^+$

(b) Curve crossings
Psuedo crossings are indicated by the dashed aidiabatic curves and crossings by the solid diabatic curves.

(c) $Ti^+ + H^+ \rightarrow Ti^{2+} + H$

(d) $H + C^{6+} \rightarrow H^+ + C^{5+}$

likely to take place at internuclear separations around R_c so that the maximum cross section will be of order

$$\hat{\sigma}_{cx} \sim \pi R_c^2. \tag{9.7}$$

At low collision energy, and hence within the aidiabatic regime, the initial (i) and final (f) states correspond to those of a steady state molecule so that the non-crossing rule (Wigner and Witmer[61]) requires that the potential curves of states with the same symmetry do not cross. Thus the aidiabatic potentials are of the form indicated by the dashed curves in Figure 25(b) and only psuedo-crossings can occur. At higher energies (in the diabatic regime) a jump is possible between the potentials curves and so the solid (diabatic) curves apply. The energy defect ΔE becomes small when $R \to R_c$ and it is evident from Eq.(9.3) that $\hat{E}$ is much smaller than in cases where no crossing or psuedo crossings take place. In the low velocity regime ($v < 1$ au, i.e. less than 2.2×10^8 cm/s) where the molecular aspects of the collision dominate, the Landau-Zener theory[62] has been applied for transitions between levels of the same symmetry. According to this approach the diabatic condition Eq.(9.2) is modified to take the form

$$\frac{a' \, \Delta E(R_c)}{h \, \hat{v}} = 1 \tag{9.8}$$

where

$$a' = \frac{\Delta E(R_c)}{\frac{d}{dR}(V_i - V_f)_{R = R_c}} \tag{9.9}$$

and

$$\Delta E(R_c) = 2 \, H_{if}. \tag{9.10}$$

Here H_{if} is the matrix element for the coupling of the diabatic states. The denominator in Eq.(9.9) is the difference in the slopes between the diabatic potential curves (V_i and V_f) at R_c. Such calculations take note of the fact that the particles must pass through R_c as they approach each other and also as they separate after the collision.

The reactions

$$Ti^+ + H^+ \to Ti^{2+} + H + (\Delta E \approx 0) \tag{36}$$

and

$$Fe^+ + H^+ \to Fe^{2+} + H + (\Delta E \approx 2.4 \text{ eV}) \tag{37}$$

which are shown in Figure 26 are relevant examples of the effects

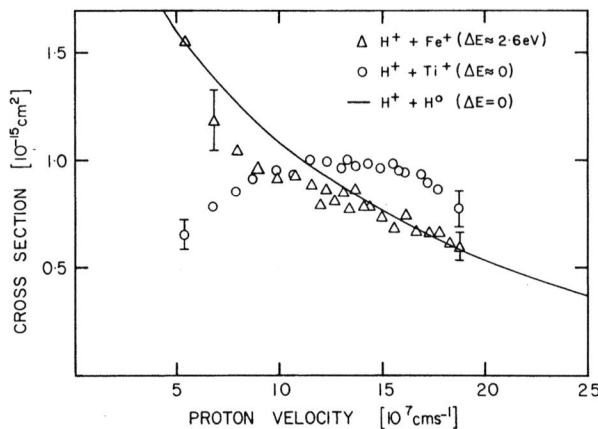

Fig. 26   Measured cross sections for the production of doubly
          charged ions in collisions between protons and Ti^+ and
          also Fe^+ ions. Data which are taken from Ref. 63 relate
          predominantly to charge capture by protons.

of variation of ΔE with R. The $Ti^+ + H^+$ collision for which
$\Delta E \approx 0$ at infinite separation exhibits the characteristic of a
non-resonant interaction because the potential energy curves are
expected to be of the form shown in Figure 25(c). In contrast,
the $Fe^+ + H^+$ collision tends to display resonant behaviour which
indicates that the value of ΔE at R_c must be very much smaller
than ΔE at infinite separation.

The ionisation potential of a multiply charged ion A^{z+} always
exceeds that of any neutral collision partner so that capture into
an excited state of the $A^{(z-1)*}$ ion is a likely event. One of the
simplest reactions is

$$He^{2+} + H(1s) \rightarrow He^+(n=2) + H^+ + (\Delta E=0).$$ (38)

However this cross section is small in the aidiabatic regime
because the Coulomb repulsion between the He^+ and H^+ ions causes
the potential curves to take a form comparable to that shown for
$Ti^+ + H^+$ in Figure 25(c). However, in the case of higher charge
states, the number of available excited states becomes larger and
multiple curve crossings, such as those shown for $H + C^{6+}$ in
Figure 25(d), can contribute significantly to charge exchange. In
this particular interaction the partners $[H + C^{5+}(n=6)]$ have
$\Delta E = 0$ at infinite separation but the charge exchange interaction
is expected to be dominated by the $[H + C^{5+}(n=4)]$ partners. The
effect of such multiple crossings is to flatten the peak of the
cross section. As the ion charge state increases more and more
curve crossings are involved and the weaker becomes the velocity

330

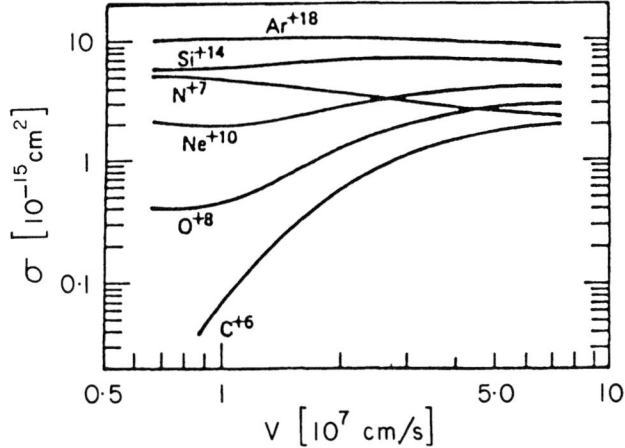

Fig. 27 Total Landau-Zener cross sections for charge exchange
between H(1s) atoms and various stripped ions.
Data taken from Salop and Olson (Ref. 64).

dependence of the cross section. The trend, as predicted by the
Landau-Zener theory, is illustrated in Figure 27 and there are
experimental data (see the reviews in Refs. 39 and 52) which
support this characteristic behaviour. Greenland[58] has compiled
and assessed data for low energy charge exchange between H atoms
and multiply charged ions of particular relevance to the plasma
edge. He also provides convenient analytical expressions for
these cross sections.

In the region of low velocity (i.e. v < 1 au) and low charge
state (z ≤ 4) there are relatively few curve crossings especially
in collisions which involve simple atomic systems such as fully
stripped ions and/or H atoms. The electron capture cross sections
are then strongly dependent upon the detailed structure of the
molecular potentials and so no simple scaling with ion charge
state can be established. However, for collisions involving non-
hydrogenic atoms and/or partially stripped ions, the number of
crossings increases very strongly with z and the collision range
[R_c in Eq. (9.6)] can be considered to vary continuously with z.
Indeed the number of crossings and hence collision channels
becomes so great that electron capture can be regarded as the
decay of the initial electronic state into the quasi-continuum of
final ionic states and so the binding energy of the initial state
can also be related to z. Janev and Hvelplund[65] propose (for ions
with z ≥ 5) a scaling relationship based upon a reduced charge
exchange cross section ($\tilde{\sigma}_{cx} = \sigma_{cx}/z$) and a reduced velocity
($\tilde{v} = vz^{-1/4}$). The scaling is

$$\tilde{\sigma}_{cx} = \sigma_{cx}(\tilde{v}) \; z^{\alpha(\tilde{v})}. \qquad\qquad (9.11)$$

Curves of $\tilde{\sigma}_{cx}$ versus reduced collision energy ($\tilde{E}/\sqrt{z}$) have been fitted to a wide range of experimental data and those for H and He target atoms are shown in Figure 28. Three regimes of the parameter α can be identified and related to the collision velocity, namely:

$0.1 \leqslant \tilde{v} \leqslant 1$:- α is weakly dependent upon $\tilde{v}$ and it has a value close to unity.

$\tilde{v} > 1$:- α increases to an asymptotic value ($\alpha = 5$) when $\tilde{v} \gg 1$.

$\tilde{v} \approx 2$ to 4 :- α becomes constant with a value close to 3.

The first of these regimes is the one most related to the plasma edge.

The preceding discussion has emphasised the role of ground state hydrogen atoms, but charge exchange may also involve excited

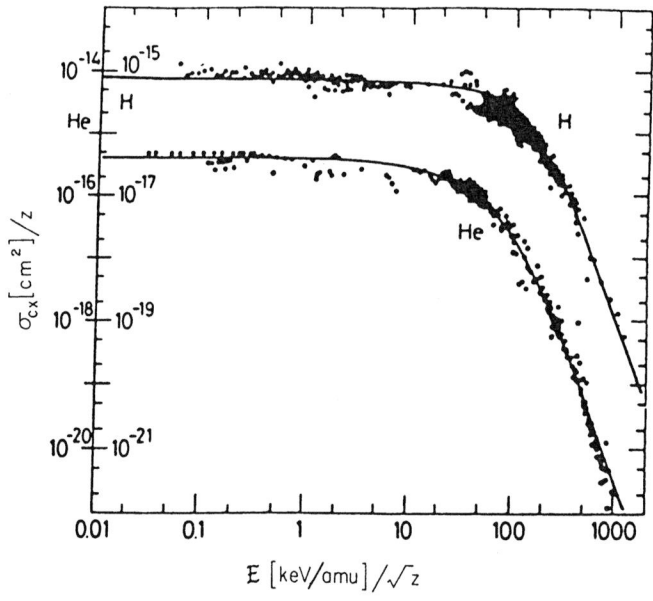

Fig. 28   Reduced electron capture cross section for ions with
          $z \geqslant 5$ in H and He atom targets.
          Data taken from Janev and Hvelplund (Ref. 65).

H atoms. Cross sections[66] for charge exchange between $H(n) + He^{2+}$ are very large $\sim 10^{-12}$ cm^{-2} when $n > 8$ but such highly excited states of the hydrogen atom are unlikely to be strongly populated in the boundary plasma (see Section 10.1).

10. INFLUENCE OF THE PLASMA ENVIRONMENT

Application of basic cross section data must take account of the distribution of collision velocities within the plasma and also of the effects of multi-step collisions of the type discussed in Section 3.2. The possible presence of non-thermalised particles is neglected in order that the velocity distribution of plasma particles can be taken to be Maxwellian. The rate coefficient for an inelastic process with threshold energy E can then be calculated using

$$\langle \sigma v \rangle = \left(\frac{8k_B T}{\pi m} \right)^{1/2} \int_{E/k_B T}^{\infty} \sigma(E)\ (E/k_B T)\exp(-E/k_B T)d(E/k_B T). \quad (10.1)$$

The influence of this integral can be appreciated by comparing the electron impact ionisation cross section for H(1s) shown in Figure 12 with the corresponding ground state rate coefficient $S_i(g)$ shown in Figure 2. When $k_B T_e < E$ [say ~ 2 eV for ionisation of H(1s) atoms] the form of the velocity distribution dominates the form of the rate coefficient so that uncertainty in the atomic cross section is of minor significance. However, when $k_B T \geqslant E$ the rate coefficient is sensitive to the cross section. It is useful when considering recycling of atomic hydrogen at boundary surfaces to note that, in a low temperture regime, the variation of the ionisation rate coefficient S_i with electron temperature is much greater than the possible variation of plasma parameters, e.g. $S_i(T_e)$ increases by a factor 10^4 over a temperature range $k_B T_e = 1$ to 10 eV. At moderate n_e, ionisation tends to take place near to the 5 eV isotherm.

10.1 Radiation Power Loses from Recycling Hydrogen

Electron collisions with hydrogen atoms and with protons give rise to radiative power losses as a consequence of the various reactions discussed in Section 3.2. The balance between these conflicting processes, which was first elucidated by Bates et al.[7], has been discussed in a number of papers, notably Bates and Kingston[67], McWhirter and Hearn[68] and Hutcheon and McWhirter[69]. Its relevance to the boundary plasma has been considered by Harrison[2,3] and by Janev et al[8]. Ionisation occurs either by a direct transition of the bound electron to the continuum (reaction 1, Section 3.2) or as the consequence of a sequence of transitions between excited levels (reaction 2) which terminates at the

continuum. The likelihood of the latter route depends upon the
balance between the lifetime t_n of the excited states and the
associated electron collision times

$$t_{en} = (n_e < \sigma_n v_e>)^{-1}.$$ (10.2)

The lifetime of the state increases with increasing principle
quantum number n and so does the collision cross section
(i.e. $\sigma_n \propto n^4$) so that the multi-step route to ionisation becomes
dominant for all but the lowest n states when $n_e \sim 10^{14}/cm^3$ or
larger. In addition to the excited state n being destroyed by
either an upward transition or direct ionisation it can also be
destroyed by a downward, super-elastic collision (reaction 4). In
effect the ability of the excited atom to radiate is reduced in
favour of (i) ionisation by a chain of non-radiative upward
transitions and (ii) by a complementary chain of non-radiative
downward transitions which populate the ground state of the atom.
In the limit of low electron density ($n_e \sim 10^{10}/cm^3$), the collision time t_{en} is much smaller than t_n so that the effects of
multi-step processes can be neglected but, at higher density
($n_e \sim 10^{16}/cm^3$) most of the radiation is suppressed. It is convenient to express the volume rate for production of protons in
the form

$$v_{H^+} = n_e \left[n_o(g) \, S_{CR} - n_i \alpha_{CR,H} \right].$$ (10.3)

Here the collisional radiative coefficients are composites of
several components which account for the multi-step routes;
$S_{CR}(T_e)$ for ionisation and $\alpha_{CR,H}(T_e)$ for electron-proton recombination. Both two body radiative processes (reaction 6) and
three body processes (reaction 7) must be included, i.e.

$$\alpha_{CR,H}(T_e) = [\alpha(\text{two-body}) + \alpha(\text{three-body})].$$ (10.4)

The ion recombination time

$$t_\alpha = (n_e \, \alpha_{CR,H})^{-1}$$ (10.5)

is generally smaller than the proton residence time in the boundary plasma so that recombination tends to be insignificant. In
contrast, the electron-atom collision times t_{en} are sufficiently
short for a quasi-steady state population of excited levels to be
established by multi-step processes. In such conditions the
average energy required to produce one proton-electron pair (E_{ion})
can be considered in a collective manner as the amount of energy
expended in ionisation plus the amount radiated from the small
number of low lying excited states which are not depopulated by
multi-step processes. Following the approach of McWhirter and
Hearn[68], this energy can be expressed as

334

$$E_{ion} = \frac{(E_i S_{CR} + P_1 \, n_e^{-1})}{S_{CR}} \qquad (10.6)$$

where P_1 is the coefficient for line radiation losses defined in relation to the density of ground state atoms [in a manner analogous to that indicated in Eq. (10.15)]. Estimated values of E_{ion} which are shown in Figure 29 are taken from a calculation by McWhirter and Hearn which relates to hydrogenic ions (e.g. He^+, Li^{2+} etc.) so that the plasma parameters can be scaled with atomic number Z in the following manner:

$$[T_e]_Z \equiv Z^2 [T_e]_H; \quad [E_{ion}]_Z \equiv Z^2 [E_{ion}]_H \text{ and } [n_e]_Z \equiv Z^{-7} [n_e]_H.$$

The plotted data are for the equivalent hydrogen "atom" and so correspond to Z = 1. These authors employ Coulomb Born ionisation cross sections for the determination of S_{CR} but the Coulomb acceleration causes such calculations to overestimate the ionisation cross section of the hydrogen atom at low electron energy (as can be seen in Figure 13). The extent of this overestimation can be seen by comparison with the data points in Figure 29 which refer specifically to hydrogen atoms. More recent detailed calculations for hydrogen atoms by

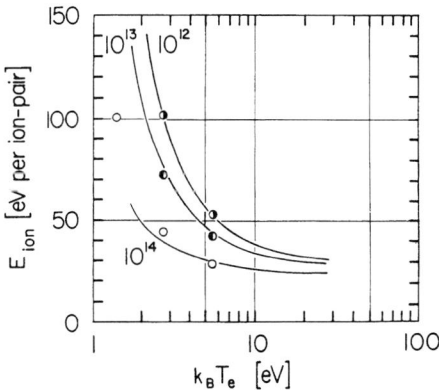

Fig.29   The average energy require to produce one electron-proton pair in atomic hydrogen.
Solid lines are data for hydrogenic ions taken from McWhirter and Hearn[69] and the circles show the data for hydrogen atoms (taken from Bates et al.[7] and Bates and Kingston (Ref. 67).

Janev et al.[8] are in close agreement with these data. In this collective concept, the amount of radiated energy associated with each ionisation event is given by

$$E_H^{rad} = E_{ion} - 13.6 \qquad [eV].\qquad\qquad (10.7)$$

It is worthwhile noting the Z dependence of the scaling relationships which imply that multi-step processes for He^+ are of but slight significance in the boundary plasma where n_e is unlikely to exceed a few $10^{14}/cm^3$.

The influence of the confining magnetic field upon the excited state population has been neglected in the preceding discussion. Its effects are two-fold. Firstly there are Lorentz forces $[e(\vec{B} \times \vec{v})]$ exerted upon the bound electron due to atom motion across the magnetic field and the associated electric field can destroy the higher n levels by field ionisation. Ionisation of highly excited Rydberg states by electric fields is discussed in detail by Brouillard[70]. Janev et al.[8] estimate for typical boundary plasma conditions that Lorentz forces lower the ionisation continuum so that it coincides with an n value of about 26. Such lowering of the ionisation continuum by Lorentz forces, and also by electric fields which arise due to statistical deviation from local charge neutrality of the plasma, are in practice insignificant because the destruction rate of these high n states by electron collisions is very large indeed. Secondly the presence of the magnetic field causes the atomic levels to be split due to the Zeeman effect (which is described, for example, in Ref. 9) so that neighbouring states become mixed. Stark splitting, which arises due to electric fields (such as the Lorentz field) also causes level mixing. In the particular case of atomic hydrogen such mixing dramatically reduces the lifetime of the metastable $2^2S_{\frac{1}{2}}$ state which mixes readily with the short lived $2^2P_{\frac{1}{2}}$ state (for a detailed discussion of the spectra of the H atom see Ref. 71). Consequently the effects of metastable atoms can be ignored. This does not conflict with the preceding calculations which account only for the principle n levels and thereby neglect contributions from sub-levels.

10.2 Charge State Distribution of Impurity Ions and Radiative Power Losses

The treatment of impurity species follows along similar lines to that described for hydrogen atoms and hydrogenic ions but it is more complicated due to the complex nature of the electron configuration of these atomic species. It is in general reasonable to accept that ionisation by electron collisions proceeds in a stepwise manner, i.e. $X^o \rightarrow X^+ \rightarrow X^{2+} \dashrightarrow$ etc., so that the steady state balance of an ionisation stage z can be expressed as

336

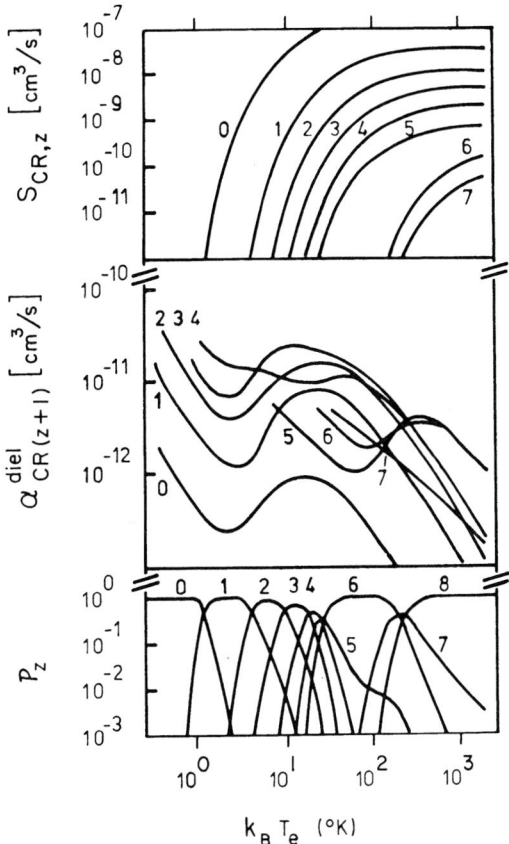

Fig. 30   Collisional radiative rates for ionisation, $S_{CR,z}(T_e)$,
dielectronic recombination $\alpha_{CR(z+1)}^{diel}(T_e)$ together with the
charge state population (P_z) for oxygen.
Data which are taken from Summers[73] are for $n_e =$
$10^{12}/cm^3$.

The ion charge state z is shown for each curve.

$$n_e n_{(z-1)} \, S_{CR(z-1)} - n_e n_z S_{CR,z} + n_e n_{(z+1)} \, \alpha_{CR(z+1)}$$

$$- n_e n_z \, \alpha_{CR,z} - n_z/\tau_z = 0. \qquad (10.8)$$

Here the ion density (n_z etc.) refers to the ground state and any
contribution from the population of excited states are accommodated
by the use of appropriate collisional radiative coefficients.
Details of the physics involved and of the modelling employed in
determining these coefficients can be found in McWhirter and
Summers[6] and also in Drawin[72]. Dielectronic recombination
(reaction 12) provides a powerful route for recombination of these
multi-electron species and so the various contributions to recombination

$$\alpha_{CR,z} = [\alpha(\text{two body radiative}) + \alpha(\text{dielectronic}) + \alpha(\text{three body})] \qquad (10.9)$$

must be included.

The time τ_z in Eq. (10.8) is the residence time of the ions in the particular plasma region under consideration. If this region is sited deeply within the closed confinement field of the plasma then

$$\tau_z \gg [n_e \, \alpha_z(T_e)]^{-1} \qquad (10.10)$$

and the charge state density population tends to be in equilibrium. It is then governed by the electron collision rates so that

$$\frac{n_{(z+1)}}{n_z} = \frac{S_{CR,z}(T_e)}{\alpha_{CR(z+1)}(T_e)}. \qquad (10.11)$$

This is the condition of "local thermal equilibrium" which is often referred to as "coronal equilibrium" and its characteristics are illustrated for the example of oxygen in Figure 30. It should be noted that the population ($P_z = n_z/\Sigma n_z$) of the charge state z is substantial when $S_{CR,z}(T_e) = \alpha_{CR(z+1)}(T_e)$ and that this equality occurs in the regime where $k_B T_e < E_{i,z}$. Thus the population of ionisation stages tends to be sensitive to the low temperature regime of the ionisation rate coefficient.

If the plasma region under consideration lies close to the edge or if the effective drift velocity of the atomic particles is large (one such example arises when substantial numbers of energetic atoms are injected) then the inequality expressed in Eq. (10.10) is no longer valid and the ion loss rate in Eq. (10.8) becomes dominant. These "non-coronal" conditions are particularly evident in the open magnetic field region of the plasma edge because here ions are lost from the system due to rapid transport along the field to the boundary surfaces so that the effects of recombination are substantially reduced. It is convenient to simplify Eq. (10.8) by assuming that the residence time of impurity ions is insensitive to their charge state (i.e. $\tau_z = \tau_{imp}$) and the results of one such calculation for oxygen by Abramov[74] are shown in Figure 31. Here the average charge state

$$\bar{z} = \frac{\Sigma z n_z}{\Sigma n_z} \qquad (10.12)$$

is plotted as a function of $k_B T_e$ for various values of $n_e \tau_{imp}$; typically $n_e \tau_{imp} \sim 5 \times 10^{10}$ cm^{-3}s in the boundary plasma.

Estimation of τ_z or τ_{imp} requires a detailed knowledge of plasma transport in the edge plasma. Neuhauser et al.[75] have

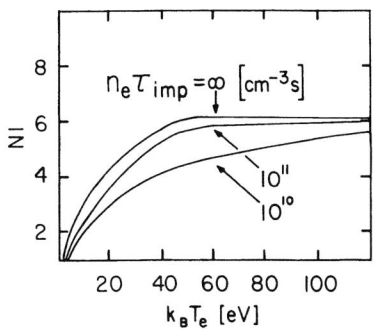

Fig. 31   The average charge state $\bar{z}$ of oxygen plotted as a
function of electron temperature.
Data are taken from Abramov[74] and the condition
$n_e\tau_{imp} = \infty$ is equivlent to the coronal equilibrium
conditions shown in Figure 30.

evolved both one dimensional and quasi-two dimensional models for
impurity ion transport parallel to the magnetic field within the
drifting hydrogen plasma of the scrape-off and divertor region of
a tokamak. They find that the peak of the charge state distribution is likely to occur at z ~ 2 to 3 when most impurity ionisation occurs within the high recycling region close to the divertor
target. A similar conclusion is reached by Harrison[2,3] but his
approach is based upon a much simpler concept; recombination is
neglected and τ_{imp} set equal to the thermalisation time of the
impurity ions within the drifting background hydrogen plasma. It
must be stressed that such low charge states are attained solely
by impurities which recycle in the immediate vicinity of a
divertor target (or limiter). The residence time of
ions which originate from the bulk of the first wall can be quite
long because of the low flow velocity within the scrape-off plasma
when high recycling occurs at a divertor target. In such cases
z ~ 10 may be encountered for medium and high z ions.

Contributions to the balance of ionisation states of impurities due to charge capture from H atoms is to be expected whenever
the local density of atomic hydrogen is significant and the charge
state of impurity ions moderately high. A discussion of the
likely significance of such effects can be found in Drawin[1].
Hydrogen atom density can be high in the recycling region adjacent
to boundary surfaces but here the average charge state of
impurities is insufficient for charge exchange to contribute
substantially. In general, the effects of charge exchange are

only likely to impact upon the ionisation stage population when
fast atoms are injected into the plasma.

The population of excited states reaches a steady state value
in a much shorter time than does the charge stage distribution.
The equilibrium balance of excited states can be conveniently
expressed in the form

$$\frac{n_{zq}}{n_{zo}} = \frac{n_e\, S_{oq}^{CR}}{\sum\limits_{p<q} A_{(q\rightarrow p)}} \qquad (10.13)$$

where n_{zq} is the density of ions (of charge stage z) in excited
level q, n_{zo} is the density of groundstate ions, S_{oq}^{CR} is the collisional radiative excitation coefficient for transitions $0 \rightarrow q$
and $(A_{q\rightarrow p})^{-1}$ is the lifetime for spontaneous decay from q to a
lower level p. The density of power radiated due to the spontaneous decay of level q to p is

$$P_{z(q\rightarrow p)} = n_{zq}\, A_{(q\rightarrow p)}\, E_{pq} \qquad [1.6 \times 10^{-19}\, W/cm^3]. \qquad (10.14)$$

The power loss due to line radiation from ionisation stage z

$$P_{zl}\,(T_e) = \sum_q P_{z(q\rightarrow p)}(T_e) \qquad (10.15)$$

is determined by summation over those q levels which (a) have a
significant excitation rate coefficient, (b) are not depopulated
by multi-step processes and (c) emit photons which carry a
significant amount of energy. Details of modelling methods can be
found in McWhirter and Summers Ref. 6. As would be expected from
the discussion of excitation given in Section 8, the summation in
Eq. (10.15) tends to be dominated by transitions in which $\Delta n = 0$.

The total power loss $P_{zt}(T_e)$ due to radiation associated with
charge state z must also include contributions from recombination
which are of the form

$$P_{\alpha z} \sim n_e\, n_z\, \alpha_{CR,z}\,(T_e)\, \Delta E \qquad (10.16)$$

where ΔE is the total amount of kinetic energy lost by the plasma
electron. These power loss components can be grouped in the
following manner

$$P_{tz}(T_e) = [P_l(T_e) + P_\alpha(T_e) + P_{br}(T_e)]_z = n_e\, n_z\, F_z(T_e) \qquad (10.17)$$

so that the radiated power function

$$F_z(T_e) = P_{tz}(T_e)/n_e\, n_z \qquad [W\ cm^3] \qquad (10.18)$$

340

can be used as a measure of the radiating efficiency of each
ionisation stage z. Summation of $F_z(T_e)$ over the population of
ionisation stages yields the radiated power loss coefficient
$F = P(T_e)/n_e n_{imp}$ which is characteristic of the particular atomic
species. The radiated power loss functions for the coronal
equilibrium charge state distributions of oxygen (shown in Figure
30) are of the form shown in Figure 32.

The total radiated power function can be strongly sensitive
to deviation of plasma conditions away from those of local thermal

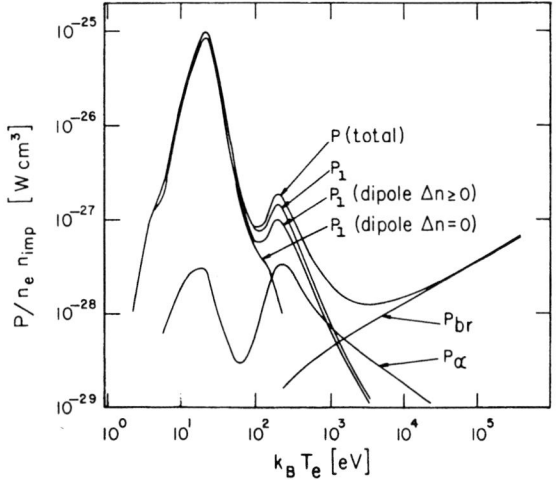

Fig. 32   Radiated power loss function for oxygen.
Data are taken from Summers and McWhirter[76].
Individual contributions from line radiation, recombination
radiation and bremsstrahlung radiation are shown. The
component due to line radiation is further resolved into
contributions from dipole transitions where
$\Delta n > 0$ and where $\Delta n = 0$.

equilibrium. This effect can be seen in Figure 33 where $P/n_e n_{imp}$,
for various values of $n_e \tau_{imp}$, is plotted as a function of $k_B T_e$ (the
associated average charge states are presented in Figure 31). A
marked sensitivity of the total radiation function to $n_e \tau_{imp}$ is
evident when $k_B T_e$ exceeds 30 eV.

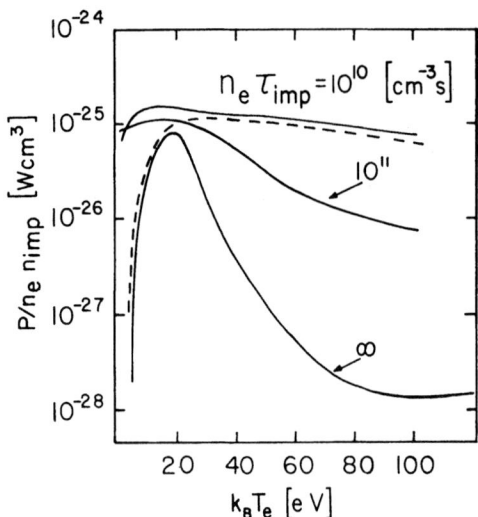

Fig. 33 Total radiated power loss functions for oxygen.
Solid curves are the data of Abramov[74] and apply to the
average charge states shown in Figure 31. The dashed curve
is the data of Shimada et al. (Ref. 77).
Note coronal equilibrium conditions correspond
to $n_e \tau_{imp} = \infty$.

Detailed calculations along the lines described by Summers and
McWhirter (and shown here in Figure 32) yield the best available
data for radiative power losses but the procedure is both complex
and time consuming. Such data are therefore restricted to a
relatively small number of atomic species. Furthermore, the collision data for neutral and lowly charged complex atomic species such
as iron and tungsten are so uncertain that such precise calculation
of their power losses is not warranted at low plasma temperatures.
Jensen et al.[78] have invoked the concept of an "average ion" model
in order to simplify calculation of radiative power losses and
thereby provide a wider base of data. The different ion charge
states of each atomic species are replaced by a single conceptual
"average ion". The populations of the ionisation stages in the real
plasma are statistically accounted for by assigning equivalent
electron populations to the principle electron shells of the average
ion. Transitions between these levels are equivalent to changes in
ionisation stage. Radiative power losses are related to transitions
between the electron levels of this average ion. A wide range of
data based upon this method has been reported by Post et al[79].

342

11. CONCLUSION

Although the preceding discussion has emphasised the physical properties of atomic and molecular collisions within the edge region it is obvious that these basic properties are valid throughout the bulk of the confined plasma. Nevertheless, the significance of a particular process is dependent upon the local plasma environment. In the hotter central region, the ion residence times are relatively long so that (a) impurity ions can be raised to much higher charge states and (b) recombination plays a powerful role in determining the population of ionisation states. In contrast, molecular processes become insignificant. Electron-proton recombination exercises a negligible effect upon the bulk plasma and charge capture due to collisions between impurity ions and hydrogen atoms become significant only if the atoms can penetrate the plasma. Typical examples are energetic neutral beams used for heating or diagnostics. Such energetic hydrogen atoms are confined within the plasma when they become ionised mainly by collisions with plasma protons. Both charge exchange and direct ionisation contribute but at high atom energies ($E > 50$ keV for H atoms) trapping due to proton impact ionisation has the most pronounced effect. Cross sections for double electron capture become quite large at very high energies (a few 100 keV/amu) and so fusion α-particles can partake in such interactions. In addition to collision processes which directly influence plasma conditions within the bulk and edge regions there are many other processes which impact upon diagnostic studies and upon the peripheral technological requirements of fusion devices. A typical example is the generation of intense beams of energetic atoms. Recent reviews of these broader issues can be found in Drawin[80], Post[81] and Harrison[82] and beam formation is reviewed by Green in Ref. 83.

The status of available data relevant to the edge plasma reflects the balance between the present limitations of theory and experiment. In the case of cross sections for electron ionisation and excitation of atomic hydrogen there is a good base of measured data and extrapolation by precise theoretical treatment is feasible for this simplest atomic system. Measurements of H atom charge transfer (and ionisation) in collisions with protons are plentiful and extend over a wide range of energy (from 1 eV to several MeV). Measured data for charge exchange between H atoms and impurity ions are rather sparse and limited to relatively low charge states (typically $z < 10$). Theoretical treatment and scaling of charge exchange is uncertain at low energies and low charge states. There are very few experimental studies of collisions between protons and impurity ions and the data do not extend below proton energies of ~ 1 keV; theoretical treatment of these collisions is uncertain. Measured cross section data for electron impact ionisation of atoms and ions of low atomic number elements is plentiful (at least for charge states less than $z = 5$), so that entrapolation can be under-

taken with a fair degree of confidence (see Ref. 40). There are but
a few measured ionisation cross sections for atoms and ions of
heavier elements and theoretical treatment is uncertain for the
lowly charged ions of complex atomic species; for lack of better
data, semi-empirical cross sections have perforce to be employed.
Apart from those atoms which can be studied in gas cell experiments,
there are virtually no measured data for electron impact excitation
cross sections of relevant atoms and ions. The main experimental
problems are associated with the precise detection of energy
resolved photons in colliding beam experiments (see the brief discussions in Section 6.3). This problem can, in principle, be
avoided by detecting the inelastically scattered electrons rather
than the photons. This solution has long been appreciated (see
comments by Harrison[84] and Dunn[33]) but such experiments are difficult and the first such study (for Zn^+) was reported by Chutjion and
Newall[85] as late as 1982. In the case of electron collisions with
neutral H_2 molecules (see the review by de Heer[39]) there is a fair
amount of measured data for the production of radiation and also of
charged particles but the cross section for dissociation of H_2 into
two ground state H atoms remains uncertain. Fast colliding beam
experiments (reviewed by Dolder and Peart[86]) have provided a sound
base of data for electron collisions with H_2^+ ions.

Recombination becomes increasingly important as the region
under consideration is sited more deeply within the confined
plasma. Concise reviews of experimental methods used to study both
radiative and dielectronic recombination under single collision conditions can be found in Refs. 86, 31 and 33 . Unfortunately such
methods do not yet yield data for multiply charged impurity ions and
so the present results are not directly applicable. Studies of moderately intense plasmas can of course, when precisely interpreted,
yield data for the rate coefficients involved in, ionisation, excitation and recombination. The techniques are reviewed by Kunze[87]
and measurements of excitation rates are discussed by Gabriel and
Jordan[88].

In addition to the valuable review articles that emerge from
the academic community the fusion researcher is also served by a
number of data centres wherein data are assessed and compiled.
Notable amongst these centres are the Oak Ridge National Laboratory
and the Bureau of Standards in the USA, the Institute of Plasma
Physics (Nagoya University, Japan), the Kurchatov Institute (Moscow,
USSR) and Queen's University (Belfast, UK). The International
Atomic Energy Agency (Vienna, Austria) provides an international
service in close collaboration with the fusion data centres. The
I.A.E.A. publishes a quarterly "International Bulletin on Atomic and
Molecular Data for Fusion" which contains up to date indexes of the
relevant literature. A comprehensive index of literature up to 1979
can be found in CIAMDA 80 (Ref. 89). Assessment of the validity of
data is a difficult and somewhat subjective task calling for the

balanced opinions of both theorists and experimentalists. Specialist workshops are organised by the I.A.E.A. and these recommend (Ref. 90) the best available data.

REFERENCES

1. H.W. Drawin, Atomic and Molecular Processes in High-Temperature, Low-Density Magnetically Confined Plasmas, in Ref. 17, p.341.
2. M.F.A. Harrison, The Plasma Boundary and the Role of Atomic and Molecular Processes, in Ref. 17, p.441.
3. M.F.A. Harrison, Boundary Plasma, in Ref. 19, Vol. 2, p.395.
4. W. Grotian "Graphische Darstellung der Spektren von Atomen und Ionen mit zwei und drei Valenzelektronen", Springer, Berlin (1928).
5. K. Takayanagi and H. Suzuki, "Cross Sections for Atomic Processes, Vol.1. Processes involving Hydrogen Isotopes, their ions, electrons and photons", English version of Report IPPJ-DT-48, Nagoya University, Institute of Plasma Physics, Nagoya (1978).
6. R.P. McWhirter and H.P. Summers, Atomic Radiation from Low Density Plasma, Ref.19, Vol.2, p.52.
7. D.R. Bates, A. Kingston and R.W. McWhirter, Proc. Roy. Soc. A267, 297, 1962.
8. R.K. Janev, D.E. Post, W.D. Langer, K. Evans, D.B. Heifetz and J.C. Weisheit, J.Nucl.Mater. 121: 10,(1984).
9. G. Herzberg, "Atomic Spectra and Atomic Structure", Dover Publications, New York (1944).
10. C. Candler, "Atomic Spectra and the Vector Model", Hilger and Watts, London (1964).
11. V.L. Jacobs, J. Davies, J.E. Rogerson and M. Blaha, Astrophys. J.230: 627 (1979).
12. A.G. Gaydon, "Dissociation Energies and Spectra of Diatomic Molecules", Chapman and Hall, London (1953).
13. C.F. Barnett, J.A. Ray and J.C. Thompson, "Atomic and Molecular Collision Cross Sections of Interest in Controlled Thermonuclear Research", Report ORNL-3113 Oak Ridge National Laboratory, Oak Ridge (1964).
14. H.S.W. Massey, E.H.S. Burhop and H.B. Gilbody, "Electronic and Ionic Impact Phenomena", Vols. 1 to 5, Oxford University Press, Oxford (1974).
15. E.W. McDaniel, "Collision Phenomenon in Ionized Gases", John Wiley and Sons, New York (1964).
16. "Atomic and Molecular Processes in Controlled Thermonuclear Fusion" ed., M.R.C. McDowell and A.M. Ferendeci, NATO ASI, Bonas, France, Plenum Press, New York and London (1980).
17. "Atomic and Molecular Physics of Controlled Thermonuclear Fusion" ed., C.J. Joachain and D.E. Post, NATO ASI, Santa Flavia, Italy, Plenum Press, New York and London (1983).

18. "Physics of Ion-Ion and Electron-Ion Collisions, ed., F. Brouillard and J.W. McGowen, NATO ASI, Baddeck, Canada, Plenum Press, New York and London (1983).
19. "Applied Atomic Collision Physics" Vols. 1 to 5, ed., H.S.W. Massey, E.W. McDaniel and B. Bederson, Academic Press, New York and London (1983).
20. J. J. Thomson, Phil. Mag. 23: 419 (1912).
21. M. Born, Z. Physik, 38: 803 (1926).
22. H.A. Bethe, Ann.Physik, 5: 325 (1930).
23. C.J. Joachain, Theoretical Methods for Atomic Collisions - A General Survey, in Ref. 16, p.147 and Recent Progress in Theoretical Models of Atomic Collisions, in Ref. 17, p.139.
24. T. Kato, "Ionisation and Excitation of Ions by Electron Impact - Review of empirical formulae", Report IPPJ-AM-2, Nagoya University, Institute of Plasma Physics, Nagoya (1978).
25. Y. Itikawa and T. Kato, "Empirical formulas for Ionisation Cross Section of Atomic Ions for Electron Collisions", Report IPPJ-AM-17, Nagoya University, Institute for Plasma Physics, Nagoya (1981).
26. W. Lotz, Z. Phys. 220: 466 (1969).
27. H. van Regemorter, Astrophys. J. 136: 906 (1962).
28. A. Burgess, Mem. Soc. Roy. Sci. Liege, 4: 299 (1961).
29. M.J. Seaton, Excitation and Ionisation by Electron Impact, p.375, in "Atomic and Molecular Processes" ed., D.R. Bates, Academic Press, New York (1962).
30. J.A. Gaunt, Phil. Trans. A229: 163 (1930).
31. D.H. Crandall, Electron Ion Collisions, in "Atomic Physics of Highly Ionised Atoms", ed. R. Marrus, NATO ASI, Cargese, Corsica, France, Plenum Press, New York and London (1982) also Report ORNL/TM-8453, Oak Ridge National Laboratory, Oak Ridge (1982).
32. D.H. Crandall, Electron Impact Excitation of Ions, in Ref. 13, p.201.
33. G.H. Dunn, Electron Ion Collisions in "Physics of Ionised Gases", ed., H. Matie and B. Kidric, Inst. Nucl. Science, Belgrade (1980).
34. K.T. Dolder, Experimental Aspects of Electron Impact Ionisation and Excitation of Positive Ions, in Ref. 17, p.213.
35. W.L. Fite and R.T. Brackman, Phys. Rev. 112: 1141 (1958).
36. K.T. Dolder, M.F.A. Harrison and P.C. Thonemann, Proc. Roy. Soc. A 264: 367 (1961).
37. M.F.A. Harrison, Colliding Beam Studies of Atomic Collision Processes, in "Low-energy Ion Beams", p.190, ed. K.G. Stephens, I.H. Wilson and J.L. Moruzzi, Inst. Phys. Conf. Ser. No.38 (1978).
38. L.J. Kieffer and G.H. Dunn, Rev. Mod. Phys. 38: 1 (1966).
39. F.J. De Heer, Experiments on Electron Capture and Ionisation by Ions, in Ref. 16, p.351 and Experiments on Electron Capture and Ionisation by Multiply Charged Ions, in Ref.17, p.269.

40 K.L. Bell, H.B. Gilbody, J.G. Hughes, A.E. Kingston and F.J. Smith, "Recommended Cross Sections and Rates for Electron Ionisation of Light Atoms and Ions", Report CLM-R216, Culham Laboratory, Culham (1982).

41. P. Defrance, W. Claeys, A. Cornet and G. Poulaert, J.Phys.B: Atom. Molec. Phys. 14: 111 (1981).

42. R.G. Montague, M.J. Diserens and M.F.A. Harrison, J. Phys. B. 17: 2085 (1984).

43. J. McGuire, Phys. Rev. A 16: 73 (1977).

44. C.F. Barnett, J.A. Ray, E. Ricci, M.I. Wilker, E.W. McDaniel, E.W. Thomas and H.B. Gilbody, "Atomic Data for Controlled Fusion Research". Reports ORNL-5260 (Vol. 1) and ORNL-5207 (Vol. 2), Oak Ridge National Laboratory, Oak Ridge (1977).

45. D.F. Dance, M.F.A. Harrison and A.C.H. Smith, Proc. Roy. Soc. A290: 74 (1966).

46. A. Burgess and M.C. Chidichimo, Mon. Not. R.A.S. 203: 1269 (1983).

47. D.F. Dance, M.F.A. Harrison, R.D. Rundel and A.C.H. Smith, Proc. Phys. Soc. 92: 577 (1967).

48. G.H. Dunn, J. Chem. Phys. 44: 2592 (1966).

49. F.J. de Heer, Physica Scripta, 23: 170 (1981).

50. B.H. Bransden, Theoretical Models for Charge Exchange, in Ref. 16, p.185 and The Theory of Charge Exchange and Ionistion by Heavy Particles, in Ref. 17, p.245.

51. R.K. Janev and B.H. Bransden, "Charge Exchange between Highly Charged Ions and Atomic Hydrogen: A Critical Review of Theoretical Data", Report INDC (NDS)-135/GA, IAEA, Vienna (1982).

52. H.B. Gilbody, Physica Scripta, 24,: 712 (1981).

53. H.S.W. Massey, Rep. Prog. Phys. 12: 248 (1949).

54. J.B. Hasted in "Advances in Eectronics and Electron Physics" Vol. XIII, p.1, ed. L. Morton, Academic Press, New York (1960).

55. R.E. Olson, A. Salop, P.A. Phaneuf and F.W. Meyer, Phys. Rev. A16: 1867 (1977).

56. P. Hvelplund and A. Andersen, Physica Scripta, 26: 370 (1982).

57. J.H. Newman, J.D. Cogan, D.L. Ziegler, D.E. Nitz, R.D. Rundel, K.A. Smith, R.F. Stebbings, Phys. Rev. A25: 2976 (1982).

58. P.T. Greenland, "Low Energy Charge Capture Cross Sections", Report AERE-R11282, Atomic Energy Research Establishment, Harwell (1984).

59. W.L. Fite, A.C.H. Smith and R.F. Stebbings, Proc. Roy. Soc. A268: 527 (1963).

60. R.F. Stebbings, A.C.H. Smith and H. Ehrhardt, in "Atomic Collision Processes", p.814, ed. M.R.C. McDowell, North Holland, Amsterdam (1964).

61. G. Herzberg, "Molecular Spectra and Molecular Structure, I. Spectra of Diatomic Molecules", D. van Nostrand Co. Inc. Princeton.

62. L. Landau, Z. Phys. Sowjet 2: 46 (1932) and C. Zener, Proc. Roy. Soc. A137: 696 (1932).

347

63. D.A. Hobbis, P. Nickolson, M.F.A. Harrison and R.G. Montague, to be submitted to J. Phys. B.
64. A. Salop and R.E. Olson, Phys. Rev. A13: 1312 (1976).
65. R.K. Janev and P. Hevelplund, Comments in Atomic and Molecular Physics, XI, 75 (1981).
66. M. Buriaux, F. Brouillard, A. Joynaux, T. Govers and S. Szücs, J. Phys. B. 10: 2421 (1977).
67. D.R. Bates and A.E. Kingston, Planet and Space Science, 11: 1 (1963).
68. R.W.P. McWhirter and A.G. Hearn, Proc. Phys. Soc. 82: 641 (1963).
69. R.J. Hutchinson and R.W.P. McWhirter, J. Phys. B. 6: 2668 (1973).
70. F. Brouillard, Rydberg States, in Ref. 17, p.313.
71. G.W. Series, "The Spectrum of Atomic Hydrogen", Oxford University Press, Oxford (1957).
72. H.W. Drawin, Phys. Rep. 37: 125 (1978).
73. H.P. Summers, "Tables and Graphs of Collisional Dielectronic Recombination and Ionisation Coefficients and Ionisation Equilibria of H-like to A-like Ions of Elements", Appleton Laboratory Memo. IM 367, Appleton Laboratory, Culham (1974).
74. V.A. Abramov, in "USSR Contributions to the INTOR Phase-Two-A Workshop", Report, Kurchatov Institute, Moscow (1982) also "International Tokamak Reactor: Phase-Two-A", p.215, International Atomic Energy Agency, Vienna (1983).
75. J. Neuhauser, W. Schneider, R. Wunderlich and K. Lackner, J. Nucl. Mater. 121: 195 (1984).
76. H.P. Summers and R.W.P. McWhirter, J. Phys. B. 12: 2387 (1979).
77. M. Shimada, M. Nagami, K. Ioki, S. Izumi, M. Maeno, H. Yokomizo, K. Shinya, H. Yoshida, N.H. Brooks, C.L. Hsieh, A. Kitsunezaki and N. Fujisawa, in "Japanese contribution to the INTOR Phase-Two-A Workshop", Japan Atomic Energy Research Institute, Tokai-mura (1982).
78. R.V. Jensen, D.E. Post, W.H. Grasberger, C.B. Tarter and W.A. Lokke, Nucl. Fusion, 17: 1187 (1977).
79. D.E. Post, R.V. Jensen, C.B. Tarter, W.H. Grasberger and W.A. Lokke, At. Data and Nuc. Data Tables, 20: 397 (1977).
80. H.W. Drawin, Physica Scripta, 24: 622 (1981).
81. D.E. Post, The Role of Atomic Collisions in Fusion, in Ref. 18, p.37.
82. M.F.A. Harrison, The Relevance of Atomic Processes to Magnetic Confinement and the Concept of a Tokamak Reactor, in Ref. 16, p.15.
83. T.S. Green, Neutral Beam Formation and Transport, in Ref. 19, Vol. 2, p.339.
84. M.F.A. Harrison, Electron Impact Ionisation and Excitation of Positive Ions, in "Methods in Experimental Physics", Vol. 7B, p. 95, ed. W.L. Fite and B. Bederson, Academic Press, New York (1968).

85. A. Chutjian and W.R. Newall, Phys. Rev. A26: 2271 (1982).
86. K.T. Dolder and B. Peart, Rep. Prog. Phys. 39: 697 (1976).
87. H.J. Kunze, Space Sci. Rev. 13: 565 (1972).
88. A.H. Gabriel and C. Jordan, in "Case Studies in Atomic Collision Physics II", p.209, ed. E.W. McDaniel and M.R.C. McDowell, North-Holland Publishing Co., Amsterdam (1972).
89. CIAMDA 80, "An Index to the Literature of Atomic and Molecular Collision Data Relevant to Fusion Reseach, International Atomic Energy Agency, Vienna, Austria (1980).
90. "Research Coordination Meetings on Atomic Collision Data for Diagnostics of Magnetic Fusion Plasmas":
    First Meeting: IAEA Report, INDC(NDS)-136/GA (1982).
    Second Meeting: IAEA Report, INDC(NDS)-150/GA (1984).
    Third Meeting: IAEA Report, INDC(NDS)-160/GA (1984).

PHYSICAL SPUTTERING OF SOLIDS AT ION BOMBARDMENT

J. Roth

Max-Planck-Institut für Plasmaphysik

EURATOM Association, D-8046 Garching

ABSTRACT

An outline is given on the current understanding of physical sputtering with emphasis of the use in fusion reactor technology. This determines the ion species H, D, He, C, O and self ions as well as the energy range from the sputtering threshold to several keV. Additionally an attempt is made to present empirical analytic formula where comprehensive theories do not exist. Finally, as an example, the impurity fluxes from charge exchange wall sputtering in ASDEX are calculated.

INTRODUCTION

The limited confinement of plasma ions and the loss of energetic neutrals lead to a constant bombardment of the first wall of a plasma device; one inevitable mechanism causing impurity production and wall erosion therefore is sputtering of metal surface atoms by energetic particles.

In this contribution the physical principles of light ion sputtering are briefly presented. As there are extensive recent reviews of light ion sputtering /1, 2/ and data collections exist which attempt to give analytic formulae for the sputtering yield /3, 4, 5/ no attempt is made here to cover the whole literature. Only the most recent experimental results on light ion sputtering are presented especially as far as the extension of the data to low ion energies is concerned.

However, the attempt is made here to demonstrate for specific applications, e.g. the sputtering due to charge exchange neutrals which have been measured extensively at ASDEX /6/, how todays knowledge of the energy and angular dependence of the sputtering yield and the energy and angular distribution of the sputtered particles can be combined to calculate the differential particle fluxes leaving the wall. Doing this reveals, where the lack of information requires future basic studies. As particles impinging from the plasma not only erode the wall but are also deposited, the wall composition changes and a surface layer may be composed of materials sputtered from different areas of the machine. This leads to necessity for understanding the behaviour of compounds under ion irradiation and the ion induced desorption of surface layers. These effects are discussed in a final part of this contribution.

A. PHYSICAL PRINCIPLES OF LIGHT ION SPUTTERING

For estimating the impurity production due to plasma wall interaction, the most important quantity is the number of wall atoms removed by one incident ion, depending on ion energy and angle of incidence, i.e. the sputtering yield $Y(E_o, \alpha)$. To determine the probability, that a sputtered atom will enter the plasma, more differential quantities have to be known, i.e. the energy and angular distribution of the sputtered atoms $d^3Y/d^2\Omega dE_1$.

In the following firstly the results given by linear analytic theory for the sputtering yield are given. Secondly, for light ion sputtering of monoelemental solids, where no comprehensive theory exist, data and empirical equations for the total and differential yields are presented. Thirdly, some special features of compound sputtering are mentioned. Finally, the importance of ion induced desorption, i.e. sputtering of adsorbed layers, is shortly emphasized.

1. Linear analytic sputtering theory

A typical sputtering event as calculated using the Monte Carlo simulation program TRIM /7/ for the case of a 1 keV Ar ion impinging at 40° to the surface normal onto a Ni-surface is shown in Fig. 1a /8/. The figure shows the trajectory of the incident ion as well as of the recoiling target atoms. It can be seen that a large collision cascade is initiated which spreads towards the surface. Eventually 5 surface atoms are knocked-on and two receive energies large enough to overcome the surface binding energy and are sputtered. The average sputtering yield for this case is calculated from a large number of incident ions and turns out to be 3.6 /9/. In contrast to the case of Ar bombardment, the cascade for He bombardment is very small and isotropy is not established.

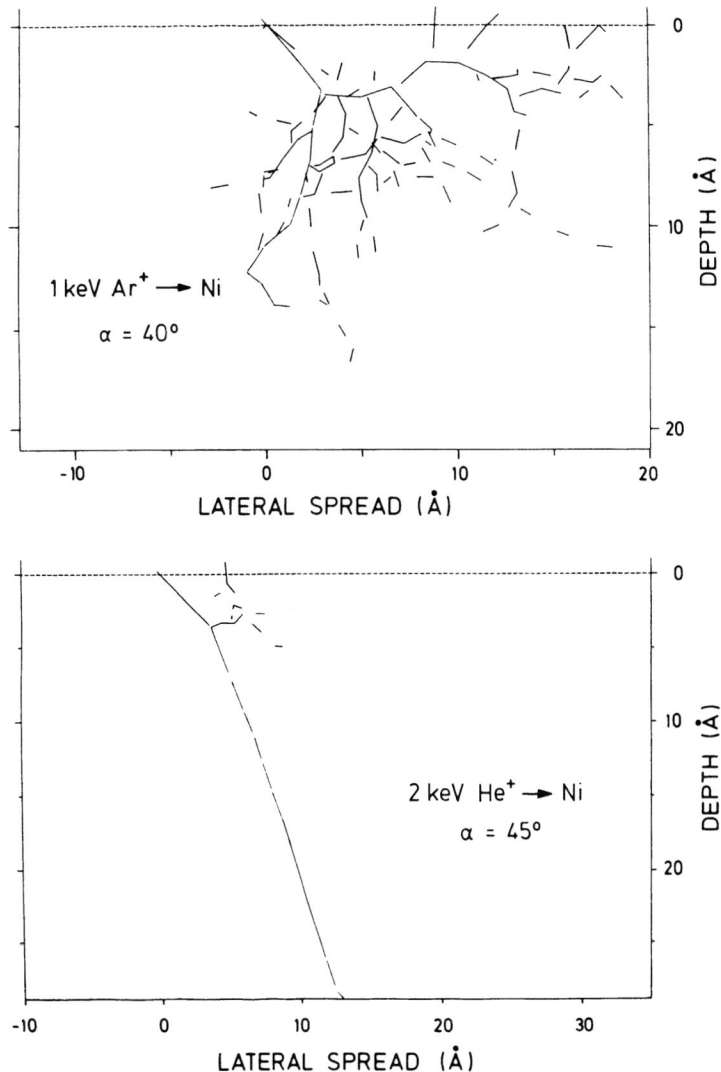

Fig. 1: Typical collision cascades initiated by a 1 keV Ar ion
and 2 keV He ion as calculated using the TRIM program.
Recoil atoms close to the surface may be ejected after
diffraction at the planar surface potential or be reflected into the bulk.

Figure 1b shows an example for a sputtering event. In this case on the average only every third incident ion leads to a sputtered atom.

The most detailed ansatz for a sputtering theory /10/ made use of the idea of a large collision cascade and used Boltzmann transport equations to calculate the flux of target atoms moving through a plane within the solid. With the assumptions that the energy transferred to the target atoms is much larger than the binding energy and that the velocity distribution within the cascade is isotropic this theory results in the well known formula for the sputtering yield

$$Y(E_o,\alpha) = \frac{0.042}{n\,U_o}\,F_D\,(E_o,\alpha,x = 0) \tag{1}$$

with U_o the surface binding energy in eV and n the atomic density in A^{-3}. $F_D\,(E_o,\alpha,x = 0)$ is the deposited energy density in the surface, i.e. at depth x=0. E_o is the projectile energy and α the angle of incidence to the surface normal.

It has been shown /11/ that the deposited energy density scales with the dimensionless energy ε as defined in Möller chapter. The function $F_D(\varepsilon)$ can be approximated by the nuclear stopping power $s_n(\varepsilon)$ times a correction term $a(m_1/m_2)$ with $a \cong 0.2$ for heavy ions, while for light ions the correction term $a(m_1/m_2,\alpha)$ increases drastically /10/. Using this energy unit, eq. (1) reads

$$Y(\varepsilon,\alpha) = \frac{3.56}{U_o}\,\frac{m_1}{(m_1+m_2)}\,\frac{Z_1 Z_2}{(Z_1^{2/3}+Z_2^{2/3})^{1/2}}\,a(\frac{m_1}{m_2},\alpha)\,s_n(\varepsilon)\;. \tag{2}$$

Figure 2 shows a comparison of the deposited energy density extracted from measured sputtering yields using eq. (2) with the deposited energy at the target surface as calculated from collision theory /12/. The agreement for heavy ions like Ar is very good for both the energy dependence and the scaling of different materials. For light ions, different materials show at high energies a similar energy dependence but differ considerably in absolute value. For low energies clearly threshold effects dominate the energy dependence.

The region of validity of eq. (2) is schematically given in Fig. 3, which is an extension towards low ion energies of a discussion of the linear analytic sputtering theory made earlier /13/. It can be seen that the assumption of isotropic cascades does not hold for particle energies below 1 keV and that at energies below 500 eV sputtering threshold effects have to be taken into account. Additionally, the theory does not treat cases, where the sputtering ions build up to large concentrations in the target material and where strong chemical interations between ions and target atoms

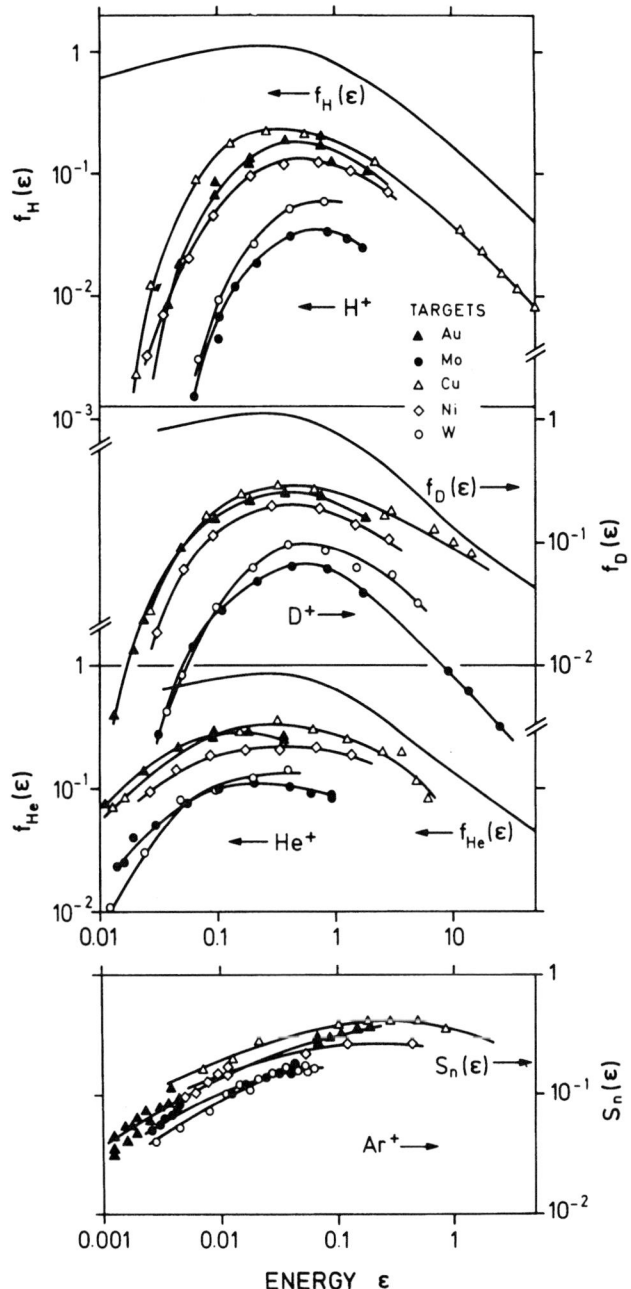

Fig. 2: Comparison of the deposited energy $f(\epsilon)$ and $s_n(\epsilon)$ extracted from sputtering data of Ar, He, D and H using eq. (2).

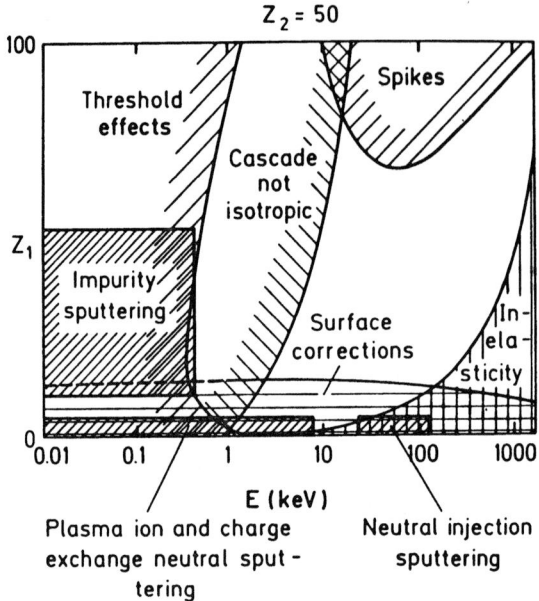

Fig. 3: Region of validity of the analytic theory of isotropic
collision cascade sputtering as a function of projectile
energy and atomic number Z_1 for a target atomic number
Z_2 = 50. Within each shaded region the main effect contributing to the breakdown of the theory is given. In
addition the regions of interest in plasma-wall interactions are given.

exist. Some of these effects are discussed in Roth chapter. In Fig. 3
also the mass and energy regimes for sputtering in fusion devices
are indicated.

2. Low energy light ion sputtering

i) Systematics and approximations for normal incidence.

The similarity in the energy dependence of the sputtering yield
for different ion/target combinations as seen in Fig. 2 has led to
the attempt to develop a universal formula for low energy light ion
sputtering. Figure 4 shows the energy dependence in the threshold
regime for a variety of ions and target materials /14, 15/. The
energy scale has been divided by a parameter E_{th}, which has been
identified as the threshold energy for sputtering. Figure 5 shows
a collection of the threshold energies divided by the surface
binding energy U_0 for different ion-target combinations. It can be
seen that these values also show a universal behaviour. The
threshold decreases with increasing ion mass m_1 as the energy
transferred to surface atoms increases. For heavy ions the momentum

356

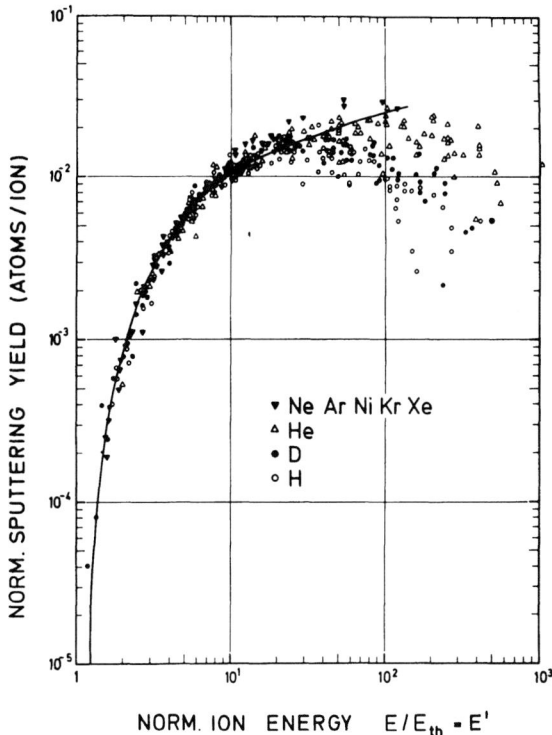

Fig. 4: Universal energy dependence in the threshold energy regime for a variety of ions and target materials. The energy dependence is best approximated by $(E_o/E_{th})^{1/4} (1-E_{th}/E_o)^{7/2}$ (solid line).

reversal necessary for sputtering can only be accomplished by multiple collisions and the threshold energy for sputtering increases again. Included in the figure are three recent analytic fitting formula /16, 17, 18/.

This universal behaviour of the sputtering yield in the threshold regime, expressed in terms of the reduced energy E_o/E_{th} /14, 15/ as well as the universal behaviour for high energies, expressed in terms of the reduced energy ε /10/, have led to various attempts to combine the two regimes, at least for normal incidence $\alpha = 0$.

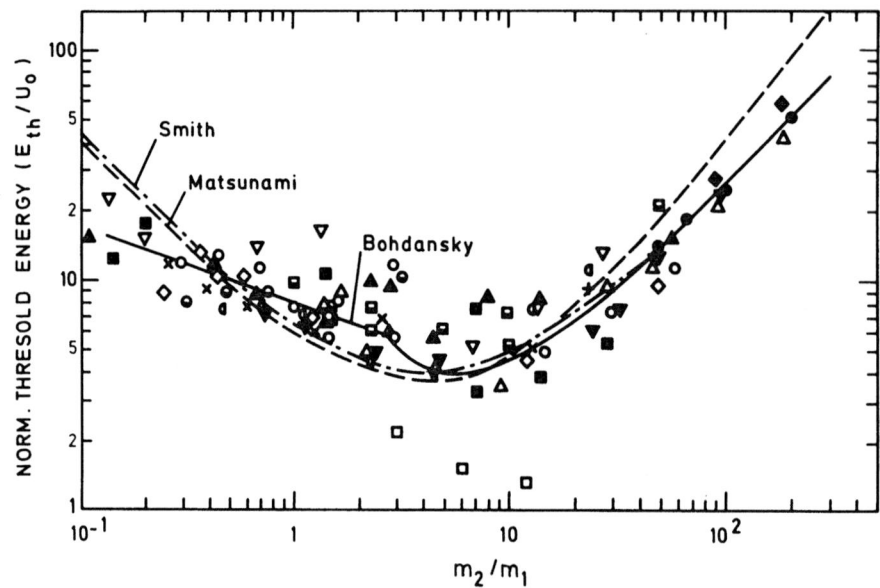

Fig. 5: Dependence of the reduced threshold energy E_{th}/U_o on the mass ratio ions and target atoms m_1/m_2. Approximations used in the different fitting formulas are also given.

Bohdansky /16/ has presented an analytic formula for the sputtering yield covering light and heavy ion bombardment at all energies. His proposed formula retains the structure of eq. (2) but adds several new correction terms:

$$Y = \frac{3.56}{U_o} \frac{m_1}{m_1+m_2} \frac{Z_1 Z_2}{(Z_1^{2/3}+Z_2^{2/3})^{1/2}} a(\frac{m_1}{m_2}) (\frac{R_p}{R_1}) s_n(\varepsilon) f(\frac{E_o}{E_{th}}) \qquad (3)$$

where

$$f(\frac{E_o}{E_{th}}) = [1 - (\frac{E_{th}}{E_o})^{2/3}] [1 - \frac{E_{th}}{E_o}]^2 \qquad (5)$$

is a good fit to the universal curve shown in Fig. 4,

$$E_{th} = \begin{cases} \dfrac{U_o}{\gamma(1-\gamma)} & \text{for} \quad \dfrac{m_1}{m_2} < 0.2 \\[3ex] 8U_o (\dfrac{m_1}{m_2})^{2/5} & \text{for} \quad \dfrac{m_1}{m_2} > 0.2 \end{cases} \qquad (6)$$

is a fit to the threshold energies shown in Fig. 5 with $\gamma = \dfrac{4m_1 m_2}{(m_1+m_2)^2}$.

$$\frac{R_p}{R_1} = k\left(\frac{m_2}{m_1}\right) + 1 \qquad (7)$$

is a measure for the number of surface crossings in an infinite medium taken as surface correction term, where k varies between 0.1 and 0.4, and

$$s_n(\varepsilon) = \frac{3.441\sqrt{\varepsilon}\ \log(\varepsilon+2.718)}{1+6.35\sqrt{\varepsilon} + \varepsilon(6.882\sqrt{\varepsilon}-1.708)} \qquad (8)$$

is a fit to the universal nuclear stopping power function.

The only fitting factor in this equation for individual materials is k. A table of best fit values is given in /1/. If no values for k are known, k is put equal to 0.2. More recently, using also E_{th} as a free fitting parameter, the overall agreement with the data could be improved for a large collection of materials /1/. For light ion and selfsputtering of C, Ni and W calculated values are compared with experimental data in Figs. 6, 7 and 8. For Ni also oxygen sputtering data are shown. Due to chemical effects they are a factor of 2 to 3 lower than calculated using the empirical formulae. Carbon ion sputtering generally leads to a build-up of carbon layers and selfsputtering of carbon prevails.

Matsunami et al. /17/ have also retained the structure of the original sputtering relation (eq. 2) for their fitting formula. With the argument, that light ion sputtering is dominated by reflected ions /19/ and that the reflection coefficient is dependent on the electronic stopping function /20/, they include the electronic stopping from LSS-theory /11/ into their sputtering equation.

$$Y = \frac{3.56\ Q}{U_o}\ \frac{m_1}{m_1+m_2}\ \frac{Z_1 Z_2}{(Z_1^{2/3}+Z_2^{2/3})^{1/2}}\ a*\left(\frac{m_1}{m_2}\right)\frac{s_n(\varepsilon)}{[1+0.35U_o\ s_e(\varepsilon)]}\ f\left(\frac{E_o}{E_{th}}\right)$$
$$(9)$$

$$s_e(\varepsilon) = \frac{0.79\ (m_1+m_2)^{2/3}}{m_1^{3/2}\ m_2^{1/2}}\ \frac{Z_1^{2/3}\ Z_2^{1/2}}{(Z_1^{2/3} + Z_2^{2/3})^{3/4}}\ \varepsilon^{1/2}\ , \qquad (10)$$

s_n as given in eq. (8),

$$f\left(\frac{E_o}{E_{th}}\right) = \left[1 - \left(\frac{E_{th}}{E_o}\right)^{1/2}\right]^{2.8}\ , \qquad (11)$$

$$a*\left(\frac{m_1}{m_2}\right) = 0.08 + 0.164\left(\frac{m_2}{m_1}\right)^{0.4} + 0.0145\left(\frac{m_2}{m_1}\right)^{1.29}\ , \qquad (12)$$

and

$$E_{th} = U_o\left[1.9 + 3.8\left(\frac{m_2}{m_1}\right)^{-1} + 0.134\left(\frac{m_2}{m_1}\right)^{1.24}\right] \qquad (13)$$

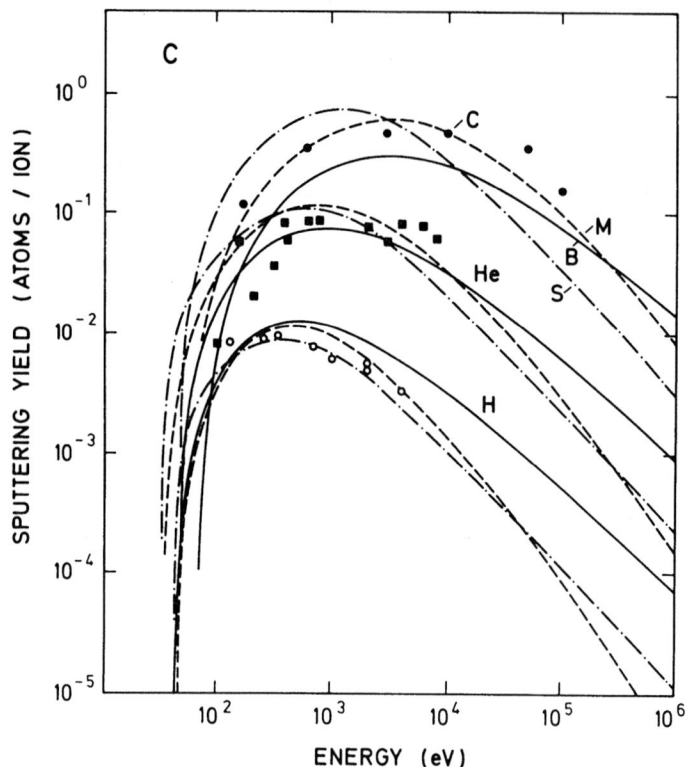

Fig. 6: Comparison of three different analytic fitting formula for normal incidence sputtering of carbon with H, He and C ions. Oxygen sputtering yield data are given in the chapter dealing with chemical sputtering.

as shown in Fig. 5, where eqs. (11) to (13) are best fits to available data.

In addition, the formula contains the fitting factor Q, which only in few cases (Be, B, C) differs largely from 1. Sputtering yields calculated from eq. (9) are also included in Figs. 6 to 8.

Finally, a very simple, completely empirical sputtering formula has been developed by D.L. Smith et al. /18/ and is also shown in Figs. 6 to 8.

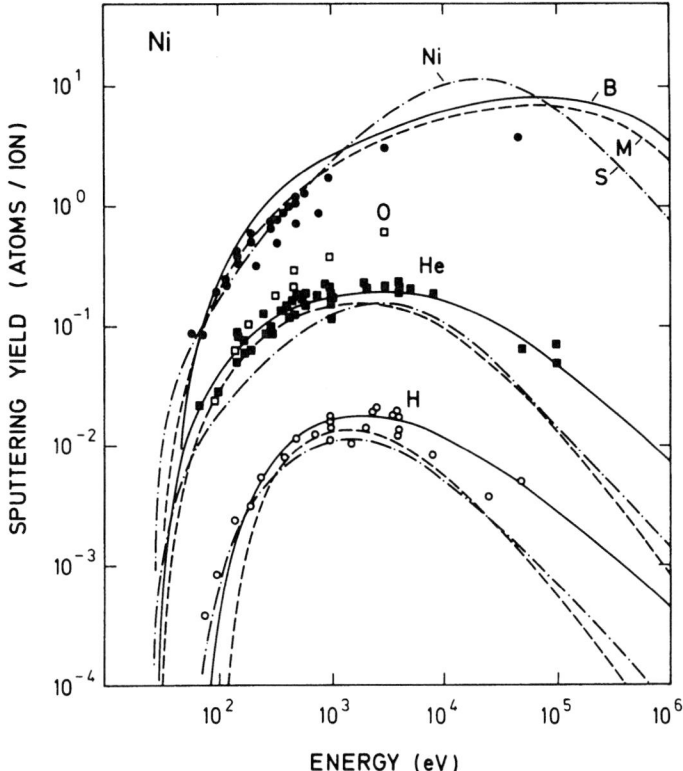

Fig. 7: Comparison of three different analytic fitting formula
for normal incidence sputtering of nickel with H, He,
and Ni ions. The data for oxygen sputtering are generally
a factor of 2 to 3 lower than given by empirical fitting
formulae.

$$Y = \frac{C}{U_o} Z_1^{0.75} (Z_2 - 1.8)^2 \; (\frac{m_1}{m_2})^{-0.8\,1.5} \; \frac{E_o - E_{th}}{(E_o - E_{th} + 50 Z_1^{0.75} Z_2)^2} \qquad (14)$$

with

$$E_{th} = U_o \frac{(4m_1 + m_2)^2}{4m_1 m_2} \qquad (15)$$

and

$$C = 2000 \text{ for } H^+ \text{ ions and } 400 \text{ for all other particles.} \qquad (16)$$

This formula contains no additional fitting parameter. All energies
are in eV units.

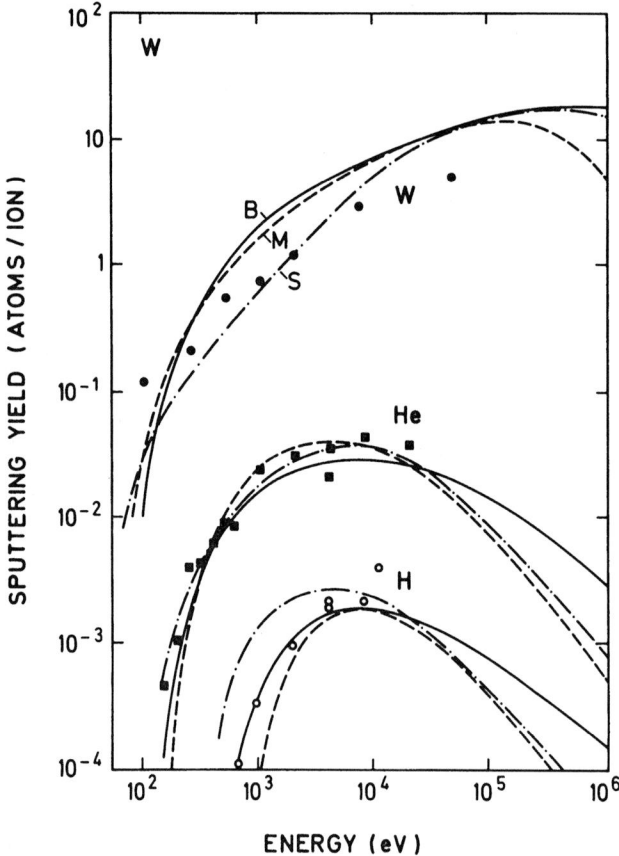

Fig. 8: Comparison of three different analytic fitting formula
for normal incidence sputtering of tungsten with H, He
and W ions.

A comparison of these three analytic formulas show that none
clearly excells the others in fitting the data. Whereas the formula
by Matsunami et al. (eq. 9) and Bohdansky (eq. 3) are based on
physical arguments about the sputtering event, D.L. Smith et al.
(eq. 14) only tries to empirically reproduce the existing data. At
high energies, the inclusion of the electronic stopping in
Matsunami's formula leads to a steeper decrease of the sputtering
yield than predicted by the energy dependence of the nuclear stopping alone, i.e. Bohdansky's formula. Wherever there are data for
light ions, this steep decrease does not seem to be justified. At
low energies, i.e. in the threshold regime, the fitting to the
existing values of the threshold energy E_{th} is important (Fig. 5)
and here Bohdansky's formula yields the best overall description.
In view of its simplicity, however, the formula by D.L.Smith et al.

(eqs. 14-16) in many cases may be preferrable. The overall accuracy
of all the formula is not better than a factor of 2 to 3.

ii) Light ion sputtering at grazing incidence

The increase of the sputtering yield for light ions with the
angle of incidence α has been shown /15/ to depend for given material
parameters on the ion energy. For high energies the increase is
more pronounced than the $1/\cos\alpha$ dependence expected from the deposited energy density in the surface, while for energies below
1 keV the increase is less pronounced. Therefore the dependence on
angle cannot be separated from the dependence on energy /1/. The
yield shows a maximum at angles of incidence α_{opt} between 70 and
85°. Figure 9 gives experimental examples for sputtering of Ni and
C with H, D and He ions. For energies below 450 eV the results have
been completed by computer simulation using the TRIM code, which
agrees nicely with experiment at higher energies. The drastic increase of the sputtering yield with angle of incidence for high
energies is assumed to be due to direct sputtering of surface
atoms in a mechanism, which is schematically shown in Fig. 10
/22/.

Again, fitting formulas have been proposed for the angular
dependence of the sputtering yield $Y(E_o,\alpha)$ normalized to the yield
for normal incidence $Y(E_o,0)$. Yamamura /5/ gives

$$\frac{Y(E_o,\alpha)}{Y(E_o,0)} = \cos\alpha^{-f}\, e^{f\,\cos\alpha_{opt}(1-\cos\alpha^{-1})} \tag{17}$$

α_{opt} is calculated analytically as result of a mechanism, where the
sputtered surface atoms are ejected in a simple direct collision
with the incident ions at grazing angles of incidence (Fig. 10).
For a power potential with $m = 1/2$ Yamamura et al. /23/ took for
α_{opt} the angle of incidence, where the maximum impact parameter S,
which leads to direct emission approaches $n^{1/3}\cdot\cos\alpha$. Then they get
the theoretical expression for α_{opt}

$$\alpha_{opt} = \frac{\pi}{2} - 0.4685\, n^{-1/3} [(z_1^{2/3} + z_2^{2/3})\, 2\varepsilon \sqrt{\frac{U_o}{\gamma E_o}}\,]^{-1/2} \tag{18}$$

The parameter f in eq. (17) is an empirical fitting factor and
tabulated values are given in /5/. The best overall fit is

$$f = \sqrt{U_o}\, (0.94 - 1.33\cdot 10^{-3}\, \frac{m_2}{m_1})\cdot \tag{19}$$

The fact that f is indepencent of energy certainly is a short-
coming in view of the data presented in Fig. 9.

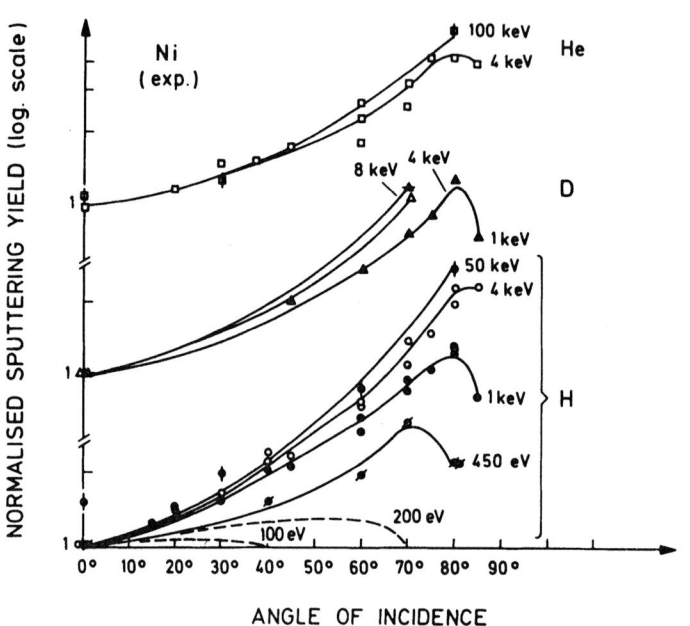

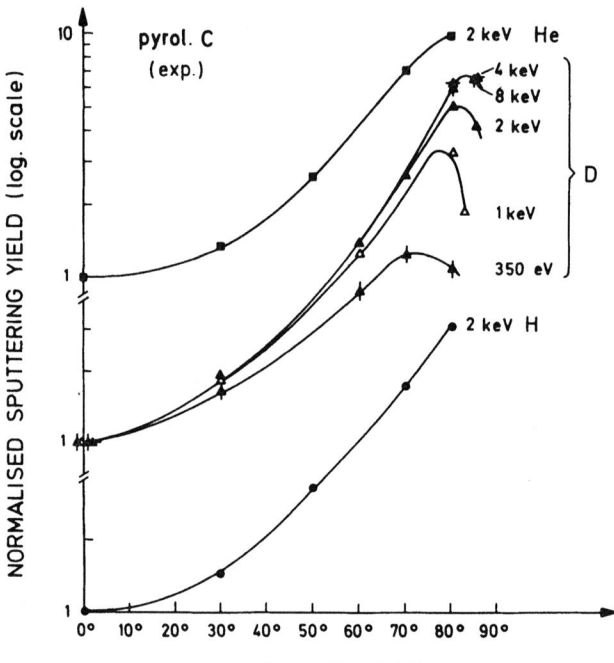

Fig. 9: Dependence of the normalised sputtering yield of C and Ni for H, D and He on the angle of incidence. Data calculated using the TRIM program, which show good agreement with experimental data, are shown as dashed curves for 100 and 200 eV H^+ sputtering of Ni.

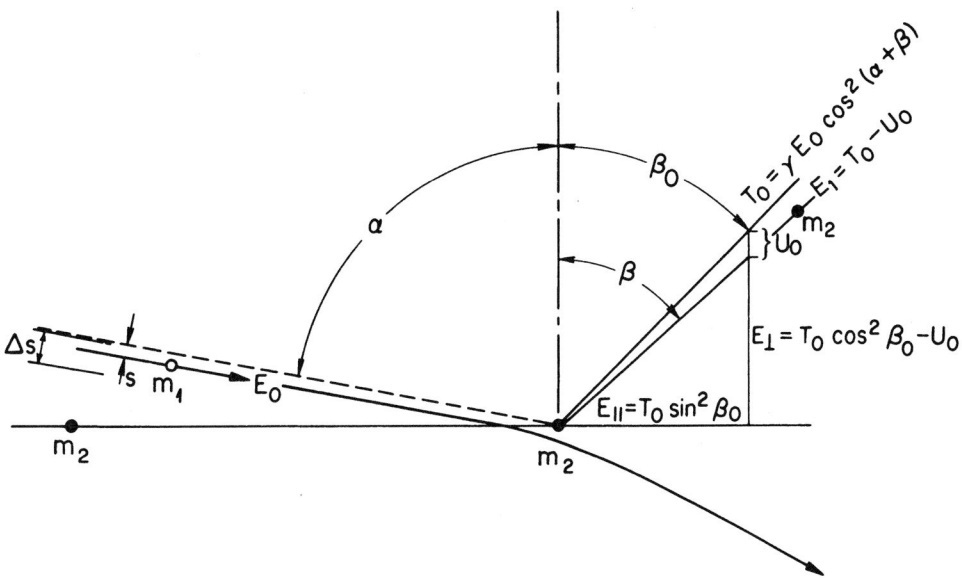

Fig. 10: Geometry of collisions leading to direct emission of a
          surface atom.

Another more empirical fit has been proposed by Smith et al.
/18/. They give

$$
\frac{Y(E_o,\alpha)}{Y(E_o,0)} = \begin{cases} \cos^{-f}\alpha & \text{for } E_o > 4\,E_{th} \\[2ex] 1 & \text{for } E_o < 4\,E_{th} \end{cases} \tag{20}
$$

with

$$
f = \frac{1}{(20Z_1)^{1/2}}\left(\frac{m_2}{m_1}\right)^{1/4}(E_o - 4\,E_{th})^{1/4} \tag{21}
$$

where E_{th} is taken from eq. (15).

The exponent f now is energy dependent. However, eq. (21) is
only defined for energies E_o larger than $4\,E_{th}$. For smaller energies Smith et al. propose in view of the weak dependence of
$Y(E_o,\alpha)$ on the angle of incidence (see Fig. 9) to put $Y(E_o,\alpha)$ =
$Y(E_o,0)$.

This latter assumption is important in view of the low particle
energies responsible for sputtering in plasma devices (see Sect. 3
of this review). As the overall fit of Yamamura's /5/ as well as
of Smith's /18/ formula is not better than a factor of 2, the
assumption of

$$
\frac{Y(E_o,\alpha)}{Y(E_o,0)} = \begin{cases} \cos^{-2}\alpha & \text{for } E_o > 4\,E_{th} \\[2ex] 1 & \text{for } E_o < 4\,E_{th} \end{cases} \tag{22}
$$

is justified within a factor of 2 for particle energies up to the keV range and angles up to 60°. Larger angles are improbable both for plasma ions /24/ and charge exchange neutrals (see Sect.3).

iii) Angular distribution of sputtered atoms

The original theory for cascade sputtering /10/ predicted a cosine distribution of sputtered particles due to the assumption of an isotropic collision cascade. This has been modified taking an anisotropic momentum distribution in the cascade into account /25/. In the regime where sputtering processes in direct single collisions dominate (see Fig. 10), i.e. for light ions and very grazing incidence, strongly anisotropic angular distributions are observed. A typical example is shown in Fig. 11 for the case of 2 keV D^+ sputtering of molybdenum.

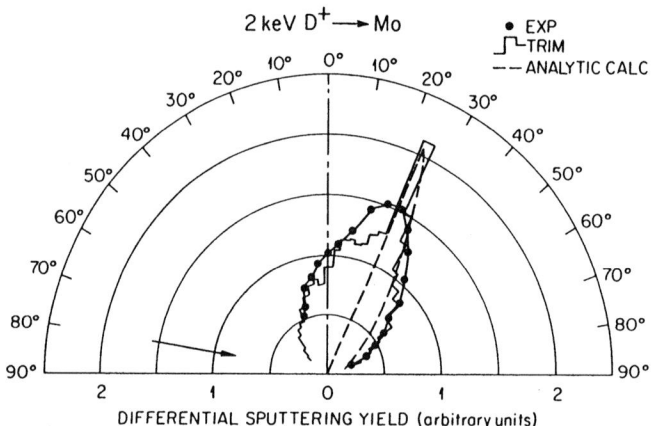

Fig. 11: Comparison of the angular distribution of sputtered Mo atoms due to 2 keV D bombardment at an angle of incidence of 80° as measured experimentally, calculated by TRIM and from the direct sputtering model shown in Fig. 10.

The experimental data are compared with TRIM calculations and an analytic calculation for processes shown in Fig. 10 assuming a Thomas-Fermi potential /22/. The agreement is very good especially in view of the limited angular resolution of ± 7° in the experiments.

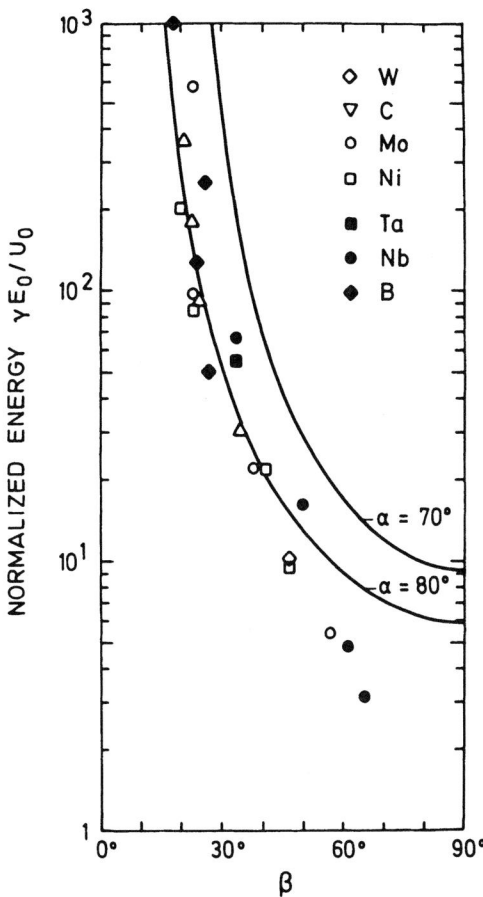

Fig. 12: Comparison of the angle of maximum differential sputtering
yields as determined experimentally for 7 materials with
the minimum angle of emission of sputtered atoms β_{min} as
calculated from the direct sputtering model shown in
Fig. 10. The experimental data are for $\alpha = 80^\circ$.

The emission angle β of maximum sputtering yield is compared in
Fig. 12 with the angle of maximum sputtering cross-section as calculated analytically /22/. It can be seen that the maximum emission
angle scales with the parameter $\gamma E_0/U_0$ for different materials and
energies. For conditions, which predict β values larger than 70°,
deviations are found, which most probably originate from scattering
at neighbouring surface atoms. Similar good agreement has been found
not only in the plane of incidence, but also for the azimuthal
distribution of sputtered particles /26/. For $\gamma E_0/U_0$ values smaller
than 3, i.e. for 100 eV, Fig. 12 shows that the direct sputtering
process is no longer possible. At lower energies the angular distributions return to an isotropic one.

Analytic expressions for these anisotropic distributions have not been developed. However, for the case of normal incidence as well as isotropic ion incidence it has been shown, that the assumption of a cosine distribution for sputtered atoms is well justified /27/. For the case of strongly nonisotropic incidence of charge exchange neutrals in Sec. 3 the distribution of sputtered atoms will be calculated.

iv) Energy distribution of sputtered atoms

The energy distribution of moving particles in an isotropic collision cascade follows an $(E_1')^2$ dependence /28/. The flux of particles through a surface plane from such a cascade has been evaluated by Thompson /29/. With the assumption of a planar surface potential U_o, which has to be overcome by the sputtered surface atoms he derived the formula

$$\frac{d^2Y}{dE_1 d\beta} = s_n(\varepsilon) \frac{0.22\ E_1}{(E_1 + U_o)^3} \cos \beta \qquad (23)$$

This distribution has a maximum at $1/2\ U_o$ and falls off at high energies as $1/E_1^2$ until the maximum transferable energy is reached. It has been frequently experimentally confirmed (for a review see /15/). Only very recently, however, data have been published for light ion sputtering in the energy regime relevant for plasma-wall interaction. Figure 13 shows measured energy distributions of Ti atoms sputtered by H, D and He atoms in the energy range of 100 eV to 8 keV for normal incidence /30/.

For sputtering with 8 keV He ions the energy distribution is still close to the theoretical one (eq. 23 including the geometric conditions of the experiment). With decreasing energy and ion mass, i.e. decreasing maximal transferable energy, the maximum is shifted towards lower energies and the high energy tail falls of more steeply. Analytical calculations /25/ indicate that the distributions in this regime may be described by the parameters U_o and γE_o, but further investigations are necessary to clarify this point.

For non-normal angle of incidence, the linear cascade theory for sputtering does not predict a different energy distribution of sputtered atoms. Taking the anisotropy of the cascade due to the momentum of the incident ions into account, a finite but small deviation from eq. (23) could be deduced depending on angle of ejection /25/.

If sputtering processes due to direct single collisions dominate (see Fig. 10), the arguments deriving eq. (23) no longer hold and a different energy distribution is expected, strongly depending on the angle of ejection. The sputtering yield is concentrated into a narrow angular interval around the emission angle

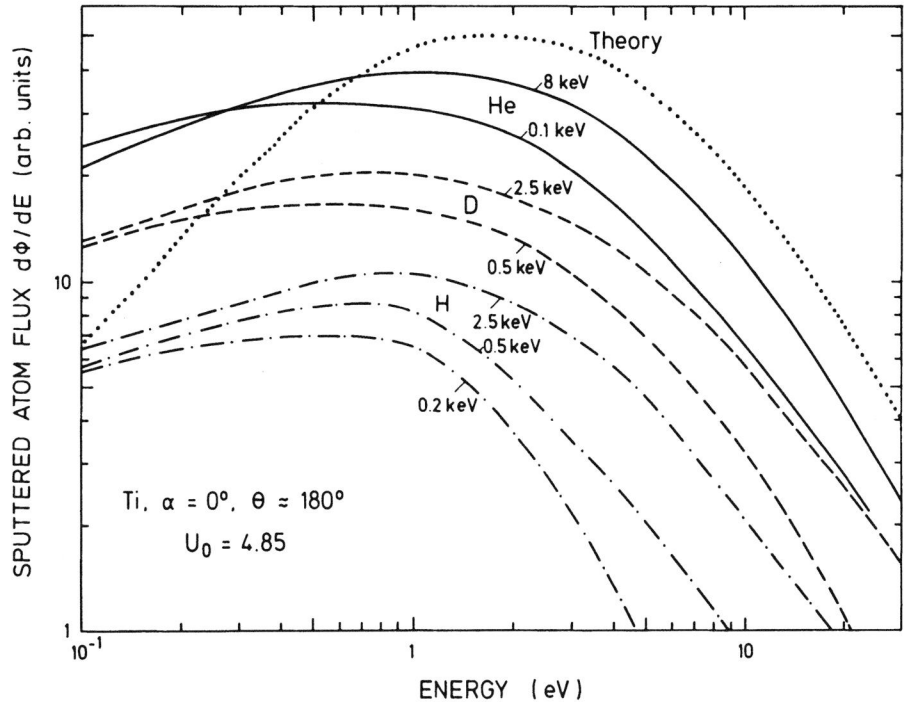

Fig. 13: Energy distribution of sputtered Ti atoms by low energy
H, D and He ions at normal incidence /30/. For comparison
the energy distribution given by collision cascade sputtering is shown.

β_{opt} for maximum differential yield. Here it can be shown that the
sputtered atoms at β_{opt} result from processes, where the transferred energies are small but large enough that the diffraction at
the surface potential is only moderate. The resulting energies E_1
after ejection are of the order of 0.5 to 1 U_0. At smaller angles β
ejection is only possible via multiple collisions or collision cascades. At larger angles β a two-fold energy distribution is expected,
which consist of a large contribution of atoms with small transferred energies and large diffraction on the surface potential and
a second, smaller peak at large transferred energies and small
diffraction. This high energetic contribution for grazing angle of
incidence has been observed recently /31, 32, 33/ and Fig. 14 shows
the position of this peak together with analytic calculation and
computer simulation.

3. Sputtering of compounds

While sputtering of monoelemental targets is most extensively
investigated, in real application sputtering occurs in most cases

369

from composite material. In plasma-wall interaction metal walls are made from alloys or carbides, the surface layer consisting from deposited material sputtered throughout the vessel. In the poor vacuum of the device this surface layer is superficially oxidised.

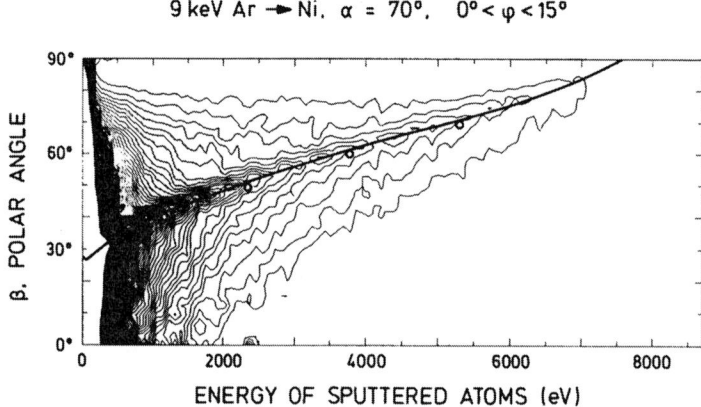

Fig. 14: Dependence of the energy of sputtered Ni atoms on the polar emission angle ß for 9 keV Ar ions at an angle of incidence of 70°. The contour lines are calculated using the TRIM program, the solid line is an analytic calculation from the direct sputtering model shown in Fig. 10. The open points are positions of peaks in the energy distribution of sputtered atoms as measured experimentally /33/.

In addition the sputtering plasma ions may accumulate to form a hydride. Thus physical sputtering as studied on elemental targets has to be applied to the case of compounds. Compound sputtering has recently been reviewed /34/ and the main emphasis here will be its application to plasma-wall interaction.

i) Prediction from linear cascade theory.

The main difference between elemental and compound target material is the possibility, that the different components have different sputtering yields. In the scope of the linear cascade theory the sputtering ratio of two components is expected to be /35/

370

$$\frac{Y_1}{Y_2} = \frac{c_1}{c_2} \left(\frac{m_2}{m_1}\right)^{2m} \left(\frac{U_o^2}{U_o^1}\right)^{1-m} \tag{24}$$

where m is the exponent for the power potential used in the inter-action of both target atoms. As m is small, of the order of $0 < m < 0.2$ for the low energies in the collision cascade mass effects are much less pronounced than effects depending on binding energy.

The yield ratio does not depend on ion mass or energy within the validity of the theory. However, eq. /24/ predicts
- the lightest component sputters preferentially,
- the least bound species sputter preferentially.

With the ratio of the sputtering yield given by eq. (24) for given concentrations c_1 and $c_2 = 1 - c_1$, in the course of the bombardment the surface concentration will change. In steady-state the surface concentrations will be arranged such that the ratio of the sputtering yields reflect the stoichiometric bulk concentrations. The ratio of surface concentrations in a surface layer Δx_o will then be

$$\left(\frac{c_1}{c_2}\right)_{\Delta x_o} = \left(\frac{c_1}{c_2}\right)_{bulk} \left(\frac{m_1}{m_2}\right)^{2m} \left(\frac{U_o^1}{U_o^2}\right)^{1-2m} \tag{25}$$

The time, i.e. the fluence until steady state is reached, depends on the range over which the composition is changed, which in turn is influenced by thermal and radiation enhanced diffusion of the components, by ion mixing and recoil implantation and surface segregation.

In Fig. 15 a collection of measured surface compositions for a number of binary compounds is given. The surface composition after steady state sputtering as taken from /34/ is plotted against the mass ratio of the target atoms. The prediction from eq. (26), assuming m = 1/10 and $U_o^1 = U_o^2$ is also shown. In most cases, as predicted, the heavier component is enriched, though wide variations of the data indicate the importance of such additional parameters as the surface binding energies. The different data points for TaC, TiC and WC are obtained for different ion masses, where the enrichment increases with decreasing ion mass.

ii) Low energy light ion sputtering.

As shown in Fig. 15, much more pronounced surface enrichments are observed for light ions especially at low ion energies /36/. This has been explained by the strong difference in sputtering yield for different target masses as the ion energy approaches the threshold regime. This is demonstrated in Fig. 16 for the materials Ta, Ti and C for He and H bombardment. While hydrogen at 500 eV does not sputter Ta any more, its sputtering yield for C is close to the maximum. At higher energies, the sputtering yields are about equal.

This has led to the attempt to correlate the observed surface concentration of the enriched heavier component as measured by Auger electron spectroscopy /36, 37/ against an energy parameter E* which

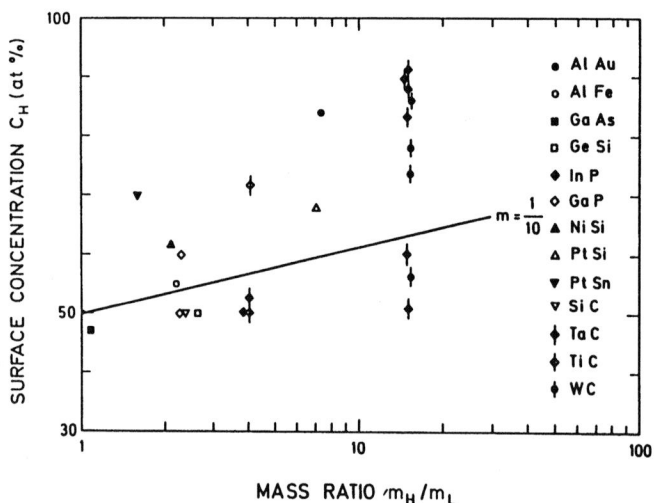

Fig. 15: Collection of the measured surface concentrations of the heavier component of binary alloys with $C_H = C_L$ after ion bombardment /34/ plotted against the mass ratio. For comparison the prediction of the linear cascade theory is shown for m = 1/10 with the assumption of equal surface binding energies $U_o^1 = U_o^2$.

gives the ratio of the difference of the energy of incidence with the threshold energy for sputtering for both materials (Fig. 17). For the carbides TiC, TaC and WC, where measurements exist, sputtering by H, He, Ar, Xe ions show good scaling with this parameter. As soon as the relative energy deviates from 1, i.e. threshold effects become measurable, the surface concentration increases and comes close to 90 % already for parameter values of 1.2. Closer to the threshold, i.e. for larger E*, no further depletion of carbon is observed. However, experimentally carbon coverages on carbides smaller than 10 % are difficult to measure. This explains the larger depletion of carbon obtained from dynamic TRIM calculations /21/ (Fig. 17). Taking the empirical energy dependence of the sputtering

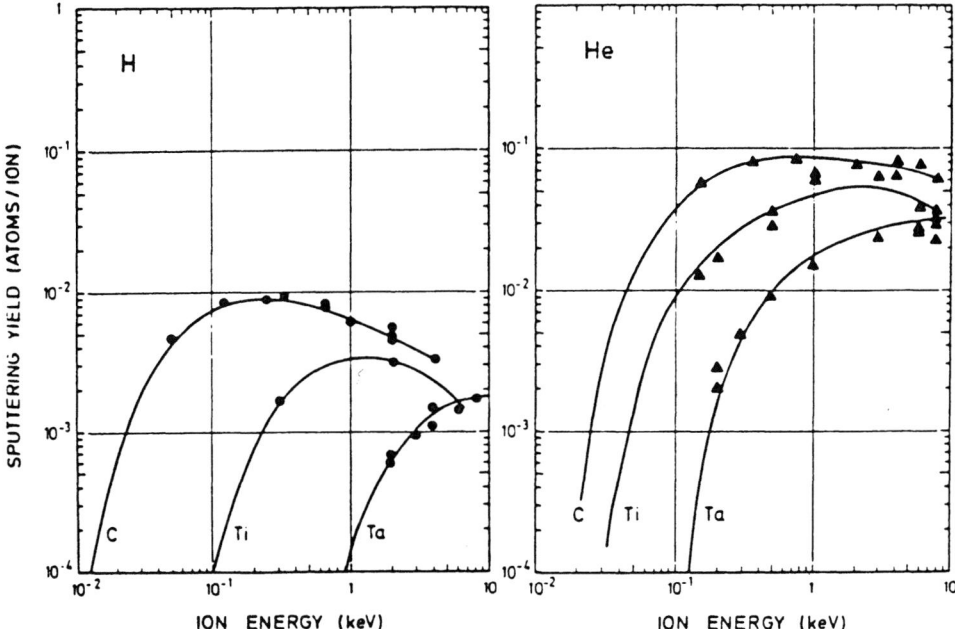

Fig. 16: Energy dependence of the sputtering yield of C, Ti and Ta
with He and H ions.

yield in the threshold regime as shown in Fig. 4 one can, neglecting
surface binding energy effects, deduce a concentration for the
heavier element

$$C_H = \frac{1}{F(\frac{E_o - E_{th}^H}{E_o - E_{th}^L})^{7/2} + 1}$$
(26)

where the factor F varies for the examples between 0.96 and 1.2.
This curve is indicated in Fig. 17. The actual data points show,
however, a much steeper increase of the enrichment approaching the
threshold regime for the heavier element. This can be understood,
as for energy parameter E* smaller than 1.3 the sputtering yield of
carbon is not yet dominated by threshold effects, but increases with
decreasing E*.

iii) Total and differential yield.

Steady state sputtering of the compound requires stoichiometric
removal of both components. The total sputtering yield is therefore
limited by the sputtering yield of the heavier component and in
general follows closely the sputtering yield for the pure heavier
element.

373

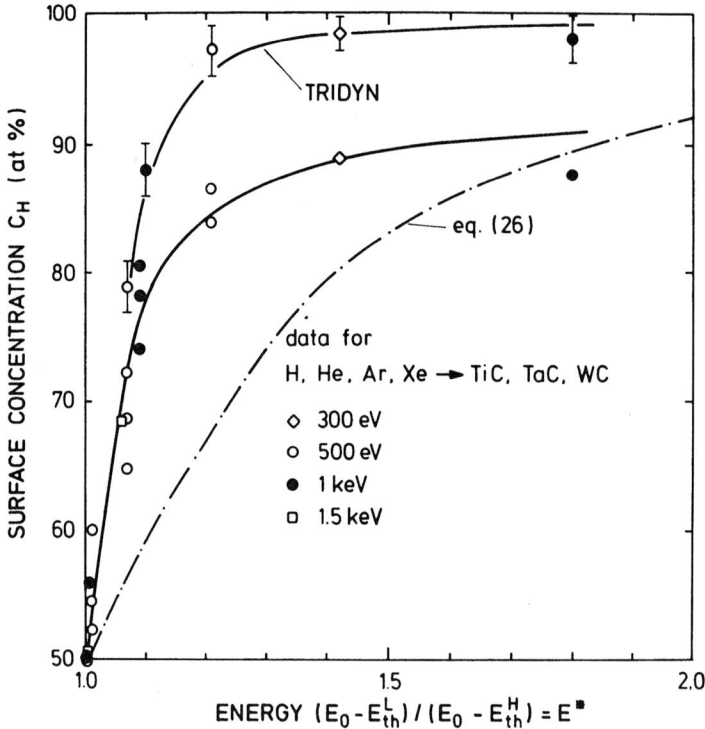

Fig. 17: Measured surface concentration of the heavier component taken from /36, 37/ plotted versus the relative difference of the ion energy E_0 from the threshold energy E_{th} for the different components.

The requirement of stoichiometric sputtering, however, does not hold for any solid angle. The angular distributions for both components, especially for non-normal incidence, may not be identical and are determined by the parameter $\gamma E_0/U_0$ as demonstrated for pure elements /22/. One example for the angular distribution of boron and niobium sputtered from NbB_2 due to deuterium bombardment at an angle of incidence of 70° is shown in Fig. 18. The measured distributions are compared with ones analytically calculated and obtained by computer simulation using the same surface binding energy for both components. Both calculation and computer simulation clearly predict the observed shift of the angle of maximum ejection β_{opt} from 35° for B to 50° for Nb.

Energy distributions from oxide layers have been measured for both the metal and the oxygen atoms /31, 33, 38/ for Ar and Ne ions at grazing angle of incidence. They show the same behaviour, as the distributions for metals (Fig. 14). The used sputtering ions and energies did not allow the extract information about the respective binding energies of the components.

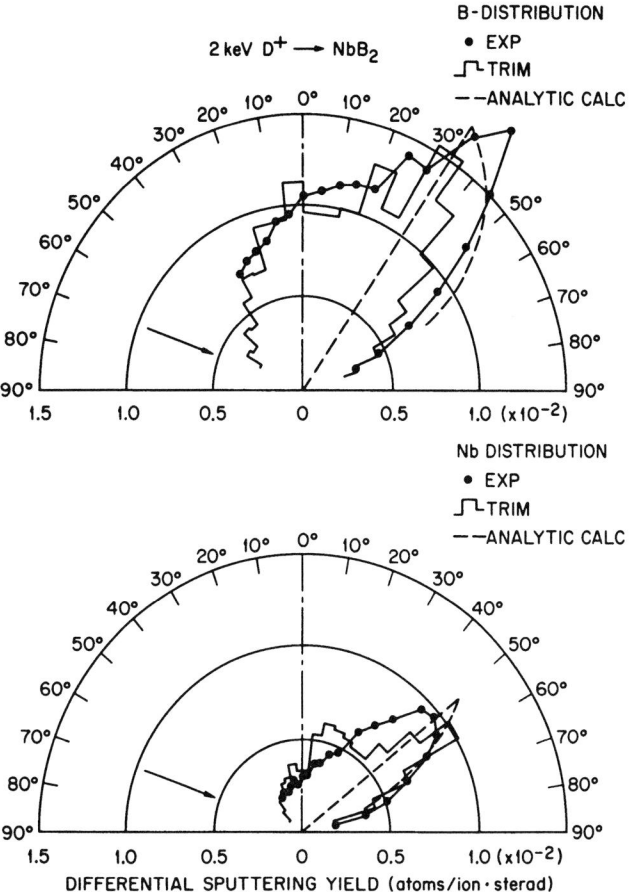

Fig. 18: Angular distributions of sputtered B and Nb from NbB_2 bombarded with 2 keV D^+ at an angle of incidence of $70°$.

4. Sputtering of adsorbed layers

Sputtering of adsorbed layers or ion induced desorption can be described /51, 1/ by the same atom-atom collision processes as sputtering. Three different processes can be distinguished (Fig.19). Figure 19a shows the sputtering of adsorbed surface atoms 3 by direct knock-on by the incoming ions 1 and subsequent reflection at matrix atoms 2. Desorption due to the flux of reflected ions and sputtered matrix atoms is shown in Fig. 19b and c respectively. For all three processes, knowing the reflection probability and matrix sputtering yield, the cross section can be calculated. The sum of these cross sections is compared in Fig. 20 with desorption data for O on Ni by different ions as a function of energy /1/.

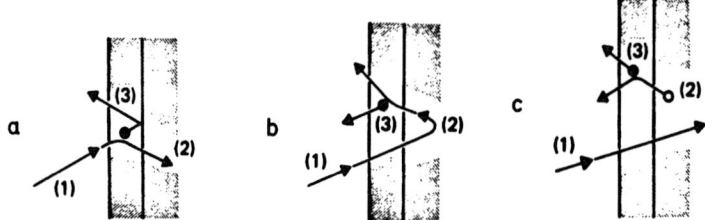

Fig. 19: Mechanisms of desorption by knock-on collisions of adsorbed atoms 3 on a substrate 2 by incident ions 1.
a) and b) knock-on by the incoming and reflected ions,
d) knock-on by sputtered matrix atoms.

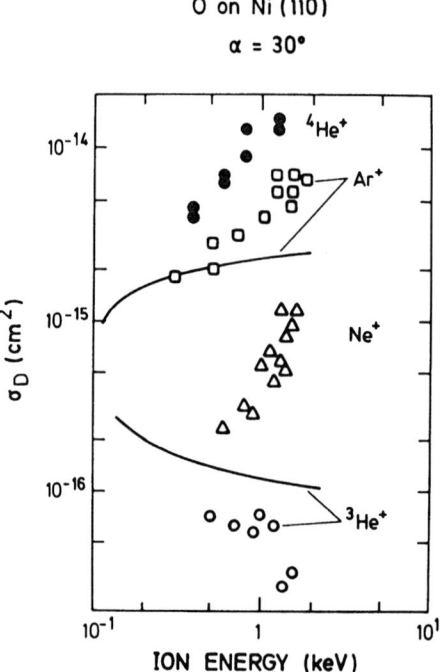

Fig. 20: Measured ion induced desorption cross sections of oxygen
on Ni by Ar, Ne and He ions vs. ion energy. The data are
compared for the case of Ar and He to calculated cross
sections for the sum of all three processes described in
Fig. 19. Open data points are for adsorbed oxygen, solid
points for adsorbed CO molecules.

376

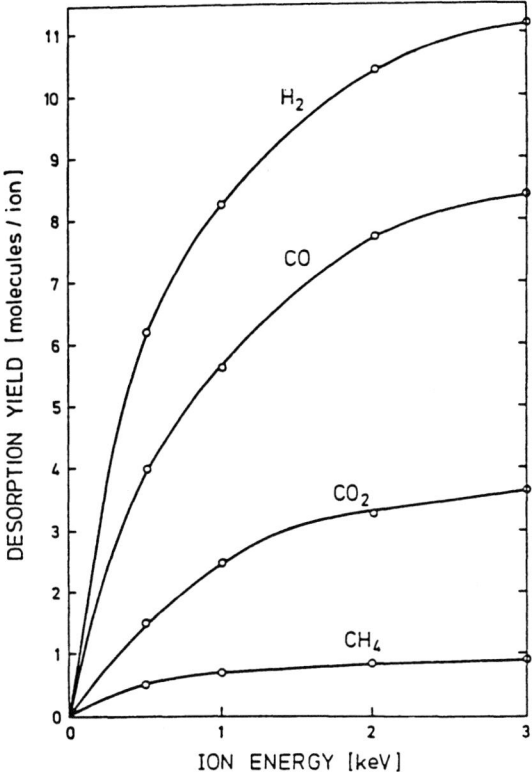

Fig. 21: Energy dependence of the desorption yield for bombardment
of unbaked stainless steel with Ar^+ ions.

The desorption cross sections, being of the order of 10^{-16} cm^2
for light ions to 10^{-14} cm^2 for heavy ions are reproduced in absolute
value within a factor of 2 as well as in their energy dependence.
They depend strongly on the surface binding energy and the mass of
the adsorbed species. This can be seen in Fig. 20 from the comparison
of the cross section for adsorbed CO, which is about two orders of
magnitude higher than for adsorbed oxygen. For oxygen desorption
from different metals a dependence of $1/U_O$ for the desorption
cross-section has been deduced /1/. The lower surface binding energies of adsorbed species compared to metal surface atoms also leads
to a lower threshold energy for desorption.

As ion induced desorption can be described by a desorption
cross section, the desorption yield depends on surface coverage with
adsorbed species. For a monolayer coverage, the desorption cross
sections shown in Fig. 20 are equivalent to desorption yields ranging
from 0.01 to 10 atoms or molecules released per incident ion. For He
desorption of CO the yield is of the order of 20, much larger than
metal sputtering yields. Desorption yields of different molecules
from unbaked stainless steel are shown in Fig. 21 as a function of
Ar ion energy.

B. APPLICATION TO SPUTTERING IN PLASMA DEVICES
    - EXAMPLE CHARGE EXCHANGE SPUTTERING

The contribution of impurity production due to sputtering of ions and charge exchange sputtering has frequently been estimated and compared to the impurity concentration in a plasma machine. For ASDEX divertor discharges in hydrogen, the metal impurity concentration in the main chamber, being predominantly Fe, showed the same time and density behaviour as the charge exchange flux normal to the wall in the midplane of the chamber /39/. Its absolute value agreed well with the total impurity influx due to sputtering and the transition from H to D yielded an increase of a factor of 5. Only about 3 % of the plasma ions hit structures in the main chamber and their contribution to impurity production is negligible for H_2 discharges /49/. In the divertor chamber, thermal and fast plasma ions are identified to be responsible for the release of Ti atoms /39, 40/ and agree quantitatively with spectroscopically determined concentrations /41/.

These estimates, however, did not take into account the detailed angular dependences and angular and energy distributions as outlined above. In the following, an attempt is made to demonstrate the influence of these details on the impurity influx. Charge exchange sputtering on the wall is considered as this is an inevitable source of impurities, while plasma ion sputtering may be removed effectively into a divertor with negligible impurity influx into the main plasma.

1. Characteristic of the wall and wall fluxes

The flux and energy distribution of charge exchange neutrals impinging onto the wall of the ASDEX-vacuum vessel is measured routinely using an electrostatic energy analyzer after ionizing the neutrals in a gas cell /42/ and by time-of-flight measurements /43/. The total flux and its energy distribution has been shown to depend strongly on the position relative to the gas valve and limiter, on the plasma density and auxiliary heating /42/.

For an ohmic divertor discharge Fig. 22 shows the angular dependence of the incident neutral hydrogen flux using the gas-cell ionisation during the plateau phase. By scanning up and down across the plasma cross-section the flux increases drastically as the analyser approaches the stagnation points of the ASDEX divertor. For an undiverted limiter discharge the hydrogen flux tangential to the plasma boundary is less strongly peaked. The scan right-left in the horizontal midplane is extrapolated from an up-down scan in a series of limiter discharges. The average flux, being of the order of 10^{14} H/cm^2 sec str may be a factor of 20 higher for a position close to the gas inlet valve or limiter. The energy distribution of

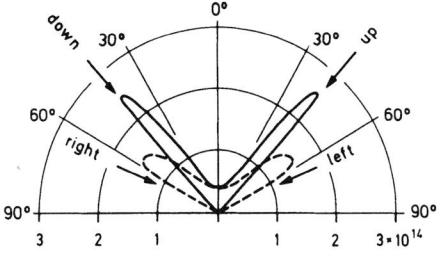

CHARGE EXCHANGE FLUX H⁰/cm² sec str.

Fig. 22: Angular distribution of the flux of charge exchange
neutral hydrogen atoms at a wall position of the divertor experiment ASDEX. The angular distribution in vertical
direction has been measured, while the distribution in
horizontal direction has been extrapolated.

the fluxes for different angles is given in Fig. 23. It can be seen
that while the low energy part of the spectrum increases with
grazing emergence from the plasma column, the high energy part of
the spectrum decreases. The lowest energies shown are at 200 eV.
For lower energies the ionization probability in the gas cell decreases drastically. TOF-measurements, however, give reliable data
for energies down to 20 eV /43/.

The data in Fig. 23 are taken on a wall position far away from
the gas valve. Measurements on the neutral particle fluxes as a
function of the distance towards the gas inlet valve /44/ or limiter /42/ show, that the increased flux close to these neutral density sources may contribute a factor of 4 higher total fluxes to
the wall.

Additional to the angular and energy distribution of the charge
exchange flux, also its time dependence during the discharge is
measured for various densities and auxiliar heating powers and can
be taken into account /42/.

The wall in ASDEX and the main wall structures as protection
plates are made from stainless steel. Analysis of the surface composition of the wall /45/ indicates, however, that the wall is
covered with an oxidised metal layer representing all elements used
in the machine and reflecting the conditions of machine operation.
In ASDEX the deposit consisted of stainless steel droplets and metal
layers in an earlier period using an SS-limiter and was covered with
a titanium oxide layer in a later period with extensive Ti-gettering

379

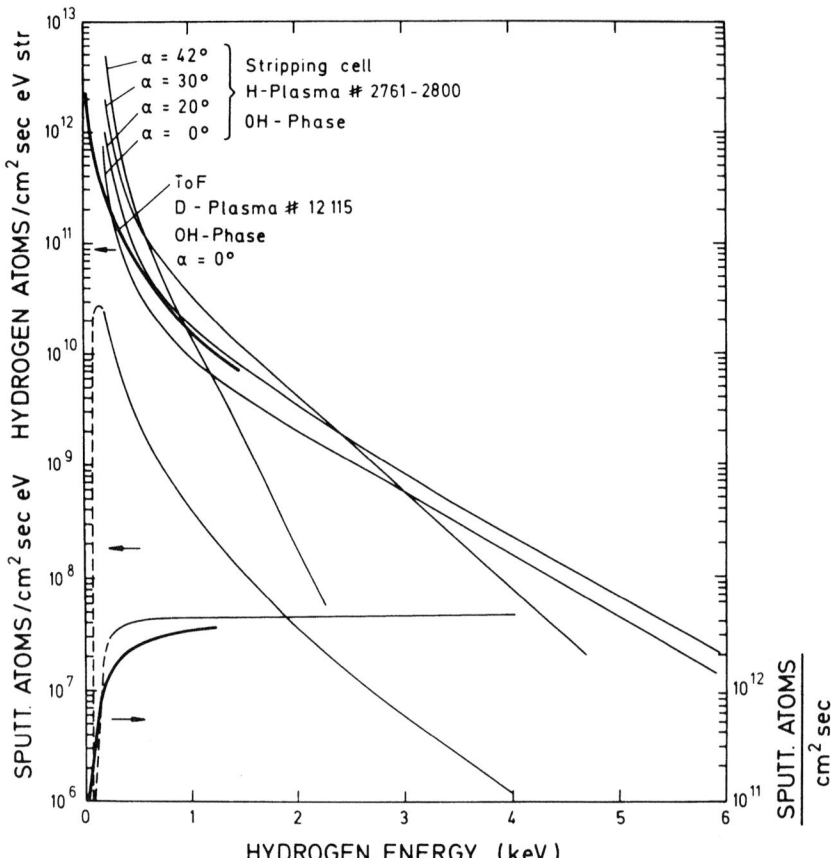

Fig. 23: Energy dependence of the charge exchange hydrogen atoms measured using a TOF-technique /43/ and after ionization in a stripping cell /42/ for different angles of incidence on a wall position of ASDEX. The sputtered Fe atom flux as a function of hydrogen energy has been calculated and the integral of the sputtered flux is shown in the bottom curve.

in the divertors. Figure 24 shows the toroidal distribution of metal layer thickness and droplet density. As the data shown in Fig. 23 were obtained in a phase before Ti gettering, the wall may be assumed to consist from oxidised SS. After the end of Ti-gettering experiments, this may be also the situation today as surface analysis of carbon limiter and protection plates indicate /46/. The Fe sputtering yield from an oxidised surface has been simulated for ASDEX conditions and results in a reduction of about a factor of 2.

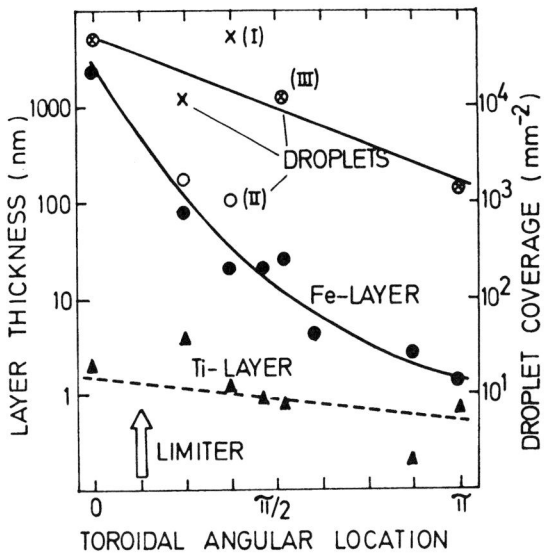

Fig. 24: Toroidal dependencies of the wall coverage with droplets
and of the layer thicknesses.

2. Impurity influx characteristic

i) Total flux

In Fig. 23 the hydrogen flux is multiplied after integration
over the angular distribution with the sputtering yield of pure
stainless steel /3/. This yields the amount of atoms sputtered per
cm^2 and sec and eV incident hydrogen energy. It can be seen, that
hydrogen atom energies around 150 eV contribute most to the production of impurities. Following the recommendation in eqs.
(20 - 22), no increase of the sputtering yield with angle of incidence was assumed. The integration of this curve (bottom curve)
shows, that a total impurity atom influx of about 5 x 10^{12} atp,s/cm^2
sec is obtained. 80 % of this influx, however, results from hydrogen
energies below 400 eV. A very similar number is obtained by applying the same procedure to charge exchange fluxes measured by the
TOF-technique. The oxidised state of the surface may reduce the total influx by a factor of 2, while the higher charge-exchange fluxes
close to the gas inlet valve may lead to an increase up to a factor
of 2. The average influx of Fe, Cr, Ni atoms may therefore be
estimated to about 5 x 10^{12} atoms/cm^2 sec.

Light impurities like O and C are present in the machine in
form of adsorbed hydrocarbon CO and water layers on the wall. From
the cross sections given in Fig. 20 and a charge exchange flux of

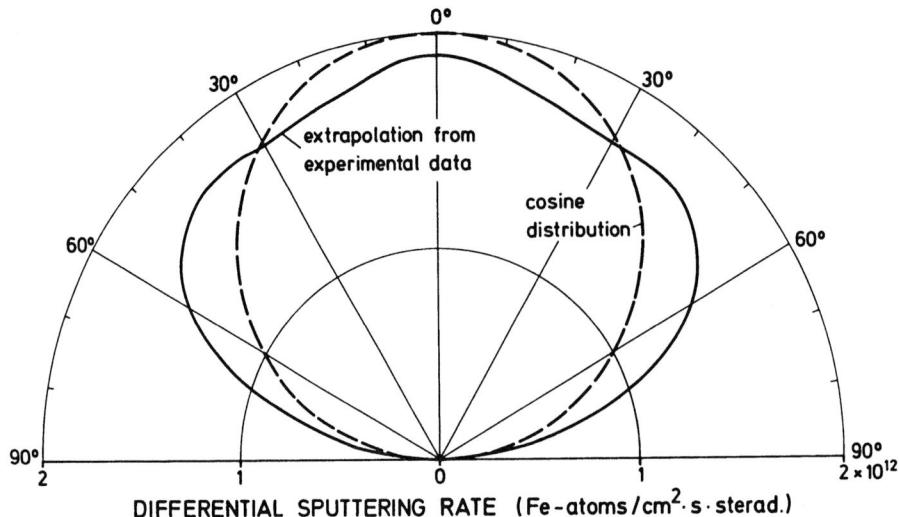

DIFFERENTIAL SPUTTERING RATE (Fe-atoms/cm^2·s·sterad.)

Fig. 25: Angular distribution of sputtered Fe atoms extrapolated
from experimental data for an incident hydrogen flux
distribution as shown in Figs. 22 and 23. For comparison
a cosine distribution is indicated.

10^{15} at/cm^2 sec one sees that a monolayer coverage is completely
desorbed within the duration of a discharge. From the annealed and
conditioned wall of ASDEX all adsorbed molecules are desorbed, leading to roughly the same amount of C and O impurities in the plasma
for H_2 and D_2 discharges /50/. The initial flux may be estimated to
10^{14} atoms/cm^2 sec.

ii) Angular distribution

The angular distribution of sputtered metal atoms is calculated
by assuming a rotational symmetry of the incident flux representing
an average of the angular distributions shown in Fig. 22. To obtain
the angular distribution of sputtered atoms the angle of preferred
emission has been calculated for each angle of incidence and different energy intervals using Fig. 12. For particles with energies below 300 eV no preferred emission is expected and the distributions
have been assumed to be isotropical.

The resulting angular distribution is shown in Fig. 25. The
distribution has a maximum normal to the surface and the predominantly oblique incidence leads only to slight preferential emission
at angles around 45. Within a factor of 1.5 the distribution can be
approximated by a cosine distribution.

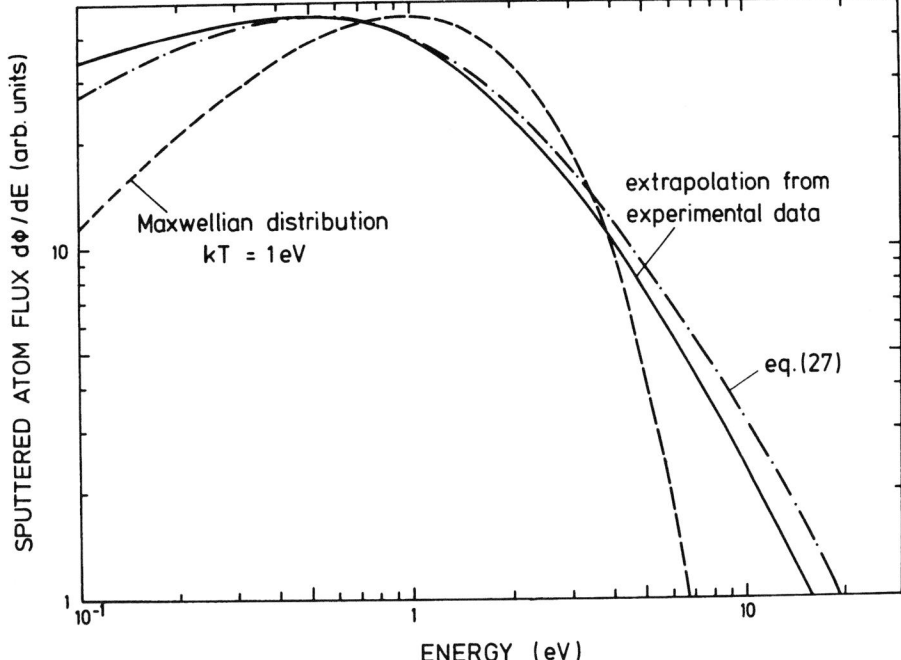

Fig. 26: Energy distribution of sputtered Fe atoms extrapolated
from experimental data for the incident hydrogen flux
distribution as shown in Fig. 23. For comparison a Maxwellian distribution with kT = 1 eV as well as the fit
from eq. (27) are indicated.

iii) Energy distribution

To obtain the energy distribution of sputtered particles the
energy distribution of incident atoms is divided into 5 energy intervals and the resulting distributions are added with the appropriate weight factors. The dependence of the energy distribution on
the energy of incidence as shown in Fig. 13 is taken into account.
The resulting energy distribution for sputtered Fe atoms, as shown
in Fig. 26, has a broad maximum at about 0.5 eV. At high energies
it decreases as $1/E_1^2$. It is compared in Fig. 26 with the energy
distribution given by kT = 1 eV, which, however, fails to describe
the high energy tail of the distribution. The best simple fit can
be obtained by using

$$\frac{dY}{dE_1} \ \alpha \ [\frac{E_1}{(E_1 + 1/3 \ U_o)^4}]^{2/3} \tag{27}$$

with $U_o = 4.5$ eV for SS.

3. Comparison with plasma impurity concentration

The calculated erosion of 5×10^{12} atoms/cm^2 sec during the plateau phase from a wall position far away from the gas valve can be compared to measured erosion rates of clean Ni samples in wall positions in ASDEX of 1.8×10^{13} Ni/cm^{2}sec /47/. This value is averaged over all different plasma phases, such as start up, plateau and end phase, neutral injection and RF-heating as well as disruptive phases. As phases with poor confinement contribute strongly to impurity production, the agreement between these values is reasonable.

For divertor discharges the interaction of plasma ions with the walls is removed into the separate divertor chamber. In the main vessel, the predominant erosion process remains the charge exchange neutral sputtering at the wall /39/. For the discharges where the data from Fig. 23 were obtained, the mean Fe-density in the plasma has been determined spectroscopically to 4.4×10^8 cm^{-3}. The influx calculated from the wall sputtering of 5×10^{12} atoms/cm^{2}sec would lead to an impurity concentration in the plasma of 6.3×10^8 cm^{-3} /48/. As the error in the sputtering yield alone is of the order of 2 due to the unknown degree of oxidation, this indicates that wall sputtering indeed is the dominant erosion process. However, a small additional contribution due to sputtering of wall structures by plasma ions cannot be ruled out. Another striking observation underlining this assumption is the similar time dependence of the impurity concentration in the plasma and the charge exchange neutral flux to the walls as shown in Fig. 27 /38/. For an ohmic divertor discharge in H_2 the gas density has been decreased abruptly by closing the gas valves. At decreasing densities the charge exchange flux increases, especially in the high energy tail of the distribution. Consequently, the sputtered flux of wall atoms increases and parallel to it the observed impurity concentration. Again the impurity concentration corresponding to the influx agrees within a factor of 2 with the measured concentrations.

CONCLUSION

Physical sputtering due to energetic light ions still lacks the theoretical foundation obtained for heavy ion sputtering. The basic processes, however, are identified and experimental data and fitting formulae are available in the region of interest of plasma-wall interaction. For ions with energies in the 100 eV range relatively little data exist on the angular and energy distribution of sputtered atoms.

Energetic atom and ion fluxes to the wall are routinely measured. Here, however, data for atoms with energies lower than 150 eV are scarce. In addition, more information about the angular distribution of incident atoms and on the dependence on toroidal and poloidal position is needed.

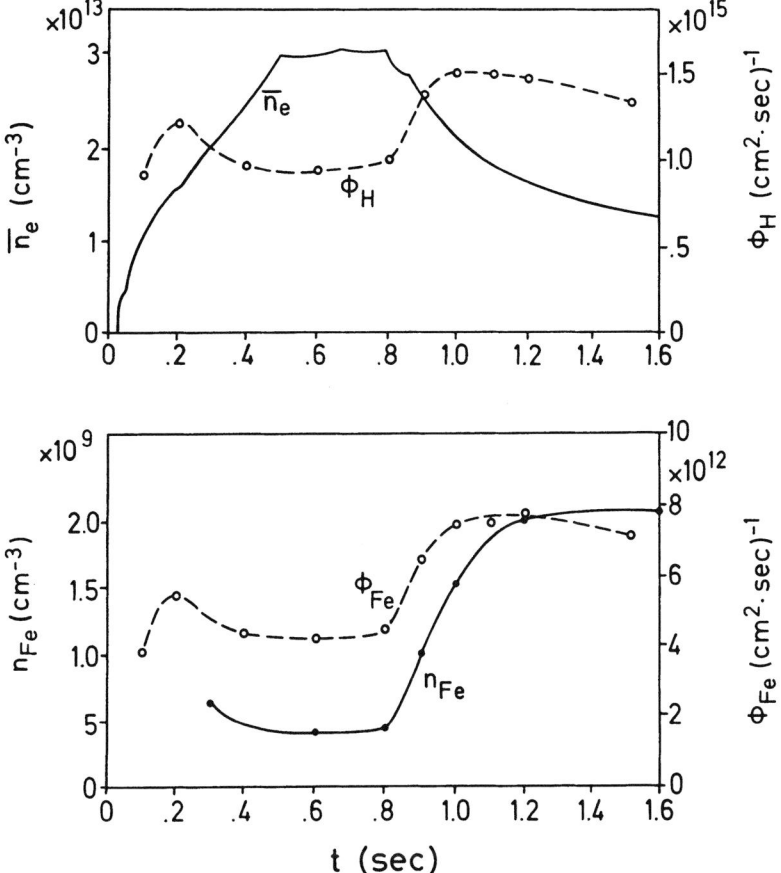

Fig. 27: Plots from an ASDEX divertor discharge of:
$\bar{n}_e$: time variation of the electron density;
ϕ_H: measured total charge exchange flux to the wall /42/;
ϕ_{Fe}: calculated sputtered iron flux from the wall;
n_{FE}: measured ion density within the discharge /48/.
The external gas valve is switched off at 0.8 s so that
the electron line density decreases /from 39/.

The largest uncertainty arises from the unknown surface composition of the wall. This surface composition consists of materials from throughout the machine and reflects the mode of operation. The degree of oxidation of deposited metal films makes estimates of impurity production rates uncertain at least within a factor of 2.

For divertor discharges in ASDEX, where direct plasma wall contact only occurs in the divertor chamber, charge exchange wall sputtering quantitatively explains the measured impurity concentrations. Light impurities like O and C are produced by ion induced desorption of CO, H_2O and hydrocarbon molecules from the wall.

Sputtering of plasma ions at wall structures may, however, still contribute to the impurity generation. In limiter discharges ion sputtering at the limiter, also due to impurity ions, may dominate over charge exchange sputtering at the wall.

REFERENCES

/1/ R.A. Langley, J. Bohdansky, W. Eckstein, P. Mioduszewski, J. Roth, E. Taglauer, E.W. Thomas, H. Verbeek, K. Wilson, Nucl. Fus., Special Issue (1984)

/2/ R. Behrisch ed., "Sputtering by particle bombardment I", TAP 47 (Springer Verlag Berlin 1981)

/3/ J. Roth, J. Bohdansky, W. Ottenberger, Report IPP 9/26 (1979)

/4/ N. Matsunami, Y. Yamamura, Y. Itikawa, N. Itoh, Y. Kazumata, S. Miyagawa, K. Morita, R. Shimizu, H. Tawara, Nagoya University, Japan, Report IPPJ-AM-32 (1983)

/5/ Y. Yamamura, Y. Itikawa, N. Itoh, Nagoya University, Japan Report IPJ-AM-26 (1983)

/6/ F. Wagner, this volume

/7/ J.P. Biersack, W. Eckstein, Appl.Phys. A34:74 (1984)

/8/ W. Eckstein, private communication

/9/ W. Eckstein, Max-Planck-Institut für Plasmaphysik, Garching, IPP-Report 9/41 (1983)

/10/ P. Sigmund, Phys.Rev. 184:383 (1969)

/11/ J. Lindhard, M. Scharff, H.E. Schiøtt, Kgl.Danske Videnskab. Selskar.Mat.-Fys.Medd. 33, No. 14 (1983)

/12/ U. Littmark, G. Maderlechner, Proc. 8th Symp.Phys.Ionized Gases, Dubrovnik, Jugoslavia, p. 139 (1976)

/13/ H.H. Andersen, H.L. Bay, J.Nucl.Mat. 93/94: 625 (1980)
H.H. Andersen, H.L. Bay, in ref. /2/, page 145

/14/ J. Bohdansky, J. Roth, H.L. Bay, J.Appl.Phys. 51:2861 (1980)

/15/ J. Roth, Proc. Symp. on Sputtering, eds. P. Varga, G. Betz, F.P. Viehböck, (Techn. University Wien, 1980) p. 773

/16/ J. Bohdansky, Nuclear Instr.Meth. B2: 587 (1984)

/17/ N. Matsunami, Y. Yamamura, Y. Itikawa, N. Itoh, Y. Kazumata, S. Mayagawa, K. Morita, R. Shimizu, H. Tawara, Atomic data and Nuclear data tables 31: 1 (1984)

/18/ D.L. Smith, J.N. Brooks, D.E. Post, D. Heifetz, Proc. 9th Symp. on Engineering Problems of Fusion Research (Chicago 1981), Institute of Electrical and Electronics Engineers, N.Y. (1982) 719

/19/ U. Littmark, S. Fedder, Nucl.Instr.Methods 194: 607 (1982)

/20/ J. Vukanic, P. Sigmund, Appl.Phys. 11: 265 (1976)

/21/ W. Eckstein, W. Möller, Nucl.Instr.Meth. (1984), to be published

/22/ J. Roth, J. Bohdansky, W. Eckstein, Nucl.Instr.Meth. 218:751 (1983)

/23/ Y. Yamamura et al., Rad. Eff. 80:57 (1984)

/24/ R. Chodura, J.Nucl.Mat. 111/112: 420 (1982)

/25/ H.C. Rosendaal, J.B. Sanders, Rad.Eff. 52:137 (1980)
/26/ R. Becerra-Acevedo, J. Bohdansky, W. Eckstein, J. Roth,
    Nucl.Instr.Meth. B2: 631 (1984)
/27/ J. Bohdansky, G.L. Chen, W. Eckstein, J. Roth,
    103/104: 339 (1981)
/28/ M.T. Robinson, Phil.Mag. 17: 639 (1968)
/29/ M.W. Thompson, Phil.Mag. 18:377 (1968)
/30/ H.L. Bay, W. Berres, E. Hintz, Nucl.Instr.Meth. 194:555 (1982)
/31/ S. Prigge, H. Niehus, E. Bauer, Proc. 7th Int.Vac.Congr.
    and 3rd Int.Conf. Solid Surfaces (Vienna 1977), p. 1381
/32/ S. Ahmad, B.W. Farmery, M.W. Thompson, Phil.Mag. A44:1387
    (1981)
/33/ H.-J. Barth, E. Mühling, W. Eckstein, to be published
/34/ G. Betz, G.K. Wehner, in: "Sputtering by particle bombard-
    ment, II", ed. R. Behrisch, TAP 52 (Springer-Verlag
    Berlin 1983), p. 11
/35/ P. Sigmund, in: "Sputtering by particle bombardment, I"
    ed. R. Behrisch, TAP 47 (Springer Verlag Berlin 1981),p.9
/36/ P. Varga, E. Taglauer, J.Nucl.Mat. 111/112: 726 (1982)
/37/ E. Taglauer, W. Heiland, Proc.Symp. on Sputtering, eds.
    P. Varga, G. Betz, F.P. Viehböck (Techn.University Wien,
    1980) p. 423
/38/ P.-J. Schneider, W. Eckstein, H. Verbeek,
    Nucl.Instr.Meth. B2: 655 (1984)
/39/ J. Roth, J.Nucl.Mat. 107/108:291 (1981)
/40/ B. Schweer, P. Bogen, E. Hintz, D. Rusbüldt, S. Goto,
    K.H. Steuer, J.Nucl.Mat. 111/112:71 (1982)
/41/ G. Fussmann, U. Ditte, W. Eckstein, T. Grave, M. Keilhacker,
    K. McCormick, H. Murmann, H. Röhr, M. ElShaer, K.H. Steuer,
    Z. Szymanski, F. Wagner, G. Becker, K. Bernhardi, A. Eber-
    hagen, O. Gehre, J. Gernhardt, G. v.Gierke, O. Glock,
    O. Gruber, G. Haas, M. Hesse, G. Janeschitz, F. Karger,
    S. Kissel, O. Klüber, M. Kornherr, G. Lisitano, H.M.Mayer,
    D. Meisel, E.R. Müller, W. Poschenrieder, F. Ryter, H.Rapp,
    F. Schneider, G. Siller, P. Smeulders, F. Söldner, E. Speth,
    A. Stäbler and O. Vollmer, J.Nucl.Mat. 128/129:350 (1984)
/42/ F. Wagner, J.Vac.Sci.Technol. 20:1211 (1982)
/43/ H. Verbeek, Report IPP 9/50, Max-Planck-Institut für Plasma-
    physik, Garching (1984)
/44/ F. Wagner, H.M. Mayer, Proc.Intern.Symp. on Plasma Wall
    Interaction (Jülich, W. Germany 1976), p. 149
/45/ P. Staib, H. Kukral, E. Glock, G. Staudenmaier,
    J.Nucl.Mat. 111/112:173 (1982)
/46/ R. Behrisch, P. Børgesen, J. Ehrenberg, B.M.U. Scherzer,
    B.D. Sawicka, J.A. Sawicki and the ASDEX-Team,
    J.Nucl.Mat. 128/129:470 (1984)
/47/ G. Staudenmaier, private communication, J.Vac.Sci.Technol.,
    to be published
/48/ G. Fussmann, private communication

/49/ J. Roth, G. Staudenmaier, G. Fussmann, unpublished
/50/ K. Behringer, G. Fussmann, W. Poschenrieder, E. Taglauer,
      K. Bernhardi, A. Eberhagen, O. Gehre, J. Gernhardt,
      G. v.Gierke, E. Glock, G. Haas, F. Karger, M. Keilhacker,
      S. Kissel, O. Klüber, M. Kornherr, G.G. Lister, G. Lisitano,
      H.-M. Mayer, K. McCormick, D. Meisel, E.R. Müller,
      H. Murmann, H. Niedermeyer, H. Rapp, B.M.U. Scherzer,
      F. Schneider, G. Siller, P. Smeulders, F. Söldner,
      E. Speth, A. Stäbler, P. Staib, G. Staudenmaier,
      K. Steinmetz, K.-H. Steuer, Z. Szymanski, O. Vollmer,
      G. Venus, and F. Wagner: Proc. 11th Europ. Conf. on Contr.
      Fusion and Plasma Physics, Aachen 1983, Eds. S. Methfessel,
      G. Thomas, EPS Geneva 1983, 7D,II: 467
/51/ H.F. Winters, P. Sigmund, J.Appl.Phys. 45:4760 (1974)

CHEMICAL SPUTTERING AND RADIATION ENHANCED SUBLIMATION OF CARBON

J. Roth

Max-Planck-Institut für Plasmaphysik
EURATOM Association, D-8046 Garching

ABSTRACT

The particularities of graphite sputtering at elevated temperatures are presented. Chemical sputtering, i.e. volatile molecule
formation with hydrogen isotopes is extensively studied at energies
above 100 eV and at thermal energies. Data on radiation enhanced
sublimation of carbon are still scarce and available only for ion
energies above 100 eV. The current understanding of both processes
is summarized and an attempt is made to quantify erosion yields for
fusion relevant conditions.

I. INTRODUCTION

Among all fusion relevant metals and alloys, carbon and its
carbides show a singular behaviour as far as erosion due to light
ion bombardment is concerned. One difference is the chemical reactivity of carbon which in case of reactive gases like hydrogen and
oxygen may lead to volatile reaction products like hydrocarbons
and carbon oxides /see 1, 2/. Another point is the large anisotropic
volume change of graphite under radiation - as experienced in
fission reactors /3/ - which is assumed to be intimately related
to the radiation enhanced sublimation at elevated temperatures /4/.
Both effects lead to the observation of strong temperature variations of the sputtering yield, which is so far unknown for metals
/5/.

The processes leading to erosion are of interest in fusion research only if they exceed the erosion due to physical sputtering
or sublimation at high temperatures. The temperature and energy

regimes, where these processes dominate, will therefore initially be discussed. Subsequently, the status of the current understanding of the processes will be presented with emphasis on possible synergistic effects in a fusion reactor environment. Eventually an attempt will be made to estimate the importance of both effects for the application of graphite in fusion devices.

II. DEFINITION

Before presenting the effect in detail, the definitions should be given which are used in the following:

In chemical sputtering molecules are formed on the surface due to a chemical reaction between the incident particles and target atoms, which have a binding energy low enough to desorb at the temperature of the solid. Processes, where target atoms or atom clusters are thermally released during irradiation in excess to the vapour pressure without molecule formation with the incident particles, will be called radiation enhanced sublimation.

A mass analysis of the sputtered species should distinguish the processes. Both processes can be distinguished from physical sputtering of molecules by an energy analysis of the sputtered particles, which should result in a distribution equivalent to the surface temperature.

The temperature dependence of chemical sputtering may be quite complicated, as chemical reactions are involved, and it may be confined to a certain temperature interval. Radiation enhanced sublimation is expected to increase with increasing target temperature. Figure 1 gives the limits in ion energy and target surface temperature for the erosion processes occurring in the interaction of hydrogen ions with carbon. The border lines are shown schematically. They are not sharp and depend among others on binding energy of atoms and molecules, ion flux and surface conditions.

III. CHEMICAL SPUTTERING

Figure 2a,b shows a collection of data for the chemical sputtering yield of hydrogen and oxygen with carbon of energetic ions and thermal atoms respectively /6-25/. From Fig. 2 it is seen that both energetic hydrogen ions and thermal hydrogen atoms show a strong temperature dependence of the erosion yield. While the yield for energetic ions lies in the range of 10^{-2} to 10^{-1} depending on ion energy and flux with the maximum occurring around 800 - 900 K, the yield for thermal atomic hydrogen with energies much smaller than 1 eV varies between 10^{-6} and 10^{-3} and the maximum may occur between 500 and 1100 K. In view of the small erosion yield

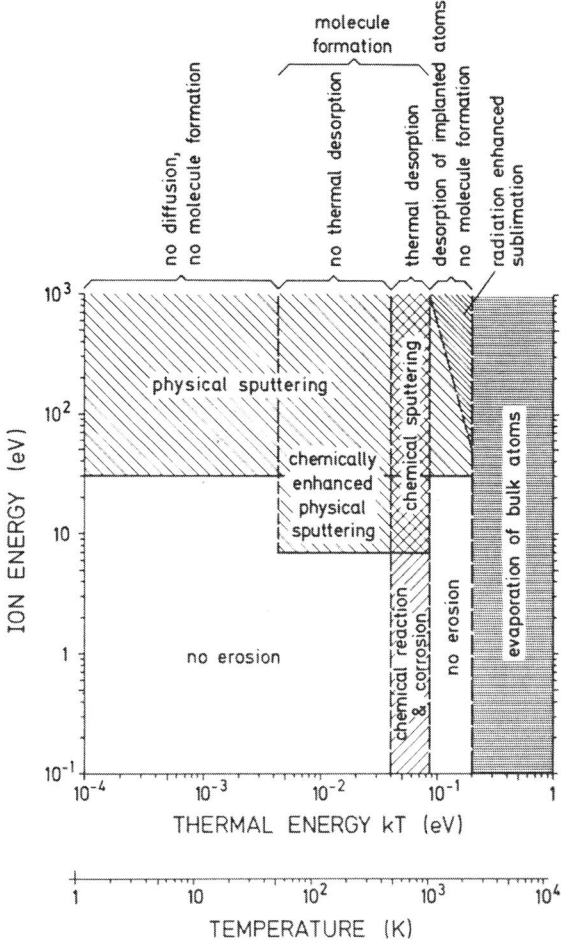

Fig. 1: Limits in ion energy and surface temperature for erosion
processes like physical sputtering, chemical sputtering,
radiation enhanced sublimation and evaporation.

of thermal atomic hydrogen and the comparable low densities and
fluxes of thermal atomic hydrogen towards carbon structures in tokamaks compared to the energetic ion and neutral fluxes, their contribution to erosion will be small. To estimate the effect of Franck-
Condon neutrals (3 eV) an understanding of the erosion mechanism
of thermal atoms may be important /33/.

Oxygen also shows enhanced erosion yields with pronounced
temperature dependences. For energetic ions the yields are close to
1 almost independent on energy and indicate CO formation. At high
temperatures ion enhanced erosion occurs as found independent of

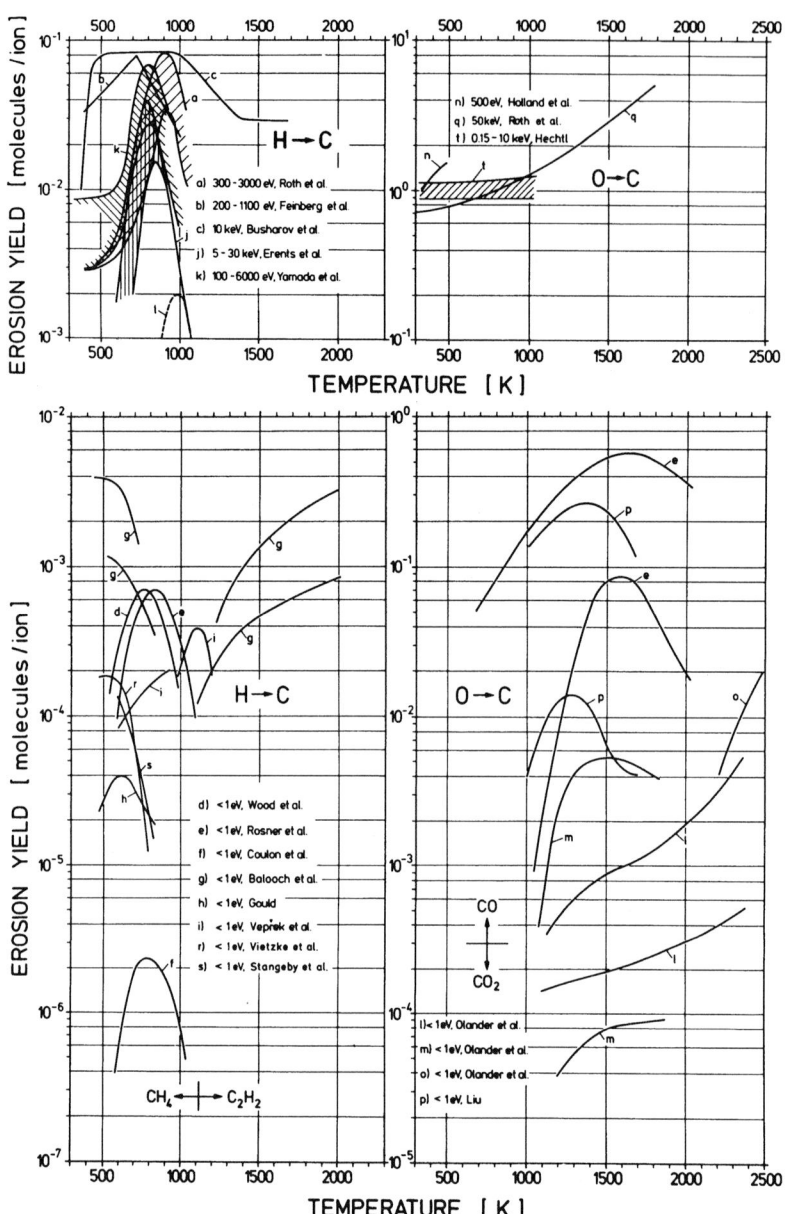

Fig. 2a: Temperature dependence of the chemical sputtering yield of carbon bombarded with H^+ and O^+ ions. Shaded areas between two curves indicate the range of measured data for different ion energies.

Fig. 2b: Temperature dependence of the chemical erosion yield for carbon with thermal hydrogen and oxygen atoms. The observed molecular species are indicated.

ion species (see Sect. IV). As little information exists on the oxygen-carbon reaction, the following discussion is limited to hydrogen.

1. Current Models for Chemical Sputtering

The interaction of hydrogen with carbon surfaces is most widely investigated. Atomistic models for the subsequent addition of hydrogen atoms to carbon surface atoms are developed /14, 17/, detailed balance equations for the reaction products are fitted to the data /26/ and global reaction rates are calculated from a hydrogen balance and the assumption of the reaction rate being proportional to the surface concentration of hydrogen /8, 9/.

The model today is shown schematically in Fig. 3. Energetic hydrogen ions penetrate the surface and accumulate within a well defined range profile. As the saturation concentration is reached, hydrogen is reemitted either through the front surface or through microchannels and cracks which are formed due to structural changes. At the surface hydrogen atoms may recombine to H_2 or D_2 molecules or form hydrocarbon complexes. Detailed measurements by Vietzke et al. /20/ indicate that eventually the radical CH_3 is released. This is most clearly seen by using thermal atomic hydrogen, which reacts at the front surface, while the reemission of CH_3 through microchannel will result in recombination of the radial with the hydrogen covered walls and the observation of CH_4. A simultaneous

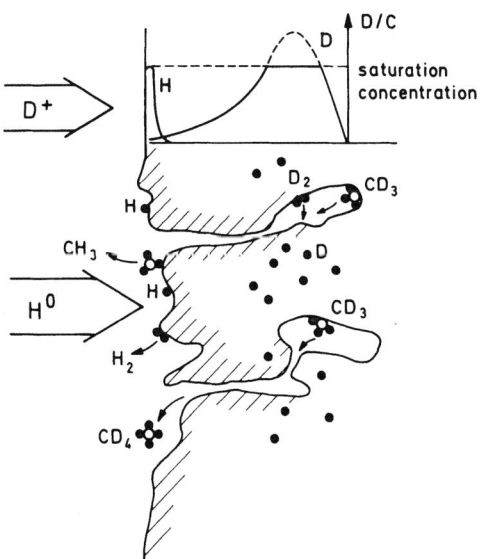

Fig. 3: Schematic model for simultaneous bombardment of carbon with hydrogen atoms and deuterium ions.

bombardment by deuterium ions and hydrogen atoms actually showed the formation of CH_3 and CD_4 molecules without much mixing of the isotopes.

2. Analytic Formula

The chemical reaction constants involved in the model as well as the sticking coefficient of atomic hydrogen on graphite has been obtained by fitting reaction yields from this model to measured data for thermal atomic hydrogen /17/. The sticking of molecules and therefore their reaction yield appear to be negligible /17, 18, 20/. With the assumption of a surface saturation of hydrogen this model allows also to explain the shift of the maximum yield in temperature for different incident atom fluxes /1/. Measurements at low atom fluences indicate, however, the importance of the state of activation of the surface /21/.

For energetic ion bombardment more global balance equations were used to explain the observed temperature dependence /8, 9/. In steady-state the reemitted hydrogen flux is equal to the incident ion flux Γ_H. The surface concentration c_{sH} can be calculated depending on the residence time of the hydrogen and is assumed to be proportional to the reaction yield. This yields

$$\Gamma_H \left(1 - \frac{c_{sH}}{c_{so}}\right) = \frac{\Gamma_H}{\tau_o} e^{-Q_2/kT} \tag{1}$$

where the residence time is equal to $\tau_o e^{Q_2/kT}$ and c_{so} is the surface saturation concentration.

Then

$$Y \alpha \frac{e^{-Q_1/kT}}{\Gamma_H/c_{so} + \frac{1}{\tau_o} e^{-Q_2/kT}} \tag{2}$$

This predicts a maximum at a temperature

$$T_m = \frac{Q_2}{k} \left(\ln \frac{(Q_2-Q_1) c_{so}}{\Gamma_H Q_1 \tau_o}\right)^{-1} \tag{3}$$

By comparing (3) with the measured maximum temperature values /a collection is shown in 27/ the activation energies Q_1 and Q_2 and c_{so} can be extracted and eq. (2) becomes

$$Y = 1.3 \cdot 10^3 \frac{c_{so} e^{-1/kT}}{\Gamma_H + 3 \cdot 10^{10} c_{so} e^{-2/kT}} \tag{4}$$

with kT in eV, Γ_H in $cm^{-2} sec^{-1}$ and c_{so} in cm^{-2}. The trapped amount in saturation becomes of the order of $1 \cdot 10^{17} cm^{-2}$ increasing with energy, which is equivalent to the saturation of carbon with hydrogen over several hundred monolayers.

However, eq. (4) does not contain an energy dependence other than c_{so}, while the yield data clearly show a maximum, correlated with the deposited energy in the carbon surface /6, 10/ (Fig. 4). Also, experiments using thermal atomic hydrogen could distinguish surfaces with different degrees of activation /2, 21/ and the influence of simultaneous bombardment with heavy inert gas ions has been demonstrated /28/. The nature of the activated state remains unknown and the existance of surface sites for the chemical reaction have been proposed /2/ as well as the reaction at radiation-induced vacancies at the surface /29/. Yamada and Sone /29/ have made an attempt to combine the model of eq. (2) with the obvious dependence on the energy deposited $F_D(E)$ into the surface layer. They derived the equation

$$Y = A \left(\frac{\Gamma_H \, F_D}{\nu_o \, e^{-Q_3/kT}} \right)^\ell \frac{(1-B) \, e^{-Q_1/kT}}{\frac{\Gamma_H}{c_{so}} + \frac{1}{\tau_o} \, e^{-Q_2/kT}} \tag{5}$$

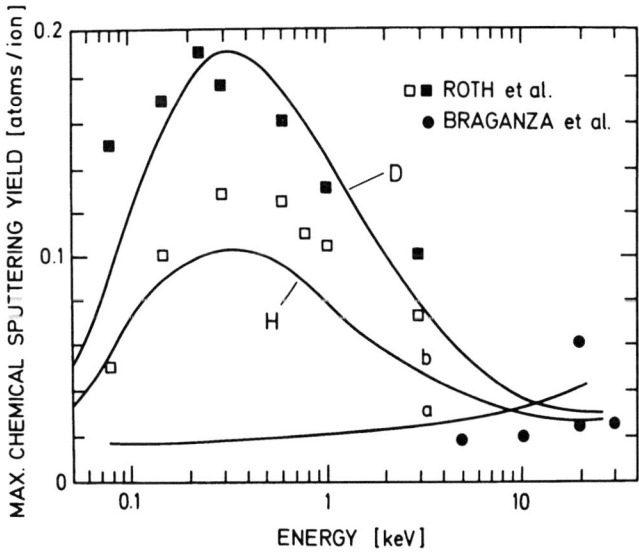

Fig. 4: Energy dependence of the maximum chemical sputtering yield for a constant hydrogen ion flux of 10^{15} H/cm² sec. Open points are for hydrogen, full points for deuterium ions. Curve a is calculated from eq. (4), curves b from eq. (6).

Here B is the energy dependent reflection coefficient of hydrogen at the carbon surface, $F_D(E)$ the deposited energy in the surface layer. $\Gamma_H \cdot F_D$ is the production rate and $\nu = \nu_0 \, e^{-Q_3/kT}$ is a jump frequency proportional to the annealing rate of radiation-incuded vacancies and interstitials. By taking $F_D(E)$ from physical sputtering calculations they could reasonably fit the measured dependence of the methane production rate on the ion energy with the exponent ℓ between 0.5 and 1 and $A(\nu)^{-\ell}$ as a fitting factor. The activation energies Q_1, Q_2 and Q_3 could be extracted from the temperature dependence of the methane yield.

This equation, however, assumes that methane formation only occurs if energy is deposited into the surface. Data for very low energy (< 1 eV) and very high energy ($\geq$ 20 keV) hydrogen bombardment show that the methane yield here does not follow the deposited energy function. It seems therefore to be appropriate to combine eqs. (4) and (5). The best fit to the data appears to be, with $c_{so} = 10^{17}$ at/cm²

$$ Y = \frac{6 \cdot 10^{19} \, e^{-1/kT}}{\Gamma_H + 3 \cdot 10^{27} \, e^{-2/kT}} \, (1 + 6 \cdot 10^3 \, m_1 \, (\frac{E_o - 35}{(E_o + 265)^2})) \tag{6} $$

where the energy dependence for energies larger than E_{th} = 35 eV has been taken from the fit by D. Smith et al. to the physical sputtering yield /30/. m_1 is the mass of the hydrogen isotope. Equation (6) fits the existing data better than within a factor of 2. The reflection coefficient B has been put to 0.5. As the energy dependence of F_D dominates, c_{so} and B have been assumed independent on energy. Unlike eq. (5) the annealing process, leading to the term $(\Gamma_H)^\ell$ has been disregarded. As a consequence, the yield decreases as $1/\Gamma_H$. There is some experimental evidence for this behaviour at high fluxes, but the data base does not allow the extrapolation to fluxes impinging onto a carbon limiter or divertor plate in a fusion device.

3. Synergistic Effects

The chemical sputtering yield of carbon with hydrogen depends on the energy deposited into nuclear damage in the surface layer, as shown in Fig. 4. Thus, it is expected that the methane yield resulting from thermal hydrogen atoms will be enhanced during simultaneous bombardment with helium or other energetic ions. For the case of argom bombardment, Vietzke et al. /28/ showed an enhancement of a factor of 50 for the reaction yield of atomic hydrogen with pyrolytic graphite to form volatile hydrocarbons (Fig. 5). The low sticking coefficient of atomic hydrogen may, however, limit the reaction yield to 10^{-2}. Simultaneous bombardment with thermal hydrogen atoms and energetic hydrogen ions results in a linear

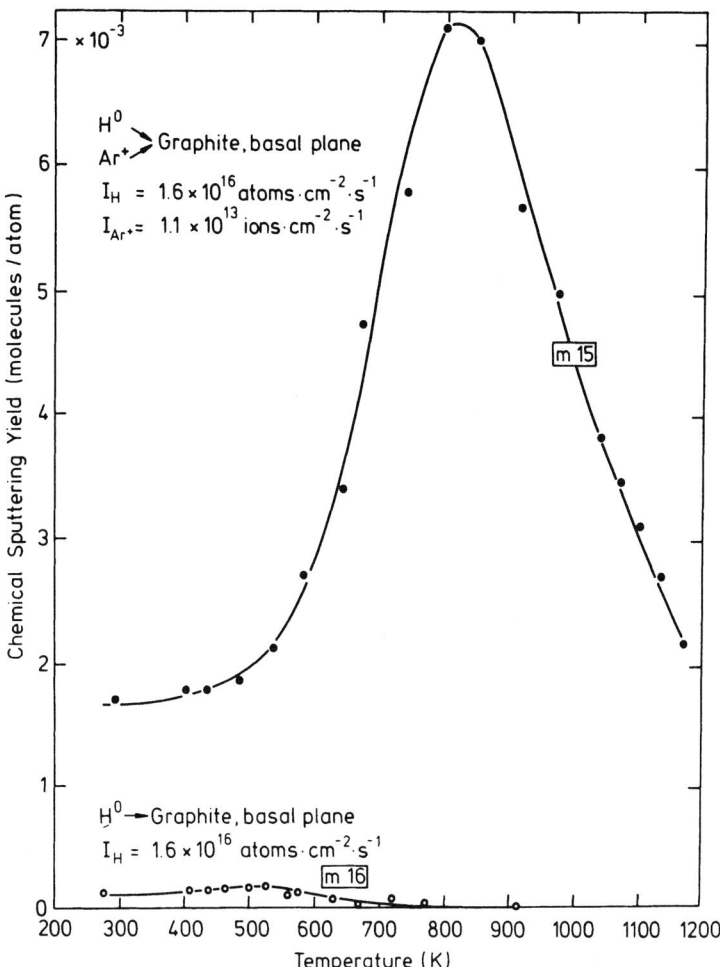

Fig. 5: Temperature dependence of the chemical sputtering yield
for mass 15 (CH_3) and mass 16 (CH_4) due to irradiation
of graphite with hydrogen atoms and simultaneously with
hydrogen atoms and Ar^+ ions, respectively.

addition of both reaction yields /31, 32/, although other authors
report a more than linear superposition of both yields /33/.

For the simultaneous interaction of energetic (> 40 eV)
electrons and thermal hydrogen atoms dissociated at a tungsten fila-
ment in front of the target, Ashby et al. /34, 35/ report an enhancement of the methane yield in the temperature range between 300 K
and 600 K. The enhancement factor, amounting roughly to a factor of
20, increases with surface temperature and depends on the relative
supply of hydrogen atoms and energetic electrons. A similar effect
is mentioned (although ion and electron beam densities and carbon

surface conditions are not given in detail) in the U.S.S.R. contribution to the International Tokamak Reactor (INTOR) Workshop in 1982 /36/. In contrast to the work of Ashby et al., the enhancement factor due to electrons is highest at room temperature (∿ 30) and is only 2.5 at the peak temperature for chemical sputtering. The enhancement shows a maximum at 400 eV and disappears at energies above 1 keV. Independent experiments with an energetic electron beam and atomic hydrogen simultaneously impinging on a graphite surface were performed by Vietzke et al. /20/ and by Haasz et al. /21/. Vietzke et al. used an atomic hydrogen beam and Haasz et al. used thermal atomic hydrogen produced using a backfilling method similar to that of Ashby /34/. In both experiments, only a small (∿ 50 - 100 %) enhancement of the methane reaction yield was found. Haasz et al. found that stringent sample and vacuum system conditioning was necessary to avoid spurious effects that can obscure true synergistic effects. Any extrapolation to fusion reactor conditions for the simultaneous effect of electrons and hydrogen atoms remains uncertain.

IV. RADIATION ENHANCED SUBLIMATION

The temperature dependences for hydrogen, deuterium, and helium sputtering of carbon show a steady increase at temperatures above 1300 K (Fig. 6) as measured for pyrolytic graphite and PAPYEX strips /37, 38/. For hydrogen sputtering, residual gas analysis, which re-

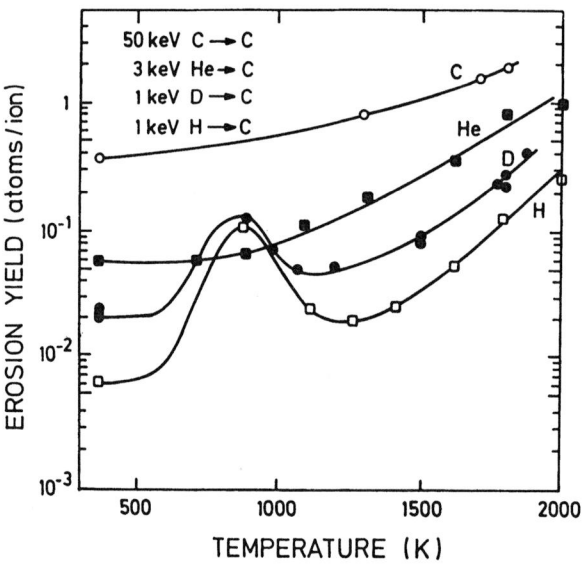

Fig. 6: Temperature dependence of the sputtering yield of graphite bombarded with H, D, He and C ions.

veals that CH_4 formation is responsible for the erosion maximum at 900 K, does not show an enhancement of hydrocarbon formation at temperatures above 1300 K. Since the same increase occurs for the inert helium ions and for bombardment with carbon, oxygen, and argon ions /38/, chemical interaction between the ions and the target atoms cannot be responsible. It can be shown that carbon atoms are sputtered /39/ with isotropic angular distribution /37/ and an energy distribution roughly equivalent to the surface temperature /40/. This process has consequently been named radiation-enhanced sublimation, although the distribution of carbon atoms clusters emitted during ion bombardment is different from sublimation (Fig.7)

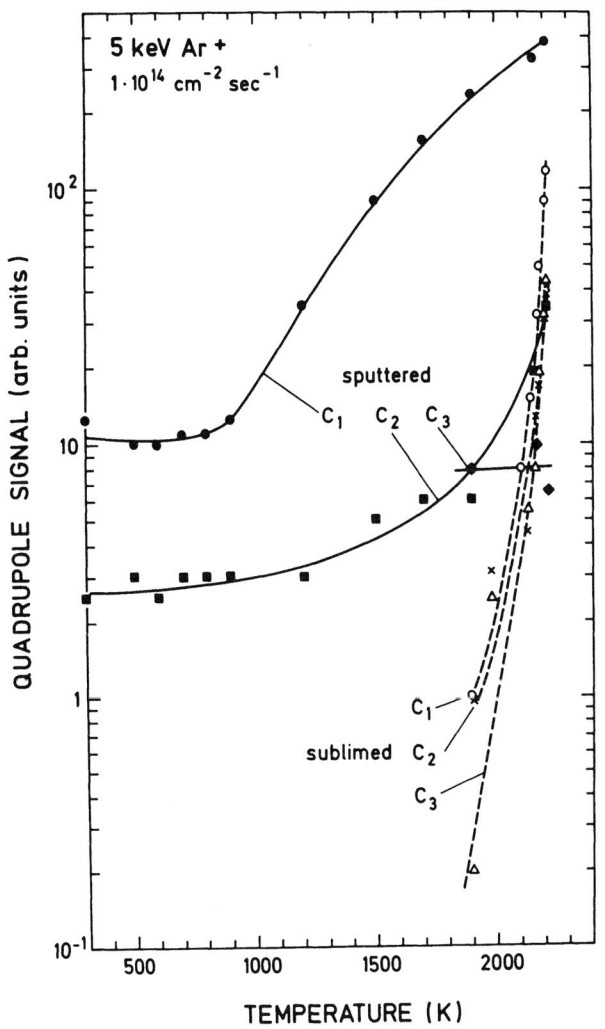

Fig. 7:

Temperature dependence of the C_1-, C_2-, C_3-components released with and without 5 keV Ar^+-ion bombardment from the basal plane of pyrolytic graphite.

and predominantly atomic species are ejected. The number of atoms ejected scales well with the energy deposited in nuclear damage by the sputtering ions /38/ at a given temperature. Figure 8, an Arrhenius plot of the data of Fig. 5 (where the peak due to chemical sputtering at 900 K has been omitted), reveals a temperature independent physical sputtering yield at low temperatures and a thermally activated sputtering yield above 1300 K. The activation energy varies between 0.6 and 1 eV for different ions.

The striking similarity of the temperature dependence of the volume swelling of graphite under neutron bombardment with the temperature dependence of the radiation enhanced sublimation yield (Fig. 8) has led to an attempt to adopt the same atomistic model /4/. The radiation damage theory of graphite assumes the creation of interstitial carbon atoms in the graphite lattice due to nuclear

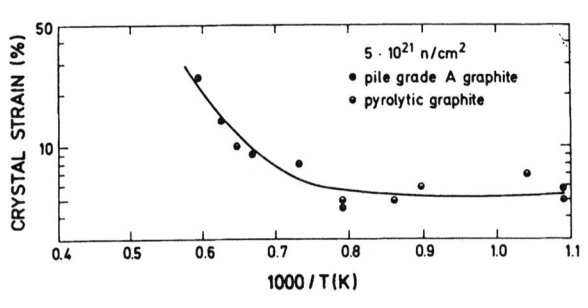

Fig. 8a:

Temperature dependence of the crystal strain parallel to the c-axis of graphite for a constant neutron fluence of $5 \cdot 10^{21}$ cm^{-2}.

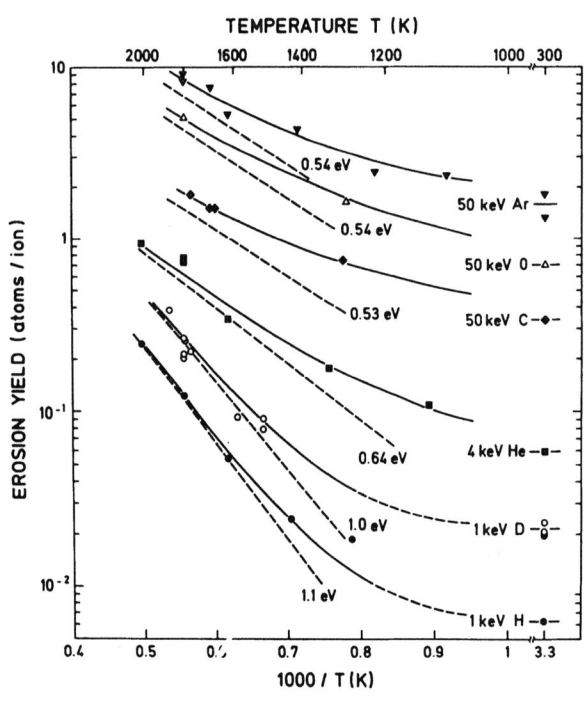

Fig. 8b:

Temperature dependence of the erosion yield of graphite due to ion bombardment collision with a displacement energy of 23 to 48 eV /3/. These interstitial atoms are very mobile between the graphite planes already at 500 K and diffuse until they are annihilated at vacancies or cluster at fixed nuclei, which are present at a density of 10^{16} to 10^{17} cm^{-3}, or arrive at the grain boundaries. Di-interstitials and small interstitials clusters are still assumed to be mobile. Larger interstitial clusters form additional lattice planes in the graphite thus leading to the observed swelling parallel to the c-axis. The vacancy migration energy is of the order of 3 to 4 eV. To explain the observed shrinkage of the graphite perpendicular to the c-axis, it has been assumed that the vacancies cluster in form of vacancy lines, which collapse and are no longer sinks for interstitials. This model could adequately explain the anisotropic volume swelling of graphite /3/.

For its application to ion bombardment, where the damage is not homogeneously distributed over the bulk, but concentrated within a damage profile close to the surface, the above model has been slightly modified. As the close surface represents a strong sink for both vacancies and interstitials, other fixed sinks distributed over the bulk have been neglected. Furthermore, free interstitials arriving at the surface are only bound to the surface by weak van-der-Waals forces and therefore assumed to sublimate at the elevated temperatures. The source distribution for interstitial and vacancy formation has been obtained from TRIM calculations /41/ assuming the modified Kinchin-Pease model with a fixed mean displacement energy E_d. The interstitial and vacancy profiles in steady state have been calculated using the diffusion code PIDAT /42, 13/. As boundary condition at the surface, zero concentrations for both interstitials and vacancies are assumed. The erosion yield is then given by the flux of interstitials through the surface, i.e. by the gradient of the interstitial profile at the surface. Table 1 summarises the important input data for the model calculations. Apart from these parameters taken from literature the model does not contain any free fitting factor. Details of the calculations are given in Ref. /4/.

Table 1: Input data for the model calculations.

Incident ion flux	10^{16}/cm² sec
damage profile	TRIM code
displacement energy	25 eV
interstitial migration energy	0.3 eV
vacancy migration energy	3.5 eV
preexponential factors of interstitial and vacancy migration	$7.1 \cdot 10^{-4}$
vacancy-interstitial and vacancy-vacancy recombination radius	0.21 nm.

Figure 9 shows a comparison of the temperature dependence of the erosion yield due to H, D, He and Ar bombardment /38/, with the interstitial flux through the surface as calculated using the present model. For better comparison the physical sputtering yield at room temperature has been added to the calculated values. It can be seen that both, the temperature dependence as well as the dependence on ion mass, are well reproduced by the model. The absolute values are in surprisingly good agreement with the data as no additional fitting is involved.

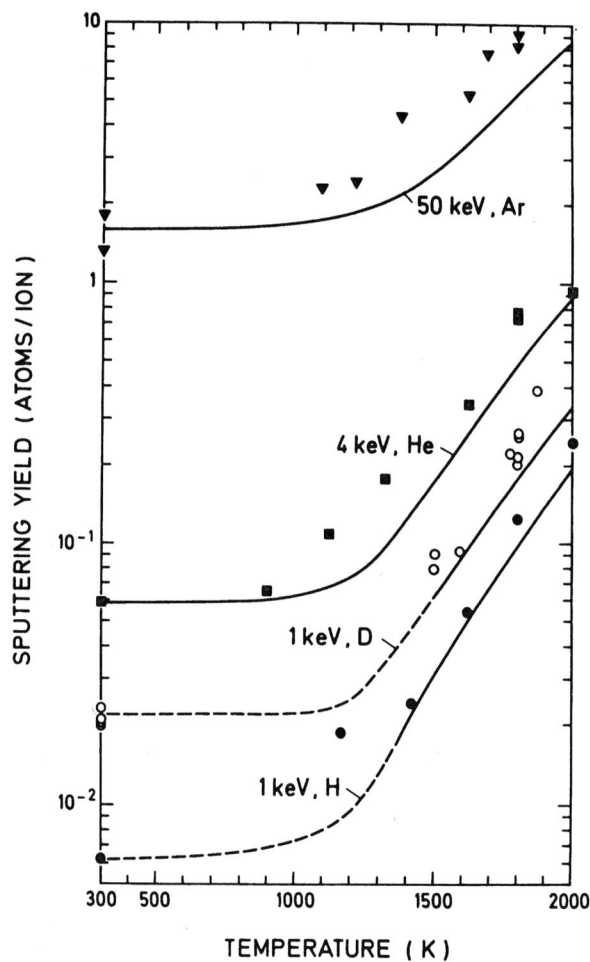

Fig. 9: Comparison of the yield of sublimating interstitials calculated from the model with the measured sputtering yields as a function of temperature.

For the case of hydrogen, where the most detailed energy dependence has been measured, Fig. 10 shows a comparison with the model calculations. Again it can be seen that the energy dependence is well reproduced and the absolute value is within a factor of 2 of the measured data. The yield shows a maximum at ion energies around 200 eV in agreement with experiments. A threshold energy is reached at an energy, where no interstitials can be produced

$$E_{th} = \frac{48\, m_1}{(m_1 + 12)^2}\, E_d \tag{7}$$

with m_1 denoting the atomic masses of the hydrogen isotope.

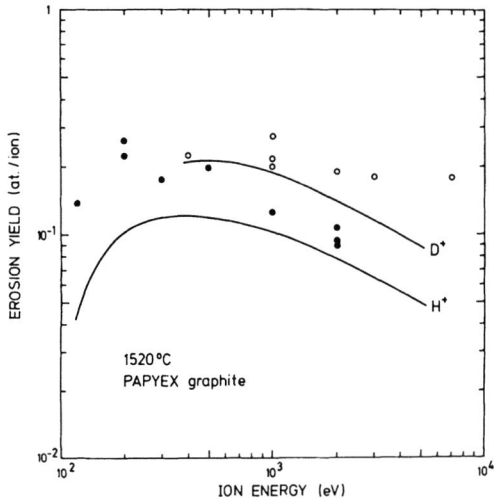

Fig. 10:

Energy dependence of the yield of sublimating interstitials for hydrogen and deuterium bombardment at 1800 K compared to experimental data.

V. CARBIDES AND INFLUENCE OF IMPURITIES

The temperature dependences of methane formation for B_4C and TiC during steady-state hydrogen or deuterium bombardment are compared with that for pyrolytic graphite in Fig. 11 /6, 43, 44/.

For SiC no CH_4 formation was observed during prolonged bombardment. However, at low fluences both SiC and TiC show high methane production yields that decrease exponentially with fluence as the surface layer is depleted in carbon throughout the ion range. In steady-state the composition of the sputtered particles should reflect the bulk composition of the compound. A comparison with the physical sputtering yield at room temperature shows that for TiC sputtering by hydrogen essentially all carbon atoms are sputtered as CH_4 molecules and that titanium sputtering is the rate-limiting process over the entire temperature range /44/. The energy dependence of the CH_4 formation during irradiation of TiC with hydrogen ions consequently reflects closely the energy dependence

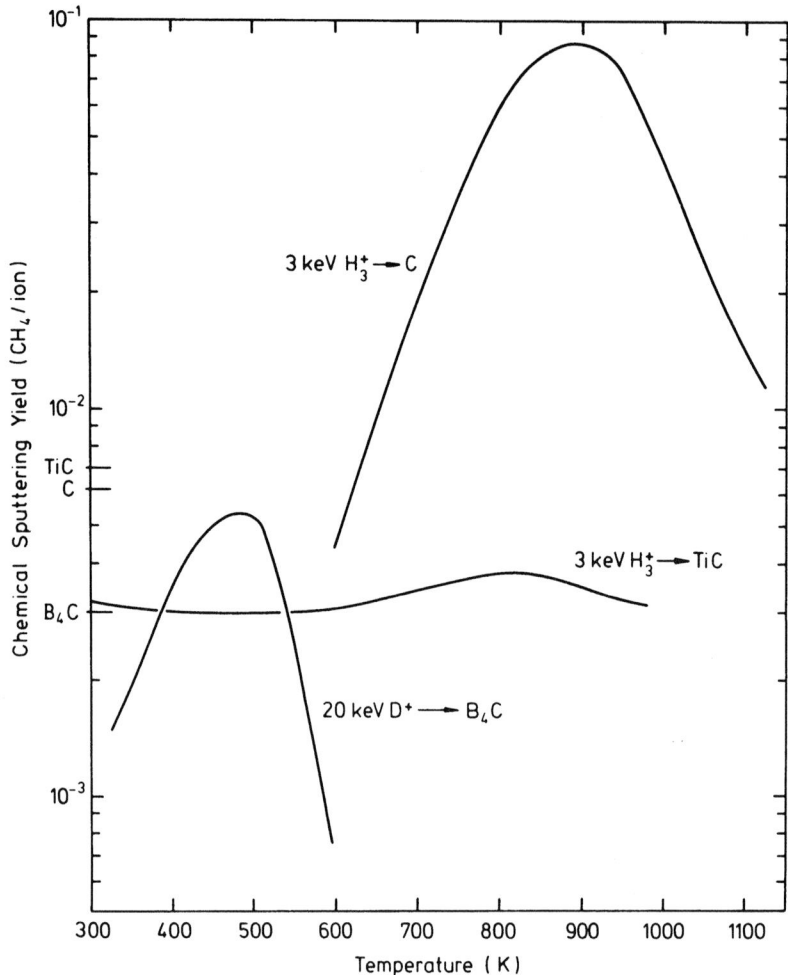

Fig. 11: Temperature dependence of the chemical sputtering yield of TiC, B$_4$C and pyrolytic graphite. Corresponding physical sputtering yield at room temperature are indicated.

of the total sputtering yield as measured by weight loss. More recently, however, the high steady-state production of CH$_4$ from TiC could not be reproduced /45/. After high initial CH$_4$ formation the carbon erosion in steady-state was mainly due to physical sputtering of carbon atoms. B$_4$C at room temperature shows physical sputtering of carbon atoms, and the total sputtering yield for 20 keV D$^+$ at 500 K should be 2.5 $\cdot$ 10^{-2}.

The behaviour of carbides, which show no or only small steady-state formation of methane, has led to a number of investigations of the influence of impurities on the chemical erosion of carbon.

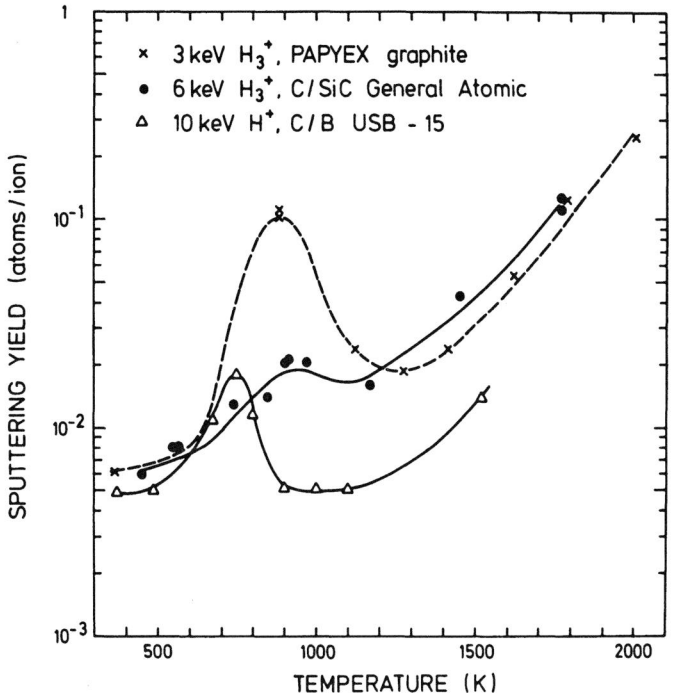

Fig. 12: Temperature dependence of the sputtering yield of C/SiC and C/B alloys compared with pure graphite for hydrogen bombardment.

A coverage of the surface with some monolayers of evaporated Ti, Ni, Si or Mo resulted in a factor of about 2 smaller CH_4 formation from energetic ions /46/ as well as for atomic hydrogen /47/. Much stronger reduction could be obtained by adding small amounts of impurities into the bulk graphite. Figure 12 shows the influence on 4 at% Si in graphite and 15 at% B in glassy carbon /46/.

For both alloys the chemical sputtering yield is reduced by almost an order of magnitude. The depth distribution of the Si in the C/SiC alloy showed a depletion of Si at temperatures above 1300 K, while around 900 K preferential removal of C occurs until Si enrichment suppresses further methane formation.

At high temperatures Si is evaporated from the C/SiC alloy and erosion proceeds as in pure graphite. Here the simultaneous evaporation of Ti during ion bombardment proved to be successful in reducing the radiation induced sublimation process /46/. The Ti is distributed over about 1500 A with a long tail into the bulk, which is of the order of the cascade extention due to the ion bombardment (Fig. 13). The width of a Ti layer evaporated without Ar bombardment and subsequently annealed to 1800 K did not exceed the 400 A depth

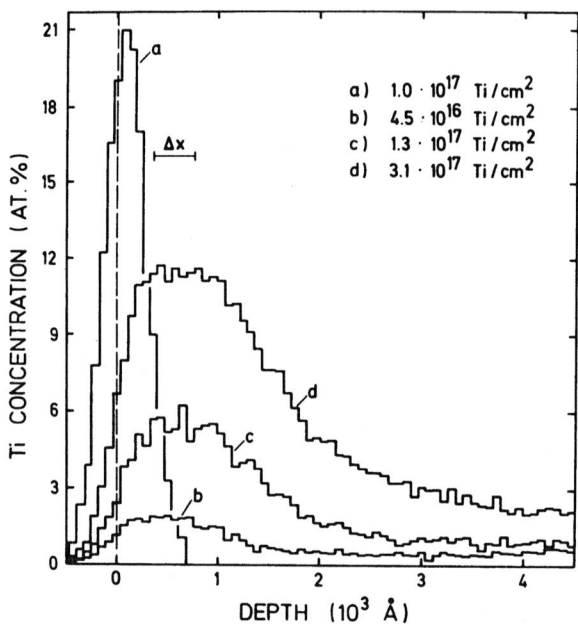

Fig. 13: Depth distribution of Ti evaporated onto graphite.
Profile a) is obtained for evaporation at 1800 K,
while b), c) and d) are obtained for simultaneous
50 keV Ar bombardment at 1800 K.

resolution of this measurement. No clustering of Ti could be observed in the SEM.

This Ti concentration has a strong effect on the ion-enhanced sublimation yield of carbon atoms (Fig. 14). Small concentrations of Ti initially increase the erosion yield while a concentration of about 10 at% in the surface layer leads to an almost complete suppression of the enhanced yield. In the model outlined in Sect.IV the initial increase may be understood by the occupation of vacancies by Ti atoms and consequently a higher density of free interstitials which diffuse to the surface. In addition, TRIM computer simulations show an increase of the deposited energy close to the surface of graphite for small additions of Ti owing to the better mass match of Ti with Ar ions. At higher Ti concentrations the erosion yield of carbon atoms approaches values observed for TiC.

Considering the suppression of methane formation at 900 K due to Si additions and the influence on Ti on the enhanced sublimation, it has been proposed /46/ to alloy a small concentration of TiC into the graphite to reduce both the chemical sputtering and the radiation-enhanced sublimation.

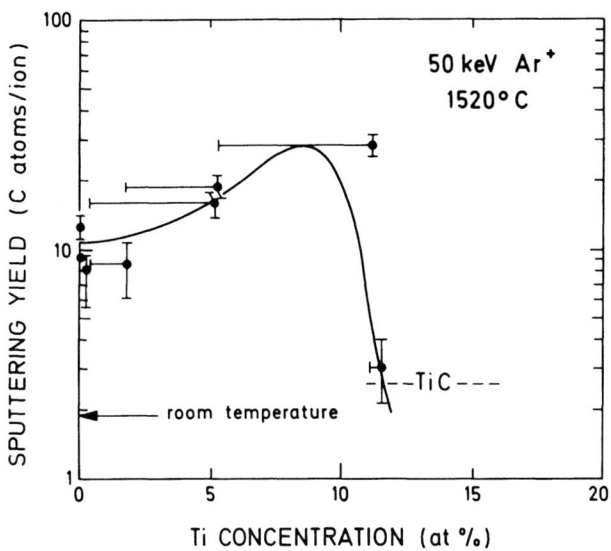

Fig. 14: Dependence of the radiation enhanced sublimation yield
on the Ti surface concentration from Fig. 13.

VI. EXTRAPOLATION TO FUSION REACTOR CONDITIONS

The irradiation conditions in a fusion device are different
from the conditions of the laboratory experiments presented here.
Hydrogen ion fluxes to a limiter surface are orders of magnitude
higher (10^{19} particles cm^{-2} sec^{-1}) and the particle energy is considerably lower (about 10 eV to 100 eV) compared to the present experiments. Experimental results show the methane yield to be flux-
dependent (Fig. 15), in rough agreement with the model for the
chemical erosion in the range of 10^{-15} - 10^{-17} H$^+$/cm^2 sec. The erosion
yield for graphite measured by the weight loss method (also shown
in Fig. 15) does not agree with the model so well. The experimental
data do not reflect a strong flux dependence of the erosion yield.
Extrapolations to fluxes of 10^{19} H$^+$/cm^2 sec remain therefore doubtful. Low ion energies and high fluxes together may result in
chemical erosion yields less or equal to physical sputtering /48/.

Though much work has been done on energetic hydrogen ions and
thermal hydrogen atoms, the effect of Franck-Condon neutrals with
about 3 eV is very difficult to estimate. At this energy the reflection coefficient is still below 70 % /51/ in contrast to a
sticking coefficient of about 2 % of thermal hydrogen atoms. Absolute yields, however, cannot be interpolated without additional
experimental data. As reported, their erosion yield may additionally be enhanced by the combined action of energetic hydrogen ions.
Actually, the only reported case of enhanced erosion of a carbon
surface in a tokamak /52/ is the result of the combined aciton of

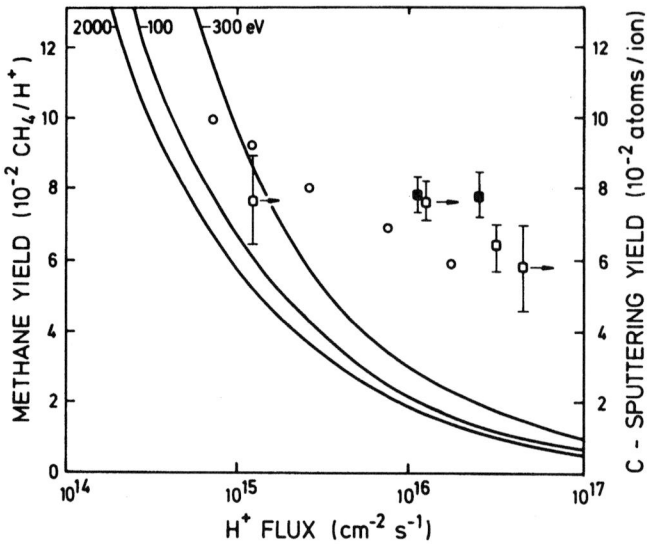

Fig. 15: Flux dependence of the maximum methane yield and the total sputtering yield of pyrolytic graphite with 2 keV H$^+$ ions. Open points are measured by residual gas analysis, squares by total weight loss. Solid squares are weight loss data at 1 keV. The data are compared with predictions using eq. 6.

nonthermal fast ions and low energetic ions or Franck-Condon neutrals.

Chemical erosion occurs at temperatures around 900 K. At temperatures above 1200 K an increase in the erosion yield was observed which can be explained by an enhanced sublimation. The model is based on the generation and diffusion of vacancies and interstitials. Computer calculations using this model show good agreement with experimental results. Such calculations also show a flux dependence of this effect /4/. In Fig. 16 calculated yield values for 1800 K and a flux of 10^{16} and 10^{18} D$^+$/cm^2 sec are compared to the physical sputtering at room temperature. The effect is reduced at higher fluxes also in the case of radiation-enhanced sublimation. Due to the higher threshold of this effect the yield below 50 eV is dominated by physical sputtering.

The calculations for chemical sputtering and radiation-enhanced sublimation together with estimations as discussed previously indicate only a moderate importance of chemical effects for the erosion of graphite limiters in fusion devices. However, all these estimates are based on extrapolations which contain a certain degree of uncertainty. Experiments at high particle flux and in the

energy range of 3 to 100 eV are needed in order to give a decisive
answer about the importance of the effects discussed here.

The influence of impurities may further reduce the enhanced
erosion processes. The analysis of carbon limiters in JET /49/ and

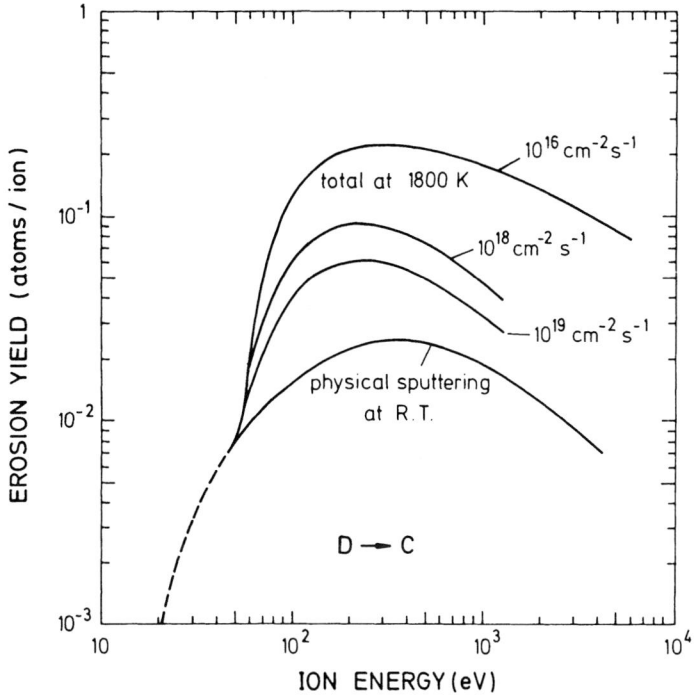

Fig. 16: Erosion yield of graphite at 1800 K due to deuterium
          bombardment as a function of energy.

ASDEX /50/ show coverages from several times 10^{16} to 5×10^{17} metal-
atoms/cm^2. These metal coverages are well within the range where
both chemical sputtering and radiation enhanced sublimation is
reduced. The distribution of the impurities - homogeneous films,
intermixed carbides, islands or droplets - are important parameters, and erosion measurements directly on these surface would
yield valuable information.

REFERENCES

/1/   J. Roth in: "Sputtering by particle bombardment, II", ed. R. Behrisch, TAP 52:91 (1983) (Springer-Verlag,Berlin)

/2/   O. Auciello, review presented at the "Workshop on Synergistic Effects", Nagoya (May 1984), Rad. Effects to be published

/3/   B.T. Kelly, Physics of Graphite, Applied Science Publishers (London 1981)

/4/   J. Roth, W. Möller, Nucl.Instr.Methods, to be published

/5/   K. Besocke, S. Berger, W.O. Hofer, U. Littmark, Rad.Effects 66:35 (1982)

/6/   J. Roth, J. Bohdansky, W. Poschenrieder, M.K. Sinha, J.Nucl.Mat. 63:222 (1976)

/7/   B. Feinberg, R.S. Post, J.Vac.Sci.Technol. 13:443 (1976)

/8/   N.B. Busharov, E.A. Gorbatov, V.M. Gusev, M.I. Guseva, Yu.V. Martynenko, Sov.J.Plasma Phys. 2:321 (1976)

/9/   S.K. Erents, C.M. Braganza, G.M. McCracken, J.Nucl.Mat. 63:399 (1976)

/10/   R. Yamada, K. Nakamura, K. Sone, M. Saidoh, J.Nucl.Mat. 95:278 (1980)

/11/   L. Holland, S.M. Ojha, Vacuum 26:53 (1976)

/12/   J. Roth, J.B. Roberto, K.L. Wilson, J.Nucl.Mat. 122/123:1447 (1984)

/13/   E. Hechtl, J. Bohdansky, J.Nucl.Mat. 122/123:1431 (1984)

/14/   B.J. Wood, H. Wise, J.Phys.Chem. 73:1348 (1969)

/15/   D.E. Rosner, H.D. Allendorf, Proc.Int.Conf. Heter.Kinetics at Elevated Temp. (Univ. Pennsylvania 1969) p. 231

/16/   M. Coulon, L. Bonnetain, J.Chim.Phys. 71:711, 717, 725 (1974)

/17/   M. Balooch, D.R. Olander, J.Chem.Phys. 63:4772 (1975)

/18/   R.K. Gould, J.Chem.Phys. 63:1825 (1975)

/19/   S. Veprek, M.R. Haque, H.R. Oswald, J.Nucl.Mat.63:405 (1976) and R. Brewer, H. Stuessi, S. Veprek, A.P. Webb, J.Nucl.Mat. 93/94:634 (1980)

/20/   E. Vietzke, K. Flaskamp, V. Philipps, J.Nucl.Mat. 128/129:545 (1984)

/21/   P.C. Stangeby, O. Auciello, A.A. Haasz, J.Vac.Sci.Technol. A1:1425 (1983)

/22/   D.R. Olander, R.H. Jones, J.A. Schwarz, W.J. Siekhaus, J.Chem.Phys. 57:421 (1982)

/23/   D.R. Olander, W. Siekhaus, R. Jones, J.A. Schwarz, J.Chem.Phys. 57:408 (1972)

/24/   D.R. Olander, T.R. Acharya, A.Z. Ullman, J.Chem.Phys. 67:3549 (1977)

/25/   G.N.-K. Liu, MIT Techn. Report 186 (1973)

/26/   M. Balooch, Jap.J.Appl.Phys. 16:1557 (1977)

/27/   J. Roth, J. Bohdansky, K. Wilson, J.Nucl.Mat. 111/112:775 (1982)

/28/   E. Vietzke, K. Flaskamp, V. Philipps, J.Nucl.Mat. 111/112:763 (1982)

/29/   R. Yamada, K. Sone, J.Nucl.Mat. 116:200 (1983)

/30/ D.L. Smith, J.N. Brooks, D.E. Post, D. Heifetz, Proc. 9th
Symp. on Engineering Problems of Fusion Research
(Chicago 1981), Institute of Electrical and Electronics
Engineers, N.Y. (1982) 719

/31/ E. Vietzke, K. Flaskamp, V. Philipps, J.Nucl.Mat. 128/129:545
(1984)

/32/ R. Yamada, K. Nakamura, M. Saidoh, J.Nucl.Mat. 98:167 (1981)

/33/ A.A. Haasz, O. Auciello, P.C. Stangeby, I.S. Youle, J.Nucl.
Mat. 128/129:593 (1984) and J.Vac.Sci.Technol. A1:1425 (1983)

/34/ C.I.H. Ashby, R.R. Rye, J.Nucl.Mat. 92:141 (1980)

/35/ C.I.H. Ashby, J.Nucl.Mat. 111/112:750 (1982)

/36/ USSR Contribution to Phase IIA of the INTOR Workshop, Vol. 2,
VII-13 (1982)

/37/ J. Roth, J. Bohdansky, K.L. Wilson, J.Nucl.Mat. 111/112:775
(1982)

/38/ J. Roth, J.B. Roberto, K.L. Wilson, J.Nucl.Mat. 111/112:1447
(1984)

/39/ V. Philipps, K. Flaskamp, E. Vietzke, J.Nucl.Mat.
111/112:781 (1982)

/40/ E. Vietzke, K. Flaskamp, M. Hermes, V. Philipps, Nucl.Instr.
Meth. B2:617 (1984)

/41/ J.P. Biersack, L.G. Haggmark, Nucl.Instr.Meth. 174:257 (1980)

/42/ W. Möller, Max-Planck-Institut für Plasmaphysik, Garching,
Report IPP 9/44 (1983)

/43/ C.M. Braganza, G.M. McCracken, S.K. Erents, Proc. Int. Symp.
on Plasma Wall Interaction (Jülich 1976) Pergamon Press,
Oxford (1977) p. 257

/44/ R. Yamada, K. Nakamura, M. Saidoh, J.Nucl.Mat. 111/112:744
(1982)

/45/ E. Vietzke, K. Flaskamp, V. Philipps, J.Nucl.Mat. 128/129:564
(1984)

/46/ J. Roth, J. Bohdansky, J.B. Roberto, J.Nucl.Mat.
128/129:534 (1984)

/47/ V. Philipps, K. Flaskamp, E. Vietzke, J.Nucl.Mat. 122/123:1440
(1984)

/48/ J. Bohdansky, J. Roth, Rad.Effects, to be published

/49/ J. Ehrenberg, R. Behrisch, Max-Planck-Institut für Plasmaphysik, Garching, Report 9/47 (1984)

/50/ R. Behrisch, P. Børgesen, J. Ehrenberg, B.M.U. Scherzer,
B.D. Sawicka, J.A. Sawicki, and the ASDEX-Team,
J.Nucl.Mat. 128/129:470 (1984)

/51/ R.A. Langley, J. Bohdansky, W. Eckstein, P. Mioduszewski,
J. Roth, E. Taglauer, E.W. Thomas, H. Verbeek, K.L.Wilson,
Nuclear Fusion, Special Issue (1984)

/52/ D. Manos, T. Bennett, R. Budny, S. Cohen, S. Kilpatrick,
J. Timberlake, J.Vac.Sci.Technol. A2:1348 (1984)

ION BACKSCATTERING FROM SOLID SURFACES

R. Behrisch and W. Eckstein

Max-Planck-Institut für Plasmaphysik
EURATOM Association, D-8046 Garching

ABSTRACT

When energetic hydrogen and helium particles impinge on the solid walls in high temperature plasma experiments, a fraction of them is immediately backscattered retaining a major part of their kinetic energy and they are thus recycled into the plasma. The particles which are not backscattered are implanted into the walls and are lost from the plasma. These hydrogen atoms are reemitted into the plasma at a later time mostly as thermal molecules. Measured data on particle and energy backscattering are still limited, they have been investigated only for primary energies above 100 eV. The results can be reasonably described with the computer simulation codes using the binary collision approximation. These codes have been used to calculate backscattering data down to the 1 eV energy range.

1. INTRODUCTION

In high temperature plasma experiments in respect to fusion, the walls of the containing vessel are bombarded from the plasma by fluxes of hydrogen ions and neutrals and some He-ions and impurity ions with energies in the eV to keV range. The most probable energies lie between 10 and 100 eV and currents range from 10^{14} to 10^{19}/cm² s. At the walls about 50 % of the incident hydrogen ions are backscattered in a time of $\lesssim 10^{-12}$ seconds predominantly as energetic neutral atoms and are thus directly recycled into the plasma. The other ions come to rest at some depth in the wall and will diffuse to the surface and into the bulk, depending on the wall temperature and the concentration of implanted atoms. At the

413

surface they will mostly be released as molecules with an energy corresponding to the surface temperature. For describing the plasma and the plasma-surface interactions it is important to have detailed data about backscattering probabilities and the distributions of the backscattered particles. Further, data about trapping, diffusion and reemission of hydrogen and helium after implantation into solids are needed. This latter topic is treated in a subsequent chapter by Möller and Roth.

Energetic ions impinging on the surface of a solid are generally scattered and slowed down by interactions with the atoms of the solid. These interactions may be described by collisions with the electrons where only energy is lost but the direction of motion is not changed (inelastic energy loss) and collisions with the nuclei where both energy is transferred and the direction of the trajectory is changed (elastic collisions) /1, 2, 3/. Those ions which are backscattered in collisions with the atoms of the solid may pick up an electron at the surface and leave the solid as a neutral atom or even as a negative ion. The other ions come to rest generally at interstitial positions in the crystal lattice of the solid and will diffuse or will be trapped generally at damage sites.

Particle and energy backscattering from solid surfaces has been investigated since about 20 years. Several review articles /4-9/ including some summaries about the data in respect to fusion research have been published /6-12/. The following review is mostly based on Refs. /6/ and /9/.

2. REFLECTION COEFFICIENTS, DISTRIBUTIONS OF BACKSCATTERED PARTICLES

If ions with an energy, E_0, bombard a surface at an angle of incidence, α, relative to the surface normal, the particles backscattered generally have a broad distribution in energy, E, a distribution in exit polar angles, β, and azimuthal angles, ϕ, (see Fig. 1) and a distribution in charge state, q_ν, which may be given by $R_N(E_0,\alpha; E, \beta, \phi, q_\nu)$. The average number of particles backscattered per incident ion, named particle reflection coefficient $R_N(E_0,\alpha)$, is obtained by integrating over the distributions.

$$R_N(E_0,\alpha) = \sum_\nu \int_o^{E_o} dE \int_o^{\pi/2} d\beta \int_o^{2\pi} d\phi \; R_N(E_0,\alpha; E,\beta,\phi,q_\nu) \sin \beta \quad (1)$$

The particle reflection coefficient depends further on the solid material, and the incident particle.

The kinetic energy carried back by the reflected particles is given by the energy reflection coefficient $R_E(E_0,\alpha)$:

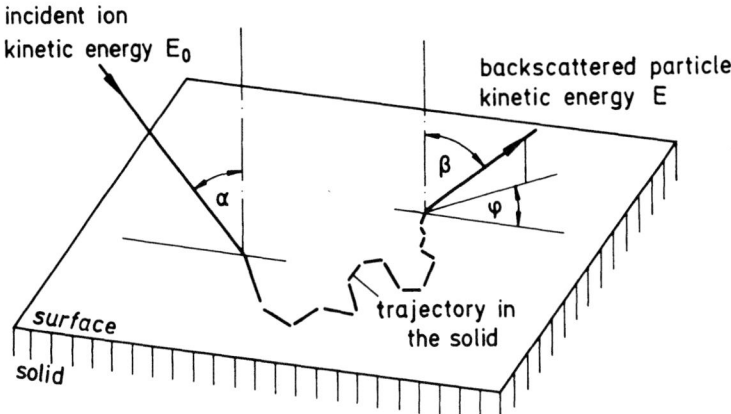

Fig. 1: Backscattering of an ion with incident energy E_0 from the surface of a solid, definition of angles.

$$R_E(E_o, \alpha) = \frac{1}{E_o} \sum_v \int_o^{E_o} dE \int_o^{\pi/2} d\beta \int_o^{2\pi} d\phi \; E \cdot R_N(E_o, \alpha; E, \beta, \phi, q_v) \; \sin \beta \quad (2)$$

In order to calculate the average energy deposited in the surface layers of the solid per incident hydrogen ion with kinetic energy E_0 the binding energies and electronic contributions have to be included. At high fluences, when the surface layers are saturated and for each incident ion on the average one hydrogen atom is released, the deposited energy is given by:

$$E_{dep} = (1-R_E)E_o + (1-R_N)E_r/2 + [1-(1-\eta^o)R_N](I_o-\phi)-\gamma_i \; \bar{E}_e \quad (3)$$

where the first term represents the energy of the kinetically reflected particles and the second term the recombination energy ($E_r \cong 4.5$ eV) of hydrogen atoms, which form molecules when they leave the surface. The third and fourth term represent the neutralization energy of the incident ions and the energy of the emitted electrons, they both are close to zero for neutral particle bombardment. I_o is the ionisation energy (13.6 eV for hydrogen), ϕ is the work function of the material ($\phi \cong 1.5$ to 5 eV), $\eta^o(E_o, \alpha)$ is the probability that an incident ion is backscattered as a neutral atom, γ_i is the ion induced electron emission coefficient, and $\bar{E}_e$ is the mean energy of the emitted electrons (0 - 5 eV). Those particles backscattered as positive ions do not deposit the ionisation energy, and the contribution of particles backscattered as negative ions to the energy deposition is small and has been neglected. Photon emission is not regarded because its contribution to the energy balance is negligible. The energies ($I_o-\phi$), $\bar{E}_e$, and E_r become important only at low incident energies, i.e. for $E_o \lesssim 100$ eV. Here most of the backscattered particles are neutrals and we may put $\eta^o \sim 1$. For low hydrogen fluences the second term in equation (3) is zero, however in addition the surface binding energy and the heat of solution enter into the energy deposition. There are no experimental data about the energy deposition of 1 to 100 eV hydrogen atoms.

The mean kinetic energy of the backscattered particles $\bar{E}$ may be calculated by

$$\bar{E}(E_o, \alpha) = E_o \frac{R_E(E_o, \alpha)}{R_N(E_o, \alpha)} \tag{4}$$

Finally the energy distributions and the angular distributions of the backscattered particles can be obtained from $R_N(E_o, \alpha; E, \beta, \phi, q_\nu)$ by appropriate integration.

In many publications backscattering data are represented not as a function of the incident energy but as a function of the reduced incident energy ε_L /13/ given by /2/

$$\varepsilon_L = 32.55 \cdot \frac{m_1}{m_1 + m_2} \cdot \frac{1}{Z_1 Z_2 (Z_1^{2/3} + Z_2^{2/3})^{1/2}} E_o \tag{5}$$

with E_o in keV and m_1, Z_1, m_2, Z_2 being the mass and nuclear charge of the incident ions and target atoms.

The use of ε_L instead of the real energy E_o allows to scale approximately backscattering data for different ion-target combinations as long as $m_2 \gg m_1$ /14, 15/. The value $\varepsilon_L = 1$ is the energy where in a head on-collision two atoms approach to a distance of the Thomas-Fermi radius given in $\overset{o}{A}$ by

$$a_{TF} = \frac{0.468}{(Z_1^{2/3} + Z_2^{2/3})^{1/2}} \quad , \tag{6}$$

which is the classical radius of the inner electron shell /2, 13/. Some values of ε_L/E_o are given in Table I.

Particle and energy reflection coefficients have been investigated both experimentally with different methods as well as theoretically. Most experiments have been performed for energies above 1 keV and the measured values can be generally reproduced by computer simulation codes /6, 9/. The codes also allow to extrapolate backscattering data to energies in the 1 to 100 eV range /16/.

416

Table I

Target atom			ε_L/E_o [1/keV]			
Element	Z_2	M_2	H	D	T	^{4}He
C	6	12.0	2.414	2.242	2.092	0.9200
Ti	22	47.90	0.4871	0.4774	0.4680	0.2222
Fe	26	55.85	0.3934	0.3866	0.3800	0.1814
Ni	28	58.69	0.3575	0.3516	0.3459	0.1655
Mo	42	95.95	0.2121	0.2099	0.2078	0.1006
W	74	183.92	0.1014	0.1008	0.1003	0.04911

3. BACKSCATTERING MEASUREMENTS

The intensity and the distributions of backscattered particles have been investigated experimentally by different techniques.

1) The first technique is based on measuring trapping of the bombarding ions in the solid below saturation at temperatures where diffusion can be neglected. The particle reflection coefficient is then given by the fraction of ions not trapped. The trapped amount is measured by subsequent thermal degassing, by weight gain, or by nuclear reaction analysis. The trapping method is not well suited if the trapping coefficents are close to 1 /17-21/.

2) In the second method the backscattered particles are measured by condensation on a surface /22/, by a large proportional counter surrounding the target /23, 24/, or by detecting the neutrals back-scattered into a given solid angle /25-27/. The neutrals are generally ionised by stripping in passing them through a gas cell and are subsequently analysed electrostatically in respect to their energy /28/, while positive and negative ions are directly energy analysed. This method allows to measure the energy and angular distributions in some detail. It works for hydrogen down to energies of about 200 eV. For lower energies time-of-flight techniques can be applied /29, 30/. Generally, the absolute accuracy of the most careful experiments is about 20 %.

3) The backscattered energy can be calculated from the backscattering distributions, but has been investigated for incident energies above 1 keV also directly by measuring calorimetrically the energy deposited on the target /31-34/.

4. BACKSCATTERING CALCULATIONS

The first model used for describing light ion backscattering is the single collision model, originally developed for higher energies, i.e. above a few 100 keV /35-40/. The incident ion penetrates the solid on a straight trajectory but looses energy to electrons. At some depth the ion is backscattered in a single elastic collision with an atom of the solid. On the trajectory towards the surface the ion moves again on a straight trajectory and looses energy to electrons before it leaves the surface. As the scattering cross section and thus the probability for the atomic collisions increases with decreasing energy, this model is re-stricted to high energies, i.e. about ε_L > 100 or E_o above a few 100 keV for hydrogen ions. It has been tried to extend the applica-bility to lower energies by including double scattering /41/.

For lower energies transport theory /42/ has been applied to calculate backscattering yields and energy distributions for light ions bombarding amorphous solids /17, 43, 44/. Compared to an in-finite medium, the presence of a surface which inhibits multiple crossing at this plane has to be included in the calculations. Backscattering distributions generally describe the experimentally observed dependencies, but absolute numbers still have some un-certainties.

With the availability of large computers Monte Carlo simulation codes have been increasingly used for describing the interaction of energetic particles with solids /45-52/. Here the classical tra-jectories are calculated for the incident ion and for those target atoms which are knocked from their lattice sites. The particles are followed until they reach the surface or they are slowed down in the solid below some threshold energy. At the surface an energy corresponding to the surface binding energy is substracted from the energy component of the particle normal to the surface before it can leave the solid. A picture of such trajectories projected on a plane normal to the surface for the particles stopped in the solid and for the particles backscattered is shown in Fig. 2 /6/.

Nearly all of the computer simulation codes used up to now are based on the binary collision approximation (BCA) /45-47, 49-56), in one case a classical dynamical code is used /48/. In the BCA-codes three different models are applied for the solid: a crystal structure in MARLOWE /45-47/ and TAVERN /46, 49/, an amorphous target structure as in a gas-like solid in SAVOY /50/ and in /51/ and a local density model in TRIM /52/. The models with the crystal structure allow to deal with crystalline, polycrystalline and amorphous targets, however, they need more computer time than the other amorphous simulations. In the simulation codes for each set of parameters the trajectories of a large number of particles has to be calculated to obtain the distributions with reasonably

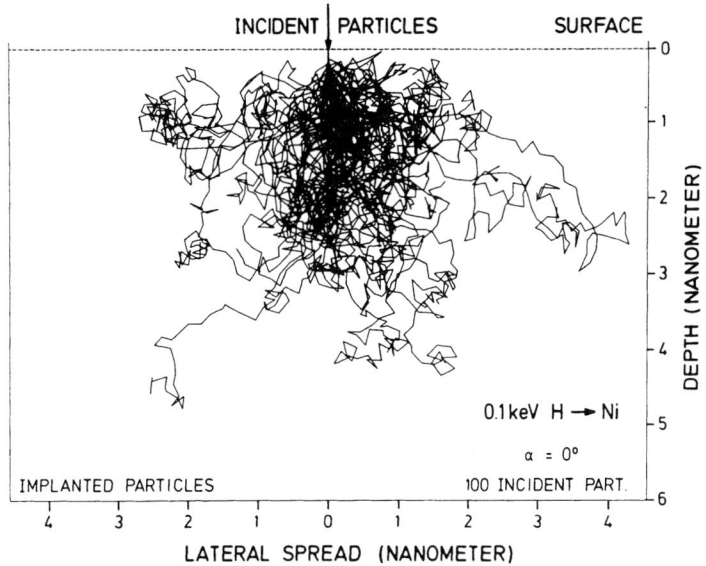

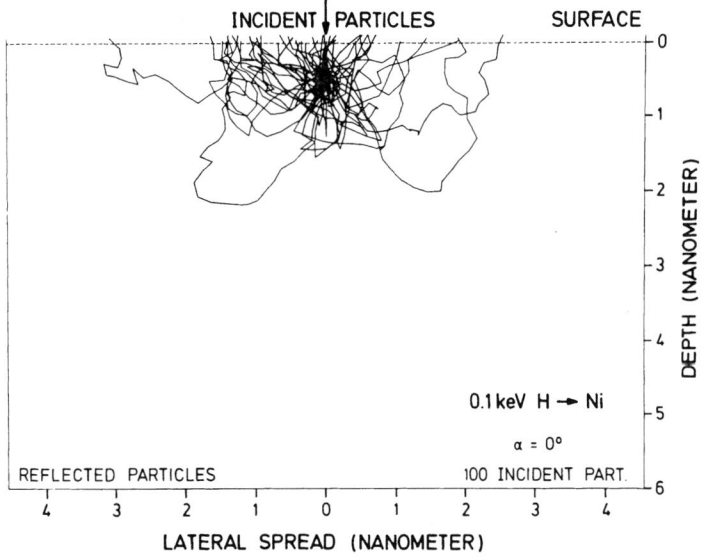

Fig. 2: Trajectories of the reflected (top) and the implanted (bottom) particles for normal incidence of 100 eV H⁺ on Ni. The trajectories have been calculated with TRIM and are projected in a plane normal to the surface /6, 102/.

small statistical errors. Most of the programs allow to describe also composite solids and layered structures /45-47, 53, 54/.

Backscattering distributions calculated with the codes and measured data generally agree within the uncertainties of the calculations and the experiments. Main uncertainties of the calculations are input parameters as the electronic energy loss, the interatomic potentials and the surface binding energies, E_s (chemical bonding), which are not sufficiently known especially for the low energies of interest here. Finally in the calculations the surface has generally been assumed to be flat within an atomic distance. The influence of the surface roughness was investigated only in two publications /55, 56/.

5. DATA ON ION BACKSCATTERING

Measured data on ion backscattering are available mostly for ion beams with incident energies of $E_0 \geq 1$ keV. Predominantly total particle and energy reflection coefficients are investigated, while detailed distributions of the backscattered particles have been measured in some detail for normal ion incidence, but only few measurements have been performed for oblique ion incidence /57/. Reflection coefficients for all incident energies of interest and details about the backscattering distributions are mostly obtained from Monte Carlo computer simulation codes. These results have, however, been checked with experimental data in the ranges where these are available. In the following some examples are given for particle reflection coefficients and energy and angular distributions of backscattered particles.

5.1 Particle and Energy Reflection Coefficients

A compilation of all measured and calculated values for the particle reflection coefficients for light ions bombarding different solids at normal incidence is shown in Fig. 3 /58-69/. The data are plotted as a function of the reduced energy ε_L, which, however, does not give a good scaling for light target atoms like carbon.

For hydrogen and helium bombardment of Fe, Ni and stainless steel all measured and calculated values for the reflection coefficients are collected in Fig. 4 /6, 9/. As the masses of Ni and Fe are very similar and the ε_L-values for the light ions differ only slightly, all data should fall on one line for each quantity. The scattering of the values of about $\pm$ 10 % is due to different measuring techniques as well as due to slightly different input data in the computer codes, mostly MARLOWE and TRIM. The lines represent the analytical formulae given in the figure /9/.

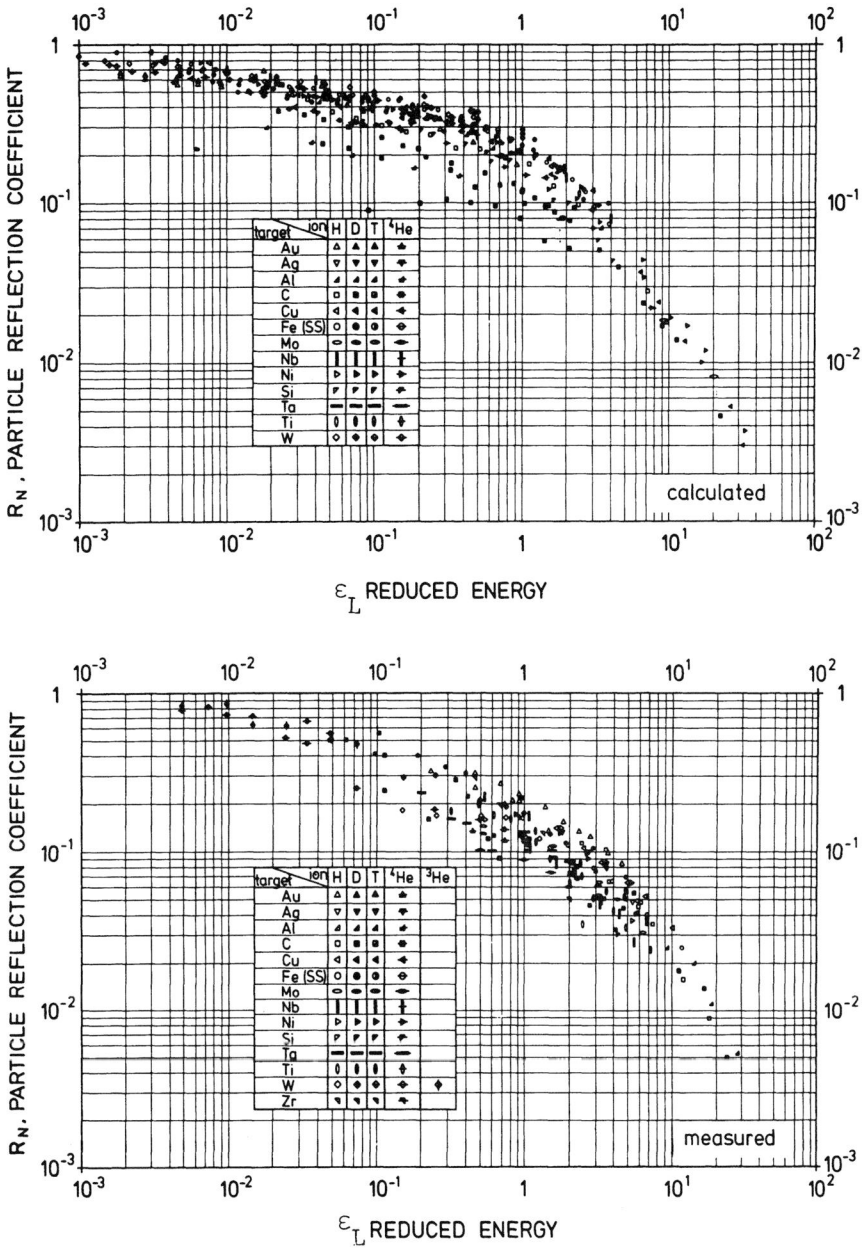

Fig. 3:   Calculated and measured particle reflection coefficients,
R_N, versus the reduced energy ε_L for many ion-target combinations /9/.

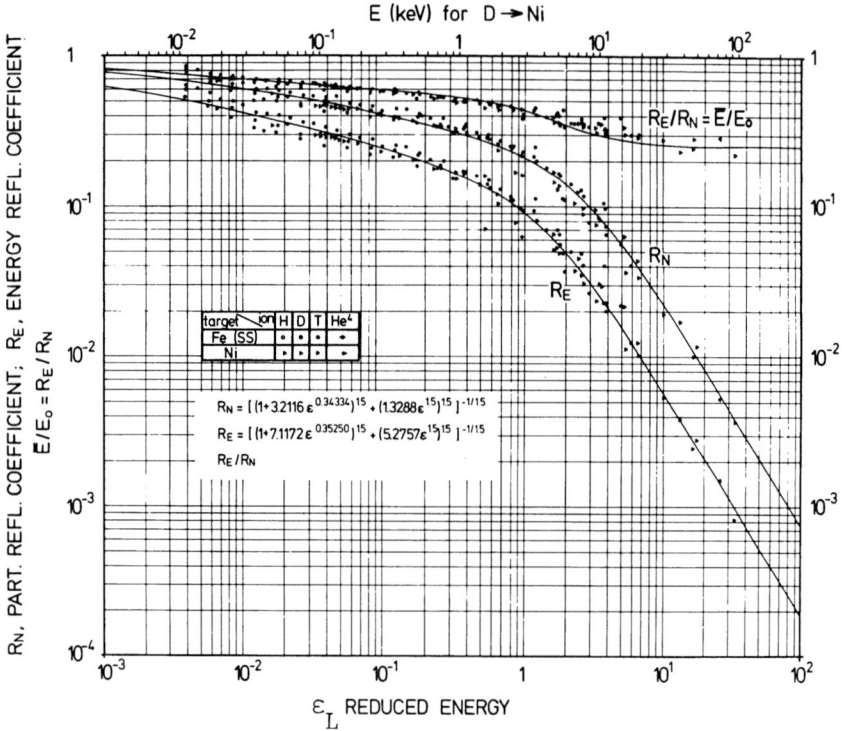

Fig. 4: Particle and energy reflection coefficients, R_N, and R_E, and the mean relative energy of reflected particles, $\bar{E}/E_O$, versus the reduced energy ε_L. Data for Ni and Fe(SS) bombarded by hydrogen isotopes and He. Analytic fitting to the data for R_N and R_E are indicated by the formulae and the solid lines /9/.

Recently some effort has been made to obtain more realistic backscattering data for primary energies E_O between 0.2 and 100 eV for hydrogen bombardment of Fe, Ni or stainless steel. Here still no measurements exist. The results for particle and energy reflection coefficients from TRIM calculations using the version described in /70/, assuming a surface binding energy $E_S = 1$ eV for hydrogen on Ni /16/, are shown in Fig. 5 together with the first results of a classical dynamical calculation /48/. The value of $E_S = 1$ eV gives a reasonable fit to the dynamical calculations and was taken also for all other calculations. For He the surface binding energy was taken to be zero. TRIM-results for hydrogen reflection from C and from W are shown in Fig. 6. In all cases the hydrogen particle reflection decreases at energies below 2 eV, which would mean that an increasing number of hydrogen atoms are thermalised on the surface and released as molecules. The amount of hydrogen adsorbed on the surface depends on temperature but it is negligible above 300° C. Adsorbed hydrogen is not taken into account in these calculations.

422

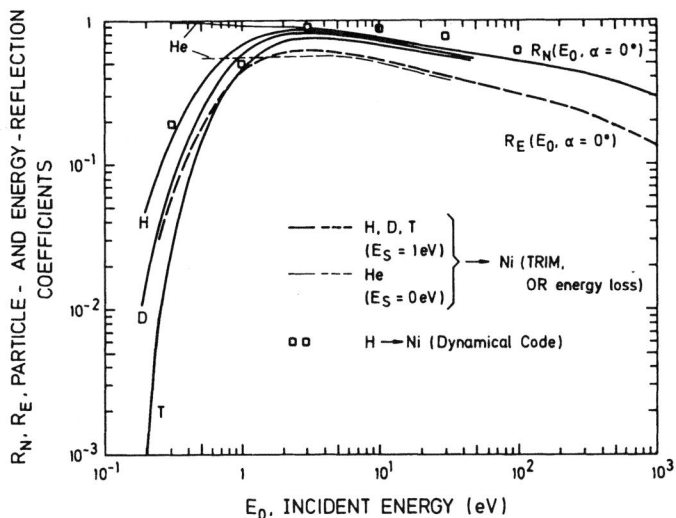

Fig. 5: Calculated particle and energy reflection coefficients,
R_N and R_E, at low energies E_O and normal incidence on Ni
for the hydrogen isotopes and He. A surface binding energy
$E_S = 1$ eV is used /16/. Further the values calculated with
a dynamical code for hydrogen on Ni are introduced /48/.

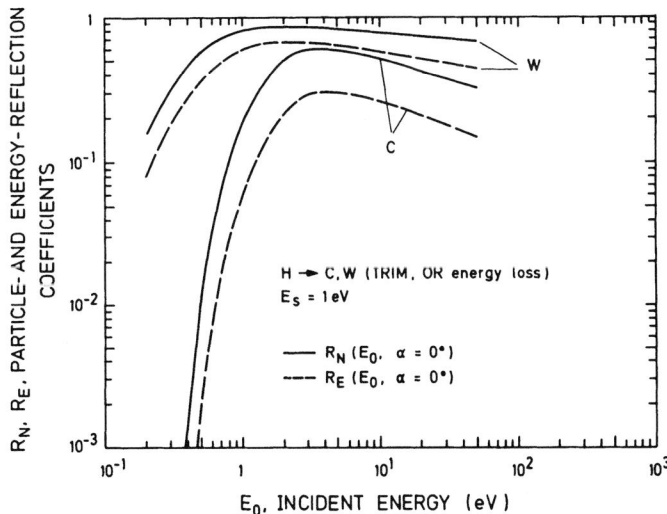

Fig. 6: Calculated particle and energy reflection coefficients,
R_N and R_E, versus the incident energy, E_O, for bombardment
of C and W at normal incidence and a surface binding
energy $E_S = 1$ eV /16/.

Particle and energy reflection coefficients are generally lower by less than 50 % for metal surfaces contaminated with low Z impurities as oxides and carbides or implanted with H or He. For rough surfaces the reflection is also reduced compared to well polished surfaces /55, 56/ and may amount up to a factor of 2.5 for a specially shaped needle-like surface structure /6/.

For ion incidence other than normal to the surface only few measurements and calculations are available /21, 33, 57, 66, 71-75/. For a limited energy range both the particle and the energy reflection coefficients $R_N(E_o,\alpha)$ and $R_E(E_o,\alpha)$ scale approximately with $\varepsilon_L \cos^2\alpha$ as was first pointed out in Ref. /42/ and is shown in Figs. 7a and 7b. The measured values for the particle reflection coefficient R_N agree reasonably well with calculated numbers while the measured energy reflection coefficients R_E tend to be lower. All computer simulations apply a flat surface partly with a roughness of one atomic distance. The measurement were mostly performed at polished surfaces or at surfaces with a structure as it develops during ion bombardment. Such surfaces are generally not flat but have not been well characterised. A rough surface may have a larger influence on the energy reflection than on the particle reflection /33/. First calculations of particle reflection coefficients for 50 eV H^+ from Cu having a

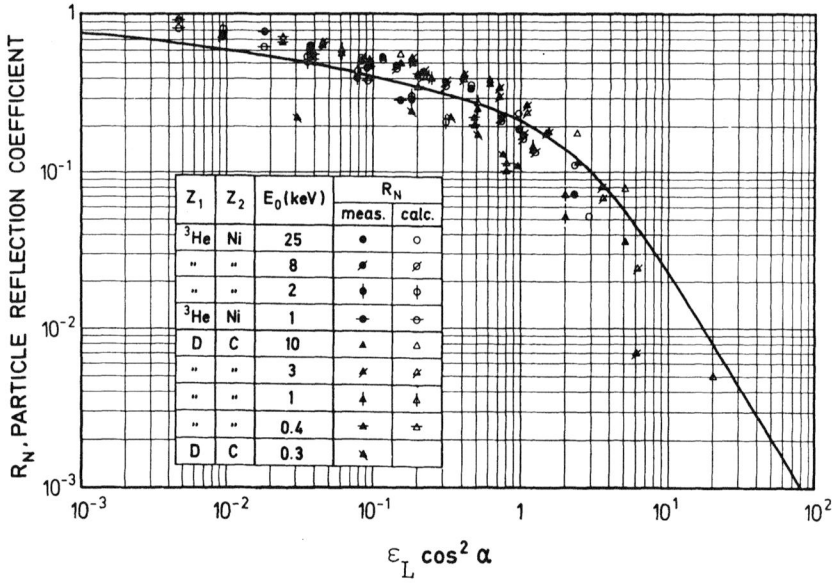

Fig. 7a: Particle reflection coefficient versus the energy and incident angle parameter $\varepsilon_L \cos^2\alpha$ for bombardment of Ni with ^{3}He and C with D. The solid curve represents the analytical fit given in Fig. 4 /21, 71/.

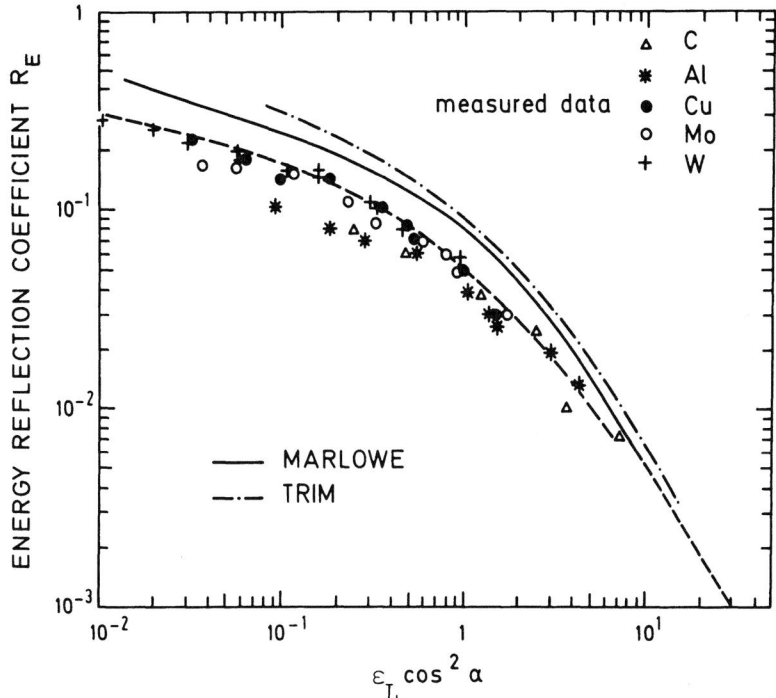

Fig. 7b: Energy reflection coefficient versus the energy and
incident angle parameter $\varepsilon_L\cos^2\alpha$ for bombardment of
C, Al, Cu, Mo, W with H together with calculated data
from MARLOWE and TRIM /34/.

special surface topography show an influence of the roughness on
the reflection coefficient, especially at large angles of incidence, which is however below a factor of 2 (Fig. 8) /55, 56/.

5.2 Distributions of Backscattered Particles

The energy distributions of the backscattered particles depend on the incident energy, E_0, the incidence angle, α, and the
target material. They further vary with the polar and azimuthal
angles of emergence ß and ϕ and charge state q_ν of the backscattered particles /6, 10, 12, 40, 44, 53, 54, 58-65, 67, 68, 76,
82/. An example of measured energy distributions for 5 keV D ions
bombarding different solids at normal incidence is shown in Fig. 9
/60/. The area below the distribution is proportional to the
particle reflection coefficient.

The dependence of the energy distribution of the backscattered
particles on the primary energy has been investigated both in experiments and with computer codes /6, 7, 9, 14, 15, 59/. Results of

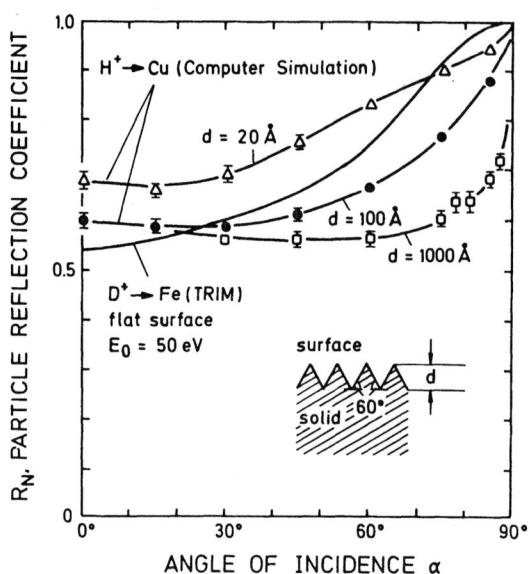

Fig. 8: Particle reflection coefficient versus the angle of incidence, α, for a special triangular surface structure
with different heights for bombardment of Cu with 50 eV H^+.
The target is rotated to get an average reflection coefficient independent on the azimuthal incident angle ϕ
/55, 56/. The reflection coefficients obtained by TRIM for
50 eV H^+ bombarding Fe having a surface roughness corresponding to one lattice distance are also included /16/.

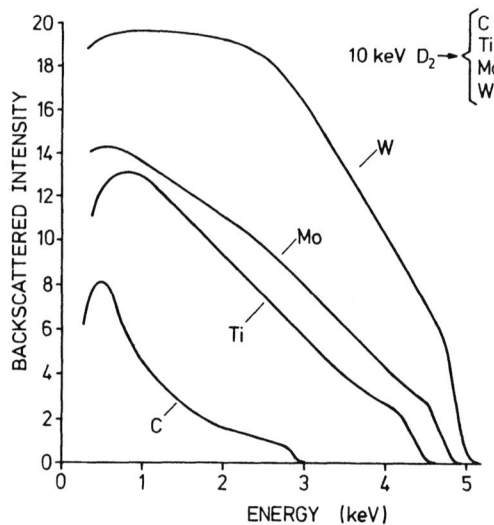

Fig. 9: Measured energy distribution of the backscattered deuterons
for bombardment of C, Ti, Mo, W with 5 keV D ions at normal
incidence /60/.

calculations using the TRIM-code are shown in Fig. 10 for D^+ ions bombarding Ni /9/. At low incident energies (10 to few 100 eV) the distribution is peaked close to the incident energy, while it is more flat at incident energies in the low keV range and is peaked at low energies for higher incident energies. The energy distri-

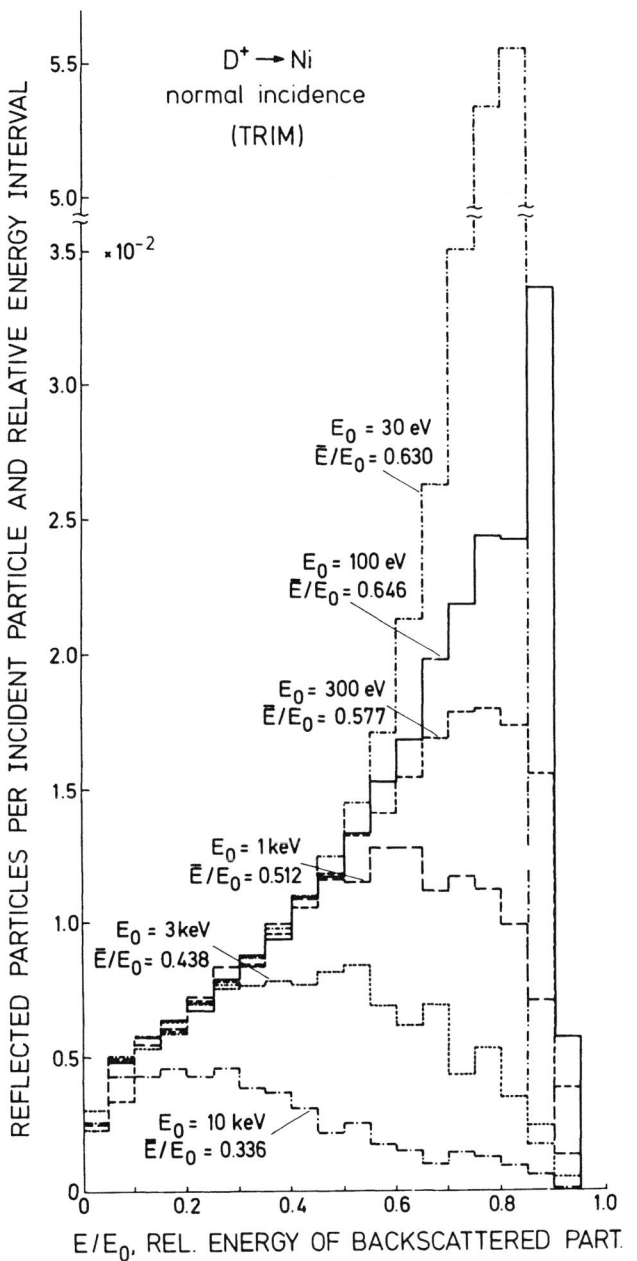

Fig. 10:

Energy distributions of D backscattered from Ni for normal incidence at several energies, E_0, as calculated with TRIM /9/.

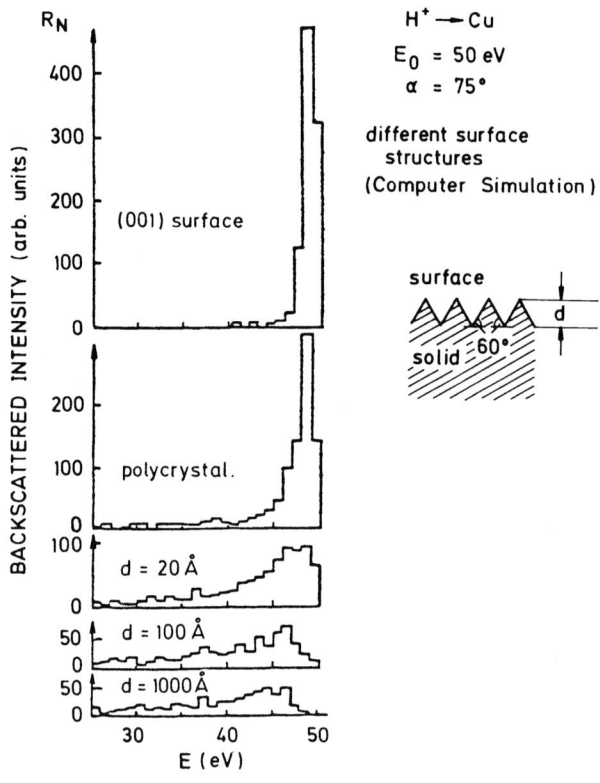

Fig. 11: Calculated energy distributions of the backscattered
hydrogen atoms 50 eV ions, bombarding a Cu-surface with
different surface roughness at an angle of incidence
$\alpha = 75°$. The target is rotated around the surface normal
/55, 56/.

bution of the backscattered particles can be largely influenced
by surface roughness especially for oblique angles of incidence.
This is demonstrated by computer calculations for a special topography as shown in Fig. 11 for an incident energy of $E_0 = 50$ eV
and an incident angle $\alpha = 75°$ /55/. At a smooth surface of a single
crystal and an amorphous solid nearly all particles are reflected
at an energy corresponding to the energy loss in a few collisions
with surface atoms. For this surface topography the energy distribution gets the broader the rougher the surface.

The angular distribution of the backscattered particles for
normal ion incidence and energies in the low keV range can be described as a first approximation by a cosine distribution /37, 62,
68, 69, 81, 81/. For non-normal incidence the angular distribution

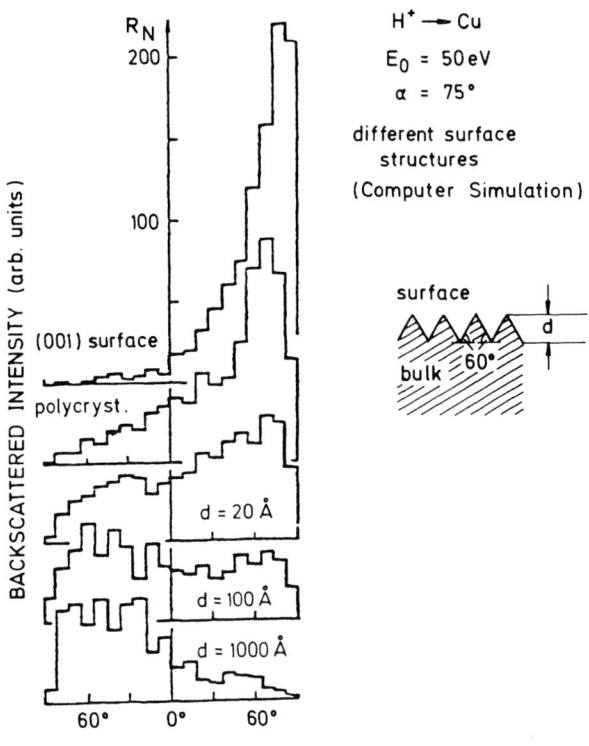

Fig. 12: Calculated angular distributions of the backscattered hydrogen atoms for 50 eV H ions bombarding a Cu surface with different surface roughness at an angle of incidence, $\alpha = 75°$. The target is rotated around the surface normal /55, 56/.

depends also on the azimuthal angle, ϕ, for large angles of incidence it gets strongly peaked around the direction of specular reflection. The angular distribution is also influenced by the surface topography. For a special surface topography this is shown in Fig. 12 for the case of 50 eV H^+ ions bombarding a Cu surface at an angle of incidence of $\alpha = 75°$ /55/. The angular distribution is the less peaked the larger the surface roughness.

The charge states of the backscattered particles depend on the surface composition and the particle velocity normal to the surface and has been investigated in some detail /6, 59, 82-85/. For energies of the backscattered hydrogen atoms below 40 keV the neutral fraction dominates. The negative fraction is typically below 5 % and is equal to the positive fraction at about 1 keV (see Fig. 13). The negative fraction is especially high if the surface is covered with alkali metal atoms /6, 84, 86-92/.

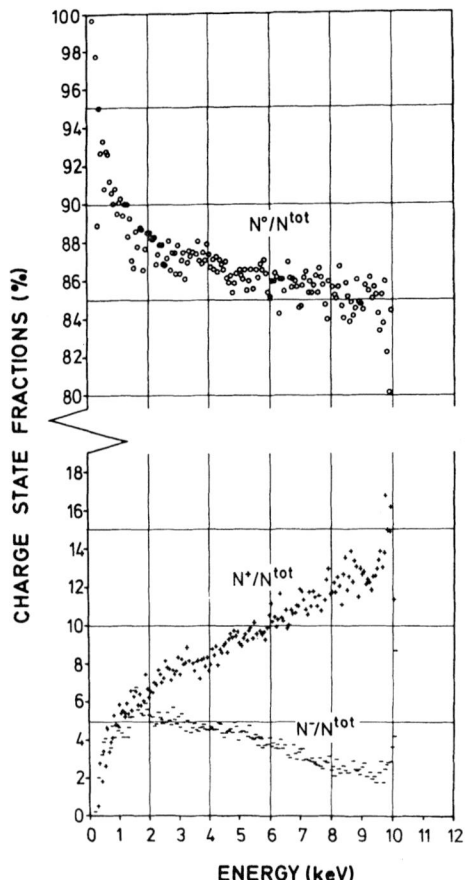

Fig. 13

Charge state fractions, η, versus the energy, E, of the particles backscattered at a polar angle, $\beta = 45°$, for bombardment of stainless steel by 10 keV H ions at normal incidence, $\alpha = 0°$ /6, 9/.

For ions with higher mass than He some information about back-scattering is known from computer simulations /92, 93/. Generally reflection coefficients for normal incidence and about equal masses are of the order of 10^{-2} and lower. They increase at grazing incidence.

5.3 Data Representation for Plasma Boundary Codes

For a correct description of the boundary plasma, especially for the neutral particles, the 6-dimensional particle- and energy reflection coefficient $R_N(E_0, \alpha, E, \beta, \phi, q) dE \sin\beta \, d\beta \, d\phi$ enters into the transport code /94, 95 and chapter by Heifetz/. These full distributions have been measured only in one case /57/, they have, however, been obtained with computer codes like MARLOWE /96, 97/ and TRIM /52/. Besides the questions about the accuracy of these results because of the not well established input data, especially the surface topography and composition, the data have

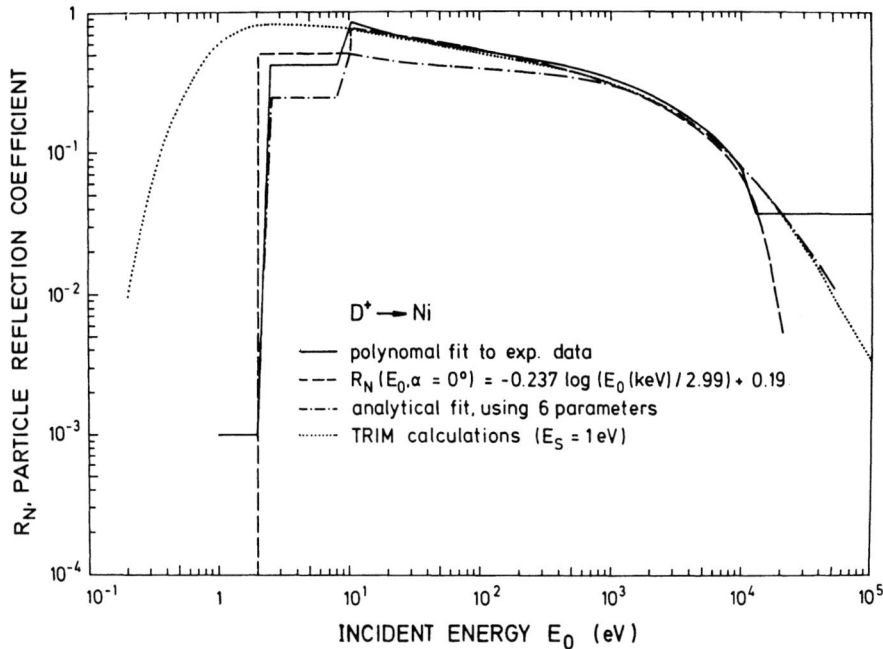

Fig. 14: The particle reflection coefficient, R_N, versus the incident energy, E_0, at normal ion incidence.

     ———— polynomal fit used as one choice in DEGAS /99/;

     ----- given by Seki et al. /97/ used also as a choice in DEGAS;

     -·-·- analytical fit proposed by Tabata et al. /103/;

     ..... results from TRIM calculations: for energies above 20 eV as given in Fig. 4 /9, formula 2.12/, for lower energies see Fig. 5 /16/.

to be provided in a way to put them easily into the transport codes. Mainly three approximations have been used:

The first approximation used is to take the global particle reflection coefficients, an average energy of the backscattered particles according to eq. (4) and cosine angular distributions. For a better approximation in addition the energy distribution of the backscattered neutrals as proposed in Ref. /7/ was included /94, 95/.

For the global particle and energy reflection coefficients several different approximations have been used which are shown in Fig. 14 /9, 10, 96, 97, 98/. These curves are mostly a best fit to experimental data, and data calculated with MARLOWE or TRIM as

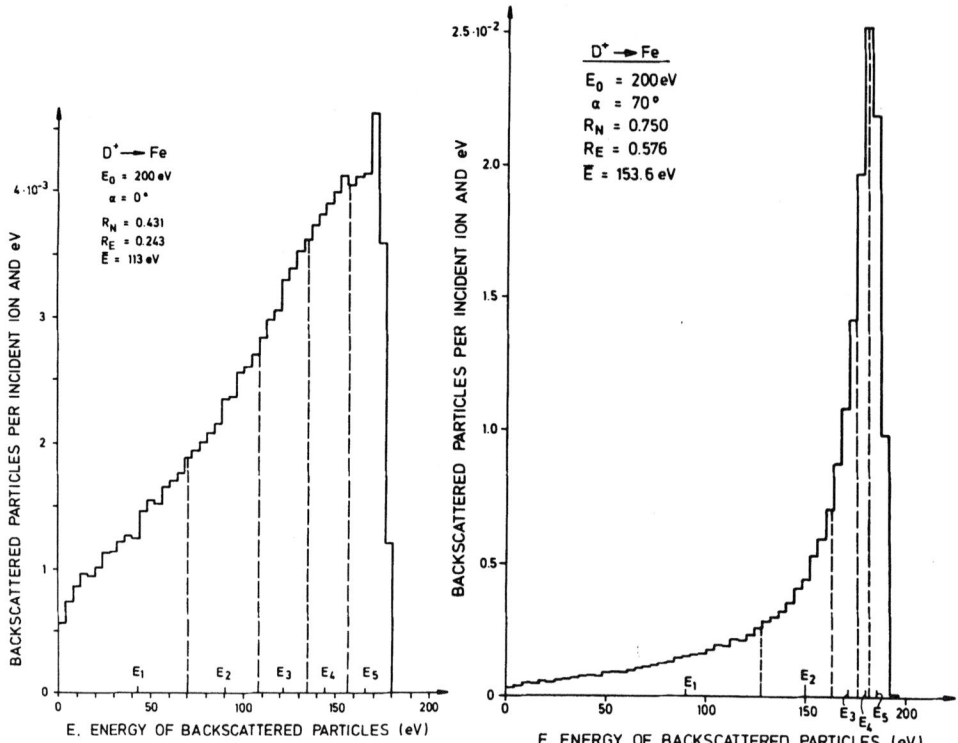

Fig. 15: Calculated energy distributions (TRIM) for the bombardment of Fe by 200 eV D. The energy regions and the mean
energies $F^1 = E_\mu$ of these regions to be used for plasma
boundary codes are indicated.
a) normal incidence, $\alpha = 0°$, b) angle of incidence, $\alpha = 70°$.

shown in Fig. 3. In the figure also the data calculated for low
incident energies are included ($E_0 \leq 10$ eV). At these low energies
the values used in most approximations are too low.

A better way to present backscattering data which was proposed by Bates et al. /96,97,99/ are inverse cumulative distributions,
which is an algorithm used to present multidimensional distribution
functions in probability theory. A given multidimensional distribution function as $R_N(E_0,\alpha; E,\beta,\phi,q_\nu)$ is firstly integrated over
all variables except one which may be the energy E.

$$f^1(E_0,\alpha;E) = \sum_\nu \int_0^{\pi/2} d\beta \int_0^{2\pi} d\phi \, R_N(E_0,\alpha;E,\beta,\phi,q_\nu) \sin \beta.$$

Such a distribution is plotted in Fig. 15 for 200 eV D^+ bombarding
Fe at normal and oblique ($\alpha = 70°$) incidence. The area of this
distribution is now divided into n (here n = 5) equal areas, and

432

for each area an average energy, E_μ, is taken giving n discrete values for the energy. Now at each backscattering event for the plasma simulation code, one of these energies E_μ is taken randomly for the backscattered energy.

The same procedure is subsequently used to calculate for each selected energy E_μ firstly n (n = 5) different angles β_λ and subsequently for each pair of (E_μ, β_λ) also n (n = 5) values of ϕ_K. This means n distribution functions are defined by

$$f^2(E_o, \alpha, E_\mu, \beta) = \sum_\nu \int_o^{2\pi} d\Phi \; R_N(E_o, \alpha; E_\mu, \beta, \phi, q_\nu) \; \sin \beta.$$

The functions $f^2 / \sin \beta$ are plotted in Fig. 16 in polar coordinates. Here again the area is divided in n (n = 5) equal areas and n average angles β_λ are calculated.

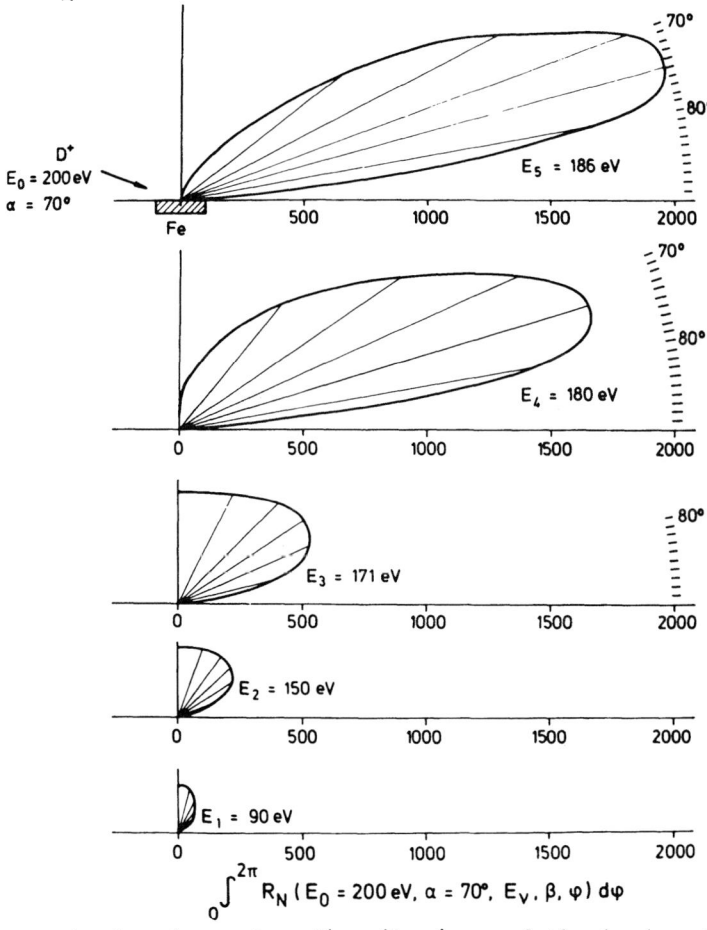

Fig. 16: Calculated angular distributions of the backscattered deuteron for the energies, $E_1 \ldots E_5$, as defined in Fig.15b.

433

Finally, the following distribution functions are defined:

$$f^3(E_o, \alpha; E_\mu, \beta_\lambda, \phi) = \sum_\nu R_N(E_o, \alpha; E_\mu, \beta_\lambda, \phi, q_\nu)$$

and for each (E_μ, β_λ) n (n = 5) average values for ϕ_K are calculated. Due to the symmetry these angles reach from 0 to 180° only.

With this procedure one value for $R_N(E_o, \alpha)$, n values for $F^1 = E_\mu$, n^2 values for β_λ or $F^2 = \cos \beta_\lambda$ and n^3 values for ϕ_K or $F^3 = \cos \phi_K$ are calculated for a given distribution. At each backscattering event first a value E_μ, then a value β_λ and finally a value of ϕ_K are chosen randomly. This procedure is a considerable improvement compared to the first scheme.

The input data for such an algorithm have first been calculated using the computer code MARLOWE /96,97,99/ and later more refined with the code TRIM /100/ for the case of D ions bombarding iron and carbon. Here flat surfaces are used and the best known data have been taken for the electronic stopping power, the atomic potentials and the surface binding energy.

A final possibility being discussed for the use of surface data, is to couple a backscattering code like MARLOWE or TRIM directly to the plasma boundary code. This would, however, be very unefficient and no more accurate than using available backscattering data. It would be necessary to calculate a very high number of particles in order to obtain reasonably small statistical errors for the results.

An introduction of the surface backscattering data into neutral particle codes as DEGAS /96,97/ developed for the boundary plasma and for describing divertors and pumped limiters has shown a definite influence of different data sets on the results /101/. However, systematic investigations have still not yet been performed. For a comparison of calculated results with those of experiments at today's plasma machines, the surface composition and topography of the walls should also be taken into account, which has only been partly investigated. Layers of low Z atoms on a metal surface decrease the particle backscattering coefficient, while a rough surface seems to give lower energies for the backscattered particles.

However, the surface conditions in plasma experiments are becoming better controlled and investigated. Computer codes are improved in order to get a better understanding of particle confinement, refuelling, particle exhaust and wall effects in running plasma experiments and for predicting these processes in future machines. In these codes realistic data for the surface processes have to be used and parametric studies would be very useful. The most correct way to introduce surface data into the codes are the use of inverse cumulative distributions and first data sets have been prepared /96, 100/.

ACKNOWLEDGEMENTS

It is a great pleasure to thank D. Post, D. Heifetz, D. Reiter, and A. De Matteis for many helpful discussions in preparing this manuscript.

REFERENCES

/1/ N. Bohr, Mat.Fys.Medd.Dan.Ved.Selsk. 18:1108 (1948)
/2/ J. Lindhard, M. Scharff, H.E. Schiøtt, Mat.Fys.Medd.Dan.Ved. Selsk. 33:No.14 (1963)
/3/ H.E. Schiøtt, Mat.Fys.Medd.Dan.Vid.Selsk.35:No.9 (1966)
/4/ E.S. Mashkova, V.A. Molchanov, Radiat.Eff. 16:143 (1972) and Radiat.Eff. 23:215 (1974)
/5/ E.S. Mashkova, V.A. Molchanov, "Medium Energy Ion Scattering from Surfaces of Solids", Moscow, Atomisdat (1980)
/6/ W. Eckstein, H. Verbeek, IPP-Report 9/32 (1979) Max-Planck-Institut für Plasmaphysik, D-8046 Garching/München, FRG.
/7/ R. Behrisch in "Physics of Plasmas close to thermonuclear conditions", Proc.Int.School on Plasmaphys., Varenna 1979, Pergamon Press 1980
/8/ K.L. Wilson, J.Nucl.Mat. 103/104:453 (1981)
/9/ W. Eckstein, H. Verbeek, in "Data compendium for plasma surface interactions", R.A. Langley, J. Bohdansky, W. Eckstein, P. Mioduszewski, J. Roth, E. Taglauer, E.W. Thomas, H. Verbeek, K.L. Wilson, Nucl.Fusion, Special Issue, IAEA, Vienna (1984) 12
/10/ T. Tabata, R. Ito, Y. Itikawa, N. Itoh, K. Morita, Institute of Plasma Physics, Nagoya University, IPPJ-AM-18 (1981) and Atomic Data and Nucl. Data Tables 28:493 (1983)

/11/ K. Morita,T. Tabata, R. Itoh, J.Nucl.Mat. 128/129:681 (1984)
/12/ T. Tabata, R. Ito, Y. Itikawa, N. Itoh, K. Morita, H. Tawara, Institute of Plasma Physics, Nagoya University, IPPJ-AM-34 (1984)
/13/ O.B. Firsov, JETP 7:308 (1958)
/14/ M.T. Robinson, Proc. 3rd National Conference on Interaction of Atomic Particles with Solids, Kiew, USSR, Oct.14-16, 1974
/15/ O.S. Oen, M.T. Robinson, Nucl.Instr.Meth. 132:647 (1976)
/16/ W. Eckstein, Appl.Physics (1985)
/17/ J. Bøttiger, K.B. Winterbon, Rad.Eff. 20:65 (1973)
/18/ J. Bohdansky, J. Roth, M.K. Sinha, W. Ottenberger, J.Nucl.Mat. 63:115 (1976)
/19/ J. Bøttiger, N.Rud.Inst.Phys.Conf.Ser. No. 28, Chapter 5:224 (1976)
/20/ E.W. Thomas, J.Appl.Phys. 51:1176 (1980)
/21/ C.K. Chen, B.M.U. Scherzer, W. Eckstein, Appl.Phys. A31:37 (1983)
/22/ J. Bøttiger, J.A. Davis, Rad.Eff. 11:61 (1971)

/23/ G. Sidenius, Phys. Lett. 49A:409 (1974)

/24/ G. Sidenius, T. Lenskjaer, Nucl.Instr.Meth. 132:673 (1976)

/25/ P. Meischner, H. Verbeek, J.Nucl.Mat. 53:276 (1974)

/26/ H. Verbeek, W. Eckstein, F.E.P. Matschke, J.Phys. E10:944 (1977)

/27/ G.I. Zhabrev, Thesis, Moscow (1979)

/28/ C.F. Barnett, J.A. Ray, Nucl. Fusion 12:65 (1972)

/29/ D.E. Voss, S.A. Cohen, J.Nucl.Mat. 93/94:405 (1980)

/30/ H. Verbeek, IPP-Report 9/50 (1984), Max-Planck-Institut für Plasmaphysik, D-8046 Garching/München, FRG.

/31/ H.H. Andersen, T. Lenskjaer, G. Sidenius, H. Sørensen, J.Appl.Phys. 47:13 (1976)

/32/ H.H. Andersen, Rad. Eff. 3:51 (1970)

/33/ H. Sørensen, Appl.Phys. 9:321 (1976)

/34/ C.K. Chen, J. Bohdansky, W. Eckstein, M.T. Robinson, J.Nucl.Mat. 128/129:687 (1984)

/35/ A.B. Brown, C.W. Snyder, A.W. Fowler, C.C. Lauritsen, Phys.Rev. 82:159 (1951)

/36/ S. Rubin, Nucl.Instr.Meth. 5:1137 (1958)

/37/ R. Behrisch, Ph.D.Thesis, Technical University Munich (1968)

/38/ G.M. McCracken, N.J. Freeman, J.Phys. B2:661 (1969)

/39/ J. Vukanic, P. Sigmund, Appl.Phys. 11:265 (1976)

/40/ W. Eckstein, J.P. Biersack, Z. Phys. A, Atoms and Nuclei 310:1 (1983)

/41/ E.S. Parilis, V.K. Verleger, J.Nucl.Mat. 93/94:512 (1980)

/42/ J. Lindhard, V. Nielsen, M. Scharff, P.V. Thompson, Mat.Fys.Medd.Dan.Vid.Selsk. 33:No.10 (1963)

/43/ R. Weißmann, P. Sigmund, Rad. Eff. 19:7 (1973)

/44/ U. Littmark, A. Gras-Marti, Appl.Phys. 16:247 (1978)

/45/ M.T. Robinson, J.M. Torrens, Phys.Rev. B9:5608 (1974)

/46/ D.P. Jackson, in Proc. of the Symp. on Sputtering; P. Varga, G. Betz, F.P. Viehböck (eds.), Institut für Allgemeine Physik, Technische Universität Wien, Austria (1980)

/47/ M.T. Robinson in "Sputtering by particle bombardment I, Chapter 3, R. Behrisch (ed.), Topics in Applied Physics 47, Springer-Verlag, Berlin-Heidelberg-New York (1981)

/48/ M.I. Baskes, J.Nucl.Mat. 128/129:676 (1984)

/49/ D.P. Jackson, Proc. VII. Int.Conf. on Atomic Collisions in Solids, Moscow (1977), Vol. II, p. 141. Moscow State University Publ. House (1980)

/50/ D.P. Jackson, J.Nucl.Mat. 93/94:507 (1980)

/51/ J.E. Robinson, S. Agany, Proc. Vth Int. Conf. on Atomic Collisions in Solids, Gatlinburg (1973); S. Datz, B.R. Appleton, G.D. Moak (eds.), Plenum Press New York, Vol. I:215 (1975)

/52/ J.P. Biersack, L.G. Haggmark, Nucl.Instr.Meth. 174:257 (1980)

/53/ J.E. Robinson, D.P. Jackson, J.Nucl.Mat. 76/77:353 (1980)

/54/ D.P. Jackson, W. Eckstein, Nucl.Instr.Meth. 194:671 (1982)

/55/ V.M. Sotnikov, Sov.J.Plasma Phys. 7:236 (1981)

/56/ N.N. Koborov, V.A. Kurnaev, V.M. Sotnikov, J.Nucl.Mat.
      128/129:691 (1984)
/57/ W. Eckstein, H. Verbeek, J.Nucl.Mat. 93/94:518 (1980)
/58/ H. Verbeek, J.Appl.Phys. 46:2981 (1975)
/59/ W. Eckstein, F.E.P. Matschke and H. Verbeek, J.Nucl.Mat.
      63:199 (1976)
/60/ W. Eckstein, H. Verbeek, J.Nucl.Mat. 76/77:365 (1978)
/61/ G. Staudenmaier, J. Roth, R. Behrisch, J. Bohdansky,
      W. Eckstein, P. Staib, S. Matteson and S.K. Erents,
      J.Nucl.Mat. 84:149 (1979)
/62/ R.S. Bhattacharya, W. Eckstein and H. Verbeek, J.Nucl.Mat.
      79:420 (1979)
/63/ J. Amano and D.N. Seidman, J.Appl.Phys. 52:6934 (1981)
/64/ H. Verbeek, W. Eckstein and R.S. Bhattacharya,
      J.Appl.Phys. 51:1783 (1980)
/65/ S.H. Overbury, P.F. Dittner, S. Datz and R.S. Thoe,
      J.Nucl.Mat. 93/94:529 (1980)
/66/ M. Braun and E.W. Thomas, J.Appl.Phys. 53:6446 (1982)
/67/ J.E. Robinson, A.A. Harms and S.K. Karapetsas,
      Appl.Phys.Lett. 27:425 (1975)
/68/ O.S. Oen and M.T. Robinson, Nucl.Instr.Meth. 132:647 (1976)
/69/ W. Eckstein, H. Verbeek and J.P. Biersack, J.Appl.Phys.
      51:1194 (1980)
/70/ J. Biersack, W. Eckstein, Appl.Phys. 34:73 (1984)
/71/ C.K. Chen, B.M.U. Scherzer and W. Eckstein, Appl.Phys.
      A33:1 (1984)
/72/ N.N. Koborov, V.A. Kurnaev, V.G. Telkovsky and G.I. Zhabrev,
      Rad.Eff. 69:135 (1983)
/73/ W. Eckstein, IPP-JET-Report No. 3 (1981), Max-Planck-Institut
      für Plasmaphysik, D-8046 Garching/München, FRG.
/74/ J.E. Robinson, Rad. Eff. 23:29 (1974)
/75/ A.F. Akkerman, Phys.Stat.Sol. (a) 48:K 47 (1978)
/76/ O.B. Firsov, E.S. Mashkova, V.A. Molchanov, and V.A. Snisar,
      Nucl.Instr.Meth. 132:695 (1976)
/77/ G.I. Zhabrev, V.A. Kurnaev and V.G. Tel'kovskii,
      Sov.Phys.Tech.Phys. 19:978 (1975)
/78/ M. Hou, W. Eckstein and H. Verbeek, Rad. Eff. 39:107 (1978)
/79/ J.E. Harris, R. Young and E.W. Thomas, J.Appl.Phys. 51:5344
      (1980)
/80/ V.A. Kurnaev and V.G. Tel'kovskii, Proc. of the 3rd All-Union
      Conf. on Interaction of Atomic Particles with Solids,
      Naukova Dunka, Kiev (1974), Vol. I:323
/81/ O.S. Oen, M.T. Robinson, J.Nucl.Mat. 111/112:789 (1982)
/82/ W. Eckstein in "Inelastic Particle Surface Collisions",
      eds.: E. Taglauer and W. Heiland, Springer Series in
      Chemical Physics, Vol. 17:157, Springer-Verlag, Berlin
      (1981)
/83/ R.S. Bhattacharya, W. Eckstein and H. Verbeek, Surf.Sci.
      93:563 (1980)
/84/ H. Verbeek, W. Eckstein and R.S. Bhattacharya, Surf.Sci.
      95:380 (1980)

/85/ R. Behrisch, W. Eckstein, P. Meischner, B.M.U. Scherzer and
     H. Verbeek, in "Atomic Collisions in Solids", Eds. S. Datz,
     B.R. Appleton and C.D. Moak, Plenum Press, New York,
     Vol. 1:315 (1975)

/86/ W. Eckstein, H. Verbeek and R.S. Bhattacharya, Surf.Sci.
     99:356 (1980)

/87/ P.J. Schneider, K.H. Berkner, W.G. Graham, R.V. Pyle and
     J.W. Stearns, Phys. Rev. B23:941 (1981)

/88/ P.J. Schneider, W. Eckstein and H. Verbeek, J.Nucl.Mat.
     111/112:795 (1982)

/89/ P.J. Schneider, W. Eckstein and H. Verbeek, Nucl.Instr.Meth.
     194:387 (1982)

/90/ J.N.M. van Wunnik, J.J.C. Geerlings, E.H.A. Grauneman and
     J. Los, Surf.Sci. 131:17 (1983)

/91/ A.R. Krauss and D.M. Gruen, J.Nucl.Mat. 93/94:686 (1980)

/92/ M.T. Robinson, J.Appl.Phys. 54:2650 (1983)

/93/ W. Eckstein, Nucl.Instr.Meth. B (1985)

/94/ D.F. Düchs, D.E. Post, P.H. Rutherford, Nucl. Fusion 17:565
     (1977)

/95/ D. Reiter, A. Nicolai, J.Nucl.Mat. 128/129:458 (1984)

/96/ D. Heifetz, D.Post, M. Petravic, J. Weisheit, G. Bateman,
     PPPL-1843 (1981), Princeton Plasmaphysics Laboratory

/97/ D. Heifetz, D. Post, M. Petravic, J. Weisheit, G. Bateman,
     J. Comp. Phys. 46:309 (1982)

/98/ Y. Seki, Y. Shimomura, K. Maki, M. Azumi, T. Takizuka,
     Nucl. Fusion 20:1213 (1980)

/99/ G. Bateman, PPPL Appl. Phys. Report No. 1, Princeton Plasma
     Physics Laboratory (1980)

/100/ W. Eckstein, to be published

/101/ D. Heifetz, Princeton Plasma Physics Laboratory, private
      communication

/102/ W. Eckstein, IPP-Report 9/43 (1983), Max-Planck-Institut für
      Plasmaphysik, D-8046 Garching/München, FRG.

/103/ T. Tabata, R. Ito, K. Morita, Y. Itikawa, Jap.J.Appl.Phys.
      20:1929 (1981)

IMPLANTATION, RETENTION AND RELEASE OF HYDROGEN ISOTOPES IN SOLIDS

W. Möller and J. Roth

Max-Planck-Institut für Plasmaphysik
EURATOM-Association
D-8046 Garching/München, FRG

ABSTRACT

A review is given of the physical mechanisms involved in the interaction of fast hydrogen ions with solids, their diffusion and trapping in the solid lattice after slowing down, and the surface recombination allowing them to get reemitted. Analytical and computational models are discussed which describe the buildup of hydrogen inventory in a vessel wall and its release through the surfaces. They are applied to characterize wall material candidates for fusion devices with respect to their interaction with hydrogen isotopes, and to compare their suitability as seen from plasma control and safety demands.

1. INTRODUCTION

Energetic hydrogen atoms or ions incident on a solid surface may either be directly reflected after suffering one or several collisions with the target atoms, or they may be implanted, i.e. slowed down to thermal velocities within the subsurface layer. Particle reflection is described in an earlier chapter of this volume /1/. The fate of an incident atom will first be determined by the physics of atomic collisions, which govern the scattering and slowing-down processes. Subsequently, however, the implanted atom will be subject to thermal-energy interactions with the solid. These, for example, determine whether the implanted atom will be retained in the material (thus modifying its composition), or migrate to one of its surfaces in order to be released.

These phenomena have been investigated with increasing activity during the past few years, because of their relevance to fusion research /2-9/. In today's plasma experiments, employing magnetic confinement, the intimate interaction of the plasma with the surrounding walls is manifested by the fact that each plasma ion leaves the plasma after an average 'particle confinement' time which is about ten times shorter than the duration of the discharge. Consequently, the buildup of hydrogen in the subsurface region of the wall and the release back into the plasma ('recycling') will influence the particle balance at the beginning of a discharge, and also the stationary plasma parameters. Further, the buildup of hydrogen isotopes throughout the thickness of the wall during a series of discharges, or even the total life of a device might create environmental problems due to the tritium 'inventory' within the wall, as well as material problems like embrittlement due to high hydrogen concentrations /10/. Finally, a fraction of deuterium and tritium might 'permeate' through the wall material and contaminate the coolant.

Final goals of investigating the above phenomena for fusion research are to encorporate them into plasma or material modelling codes, in order to understand the particle balance in plasma experiments. A large range of parameters may be of influence: Different materials (metals, metallic alloys and compounds, graphite) with varying surface contaminations may be employed in a wide range of temperatures (from below room temperature to the melting point). The first wall will be exposed to a wide spectrum of incident energies ($\sim$ 50 eV to $\sim$ 100 keV), at fluxes which vary by orders of magnitude ($\sim 10^{15}$ cm^{-2} s^{-1} to $\sim 10^{19}$ cm^{-2} s^{-1}). Therefore it appears impossible to provide all necessary data by individual simulation experiments. The aim of any investigations should rather be to understand the basic mechanisms and to formulate model descriptions, which are supported by specific experimental informations and can be used for extrapolation.

The present lecture will review the current understanding of the processes involved in the implantation and post-implantation behavior of hydrogen atoms in solids, and report on analytical and computational models based on these processes. First we will treat the atomic collisions slowing down the implanted atoms, and the collisional damage in the implanted region. The retention and release of the slowed-down atoms will then be shown to be governed by atomic bulk diffusion, by trapping at damage sites, and by the recombination to molecules at the surfaces. Based on these mechanisms, simple analytical and more detailed numerical models will be presented. A final chapter will describe and compare the recycling, inventory and permeation properties of some materials which are of relevance for fusion reactor research.

2. RANGE AND DAMAGE DISTRIBUTIONS

2.1 Stopping power

In the following, only those incident atoms are considered which penetrate the surface and are slowed down in the target substance. The loss of kinetic energy E of an incident atom per unit pathlength x is described by the 'stopping power' -dE/dx or the 'stopping cross section'

$$s = - \frac{1}{n} \frac{dE}{dx} \qquad (1)$$

where n denotes the target atomic density.

Energy can be transferred to the target electrons, leading to ionization, atomic excitations, or collective electron excitations. Due to the small relative mass of the electron, angular deflections of the atom occuring during these collisions are negligibly small. Alternatively, the incident particle may interact with the target atom cores, i.e. be scattered by the screened electrostatic potential of a target nucleus, and be deflected significantly. Consequently, the total stopping cross section can be subdivided into an 'electronic' or 'inelastic' fraction and a 'nuclear' or 'elastic' one:

$$s = s_e + s_n \qquad (2)$$

Both contributions can, in principle, be calculated from the differential cross section of the single scattering event,

$$s = \int T_c d\sigma \qquad (3)$$

with T_c denoting the energy transfer.

2.1.1 Nuclear Stopping

For the elastic contribution, s_n can be derived using classical mechanics for any interatomic interaction potential (for a review on interatomic potentials, see ref. /11/; more recent considerations are found in, e.g., ref. /12/). For practical applications, it is often desirable to derive universal approximations which can be used for all projectile-target combinations. Lindhard and coworkers /13/ have given a universal scattering cross section and a corresponding elastic stopping power using reduced units

$$\varepsilon_L = \frac{4\pi \, \varepsilon_o \, a_{TF} m_2}{Z_1 Z_2 \, e^2 (m_1 + m_2)} \cdot E \qquad (4)$$

for the energy and

$$\rho_L = \pi a_{TF}^2 n \frac{4\ m_1 m_2}{(m_1 + m_2)^2} \cdot x \tag{5}$$

for the pathlength. In eqs. (4) and (5), Z_i and m_i denote the atomic numbers and masses of the projectile (i=1) and the target (i=2), and a_{TF} the Thomas-Fermi screening parameter /13/

$$a_{TF} = \frac{0.8853\ a_0}{(Z_1^{2/3} + Z_2^{2/3})^{1/2}} \tag{6}$$

with a_0 representing the radius of the first Bohr orbit ($a_0 = 5.29 \times 10^{-9}$ cm). The reduced nuclear stopping power

$$s_n^L = - \left(\frac{d\varepsilon_L}{d\rho_L}\right)_n \tag{7}$$

has been calculated and tabulated for different interatomic potentials /13-15/. Good approximation formulae based on the Thomas-Fermi potential are /16/

$$s_n^L(\varepsilon_L) = \begin{cases} 39.5\ \varepsilon_L^{0.73}\ \exp(-4.7\ \varepsilon_L^{0.2}), & 10^{-2} \underset{\sim}{<} \varepsilon_L \underset{\sim}{<} 10 \\[2mm] (2\varepsilon_L)^{-1}\ \log(1.29\varepsilon_L), & \varepsilon_L \underset{\sim}{>} 10 \end{cases} \tag{8}$$

A recent alternative analytical fit to $s_n^L(\varepsilon_L)$ is of similar quality /17/.

2.1.2 Electronic Stopping

Simple expressions for the electronic stopping have been derived both for high (here: still non-relativistic) and low energies. From quantum mechanical calculations using the first Born approximation, the Bethe formula results as

$$s_e = \frac{e^4}{4\pi\ \varepsilon_0^2\ m_e}\ \frac{Z_1^2 Z_2}{v^2}\ \log \frac{2\ m_e\ v^2}{I_0} \tag{9}$$

with the electron rest mass m_e, the mean ionization potential I_0, and the projectile velocity v. (A detailed discussion is found in, e.g., ref. /18/). For hydrogen isotopes, eq. (9) is valid for energies well above 100 keV/amu.

For low energies (below $\sim$ 25 keV/amu for hydrogen isotopes), classical scattering theories predict the electronic stopping to be proportional to the projectile velocity. Again to be understood as an universal approximation for different projectile-target combinations, Lindhard and Scharff /19/ write

442

$$\left(\frac{d\varepsilon_L}{d\rho_L}\right) \cong \frac{0.0793 \, z_1^{2/3} \, z_2^{1/2} \, (m_1+m_2)^{3/2}}{(z_1^{2/3} + z_2^{2/3})^{3/4} \, m_1^{3/2} \, m_2^{1/2}} \, \sqrt{\varepsilon_L} . \qquad (10)$$

At intermediate energies, the theoretical understanding is still rather incomplete to date. In order to provide reliable data, one may fit experimental data using an interpolation formula

$$\frac{1}{s_e} = \frac{1}{s_e^{low}} + \frac{1}{s_e^{high}} \quad , \qquad (11)$$

with $s_e^{low} \sim \sqrt{E}$ and $s_e^{high} \sim E^{-1} \log E$, as proposed by eqs. (9) and (10), respectively. It is seen that both limits are correctly reproduced by eq. (11). Semiempirical stopping cross sections obtained in this way are given by Andersen and Ziegler [20]. Their data set may be regarded as today's standard compilation of hydrogen stopping powers. Nevertheless, it should be pointed out that the stopping cross sections at the lowest energies ($E \stackrel{\scriptstyle <}{\scriptstyle \sim} 1$ keV/amu) are generally not supported by experimental information and are therefore doubtful to some extent.

Figure 1 shows both nuclear and electronic stopping cross sections for the representative case of D in Ni from eqs. (8-10) and the Andersen/Ziegler [20] tabulations. Above 1 keV, the inelastic stopping is largely dominating. Nevertheless, the nuclear interaction cannot be neglected for the construction of range distributions up to $\sim$ 100 keV energies, as - in contrast to the electronic collisions - it is accompanied by angular deflections of the projectile.

2.2 Ranges of Implanted Ions

Knowing the total stopping cross section, the mean total pathlength of a beam of implanted ions may directly be calculated according to

$$\overline{R}_t = \frac{1}{n} \int_0^{E_O} \frac{dE}{S(E)} \quad , \qquad (12)$$

where E_O is the incident energy. The total pathlength is a measurable quantity only in the limit of very high energies ($E_O \stackrel{\scriptstyle >}{\scriptstyle \sim} 1$ MeV/amu) where angular deflections resulting from nuclear collisions are small over a large fraction of the total range. Generally and especially for light ions, the elastic scattering events will create polygonal trajectories with the mean range along these paths given by eq. (12). An example is given in Fig. 2, where the trajectories of implanted hydrogen atoms have been generated in a computer simulation [21] (see below). It is seen that

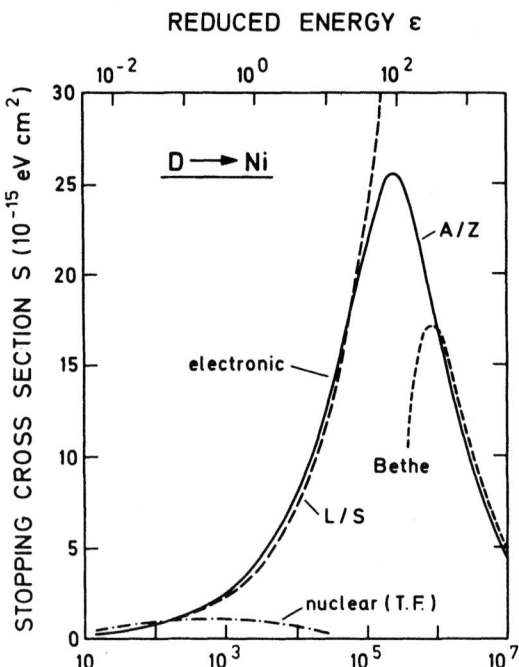

Fig. 1: Stopping cross-section for deuterium in nickel from the
semiempirical tabulations of Andersen and Ziegler /20/
(eq. (11)) and theory: Thomas-Fermi nuclear cross-section
(eq. (8)), Lindhard-Scharff formula (eq. (10)) and Bethe
formula (eq. (9)) with a mean ionization energy of
I_o = 310 eV.

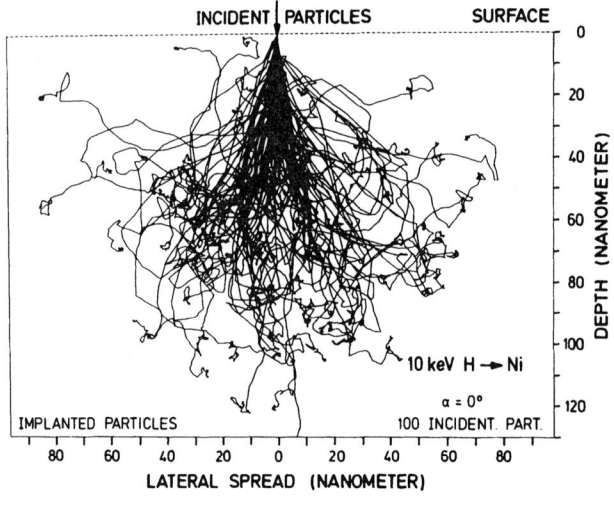

Fig. 2:

Planar projection
of ion trajectories
generated by a TRIM
simulation /21/ for
hydrogen in Ni. Of
the 100 incident
particles, 8 have
been reflected
kinematically. Their
trajectories have
been discarded for
this plot.

the total pathlength of an individual ion is significantly greater
than the depth of its final position. The distribution of the final
depths, which is constituted from the 'projected' ranges of the individual atoms, represents the source distribution of the implanted
atoms, which we need for further discussions.

Figure 2 also shows a distribution of lateral spreads of the
final depth; at normal incidence this has no physical meaning as
the extension of the bombarded area is in reality much larger than
the implantation depth. At oblique incidence, however, the lateral
spread will cause a broadening of the depth distribution in addition
to the one along the incident direction.

The calculation of projected range distributions requires
more implicate procedures. From a simple evaluation of the Boltzmann transport equation, Schiøtt /14/ obtained an approximate
correction factor between mean projected and mean total range

$$\frac{\bar{R}_p}{\bar{R}_t} \cong \lambda(1-2\lambda); \qquad \lambda = \frac{m_1}{m_2} \frac{s_e(E_o)}{s_n(E_o)} , \qquad (13)$$

which is valid for $\lambda \ll 1$ and $s_n < s_e$, i.e. for hydrogen ions of
about 1...10 keV.

More elaborate evaluations of transport theory /22, 23/ have
been performed successfully, but suffer from certain problems,
which become especially serious at low (< 1 keV) energies which we
need to cover for our subject. As the theory is normally set up
for infinite media, scattering phenomena close to the surface are
handled incorrectly. (However, more recently a formalism has been
developed for semi-infinite targets /24/.) In addition, transport-
theoretical analytical solutions provide in the first step the
moments of the distribution, from which the distribution functions
have to be reconstructed. This procedure only works well for distributions which are of approximately symmetric (gaussian) shape,
which is not the case for either very low (<< 1 keV) or very high
(>> 10 keV) energies.

An alternative approach in particular for plasma-wall interaction applications is the use of Monte Carlo computer simulations
/25-27/ to generate range distributions of implanted hydrogen isotopes. In these simulations, the trajectories of a large number of
'pseudoprojectiles' are traced by a sequence of binary elastic
collisions taking into account the electronic energy loss along
the path, with the scattering parameters chosen from a sequence of
random numbers. The range distribution then arises naturally from
the distribution of the endpoints of the pseudoprojectile paths.
As a further advantage of such simulations, one may treat a
spectrum of incident energies (e.g., a maxwellian one) directly,

by choosing the start energy of each pseudoprojectile statistically. A code being frequently used for these purposes is TRIM /27/.

In the example of Fig. 2, the ratio of the average projected
range and the average pathlength amounts to 0.24 compared to 0.10
according to the simple analytical approximation of eq. (13).

Reasonable agreement, however, is found for keV energies when
comparing the simulation results to numerical evaluations of the
analytical theory. Figure 3 shows a TRIM result and the first two
moments obtained from an analytical calculation /20/. Furthermore,
the theoretical predictions are compared to an experimental profile
which has been obtained by a nuclear reaction depth profiling technique /28/. The mean projected ranges from theory and experiment
agree well within ∿ 10 %, whereas the experimental width is about
30 % larger compared to theory. This has been attributed to effects
of target crystallinity /29/. Better agreement also for the widths
is obtained at higher energies of 100 ... 400 keV /30/.

Experimental depth profile studies for lower implantation
energies have been performed in C using the SIMS technique /31/.
For both 1.5 keV H and D, these studies show also a significantly
(∿ 50 %) larger width than the one predicted by TRIM, while the mean
ranges agree reasonably well.

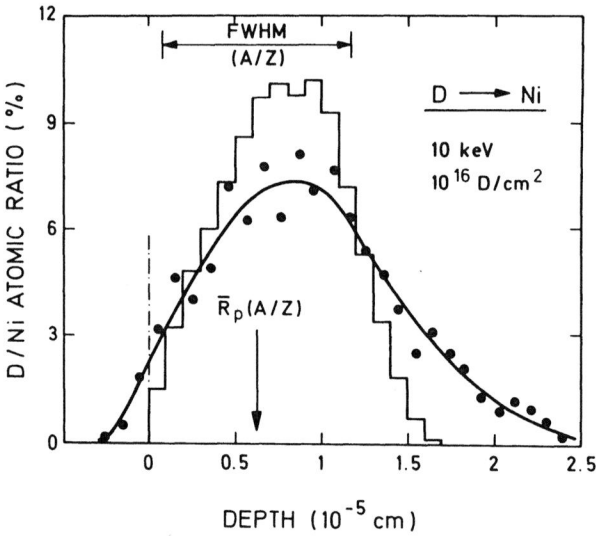

Fig. 3: Range distributions for 10 keV D implanted into Ni from
experiment /28/ and a TRIM computer simulation. $\overline{R_p}$ and
FWHM denote the mean projected range and the full width
at half maximum taken from the analytical prediction in
Ref. /20/.

446

Summarizing, we may conclude that investigations on the behaviour of implanted hydrogen isotopes may rely on rather well established mean range data, whereas the profile shape as taken from theoretical predictions is subject to some uncertainties. However, many of the phenomena to be discussed in the following sections do not depend on the implantation profile shape in a critical way. This might not be true for very low (10...100 eV) energies, where a large fraction of the incoming beam ($\sim > 50$ %) will be reflected /1/, and the maximum of the range distribution is very close to the surface. Due to the lack of experimental information, one must rely entirely on computer simulation studies in this energy range. There is some hope, however, that those are not too far from reality, as also sputtering yields, which result from collisional processes at down to ~ 5 eV energies, are quite well reproduced by Monte Carlo calculations /32/.

2.3 Damage

2.3.1 Frenkel Defects

Each nuclear collision during the slowing-down of a projectile will transfer energy to a target atom according to

$$T_n = \frac{4\, m_1 m_2}{(m_1 + m_2)^2}\, E \sin^2 \frac{\theta_{CM}}{2} = \gamma E \sin^2 \frac{\theta_{CM}}{2}\ , \tag{14}$$

where the probability for a scattering process with the deflection angle θ_{CM} in the center-of-mass system is given by its differential cross-section. As the binding energy of a target atom in its lattice site is of the order of a few eV only, quite soft collisions will remove the atom from that site. However, a relatively large energy transfer of $\sim 10...50$ eV is required to move the atom far enough from its original site in order to create a stable displacement ('Frenkel pair'). For each material, this threshold depends on the knock-on direction. Its average value is called the 'effective displacement energy' E_d. (For reviews, see e.g. Refs. /33-35/.)

At sufficiently high energy transfer, the 'primary' knock-on atom may create a collision cascade and further stable Frenkel pairs. According to the 'modified Kinchin/Pease' model /36/, the average number of Frenkel pairs created in a cascade with the primary energy transfer T_n amounts to

$$\nu_F(T_n) = \begin{cases} 0 & \text{for } T_n < E_d \\[2mm] 1 & \text{for } E_d < T_n,\ \ T_n^* < 2.5\, E_d \\[2mm] \dfrac{0.8}{2E_d}\, T_n^* & \text{for } T_n^* > 2.5\, E_d \end{cases} \tag{15}$$

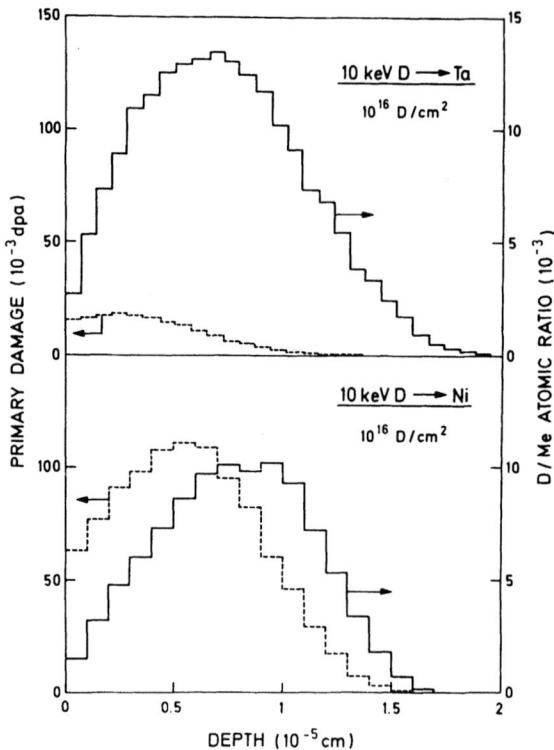

Fig. 4: Range and primary damage distributions for 10 keV deuterons implanted into Ni and Ta generated by TRIM simulations with 10000 (Ni) and 15000 (Ta) projectile histories.

In eq. (15), T_n^* is the energy transfer corrected for the inelastic energy losses in the cascade. For small relative energy transfers, $T_n^* = 0.9 \, T_n$ is a good approximation /37/. Experimentally, one finds the number of Frenkel pairs created by ion bombardment at low temperature in rough agreement with eq. (15). However, it tends to be overestimated at high energy transfers (>> 100 eV) by up to a factor of 2 /33, 38/ and underestimated at low energy transfer by about a factor of 3 for keV light ion bombardment /39/.

By applying the modified Kinchin-Pease formula for each nuclear collision, the depth distribution of Frenkel pairs can be generated in a computer simulation simultaneously with the range distribution of implants. Two results comparing a medium Z and a high Z material are shown in Fig. 4: The damage distributions peak closer to the surface than the range distributions as the projectiles may still move at low residual energy, while no longer creating any damage. Considerably less damage is created in Ta than in Ni because of the small energy transfer factor γ (eq.14) and the large effective displacement energy (90 eV compared to 33 eV in

Ni /34/). The present type of simulations does not take into account any annealing of Frenkel pairs (see Sect. 2.3.2).

Apparently, no Frenkel pairs at all will be created at energy transfers $T_n < E_d^{min}$, where E_d^{min} is the minimum displacement energy for any knock-on direction, which varies from ~ 10 eV (Pb) to ~ 40 eV (Au). This imposes the condition

$$E_o > \frac{E_d^{min}}{\gamma} \tag{16}$$

on the creation of stable Frenkel defects. For graphite with $E_d^{min} \cong 20$ eV, no damage will be created by hydrogen bombardment below 70 eV, for gold below 1.8 keV.

On the other hand, lattice damage might result from other than collisional processes. For semiconductors and insulators, it is well known that the conversion of electronic excitations into kinetic energy can lead to damage /40, 41/. However, any systematic investigations on hydrogen bombardment are missing up to date. Furthermore, it has been found that thermal He atoms in metals might cluster and thereby eject lattice atoms from their sites /42/. It is unknown whether such a mechanism could also exist for hydrogen at high concentrations.

2.3.2 Annealing and Extended Defects

The above considerations are only valid for low defect concentrations. Due to spontaneous recombination of close Frenkel pairs and the 'subthreshold' recombination of vacancies with energetic interstitial atoms, the maximum atomic concentration of Frenkel defects is limited to about 0.2 % for fcc metals and 0.5 % for bcc metals /43/. For the examples of Fig. 4, a saturation of Frenkel defects would occur at fluences well below a fluence of 10^{16} D/cm².

Moreover, Frenkel defects only survive at sufficiently low temperatures where both interstitial atoms and vacancies are immobile. Interstitial atoms become mobile generally below 100 K. They may then be annihilated at vacancies, or form clusters like dislocation loops /44/. Therefore, only vacancies will remain, but to an uncertain concentration. In addition, recent interpretations of experimentally observed hydrogen-defect interactions /45/ indicate that interstitial atoms might get stabilized by the presence of implanted hydrogen.

Vacancies become mobile at an absolute temperature which is roughly a quarter of the melting temperature T_m. They may then agglomerate into clusters (loops, voids), which at even higher temperature ($T > T_m/2$) dissolve to complete annealing.

The above indicates that the information on damage introduced by hydrogen implantation is by no means sufficient to formulate any models which would describe the presence of defects as function of implantation fluence, flux density and temperature, or even their interaction with the implanted hydrogen atoms. Therefore, defects, which arise from irradiation damage, will be treated in the following sections rather phenomenologically, based on the experimental observation of hydrogen trapping at defects /30, 45, 46/.

3. MECHANISM GOVERNING THE BEHAVIOUR OF IMPLANTED ATOMS

In the preceding paragraph, the slowing-down of energetic atoms implanted into a solid was treated. Now we ask how they will behave after coming to rest. It turns out that the knowledge from investigations on the behaviour of hydrogen in unimplanted systems alone is not sufficient to describe these phenomena. Implanted systems expose new characteristics: Besides the creation of radiation defects, it is possible to achieve extremely high local hydrogen concentrations, when diffusion away from the implanted layer is hindered at low temperatures, or when the implantation flux densities are sufficiently high. In the following, we will first list the mechanisms, as far as they are known from solid state and ion-implantation physics. This information will then be used in a subsequent chapter to deduce model descriptions on the inventory and release of the implanted hydrogen atoms.

The different interactions of injected hydrogen atoms with the solid are sketched in the energy diagram of Fig. 5. Once implanted into the solid, the atoms may reside on interstitial solution sites, from which they may diffuse interstitially, subject to the activation energy of diffusion, U_d. Traps may be present (e.g., from radiation damage, impurities, or mechanical cold-work) with additional binding enthalpies U_t with respect to the normal interstitial solution sites. At high concentrations, it may be energetically favourable for the hydrogen atoms to precipitate into hydride phases or gas bubbles with a binding energy U_p to precipitates. Other processes which influence the release of the implants are associated with the surface; on most metal surface hydrogen atoms can be chemisorbed with an enthalpy U_c with respect to a molecule in the gas phase. (The dotted line in Fig. 5 indicates the high dissociation energy $U_m = 2.3$ eV/atom; the threshold for hydrogen chemisorption at solid surfaces is generally found to be much less, excluding a two-step process of molecular dissociation and atomic adsorption.)

A hydrogen atom will preferentially leave the surface after recombination with a second atom to form a molecule. (At sufficiently high temperatures, also atomic desorption should occur; also, in the presence of atomic hydrogen in the gas above the surface, direct

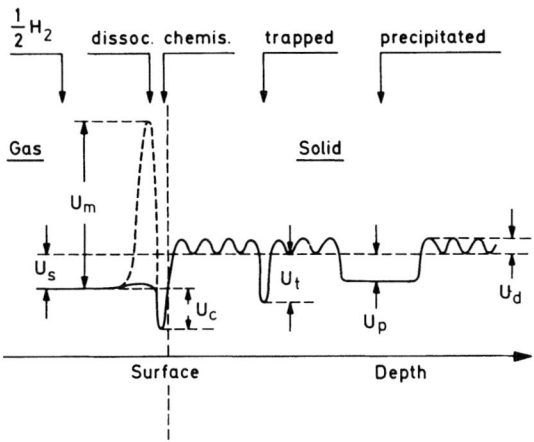

Fig. 5: Schematic energy diagram for hydrogen in metals. The different energies are: U_s enthalpy of solution, U_m dissociation energy of the hydrogen molecule, U_d activation energy of diffusion, U_c, U_t and U_p enthalpies of chemisorption, trapping and precipitation.

recombination with a gas atom will contribute. Both processes are neglected in the following.) The difference between the molecular gas level and the atomic solution level in the bulk is the solution enthalpy U_s which may be positive or negative for endothermal or exothermal hydrogen-solid systems, respectively.

The bulk processes occuring in unimplanted systems have been studied in great detail for many systems (see, e.g., Refs. /47-49/). Here, we only recall the main findings briefly and as far as they are needed for our further discussions.

3.1 Solubility

Hydrogen isotopes have been found to be soluble in most solid elements, except for the groups V_B, VI_B, VII_B and 0 of the periodic system. In particular, high concentrations of hydrogen may be dissolved exothermically in the group III_A – V_A elements including the 4f and 5f transition metals, whereas the VI_A – II_B transition metals dissolve hydrogen endothermically (with the exception of the slightly exothermal palladium-hydrogen system). For the equilibrium between hydrogen dissolved in the solid and the hydrogen gas with pressure p, Sievert's law holds for sufficiently low bulk concentrations c_{bs} of hydrogen in solution:

$$c_{bs} = S_s \cdot p^{1/2} \tag{17}$$

with S_s denoting the solubility given by

$$S_s = S_o \ e^{-\frac{U_s}{kT}} \ . \tag{18}$$

Solubility data can be found in the literature for many metal-hydrogen systems /48/ and are also available for some nonmetals like carbon and carbides /9/.

At higher concentrations, metal-hydrogen systems tend to convert from the so-called 'α-phase', in which hydrogen atoms are statistically dissolved, to hydride phases with an ordered occupation of solution sites. Such phenomena have been investigated to some detail only in exothermal systems. There is, however, also evidence for hydride formation in endothermal ones like Ni at extreme conditions, e.g., very high pressures /49/.

For any detailed information on the solubilities at higher concentrations one has to look up the phase diagram for a specific system. An example is given in Fig. 6 for hydrogen in Ti. In the (T,c) diagram, several mixed-phase regimes are seen with plateau ('decomposition') pressures as function of temperature. In these

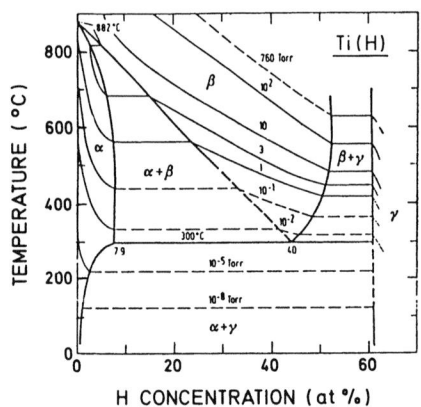

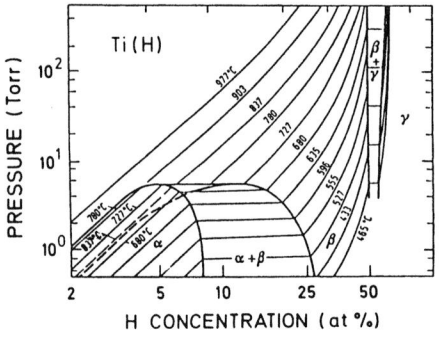

Fig. 6:

Phase diagram of hydrogen in Ti in (T,c) and (p,c) representation. Note that the concentration scale is given in at% (i.e., 66.7 at% corresponds to stoichiometric hydride, TiH_2).

(From Ref. /48/.)

regimes, precipitates of a higher-concentration phase are formed
within the lower-concentration one. As seen from the (p,c) plot,
Sievert's law (eq. 17) is valid in the α-phase, and, within certain
limits, also in the ß-phase, however with different constants for
the different phases.

The progress of theoretical approaches to calculate solubilities, or even to describe phase diagrams quantitatively, is still
rather limited. However, solution enthalpies have recently been
calculated with satisfactory agreement with experimental values for
several metals /50, 51/.

3.2 Diffusion

At elevated temperatures, the dissolved hydrogen atoms may
migrate in the host lattice. In the case of a concentration gradient, a net flux j_s of diffusing particles then results according
to Fick's first law (for reference, see e.g. /52/):

$$j_s = - D \frac{\partial c_{bs}}{\partial x} \tag{19}$$

For most problems associated with ion implantation, eq. (19)
can be written in its 1-dimensional form, as lateral extensions are
typically large compared to the considered depths. For a diffusion
coefficient D which is independent of the depth, the continuity
equation transforms eq. (19) into Fick's second law

$$\frac{\partial c_{bs}}{\partial t} = D \frac{\partial^2 c_{bs}}{\partial x^2} \; . \tag{20}$$

In the simplest classical picture, D can be derived from a
model of "random walk", i.e. random jumps from one solution site to
an adjacent one after thermal activation. In this model, the diffusion coefficient results as

$$D = \frac{1}{6} \nu_o \, e^{-\frac{U_d}{kT}} \, a_d^2 \; , \tag{21}$$

with the jump distance a_d, the attempt frequency ν_o, and the activation energy U_d of diffusion. Writing instead

$$D = D_o \, e^{-\frac{U_d}{kT}} \; , \tag{22}$$

one would expect $D_o \cong 10^{-3}$ cm²/s for vibration frequencies
$\nu_o \cong 10^{13}$ s⁻¹ and $a_d \cong 2 \times 10^{-8}$ cm.

Any more elaborate predictions for D_o or U_d cannot be given up
to date. However, a broad data base exists for diffusion of hydrogen
atoms in metals /49, 53/. An example for the case of H in Ni is

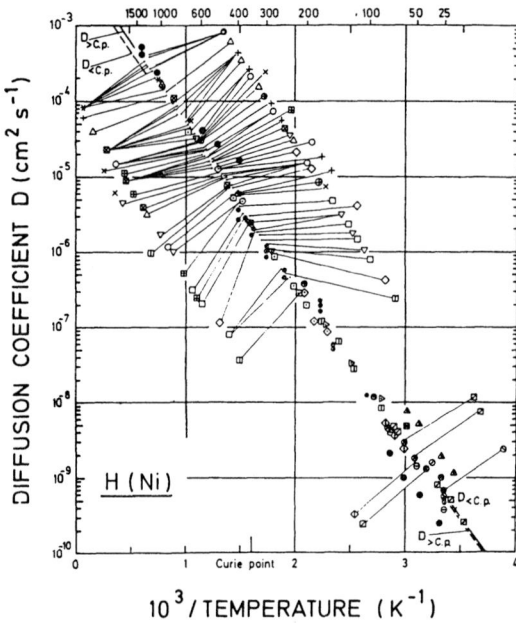

TEMPERATURE (°C)

DIFFUSION COEFFICIENT D (cm²s⁻¹)

H (Ni)

10^3 / TEMPERATURE (K⁻¹)

Fig. 7: Diffusion coefficient of hydrogen in Ni ($D_{<C.p.}$ and $D_{>C.p.}$ denote the best Arrhenius fits below and above the Curie point, respectively). (From Ref. /49/.)

given in Fig. 7: Over a large range of temperatures, the diffusion coefficient obeys the Arrhenius relation (eq. 22). For metals, this is valid above ∿ 200 K, below which quantum effects might play a role (see, e.g. /54/).

The availability of diffusion data is less satisfactory for nonmetals. In carbon, it is unclear whether the activation energy of diffusion is ∿ 4 eV or ∿ 0.5 eV as it is difficult to distinguish between diffusion and trapping processes (see Sect. 3.5). For carbides, only a few data are available /9/.

3.3 Diffusion in a Temperature Gradient

Even in the absence of a concentration gradient, hydrogen can diffuse due to a temperature gradient ('Soret effect'). Intuitively, one might expect that thermal motion would establish a net flux from the hot side of a membrane to the cold one. This is indeed observed for metals which dissolve hydrogen exothermally; the temperature dependence of the solubility, however, will counteract for the endothermal case, thus leading to a diffusion to the hot side, as, e.g.,for Ni or Fe. The flux arising from thermal diffusion can be written as /52/:

454

$$j_s = c_{bs} \frac{U_{SE}}{kT^2} \frac{\partial T}{\partial x} \tag{23}$$

with the 'heat of transport' U_{SE}. Recent data for U_{SE} are discussed in Refs. /55, 56/. The superposition of eqs. (19) and (23) yields then for the diffusional transport, including the thermal one

$$\frac{\partial c_{bs}}{\partial t} = D \frac{\partial^2 c_{bs}}{\partial x^2} - \frac{U_{SE}}{kT^2} \frac{\partial}{\partial x} \left(c_{bs} \frac{\partial T}{\partial x} \right). \tag{24}$$

3.4 Surface Recombination

The release of hydrogen at the surface of a solid into the vacuum proceeds for most materials via the recombination of dissolved or chemisorbed atoms to a physisorbed hydrogen molecule. This subject has attracted interest for a number of a decades /57, 58/. Nevertheless, little progress has been achieved in its theoretical understanding and the formulation of models.

For our problem, we have to describe a nonequilibrium situation where implanted hydrogen atoms diffuse to the surface, recombine there and are emitted into the vacuum. Models available to date result from equilibrium considerations: In equilibrium with a gas pressure present, all individual fluxes (gas to surface, surface to bulk, gas to bulk and vice versa) will balance. All these individual fluxes depend on the concentrations (gas pressure, surface coverage, and bulk concentrations). If now the gas is replaced by a vacuum, the dependence of the remaining fluxes on the concentrations is expected still to be identical with the one in equilibrium.

The simplest consideration might neglect the surface and take into account the gas to bulk flux and the reverse one only. In equilibrium, the flux of atoms from the gas onto the surface is given by kinetic theory:

$$j_g = 2 \cdot \frac{p \cdot v_t}{4kT} \quad , \tag{25}$$

where p denotes the gas pressure and v_t the mean thermal velocity,

$$v_t = \sqrt{\frac{4kT}{\pi m}} \tag{26}$$

(m mass of hydrogen atom).

As the recombination is a second-order process, one might expect the outgoing flux of atoms to be proportional to the square of the bulk concentration close to the surface:

$$j_- = K_r \cdot [c_{bs}(x = 0)]^2 \tag{27}$$

with the 'recombination (rate) constant' K_r. By combining eqs. (25-27) with Sievert's law (eqs. 17-18), one obtains

$$K_r = (S_o^2 \sqrt{\pi m k T})^{-1} e^{\frac{2U_s}{kT}} . \tag{28}$$

Taking into account a sticking coefficient α_s of the incoming molecules and an activation threshold for their atomic solution (see Fig. 5)

$$U* = \max (0, U_s + U_d), \tag{29}$$

thus modifying eq. (25) to yield the deposited flux from the gas

$$j_g^* = \alpha_s e^{-\frac{U*}{kT}} \cdot j_g, \tag{30}$$

eq. (26) becomes modified to

$$K_r = \frac{\alpha_s}{S_o^2 \sqrt{\pi m k T}} \cdot \begin{cases} e^{\frac{2U_s}{kT}} & \text{for } U_s + U_d \leq 0 \\[2mm] e^{\frac{U_s - U_d}{kT}} & \text{for } U_s + U_d > 0 . \end{cases} \tag{31}$$

According to eqs. (28) and (31), the recombination constant will later be given in the form

$$K_r = \frac{K_r^o}{\sqrt{T}} e^{-\frac{U_r}{kT}} . \tag{32}$$

A correct derivation of the second-order form of eq. (27) and the corresponding recombination constant requires information on the individual processes at the surface. In addition to the idealised gas-bulk transition leading to the above equations, one has to include the presence of chemisorbed atoms on the surface (see Fig. 5). The sticking coefficient of arriving molecules from the gas, even in the limit of low surface coverage, depends on the presence of an activation threshold for chemisorption (which, however, is understood to be much lower than the molecular dissociation energy $U_m = 2.3$ eV per atom). Very little is known on the energy accommodation during chemisorption. For the desorption process, there is no quantitative picture of the influence of surface diffusion.

Thus the extended models remain mainly heuristic at present. Baskes /59/ has verified eq. (27) by describing the individual fluxes, including the direct splitting of an arriving molecule into a chemisorbed one and a dissolved one, in an approximation. His calculation yields a recombination constant which exposes the temperature dependence of eq. (31), but is a (for practical purposes negligible) factor of 2 larger than given here. Neglecting direct gas-bulk transitions, Pick and Sonnenberg /60/ conclude that eq. (28) is generally valid for clean metal surfaces. This formula has been used earlier by Hotston and McCracken /61/ for exothermal systems.

A further complication arises from the fact that all individual surface processes are influenced by surface contamination. This is also indicated by experimental results, which are available mainly for stainless steel (Fig. 8). Probably depending on the presence of surface impurities, the experimental values span a range of about four orders of magnitude, with the highest results given by eq. (31). The influence of different surface treatments has been proven in one of these experiments /62/.

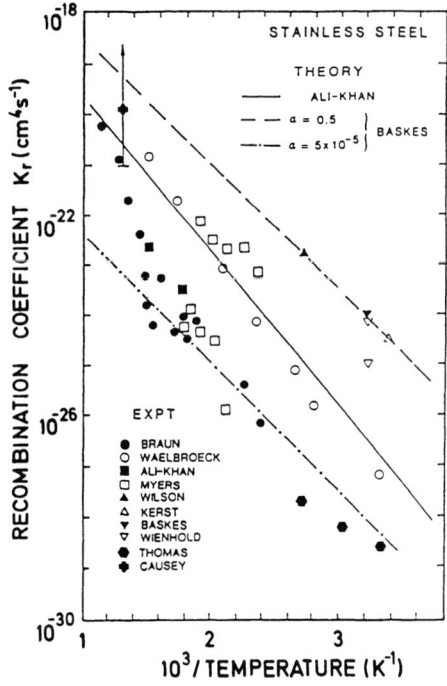

Fig. 8:

The surface recombination coefficient for hydrogen in stainless steel measured by a variety of techniques compared to theoretical estimates. (From Ref. /9/.)

Recent results for Ni /63, 64/ also document the lack of knowledge in this area. Figure 9 shows the temperature dependence of the data to be in better agreement with eq. (28), whereas the absolute values are 6 orders of magnitude lower and close to the

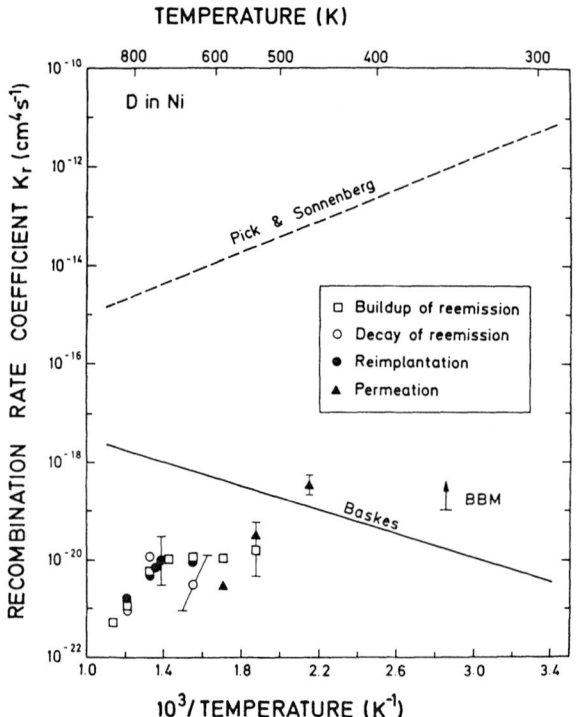

Fig. 9: Surface recombination coefficients for deuterium on Ni obtained from permeation and reemission measurements /63, 64/ compared with results of model calculations /59, 60/. An experimental lower limit estimate, which has been given by Besenbacher et al. /75/, is denoted by 'BBM'. (From Ref. /64/.)

prediction of Baskes /59/ or eq. (31). Note that both Ni and SS dissolve hydrogen endothermally so that no qualitative differences should be expected.

Much more experimental information, including surface characterization, is required to provide reliable data. For nonmetals, no information on the surface recombination is available at all.

3.5 Trapping

The influence of traps, i.e. sites exposing a higher binding energy than a regular solution site, on the diffusion of implanted hydrogen has been observed in many hydrogen-metal systems. Experimentally, both a delayed thermal release of previously implanted hydrogen (see Ref. /30/ and references therein), and a delayed onset of reemission or permeation during implantation /65, 67/ have been observed. The trap concentrations involved in these observations are typically in the order of per cent or less of the host atomic concentration. Often the results cannot be explained by a single type of traps, but indicate the presence of traps with different binding energies.

Traps arise from any defect or impurity in the host lattice. They may also be present throughout the whole bulk of the material in the absence of any irradiation effects, e.g. due to bulk impurities or cold-work generated dislocation networks (see, e.g., Refs. /67-69/).

The knowledge on the nature of ion-induced traps (see Sect. 2.2) is still rather limited, especially at larger implantation fluences ($\gg 10^{15}$ cm^{-2} for $\sim$ keV hydrogen). However, for low fluences ($\lesssim 10^{15}$ cm^{-2} at keV energies), recent experimental and theoretical work has improved the understanding considerably (see Ref. /45/ and references therein):
One finds several types of traps due to the interaction with vacancies, possibly interstitials, and voids. Of these, the trapping at vacancies is best established by the comparison of channeling investigations /45/ and theoretical calculations obtained by means of the effective medium theory /70/. As an interesting feature, the hydrogen atom does not occupy the center of a vacancy, but is slightly shifted away due to electronic interactions. This is also corroborated by quantum-mechanical cluster calculations /71, 72/. Figure 10 compares potential contour plots for hydrogen in the interstitial solution site and at the minimum energy trapping site near a vacancy.

Trapping of implanted hydrogen atoms occurs above a certain critical temperature where short-range diffusion sets in which is necessary to reach the traps /30/. For metals, this temperature is far below room temperature. The rate of capture into traps can be written according to (c_{bt} concentration of trapped atoms)

$$\left.\frac{\partial c_{bt}}{\partial t}\right|_c = 4\pi \, r_T \, D \, c_{bs}(c_T - c_{bt}), \tag{33}$$

which is valid for low trap concentrations c_T and a diffusion controlled flux to the traps /73/. r_T is denoted as effective trap radius and is normally taken equal to the lattice constant.

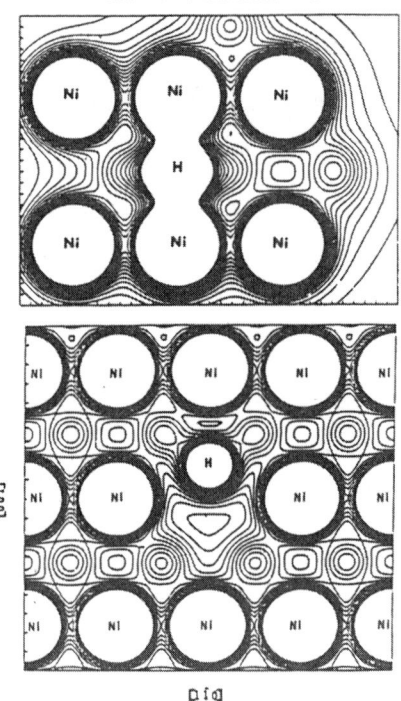

(110) PLANE CONTAINING HYDROGEN

Fig. 10: Contour plots of the electronic charge densities obtained
by means of quantum-mechanical cluster calculations, for
hydrogen in Ni dissolved interstitially at an octahedral
site (upper) and trapped at a vacancy (lower figure).
(From Refs. /71, 72/.)

Equation (33) assumes the traps to be saturable at an occupation
of one atom per trap.

At sufficiently high temperature, the trapped atoms can
escape again from the traps by thermal activation (see Fig. 5).
Assuming the release rate again to be controlled by diffusion,
one may write

$$\left.\frac{\partial c_{bt}}{\partial t}\right|_R = -\,4\pi\ r_T\ D\ n\ c_{bt}\ Z_t\ e^{-\frac{U_t}{kT}} \tag{34}$$

with a coordination number Z_t (often taken as 1). More precisely,
Z_t can be defined as the ratio of the partition functions of the
vibrational energy states of hydrogen atoms at trapping and solution sites /74/.

Trap binding enthalpies U_t have been derived experimentally to
quite high precision for different hydrogen-metal systems /30,45/.
However, some ambiguity exists in the literature for the depth

460

distribution of traps, $c_T(x)$, in the case of implantation-generated traps, due to the uncertain mechanism of their formation (see Sect. 2.2): The trap profiles have been taken as rectangular-shaped /65/ or from TRIM damage distributions /75/. Some depth-profiling experiments, however, indicate that the depth distributions of trapped atoms are rather close to the range distributions /30, 66/, which is consistent with the picture that the traps might be stabilized by the implanted atoms.

For hydrogen in nickel, table 1 gives a compilation of trap binding enthalpies typical of different kinds of traps. It should be noted that basically no information is available for trapping at very low implantation energies, e.g., below the damage threshold (Sect. 2.2). However, one might also conceive mechanisms of trap generation other than radiation damage, as for example by the distortion of the lattice by the precipitation of hydrides /49/.

For carbon and other nonmetals, basically no information is available on the role of low-concentration traps. In carbon, it is uncertain whether the low mobility of implanted hydrogen atoms is due to low diffusivity or to deep traps /9/ (see also Sect. 3.2). Generally, trapping effects will be most important for fusion reactor applications, if bulk traps were present with high binding energies. These may be generated, for example, by 14 MeV neutron irradiation damage, and contribute to excessively high tritium inventories /9/.

It is well known further than He preimplantation leads to an enhanced trapping of hydrogen /76, 77/. Although the fluxes of energetic He ions in fusion devices are expected to be rather low (10^{12} cm^{-2} s^{-1} or less /78/), the implanted He may accumulate during the lifetime of the first wall and lead to significant additional hydrogen trapping.

Table 1: Binding Enthalpies U_t of different types of traps for hydrogen in nickel. (From Ref. /45/)

Solution site	0
Substitutional impurities	$\lesssim 0.1$ eV
Dislocations	~ 0.1 eV
Interstitial impurities	$\lesssim 0.1$ eV
Self-interstitials (?)	0.24 eV
Vacancies	0.43 eV
Voids	0.52 eV

3.6 Local Saturation

In this subsection we will treat a special situation which is known only for ion-implanted systems: At low temperatures, the diffusivity or the detrapping rate may become so small that very high concentrations of the implants can be built up, largely exceeding solubility limits or even the concentrations of stoichiometric hydrides. It seems obvious that the material cannot accomodate an arbitrarily high concentration without significant modifications. In the case of gaseous ions, one observes gas release above a certain, energy-dependent fluence. The microscopic mechanisms for these saturation effects are poorly understood, nevertheless there have been several investigations with different materials which allow the phenomena to be described.

Saturation under hydrogen bombardment is probably accompanied by a highly damaged state of the material. Electron micrographs obtained after helium bombardment of metals show that the material has disintegrated into small grains, with microchannel networks arising from the interconnection of bubbles /79, 80/. This picture is likely to be applicable also for hydrogen implantations at low temperature, which is corroborated by the fact that saturation occurs simultaneously with blister formation for both hydrogen implantation at low temperature (< 100 K) and helium at room temperature in several metals /81, 82/.

Measurements of the retained amount and depth profiles of implanted hydrogen show that the saturated state is given by a maximum concentration which is nearly independent of depth (Ref. /30/ and references therein). Mixing experiments where protium and deuterium were implanted consecutively, proved to be especially valuable in establishing this effect for metals at low temperature and keV energies. Similar studies have been carried out with carbon at lower energies. The SIMS profiles of Fig. 11 /31/ show that the saturation level of D implanted into C is practically independent on depth and implantation energy.

From the above knowledge, it is straightforward to introduce a model of 'local saturation' in which the total bulk concentration, c_b, of implanted atoms is limited to a maximum value, c_M so that

$$\frac{\partial c_b(x)}{\partial t} = \begin{cases} S(x) & \text{if } c_b(x) < c_M \\ \\ 0 & \text{if } c_b(x) = c_M \end{cases} , \qquad (35)$$

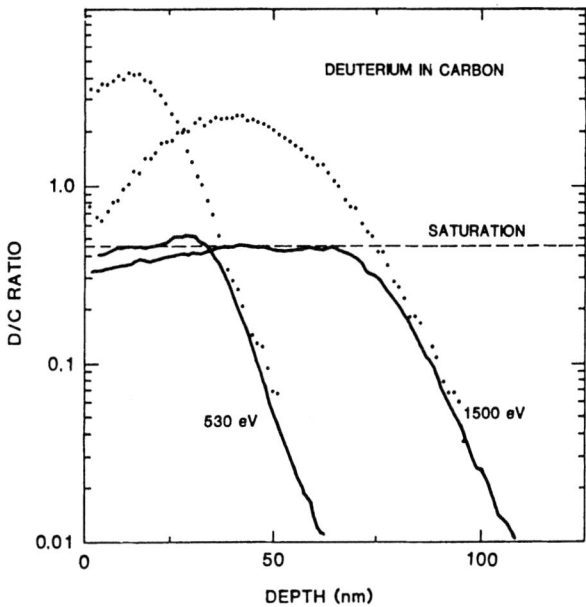

Fig. 11: Depth profiles obtained from Secondary Ion Mass Spectrometry for deuterium implanted into carbon at room temperature with fluences of 10^{18} D/cm² (solid lines). For
comparison the dotted curves show the profiles of
10^{16} D/cm² which have been scaled by a factor of 100.
(From Ref. /31/)

where $S(x)$ denotes the number of deposited particles per unit
volume and time, as given by the range distribution of the implanted atoms. This simple description of local saturation was
originally developed for hydrogen in carbon and silicon /83, 84/
and later reformulated and applied to isotope mixing in metals
/85, 86/. The model implies that in the saturated depth region
for each atom coming to rest one atom is released directly into
the vacuum, without the possibility of getting trapped in adjacent
non-saturated regions. This is qualitatively consistent with a
picture of a release through microchannels, as mentioned above.

For postimplantation of an isotope of type 2 after implantation with type 1, one obtains /86/

$$\frac{\partial c_1}{\partial t} = \begin{cases} 0 & \text{if } c_1 + c_2 < c_M \\[2ex] -\dfrac{S_2(x)\, c_1\, \gamma_r}{(\gamma_r - 1)c_1 + c_M} & \text{if } c_1 + c_2 = c_M \end{cases} \tag{36a}$$

463

and simultaneously

$$\frac{\partial c_2}{\partial t} = \begin{cases} S_2(x) & \text{if } c_1 + c_2 < c_M \\ \\ -\frac{\partial c_1}{\partial t} & \text{if } c_1 + c_2 = c_M . \end{cases} \tag{36b}$$

Here $\gamma_r = 1$ normally, but $\gamma_r \neq 1$ accounts for the possibility that the replacement probabilities for both isotopes are not symmetric (which has only been found for Pd).

Various saturation and isotope exchange experiments show the results in surprisingly good agreement with this simple picture of 'local mixing'. The maximum concentration c_M varies from 0.4 (H in C at room temperature) to $\sim$ 5 (D in Zr at 70 K) /86/. In metals, it exceeds the concentrations of the stoichiometric hydrides. There are indications, however, that a large fraction of the retained hydrogen is present in the form of these hydrides /86/.

Whereas for metals systematic investigations of the temperature dependence of these saturation effects are not available, this has been studied for TiC, and in some detail for graphite /87/. In these materials, saturation effects occur far above room temperature due to the low mobility or deep traps (see Sects. 3.2 and 3.5). To explain the observed temperature dependence of the saturation concentration, other mechanisms have to be invoked than the replacement of atoms at a fixed maximum concentration only. Competitive mechanisms can be the thermal detrapping (Sect. 3.5) or beam-enhanced transport processes, which will be briefly discussed in the following subsection.

3.7 Beam-Enhanced Processes

There are few experiments which demonstrate that some of the elementary processes described in Sects. 3.1 - 3.6 are influenced by the presence of the implantation beam, probably due its energy deposition or due to the generation of defects, which may migrate at elevated temperatures. Before readressing the case of carbon and similar materials exposing local saturation behaviour, we will first mention experiments around room temperature for which diffusion and low-concentration traps dominate /88, 89/. For example, Braganza et al. /88/ observed a strong reemission of preimplanted deuterium after switching to proton bombardment for stainless steel in a temperature regime, where thermal detrapping (eq. 34) is negligible. This indicates that during irradiation previously trapped and incident atoms may compete for trapping sites, which is in contradiction to the picture given in Sect. 3.5, which describes statically occupied traps at sufficiently low temperature.

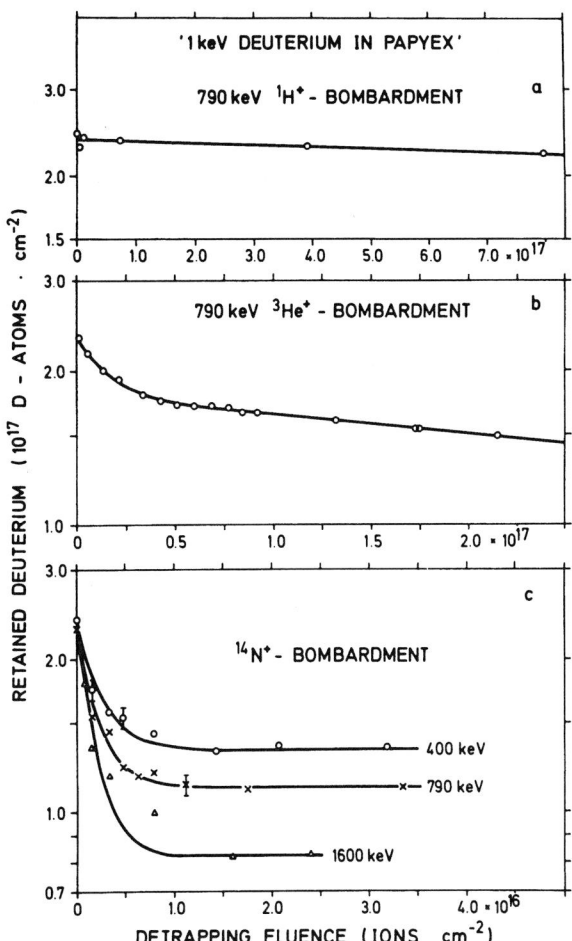

Fig. 12: Beam-enhanced release of deuterium in carbon after implantation to saturation, during bombardment with different ions and energies. (From Ref. /90/)

Another example of radiation-induced effects has been observed for carbon /90, 91/ and BeO /92/ in the local saturation regime. Figure 12 shows that deuterium is released by bombardment with ions the ranges of which considerably exceed the preimplanted layer, so that any replacement effects can be excluded.

In order to describe the above phenomena, one might introduce a cross-section of beam-induced release, σ_B, so that

$$\frac{\partial c_b}{\partial t}\bigg|_B = - j_o \cdot \sigma_B \cdot c_b , \qquad (37)$$

where j_o denotes the incident (nonreflected) ion flux. Assuming σ_B

to be independent of depth, one can then derive it by measuring the total areal density of retained hydrogen, or the amount of released hydrogen, as function of the post-irradiation fluence (see, e.g., Refs. /90/ and /88/, respectively). One might also propose a depth-dependent cross-section which is proportional, e.g., to the energy deposition of the implanted particles.

In an elaboration of the local mixing model, Brice et al. /87/ introduce a release term

$$\left.\frac{\partial c_b}{\partial t}\right|_B = - V_D \, S(x) \, c_b \qquad (38)$$

with a 'disturbed' volume per deposited particle, V_D. This model cannot explain the release induced by long-range ions mentioned before, but can be used to explain the temperature dependence of local mixing in C and TiC (see Sect. 4.3). The physical meaning of the above parameters σ_B, or V_D, is rather nebulous, and they can only be understood as fitting parameters deduced from experiments, which may be useful for model calculations in some specific cases.

It has been proposed that the diffusion of implanted hydrogen atoms is enhanced by their interaction with moving defects /93/. The existence of such a mechanism is doubtful and seems to be disproven by recent experiments /94/.

4. MODELS OF INVENTORY AND RELEASE

With the partly well-established, but partly rather rudimentary information on the basic mechanisms we can now discuss the models which describe the hydrogen inventory and release. At present, there is no universal model available which covers all these mechanisms; the existing models are thus limited to describe specific materials in certain parameter ranges.

It is also evident from the above classification that a crucial parameter is the temperature, which also controls the maximum concentration in a continuous implantation experiment. The available models are thus valid only for certain temperature regimes, which may be considerably different on an absolute scale for different materials (see table 2).

4.1 High Temperature: Diffusion and Recombination

At sufficiently high temperatures (typically above 400 K for metals except Be, probably above 1000 K for C - see table 2), diffusion becomes very fast, and trapping effects can be neglected if the concentration of traps is small. Due to the fast diffusion at elevated temperature, the maximum concentration of

Table 2. Approximate temperature limits for the regimes of local saturation and free bulk diffusion. In the intermediate regime, trapping effects have to be taken into account. The data have been estimated partly from diffusion/trapping data (see Ref. /30/ and Sect. 5.1), and partly from experimental information on Be /95, 96/ and C /87, 97/.

Material	Local Saturation Below:	Free Diffusion Above:
Be	450 K	750 K
C	650 K	1000 K ?
Ti	200 K	500 K
α-Fe	70 K	400 K
SS	220 K	400 K
Ni	160 K	380 K
Inconel	200 K	400 K

implanted hydrogen atoms will also be small, so that phase precipitation can be excluded. Neglecting any radiation-enhancement, diffusion and surface recombination remain as the only mechanisms governing the system.

4.1.1 Stationary State

For simple estimates, we consider the stationary state and assume that all nonreflected atoms are deposited according to a δ-function at $x=R$. Then, the solute concentration will decrease linearly towards both surfaces (the front or 'upstream' surface at $x=0$ and the rear or 'downstream' one at $x=L$) of the sample. A schematic representation is given in Fig. 13. The partial fluxes of implantation, reemission and permeation must then balance:

$$j_o = j_- + j_+ , \tag{39}$$

with (according to eqs. (19) and (27))

$$j_- = K_o c_o^2 = D \frac{c_R - c_o}{R} \tag{40}$$

and

$$j_+ = K_L c_L^2 = D \frac{c_R - c_L}{L-R} , \tag{41}$$

with the abbreviations $c_o = c_{bs}(0)$, $c_L = c_{bs}(L)$, and $c_R = c_{bs}(R)$, and with K_o and K_L denoting the possibly different recombination

467

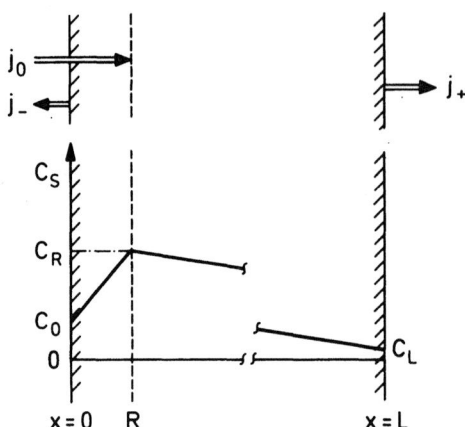

Fig. 13: Schematic plot of solute concentration versus depth and partial fluxes in the stationary case. j_0 denotes implanted flux of nonreflected atoms, j_- and j_+ the reemitted and permeating fluxes.

constants for the upstream and downstream surfaces, respectively. One can then solve eqs. (39-41) to obtain c_0, c_L and c_R. A general solution, however, cannot be obtained analytically. Nevertheless, two extreme situations are easy to cover: The release through either surface may either be controlled by diffusion or by surface recombination. If diffusion is slow compared to the recombination at both surfaces ('DD' regime), one may approximate the solute profile by a triangular-shaped one with vanishing surface concentrations (see table 3). With $R \ll L$, we obtain

$$c_R(DD) = \frac{R}{D} \, j_0 \qquad (42)$$

and from this the fractional fluxes, j_+ and j_-, listed in table 3. which are related according to the distances to the surfaces

$$\frac{j_+}{j_-} \, (DD) = \frac{R}{L-R} \, . \qquad (43)$$

In the other extreme, surface recombination limits the release through both surfaces ('RR' regime). Then the profiles can be approximated by a rectangular one with

$$c_{bs}(RR) = (\frac{j_0}{K_0 + K_L})^{1/2} \, . \qquad (44)$$

As in the DD regime, the released fluxes are proportional to the incident one, with

468

Table 3. Transport regimes for hydrogen atoms implanted into solids under the influence of diffusion and surface recombination. Approximate expressions for the stationary up- and downstream release fluxes, j_- and j_+, and the inventory I are given as derived in Sect. 4.1. Time constants estimated from diffusion theory are also included. See text for further explanation of other quantities.

Regime	DD	RD	RR
Schematic profile			
Criteria	$W_L \gg W_o \gg 1$	$W_L \gg 1 \gg W_o$	$1 \gtrsim W_L \gg W_o$
Main Release	Upstream	Upstream / Downstream ($K_L \gtrsim 10^3 K_o$)	Upstream ($K_L \ll K_o$) / Downstream ($K_L \gg K_o$)
j_-/j_o	$\cong 1$	$\cong 1$ / $\left(\dfrac{L}{D}\right)^2 K_o j_o$	$\dfrac{K_o}{K_o + K_L}$
j_+/j_o	$\dfrac{R}{L}$	$\dfrac{D}{L\sqrt{K_o j_o}}$ / $\cong 1$	$\dfrac{K_L}{K_o + K_L}$
I/j_o	$\dfrac{1}{2}\dfrac{RL}{D}$	$\dfrac{L}{2\sqrt{K_o j_o}}$ / $\dfrac{1}{2}\dfrac{L^2}{D}$	$\dfrac{L}{\sqrt{(K_o + K_L)j_o}}$
τ_-	$\dfrac{R^2}{D}$	$\dfrac{D}{K_o j_o}$	$\dfrac{L}{\sqrt{(K_o + K_L)j_o}}$
τ_+	$\dfrac{L^2}{6D}$	$\dfrac{L^2}{6D}$	$\dfrac{L}{\sqrt{(K_o + K_L)j_o}}$
τ_I	$\dfrac{L^2}{10D}$	$\dfrac{L^2}{10D}$	$\dfrac{L}{2\sqrt{(K_o + K_L)j_o}}$

$$\frac{j_+}{j_-} \ (RR) = \frac{K_L}{K_o} \ . \tag{45}$$

For the intermediate situation between the DD and RR regimes, only an approximate solution can be obtained, which however will enable us to formulate quantitative criteria for the regime limits. Let us first extrapolate from the DD regime by assuming that $j_- \cong j_o$ /6/. Then, from eqs. (39-41):

$$\bar{c}_o = (\frac{j_o}{K_o})^{1/2} \tag{46}$$

$$\bar{c}_R = (\frac{j_o}{K_o})^{1/2} \ (1 + W_o) \tag{47}$$

$$\bar{c}_L = (\frac{j_o}{K_L})^{1/2} \ \frac{1}{2W_L} \ [(1+4 \ (\frac{K_L}{K_o})^{1/2} \ W_L (1+W_o))^{1/2} - 1] \tag{48}$$

with the 'transport parameters' /8, 98, 99/

$$W_o = \frac{R}{D} \ (j_o K_o)^{1/2} \tag{49}$$

and

$$W_L = \frac{L-R}{D} \ (j_o K_L)^{1/2} \ . \tag{50}$$

L and R differ typically by 4 orders of magnitude or more. We therefore can first assume $W_L \gg W_o$. The physical significance of the transport parameters is seen as follows: If $W_o \gg 1$, then $\bar{c}_R \gg \bar{c}_o$, i.e. the release is diffusion-controlled at the upstream surface. For $W_o \ll 1$, the upstream release becomes recombination-limited. Similarly, eq. (48) can be approximated to show $\bar{c}_R \gg \bar{c}_L$ or $\bar{c}_R \cong \bar{c}_L$ for $W_L \gg 1$ or $W_L \ll 1$, respectively, for the downstream release. It should be noted that eq. (48) is consistent with eqs. (43) and (44) in the limits of the DD and RR regimes, respectively, provided that $K_L \ll K_o$ (i.e. $j_+ \ll j_o$) in the RR regime.

Of the two possible intermediate regimes, the 'DR' regime can be excluded as the conditions $W_o \gg 1$, and $W_L \ll 1$ would conflict with $W_L \gg W_o$. For the 'RD' regime, we obtain from eq. (48)

$$\bar{c}_L (RD) = [\frac{j_o}{K_L} \ \frac{1}{W_L} \ (\frac{K_L}{K_o})^{1/2}]^{1/2} \ . \tag{51}$$

In this case, the permeation flux (see table 3) becomes proportional to the square root of the incoming flux.

We might further ask if a permeation-dominated solution exists in the RD regime, which is apparently possible for $K_L \gg K_O$ in the RR regime. Assuming $j_+ \cong j_0$, a similar procedure as above leads to

$$c_o^+ = (\frac{j_o}{K_o})^{1/2} \frac{1}{2W_o} [(1+4 (\frac{K_o}{K_L})^{1/2} W_o (1+W_L))^{1/2}-1]. \qquad (52)$$

A consistent result (i.e. fulfilling $j_- \ll j_o$) is obtained for $W_L \gg 1 \gg W_o$, i.e. the RD regime, if $K_L \gg K_o$. Then

$$c_o^+ = (\frac{j_o}{K_L})^{1/2} W_L \qquad (53)$$

and

$$\frac{j_-}{j_o} = \frac{K_o}{K_L} W_L^2 \quad , \qquad (54)$$

which fulfills the assumption of $j_- \ll j_o$ only for $K_L \gtrsim 10^3 K_o$. Thus, we conclude that a permeation-dominated release is only possible for strongly different surface conditions on both sides.

Figure 14 shows the transport parameters with the different transport regimes for the case of deuterium implantation into a stainless steel foil. It is seen that the RR regime is hardly reached below the melting point, even for a very thin membrane.

In Fig. 15, a result is displayed from an experiment /63/ which verifies the conditions for the permeation-dominated RD regime. A getter layer applied to the downstream surface guarantees

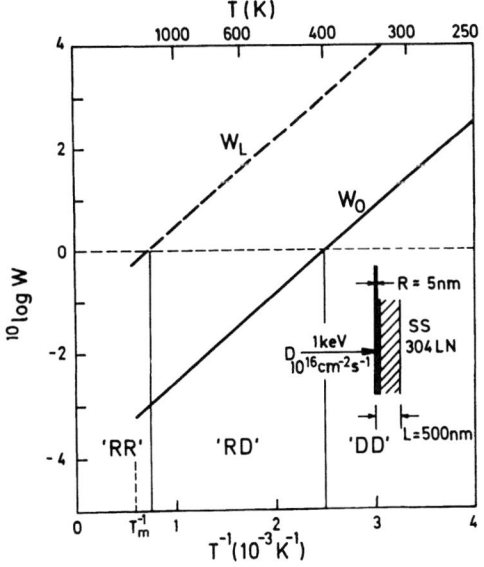

Fig. 14:

Upstream and downstream transport parameters, W_O and W_L, for deuterium implanted into a stainless steel membrane. The underlying diffusion and surface recombination data have been taken from Ref. /65/.

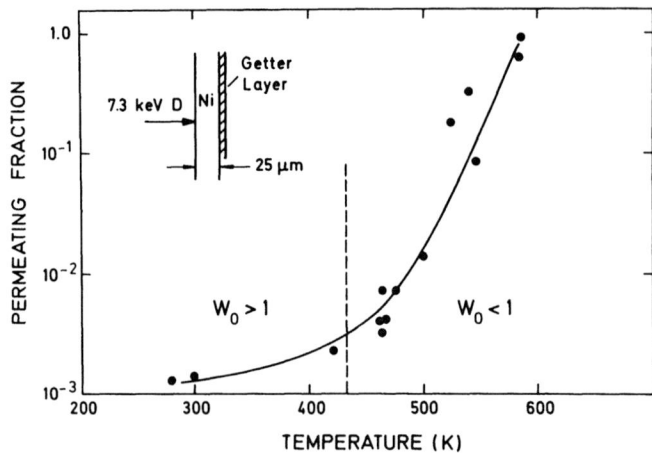

Fig. 15: Permeating fraction of implanted deuterium through a
nickel foil equipped with a downstream getter layer.
The plot is only qualitative as different ion fluxes
have been employed for the individual experimental points.
The solid line is drawn to guide the eye. The dashed line
indicating $W_0 = 1$ is valid for a flux of $1 \times 10^{15}\,D\,cm^{-2}\,s^{-1}$
and derived from diffusion and solubility data from
Refs. /53/ and /48/, using $K_0 = 10^{-6}\,K_r$ (eq. (28) - see
Fig. 9). (From Ref. /63/)

$K_L \gg K_o$, with K_o decreasing with temperature as seen in Fig. 9.
The getter layer simultaneously allows the permeating
fraction to be measured in stationary state. Whereas this fraction
is small and independent of temperature in the DD regime, it increases strongly as function of temperature in the RD regime.

A summary of the results obtained using the above simple
approximations is presented in table 3. It also contains the
corresponding approximate results for the hydrogen inventory in
stationary state given by

$$I = \int_o^L c_{bs}(x)\,dx \; . \tag{55}$$

4.1.2 Transients

The time constants characteristic for the buildup of up- and
downstream release and inventory require a more rigorous treatment
by diffusion theory (see, e.g., Refs. /99-103/, except for
the RR regime where a simple hyperbolic time-dependent solution

for the constant concentration can be obtained. For diffusion-controlled release, the time to establish half of the stationary quantity when starting from a zero concentration is approximately given by

$$t \sim \frac{d^2}{D} , \tag{56}$$

where d is a 'diffusion length' (R or L-R in the above treatment). In the RD regime, diffusion and recombinative release compete yielding an upstream time constant ('recycle time') of

$$\tau_-(RD) = \frac{D}{K_o j_o} . \tag{57}$$

It is obvious that in the DD and RD regimes, the time constants for the inventory are similar to those for the downstream release. All time constants in table 3 are understood to represent approximate values as their definition depends on the functional shape of the specific solutions.

To obtain predictions without the above simplifying assumptions or even the complete time-dependent solutions, one can either employ numerical methods or analytical approximations to solve the diffusion equation including the source term:

$$\frac{\partial c_{bs}}{\partial t} = D \frac{\partial^2 c_{bs}}{\partial x^2} + S(x), \tag{58}$$

with the boundary condition (from eqs. (19) and (27))

$$\left. \frac{\partial c_{bs}}{\partial x} \right|_{x=0} = \frac{1}{D} K_o [c_{bs}(0)]^2 \tag{59}$$

for the upstream surface and a corresponding one downstream. Analytical approximations have been treated in a series of papers by Doyle and Brice /74, 99, 102, 104, 105/ including realistic source distributions, diffusion in a thermal gradient, and the influence of kinetic orders of surface recombination different from 2. Recently, Stangeby /106/ published some approximate time-dependent solutions.

Although analytical approximations are very useful for estimates and parameter studies, it is often more convenient to employ numerical solutions. Several codes have been developed to meet the special requirements of implantation problems, as DIFFUSE /107/, PIDAT /108/, and PERI /109/. Of these, PERI does not take into account the particle implantation with realistic range distributions which are instead approximated by a δ-function at the surface; it is thus not applicable for the DD regime. It also does not treat any trapping effects (see Sect. 4.2), whilst DIFFUSE and PIDAT have included both.

473

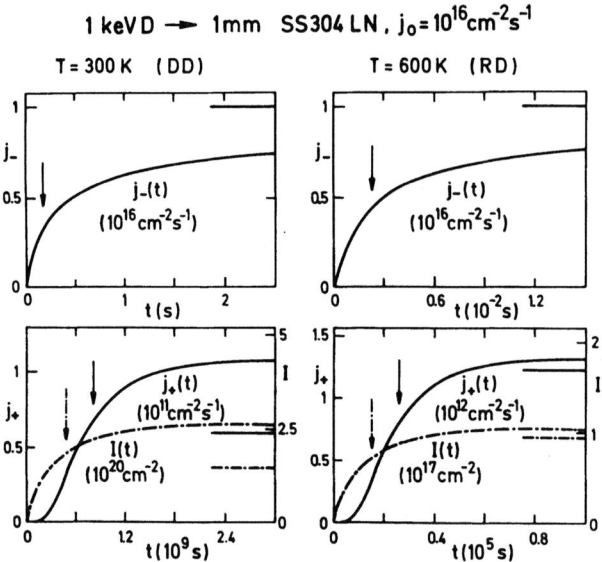

Fig. 16: Time-dependent up- and downstream release fluxes
(j_-, j_+) and inventory (I) for 1 keV deuterium implanted
into stainless steel obtained from PIDAT calculations.
The horizontal bars and arrows denote the simple
estimates for the stationary values and the time constants,
respectively, given in table 3. (A gaussian shape of
the range distribution has been assumed with a most
probable range of R = 6 nm and a standard deviation of
σ = 10 nm /20/; the analytical estimates were done for
R = 6 nm).

Figure 16 displays some results from a numerical PIDAT calculation for deuterium implanted into stainless steel, for both the
DD and the RD regimes (see Fig. 14). It is seen that the simple
estimates (table 3) for the stationary quantities and time constants agree surprisingly well with those obtained from the exact
calculations, except for the permeation flux and inventory in the
DD regime, where larger values are obtained in the numerical calculation due to the realistic source distribution inserted there.

It is also evident from Fig. 16 that the setup of the stationary situation requires unrealistically long times in many cases.
In the present example at room temperature, the downstream quantities j_+ and I do not become stationary in less than 50 years, and
even at 600 K this requires more than 1 day. Therefore, for

practical applications in a fusion device, time-dependent predictions are often needed far below the times required to reach the stationary state; they are required also for cyclic implantations. These can only be obtained numerically.

4.2 Intermediate Temperature: Trapping and Precipitation

The influence of traps becomes most obvious in an intermediate temperature range, where diffusion is still fast but the detrapping rate small. The approximate temperature limits are given in table 2 for some materials. From eqs. (33) and (34), it is seen that the capture rate increases faster than the detrapping rate at low temperature, so that the fraction of trapped atoms increases towards lower temperature. The diffusion equation has then to be extended according to

$$\frac{\partial c_{bs}}{\partial t} = D \frac{\partial^2 c_{bs}}{\partial x^2} - \frac{\partial c_{bt}}{\partial t} + S(x) \tag{60}$$

with the trapping term

$$\frac{\partial c_{bt}}{\partial t} = 4\pi r_T D \left[c_{bs}(c_T - c_{bt}) - n\, c_{bt}\, Z_t\, e^{-\frac{U_t}{kT}} \right]. \tag{61}$$

Equation (61) has first been used in Ref. /110/. Other, but equivalent formulations have been employed in several other treatments of diffusion under the influence of traps (e.g., Refs. /107, 111/).

For the stationary situation, the model calculations described in the previous subsection apply directly to the solute concentration, as $\partial c_{bt}/\partial t = 0$. Then, setting eq. (61) equal to zero, also the trapped concentration c_{bt} can be calculated, and thereby the total inventory. This has been carried out also in the approximate analytical solutions given by Brice and Doyle /74/.

Transients can also be incorporated into a simple analytical approximation in the case of small trap occupation, $c_{bt} \ll c_T$, i.e. at sufficiently high temperature. Assuming local equilibrium, which is fulfilled over a wide range of conditions /69/, eq. (61) becomes locally

$$c_{bt} = \frac{c_{bs}\, c_T}{n\, Z_t\, e^{-U_t/kT}} \tag{62}$$

By setting

$$\frac{\partial c_{bs}}{\partial t} + \frac{\partial c_{bt}}{\partial t} = \frac{\partial c_{bs}}{\partial t} \left(1 + \frac{dc_{bt}}{dc_{bs}} \right), \tag{63}$$

we obtain from eq. (60):

475

$$\frac{\partial c_{bs}}{\partial t} = \frac{D}{1 + \dfrac{c_T}{nZ_t} e^{U_t/kT}} \frac{\partial^2 c_{bs}}{\partial x^2} + \frac{S(x)}{1 + \dfrac{c_T}{nZ_t} e^{U_t/kT}} \qquad (64)$$

This is the so-called 'effective diffusion' approach /112/. It should be noted that the effective diffusion coefficient

$$D_{eff} = \frac{D}{1 + \dfrac{c_T}{nZ_t} e^{U_t/kT}} \qquad (65)$$

only enters transient phenomena; stationary fluxes are still given by the undisturbed bulk diffusion.

For nonstationary problems with higher trap occupation, numerical solutions remain as the only possible ones. The numerical codes DIFFUSE and PIDAT mentioned above include trapping effects, even for different types of simultaneously occurring traps, which can easily be taken into account according to eq. (61).

In Fig. 17, the result of such a model calculation (PIDAT) is compared to experimental permeation data obtained by the technique indicated in the insert of Fig. 15 /63/. At room temperature, pure diffusion in the DD regime would yield a permeation time constant $\tau_+ \cong 0.5$ h. Instead, $\tau_+ \cong 5$ h is observed indicating that atoms diffusing downstream first fill traps before arriving at the downstream surface. The comparison with the model calculations shows that the bulk atomic concentration of traps is $c_T/n \cong 3 \times 10^{-4}$, at a binding energy of $U_t = 0.27$ eV /113/. These traps probably stem from the cold-work of the foils. As expected, the stationary permeation flux remains unchanged in the presence of traps.

Another advantage of numerical calculations is their ability to predict time-dependent hydrogen profiles, which is especially important for traps which are generated by radiation damage and thus occur in the implanted region only. Figure 18 (a) represents an example /114/ in which the model calculation reproduces a peak of trapped atoms and a tail of diffusing ones. Similar calculations have been employed to model the thermal release after implantation and to deduce trap binding energies. (In the present example, Fig. 18 (b), two types of traps have been assumed with $U_t = 0.20$ (0.34) eV at homogeneous concentrations $c_T/n = 0.05$ (0.075) in the implanted region.)

Lowering the temperature will increase the implant concentration due to the slowing down of diffusion, and also shift the α-phase solubility limit towards lower concentrations (see, e.g.,

476

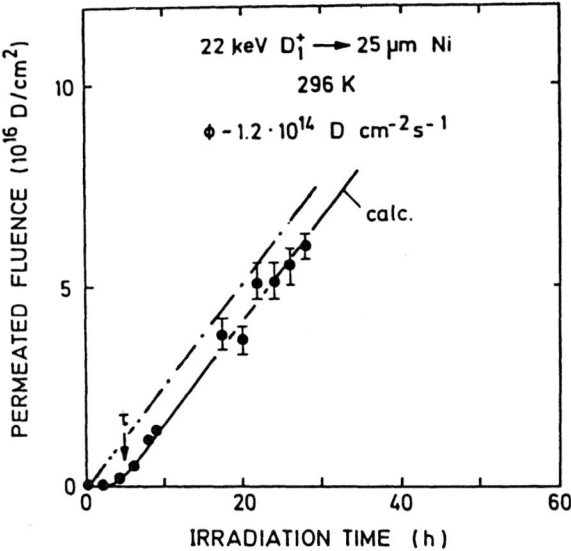

Fig. 17: Permeated fluence (integrated permeation flux) during
deuterium implantation into Ni from experiment and
numerical PIDAT calculation, assuming 3×10^{-2} % traps
with $U_t = 0.27$ eV throughout the foil. The experimental
arrangement was as for Fig. 15. The dashed-dotted line
was calculated without trapping effects. (From Ref./113/)

Fig. 6), so that precipitation of hydrides can be expected in many
cases. The formation of small hydride precipitates can also be
modelled by numerical solutions of eq. (60), employing different
trapping terms /30, 115/. However, in implantation experiments
with Ti /116, 117/ and Zr /118/ deuterium concentrations approaching the stoichiometric dihydrides have been achieved around room
temperature. These results can be satisfactorily modelled by
assuming the surface to represent an ideal sink for the diffusing
implants at which hydride formation is initiated; at increasing
fluence, the interface between hydride layer and underlying bulk is
assumed to move according to the deposition in this layer and the
arrival rate of diffusing atoms. With this boundary conditions,
PIDAT calculations /30/ were found to describe the experimental
results properly (see Fig. 19).

4.3 Low Temperature: Local Saturation

At sufficiently low temperature (see table 2), the diffusion
coefficient becomes so small that implanted hydrogen atoms may be
regarded as immobile. Then, the rate equations (35, 36) for the
local saturation and mixing model are to be integrated to calculate the local concentration as function of the incident fluence.
For the implantation with one isotope only, one obtains directly

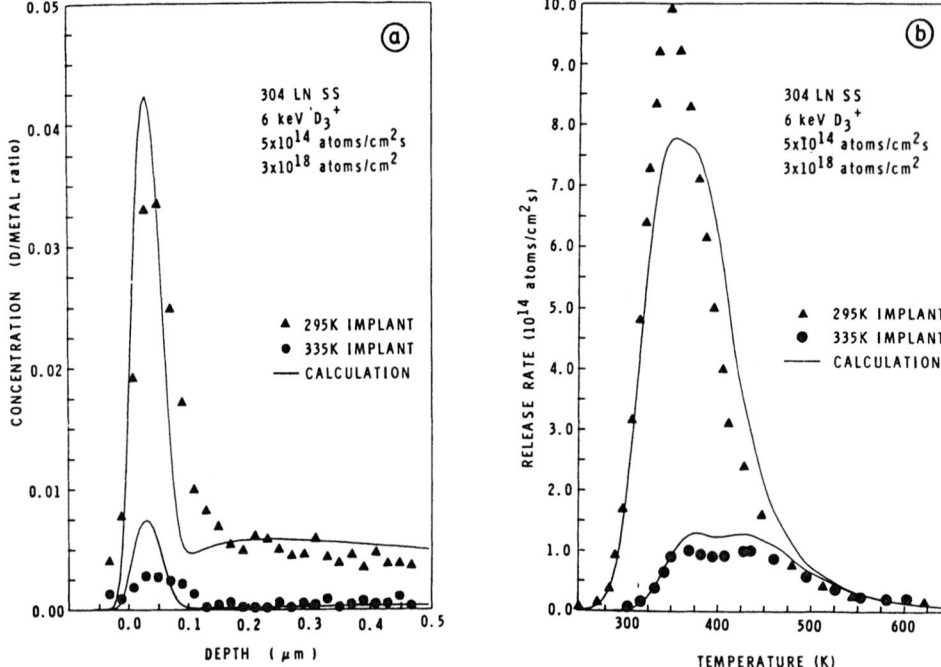

Fig. 18: Depth distributions of deuterium implanted into stainless steel at two different temperatures (a), and corresponding thermal release curves obtained during 1.2 K/s linear ramp annealing (b). The DIFFUSE numerical calculations (solid lines) have been obtained assuming two types of different traps (see text). (From Ref. /114/)

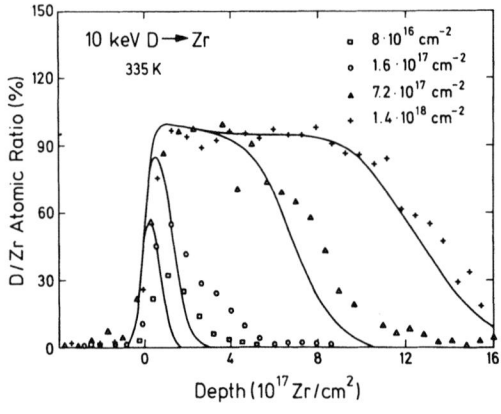

Fig. 19:

Experimental depth profiles and PIDAT numerical results (convoluted with the experimental depth resolution) for deuterium implantations into Zr around room temperature. The model calculation assumes the formation of a plane hydride layer, with a maximum atomic ratio of 100 %. (This concentration, however, is different for different types of Zr samples.) (From Ref. /30/)

$$c_b(x,t) = \begin{cases} j_o \, S(x) \cdot t & \text{if } t < c_M/(j_o S(x)) \\ c_M & \text{in saturation.} \end{cases} \tag{66}$$

Also for an isotope replacement experiment (see Sect. 3.6), where isotope 2 is implanted from t = o after prebombardment with isotope 1, a relatively simple analytical expression can be obtained /85/ for identical replacement probabilities (γ_r = 1):

$$c_1(x,t) = \begin{cases} c_1(x,0) & \text{if } t < \dfrac{c_M - c_1(x,0)}{j_o \, S_2(x)} \\[4mm] c_1(x,0) \, \exp\left(-\dfrac{j_o \, S_2(x)t + c_1(x,0) - c_M}{c_M}\right) & \text{in saturation} \end{cases} \tag{67a}$$

and

$$c_2(x,t) = \begin{cases} j_o \, S_2(x) \cdot t & \text{if } t < \dfrac{c_M - c_1(x,0)}{j_o \, S_2(x)} \\[4mm] c_M - c_1(x,t) & \text{in saturation.} \end{cases} \tag{67b}$$

Using gaussian approximations /84/ or TRIM distributions /83, 85, 119/, the above expressions have been evaluated for several cases. The depth integration required to obtain inventory and upstream release data is rather unwieldy, so that eqs. (35, 36) have been solved numerically as well /86/. An advantage of the local mixing model is the small number of additional parameters. For γ_r = 1, which has been established for most cases /86/, only c_M remains, which can even be determined experimentally in depth-profiling experiments.

For the low-temperature retention of deuterium during implantation and protium postbombardment in stainless steel, an example is given in Fig. 20 /120/. It confirms a rather good agreement of experimental data and local mixing model calculations for different implantation energies.

For carbon at room temperature, a study by Sone and McCracken /121/ indicates that traps may influence the reemission transients during buildup of local saturation. A satisfactory modelling of these phenomena is not available up to date. As a further complication, c_M has been found to depend on temperature for carbon and some other non-metallic materials /122/. Such a temperature dependence can be incorporated into the local mixing model in an empirical way; however, it would be desirable to predict a temperature-dependent saturation on physical grounds. Brice et al. /87/ have attempted this in their 'extended local mixing' model, which, in addition to the local mixing model, contains a beam-enhanced activation of deposited atoms according to eq. (38), and the possibility of thermal release. With this extension, several

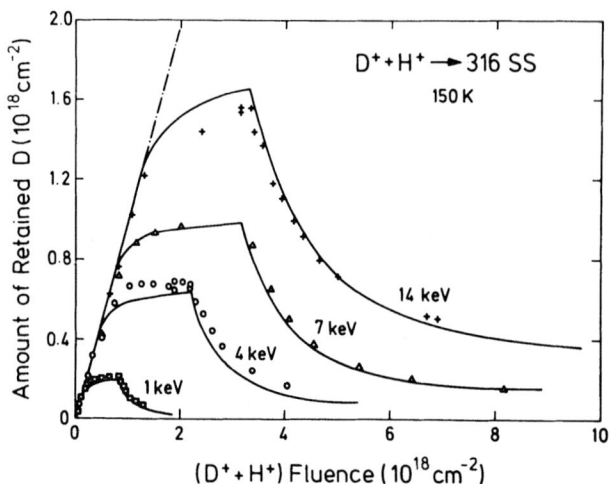

Fig. 20: A replacement experiment for hydrogen isotopes implanted
into stainless steel at low temperature. The experimental
data /120/ are compared to results from local mixing
model calculations, based on TRIM range distributions.
(From Ref. /30/)

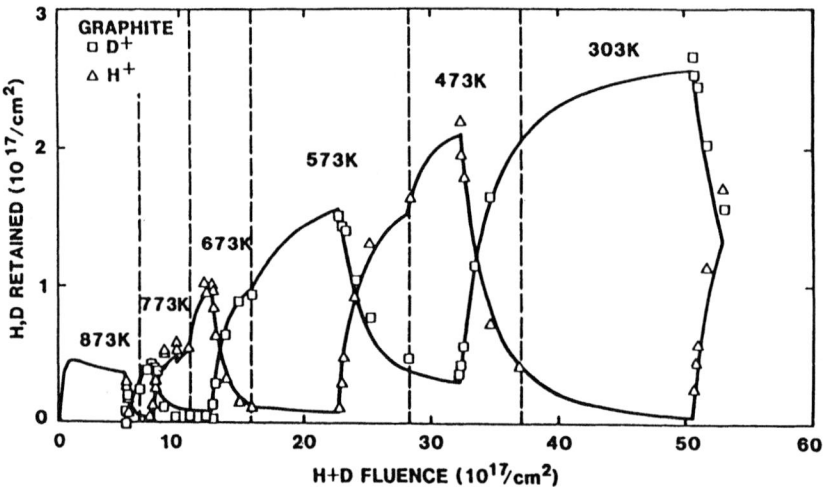

Fig. 21: Retained amounts of hydrogen isotopes in graphite during
consecutive implantations of protium and deuterium at 1.5 keV
energy. In the course of the experiment, the temperature
was decreased stepwise. Solid lines have been calculated
using the extended local mixing model. (From Ref. /87/)

additional free parameters enter the model, which can only be determined by curve fitting. An excellent fit, however, is obtained
for a temperature-dependent isotope mixing experiment with graphite
(see Fig. 21).

The local mixing model with thermal release included can be formulated within the above framework for the purpose of numerical solution. With eqs. (35, 36), and assuming a first-order thermal release, one obtains for the implantation of one species only:

$$\frac{\partial c_b}{\partial t} = - c_b \nu_o e^{-\frac{U_a}{kT}} + \begin{cases} S(x) & \text{if } c_b < c_M \\ \\ 0 & \text{in saturation} \end{cases} \quad (68)$$

and for the postimplantation with a different isotope:

$$\frac{\partial c_1}{\partial t} = - c_1 \nu_o e^{-\frac{U_a}{kT}} - \begin{cases} 0 & \text{if } c_1 + c_2 < c_M \\ \\ \frac{S_2(x) c_1 \gamma_r}{(\gamma_r - 1) c_1 + c_M} & \text{in saturation} \end{cases} \quad (69a)$$

and

$$\frac{\partial c_2}{\partial t} = - c_2 \nu_o e^{-\frac{U_a}{kT}} + \begin{cases} S_2(x) & \text{if } c_1 + c_2 < c_M \\ \\ \frac{S_2(x) c_1 \gamma_r}{(\gamma_r - 1) c_1 + c_M} & \text{in saturation} \end{cases} \quad (69b)$$

In eqs. (68, 69), ν_o denotes the attempt frequency of thermal release (in the order of the lattice vibration frequency: $\nu_o \cong 10^{13}$ s^{-1}), and U_a the activation energy of release.

5. COMPARISON OF FUSION-DEVICE MATERIALS

With the knowledge on the different processes involved in the interaction of implanted hydrogen atoms with materials, and the ability to formulate model descriptions, we will now finally address the behaviour of materials in fusion machines. One objective is to select materials the properties of which are favourable with respect to hydrogen retention and release. The main considerations are:

(i) One would like to keep the hydrogen wall inventory small, which, in the case of tritium, substantially adds to the total amount of radioactive materials being present in a fusion plant.

(ii) Similarly, the permeation to the outer wall surface should be small in order to avoid contamination of the coolant by tritium.

(iii) At the first glance, the reemission behaviour ('recycling') seems to play a minor role for the selection of materials, as any reemission coefficient smaller than 1 can be balanced by the admission of gas to the tokamak. This might, however, introduce toroidal asymmetries. Especially for the establishment of a laterally uniform high-density, low-temperature layer (the 'cold gas mantle') it may thus be desirable to establish a high recycling quickly from the beginning of the discharge.

(iv)   Some concern has to be given to the possibility that the re-emitted flux could exceed the incident one due to thermal release of previously retained hydrogen, when the wall temperature is raised during a discharge. This would cause problems associated with the density control of the plasma.

The following subsections will treat these problems on the basis of the transport parameter formalism, so that a compilation of the transport parameters for fusion-relevant conditions and several materials will first be given.

5.1 Transport Parameters

For later purposes, we shall first compare the transport parameters W_O and W_L (see 4.1.1) for different materials. Figure 22 shows a recent compilation /104/ for a mean range of R = 10 nm (corresponding to a mean energy of $\sim$ 1 keV). (A more realistic mean range for $\sim$ 100 eV mean energies changes the regime limit very little on the scale of Fig. 22.) The data are based on the para-

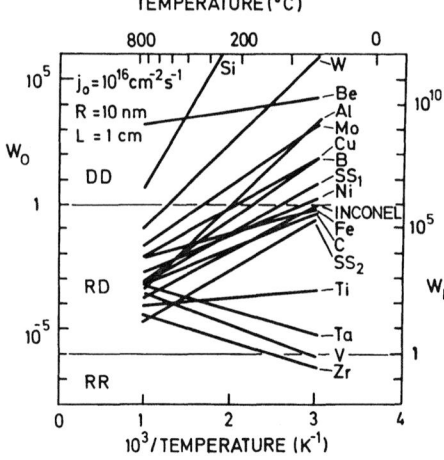

Fig. 22:

Transport parameters for different materials for hydrogen implantation at a mean depth of 10 nm into a wall of 1 cm thickness. The data are taken from Ref./104/. DD, RD and RR indicate the different transport regimes (see Sect. 4.1).

meters given in table 4 /104/. For stainless steel, two curves are shown, according to the uncertainty of surface recombination data. It is seen that for most materials the conditions of the RD regime govern the hydrogen transport at relevant temperatures of operation (200°C < T < 800°C).

482

Table 4. Parameters used for the calculation of data sets in
Figs. 22-24. Included are also the activation energy for
the Soret effect and approximate values for the trap concentrations and trap binding energies, which are only
valid for irradiation-induced traps. The pre-exponential
K_r° and activation energy U_r for surface recombination
are taken from the model of Baskes /59/ assuming a
sticking coefficient (αs = 1 (see Sect. 3.4).
Values in brackets indicate the decimal exponent. The
data are taken from ref. /104/. Literature sources are
found therein.

Mat.	n (cm^{-3})	D_o $(cm^2 s^{-1})$	U_d (eV)	K_r° $(cm^4 s^{-1} K^{\frac{1}{2}})$	U_r (eV)	U_{SE} (eV)	$\frac{c_T}{n}$	U_t (eV)
SS	8.5(22)	3.8(-3)	.56	2.1(-18)	.34	-.065	.01	.22
Fe-α	8.4(22)	7.8(-4)	.08	1.3(-15)	-.20	-.35	.01	.62
INCONEL	8.6(22)	1.7(-2)	.54	6.3(-18)	.37	-.075	.01	.34
Al	6.0(22)	2.1(-1)	.47	5.4(-18)	-.37	(0)	.04	.8
Mo	6.4(22)	4.8(-3)	.39	6.9(-16)	-.15	–	.05	.8
Cu	8.5(22)	1.1(-2)	.40	2.9(-15)	.03	–	.01	.6
Ni	9.1(22)	6.9(-3)	.41	5.4(-16)	.25	-.075	.01	.34
W	6.3(22)	4.1(-3)	.39	1.2(-16)	-.56	–	.01	.8
Ti	5.6(22)	1.8(-2)	.54	2.2(-15)	.98	.26	–	–
V	7.2(22)	5.2(-4)	.08	2.7(-14)	.68	.06	–	–
Ta	5.5(22)	5.0(-4)	.16	2.2(-14)	.72	.33	–	–
Zr	4.3(22)	4.2(-3)	.41	1.3(-14)	1.26	.26	–	–
Be	12.3(22)	3.0(-7)	.19	9.4(-11)	.19	–	.25	1.5
Si	5.0(22)	9.4(-3)	.48	9.8(-18)	-1.39	–	.5	1.5
C	11.3(22)	7.0(-3)	.50	8.5(-17)	.45	–	.4	2.35
B	13.0(22)	1.0(-3)	.85	3.4(-20)	.75	–	.45	1.15

The RR regime is never reached for wall thicknesses of a structural component, (d > 1 mm), except for the reactive metals V and Zr at comparatively low temperature. However, for the example of Zr at 50°C it would take about 50 years to establish the stationary state for the RR regime.

It should be noted that some of the parameters given in table 4 are not well established. For diffusion and solubility, this is particularly true for Be, C, and Si. Furthermore, the recombination constants and thereby the transport parameters may change strongly with different surface contamination. For stainless steel, some useful investigations on this subject are available /62, 123/ with artificial surface modifications. Knowledge of the (probably time-dependent) surface composition of real tokamak walls is lacking and is a main reason for uncertainty in this area.

5.2 Permeation and Inventory

A comparison of the stationary hydrogen isotope inventories and permeation rates for different materials is given in Fig. 23 /104/. As mostly the RD regime is of concern, one should select materials with large surface recombination coefficients to minimize the inventory, and large surface recombination coefficients combined with small diffusion coefficient to minimize the permeation rate (see table 4). This is reflected in Fig. 23.

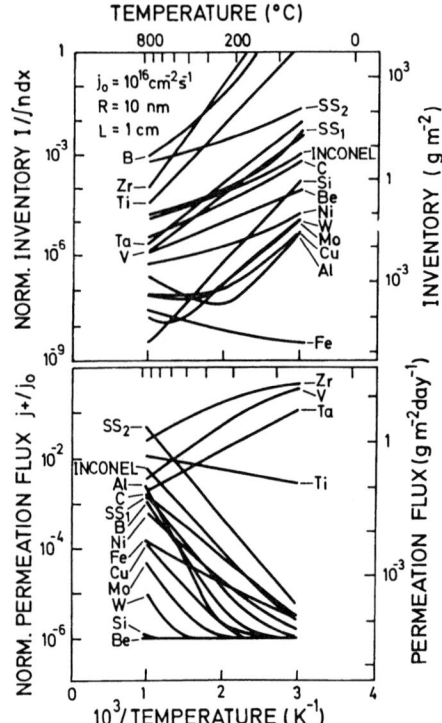

Fig. 23 :

Stationary hydrogen inventory and permeation rate for different wall materials as a function of temperature, assuming a mean implantation depth of 10 nm into a wall of 1 cm thickness /104/. The right-hand scale quantities are the total amounts retained and the mass permeation rates assuming tritium transport through a wall of 100 m² area.

484

The stationary quantities, however, must be applied with some care, as they may not be established within reasonable time intervals. For example, in a material showing high permeation and inventory like Zr the time constant to establish the stationary state at 200°C is about 10 days. Nevertheless, tritium inventory and permeation remains as a serious problem, which is evident from the right-hand scales of Fig. 23. With 1 g of T corresponding to a radioactive decay rate of 10^4 Ci, permeation rates and inventories even for the best materials are excessively high. Moreover, any trapping effects which may occur at lower temperatures are not included in the above calculations. Trapping effects are beneficial to reduce permeation in the nonstationary phase, but may add considerably to the inventory. For example, a concentration of 10^{-3} tritium atoms per wall atom being trapped at bulk traps from, e.g., neutron damage, would form an additional amount of ∿ 1 kg T per 100 m² wall area.

In view of these excessive problems associated with T transport, one has to look for additional means to reduce the hydrogen permeation and inventory. A straightforward approach would be the creation of different surface conditions at the upstream and downstream surfaces. Several possibilities are available: One could try to push the system into the DD regime with its much lower inventory and permeation rate by increasing the surface recombination coefficient. Whether this could be achieved by the deposition of special surface layers, which would have to be continuously renewed during prolonged machine operation, has not yet been investigated to our knowledge. In the DD regime, an enhanced diffusion in the implanted layer would be beneficial. Recent results for deuterium bombardment of Ni show that the permeation rate can be strongly reduced even in the DD regime, after the sample has been preirradiated with He ions to create high surface damage and blistering /67/. This finding can probably be explained by the formation of microchannels (see Sect. 3.6) which enhance the outdiffusion at the upstream surface. Whether the effect can be utilized for practical application, remains yet to be investigated.

Similarly, materials with an extremely low diffusivity could be employed in order to confine the implanted hydrogen atoms in a region close to the upstream surface. There, local saturation would establish quickly a nondiffusional release through the upstream surface. Possible candidates would be metal oxides or carbides. In carbon, for which this behaviour is also expected, the bulk porosity of technical graphites might make it less effective. Other difficulties arise from practical considerations: As the above materials are not suitable for structural components, they would have to be deposited as coatings thick enough to cover the ranges of implanted ions at the highest energies within the thermal distribution (ions penetrating through the coating would build up an enhanced inventory and permeation as the coating would present an outdiffusion barrier). Such coatings may be damaged and flake off under tokamak conditions or due to thermal fatigue (see, e.g., Refs. /124, 125/), so that practical problems might arise. Certainly, laboratory simulations should be performed to study the possible reduction of permeation under local saturation conditions at the upstream side.

Downstream permeation barriers could be employed to reduce the permeation rate /8/, however, only at the cost of increasing the inventory. Indeed it has been observed that thin oxide layers ($\sim$ 10 nm) applied to both sides of austenitic alloy membranes can reduce the gas-driven permeation rate by factors of up to 1000 or more /126, 127/.

A further possibility of enhancing the outdiffusion towards the upstream side is the establishment of a thermal gradient across the wall thickness. A reduction of inventory and permeation, however, cannot be achieved for all materials, as the Soret effect (see Sect. 3.3) may counteract the increase of the diffusion coefficient at the hot side. For stainless steel, Doyle and Brice /104/ find a reduction of permeation by one order of magnitude for a temperature gradient of $\sim$ 600 K across a wall of 1 cm thickness. Such a thermal gradient is probably difficult to establish. It requires a thermal load to the wall of $\sim$ 100 W/cm², which is realistic only for those parts of the wall, which are exposed to the plasma (e.g. limiters), and which do normally not form a structural part of the wall.

5.3 Recycling

As mentioned above, a short recycle time might be desirable for tokamak operation. Figure 24 /102/ shows the recycle time to be sufficiently small for a large number of materials. Problems

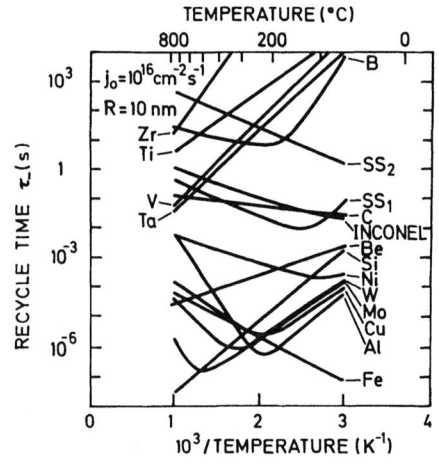

Fig. 24:

Recycle time (time at which the upstream release flux reaches 50 %) for different materials as function of temperature for a flux of 10^{16} cm^{-2} s^{-1}. (From Ref. /102/)

may arise for the reactive metals and for a 'dirty' stainless steel surface. This again confirms the necessity of detailed investigations of the influence of the surface state on the hydrogen transport behaviour.

The recycling behaviour for three typically different materials is furthermore characterized by the model calculations displayed in Fig. 25, which have been performed for the implantation at constant temperature (a), thermal release after implantation (b) and an implantation during a fast linear ramp of the temperature (c). The numerical calculations for Ti were of pure diffusional type (eqs. 58 and 59), i.e. neglecting traps. For SS, radiation-induced traps with a concentration of 4 % to a depth of $\sim$ 50 nm have been assumed for the higher energy case (a and b), whereas for the low energy linear ramp implantation (c) 4 % traps throughout the whole bulk have been inserted, which could be generated, e.g., by neutron damage. For C, the formalism of eq. (68) has been employed using five different types of traps with gaussian distribution concentrations centered around a mean activation energy $U_a = 2.85$ eV with a standard deviation $\Delta U_a = 0.54$ eV /87/, with a temperature-dependent total concentration $c_M(T)$ /122/. However, both the model and the parameters chosen for C are subject to some uncertainties.

At a bombardment temperature of 375 K (Fig. 25 (a)), 'clean' (i.e., $\alpha_s = 1$) stainless steel is found in the transition regime between diffusion- and recombination-controlled upstream release, whereas the reemission from Ti is extremely recombination-limited. C exposes local saturation behaviour. Immediate reemission and a fast establishment of the stationary state is found for stainless steel, a delayed one due to the buildup of local saturation in C, and no reemission in addition to the kinematical reflection for Ti, which, for the present fluences, is far from the stationary situation ($\tau_- = 1.3 \cdot 10^4$ s from table 3).

Starting a slow linear temperature ramp with the final depth profiles from the runs for Fig. 25 (a), stainless steel is found to release the retained deuterium promptly, whereas C and Ti retain it up to rather high temperatures, as shown in Fig. 25 (b). The total amount of retained hydrogen is larger for Ti than for C despite of the larger kinematical reflection for Ti, due to the ion-induced release for the local saturation behaviour of carbon.

Figure 25 (c) finally shows the release occurring for a linear temperature ramp during implantation, with parameters close to those expected for fusion machine operation. It also addresses the question of whether the thermal release of hydrogen implanted at lower temperature may lead to difficulties with the density control of the plasma. The SS result shows a recycling factor which

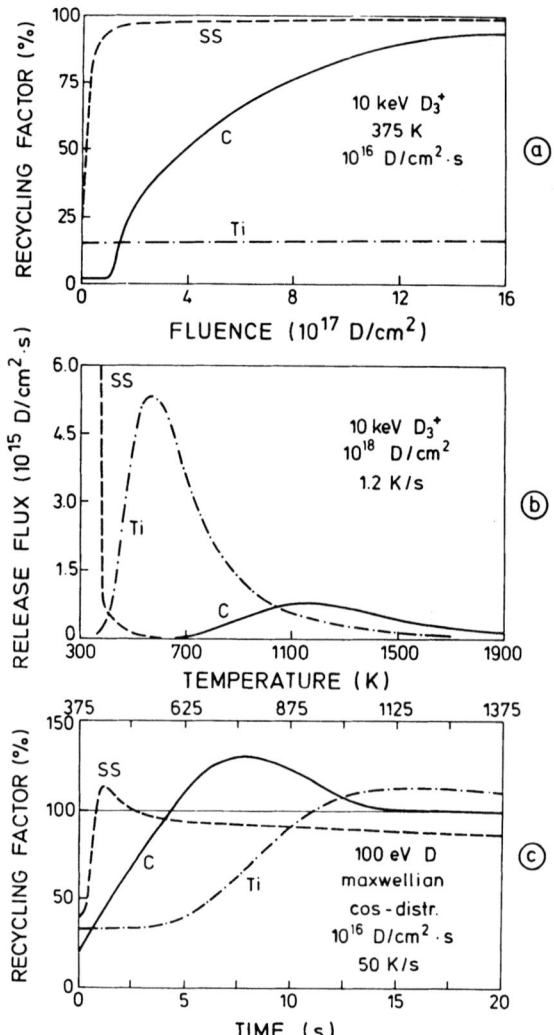

Fig. 25: Recycling behaviour for SS, C, and Ti during deuterium
bombardment under different experimental conditions:
(a) Implantation to steady-state at 375 K,
(b) Thermal release after implantation at 375 K,
(c) Implantation during a linear ramp of temperature.
The recycle coefficient accounts for both kinematical
reflection and reemission. The calculations are performed
by PIDAT-calculations according to eqs. (58, 59) for Ti,
eqs. (59-61) for stainless steel, and eq. (68) for carbon,
based on TRIM range distributions (see text for further
details).

exceeds 1 shortly after the beginning of the pulse, which is attributed to detrapping from bulk traps. Then, it keeps slightly below 100 % due to the buildup of the bulk inventory in the RD regime. Ti exposes a slow increase of recycling, with a recycling factor exceeding 100 % above 900 K. C shows the strongest excess recycling. However, according to the local saturation mechanism this is reduced considerably for higher fluxes (up to 10^{19} cm^{-2} s^{-1}), which are typical for the most exposed parts of the machine walls (e.g., limiters). The excess recycling factor from the locally saturated surfaces would then become 10^{-3} or less, meaning that the use of carbon limiters would not be in conflict with density control requirements. (This conclusion has also been drawn recently by Vernickel /128/ for specific wall and bombardment conditions.) Further, a recycling factor deviating from 100 % for wall components other than limiters would not influence the density control as well, as the integrated fluxes there are negligible compared to the fluxes to the limiters. Therefore, we can expect a tokamak vessel from, e.g., stainless steel, which is equipped with carbon limiters, not to expose any problems with respect to density control, if the mechanisms described here are the only ones for hydrogen desorption.

6. CONCLUSION

The present paper shows that the understanding of hydrogen implantation and transport phenomena in solids has increased considerably due to simulation experiments and accompanying model calculations during the past few years.

With respect to fusion reactors, it is demonstrated that the tritium inventory and its permeation represent a serious safety problem, particularly in connection with materials which are preferably envisaged at present, e.g., austenitic stainless steel or inconel. If no effective way to reduce permeation and inventory can be developed for these, other structural materials, like Mo or W to be operated at a temperature of about 900 K, might be envisaged. This, however, might conflict with other demands, like for low material activation and low impurity radiation.

Many additional investigations are needed to obtain predictions which are more relevant to the processes in a real fusion device. Specific areas to be addressed are:

(i)   Extension of experimental simulations to lower implantation energies, as most of the above findings are derived from experiments at energies above 1 keV, whereas data down to ∿ 50 eV are needed.

(ii) Experiments at higher temperatures (up to ∿ 1300 K) in order to check the concepts of inventory and permeation exposed above.

(iii) Investigation of the influence of surface coverages and compositions on the different phenomena. This also includes temporal changes due to ion-induced surface alterations by, e.g., precipitation.

(iv) Studies of the influence of bulk traps generated by neutron irradiation.

(v) Properties of realistic materials and surfaces. These could be investigated either by laboratory investigations on materials which have been exposed in a plasma device, or by direct experiments at these machines.

On-site experiments at plasma machines generally require rather difficult technical procedures, which make the unique determination of all experimental parameters difficult or even impossible. Thus, in future one will have to find a reasonable balance between such experiments and laboratory investigations to learn more about hydrogen transport and to provide reliable data for realistic materials.

ACKNOWLEDGEMENT

The authors would like to thank R. Behrisch and G.M. McCracken for their critical comments on the manuscript.

REFERENCES

/1/ R. Behrisch and W. Eckstein, these proceedings
/2/ E.S. Marmar, J.Nucl.Mat. 76&77:59 (1978)
/3/ G.M. McCracken and P. Stott, Nucl.Fus. 19:889 (1979)
/4/ R. Behrisch, in Physics of Plasmas Close to Thermonuclear Conditions, Pergamon Press, Oxford (1980)
/5/ K.L. Wilson, J.Nucl.Mat. 103/104:453 (1981)
/6/ J. Bohdansky, F. Pohl, F. Waelbroeck, R. Wienhold, and J. Winter, INTOR Phase IIA Critical Issues, European Contributions, EURFUBRU/XII-132/82EDV30, Vol. III Ch. VII 2.1 (1982)
/7/ F. Waelbroeck, J.Nucl.Mat. 111/112:185 (1982)
/8/ R.A. Kerst and W.A. Swansiger, J.Nucl.Mat. 122/123:1499 (1984)
/9/ R.A. Langley, J. Bohdansky, W. Eckstein, P. Mioduszewski, J. Roth, E. Taglauer, E.W. Thomas, H. Verbeek, and K.L. Wilson: Data Compendium for Plasma-Surface Interactions, Nucl.Fus. Special Issue (1984)
/10/ K. Sonnenberg and H. Ullmaier, J.Nucl.Mat. 103/104:859 (1981)
/11/ P. Sigmund, Rev.Roum.Phys. 17:823 (1972)
/12/ J.P. Biersack and J.F. Ziegler, Nucl.Instr.Meth. 194:93 (1982)
/13/ J. Lindhard, V. Nielsen, and M. Scharff, Kgl.D.Vid.Selsk.Mat. Fys.Medd. 36:no.10 (1968)

/14/ H.E. Schiøtt, Kgl.D.Vid.Selsk.Mat.Fys.Medd. 35:no.9 (1966)

/15/ K.B. Winterbon, Report AECL-3194 (1968), Chalk River Nat.
     Laboratories

/16/ K. Shimizu and H. Kawakatsu, Jap.J.Appl.Phys. 13:1161 (1974)

/17/ N. Matsunami, Y. Yamamura, Y. Hikawa, N. Itoh, Y. Kazumata,
     S. Miyagawa, K. Morita, R. Shimizu, and H. Tawara, Report
     IPPJ-AM-32 (1983), Institute of Plasma Physics, Nagoya
     University

/18/ U. Fano, Ann.Rev.Nucl.Sci. 13:1 (1963)

/19/ J. Lindhard and M. Scharff, Phys.Rev. 124:128 (1961)

/20/ H.H. Andersen and J.F. Ziegler, Hydrogen Stopping Power and
     Ranges in All Elements (Pergamon, New York 1977)

/21/ W. Eckstein, Report IPP 9/43, Max-Planck-Institut für
     Plasmaphysik, Garching (1983)

/22/ R. Weissmann and P. Sigmund, Rad.Eff. 19:7 (1973)

/23/ U. Littmark, G. Maderlechner, R. Behrisch, B.M.U. Scherzer,
     and M.T. Robinson, Nucl.Instr.Meth. 132:661 (1976)

/24/ S. Fedder and U. Littmark, J.Appl.Phys. 52:4259 (1981)

/25/ M.T. Robinson and I.M. Torrens, Phys.Rev. B9:5008 (1974)

/26/ W. Möller, G. Pospiech, and G. Schrieder, Nucl.Instr.Meth.
     130:265 (1975)

/27/ J.P. Biersack and L.G. Haggmark, Nucl.Instr.Meth. 174:257 (1980)

/28/ P. Børgesen, J. Bøttiger, and W. Möller, J.Appl.Phys. 49:4401
     (1978)

/29/ W. Möller and W. Eckstein, Nucl.Instr.Meth. 194:121 (1982)

/30/ W. Möller, Nucl.Instr.Meth. 209/210:773 (1983)

/31/ W. Wampler and C.W. Magee, J.Nucl.Mat.103/104:509 (1981)

/32/ J.P. Biersack and W. Eckstein, Appl.Phys. 34:73 (1984)

/33/ K.L. Merkle, in Radiation Damage in Metals, eds. N.L.Petersen
     and S.D. Harkness (Am.Soc. for Metals, Metals Park,
     Ohio 1976)

/34/ P. Lucasson, in Fundamental Aspects of Radiation Damage,
     eds. M.T. Robinson and F.W. Young (ERDA CONF-751006-P1,
     1975)

/35/ W. Schilling, J.Nucl.Mat. 72:1 (1978)

/36/ M.T. Robinson and O.S. Oen, J.Nucl.Mat. 110:147 (1982)

/37/ M.T. Robinson, in Nuclear Fusion Reactors (British Nuclear
     Energy Soc., London 1970)

/38/ R.S. Averback, R. Benedek, K.L. Merkle, J. Sprinkle, and
     L.J. Thompson, J.Nucl.Mat. 113:211 (1983)

/39/ H. Bernas and A. Traverse, Appl.Phys. Lett. 41:245 (1982)

/40/ J.W. Corbett, J.P. Karins, and T.Y. Tau, Nucl.Instr.Meth.
     182/183:457 (1981)

/41/ P.D. Townsend, in Sputtering by Particle Bombardment II,
     ed. R. Behrisch, Springer-Verlag (1983)

/42/ W.D. Wilson, C.L. Bisson, and M.I. Baskes, Phys.Rev. B24:5616
     (1981)

/43/ R.C. Birtcher, R.S. Averback, and T.H. Blewitt, J.Nucl.Mat.
     75:167 (1978)

/44/ G.J. Thomas and K.L. Wilson, J.Nucl.Mat. 76/77:332 (1978)

/45/ F. Besenbacher, S.M. Myers and J.K. Nørskov, Proc. IBMM'84,
     to be published in Nucl.Instr.Meth. B
/46/ S.T. Picraux, Nucl.Instr.Meth. 182/183:413 (1981)
/47/ W.M. Mueller, J.P. Blackledge, and G.G. Libowitz (eds.),
     Metal hydrides (Academic Press, New York 1968)
/48/ E. Fromm and E. Gebhard (eds.), Gase und Kohlenstoff in
     Metallen (Springer-Verlag, Berlin 1976)
/49/ G. Alefeld and J. Völkl (eds.), Hydrogen in Metals I+II
     (Topics in Applied Physics, Vol. 28+29; Springer-Verlag,
     Berlin 1978)
/50/ A.R. Williams, C.D. Gelatt, and V.L. Moruzzi, Phys.Rev.Lett.
     44:429 (1980)
/51/ J.K. Nørskov, Phys.Rev. B26:2875 (1982)
/52/ P.G. Shewmon, Diffusion in Solids (McGraw-Hill, New York
     1963)
/53/ J. Völkl and G. Alefeld, in Diffusion in Solids, eds.
     A.S. Nowick and J.J. Burton (Academic Press, New York
     1975)
/54/ B. Hohler and H. Schreyer, J.Phys. F12:857 (1982)
/55/ D.T. Peterson and M.F. Smith, Metall.Trans.14A:871 (1983)
/56/ M. Sugisaki, S. Mukai, K. Idemitsu, and H. Furaya,
     J.Nucl.Mat. 115:91 (1983)
/57/ G. Ehrlich, J.Chem.Phys. 31:4 (1959)
/58/ R. Ash and R.M. Barrer, Phil.Mag. 8:1197 (1959)
/59/ M.I. Baskes, J.Nucl.Mat. 92:318 (1980)
/60/ M. Pick and K. Sonnenberg, J.Nucl.Mat. (in press)
/61/ E.S. Hotston and G.M. McCracken, J.Nucl.Mat. 68:277 (1977)
/62/ S.M. Myers and W.R. Wampler, J.Nucl.Mat. 111/112:579 (1982)
/63/ P. Børgesen, B.M.U. Scherzer, and W. Möller, Nucl.Instr.Meth.
     B (in press)
/64/ D. Presinger, P. Børgesen, W. Möller, and B.M.U. Scherzer,
     to be published
/65/ K.L. Wilson and M.I. Baskes, J.Nucl.Mat. 111/112:622 (1982)
/66/ D. Presinger, PhD thesis (in preparation)
/67/ P. Børgesen, B.M.U. Scherzer, and W. Möller, J.Appl.Phys.
     (in press)
/68/ G.J. Thomas, Report SAND80-8656, Sandia Nat.Lab., Livermore
     (1980)
/69/ R.A. Oriani, Acta Metall. 18:147 (1970)
/70/ J.K. Nørskov, Phys.Rev. B26:2875 (1982)
/71/ C.F. Melius, C.L. Bisson, and W.D. Wilson, Report
     SAND78-8610, Sandia Nat.Lab., Livermore (1978)
/72/ M.I. Baskes, C.F. Melius, and W.D. Wilson, Report
     SAND80-8626, Sandia Nat.Lab., Livermore (1981)
/73/ F.S. Ham, J.Phys.Chem. Sol. 6:335 (1958)
/74/ D.K. Brice and B.L. Doyle, J.Nucl.Mat. 120:230 (1984)
/75/ F. Besenbacher, J. Bøttiger, and S.M. Myers, J.Appl.Phys.
     53:3536 (1982)
/76/ F. Besenbacher, J. Bøttiger, T. Laursen, and W. Möller,
     J.Nucl.Mat. 93/94:617 (1980)

/77/ K.L. Wilson, A.E. Pontau, L.G. Haggmark, M.I. Baskes, J. Bohdansky, and J. Roth, J.Nucl.Mat. 103/104:493 (1981)

/78/ R. Behrisch and B.M.U. Scherzer, Rad.Eff. 78:393 (1983)

/79/ G.J. Thomas and K.L. Wilson, ANS-Trans. 27:273 (1977)

/80/ W. Jäger and J. Roth, J.Nucl.Mat. 93/94:756 (1980)

/81/ W. Möller, F. Besenbacher and T. Laursen, J.Nucl.Mat. 93/94:750 (1980)

/82/ B.M.U. Scherzer, J. Ehrenberg, and R. Behrisch, Rad. Eff. 78:417 (1983)

/83/ G. Staudenmaier, J. Roth, R. Behrisch, J. Bohdansky, W. Eckstein, Ph. Staib, S. Matteson,and S.K. Erents, J.Nucl.Mat. 84:149 (1979)

/84/ S.A. Cohen and G.M. McCracken, J.Nucl.Mat. 84:157 (1979)

/85/ B.L. Doyle, W.R. Wampler, D.K. Brice, and S.T. Picraux, J.Nucl.Mat. 93/94:551 (1980)

/86/ W. Möller, F. Besenbacher and J. Bøttiger, Appl.Phys. A27:19 (1982)

/87/ D.K. Brice, B.L. Doyle, and W.R. Wampler, J.Nucl.Mat. 111/112:598 (1982)

/88/ C. Braganza, S.K. Erents, E.S. Hotston, and G.M. McCracken, J.Nucl.Mat. 76/77:298 (1978)

/89/ R. Yamada, K. Nakamura, K. Sone, and M. Saidoh, J.Nucl.Mat. 101:100 (1981)

/90/ J. Roth, B.M.U. Scherzer, R.S. Blewer, D.K. Brice, S.T.Picraux, and W.R. Wampler, J.Nucl.Mat. 93/94:601 (1980)

/91/ W.R. Wampler and S.M. Myers, J.Nucl.Mat. 111/112:616 (1982)

/92/ B.M.U. Scherzer, R.S. Blewer, R. Behrisch, R. Schulz, J. Roth, J. Borders, and R.A. Langley, J.Nucl.Mat. 85/86:1025 (1979)

/93/ A.E. Gorodetsky, A.P. Zakharov, V.M. Sharapov, and V.Kh. Alimov, J.Nucl.Mat. 93/94:588 (1980)

/94/ B.M.U. Scherzer and P. Børgesen, Rad.Eff. 76:169 (1983)

/95/ W.R. Wampler, J.Nucl.Mat. 122/123:1598 (1984)

/96/ W. Möller, B.M.U. Scherzer, and P. Børgesen, to be published

/97/ B.M.U. Scherzer, R.A. Langley, W. Möller, J. Roth, and R. Schulz, Nucl.Instr.Meth. 194:497 (1982)

/98/ I. Ali-Khan, K.J. Dietz, F.G. Waelbroeck, and P. Wienhold, J.Nucl.Mat. 76/77:337 (1978)

/99/ B.L. Doyle, J.Nucl.Mat. 111/112:628 (1982)

/100/ W. Jost: Diffusion (Academic Press, New York 1952)

/101/ K. Erents and G.M. McCracken, Brit.J.Appl. Phys. (J.Phys. D) 2:1397 (1969)

/102/ B.L. Doyle and D.K. Brice, Proc. of the Int. Workshop on 'Synergistic Effects in Surface Phenomena Related to Plasma-Wall Interactions', Nagoya, Japan, May 1984

/103/ J. Crank, The Mathematics of Diffusion, Oxford University Press (1975)

/104/ B.L. Doyle and D.K. Brice, J.Nucl.Mat. 122/123:1523 (1984)

/105/ D.K. Brice, J.Nucl.Mat. 122/123:1531 (1984)
/106/ P.C. Stangeby, J.Nucl.Mat. 126:190 (1984)
/107/ M.I. Baskes, Sandia Nat. Laboratories, Report SAND83-8231
      (1983)
/108/ W. Möller, Max-Planck-Institut für Plasmaphysik, Garching,
      Report IPP 9/44 (1983)
/109/ P. Wienhold, M. Profant, F. Waelbroeck, and J. Winter,
      J.Nucl.Mat. 93/94:866 (1980)
/110/ S.M. Myers, S.T. Picraux and R.E. Stoltz, J.Appl.Phys.
      50:5710 (1979)
/111/ A. McNabb and P.K. Foster, Trans.Metall.Soc. AIME
      227:618 (1963)
/112/ H.H. Johnson and R.W. Lin, in: Hydrogen Effects in Metals
      (AIME 1981)
/113/ P. Børgesen, B.M.U. Scherzer, and W. Möller, Nucl.Instr.
      Meth. B (in press)
/114/ J. Bohdansky, K.L. Wilson, A.E. Pontau, L.G. Haggmark,
      and M.I. Baskes, J.Nucl.Mat. 93/94:594 (1980)
/115/ S.M. Myers and H.J. Rack, J.Appl.Phys. 49:3246 (1978)
/116/ J. Roth, W. Eckstein, and J. Bohdansky, Rad.Eff. 48:231 (1980)
/117/ A.E. Pontau, K.L. Wilson, F. Greulich, and L.G. Haggmark,
      J.Nucl.Mat. 91:343 (1980)
/118/ W. Möller and J. Bøttiger, J.Nucl.Mat. 88:95 (1980)
/119/ W. Eckstein, Report IPP 9/33, Max-Planck-Institut für Plasma-
      physik, Garching (1980)
/120/ R.S. Blewer, R. Behrisch, B.M.U. Scherzer and R. Schulz,
      J.Nucl.Mat. 76/77:305 (1978)
/121/ K. Sone and G.M. McCracken, J.Nucl.Mat. 111/112:606 (1982)
/122/ B.L. Doyle, W.R. Wampler, and D.K. Brice, J.Nucl.Mat.
      103/104:513 (1981)
/123/ R.A. Causey, R.A. Kerst, and B.E. Mills, J.Nucl.Mat.
      122/123:1547 (1984)
/124/ R. Yamada, K. Nakamura, M. Saidoh, and Y. Murakami,
      J.Nucl.Mat. 111/112:856 (1982)
/125/ Y. Hirohata, S. Adachi, S. Fukuda, M. Mohri, T. Yamashina,
      N. Noda, S. Tanehashi, J. Fujita, Y. Gomay, J.Nucl.Mat.
      122/123:1160 (1984)
/126/ D. Stöver, H.P. Buchkremer, R. Hecker, and H.J. Leyers,
      J.Nucl.Mat. 122/123:1541 (1984)
/127/ W.A. Swansiger and R. Bastasz, J.Nucl.Mat. 85/86:335 (1979)
/128/ H. Vernickel, J.Nucl.Mat. 128/129:708 (1984)

SURFACE EROSION BY ELECTRICAL ARCS

R. Behrisch

Max-Planck-Institut für Plasmaphysik, EURATOM Association

D-8046 Garching/München, Federal Republic of Germany

ABSTRACT

On the vessel walls of high temperature plasma experiments traces of cathode spots from electrical arcs have been observed. They originate from short metal plasma discharges which can be ignited at the walls due to the Langmuir sheath potential in front of the walls and thus the hydrogen plasma acts as the anode. The arcs may also be caused by electrical potentials which are created by gradients in the plasma, by plasma motion or during disruptive plasma phases. The material eroded from the walls at the cathode spots contributes to impurity introduction into the hydrogen plasma. Further arcing is a critical issue in high intensity ion sources and in RF transmitters and antennas for plasma heating.

1. INTRODUCTION

Surface erosion and impurity introduction due to electrical arcs at the solid walls in high temperature plasma experiments have been a matter of major concern already very early in fusion research in the pinch experiments ZETA and T-2 /1, 2/. In tokamaks the observation of arc tracks on the first wall has first been reported for PULSATOR /3/ and arcing phenomena have been explored since in some detail at several fusion experiments /4-34/ The ignition of such arcs could be largely reduced in today's tokamaks by careful discharge cleaning of the limiters and vessel walls /13, 17/. However, arcs are still observed during start-up and shut-down and during unstable plasma phases or during additional heating, and they may get again of importance for an ignited d,t-plasma.

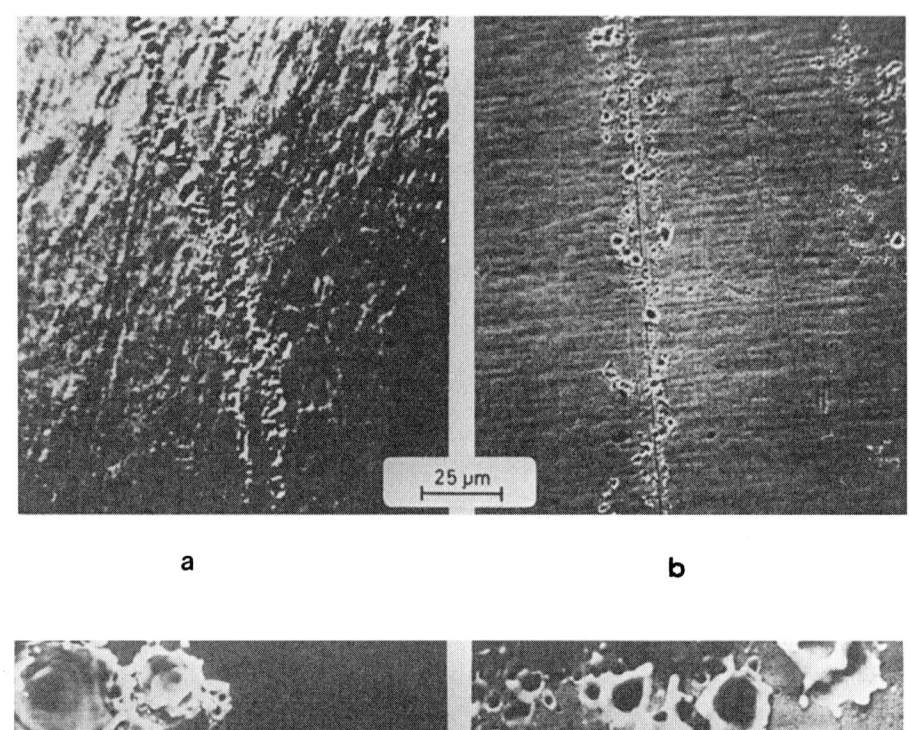

a   b

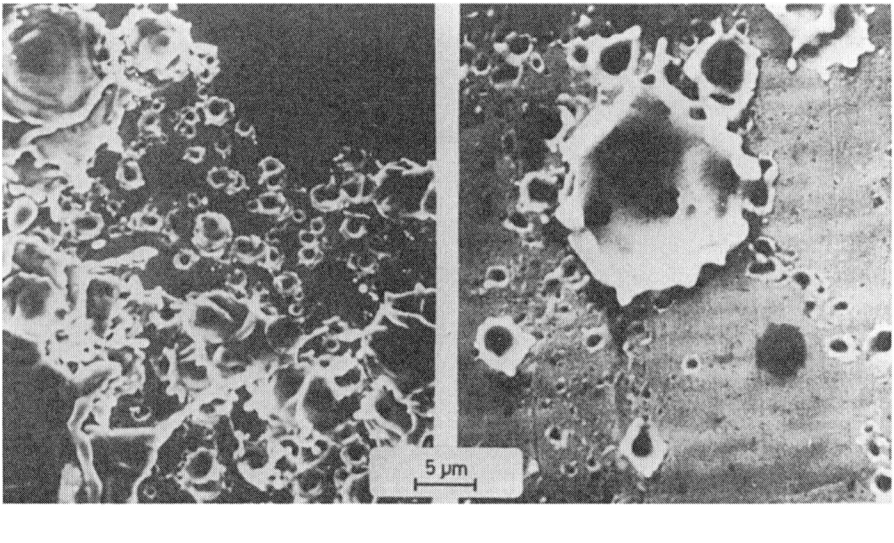

c   d

Fig. 1: Cathode spots from electrical arcs on stainless steel 316,
         a) and c) found on the first wall of tokamak TFR 600,
         b) and d) on the cathode of a vacuum arc with 11.6 A /44/.

496

In high temperature plasma experiments the arcs mostly burn
between cathode spots on the vessel walls which supply also the
metal atoms for the arc plasma and the hydrogen plasma acting as
an anode. They are named unipolar arcs because generally only one
solid electrode is present. However, arcing has also been found
between adjacent wall areas. The electromotive force for the
initiation and the burn of such arcs can be the Langmuir sheath
potential between the plasma and the walls /35-40/ or electrical
potentials induced by plasma gradients, plasma motion or plasma
disruptions /23, 24, 39-43/.

Electrical arcs always ignite if the electrical field at an
area of the vessel surface is sufficiently high, i.e. larger than
a critical value which depends on the surface conditions /46/. The
arcs are a genuine phenomenon in plasma surface interactions for
high temperature dense plasmas and can be regarded as an instability leading to a local concentration in the interaction of the
plasma with the surfaces of the vessel /37-40/.

The cathode phenomena of unipolar arcs are very similar to
those of vacuum arcs (Fig. 1) /44/. Thus in the following today's
knowledge about such cathode spots, their initiation, movement and
surface erosion is reviewed. Further ignition mechanisms and the
burn conditions of unipolar arcs are described and their occurrence
in plasma experiments is discussed.

2. CATHODE PHENOMENA IN ELECTRICAL ARCS

A vacuum arc is a very short and concentrated metal plasma
discharge between two electrodes in a vacuum. At the negative
electrode electrons are emitted from a cathode spot which is typically a few μm in diameter. Desorption and evaporation from this
cathode spot provides further the atoms and ions for the plasma
which forms the channel for the arc current of typically 10 to
100 A. The burning conditions of such a cathode spot plasma have
been investigated in detail /38, 40, 46-49/ and existance diagrams
have been calculated. The anode has mostly a passive role in
vacuum arcs, however, desorption at the anode due to electron bombardment may also contribute to arc initiation.

2.1 Arc Initiation

The initiation of a cathode spot leading to a vacuum arc has
been the subject of very numerous publications, especially in respect to prevent breakdown in vacuum. However, the details are
still not fully understood /16, 45, 46, 50-57/.

For igniting an arc between two clean metal surfaces in high
vacuum generally voltages in the 10 keV range are needed. Field
electron emission occurs at the cathode at electric fields of
about $5 \cdot 10^7$ V/cm /40/. Such fields may be reached at the tips of
protrusions which are always present. The emitting tips are subsequently heated by Joule heating and the Nottingham effect. The
electron emission current is increased by thermal field emission
and will finally result in an explosion of the tip, releasing a
sufficient amount of metal atoms to build up the arc plasma /45-50,
52/. During the burn of the arc mainly the voltage of the cathode
drop, i.e. about 15 to 25 V are necessary. Generally vacuum arcs
are started by touching the electrodes and subsequent separation
or by injecting a pulse of dense plasma.

At contaminated surfaces arcs can ignite at a much lower
voltage, even as low as the burning voltage. Such surfaces can have
a considerably reduced workfunction, which depends on the degree
of coverage, the contaminating material and its binding to the surface /58/. Electron emission can occur at much lower electric fields
than for clean surfaces, and atoms are provided by desorption to
form the arc plasma /53-55/. Such contaminated surfaces get generally cleaned by igniting arcs. In spite of the surface becoming
rough, the ignition probability for further arcs at the same
applied external voltage is decreased, i.e. the surfaces get conditioned /57, 58/.

2.2 Burning of Arcs

The development of a cathode spot and the burning of an arc is
shown schematically in Fig. 2 /59/. After ignition the current
channel of the metal plasma arc contracts by the self magnetic
field yielding current densities of the order of 10^8 A/cm² /60, 61/.
During the burning of the arc the metal plasma has a density up to
10^{18}/cm² and electron temperatures of a few eV. The energy balance
at the cathode spot is given by ion impact heating from the arc
plasma and by Joule heating due to the concentric electron current
in the solid toward the cathode spot. The arc spot is cooled by
melting and extrusion of the melt layer, by heat conduction to the
bulk of the solid, by evaporation and due to the inversion of the
Nottingham effect at higher tempratures /45-47, 56, 57/. Generally
ion impact heating and melt layer extrusion are likely the dominant
effects /61/.

Due to the loss of cathode material the arc spot diameter
grows and the current density decreases until the cathode spot
heating is no more sufficient. The arc extinguishes typically
after 10 to a few 100 ns and starts at a new, more favourable site
as on the rim of the old crater or, for contaminated surfaces, at
another area with very reduced work function. Thus the arc spot
moves randomly on the surface of the cathode /62/.

498

2.3 Arc Spot Movement

If an external magnetic field is applied normal to the arc current the random movement of the arc becomes directed normal to the $\vec{B}$ field. However, the vacuum arc moves opposite to the Lorentz

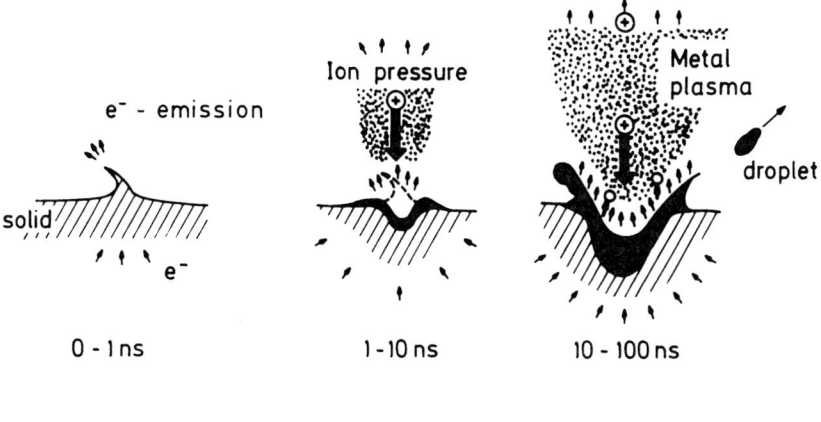

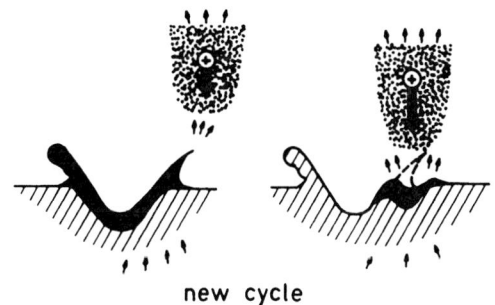

new cycle

Fig. 2: Schematic of the cathode phenomena of an electrical arc /59/.

force, $\vec{J} \times \vec{B}$, which acts on the current channel, i.e. the arc moves in the retrograde direction. Several mechanisms have been proposed for the retrograde motion. A plausible explanation is that the direction is determined by the movement of the cathode arc spot rather than by the plasma /63/. The overlap between the external magnetic field and the magnetic field of the arc current result in an increase of the magnetic field on that side from where the Lorentz force acts on the plasma channel and a decrease of the magnetic field on the opposite side. The density of the arc plasma is increased on the higher field side and decreased on the lower field side. As the probability for starting a new cathode spot is higher for a higher plasma density, new cathode spots always ignite on the high field side and thus the arc spot moves opposite to the Lorentz force /63/.

The velocity of the arc spot movement depends on the time the arc remains at one cathode spot and the displacement distance to the new spot /62/. For clean metal surfaces the burn time on one spot is in the 100 ns range, the crater radius is about 5 μm, and a new cathode spot generally ignites on the rim of the previous crater. Thus the arc velocity can be estimated to be about 10 m/s which is in agreement with measurements /65/. For high cathode temperatures the crater diameter and the burn time per cathode spot increase and the velocity of the arc decreases. Figure 3 shows traces of such arc spots on Inconel at 500° to 1000°C. For contaminated surfaces the arc spots are generally smaller and the arcs preferentially ignite at some distance from the previous spot. This gives larger arc velocities, i.e. in the 100 m/s range /65-68/.

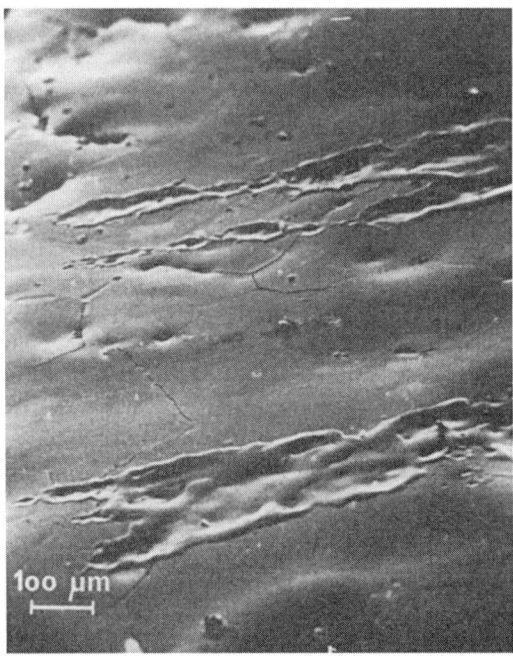

Fig. 3:

Traces of cathode spots on Inconel at temperatures between 500° and 1000°C. The SEM picture was taken at a below protection plate of the Joint European Torus (JET, 1984).

500

2.4 Cathode Erosion

The electrical arc causes an erosion of the cathode surface.
The material is released from the cathode spot predominantly in the
form of ions and of small particles /69-78/. The ions originate
from the arc plasma. They may have entered the plasma by evaporation or sputtering from the cathode spot. Part of the ions are
doubly charged, their energies ly in the eV to 100 eV range.
Generally the ion current is about 5 % to 10 % of the electron
current in the arc /70-74/.

The particles emitted from the cathode spot are predominantly
small metal droplets with 1 μm to 30 μm diameter. They are emitted
mostly tangentially to the surface, i.e. at an angle of 70° to 80°
to the normal /75/ and velocities in the 100 m/s range have been
measured /74/. Such metal droplets are observed in the form of
splashes on collectors positioned around the arc spot /74-76, 78/.
For Carbon cathodes a melt layer cannot form on the surface, however, part of the material is likely released as small carbon
pieces. Such arc traces are much more difficult to identify than for
metals (Fig. 4) /77/.

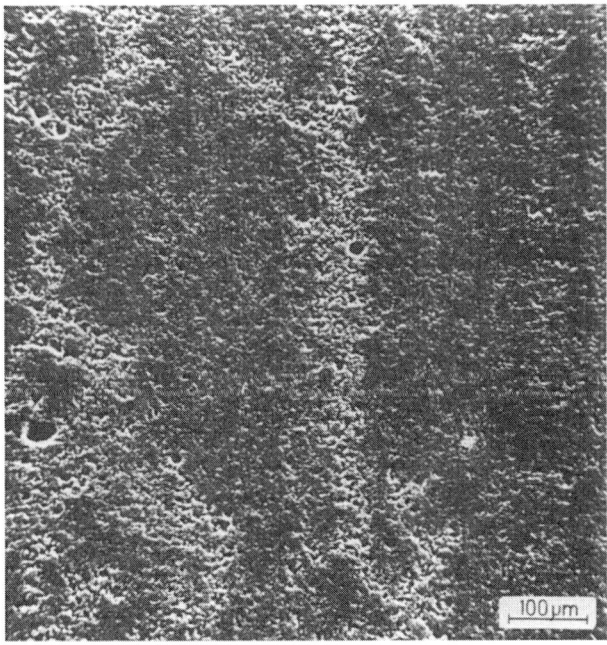

Fig. 4: Traces of cathode spots on the surface of Poco-Graphite
AXF-5Q after being sujected to vacuum arcs with a current
of 8,75 A at room temperature /77/.

501

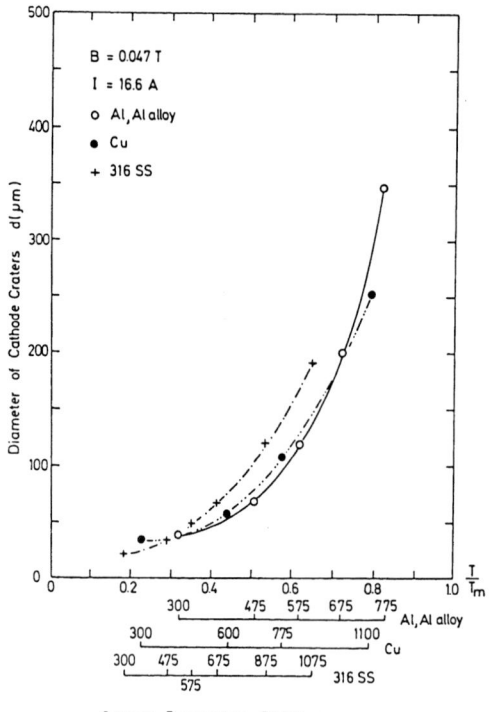

Fig. 5:

Average cathode diameter found on different metals and its dependence on the cathode temperature /66/.

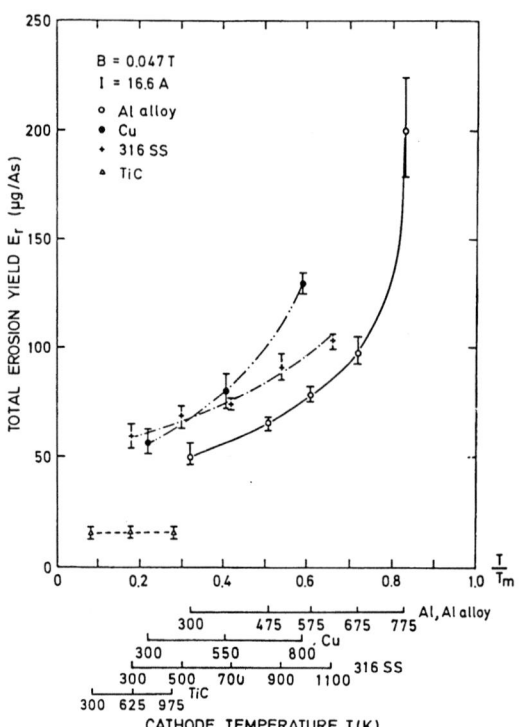

Fig. 6:

Temperature dependence of the erosion yields at the cathode for different metals /78/.

There are only few investigations about cathode erosion at elevated target temperatures. They have shown that for temperatures, T, larger than half the melting point T_m, i.e. $T \gtrsim 0.5\ T_m$ parallel to the increase of the size of the cathode spots (Fig.5), the erosion yields (Fig.6) and the size of the released droplets increase /64, 78, 79/.

Total cathode erosion yields due to vacuum arcs have been measured by different authors /72-79/. The values for fusion relevant materials are summarized in Table 1. At low temperatures the eroded material is predominantly in the form of ions corresponding to erosion yields of 0.05 to 0.1 atoms/electron. While at higher temperatures erosion in the form of particles or droplets dominates.

Table 1: Cathode erosion by electr. arcs for fusion relevant materials:

Ele-ment	Mass number	Erosion yield	
		μg/C Room temp. (heated)	atoms/electron Room temp. (heated)
Be	9.01	10-15	0.1 - 0.15
C	12.01	13-17 (5-25)	0.1 - 0.14 (0.04-0.2)
Al	26.98	22-25 (50-200)	0.08 - 0.09 (0.17-0.72)
Ti	47.9	45-52	0.09 - 0.11
Cr	52	22-40	0.04 - 0.07
Fe	55.85	40-50 (60-100)	0.07 - 0.9 (0.1-0.17)
Ni	58.71	44-50	0.07 - 0.08
Cu	63.54	115-130 (50-150)	0.18 - 0.2 (0.08-0.23)
Mo	95.94	47-55	0.05 - 0.06
W	183.85	62-90	0.03 - 0.05

3. UNIPOLAR ARCS

If a solid surface is brought in contact with a plasma a Langmuir sheath potential builds up (Fig. 7) /35/. For a hydrogen plasma this is of the order of 3 kT_e, /see chapters by Stangeby and by Chodura/. Typical electron temperatures in the boundary plasma of fusion experiments are in the 10 to 100 eV range giving sheath potentials of 30 to 300 V. These voltages are generally too low to ignite an arc, except for contaminated surfaces. They can, however, be sufficient to keep the arc burning, when it is ignited. After ignition the Langmuir sheath potential will drop around the cathode spot and more electrons can hit the surrounding surfaces. During

the burning the electron current circulates as indicated in
Fig. 7. As the driving force for such an unipolar arc is the
Langmuir sheath potential, the probe must be sufficiently large
to provide the necessary surface area for the current backflow.

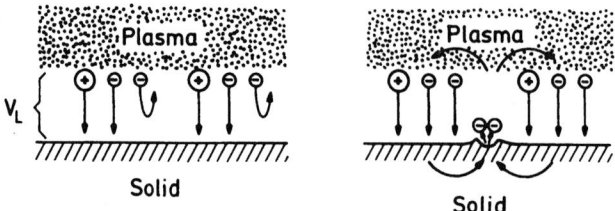

Fig. 7:   Schematic picture of the Langmuir sheath (stable plasma
          surface transition) and of an electrical arc in the
          Langmuir sheath (instability in the plasma surface
          transition) /35/.

Unipolar arcs have been shown to exist in laboratory experiments with RF-plasma discharges /35, 36/. Unipolar arcs are also
observed if a plasma cloud hits a surface /80/ or if a plasma is
created in front of a surface by laser irradiation /81/.

4.   ARCING IN FUSION DEVICES

Traces of cathode spots from electrical arcs are observed on
nearly all solid surfaces exposed to the hot plasma in fusion experiments, especially on limiters, divertor plates, protection
plates and probes inserted into the boundary plasma /1-34/. The ignition of the arcs, the current flow during the burning of the arc
and their contribution to impurities in the plasma, have been investigated at most tokamaks /5-8, 10, 12-24, 27-29, 33, 34/, however, still several questions are open.

4.1 Arcing during a Plasma Discharge

Time resolved measurements of the limiter released impurities
during the start of a plasma discharge have been performed already
10 years ago. They show maxima in the impurity lines of the limiter
material, when magnetic surfaces with rational values for the safety
factor q touched the limiter. The processes for the increased
erosion have not been identified /82/.

504

More detailed investigations have been performed with probes inserted into the boundary plasma in order to diagnose the ignition of arcs /6-8/. For identifying the arcs, the current to the probe as well as the optical emission from probe material in front of the probe was measured together with the loop voltage. Electrical arcs are found to occur predominantly during the rise of the plasma current after the start of the discharge (Fig. 8). Arcs occur further during plasma instabilities and during the current ramp down /6-10, 12, 14/.

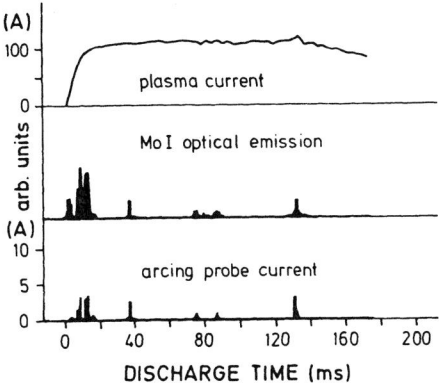

Fig. 8: Arcing as a function of time in a DITE discharge. Arcing is indicated by the correlation between the current flowing between two concentric cylinders in a probe and the optical emission of neutral Mo lines also from the probe Typical arcing periods at the beginning of the discharge and at a minor disruption near the end of the discharge.

Measurements with better time resolution confirmed the early observations that the peaks in the optical radiation of limiter material, which indicate arcing during the current built, always occur when magnetic surfaces with rational values for the safety factor q touch the limiter. They are further accompanied by spikes in the loop voltage (Fig. 9) /9, 10, 83, 84/.

Similar investigations with an arcing- and a Langmuir probe have been performed in the divertor tokamak ASDEX. Arcing at the divertor plates was found primarily during neutral beam injection. During the current rise arcing could not be observed /33, 34/.

505

In arcing investigations at the TM1 tokamak in Prague the surfaces of the arcing probe had been examined after exposure. For extremely clean probe surfaces the measured current spikes to the probe during the plasma discharges could not be correlated with arc spots on the probe surface. However for slightly contamined surfaces arc traces were found /15, 16/.

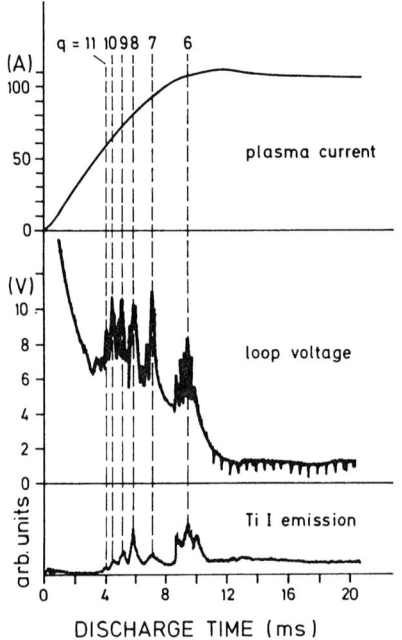

Fig. 9:

Correlation between the occurence of arcing, indicated by the optical emission of Ti (481 Å), the spikes in the loop volts and the rational q surfaces. A 110 kA discharge in DITE /9, 10/.

4.2 Arc Ignition

The ignition of the arc on the limiters or divertor plates has been investigated mostly at smaller tokamaks showing that different effects may contribute.

Measurements at the tokamak JFT-2 /20, 21/ show that the ignition of arcs is preceeded by run-away bombardment of the limiter as indicated by X-ray emission. When run-away electrons hit the limiter, it may become charged to a sufficient high negative voltage to ignite the arc. The ignition was identified by the current measured to the probe, by optical emission and by a subsequent observation of the surface topography. Cathode traces have generally been found on both the ion drift and the electron drift side of the limiter. However, if two electrically insulated plates are used for the electron- and ion drift sides, the cathode traces are only seen on the electron drift side, emphasising the role of the run-away electrons.

Measurements performed at the divertor plates of ASDEX had shown that arcing occurs predominantly during neutral beam injection, where the plasma is slightly unstable /33, 34/. Before the arc ignites the Langmuir probe situated close to the arcing probe show a large rise of the electron temperature, well above 30 eV. Arcs on the Mo-probe in front of the divertor plates have further been triggered by the application of an external potential to the probe. During plasma instabilities arcs are ignited even when the potential was positive, while during quiescent discharges negative potentials of the order of several 100 V were necessary to initiate arcs.

At the T-3M tokamak arcing was investigated using a graphite probe which was made out of two separated graphite plates being electrically insulated. The electrical current between the plates, the C II line emission, the loop voltage and the surface topography have been measured as an indication for arcing to occur. It was found that arcs ignite predominantly during tearing instabilities, when magnetic island touch the limiter. Contrary to the investigations at JFT-2, arc traces have nearly exclusively been found on the ion drift side, together with a metal deposition.

4.3 Electrical Circuit of Arcs

The current flow for an electrical arc ignited at a limiter of a magnetically confined plasma is shown schematically in Fig. 10. The arc current is restricted predominantly to the magnetic flux tube connected to the arc spot. If a magnetic island hits the limiter the flux tubes are partly destroyed so that the current backflow may go directly to the limiter. In stable discharges the arc current circuit must be closed at the walls by electron diffusion perpendicular to the field lines or at some distance where the flux tube hits a surface. In this case the flux tubes may reach several meters and represent a high inductivity /34/. Because electrical arcs deal with short current pulses, the inductivities in the circuit play a major role and may prevent the built-up of large currents /34/.

4.4 Motion of the Arc Spot on the First Wall

The arc traces show that the arcs generally move in the retrograde direction in respect to the magnetic field $\vec{B} = \vec{B}_{tor} + \vec{B}_{pol}$. At most wall areas the magnetic field lines are not parallel but intersect the surface at some angle. In this case the field compo-

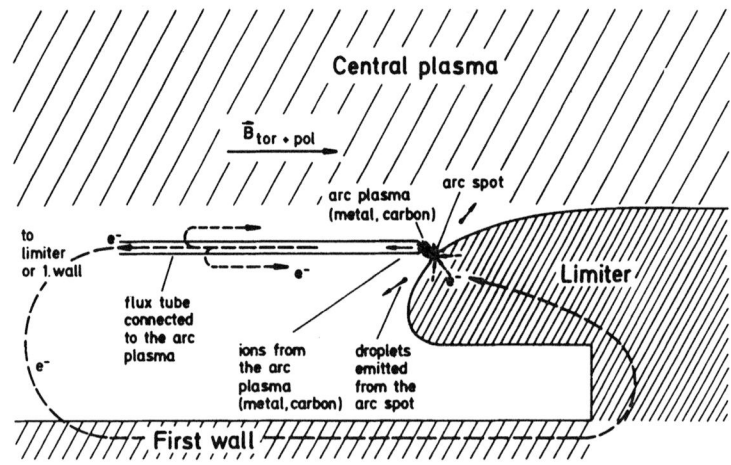

Unipolar arc at a limiter

Fig. 10: Electrical circuit for an arc at a limiter /32/.

nent, $\vec{B}_{\shortparallel}$, parallel to the surface is the major force for driving the arc in the retrograde direction. Due to the $\vec{J} \times \vec{B}$ force the arc column is, however, bent and the resulting component of the arc current, $\vec{J}_{\shortparallel}$, parallel to the surface interacts with the field component, $\vec{B}_{\perp}$, normal to the surface. This force causes the arc to move at an angle relative to its original direction. The extra motion is not in the retrograde direction but in the direction of the corresponding Lorentz force /86-87/.

4.5 Wall Erosion and Impurity Introduction

Arcs have been found to occur predominantly on the front part of limiters and on protection plates close to the central plasma /6-9, 14/, thus the major erosion takes place at these areas. From the investigations of vacuum arc it is known that the material released from the cathode spots at the vessel walls are

ions in the energy range up to 100 eV and particles, predominantly
small droplets in the 1 to 30 μm range.

The ions will mostly be trapped at the magnetic field lines.
The major part of them may be guided directly back to the sides of
limiter and/or to protection plates and only a small fraction may
enter the central plasma. Peaks in the metal deposition have been
found on time resolving collector probes in the scrape-off plasma
during the start-up phase and during plasmainstabilities in most
tokamaks /86, 88, 89/.

The material released as small particles or droplets has a
higher probability to enter the central plasma. When a tokamak discharge is filmed with a high speed camera generally bright objects
(UFO's) are found to move through the plasma /32/. These can be
flakes from metal films which had been poorly attached to the walls.
However part of the bright objects are likely to be droplets released at the walls by electrical arcs. These bright objects are
found predominantly at low density discharges, where high
electric fields may occur near the walls. They are further observed
as being released in large numbers from the limiter surface during
disruptions when the limiter is heated close to the melting poing
(Fig. 11) /32/. This is consistent with the observed large erosion
and emission of big droplets from hot cathodes in vacuum arcs.

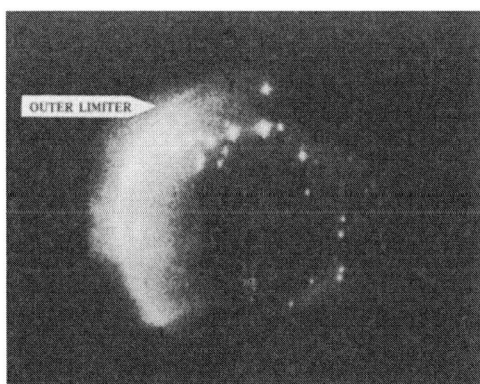

Fig. 11:

Emission of bright objects
(UFO's), being likely metal
droplets from the outer poloidal limiter of ASDEX during a
disruptive discharge for a
plasma current of 225 kA and
a plasma density of $1.9 \cdot 10^{19}$
m^{-3} /32/.

The droplets which enter the central plasma will partly be
eroded and thus introduce metal atoms. A major part of the droplets
impinges on the vessel walls. Wall analysis of most tokamaks have
shown that they are covered with large amounts of metal splashes
/90/.

5. CONCLUSIONS

Arcing is observed in today's tokamaks predominantly during unstable plasma conditions and/or at wall areas which are contaminated with surface layers. Arcing can be largely reduced for well conditioned walls, this is likely one of the major reasons for the necessity of discharge conditioning especially in limiter tokamaks. For clean walls the voltage for ignition of arcs is several 100 V.

In order to avoid the ignition and the burn of electrical arcs the plasma must be operated in a stable mode and the boundary plasma should stay at a low temperature. Arcing will be a severe process of wall damage and a source of metal droplets at unstable plasma conditions.

REFERENCES

/1/   J.L. Craston, R. Hancox, A.E. Robson, S. Kaufmann, H.T. Miles, A. Ware, J.A. Wesson, Proc. 2nd Int. Conf. Atomic Energy 32:414 (1958)

/2/   V.A. Simonov, B.N. Shvilkin, G.P. Katukov, Proc. Int. Conf. Plasmaphys., Contr. Nucl. Fusion Research, Salzburg (1961) Nucl. Fus. Suppl., p. 325 (1962) Engl. trans. AEC-No. 5589

/3/   Ph. Staib, G. Staudenmaier, J. Nucl. Mat. 63:37 (1976)

/4/   Ph. Staib, G. Staudenmaier, J. Nucl. Mat. 76/77:405 (1978)

/5/   G.M. McCracken, G. Dearnaley, S.F. Fielding, D.H.J. Goodall, J. Hugill, J.W.M. Paul, P.E. Stott, J.F. Turner and J. Vince, Proc. 8th Europ. Conf. Controlled Fusion and Plasmaphysics, Prague 1977, Vol. p. 40

/6/   G.M. McCracken, D.H. Goodall, Nucl. Fusion 18:537 (1978)

/7/   D.H.J. Goodall, T.W. Conlon, C. Sofield and G.M. McCracken, J. Nucl. Mat. 76/77:492 (1978)

/8/   D.H.J. Goodall and G.M. McCracken, Nucl. Fusion 19:1396 (1979)

/9/   G.M. McCracken, J.Nucl. Mat. 93/94:3 (1980)

/10/ D.H.J. Goodall, J. Nucl. Mat. 93/94 (1980) 154

/11/ S.A. Cohen, H.F. Dylla, S.M. Rossnagel, S.T. Picraux, J.A. Borders, C.W. Magee, J. Nucl. Mat. 76/77:459 (1978)

/12/ M. Mioduszewsky, R. Clausing, L. Heatherly, J. Nucl. Mat. (5/86:968 (1979)

/13/ P. Mioduszewsky, R. Clausing, L. Heatherly, J. Nucl. Mat. 91:297 (1980)

/14/ B. Jüttner, M. Laux, J. Lingertat, P. Pech, P. Siemroth, H. Wolff, Nucl. Fusion 20:497 (1980)

/15/ J. Jakubka, B. Jüttner, Proc. 10th Europhys. Conf. on Contr. Nucl. Fusion and Plasmaphysics, Moscow (1981), Vol. I, paper J5

/16/ J. Jakubka, B. Jüttner, J. Nucl. Mat. 102:259 (1981)

/17/ K. Ohasa, H. Maeda, S. Yamamoto, M. Nagami, H. Ohtsuka,
S. Kasai, K. Odajima, H. Kimura, S. Sengoku and
Y. Shimomura, J. Nucl. Mat. 76/77:489 (1978) and
Nucl. Fusion 18:872 (1978)

/18/ Y. Gomay, N. Fujisawa, M. Maeno, J. Nucl. Mater. 85/86:967
(1979)

/19/ H. Ohtsuka, N. Ogiwara and M. Maeno, J. Nucl. Mat. 93/94:161
(1980)

/20/ M. Maeno, H. Ohtsuka, S. Yamamoto, T. Yamamoto, N. Suzuki,
N. Fujisawa, Nucl. Fusion 20:1415 (1980)

/21/ H. Ohtsuka, M. Maeno, N. Suzuki, S. Konoshima, S. Yamamoto,
N. Ogiwara, Nucl. Fusion 22:823 (1982)

/22/ J.N. Smith, C.H. Meyer,Jr., General Atomic Co. Report
GA-A16045 (1980)

/23/ D.G. Baratov, V.N. Demyanenko, S.V. Mirnov, V.V. Myalton,
L.E. Demyanenko, T.B. Myalton, V.S. Semenov, V.P. Fokin,
Proc. 10th Europ. Conf. on Controlled Nuclear Fusion and
Plasmaphysics, Moscow (1981), Vol. I, paper J11

/24/ D.G. Baratov, G.V. Gordeeva, M.I. Guseva, V.N. Demyanenko,
N.N. Mansurova, S.V. Mirnov, V.A. Stepanchikov,
V.P. Fokin, Atomnaya Energya 53:396 (1982)

/25/ J.K. Tien, N.F. Panayotou, R.D. Stevenson, and R.A. Gross,
J.Nucl.Mat. 76/77:482 (1978)

/26/ G.M. McCracken, L. Firth, D.H.J. Goodall, R.E. King,
K.E. Lavender, A.A. Newton and V.K. Thompson, B.C. Edwards
and J. Titchmarsh, J. Nucl. Mat. 111/112:522 (1982)

/27/ J. Cao, J. Chen, W. Duan et al., J. Nucl. Mater. 93/94: 343
(1980)

/28/ W. Duan, J. Chen, S. Yang, J. Nucl. Mat. 111/112:502 (1982)

/29/ V.S. Voitsenya, E.D. Volkov, Yu.A. Grivanov, A.G. Dikii,
E.M. Latsko, V.F. Rybalko, S.I. Solodovchenko, Sov. Phys.
Techn. Phys. 25:248 (1980)

/30/ P. Mioduszewsky, "Data compendium for Plasma Surface Interactions",Nuclear Fusion Special Issue 105 (1984)

/31/ B. Jüttner, Plasmaphysics and Controlled Fusion 26:249 (1984)

/32/ D.H.J. Goodall, J. Nucl. Mat. 111/112:11 (1982)

/33/ K. Ertl, R. Behrisch, B. Jüttner, Proc. 11th Europ. Conf. on
Controlled Fusion and Plasmaphysics, Aachen 1983, Vol. II,
p. 385

/34/ K. Ertl, B. Jüttner, Nucl. Fusion 25 (1985)

/35/ A.E. Robson, P.C. Thoneman, Proc. Phys. Soc. 73:508 (1959)

/36/ K. Höthker, W. Bieger, H. Hartwig, E. Hintz and K. Koizlik,
J. Nucl. Mat. 93/94:785 (1980)

/37/ E. Hantzsche, Beiträge zur Plasmaphysik 24:329 (1980)

/38/ G. Ecker, Proc. Int. Symp. Plasma Wall Interaction, Jülich
1976, Pergamon Press 245 (1977)

/39/ C. Wiekert, J. Nucl. Mat. 76/77:499 (1978)

/40/ G. Ecker, Proc. ICPIG Düsseldorf (1983)

/41/ A.V. Nedospasov, V.G. Petrov, J.Nucl. Mat. 76/77:504 (1978)

/42/ A.V. Nedospasov, V.G. Petrov, J.Nucl.Mat. 93/94:775 (1980)

/43/ L. Oren, R.J. Taylor, F. Schwirtzke, J. Nucl. Mat. 76/77:412
(1978)
/44/ A.W. Nürnberg, U.H. Bauder, C. Mooser, R. Behrisch,
Beitr. Plasmaphysik 21:127 (1981)
/45/ W.P. Dyke, W.W. Dolan, "Field Emission", in Adv. in Electr.
and Electron Phys. 8:89 (1956)
/46/ J.J. Lafferty (ed.) "Vacuum arcs, theory and application",
John Wiley & Sons, New York, Chichester, Brisbane,
Toronto (1980)
/47/ G. Ecker in Ref. /46/
/48/ G. Ecker, Beitr. Plasmaphysik 11:405 (1971)
/49/ G. Ecker, Z. Naturforschung 28a:228 und 417 (1973)
/50/ E.E. Kunhardt, L.H. Luessen, "Electrical breakdown and discharges in gases", NATO ASI Series, Physics Vol. 89 b.
Plenum Press, New York & London (1983)
/51/ A.E. Guile, IEE Proc. 131:450 (1984)
/52/ E. Hantzsche (ed.), Proc. XI. Int. Symp. on Discharges and
Electric Insulation in Vacuum Berlin, GDR, Central Institute of Electron Physics (ZIE) (1984),Reviews by
G. Fursey, by G.A. Mesyats and by V.Z. Rakhovsky
/53/ R. Hancox, Brit. J. of Appl. Phys. 11:468 (1960)
/54/ J.P. Daltov, A.E. Guile, B. Jüttner, Beitr. Plasmaphysik
21:135 (1981)
/55/ J. Achtert, B. Altrichter, B. Jüttner, P. Pech, H. Pursch,
H.D. Reiner, W. Rohrbeck, P. Siemroth, H. Wolff,
Beiträge Plasmaphysik 17:419 (1977)
/56/ J. Daalder, PhD Thesis, Technical University Eindhoven 1978
/57/ B. Jüttner, Thesis B, Central Institute of Electron Physics
(ZIE) Berlin, GDR (1982)
/58/ L. Malter, Phys. Rev. 50:48 (1936)
/59/ B. Jüttner, Beitr. Plasmaphys. 19:25 (1979)
/60/ E. Hantzsche, B. Jüttner, /Ref. 52/
/61/ B. Jüttner, H. Pursch, S. Anders, J. Phys. D. Appl. Phys.
17:L111 (1984)
/62/ B. Jüttner, W. Rohrbeck, Beitr. Plasmaphys. 17:229 (1977)
/63/ M.G. Drouet, Jap. Journal of Applied Phys. 20:1027 (1981)
and Ref. /52/
/64/ B. Jüttner, Physica 114C:255 (1982)
/65/ D.Y. Fang, A. Nürnberg, U.H. Bauder, R. Behrisch,
J. Nucl. Mater. 111/112:517 (1982)
/66/ W.H. Zhao, A. Koch, U.H. Bauder, R. Behrisch,
J. Nucl. Mater. 128/129:613 (1984)
/67/ S.K. Sethuraman, M.R. Barrault, J.Nucl.Mat. 93/94:791 (1980)
/68/ B. Jüttner, J. Phys. D, Appl. Phys. 14:1265 (1981)
/69/ E. Hantzsche, Beitr. Plasmaphysik 20:61 (1980)
/70/ E. Hantzsche, B. Jüttner, V.F. Puchkarov, W. Rohrbeck and
H. Wolff, J. Phys. D, Appl. Phys. 9:1771 (1976)
/71/ Ya. Udris, Radio Engin. Electron. Phys. 8:1050 (1963)
/72/ C.W. Kimblin, J. Appl. Phys. 44:3074 (1973)
/73/ C.W. Kimblin, J. Appl. Phys. 45:5235 (1974)

/74/ H.C. Miller, J. Appl. Phys. 52:4523 (1981)

/75/ T. Utsumi, J.H. English, J. Appl. Phys. 46:126 (1975)

/76/ J.E. Daalder, J. Phys. D, Appl. Phys. 9:2379 (1976)

/77/ A.W. Koch, A.W. Nürnberg, R. Behrisch, J. Nucl. Mat.
     122/123:1437 (1984)

/78/ A.W. Nürnberg, D.Y. Fang, U.H. Bauder, R. Behrisch, F. Brossa,
     J.Nucl. Mat. 103/104 (1981)

/79/ S.K. Sethuraman, P.A. Chatterton, M.R. Barrault,
     J.Nucl. Mat. 111/112:510 (1982)

/80/ H. Ehrich, J. Kartan, K.G. Müller, J. Nucl. Mat.
     111/112:526 (1982)

/81/ F. Schwirzke, "Unipolar Arcing, a basic laser damage mechanism"
     Technical Report, Naval Postgraduate School, Monterey,
     Cal. (1983)

/82/ D.A. Shcheglov, JETP Letters 22:114 (1976)

/83/ G.A. Bobrovskii, A.A. Kondratev, Sov. J. Plasma Phys.
     3:115 (1977)

/84/ S.V. Mirnov, I.B. Samenov, Sov. J. Plasma Phys. 4:27 (1978)

/85/ A.E. Robson, Proc. 4th Int. Conf. on Ion Phen. in Gases,
     Uppsala 1:346 (1959)

/86/ G.M. McCracken, P. Stott, Nucl. Fusion 19:889 (1979)

/87/ W. Hintze, M. Laux, Beitr. Plasmaphysik 21:247 (1981)

/88/ G.M. McCracken, G. Dearnaley, R.D. Gill, J. Hugill,
     J.W.M. Paul, B.A. Powell, P.E. Stott, J.F. Turner and
     J.E. Vince, J. Nucl. Mat. 76/77:431 (1978)

/89/ E. Taglauer, B.M.U. Scherzer, P. Varga, R. Behrisch,
     Chen Cheng Kai and ASDEX Team, J. Nucl. Mat. 111/112:142
     (1982)

/90/ R. Behrisch, R.S. Blewer, H. Kukral, B.M.U. Scherzer,
     H. Schmidl, P. Staib, G. Staudenmaier, J. Nucl. Mat.
     1976/77:437 (1978)

ELECTRON EMISSION FROM SOLID SURFACES

K. Ertl and R. Behrisch

Max-Planck-Institut für Plasmaphysik, EURATOM Association

D-8046 Garching/München, Federal Republic of Germany

ABSTRACT

Electrons are emitted from solid surfaces as a result of heating, of ion, electron and photon bombardment from the plasma, and if a sufficiently high electric field is applied. Thermal electron emission only becomes important at temperatures above 2000° C. With particle bombardment at energies in the 10 to 100 eV range, the yields are typically of the order of 0.1 to 1 except in the case of highly charged ions, where potential emission yields may be well above 1. Field electron emission may be important for the ignition of electrical arcs.

1. INTRODUCTION

The emission of electrons from the solid walls in high-temperature plasma experiments due to heating, to particle and photon bombardment and field emission reduces the sheath potential between the plasma and the walls and thus influences the energy and particle fluxes to the surfaces, which is important especially at limiters and divertor plates /1 to 4/. In power transmission by waveguides for RF-heating of the plasma, electron emission at the walls of the waveguides leads to a reduction of the transmitted power due to multipactor effects. In addition, electron emission is an important parameter for the interpretation of probe measurements in the plasma scrape-off layer, especially for Langmuir probes and energy flux probes (see chapter by Manos and McCracken). Reliable data on the particle- and photon-induced electron emission yields from solids are needed. For more sophisticated calculations, the energy and angular distributions of the emitted electrons are required too.

515

With respect to plasma-wall interactions, mostly it is the
total secondary electron emission coefficient γ that is of primary
interest, this being defined as the average number of electrons
emitted from a solid per incident electron. To consider individual
physical details, however, one needs more refined definitions such
as the electron emission yields due to the different processes.
As shown in the following, only very few data measured under well
defined conditions are known in the energy range most relevant for
fusion research.

2. PHYSICAL PROCESSES

For the emission of electrons from the surface of a solid,
energies larger than the surface potential or work function ϕ have
to be transferred to an electron in the conduction or valence band
(Fig. 1). This energy is of the order of a few eV. It may be provided by heating the solid or bombarding it with energetic particles, or by applying an electric field so that the electrons can
tunnel through the resulting potential barrier from the conduction
band into the vacuum.

2.1 Field Electron Emission

The field emission current density is given by the Fowler
Nordheim equation /5, 6/

$$j = \frac{1.54 \times 10^{-6} \; F^2}{\phi t^2(y)} \; e^{-6.83 \times 10^7 \frac{\phi^{3/2}}{F} v(y)} \quad A/cm^2 \qquad (1)$$

with $y = 3.79 \times 10^{-4} \; F^{1/2}/\phi$, F = electric field in V/cm, ϕ = work
function in eV, $t \simeq 1$ and v = elliptic functions (Table in /6/).

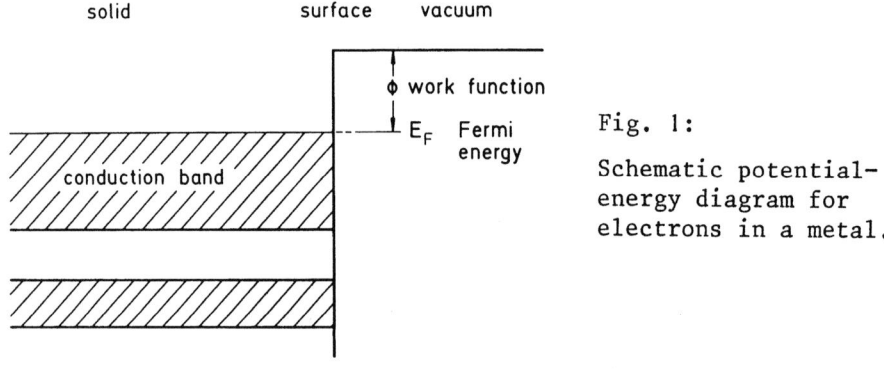

solid   surface   vacuum

ϕ work function

E_F Fermi energy

conduction band

Fig. 1:

Schematic potential-
energy diagram for
electrons in a metal.

Field emission of electrons requires electric fields in the order of $(3-6) \times 10^7$ V/cm and is thus limited to micro tips of rough surfaces. In fusion research field emission is not likely to play a role in normal Ohmic discharges, but it may contribute to the ignition of electric arcs; see chapter by R. Behrisch.

2.2 Thermal Electron Emission

At temperatures T >> 0 K some of the electrons in the conduction band of a solid have an energy large enough to overcome the potential barrier when they reach the surface. This fraction gives an emitted electron current density which can be described by the Richardson-Dushman equation |5-9|

$$j = AT^2 e^{-\frac{\phi}{kT}} \quad A/cm^2 \qquad (2)$$

with $A = 4 \pi m_e e_0 k_B h^{-3} = 120$ A/cm² deg², where m_e is the electron mass ($m_e = 9.110 \times 10^{-31}$ kg), e_0 the elementary charge ($e_0 = 1.602 \times 10^{-19}$ As), k_B the Boltzmann constant ($k_B = 1.381 \times 10^{23}$ J K^{-1}), h Planck's constant (h = 6.626×10^{-34} Js), and T the absolute temperature.

Experimentally determined values for A differ from the calculated ones owing to modifications caused by the temperature dependence of the work function, surface topography and surface composition.

Table 1: Measured parameters for thermal electron emission /7, 8, 9/

Metal	A (A/cm² deg²)	Work function (eV)
W	15–156	4.5
Ni	30–1380	4.6–5.2
Cr	48	4.6
Fe	1–26	4.2–4.5
C	15–30	4.4–5
Mo	25–338	4.3

These data show that thermal electron emission only becomes important at temperatures above 2000° C.

2.3 Ion-induced Electron Emission

The electron emission due to ion bombardment is generally described by the electron emission coefficient γ_i, giving the average number of electrons released per incident ion. Depending on the energy of the incident ions, two different processes can dominate the electron emission:

- at low particle velocities ($v_0 < 10^7$ cm/sec) electron ejection is mainly due to <u>potential emission</u>,

- at higher impact energies <u>kinetic emission</u> is the dominant mechanism.

Besides the energy being one of the major parameters, the emission yields depend on the projectile-target combination, the mass, the charge state, the excitation state, the angle of incidence of the projectile, and the atomic number, composition, topography and surface contamination of the target.

2.3.1 Potential emission

In the case of potential emission /10 to 14/, the transfer of potential energy of the incoming ions or metastables to electrons of the target atoms during neutralization or de-excitation of the incident projectile is responsible for the electron emission. This process is nearly independent of the kinetic energy of the incident particle and dominates the electron yield at low velocities. Quantum mechanical theories for potential electron emission are based mainly on the theory by Hagstrum /15/. Some of the dominant processes involved in potential emission are shown in Fig. 2.

- In the process of Auger neutralization of an incident ion (mechanism a) an electron from an electronic state in the solid tunnels into the atomic ground state of the incident ion. If the energy difference is $E_i > 2\phi$, this energy is sufficient to eject another valence electron. Radiative transition can be neglected for this process.

- In the process of Auger de-excitation of a metastable atom into the ground state near the surface (mechanism b) the energy is transferred to an electron of the valence band of the solid that is then emitted. The process can only take place if the condition $E_m > \phi$ is valid. The energy distribution of the emitted electrons is directly correlated with the structure of the valence band and its occupation density.

- In the case of resonance neutralization (mechanisms c and d) an electron from an electronic state below the Fermi level of the solid (valence band or core level) tunnels into an empty state of the ion having the same energy. Emission of an electron is possible if $E_h = E_i > 2\phi$.

- Electron emission may be the result of a multi-step process such as resonance neutralization of an ion (process e) followed by Auger de-excitation or resonance ionization of an excited atom (process f) followed by Auger neutralization.

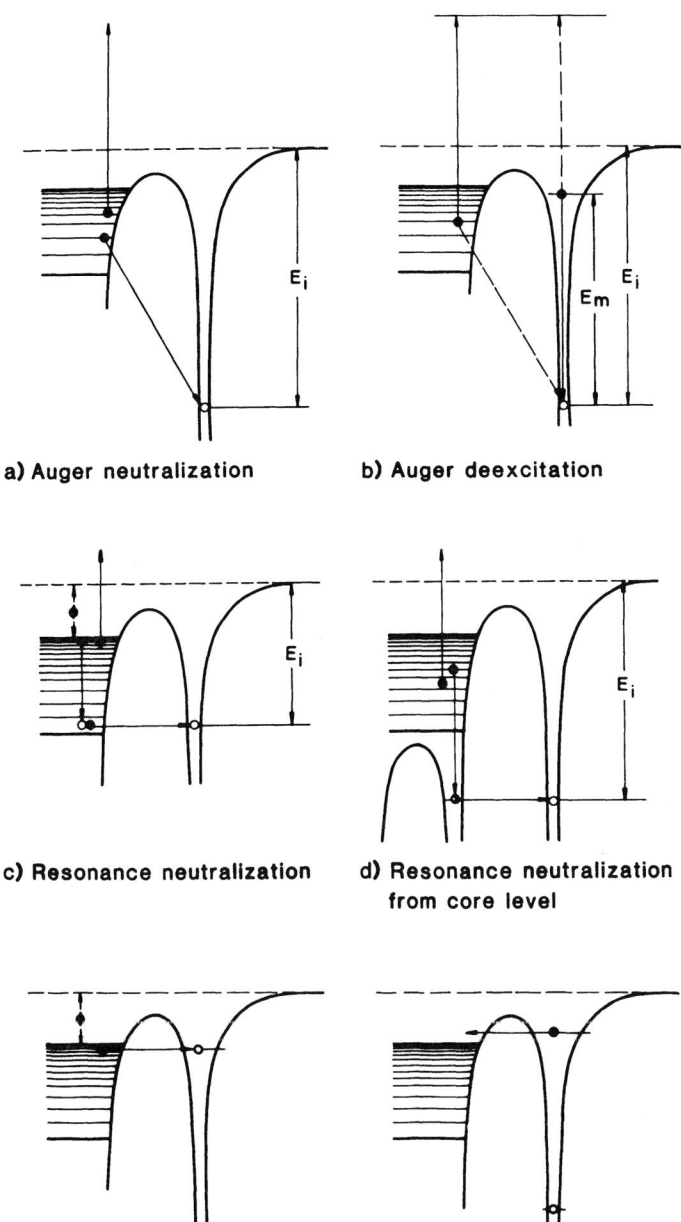

a) Auger neutralization   b) Auger deexcitation

c) Resonance neutralization   d) Resonance neutralization
from core level

e) Resonance neutralization   f) Resonance ionization

Fig. 2:   Processes leading to potential emission of electrons at
ion - or excited atom - surface interaction.

In all these processes simultaneous effects such as the image potential, polarization, and nuclear repulsion cause shifts of the energy levels compared with the free particle. Transition probabilities are calculated with quantum mechanical perturbation theory. There is no simple theoretical formula for the emission yield, but several experimental data are available.
Emission yields for singly charged ions are typically of the order of 0.1, but can amount to values well above 1 for highly charged ions /11/. Recent measurements show a nearly linear dependence of the emission yield on the potential energy of the impinging ion, the factor of proportionality being 0.01 - 0.02 (electrons/eV), depending on the surface conditions. N^{5+} (pot. energy 270 eV) impinging on non degassed W with a kinetic energy of 300 eV gives an emission yield $\gamma_i = 2$, Kr^{9+} (pot. energy 740 eV) gives $\gamma_i = 7$ /16/.

The energy distribution of the emitted electrons depends on the electronic states of the projectile and the target and extends from the eV to the 15 eV range. An example is given in Fig. 3.

2.3.2 Kinetic emission

In the case of kinetic emission /18, 19/ the kinetic energy of the impinging projectile is transferred to target electrons. This may occur in a large number of scattering processes with ionization of bound electrons, single and multiple excitation,

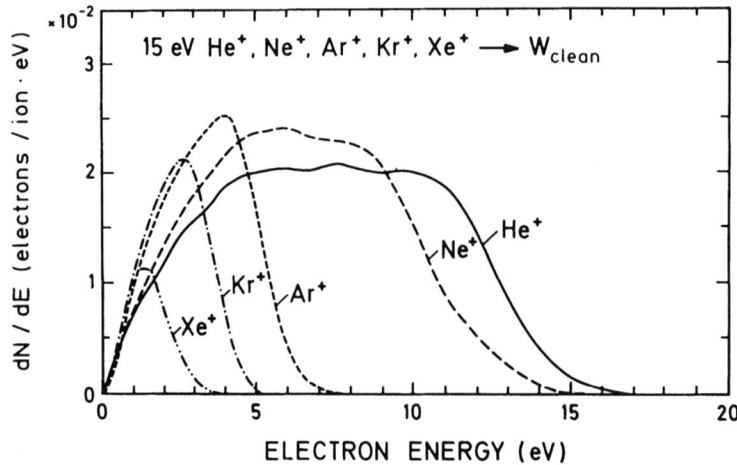

Fig. 3: Energy distribution of electrons, emitted by potential emission at bombardment of a clean W-surface with different noble gas ions having an incident energy of 15 eV /17/.

energy and charge transfer processes, scattering of the deliberated electrons at other electrons or target nuclei, until finally one or several electrons are released at the surface /19a/. The minimum energy transfer necessary for electron emission to be possible leads to a well-defined threshold energy for electron emission which depends on the projectile-target combination and has an amount between (0.6 - 0.7) x 10^7 cm/sec /19/.

For high incident energies (> 100 keV) and low atomic numbers of the projectile, theoretical formulae were derived and tested in experiments. In ionization cascade theories transport methods are used /20/, which have also been successfully applied for analyzing sputtering. The electron emission is closely related to the electronic stopping power, i.e. the energy deposited into the electrons of the target.

At low and intermediate energies theories are not completely satisfying. One of the most frequently used, but not undisputed approaches in the energy range 1-100 keV involves the formulae of Parilis and Kishinevskii /21/. Here the interaction of the projectile and the solid is described as a bi-particle collision based on the theory of Firsov /22, 23/. Ionization of an atom of the solid produces a hole in the valence band. Auger recombination with an electron from the conduction band causes emission of another electron from the conduction band (Fig. 4). The threshold velocity in this theory is determined by the minimum energy transfer, where excitation of an electron from the valence band into the conduction band is still possible /21/.

With respect to the energy dependence of the kinetic electron emission yield three ranges of the particle velocity $v_o = \sqrt{2E_o/m}$ result from this theory being in good agreement with experiments /19, 21/.

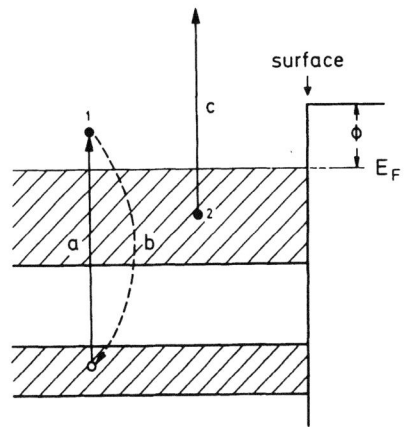

Fig. 4:
Model of kinetic electron ejection according to Parilis and Kishinevskii /21/:

a) Ionization of an inner shell (valence band) by the incoming ion or atom;

b) Hole-electron recombination, the energy is coupled to electron 2;

c) Electron 2 is excited to the emission level.

521

- Near threshold ($v_o \lesssim 10^7$ cm/sec): $\gamma_i \sim E_o^2$
- Linear range ($10^7 < v_o < 3 \times 10^7$ cm/sec): $\gamma_i \sim E_o$
- Square root range ($v_o > 3 \times 10^7$ cm/sec): $\gamma_i \sim E_o^{1/2}$,

with E_o being the kinetic energy of the ion, and m the ion mass. The kinetic emission coefficient is independent of the ion charge state, but it depends on the atomic number of the projectile and the atoms of the solid.

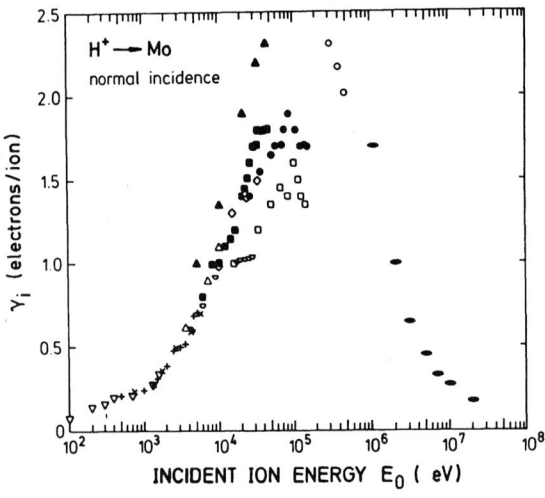

Fig. 5:
Electron emission yield for Mo at bombardment with H^+-ions at normal incidence. Data from references given in /18, 19, 24/.

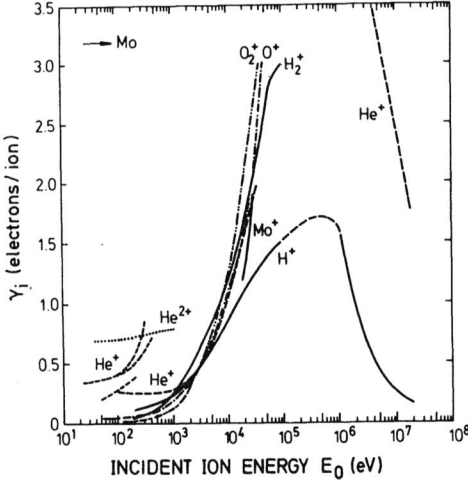

Fig. 6:
Electron emission yield for Mo at bombardment with different ions and different charge states. Data from references given in /18, 19, 24/.

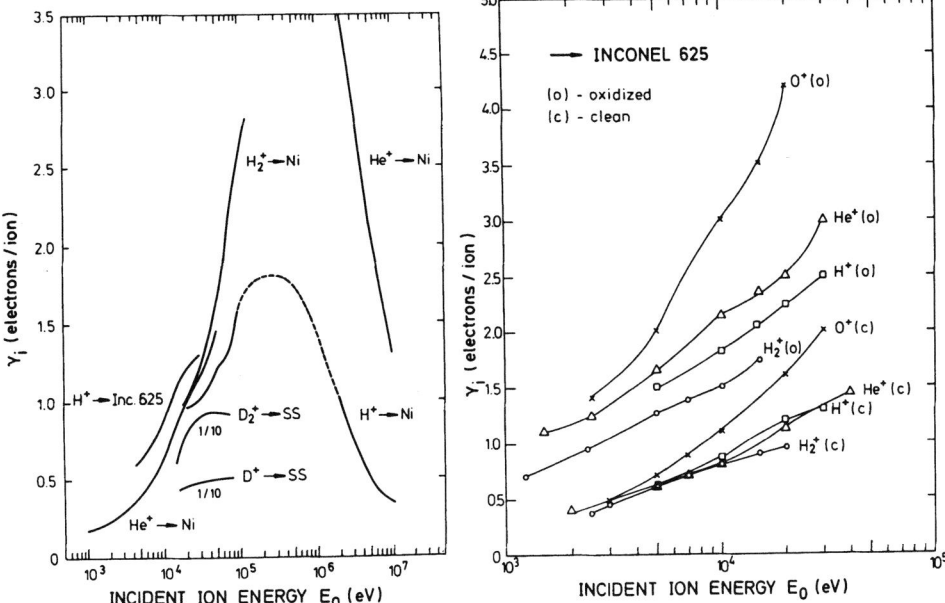

Fig.7: Electron emission yields of Ni, Inconel and stainless steel (surfaces not specially prepared) at H^+, H_2^+, D^+, D_2^+ and He^+ bombardment. The yields for stainless steel are reduced by a factor of 10. Data from references given in /24/.

Fig.8: Electron emission yield of Inconel at H^+, H_2^+, He^+ and O^+ bombardment for differently prepared surfaces /25/.

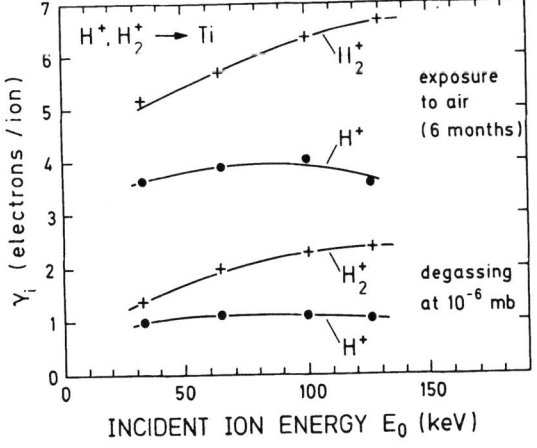

Fig.9:

Electron emission yield at H^+ and H_2^+ bombardment of Ti for different surface conditions /26/.

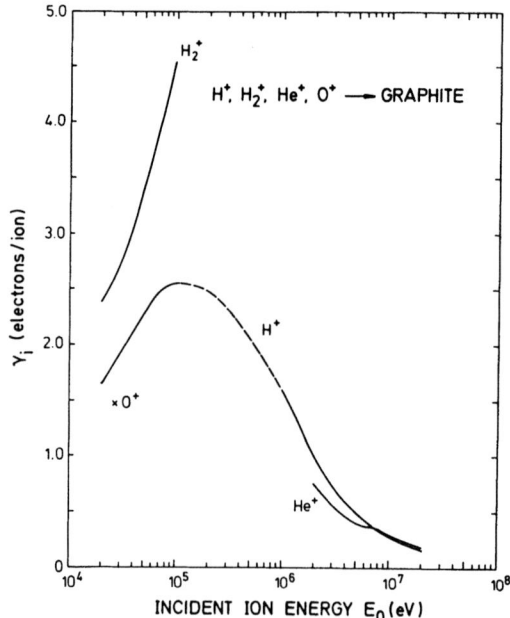

Fig. 10:

Electron emission yield
of graphite at H^+, H_2^+, He^+
and O^+ bombardment. Data from
references given in /24/.

The data existing for relevant projectile-target combinations
are collected in Figs. 5 to 10. They are put together from the references given in several reviews /18, 19 and 24/. Many of the data
were measured for not well defined surface conditions of the materials investigated. As can be seen in Figs. 8 and 9, the emission coefficients for clean and oxidized surfaces may differ by a factor of
up to 4. No data have been found for low-energy ions on carbon.

The dependence of the electron emission yield on the angle of
incidence is generally correlated to the escape depth of the produced
electrons and varies for light ions as /27/

$$\gamma_i(\alpha) = \gamma_i(0)/\cos\alpha \tag{3}$$

The experimental energy distributions of the emitted electrons
peak at a few eV as shown in Fig. 11 and are in good agreement with
theoretical calculations by Parilis and Kishinevskii /21/.

There are only few measurements on the angular distribution
of the ejected electrons. In general an isotropic cosine distribution
has been found for clean polycristalline materials, approaching a
$\cos^2$ distribution for lower ion energies /29/. Surface contaminations
may cause anisotropic electron emission /30/.

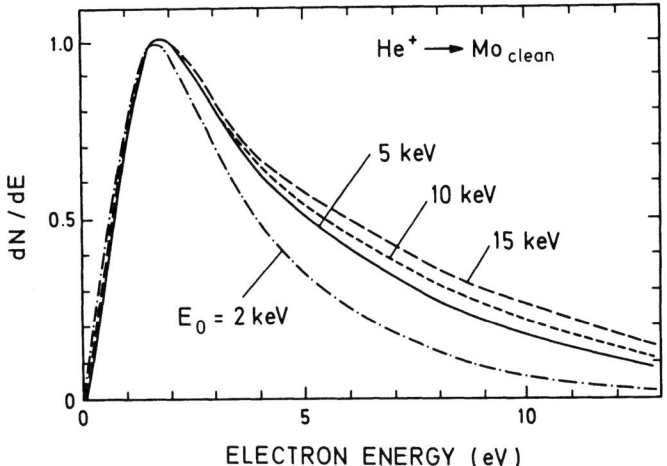

Fig. 11: Energy distribution of electrons emitted by kinetic
emission at bombardment of a clean Mo surface with
He$^+$ ions /28/.

2.4 Electron-induced Electron Emission

The electron emission during electron bombardment of a solid
surface has been investigated in some detail in respect of scanning
electron microscopy /31/. The emitted electrons consist of elastically and inelastically reflected electrons (RE) and of true secondary
electrons peaking at low energies (SE) (Fig. 12), for review see /32
to 35/. The separation between these two components SE and RE is arbitrari-

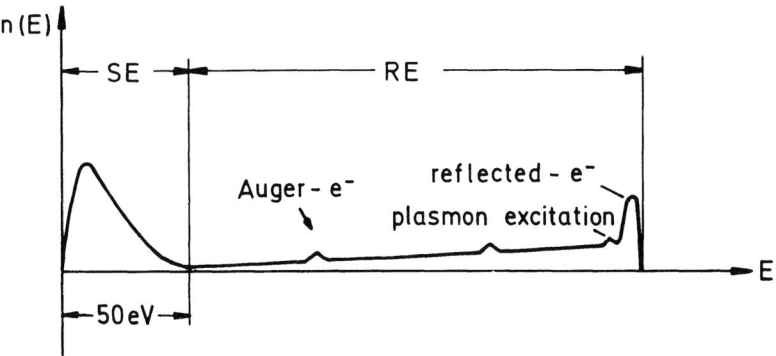

Fig. 12: Schematic energy distributions of electrons emitted
from a surface at electron bombardment.

ly drawn in the literature at an emergent energy of about 50 eV. In addition, peaks are found which are superimposed at discrete energies on the energy spectrum and which are due to Auger electrons or due to backscattered electrons which have suffered energy losses by excitation of plasmons. The total electron emission yield γ_e is the sum of the electron backscattering coefficient η_e, defined as the number of reflected electrons per incident primary electron, and the secondary electron emission yield δ_s, defined as the number of secondary electrons per incident primary electron:

$$\gamma_e = \delta_s + \eta_e \qquad\qquad (4)$$

2.4.1 Backscattered electrons

Measured electron backscattering coefficients η_e and their dependence on the incident energy as well as on the target mass are

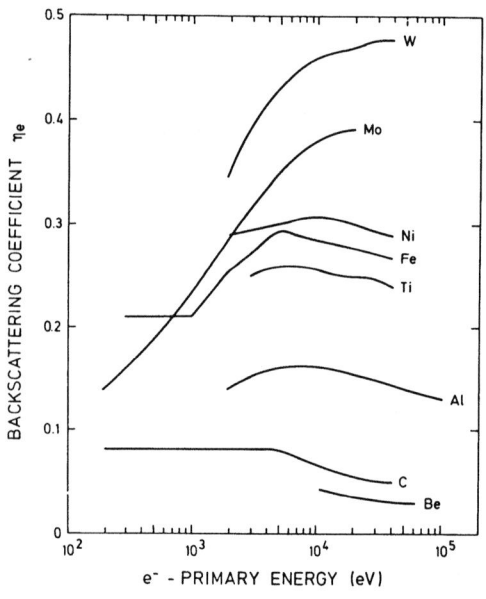

Fig. 13:

Electron backscattering coefficient η_e for different materials and different incident energies at normal incidence.
Data from references given in /24/.

shown in Figs. 13 and 14. They can be described by an empirical formula which was given by Hunger and Kuechler for normal electron incidence ($\alpha = 0$) /36/, and by Darlington for oblique electron incidence ($\alpha \neq 0$) /37/.

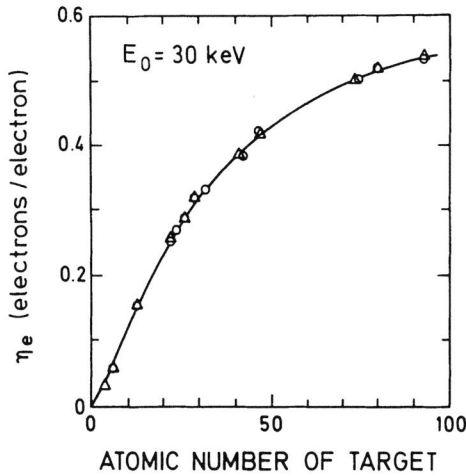

Fig. 14:

Dependence of the electron backscattering coefficient η_e on the atomic number of the target /31/.

For $\alpha = 0$ and incident energies 4 keV < E_o < 40 keV on targets with atomic number 5 < Z < 92:

$$\eta_e (Z,E) = E^{m(Z)} \exp (C(Z)) \qquad (5)$$

with $\quad m(Z) = 0.1382 - 0.9211 \; Z^{-0.5}$ and
$\qquad C(Z) = 0.1904 - 0.2236 \; \ln Z + 0.1292 \; \ln^2 Z - 0.01491 \; \ln^3 Z,$

and for $\alpha \neq 0$

$$\eta_e(\alpha) = 0.891 \; (n_e(0)/0.891)^{\cos\alpha}. \qquad (6)$$

There are hardly any data on electron backscattering yields at low incident energies and for well-defined surfaces.

2.4.2 Secondary electrons

In secondary electron emission several processes of the electrons moving in the solid have to be considered, as deceleration, scattering and energy loss of the primaries, the excitation and the scattering and energy loss of the secondary electrons and finally the emission process at the surface. There are many attempts to treat the secondary electron production on a classical or quantum mechanical basis /38/, for references see also /33/. For electrons with energies in the keV range the transport theory by Schou /20/ may be applied successfully. Experimental data were obtained partly for contaminated surfaces, owing to insufficient vacuum conditions. Only within the last years yields have been measured at well defined surfaces in UHV.

It was found that the dependence of the secondary electron emission coefficients on the primary electron energy has the same

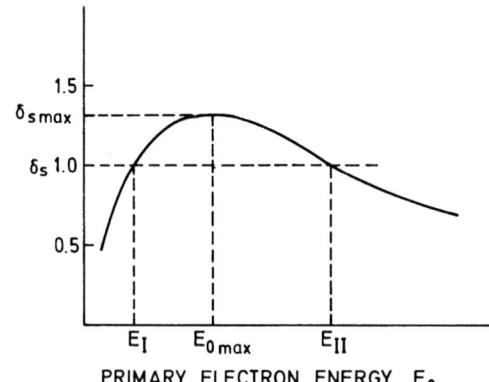

Fig. 15:

Schematic diagram of the secondary electron yield δ_s depending on the incident energy. E_I and E_{II} are those energies, where the yield drops to 1 if $\delta_{s\ max} > 1$.

shape for different materials. It is usual to characterize these curves by the energy $E_{0\ max}$ where the emission coefficient δ_s has its maximum yield $\delta_{s\ max}$ (Fig. 15). Table II presents some data on relevant materials.

Table 2: Maximum of the secondary electron emission yield for different materials /39/

Z	Atom	$\delta_{s\ max}$	$E_{0\ max}$
4	Be	0.5 - 0.9	200 - 300
6	C	0.9 - 1.0	300 - 1100
13	Al	0.9 - 1.0	250 - 300
22	Ti	0.7 - 0.9	300
26	Fe	1.1 - 1.3	400
28	Ni	1.0 - 1.3	500 - 550
42	Mo	1.0 - 1.2	400
74	W	1.0 - 1.4	700

The normalized yield curves ($\delta_s/\delta_{s\ max}$ versus $E_0/E_{0\ max}$) coincide for most of the different materials as seen in Fig. 16. Following up on a theory of Sternglass /39/, Kollath /33/ proposes the following semi-empirical formula for the normalized yield curve:

$$\frac{\delta_s}{\delta_{s\ max}} = (2.72)^2 \frac{E_0}{E_{0\ max}} \exp\left(-2\left(\frac{E_0}{E_{0\ max}}\right)^{1/2}\right). \tag{7}$$

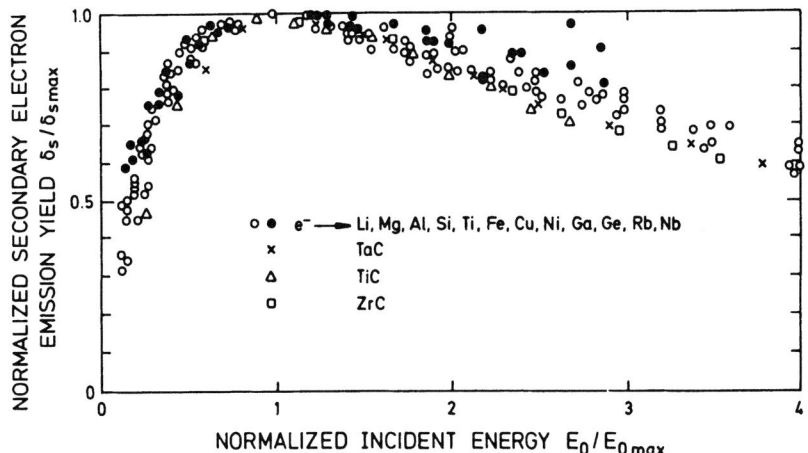

Fig. 16: Normalized secondary electron emission yield curves
for different materials /33, 43/.

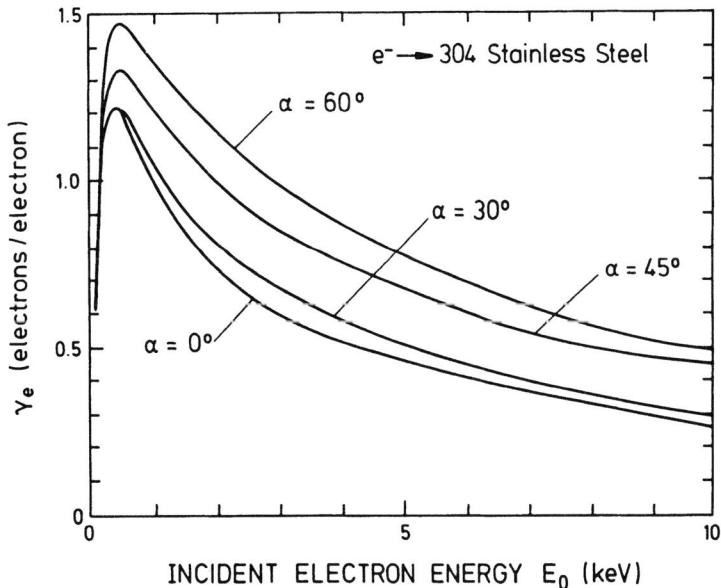

Fig. 17: Secondary electron emission coefficient for stainless
steel at different angles of incidence /42/.

The maximum secondary electron emission coefficient increases with the work function of the solid according to an empirical equation /41/

$$\delta_{s\ max} = (0.35\ \phi)^{1/2} \tag{8}$$

This shows that the secondary emission coefficient depends on the surface contamination. Data for different materials, surface conditions and angles of incidence are summarized in the Figs. 17 - 22. For very high incident energies such as typically occur with runaway electrons there are very few data available on secondary electron emission coefficients. Figure 21 gives an example for Al and Cu.

The dependence of the secondary electron emission yield on the angle of incidence α is determined like the ion induced electron emission by the changing penetration and escape depth of the electrons and varies as follows /44/:

$$\delta_s(\alpha) = \delta_s(0)/\cos\alpha \tag{9}$$

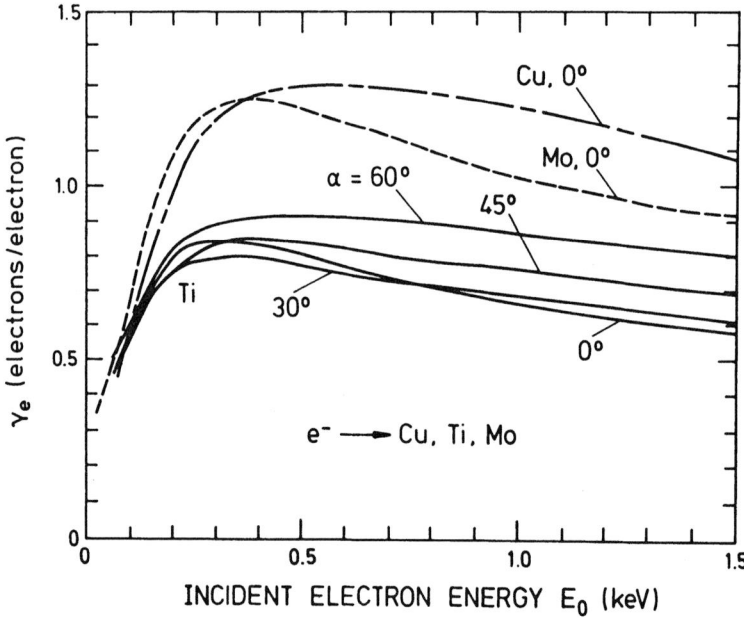

Fig. 18: Secondary electron yield for Ti at different angles of incidence and Cu and Mo at normal incidence /42/.

530

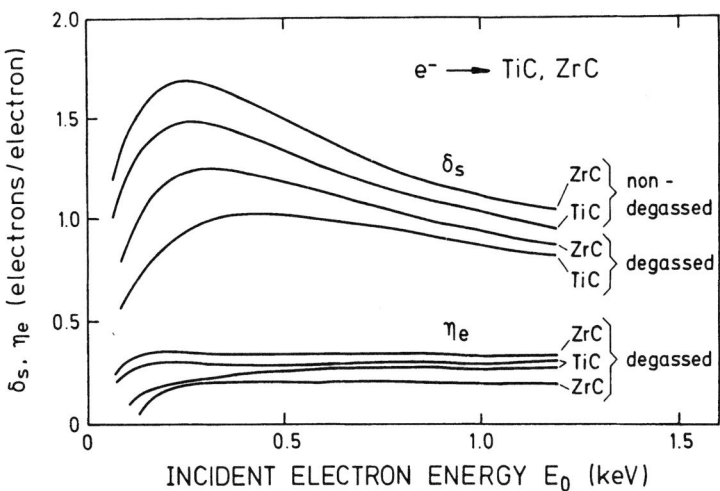

Fig. 19: Reflection coefficient η_e and secondary emission coefficient δ_s from ZrC and TiC at normal incidence and
for different surface conditions /49/.

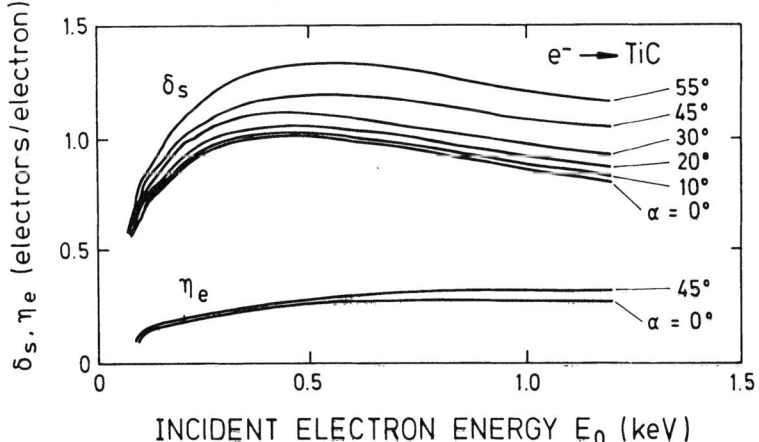

Fig. 20: Electron reflection coefficient η_e and secondary
electron emission coefficient δ_s from TiC for
different angles of incidence /49/.

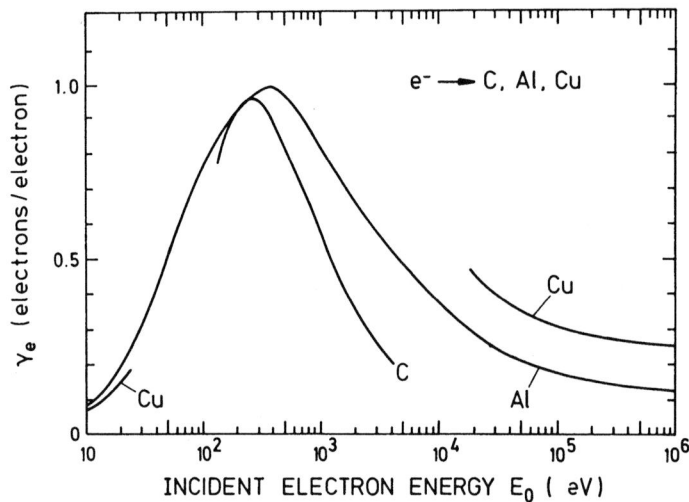

Fig. 21: Secondary electron emission yield for Al, C and Cu.
Data from references given in /25/.

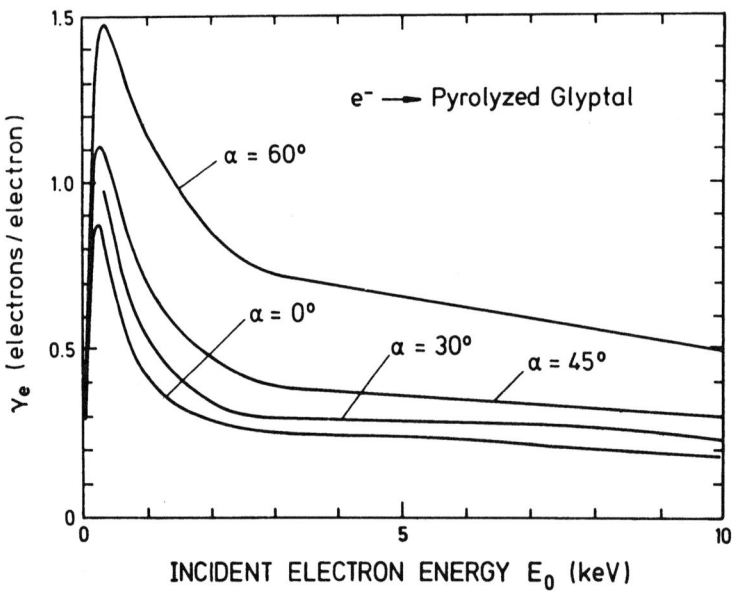

Fig. 22: Secondary electron emission yield of C (pyrolized Glyptal)
for different angles of incidence /42/.

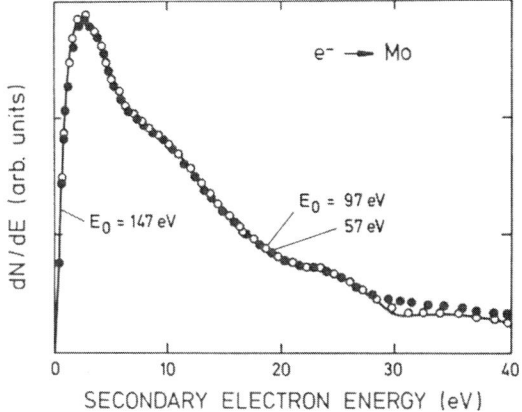

Fig. 23:

Energy distribution of secondary electrons emitted from degassed Mo /45/.

The energy distribution of the secondary electrons peaks typically at a few eV, as shown in Fig. 23. The angular distribution of the ejected secondary electrons is shown to vary a $\cos\theta$, where θ is the angle between the surface normal and the emitted electron /41/.

For waveguide tubes it is essential to have surfaces with a secondary electron emission coefficient well below 1 in order to suppress multipactor discharges. This was achieved either by using a low-Z coating such as carbon /42/ (Fig. 22) or by depositing a very rough surface coating (Fig. 24). In the latter case, the ratio

Fig. 24:

SEM-picture of an electrolytically deposited rough Au-coating /46/.

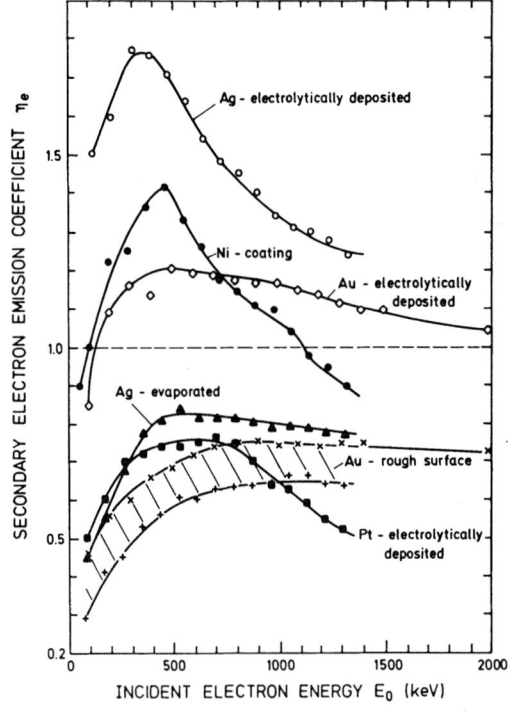

Fig. 25:

Secondary electron emission
yield for different sur-
face coatings /46/.

of the mean free depth to the mean pitch of the coarse-grained
surface has to exceed a critical value, depending on the material,
e.g. greater than 1 for gold. In high magnetic fields the mean pitch
has to be less than the gyroradius of the secondaries. Successes
were achieved with Ag-coated materials, best results being obtained with special Au coatings /46, 47/ (Fig. 25).

2.5 Photon-induced Electron Emission

When surfaces are bombarded with photons having energies $h\nu$
larger than the work function ϕ of the solid, electrons from the
conduction band can be directly emitted upon photon absorption
(photo effect, reviews /48 to 50/). At higher photon energies high-
energy electrons may be produced and also inner-shell electrons may
be emitted. Further electrons may be produced and emitted by
multi-step processes. For a theoretical treatment of the phenomenon
quantum-mechanical approaches as for secondary electron emission
are necessary.

The photoelectron emission yield depends strongly on the
material and the work function and is very sensitive to surface
conditions and impurities (Fig. 26). Relatively old experimental

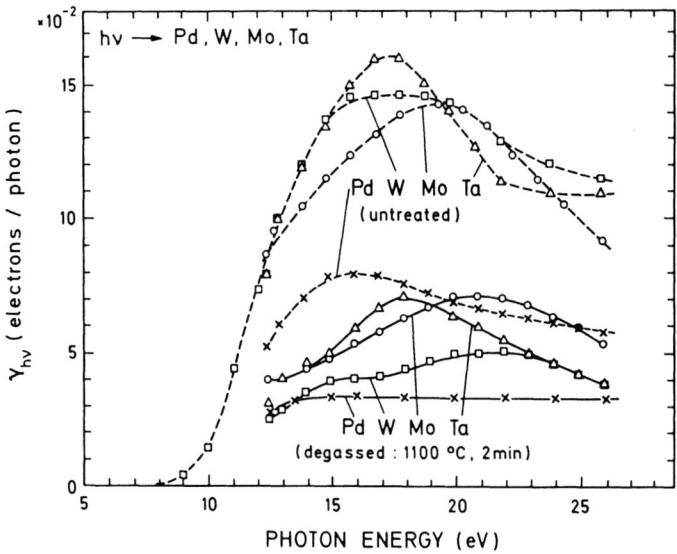

Fig. 26: Electron emission yield at photon bombardment of Pd, W, Mo and Ta for different surface conditions /48/.

data mainly on not well-defined surfaces are available for small photon energies up to the UV-range from photoelectric investigations for photo-cathodes in multiplier tubes. For higher photon energies there are only few experimental results for the absolute electron yield available. It was found, however, that the curve of the energy dependence of the photoelectron emission yield is very similar to the photo-absorption curve of the material considered /51/, and thus some qualitative data can be estimated from mostly well-investigated photon absorption curves (Figs. 27, 28).

The energy distribution of the emitted electrons peaks similar to the ion- and electron-induced emission at a few eV. The structured high energy tails for higher photon energies have been measured in some detail with respect to surface analysis with UPS and XPS /53, 54/ (ESCA), but the main intensity is contained in the low energy peak of the distribution.

3. CONCLUSION

The data available on electron emission from solid surfaces for fusion-device-relevant materials and projectiles are still incomplete especially for low ion energies and highly charged ions as well as for well-defined and clean surfaces. On the other hand, in plasma devices the surface composition of the vessel walls is

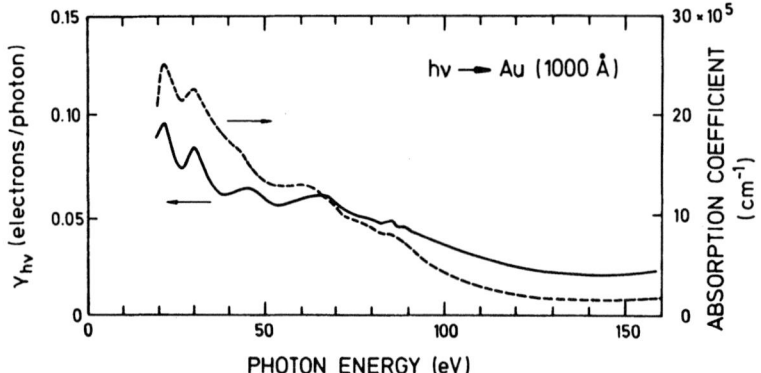

Fig. 27: Electron emission yield and absorption coefficient at photon bombardment of Au /52/.

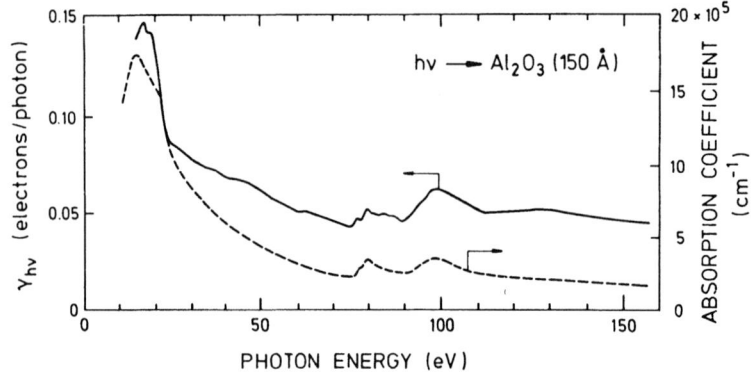

Fig. 28: Electron emission yield and absorption coefficient at photon bombardment of Al_2O_3 /52/.

generally not well defined. It represents a mixture of all materials once used in the plasma vessel. Also a better knowledge of wall fluxes and of the energy distributions of ions, electrons and photons is necessary.

ACKNOWLEDGEMENTS
    We would like to thank Dr. P. Varga and Dr. D. Ruzic for helpful comments in preparing this manuscript.

4. REFERENCES

1. G.D. Hobbs, J.A. Wesson, Culham Laboratory Report CLM-R57 (1966)
2. G.D. Hobbs, J.A. Wesson, Plasma Phys. 9:85 (1967)
3. P.J. Harbour, M.F.A. Harrison, Nucl. Fusion 19:695 (1979)
4. G. Fuchs, A. Nicolai, Nucl. Fusion 20:1247 (1980)
5. R. Gomer, Field Emission and Field Ionization, Harvard University Press (1961)
6. R.H. Good, E.W. Müller, Handbuch der Physik 21, Springer Verlag, Berlin (1956)
7. W.B. Nottingham, Handbuch der Physik 21, Springer Verlag, Berlin (1956)
8. F. Kohlrausch, Praktische Physik 3, B.G. Teubner Verlag, Stuttgart (1968)
9. C. Kittel, Introduction to Solid State Physics; 3rd edition, J. Wiley & Sons, New York (1967)
10. M. Kaminsky, Atomic and Ionic Impact Phenomena on Metal Surfaces, Springer Verlag, Berlin (1965)
11. I.A. Abroyan, M.A. Eremeev, N.N. Petrov, Soviet Physics Uspekhi 10:332 (1967)
12. V.A. Arifov, L.M. Kishinevskii, E.S. Mukhamadiev, E.S. Parilis, Soviet Phys.-Tech. Phys. 18:118 (1973)
13. G. Carter, J.S. Colligon, Ion Bombardment of Solids, Heinemann Educational Books, London (1969)
14. R.A. Baragiola, Rad. Effects 61:47 (1982)
15. H.D. Hagstrum, Phys. Rev. 96:336 (1954)
16. M. Delaunay, S. Dousson, M. Fehringer, R. Geller, P. Varga, H. Winter, Verh. DPG (1985) and private communication
17. W. Hofer, thesis Technische Universität Wien (1983)
18. D.B. Medved, Y.E. Strausser, Advances in Electronics and Electron Physics 21:101 (1965)
19. K.H. Krebs, Fortschritte der Physik, 16:419 (1968)
19a. P. Sigmund, S. Tougaard in: Inelastic Particle Surface Collisions, Springer Series in Chemical Physics 17:2, Springer Verlag, Berlin (1981)
20. J. Schou, Phys. Rev. 22:2141 (1980)
21. E.S. Parilis, L.M. Kishinevskii, Soviet Phys. Solid State 3:885 (1960)
22. O.B. Firsov, Soviet Phys. JETP 7:308 (1958)
23. O.B. Firsov, Soviet Phys. JETP 9:1076 (1959)
24. E.W. Thomas, in Data Compendium for Plasma Surface Interactions, Nucl. Fusion, special issue (1984)
25. E.V. Alonso, R.A. Baragiola, J. Ferron, A. Oliva-Florio, Rad. Effects 45:119 (1979)
26. L.N. Large, Proc. Phys. Soc. 81:175 (1963)
27. R.A. Baragiola, E.V. Alonso, A. Olivia-Florio, Phys. Rev. B 19:121 (1979)
28. G.K. Wehner, Z. Phys. 193:439 (1966)
29. H.J. Klein, Z. Phys. 188:78 (1965)

30. W.H.P. Losch, phys. stat. sol. (a) 2:123 (1970)
31. H. Bethge, J. Heydenreich, Elektronenmikroskopie, Springer Verlag, Berlin (1982)
32. K.G. McKay, Advances in Electronics and Electron Physics 1 (1948)
33. R. Kollath, Handbuch der Physik 21, Springer Verlag Berlin (1956)
34. A.J. Dekker, Sol. State Phys. 6:251 (1958)
35. O. Hachenberg, W. Braun, Advances in Electronics and Electron Physics 11 (1959)
36. H.J. Hunger, L. Kuechler, phys. stat. sol. (a) 56:K 45 (1979)
37. E.H. Darlington, J. Appl. Phys. (J. Phys. D) 8:85 (1975)
38. M. Rösler, W. Brauer, phys. stat. sol. (b) 104:161 (1981) and phys. stat. sol. (b) 104:575 (1981)
39. H. Seiler, J. Appl. Phys. 54:R1 (1983)
40. E. Sternglass, Westinghouse Res. Sci. Paper 1772 (1954)
41. E.M. Baroody, Phys. Rev. 78:780 (1950)
42. D. Ruzic, R. Moore, D. Manos, S. Cohen, PPPL report 1874 (1982) and J. Vac. Sci. Techn. 20:1313 (1982)
43. S. Thomas, E.B. Pattinson, J. Appl. Phys. (J. Phys. D) 2:1539 (1969)
44. I.M. Bronshtein, S.S. Denisov, Soviet Physics, Solid State 9:731 (1967)
45. L.J. Haworth, Phys. Rev. 48:88 (1935)
46. H. Derfler, J. Perchermeier, H. Spitzer, Max-Planck-Institut für Plasmaphysik (IPP), 8046 Garching, german patent # 3247268
47. J. Brunnhuber, Max-Planck-Institut für Plasmaphysik (IPP), 8046 Garching, unpublished data, and IPP annual reports (1982 und 1983)
48. G.L. Weissler, Handbuch der Physik 21, Springer Verlag Berlin (1956) 342
49. P. Görlich, Advances in Electronics and Electron Physics 11 (1959)
50. A.H. Sommer, Photoemissive Materials, Wiley New York (1968)
51. W. Gudat, C. Kunz, Phys. Rev. Lett. 29:169 (1972)
52. W. Lenth, DESY Internal Report 1F 41/75/07 (1975)
53. M. Cardona, L. Ley, Topics in Applied Physics 26 (1978)
54. L. Ley, M. Cardona, Topics in Applied Physics 27 (1978).

PROPERTIES OF MATERIALS

M.F. Smith and J.B. Whitley

Sandia National Laboratories
Albuquerque, N.M.   87185 U.S.A.

INTRODUCTION

The selection of materials for plasma-interactive components in
a magnetic confinement fusion device is a complex process that
usually involves numerous trade-offs. In general, the selection
criteria are heavily weighted according to specific design objectives
or constraints, and it is difficult to make broad statements
regarding the properties of "good" versus "bad" materials. Accordingly, no attempt is made in this chapter to describe a set of
properties for an optimum material, but rather, the purpose here is
to review some important fundamental properties that can be used to
guide the selection or development of materials for a specific
first-surface application.

The following discussion is divided into four sections. The
first section describes major differences in the basic structures of
metals, ceramics, and graphites. In the second section, physical
properties which strongly influence materials performance in a fusion
plasma environment are discussed. The third section treats the
response of materials to severe thermal loads that can be imposed by
contact with a plasma. This section includes several figures-of-
merit and materials comparisons relating to thermal stress. The
final section deals with radiation effects on materials.

STRUCTURE OF MATERIALS

Candidate materials for plasma-interactive surfaces can be
broadly classified as metals, ceramics, or graphites. Important
generic differences among the properties of these three classes of

539

materials are fundamentally related to differences in inter-atomic bonding. Thus, a brief review of the basic structures of metals, ceramics, and graphites will provide a good foundation for the subsequent discussions of materials properties and responses to the fusion environment.

Metals

Metallic bonding, which distinguishes metals from other materials, is characterized by partially filled valence bands that contain empty electron energy states immediately above the Fermi level. Thus, very little energy is required to excite electrons near the Fermi level into higher available states. In metals that have a high degree of crystalline perfection, the mean free path for these excited electrons is long, and they act as very effective carriers of both electric charge and thermal energy. For this reason many pure metals are very good electrical and thermal conductors. Factors which decrease the effective density of conduction electrons or reduce the electron mean free path (i.e., disrupt the periodicity of the lattice) impede electronic conduction. Impurities, alloying, phase changes, etc. can, therefore, strongly influence conductivity. For this reason, stainless steels and other highly alloyed metals tend to be relatively poor electrical and thermal conductors.

The excellent ductility of many metals (i.e., the ability to deform permanently without fracturing) is also fundamentally related to the delocalization of electrons in metallic bonding. However, in order to gain some intuitive insight regarding the effects of crystal structure, impurities, etc., it is more instructive to discuss ductility in terms of mobile crystal defects called dislocations*. There are several types of dislocations, however, they all influence ductility in essentially the same way. The edge dislocation, shown in Fig. 1, is the easiest to visualize and can be thought of as a line extending along the edge of an extra half-plane of atoms in a crystal. The movement of dislocations in response to an applied stress facilitates deformation of a crystal. The shear stress shown in Fig. 1 causes the edge dislocation to migrate across the crystal from left to right in wave-like fashion. This dislocation motion results in a permanent displacement of the top half of the crystal relative to the bottom half. If the crystal in Fig. 1 did not contain a dislocation, all atoms above the midplane of the crystal would have to shift in unison in order to produce this same net displacement. This motion would require a much higher stress since

---

*Although ductility is usually controlled by dislocations, other mechanisms such as twinning and grain boundary sliding are important in some cases. For a more comprehensive discussion of factors that influence the mechanical properties of metals see Hertzberg [1].

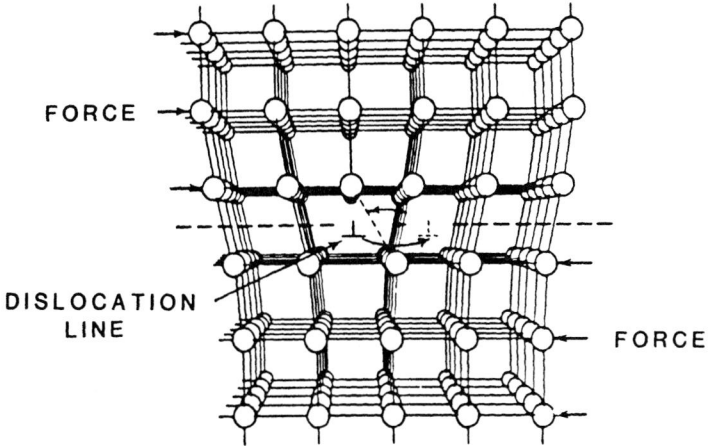

Figure 1.   The edge dislocation shown above is a linear defect that
extends into the page along the lower edge of an extra
half-plane of atoms. As the edge dislocation migrates
through the crystal, the top half of the crystal is
permanently displaced relative to the bottom half.

many atomic bonds must be broken simultaneously, and it is likely
that the crystal would fracture. Consider by analogy a person trying
to move a heavy rug across a floor. By flicking the rug so as to
introduce a large wrinkle that passes across the rug as a running
wave, the rug can be moved in short jerks with a modest level of
effort. But, if one attempts to move the rug by exerting a steady
pull, it is necessary to pull very hard, and the rug may tear. Thus,
the concept of dislocation motion provides a somewhat intuitive model
that can be used to explain differences in the strength and ductility
of metals.

Dislocations occur naturally in metals in concentrations ranging
from 10^2 to 10^{11} per square centimeter. There are many different
ways to modify the concentration and mobility of dislocations in a
metal in order to alter its strength and ductility. A simple example
is found in plain carbon steel. Plain carbon steel is made by
alloying carefully controlled small amounts of carbon with iron. By
proper heat treatment, a two phase microstructure consisting of small
particles of iron carbide in an iron matrix can then be formed. The
small carbide particles set up local stress fields that impede the
motion of dislocations and thus increase the strength of the metal.
However, this increase in strength is achieved at the expense of a
corresponding decrease in ductility and, compared to pure iron, steel
has a greater tendency to fracture rather than bend or deform when a
high stress is applied.

Many structural metals, such as austenitic steels, aluminum, and copper, have a face-centered cubic (fcc) crystal structure. The fcc structure is highly symmetric with essentially equivalent inter-atomic bond strengths in several directions. Thus, dislocations can move with the same activation energy along equivalent planes of varying orientations, and the crystal can readily deform in response to an applied stress acting in almost any direction.

Body-centered cubic (bcc) metals, such as ferritic steels, molybdenum, and tungsten, do not have as many directions of equivalent bond strength, and deformation is more difficult in some cases. The force required to move a dislocation through the crystal (called the Peierls-Nabarro or Peierls force) also has a significant temperature dependence in bcc metals. In general, the Peierls force decreases rapidly with temperature, and bcc metals typically exhibit a marked increase in ductility above a minimum temperature known as the ductile-brittle transition temperature (DBTT). It is noteworthy that the DBTT is not a fundamental material constant, but varies according to factors which influence the mobility of dislocations in the metal, e.g., impurities, alloying, process history, radiation damage, etc. The DBTT also tends to increase with increasing strain rate.

Beryllium, alpha phase titanium, and a few other metals have non-cubic crystal structures. The room temperature ductility of non-cubic metals is usually low because dislocation motion is restricted to only a few planes that have the lowest activation energies for dislocation migration. This inherent limitation can be overcome to some degree by minimizing the crystal size and maximizing the randomness of crystal orientation in a polycrystalline metal, so that the bulk material is more isotropic and less brittle than a single crystal. The ductility of some non-cubic metals also improves dramatically with increasing temperature, as more energetic deformation modes are activated. For example, the ductility of one grade of polycrystalline beryllium increases from roughly 5% elongation at room temperature to nearly 50% at 600 Kelvin [2].

Ceramics

Simply stated, ceramics are compounds that contain metallic and non-metallic elements. Inter-atomic bonding in most ceramic compounds has a partly ionic, partly covalent character. The alkali halide compounds tend to be predominantly ionically bonded, as do compounds of group II and group VI elements. Covalent bonding predominates in a few compounds that contain group IV elements, e.g., SiC. The high melting points of most ceramics are a consequence of these strong ionic and covalent bonds. Unlike metals, valence electrons in many ceramics are highly localized with band gaps of 7 eV or more. Therefore, ceramics are generally poor electrical conductors and thermal conduction is limited to phonons in most cases. Because phonon conduction involves lattice vibrations, it is influenced by factors such as inter-atomic bonding forces and atomic masses. As explained in the subsequent discussion of thermal properties, this leads to relatively high thermal conductivities in some low atomic number (low atomic mass) ceramics such as BeO and SiC.

Due to strong electrostatic interactions in ionic ceramics and directionality of bonding in covalent ceramics, the energy required for dislocation motion is normally much higher in ceramics than in metals. For this reason, dislocation motion in ceramics is usually insignificant except at very high temperatures. With a few exceptions, ceramics are therefore relatively brittle, and their use in structural applications is limited.

Graphites

As shown in Fig. 2, single crystal graphite has a layered structure composed of parallel sheets of carbon atoms that are covalently bonded in hexagonal arrays. The bonds between the sheets are much weaker resonating bonds known as π bonds. These large directional variations in bonding result in highly anisotropic properties for graphite crystals. For example, the coefficient of thermal linear expansion for a single crystal parallel to the layered planes is approximately $1 \times 10^{-6} \text{ K}^{-1}$, whereas, in the direction normal to the layered planes it is $27 \times 10^{-6} \text{ K}^{-1}$ [3]. The interested reader is referred to a book by Kelley [3] for a comprehensive discussion of the fundamental properties of graphite.

There are many different processes for producing polycrystalline bulk graphites. By manipulating the degree of crystallinity, particle size, orientation, and void texture, many different combinations of physical and chemical properties can be achieved in the bulk material. Hence, there are wide variations in properties among commercially available graphites. The graphites that are potentially interesting for fusion applications can be broadly classified into three major types. The first type is produced by blending fine coke or graphite particles with an organic binder such as pitch or resin. The resulting mixture is shaped by pressing or extruding and is then heated in several stages to a maximum temperature of approximately 3000 to 3300 K in order to drive off volatiles and graphitize the carbon. Multiple cycles of impregnation with an organic fluid followed by a furnace treatment are usually required in order to reduce the porosity in the final product. The bulk properties of polycrystalline graphite produced in this way are usually somewhat anisotropic due to preferential alignment of particles during the pressing or extruding operation.

The second type of graphite, pyrolytic graphite, is produced by decomposing a gaseous hydrocarbon at the heated surface of a substrate in order to produce a carbonaceous deposit. By adjusting

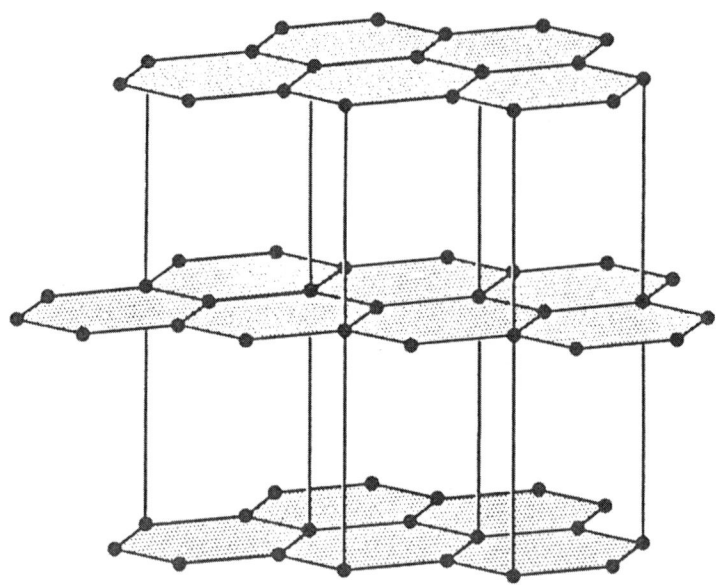

Figure 2. Crystal structure of graphite. Strong covalent bonding
within the planar hexagonal arrays of atoms and much
weaker π bonding between these planes produce highly
anisotropic properties in single crystal graphite (taken
from [4]).

process variables such as the initial hydrocarbon, gas flow rate, hot
zone geometry, and substrate surface temperature, a wide range of
structures can be achieved in the final material. At one extreme the
polycrystalline product is highly oriented with anisotropic
properties that closely resemble a large single crystal; and at the
opposite extreme the deposited material is highly disordered with
relatively isotropic properties. By introducing appropriate
compounds in the gas phase, it is also possible to co-deposit carbon
with other elements such as silicon.

A third type of graphite for possible fusion applications is the
carbon-carbon composite. Carbon-carbon composites are produced by
weaving a felt of carbon fibers which is then infiltrated with an
organic liquid, such as pitch or resin, and subsequently heated to
graphitization temperatures. Alternatively, the felt material can be
infiltrated with carbon by chemical vapor deposition (CVD) and then
graphitized. In either case, multiple infiltration and heating
cycles are normally required in order to reduce the porosity of the
final material. The carbon fibers are commonly produced by a complex
heat treatment of a rayon or acrylic fiber precursor. Because carbon

544

fibers have very small diameters, they are relatively free of the defects which limit tensile strength in most brittle materials. In addition, these fibers have a highly oriented structure with the high-strength direction parallel to the length of the fiber. Hence, these fibers are exceptionally strong and tend to reinforce the bulk composite. Carbon-carbon composites are usually stronger than other forms of bulk graphite with tensile strengths on the order of 35 to 100 MPa (5,000 to 15,000 psi). However, the properties of most carbon-carbon composites are very anisotropic due to preferred orientation of the carbon fibers.

PHYSICAL PROPERTIES

The physical properties of a material have a critical influence on its performance in a fusion plasma environment. Because the macroscopic properties of a material are largely determined by its basic physical and chemical structure, some understanding of the relationships between properties and structure is important in order to quickly identify promising general types of materials for a specific first-surface application. This section describes some of the mechanical, electrical, and thermal properties which impact the performance of materials that are exposed to a plasma and discusses structural influences on these properties. General trends in the properties of metals, ceramics, and graphites are also described to illustrate how structure can be used as a guide in the selection and development of materials.

Mechanical Properties

Four fundamental concepts of mechanical behavior will be discussed. These are: 1) elastic deformation, 2) plastic deformation, 3) fatigue, and 4) toughness.

Elastic Deformation. When a force acts on a material, deformation occurs. The force per unit area of the material is termed stress (σ), and the resulting deformation per unit length of the material is termed strain (ε). Consider the stress-strain diagram in Fig. 3 for a bar of brittle material pulled in tension. Note that the strain is linearly proportional to the stress up to the point of fracture. Furthermore, if the stress were removed at any point below the point of fracture, the strain would be completely recovered and the bar would return to its original length. This occurs because atomic bonds are not broken, but merely stretched, and there is no permanent shift in the relative positions of atoms. This type of stress-strain response is termed linear-elastic. The simple equation that relates stress and strain in the linear-elastic regime is Hooke's law:

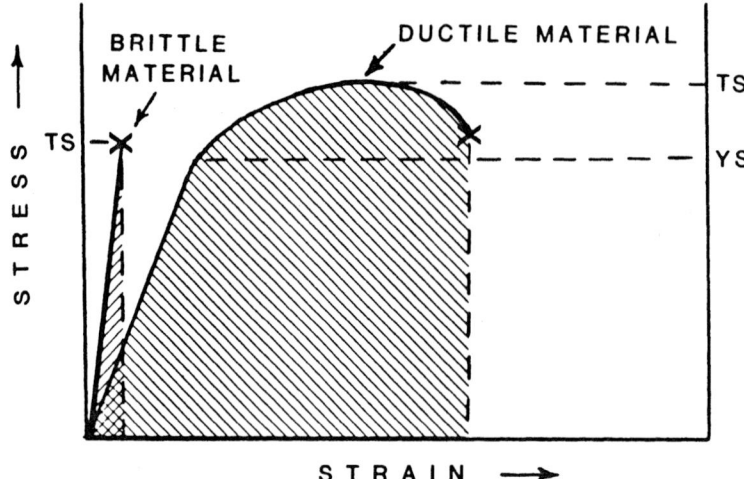

Figure 3. Comparison of engineering stress-strain relationships for
typical brittle and ductile materials. Plastic
deformation of the ductile material results in a much
larger area under the stress-strain (force-displacement)
curve. Therefore, much more energy is required to
fracture the ductile material. TS = Tensile Strength,
YS = Yield Strength, X = Fracture.

$$\sigma = E_Y \, \epsilon \tag{1}$$

The constant of proportionality between stress and strain, E_Y, is the
modulus of elasticity or Young's modulus. The modulus of elasticity
is closely related to inter-atomic bonding forces and has dimensions
of force per unit area. Ceramics, due to their strong ionic/covalent
bonds, typically have high moduli (200,000 to 350,000 MPa). Metals,
with somewhat weaker metallic bonding, tend to have slightly lower
moduli (50,000 to 200,000 MPa). Polycrystalline graphites have very
low moduli (5,000 to 15,000 MPa) because of the weak inter-planar
bonds in the individual graphite crystals.

The modulus of elasticity varies inversely with the distance of
atom or ion separation to the fourth (or higher) power. Since the
equilibrium distance of atom separation increases with increasing
temperature (thermal expansion) the modulus of elasticity decreases
as the temperature of the material increases. However, this decrease
is relatively gradual for most materials, with typically only a few
percent change in modulus per 100 K change in temperature.

From an engineering viewpoint, the modulus of elasticity is an
important indicator of materials response. For example, a high-
modulus material deflects less in response to a given applied stress

than a low-modulus material. Conversely, stresses induced by dimensional changes resulting from thermal expansion, radiation swelling, etc. would be lower in a low-modulus material. It is noteworthy that a few materials exhibit a small non-linear region of elastic (fully recoverable) deformation at the upper end of the elastic regime. However, this slight departure from linearity is seldom significant for design purposes.

Plastic Deformation. As shown in Fig. 3, a ductile material initially exhibits a linear-elastic response, but a transition to a much more complex stress-strain relationship occurs above a critical stress called the yield stress* or yield strength. This transition corresponds to a change in the atomic mechanism of deformation. As the stress increases, a point is reached where a significant number of atomic bonds begin to break and reform, permanently shifting the relative positions of atoms. Therefore, a material that is stressed beyond its yield strength will not return to its original state after the stress is removed because only the elastic portion of the deformation is recovered. The non-recoverable or permanent portion of the total deformation is called plastic deformation or plastic strain.

Dislocation migration, which was explained in the preceding discussion of the structure of metals, is the primary mechanism of plastic deformation in most ductile materials. The yield stress is therefore related to the onset of significant dislocation movement. For this reason, the yield strength of a material may have a strong temperature dependence, and it is also influenced by microstructural changes which alter the energy requirements for dislocation migration. (In contrast, the modulus of elasticity, which is fundamentally related to inter-atomic bonding forces, has only a slight temperature dependence and is relatively insensitive to microstructural variations.) Examples of some common microstructural variables that can influence the yield strength of a material are grain size (the size of individual crystals in a polycrystalline material), foreign atoms (impurities or alloy elements), particles of a second phase, and defects due to mechanical deformation (cold working). In addition, factors such as elevated temperatures, radiation, etc. in the fusion environment can alter the microstructure of a material, and significant changes in strength may therefore occur during the service life of a fusion component. Thus, yield

---

*Technically, the stress at the precise point of transition from elastic to plastic deformation is called the elastic limit. Because the elastic limit is very difficult to measure, the yield stress, which is operationally defined as a measurable point on the stress-strain curve, is used instead. For engineering purposes, the difference between the elastic limit and the yield stress is negligibly small.

strengths for design calculations must be carefully selected to assure that they are representative of the condition of the material, and possible changes in strength due to the service environment should be considered.

The maximum stress on each of the engineering stress-strain curves in Fig. 3 is termed the ultimate tensile stress or simply the tensile strength. An applied tensile stress equal to the tensile strength will result in fracturing of the material. Tensile strength is influenced by the same variables that affect yield strength. In addition to the variables that have already been discussed, the strengths of both brittle and ductile materials sometimes vary according to the rate at which a load is applied. Such materials are said to be strain rate sensitive. Strain rate sensitivity varies widely among different materials, however, it is generally less important in face-centered cubic (fcc) metals than in other metals, ceramics, or graphites.

There is an apparent decrease in stress between the ultimate stress and the point of fracture for the ductile material in Fig. 3. However, this decrease is really an artifact of a convention that is used to calculate engineering stresses. Engineering stresses are computed on the basis of the initial cross sectional area of a test specimen and do not account for a reduction in the cross sectional area of a tensile test specimen as it plastically deforms before fracturing. Adjustments for this reduction in area are omitted from engineering strengths for two reasons: 1) design calculations are based on original (undeformed) cross sectional areas, and 2) in most engineering applications the range of permissible loads is restricted to values below the yield strength of the material. If a correction is made for the reduction in area due to plastic deformation, a quantity called true stress is obtained. The true stress for most ductile material continues to increase up to the point of fracture.

In general, fusion components are designed so that the yield stress will never be exceeded and the materials always remain within the elastic strain regime. Nevertheless, in special cases it may be possible to design into the plastic strain regime, and thus extend the useful stress range of a material. For example, a water cooled copper test target was thermally cycled in the plastic regime by electron beam heating for more than 10,000 cycles at a power density of 4 kW/cm^2. The result was extensive cracking of the surface due to cyclic plastic deformation, but the cracks were too shallow to significantly degrade the structural integrity or performance of the target [5]. Despite successes such as this, it must be strongly emphasized that designs which permit plastic deformation are outside the realm of conventional engineering experience, and thorough testing of the proposed component is extremely important.

Brittle materials (i.e., most ceramics and graphites as well as a few metals) fracture under tensile loading by the rapid propagation of a crack through the material. As a crack grows, the load bearing area in the plane of the crack is reduced, and the stress (force per unit area) therefore increases. In addition, opening of the crack greatly increases the stress in the material immediately ahead of the advancing crack front. This process is analogous to the amplification of an applied load by means of a simple lever. The fulcrum of the "lever" in this case is near the advancing crack front. As the crack advances, this localized region of higher stress sequentially "unzips" the inter-atomic bonds, and fracturing of the material therefore occurs at a much lower applied load than would be required to instantaneously break all of the inter-atomic bonds across the plane of fracture. This mechanism to locally amplify the applied load does not apply to compressive loading because the applied forces tend to close rather than open cracks. Hence, the compressive strength of a brittle material is generally much higher than the tensile strength, and it is good engineering practice to load brittle materials only in compression wherever possible. Because yielding occurs before fracturing in ductile materials, the useful strength of a ductile material is normally established by the yield stress, which is essentially the same for tensile or compressive loading.

Fatigue. The stress that a material can withstand is much lower for cyclic loading (either thermal or mechanical) than for static loading. In the early days it was said that metals subjected to cyclic stresses became "tired" and failed due to fatigue. It is now understood that fatigue failures result from localized micro-structural changes that produce cracks. The number of cycles that a material will endure decreases as the magnitude of the stress increases. Figure 4 shows a typical S-N curve (cyclic stress versus number of cycles to failure) for fatigue fracture. It is apparent from Fig. 4 that tensile strength and yield strength may be quite misleading if used to guide the design of a component that will be subjected to cyclic stresses. When designing for unlimited cyclic loading, it is necessary to limit stresses to values below the endurance limit, which is the level of stress below which the number of cycles to failure becomes indeterminately large. It is difficult to make general statements or provide "rules-of-thumb" regarding specific reductions in strength due to fatigue. However, it should be apparent that fatigue can substantially reduce the useful strength of a material that is subjected to cyclic loading in a pulsed fusion device.

Toughness. Unlike strength, which refers to the stress that is required to deform or fracture a material, toughness is a measure of the energy that is required to fracture a material under carefully controlled test conditions. Toughness is usually measured with one of several standardized tests. Two widely used tests are the Charpy and Izod impact tests. In these tests a specimen with a carefully

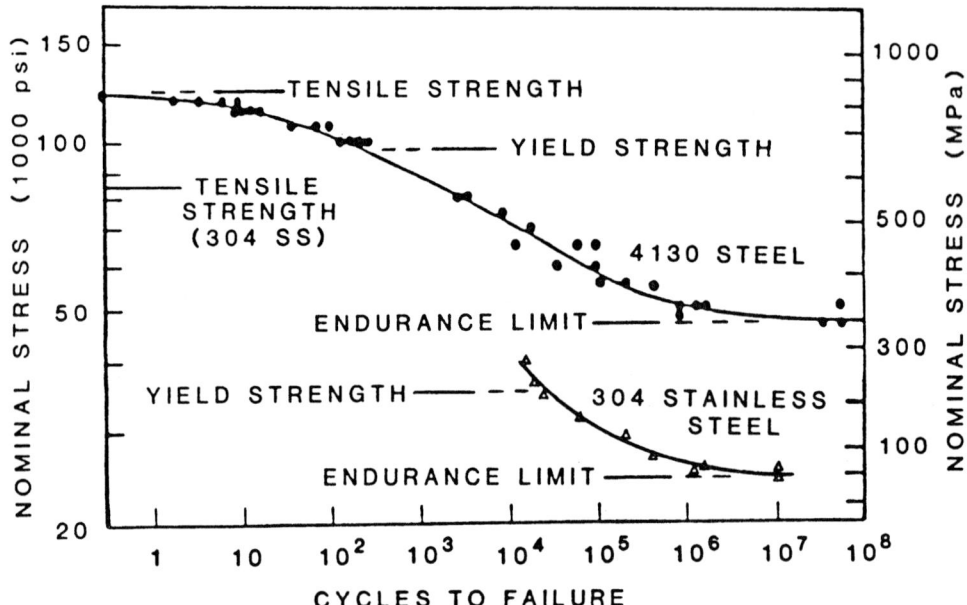

Figure 4. Cyclic stress versus number of cycles to failure (S-N
curves) for 304 stainless steel and 4130 chromium-
molybdenum steel (data from [6,7]). In order to achieve
reliability over an essentially unlimited number of
cycles, cyclic stresses in the material must not exceed
the endurance limit. Hence, cyclic loading greatly
reduces the useful engineering strengths of these metals.

specified geometry is broken by the impact of a large swinging
pendulum. The energy absorbed is then calculated from the arc height
of the follow through swing of the pendulum. If the energy absorbed
is large, the material is said to be tough or to possess high
fracture toughness. Conversely, if the energy absorbed is low, the
material would be described as brittle (low fracture toughness).

Toughness is closely related to the area under a stress-strain
(force-displacement) curve,

$$energy = \int_{0}^{\varepsilon_{fracture}} \sigma \, d\varepsilon \qquad (2)$$

It is apparent from the areas under the stress-strain curves in
Fig. 3 that much larger amounts of energy are absorbed during plastic
deformation than during elastic deformation. Therefore, ductility,
which is a measure of the total plastic strain up to the point of

fracture, can be used as a rough guide concerning the toughness of a material. With some exceptions, metals usually have much higher ductility and therefore superior toughness as compared to more brittle ceramics or graphites. However, extreme ductility alone (e.g., taffy) does not necessarily impart high toughness. Maximum toughness (maximum area under the $\sigma - \varepsilon$ curve) is achieved with an optimum combination of both strength and ductility.

Toughness is an important parameter in engineering design because the small amounts of energy that are required to fracture brittle materials result in a high probability of catastrophic failure in many design applications. The science of fracture mechanics utilizes the concept of toughness to calculate the effects of common flaws (e.g., cracks, holes, grooves, etc.) on the load carrying capability of a material. To present a highly simplified description, the energy required to propagate a crack through a material must be provided by mechanisms such as the release of stored elastic energy or work done by movement of external loads. Hence, for a specified load, flaw size, geometry, etc., equations based on an energy balance can be used to determine whether or not the flaw will cause unstable crack growth resulting in fracture of the material. Similarly, the maximum permissible (critical) flaw size can be calculated for a specified set of conditions. An excellent introduction to the complex subject of fracture mechanics is presented in a book by Hertzberg [1].

Electrical Properties

Ohm's law of electrical conduction is one of the most widely known and practically useful laws of physics. This law expresses a linear relationship between the electric current density $\vec{j}$ and the electric field vector $\vec{E}$

$$\vec{j} = \sigma_{el}\vec{E} \tag{3}$$

The constant of proportionality σ_{el} is the electrical conductivity, which is a scalar for most common materials. Ohm's law applies to both good and poor electrical conductors. The difference between good conductors and insulators is simply the magnitude of σ_{el}. As shown in Fig. 5, σ_{el} is of the order of 10^5 ohm^{-1}cm^{-1} for the better metallic conductors and 10^{-12} ohm^{-1}cm^{-1} for good insulators.

The electrical conductivities of materials that are used in plasma interactive components can be very important. For example, limiter blades or other plasma-interactive components that contain materials with high electrical conductivities can experience large Lorentz forces during off-normal events such as disruptions. Similarly, differences in electrical potential can develop across an insulating material and may cause arcing in some applications. Differences in the basic structures of metals, ceramics, and

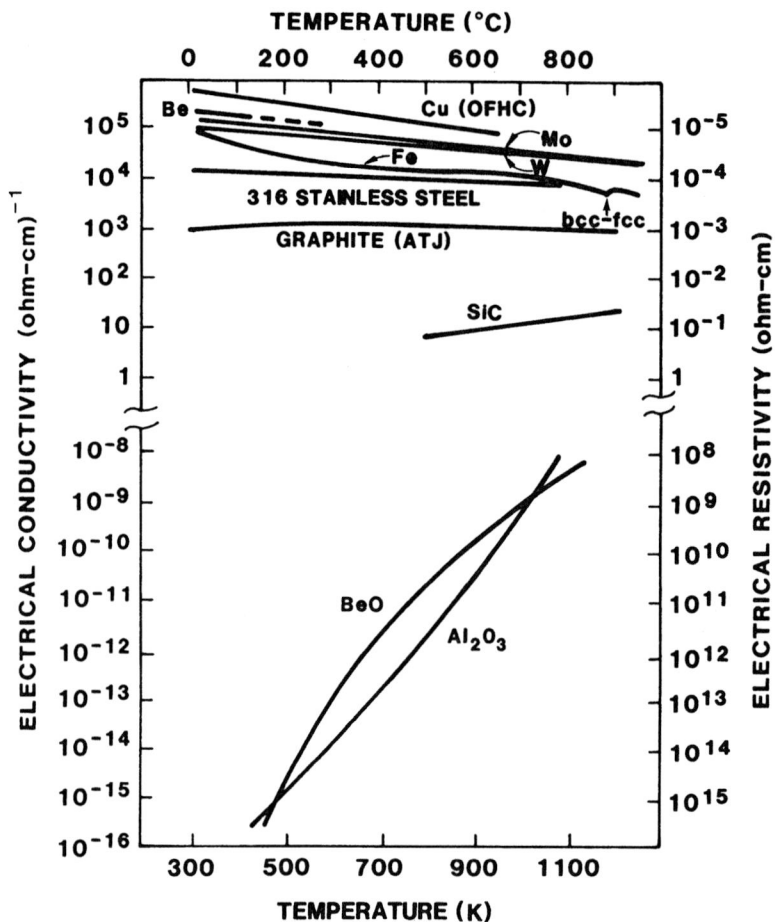

Figure 5. Electrical conductivities of representative metals,
ceramics, and a polycrystalline graphite (data from
[8-14]). Materials such as Al_2O_3 and copper are not
promising candidates for plasma-interactive surfaces, but
they have been included in here and in other figures
because they provide familiar benchmarks that can be used
to place the properties of less familiar materials in
perspective.

graphites strongly influence electrical conductivity. We will now examine electrical conduction in these three general classes of materials.

Metals. As noted in the preceding discussion of the structure of metals, metallically bonded materials have partially filled valence bands that contain empty electron energy states immediately above the Fermi level. Thus, electrons influenced by an electric field can be excited to higher energy states with small energy absorption. These excited electrons have a mobility in the solid which arises from an overlap of atomic state functions and could, in principle, traverse a perfect crystal without suffering any collisions with lattice ions. Because an electric field biases the velocity distribution of these mobile electrons, there is a net charge transport and hence an electric current.

Factors which influence electrical conduction in a metal can be explained in terms of the following expression for electrical conductivity

$$\sigma_{el} = n_{ce} e^2 \tau_{er}/m^* \tag{4}$$

where n_{ce} is the number of conduction electrons, e is the electron charge, τ_{er} is the relaxation time for conduction electrons, and m^* is the effective electron mass. Both n_{ce} and m^* are influenced by the band structure of a metal. Small but important differences in the band structures of pure metals therefore cause systematic variations in electrical conductivity across the periodic chart.

The monovalent metals, Li, Na, K, etc., are not candidates for first wall surfaces, however, they are the simplest metals and will serve as a useful introduction to metals with more complex band structures. The single valence electron in a monovalent metal fills only half of an allowed energy band. This band structure favors electrical conduction for two reasons. First, because the s state band is only half filled, the Fermi level falls near the maximum in the roughly bell-shaped distribution curve for the density of s electron states. Consequently, the number of electrons with energies just below the Fermi level is large, and there are also many unoccupied states with energies slightly above the Fermi level. When an electric field is applied, a large number of electrons can therefore be easily excited to higher energy states (i.e., large n_{ce}). Second, the value of m^* near the Fermi level is only slightly larger than the free-electron mass. Thus, the electrons are readily accelerated by an electric field.

Electrical conductivities for divalent metals, Be, Mg, etc., tend to be lower than those of the monovalent metals. The two valence electrons nearly fill the s state band and only partially fill the higher p state band. Hence, the Fermi level falls near a

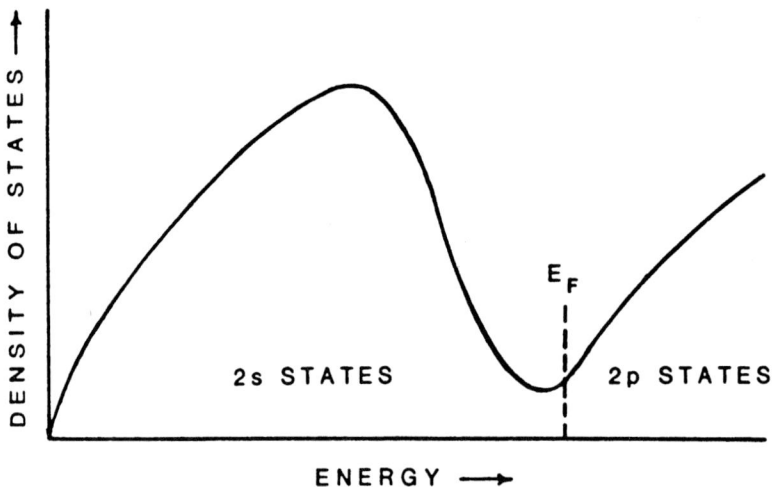

Figure 6.   Density of electron energy states for beryllium. Note
that the density of states at the Fermi level (E_F) is low
due to the small overlap of the 2s and 2p states (taken
from [15]).

minimum in the density of states distribution curve as shown in
Fig. 6. In comparison to the monovalent metals, the density of
states near the Fermi level in a divalent metal is therefore lower,
and fewer electrons can be easily promoted to higher energy states
when an electric field is applied (i.e., n_{ce} is reduced). In
addition, m* near the high energy limit of the s state band is much
larger than the free-electron mass. An increase in m* tends to
reduce electron mobility and therefore decreases σ_{el} as indicated in
Eq. 4.

In general, transition metals (metals such as Ti, V, etc., that
have partially filled d states) have poor electrical conductivities
in comparison to simpler metals. Transition metals are characterized
by extensive overlap of s and d state bands. Details of the band
structure are not fully understood, but it is thought that the s
state band is very broad with an average effective mass that is close
to the free-electron mass. The d state wave functions are highly
localized, with relatively little overlap from one atom to another.
Thus, the d state bands are probably narrow with a large effective
mass. Because the high effective mass of the d electrons drastically
reduces their response to an electric field, it is assumed that s
electrons are the primary current carriers in transition metals.
However, s electrons that have been accelerated by the electric field
are easily scattered by crystal defects, lattice vibrations, etc.
into empty d states. Therefore, the conductivities of these metals
are expected to be lower than the conductivities of similar metals in
which all d states are completely filled.

554

The group IB metals, Cu, Ag, and Au, are excellent electrical conductors. These metals have a single s electron outside completely filled inner shells. Thus, they are similar in many respects to the high-conductivity monovalent metals that were discussed earlier. Unlike the transition metals, the d states in the group IB metals are completely filled. Therefore, electrical currents carried by s electrons are not diminished by scattering into empty d states.

The relaxation time τ_{er} in Eq. 4 is defined as half of the average time between electron collisions or scattering events. Conduction electrons can be scattered by physical lattice defects, such as foreign atoms, vacancies, dislocations, etc., and by thermal vibrations of the lattice. Mathiesson's rule states that the total electrical resistivity R (defined as $1/\sigma_{el}$) is simply a linear sum of contributions from each scattering mechanism

$$R_{Total} = R_{Defect} + R_{Vibration} \tag{5}$$

Foreign atoms, both impurities and alloy elements, generally cause a much greater increase in resistivity than other lattice defects such as vacancies, dislocations, etc. For this reason highly alloyed metals, such as stainless steels and Inconel, tend to be poor electrical conductors, and metals that are to be used as electrical conductors are strengthened by mechanical deformation (cold working) rather than alloying whenever possible. Radiation damage, which is discussed in a subsequent section of this chapter, can produce impurities (e.g., transmutation products) as well as other types of lattice defects. Thus, radiation damage can also degrade the electrical conductivity of a metal.

The R_{Defect} term in Eq. 5 is normally considered to be independent of temperature. But, diffusion controlled processes in metals held at elevated temperatures can produce gradual micro-structural changes (e.g., healing of lattice defects) that alter the electrical resistivity. Since the rates of diffusion controlled processes are highly temperature dependent, such changes seldom occur at a significant rate at temperatures below roughly one-third of the absolute melting point of the metal.

Above room temperature, electrical resistivity due to the vibrational term in Eq. 5 increases almost linearly with increasing temperature in most metals. Note that the electrical conductivities (1/R) of all of the metals except iron in Fig. 5 decrease essentially linearly with increasing temperature. Even in the case of iron, linear approximations could be used without serious error over limited temperature intervals. Hence, the effect of temperature on the resistivity of a metal can often be estimated with sufficient accuracy by means of the following simple equation

$$R_2 = R_1[1+\alpha_{el}(T_2-T_1)] \qquad\qquad (6)$$

where α_{el} is the temperature coefficient of the resistivity, and the numerical subscripts denote corresponding values of temperature and resistivity. Values of α_{el} for various metals over specified temperature ranges are tabulated in many materials handbooks along with appropriate values of R_1.

Ceramics. As shown in Fig. 5, the electrical conductivities of ceramic materials span a range of roughly twenty orders of magnitude. Many different mechanisms contribute to this wide variation in conductivity, and a thorough treatment of this subject is beyond the scope of this chapter. We will, however, discuss the basic characteristics of electrical conduction in ceramics and point out some important differences between ionically bonded versus covalently bonded materials.

Ionically bonded ceramics tend to be poor electrical conductors. In fact, many ionic ceramics are used as electrical insulators. The electronic band structure for a typical electrical insulator is shown in Fig. 7. This band structure consists of a filled band that is separated from an empty, higher energy band by a region of forbidden energy levels. In ionic ceramics the full band (called the valence band) corresponds to the filled shells of the negative ions, and the empty band (called the conduction band) corresponds to the empty valence levels of the positive ions. Electrons in the filled shell are tightly bound and do not move appreciably under the influence of an external field. Therefore, these electrons are not effective charge carriers for electrical conduction.

Electrons excited into the conduction band would have much greater mobility and could act as effective charge carriers. However, in a good electrical insulator the energy gap separating the valence band from the conduction band can be 7 eV or more. Thus, the energy required to promote an electron into the conduction band is much greater than the energy available from thermal excitation (i.e., $E_{gap} \gg k_B T$). In addition, because the density of states at the upper limit of the valence band is low, only a small number of electrons could be promoted into the conduction band by absorbing energies close to the band gap energy. Even higher energies would be required in order to promote electrons from lower energy levels in the valence band. For these reasons, ionic solids with large band gaps have essentially no mobile electrons to act as charge carriers for electrical conduction.

Ions can of course act as charge carriers, and significant ionic conduction does occur in some materials. It is noteworthy that ionic electrical conduction necessarily results in a simultaneous transfer of mass. Also, since ionic conduction is basically a diffusion process, electrical conductivity in ionic conductors tends to

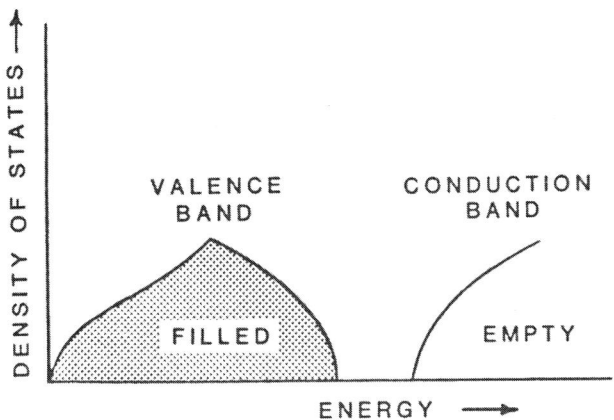

Figure 7.   Schematic representation of the electronic density of
states in an insulator. The forbidden energy region
between the filled valence band and the empty conduction
band (i.e., the band gap) increases the minimum energy
required to excite an electron into a higher available
state. In good electrical insulators the band gap is
often 7 eV or more (taken from [15]).

increase exponentially with increasing temperature. This behavior is
in contrast to electrical conductivity in metals, which usually
decreases as temperature increases.

The band structure of covalently bonded ceramics is similar to
that just described for ionic ceramics. Nevertheless, some covalent
ceramics are relatively good electrical conductors and a few have
conductivities comparable to some metals. This increased conductivity is possible because the energy gap between the valence and
conduction bands in these conductive ceramics is much smaller than
the band gaps in electrical insulators. At temperatures where
$E_{gap} < k_B T$ electrons can be thermally excited into the conduction
band, causing what is termed intrinsic conduction. Although
electrical conduction in such materials usually increases with
increasing temperature, other mechanisms can contribute to the
temperature dependence of σ_{el} and an inverse temperature relationship
occurs in some covalent ceramics over certain temperature ranges.

Impurities and off-stoichiometric compositions can also produce
discrete, allowed energy levels within the otherwise forbidden energy
region that separates the valence and conduction bands. If these
defect energy levels lie immediately below the conduction band,
electrons can be easily promoted from the defect levels into the
conduction band. Similarly, unoccupied defect energy levels
immediately above the valence band can accept electrons from the

valence band, producing so-called "holes" in the valence band. Because defect energy levels can contribute greatly to the production of mobile charge carriers (electrons or holes), electrical conductivity can be strongly influenced by changes in chemical composition. In contrast to metals, where foreign atoms normally decrease electrical conductivity by scattering conduction electrons, foreign atoms in a covalent ceramic can sometimes increase electrical conductivity by increasing the concentration of mobile charge carriers.

In summary, covalent ceramics tend to be better electrical conductors than ionic ceramics, but conductivity may vary greatly depending upon factors such as temperature and chemical composition. Most ceramics have a partly ionic, partly covalent character, and electrical conductivities of ceramic materials therefore span an extremely wide range.

Graphites. Measured values of electrical conductivity parallel to the covalently bonded basal planes in single crystal graphite fall in the range of $1-3 \times 10^4$ ohm^{-1} cm^{-1} at room temperature [16, 17]. Electrical conductivities perpendicular to the basal planes are reported to be two to three orders of magnitude lower [18]. This highly anisotropic conductivity is not surprising considering the large differences in intra-planar versus inter-planar bonding in graphite. The electrical conductivities of most polycrystalline graphites fall between the maximum and minimum values for single crystals. With the exception of highly oriented pyrolytic graphites, electrical conductivity in most polycrystalline graphites is more isotropic than in a single crystal due to the randomization of crystal orientations in a polycrystalline material.

Most bulk graphites are not fully dense, and porosity can significantly reduce electrical conductivity in some cases. Decreases in electrical conductivity due to porosity are proportional to the reduction in cross sectional area of conducting material in a plane normal to the direction of current flow. Densification of bulk graphites upon heating can increase electrical conduction. Nevertheless, electrical conductivity in most high-density bulk graphites tends to gradually decrease with increasing temperature. A representative curve for the electrical conductivity of a high quality commercial graphite (ATJ) is shown in Fig. 5. Electrical conductivity in this particular graphite is somewhat anisotropic due to a slight preferred orientation of graphite crystals in the bulk material. However, these small directional variations in conductivity are insignificant on the scale of Fig. 5.

Thermal Properties

Many different thermal properties impact the performance of materials that are exposed to a fusion plasma environment. In order to limit the present discussion to a reasonable length, we will focus on thermal conductivity, heat capacity, and thermal expansion. Because these properties strongly influence surface temperatures and thermal stresses, they affect materials selection and component design for virtually any first-surface application. These three properties are also important parameters in several figures-of-merit that will be discussed later in this chapter.

Additional thermal properties should normally be considered when selecting materials for a particular application in a fusion device. For example, the heats of fusion and vaporization affect melting and vaporization of first-surface materials during off-normal events such as disruptions or arcs. Similarly, temperature limitations associated with phase transitions, changes in microstructure, diffusion or chemical reaction rates, etc. can be critically important.

Thermal conductivity. The principal carriers of heat in metals are electrons and phonons (quantized lattice vibrations). It can be shown that electronic thermal conductivity k_{el} and phonon thermal conductivity k_{ph} are linearly additive. Hence, the total thermal conductivity is simply

$$k = k_{el} + k_{ph} \tag{7}$$

The magnitude of k_{el} relative to k_{ph} varies greatly. As expected, k_{el} is large in metals that have high electrical conductivities. Therefore, good electrical conductors such as copper, silver, and gold are also excellent thermal conductors. As a rough generalization, k_{el} usually dominates thermal conduction at all or nearly all temperatures in metals with $\sigma_{el} > 10^5$ ohm^{-1} cm^{-1} at room temperature. In metals with lower electrical conductivities, the contribution from k_{ph} is often important at the temperatures of interest for fusion related hardware.

The relationship between k_{el} and σ_{el} has been formalized as the law of Wiedemann-Franz-Lorentz

$$k_{el}/\sigma_{el}T = \pi^2 k_B^2/3e^2 = 2.44 \times 10^{-8} \text{ J Ohm/sec K}^2 \tag{8}$$

where T is the absolute temperature, k_B is the Boltzmann constant, and e is the electron charge. The constant shown on the right side of Eq. 8 is called the Lorentz number. Because k_{el} and σ_{el} are related, the previous discussion of factors that tend to enhance or reduce σ_{el} also applies to k_{el}. Hence, k_{el} is sensitive to the band structure of a metal and is reduced by structural imperfections in the solid (foreign atoms, vacancies, etc.) as well as dynamic imperfections due to lattice vibrations.

The phonon contribution to thermal conductivity k_{ph} is also influenced by imperfections in real solids. In a perfectly periodic crystal which exactly obeys Hooke's law of linear elasticity (i.e.,

559

all inter-atomic restoring forces are linear functions of relative displacement, and the elastic energy is a quadratic function of relative displacement), each elastic wave would be completely independent of all other waves and would maintain a constant level of energy as it traversed the crystal. In such a crystal, heat could be conducted as a net flow of elastic energy without a driving force, and the thermal conductivity would therefore be infinite. However, structural defects and deviations from linear elasticity in real crystals cause scattering and interchanges of energy among lattice waves. Thus, k_{ph} is reduced by crystal defects in a manner analogous to the reductions in electronic conduction that have already been discussed.

In a pure, defect-free crystal the Debye temperature θ_D is related to a characteristic maximum vibrational frequency ν_{max} by the following expression

$$\theta_D = h\nu_{max}/k_B \tag{9}$$

where h and k_B are the Planck and Boltzmann constants, respectively. At temperatures higher than the Debye temperature, the intrinsic thermal resistivity of a crystal lattice is dominated by phonon scattering due to anharmonic effects (i.e., departures from Hooke's law). In this temperature regime the intrinsic contribution to the lattice thermal conductivity falls off roughly as 1/T with increasing temperature. It can also be shown [19, 20] that the overall magnitude of the intrinsic thermal resistivity is qualitatively proportional to $1/\theta_D^3$. Therefore, metals with high Debye temperatures tend to have higher intrinsic values of k_{ph}.

Common structural metals such as iron, copper, and aluminum have Debye temperatures on the order of 300 to 400 K. Metals with higher atomic masses (tungsten, gold, etc.) tend to have Debye temperatures below 300 K. Conversely, beryllium, one of the lightest metals, has an exceptionally high Debye temperature of 1200 K. The high θ_D of beryllium can be understood in terms of the direct proportionality between θ_D and ν_{max} in Eq. 9. To a first approximation, the maximum vibrational frequency of a lattice increases as the strength of the inter-atomic bonding increases (which usually corresponds to higher melting temperatures) and as the mass of the vibrating atoms decreases. Hence, the unusually high Debye temperature of beryllium is consistent with its moderately high melting temperature (1550 K) and its very low atomic mass (9 amu).

Thermal conductivities for several pure metals and alloys are plotted over a range of temperatures in Fig. 8(a). Due to the importance of the electronic contribution to thermal conductivity in metals, the relative ordering of metals in Fig. 8(a) is similar to the order shown in Fig. 5 for electrical conductivities. Copper, which has a very favorable electronic structure, has the highest

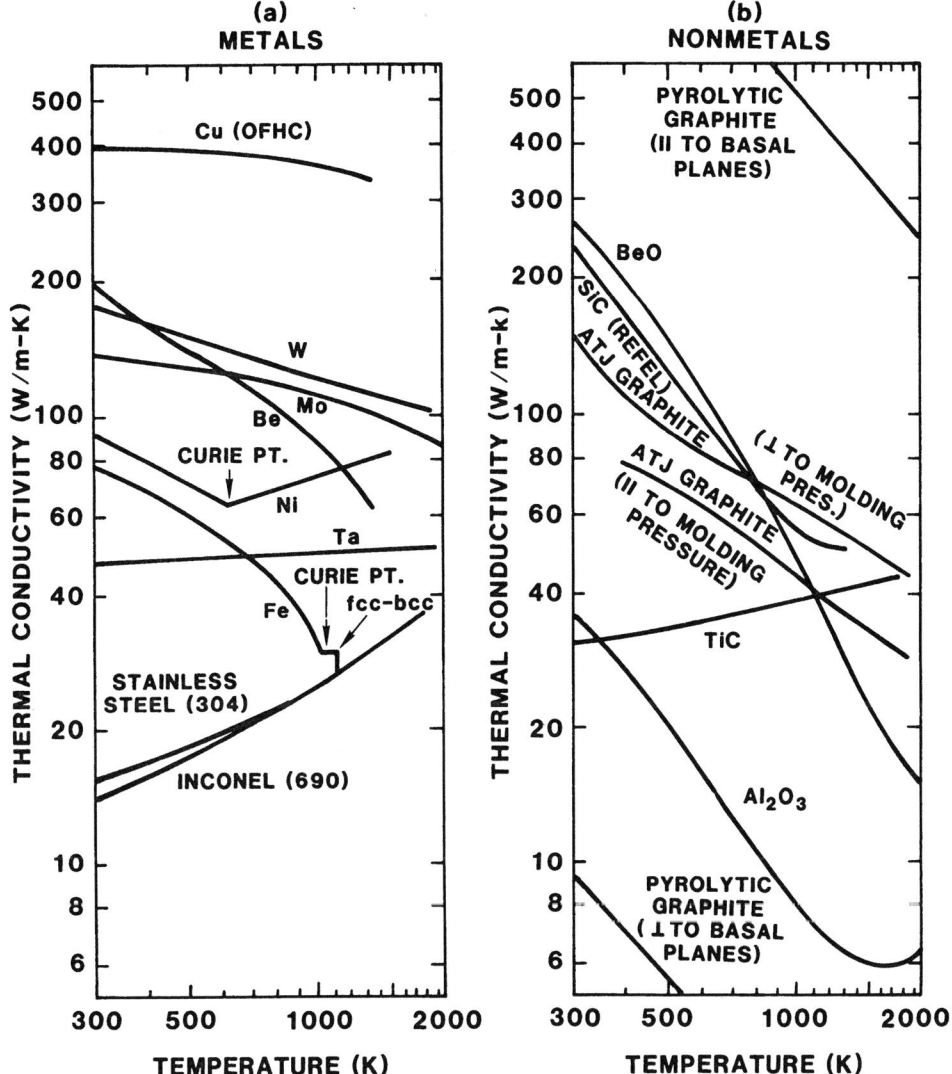

Figure 8. Thermal conductivity versus temperature for representative materials (compiled from data in [8,9,21,22]).

conductivity in both figures; beryllium and the pure transition metals have intermediate conductivities; and highly alloyed metals (stainless steel and Inconel) have the lowest conductivities.

The temperature dependence of the thermal conductivities in Fig. 8(a) varies greatly because there are at least two types of carriers responsible for heat transfer and several different mechanisms can limit the overall thermal conductivity. In addition to the electron and phonon processes that have already been discussed, other mechanisms of heat transfer are important in some cases. For example, cooperative effects between magnetic moments arranged in a regular lattice can result in phenomena known as spin waves or magnons. This mechanism can provide additional heat transfer, but it may also decrease heat conduction by electrons or phonons. These magnetic effects can be important in ferromagnetic or nearly ferromagnetic transition metals. Note that the thermal conductivity curves for iron and nickel in Fig. 8(a) show abrupt changes at their respective Curie temperatures.

Electronic heat conduction in ionically bonded ceramics is usually negligible due to the large band gaps that are characteristic of these materials. Lattice vibrations (phonons) are the primary carriers of thermal energy. Phonon conduction also dominates heat transfer in most covalently bonded ceramics, however, a few covalent ceramics can have a significant electronic contribution. For example, the electronic contribution can become significant at high temperatures in compounds that contain silicon or other semiconductors. Electronic heat conduction in such materials is also sensitive to the stoichiometry of the compound and to the presence of certain impurities.

Thermal conductivities for several ceramic materials over a range of temperatures are shown in Fig. 8(b). As a rough generalization, the ceramics with the highest thermal conductivities (e.g., BeO and SiC) are simple compounds with strong inter-atomic bonds between elements that have low atomic weights. This observation is primarily attributable to increases in k_{ph} with increasing θ_D or ν_{max} as discussed earlier for the case of beryllium metal.

The preceding discussion of phonon conduction in metals applies equally well to ceramics. However, the physical and chemical structures of ceramic materials are usually more complex, and there are often additional factors that influence phonon conduction. For example, materials with complex structures have a greater tendency to scatter lattice waves and therefore tend to have lower thermal conductivities than materials with simple structures. Many ceramics also have anisotropic crystal structures, and this can cause significant directional variations in thermal conductivity. Finally, differences in the atomic masses of the constituent elements in a ceramic compound can produce anharmonicities in the lattice vibra-

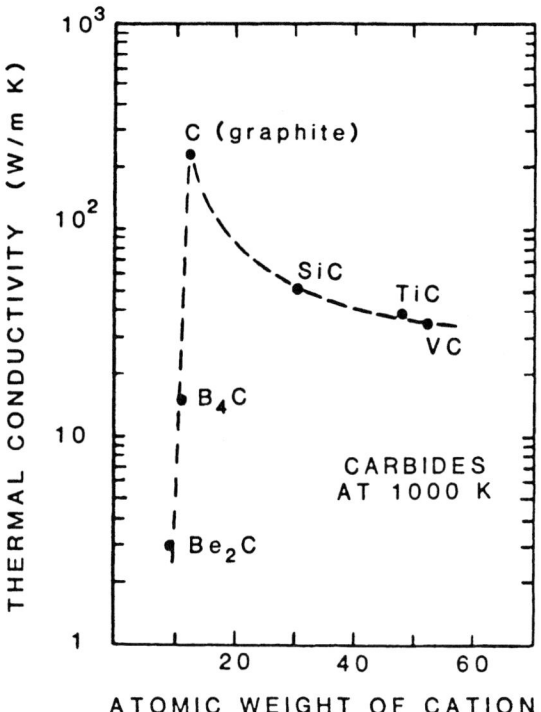

Figure 9. Increasing differences in the atomic masses of constituent
elements in a ceramic compound can produce vibrational
anharmonicities which decrease thermal conductivity. For
this group of carbides, thermal conductivity decreases
both with progressively lighter and heavier cations (The
value shown for pure carbon is an average for pyrolytic
graphite parallel and normal to the basal plane).

tions, and thermal conductivity is therefore reduced. This effect is
illustrated for a series of carbides in Fig. 9. Compared to an
average thermal conductivity for pyrolytic graphite (essentially pure
carbon), the thermal conductivities of the carbide compounds in
Fig. 9 systematically decrease for progressively lighter elements (B,
Be) and also for increasingly heavier elements (Si, Ti, V).

Many ceramics are mixtures of one or more solid phases together
with a pore "phase". The thermal conductivity of the bulk material
therefore depends upon the relative amount and geometry of each phase
as well as their individual conductivities. Most two-phase ceramic
microstructures can be modeled either as a continuous major phase
with a minor amount of dispersed second phase, or as a discontinuous
major phase with the minor phase appearing as a continuous layer
between the major phase particles. For these two cases, the thermal

conductivity of the bulk material k is given by [4]

$$k = k_c \frac{1+2v_d(1-k_c/k_d)/(2k_c/k_d+1)}{1-v_d(1-k_c/k_d)/(k_c/k_d+1)}$$ (10)

where k_c and k_d are the thermal conductivities of the continuous and dispersed phases, respectively, and v_d is the volume fraction of the dispersed phase. If $k_c > k_d$, the conductivity is approximately $k \cong k_c(1-v_d)/(1+v_d)$, and when $k_c < k_d$, then $k \cong k_c(1+2v_d)/(1-v_d)$.

Equation 10 can be used to estimate the effects of porosity on the thermal conductivity of a ceramic, however, the result will vary according to the assumed value for the effective conductivity of the pores. At low temperatures, pores usually have a low thermal conductivity relative to any solid phase. In this case, the thermal conductivity of the bulk material decreases essentially linearly with increasing porosity. However, at higher temperatures radiative heat transfer across the pores may become significant. If the material surrounding the pores is opaque to infrared radiation and the temperature gradient is small, an effective radiation conductivity for a pore can be derived based on an assumption that the radiant energy transfer across a pore is proportional to the temperature difference between pore surfaces [4]. The effective conductivity developed in this way is

$$k_{eff} = 4d_p \eta^2 k_{SB} e_{eff} T_m^3$$ (11)

where d_p is the width of the pore parallel to the direction of heat flow, η is the refractive index, k_{SB} is the Stefan-Boltzmann constant, e_{eff} is the emissivity, and T_m is the mean temperature across the pore. Equation 11 indicates that radiative conduction across pores increases linearly with increasing pore size and with increasing temperature raised to the third power. Consequently, small pores are more effective barriers to thermal conduction except at low temperatures (where radiative conduction is negligible) and at very high temperatures (where the T_m^3 term dominates k_{eff}).

Due to the anisotropic crystal structure of graphite, thermal conductivity parallel to the covalently bonded basal planes is much higher than it is normal to these planes. This difference is dramatically illustrated in Fig. 8(b) by the two curves for thermal conductivity measured parallel and perpendicular to the basal (layer) planes in a highly oriented pyrolytic graphite. It is noteworthy that the thermal conductivity parallel to the basal planes exceeds the conductivity of oxygen-free high-conductivity (OFHC) copper from room temperature up to the melting point of copper (1356 K).

Figure 8(b) also shows two thermal conductivity curves for a commercial hot-pressed graphite (ATJ), which has a more random crystallite orientation. As expected, these curves fall between the

564

two extremes of the pyrolytic graphite. The residual anisotropy in the thermal conductivity of the hot-pressed graphite results from a slight preferential orientation of graphite platelets with the basal planes normal to the direction of the applied molding pressure. Thus, conductivity normal to the direction of molding pressure (parallel to the preferred orientation of the basal planes) is somewhat higher. Thermal conductivities for other commercial graphites are similarly influenced by the degree of preferred orientation. Most commercial graphites also contain some porosity, which reduces thermal conductivity. In cases where the graphitization process has not been carried to completion, residual ungraphitized material can further degrade thermal conductivity.

In comparison to metals and ceramics, there has been relatively little experimental investigation of thermal conductivity in graphite at temperatures above 300 K. In addition, the anisotropic structure of graphite greatly complicates modeling of thermal conduction processes. Consequently, thermal conduction in graphite is not presently understood in detail. The current consensus seems to be that electronic thermal conduction may be important in some materials in some temperature ranges, but phonon conduction appears to be the dominant mechanism of heat transfer at temperatures above 300 K. Calculated thermal conductivities based upon a phonon-phonon scattering model [3] are higher than experimental values over the range from 300 to 1000 K, but the temperature dependence is nearly correct.

We have seen that electrons and phonons are the primary carriers of thermal energy in metals, ceramics, and graphites. Nevertheless, the processes that influence heat conduction in these materials can be quite complicated, as evidenced by the wide variations in magnitude and temperature dependence among the curves shown in Fig. 8. Although a comprehensive treatment of this complex subject has not been attempted here, the preceding discussions of important variables such as band structure, lattice defects, atomic mass, crystal structure, etc. should provide useful guidance for materials selection and insight concerning the qualitative effects of various environmental factors on the thermal conductivity of a material. Additional information concerning changes in thermal properties due to radiation damage is presented in the radiation effects section of this chapter.

Heat capacity -- Heat capacity is a measure of the energy required to raise the temperature of a material; or from a slightly different perspective, it is the increase in energy per degree temperature rise. Based on the concepts of thermodynamics, heat capacity can be more formally defined as the temperature derivative of the internal energy taken at constant pressure C_p or at constant volume C_v. For condensed phases, differences between C_p and C_v are usually small at low temperatures, but significant differences may develop as the temperature approaches or exceeds the Debye temperature. In general, C_v remains essentially constant above the Debye temperature, but C_p gradually increases with increasing temperature. Although increases in C_p at temperatures above θ_D can result from more than one mechanism, a linear increase in C_p with increasing temperature above θ_D is often a satisfactory approximation.

There are several mechanisms by which a solid can absorb thermal energy. For example, energy can be absorbed by increased atomic vibration, increased rotational energy in crystals with rotational degrees of freedom, electronic excitation, and changes in atomic arrangement such as the formation of lattice defects. Above room temperature most of the heat capacity in solids can be attributed to increased atomic vibration. Based on a very simple classical model, each atom in a solid should have an average of $1/2\ k_B T$ kinetic energy plus $1/2\ k_B T$ potential energy for each degree of vibrational freedom. As a three-dimensional oscillator which has six degrees of freedom, each atom should therefore have an energy of $3k_B T$ at thermal equilibrium. A gram-atom of an element contains Avogadro's number N_A of atoms; hence, the total internal energy U is $3N_A k_B T$. The specific heat at constant volume is therefore

$$C_v = (\partial U/\partial T)_v = 3N_A k_B = 3R_g \qquad (12)$$

where R_g is the gas constant. This simple equation is known as the law of Dulong and Petit. With SI units for the value of R_g, Eq. 12 predicts that $C_v = 25$ J/g-atom K.

The predicted value of 25 J/g-atom K for C_v is remarkably close to measured values of C_p in many solids at temperatures above the Debye temperature. The heat capacities for various metals in Fig. 10(a) can be readily converted to a gram-atomic basis by multiplying by the appropriate atomic weights. A comparison of the resulting gram-atomic heat capacities at 500 K is shown in Table 1. Note that the heat capacities for Cu, Mo, Ta, and W, which all have Debye temperatures below 500 K, are very close to 25 J/g-atom K. Although measured values of C_p continue to increase at temperatures above 500 K, the 25 J/g-atom K value is a reasonable approximation up to fairly high temperatures.

The heat capacity for beryllium is significantly lower than the other metals in Table 1. This anomaly can be attributed to its exceptionally high Debye temperature of 1200 K. Below the Debye temperature, the heat capacities of nearly all solids tend to decrease rapidly with a complex temperature dependence. The comparatively high molar heat capacities of iron and nickel in Table 1 are also noteworthy. These elements are ferromagnetic and, as shown in Fig. 10(a), their heat capacities exhibit sharp increases at their respective Curie temperatures. Thus, there is an important additional mechanism of energy absorption in these magnetic materials (i.e., randomization of adjacent spins).

566

Table 1. Comparison of Heat Capacities

Metals	Be	Fe	Ni	Cu	Mo	Ta	W
C_p @ 500 K (J/g-atom K)	21.1	29.9	31.0	25.7	25.3	25.9	25.4
Non-metals	Be_2C	BeO	BN	Al_2O_3	SiC	TiC	Graphite
C_p @ 1000 K (J/g-atom K)	24.4	24.7	22.7	25.0	24.1	25.5	20.7

It might be expected that the valence electrons in metals would make a large contribution to the heat capacity. Based upon a simple free electron approximation, it can be shown that the electronic contribution to the heat capacity C_{el} is [15]

$$C_{el} = (\pi^2/3)\,(3R_g/2)\,(k_BT/E_F) \tag{13}$$

where E_F denotes the Fermi level. Since $k_BT \ll E_F$ at normal temperatures, the electronic contribution is only a small addition to the lattice heat capacity except at very low temperatures (where the lattice heat capacity is small) or at very high temperatures (where the lattice heat capacity is essentially constant). In actuality, many metals melt before C_{el} becomes significant. The relatively small magnitude of C_{el} is consistent with the observation that only those electrons within approximately k_BT of the Fermi level are affected by changes in temperature. Therefore, the vast majority of the electrons in a metal make no contribution to C_{el}.

The molar specific heat of an alloy or compound can often be closely approximated as the sum of the atomic specific heats of its constituent elements weighted according to the atomic fraction X_i of each constituent

$$C_p = \sum_{i=1}^{n} X_i C_{p,i} \tag{14}$$

This equation, known as the Kopp-Neumann law, can be very useful; but it should be applied with caution at temperatures near phase transitions or magnetic transformations.

As shown in Table 1, if C_p values for ceramic compounds in Fig. 10(b) are converted to a gram-atomic basis (i.e., multiply by the formula weight and divide by the number of atoms per formula unit), they all fall close to 25 J/g-atom K at 1000 K. Thus, these

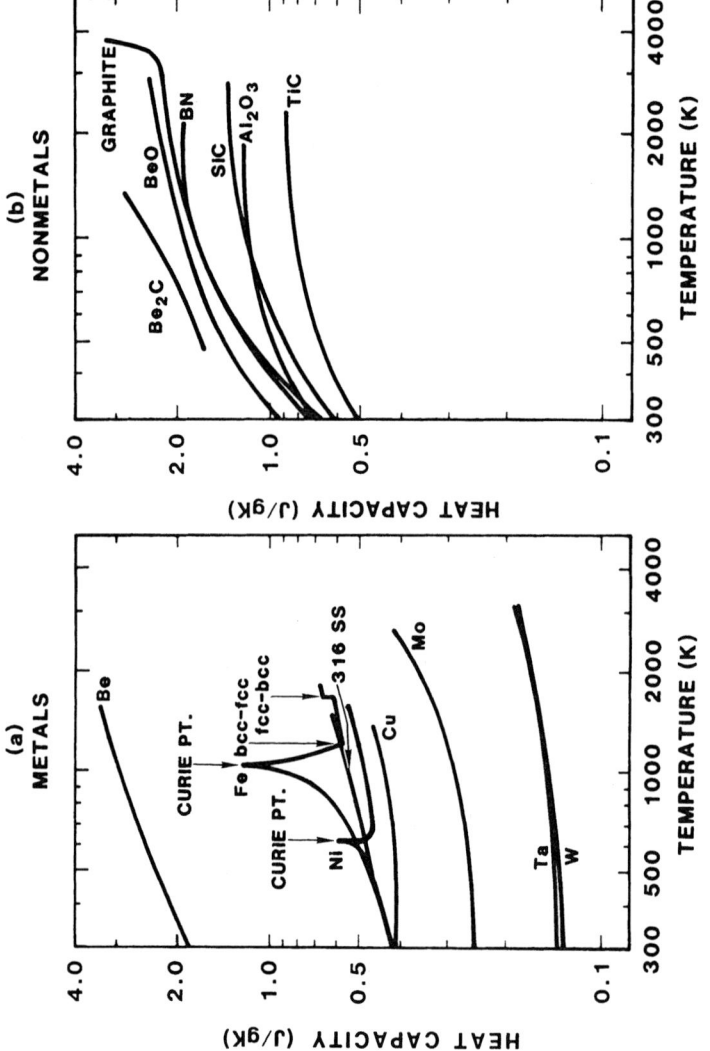

Figure 10. Heat capacity versus temperature for representative materials (compiled from data in [23,24]).

ceramics behave very much like the metals in Fig. 10(a) with the
exception that the Debye temperatures of the ceramics tend to be
somewhat higher. These higher Debye temperatures are apparent from
the rapid decreases in the C_p curves at temperatures below roughly
1000 K. Ceramic compounds that undergo order-disorder transformations can exhibit sharp peaks in their C_p curves. These peaks are
similar in appearance to the Curie point peaks for iron and nickel in
Fig. 10(a), however, the mechanism is different. Like metals,
ceramics may also exhibit abrupt changes in heat capacity as a result
of phase changes. One difference between ceramics and metals is
related to porosity, which is more prevalent in ceramics. Porosity
can significantly affect the heat capacity per unit volume of a bulk
ceramic material.

The Debye temperature of graphite is very high (1500 K), and the
heat capacity of graphite therefore remains well below 25 J/g-atom K
up to relatively high temperatures. As shown in Fig. 10(b), the heat
capacity of graphite also exhibits a sharp, reproducible increase at
temperatures above 3500 K. Hove [27] rejects the possibility that
this increase is related to vaporization based on measurements
carried out in a chamber that was pressurized to 10 atmospheres.
Instead, the exponential rise in C_p is attributed to the reversible
formation of an equilibrium concentration of lattice defects,
specifically vacancies.

Thermal expansion -- Thermal expansion can be readily understood
by considering the plots of cohesive energy versus interatomic
spacing shown in Fig. 11. Note that the potential well in each of
the two cohesive energy curves is non-symmetrical, with a more
shallow curvature at larger atomic spacings. Hence, as the temperature increases (i.e., a shift to higher energy levels in the potential well) the mean atomic separation becomes larger, corresponding
to expansion of the lattice. Consideration of the potential wells in
Figs. 11(a) and 11(b) suggests that there should be less thermal
expansion in a strongly bonded (high melting) material with a deep,
steep-sided potential well, as compared to a weakly bonded material
that has a shallow, more asymmetrical well. This generalization
neglects other important variables such as the crystal structure of
the material, however, it holds reasonably well provided that the
comparison is limited to structurally similar materials.

The change in lattice volume due to thermal expansion is closely
related to the amount of thermal energy that is absorbed. Hence, the
coefficient of thermal linear expansion α_l has a temperature dependence that tends to parallel the temperature dependence of the heat
capacity, i.e., α_l rapidly increases up to the Debye temperature,
with comparatively little additional increase at higher temperatures
(Fig. 12). Two practical consequences of this temperature dependence
are noteworthy. First, it is risky to neglect the temperature
dependence of the thermal expansion coefficient unless the tempera-

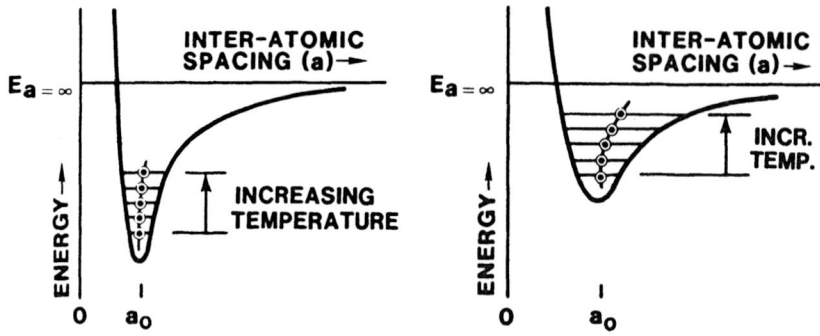

Figure 11. Cohesive energy versus inter-atomic spacing. Due to the
assymetry of the energy well, the equilibrium
inter-atomic spacing (denoted by dots in the diagrams
above) increases as the thermal energy is added to the
system. (a) Strongly bonded (high-melting) material with
low thermal expansion. (b) Weakly bonded (lower melting)
material with higher thermal expansion.

tures of interest are above θ_D for the material, and a value of α_1 at
or above θ_D is used. Second, materials comparisons based solely upon
room temperature values of α_1 (those most commonly tabulated) can be
misleading. In some cases, it may be preferable to compare values of
α_1 for the respective Debye temperature of each material.

Thermal linear expansion coefficients along different crystal
axes in cubic crystals are all equal, and dimensional changes with
temperature are therefore symmetrical. In such systems, the volume
expansion coefficient is simply three times the linear expansion
coefficient. For other, non-symmetrical crystal structures, thermal
expansion is anisotropic and varies along different cystal axes. In
this case, the volume expansion coefficient is the sum of the linear
expansion coefficients in three mutually orthogonal directions.
Common structural metals such as steels, aluminum alloys, and copper
alloys have essentially isotropic thermal expansion. Graphite, many
ceramics, and a few metals such as beryllium exhibit varying degrees
of anisotropic thermal expansion.

It is noteworthy that there are other mechanisms for temperature related volume change. For example, solid state phase transitions often involve substantial volume changes that occur over small
temperature intervals. Such phase changes can also cause abrupt

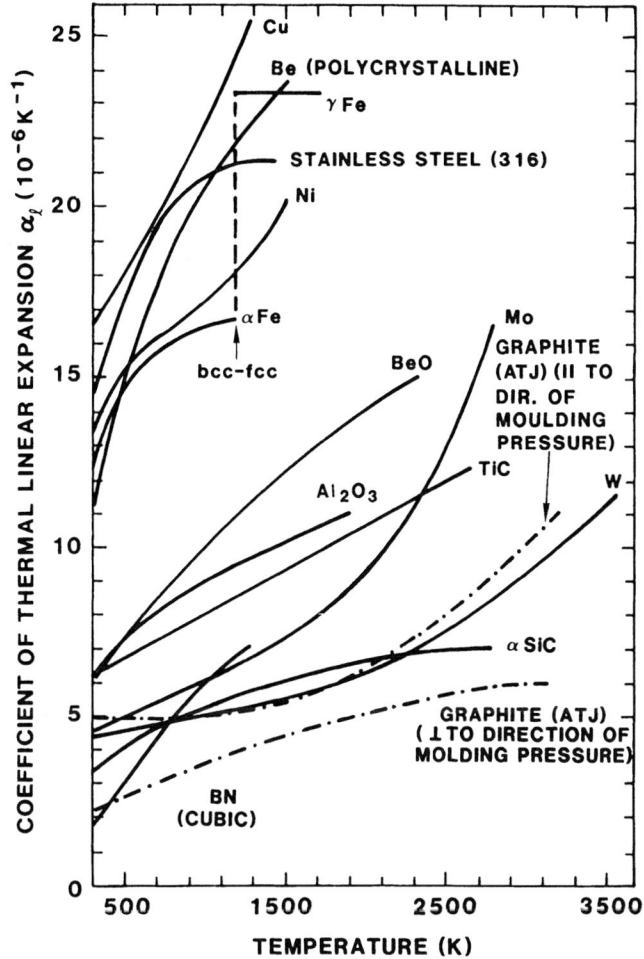

Figure 12. Coefficient of thermal linear expansion versus temperature (compiled from data in [25,26]).

shifts in the thermal expansion coefficient due to the change in
crystal structure. Volume changes associated with phase transitions
and a few other mechanisms can produce large internal stresses which
may cause cracking or fragmentation of the material.

At first glance, some of the information that has been presented
in this physical properties section may seem far removed from the
problem of selecting or developing a material for a particular
first-surface application. Certainly, a comparison of tabulated
values for a particular property can be made without a detailed
knowledge of the underlying variables which control that property.
However, if these variables are clearly understood, the most promising general types of materials can be identified more easily, and the
risk of inadvertently using literature data that are inappropriate to
a particular material condition, application, or service environment is greatly reduced. In addition, there are many instances when
the specific data of interest are not available, or when none of the
candidate materials have acceptable properties. In such cases, the
information that has been presented here should provide a basis for
making reasonable estimates of unknown properties or trends, and may
also suggest ways to develop a better material for a particular
application by methods such as impurity control, alloying, grain
refinement, and so forth.

THERMAL RESPONSE AND FIGURES-OF-MERIT

Plasma interactive materials are, by definition, exposed to
plasma bombardment. This bombardment delivers a heat load to the
component surface, a heat load that is often quite large. This
thermal load will alter the component's temperature distribution and
may often dictate the choice of materials and the design (see [28-34]
for further discussions of this subject). This section will give a
brief overview of the thermomechanical considerations involved in
component design and give a brief description of several figures-of-
merit used to rank high heat flux materials.

Thermal Response

The flow of heat through solids has been found by empirical
observation to be given by a relationship called the Fourier
conduction law, generally written as:

$$q = -k \frac{\partial T}{\partial n} \qquad (15)$$

where q is the flux of heat (energy per unit time per unit area) in
the x direction and k is a constant of proportionality called the
thermal conductivity [35,36]. Note that the thermal conductivity is

a material property and does not depend on the object's configuration. This relationship can be used to develop a general equation that describes the temperature distribution throughout a heated body. By requiring that all points within the body satisfy a volumetric heat balance regulated by Eq. 15 and a similar relationship between the temperature rise and the materials specific heat, a general heat conduction equation can be derived:

$$\rho C_p \frac{\partial T}{\partial t} = \nabla \cdot (k\nabla T) + q* \tag{16}$$

where ∇ is the vector operator, ρ the material density, C_p the material specific heat, t the time, and $q*$ the volumetric heat generation rate.

In general, the physical properties ρ, C_p, and k are functions of temperature and the volumetric heat generation term $q*$ can be a function of both position and time. Internal heat generation is usually included in high heat flux component calculations only when performing detailed calculations on components operating in the presence of large neutron or gamma fluxes or in cases where significant joule heating is generated.

Various simplifications of Eq. 16 are often encountered. If the volumetric heat generation term is neglected and k is assumed to be a constant independent of temperature, then the Fourier equation is derived:

$$\frac{\partial T}{\partial t} = D \nabla^2 T \tag{17}$$

with $\quad D \equiv \frac{k}{\rho C_p}$. $\tag{18}$

where D is called the thermal diffusivity and is the ratio of the thermal conductivity to the specific heat per unit volume. For the special case of steady state heat transfer, the general equation further simplifies to the Laplace equation:

$$\nabla^2 T = 0 \tag{19}$$

When the appropriate boundary conditions are applied, the solutions to the above equations will fully and uniquely describe the temperature distribution development in an object. Full solutions to Eq. 16 usually require the use of numerical methods, and several general computer codes exist that allow three dimensional solutions with temperature dependent materials properties and various boundary conditions such as thermal radiation and convective heat transfer to fluids [37]. An example of a two dimensional temperature profile generated by the ABACUS code is shown in Fig. 13.

To help understand the development of temperature profiles in materials exposed to surface heat fluxes, it is instructive to solve the Fourier equation (Eq. 17) for the case of a semi-infinite slab subjected to a uniform and constant heat flux q_o starting at time t=0

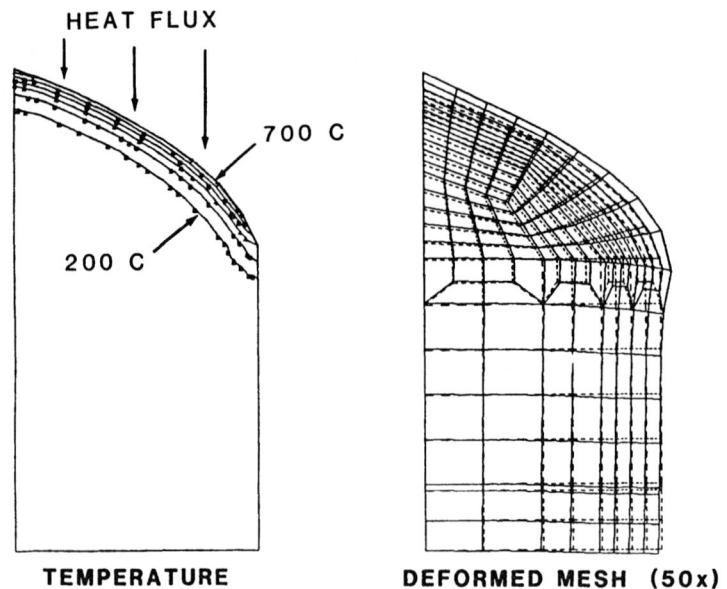

Figure 13.   A two dimensional temperature profile generated by a finite element thermal code. This case is for a beryllium limiter after exposure to a heat pulse of 25 MW/cm^2 for 0.3 seconds.

and ending to time $t=t_1$. The initial starting temperature is defined as T(0,0) and neglecting any cooling by radiation, the solution is found to be [38, 39]:

$$T(x,t) = T(0,0) + \frac{2q_o}{\sqrt{\pi}} \sqrt{t} \; \frac{1}{\sqrt{\rho C_p k}} \left\{ e^{-u^2} - 2u \int_u^{\infty} e^{-\xi^2} d\xi \right\} \quad (20)$$

$$\text{with } u = x/L_{th} = \frac{x}{2\sqrt{Dt}} \quad (21)$$

where L_{th} is called the thermal·diffusion length and is a measure of the depth of penetration of a heat pulse in a time period t. The semi-infinite slab solution is appropriate for cases where the

574

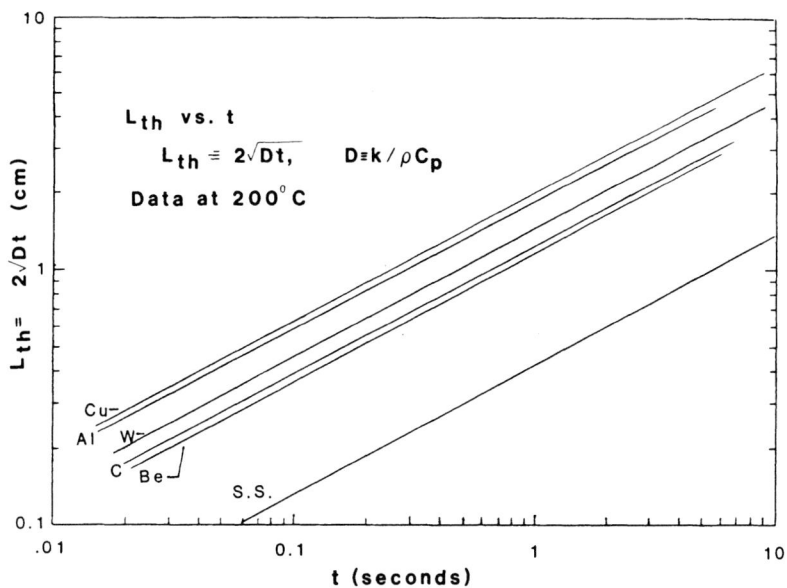

Figure 14. The thermal diffusion length L_{th} vs. time for several different materials.

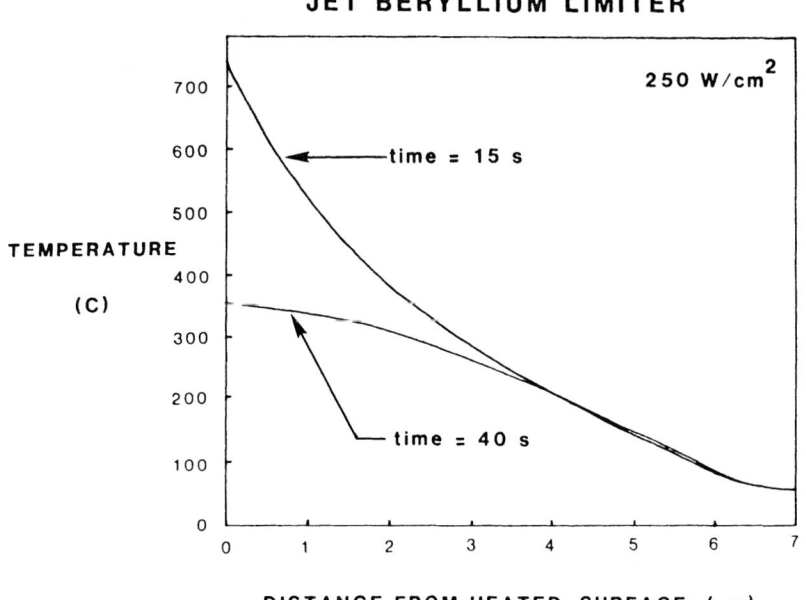

Figure 15. The one dimensional temperature profiles generated by exposing a beryllium surface to a heat flux of 250 W/cm^2 for 15 seconds followed by a cooling period of 25 seconds.

thickness of the solid is greater than a few times L_{th}. Values of L_{th} for several materials as a function of time are shown in Fig. 14. Temperature profiles for the case of a beryllium limiter exposed to a surface heat flux of 250 W/cm^2 for 15 seconds are shown in Fig. 15 for times of 15 seconds (i.e., at the end of the heating pulse) and at 40 seconds (after a 25 second cooling period). The temperature rise at the surface can be found by evaluating Eq. 20 at u=0 and is found to be.

$$T(0,t) = T(0,0) + \frac{2}{\sqrt{\pi}} \; q_o \; \frac{\sqrt{t}}{\sqrt{\rho C_p k}} \qquad (22)$$

Notice that the surface temperature rise is determined by the heat flux q_o, the pulse length t, and by the material's density, specific heat and thermal conductivity, with high values of these material properties desirable to reduce the transient temperature rise.

Many components in long pulse fusion devices are actively cooled. In these cases, the thickness of material between the coolant and the heated surface is often much less than the thermal diffusion length. When a heat pulse is applied to these types of components, the initial temperature behavior is still described by equations 21 and 22, but as the heat pulse reaches the coolant channel, the temperature distribution begins to flatten and eventually reaches a linear profile found by solving the conduction law (Eq. 15); i.e.,

$$T(x) = T_d + \frac{q_o (d - x)}{k} \qquad (23)$$

where T_d is the temperature of the solid at the coolant channel (x=d) with the front surface at x=0. The steady state temperature difference is proportional to k^{-1} and the thickness d, indicating that the material's thermal conductivity is the only material property that influences the temperature gradient. Note that this thermal profile will actually be slightly non-linear in most materials due to the temperature dependence of the thermal conductivity.

Evaporation and Melting

Under some conditions, such as during plasma disruptions, the heat flux and pulse length can be such that the surface temperature exceeds the material's melting or even vaporization temperature. In these cases, the determination of the amount of material lost by evaporation can be a critical factor in estimating the component lifetime. While the amount of material lost by evaporation can be estimated using relatively simple models [33,40], more recent calculations have been performed which carry out a detailed energy balance including surface cooling by evaporation and radiation.

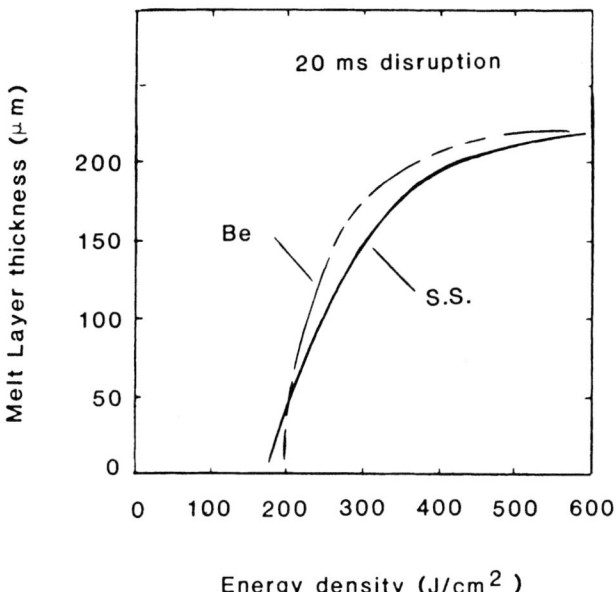

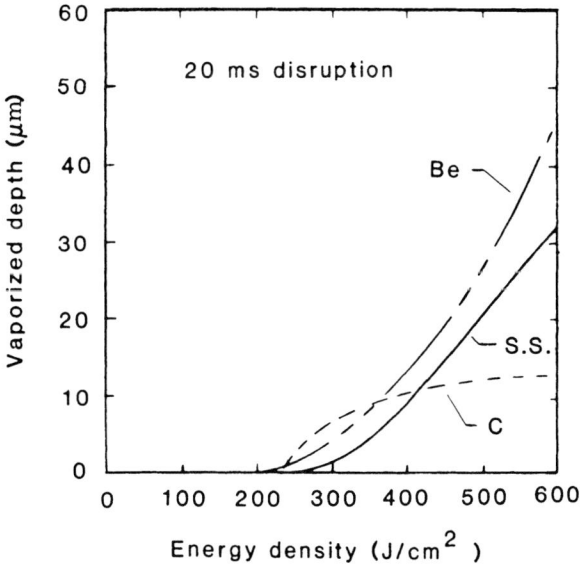

Figure 16. Calculated melt layer thicknesses and vaporization thicknesses for several materials exposed to disruption type heat loads [41,42].

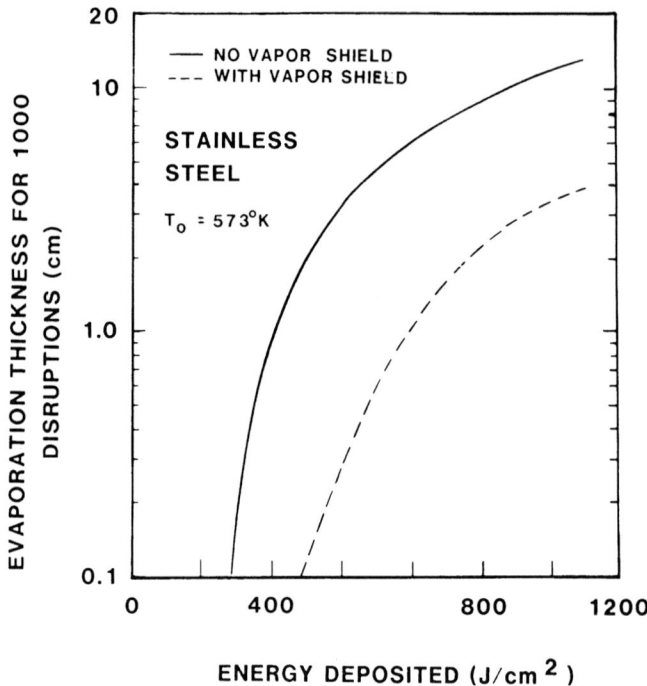

Figure 17. Calculated vaporization losses during plasma disruptions
with and without vapor shielding [42].

These computer codes solve this two moving boundary problem (i.e.,
the liquid-vapor and the solid-liquid interfaces) and predict the
surface temperature rise, the melt layer depth, and the amount of
material lost by vaporization [41,42,43]. A typical result from
these calculations is shown in Fig. 16. In this calculation, all of
the incident energy was assumed to be delivered to the first solid or
liquid surface. The actual heat flux delivered when evaporation is
occuring will usually not be a simple function of the plasma power
incident on the surface but will be reduced by the shielding that
occurs when a vapor cloud forms. This vapor shield can be roughly
modeled by calculating the energy deposition in the vapor cloud and
allowing this energy to radiate isotropically. This often results in
a significant reduction in the incident heat flux as is shown in Fig.
17. One area of uncertainty in these calculations is the stability of
the melt layer. As can be seen from the results of Fig. 16, the
depth of melted material resulting from a single disruption can be
quite large. If the melt layer resolidifies without significant

movement, then the problem of material loss from disruption heat loads seems manageable. If, however, the molten material is redistributed before it solidifies, then the use of many materials such as stainless steel is questionable unless the expected disruption frequency can be significantly reduced [44].

Thermal Stress

Thermal stresses develop when temperature differences induce differential thermal expansion in a constrained material. If a material is unconstrained and the temperature gradient is a linear function of a rectangular coordinate, then no thermal stresses will be generated and the object will assume a curved, stress free shape. If, however, the material is physically constrained from achieving its equilibrium shape or if the temperature gradient is non-linear (such as during transient heating) then stresses will develop [45]. Most high heat flux component designs attempt to minimize external constraints and hence the major stress is often due to the non-linear temperature profile or to constraints imposed by the joining of materials with different thermal expansion properties. Thermal stresses generated by non-linear temperature profiles can be pictured by visualizing a thin heated layer of material near the surface which is constrained from undergoing its normal thermal expansion by the massive underlayer of material at a lower temperature. It is quite common for the thermal stresses generated in high heat flux components to exceed the material's yield strength and cause extensive plastic deformation of the surface. The thermal stress that develops in a constrained body subjected to a heat flux producing a temperature gradient ΔT can be obtained by applying Hooke's law assuming that the strain introduced is given by the product of the materials thermal expansion and the temperature difference in the material. For a material constrained in only one dimension, the stress is described by [46]:

$$\sigma_{th, \ 1D} = -E_Y \alpha \Delta T \tag{24}$$

or, for materials constrained in two dimensions, by:

$$\sigma_{th, \ 2D} = - E_Y \alpha \Delta T \tag{25}$$

where E_Y is Young's Modulus, α the coefficient of thermal expansion, and ν is Poisson's ratio. Note that the thermal stresses are less for the one-dimensional case where the constraint is applied only in one direction and the body is free to expand in the other two directions. In the two-dimensional case, the body is only free to expand in one direction (i.e., perpendicular to the surface). The thermal stress that actually develops in a high heat flux component will be somewhat lower than these values since some deflection of the object will occur and hence relieve the peak stress.

Actively cooled components are often designed with a protective tile rigidly bonded to a substrate to enhance the interface thermal conductance the hence lower the surface temperature. The stresses that develop in these cases are more complex and design specific. Depending on the materials used and the design, these stresses can be quite large and can only be modeled using sophisticated elastic-plastic stress codes [37].

Figures-of-Merit

To aid in the materials selection and screening process, designers often use a "figure-of-merit". A figure-of-merit is a combination of fundamental material properties that attempts to weigh each factor according to the end use environment. A figure-of-merit may be a dimensionless ratio or may be related to a given material response such as surface melting. Caution must be used when studying figures-of-merit, for a given figure-of-merit will present no more information than is contained in its derivation. A figure-of-merit could probably be derived that would highly rank any given material, but that does not mean that this material would be the best choice for the task at hand. The key questions are: 1) is the figure-of-merit closely tied to a critical material response, and 2) does the actual use of the material correspond to the comparison criteria. For example, a figure-of-merit derived on a comparison of the materials elastic behavior should not be used if the components are plastically deformed during use. In general, the use of figures-of-merit is no substitute for extensive screening tests using a representative environment. With these limitations in mind, a few examples of figures-of-merit will now be given.

Transient Heating Figure-of-Merit. Many components in use in operating devices are passively cooled in that very little of the incident energy is removed during the heat pulse. These components are thermally "thick" and simply heat up during the heat pulse and cool during the long rest period between shots. During disruptions, the heat pulse is of such short duration that most components, whether actively or passively cooled, will look thermally thick and hence a figure-of-merit that describes the ability of a material to absorb intense heat pulses without surface melting is useful. Such a relationship can be approximated for a slab geometry by setting the surface temperature rise of Eq. 22 to the value necessary to raise the surface temperature to the materials melting point. That is, for a material with a melting point T_m with a surface temperature of T_o before the pulse, the limiting heat flux is given by:

$$q_o \sqrt{t} = \frac{\sqrt{\pi}}{2} \sqrt{\rho C_p k} \, (T_m - T_o)$$

(26)

580

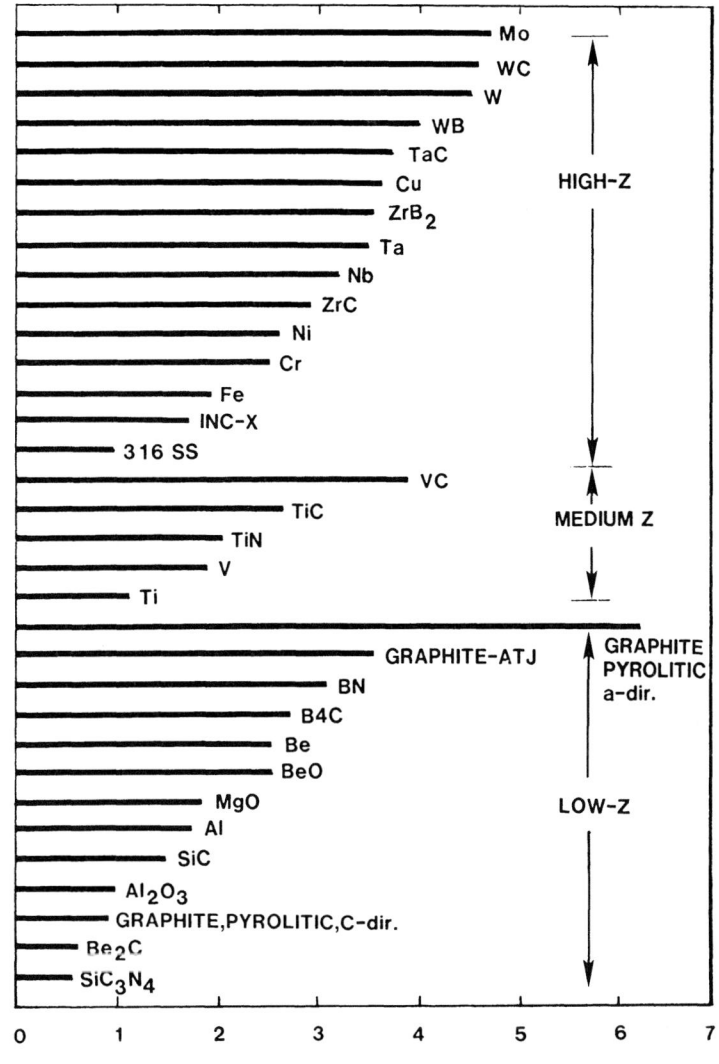

HEAT CAPACITY FIGURE-OF-MERIT $(T_m - T_o)\sqrt{\rho C_p K}$ KJ/cm$^2\sqrt[2]{\text{sec}}$

Figure 18. The transient heating figure-of-merit. Larger numbers imply that more heat can be absorbed in a given amount of time without melting.

581

This factor has been plotted for a variety of materials in Fig. 18. Remember that this solution assumed a uniformly heated, semi-infinite slab and hence the absolute values shown for the limiting heat flux cannot be directed applied to other cases. Note that this comparison is based only on the surface melting criteria, and says nothing about mechanical behavior, vaporization losses, etc.. As can be seen, the refractory metals molybdenum and tungsten rank the highest, with ATJ graphite and copper ranking quite well and stainless steel one of the lowest.

Evaporation Flux Figure-of-Merit. Most next generation machines call for pulse lengths approaching steady state operation. These devices will require high heat flux components that can carry off in some form of ^oolant an amount of energy equal to the incident energy without adversely affecting plasma operation. While a figure-of-merit that compares the relative heat fluxes necessary to melt the material surface is a useful guide in ranking a materials ability to withstand disruption damage, it is not a good comparison to use for steady state operation. Since the purpose of a limiter is to define the plasma edge without contaminating the plasma, a figure-of-merit will be derived which will rank materials by the maximum heat flux they can withstand without exceeding an impurity flux limit based on plasma performance considerations. For this analysis, impurity fluxes generated by evaporation will be considered.

The first step in this derivation will be to define a maximum allowable impurity flux level. An analysis by Cecchi [47,48] defines this level by limiting the plasma impurity concentration to a level which will keep the reduction in Q (i.e., the fusion power gain) due to the impurities to below 10%. The impurities were assumed to influence Q by two mechanisms, namely by impurity radiation reducing the plasma temperature and by impurities reducing the fuel density. The radiation limit for conditions relevant to TFTR was found to require:

$$n_i/n_e < 10/Z^3 \tag{27}$$

and the dilution limit to require:

$$n_i/n_e < 0.05/Z \tag{28}$$

where n_i is the impurity concentration in the plasma and n_e the average plasma electron density. The maximum impurity concentration as a function of the atomic number of the impurity is shown plotted in Fig. 19. Note that at high Z the limit is due to radiation while at low Z the fuel dilution limits the allowable concentration. For a particular combination of limiter area, plasma volume and particle confinement time, a relationship between the maximum allowable impurity flux and the maximum allowable impurity concentration can be

582

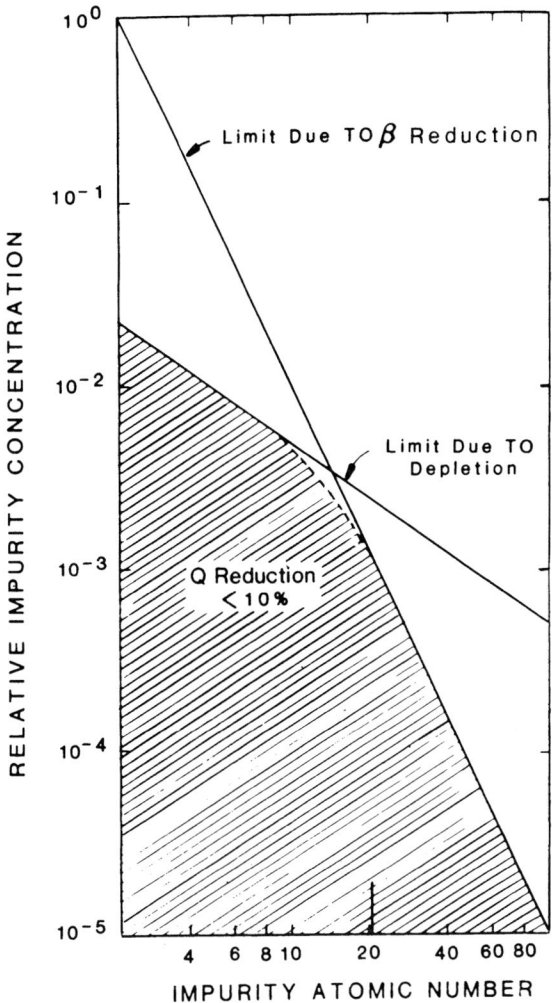

Figure 19. The concentration of impurities that will reduce the fusion power gain (Q) by 10% in a TFTR type plasma.

derived and is found to be given by [47]:

$$\Gamma^{max}_{imp} = \frac{(n_i/n_e)^{max} \, V \, \langle n_e \rangle}{A \, \tau_p} \tag{29}$$

where V is the plasma volume, $\langle n_e \rangle$ the average electron density, A the active limiter area, and τ_p the impurity particle confinement time. For TFTR these values were taken to be $V=4.4\times10^7$ cm^3, $\langle n_e \rangle = 8\times10^{13}$ cm^{-3}, $A=10^4$ cm^2 and $\tau_p=0.05$ seconds. The calculated maximum allowable fluxes for several materials are given in Table 2.

583

Table 2. Maximum allowable plasma concentrations and maximum allowable impurity fluxes for a TFTR type plasma.

Material	n_i/n_e	Maximum flux $(cm^{-2}s^{-1})$
Be	1.2×10^{-2}	8.8×10^{16}
C	8.3×10^{-3}	5.9×10^{16}
Al	3.9×10^{-3}	2.7×10^{16}
Steel	5.7×10^{-4}	4.0×10^{15}
Cu	4.1×10^{-4}	2.9×10^{15}
Mo	1.4×10^{-4}	9.5×10^{14}
W	2.5×10^{-5}	1.7×10^{14}

Having completed the calculation of the maximum impurity flux for the materials of interest, an attempt to compare these materials can be made. For this analysis, the materials will be compared by considering only impurity introduction by evaporation. The rate at which material evaporates from a surface at a temperature T_o is given by [39]:

$$\Gamma_{evap} = \tilde{\alpha} \, 3.5 \times 10^{22} \frac{P}{\sqrt{mT}} \tag{30}$$

where P is the materials vapor pressure (in torr) at the temperature T_o, $\tilde{\alpha}$ is a sticking coefficient taken to be 1, and m the mass of the material. Vapor pressures for several materials are plotted as a function of their homologous melting temperatures in Fig. 20. From this calculation, the surface temperature that corresponds to the maximum allowable impurity flux can be found for each material. This surface temperature can then be related to a surface heat flux using the steady state solution of Eq. 23. The back surface temperature will be defined assuming heat transfer to a liquid coolant governed by Newton's Law:

$$q_o = h(T_d - T_f) \tag{31}$$

where h is the heat transfer coefficient to a fluid, T_d the wall temperature at the coolant channel, and T_f the mean coolant

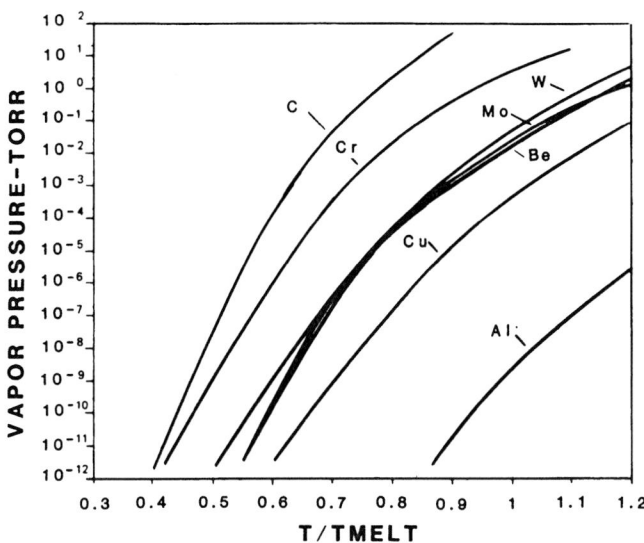

Figure 20. Vapor pressures vs. homologous melting temperature (T/T_m) for several materials. For graphite a T_m temperature of 3200 K was used.

temperature. Assuming that the effective coolant channel area is equal to the front surface area, one then finds that the front surface temperature is related to the incident heat flux by the following relationship:

$$q_o = \frac{T_o - T_f}{\frac{1}{h} + \frac{d}{k}}$$

(32)

where d is the distance below the heated surface, i.e., the thickness.

By setting T_o equal to the maximum allowable surface temperature, the limiting heat flux q_o^{vap} can be calculated as a function of d and h. Plots of these values for the case of perfect heat transfer to the coolant ($h = \infty$) and a typical heat transfer coefficient of $h = 5 W/cm^2 s$ are plotted in Fig. 21. Note that for the limiting case of $h = \infty$, copper is the highest ranking material, with stainless steel the lowest. For cases with an h typical of water cooling, the curves cross with the refractory metals being superior for thin structures but with copper surpassing them for thicker structures.

585

Heat Flux Limit from Evaporation

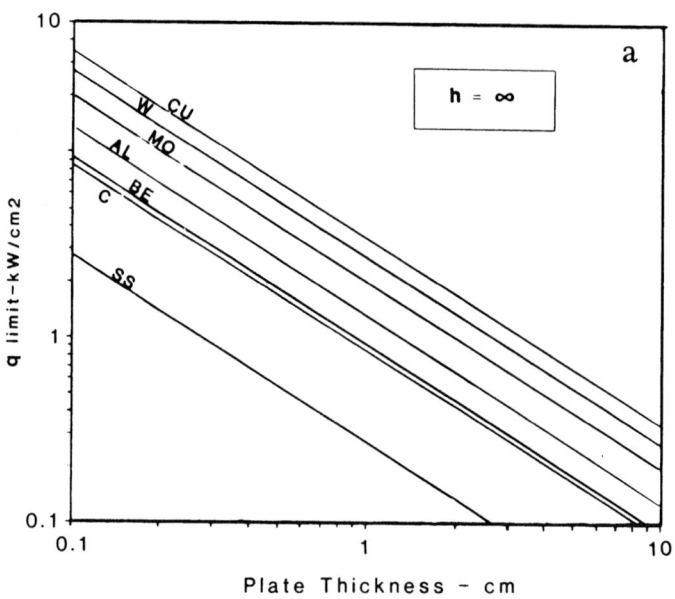

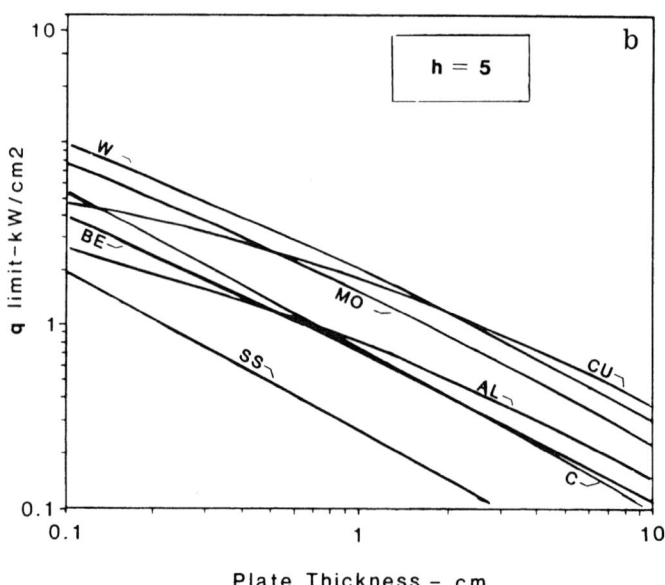

Figure 21. Maximum heat flux limits based on plasma contamination by limiter impurities generated by vaporization. Figure 21a is for perfect heat transfer to a coolant while Figure 21b is for a typical heat transfer value.

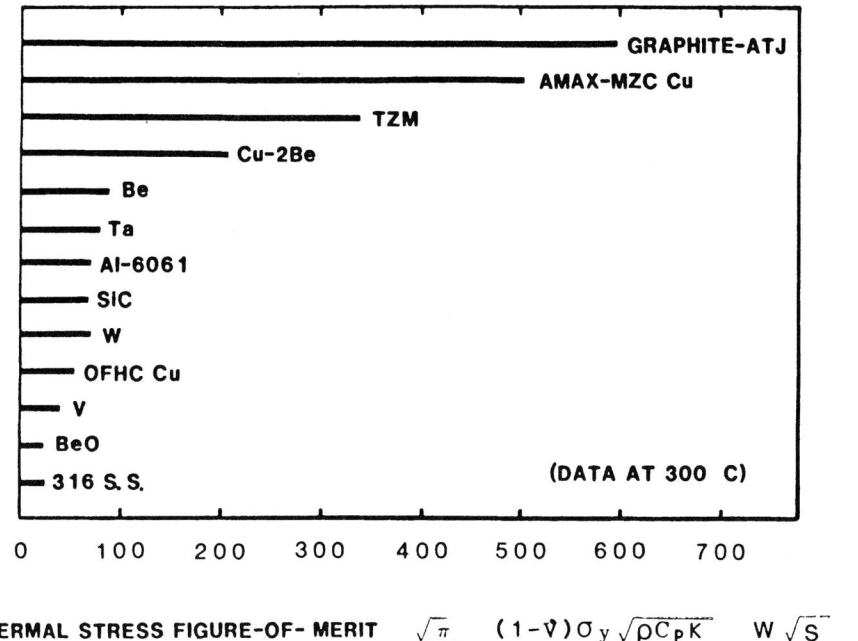

THERMAL STRESS FIGURE-OF- MERIT $\quad\dfrac{\sqrt{\pi}}{2}\quad\dfrac{(1-\nu)\sigma_y\sqrt{\rho C_p K}}{E_y\alpha}\quad\dfrac{W\sqrt{s}}{cm^2}$
(TRANSIENT)

Figure 22. Thermal stress figure-of-merit for transient heating.
Higher values of the figure-of-merit imply that more heat
can be absorbed during a given amount of time without
yielding the material.

Thermal Stress Figure-of-Merit. The thermal stress at the
surface of a heated component constrained from expanding in all
directions except perpendicular to the surface was given by Eq. 25.
By setting this stress equal to the materials yield stress σ_y, a
thermal stress figure-of-merit that defines the maximum heat flux
allowable before this elastic limit is exceeded can be derived for
both transient and steady state heating. For the transient case, the
surface temperature rise solution of Eq. 22 will be substituted for
the temperature difference of Eq. 25, giving:

$$q_o\sqrt{t} = \frac{\sqrt{\pi}}{2}\ \sigma_y\ (1-\nu)\ \frac{\sqrt{\rho C_p k}}{E_y\alpha} \tag{33}$$

The solution for the steady state case is dependent on the type of
constraints applied to the component. If the component is
constrained in such a manner that bending is not allowed but the
component is free to expand in all directions, then the thermal
stress profile that develops will be compressive in the heated half
and tensile in the half near the cooled surface. This implies that
the effective temperature difference generating the surface stress is

587

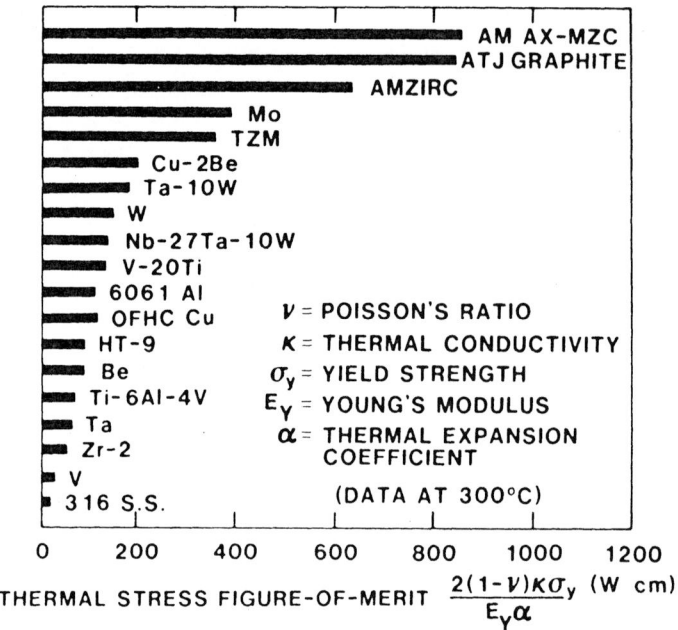

Figure 23. Thermal stress figure-of-merit for steady-state heating. Larger values imply that greater heat fluxes can be removed steady-state without yielding the material.

only half the actual temperature difference given by Eq. 23. Hence the solution for this case is given by:

$$q_o d = \frac{2\,\sigma_y(1 - \nu)k}{E_Y \alpha} \tag{34}$$

These values are shown plotted in Fig. 22 and 23 respectively. In both cases, graphite, a copper alloy and molybdenum show the best performance. Note that for these calculations, unirradiated material properties were used. If a component is to be used in a device that will accumulate a significant neutron exposure, then the effect of the irradiation on these factors must be considered. A brief review of the radiation damage process and its effects on material properties will now be given.

RADIATION EFFECTS

The intense radiation environment present in the first wall region of any D-T or even D-D or D-^{3}He burning fusion reactor will dramatically influence almost all material properties (see, for example, [49-51]). The obtaining of an improved understanding of these changes and the design of materials resistant to these property degradations is a large world-wide effort, the success of which may determine the economic viability of fusion as a power source. This section will give only a very brief introduction to this subject, with numerous references supplied for those interested in learning more about this critical technological problem.

Radiation Damage Production

There are basically two different mechanisms by which neutron irradiation can affect a metal, namely by the creation of impurities by transmutation reactions and by the atoms being physically ejected from their crystal lattice sites by momentum transfer from elastic and inelastic neutron interactions. The calculation of the buildup of transmutation impurities is straightforward and is given by [52]:

$$c_{imp} = x \, \sigma_{trans} \, \phi t \tag{35}$$

where c_{imp} is the impurity concentration, x the number of impurities atoms formed by each interaction, σ_{trans} the effective neutron cross-section, and ϕt the integrated neutron fluence.

With a knowledge of the relevant cross-sections, the impurity buildup can be calculated for any neutron spectrum. From a materials standpoint, the primary concern with these reactions comes from the large amounts of gaseous products formed by (n,p) and (n, α) reactions and to a lesser degree the modification of the alloy composition by transmutation products such as occurs when aluminum absorbs a neutron and subsequently decays to form silicon. The gaseous elements in particular will embrittle many commercial alloys and play an critical role in the formation of defect clusters (see, for example, numerous papers in [53-55]). Transmutation reactions are also the source of the induced radioactivity in fusion components. Most proposed first wall materials (such as stainless steel) will reach activation levels that will require long term radioactive disposal facilities. A requirement that only "low" activation materials be used will severely limit the materials choices available since only a few materials (such as beryllium, carbon, silicon, aluminum, and vanadium) meet this requirement and then only if the levels of impurities are carefully controlled.

The calculation of the total number of atomic displacements caused by neutron irradiation is more complicated since the initial neutron interaction usually initiates a large cascade which in turn produces many subsequent displacements [52,56,57]. The displacement of a lattice atom will form a point defect pair consisting of an interstitial atom and a vacancy, where the interstitial consists of a lattice atom which has been literally "stuffed" into the middle of a set of occupied lattice sites and the vacancy is the hole left in the lattice by the removal of an atom. These defects can move, cluster, be absorbed at sinks, or annihilate each other.

589

The accepted unit of radiation damage is the displacement per atom (dpa) which is equivalent to the average number of times a lattice atom is energetically forced from its lattice site to an interstitial position during the irradiation. Note that this definition includes only those events that produce point defects, and does not include the events where an energetic atom collides and ejects the host atom from its lattice site, but in the collision loses enough energy to become trapped in the ejected atom's site. These replacement events are only important in materials with a specific ordered structure such as ordered alloys.

The rate at which displacements are produced by an arbitrary flux of particles $\phi(E)$ is given by:

$$R_d = N \int_{E_{d/\gamma}}^{\infty} \phi(E) \, dE \int_{E_d}^{\gamma E} \sigma(E,T)\nu(T)dT \qquad (36)$$

where N is the atom density of the target, $\sigma(E,T_o)$ the differential cross section for an incoming particle of energy E to transfer energy T_o to a recoil atom, $\nu(T_o)$ the number of displacements caused by a recoil of energy T_o, with

$$\gamma = \frac{4 \, m_1 m_2}{(m_1 + m_2)^2} \qquad (37)$$

where m_1 and m_2 are the masses of the incoming and target atoms respectively.

The measurement or calculation of $\phi(E)$ and $\sigma(E,T)$ for neutrons is a well developed field and will not be discussed here (see, for example, [57]). In a fusion neutron spectrum, the average energy transfered to a lattice atom will be hundreds of times larger than the energy necessary to move an atom from its lattice site. This energetic recoil atom, called the primary knock-on atom or PKA, will in turn interact with other lattice atoms, producing many more secondary displacements (shown schematically in Fig. 24). This secondary effect is the actual source of the vast majority of the displacement damage under high energy neutron irradiation. The calculation of $\nu(T_o)$ is therefore one of calculating the displacements produced by an energetic, partially ionized projectile that is interacting with both the electronic and nuclear charge distributions within the target. In metals, the interactions of the PKA with the electronic charge distribution of the target do not result in displacements but only dissipate the ion energy, while the nuclear collisions result in momentum transfer from the PKA to the target atoms and hence are capable of generating displacements. The production of displacements in non-metals is complicated by the reduced ability of the crystal to neutralize electronic charge disturbances. For example, displacements can be produced in ionic bonded crystals using low energy X-rays, where the apparent

Point Defects

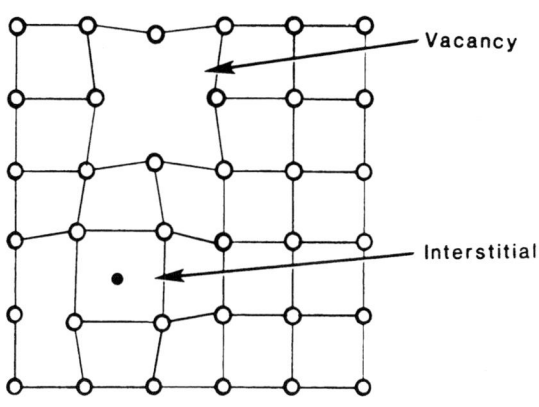

Vacancy

Interstitial

The Displacement Spike

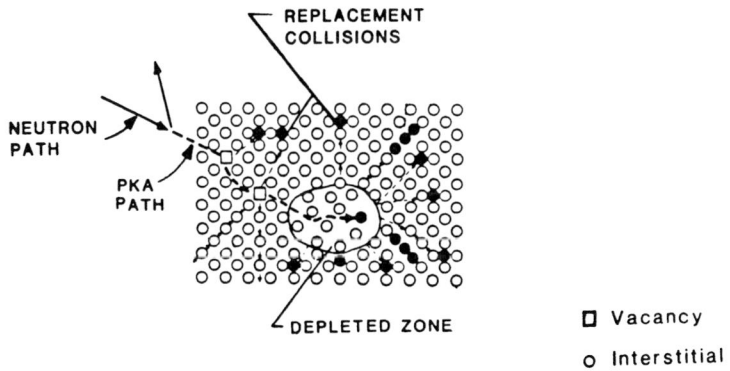

REPLACEMENT
COLLISIONS

NEUTRON
PATH

PKA
PATH

DEPLETED ZONE

□ Vacancy

o Interstitial

Figure 24. A qualitative picture of a point defect pair and a displacement spike.

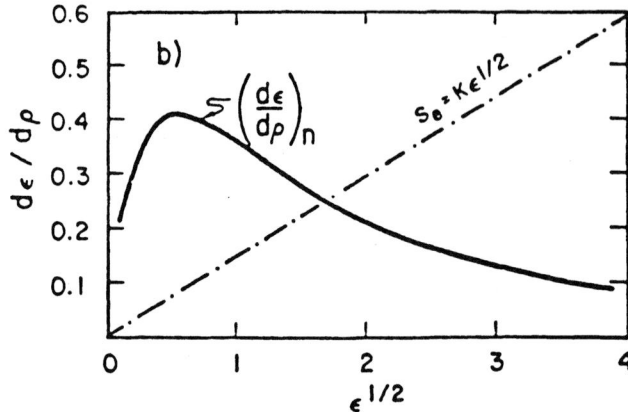

Figure 25. The LSS result for the nuclear stopping cross section.
The abscissa is proportional to the ion velocity. The
full drawn curve is the nuclear portion and the broken
curve the electronic portion [58].

displacement mechanism is the ionization of a lattice atom by the
photon, creating a local charge imbalance which ejects a lattice
atom from its site.

A precise description of the displacement process would involve
the solution of a many bodied problem requiring detailed knowledge of
the interaction potentials for both the incident PKA and the target
atoms. In the absence of such detailed knowledge, various
semi-empirical formulations have been developed to calculate the
damage function. These formulations first partition the recoil
energy into electronic and nuclear energy losses and then convert the
nuclear energy losses into dpa.

The energy partition model developed by Lindhard, Scharff, and
Schiott [58] has proven to be useful in determining the energy loss
for charged particles. This model assumes the electronic and nuclear
energy loss mechanisms can be totally separated. Using a Thomas-
Fermi model for the atom, and assuming an amorphous solid, they
developed an universal range theory that can describe, to a good
approximation, the nuclear energy loss for many target/projectile
combinations by a single curve. This curve is shown in Fig. 25,
where ε is a dimensionless energy ratio. Note that the electronic loss
portion increases at a rate proportional to the ion velocity, while
the nuclear portion peaks at a relatively low ion energy and

592

subsequently drops. The amount of energy deposited into nuclear collisions S_D can now be calculated by integrating over the PKA path.

Once the nuclear energy deposition is known, the number of displacements can be calculated using a modified Kinchen and Pease model:

$$N_d = \frac{\kappa S_D}{2E_d} \tag{38}$$

The factor κ is called the displacement efficiency and is used to account for the displacement events that are not elastic and has been found to vary from 0.3 to 1.0, depending on the primary recoil energy spectrum [59]. E_d is the mean displacement energy and is the average energy that must be transferred to an atom to displace it from its lattice site. E_d is actually strongly dependent on the crystallographic direction in which the atom is displaced, and the use of a single value for E_d is a spherical average of the saddle points in the potential barrier surrounding the equilibrium site [57]. Values for E_d are normally measured using high energy electrons to find the minimum electron energy that will produce evidence of displacement damage such as an increase in the low temperature electrical resistivity. For example, in nickel this value corresponds to an energy transfer of 24 eV. This measured threshold value is then multiplied by 1.66 to correct for such things as the directional dependence and spontaneous recombination events. This gives the commonly used displacement value for nickel of 40 eV [60].

Point Defect Behavior

The vacancy/interstitial pair is the simplest type of damage produced by irradiation. The thermodynamics of the crystal determines the behavior of these defects, and from a thermodynamic argument one can determine the number of these defects that will exist in a crystal under thermal equilibrium. This thermal concentration of defects is found by minimizing the Helmholtz free energy and is given by [52]:

$$C_{v,i}{}^{eq} = f_{v,i} \exp(-E_{fv,i}/kT) \tag{39}$$

where v,i subscripts refer to vacancy or interstitial, $f_{v,i}$ is a frequency term, E_f the defect formation energy, and k the Boltzmann constant. The formation energy E_f is the amount by which the energy of the crystal is raised when the defect is introduced. Another exponential term describes the defect jump frequency:

$$\nu_{v,i} = \nu_o \exp(-E_{mv,i}/kT) \tag{40}$$

593

where $\nu_{i,v}$ is the defect jump frequency, ν_o the jump attempt frequency and $E_{mv,i}$ the defect migration energy.

The formation of an interstitial introduces a large amount of strain energy into the crystal and hence the formation energy for interstitials, E_{fi}, is generally several times larger than that for vacancies, E_{fv}. Once formed, the interstitial's strain field is relatively unstable and is easily freed from its potential well by thermal vibrations and hence has a lower migration energy, E_{mi}, than that of a vacancy, E_{mv}. For metals, values for these parameters are the order of:

$$E_{fv} \sim 1eV \qquad E_{fi} \sim 5eV$$

$$E_{mv} \sim 1eV \qquad E_{mi} \sim 0.1eV$$

An unirradiated crystal in thermal equilibrium at $1000^{o}K$ will contain a concentration of vacancies of about 10^{-5} and a concentration of interstitials several orders of magnitude lower.

The concentration of point defects during irradiation is determined by the defect production and loss rates in the crystal. The defects are lost by either annihilation of an opposite defect or by migration to a sink such as a dislocation, void, precipitate particle, or grain boundary. During irradiation, the defect concentration will be supersaturated and hence the crystal can lower its free energy by clustering the defects and forming new sinks. These clusters are either two-dimensional objects which form plate-like dislocation loops or voids which are roughly spherical vacancy clusters that may contain gas atoms. Voids are differentiated from bubbles in that bubbles contain enough gas to equilibriate the interior gas pressure with the cavity surface energy while voids contain less than than equilibrium gas amounts. The total bubble volume of a material containing bubbles is determined by the gaseous product production rate and therefore this bubble driven swelling rate is usually small.

The formation of voids leads to a reduction of the material density, and since material is conserved, to an increase in volume. The magnitude of this void induced swelling can become quite large at fluences typical of that expected in power reactors. For example, swelling values greater than 50% have been observed in 304 stainless steel under fast breeder irradiation [61]. Since some materials exhibit little swelling under similar conditions, a brief description of swelling theory and the direction materials designers are taking will be presented.

There are four general requirements necessary for void swelling to develop in a material. They are: 1) the crystal must have a vacancy supersaturation, 2) the vacancies must be mobile, 3) voids must nucleate, and 4) more vacancies than interstitials must enter the void (see [62 and 63] for dated but good reviews of this subject). The first requirement places an upper limit on the temperature for void formation since the thermal vacancy concentration increases rapidly with temperature thereby reducing the radiation induced supersaturation. The second requirement establishes a lower temperature limit since vacancy mobility is negligible below about 0.2 times the absolute melting temperature T_m and voids cannot grow if vacancies cannot diffuse to them. The third requirement is stated simply because voids can only form by nucleation. Nucleation theories have not been particularly successful to date, and will not be discussed further except to say that most materials of interest do nucleate voids under irradiation. The fourth requirement is necessary since if a void is to grow, it must receive a net flux of vacancies. This net influx of vacancies implies that another sink with a net influx of interstitials must exist since both defects are being produced at equal rates. Since a vacancy possesses a small long-range stress field, it is generally believed that the strain field around the interstitial is interacting with the strain field of dislocations, causing the interstitials to have a preferred drift toward dislocation sinks. It was the failure to recognize the existence of this biased sink that lead to the lack of a prediction of void swelling as a problem under high energy neutron irradiation.

Under irradiation, the first major microstructural feature to develop is the dislocation structure. In a well annealed material, dislocation loops will nucleate and grow into a dense dislocation network, while the dislocation structure of a cold worked material may increase or decrease in density depending on the temperature, initial dislocation density and/or irradiation conditions. After a few dpa, voids will begin to form and swelling will start. At about the same time, the irradiation may induce precipitate formation or modify the material's existing precipitate structure. A typical swelling versus fluence curve is shown in Fig. 26. Note that swelling does not begin in this case until a dose level of about 20 dpa. Most materials will exhibit this "incubation dose" before swelling begins, and then swell at a fairly constant rate. In some materials, the swelling has been observed to rise to some level (typically a few percent) and then saturate, with no subsequent increase in swelling with increasing dose.

Gas atoms, especially helium, are believed to play an important role in the nucleation of voids. Helium is believed to stabilize small vacancy clusters until they can grow above the critical size necessary to sustain growth. Soluble impurities such as hydrogen can assist void nucleation by lowering the surface energy and stabilizing the cluster. Understanding the effect of gas production on void formation is particularly important for fusion since the high energy neutron spectrum will produce much larger amounts of gas than are

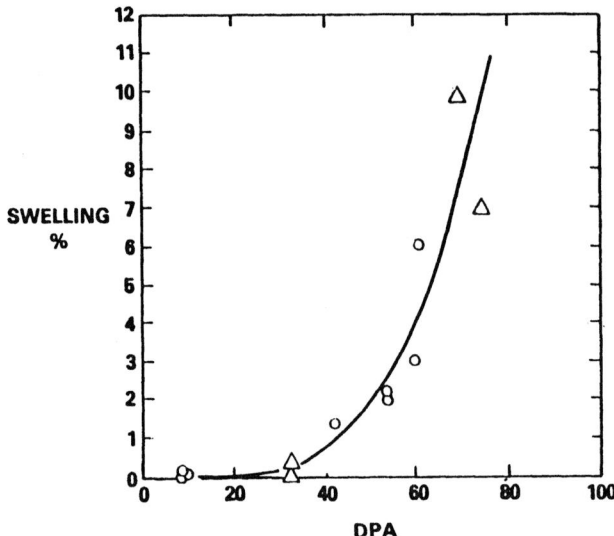

Figure 26. A typical swelling versus neutron damage level for 316
stainless steel. Note the presence of an incubation
period and the roughly linear swelling region [61].

normally produced in fission reactor irradiations. Virtually all
existing neutron induced swelling data has been collected from
fission irradiations since the only operational "fusion" neutron
sources have limited irradiated volumes and relatively low neutron
fluxes.

One of the difficulties involved in developing alloys resistant
to swelling lies in producing low swelling rates without adversely
affecting the materials other mechanical properties. Many alloy
modifications which reduced the swelling tendencies also reduced the
alloy ductility to unacceptable levels. Most alloy modification
programs try to form a microstructure with a high density of sinks,
such as precipitate particles, which will reduce the defect
concentration and/or trap gas atoms, thereby reducing swelling.
Another approach is to take an inherently low swelling alloy, such as
a ferritic steel, and attempt to modify its other properties to make
its use suitable for fusion. At the present time it is the ferritic
steels which seem to show the most promise as low swelling
structural materials.

Property Changes from Irradiation

Neutron irradiation will affect virtually all of the properties
important in the design of plasma-interactive components. The

596

changes in material properties observed after irradiation can be
attributed to five mechanisms, namely: 1) a high concentration of
radiation produced point defects "frozen" into the crystal, 2) the
presence of defect clusters, 3) the presence of impurity atoms, 4)
the presence of metastable precipitate phases formed by the
irradiation, and 5) the net flux of interstitials to dislocations.
The first mechanism is only important at very low irradiation
temperatures and will generally not contribute to property
degradation for plasma-interactive materials. All of the other
mechanisms can operate at temperatures typical for these components.
The major mechanical properties pertinent to these materials that are
strongly affected by irradiation will be briefly discussed.

Radiation hardening and embrittlement. Radiation hardening
usually refers to the increase in the yield stress and ultimate
tensile strength that occurs as a function of fast neutron fluence
and irradiation temperature. Embrittlement refers to the decrease in
the amount of plastic deformation (i.e., ductility) observed prior to
failure. A schematic representation of a typical engineering
stress-strain curve for stainless steel before and after irradiation
is shown in Fig. 27. Irradiation at low temperature (less than about
half the absolute melting temperature) leads to a significant
increase in the yield stress and to a decrease in ductility, while
irradiation at high temperature reduces the total ductility without
increasing the yield point. The low temperature behavior shown in
Fig. 27 is observed for virtually all metals, and is attributed to
the high density of dislocation pinning points produced by the
irradiation. Virtually all irradiation produced structures can pin
or increase the amount of force necessary to move dislocations. This
increased resistance to dislocation motion increases the materials
yield point, but does not increase the theoretical fracture stress.
Hence a material will become harder at the expense of reduced
toughness.

The mechanical property changes observed after high temperature
irradiation vary somewhat from material to material. For example,
the unirradiated properties of ferritic steels can be recovered by a
high temperature anneal, whereas an austenitic steel will exhibit a
sharply reduced ductility after irradiation that cannot be recovered
by annealing. The difference seems to lie in the behavior of the
transmutation produced helium atoms. In the austenitic steels, the
helium migrates to the grain boundary region where grain boundary
bubbles form. Under an applied stress, these bubbles will eventually
link up and cause failure. This migration does not seem to occur
with the ferritic steels.

A related phenomena that is critical for the ferritic steels and
other body centered cubic materials such as molybdenum is the shift
in the ductile to brittle transition temperature observed under
irradiation. The ductile to brittle transition temperature, or DBTT,

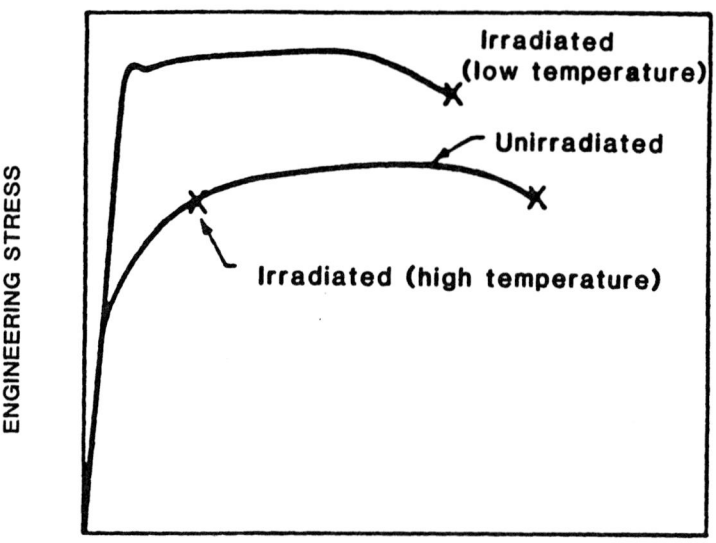

Figure 27. A typical stress-strain curve for a face centered cubic
material in the unirradiated state and after neutron
irradiation at low and high temperature.

is the temperature at which the observed failure mode for these
materials switches from a brittle failure mode at low temperatures to
a ductile failure mode at high temperatures. Since ductile materials
can absorb much higher shock loads without catastrophic failure than
can brittle materials, designers like to avoid the use of brittle
materials whenever possible. By proper alloying and thermomechanical
treatment, the DBTT in unirradiated alloys can usually be brought
below room temperature and hence presents no major design problem.
Under irradiation, however, the DBTT gradually increases and can
approach the service temperature in some cases. Current research
efforts to limit this increase are meeting with some success and
hence it appears this problem may be manageable. (for further
information, see the numerous papers on this subject in [53-55]).

Irradiation Creep. Irradiation creep refers to the augmentation
of the thermal creep process by irradiation. While thermal creep is
very temperature dependent, irradiation creep is much more
temperature independent and is instead related to the neutron flux.
Irradiation creep is due basically to the existence of the large
fluxes of point defects in a material under irradiation and hence is

598

closely related to the swelling phenomenon. There are basically two mechanisms that have been proposed for the irradiation creep process, either 1) the growth of stress oriented dislocation loops or 2) accelerated climb of dislocations past obstacles. Under the first mechanism, the dislocation loops are assumed to preferentially nucleate and grow in planes perpendicular to the applied stress, leading to an enhanced dimensional change (creep) in that direction. With the second mechanism, the defects are assumed to diffuse and nucleate loops with random orientation. Due to the the net bias of dislocations for interstitials, there will be a net flux of interstitials to the dislocations present in the materials. The continual flux of interstitials will allow the dislocation to continually climb to new locations in the crystal, thereby freeing itself from pinning sites. Creep would therefore be enhanced since the blocking of dislocation movement has been reduced. Radiation enhanced creep can lead to premature component failure in some cases, but in other instances, it may serve to relieve the stresses generated by swelling. Radiation creep is not usually the life limiting phenomena.

Thermal and Electrical Properties. Changes in the thermal or electrical properties of metals at the temperatures relevant to plasma-interactive components are expected to be small. At very low temperatures, large atomic concentrations of vacancies can be trapped in a metal lattice and can significantly reduce the electrical conductivity. At higher temperatures (roughly above $200^{\circ}C$) these defects have migrated to sinks or clustered, where they have a relatively small affect on either the thermal or electrical conductivity of metals. For non-metals, the picture is quite different. For example, in insulators the presence of a radiation field can increase the electrical conductivity by changing the band gap structure or by exciting electrons into normally empty bands [64]. The thermal conductivity of materials that conduct by phonon processes (such as SiC) can also be significantly reduced by the presence of radiation induced scattering centers [65].

Radiation Damage in Graphite. The preceeding discussion has concentrated on radiation effects in metals. Graphite is a material that is often considered for use in high heat flux components in power reactors, and since its response to radiation is quite different, it will be briefly discussed.

The radiation response of graphite is related to its anisotropic, layered structure [66]. The presence of the widely spaced basal planes creates a structure that induces a preferred collection area for radiation induced defects. This effect is most noticeable when single crystal graphite is irradiated and large anisotropic dimensional changes are measured. The crystal is observed to grow in the direction parallel to the normal of the basal planes (c-axis) and contract along the basal planes. This behavior is due to the preferential formation of interstitial loops inbetween the widely spaced layer planes, with the vacancies preferentially forming loops perpendicular to the layer planes. If the graphite is polycrystalline, then each crystallite will behave in the above manner, but the bulk response will be dictated by the behavior of the

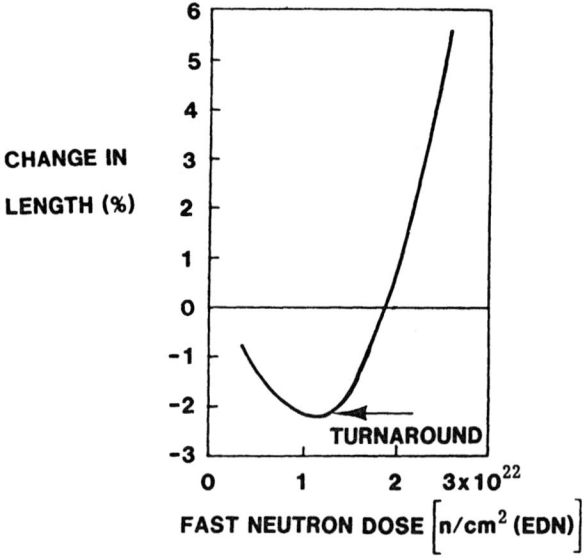

Figure 28. Typical dimensional change of polycrystalline graphite after fast neutron irradiation. 1×10^{23} n/cm^2 corresponds roughly to 1.3 dpa [66].

grain boundaries. The initial expansion of the c-axis is absorbed by intergranular cracks and hence the contraction of the other directions dominates the bulk response. This leads to an increase in density. When the accommodating cracks are eliminated, the behavior will turn around and the material will begin to swell. This behavior is shown schematically in Fig. 28. The rapid swelling will then start to degrade the graphite's mechanical and thermal properties. The neutron fluence at which this turnaround occurs and the speed at which the properties degrade varies depending on the type of graphite. In general, the isotropic "reactor" grade graphites are the most resistant to radiation damage. However, the present data indicates that if graphite is to be used in power reactors in regions with high neutron fluxes, that it should only be used in components that have a relatively short expected lifetime and can be easily replaced.

CONCLUSION

It should now be evident that the selection of materials for plasma-interactive components in a magnetic confinement fusion device is a complex problem that normally involves many trade-offs. Fortunately, the responses of a material to many of the loads imposed by a fusion device can be predicted on the basis of appropriate materials properties, and a careful analysis of the anticipated loads (thermal, mechanical, radiation, etc.) can be used to identify the critical properties for a particular application. Appropriately weighted figures-of-merit can then be developed and used to compare candidate materials. Although it is often expedient to use published figures-of-merit to rank materials according to a particular response, it must again be emphasized that a figure-of-merit is a highly specific comparison criterion which provides useful information only when it is applied to materials and conditions that are consistent with the assumptions and priorities established in its derivation. Once suitable figures-of-merit and/or critical materials properties have been identified, generic differences and trends among the properties of metals, ceramics, and graphites can then be used to help focus the search on the most promising general types of materials. If necessary, a careful examination of the basic structural features which control important macroscopic properties can also provide useful guidance for possible materials modifications (e.g., alloying, microstructural changes, etc.) to improve performance. After the field has been narrowed to the most promising materials, a series of carefully designed tests of critical responses is often required in order to make the final evaluation with a reasonable degree of confidence.

At the present time, graphite (with and possibly without coatings) and beryllium appear to be the leading candidates for surfaces that directly contact the plasma in major fusion devices of the future; however, many alternative materials including refractory metals and various ceramic compounds are also being considered. Potentially attractive attributes of graphites include a high temperature capability, good resistance to thermal shock, relatively low atomic mass, low activation, prior operational experience in fission and fusion devices, lack of toxicity, ready availability, comparatively low cost, and relative ease of fabricating complex shapes. Areas which presently require further study to determine if serious problems exist include questions regarding chemical and/or radiation-enhanced sputtering, possible problems with tritium retention and recycling, radiation induced loss of thermal conductivity, and radiation swelling.

Potential advantages of beryllium for plasma-interactive surfaces include the lowest atomic number (Z=4) of any viable first-surface material, a self-sputtering yield that is predicted to be less that unity at all plasma edge temperatures, no reported evidence of chemical sputtering, low hydrogen solubility and permeability, and a relatively high thermal conductivity which may not be seriously degraded upon irradiation. In near-term (i.e., no tritium) devices, the toxicity of respirable beryllium particulate is a serious disadvantage due to the added time and expense that will be necessary to safely control worker exposure. However, this limitation should become much less important in future tritium burning devices because similar control measures will already be required to limit worker exposure to tritium. For applications such as limiters in tokamaks there is also concern regarding possible degradation of beryllium surfaces due to disruptions and thermal fatigue. In an intense neutron environment helium production due to an (n,2n) reaction may also cause swelling problems at temperatures above 600 $^{\circ}$C unless countermeasures (e.g., controlled introduction of interconnected porosity) can be successfully developed.

In conclusion, an "ideal" material for plasma-interactive surfaces is not presently available and probably cannot be developed. The selection of a material for a specific application must be based upon compromise according to the relative importance of various competing requirements. In actual practice, factors other that those discussed here are often important. For example, limitations imposed by inadequate fabrication technology, lack of prior experience, high cost, or limited availability often play a major role in "real world" situations.

ACKNOWLEDGMENTS

The authors wish to thank R. D. Watson, W. B. Gauster, and B. W. Tafoya for their assistance in the preparation of this chapter. This work was performed at Sandia National Laboratories, supported by the U.S. Department of Energy under contract number DE-AC04-76DP00789.

REFERENCES

1. R. W. Hertzberg, "Deformation and Fracture Mechanics of Engineering Materials," Wiley and Sons, New York (1976).
2. R. D. Watson, M. F. Smith, J. B. Whitley, and J. M. McDonald, Thermomechanical testing of beryllium for the JET/ISX-B beryllium limiter experiment, in: "Proceedings of the 13th Symposium on Fusion Technology," Varese, Italy, Sept. 24-28, 1984.
3. B. T. Kelley, "Physics of Graphite," Applied Science, London (1981).
4. W. P. Kingery, H. K. Bowen, and D. R. Ahlman, "Introduction to Ceramics," Wiley and Sons (1976).

5. J. B. Whitley, A. W. Mullendore, R. D. Watson, M. F. Smith, and R. S. Blewer, J. Nucl. Materials 111&112:866 (1982).

6. R. E. Peterson, Materials Res. and Standards 3(2):122 (1963).

7. ASM Committee on Wrought Stainless Steels, Section I: Stainless steel types and their characteristics, in: "Sourcebook on Stainless Steels," A. G. Gray, ed., Am. Soc. for Metals, Metals Park, Ohio (1976).

8. Anonymous, Reaction Bonded Silicon Carbide, Technical Bull. No. 811, Pure Carbon Co., St. Mary's, Pennsylvania.

9. Anonymous, UCAR Premium Grade ATJ, Technical Bull. No. 463-205, Union Carbide Corp., New York.

10. E. J. Stenfanides, Design News, Jan. 22, 1979.

11. R.F. Mattas, Austenitic Stainless Steel Bulk Property Considerations for Fusion Reactors, Report No. ANL/FPP/TM-86, Argonne Nat. Lab., Argonne, Illinois (1979).

12. L. R. Smith and W. C. Leslie, Properties of pure metals - iron, in: "Metals Handbook," 9th edition, vol. 2, D. Benjamin, ed., Am. Soc. for Metals, Metals Park, Ohio (1979).

13. T. C. Chi, J. Phys. Chem Ref. Data, 8(2):439 (1979).

14. G. T. Meaden, "Electrical Resistance in Metals," Plenum, New York (1965).

15. C. A. Wert and R. M. Thompson, "Physics of Solids," 2nd edition, McGraw-Hill, New York (1964).

16. D. E. Roberts, Ann. Phys., 40:453 (1963).

17. W. Meissner, H. Franz, and H. Westerhof, Ann. Phys. Lpz., 13:555 (1932).

18. N. Ganguli and K. S. Krishnan, Nature, 144:67 (1939).

19. P. G. Klemens, Thermal conductivity and vibrational modes, in: "Solid State Physics," vol. 7, Academy Press, New York (1958).

20. G. Leibfried and E. Schlomann, Nachr. Akad. Wiss. Gottingen, Math-Physik. Kl., 2a(4):71 (1954); English translation AEC-TR-5892, 1-36 (1963).

21. Y. S. Touloukian, R. W. Powell, C. Y. Ho, and P. G. Klemens, "Thermophysical Properties of Matter, Volume 1, Thermal Conductivity-Metallic Elements and Alloys," Plenum, New York (1970).

22. Y. S. Touloukian, R. W. Powell, C. Y. Ho, and P. G. Klemens, "Thermophysical Properties of Matter, Volume 2, Thermal Conductivity-Nonmetallic Solids," Plenum, New York (1970).

23. Y.S. Touloukian and E. H. Buyco, "Thermophysical Properties of Matter, Volume 4, Specific Heat-Metallic Elements and Alloys," Plenum, New York (1970).

24. Y.S. Touloukian and E. H. Buyco, "Thermophysical Properties of Matter, Volume 5, Specific Heat-Nonmetallic Solids," Plenum, New York (1970).

25. Y. S. Touloukian, R. K. Kirby, R.E.Taylor, and P. D. Desai, "Thermophysical Properties of Matter, Volume 12, Thermal Expansion-Metallic Elements and Alloys," Plenum, New York (1976).

26. Y. S. Touloukian, R. K. Kirby, R.E.Taylor, and T. Y. R. Lee, "Thermophysical Properties of Matter, Volume 13, Thermal Expansion-Nonmetallic Solids," Plenum, New York (1977).

27. J. E. Hove, "Proc. First Conference on Industrial Carbon and Graphites," SCI, London (1958).

28. W.B. Gauster, J.A. Koski and R.D. Watson, J. Nucl. Materials 122&123:80 (1984).

29. G.L. Kulcinski, Plasma Physics and Controlled Nuclear Fusion, Vol. II:251 (1974), IAEA-CN-33/S3-1.

30. R.W. Conn, J. Nucl. Materials 103&104:7 (1981).

31. D.L. Smith, J. Nucl. Materials 103&104:19 (1981).

32. R.E. Nygren, J. Nucl. Materials 103&104:31 (1981).

33. R. Behrisch, J. Nucl. Materials 85&86:1047 (1979).

34. G.M. McCracken and P.E. Stott, Nucl. Fusion, Vol. 19, no. 7:889 (1979).

35. A.J. Chapman, "Heat Transfer," Macmillan Pub. Co., N.Y., 1984.

36. "Handbook of Heat Transfer," edited by W.M. Rohsenow and J.P. Hartnett, McGraw Hill (1973).

37. Hibbit, Karlsoon and Sorenson, Inc., "ABAQUS Users Manual," (1984).

38. M. Ulrickson, in this volume.

39. R. Behrisch, J. Nucl. Material 93&94:498 (1980).

40. R. Behrisch, Nuc. Fusion 12:695 (1972).

41. B.J. Merrill, in "Proceedings of the 9th Sym. on Eng. Problems of Fusion Research," Chicago, Ill. (1981), IEEE 81CH1715-2:1621.

42. A.M. Hassanein, G.L. Kulcinski and W.G. Wolfer, Surface Melting and Evaporation During Plasma Disruptions in Magnetic Fusion Devices, to be published in Nucl. Eng. Design/Fusion (also UWFDM-494, 1982).

43. C.D. Croessmann, G.L. Kulcinski and J.B. Whitley, J. Nucl. Materials 128&129:816 (1984).

44. W.G. Wolfer and A.M. Hassanein, J. Nucl. Materials 111&112:560 (1982).

45. J.F. Schivell, D.J. Grove, J. Nucl. Materials 53:107 (1974).

46. S. Timoshenko and J. N. Goodier, "Theory of Elasticity," McGraw Hill, N.Y. (1951).

47. J.L. Cecchi, Impurity Control in TFTR, PPPL-1668, June (1980).

48. J.L. Cecchi, "9th Sym. on Eng. Problems of Fusion Research," Chicago, Ill. (1981), IEEE 81-CH1715-2:1378.

49. G.L. Kulcinski, Contemp. Phys, Vol. 20, no. 4:17 (1979).

50. R.E. Gold, E.E. Bloom, R.W. Clinard, D.L. Smith, R.D. Stevenson and W.G. Wolfer, Nucl. Tech. Fusion, Vol. 1:169 (1981).

51. D.I. Roberts, S.N. Rosenwasser and J.F. Watson, J. Materials for Energy Systems, Vol. 3:54 (1981).

52. M.W. Thompson, "Defects and Radiation Damage in Metals," Cambridge Press (1969).

53. "Proceedings of the Third Topical Meeting on Fusion Reactor Materials," J.B. Whitley, K.L. Wilson and F.W. Clinard, editors. Published in J. Nucl. Materials 122&123 (1984).

54. "Proceeding of the Second Topical Meeting on Fusion Reactor Materials," R.E. Nygren, R.E. Gold and R.H. Jones, editors. Published in J. Nucl. Materials 103&104 (1981).

55. "Proceedings of the First Topical Meeting on Fusion Reactor Materials," F.W. Wiffen, J.H. Devan and J.O. Stiegler, editors. Published in J. Nucl. Materials 85&86 (1979).

56. B.T. Kelly, "Irradiation Damage to Solids," Pergamon Press (1966).

57. D.R. Olander, "Fundamental Aspects of Nuclear Reactor Fuel Elements," National Technical Information Center, TID-26711-P1 (1976).

58. J. Lindhard, M. Scharff and H.E. Schiott, Mat. Fys. Medd. Dan. Vid. Selsk. 33, no. 14 (1963).

59. J.H. Kinney, M.W. Guinan, Z.A. Munir, J. Nucl. Materials 122&123:1028 (1984).

60. P.G. Lucasson and R.M. Walker, Phy. Rev. 127:485 (1962).

61. F.A. Garner, J. Nucl. Materials 122&123:459 (1984).

62. D.I.R. Norris, Rad. Effects 14, no. 1 and Rad. Effects 15, no. 1 (1972).

63. B.L. Eyre, in "Conf. on Fund. Aspects of Rad. Damage in Metals," Gatlinburg, Tenn. (1975) :729.

64. F.W. Clinard, Jr. and G.F. Hurley, J. Nucl. Materials 103&104:705 (1981).

65. R.J. Price, J. Nucl. Materials 46:268 (1973).

66. G.B. Engle and B.T. Kelly, J. Nucl. Materials 122&123:122 (1984).

PLASMA TRANSPORT NEAR MATERIAL BOUNDARIES

C. E. Singer
Plasma Physics Laboratory, Princeton University
Princeton, New Jersey 08544

ABSTRACT

The fluid theory of two-dimensional (2-d) plasma transport
in axisymmetric devices is reviewed. The forces which produce
flow across the magnetic field in a collisional plasma are
described. These flows may lead to up-down asymmetries in the
poloidal rotation and radial fluxes. Emphasis is placed on
understanding the conditions under which the known 2-d plasma
fluid equations provide a valid description of these
processes. Attempts to extend the fluid treatment to less
collisional, turbulent plasmas are discussed. A reduction to
the 1-d fluid equations used in many computer simulations is
possible when sources or boundary conditions provide a large
enough radial scale length. The complete 1-d fluid equations
are given in the text, and 2-d fluid equations are given in the
Appendix.

INTRODUCTION

Understanding the interaction of plasma flows with material
boundaries is of fundamental importance in experimental plasma
physics. Several of the lectures in this series refer to the
development and application of fluid models of these plasma
flows. These models are beginning to play an important role in
the design and interpretation of tokamak experiments. The
physicist interested in this area therefore needs at least a
qualitative understanding of the physics used in these models.
Where do the model equations come from? When is the general
approach used adequate, and when does it have to be abandoned or
supplemented with a more sophisticated treatment? The purpose

607

of this review is to provide some answers to these questions. The discussion will be kept intuitive and mostly qualitative. For those interested in the complete transport equations, however, an Appendix is provided.

CONSERVATION LAWS

The only equations which give a complete description of boundary plasma flows are the collisional Vlasov equations.[1] These consist of the Fokker-Planck equation for each type of particle

$$\frac{\partial f}{\partial t} + v_\alpha \frac{\partial f}{\partial x_\alpha} + a_\alpha \frac{\partial f}{\partial v_\alpha} = C + \Sigma \ , \qquad (1)$$

and Maxwell's equations. f represents the number of particles in a unit cube centered on physical location x_α with velocities in a unit cube centered on v_α. In one dimension, $\partial f/\partial t$ depends on the difference in the number of particles advecting towards point x (and, hence, on $\partial f/\partial x$). $\partial f/\partial t$ similarly depends on the acceleration, a, of particles up to velocity, v, and the acceleration away from v (and, hence, on $\partial f/\partial v$). In three cartesian coordinates, we sum over the index α and note that $\vec{a} = (e/m)(\vec{E} + \vec{v} \times \vec{B}/C)$. Coulomb collisions between particles separated by small distances can also change the distribution function f. The effect of $C_a = \Sigma_b\, C_{ab}(f_a, f_b)$ of all collisions on species a results in a contribution roughly proportional to the Coulomb collision frequencies

$$\nu_{aa} = \frac{4\pi^{1/2} n_a e^4 \lambda_{aa}}{3m_a^{1/2} T_a^{3/2}} \ , \qquad (2)$$

for like-particle collisions and, e.g. $\nu_{ei} = 4(2\pi\, m_e)^{1/2} n_e e^4 \lambda_{ei}/(3m\, T_e^{3/2})$ for collisions between hydrogen ions of mass m, electrons of mass m_e, and temperature T_e, where λ_{ab} is the Coulomb logarithm.[2] The rate, Σ, of creation and scattering of particles by neutral and inelastic atomic collisions depends on the cross sections described by M. Harrison in this lecture series. Collisions with neutrals are generally computed using the Monte Carlo methods described by D. Heifetz in this lecture series.

Now we make three enormous simplifications. First, we assume that the magnetic field in the region of interest is externally determined. For 2-d calculations this is generally the case, since currents in the poloidal field coils and core plasma are much larger than in the boundary plasma. (With

608

extreme gradients as in so-called edge relaxation phenomena (ERPS) and in high confinement (H-mode) divertor plasmas this condition may be violated.

Second, we assume $\partial/\partial t = 0$, which eliminates all forms of microturbulence. We therefore cannot generally treat low-density plasmas, which are typically microturbulent. The trend towards higher density scrape-off plasmas in future experiments gives hope that higher collisionality will rapidly damp out any microturbulence driven by core plasma fluctuations. Later, we will discuss attempts to reintroduce turbulence effects phenomenologically into the fluid equations.

Third, we assume collisions and gyromotion are sufficient to maintain nearly Maxwellian ion distribution functions

$$f = f_o(1 + \Phi) , \qquad (3)$$

where $\Phi \ll 1$ and $f_o = n\pi^{-3/2} v_{th}^{-3} \exp[(\vec{v} - \vec{u})^2 / v_{th}^2]$ with $v_{th} = (2T/m)^{1/2}$. (From now on we suppress the species index when referring to ions.) A similar equation is used for the electron distribution function f_e. This places three restrictions on the validity of our treatment. First, inspection of the Fokker-Planck equation shows that advection parallel to the magnetic field will overly distort the nearly Maxwellian distribution unless $v_{th} f/L_\parallel \ll \nu f$, where $L_\parallel = Bf/\vec{B}\cdot\nabla f$ is the parallel scale height. This requires a small "inverse collisionality," $\Delta = \lambda_{mfp}/L_\parallel = (v_{th}/\nu)/L_\parallel$. Second, local gyromotion will also overly distort the ion Maxwellian distribution unless $\delta = \rho/L_\psi \ll 1$ (where $\rho = v_{th}/\Omega = v_{th}/[eB/(mc)]$ is the ion gyroradius and $L_\psi = |\nabla\psi|f/(\nabla\psi\cdot\nabla f)$, where ψ is the magnetic flux surface label in orthogonal coordinates.) The contributions to f of order Δ and δ have been computed in works on classical transport theory[3,4] and pose no problem.

However, there is also a distortion of the distribution function due to the radial drifts experienced by particles in a curved magnetic field. In the absence of collisions, there is a contribution of order δ_p to the correction function Φ, where $\delta_p = \delta B/B_\theta$, and B_θ is the poloidal magnetic field. But particles can only execute a fraction Δ of their poloidal drift before being scattered by collisions, so a collisional plasma experiences a resulting distortion of order $\Delta\delta_p$. Suppose we rely on existing work for computation of the perturbation, Φ. Then the unknown correction of order $\Delta\delta_p$ should be small compared to the know corrections of order Δ and δ. This requires $\Delta\delta_p \ll \Delta$ and $\Delta\delta_p \ll \delta$, i.e.,

$$\delta_p \ll 1 , \tag{4}$$

$$\Delta \ll B_\theta/B, \tag{5}$$

where B_θ is understood to be averaged over a region significantly smaller than $L_\psi \times L_\parallel$. The first of these requirements is in general more severe. For typical large tokamak parameters, $(n/10^{14} cm^{-3})/(T_e/10 \; eV)^{3/2} \gtrsim 1$ is required.[5] This is achieved only in high-density divertors.

When the above conditions are satisfied, we can usefully integrate the Fokker-Planck equations over velocity space[4] after multiplying by various powers of v_α. This gives the conservation equations

$$\nabla \cdot (n\vec{u}) = S, \tag{6}$$

$$\nabla : (m\vec{u}\vec{u}) + \nabla(nT) + \nabla : \overset{\leftrightarrow}{\Pi}_{vis} + en(\vec{\nabla}\phi - \vec{u} \times \vec{B}/c) = \vec{F}_{Coul} + \vec{X}_{Non-Coul}, \tag{7}$$

$$\nabla \cdot (\frac{1}{2} mu^2 \vec{u} + \frac{5}{2} nT\vec{u} + \vec{u} : \overset{\leftrightarrow}{\Pi}_{vis} + \vec{q}_{cond}) + en\vec{u} \cdot \nabla\phi,$$

$$= Q_{coul} + \vec{u} \cdot \vec{F}_{coul} + Y_{Non-Coul}, \tag{8}$$

for each plasma species, with similar equations for higher velocity moments. The Coulomb and non-Coulomb friction and energy sources are obtained by integrating over the collision terms C (using polynomial expansion techniques) and Σ (by summing over Monte Carlo calculations). Here we have used $\partial/\partial t = 0$ to write $\vec{E} = -\nabla\phi$; this requires the usual situation of a small externally applied electric field, $e\phi_{ext} \ll kT_e$.

In general these equations should be supplemented by Poisson's equations, $\nabla^2\phi = n_i - n_e$, but an excellent approximation in practical pure hydrogen collisional plasma regions is $n_e \approx n_i \equiv n$ (cf. Stangeby's lecture in this series). We can therefore replace Poisson's equation by this approximation. We are left with ten equations for the ten unknowns, n, $\vec{u}$, T, $\vec{u}_e$, T_e, ϕ.

CROSS-FIELD TRANSPORT

In toroidally symmetric situations, the above equations

610

become tractable if we retain only the dominant terms. Since the mathematical details are outlined elsewhere for the case of tokamak boundary plasmas,[5] we present here only the most interesting physics.

Ion continuity provides a useful starting point. The continuity equation is

$$g^{-1/2} \left[\frac{\partial}{\partial \psi} \left(nu_\psi |\vec{\nabla}\psi| g^{1/2} \right) + \frac{\partial}{\partial \theta} \left(nu_\theta |\vec{\nabla}\theta| g^{1/2} \right) \right] = S, \qquad (9)$$

where $g^{1/2} = 1/|\vec{\nabla}\psi \times \vec{\nabla}\theta \cdot \vec{\nabla}\xi|$ is the Jacobian, and ξ is the toroidal symmetry angle. The orthogonal coordinates (ψ,θ,ξ) have the advantage of geometric simplicity. However, we have to resolve the poloidal flow,

$$u_\theta = \left(B_\theta/B \right) u_\parallel + \left(B_\xi/B \right) u_d , \qquad (10)$$

into components in the "parallel" direction $\hat{B} = \vec{B}/B$, and the electron "diamagnetic" direction $\hat{B} \times \vec{\nabla}\psi/|\vec{\nabla}\psi|$ perpendicular to the magnetic field within the magnetic flux surfaces. We start our discussion of cross-field transport by analyzing the diamagnetic ion flow velocity, u_d.

Diamagnetic Ion Flow Velocity

Due to the large $\vec{u} \times \vec{B}$ (Lorentz) term in the momentum balance, cross-field transport in one direction is determined by the momentum balance in the orthogonal direction across the magnetic field. This is a reflection of the fact that a cross-field force tends primarily to make particles drift in the picture direction perpendicular to that force. Thus, the diamagnetic flow is determined from the radial momentum balance. The only significant radial forces are due to the radial gradients of the pressure and potential, so

$$nu_d = \frac{|\vec{\nabla}\psi|}{m\Omega} \left(\frac{\partial (nT)}{\partial \psi} + en \frac{\partial \phi}{\partial \psi} \right) . \qquad (11)$$

Normally nT and ϕ decrease radially outward, so the poloidal projection $(B_\xi/B)u_d$ corresponds to a unidirectional poloidal rotation. This rotation tends to compress the plasma against one side of the limiter or divertor(s). Even in an up-down symmetric geometry, the flow will not be up-down symmetric. The magnitude of the rotation is comparable to that needed to explain observed up-down asymmetries in the ASDEX tokamak.[6] The sign of the rotation depends on the sign of the toroidal magnetic field,[7] as do experimentally observed up-down asymmetries.[8]

Radial Ion Flow

The cross-field transport perpendicular to magnetic flux surfaces is more complicated. This "radial" flux is driven by forces in the diamagnetic direction. For very subsonic flows, only the gradients of nT and ϕ are significant. Since $\partial/\partial\xi = 0$, only $\partial(nT)/\partial\theta$ and $en\partial\phi/\partial\theta$ contribute. The radial flux can be expressed directly in terms of these gradients, as illustrated in Fig. 1. Alternatively, the electron momentum balance parallel to the magnetic field

$$\frac{B_\theta}{B}\left[\frac{\partial(nT)}{\partial\theta} + en\frac{\partial\phi}{\partial\theta}\,|\nabla\theta|\right] = F_\parallel = -F_\parallel^e = -\frac{\alpha_o m\nu_{ei}J_\parallel}{e} + \beta_o n\frac{B_\theta}{B}\frac{\partial T_e}{\partial\theta}\,|\nabla\theta|,$$

(12)

can be used to eliminate the poloidal gradients. Choosing this alternative, we find two contributions to the radial flux

$$\Gamma_{PS} = \frac{1}{m\Omega_p}\frac{B_\xi}{B}\frac{\alpha_o m\nu_{ei}J_\parallel}{e},$$

(13)

and

$$\Gamma_F = \frac{1}{m\Omega_p}\frac{B_\xi}{B}n\nabla_\parallel T_e,$$

(14)

where $\alpha_o = 0.5129$, $\beta_o = 0.711$, $\nabla_\parallel = (B_\theta/B)\partial/\partial\theta$, and $J_\parallel$ is the current parallel to the magnetic field.

The Scrape-off Thickness as a Function of Coulomb Collisionality

The usual Pfirsch-Schlüter (PS) term is significant only at very high collisionality, $[\Delta^{-1} \gtrsim (m/m_e)^{1/2}]$. However, the $\nabla_\parallel T_e$ term is independent of collisionality as pointed out by Feneberg.[9] This so-called Neinst is intimately related to the poloidal rotation, as is evident from Eqs. (9-12). The net effect of these two terms is to map particles onto the material boundary not from a magnetic flux surface, but rather from a surface shifted downwards (for the configuration shown in Fig. 2). There may therefore be a "superplateau regime" where the scrape-off flows are dominated by forces independent of the

612

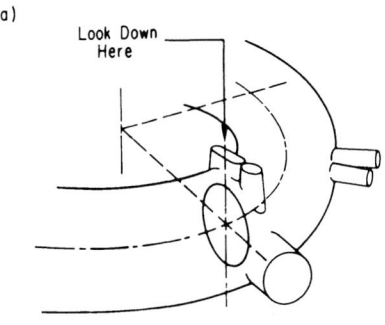

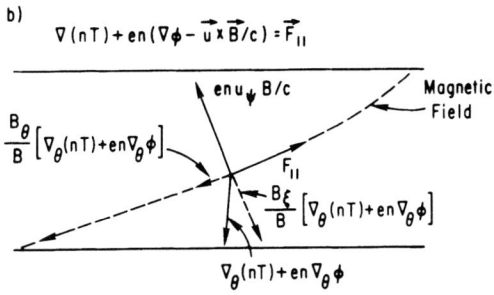

Fig. 1.   Force balance which describes neoclassical radial
particle flows. (a) Overall geometry; looking down
through the indicated port, one views from above the
tangent plane shown in Fig. 1b. (b) Force balance in
a plane tangent to the magnetic flux surface (solid
vectors). The Lorentz force $enu_\phi B/C$ is balanced by
the projection (long dashed vector) of the poloidal
force, $\nabla_\theta(nT) + en\nabla_\theta\phi$. For subsonic flows, the
poloidal force can in turn be related to the dominant
Coulomb frictional force, $F_\parallel = (B_\theta/B)[\nabla_\theta(nT) +$
$en\nabla_\theta\phi]$. (For near-sonic flows, additional inertial
and other contributions to $F_\parallel$ make the radial flux
different for ions and electrons, cf. Sec. 3.4.)

collisionality. This superplateau regime would lie in the
collisionality regime $1 \ll \Delta^{-1} \lesssim (m/m_e)^{1/2}$. At lower
collisionality, the thermal friction is no longer independent of
collisionality. At very high collisionality, the usual Pfirsch-
Schluter flux produces a thicker scrapeoff. In the putative
superplateau regime, the poloidal rotation adjusts to be
competitive with the radial flux due to thermal friction. The
actual scrape-off thickness then depends on the other poloidal
and radial flows. Flow along the magnetic field will tend to
cancel part of the poloidal rotation. In addition there are
radial Pfirsch-Schluter flows, which always go towards
increasing major radius. At very high collisionality, the

613

Pfirsch-Schluter flux overwhelms the other radial fluxes, leading to a different flow pattern. In the superplateau regime, the poloidal gyroradius is a consistent, but not unique possibility for the scrape-off thickness. Only in the subsonic supercollisional regime is the scrape-off thickness known for certain to be larger than the poloidal gyroradius.

Breakdown of Local Ambipolarity

In the subsonic flow regime upstream from an intense recycling region, the radial ion and electron flows are equal. This local ambipolarity results from the fact that the friction force is equal and opposite for electrons and ions, as is the Lorentz force. (Thus the same velocity $\vec{u}$ balances the Lorentz force against the poloidal forces held off by parallel friction.) For sonic flows, however, the larger ion mass leads to Coriolis and viscous forces which act almost exclusively on the ions. This complicates the flow pattern and destroys the ambipolarity of the radial particle flows. The poloidal flow of the resulting radial electric current can be returned through the plasma and/or an electrically conducting limiter.

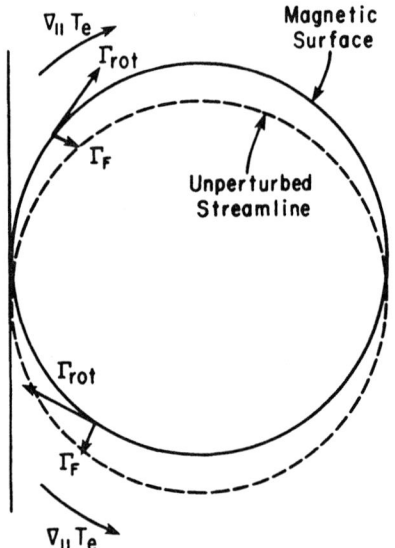

Fig. 2.   Poloidal streamlines which would result from poloidal rotation, $\Gamma_{rot} = (B_\xi/B)nu_d$, and from thermal friction force, Γ_F if these fluxes were not perturbed by parallel flows and by other radial flows (dashed line). These unperturbed streamlines are shifted downwards from the magnetic flux surfaces by about a poloidal gyroradius. The axis of symmetry is to the left and the toroidal magnetic field comes out of the page. (84P0215)

Terminology for Radial Flows

According to the nomenclature used in the authoritative review by Hinton and Hazeltine,[10] all of the above-mentioned radial particle fluxes are neoclassical, in the sense that they are absent in a rectilinear magnetic field geometry. There are also classical radial particle fluxes resulting from resistance to the diamagnetic current, $J_d = (c/B)|\vec{\nabla}\psi| \partial [n(T_e + T_i)]/\partial\psi$. However, the classical radial particle flux is small compared to the neoclassical fluxes.

PARALLEL FLOWS

When transport across the magnetic field is negligible, the plasma fluid equations become much simpler. One way this can happen is that reionization of charge-exchange neutrals may dominate cross-field energy transport. This may occur in some high-recycling divertor geometries. An example of such broad scrape-off thickness is given in the D-III probe data given by D. Post in this lecture series. Alternatively, anomalous transport upstream from a collision-dominated region can produce a broad scrape-off plasma. This may occur near the tokamak midplane in divertor experiments. If the midplane turbulence damps away as the plasma approaches a collision-dominated divertor throat, then 1-d fluid equations can be used to integrate the flows through the divertor throat. Under these circumstances, it is traditional to use the distance x along the magnetic field as a coordinate. The fluid equations then become

$$B \frac{d(nu_\parallel/B)}{dx} = S, \tag{15}$$

$$mnu_\parallel \frac{du_\parallel}{dx} = - \frac{dp_{tot}}{dx} + \left(X_\parallel - mu_\parallel S\right), \tag{16}$$

$$\frac{3}{2} nT_j + nT_j B \frac{d(u_\parallel/B)}{dx} = B \frac{d}{dx} \left(\frac{\kappa_\parallel^j}{B} \frac{dT_j}{dx}\right) + W_j , \tag{17}$$

for j = ions or electrons, where the source terms S, $X_\parallel$, and W_j can be gleaned from the Appendix. In numerical simulations, the small variation of field strength B is sometimes neglected. Only the sum of the ion and electron momentum balances is needed, so only the total pressure $p_{tot} = n (T_e + T)$ enters in the momentum balance. This is adequate if no significant current flows through the region, an assumption which is valid for 1-d flows unless a potential difference $e\Delta\phi \gg T_e$ is deliberately applied across the region of interest. The ion thermal conductivity $\kappa_\parallel \ll \kappa_\parallel^e$ is formally small in the fluid regime, as is the bulk viscosity η_0, but these effects are sometimes nevertheless included in numerical simulations.

Since the solution of these equations is discussed elsewhere in these lectures, comments here are restricted to collecting the above-mentioned constraints on their applicability. These are summarized in Table 1.

Table 1. Restrictions on 1-D Fluid Equations.

REQUIREMENT	FORMULA	WHERE VIOLATED
Collision-dominated	$\Delta \ll 1$	Near plate; usually near midplane
Broad scrapeoff	$\delta_p \ll 1$	Usually marginal
$\vec{B}$ determined externally	$\varepsilon d\ell n\beta_p/d\ell n\psi < 1$	Steep gradients
Steady State	$\partial\ell nf/\partial t \ll v_{th}/L_\parallel$ observable	When fluctuations are observable
Axisymmetry	$\partial/\partial\xi = 0$	Non-axisymmetric limiters·
No excess biasing	$e\phi_{applied} \lesssim T_e$	RF heating?
Pure hydrogen	$Z_{eff}-1 \ll 1$	Many experiments

Given the many limitations listed in Table 1, one might wonder why 1-d plasma transport calculations have been so successful in helping to explain the difference between low- and high-recycling tokamak divertors. The basic reason is that a qualitative treatment requires primarily a careful analysis of neutral gas transport and is relatively insensitive to the details of plasma transport. A quantitative comparison with detailed experimental observations will require more careful analysis of several problems, however. The most difficult problems are a phenomenological treatment of turbulence and inclusion of impurities with appropriate boundary conditions.

TURBULENCE, IMPURITIES, AND BOUNDARY CONDITIONS

Turbulence

There are a number of microinstabilities which occur when the collisionality is insufficient to suppress them. In some cases, it may be possible to obtain stationary conservation equations for the turbulence- averaged properties of the plasma. It is therefore worth examining possible types of turbulence for hints as to how the fluid equations could be modified.

The simplest instability occurs when radial gradients are relatively flat but the parallel collisionality approaches unity, $\Delta \sim 1$. Then the electron distribution is highly distorted along field lines. Energetic electrons can then escape from hot regions unimpeded by collisions. This sets up a positive potential which draws back a current of cold electrons. Ion cyclotron waves resonant with the cold electron current can be driven unstable. These waves propagate obliquely across the magnetic field and damp on the ions.[10] The energy they extract from the parallel electron flow should therefore result in anomalous ion heating.

Even at somewhat higher collisionality (up to $\Delta^{-1} \gtrsim (m/m_e)^{1/2}$), the known neoclassical effects may lead to a scrapeoff thickness comparable to the poloidal gyroradius, $\rho_p = \rho B/B_\theta$ (where $\rho = v_{th}/\Omega$ is the ion Larmor radius). Loss of poloidally drifting ions may then generate loss-cone instabilities,[11] which mix ions from different magnetic flux surfaces.

However, both of these instabilities may be overshadowed by resonance of waves with poloidal phase velocity near the diamagnetic drift velocity, u_d. While such instabilities have been studied extensively for core-plasma conditions, strong poloidal and radial scrape-off gradients make most of these studies inapplicable. The only method available so far for studying the possible 2-d effects of such instabilities has been to extrapolate empirical enhancements of some cross-field transport coefficients from core-plasma transport studies. Since the 2-d fluid equations are much more complicated than their flux-surface-averaged counterparts, this can be dangerous. For example, enhancing the cross-field particle transport alone can force the ions in a two fluid model to do more work than is available in the stored ion energy. The resulting negative ion temperatures may be an indication that the ions have been told to flow into a region where the high electrostatic potential ought to exclude them. Unless Ohm's law is modified consistently with the cross-field diffusion, such paradoxes will occasionally arise. One way to avoid such problems is to modify the underlying expressions for the electron-ion, ion-ion, and/or electron-electron collisionality. This will avoid such inconsistency problems, and it may give some insight into the action of different kinds of turbulence. However, the actual instabilities are localized and anisotropic in velocity space, so enhancement of the collision frequencies is only a first crude step towards a useful qualitative picture of turbulence effects. In the equations in the Appendix, the various collisionalities are nevertheless kept separate to facilitate this approach.

Impurities

For a multispecies plasma, the neoclassical radial transport of the various ions is driven by the same type of forces which determine hydrogen transport. However, the Coulomb friction between various ions is much stronger than between ions and electrons. Small differences between the parallel flow velocities of the various ions can therefore contribute strongly to radial transport.

For a broad scrapeoff, the 1-d transport of impurities along the magnetic field is trivial in a collisional plasma: collisions force the impurities to flow with the bulk plasma flow. In the sonic flow regime, other effects such as thermal friction only become significant when the collisionality is of order one and the fluid approximation breaks down. Numerical fluid studies including such terms, discussed by K. Lackner in these lectures, give quantitative insight to the extent the results are independent of the detailed particle kinetics, or the computation is confined to an intense recycling regime with subsonic flow throughout.

The lowest order approximation of equal parallel ion flow velocities is unfortunately inadequate for determining the cross-field transport. In principle, there is no difficulty in numerically solving directly for the difference between the various flow velocities in the fluid regime. The major difficulty lies in determining the boundary condition on the relative velocities.

Boundary Conditions

A scrape-off fluid is always bounded by a region near the material boundary, where fluid theory breaks down. We must therefore turn to kinetic solutions of the Fokker-Planck equation for boundary conditions on the fluid equations. There is at present no general solution relevant to scrape-off problems which includes a correct quantitative treatment of the

618

transition from a realistic sheath plasma to a collision-dominated fluid. However, there are a number of treatments of this interface which contain qualitatively reasonable assumptions about the sheath and fluid regions as discussed elsewhere in these lectures.[*] Our understanding of the qualitative behavior of hydrogen plasma flows is therefore not necessarily compromised by uncertainties about the boundary conditions.

The same cannot be said about impurity flows. At present there is to our knowledge not even a qualitative theory which adequately predicts the <u>relative</u> velocities of impurities in the kinetic regime. Since this is essential for studying radial impurity transport, a careful study of the multispecies kinetic theory is a high priority.

CONCLUSIONS

Considerable progress has been made in the past few years in understanding the physics of axisymmetric plasma flows near material boundaries. Despite the above-mentioned limitations of the present fluid theory of these flows, at least four significant results have emerged. First, 1-d analytic and computational work has pointed out the importance of localized intense recycling near the material boundary (cf. lecture by D. Post in this series.) Second, up-down asymmetry in the plasma flow can be related to poloidal rotation and up-down asymmetry in the radial transport. Third, the poloidal gyroradius is identified as a possible minimum radial thickness of a collisional scrape-off plasma. Fourth, the very high collisionalities envisioned for advanced divertor plasmas such as ASDEX-Upgrade and INTOR should have a larger neoclassical scrape-off thickness due to the radial Pfirsch-Schluter fluxes. For such cases, a complete fluid treatment is relatively straightforward. Since the very high collisionality in such plasmas should help prevent kinetic microturbulence, detailed numerical solution of the 2-d neoclassical transport

---

[*]Truly quantitative treatments of the sheath boundary conditions may soon be forthcoming. To be computationally feasible, this will require matching a kinetic computation near the material boundary to a fluid computation away from the boundary. The fluid calculation will have to have transport coefficients accurate to at least second order in the small parameter $\Delta^{-1} = L/L_{mfp}$ (e.g., parallel electron heat conduction $q^e = -\kappa_1 \nabla T_e + \kappa_2 \nabla (\nabla T_e) + ...$). (From the point of view of numerical efficiency alone, third or fourth order accuracy in Δ^{-1} is optimal, but this is not likely to be practical.)

equations for such devices may provide considerable insight. It should be possible to discover which transport equilibria lead to acceptable rates of plasma pumping and material boundary erosion. It should also be possible to perform accurate fluid and kinetic stability analyses on the computational equilibria to investigate limits on the achievable boundary gradients of pressure, current, and resistivity.

ACKNOWLEDGMENTS

The author is grateful to A. Boozer, W. Feneberg, K. Lackner, J. Neuhauser, and D. Post for useful discussions and to B. Braams for pointing out the problem of negative ion temperatures in 2-d transport simulations.

This work was supported by the U.S. Department of Energy Contract No. DE-AC02-76-CHO-3073.

APPENDIX

Here we outline a complete set of axisymmetric steady-state transport equations for a quasineutral, pure hydrogen plasma. The plasma is assumed to be collision dominated. The applied electrostatic potential is assumed to be of no greater order of magnitude than that arising from gradients in the plasma.

The transport equations are derived from the lowest four velocity moments of the Fokker-Planck equation. The zeroth velocity moments yield the electric current and ion conservation laws

$$\vec{\nabla} \cdot \vec{J} = 0, \tag{A1}$$

$$\vec{\nabla} \cdot \vec{\Gamma} = S, \tag{A2}$$

where $\vec{\nabla} \cdot \vec{A} = g^{-1/2}[\partial(A_\psi|\nabla\psi|g^{1/2}) + \partial(A_\theta|\nabla\theta|g^{1/2})/\partial\theta]$, $g^{1/2} = 1/|\nabla\psi \times \nabla\theta \cdot \nabla\xi|$, and (ψ,θ,ξ) are right-handed orthogonal toroidal coordinates with toroidal symmetry angle ξ. The poloidal current and the poloidal flux $\Gamma_\theta = nu_\theta$ are decomposed into components perpendicular and parallel to the magnetic field, $\vec{B}$, to obtain $J_\theta = (B_\xi/B)J_d + (B_\theta/B)J_\parallel$ and $\Gamma_\theta = (B_\xi/B)\Gamma_d + (B_\theta/B)\Gamma_\parallel$.

The dominant terms in the radial total and ion momentum balances give the diamagnetic flows:

620

$$J_d = (c/B) \, \nabla_\psi [n(T + T_e)], \tag{A3}$$

$$\Gamma_d = \frac{1}{m\Omega} \left[\nabla_\psi (nT) + en\nabla_\psi \phi \right], \tag{A4}$$

where $n = n_e$ is the ion density, T the ion temperature, T_e the electron temperature, m the ion mass, $\Omega = eB/(mc)$ the ion gyroradius, e the ion charge, ϕ the electrostatic potential, and $\nabla_\psi = |\nabla\psi| \, \partial/\partial\psi$. The total and ion toroidal momentum balances give the radial current and radial ion flow

$$J_\psi = \frac{e}{m\Omega_p} \, F_{Mach} \, , \tag{A5}$$

$$\Gamma_\psi = \frac{J_\psi}{e} - \frac{1}{m\Omega_p} \frac{B_\xi}{B} \, F_{\parallel} \, , \tag{A6}$$

where

$$F_{\parallel} = -\alpha_o \, m\nu_{ei} J_{\parallel}/e \tag{A7}$$

is the Coulomb friction on the ions parallel to the magnetic field, $\alpha_o = 0.5129$, $\beta_o = 0.711$, m_e is the electron mass, $\nu_{ei} = 4(2\pi)^{1/2} m_e^{1/2} e^4 \lambda_{ei}/(3m \, T_e^{3/2})$ is the electron-ion collision frequency, λ_{ei} is the Coulomb logarithm, $\nabla_{\parallel} = (B_\theta/B)\partial/\partial\theta$ and $\Omega_p = (B_\theta/B)\Omega$ is the ion poloidal gyrofrequency. We have neglected the small classical contribution to Γ_ψ resulting from resistance to the diamagnetic current. For nearly sonic flows, there can be several forces producing radial transport which act primarily on the ions

$$F_{Mach} = F_{Cor} + F_{Vis} + F_{CX} \, . \tag{A9}$$

(We call the sum of these forces "F_{Mach}" since they are only significant when the parallel Mach number is of order 1.) The Coriolis force is

$$F_{Cor} = -\left(m\Gamma_\psi \nabla_\psi u_\xi + m\Gamma_\theta \nabla_\theta u_\xi \right), \tag{A10}$$

where u_ξ can be replaced by $u_{\parallel}$, and an accurate solution is retained. The classical viscous force can be adequately approximated by

$$F_{Vis} = \frac{B_\xi}{B} \left[\nabla_\phi \left(-\eta_2 \nabla_\phi u_\parallel + \eta_4 \nabla_\theta u_\parallel \right) - \nabla_\theta \left(\eta_4 \nabla_\phi u_\parallel \right) \right], \qquad (A11)$$

where $\eta_4 = nT/\Omega$ and $\eta_2 = 6\nu_{ii}\eta_4/(5\Omega)$ are viscosity coefficients, $\nu_{ii} = (4\pi^{1/2} ne^4 \lambda_{ii}/3m^{1/2} T^{3/2})$ is the ion-ion collision time, and $\nabla_\theta = |\nabla\theta| \partial/\partial\theta$. The toroidal charge-exchange friction

$$F_{CX} = X_\xi - mu_\xi S \qquad (A12)$$

is obtained by integrating the non-Coulomb collision operator over a Maxwellian approximation to the distribution functions

$$X_\xi = \int d^3v \, mv_\xi \, \Sigma\left(f_o, f_o^e\right), \qquad (A13)$$

and subtracting out the momentum dilution term, $mu_\xi S$.

The total and electron momentum balances parallel to the magnetic field help determine the parallel ion flux and the electrostatic potential. An adequate approximation of the total parallel momentum balance is

$$\frac{B_\xi}{B} \nabla_\parallel \left[n\left(T + T_e\right) \right] = F_{Mach} .$$

The parallel electron momentum balance (Ohm's law) can be approximated by

$$en\nabla_\parallel \phi = -F_\parallel - \nabla_\parallel \left(nT_e\right) . \qquad (A15)$$

Finally, the ion and electron heat balances are

$$\frac{3}{2} \left(\Gamma_\phi \nabla_\phi T + \Gamma_\theta \nabla_\theta T \right) + nT\nabla \cdot \vec{u} = -Q_\Delta + W_{Vis}$$

$$+ \left(Y - \vec{u} \cdot \vec{X} - \frac{3}{2} TS + \frac{m}{2} u^2 S \right) - \nabla \cdot \vec{q} , \qquad (A16)$$

$$\frac{3}{2}\left(\Gamma^e_\phi \nabla_\phi T_e + \Gamma^e_\theta \nabla_\theta T_e\right) + nT_e \nabla\cdot\vec{u}_e = Q_\Delta - F_\| J_\|/(en)$$

$$+ \left(Y^e - \frac{3}{2}T_e S\right) - \nabla\cdot\vec{q}^e . \tag{A17}$$

Here $Q_\Delta = 3n\nu_{ei}(T - T_e)$ is the Coulomb interchange ion heating, $Y = \int d^3v\,(1/2)\,mv^2\,\Sigma(f_o, f^e_o)$ is the non-Coulomb ion energy source, $\vec{\Gamma}^e = n\vec{u}^e = \vec{\Gamma} - \vec{J}/e$ is the electron flux. The poloidal components of the heat conduction fluxes, $\vec{q}$ and $\vec{q}^e$ are decomposed into parallel and diamagnetic components, e.g.,

$$q^e_\theta = \left(B_\xi/B\right)q^e_d + \left(B_\theta/B\right)q^e_\| . \tag{A18}$$

The classical diamagnetic heat fluxes

$$q^e_d = -\frac{5}{2}\frac{nT_e}{m\Omega}\nabla_\phi T_e , \tag{A19}$$

$$q_d = \frac{5}{2}\frac{nT}{m\Omega}\nabla_\phi T , \tag{A20}$$

cancel each other for equal ion and electron temperatures. The classical parallel electron heat flux is

$$q^e_\| = -\left(\kappa^e_\| \nabla_\| T_e + \beta_o TJ_\|/e\right), \tag{A21}$$

where $\kappa^e_\| = 3.1616\, nT/(2^{1/2}\, m_e \nu_{ee})$, $\nu_{ee} = 4\pi^{1/2}\, n\, e^4$ $\lambda/(3m_e^{1/2}T_e^{3/2})$, and $\beta_o = 0.711$. The parallel ion heat conduction is formally small in the fluid regime. The radial electron heat flux is dominated by the neoclassical contribution

$$q^e_\phi = -\frac{B_\xi}{B}\frac{5nT_e}{2m\Omega_p}\nabla_\| T_e . \tag{A22}$$

Due to the larger ion mass, the radial ion heat flux, q_ψ, is considerably more complicated. An exact expression for q_ψ is obtained from the diamagnetic component of the $(1/2)mv^2\vec{v}$ moment of the Fokker-Planck equation

$$q_\psi = -\frac{1}{\Omega_p} \left\{ \frac{B_\theta}{B} \left[\left(V:\vec{R} - \frac{5T}{2m} V:\vec{P} \right)_d + \frac{e}{m} \left(\nabla\phi:\vec{\Pi} \right)_d - enu_d \vec{u}\cdot\nabla\phi \right] \right.$$

$$\left. + \frac{B_\xi}{B} \frac{enu^2}{2} \nabla_\parallel\phi + \frac{B_\theta}{B} \left[(G_d + z_d) - \frac{5T}{2m} (F_d + x_d) \right] \right\}$$

$$- \left(u:\vec{\Pi} \right)_\psi - \frac{m}{2} nu^2 u_\psi . \tag{A23}$$

For very subsonic flows, this reduces to $q_\psi = (B_\psi/B)[(5nT)/(2m\Omega_p)]\nabla_\parallel T + q_\psi^{classical}$ where

$$q_\psi^{classical} = -\frac{G_d}{\Omega} = \frac{2nT}{m\nu_{ii}} \left(\frac{\nu_{ii}}{\Omega} \right)^2 \nabla_\psi T. \tag{A24}$$

For near sonic flows, one can straightforwardly obtain an expression valid for q_ψ at very high collisionality [$\Delta^{-1} \gg (m/m_e)^{1/2}$] by using the method of Braginskii (cf. Wong[12]) to obtain the correction Φ to the distribution function $f = f_o(1 + \Phi)$. One then determines the heat flux using the definitions $\vec{R} = \int d^3v(mv^2/2) \vec{v}\vec{v}f$, $P = \int d^3v \vec{v}\vec{v}f$. (Usually, the diamagnetic contribution from $\vec{Z} = \int d^3v(mv^2/2)\vec{v}\Sigma$ should be negligible.) The viscous heating $Q_{Vis} = (\nabla\vec{u}:\vec{\Pi}_{Vis})$ is also potentially important for near-sonic flows and can be similarly computed. While this computation of q_ψ is straightforward, we have not carried it through because the results are valid only at very high collisionality. Note that in this "super-Pfirsch-Schluter" regime, the electron and ion temperatures are very nearly equal, and the diamagnetic and neoclassical electron heat fluxes are cancelled by similar ion heat flux terms. It is therefore highly desirable to add ion and electron heat balance equations to obtain a simpler total heat balance.

For lower collisionality, where one cannot generally exclude the condition $\delta_p \sim 1$, drift effects must be included in computing the transport coefficients. After this is done, one could include one of the ion or electron heat balances along with the total heat balance to obtain a complete set of ten differential plasma transport equations. It might then be computationally convenient to add the total mechanical energy to the total heat balance to obtain the total energy transport equation

$$\nabla\cdot\left(\frac{1}{2} mu^2 \vec{u} + \frac{5}{2} nT\vec{u} + \frac{5}{2} nT_e\vec{u}_e + \vec{u}:\vec{\Pi}_{Vis} + \vec{q}_\theta + \vec{q}_\theta^e + \vec{q}_\psi + \vec{q}_\psi^e \right)$$

$$= -F_\parallel J_\parallel/(en) + \vec{u}\cdot\vec{F}_{Coul} + Y + Y^e . \tag{A25}$$

624

Use of this "conservative" form has some computational advantages for determining the total energy density. Due to the large mechanical and other energy interchanges between the ions and electrons, however, such conservative forms are unlikely to be useful for separately determining the electron and ion temperatures. Rather, one should combine Eq. (A25) with either Eq. (A16) or Eq. (A17) to determine the difference between the ion and electron temperatures.

REFERENCES

1. D. Sharkovsky, T. W. Johnston, and M. P. Bachyuski, The Particle Kinetics of Plasmas (Addision-Wesley, London, 1966).
2. D. L. Book, NRL Plasma Formulary, Naval Research Laboratory (Washington, DC, 1983).
3. S. I. Braginskii, ZHETF 33, 4591 (1957); Sov. Phys. JETP 6, 358 (1958).
4. S. I. Braginskii, in Reviews of Plasma Physics, edited by M. A. Leontovich (Consultants Bureau, NY, 1965).
5. C. E. Singer and W. Langer, Phys. Rev. A28, 994 (1983).
6. J. Neuhauser, Max Planck Institut fur Plasmaphysik, private communication (1984).
7. U. Daybelge, Nucl. Fusion 21, 1589 (1981).
8. G. Fussmann, K. Behringer, K. Bernhardi, G. Haas, W. Poschenrieder, et al., "Impurity Retainment in the Divertor of ASDEX," in Proceedings of the Eleventh European Conference on Controlled Fusion and Plasma Physics (Aachen, Germany, 1983) European Physical Society (1983), Vol. 7D, Part II, p. 373.
9. W. Feneberg, JET Joint Undertaking, Abingdon, Oxfordshire, England, private communication (Feb. 1984).
10. C. E. Singer, J. Geophys. Res. 82, 2686 (1977).
11. F. H. Hinton and R. D. Hazeltine, Rev. Mod. Phys. 48, 239 (1976).
12. S. K. Wong, Matrix Elements of the linearized Fokker-Planck Operator, GA Technologies, Inc. Report No. GA-A1749 (1984), submitted to Phys. Fluids.

PLASMA MODELS FOR IMPURITY CONTROL EXPERIMENTS

D. E. Post
Plasma Physics Laboratory, Princeton University
Princeton, N.J. 08544

K. Lackner
Max Planck Institut für Plasmaphysik
Garching bei München, Federal Republic of Germany

INTRODUCTION

There have been significant advances in impurity control modeling in the last five years. The models for impurity control has grown in sophistication from simple, almost heuristic models to two and three-dimensional codes embodying realistic geometries and many of the important plasma wall and atomic physics processes. Perhaps more importantly, the growth in the models has helped to increase our understanding of how divertors and limiters work as impurity control devices and has aided in the design of improved divertors and limiters. The greatest success of the models has been in helping to identify the key role that local recycling can play in producing a "high recycling" divertor in which a cool, dense plasma is produced next to neutralizer plate. This cool plasma allows the possibility of removing the heating power (neutral beams, ICRF, alpha particles, etc.) from the plasma without the generation of large amounts of impurities. The low temperature reduces the sheath potential and thus reduces the energy of the ions and neutrals that strike the neutralizer plate to below the sputtering threshold of a number of candidate wall materials, eliminating the divertor plate as a source of impurities. In addition, the high density means that adequate pumping of helium will be possible with relatively modest pumping systems.

The level of confidence in these models has reached the point where they can be used with a reasonable degree of confidence to design a divertor system for a large, high power fusion experiment.

627

As described in the other chapters of this book, a model
for the plasma conditions and performance of impurity control
and particle control systems for large experiments with high
power heating involves a large number of diverse ingredients
(Fig.1). First, one must have a set of equations to solve for
the plasma transport, such as described in the chapter by C.
Singer. Then one must have boundary conditions. The sheath
conditions, such as described in the chapter by Stangeby and
Chodura, are usually appropriate. The source terms must also be
calculable. The source terms for the recycling and injected
neutral gas are usually obtained from an auxiliary calculation
of the neutral gas transport, such as discussed in the chapter
by Heifetz. These neutral gas transport calculations involve
detailed descriptions of the neutral and ion reflection from
surfaces and of the atomic and molecular processes involved in
neutral-plasma collisions. In addition, impurity production and
source rates are needed. A fast, convenient model for
sputtering, for example, is required. Ionization,
recombination, and excitation rates for impurity ions of
interest are required. In addition, some model for the plasma
and impurity cross-field transport is necessary. Such transport
is usually anomalous, so that empirical models are used. These
calculations need to be carried out in realistic geometries.
Finally, all of these effects have to be combined in a model
where they can interact with each other.

HIGH BETA PLASMA

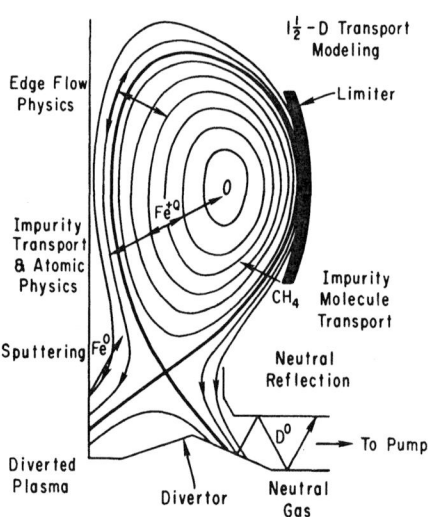

Fig. 1. Schematic illustration of the physics ingredients of
impurity and particle control systems for a large, high
power fusion experiment.

Once a system of equations, boundary conditions and sources is defined, one must be able to obtain a solution to the problem. The equations are highly nonlinear and multidimensional, so that analytic methods are difficult and numerical techniques are often applied. This chapter will outline some of the various models used for calculating plasma edge conditions, and will discuss a few of the numerical techniques used for both hydrogen and impurity transport. The first half of the chapter will concentrate on the transport of hydrogen plasmas, and the second half will concentrate on impurity transport.

1. HYDROGEN DYNAMICS

1.1. Formulation of the problem

The calculation of the plasma parameters at the edge begins with a set of equations. The typical starting point is the fluid approximation, which may not be adequate in some circumstances.

The plasma edge region is characterized by plasma densities of 10^{11} to 10^{15} cm^{-3} and temperatures of 1 eV to 300-400 eV. If we define a characteristic length L which for axisymmetric toroidal devices is the connection length $L = \pi Rq$, where R is the major radius, and q is the safety factor, and the mean free path of ions and electrons for a 90° scatter as λ, we can then define an effective collisionality $\nu = \lambda/L$. When $\nu \ll 1$, the fluid approximation is appropriate. When $\nu \gg 1$, a kinetic approach is preferable. In between, the fluid approximation can be used, but with caution. The mean free path for a cumulative 90° scattering can be written as [1]:

$$\nu = \frac{\lambda}{L} = \frac{108 \ \pi^{1/2} (\varepsilon_o kT)^2}{Lnc^4 \ \ell n \ \Lambda} \approx 5 \times 10^{16} \ \frac{T^2 (eV)}{nL} \ . \tag{1}$$

Far out in a scrape-off layer, where $n \sim 10^{18}$ m^{-3} and $T \sim 10$ eV, $\lambda \sim 5$ (Fig. 2). In a high recycling divertor where $n \sim 10^{20}$ m^{-3} and $T \sim 5$ eV, $\lambda \sim 1 \times 10^{-2}$ m. For typical L's of 10 to 50 m, this implies that the fluid approximation is good at low temperatures, and not very good when the temperature is in the 50-100 eV range and the density is low. Nonetheless, the fluid approximation gives us tractable equations, and is usually better than would be expected from simple collision arguments. The fluid model is reasonably valid for the high-density and low-temperature plasmas needed for a practical impurity control system. At the end of this section, we will return to a discussion of kinetic treatments.

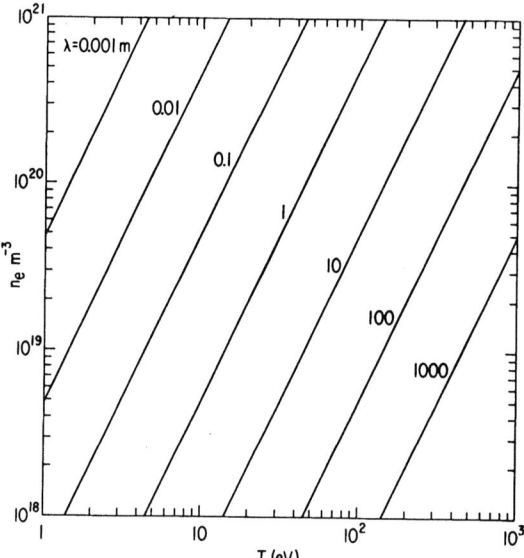

Fig. 2. The mean free path for a 90° scattering as a function
         of n and T [1].

        The fluid equations are generally derived from the
equations formulated by Braginskii [2]. Neglecting a few terms,
they have the form of conservation equations for particles,
momentum, and energy:

$$-\nabla \cdot \vec{\Gamma} + S_n = 0, \quad -\nabla \cdot \vec{R} + S_p = 0,$$
and
$$-\nabla \cdot \vec{Q} + S_E = 0, \tag{2}$$

where $\vec{\Gamma}, \vec{R}$, and $\vec{Q}$ are the particle, momentum, and energy fluxes,
respectively, and S_n, S_p, and S_E are the appropriate sources.
Dr. Singer in his lecture presents a detailed set of equations
including all of the neoclassical effects (see also Ref. [3]).
As discussed by Dr. Singer, these equations are obtained from
the Boltzmann equation by taking the first three velocity
moments of the Boltzmann equation.

        It is often convenient to have these equations expressed in
a "semi-conservative" form for use in plasma modeling. By
"semi-conservative" we mean that as many terms as possible are
expressed as the divergence of a flux. For circular magnetic
flux surfaces, the equations become [4,5]

630

density:

$$\frac{\partial(nv_\parallel)}{\partial\zeta} = \frac{1}{r}\frac{\partial}{\partial r}\, r\Big(D_\perp \frac{\partial n}{\partial r} - nv_r\Big) + S_n \; , \tag{3}$$

momentum:

$$\frac{\partial}{\partial\zeta}\Big(p_i + p_e + mnv_\parallel^2 + 1.28\ \tau_i nT_i \frac{\partial v_\parallel}{\partial\zeta}\Big)$$

$$= S_p + \frac{1}{r}\frac{\partial}{\partial r}\, r\Big[mv_\parallel\Big(-nv_r + D_\perp \frac{\partial n}{\partial r}\Big)\Big], \tag{4}$$

energy:

$$\frac{\partial}{\partial\zeta}\Big[\Big(\frac{5}{2}T_i + \frac{1}{2}nmv_\parallel^2\Big)nv_\parallel - n\chi_\parallel^i \frac{\partial T_i}{\partial\zeta}\Big] = -v_\parallel \frac{\partial}{\partial\zeta}(nT_e)$$

$$+ \frac{1}{r}\frac{\partial}{\partial r}\, r\Big[n\chi_\perp^i \frac{\partial T_i}{\partial r} + \Big(\frac{5}{2}T_i + \frac{1}{2}mv_\parallel^2\Big)\Big(D_\perp \frac{\partial n}{\partial r} - nv_r\Big)\Big] + Q_\Delta + S_E^i \; ,$$
$$\tag{5}$$

and

$$\frac{\partial}{\partial\zeta}\Big(\frac{5}{2}Tnv_\parallel - n\chi_\parallel^e \frac{\partial T_e}{\partial\zeta}\Big) = v_\parallel \frac{\partial}{\partial\zeta}(nT_e)$$

$$+ \frac{1}{r}\frac{\partial}{\partial r}\, r\Big[\frac{5}{2}T_e\Big(D_\perp \frac{\partial n}{\partial r} - nv_r\Big) + n\chi_\perp^e \frac{\partial T_e}{\partial r}\Big] - Q_\Delta + S_E^e \; . \tag{6}$$

Viscous effects have been neglected in the energy equations. ζ is the coordinate along the field lines, the subscript r denotes the radial direction (across flux surfaces), $v_\parallel$ is the velocity along the field line, n is the particle density, $T_{i,e}$ are the ion, electron temperatures, m is the ion mass, τ_i is the ion collision time, $D_\perp$ is the radial density diffusion coefficient (usually anomalous [6]), χ^e and χ^i are the electron and ion radial heat diffusivities (also anomalous [6]), Q_Δ is the classical ion-electron energy equipartition rate, S_n, S_p, S_E^i, and S_E^e are the density momentum, ion energy, and electron energy sources from neutral atom and impurity collisions (ionization, charge exchange, radiation, etc.). $p_{i,e}$ is the ion, electron pressure.

These are four highly nonlinear, two-dimensional equations for n, $v_\parallel$, T_i, and $T_e(p = nT)$. Many of the terms involve the product of three variables, and one term is the product of four variables ($nv_\parallel^3$). The source terms from the neutral plasma collisions will be large, and will also depend nonlinearly on the plasma parameters.

The first step in solving these equations is to determine a set of boundary conditions. One can specify $\Gamma = \Gamma_o$, $Q_i = Q_i^o$, and $Q_e = Q_e^o$, where Γ, Q are the parallel particle/energy fluxes for ions/electrons, and the superscript "o" denotes a specification. There are four boundaries (Fig. 3) for the computational mesh, the main plasma, the two "collector plates" where the field lines strike the wall, and the wall region where the field lines are parallel to the wall. The main plasma boundary conditions can be set either by specifying a set of edge plasma conditions, such as n_e, T_e, and T_i, or as a set of fluxes from the main plasma. Similar specifications can be made at the wall parallel to the field lines. For the field lines that strike the collector plates, one can either specify the fluxes at each plate, or take advantage of the symmetry of the problem to note that since the plasma flows to the plates at each end of the field lines where the field lines strike the plates and integrates from the "stagnation" point where the net fluxes along the field lines are zero to the plates.

As discussed in the chapters by Stangeby and Chodura, an electrostatic potential forms (at the point the plasma strikes the plates) to retard the electron flow so that the electron and ion currents to the plate are equal. This requirement leads to boundary conditions for the particle and heat fluxes. From the continuity and momentum equations, Stangeby in his chapter has shown that, in the fluid picture, the parallel flow velocity cannot exceed the sound speed as the plate is approached. On the other hand, the "Bohm" criterion states that the flow velocity must be at least sonic at the plate for a stable sheath

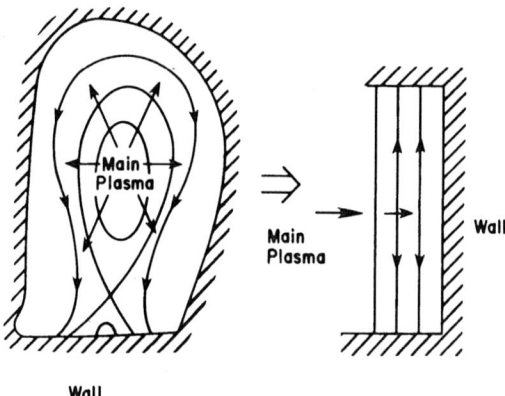

Wall

Fig. 3. Schematic outline of the field line geometry at the plasma edge showing parallel flow to the divertor plates and perpendicular flow from the main plasma to the wall.

632

to form. Thus, the condition that $v_{\parallel} = c_s$, the ion sound speed, is a reasonable approximation at the sheath boundary. Similar considerations apply for the heat flux. The result from Stangeby's chapter is that

$$Q_e = \delta \, kT_e \, \Gamma, \tag{7}$$

$$Q_i = 2 \, kT_i \, \Gamma, \tag{8}$$

and

$$\Gamma = nc_s \, , \tag{9}$$

where $\delta_0 \sim 4.8$ for equal ion (T_i) and electron (T_e) temperatures, n and T are the densities and temperatures just in front of the surface, and c_s is the ion sound speed, $[(T_i + T_e)/m]^{1/2}$. Secondary electron emission effects are shown to modify δ_e giving

$$\delta_e = \frac{2}{1 - \lambda} - \frac{1}{2} \, \ell n \left[2\pi \, \frac{m_e}{m_i} \left(1 + \frac{T_i}{T_e}\right)(1 - \gamma)^{-2} \right] \, , \tag{10}$$

where λ is the secondary electron yield per incident ion-electron pair.

The boundary condition (8) is not quite correct, due to the neglect of the plasma before the sheath. A better condition [4] can be derived from Eqs. (3) and (4) by combining them to obtain

$$\left(\frac{T_i + T_e}{2u} - 1\right) \frac{\partial u}{\partial \zeta} = \left(2u + T_i + T_e\right)[s_n$$

$$+ \frac{1}{r} \frac{\partial}{\partial r} \, r \left(D_{\perp} \frac{\partial n}{\partial r} - nv_r\right)]/\Gamma - \frac{\partial T_i}{\partial \zeta} + \frac{\partial T_e}{\partial \zeta} \, , \tag{11}$$

where we let $u = mv_{\parallel}^2/2$. As the sheath is approached, the coefficient of $\partial u/\partial \zeta$ approaches zero. Thus in order for $\partial u/\partial \zeta$ to stay finite, the terms on the right-hand side of (11) must go to zero. Thus, for self-consistency, we have

$$\left.\frac{\partial T_i}{\partial s}\right|_{sh} = - \left\{\left(2u + T_i + T_e\right)[s_n\right.$$

$$\left. + \frac{1}{r} \frac{\partial}{\partial r} \, r\left(D_{\perp} \frac{\partial n}{\partial r} - nv_r\right)]/\Gamma\right\}_{sh} - \left.\frac{\partial T_e}{\partial \zeta}\right|_{sh} \, , \tag{12}$$

which should be used as a boundary condition for Eq. (5).

To return to the question of the fluid versus kinetic picture, we note that in many cases (Fig. 2), the fluid approximation is marginal at best. As discussed in the sheath theory lectures, the sheath is essentially collisionless. A number of attempts have been made to treat the plasma edge with kinetic models [7,8]. In general, these models were for one spatial dimension and one to two velocity dimensions. The major result of these models was that the potential distribution along the field lines can be substantially modified by the ionization of cold gas. An analytic model was developed by Bailey and Emmert [9] which obtained solutions close to the numerical ones. However, similar behavior is present in the fluid simulation. A recent particle simulation of the sheath and presheath region by Takizuka et al. [10] shows that kinetic effects can substantially modify the heat flux. Indeed, on ASDEX [11], there is some evidence that hot electrons from the main plasma edge can penetrate into the divertor plasma. There is clearly a need for improved kinetic treatments which include as many as possible of the important atomic processes, effects of real geometry, and surface processes.

The fluid approximation breaks down when the ratio of the collisional mean free path to the scale lengths of the system is near one or greater. One effect in this regime is that the electron heat flux can no longer be written as $n\chi_\parallel \, \partial T_e/\partial \zeta$. Since $n\chi_\parallel \propto T_e^{5/2}$ and $\lambda \propto T_e^2$, the low collisionality regime occurs when T_e is large, and the heat flux needs modification. One such modification is the use of a "flux-limiter" for the electron heat transport. The concept of a flux limiter was developed at Livermore in the mid-sixties by Wilson and LeBlanc [12]. It has been used extensively in laser fusion studies [13]. The basic idea is that the electron energy cannot flow any faster than the electron sound speed. Thus, the heat flux is "limited" to

$$Q_e \lesssim v_e n T_e \; ,$$

where v_e is the electron thermal speed. A suggested form for the heat flux might be

$$Q_e = \frac{1}{(1/n\chi_\parallel^e) + (|\nabla T_e|/\alpha v_e n T_e)} \, \nabla T_e \; , \qquad (13)$$

where α is chosen to match experiments ($\alpha \sim 0.03-0.1$). Where λ

634

is small, $\chi_{e\parallel}$ is small, and the heat flux is the usual value of $n\chi_{\parallel}^e \nabla T_e$. When λ is large, $\chi_{\parallel}^e$ is large, and the heat flux becomes $\alpha v_e n T_e$. The first application of this limiter to edge transport has been by Singer and Langer [14,15].

In general, the fluid approach is better that one might expect based on collision rate and mean free path arguments. It does, after all, conserve particles, momentum, and energy, and as long as some type of average description of the plasma is appropriate, it will not be too far wrong, especially if some corrections can be made for the nonlocal effects.

1.2. Zero-dimensional models

Real tokamaks and other fusion experiments usually have complex geometries requiring two or three dimensions for a model. However, two-dimensional models are large and complex, and often a great deal of insight can be gained from simpler models. In many cases, a zero-dimensional model can be quite useful. In these models one approximates the full transport equations

$$\frac{\partial n}{\partial t} = -\nabla \cdot \vec{\Gamma} + S_n ; \quad \frac{\partial (m n v_\parallel)}{\partial t} = -\nabla \cdot \vec{R} + S_p ;$$

$$\frac{\partial (3/2 \; nT)}{\partial t} = -\nabla \cdot Q + S_E , \tag{14}$$

where the terms are defined as in Eq. (2). In the zero-dimensional problem, one replaces the divergence of the flux with a density divided by a confinement time (a la Gauss's theorem) and obtains equations of the form

$$\frac{\partial n}{\partial t} = S - \frac{n}{\tau_\parallel} - \frac{n}{\tau_\perp} , \; \text{etc.} \tag{15}$$

where $\tau_\parallel$ and $\tau_\perp$ are the parallel and perpendicular loss times, and S is the sum of the volume sources and sinks due to ionization, charge exchange, etc. The advantage of these models is that they can include quite complex treatments for the source terms in a code with a simple structure. Simple predictor-corrector, or semi-implicit techniques are quite adequate [16]. This type of model has been used for plasma start-up calculations [17,18] and for modeling ion sources for neutral beams [19].

1.3. One-dimensional models

The next level of complication involves keeping either the parallel or perpendicular part of the fluxes in the divergence term, and integrating the one-dimensional part of the equation. Thus, the transport would be of the form

$$\frac{\partial f}{\partial t} = -\nabla_\perp \cdot \left(f v_\perp \right) - \frac{f}{\tau_\parallel} + S_f \ , \tag{16}$$

or

$$\frac{\partial f}{\partial t} = -\nabla_\parallel \cdot \left(f v_\parallel \right) - \frac{f}{\tau_\perp} + S_f \ . \tag{17}$$

Models of the form of Eq. (16) are typically used in one-dimensional tokamak transport codes [20-23], or in a model just for the plasma edge [24]. A typical set of equations in the form of Eq. (16) for the plasma edge region is [24]

$$\frac{d}{dx} \left(D_\perp \frac{dn}{dx} \right) - \frac{n}{\tau_\parallel} + S_e = 0, \tag{18}$$

$$\frac{d}{dx} \left(n \chi_\perp^e \frac{dT_e}{dx} + \frac{3}{2} T_e D_\perp \frac{dn}{dx} \right) - \frac{2\gamma n T_e}{\tau_\parallel} - Q_\Delta + S_E^e = 0, \tag{19}$$

and

$$\frac{d}{dx} \left(n \chi_\perp^i \frac{dT_i}{dx} + \frac{3}{2} T_i D_\perp \frac{dn}{dx} \right) - \frac{2n T_i}{\tau_\parallel} + Q_\Delta + S_E^i = 0, \tag{20}$$

where the terms are defined as in Eqs. (3-6). The coordinate x is the distance into the scrapeoff from the last closed flux surface. The parallel confinement time is given by a connection length divided by a parallel velocity, which for an axisymmetric divertor or limiter would be

$$\tau_\parallel = \frac{L}{V_\parallel} = \frac{\pi q R}{v_\parallel} \ , \tag{21}$$

where $v_\parallel = 0.5(T_i + T_e)^{1/2}$, the ion sound speed. A code of this type was used to provide some estimate of the perpendicular transport in the plasma edge [24]. Ulrickson [24] used $\chi = \chi_0 T^a n^b$ and $D = D_0 T^c n^d$, and determined the coefficients by fitting probe data from PLT and PDX. The best fits were

$$\chi_\perp^e = (9 \pm 3) \times 10^{16} \ n^{(-1 \pm 0.5)} T^{(0 \pm 0.25)} \ \mathrm{cm}^2/\mathrm{sec} \ , \tag{22}$$

and

$$D = (1\pm0.3) \times 10^4 \; n^{(0\pm0.25)} \; T^{(0\pm0.25)} \; cm^2/sec \; . \qquad (23)$$

The quality of the fit is reasonable (Fig. 4). Models of this type have been used in conventional one-dimensional tokamak transport codes [25,26]. They form a natural set of boundary conditions at the edge for such codes. Applying these models to examine the expected edge conditions for INTOR produces estimates of $T_i \sim$ 100-200 eV, $T_e \sim$ 100-300 eV, and $n_e \sim 2 \times 10^{12}$ to $10^{13} cm^{-3}$ (Fig. 5) for the edge parameters.

An alternative approach to one-dimensional models has been to replace the parallel loss terms with transport terms to describe the parallel flow for particles and energy and to either ignore or keep only a simple model like that of Eq. (15) for the perpendicular transport [27-30]. One of the major uses of these models has been in the analysis of very cool, diverted plasmas, where large temperature and density gradients are observed along the field lines near the divertor. Many of the

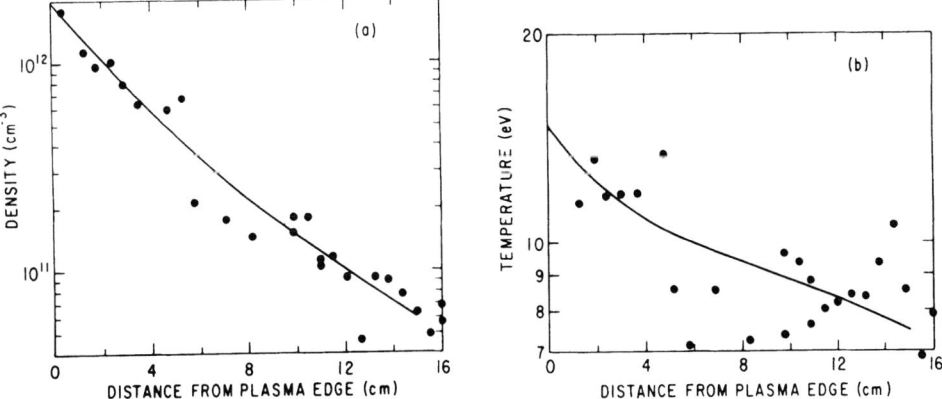

Fig. 4. (a) Fit to the PDX density data. The solid curve is the best fit to the data shown. (b) Fit to the PDX temperature data. The solid curve is the best fit to the data shown. The fit parameters are given in Eqs. (22) and (23) (taken from [24]).

637

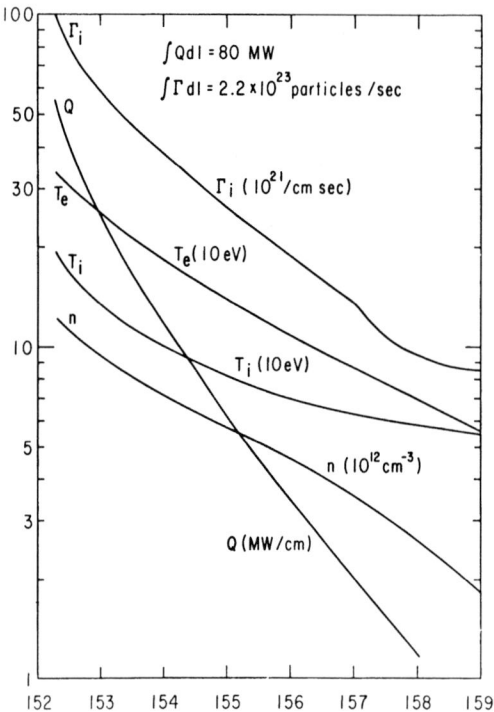

Fig. 5. Parameters in the scrape-off plasma for INTOR
calculated with the model of Ref. [21]. The separatrix
is at r = 152 cm.

scalings observed experimentally can be obtained this way.
Following Mahdavi et al. [31], the diverted plasma can be
represented as a tube of cross sectional area $A = 4\pi R\delta B_p/B_T$ (two
divertor plates) where δ is the thickness of the scrape-off
layer, B_T is the toroidal field, B_p is the poloidal field, and R
is the major radius. The tube has length ℓ. The power, P_o, is
input at $x = 0$, and the divertor plate is located at $x = \ell$. It
is assumed that, at the plasma edge, $n_e = \alpha \bar{n}_e$; i.e., that the
midplane edge plasma density is proportional to the line average
plasma density. One then neglects radiation, perpendicular
transport, convection, etc., and reduces Eq. (6) to
$\partial/\partial x[K_{\parallel} (\partial/\partial x) T_e] = 0$ where $k_{\parallel} = k_o T_e^{5/2}$. The boundary
conditions are at $x = 0$ are:

$$n_e(0) = \alpha \bar{n}_e ,\qquad\qquad (24)$$

$$-K_{\parallel} \frac{\partial}{\partial x} T_e = P_o/A ,\qquad\qquad (25)$$

and at $x = \ell$ are:

$$-K_\| \frac{\partial}{\partial x} T_e = 8nT_e \left(2T_e/m\right)^{1/2} . \tag{26}$$

Integrating the heat conduction equation one obtains

$$T_e^{7/2}(x) - T_e^{7/2}(0) = \frac{7P_o}{2k_o A} x . \tag{27}$$

Using pressure balance, $n(0)T(0) = n(\ell)T(\ell)$, and flux conservation $P_o/A = 8n(\ell)T_e^{3/2}(\ell) (2/m)^{1/2}$, one can obtain

$$T_e^{21/2}(0) = \left(\frac{P_o (m/2)^{1/2}}{8 \, \alpha \bar{n}_e}\right)^7 + \frac{P_o}{A} \frac{7\ell}{2k_o} T_e^7(0) . \tag{28}$$

For large $\bar{n}_e$, the middle term drops out, and one obtains

$$T_e(0) = \left(\frac{7\ell P_o}{2Ak_o}\right)^{2/7} . \tag{29}$$

This is independent of $\bar{n}$, and only weakly dependent on the other parameters such as ℓ, P_o, δ, etc. Using this result in the flux conservation equation, one can then obtain

$$T_e(\ell) = \frac{m}{2} \left(\frac{2k_o}{7\ell}\right)^{4/7} \left(\frac{P_o}{A}\right)^{10/7} \left(\frac{1}{8\alpha}\right)^2 \frac{1}{\bar{n}^{-2}} , \tag{30}$$

and

$$n_e(\ell) = \frac{128\alpha^3}{m} \left(\frac{A}{P_o}\right)^{8/7} \left(\frac{7\ell}{2k_o}\right)^{6/7} \left(\bar{n}\right)^3 . \tag{31}$$

Using flux conservation for the plasma and neutrals $n_n v_n = n_e v_s$, we have

$$n_n \propto \frac{n_e \left[T(\ell)\right]^{1/2}}{(T_n)^{1/2}} \propto \frac{\bar{n}_e^{-2}}{(T_n)^{1/2}} . \tag{32}$$

These scaling results fit the D-III data rather well for ohmic heating (Fig. 6). The absolute magnitude of the coefficients is not quite right, but the general trends do indicate the right scaling.

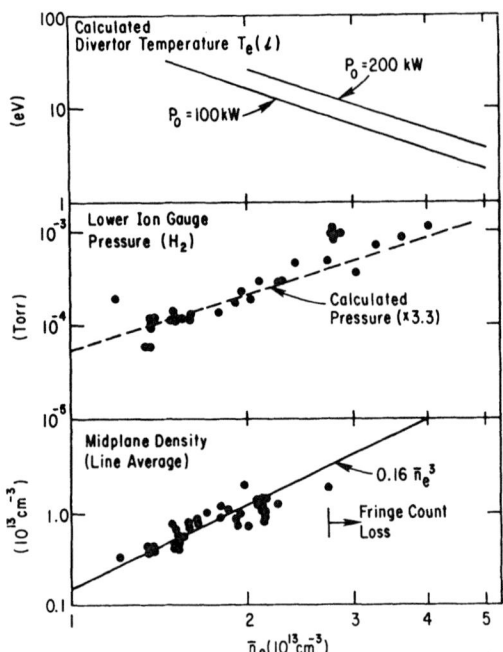

Fig. 6. Density dependence of divertor plasma parameters: (a) midplane line averaged density; the solid line shows the best fit of experimental data to the model; (b) ion gauge pressure (lower chamber); calculated divertor temperature boundary (taken from [31]).

Still missing is an identification of the detailed physics that might be going on near the sheath boundary, or, put another way, under what conditions do T(0) and T(ℓ) become different? Writing the heat flux as

$$\frac{P_o}{A} = 8\ T^{1/2}\ \Gamma, \tag{33}$$

clearly indicates that lowering T requires raising Γ, the particle flux. Since nT is constant, lowering T also raises n. This suggests that very high fluxes at high densities will be required to produce a cool plasma.

These considerations indicate that an examination of the particle transport might help identify the mechanism for cooling the diverted plasma. Following Ref. [32], we can construct a simple model based on only the continuity equations for plasmas and neutrals. Consider plasma incident on a wall at x = a (Fig. 7). The continuity equation for the plasma is

640

$$\frac{\partial(nv)}{\partial x} = S = n_e n_o \langle \sigma v \rangle_{ionization} \qquad (34)$$

Integrating this from the divertor entrance (x = 0) to the neutralizer plate (x = a), we obtain

$$\Gamma_a = \Gamma_o + \int_o^a n_e n_o \langle \sigma v \rangle \, dx, \qquad (35)$$

where Γ^a = nv at the divertor plate, and Γ_o is the input particle flux at the divertor throat. We see from this equation that the flux increases as we approach the plate. We can define $R = \Gamma^a/\Gamma_o$ as the flux amplification factor. Then, the equation becomes

$$R = 1 + \frac{1}{\Gamma_o} \int_o^a n_e n_o \langle \sigma v \rangle \, dx > 1. \qquad (36)$$

If we apply the sheath boundary conditions $Q_e^a = \delta \, T_e^a \, \Gamma^a$ (Eq. 7), and substituting $\Gamma^a = \Gamma_o R$, we have

$$T_e^a = \frac{Q_e^a}{\delta \Gamma^a} \approx \frac{Q_e^o}{\delta \Gamma_o} \frac{1}{R}. \qquad (37)$$

Raising R lowers T_e^a since $Q_e^a \lesssim Q_e^o$. The density is increased by raising R. Using $v^a \propto (T^a)^{1/2}$, $Q_e^a \sim n_a(T^a)^{3/2}$, or

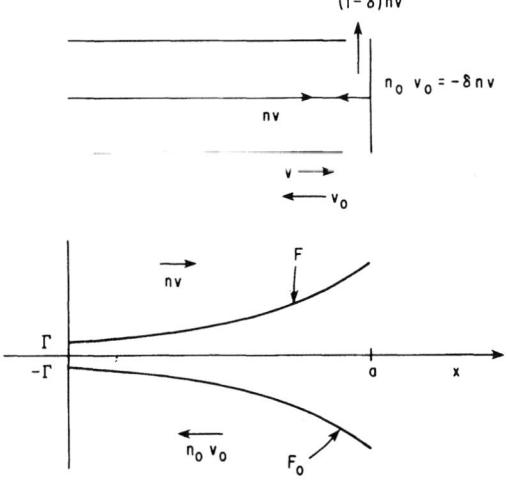

Fig. 7. Schematic illustration of a divertor model (taken from Ref. [32]). 81P0101

$$n_a \propto \frac{Q_e^a}{(T^a)^{3/2}} \propto R^{3/2} . \qquad (38)$$

Thus increasing the particle recycling raises n and lowers T. R > 1 implies that each ion-electron pair that enters the divertor has more than one chance to carry its energy to the plate, and thus the average energy per particle can be lowered. Since the flux must increase as R, the density must increase even faster than linearly in R to compensate for the lowering of the velocity by the lowering of T_e. The result of all of this is that recycling can produce a cool, dense diverted plasma. This is illustrated in Fig. 8 which shows T versus Q/Γ (energy per ion) for a variety of recycling coefficients. Typical heat fluxes for several plasma experiments are indicated in Fig. 8. Ohmic heated PDX should be able to achieve low divertor T_e with modest recycling coefficients. High power experiments require larger recycling coefficients. INTOR, for instance, may require recycling coefficients of 30 or more to achieve temperatures of 10 to 20 eV.

Here it is worth noting that recycling is a very nonlinear process. Since $1/\lambda_o = n_e \langle \sigma v \rangle / v_o$ where λ_o is the mean free path for electron impact ionization, and v_o is the neutral velocity, the formula for R can be written as follows

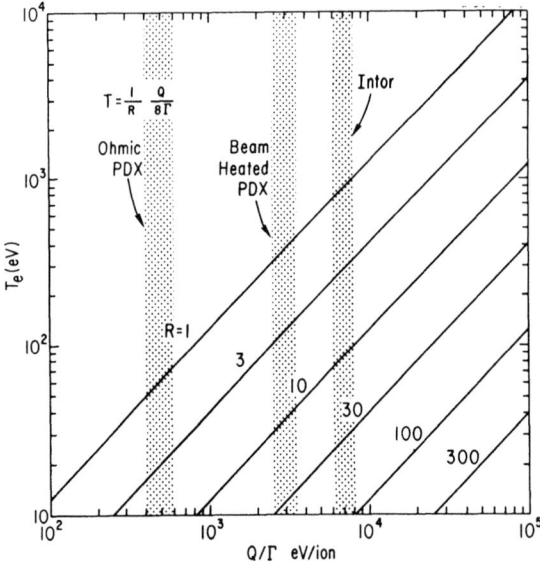

Fig. 8. Electron temperature at the divertor plate as a function of the energy per ion (ratio of heat flux to particle flux) for a variety of recycling coefficients (taken from Ref. [32]).

$$R = 1 + \frac{1}{\Gamma_o} \int_o^a \frac{v_o n_o}{\lambda_o} \, dx. \tag{39}$$

Thus, the smaller λ_o becomes, the larger R becomes. Since $n_e \propto R^{3/2}$, $\lambda_o \sim 1/R^{3/2}$. The stable operating point for a divertor exists when the continuity equation for the plasma [Eq. (34)] can be satisfied simultaneously with a continuity equation for the neutrals. In any divertor geometry, the net plasma influx must be balanced by the neutral flow back into the main chamber and down the pumping chamber. At high enough plasma densities the neutral mean free path can become too small for a solution to exist.

These considerations are moderated by the fact that as T_e drops below 10 eV, the electron impact ionization rate decreases rapidly with decreasing T_e. Nonetheless, we expect to find at least two stable operating points for divertors. One solution is where $R \sim 1$, and the neutrals stream with little ionization back to the main chamber and down the pump, and the second is a cooler and denser regime with $R > 1$, where the maximum R is limited by the requirement that the density must be low enough to allow the neutral outflux to match the plasma influx. The authors of Ref. [32] constructed an analytic model embodying these concepts. In the model, they solved two continuity equations of the form

plasma:
$$\frac{\partial(nv)}{\partial x} = n_o n_e \langle \sigma v \rangle = \frac{n_o v_o}{\lambda_o} , \tag{40}$$

neutrals:
$$\frac{\partial(n_o v_o)}{\partial x} = -n_c n_o \langle \sigma v \rangle = -\frac{n_o v_o}{\lambda_o} . \tag{41}$$

Matching the fluxes at the plate, using sheath boundary conditions, and making a barely reasonable set of assumptions, they were able to derive an existence condition of the form

$$f(x) = \frac{x}{1 - [1 - \exp(-\Delta/x^2)][1 - \exp(-1/x^2)]} , \tag{42}$$

where $\Delta = b/a$, the ratio of the divertor plasma width to the length (Fig. 7), $x = T(a)/T_2$, and $f(x) = T_1/T_2$. T_1 is the temperature with no recycling [$R = 1$, $T_1 = Q/(8\Gamma)$], and T_2 is the temperature at which the divertor is exactly one neutral mean free path long $[T_2^2 \approx (aQ_e<\sigma v>m/8)]$. The solutions to this transcendental equation are the intercepts of $f(x) = T_1/T_2$. For a given Δ, there are in general two solutions (Fig. 9). Since $T_1/T_2 \propto (Q)^{1/2}/\Gamma_o$, if the heat flux is too low or the particle flux too high, no solutions exists. In the real world, the flow would reduce until a solution is found. Plugging in numbers for a variety of divertor machines, one finds that the model predicts two solutions: a high-density, low-temperature solution with a lot of recycling, and a low-density, high-temperature solution with very little recycling. For D-III, for instance, the model predicts $n \sim 3 \times 10^{14} cm^{-3}$, $T \sim 5$ eV, and $n \sim 3.4 \times 10^{11} cm^{-3}$, and $T \sim 289$ eV for the two solutions.

The next step beyond these simple models is to actually integrate the full set of one-dimensional equations. A partial step in this direction has been the construction of "two-point" models [33,14-15]. The idea is to divide the plasma along the field lines into two regions, one at the midplane and one at the divertor target. Then one writes a simplified set of equations derived from Eqs. (3-6) of the form [15]:

$$\Gamma_2 - \Gamma_1 = \int_1^2 S ds , \qquad\qquad (43)$$

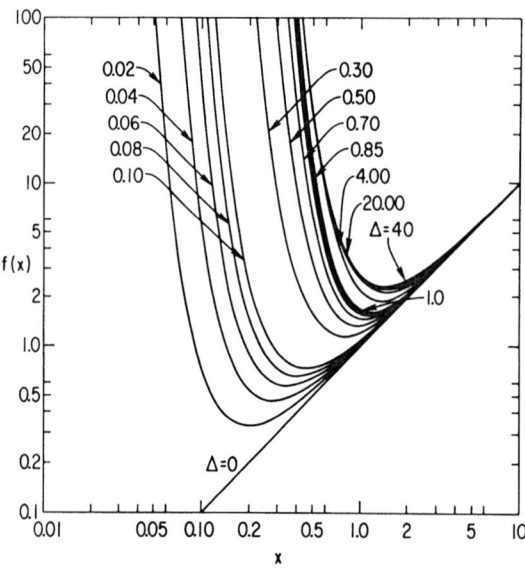

Fig. 9. Temperature equation with variable δ, $f(x) = x/\{1 - [1 - \exp(-\Delta/x^2)][1 - \exp(1/x^2)]\}$ where $x = T/T_2 \propto T/(Q)^{1/2}$. $f(x) = T_1/T_2 \propto (Q)^{1/2}/\Gamma_o$, and $\Delta = b/a$ ratio of divertor width to height (taken from Ref. [32]).

$$(mu_2^2 + 2T_2)n_2 = (mu_1^2 + 2T_1)n_1 , \tag{44}$$

and

$$Q_2 - (5T_1\Gamma_1 + q_1^e) = \Delta E \int_1^2 Sds , \tag{45}$$

where 1 is at the midplane, 2 is at the divertor target, u is the parallel velocity, ΔE is the energy radiated per ionization, S is the local ionization rate, and q_1^e is the conducted electron heat flux. Q_2 is the heat flow across the sheath ($Q_2 \sim 8T_2\Gamma_2$). This model can be combined with a simple neutral transport model which determines S. Properly calibrated, this model can be used in one-dimensional transport codes as an improvement over the model with $u = c_s$, the sound speed.

The next step is to integrate numerically the one-dimensional transport equations for the parallel flow [28,30]. The full set of equations is given in [28]. Many of the properties discussed above apply. The solution of these equations is a difficult numerical problem due to the high degree of nonlinearity. In addition, the sheath boundary conditions are nonlinear. Thus many standard solution techniques are inadequate. As an illustration of how the set of equations can be solved in one-dimension, we consider the following, slightly reduced equations:

density:

$$\frac{\partial(nv_\parallel)}{\partial x} = S , \tag{46}$$

momentum:

$$nmv_\parallel^2 + nT_e + nT_i = P_o , \quad \text{a constant} \tag{47}$$

energy:

$$\frac{\partial}{\partial x}\left(5nv_\parallel T + \frac{1}{2}mv_\parallel^2(nv) - n_e\chi_e \frac{\partial T}{\partial x}\right) - W_L = 0 , \tag{48}$$

where we have combined Eqs. (5) and (6) assuming $T_e = T_i$. W_L is the energy loss or gain due to atomic processes and heat sources.

The first step is assume a plasma from the boundary conditions at $x = 0$, $Q_\parallel = \Gamma_\parallel = 0$, $P = P_o$, and at $x = \ell$, $Q_\parallel = 8T\Gamma$, $\Gamma = nv_s$, $v_s = (2T/m)^{1/2}$. The particle and energy sources are added as line sources (S, W_L) from $x = 0$ to x_D where x_D is the entrance to divertor or pump limiter. These sources are due to perpendicular diffusion and conduction from the main plasma. Once one has a plasma, then S can be calculated. Once S is known, then compute the flux $\Gamma(x)$ as:

$$\Gamma(x) = nv = \int_0^x S dx' \; . \tag{49}$$

Then using $P^o = 4nT$ at $x = \ell$ and the sheath boundary condition, one can determine $n(x)$ from Eq. (47) as

$$n(x) = \frac{P_o}{2T} \left[1 + \left(1 - 8\Gamma^2 mT/P_o^2 \right)^{1/2} \right], \tag{50}$$

and determine $v(x)$ from

$$v(x) = \Gamma(x)/n(x) .$$

Given $n(x)$, $v(x)$, and $\Gamma(x)$, Eq. (48) can be solved for T using a fully implicit tridiagonal scheme [34]. Some care must be taken to properly center the difference equations for Eq. (48). In particular, the fluxes must be computed with "upstream" values so that negative densities, etc., are avoided. It is also important to make the sheath boundary conditions in the same form as the flux in the zone next to the wall. Then, once one has a new set of n, v, and T, one computes P_o again as $P_o = 4nT$ at $x = \ell$, and recomputes the neutral transport with the new plasma. Then the cycle can be repeated, and the iteration continued until convergence is reached. This scheme is reasonably robust and converges quickly.

Similar models are used by many groups [5,35-37] for both data analysis and divertor design calculations. As one example the JAERI group used such a model to analyze their probe measurements of an expanded boundary divertor on D-III [38]. Using probes they measured the electron temperature and density (Fig. 10) along the wall and through part of the diverted plasma. These results were matched with a one-dimensional calculation (Fig. 11) [38] which also included impurity radiation. Two points especially worth noting are that the flux, the flow velocity, and density all rise near the plate, and that the temperature falls near the plate. The temperature drops from 40 eV to ~ 8 eV. The rise in density and drop in temperature all take place over a very small distance (< 10 cm in the poloidal plane).

646

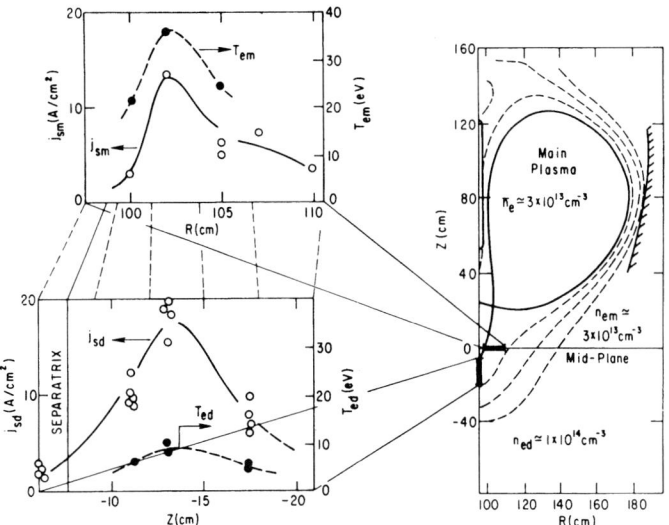

Fig. 10. The horizontal profile of the electron temperature, T_{em}, and ion saturation current, j_{sm}, across the lower divertor channel (midplane: z = 0 cm) and corresponding vertical profiles of T_{ed} and n_{ed} on the divertor plate at t = 800 msec. The connection of the field lines between both profiles is shown with dotted lines (taken from [38]). 84P0010

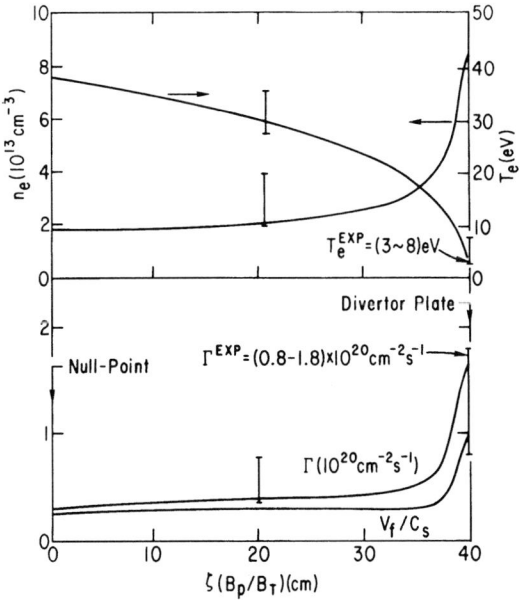

Fig. 11. Model calculations of the electron temperature, density, particle flux density, and Mach number along the field line. ζ is the coordinate along the field line. Experimental values of T_e, n_e, and Γ are shown with the error bars (taken from [38]). 8400165

647

1.4 Two-dimensional models

Realistic simulations of divertors and limiters require a two-dimensional model. The transport of the neutral gas is certainly a two or even three-dimensional problem. However, as we have seen, many of the plasma physics properties can be determined from one-dimensional considerations. Much of the coupling between flux surfaces is via the neutral ionization and charge exchange. An early example of these calculations was given by M. Petravic et al. [39], who solved a series of one-dimensional equations coupled to the DEGAS [40, also in the chapter by D. Heifetz] code for a PDX-like geometry (Fig. 12). The equations used were a subset of Eqs. (3-6). The electron energy equation was simplified by assuming that χ_e was sufficiently large and that T_e was constant along the field line. A rectangular grid was used. The effects of varying the neutral recycling were studied by varying the size of the pump opening. The results are similar to the later work of [38] and others except that the electron temperature was constant along the field line (Fig. 13). The ionization source is localized near the plate. The density rises and the temperature drops near the plate. The temperature drops along the entire profile (Fig. 14). As the recycling was increased by reducing the pump opening and pump speed, the temperature at the plate drops and the density increases (Fig. 15). The temperature at the throat and all along the field lines drops also, and the density rises. Thus many of the points made with the simple models are borne out by the more sophisticated models. In fact, the more sophisticated calculations [e.g., Eq. (9)] preceded the simpler models.

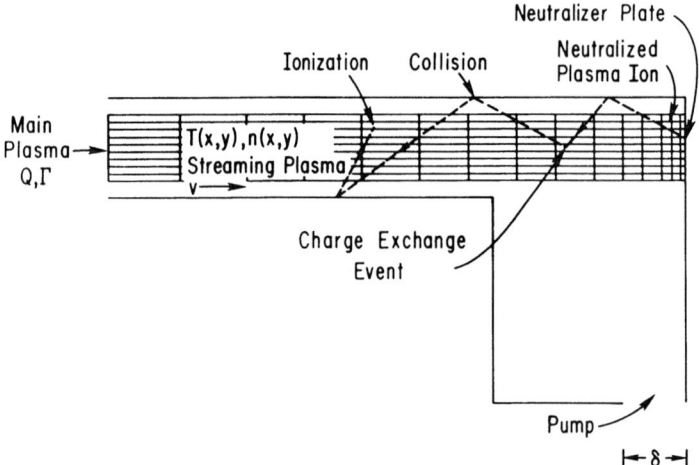

Fig. 12. Model divertor chamber for the PDX divertor (taken from [39]).

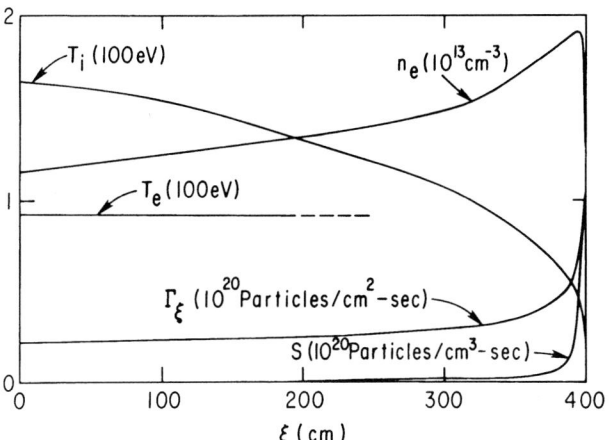

Fig. 13. Calculated plasma parameters along the separatrix in
the modified PDX divertor for a pump opening of 4 cm
(taken from [39]).

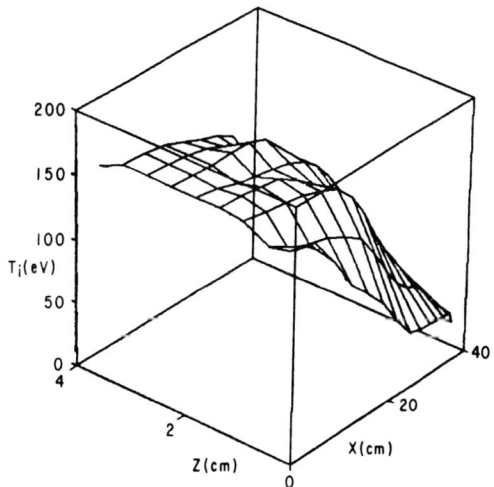

Fig. 14. Ion temperature profile for the modified PDX divertor.
X is the distance along the channel, and Z is the
distance across the plasma (taken from [39]).

649

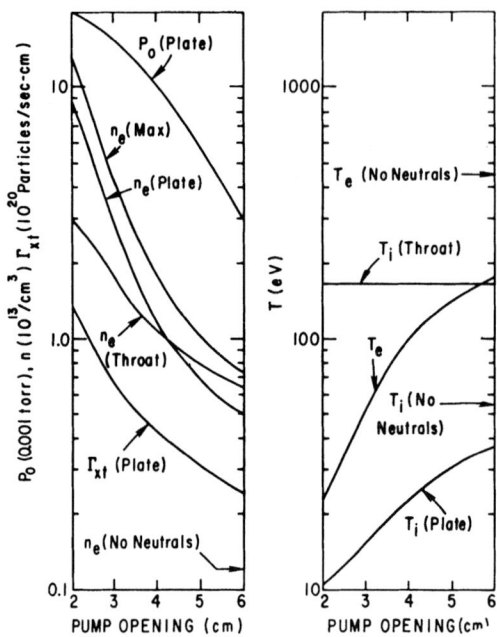

Fig. 15. The neutral pressure P_o, the plasma density at the throat and at the plate n_e, the ion temperature at the plate T_i, and the electron temperature T_e, and the total particle flux Γ at the plate as a function of the pump opening for the modified PDX divertor (taken from [39]). 85P0091

The initial calculations were improved by the addition of a finite electron heat conductivity. The D-III expanded boundary divertor experiments [29] were studied, and predictions made that D-III with high power neutral injection heating should be able to still have a high recycling divertor [41]. In these simulations (Fig. 16), the density rises along the field lines from a few $\times 10^{13}$ cm^{-3} to ~1.7 $\times 10^{14}$ cm^{-3} near the plate. The temperatures fall from 45-50 eV at the divertor entrance to ~5-10 eV at the plate. Similarly, the flux and the velocity rise near the plate. The relatively short distance (10-15 cm) over which the density rises and the temperature falls implies that a "high-recycling" divertor can be quite compact.

Calculations were done for INTOR parameters with similar results [41] (Fig. 17). With 80 MW of power to exhaust, the peak density at the plate was 1.5 $\times 10^{14}$ cm^{-3} and the peak temperature was 20 eV (Fig. 18). Similar calculations have been performed by several other groups with basically similar results [42-44] (notably the USSR, Japanese, and European INTOR delegations).

650

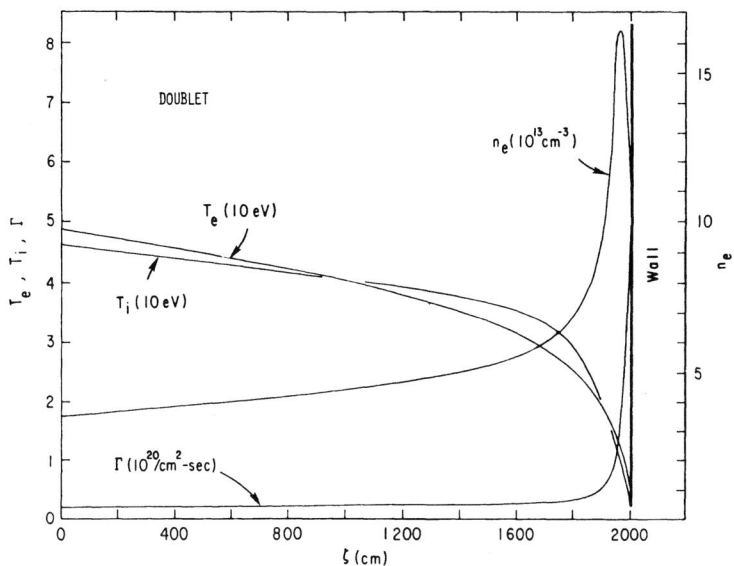

Fig. 16. Plasma density along a field line for a D-III divertor
configuration (50 cm long, 10 cm wide plasma), with
$\Gamma_{input} = 2 \times 10^{19}$ exp($-|Z|/2.5$ cm) cm^{-2}sec^{-1}, $Q^i_{input} =$
390 exp($-|Z|2.5$ cm) watts cm^{-2}, and $Q^e_{input} = 3900$ exp($-$
$|Z|2.5$ cm) watts cm^{-2} corresponding to 0.5 MW of heat
flowing into the divertor. Γ is the particle flux, n_e
is the electron density, T_e is the electron
temperature, and T_i is the ion temperature. Z is the
distance across the divertor chamber (Z=0 defines the
separatrix), and ζ cm is the coordinate along the field
line. All the fluxes are taken through a surface
normal to the field lines (taken from [41]).

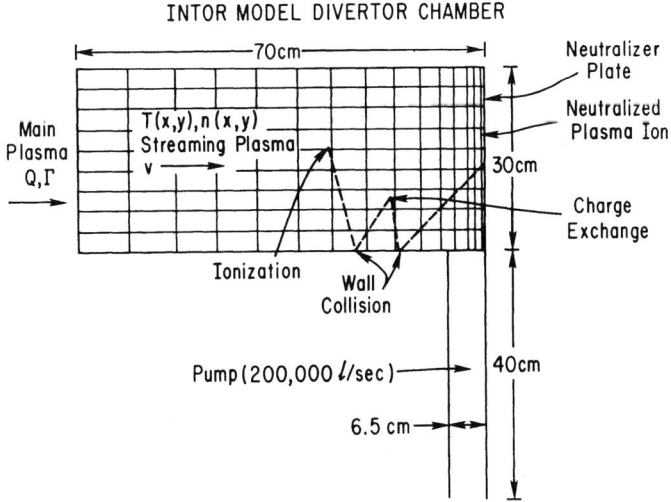

Fig. 17. Schematic of INTOR poloidal divertor (taken from [41]).

The two-dimensional calculations described so far use a
rectangular geometry, and usually neglect cross field
transport. They are primarily models for just the divertor
chamber. During the last year, progress has been made in both

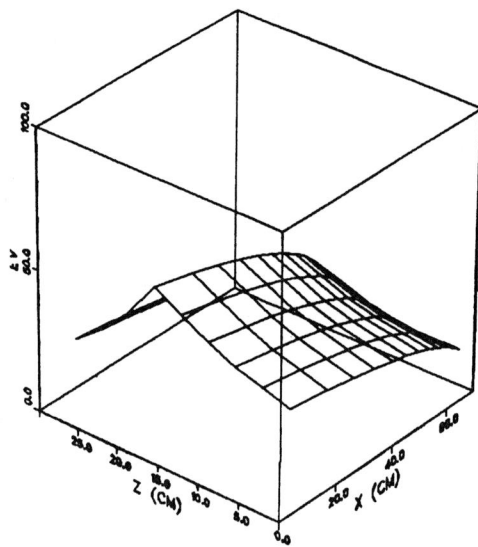

FIG. 18. Electron temperature profile for an INTOR simulation
         with the geometry of Fig. 17. Z is across the divertor
         channel, and x is along the channel. The wall is at x
         = 70 cm (taken from [41]).

areas. Petravic et al. [45] have generalized their code to
perform calculations in a realistic geometry (Fig. 19). A key
question for the open divertor configuration for INTOR is the
degree to which a high recycling divertor can be produced in the
open geometry. In these calculations, the neutrals were
sufficiently well-confined to produce a high-recycling divertor,
with densities of greater than $10^{14}cm^{-3}$ at the plate (Fig. 20).

Petravic et al. [46] also modeled the PDX divertor with a
realistic geometry (Fig. 21). There were probe meeasurements at
the indicated position (Fig. 21) of the electron temperature and
density. The gas puffing rate in the dome was about 1.3×10^{21}
atoms/sec. The measurements were made during ohmic heating, and
during beam heating with both the L-mode and H-mode. The input
heat and particle fluxes were varied to match the probe
measurements. The calculated parameters depend sensitively on
input. In general, matching both the temperature and density

652

Table 1. PDX Simulation Results (taken from [46]).

MODE	O-H		H		L	
	EXPERIMENT	MODEL	EXPERIMENTAL	MODEL	EXPERIMENT	MODEL
T_e	20 eV	20 eV	31 eV	31 eV	40 eV	40 eV
n_{ez}	6.5×10^{12}	6.5×10^{12}	1.0×10^{1}	1.0×10^{13}	7.5×10^{12}	7.5×10^{12}
$v_{||}$ (cm/s)	0.75×10^{6}	1.0×10^{6}	0.8×10^{6}	1.0×10^{6}	0.3×10^{6}	1.05×10^{6}
Power	50 kW	50 kW	200 kW	150 kW	250 kW	170 kW
Neutral Pressure (torr)	1.8×10^{-4}		2.3×10^{-4}	1.6×10^{-4}	2.9×10^{-4}	1.7×10^{-4}
Total Input Particle Flux		8.0×10^{20}/s		1.1×10^{21}/s		1.0×10^{21}/s

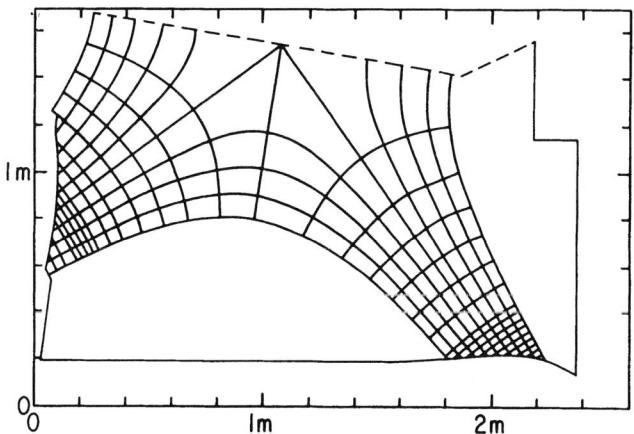

Fig. 19. The model geometry of the INTOR poloidal divertor
region. The walls closely follow the engineering
design, except that in some design versions the outer
(bottom) plate is more angled. The wall at the bottom
right-hand corner is gettered to simulate the pump
opening. The slashed line at the top represents an
opening for the neutrals (taken from [45]).

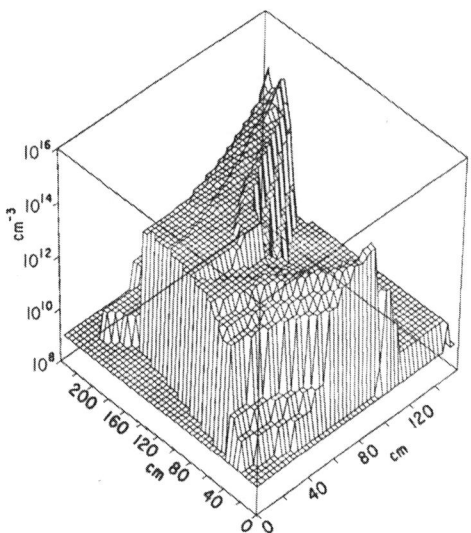

Fig. 20. A 3-D review of the plasma density looking down into
the divertor chamber. The inner divertor plate is on
the right, while the outer one is at the top of the
figure. The two divertor branches are clearly seen.
The peak electron density in both branches reaches
above 10^{14}cm^{-3}, and the density at the x-point is about
6×10^{13}cm^{-3}.

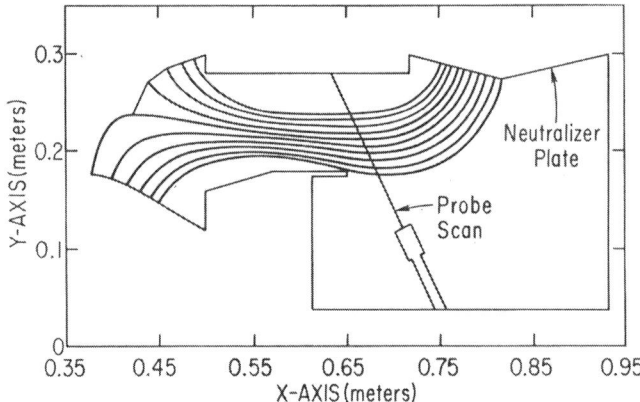

FIG. 21. The model geometry of the PDX divertor. The magnetic
geometry is faithfully represented, and so are the
walls at the divertor throat (taken from [46]).

654

required quite a narrow range of heat and particle fluxes. The best particle input flux was only slightly larger than the gas feed rate. The power could also be determined quite sensitively (Fig. 22). The general level of agreement is given in Table I. It is quite reasonable, (within 30%), except for the neutral pressure which is about a factor of 2 too low, and $v_{\parallel}$ in the L-mode where the experimental value is much lower. A better match awaits better data. A key question is the sensitivity of the calculation to neutral reflection models, which may be responsible for the pressure discrepancy. Calculations similar to these have also been performed by Schneider et al. [47] and Igitkhanov et al. [44] for ASDEX, with quite reasonable agreement.

The second improvement that has been made in the models is to include the effects of the perpendicular transport in the scrape-off layer near the main plasma. This has been done by Braams et al. [42], with a simple model for the neutrals. Igitkhanov et al. [44] have modeled ASDEX and INTOR with a rectangular geometry in which the whole scrapeoff was included. Recently, Petravic et al. [46] have modeled INTOR

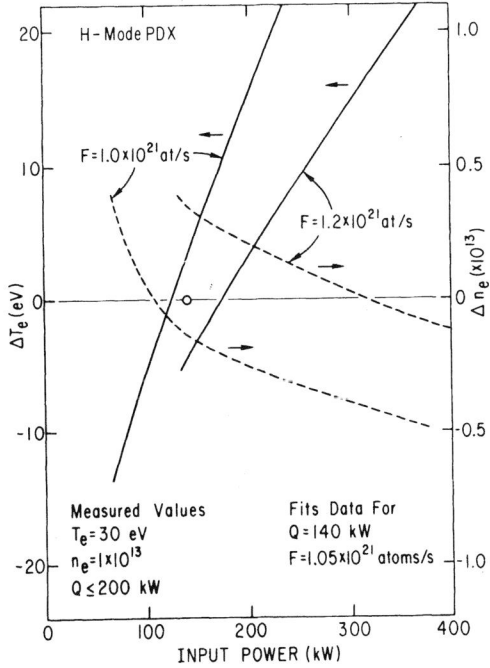

FIG. 22. The difference between the computed and the measured electron temperature and density versus the input power flux for the PDX simulation. The derivation from the measured values is a sensitive function of both the power and the particle flux (taken from [46]).

with a similar model (Fig. 23). It is interesting to note the behavior of the parallel velocity and temperature along the separatrix (Fig. 23). Referring back to Eq. (11), we note that the total particle source, $S_N + \partial/\partial y[D_\perp (\partial n/\partial y) - nv_r]$, is positive as one goes from the stagnation point toward the divertor due to diffusion from the main chamber. After passing the x-point, the separatrix becomes the point of maximum density, and the diffusion term becomes negative. Also, the temperature gradients are negative. Thus $v_\parallel$ begins to decrease after the x-point until one is close to the plate again, when S becomes large and $v_\parallel$ increases to v_s, the sound speed. The temperature drop has the same behavior as previous models.

In the INTOR simulation [46], a thermal diffusivity of $\chi_\perp^e + \chi_\perp^i = 4 \times 10^4$ cm^2/sec was used with $D_\perp = 5 \times 10^3$ cm^2/sec. Since high plasma densities were expected, $T_i = T_e$ was assumed for simplicity. The major result of these calculations is that despite the large thermal conduction, the scrape-off distances are smaller (~3 cm) than previous estimates (~6 cm) leading to high heat flux problems. However, even with shorter scrape-off distances, a low-temperature, high-density, high-recycling divertor can be achieved. While the power profile is quite peaked, the density profile is broad. A key question of interest is the two-dimensional particle flow patterns in the divertor (Figs. 24,25). Recalling the continuity equation [Eq. (3)], we see that

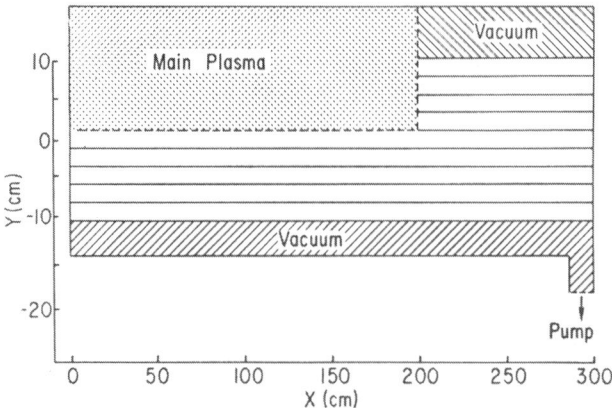

Fig. 23. A tokamak scrapeoff mapped on to a rectangular mesh. The essential topological features of the real geometry are preserved. The stagnation line with $v_\parallel$ =0 is at the left, and the x-point is 200 cm to the right. The transport from the main plasma is only across the line connecting the stagnation point with the x-point. The neutrals can escape either to the pump or into the main plasma (taken from [46]).

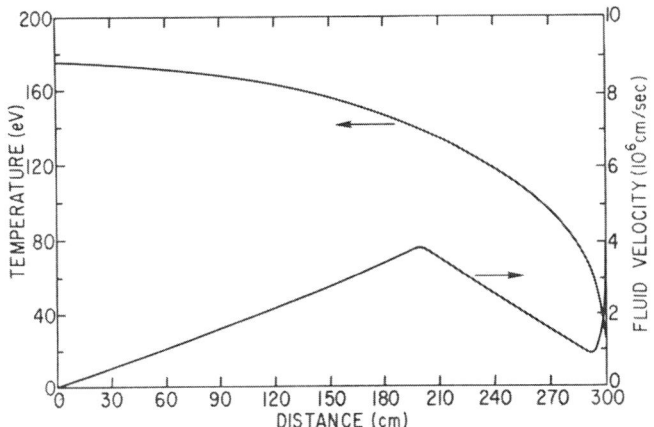

Fig. 24. The variation of the plasma temperature and the
parallel flow velocity along the separatrix from the
stagnation point to the plate (taken from [46]).

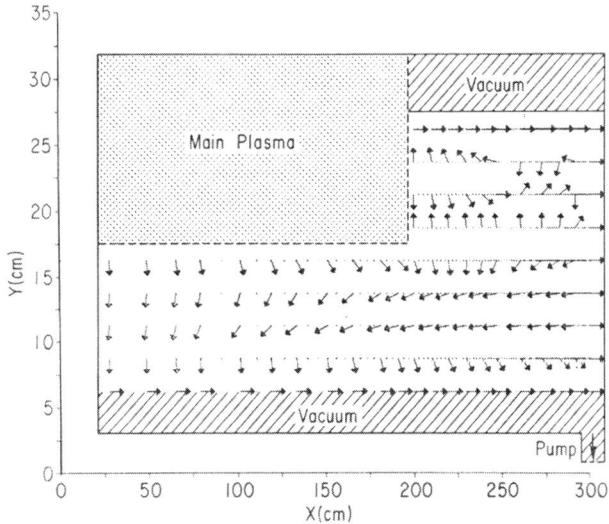

Fig. 25. The 2-D particle flow pattern for an INTOR
simulation. Due to the large variation in magnitude
only the directions of the particle flux are shown.
This pattern is due to the convex shape of the (radial)
density profile, combined with the lack of recycling
upstream of the x-point (taken from [46]).

$$v_\parallel(x,y) = \frac{1}{n(k,y)} \int_o^x \left[S_n(x',y) + \frac{\partial}{\partial y} \left(D_\perp \frac{\partial n(x,y)}{\partial y} \right) \right] dx'.$$

In the region away from the plate, $S \approx 0$, so the radial diffusion term can dominate $v_\parallel$. In particular, net radial outward diffusion can lead to negative $v_\parallel$. The details of the $v_\parallel$ profile will not only affect the recycling picture, but may also have a bearing on impurity transport. Two dimensional flow patterns have also been computed by Braams et al. [42] and Igitkhanov et al. [44]. Resolving some of the questions of the two-dimensional flows is obviously a key next step in plasma edge modeling.

Continued development of these models will probably concentrate on combining the two-dimensional transport with realistic geometries. A further step would then be the inclusion of a model for impurity transport. A code including all of these effects plus some sort of treatment for the main plasma transport would have the promise of being a predictive model for impurity production that could be calibrated on current experiments, and used for the design of future experiments.

A second area for plasma modeling is the application of these models to limiters. In a sense, limiter modeling is more difficult than divertor modeling in that the main plasma is coupled quite strongly. Key questions that could be addressed are:

(1) to what extent does local recycling in pumped limiters enhance the neutral pressure, and
(2) can a "high-recycling" condition exist with a limiter in which a high-density, low-temperature plasma be produced on the limiter surface?

The general level of agreement between the divertor models and the experiments is quite good. In fact, in the instance of the "high-recycling" divertor, the calculations predicted the importance of intense, localized recycling before it was identified experimentally. Perhaps one reason that the models have been successful is that the atomic processes, wall processes, and classical transport can all be understood individually. These processes dominate the plasma edge. Thus models which include these processes include the dominant effects, in contrast to models for the central plasma, where the dominate process is anomalous transport which is not understood as an individual process. It is thus that we can have a reasonable degree of confidence that the models will be useful and credible tools to use for not only analyzing current experiments but also for the design of future experiments.

2.   IMPURITY DYNAMICS

Most of the motivation for quantitative modeling of the hydrogen plasma and neutral particle behavior in the plasma edge region derives from the effect on the production and transport of impurities. In this section, we give first an introductory overview of the physics aspects of the impurity problem, followed by a description of assumptions and results of different existing modeling calculations, an outline of numerical techniques employed, and a summary of the deficiencies of past efforts, crucial open questions, and present lines of development.

2.1.   Physics aspects of the impurity question

2.1.1.   Impurities in the bulk plasma

Depending on their atomic number Z, impurities in the bulk plasma will have different detrimental effects on discharge performance. The contribution of a given impurity to the radiative energy losses per unit volume p_{rad} can be expressed as

$$p_{rad,Z} = \ell_Z \, n_e n_Z \tag{51}$$

with n_Z the density of ions of the species Z.

Under the assumption of coronal equilibrium (discussed in Sec. 2.2.1) the radiative cooling rate ℓ_Z is a function only of electron temperature. It is tabulated in Ref. [48] for materials of possible interest in fusion research and shown in Fig. 26 for a representative low (oxygen), medium (iron) and high Z (tungsten) impurity.

Low-Z impurities (e.g., oxygen and carbon) will typically be fully stripped above several 100 eV, and contribute to radiation losses from the interior region of a tokamak only weakly through bremsstrahlung

$$P_{br} = 1.69 \times 10^{-32} n_e^2 \left(T_e\right)^{1/2} Z_{eff} \text{ W cm}^{-3}, \ \left(T_e \text{ in eV}\right), \tag{52}$$

via their contribution to increasing Z_{eff}:

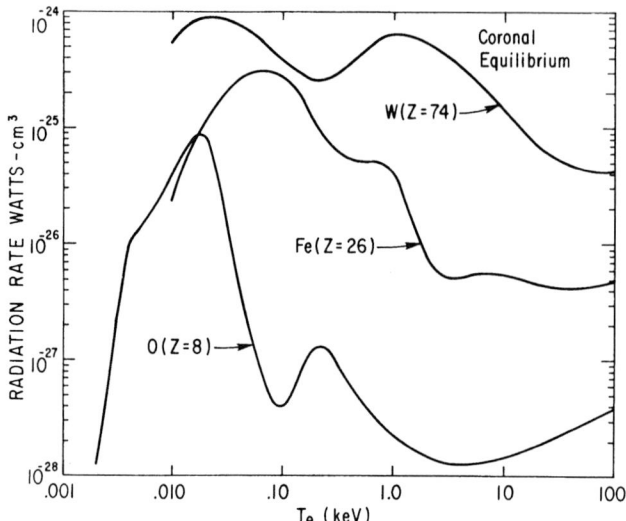

Fig. 26. Radiative cooling rates, computed on the basis of the
average ion model for oxygen, iron and tungsten as
representative low, medium and high-Z tokamak
impurities (from Ref. [48]). Cooling rates $\ell_Z(T_e)$ are
given in ergs cm^3/sec, temperatures in keV.

$$Z_{eff} = \sum_Z n_Z z^2/n_e.$$ \hfill (53)

Their main effect on the bulk plasma is the enhancement of
the plasma resistivity for ohmically heated discharges, and fuel
dilution in the reactor. The fusion power density over a wide
range of temperatures is proportional to $n_{i,D} n_{i,T} T_i^2$ and
therefore to the square of the partial pressure of the hydrogen
isotopes. The total plasma pressure

$$p = k\left[\left(n_{i,D} + n_{i,T} + \sum_Z n_Z\right)T_i + \left(n_{i,D} + n_{i,T} + \sum_Z n_Z z\right)T_e\right]$$

$$+ p_{fast\ particles} = \beta_t B_t^2/(2\mu_o) .$$ \hfill (54)

will be limited however through constraints on the achievable β_t
from MHD stability and on the magnetic field strength B_t through
superconductor technology. In particular the free electrons
associated (because of charge neutrality) with the presence of
impurities will therefore make a parasitic contribution to the
pressure, giving rise, for fixed p, to a fusion power reduction

$$\delta p_{fus}/p_{fus} \sim \sum_Z f_Z(1 + z).$$ \hfill (55)

660

For impurities like carbon or oxygen this leads to tolerable concentrations f_Z in the percent range. A particular example of light impurities are, of course, the thermalized alpha particles, for which typical central concentrations of about 5% are considered acceptable. (The fusion-produced helium ions will make a further contribution to p in the 10-20% range during their fast particle phase; however, this is unavoidable, unless an anomalous process leads to their accelerated slowing-down.)

Line radiation from light impurities can play a role at low temperatures. During the start-up phase of a tokamak discharge, for too large an impurity concentration, radiative losses from light impurities can exceed the ohmic power input and clamp T_e in the few 10 eV range. During steady state in the plasma boundary, a dominate fraction of the outflowing energy can be converted into low-Z impurity radiation in ohmically heated experiments (e.g., illustration in Ref. [49]). Thermal collapse at the edge because of excessively large oxygen or carbon concentrations is believed to be one reason for disruptions in tokamaks. For the high energy flux densities of INTOR-like devices or reactors (40 W/cm^2 compared to ≈3 W/cm^2 for ohmically heated mid-size tokamaks), the low-Z material concentrations needed to radiate a substantial fraction of the heating power would, however, be far in excess of those tolerable from the fuel-dilution point of view.

Medium-Z materials, like iron, will radiate strongly only for temperatures below 2 keV, and therefore will be a significant energy loss also in the interior regions of ohmically heated devices. However, they will be important only in the boundary zone of tokamaks with strong auxiliary heating or in ignited operation.

The latter regime has been suggested as a self-consistent solution to the problem of energy transfer to the walls and impurity generation. As the intensity of all the impurity production mechanisms discussed in this book depends ultimately upon the energy transported by particles to the wall, a situation can be envisaged in which the plasma contamination is self-limited by the conversion of a dominate fraction of the power deposited or generated in the plasma into impurity radiation. Such equilibria would be self-regulating, as an increase in impurity content would reduce the nonradiative power transfer to the walls and consequently would reduce also the further impurity production. In order not to affect strongly the temperature profiles in the interior, and therefore the confinement times and ignition probabilities, the conversion from heat transport by particle convection and conduction to heat transport by radiation has to proceed in a relatively

narrow boundary layer and requires therefore radiation
characteristics typical of medium-Z impurities [50].

The concentration leading to the conversion of the total
input power into radiation in the boundary zone can therefore be
viewed either as approximately the one established automatically
in such self-regulating situations, or as an upper limit to the
tolerable one, as larger values of f_Z would lead to the thermal
collapse of the discharge. Actual values of the limiting f_Z
depend, among other things, on energy and particle transport
rates; for the so-called Alcator-Intor heat conduction law ($n\chi_e$
$= 5 \times 10^{17} cm^{-2}$ s) and a number of further simplifying
assumptions (plane geometry, constant n_e and f_Z, coronal
equilibrium) one obtains approximately [51]

$$f_Z \sim 10^{-5} \left(\frac{q_\perp (W/cm^2)}{n_e [10^{14} cm^{-3}]} \right)^2 \tag{56}$$

for the iron concentration converting an energy flux
density $q_\perp$ into radiation. For the assumption of the JET
simulation calculations in which these concepts were
examined ($q_\perp \approx 10$ W/cm^2, $n_e \approx 10^{14} cm^{-3}$), the corresponding value
of f_Z is 10^{-3}. As f_Z is proportional to $q_\perp^2$, [Eq. (56)],
for sizeably larger power fluxes (like the 40 W/cm^2 quoted for
INTOR) these f_Z-limits might actually become larger than those
tolerable from the point of view of fuel dilution or central
radiation losses. A significant correction to the high
temperature radiation characteristics outlined before arises due
to the presence of hydrogen neutrals in the plasma interior,
which can reduce the ionization state of the impurities by
charge-exchange recombination; this obviously is the case in the
presence of neutral injection heating, where the resulting
enhancement in radiation efficiency has been clearly
demonstrated [52,53].

Heavy impurities, like tungsten, finally will radiate
strongly even from the core of a fusion plasma: from the point
of view of energy transport in the plasma they act therefore
like a reduction of local fusion power density. Tungsten
concentrations in excess of a few times 10^{-4} would therefore
make ignition impossible at the plasma burn temperatures
typically considered [54].

Present day experiments can be strongly affected by the
fact that the radiation efficiency of tungsten peaks around 1
keV: particularly with ohmic heating, the central radiation
losses can exceed the local power deposition, even though the
global power input remains larger than the total radiation. In

this case a conductive inward power flow is needed and hollow temperature profiles are observed.

The above introduction shows that impurities have different effects in the central and boundary region and underlines thereby the importance of the impurity profiles, and hence the importance of transport of impurities in the bulk plasma. This is enforced also by the fact that the methods of plasma boundary engineering discussed in this book can only control the impurity density in the edge region, leaving their global concentration determined by interior transport.

With the notable exception of helium and some nonstandard situations (injection of impurity pellets, laser blow-off experiments or contaminated neutral injection beams), there is no source of ionized impurities in the bulk plasma region. The most simple transport law ($\Gamma_{Zj,1} = -D_Z \times \partial n_{Zj}/\partial n$; Γ_{Zj} = radial flux of ionic stage j of species Z; D_Z independent of j) would then give rise to a homogeneous distribution of the total impurity number $n_Z = \sum_j n_{Zj}$. The dominate effects according to collisional impurity transport theory [55] are, however, due to drift-type terms, driven by gradients in hydrogen ion density n_i

$$\Gamma_{Zj,2} = D_o \times j \times \frac{n_{Zj}}{n_H} \frac{\partial n_i}{\partial r} , \qquad (57)$$

and in the temperature T_i

$$\Gamma_{Zj,3} = D_o \times j \times H \times \frac{n_{Zj}}{T_i} \frac{\partial T_i}{\partial r} . \qquad (58)$$

In these expressions $D_o = \rho_i^2 q^2/\tau_i$, with ρ_i, the gyroradius and τ_i, the ion-ion collision time for hydrogen ions; q is the usual safety-factor. H is a function of ion-collisionality; which is positive in the strongly collisional regime, but tends to – 1/2 for $Rq/(\tau_i \times v_{th,i}) \ll 1$ ($v_{th,i}$ being the thermal ion velocity). Collisional contributions to the diagonal terms $\Gamma_{Zj,1}$ are small, and have to be supplemented by anomalous ones to give the values of D_Z in the range of the few $10^3 cm^2/s$ needed to fit experiments. The drift terms derived from collisional theory on the other hand are significant and can compete with anomalous diffusion of the above magnitude to yield total impurity density profiles either peaked centrally or in the periphery, depending on the regime of H and the profiles of n_i and T_i.

Even though detailed attempts to fit radial scans of impurity spectra have frequently been successful using such models, we certainly cannot yet predict with confidence the expected impurity profiles in a reactor. This is in part due to the fact that these fits use different contributions in combinations which are not necessarily unique and could give vastly different predictions when extrapolated. In particular the anomalous transport might vary strongly with plasma conditions, being driven partly by sawtooth oscillations and other MHD manifestations. Finally, it should be mentioned, that even based on collisional theory, additional contributions to impurity transport could come from plasma rotation or momentum exchange with fast circulating beam-injection particles, which have not yet been fully explored in experiments [56]. Rather than to treat these effects in detail, an empirical inward pinch velocity of magnitude $|v| \sim D_Z/a$ is often used.

2.1.2.   Sources and sinks of impurities

The physical processes giving rise to the production of impurities are subject of the major part of this book, and we will restrict ourselves here to describe the way and the extent in which they are included in scrape-off model calculations.

Production processes that have been taken into account in models in partly self-consistent fashion are sputtering by ions and by hydrogen neutrals (produced from charge exchange), and by impurity self-sputtering. In addition, the fusion production of helium ions, important in a reactor experiment has sometimes been studied.

Impurity production by charge-exchange neutrals must generally be treated in simulation calculations of the bulk plasma, as knowledge of the neutral particle, electron density, and ion temperature profiles is needed over a layer of sufficient depth to determine the flux and energy distribution of the fast outflowing neutrals. Given these quantities (computed by means of a neutral particle model accompanying the radial transport calculations), the surface average sputtering rate can be calculated on the basis of the coefficients given, e.g., in [57]. As these results evidently depend on the edge neutral particle density, which at the same time determines refuelling, they are sensitive to the particle confinement assumptions.

In an idealized tokamak, significant fluxes of charged ions with sufficient energy to cause sputtering arrive only at divertor target plates or limiters. There they are further accelerated by the sheath potential, which adds a constant energy of the order $3kT_e$ to each particle. As target plate geometries capable of sustaining large power flows inevitably are associated with a near glancing incidence of field lines, the true angular distribution of the impacting ions has also to be taken into account for evaluating the sputtering yield. The required information on the hydrogen distribution function can only be calculated from a sheath model like the one described in the chapter by Chodura, which was used to evaluate the properly averaged sputtering rates for a hydrogen plasma in contact with an iron target plate shown in Fig. 27.

Friction with the plasma flow and electric fields in the presheath region will also immediately carry back a fraction f of the sputtered impurities ionized in the scrape-off fan. These impurities will both redeposit and cause sputtering with a rate Y_s. The impurities produced by this self-sputtering can return to the wall and deposit and sputter, and so on (Fig. 28). Summing the successive generations of this self-sputtering and redeposition shows that, if the product of the self-sputtering yield and f is less than one ion per incident ion, then the net sputtering is given by

$$\Gamma_H + Y_H + \frac{1 - f}{1 - fY_s} \ .$$

Since f < 1, self-sputtering and redeposition reduces the net sputtering rate when Y_s is less than one. If $fY_s >1$, the self-sputtering does not converge. The self-sputtering coefficient Y_s depends, of course, critically on the impact energy of the impurity, and therefore on the ionization state it has reached before entering the electrostatic sheath. Because of this, an actual evaluation of the self-sputtering contribution requires a model for the impurity dynamics along the scrape-off layer field lines like to be described in Sec. 2.2.2.

As f can be close to 1, and Y_s exceeds 1 for most materials at sufficiently high impact energy, the self-sputtering contributions can form an avalanche, precluding a quasistationary solution for the plasma impurity content. Obviously this situation will self-limit as additional energy losses caused by these impurities will ultimately lower the near-target plate electron temperature. In radial transport calculations for the bulk plasma, this behavior is sometimes simulated by using an edge temperature dependent impurity inflow which increases drastically above a given threshold electron temperature. As a consequence, the impurity content and the radiative power losses will adjust so as to clamp T_e at the

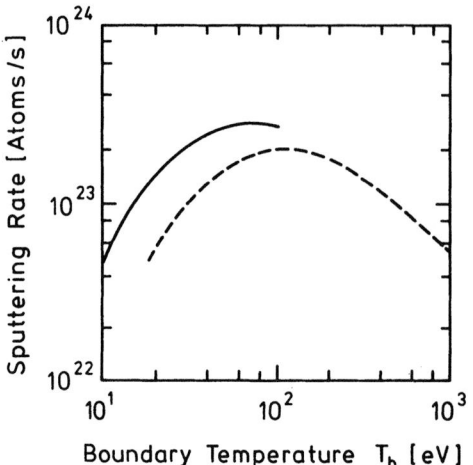

Fig. 27. Effect of sheath model onto sputtering, illustrated at hand of the sputtering rate predicted for a 3 GW thermal reactor as a function of boundary temperature. Dashed curve assuming no sheath potential, solid line taking into account electrostatic sheath structure (for $T_e = T_i$) and the angular distribution of impacting ions as following from the Monte Carlo calculations described in the chapter by Chodura.

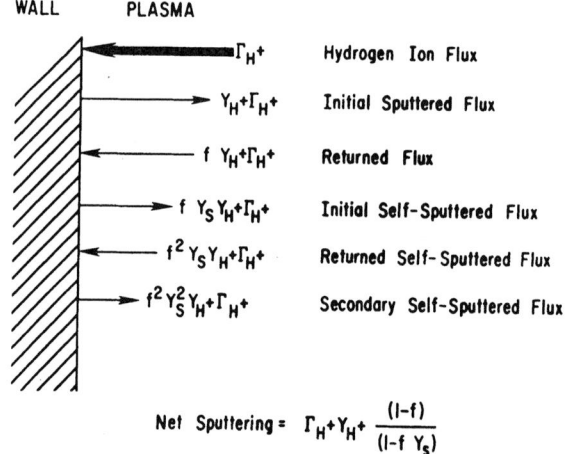

Fig. 28. Schematic illustration of the role of self-sputtering and redeposition showing the successive generations of each process. Γ_{H+} is the incoming hydrogen flux, Y_{H+} is the sputtering yield of hydrogen for the wall, f is the fraction of the sputtered wall material that returns to the wall, and Y_s is the self-sputtering yield of the wall material.

boundary at a value just marginally above this threshold, with little dependence of other results on the chosen functional form of the impurity influx.

No attempts have been made so far in plasma modeling calculations to determine self-consistently influx rates for gaseous impurities, as they depend primarily on wall conditions and vary from one discharge to the next and during the pulse and from location to location. When gaseous impurities have been taken into account; e.g., in bulk plasma simulations, the influx rates or global content have been fixed a priori.

2.1.3. The role of the scrape-off layer

The scrape-off layer affects the impurity content of the bulk plasma in two essentially different ways. On one hand, the edge plasma determines the impurity production at the target plate by charge particle sputtering. The edge plasma and the plasma in the region just inside of the separatrix corresponding to a total radially integrated electron line density of a few 10^{14} cm^{-2} determine the target plate and first wall sputtering by charge-exchange neutrals. On the other hand, for a given impurity production rate, the impurity content of the bulk plasma is largely determined by the impurity transport in the scrape-off layer.

To make the second point clearer, we first discuss the rather unphysical case of a tokamak in which the first wall at $r = a$ coincides completely with the outermost flux surface and has no protruding limiter. Impurities produced by sputtering leave the wall in a neutral state with an energy of the order of 1 eV. To arrive at a simple estimate, we assume them all to become ionized at the same radius a_i. The simplest diffusion model ($\Gamma_{Zj,1}$ of Sec. 2.1.1), summed over all the charge states j $\geqslant$ 1 of the impurity, gives a steady-state solution

$$n_Z(r) = \frac{\phi_Z}{D_Z} \left(a - a_i\right) + n_Z(a) \quad \text{for } r < a_i \qquad (59a)$$

$$n_Z(r) = \frac{\phi_Z}{D_Z} \left(a - r\right) + n_Z(a) \quad \text{for } a_i < r < a \qquad (59b)$$

in terms of the flux density ϕ_Z of the incoming neutral impurities, where we have also assumed $(a - a_i) \ll a$. The boundary density $n_Z(a)$ should be proprly determined from a transport model for the immediate wall vicinity region, where the diffusion picture is not valid. Typically, this will yield a boundary condition of the form

$$D_Z \frac{dn_Z}{dr}\bigg|_{r=a} = -v_{eff} \times n_Z(a)$$

in terms of an effective radial outflow velocity v_{eff}. For the present estimate, we assume $n_Z(a) = 0$, which is justified for $v_{eff} \gg D_Z/(a - a_i)$. (Simple arguments give v_{eff} of the order of the thermal velocity, which would satisfy this relation.)

The important point evidenced by this model is that for a given influx of impurities, their concentration in the plasma will be inversely proportional to their diffusion coefficient: this forms part of the motivation for ergodic limiter concepts, which aim at an artificial enhancement of D_Z in the boundary zone. Adding to this transport model an inward impurity drift term of order D_Z/a would change the solution in the boundary region very little (simply as in order to compete with diffusion driven by gradients with a scale length $a - a_i$ we would need drift velocities of order $D_Z/(a - a_i) \gg D_Z/a$) but would raise the central density $n_Z(0)$ above $n_Z(a - a_i)$.

The above model does not actually describe a scrape-off layer situation, in which field lines beyond a certain radius a_s (either defined by a tip of a limiter or a separatrix) intersect a material wall acting like a pump for charged particles. Again to understand qualitative trends, we describe the loss of impurities along open field lines by a volumetric sink $n_Z/\tau_\parallel$, with $\tau_\parallel$ being a properly averaged characteristic outflow time.

In the simple model, where all impurities are ionized at the same radius, we have to distinguish two cases, depending on whether the ionizing layer lies inside or outside the separatrix. The first (Fig. 29b, Case II), is relevant to a thin scrape-off layer; it bears similarity to the situation treated above (Fig. 29a, Case I), except that the charged impurity density is not forced to zero abruptly at the limiter tip or separatrix as it was before at the wall, but starts an exponential decay with a characteristic fall-off length $(\tau_\parallel D_Z)^{1/2}$. Consequently, the central density

$$n_Z(0) = \frac{\phi_Z}{D_Z} \left\{ (a_s - a_i) + (\tau_\parallel D_Z)^{1/2} \times \tanh\left[(a_w - a_s)/(\tau_\parallel D_Z)^{1/2} \right] \right\} \quad (60)$$

is somewhat higher than predicted by the previous model, provided we identify $(a_s - a_i)$ with $(a - a_i)$ of Eq. (59a). Equation (60) also shows that the actual wall distance becomes irrelevant, once that $(a_w - a_s)$ is large compared to the fall-off length $(\tau_\parallel D_Z)^{1/2}$.

668

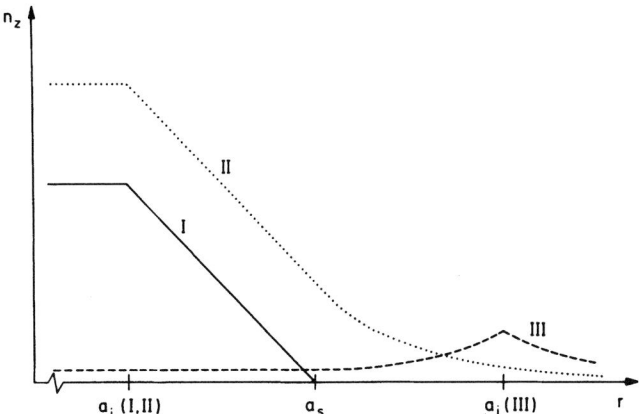

Fig. 29. Radial edge impurity profiles according to the three
simplified models of Sec. 2.1.3. Model I assumes an
absorbing wall at $r = a_s$, models II and III identify r
$= a_s$ with the inner edge of the scrape-off layer,
defined by either the limiter tip or a separatrix. The
impurity ion source is assumed to be located at $r = a_i$,
$a_i < a_s$ for models I and II, and $a_i > a_s$ for model
III. Results for cases II and III refer to a situation
with $|a_i - a_s| = 2 (\tau_\| D)^{1/2}$ and are normalized so as
to correspond for fixed D, $\tau_\|$, and $|a_i - a_s|$ to the
same impurity influx ϕ_Z.

For the parameters of present divertor tokamaks and their
extrapolation to INTOR or reactor type devices, however, a
significant or even dominating part of impurities coming from
the walls will be ionized already in the scrape-off layer. For
such a situation, a model with $a_i > a_s$ gives a more relevant
picture, yielding impurity density profiles like in Fig. 29,
case III, predicting a central density

$$n_Z(0) = \frac{\phi_Z}{D_Z} (\tau_\| D_Z)^{1/2} \sinh\left(\frac{a_w - a_i}{(\tau_\| D_Z)^{1/2}}\right)/\cosh\left(\frac{a_w - a_s}{(\tau_\| D_Z)^{1/2}}\right), \quad (61a)$$

which, for $a_w - a_s \gg (\tau_\| D_Z)^{1/2}$ reduces to

$$n_Z(0) = \frac{\phi_Z}{D_Z} (\tau_\| D_Z)^{1/2} \exp\left[-(a_i - a_s)/(\tau_\| D_Z)^{1/2}\right]. \quad (61b)$$

Keeping fixed $\tau_\|$ and D, this estimate predicts a smaller value
than for the case $a_i < a_s$, as the flow to the target plates now
constitutes a sink of particles between the ionization layer and
the radius a_s marking the limiter tip or the separatrix. For
the present transport and deposition model, the impurity density

has a maximum in the scrape-off layer, which would persist in attenuated form also for a more realistic model of the ionization of incoming impurities, and at least as a local maximum for transport models including inward drift velocities of order $D/a_s \ll (D/\tau_\parallel)^{1/2}$.

Besides giving some qualitative tendencies, the above crude models serve to show the role of the scrape-off layer in relating the production of impurities to their density at the edge of the bulk plasma. To obtain these analytic solutions we assumed constant values of $\tau_\parallel$, D_Z, and simplified deposition laws. These assumptions can be readily dropped when going over to a numerical solution of the radial transport equations. More fundamental are criticisms against the 1-D formulation, which is strictly not justified, as the sinks are localized, and the characteristic time for parallel transport $(\tau_\parallel)$ is not much shorter, but in fact equal to that for perpendicular transport $[d(\log n_Z)/dr]^{-2}/D_Z$. Purely one-dimensional relations for radial transport can still be derived by averaging, but better and more detailed models for impurity behavior along the field lines are needed to define the proper averaging procedure.

For numerical models we will frequently also drop the assumption of steady state made in the above estimates. This is generally done for numerical convenience only. Even in cases where the time-development of the impurity content of the plasma, which proceeds on a typical time scale a^2/D_Z, is analyzed, the scrape-off parameters, which equilibrate on a much shorter time scale $\tau_\parallel$, can be considered as passing through a sequence of equilibrium states.

2.2. Modeling of impurity transport

2.2.1. Impurity reaction dynamics

The spatial distribution of the j-th ionization stage of a given impurity is determined by the competition of drifts, diffusion, ionization and recombination which in a 1-D fluid model is described by an equation of the type

$$\frac{\partial n_j}{\partial t} - \frac{\partial}{\partial x}\left(D_j \frac{\partial}{\partial x} n_j\right) + \frac{\partial}{\partial x} v_j n_j + n_j/\tau_\parallel$$

$$= n_e\left[s_{j-1}n_{j-1} - \left(s_j + r_j\right)n_j + r_{j+1}n_{j+1}\right]. \tag{62}$$

The terms on the right-hand side describe ionization and recombination, which are dominated by two-body processes. The

670

ionization (s_j) and recombination rate coefficients (r_j) are therefore primarily functions of electron temperature, with a weak dependence on electron density n_e arising from correction terms in the dielectronic recombination and a possible strong one on hydrogen neutral density n_H through charge exchange recombination. These atomic physics aspects are treated in the chapter by Harrison.

For impurity modeling, a crucial question is whether the relative magnitude of the terms in Eq. (62) allows a further simplification, the "coronal equilibrium" model. It is formally obtained by a two timescale expansion and is valid if the atomic physics reaction processes described by the right-hand side terms proceed much faster than the characteristic time for diffusion or drift of a particular impurity stage. In that case, on the diffusion-drift time scale, the atomic reaction processes have to be in instantaneous equilibrium described by the system

$$s_{j-1} n_{j-1} - s_j n_j + r_j n_j + r_{j+1} n_{j+1} = 0; \quad j = 1, \ldots, Z.$$

Applying this relation to the fully-stripped ions, for which the terms $s_j n_j$ and $r_{j+1} n_{j+1}$ do not exist, we obtain first for $j = Z$, and subsequently by reinsertion for all j the coronal equilibrium relations

$$s_{j-1} n_{j-1} = r_j n_j. \tag{63}$$

Accordingly, a given ionization stage j dominates over a parameter regime where the inequality

$$\frac{s_j}{r_{j+1}} < 1 < \frac{s_{j-1}}{r_j} \tag{64}$$

is satisfied which, due to the strong temperature dependence of s_j, defines primarily a temperature interval.

Using the coronal equilibrium relations, the density of a particular impurity ionization stage can be written as $n_{Z,j} = f_{Z,j} \times n_Z$, where n_Z is the total density of the particular impurity and $f_{Z,j}$ is essentially a function of T_e only, without any explicit space or time dependence. A quantitative criterion for the validity of the model can be obtained by inserting the coronal equilibrium results into the diffusion term (neglecting possible space- and ionization-stage dependences of the diffusion coefficients)

$$D \frac{\partial^2}{\partial x^2} n_{z,j} = Df_{z,j} \frac{\partial n_z}{\partial x^2} + 2D \frac{\partial n_z}{\partial n} \left(\frac{df_{z,j}}{dT}\right) \frac{\partial T}{\partial x} + n_z D \frac{d^2 f_{z,j}}{dT^2} \left(\frac{\partial T}{\partial x}\right)^2 \,,$$

and comparing the time scale of the largest of these terms, typically the last one, with that of the atomic reactions. Coronal equilibrium is satisfied for a particular stage, if

$$\frac{r_j \times n_e}{D} \left(\frac{(\delta T)_j}{\partial T/\partial x}\right)^2 \gg 1$$

holds. Here $(\delta T)_j$ is the temperature interval over which the j stage dominates according to inequality (64), and r_j has to be taken at a temperature in this range. Among experiments with similar peak temperatures (e.g., ohmically heated ones), coronal equilibrium will therefore be better satisfied for larger devices and at higher densities. Generally for the bulk plasma, coronal equilibrium is a good approximation for medium to high-Z impurities in the ionization stages making major contributions to radiative losses, but a poor one for materials like oxygen or cargon.

In the scrape-off layer, a better estimate is gained from comparing characteristic recombination times (which for stages with j < 8 are of the order of a few ms at $n_e = 10^{14}$) with the flow time of impurities to the target plates. As the latter will tend to move with the speed of the bulk plasma, coronal equilibrium would hold for

$$10^{-17} \frac{n_e L}{M (T)^{1/2}} \gg 1 \,,$$

(where M is the Mach number) which we would just start to satisfy under parameters of the high recycling solution to the INTOR scrapeoff ($n_e = 10^{14}$ cm^{-3}, M $\approx$ 0.1, T $\approx$ 100 eV, L $\approx$ 5 $\times$ 10^3 cm), but violate under all conditions of lower recycling or for smaller values of L than corresponding to this midplane to target-plate field line connection length of INTOR.

2.2.2. Fluid models for impurity transport in the scrape-off layer

As discussed in 2.1.3, transport phenomena in the scrape-off layer form the link between the production rate and the bulk plasma concentration of impurities. In this section, we describe models which, with differing degrees of sophistication, treat the impurities as one or several fluids diffusing in a hydrogen background plasma.

Presently, common to all of them is that for the cross-field transport they make a diagonal, diffusion type ansatz for the radial charged particle flux

$$\Gamma_{z,j} = -D \frac{\partial}{\partial r} n_{z,j}$$

with a diffusion constant D assumed to be due to unspecified anomalous processes, and of comparable magnitude to that used in hydrogen diffusion. Classical, collision-dominated transport processes for scrape-off plasma have not so far been formulated in a way ready for inclusion into modeling calculations.

2.2.2.1. Analytic solutions to scrape-off layer transport. For the above reason, some model of scrape-off impurity transport is also needed in simulation calculations of bulk plasma performance which aim at a consistent inclusion of impurity effects. In the simplest version such a model, in conjunction with the coronal approximation, takes the form of boundary conditions for the bulk plasma impurity diffusion equations at radius r = a corresponding to the separatrix of a divertor or the limiter tip. They are derived from analytic solutions to the 1-D radial scrape-off transport equation

$$\frac{d}{dr} \left(D \frac{d}{dr} n_z \right) = \frac{n_z}{\tau_{\parallel}} - s_z(r)$$

for constant $D, \tau_{\parallel}$ and simple spatial distributions $s_z(x)$. Useful solutions exist [Eq. (58)] for the two situations also considered in 2.1.3

(a) $\frac{1}{s_z(r)} \frac{d}{dr} s_z(r) \gg \left(\tau_{\parallel} D \right)^{1/2}$ (light impurities in a thin scrapeoff),

and

(b) a source distribution strongly peaked in the scrapeoff at $r = a_i$ (heavy impurities, thick scrapeoff).

They can be evaluated at r = a to link the density n_z and radial outward flux $D(dn_z/dr)$ in the form

Case (a) $\left[D \dfrac{dn_Z}{dr} + \left(\dfrac{D}{\tau_\parallel} \right)^{1/2} n_Z \right]_{r=a} = \left(D\tau_\parallel \right)^{1/2} s_Z(a)$ (65)

and

Case (b) $\left[D \dfrac{dn_Z}{dt} + \left(\dfrac{D}{\tau_\parallel} \right)^{1/2} n_Z \right]_{r=a} = \phi_Z \exp\left(-\dfrac{(a_i - a)}{(D\tau_\parallel)^{1/2}} \right).$ (66)

Although obtained as solutions to steady-state scrape-off models, these relations can be used as boundary conditions to time-dependent bulk plasma models, as the characteristic transport time in the scrape-off layer ($\tau_\parallel$) is much shorter than that in the bulk plasma ($\approx a^2/D$).

   2.2.2.2. Numerical models for radial transport. The restrictions made above on D, $\tau_\parallel$, $s_Z(x)$ and the coronal assumption can readily be dropped when considering numerical solutions to the system of equations in the scrape-off layer, as is by now commonly done also for the energy and particle transport equation in general 1-D plasma simulation codes (e.g., Mense [59], Ogden et al. [60]).

   Impurity transport in the full finite rate formulation has been studied both in stand-alone calculations and as part of predictive full plasma simulations. In the former case, emphasis has always been on a close fit of experimentally measured line profiles by the appropriate adjustment of the transport models, and, when otherwise poorly known, also the reaction rate coefficients (Breton [61], Behringer [62], Hawryluk [63]). Even when including a scrape-off model, the results have generally been normalized to some density either by appropriately adjusting the particle influx, or by assuming perfect impurity recycling and renormalizing the total impurity content of the plasma. The relative intensity of different impurity lines apparently show little sensitivity to the employed scrape-off model. The main effect of different scrape-off modes is in determining the total impurity concentration for a given neutral impurity influx ϕ_Z.

   Detailed attempts for an absolute calibration of the edge transport model have been made by Behringer et al. [62] on ASDEX, using both gas in blow (silane) and naturally produced impurities (Fe and O). The impurity influx ϕ_Z was determined in the former case by the known inflow rate (and the assumption of negligible recycling of silicon); in the latter case it was determined from deposition measurements in the target plate vicinity (arguing that in steady state the neutral impurity inflow from the wall will be equal to the ionized impurity outflow) and intensity measurements of neutral atom lines.

The calculations used measured scrape-off profiles for the density and electron temperature of the background plasma; assuming $\tau_\| = L/(0.2 \times v_{th,i})$ then gave a good fit to the measured containment times illustrated by the numbers in Table 2. The strongly different magnitudes of τ_p reflect the relative penetration of the different neutral atoms. Whereas Si is produced by dissociation of silane with a low initial energy, Fe is sputtered from the walls with an initial energy of some eV. An energy similar to the latter has also to be assumed for the incoming O-atoms, but their obviously good penetration might possibly also be due to their nearly resonant charge transfer reactions with hydrogen [64].

The principle problem in applying this model is the proper choice of $\tau_\|$, for which the flow velocity of the background plasma (i.e., T_e and T_i and the Mach number M) and the relative drift of impurities to the plasma has to be known. Furthermore, the velocity field of the hydrogen scrape-off plasma is inhomogeneous (even in the case of negligible recycling it will contain a line of stagnation somewhere on each flux surface) so that differences in the location of the impurity injection may lead to changes in $\tau_\|$ not just simply proportional to the variation in the geometrical connection length L to the target plates.

Table 2. Comparison of measured and computed impurity confinement times on ASDEX [62].

Divertor/ Limiter	Working Gas	Heating	$\bar{n}_e$ [10^{13} cm^{-3}]	Impurity	τ_p meas.	τ_p calc.
D	H_2	OH	4.4	Si	1.6	1.8
D	D_2	OH	4.4	Si	1.8	1.8
tLC	H_2	OH	3.0	Si	1.2	2.2
tLC	H_2	NI	3.0	Si	0.5	1.4
tLFe	H_2	OH	1.5	O	20	35
tLFe	H_2	OH	1.5	Fe	5	5

D=divertor, tLC=toroidal carbon limiter, tLFe=toroidal stainless steel limiter. The confinement time τ_p is defined as total impurity particle number in the main plasma divided by total impurity in or outflux (whichever is better known).

These uncertainties get particularly severe when impurities produced at the limiter target plate are considered. Those produced at the limiter tip can reach a certain distance into the bulk plasma as neutrals; their rate of back diffusion to the limiter will depend on how fast they spread over the closed flux surfaces. Those ionized in the limiter shadow will have a very short connection length L and should therefore make [according to Eq. (66)] only a small contribution per unit source strength to the bulk plasma impurity content. On the other hand, their source intensity might become quite large, particularly if the threshold to self-sputtering is approached, making it difficult to draw quantitative conclusions out of these considerations.

For a divertor tokamak the situation is easier, as target plate produced impurities can generally not reach the bulk plasma as neutrals, and would have to stream against the general flow direction to reach the bulk plasma as ions. These considerations, supported by the models of Secs. 2.2.2.3 and 2.2.3, and not contradicted by experiments, suggest neglecting the target plates as sources of impurities in the bulk plasma of divertor tokamaks as long as plasma parameters remain below the self-sputtering threshold.

For the above reasons full plasma simulations can use a more convincing model of CX sputtering at the walls (although also here the concentration of recycling in the limiter vicinity has to be assessed and taken into account in $\tau_{\|}$) than for ion sputtering at the limiter. The effect of this uncertainty is attenuated only in self-limiting situations like the "radiating edge" solutions, where the impurity concentration is primarily fixed by the requirement to radiate a dominating power fraction.

The most elaborate of these simulations (e.g., [65]) determine profiles of plasma parameters in the scrape-off layer by a model like Ref. [60], and solve rate equations for the impurities in the form of Eq. (62). CX sputtering is taken into account by using the results of a Monte Carlo code for the neutral hydrogen flux and energy distribution at the walls and the tabulated sputtering yields [50]; the impurities are started as a flow of neutrals at the wall, whose attenuation by ionization determines at the same time the source term in the continuity equation for the first stage. To the latter is added, in the limiter shadow, a contribution from hydrogen ion sputtering, evaluated from the computed charged particle flux and the temperatures at the limiter, and reduced by a factor λ_s to take into account the enhanced probability of the impurities, compared to those produced by CX, to be immediately swept back to the target plates. To account for self-sputtering, the ion sputtering yield is multiplied by a function of electron temperature which is one far below the self-sputtering

676

threshold, but rises rapidly as a function of T_e near the threshold. Obviously the uncertainty in the effect of limiter-produced impurities is concentrated in the factor λ_s.

Results of such models are exemplified by Figs. 30 and 31, which show the radial distribution of the radiated power losses for JET simulation calculations using carbon, iron, and tungsten, respectively, as both wall and limiter material, as well as the distribution of ionization stages found in the calculation for the iron case. The confinement law (Alcator-INTOR) and additional heating power (29 MW of a "magic" heating method depositing 50% of the power in the central half of the volume) were chosen so as to give ignition in the iron case; the profiles describe conditions one second after switching off the additional heating. The salient points are the qualitatively different distribution of radiation losses (fairly central for tungsten, in a "ideal" radiating layer for iron, and primarily in the scrape-off layer for carbon) and the close agreement in the location of the maxima in the distribution of the iron charge stages between the finite rate model and the coronal prediction.

2.2.2.3. Impurity transport along field lines. For the reasons described in 2.1.3 a flux-surface averaged transport model for the scrape-off layer is basically not a well justified approximation, and requires, moreover, a model or an assumption for the transport along field lines to define the value of $\tau_\parallel$. One-dimensional models for impurity transport along field lines, discussed in this section, are equally limited in their applicability, but important for our qualitative understanding, as they illustrate physics aspects complementary to those treated in radial transport models.

As in the corresponding transport models for the background plasmas, cross-field diffusion or drifts out of the flux surface considered are approximated by sources and sinks. Additional sources in the transport equations for the charged impurities will be ionization of atoms coming either from the vessel walls (CX sputtered materials, oxygen), or recycling (e.g., fusion produced helium atoms), or sputtering from the target plates. Conditions treated so far assume the impurities to be a minority, with collisions among themselves, and a back reaction onto the (hydrogen) scrape-off plasma neglected. (The latter assumption has been dropped test-wise in some calculations, where radiative losses, based on computed impurity concentrations, were taken into account consistently in the energy balance of the background plasma. No particular difficulties were encountered in iterating between the two systems of equations). In this case, the single ionization

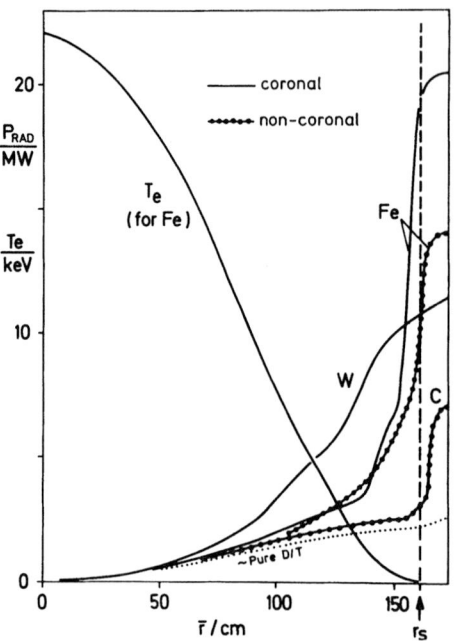

Fig. 30. Results of simulation calculations for the radially
integrated radiation loss power as a function of minor
radius for tungsten (W), iron (Fe) and carbon (C) as
limiter and wall material (from Ref. [65]).

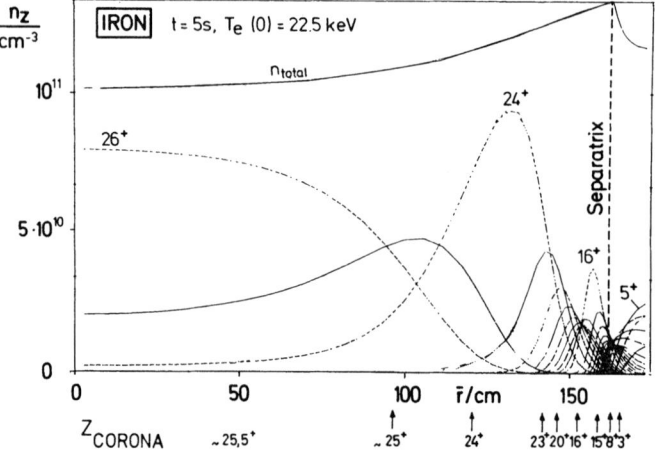

Fig. 31. Radial distribution of ionization stages for iron in a
JET-type plasma at thermonuclear condition. The
position of charge states calculated from the coronal
model is indicated by arrows (Ref. [65]).

states of the impurity Z considered are treated as test fluids, described by systems of equations [66]

$$\frac{\partial}{\partial t}\, n_j + \frac{\partial}{\partial s}\,\left(n_j v_j\right) = n_e\left[s_{j-1}n_{j-1}\right.$$

$$\left. - \left(s_j + r_j\right)n_j + r_{j+1}n_{j+1}\right] + d_j \;\cdots\;, \tag{67}$$

$$\rho_j\left(\frac{\partial}{\partial t}\, v_j + \frac{1}{2}\,\frac{\partial}{\partial s}\, v_j^2\right) + \frac{\partial}{\partial s}\, p_j + n_j\cdot j\cdot e\cdot E$$

$$- \rho_j\left(\frac{v - v_j}{\tau_j}\right) - a_j n_j\,\frac{\partial}{\partial s}\left(kT_e\right)$$

$$- \beta_j n_j\,\frac{\partial}{\partial s}\left(kT_i\right) = n_e\left[s_{j-1}\rho_{j-1}v_{j-1}\right.$$

$$\left. - \left(s_j + r_j\right)\rho_j v_j + r_{j+1}\rho_{j+1}v_{j+1}\right] - m_j v_j d_j \;\cdots\;, \tag{68}$$

$$\rho_j = m_j n_j,\; p_j = n_j kT_i,\; 1 < j < Z.$$

The sources d_j represent ionization of neutrals (for $j = 1$), or cross-diffusion into or out of the flux surface considered. (The time-dependent formulation is essentially for numerical convenience only.) The expressions for the Spitzer slowing-down time τ_j and the coefficients a_j, β_j are given in Ref. [66]. The temperature of the impurities is assumed to be tightly coupled to that of the ions, eliminating the need for separate energy equations for each ionization stage.

The background plasma properties, determined a priori by solution of a model like described in the first section of this paper, enter this system through n_e, T_e, T_i, and the background plasma velocity v, and determine also the electric field via

$$n_e eE = -\frac{\partial}{\partial s}\, p_e - 0.71\, n_e\,\frac{\partial}{\partial s}\, kT_e\,. \tag{69}$$

(The last expression can be viewed as the equation of motion of the electrons, or Ohm's law, under the condition of vanishing net current flow.) Profiles of these background plasma parameters and of the electric potential ϕ are shown in Fig. 32 for a model case, corresponding to ASDEX geometry, with

679

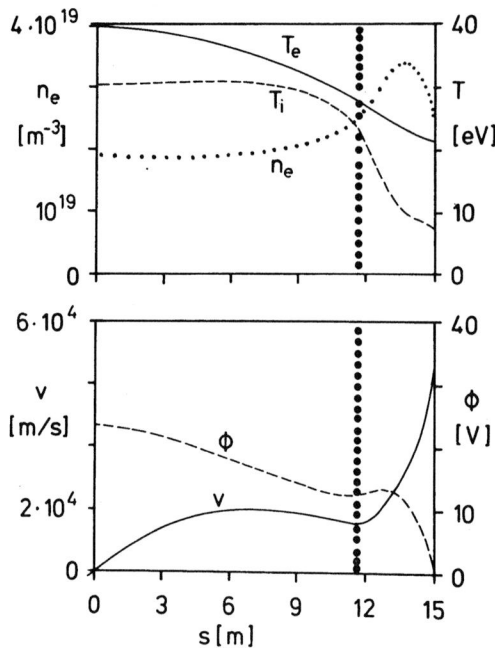

Fig. 32. Variation of scrape-off parameters along magnetic field
lines, defining the background plasma for impurity
transport studies along the field lines.

relatively high assumed particle and energy flows across the
separatrix (10^{22} s^{-1} and 1 MW, respectively) and intensive
divertor recycling.

Not trivially expected, but potentially very important, are
the thermal force terms in Eq. (68), which arise in this form
from the energy dependence of the Coulomb collision cross-
section. (Consider, in the presence of an ion temperature
gradient, an impurity test particle moving with the speed of the
plasma background. By definition of this speed, an equal number
of hydrogen ions will therefore pass it within a given distance
coming both from the upstream and downstream direction.
However, collisions will be more frequent and momentum exchange
more intensive with those coming from the cold side, resulting
in a net force onto the particle in the direction to higher
temperature).

This thermal force can balance flow friction if the
hydrogen mean free path λ_i, the temperature gradient length $\lambda_T = $
$[d(\ln T_i)/ds]^{-1}$ and the Mach number satisfy

680

$$M \lesssim \lambda_i / \lambda_T$$

which tends to happen in high power flow, high recycling situations, at a position corresponding to the beginning of the zone of strong hydrogen ionization. The electric field term, which, in contrast to the quadratic charge-state dependence of Coulomb-collision terms, depends only linearly on j, will have a strong effect only on low ionization stages, particularly for helium.

The examples given in Figs. 33-35 treat three practically relevant situations, differing in impurity species and source distribution. All three are based on an ASDEX-like geometrical situation.

The first case refers to helium, which is assumed to enter the scrapeoff from the bulk plasma at a fixed rate (like will be the case in a reactor for the fusion ash). Impinging and neutralized at the target plates, a relatively small fraction of these atoms (2%) is assumed to be pumped, the rest added as singly ionized atoms to the plasma fan in the divertor region. For the background plasma parameters considered, the effect of the thermal force shows essentially in the local peak in the density of fully-ionized atoms, and the finite fraction of singly-ionized He-atoms (labelled HeII in the figure and figure captions, where spectroscopic notation is used) which appear outside their source region, the divertor chamber. The total helium density, for these conditions, shows however an

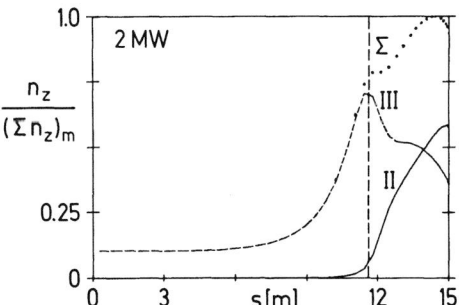

Fig. 33. Distribution of helium II and III along field lines. A helium-III source at the midplane and a helium-II recycling source in the divertor are assumed (98% recycling) (Ref. [65]).

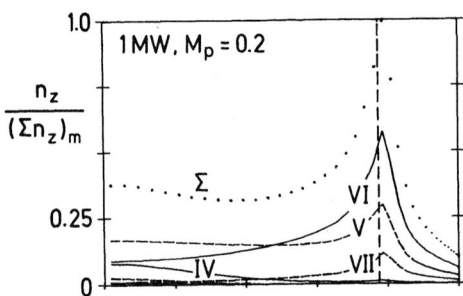

Fig. 34. Distribution of oxygen states along the field line.
The background plasma profiles are those of Fig. 7. A
Gaussian OII source centered at the midplane (s = 0) is
assumed. M_p is the Mach number of the hydrogen
background plasma outside the divertor.

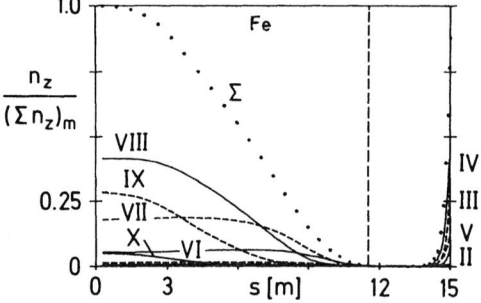

Fig. 35. Distribution of iron states along the field lines. A
homogeneous FeII source is chosen for 0 < s < 10 m. A
self-sputtering avalanche is just starting at the
target plate. The loss time constant is chosen as $\tau_\perp$ =
1 ms.

acceptable behavior, with accumulation in the divertor chamber. Due to the above-mentioned dependence of all frictional terms on j, He will in fact be the impurity least affected by thermal force effects, but not immune against them for arbitrarily strong recycling.

Example 2 treats the case of oxygen, which is assumed to be released from the walls of the main plasma vessel (appearing in our system of equations as a source of singly ionized atoms) and not to recycle from the target plates. Locally dominating retarding thermal forces cause the buildup of impurity density (and therefore pressure) gradients, which in stationary state ultimately will have to overcome the former, to allow removal of the impurities at the same rate with which they are produced. The negative effect of the thermal force will show up, as for a given impurity production the flow retardation results in an increased impurity density at the periphery, and consequently also in the interior, of the bulk plasma.

The last example concerns iron, and addresses also the question of self-sputtering. The last effect depends strongly on the energy which impurities released at the target plates have gained after ionization before re-impinging again. If acceleration in the electrostatic sheath is considered their main energy source, then the presheath electron temperature and the ionization stage of the impurity are the crucial parameters. Calculations to this rather complex point have been reported in [66] and [67]; they are illustrated here in Fig. 35 for one scenario, in which impurities from the primary source (CX) enter the scrape-off layer in the midplane. On their long way to the target plates they reach relatively high ionization stages (reflecting roughly the midplane separatrix temperatures), and are therefore strongly affected by friction forces. In the immediate target plate vicinity, where the plasma flow Mach number approaches 1 and flow friction is the clearly winning process, the impurity ions acquire a flow velocity close to that of the background plasma, and impinge onto the target plate with a nearly monoenergetic energy distribution around (for $T_e = T_i = T$)

$$\varepsilon_Z = \frac{m_Z}{m_i} kT + 3jkT$$

easily giving, particularly because of the first term, rise to a self-sputtering yield Y_s above 1 even for relatively modest T-values. To start an avalanche however, these secondary impurity ions would also have to have a self-sputtering yield above 1. They will however be ionized in the much colder region close to the target plate, intersecting it while in a much lower ionization stage. By the same token, they will also be more weakly coupled to the hydrogen plasma flow and enter the electrostatic sheath with an energy much below $(m_z/m_i) \times kT_e$; both effects contribute to make their self-sputtering yield considerably lower than that of the primary particles. Both components, properly weighted, have to be taken into account to define an effective average sputtering coefficient, which furthermore will be different, depending on whether the primary impurity atom is produced by CX in the midplane, or hydrogen ion sputtering at the target plates.

A further effect included in Fig. 35, is a simple model for cross-diffusion of impurities, taking into account the fact that the low impurity velocity in the vicinity of the point of dominating thermal forces causes a long local residence time for the ions, and therefore gives time for any anomalous cross-diffusion to compete with parallel flow. In the calculations of Ref. 67 this was taken into account by a contribution to d_j of the form of a loss term: $-n_j/\tau_\perp$. This gives rise to an impurity density profile along the field lines with two maxima, corresponding to the location of the two source contributions (the midplane region for CX and the target plate vicinity for self-sputtering). The large qualitative difference between the oxygen and the iron distributions in Figs. 34 and 35 is primarily due to this difference in the models indicating that any quantitative conclusions about transport effects along the scrape-off layer are strongly affected by cross-diffusion and the variations in the hydrogen plasma flow pattern between adjacent flux surfaces. This is particularly true as 2-D scrape-off models for the background plasma indicate the possibility even of hydrogen ion flow reversal. A valid conclusion from the present model, however, is that there is a limit to the intensity of hydrogen recycling in the divertor chamber desirable: a minimum net plasma flow from the bulk plasma proximity to the target plates will be required to sweep out impurities born in the bulk plasma (helium ash) or deposited there following CX sputtering.

The above examples illustrating the possible importance of the thermal forces all refer to primary impurity sources in the main vessel region. In the region close to the target plates, the flow Mach number will always approach one, so that impurities produced by charged particle sputtering will always be ionized in a region of weak thermal force effects and with a background plasma flow clearly directed to the target plates. At least as long as the criteria for self-sputtering avalanche formation are not violated, target plate produced impurities should have no chance to reach the bulk plasma.

2.2.3. Other models for impurity transport

Besides fluid models, other approaches have been used to describe impurity behavior in the plasma edge. Not competing, but, of course, complementary are the neutral particle Monte Carlo calculations (treated in the chapter by D. Heifetz), which for recycling gaseous impurities, particularly helium, are important to determine quantitatively the pumping rate, and the flux escaping out of the divertor chamber into the main vessel region.

Combined Monte Carlo approaches to both neutral and charged impurity transport have also been carried out. Prior to the discovery of the high recycling divertor regime and its reactor potential, expected divertor parameters for large devices were in a high temperature, low density regime, suggesting such a model in fact as a valid description for the impurity dynamics. It was developed by the authors of Ref. [68], who studied the motion of trace impurities in a given 2-D background, including a section of the bulk plasma and a topological model for the divertor region. Interaction of impurities with the background was through ionization and Coulomb collisions, and via a simple electric field simulating the presheath field of the scrape-off plasma [Eq. (69) of Sec. 2.2.2.3); additional random displacements were added to the particle trajectories so as to simulate anomalous perpendicular diffusion of magnitude $1/2 \, D_{BOHM}$. Upon impact at the target plates, the impurities gave rise to self-sputtering, with the sheath potential and the charge state of the impinging impurity taken into account in the evaluation of the yield. The principal, nontrivial outcome of these calculations was supression of formation of self-sputtering avalanches in divertor (in contrast to limiter-) geometries. This effect was due to the loss of a fraction of the sputtered impurities (either through escape as neutrals or via cross diffusion) into such regions of the divertor chamber not connected along field lines to the bulk plasma, and therefore cold.

In the high recycling regime, however, this numerical method would require taking into account an excessive number of collisions, and yield bad statistics for the very small fraction of particles actually reaching the bulk plasma.

So far we have looked at impurity modeling in order to assess the expected effect on plasma performance. For future long pulse experiments and the fusion reactor, however, material erosion is itself a crucial problem. In this case, we are not concerned about the relatively minor fraction of particles actually reaching the bulk plasma, but primarily in the magnitude of erosion and the distribution with which the particles are again deposited. This is the scope of models like the REDEP code [69], where assumed distributions of particle and energy fluxes are used to evaluate charged particle sputtering by hydrogen isotopes and helium. The angular and energy distribution $f_1(\alpha)$ and $f_2(E)$ of the sputtered particle flux Γ_Z^s and the angle between field lines and target plates δ are taken into account to evaluate the impurity backflux distribution Γ_Z via the integral relation

$$\Gamma_Z(x) = \int_\alpha \int_E \int_{x'} \frac{\Gamma_Z^s(x')}{\lambda} \, f_1(\alpha) \, f_2(E) \, \sin \delta$$

$$\exp\left(-\int_0^\infty dz'/\lambda\right) \, d\alpha \, dE \, dL_x \, ,$$

which assumes that particles flow back along the field line on which they got ionized. Here x is a coordinate along the target plate surface in a poloidal cross section (only axisymmetric configurations are considered) and dL_x the associated arc length. As self-sputtering giving a contribution to $\Gamma_Z^s(x')$ proportional to the local backflux $\Gamma_Z(x')$ is also included, the above expression becomes on inhomogeneous integral equation. The angular distribution of the sputtered impurities, together with the finite distance traversed by them before being ionized and frozen to a particular field line contribute an effective diffusion of materials over the target plates, to which is added a preferential drift in the case of an oblique angle between target plates and poloidal flux surfaces. To take into account the effect of wall produced impurities, a contribution is added to $\Gamma_Z(x)$, determined in magnitude from CX sputtering calculations using the DEGAS-code and taken with a spatial distribution at present assumed equal to that of the hydrogen ion flux at the target plates.

Results of such REDEP calculations, giving also a good general idea of the magnitude of this technical reactor problem, are shown in Table 3. It gives the total sputtering and the net erosion rates, expressed in cm/year, for the point of intersection of the separatrix with the outer divertor collector plate of INTOR, for a power flow of 40 MW onto it and different assumed presheath temperatures and target plate materials. CX sputtering was omitted for these calculations. Compared to Mo and W, Be and Ti show large sputtering rates. Moreover, heavy materials (and those with low ionization potential) have a short mean free path as neutrals; being redeposited in the nearly immediate vicinity of their location of birth will therefore lead to very small net material migration. From such results a strong preference emerges for the use of heavy materials, preferentially tungsten, as target plate materials, which for a

Table 3. INTOR Divertor Plate Erosion/Redeposition Analysis
(from Ref. [70]), using the code described in Ref.
[69].

Plate Coating Material	Plasma Edge Temperature (at Separatrix) (eV)	Total Gross Sputtering Rate at Divertor Center (cm/yr)	Net Erosion Rate of Divertor Center (cm/yr)
Beryllium	20	680	14.0
	30	688	9.1
	40	666	5.4
	50	624	4.1
Titanium	20	266	0.025
	30	658	0.056
	40	2750	0.42
Molybdenu	20	21	0
	30	64	0
	40	142	0
Tungsten	20	0.07	0
	30	1.1	0
	40	3.1	0
	45	13.7	0

divertor might seem acceptable in spite of the small tolerable
concentration in the main plasma, as all models suggest a
negligible probability for target plate material to reach the
bulk plasma.

Calculations including CX sputtering from vessel walls, and
redeposition of these impurities at these target plates,
illustrate, however, a severe problem as the latter will become
covered by the wall materials. With the model in the form
discussed above, REDEP calculations therefore lead to the
suggestion to construct the vessel walls from tungsten also. As
the fluid model of Sec. 2.2.2.3 predicts however that a
significant probability for wall-produced impurities will reach
the bulk plasma, this suggestion and the underlying calculations
have to be critically scrutinized. For one point, one expects
CX produced impurities to flow to the target plates more
predominantly in the outer fringes of the scrape-off layer than
the hydrogen ions. This, together with the effect of the
poloidal angle between target plates and poloidal flux surfaces,
which for the divertor geometry of INTOR, should contribute a
net drift of materials over the target plates towards a region
closer to the walls [70] and, hopefully, may make the use of
lower-Z wall materials still feasible.

2.3. Numerical methods for multi-species diffusion equations

A finite rate formulation of the time-dependent cross-field diffusion equations leads to a system of the type

$$\frac{\partial}{\partial t}\vec{n} - \overset{\leftrightarrow}{D}\frac{\partial^2}{\partial x^2}\vec{n} + \overset{\leftrightarrow}{v}\frac{\partial \vec{n}}{\partial x} + \overset{\leftrightarrow}{c}\vec{n} = -\overset{\leftrightarrow}{S}\vec{n} - \overset{\leftrightarrow}{R}\vec{n} + \vec{d} \quad ,$$

with $\vec{n}$ the vector of impurity ionization stages. Neglecting inertial terms, the continuity and momentum equations for the fluid dynamics along field lines of Sec. 2.2.2.2 can be cast in the same form. In the following we give a brief overview of methods of solutions available for such systems, assuming for this $\overset{\leftrightarrow}{D}$, $\overset{\leftrightarrow}{v}$, and $\overset{\leftrightarrow}{c}$ to be diagonal matrices. (When this is not satisfied, the diagonal terms generally still remain dominant, justifying a simple explicit treatment of the other terms.) The matrices $\overset{\leftrightarrow}{S}$ and $\overset{\leftrightarrow}{R}$ describe ionization and recombination, and consist, in the case of only single-stage processes, of terms only along the diagonal and one lower or upper co-diagonal, respectively.

To allow time steps $\Delta t > 1/2 \ [(\Delta x)^2/\|\overset{\leftrightarrow}{D}\|]$, $1/\|\overset{\leftrightarrow}{S}\|$, an implicit formulation has to be chosen for the diffusion and reaction rate terms. The standard way of treating the system of algebraic equations resulting from the finite differencing of such systems with full matrices $\overset{\leftrightarrow}{S} + \overset{\leftrightarrow}{R}$ is to arrange variables like $[n_1(x_1), n_2(x_1) \ .. \ n_Z(x_1), n_1(x_2) \ ... \ n_{Z-1}(x_k), n_Z(x_k)]$, making the coefficients of the densities at the new time step form a block-tridiagonal matrix. Such systems can then be solved by a vector analogue to the usual method for tridiagonal systems which involves, however, inversion of the single submatrices. During the latter process, the tridiagonal structure of $\overset{\leftrightarrow}{R} + \overset{\leftrightarrow}{S}$ is lost, so that no benefit can be drawn from it, and the total computational effort scales linearly with the number of space steps K, but quadratically with the number of ionization stages Z. This scaling can be inverted to become $\sim Z \times K^3$, by rearranging variables like $[n_1(x_1), n_1(x_2), n_1(x_3) \ ... \ n_1(x_k), n_2(x_1) \ ..., n_Z(x_{K-1}), n_Z(x_K)$ and considering the coefficients arising from the spatial differentiation as "inner" submatrices (Hulse [72]).

An alternative method, with a similarly scaling computational effort has been developed by Okamoto and Amano [73] who use a time splitting technique, solving the diffusion part of the equations by a standard Crank-Nicholson scheme and the rate equation as an eigenvalue problem. The computational effort involved in the latter step is quite large and increases again quadratically with the number of ionization stages.

Two methods have been developed using the tridiagonal structures of both the $(\tilde{R} + \tilde{S})$-matrix, and of the finite differenced diffusion terms, to produce schemes with a numerical effort $\sim Z \times K$. One consists in application of an SOR technique making full use of the sparseness of the system [74]. The second one derives originally from the observation that in many problems far from coronal equilibrium, ionization is much more important than recombination, suggesting an implicit treatment of the former, but only an explicit one of the latter terms. In this way, at each time step, the diffusion equations for the single ionization stages can be solved one after the other, starting with the singly ionized one, with an effort scaling $\sim Z \times K$. This procedure should become unstable if (somewhere) recombination rates exceed the ionization rates, a problem which has in practice generally been solved [62] by iterating several times during one time step. Another stabilization method, utilized in [75] and the applications of [65], consists in alternating between implicit treatment of ionization and recombination, with the other process taken into account in explicit formulation. This method has also been successfully used in tests in a 2-D form, by a simple time-splitting between the transport processes in the two directions.

2.4. Summary and outlook

Comparing Secs. 1 and 2 of this chapter it should become evident that the present state of scrape-off layer modeling is far less satisfactory for impurities than for the background plasma, an impression that is further enhanced when a comparison with experiments is attempted. This is partly due to the fact that the motion of impurities is primarily determined by the hydrogen background velocity distribution, which in the high recycling regime becomes a second order quantity. This implies that we could very well reproduce correctly features of the background scrape-off plasma like density and temperature distributions with a model giving large relative errors in the velocity distribution. Such errors would on the other hand immediately affect the impurity flow dynamics, for which the bulk plasma velocity is a first order quantity. In experimental terms this implies that in the high recycling regime we should be able to affect significantly the impurity transport by adjusting the refueling of the bulk plasma, possibly by an appropriate mix of deep (pellet) and shallow (gas puff) refueling [76].

Also, contrary to the situation for the hydrogen plasma component, we do not yet have a satisfactory formulation for neoclassical, collision dominated cross-field impurity transport in the scrape-off layer. This is a serious restriction as neoclassical contributions to cross-field transport could be large, and, as the discussion of the thermal force effects in Sec. 2.2.2.3 shows, vary qualitatively with recycling regime. Moreover, ASDEX experiments to the impurity flow into divertor chambers, in which top and bottom single null configurations were compared, showed significant asymmetries, which changed sign after inversion of the toroidal field direction: again effects one suspects could be explained by a properly formulated neoclassical theory [77].

Both criticized points are however under active study. Combined 2-D plasma and neutral particle calculations, fitted to satisfy experimental measurements of both plasma parameters and neutral pressures in the divertor chamber are about to come forth, and the numerical tools to study 2-d impurity transport, albeit with formally simple, anomalous diffusion models, have partly already been tested. Inclusion of the neoclassical impurity cross-field transport in the scrape-off layer finally requires the relatively straightforward application of the parallel friction coefficients to the neoclassical formulation summarized in the chapter by Singer.

ACKNOWLEDGMENTS

The authors are grateful for discussions and contributions from B. Braams, M. Harrison, D. Heifetz, R. Hulse, J. Neuhauser, M. Petravic, J. Roberto, W. Schneider, and C. Singer. They are especially grateful to M. Petravic and C. Singer for proofreading and to Ms. E. Carey for preparing the manuscript.

This work was partially supported by the U.S. Department of Energy Contract No. DE-AC02-76-CHO-3073.

REFERENCES

[1] W. Stacey, Fusion Power Analysis, (John Wiley and Sons, NY, 1981).
[2] S. Braginskii in Reviews of Plasma Physics, (Consultants Bureau, NY, 1965), Vol. 1.
[3] C. Singer and W. Langer, Phys. Rev. A 28, 994 (1983).
[4] M. Petravic et al., J. Nucl. Mater. 128&129, 91 (1984).
[5] P. Harbour, J. Morgan, Culham Laboratory, Abingdon, UK, Report No. CLM-R234 (1982).
[6] M. Ulrickson and D. Post, J. Vac. Sci./Technol. A1 907 (1983).
[7] P. Gierszewski et al., Phys. Rev. Lett. 49, 650 (1982).
[8] U. Daybelge, Nucl. Fusion 21 1589 (1981).

[9] A. Bailey and G. Emmert, J. Nucl. Mater. 111&112, 403 (1983).

[10] T. Takizuka et al., J. Nucl. Mater. 128&129, 104 (1984).

[11] G. Fussmann et al., J. Nucl. Mater. 128&129, 350 (1984).

[12] J. W. Wilson and J. LeBlanc, private communication.

[13] A. Bell, R. Evans, and D. Nicholas, Phys. Rev. Lett 46, 243 (1981).

[14] C. Singer, J. Fus. Eng. 3, 231 (1983).

[15] W. Langer and C. Singer, "Two Chamber Model for Divertors with Plasma Recycling," submitted to IEEE Trans. Plas. Sci.

[16] G. Dahlquist and A. Bjorck, Numerical Methods, (Prentice-Hall, Inc., NJ), 1984.

[17] R. Hawryluk and J. Schmidt, Nucl. Fusion 16, 775 (1976).

[18] F. Marcus, private communication (1976).

[19] P. Raimbrult and J. Girard, Phys. Scrip. 23, 107 (1981).

[20] A. Mense and G. Emmert, Nucl. Fusion 19, 361 (1979).

[21] J. Ogden et al., IEEE Trans. Plas. Sci., PS-9, 274 (1981).

[22] G. Becker and C. Singer, Max Planck Institut fur Plasmaphysik, Report No. IPP-III/75, Garching (November 1981).

[23] H. Howe, Oak Ridge National Laboratory Report ORNL/TM-7803, 1981.

[24] M. Ulrickson and D. Post, J. Vac. Sci./Technol. A1, 907 (1983).

[25] J. Hogan, Methods in Computational Physics, Controlled Fusion, (Academic Press, London, 1976). Vol. 16, 131.

[26] D. Duchs, D. Post, and P. Rutherford, Nucl. Fusion 17, 565 (1977).

[27] M. Shimada et al., Japan Atomic Research Institute Report No. JAERI M-9470.

[28] J. Morgan and P. Harbour, Fusion Technology (Pergamon, Oxford, 1980), Vol. 2, 1187.

[29] M. Ali Mahdavi et al., Phys. Rev. Lett. 47, 1602 (1981).

[30] K. Lackner and W. Schneider, in International Conference on Plasma Physics (Goteborg, Sweden, June 1982), to be published.

[31] M. A. Madhavi et al., J. Nucl. Mater. 111&112, 355 (1982).

[32] D. Post, W. Langer, and M. Petravic, J. Nucl. Mater. 121, 171 (1984)

[33] J. Galambos et al., J. Nucl. Mater. 121, 205 (1984).

[34] D. Potter, Computational Physics (Wiley Interscience, London, 1973) p. 88.

[35] M. Shimada et al., J. Nucl. Mater. 121, 184 (1984).

[36] W. Schneider et al., J. Nucl. Mater. 121, 178 (1984).

[37] B. Lipschultz et al., J. Nucl. Mater. 121, 441 (1984).

[38] S. Sengoku et al., Nucl. Fusion 24, 415 (1984).

[39] M. Petravic et al., Phys. Rev. Lett. 48, 326 (1982).

[40] D. Heifetz et al., J. Comp. Phys. 42, 309 (1982).

[41] M. Petravic et al., in Plasma Physics and Controlled Nuclear Fusion Research 1982, (IAEA, Vienna, 1983) Vol. 1, 323.

[42] R. Braams et al., J. Nucl. Mater. 121, 75 (1984).

[43] S. Saito et al., J. Nucl. Mater. 121, 199 (1984).

[44] Yu. L. Igitkhanov et al., in Eleventh European Conference on Controlled Fusion and Plasma Physics, (Aachen, 1983) Vol. 3, 397.

[45] M. Petravic, D. Heifetz, G. Kuo-Petravic, and D. Post, J. Nucl. Mater. 128&129, 111 (1984).

[46] M. Petravic, D. Heifetz, S. Heifetz, and D. Post, J. Nucl. Mater. 128&129, 91 (1984).

[47] W. Schneider and K. Lackner, in International Conference on Plasma Physics, (Goteberg, June 1982).

[48] D. E. Post et al., Atomic Data and Nuclear Tables 20, 397 (1977).

[49] K. Bol et al., in Plasma Physics and Controlled Nuclear Fusion Research 1978, (IAEA, Vienna) Vol. 1, 11.

[50] A. Gibson, M. L. Watkins, in Ninth European Controlled Fusion and Plasma Physics (Prague, 1977) Vol. 1, 31.

[51] K. Lackner, J. Neuhauser, in Divertor and Impurity Control (Proceedings of IAEA Technical Committee Meeting, Garching 1981), 58.

[52] R. A. Hulse, D. E. Post, D. R. Mikkelsen, J. Phys. B. At. Mol. Phys. 13, 3895 (1980).

[53] R. C. Isler, C. Crume, Phys. Rev. Lett. 41, 1296 (1978); W. H. M. Clark, et al., Nucl. Fusion 22 333 (1982).

[54] R. J. Jensen et al., Nucl. Sci./Eng. 65, 282 (1978).

[55] S. P. Hirschman, D. J. Sigmar, Nucl. Fusion 21, 1079 (1979).

[56] W. M. Stacey, D. J. Sigmar, Nucl. Fusion 19, 1665 (1979).

[57] J. Roth, J. Bohdansky, A. P. Martinelli, Radiation Effects 48, 213 (1980) and references cited therein.

[58] K. Borrass, Max Planck Institut fur Plasmaphysik, Report No. IPP 4/191, Garching (1980).

[59] A. T. Mense, PhD. Thesis, University of Wisconsin, UEFDM-219 (1977).

[60] J. Odgen et al., IEEE Trans. Plasma Sci. PS-9, 274 (1981).

[61] TFR Group, Nucl. Fusion 22 1173 (1982); Equipe TFT, in Plasma Physics and Controlled Nuclear Fusion Research 1982 (IAEA, Vienna), Vol. 3, 219.

[62] K. Behringer et al., in Eleventh European Conference on Controlled Fusion and Plasma Physics (Aachen, 1983) Vol. 2, 467.

[63] R. J. Hawryluk, S. Suckewer, S. P. Hirshman, Nucl. Fusion 19, 607 (1979).

[64] M. Tendler, J. Neuhauser, R. Wunderlich, Nucl. Fusion (to be published).

[65] R. Chodura, et al., in Plasma Physics and Controlled Nuclear Fusion Research 1982 (IAEA, Vienna) Vol. 1, 313.

[66] J. Neuhauser et al., Nucl. Fusion 24, 39 (1984).

[67] J. Neuhauser et al., J. Nucl. Mater. 121, 194 (1984).

[68] S. Sengoku et al., Nucl. Fusion 19, 1327 (1979).

692

[69] J. Brooks, J. Nucl. Mater. 111&112, 457 (1982).

[70] Impurity Control Engineering, in INTOR, Phase IIA, USA Contributions to the 9th Workshop-Meeting, Vol. 1 (1984).

[71] M. Harrison, J. Neuhauser, private communication.

[72] R. A. Hulse, private communication.

[73] M. Okamoto, T. Amano, J. Comp. Phys. 26, 80 (1978).

[74] C. Mercier et al., Nucl. Fusion 21, 291 (1981).

[75] K. Lackner et al., Z. Naturforsch. 37a, 931 (1982).

[76] M. Kaufmann et al., submitted to Nucl. Fusion.

[77] G. Fussmann et al., J. Nucl. Mater. 121, 164 (1984).

NEUTRAL PARTICLE TRANSPORT

D.B. Heifetz

Princeton Plasma Physics Laboratory
Princeton, New Jersey, 08544

1. Effects of Neutral Transport

Particle confinement times in present day controlled fusion devices are typically less than a tenth of the discharges' duration, and this fraction will decrease in future machines with larger discharge times. Thus it is important to study the mechanism of the recycling of neutral atoms and molecules formed as plasma ions strike device walls at the plasma edge.

By neutral transport we refer to the transport of mass, momentum, and energy in a plasma by neutral atoms and molecules. Global phenomena affected by this transport include:

1. Power and Particle Balance. Neutral particles refuel the plasma as they ionize, adding also to the plasma's momentum and energy. The process of charge-exchange alters the plasma momentum and energy as the plasma ions' momentum and energy are replaced with those of the neutrals. Plasma electrons lose energy as they excite and ionize neutral particles and dissociate molecules, and, in very hot plasmas, ions lose energy through the same processes. Neutral particles also transmit energy to the confinement device's walls in the form of radiation and, as they collide with the wall, heat. Walls absorbing the atom's momenta may suffer damage as material is sputtered off. Neutral particles may stay in walls for times which vary from a few milliseconds to a few years, so that, for example, the wall acts as a pump as well as a repository of tritium in reactors.

2. Plasma Confinement Quality. The distribution of the particle and energy refueling may be a factor in creating the increase in confinement seen during divertor operation (the H-mode) in the ASDEX, PDX, and D-III tokamaks and on ISX during impurity feeding (the Z-mode).

3. Particle and Energy Fluxes to Device Walls. Some components in tokamaks receive high heat loads due to charge-exchanging neutrals. This has not greatly restricted the designs of limiters and divertors in present-day machines, with relatively short duty cycles, but it will impact the next generation of machines.

4. Erosion of Device Walls Due to Sputtering. Impact sputtering by charge-exchanging neutrals is a major source of impurities in the plasma, and an important factor in the design of high heat and particle flux machine components.

5. Level of Tritium Contamination in Device Walls. The charge-exchange flux is a major mechanism in tritium contamination of the first wall in a $D + T$ machine. For example, it is necessary to know this flux in order to schedule tritium discharges in TFTR.

6. Performance of Pumping Systems. Future reactors will have to be pumped to remove impurities, in particular helium, in order to sustain long pulse operation. Ideally the impurities would be preferentially removed leaving relatively more tritium fuel in the system. The physics of how to do this is, at this moment, highly speculative. Among the unknowns are the relative confinement in the main discharge of hydrogen versus helium, the difference in how hydrogen and helium flow along field lines in, say, a divertor, and once neutralized, the difference between the two species in the pumping efficiency of proposed pump designs and the difference in reionization in the plasma.

7. Interpretation and Design of Diagnostics. Knowledge of neutral transport behavior is also necessary for the design and analysis of such widely used experiment diagnostics as neutral particle emission analysers [Parsons and Medley], H_α emission detectors, and pressure gauges. The atomic density also affects the relative populations of ionized states of impurities such as oxygen and carbon, hence must be known when interpreting spectroscopic measurements of lines from these elements.

In the next section we will review the physics determining the trajectories of neutral particle flights. Section 3 formulates the Boltzmann equation for neutral transport problem, and analytic and Monte-Carlo solutions of the equation are given in Secs. 4. and 5. In Sec. 6 applications of the theory are described, and suggestions for future model development are given in Sec. 7.

2. The Physics of Neutral Transport

The original outline of a basic model for neutral transport in confined plasmas was given in [Sakharov]. Atoms and molecules are introduced into the plasma through recombination at the walls or by external fueling (Fig. 1a). The atoms may charge-exchange, creating a new atom with new velocity which will continue on to either charge-exchange, be ionized (Fig. 1b), or strike a wall (Fig. 1c). Through charge-exchange and reflection off walls, generations of atoms are formed and move through the interior of the plasma. The molecules will dissociate upon entering the plasma creating atoms and ions with 5 eV or so of energy (Fig. 1b). The resultant atoms now begin their charge-exchange cascade. A graphical depiction of the movement of particles in the plasma as they charge-exchange and reflect plasma is shown in Fig. 2.

In this section we describe the processes involved in the movement of neutral particles, and the uncertainties in the existing data for these processes.

2.1. The Complete Model

A complete neutral transport model must include the following physical processes:

1. All important neutral-plasma reactions. Such a catalogue of reactions would include those for hydrogenic atoms and molecules, as well as for those light neutral impurities which commonly occur in controlled plasmas, such as He^0, Ne^0, H_2O, CO, CH_n, and Ar^0, over a temperature range from 1-10^4 eV. A detailed discussion of this atomic physics can be found in [Harrison].

2. The important neutral-neutral and neutral-ion elastic scattering processes. When the ambient neutral density rises above 10^{14} cm^{-3} elastic scattering becomes important. Such high pressures of a few tenths of a Torr occur rarely in tokamaks. However they have been routinely reached in the DITE bundle divertor [Fielding et al], and may be present in future, high-recycling, divertors [Petravic et al (1982), Hsu].

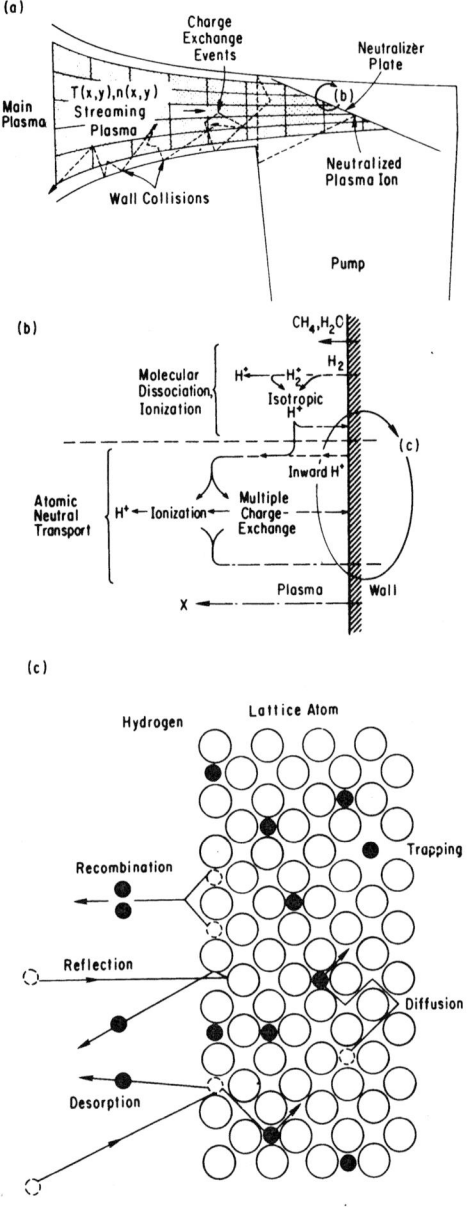

Fig. 1. We can separate into three levels the physical processes involved in neutral particle transport at the plasma edge. At the widest level (a), atoms interact with a streaming plasma, charge-exchanging, ionizing, and colliding with the device wall. Closer in to the wall (b), molecules form at a wall and dissociate into the near-wall plasma. In the wall itself (c), particles interact with the wall recombining, reflecting, and desorbing.

698

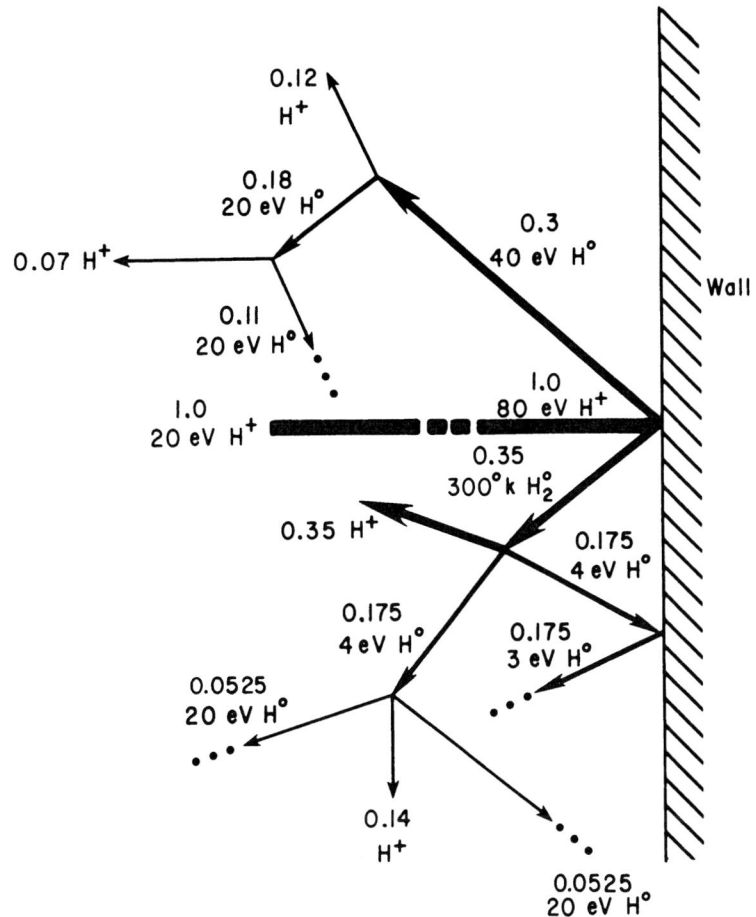

Fig. 2. A sample case showing the possible outcomes of a neutral particle flight. A single 20 eV H$^+$ crosses a sheath at the wall, striking it with 80 cV of cvcrgy. The reflection probability is then approximately 0.3, producing on the average, 0.3 H^o and $0.7/2 = 0.35$ H$_2^o$ particles. The reflected atom may have an energy of 40 eV, and the molecule desorbs at the wall temperature. The atom now goes on to a collision with the plasma where it charge-exchanges (60% probability), or ionizes (40% probability), leaving $0.6 \times 0.3 = 0.18$ atoms, which go on to be eventually ionized or pumped. The molecule ionizes and dissociates near the wall into 0.35 H$^+$ and 0.35 4 eV H^o. Assume that half of the H^o travels toward the wall and half away from the wall. Then 0.175 of the 4 eV H^o strikes the wall and reflects, for example, with 3 eV of energy. The remaining 0.175 H^o collides with plasma, splitting into 0.105 H^o and 0.07 H$^+$, and so on.

699

3. A realistic wall reflection model. The physics of wall reflection is being covered elsewhere in this school by [Behrisch and Eckstein, Roth]. The ideal reflection model we require would include data on the reflection probability, reflected energy, and reflected angle, as a function of incident particle species, energy, and incident polar angle, as well as target composition, surface roughness, and surface contamination. A model of wall desorption is also necessary, which such include information on the distribution of particles absorbed in the wall. The model will be valid over the range of incident energies of $1 - 10^4$ eV.

4. A model of physical wall sputtering. When an energetic particle strikes a wall, material from the wall may be sputtered off. To be consistent, this process should be included as part of the wall reflection model, since it involves the same physics.

5. Realistic geometries. Since the walls influence the neutral transport as much as the plasma in relatively closed components such as divertors, pump-limiters, and pump ducts, the model must not only include a realistic wall reflection model but it must also accurately know where the walls are. It is true that for some applications a schematic geometry is adequate to describe the major effects. Modeling actual experiments often does require, however, a realistic geometry in order to compare results with spatially resolved data such as for example H_α emission strength.

2.2. Physical Uncertainties

Recent reviews of existing atomic and wall reflection experimental and theoretical data can be found in [Post and Singer, Sagara and Kamada, Thomas et al, Langley et al, Conn et al (1984A), Oshiyama et al, Ozawa et al, Terasawa et al, Kato et al]. While much of the basic relevant atomic physics can be described in theory, there remain outstanding problems whose solutions are important in our modeling.

One of these is that in order to compute the rate of H_α emission, we must be able to determine the excitation states of hydrogen in systems where the atom will not be in equilibrium with the ambient plasma. For example, an atom in the throat of a narrow pump limiter may cross the throat in 10^{-7} sec, strike a wall and change its excited state.

Another unknown is a description of the vibrational states of molecular hydrogen, a process which may absorb relatively large amounts of energy in a divertor.

There are also large gaps in the measured data for reaction rates, and where the data exist, there are uncertainties in them. One difficulty is that

700

the range of temperatures found in controlled fusion devices is large, from 1 to 10^4 eV, and we may need the reaction rates over that whole range in particular calculations.

As mentioned above, neutral-neutral interaction may become an important process at pressures of a few mTorr. At present, however, there is little data on total cross-sections, σ, and $d\sigma/d\theta$, for H^0 and D^0 onto anything at energies above 2 eV and below 300 eV. Some data for 25 to 1800 eV D^0 and He^0 onto He, Ne, Ar, and Kr, measured utilizing PLT as a neutral particle source and the Low Energy Neutral Spectrometer as a detector appears in [Ruzic (1984)]. However, in pump ducts where we may find high neutral pressures, the atomic energies would be below 25 eV for which there is no data.

An even less understood phenomenon is wall reflection. Most existing data is for well characterized wall material with a clean and smooth surface. Few projectile/target pairs have been treated, and the incident energies are typically above 100 eV. This restriction in energy is due both the difficulty in generating neutral beams of low energy (less than 50 eV), and the theoretical problem of treating the multi-body interactions which become important at energies of a few eV [Behrisch and Eckstein]. A key question is how different from the ideal case is wall reflection off the rough and dirty walls actually found in fusion devices.

We must also know the process of particle desorption from a wall better. This determines in part the source rate for neutral particles, and is an important factor in the overall particle balance in the machine.

If we extend our interests to the study of the transport of the heavier neutral molecules, a whole area opens up which is largely unexplored by experimentalists. For example, methane and aceylene form on a carbon surface when exposed to a hydrogen plasma, and methane and water are produced in stainless steel when it is struck by carbon and oxygen impurities. There may be great differences between the transport in plasma of methane and water and that of atomic carbon and oxygen [Langer, Tendler]. In order to model neutral molecular transport it will be necessary to have data on the production and type of molecular impurities produced at the wall and on the molecular/plasma interaction.

A final unknown is the properties of the sheath created when the plasma strikes a wall [Stangeby, Chodura]. This determines one of the boundary conditions for our model, the distribution of energies of the neutral particles produced at a neutralizer plate, which depends on the distribution of incident energies and angles of the ions.

3. The Boltzmann Equation

The lifetimes of neutral particles in a fusion device are a function of their velocities, the mean free path lengths of interaction with the plasma, λ_p, and the distances to the device walls, ℓ_w. Assume, for example, that a deuterium atom has an energy of 75 eV, hence a velocity, v, of about 8.7×10^6 cm/sec, the electron density, n_e, of the plasma is $1 \times 10^{13}\,\mathrm{cm}^{-3}$, the electron temperature, T_e, is 50 eV, and the diameter of the device is 100 cm. Then, including only the charge-exchange, $<\sigma_{cx}v>$, and electron impact ionization, $<\sigma_e v>$, reaction rates, and equating the ion density with the electron density,

$$\lambda_p = \frac{v}{n_e(<\sigma_e v> + <\sigma_{cx}v>)} \approx 17.5\,\mathrm{cm}$$

and $\ell_w = 50\,\mathrm{cm}$. Then the particle "mean free path length", ℓ_{tot}, is

$$\ell_{tot} = \frac{\lambda_p \ell_w}{\lambda_p + \ell_w} \approx 13\,\mathrm{cm}.$$

Thus the first generation atom survives approximately 1.5×10^{-6} sec. Since the probability of charge-exchanging is

$$\frac{<\sigma_{cx}v>}{<\sigma_{cx}v> + <\sigma_e v>} \approx 0.6,$$

the next generation of atoms has half the population of the previous, and similarly for subsequent generations. Summing over all generations, we have an average total lifetime of $1.5 \times 10^{-6}(1 + 0.6 + 0.6^2 + \cdots) = 3.75 \times 10^{-6}$ sec. This is much shorter than the timescale for radial transport in a tokamak.

In this case we may then assume that the Boltzmann equation describing the neutral particle distribution $f(\vec{x}, \vec{v})$ over position, $\vec{x}$, and velocity, $\vec{v}$, space is time-independent:

$$\vec{v} \cdot \nabla f(\vec{x}, \vec{v}) = C(f)(\vec{x}, \vec{v}), \tag{1}$$

where $C(f)$ is the neutral/plasma collision term.

We also note, however, that the timescales for parallel plasma transport and neutral transport are similar. In this case the usual approach is to iterate the neutral transport calculation with a calculation of the plasma flow to determine an equilibrium solution. We will describe this more fully in the section below on neutral and plasma transport codes, and confine the discussion now to the time independent problem.

There is a large literature on solutions of Eq. (1) in the context of neutron transport, where $C(f)$ includes the processes of absorption and scattering. As mentioned before, it has been suggested [Greenspan] that existing codes for neutron transport be adopted to the neutral transport problem, and some early work followed this approach [El-Derini and Gelbard, Marable and Oblow, Pfeiffer, Burrrell, Gilligan et al]. Since the original neutronics codes were large and slow, including processes more complicated than those involved in neutral transport, the codes based on them have served best to calibrate the faster, smaller routines written especially for the neutral transport problem.

4. Analytic Solutions

4.1. The One-Dimensional Boltzmann Equation

Solving Eq. (1) in three-dimensions is a formidable task, so we simplify the assumptions to reduce the problem to a one-dimensional one.

Define the Knudsen number, K, by $K \stackrel{\text{def}}{=} \lambda/L$, where λ is the total neutral mean free path length, and L is the characteristic size of the problem. We can then classify problems as either impermeable, when $K \ll 1$, or, if otherwise, permeable [Lehnert]. Most machines in the past were permeable and TFTR/Jet/JT-60 is the first generation of impermeable devices. (We leave the divertor/pump-limiter regions out of the discussion for the moment, and consider only the main discharge.)

We follow the discussion in [Conner] for the impermeable case. In this case we may assume that the plasma is a slab bounded on one side by a flat wall, and we need solve Eq. (1) in only one-dimension:

$$v_x \frac{\partial f}{\partial x} = C(f), \tag{2}$$

for $x > 0$, with one boundary condition on f, at the wall at $x = 0$:

$$f(0, \vec{v}) = f_+(\vec{v}). \tag{3}$$

In the collision term for atomic hydrogen we may include only the electron impact ionization and charge-exchange processes, as proton impact ionization should be negligible at the plasma edge. The ionization term, $S(f)$, in $C(f)$ is then

$$S(f)(x, \vec{v}) = -f(x, \vec{v}) \int_{\vec{w}} \|\vec{v} - \vec{w}\| \sigma_e f_e(x, \vec{w}) \, d\vec{w} \tag{4}$$

where $f_e(x, \vec{w})$ is the electron distribution function, and $\|\vec{a}\|$ denotes the length of the vector $\vec{a}$. Now in general $\|\vec{w}\| \gg \|\vec{v}\|$, so that, assuming

$f_e(x, \vec{w}) = n_e(x) f_M(\vec{w})$, where f_M is a Maxwellian velocity distribution, Eq. (4) can be rewritten as

$$S(f)(x, \vec{v}) = -f(x, \vec{v}) n_e(x) \int_{\vec{w}} \|\vec{w}\| \sigma_e f_M(\vec{w}) \, d\vec{w}$$
$$= -f(x, \vec{v}) n_e(x) s, \tag{5}$$

where

$$s = \int_{\vec{w}} \|\vec{w}\| \sigma_e f_M(\vec{w}) \, d\vec{w}.$$

The charge-exchange term, $X(f)$, in $C(f)$ is

$$X(f)(x, \vec{v}) = \int_{\vec{w}} \|\vec{v} - \vec{w}\| \sigma_{cx} [f(x, \vec{w}) f_i(x, \vec{v}) - f(x, \vec{v}) f_i(x, \vec{w})] \, d\vec{w} \tag{6}$$

Simplifying Eq. (6) is the fact that $\sigma_{cx} \|\vec{v} - \vec{w}\|$ is a slowly varying function of the relative velocity, which can then be assumed to be a constant c. Taking $f_i(x, \vec{v}) = n_i(x) f_M(\vec{v})$, Eq. (6) then becomes

$$X(f)(x, \vec{v}) = n_i(x) c [f_M(\vec{v}) \int_{\vec{w}} f(x, \vec{w}) \, d\vec{w} - f(x, \vec{v})]. \tag{7}$$

Returning to Eq. (2), assuming $n_e = n_i = n$ we have

$$v_x \frac{\partial f}{\partial x} = S(f) + X(f)$$
$$= -fns + cn[f_M \int_{\vec{w}} f(x, \vec{w}) \, d\vec{w} - f]$$
$$= -fn(c + s) + ncf_M \int_{\vec{w}} f(x, \vec{w}) \, d\vec{w}. \tag{8}$$

It is useful to reparametrize the distance variable to units, z, of mean free path lengths (the optical depth):

$$z(x) = \int_0^x \frac{dy}{\lambda_{tot}(y)} = \int_0^x \frac{n(y)(c + s) \, dy}{v_{T_i}} \tag{9}$$

where v_{T_i} is the ion thermal velocity. With this change of variables, integrating Eq. (8) over v_y and v_z gives

$$u \frac{\partial F}{\partial z} + F(u, z) = \frac{c}{s + c} g(u) \int_{-\infty}^{\infty} F(u, z) \, du \tag{10}$$

where $u = v_x/v_{T_i}$, $F(u,z) = \int\int f(x,v)dv_y dv_z$, and $g(u)$ is the ion velocity distribution.

A number of solutions for Eq. (10) have been found for constant T_i, dependent on different assumptions on g. Both [Williams] and [Dnestrovskii et al] solve Eq. (10) assuming u has equal probability of being either negative or positive:

$$g = \frac{1}{2}[\delta(u-1) + \delta(u+1)].$$

A more realistic expression for g is

$$g = \begin{cases} 1/2, & -1 \le u \le 1 \\ 0, & u < -1 \text{ or } u > 1, \end{cases}$$

which leads to a transport equation which has been widely treated in the literature on neutron transport [Williams, Lewis and Miller]. Finally, a solution expressible in explicit algebraic form using the method of singular eigenfunctions for $g(u) = \pi^{-1/2}exp(-u^2)$ has been developed in [Conner].

4.2. The Integral Form of the Boltzmann Equation

If the neutral transport is important only at the plasma edge, then the case may be assumed impermeable. Note, however, that beam particles can penetrate impermeable plasmas, creating thermal neutral particles through charge-exchange whose transport may be of interest. Furthermore, charge-exchange flux measurements provide a measure of the neutral central density, so that a method which can accurately predict the central neutral density can be calibrated.

Both [Audenaerde et al] and [Tamor] have reformulated the problem in a way which can treat transport in permeable plasmas. Let ϕ be the charge-exchange rate density:

$$\phi(\vec{x}) = \int_{\vec{v}} n <\sigma_{cx}v> f(\vec{x},\vec{v})d\vec{v} + \phi^0(\vec{x}) \tag{11}$$

where ϕ^0 is an external source. For a given velocity direction, $\vec{v}_1$, choose a coordinate system with the z-axis parallel to $\vec{v}_1$. Then Eq. (1) can be written

$$\vec{v}_1 \cdot \nabla f(\vec{x},t\vec{v}_1) = \frac{\partial f}{\partial z}(\vec{x},t\vec{v}_1)$$

$$= \frac{\phi(\vec{x})F_M(\vec{x},t\vec{v}_1)}{t} - \frac{1}{\lambda_{tot}(\vec{x},t)}f(\vec{x},t\vec{v}_1) \tag{12}$$

Eq. (12) has the solution

$$f(\vec{x},t\vec{v}_1) = \int_{z_0}^{z} exp\left[-\int_{z'}^{z} \frac{dw}{\lambda_{tot}(w,t)}\right] \times \frac{\phi(z)F_M(z,t)}{t}dz$$

705

which can be rewritten as

$$f(\vec{x}, \vec{v}) = \int_0^\infty exp\left[-\int_0^s \frac{dw}{\lambda_{tot}(\vec{x} - \vec{v}_1 w, \|\vec{v}\|)}\right] \frac{\phi(\vec{x} - \vec{v}_1 s)g(\vec{x} - \vec{v}_1 s, \|\vec{v}\|)}{\|\vec{v}\|}\, ds$$

$$(13)$$

Substituting Eq. (13) into Eq. (11) gives

$$\phi(\vec{x}) = \int G(\vec{x}, \vec{y})\phi(\vec{y})d\vec{y} + \phi^0(\vec{x}), \tag{14}$$

where

$$G(\vec{x}, \vec{y}) = \int \frac{1}{\lambda_{cx}(\vec{x}, t)} \frac{e^{-\tau(\vec{x}, \vec{y}, t)}}{\|\vec{x} - \vec{y}\|^2} F_M(\vec{x}, t)\, dt,$$

and τ is the optical depth from $\vec{x}$ to $\vec{y}$:

$$\tau(\vec{x}, \vec{y}, t) = \int_{\vec{x}}^{\vec{y}} \frac{dw}{\lambda_{tot}(w, t)}.$$

Intuitively, $G(\vec{x}, \vec{y})$ is the probability that a neutral from point $\vec{y}$ charge-exchanges at point $\vec{x}$. Thus if we define

$$\Gamma(\phi)(\vec{x}) \stackrel{\text{def}}{=} \int G(\vec{x}, \vec{y})\phi(\vec{y})\, d\vec{y},$$

then $\phi^1 = \Gamma(\phi^0)$ is the density of first generation charge-exchanges for the source ϕ^0, $\phi^2 = \Gamma(\phi^1) = \Gamma^2(\phi^0)$ the second generation density, and so on, and the sum of all generations is

$$\phi = \phi^0 + \phi^1 + \cdots$$
$$= \sum_{i=0}^\infty \Gamma^i \phi^0$$
$$= (1 - \Gamma)^{-1}\phi^0, \tag{15}$$

where the second equality is meant formally. If, for example, we discretize the system so that Γ is matrix multiplication of a source vector ϕ by a matrix G, then it can be shown that the infinite sum in Eq. (15) converges to $(1 - G)^{-1}$, providing that $\|G\| < 1$, which holds in this case. Thus $(1 - G)^{-1}\phi^0$ is the sum over all $G^i\phi^0$. This is how [Audenaerde et al, Tamor] numerically solve Eq. (14). While this solution is not explicit in the dependent parameters, as was case of the first solution described above, it is computationally very fast, even when the charge-exchange matrix is calculated for a cylindrical geometry.

5. Monte Carlo Solutions

The analytic solutions described in Sec. 4 used geometric symmetries and simplified plasma parameters in order to make the problem tractable. An alternative approach is to use Monte Carlo methods to integrate Eq. (1). This is a very powerful tool for including the complicating effects of asymmetries, Maxwellian (and non-Maxwellian) plasmas, and detailed atomic and wall physical processes.

We refer the reader elsewhere for general introductions to the Monte Carlo method, and its applications to neutron transport problems [Spanier and Gelbard, James, Cashwell and Everett, Sobol, Hammersley and Handscomb, Kalos et al, Carter and Cashwell, McGrath and Irving].

Applications of the Monte Carlo method to the neutral transport problem [Parsons and Medley, Cupini et al (1983A), Hogan, Hughes and Post, Reiter and Nicolai, Seki et al, Heifetz et al (1982), Heifetz and Post] have all made use of techniques from neutronics codes.

To illustrate how the Monte Carlo method can be used to solve Eq. (1), simplify Eq. (1), for argument's sake, by assuming that no charge-exchange occurs, so that the one-dimensional version [Eq. (10)] of Eq. 1 is

$$u\frac{\partial F}{\partial z} = -F(u, z) \tag{16}$$

Making the further assumption that $|u| = 1$, Eq. (16) has the solution

$$F(x) = exp[-z(x)] = exp\left[-\int_0^x \frac{dy}{\lambda_{tot}(y)}\right]. \tag{17}$$

We now derive Eq. (17) from a different point of view. Let $\xi(x)$ be the probability that a particle travels at least a distance x without making a collision. The the probability of traveling at least a distance $x + dx$ without making a collision is

$$\xi(x + dx) - \xi(x) - \xi(x)dx/\lambda_{tot}(x),$$

and solving for ξ gives

$$\xi(x) = exp\left[-\int_0^x \frac{ds}{\lambda_{tot}(s)}\right], \tag{18}$$

which is just Eq. (17).

The Monte Carlo method of solving Eq. (16) is to track test flights using Eq. (18) to determine points of collision. Take a uniform random variate (URV), ξ, between 0 and 1, and solve

$$\xi = exp\left[-\int_0^{\bar{x}(\xi)} \frac{ds}{\lambda_{tot}(s)}\right] \tag{19}$$

for the collision point $\vec{x}(\xi)$. By running a sequence of test flights a distribution of collision points is determined and, knowing the remaining terms in $C(f)$, such as plasma density and temperature and the reaction cross-sections, we solve for f.

5.1. Tracking

To make the disscussion concrete we describe two standard methods of solving Eq. (19) for $\vec{x}(\xi)$. The first integrates

$$\int_0^{\vec{x}(\xi)} \frac{ds}{\lambda_{tot}(s)}, \tag{20}$$

by elementary calculus techniques, and the second, called the pseudo-collision algorithm (PCA), integrates Eq. (20) by the Monte-Carlo rejection method.

In the first method, assume that a test flight begins at $\vec{x}_o$ with a velocity $\vec{v}$, and that the plasma region has been divided into, say, two-dimensional zones with constant plasma parameters (Fig. 3). Choose a URV ξ, $0 < \xi < 1$, and compute the first intersection, $\vec{x}_1$, of the flight path with a zone boundary, and set

$$d_1 = exp\left[-\int_{\vec{x}_o}^{\vec{x}_1} \frac{ds}{\lambda_{tot}(s)} \right].$$

if $d_1 \geq \xi$, find the next intersection, $\vec{x}_2$, and set

$$d_2 = exp\left[-\int_{\vec{x}_1}^{\vec{x}_2} \frac{ds}{\lambda_{tot}(s)} \right].$$

If $d_1 \times d_2 \geq \xi$, continue until a point, $\vec{x}_n$, is reached where

$$d_1 \times \cdots \times d_n < \xi.$$

The solution, $\vec{x}_c = \vec{x}(\xi)$, is then between $\vec{x}_{n-1}$ and $\vec{x}_n$. Since λ_{tot} is constant in the zone between $\vec{x}_{n-1}$ and $\vec{x}_n$,

$$d_c = \xi/(d_1 \times \cdots \times d_{n-1}) = exp\left[-\frac{\vec{x}_c - \vec{x}_{n-1}}{\lambda_{tot}} \right],$$

and $\vec{x}_c = \vec{x}_{n-1} - \lambda_{tot} ln(d_c)$.

The PCA appears to be very different. Given the initial $\vec{x}_o$ and $\vec{v}$, let λ_{min} be the shortest mean free path length in the entire plasma for the test flight (Fig. 4). Move the test flight to

$$\vec{x}_1 = \vec{x}_o - ln(\xi)\lambda_{min}\vec{v}_1.$$

708

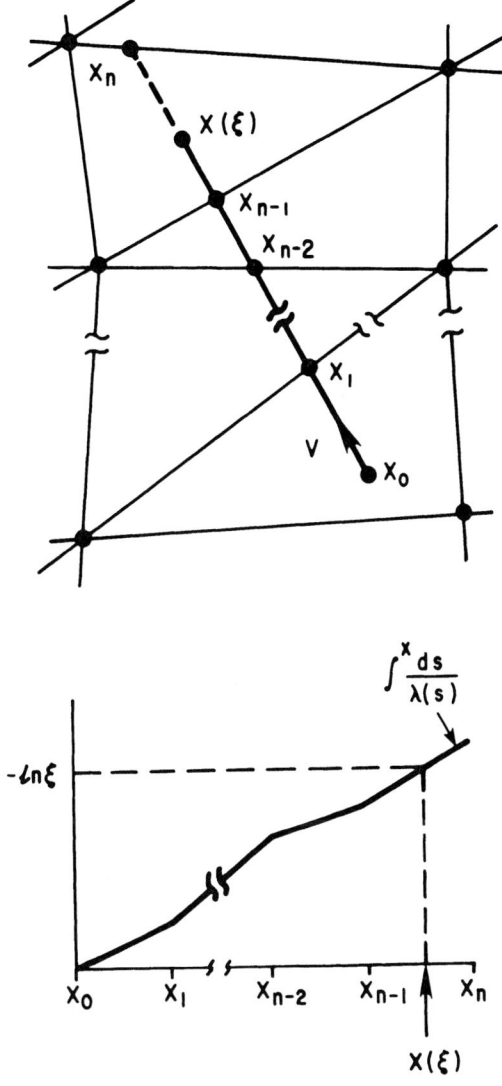

Fig. 3. Tracking in two dimensions by integrating $\int ds/\lambda(s)$ directly. Starting at $\vec{X}_o$ with velocity $\vec{V}$, choose a uniform random variable ξ, $0 \le \xi < 1$. The first intersection with the triangularized two-dimensional grid is at the point $\vec{X}_1$. If $\int_{X_o}^{X_1} ds/\lambda(s) < -\ell n\xi$, the intersections $\vec{X}_2, \ldots$, are computed, until $\int_{X_{n-1}}^{X_n} ds/\lambda(s) > -\ell n\xi$, at which point the algorithm can backtrack linearly to the point of collision $\vec{X}(\xi)$. Tracking in three-dimensions is done analogously, with tetrahedrons taking the place of triangles.

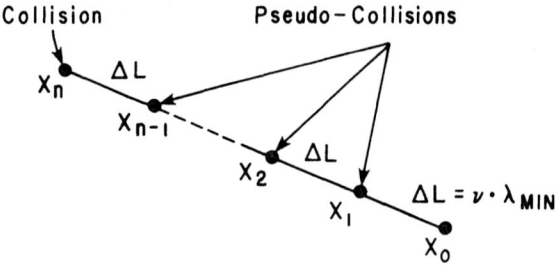

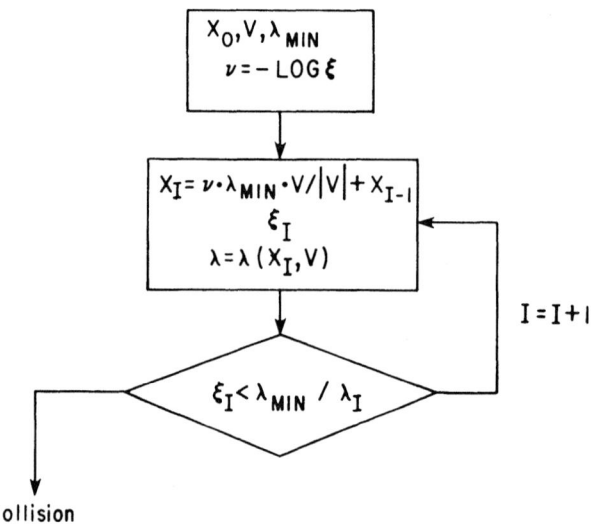

Fig. 4. Tracking using the pseudo-collision algorithm. Starting with the test flight at the position $\vec{X}_o$ with velocity $\vec{V}$, choose a uniform random variate ξ, $0 \le \xi < 1$, and let $\nu = -\ell n \xi$. With λ_{min} being the smallest mean free path length in the entire plasma of a particle that the test flight represents, move the flight to the point $\vec{X}_1 = \nu \lambda_{min}(\vec{V}/\|\vec{V}\|) + \vec{X}_o$. Choose a second uniform random variate, ξ_1, and test it against the ratio λ_{min}/λ_1, where λ_1 is the local mean free path at the point $\vec{X}_1$: If $\xi < \lambda_{min}/\lambda_1$, then a collision has occurred, but if not, move the test flight again a distance ν to $\vec{X}_2$, test again against a ratio λ_{min}/λ_2, and so on, until a point of collision, $\vec{X}_n$, is reached.

710

where ξ is a URV, $0 < \xi < 1$, and $\vec{v}_1 = \vec{v}/\|v\|$. Let $\rho = \lambda_{min}/\lambda_{tot}(\vec{x}_1)$, and test for a collision: Choose a new URV, ξ_I. If $\xi_I \leq \rho$ then a collision occurs. If $\xi_I > \rho$ then we say a pseudo-collision occurs and the procedure is repeated until a collision occurs.

The main computational work in tracking by the PCA is computing λ_{min} and λ. To compute λ_{min} quickly, one can construct a reference array of λ_{min} as a function of a set of neutral velocities before any flights are flown. This reduces computing λ_{min} to a table look-up and a linear interpolation.

Computing λ requires knowledge of local plasma temperatures and densities. If geometric zones of constant plasma density and temperature are used, the problem reduces to knowing the plasma zone where the test for collision occurs. This is done using a two-dimensional rectilinear reference array containing plasma zone indices. Indices of the array represent retangular plasma cells of dimension Δx by Δy. A point (x,y) then lies in the cell represented by the reference array

$$([x/\Delta x] + 1, [y/\Delta y] + 1)$$

where [x] = the greates integer less than or equal to x. This effectively rectilinearizes the plasma zone boundaries, thus the test flights actually travel through a rectilinearized approximation of the plasma zones. However, our scheme for computing the reference array is economical enough for sufficient resolution to be obtained by using a fine enough rectangular grid.

Note that it is easy to include continuously varying plasma parameters in the PCA calculation. In the direct integration method, the integral (Eq. 20) over optical depth was made with a step function integrand, and the simple backing up to a solution after the final step's overshoot worked because the plasma quantities were assumed to be constant over that last step. In the PCA, however, we can substitute the zone lookup described above by using polynomial fits, for example, for the plasma conditions, and use them directly to compute λ_{min} and λ as a function of position.

5.2. Test Flight Weighting

Each test flight is initially given a weight of $\omega = 1$. If a total source of J atoms/s is being modeled with N test flights, each test flight then represents $\gamma = J/N$ atoms/s at its start.

As the test flight progresses through a sequence of collisions the weight is reduced to account for attenuation by ionization. This can be done in two ways. The first is to continue the flight until the first ionization occurs, as determined for hydrogen, for example, by the ratio

$$\rho_{ion} = \frac{n_e <\sigma_e v> + n_i <\sigma_i v>}{n_i <\sigma_{cx} v> + n_i <\sigma_e v> + n_e <\sigma_i v>},$$

where σ_i is the ion-impact ionization cross-section. The weight is then set to zero, and the flight ends.

The second method is called suppressed absorption. Here ω is multiplied at each collision by the probability of charge-exchange, $\rho_{cx} = 1 - \rho_{ion}$. For example, after the first collision a test flight would represent $\rho_{cx}(\gamma\omega)$ atoms/sec, with $(1 - \rho_{cx})(\gamma\omega)$ atoms/sec having been ionized. This weight attenuation process continues until ω is less than some pre-assigned minimum weight, ω_{min}. At this point we choose between a pure charge-exchange in which case ω remains unchanged and the flight is continued, or a pure ionization, when ω is set to 0 and the flight is ended. Clearly the method of killing a flight at the first ionization is a special case of suppressed absorption, with $\omega_{min} = 1$.

Quantitatively, it can be shown that suppressed absorption can on the average reduce the standard deviations in a calculation for a fixed number of test flights by a factor

$$\left[\frac{\log(\omega_1/\omega_2)}{\omega_1/\omega_2} \right]^{\frac{1}{2}}.$$

One subtlety of a charge-exchange process is that its cross section varies over the Maxwellian distribution of ion velocities [Freeman and Jones, Cupini et al (1983B)]. It is enough to know $<\sigma_{cx}v>$ to determine the points of collision. However the choice of the velocity of the charge-exchanging ion must take into account the dependence of the reaction rate on the relative velocity between the ion and the neutral.

This variation can be included in the calculation by choosing a velocity, $\vec{v}_{new}$, from a Maxwellian at the local ion temperature for the atom created in a charge-exchange, and multiplying the test flight weight by

$$\kappa = \frac{\sigma_{cx}\|\vec{v}\|}{<\sigma_{cx}v>}$$

where σ_{cx} is the charge-exchange cross section, and $\vec{v} = \vec{v}_{new} - \vec{v}_{old}$, with v_{old} the incoming test flight velocity. On the average $\kappa = 1$.

5.3. Scoring

At a collision, a weight of $(1 - \rho_{cx})\omega$ is ionized. The sum

$$\tau_j = \sum(1 - \rho_{cx})\omega, \tag{21}$$

taken over all collisions in plasma zone j of all test flights, represents the ionization rate in zone j, in units of test flight weight. This is converted to S_j ionizations/cm^{-3} by multiplying by γ and dividing by the volume, vol_j, of zone j:

$$S_j = \gamma \cdot \tau_j / vol_j.$$

The charge-exchange ion power loss in zone j is equal to

$$\gamma \cdot \sum \rho_{cx}\omega(\kappa E^+ - E^0)/vol_j,$$

where the sum is over all collisions in zone j, κ is the charge-exchange weight correction coefficient described above, E^0 is the energy of the incoming neutral, and E^+ is the energy of the incoming ion. The power gained by ions due to ionization processes equals

$$\gamma \cdot \sum (1 - \rho_{cx}) \cdot \omega(E^0 - \rho_{ion}E^{ion})/vol_j,$$

where

$$\rho_{ion} = \frac{n_i <\sigma_i v>}{n_i <\sigma_i v> + n_e <\sigma_e v>}$$

is the ion impact ionization probability, and E^{ion} is the energy necessary for ionization.

The neutral density, n_{0_j}, in zone j, is derived from the relation

$$S_j = n_{0_j}(n_{e_j} <\sigma_e v> + n_{i_j} <\sigma_i v>),$$

where the only unknown is n_{0_j}. Note however that the rate coefficients for both electron and ion impact ionization depend on both the ion temperature, which is known, and the neutral energy (cf. Sec. 6). We can independently compute the average energy, $\overline{E_j^o}$, of neutrals ionizing in zone j is given by

$$\overline{E_j^o} = \gamma \cdot \sum (1 - \rho_{cx})\omega E^0/S_j, \qquad (22)$$

where the sum is taken as in Eq.(21), and E^0 is the neutral energy at a collision.

Equations (21) and (22) cannot be used in zones without plasma, because the sums are empty there. In plasmaless regions we posit a fictitious plasma of pseudo-ions, ψ, whose reaction

$$\psi + H^0 \rightarrow \psi + H^0, \qquad (23)$$

leaves H^0 unchanged in every way. Taking the sum

$$\tau_j^\psi = \sum \omega,$$

over all collisions with pseudo-ions in zone j, the neutral density, n_{0_j}, then obeys

$$S_j^\psi = \gamma\tau_j^\psi = n_{0_j}n_{\psi_j} <\sigma_\psi v>,$$

where n_{ψ_j} is the pseudo-ion density in zone j, and σ_ψ is the cross section of reaction (23). We are free to choose the value of $n_{\psi_j} <\sigma_\psi v>$, but we must do so carefully. Too large a value will make λ_{min} too small for an efficient use of PCA, and too small a value will result in too few collisions with pseudo-ions.

5.4. Variance Reduction Techniques

In Sec. 5.2 the method of suppressed absorption was discussed. We now describe some other widely used variance reduction techniques.

The first is the method of geometric splitting [Hammersley and Handscomb], which is used to amplify the number of scorings in selected regions. The entire geometry is divided into areas which are given weights corresponding roughly to the inverse of the expected probability that a test flight would reach them. For example, zones which are small in area, or isolated, or farthest away from the start of a test flight would be given higher weights than larger zones or zones closer to the initial flight position. Flights which go from regions of lower weight into regions of higher weight are then split as follows by a predetermined factor $\varrho > 1$: If ϱ is an integer, then the original test flight is replaced by ϱ flights, each which a Monte Carlo weight $1/\varrho$ that of the orginal. The offspring flights then proceed independently, increasing the number of scorings in the more highly weighted region.

More generally, in order to increase efficiency, the ascribing of weights to areas may be done according to the relative importance given to the areas. Scorings may be increased through splitting in regions which would catch many flights even without splitting, if they are of the most interest.

Typically ϱ is an integer, equal to 2 or 3. Different regions can be given different values of ϱ. Noninteger values of ϱ can also used. Splitting, for example, on the average 60% of the flights two ways and 40% three ways results in a $\varrho = 2.4$.

Geometric splitting is a powerful method of focusing a Monte Carlo algorithm at selected regions of the problem. However, it is most efficiently used when combined with a method for killing off flights coming from a highly weighted region into a region of less interest. Such a method is picturesquely known as Russian Roulette. Flights entering a region of lower importance are killed off, by choosing random numbers, with a probability of $1 - 1/\varrho$. Without something like Russian Roulette, the split flights will eventually wander into regions of lesser interest, reducing the calculation's efficiency.

Another variance reduction technique useful when tracking using the PCA, is to make scorings not just at each collision but at each pseudo-collision. For example, the sum in Eq. (21) is replaced by the sum over all pseudo-collisions in zone j,

$$\tau'_j = \sum \frac{\lambda_{min}}{\lambda} \cdot (1 - \rho_{cx})\omega, \qquad (24)$$

where λ is the mean free path length at the point of pseudo-collision. On the average $\tau_j' = \tau_j$, but the sum in Eq. (24) has more terms than the sum in Eq. (21), reducing in many cases the variance of the calculation for a fixed CPU time.

5.5. A Storage Algorithm for Wall Reflection Data

As mentioned in Sec. 2.1, the wall reflection model in an ideal neutral transport calculation would supply values for the reflection coefficients, backscattered velocity distributions, and sputtering yields for particles incident with a given energy and angle on a surface of arbitrary roughness and composition.

Three approaches to this ideal model are:

(1.) Using simple energy and angle dependent expressions such as the one found in [Heifetz et al(1982)] after [Seki et al]:

$$
R_N(E_{in}, \theta_{in}) = \begin{cases} -0.237 \ell n(E_{in}/2990.0) + 0.19 & \theta_{in} \leq 40° \\ 1 & \theta_{in} > 40° \end{cases} .
$$

This approach is computationally fast but simplistic; in particular it does not fully exploit the power of the Monte Carlo method, where sampling from complicated distributions can be easily included. However analytic treatments of neutral transport would benefit from realistic approximations expressed in closed form.

(2.) Using reflection simulation codes such as TRIM [Biersack and Haggmark] and MARLOWE [Robinson] to create data sets of reflection coefficients and angular distributions for use in Monte Carlo codes. The drawbacks to this method are: (a) the more distributions sampled in a Monte Carlo calculation the larger, in general, the standard deviations for a given number of test flights, (b) additional work in linearly interpolating the distributions is required, and (c) statistical inaccuracies in the original data (both TRIM and MARLOWE themselves use Monte Carlo algorithms) will be forever present in the data compilation.

(3.) Incorporating a reflection code, such as TRIM, directly into a Monte Carlo neutral transport code. This avoids problems (b) and (c) of method (2). An additional advantage is that the projectile species and wall structure can be easily varied. However difficulty (a) is still present, and the greatest drawback is now that the combined codes may be too costly to run.

At this time the second approach seems best for Monte Carlo calculations, being a compromise between code running efficiency and the inclusion of realistic data. As computers become faster in the next few years, however, the third approach may become more practical and favorable.

We now describe an implementation of the second approach which we have found useful. As stated in Sec. 2.2, existing wall reflection data is almost exclusively for smooth, clean walls, whereas walls in tokamaks are not smooth and not clean. One approach to deal with this gap in our knowledge is to use the existing reflection data, together with a roughness factor, ε, equal to the number of additional reflections a particle would have to undergo before leaving the wall. This factor would increase with greater roughness. In a Monte Carlo algorithm a test flight would be tested for reflection ε times making the effective reflection coefficient roughly equal to the measured one to the power ε. This reduces the problem to obtaining accurate and complete reflection data for smooth walls.

In general, for each projectile/target combination, experiment or computation will give a five-dimensional differential scattering distribution

$$P(E, \alpha, v, \theta, \phi) v^2 \, dv \, \sin\theta \, d\theta \, d\phi$$

where E and α are the incident energy and polar angle respectively, and v, θ, and ϕ are the velocity, polar, and azimuthal angles of the reflected particle, in that order (Fig. 5)[Behrisch and Eckstein]. (The azimuthal angle is measured from the plane of the incident trajectory).

To each incident E and α is associated the three-dimensional conditional distribution

$$P_{E,\alpha}(v, \theta, \phi) \, v^2 \, dv \, \sin\theta \, d\theta \, d\phi$$

We now describe an efficient method which can be conveniently used in Monte Carlo models for sampling from any multi-dimentional distribution [Bateman]. This method is fast and economical in storage, and avoids the pitfall of using the rejection method for this distribution, which can be highly peaked in certain regions.

The algorithm reduces sampling $P_{E,\alpha} v^2 \, dv \, \sin\theta \, d\theta \, d\phi$ to sampling consecutively from three one-dimensional distributions as follows. First v_0 is chosen from

$$f_{E,\alpha}^1(v) = \int \int P_{E,\alpha}(v, \theta, \phi) \, \sin\theta \, d\theta \, d\phi.$$

Given v_0 then θ_0 is picked from

$$f_{E,\alpha}^2(\theta) = \int P_{E,\alpha}(v_0, \theta, \phi) \, d\phi,$$

and finally, with v_0 and θ_0 known, ϕ_0 can be chosen from the distribution

$$f_{E,\alpha}^3(\phi) = P_{E,\alpha}(v_0, \theta_0, \phi),$$

716

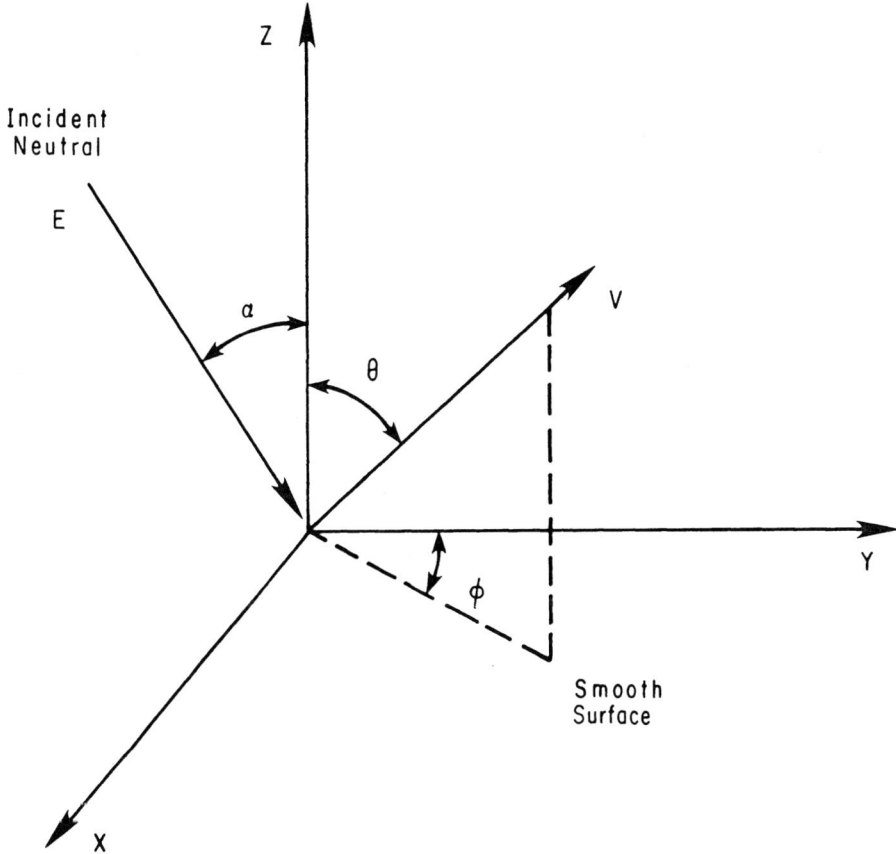

Fig. 5. Coordinate system of the wall reflection scattering distribution [Heifetz et al (1982)].

where $P_{E,\alpha}$ has been suitably normalized.

The problem now simplifies to sampling a one-dimensional distribution f. Assuming that $\int_0^a f(y)\,dy = 1$, this is done by computing the inverse, $F(\varpi) = G^{-1}(\varpi)$ of the cumulative distribution, G, of f:

$$G(x) = \varpi = \int_0^x f(y)\,dy.$$

Hence a random number ϖ between 0 and 1 corresponds by f to the point x.

Suppose that $F_{E,\alpha}^1(\xi)$, $F_{E,\alpha}^2(\eta, \xi)$, and $F_{E,\alpha}^3(\varsigma, \eta, \xi)$ are the inverse cumulative distributions of $f_{E,\alpha}^1$, $f_{E,\alpha}^2$, and $f_{E,\alpha}^3$ respectively. Choose three random numbers ς_0, ν_0, and ξ_0 all between 0 and 1. Then the corresponding velocity of the reflected particle is $F_{E,\alpha}^1(\xi_0)$, the polar angle of reflection is $F_{E,\alpha}^2(\nu_0, \xi_0)$, and the azimuthal angle of reflection is $F_{E,\alpha}^3(\varsigma_0, \nu_0, \xi_0)$.

We may conveniently tabulate the inverse distributions F by storing the values of F for a few values of ϖ, say for $\varpi = 0.1, 0.3, 0.5, 0.7$, and 0.9, and approximating F elesewhere by linearly interpolation. By computing F^1, F^2, and F^3 for a small number of incident energies, say $E=50, 100, 200, 500$, and 1000 eV, and incident polar angle, say $\alpha = 0, 20, 40, 60$, and 80 degrees, the total computer storage needed for F^1 would be $5^3 = 125$ words, $5^4 = 625$ words for F^2, and $5^5 = 3125$ words for F^3.

Table 1 gives a set of values F^1, F^2, and F^3 for $E = 200$ eV and $\alpha = 0°$. Suppose, for example, that a reflection occurs, and that, for simplicity, the three choices of random numbers between 0 and 1 are $\xi = 0.3$, $\eta = 0.5$, and $\varsigma = 0.9$. Then using Table 1, $v = F^1(2) = 100$ eV, $\cos\theta = F^2(3,2) = 0.6497$, and $\cos\phi = F^3(5,3,2) = -0.5448$. (It is more convenient to store the directional cosines $\cos\theta$ and $\cos\phi$ instead of θ and ϕ.)

5.6. Monte Carlo Algorithms on Vectorizing or Multi-Processing Computers

The measure of confidence in a result from a Monte Carlo calculation is the standard deviation of that result divided by the $\sqrt{N}$, where N is the number of test flights. Algorithmic methods for minimizing the standard deviations of results from individual test flights were described in Secs. 5.2 and 5.4. In this section we will describe some approaches to reducing the computer running time needed to track each flight for a given Monte Carlo tracking/scoring algorithm. These approaches are dependent on the computer hardware.

Efforts are being made, for example under M. Kalos at New York University, to design computer hardware especially suited for Monte Carlo calculations. It is not likely, however, that the demand for computers with such a specialized architecture will be great enough for them to become commercially viable and widely available. However two new computer architectures

718

Table 1

Results from the MARLOWE code [Robinson et al] for an incident energy of 300 eV and polar angle of 0° [Heifetz et al (1982)]

$$E = F^1(\xi)\,(\text{eV})$$

57.3	100.	127.5	150.8	168.5

$$\cos\theta = F^2(\eta,\xi)$$

57.3	100.	127.5	150.8	168.5
0.5104	0.6931	0.8096	0.8926	0.9672
0.4843	0.6817	0.8009	0.8869	0.9654
0.4753	0.6497	0.7727	0.8681	0.9609
0.4646	0.6475	0.7642	0.8642	0.9522
0.3343	0.5088	0.6313	0.7533	0.8867

$$\cos\psi = F^3(\zeta,\eta,\xi)$$
$$F^3(1,\eta,\xi)$$

57.3	100.	127.5	150.8	168.5
-0.9460	-0.5096	-0.0494	0.5299	0.9293
-0.9224	-0.5758	0.0270	0.5498	0.9519
-0.9481	-0.6291	0.0147	0.5343	0.9507
-0.9501	-0.5564	0.0380	0.5548	0.9442
-0.9594	-0.5536	0.1303	0.6682	0.9567

$$F^3(2,\eta,\xi)$$

57.3	100.	127.5	150.8	168.5
-0.9510	-0.6303	-0.1349	0.5591	0.9501
-0.9448	-0.6107	-0.0077	0.4643	0.9378
-0.9362	-0.4654	0.0031	0.5703	0.9513
-0.9322	-0.5519	-0.0030	0.5741	0.9734
-0.9388	-0.5297	0.1150	0.5879	0.9400

$$F^3(3,\eta,\xi)$$

57.3	100.	127.5	150.8	168.5
-0.9308	-0.5249	0.1161	0.6025	0.9511
-0.9346	-0.6454	-0.0176	0.4700	0.9325
-0.9448	-0.5862	-0.0408	0.6021	0.9694
-0.9341	-0.5364	-0.0625	0.5133	0.9494
-0.9474	-0.5254	-0.0254	0.5714	0.9646

$$F^3(4,\eta,\xi)$$

57.3	100.	127.5	150.8	168.5
-0.9095	-0.5492	0.0054	0.5960	0.9142
-0.9395	-0.6243	-0.0676	0.6596	0.9630
-0.9670	-0.5826	-0.0809	0.5839	0.9256
-0.9401	-0.5702	0.0194	0.5559	0.9450
-0.9376	-0.4852	0.0458	0.6537	0.9589

$$F^3(5,\eta,\xi)$$

57.3	100.	127.5	150.8	168.5
-0.9343	-0.6085	-0.0343	0.5889	0.9491
-0.9199	-0.5429	0.0072	0.6042	0.9324
-0.9343	-0.5448	0.0822	0.6140	0.9716
-0.9217	-0.4109	0.1232	0.6503	0.9696
-0.9243	-0.4775	0.0642	0.6172	0.9532

have been introduced commercially in the past few years designed for general large scale scientific calculations: pipeline/chaining computers, such as the American CRAY-1 and CDC Cyber 205, and the Japanese Fujitsu M-380 and NAS 9060 machines, and multi-processing computers, such as the CRAY X-MP, CRAY-2 and the Fujitsu VP-200.

Pipeline/chaining, popularly known as vectorizing, machines such as the CRAY computers use banks of registers to make the same calculation on sets of numbers using an assembly-line architecture in and between their arithmetic units [Russell]. This architecture is optimal for computations which perform the same numerical operations repeatedly.

Monte Carlo tracking schemes, as described in Sec. 5.1, however, use many logical tests as they branch through the set of all possible flight paths. Thus, in order to take advantage of vectorizing one must reorgranize the calculation somewhat. A flow chart of one Monte Carlo code which has been vectorized is given in Fig. 6. This program, called MONET, solves Eq. 14 by evaluating G by Monte Carlo methods using pseudo-collision tracking (Sec. 5.1) , while including the effects of wall reflection. It tracks each test flight only until its first charge-exchange, so that the test-sequence for following a flight is relatively short.

The basic idea used in vectorizing MONET is to follow 128 test flights together. (There is nothing magic in the number 128; each vector register on the CRAY-1 holds 64 words, and so it would probably be more appropriate to say that MONET follows two vector registers of test flights together.) The 128 initial positions and velocities are set in one loop, and subsequent loops handle pseudo-collision steps, make wall reflections, and so on. All but one of these loops are vectorized. The bookkeeping necessary to compress arrays of flight information after flights finish and leave the flight loop is confined to one loop which is unvectorized in the present version. Using the gather/scatter hardware and software techniques which have recently become available, however, this loop could also be vectorized.

The vectorized loops contain all of the operations which are expensive to compute, such as exponentials and trigonometric funcions. Thus the vectorized version of MONET runs in one-third of the time of the unvectorized version. The principal of tracking test flights in loops can be and has been applied to other Monte Carlo algorithms which involve longer, more complicated test flight histories.

Vectorizing involves a localized processing of events in parallel. A computer architecture which is more immediately applicable to Monte Carlo calculations is one containing multiple processors, each using a nonoverlapping copy of the flight tracking subprogram to independently track the flights (Fig. 7). This applies, of course, only to linear calculations. Thus on a

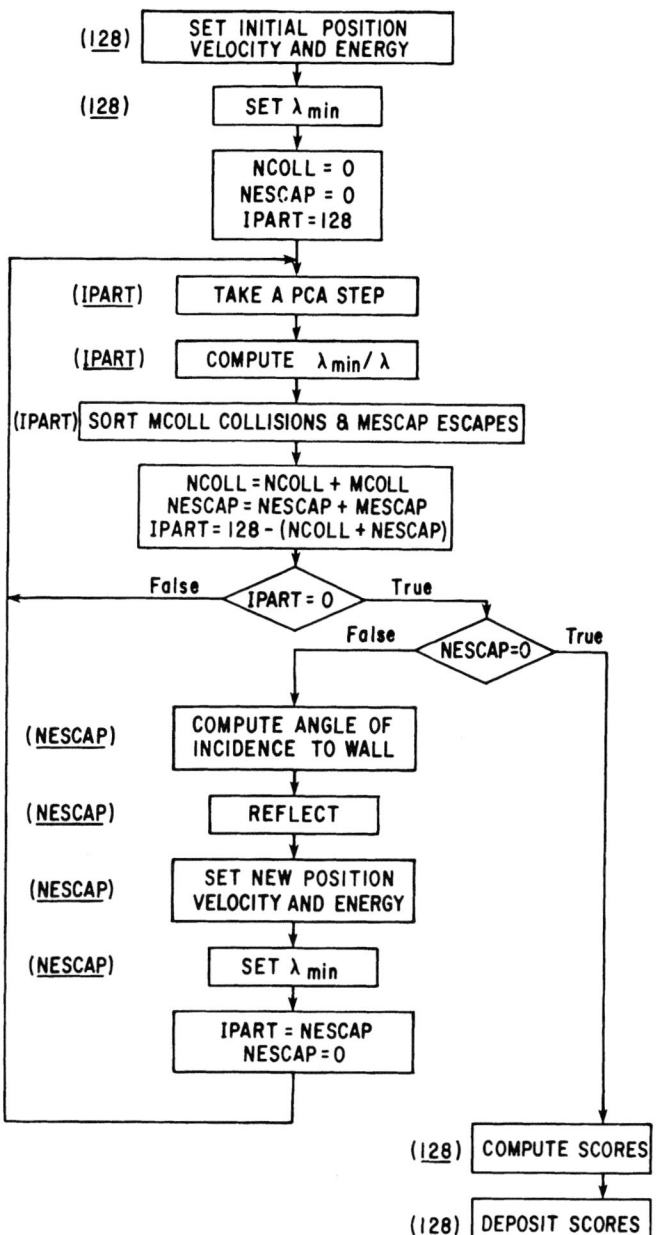

Fig. 6. Flow chart of the MONET code. This Monte Carlo algorithm has been vectorized on the CRAY-1 computer by dividing the algorithm into loops over at most 128 test flights. The loop sizes are given in parentheses. Loops whose sizes are underlined are vectorized.

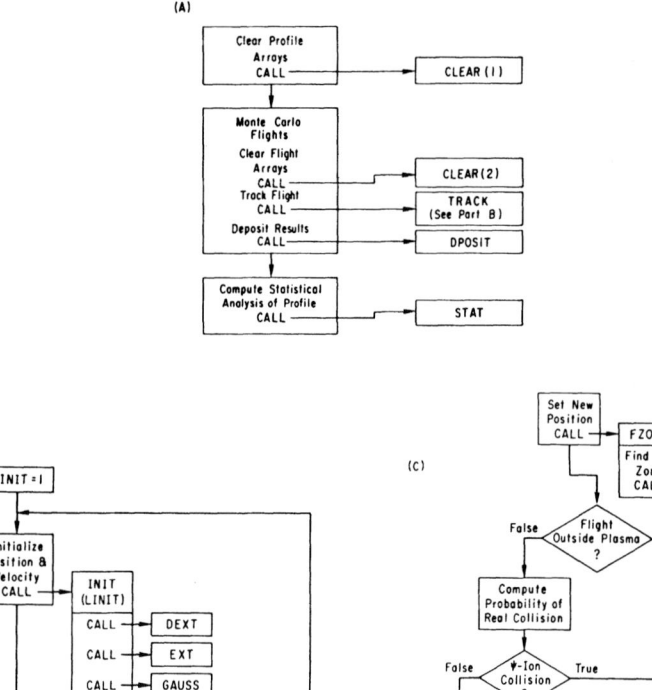

Fig. 7. Flow chart of the PCA tracking algorithm and scoring method in the SEURAT code [Heifetz and Post]. This code can exploit the architecture of a multiprocessing computer by dividing the loop (a) of Monte Carlo flights into pieces, with each processor computing part of the loop using the re-entrant subroutine TRACK (b) and its subordinate routines (c).

computer with N processors, the flight tracking loop would be divided into N smaller loops, reducing the clock-time required for the calculation by as much as $1/N$.

6. Sample Applications

To appreciate the flavor of neutral transport modeling, it would be useful for the interested reader to review recent papers devoted to a particular aspect of neutral transport, for example the modeling of divertor recycling in ASDEX [Schneider et al], PDX [Petravic et al (1984A)], DITE [Fielding et al], and D-III [Shimada et al, Petravic et al (1984B)], modeling of the behavior of the ISX pump limiter [Evans et al], the calculation of tritium inventories in TFTR [Baskes et al], the modeling of proposed experiment designs, such as the pump limiter ALT-I [Boley et al] computation of neutral transport in ASDEX-Upgrade [Vernickel et al], and the proposed reactor divertors and pump limiter for TFCX/INTOR [Pet et al (1984A), Boley].

In this section we describe recent work using the three-dimensional Monte Carlo DEGAS [Heifetz et al (1982)] which for the most part has not been published elsewhere. The DEGAS code includes the following pieces from the ideal model described in Sec. 2:

1. The set of neutral-plasma reactions listed in Table 2 for hydrogenic atoms and molecules, and electron impact ionization of four impurities, atomic helium, copper, iron, and argon. Cross sections for these reactions are taken from the data and formulas in [Janev et al(1985), Weisheit]. Neutral-ion reaction rates are computed as functions of ion temperature, density, and neutral energy (Fig. 8). The model of electron impact ionization of hydrogen includes the effects of multi-step ionization, a function of both electron density and temperature (Fig. 9). Reaction rates for these reactions are taken from the data and formulas in [Janev et al(1985), Weisheit].

2. No neutral-neutral or neutral-ion elastic scattering processes.

3. Reflection models for hydrogen and helium atoms onto various targets using both experimental data and the results of numerical simulation from: (a) the TRIM [Biersack and Haggmark] code for normal particle incidence and incident energy, E_{in}, greater than 100 eV onto C, Si, Fe, W, and Au [Eckstein and Verbeek], (b) empirical fits for normal incidence and $E_{in} > 1$ keV onto Be, C, Al, Si, V, Fe, Cu, Mo, W, and Au from [Tabata et al], (c) theoretical fits for $E_{in} < 100\,\mathrm{eV}$ from [Singer et al (1985), Baskes] onto Fe and W, and (d) the MARLOWE code [Robinson

Table 2
The Dominant Neutral/Plasma Reactions
Included in the DEGAS Code

Charge-exchange (H is any hydrogenic species):

$$H^0 + H^+ \rightarrow H^+ + H^0$$

$$H_2^0 + H_2^+ \rightarrow H_2^+ + H_2^0$$

$$He^0 + He^{+(+)} \rightarrow He^{+(+)} + He^0$$

$$He^0 + H^+ \rightarrow He^+ + H^0$$

$$H_2^0 + H^+ \rightarrow H_2^+ + H^0$$

Electron-impact ionization:

$$e^- + H^0 \rightarrow H^+ + 2e^-$$

$$e^- + He^0 \rightarrow He^+ + 2e^-$$

$$e^- + H_2^0 \rightarrow H_2^+ + 2e^-$$

Electron dissociation:

$$e^- + H_2^0 \rightarrow 2H^0 + e^-$$

$$e^- + H_2^0 \rightarrow H^0 + H^+ + 2e^-$$

$$e^- + H_2^+ \rightarrow 2H^0$$

$$e^- + H_2^+ \rightarrow H^0 + H^+ + e^-$$

$$e^- + H_2^+ \rightarrow 2H^+ + 2e^-$$

Recombination:
$$H^+ + e^- \rightarrow H^0$$

Ion-impact ionization:

$$H^0 + H^+ \rightarrow 2H^+ + e^-$$

$$He^0 + H^+ \rightarrow He^+ + H^+ + e^-$$

$$H_2^0 + H^+ \rightarrow H_2^+ + H^+ + e^-$$

$$H^0 + He^{+(+)} \rightarrow H^+ + He^{+(+)} + e^-$$

$$He^0 + He^{+(+)} \rightarrow He^+ + He^{+(+)} + e^-$$

$$H_2^0 + He^{+(+)} \rightarrow H_2^0 + He^{+(+)} + e^-$$

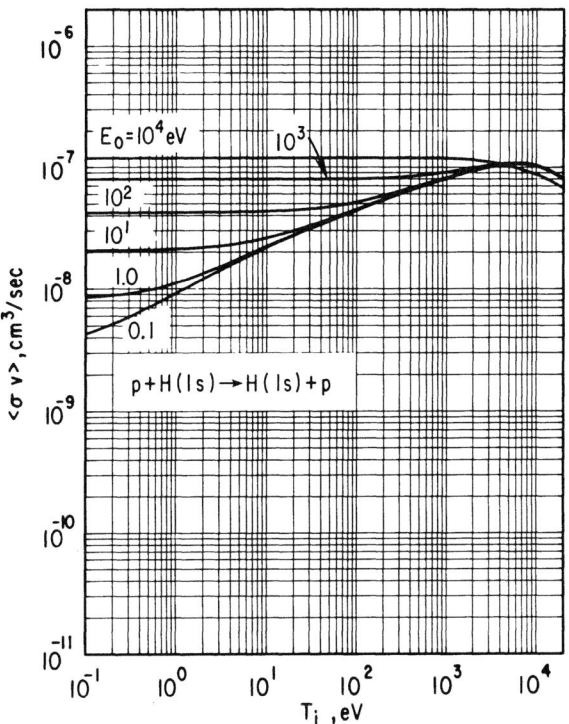

Fig. 8. Rate coefficient for $II^0 + II^+$ charge-exchange as a function of ion temperature, T_i and neutral energy, E_0 [Janev et al(1985)].

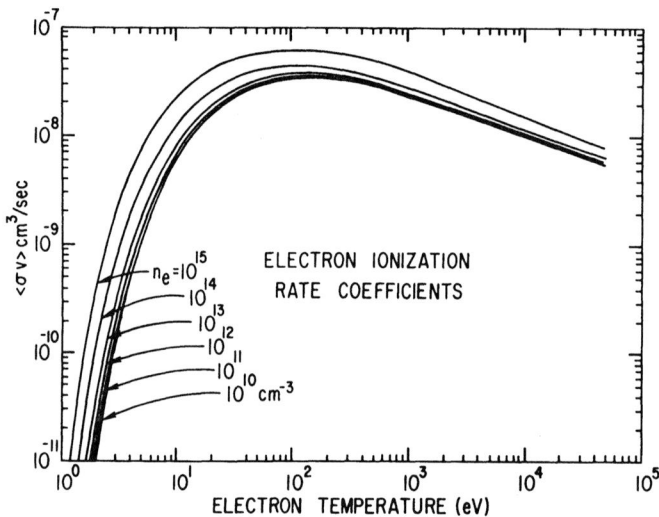

Fig. 9. Rate coefficient for electron impact ionization of hydrogen, including excitation and de-excitation processes. [Weisheit, Janev et al(1984)].

et al] and TRIM for H, D, T, and He onto Fe, for arbitrary incident angles and E_{in} from 10 to 1000 eV. A simplified time independent wall desorption model is used which assumes that atoms and molecules which stick to the wall desorb immediately as molecules (hydrogen species), or atoms (all other species) at the wall temperature.

4. Impact sputtering is computed as a function of target species and incident particle species, energy, and polar angle, using the empirical fitting routine DSPUT [Smith et al].

5. Fully three-dimensional geometries.

DEGAS uses the pseudo-collision tracking method described in Sec. 5. It has been combined with a number of plasma codes, including a one-dimensional one-fluid model [Post and Lackner], the SOLID one-dimensional, two-fluid code [Schneider and Lackner], the PLANET two-dimensional, two-fluid code [Petravic et al (1982)], and the BALDUR one-dimensional radial transport code [Singer et al (1985B)].

6.1. Radial Plasma Profiles in PDX Scoop-Limited Discharges

Most flux-averaged toroidal plasma transport codes compute as a sub-calculation the neutral density and temperature as a function of flux surface, in order to determine summands, S_F^o, in the source terms, S_F, for their transport equations:

$$\frac{\partial F}{\partial t} = \nabla \Gamma_F + S_F.$$

Here F is the particle, n, or energy, E, density. In each zone the ion source rate, $S_{n_s}^o$, for ion species s equals

$$S_{n_s}^e + S_{n_s}^i + S_{n_s}^{c-x} + S_{n_s}^{md},$$

where

$$S_{n_s}^e = n_s^o n_e <\sigma_e> \qquad \text{(electron impact ionization)}$$

$$S_{n_s}^i = n_s^o \Big(\sum_{\substack{ion \\ species \ t}} n_t <\sigma_i> \Big) \qquad \text{(ion impact ionization)}$$

$$S_{n_s}^{c-x} = n_s^o \Big(\sum_{\substack{ion \\ species \ t}} n_t <\sigma_{c-x}> \Big) \qquad \text{(charge - exchange)}$$

and $S_{n_s}^{md}$ is the ion creation rate from all the molecular dissociation processes (Fig. 10). The ion energy source, $S_{E_s}^o$ equals,

$$S_{E_s}^i + S_{E_s}^{ion} + S_{E_s}^{c-x} + S_{E_s}^{md},$$

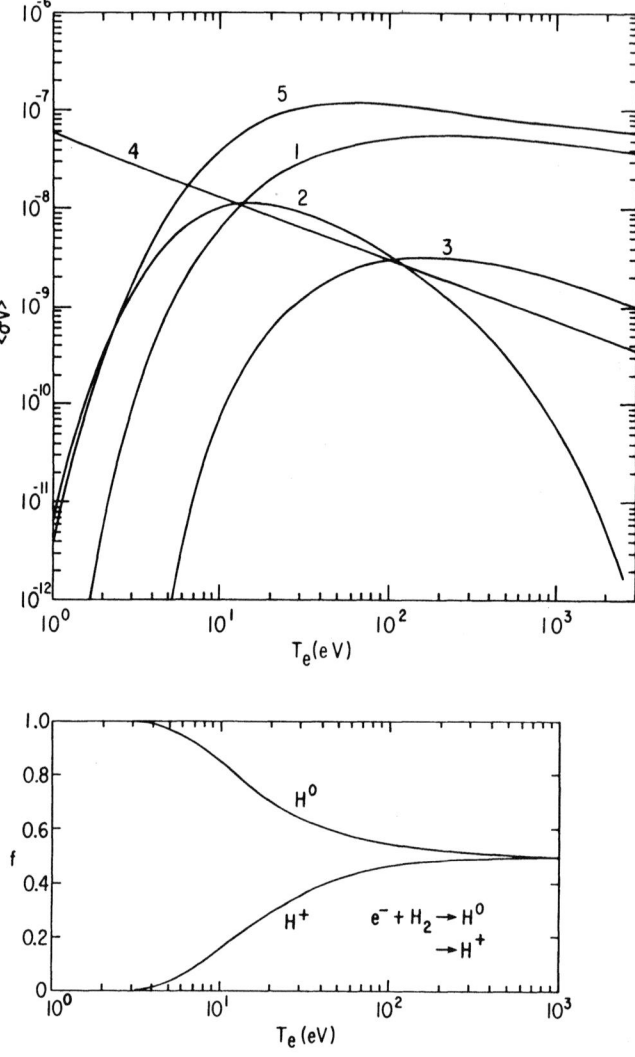

Fig. 10.a) Rate coefficients as functions of electron temperature, T_e, for the dominant H_2^0 dissociation processes:(1) $e^- + H_2^0 \rightarrow H_2^+ + 2e^-$, (2) $e^- + H_2^0 \rightarrow 2H^0 + e^-$, (3) $e^- + H_2^0 \rightarrow H^0 + H^+ + 2e^-$, (4) $e^- + H_2^+ \rightarrow 2H^0$, and (5) $e^- + H_2^+ \rightarrow H^0 + H^+ + e^-$ [Freeman and Jones, Jones] . The average H^0 and H^+ yields due to these processes are plotted in (b) [Post et al(1982)] . For $T_e < 5\,eV$ the rate coefficient for reaction (2) is larger than the coefficients for the competing $e^- + H_2^0$ reactions (1) and (3) hence the yield fraction there of H^0 approaches 1.0. Above $T_e = 30 - 40\,eV$ the rate coefficient of reaction (1) becomes larger than those for reactions (2) and (3), and at those temperatures, the dominant $e^- + H_0^+$ reaction is (5). Hence the yield fraction for H^0 for $T_e > 40\,eV$ is approximately 0.5.

728

where

$$S^i_{E_s} = -E^{ion}n_s\Big(\sum_{\substack{neutral \\ species\ t}} n^o_t <\sigma_i>\Big) \quad \text{(energy lost ionizing)}$$

$$S^{ion}_{E_s} = E^o_s(S^e_{n_s} + S^i_{n_s}) \quad \text{(energy gained from ionizing neutrals)}$$

$$S^{c-x}_{E_s} = n^o_s E^o_s\Big(\sum_{\substack{ion \\ species\ t}} n_t <\sigma_{c-x}>\Big) - \frac{3}{2}n_s\Big(\sum_{\substack{neutral \\ species\ t}} n^o_t T_t <\sigma_{c-x}>\Big)$$

$$\text{(charge} - \text{exchange)}$$

and $S^{md}_{E_s}$ is the energy gained from the products of molecular dissociation. Here E^{ion} is the ionization potential of species s, and E^o_s is the average energy of an atom of species s. Finally the electron energy source, $S^o_{E_e}$, equals

$$S^e_e + S^{ex}_e + S^{md}_e,$$

where

$$S^e_e = -E^{ion}\Big(\sum_{\substack{ion \\ species\ t}} S^e_{n_t}\Big) \quad \text{(energy lost ionizing)},$$

S^{ex}_e is the energy lost due to electron excitation (Fig. 11), and S^{md}_e is the energy lost from the process of molecular dissociation.

One widely used routine for cirular concentric flux surface geometries is ANTIC [Tamor], which uses the numerical method described in Sec. 4.2. ANTIC is fast enough to be used in transport codes run on VAX-sized computers. The Monte Carlo code SEURAT [Heifetz and Post], models general two-dimensional geometries, but requires a more powerful computer such as the CRAY-I to be used on a production basis.

The question in deciding which method should be used is whether the added expense of a more complete model is necessary in order to effectively answer the relevant questions. That is, how sensitive are the results for neutral density and temperature, and hence the source terms, to the atomic physics, wall reflection, and geometric models used?

In this section we give an example showing the importance of geometry to the calculation. The DEGAS code was used together with the BALDUR code to model both scoop and rail limited discharges in PDX [Heifetz and Budny].

The PDX scoop consisted of a broad 30×40 cm graphite limiter blade backed by a plenum for the collection of particles, mounted on the outside

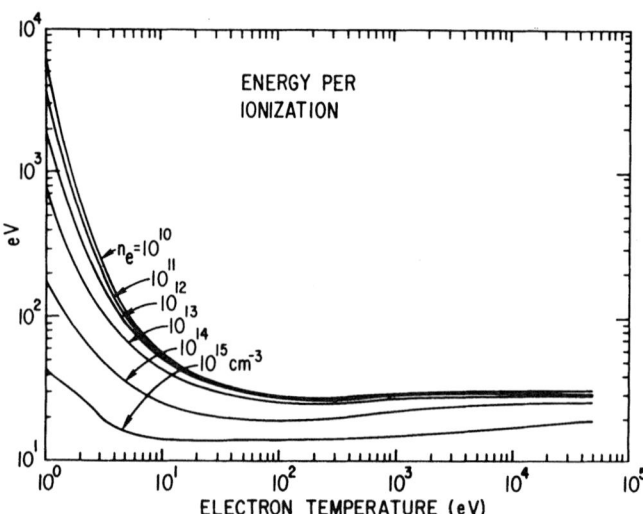

Fig. 11. The average electron energy loss for each electron-impact ionization of hydrogen including excitation/de-excitation processes, as a function
of electron temperature and density [Weisheit, Janev et al(1984)].

midplane of the machine [Budny et al.(1984A). It was the primary limiter for a brief experiment in 1983 with up to 5 MW of neutral beam heating.

Interest in modeling scoop discharges came from a comparison of results from scoop limited, rail limited, and diverted beam heated discharges. As reported in [Budny et al(1984B)], the energy confinement times, τ_E, during beam heating were 40 ms for the scoop and 50 ms for H-mode diverted discharges, as opposed to only 20 ms for rail limited discharges. Thus the scoop appeared to significantly increase τ_E above previous results for limiters, almost to the level of H-mode values.

A number of possible causes for this increase were considered using the BALDUR code and its standard one-dimensional Monte Carlo calculation, AURORA, for the neutral transport [Hughes and Post]. In order to include in the calculation any geometric effects on neutral recycling due to the position and large limiter face of the scoop it was decided to run this Monte Carlo treatment concurrently with DEGAS calculations. We note as an aside that the BALDUR+DEGAS combination took only twice as long as the standard calculation, eight CRAY-1 CPU minutes versus four minutes. Thus it could be functionally used in a production mode.

The model geometry for the neutral transport calculation is shown in Fig. 12. While the scoop was not described in full three-dimensions the three-dimensional features most important to the neutral transport calculation were included. The scoop face had the correct dimensions. The scoop channel was also included, together with a section representing the sides of the scoop box.

A plot of the toroidally and poloidally averaged ionization rates due to each neutral source 200 ms into the beam heating phase is also shown in Fig. 12. It shows that almost all the recycling occurs at the minor radius of the scoop's face. The complex Monte Carlo computation of particle recycling could almost have been replaced in the plasma transport calculation by a delta function there. This contrasted with results from the standard BALDUR calculation, where the curve for reionization had a much broader peak.

One consequence of this spike in reionization is shown in Fig. 13, where the radial plasma electron density is plotted as a function of time. The density profile computed with DEGAS is flatter than that computed by AURORA, as much as 20% lower at the center and 20% higher at a minor radius of 35 cm during beam heating. One can also get such a result by adjusting diffusion coefficients and anomalous pinch velocities, which are poorly known anyway. However the lesson to be learned here is that a good neutral calculation alone can make for significant effects in time dependent plasma transport calculations.

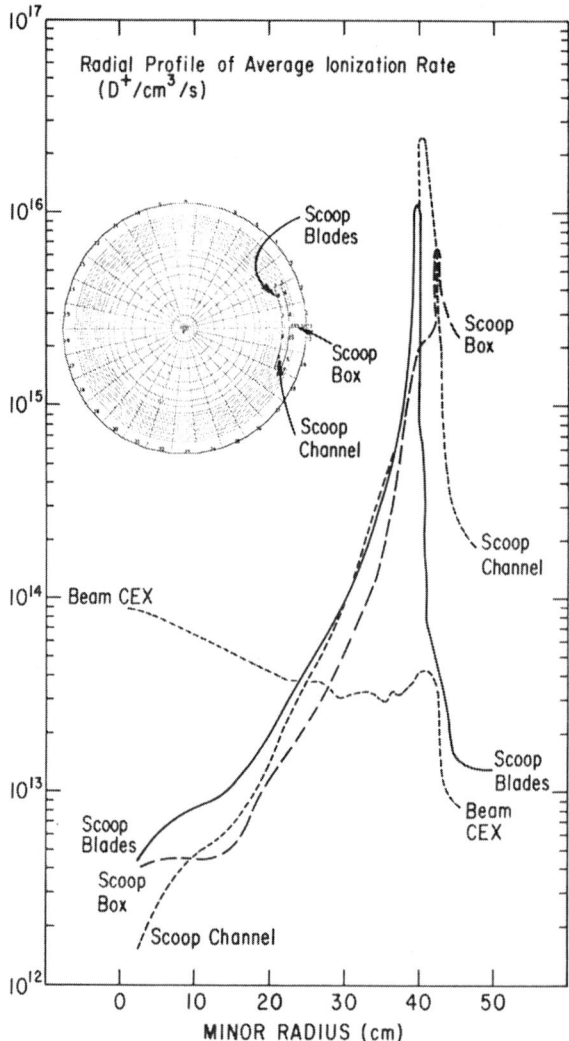

Fig. 12. Model geometry for the PDX scoop-limiter experiment (inset) and poloidally and toroidally averaged ionization rates due to various neutral sources. The model scoop was of the correct toroidal extent. A recombination source (not shown) was also included. The neutral density due to it was about one-tenth that due to beam charge-exchange.

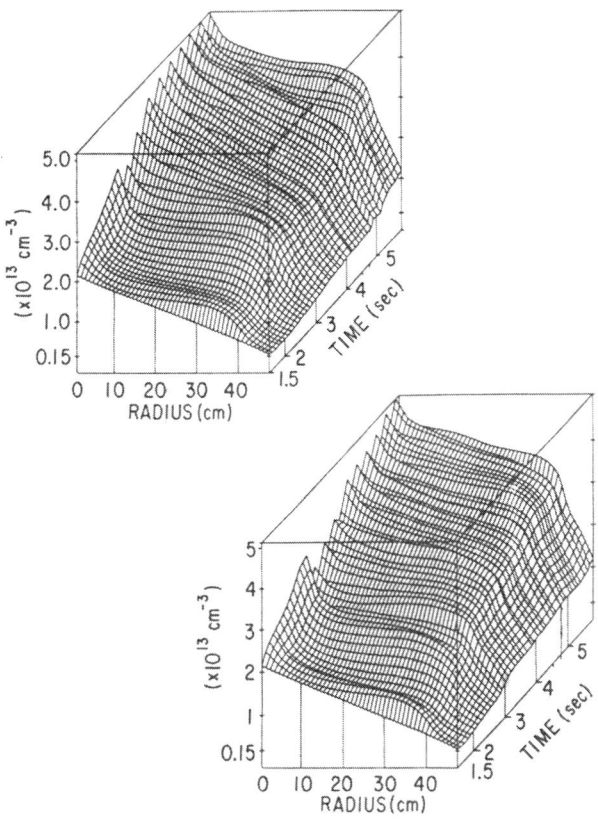

Fig. 13. Radial versus time plots of ion density as computed by the BALDUR code [Singer et al] for a PDX scoop-limiter discharge. Neutral transport was computed using both the AURORA code [Hughes and Post] (above), and the DEGAS code [Heifetz et al (1982)] (below). The effect of the more realistic treatment of the geometry in the DEGAS code is seen in the flattening of the radial density profiles.

No strong conclusions can be made yet as to what caused the "S-Mode," as the beam-heated scoop discharges have come to be known. Modeling with the combination code is still being done, including a repeat of the calculations for a rail limiter configuration, a repeat using a realistic description of the vacuum vessel geometry to test the effect on recycling of shifting the plasma major radius, and a comparison of computed three-dimensional neutral particle distributions with spatially resolved measurements.

6.2. Pumping Performance of the PDX Zr-Al Getters

Two arrays of Zr-Al alloy bulk getters were installed adjacent to the PDX outer moveable limiter in 1982 to affect the pumping of neutrals formed near the limiter (Fig. 14) [Cecchi et al]. The arrays had a projected area of 0.8% of the plasma surface, and had a measured H_2 pumping speed of 16,000 liters/sec.

Data on the performance of the limiter/getter system were obtained from circular ohmically heated discharges, with constant minor radii of 41 cm, and major radii varying from 143 cm to 154 cm. Thus the outer limiter radius, R_L, varied from 184 cm to 195 cm (Fig. 15). The line average density was feedback controlled to 1.5×10^{13} cm^{-3}. At t=0.4 sec the gas feed was terminated to facilitate the measurement of the density decay time.

It was found that the density exponential decay time at t=0.4 sec,

$$\tau_p^* = \frac{n_e}{dn_e/dt} \Big|_{t=0.4\,sec}$$

had little dependence on R_L for $184 < R_L < 189$ cm, and decreased by a factor of approximately 1.9 as R_L increased from 189 cm to 195 cm. If β_R is the recycling coefficient, then we can relate τ_p^* to τ_p, the global plasma particle lifetime, by

$$\tau_p^* = \frac{\tau_p}{1 - \beta_R}.$$

Thus $1 - \beta_R$ increased by a factor of 1.9 as R_L increased from 184 to 195 cm.

We used the DEGAS code to simulate the performance of the getter system. The principal question was to understand where the neutrals that were pumped came from. At first it was thought that the limiter was the only source. However it alone could not account for the behavior of the pumping rate as a function of limiter major radius. A second possible source of neutrals was from plasma striking the support structure of the getter panels. It was necessary to add this source to the calculation to match the experimental results.

The plasma parameters were taken from the experiment, with the values in the edge as reported in [Budny et al]. Two neutral sources were

Fig. 14. Photograph of the PDX Zr-Al getter arrays installed adjacent to the outer moveable limiter [Cecchi et al].

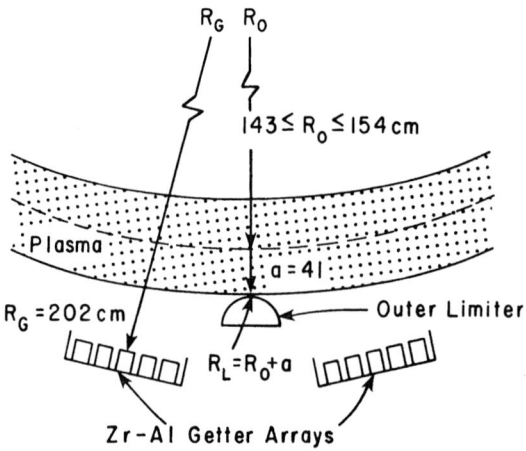

Fig. 15. Top view of the PDX Zr-Al getter array/moveable limiter system. The arrays are at a major radius of 202 cm, and the radius of the limiter can be adjusted from 184 cm to 195 cm [Cecchi et al].

included in the calculation, one, S_L, at the limiter, and one at the two getter vertical support sides, S_G. The model geometry used to track the limiter source is shown in Fig. 16. A three-dimensional model of the getter panel, its support structure, and the neighboring plasma and vacuum vessel wall was used in the calculation of the getter source transport.

Assuming a particle scrapeoff length of 3 cm, the flux at the getter, Γ_G, is related to that at the limiter, Γ_L, by

$$\Gamma_G = \Gamma_L exp\left((R_L - 198)/3\right).$$

The projected area, A_L, of the limiter was $\sim 80\,\mathrm{cm}^2$, and the exposed area of the two vertical sides, A_G, was $\sim 90\,\mathrm{cm}^2$. Thus

$$S_G = A_G\Gamma_G = 90\Gamma_L exp\left((R_L - 198)/3\right),$$

and

$$S_L = A_L\Gamma_L = 80\Gamma_L.$$

Let $S = S_G + S_L$, ϵ_L^G equal the fraction of S_L absorbed by the getter, ϵ_L^D equal to the fraction of S_L lost to the two divertor domes, and ϵ_G^G be the fraction of S_G absorbed by the getter. Then

$$(1 - \beta_R)S = \epsilon_L^G S_L + \epsilon_L^D S_L + \epsilon_G^G S_G,$$

or

$$1 - \beta_R = \epsilon_L^G \overline{S}_L + \epsilon_L^D \overline{S}_L + \epsilon_G^G \overline{S}_G,$$

where $\overline{S}_L = S_L/S$ and $\overline{S}_G = S_G/S$.

The results of the calculation are shown in Table 3. They show an increase in $1-\beta_R$ by a factor of 1.73 as R_L increases from 184 to 195 cm, versus the experimental result of 1.9. Note that for $R_L = 184\,\mathrm{cm}$, the gettering was almost all of limiter born neutrals. However for $R_L = 195\,\mathrm{cm}$ 62.8% of the gettering is of the vertical side source.

This calculation agreed well qualitatively with experiment, and showed that when the plasma major radius was largest, the getter support structure contributed the majority of the pumped neutrals.

6.3. Recycling in TFTR

We have recently begun modeling neutral particle recycling in TFTR and have reportedly our first results in [Heifetz et al (1985)]. We give here some of these results in more detail, in particular the calculations of wall particle and energy fluxes, and the variation of results as physical models for wall

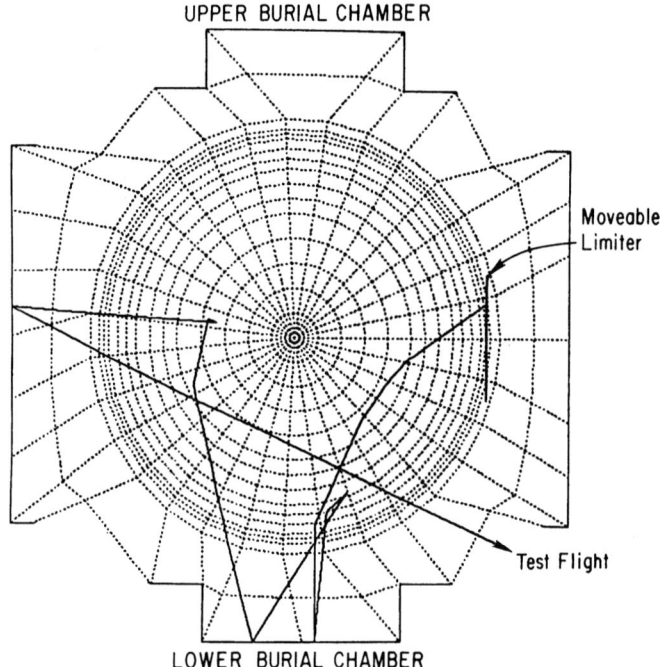

UPPER BURIAL CHAMBER

Moveable Limiter

Test Flight

LOWER BURIAL CHAMBER

Fig. 16. DEGAS model geometry for PDX including the moveable limiter and Zr-Al getter arrays. The major radius of the plasma can be adjusted from 143 cm to 154 cm, with the limiter moving along with it. The getter panels were simulated using a collection efficiency of 50% for atoms.

Table 3
PDX Zr-Al getter performance

R_L	ϵ_L^G	ϵ_L^{D*}	ϵ_G^G	$\overline{S}_L$	$\overline{S}_G$
184	0.043	0.086	0.302	0.993	0.007
195	0.086	0.086	0.353	0.707	0.293

R_L	$\epsilon_L^G \overline{S}_L$	$\epsilon_L^D \overline{S}_L$	$\epsilon_G^G \overline{S}_G$	$1 - \beta_R$
184	0.043	0.085	0.002	0.130
195	0.061	0.061	0.103	0.225

*from experiment

Measured and calculated results on the pumping performance of the PDX Zr-Al getter panels. R_L is the major radius to the moveable limiter face, ϵ_L^G is the fraction of the limiter neutral source, S_L, pumped by the getters, ϵ_L^D is the fraction of S_L lost to the divertor domes, ϵ_G^G is the fraction of the neutral source, S_G, due to ions striking the getter supports pumped by the getters, $\overline{S}_L = S_L/(S_L + S_G)$, $\overline{S}_G = S_G/(S_L + S_G)$, and β_R is the overall recycling coefficient.

reflection and atomic were changed.

The model geometry included the complete three-dimensional torus of the vacuum vessel, containing a section of a torus of the approximate dimensions of the movable limiter to simulate the main geometric effects of the limiter [Doll et al]. The plasma parameters remained fixed in all the calculations, with profiles of plasma density and temperature taken from fits to experimental data.

Three sources of neutrals were included: plasma recycling off the limiter, plasma recycling off the vessel wall, and recombination. One uncertainty was the size of the recycling sources. Knowing the limiter ion current is equivalent to knowing a particle confinement time, τ_p, which is not independently measured. We assumed that $\tau_p = 200$ msec. The wall ion current is also unmeasured, and we assumed that it was one-tenth that of the limiter's, based on PLT experience [Ruzic et al].

As a base case we took an ohmically heated purely deuterium plasma of minor radius 85 cm, with a volume averaged density of 4.2×10^{13} cm^{-3}, equal ion and eletron temperatures T, $T(0) = 2300$ eV, $T(85) = 75$ eV, and a density scrapeoff length of 2.5 cm [Budny]. The ion current onto the limiter was 7.4×10^{21} D$^+$/sec. The reflection model used reflection coefficients for normal incidence together with a model for a rough limiter and wall, where for each wall collision an atom was made to undergo not one but two to three collisions with the wall. This effectively reduced the reflection coefficient by a power of two to three (cf. Sec. 5.5). All sticking D^0 was immediately desorbed in the calculation as D_2^0, and there was no getter pumping.

As expected from the PDX scoop calculation (Sec. 6.1), the D^0 and D_2^0 populations were localized near the limiter. This localization appears in both the poloidal (Fig. 17) and toroidal directions (Fig. 18). The peak D^0 density, in front of the limiter recycling source, was a factor of 100 times greater than D^0 density anywhere else in the machine.

It is instructive to look at the individual contributions to the neutral density from each of the three sources. The poloidally averaged densities of D^0 due to each source are plotted in Fig. 18, along with the total D^0 density. Note that atoms in the outer 15 cm were due primarily to the wall source, and those in the central half of the plasma were from recombination. Since charge-exchange anlyzers can potentially measure absolute fluxes of neutrals originating from the plasma center (see below), and hence measure the neutral density, the recombination source could not be left out of our calculation.

A simple estimate of central atomic density due to recombination is gotten by assuming that the conditions in the center are in steady state and that the center is isolated from the outer plasma, in the sense that the creation rate of neutrals in the center will equal the rate of ionization of atoms there from electron impact ionization. Thus set

$$n_e(0)n_{D^0}(0) <\sigma v_e> = n_e(0)n_{D^+}(0)K_{recomb}$$

where K_{recomb} is the local recombination rate coefficient, from which

$$n_{D^0}(0) = n_{D^+}(0)\frac{K_{recomb}}{<\sigma v_e>}.$$

In our base case, the central temperature is 2300 eV, and $n_e(0) = n_{D^+}(0) = 5.6\times10^{13}$ cm^{-3}, so $<\sigma_e v> = 2.0\times10^{-8}$ cm^3/ sec, $K_{recomb} = 1.2^{-16}$ cm^3/ sec (Fig. 19), and $n_{D^0} = 3.36 \times 10^5$ cm^{-3}. The density calculated by DEGAS was 3.44×10^5 cm^{-3}.

Wall particle fluxes, again not surprisingly, peaked on the moveable limiter in the area of plasma/limiter contact at 3.5×10^{18}cm^{-2}sec^{-1} (Fig. 20). The toroidally averaged particle flux was approximately 6×10^{14} cm^{-2}sec^{-1}, and the contribution to the wall flux by the limiter source was negligible.

The sensitivity of the results for the edge neutral densities to the physical models used is shown in Fig. 21. This figure shows the variations of

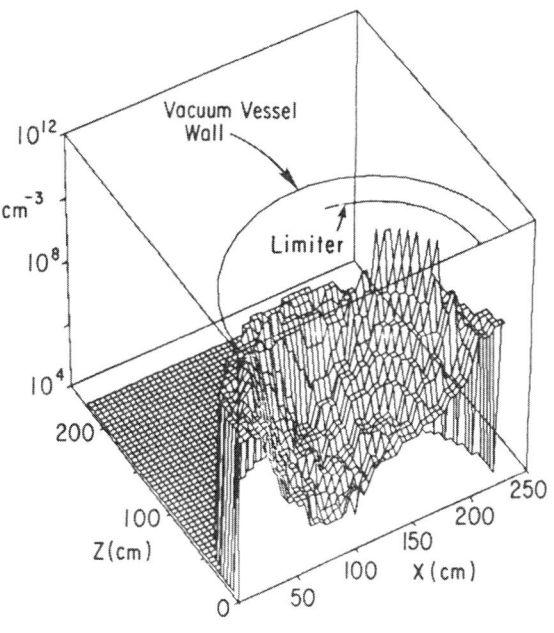

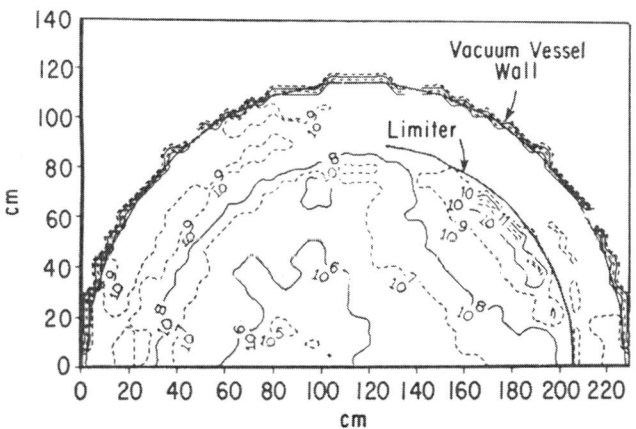

Fig. 17. Surface and contour plots of the D^0 density for the TFTR base case in a poloidal slice directly in front of the moveable limiter.

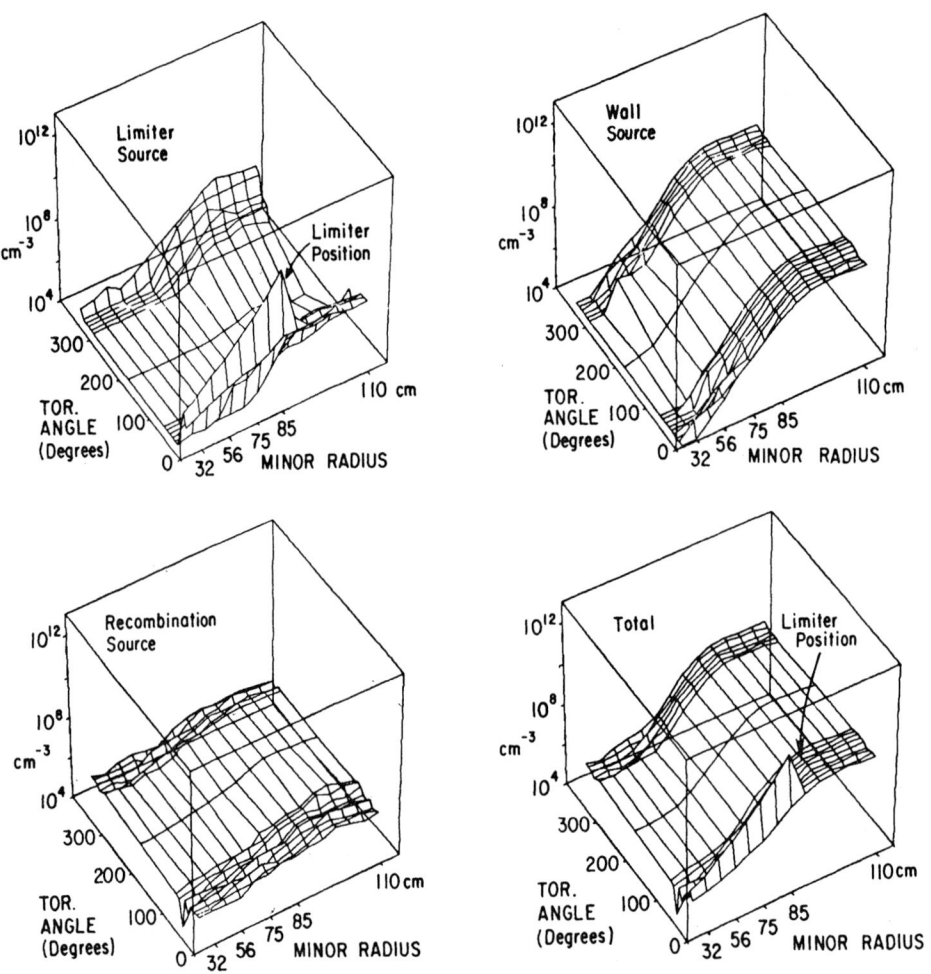

Fig. 18. Contributions to the poloidally averaged D^0 density in TFTR from three sources, and the total D^0 density [Heifetz et al (1985)].

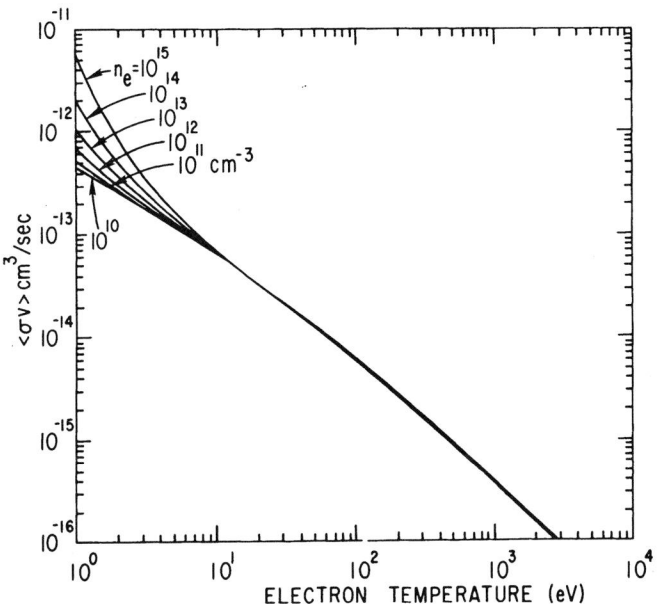

Fig. 19. Rate coefficients for e + H$^+$ recombination [Weisheit, Janev et al(1984)].

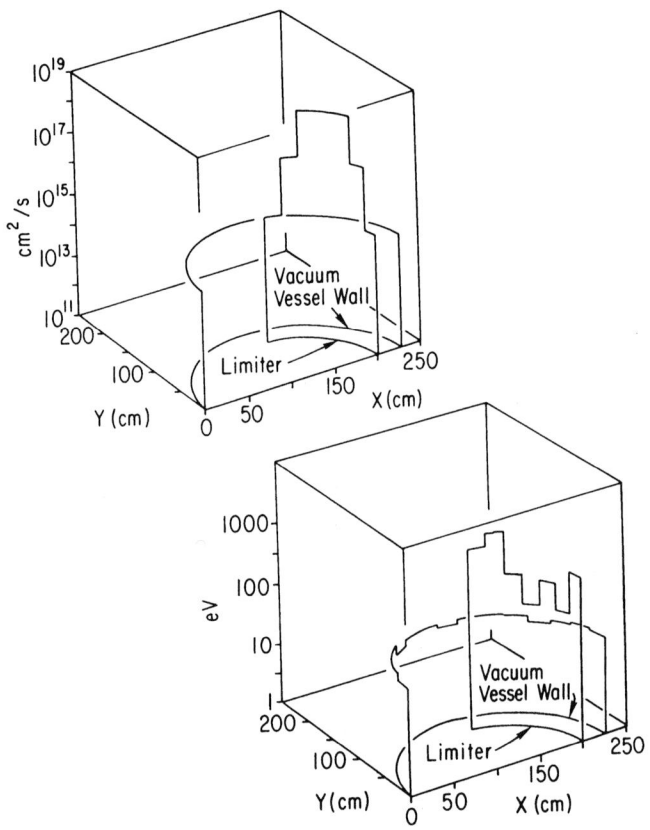

Fig. 20. The D^0 wall flux (a) and the average energy of the D^0 striking the wall (b) in TFTR.

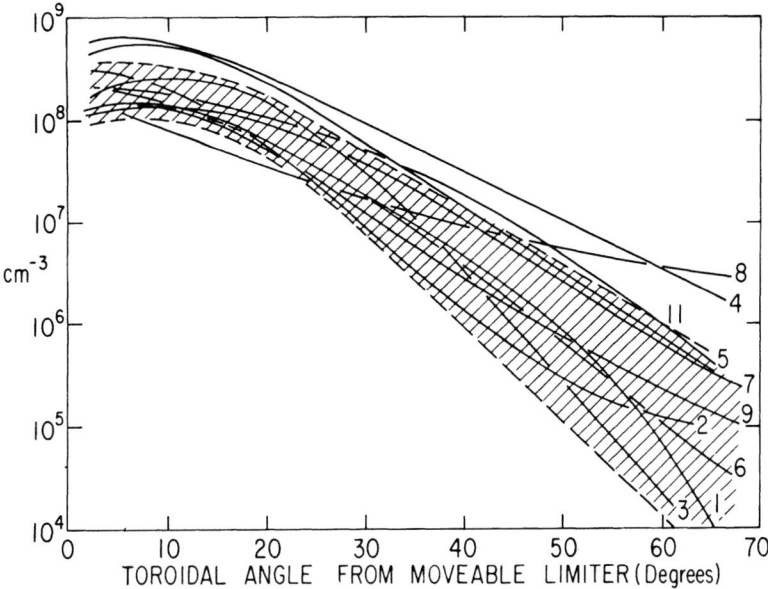

Fig. 21. Midplane outboard D_2^0 density at the vacuum vessel wall due to a limiter source in TFTR as a function of toroidal angle from the moveable limiter, computed using different wall reflection and atomic physics models and plasmas. The base case (1) was for a 100% deuterium plasma with $Z_{eff} = 1$, $<n>= 4.2 \times 10^{13}$ cm^{-3}, n(limiter)=10^{13} cm^{-3}, a density scrapeoff length of 2.5 cm, $T_i = T_e = T$, T(0) = 2300 eV, T(limiter) = 75 eV, and T(edge) = 10 eV. Wall reflection coefficients for normal incidence were used, and the distribution in reflected direction was cosine. All unreflected D^0 was assumed to immediately desorb as wall temperature D_2^0. Variations on this base case included: (2) the same as (1) but with T(edge) = 5 eV, (3) same as (1) with T(edge) = 20 eV, (4) the same as (1) with T(0) = 6900 eV, T(limiter) = 150 eV, and T(edge) = 10 eV, (5) same as (4) but with T(edge) = 20 eV, (6) same as (1) with 10 % of the unreflected D^0 absorbed into the wall, (7) same as (6) with 50 % of unreflected D^0 absorbed into the wall, (8) same as (1) but with the reflection coefficient tending to unity as the incident normal approached 90°, and the reflected direction distribution varying from a cosine distribution for normal incidence to a specular one at shallow incident directions, (9) same as (1) but using MARLOWE/TRIM wall reflection data, (10) same as (1) but assuming all D^0 desorbs off the walls as atoms, and (11) same as (1) but with $<\sigma v>$ independent of n_e (so no multistep ionization processes are included) and $<\sigma v_{cx}>$ is independent of the neutral's energy. The shaded region indicates the range within the correct density is most likely to be.

the midplane outboard edge D_2^0 density due to the limiter source in the toroidal direction, as computed using different wall reflection and atomic physics models. Because of the uncertainties in our data for these processes, the best that can be done is to vary the models over a resonable range and produce a range in which the actual answer most probably lies. In this case, the resulting range in densities at 50-60° from the limiter center was one to two orders of magnitude.

The neutral central density changed relatively little with model, probably because, as noted above, it results mainly from the one process of recombination, which occurs in isolation in the core. As also mentioned before, the neutrals in the center can be sampled by charge-exchange analyzers to diagnose the ion temperature, and, in the case of absolutely calibrated analyzers, the neutral central density can be determined absolutely.

In general, an analyzer looking through a plasma along a chord of length L will see the energy spectral distribution

$$\frac{dn}{d\epsilon} \sim \int_0^L n_i n_{H^0} <\sigma v_{cx}> \frac{exp(-\epsilon/T_i)}{T_i^{3/2}} \epsilon^{1/2} exp\left(-\int_0^x \frac{ds}{\lambda_{tot}(s)}\right) dx.$$

This integrand consists basically of three factors. The first, $n_i n_{H^0} <\sigma v_{cx}>$, is the rate at which charge-exchanging with ions of energy ϵ is occuring. This factor times the second, $exp(-\epsilon/T_i)\epsilon^{1/2}T^{-3/2}$, the Maxwellian energy distribution of the ions, gives the value for $dn/d\epsilon$ if there was no attenuation of the neutrals along the line of sight. The third factor, $exp(-\int_0^x ds/\lambda_{tot}(s))$, accounts for this attenuation.

We can simplify the expression for $dn/d\epsilon$ when $\epsilon \gg T_i$. Due to the attenuation by the Maxwellian distribution, we can expect the main contribution to the integral for $dn/d\epsilon$ to come from the chordal segment, say from $x = a$ to $x = b$, were the plasma is hottest. Also at such energies λ_{tot} will be much greater than this chord segment length, so that attenuation by ionization and charge-exchange will be minimal, that is $exp(-\int_0^x ds/\lambda_{tot}(s)) = 1$. Hence we can write

$$\frac{dn}{d\epsilon} \sim \int_a^b n_i n_{H+} <\sigma v_{cx}> \frac{exp(-\epsilon/T_i)}{T^{3/2}} \epsilon^{1/2} exp\left(-\int_0^x \frac{ds}{\lambda_{tot}(s)}\right) dx$$

$$= \|b - a\| n_i n_{H+} <\sigma_{cx} v> \frac{exp(-\epsilon/T_i)}{T^{3/2}} \epsilon^{1/2}$$

$$= C exp(-\epsilon/T_i)\epsilon^{1/2},$$

assuming that n_i, n_{H+}, $<\sigma_{cx} v>$, and T_i do not vary from $x = a$ to $x = b$. Taking the logarithms of both sides we get

$$ln\left(\epsilon^{-1/2}\frac{dn}{d\epsilon}\right) = \left(-\frac{1}{T_i}\right)\epsilon + C'.$$

This states that for large enough ϵ, the curve for $ln(\epsilon^{-1/2}dn/d\epsilon)$ as a function of ϵ will approach a straight line of slope of $-T_i^{-1}$.

As an example from our calculations, a graph of $ln(\epsilon^{-1/2}dn/d\epsilon)$ as a function of ϵ is given in Fig. 22 for a TFTR high temperature, low density (few $\times 10^{12}$ cm^{-3}) beam heated discharge. The central ion temperature was set at 10 keV. The curve for $ln(\epsilon^{-1/2}dn/d\epsilon)$ becomes roughly linear above 20 keV. As a rule of thumb this will happen at $\epsilon = 2$ to $3 \times T_i$. From the slope of the curve there we computed that $T_i = 6100$ eV. This is less than two-thirds of the central T_i in the calculation, but this is typical for a central temperature measurement. However, as we viewed along chords moving away from the center all the way out to the plasma edge, (something the experimentalists cannot do on TFTR) the computed ion temperatures, though usually lower, did approach the given values.

The reason for the bad result for central temperature is that in this case T_i did not fall off sharply from the center, so that major contributions to the high energy end of the spectrum were being made by regions with lower T_i. Because of this, experimentalists typically make a correction upward in the T_i when measured this way. Views through the plasma edge do not see such a wide variation in the temperature, thus there is less mixture of different Maxwellians of energies and the measurement tends to be more accurate.

In light of this experience, a seeming paradox can occur in calculations for very dense, large plasmas. For example, [Reiter] has modeled JET discharges with $<n> =$ few $\times 10^{13}$ cm^{-3}, and found that for the case of a given central ion temperature of 10 keV, the computed temperature was over 13 keV. The plasma is acting here as a speed filter on the atoms, with only the very hottest atoms being able to escape from the center due to the high attenuation of the dense plasma. The assumption in our derivation above that λ_{tot} was much larger than the length of the chord of integration was incorrect.

6.4 Recycling in the PDX Divertor

Recent studies of the existence of H-mode discharges in ASDEX, PDX, and D-III [Wagner et al, Kaye et al, Ohyabu et al] have raised questions about particle refueling through divertors. Two of the more important questions concern the degree of multiple recycling of particles in the divertor, that is, determining the divertor particle confinement time, and the related question of the conductance of the plasma filled channels in the ASDEX and PDX divertors.

Modeling of the H-mode has been done using fluid treatments of varying sophistication [Morgan and Harbour, Pet et al (1982), Schneider and Lackner, Shimada et al, Post et al, Saito et al, Galambos and Peng]. The calculations described here were of neutral transport in the PDX divertor

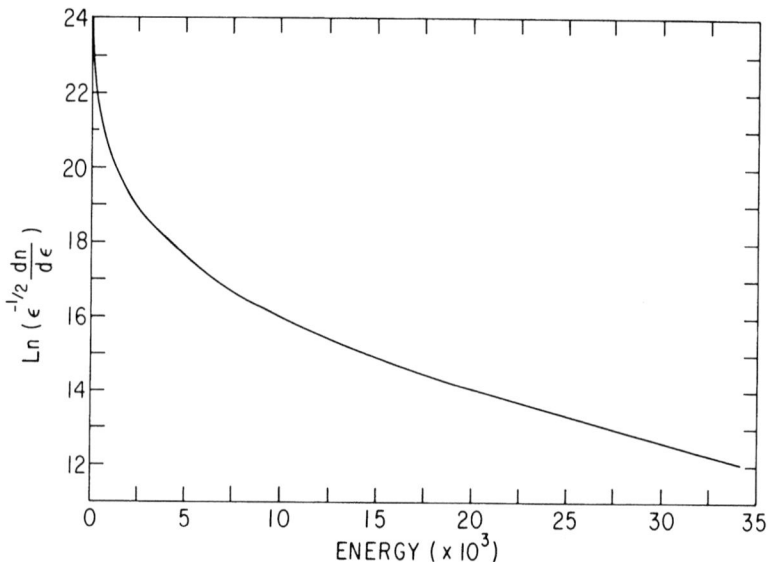

Fig. 22. Computed $ln(\epsilon^{-1/2}dn/d\epsilon)$ versus energy, ϵ, for a high temperature $(T_i(0) = 10\,\text{keV})$, low density $(n_e = few\,10^{12}\,\text{cm}^{-3})$ TFTR beam heated discharge. The six curves shown correspond to the six views of the charge-analyzer in TFTR. The curves lie close together since all the views are in the mid-plane, pass through the center, and no beam neutral sources are included.

(Fig. 23), where the plasma was described from experimental data, and assumed to remain constant during the calculation. This approach was designed to focus on issues involving the neutral transport alone, such as atomic physics, wall reflection, and wall desorption physics.

We began by constructing a consistent picture of the gas pressure, H_α radiation, and line average density measured in the divertor dome, for a series of ohmically heated discharges with varying main chamber line-average densities [Dylla et al]. In particular we hoped to explain the variation of dome pressure (Fig. 24), which increased exponentially as the mid-plane line-averaged density, $\overline{n}_e(m)$, increased from 1 to $3 \times 10^{13}\,\mathrm{cm}^{-3}$, and then saturated for $\overline{n}_e(m)$ above $3 \times 10^{13}\,\mathrm{cm}^{-3}$.

The key unknown from the viewpoint of modeling the neutral transport was the dome electron temperature, T_e, because there existed in the data only one Langmuir probe measurement for T_e in the relevant parameter range. We estimated T_e as follows. In the electron density and temperature ranges expected in the divertor, the H_α light emission, Γ, varies approximately (Fig. 25) as

$$\Gamma \propto n_e n_o (T_e)^\alpha,$$

where

$$\alpha = \begin{cases} 5.3 & T_e \leq 5\,\mathrm{eV} \\ 1.0 & 5 < T_e \leq 60\,\mathrm{eV} \end{cases}$$

where n_o is the density of atomic hydrogen. Assuming n_o is proportional to the neutral pressure (Fig. 24), we can then determine the variation of dome T_e since we know that of the dome line-averaged density $\overline{n}_e(d)$, and the dome H_α light emission.

The resulting plot for T_e, scaled at one point with the data from a Langmuir probe, is shown in Fig. 26. Note that T_e falls below $10\,\mathrm{eV}$ for $\overline{n}_e(m)$ above $3 \times 10^{13}\,\mathrm{cm}^{-3}$.

Using this data for $\overline{n}_e(d)$ and T_e, neutral transport was computed. Since the dome containes a population of hot atoms with energies above $1\,\mathrm{eV}$ as well as wall temperature molecules, the definition of pressure must made carefully. Pressure measurements in PDX were made at the end of a long thin tube, where we may assume that the entire population is molecular. Thus we can use particle continuity at the tube's entrance to compute the number of molecules in the tube produced by atoms sticking to the tube's walls. If $\overline{E}_{D^o}$ is the average atomic energy, n_{D^o} the atomic denisty, and T_{wall} is the wall temperature, then the molecular density in the tube, $n'_{D_2^o}$, produced by the sticking atoms is determined by

$$n_{D^o}\sqrt{2\overline{E}_{D^o}/3m_{D^o}} = 2n'_{D_2^o}\sqrt{T_{wall}/2m_{D^o}}$$

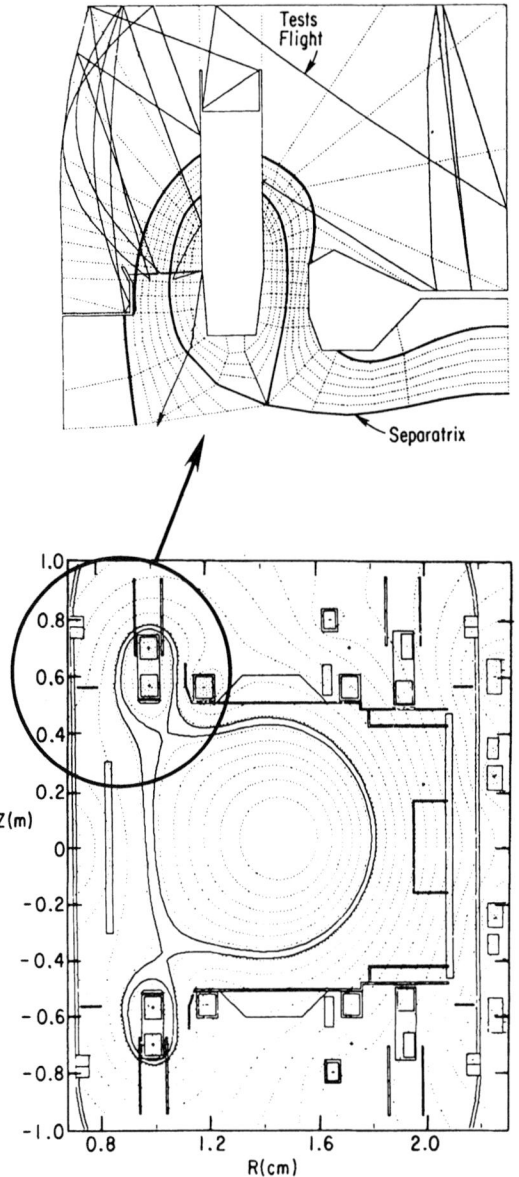

Fig. 23. Poloidal cross-section of the closed PDX divertor (bottom), and the divertor as modeled by DEGAS (top).

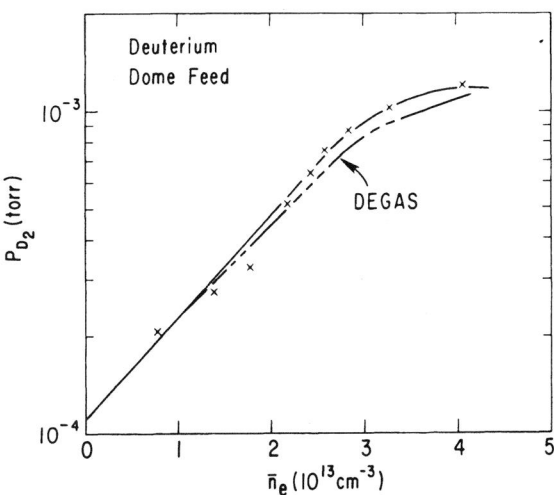

Fig. 24. Measured dome pressure as a function of the line averaged density in the main chamber (solid line), and computed pressure (dashed line).

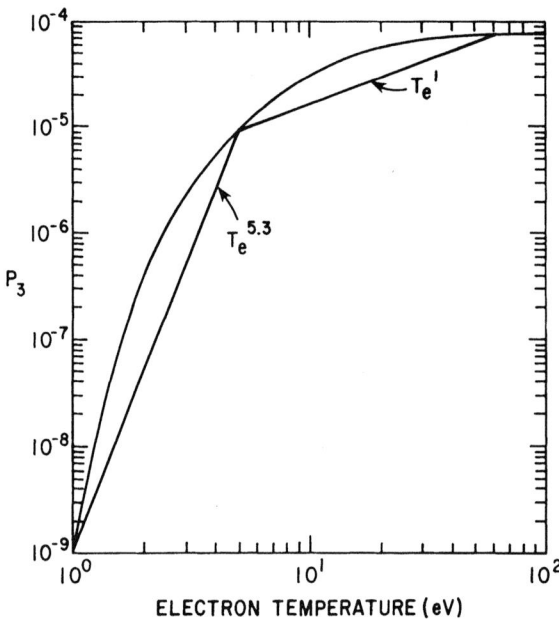

Fig. 25. The fraction of D^0 excited to the n=3 level, as a function of electron temperature, T_e, for $n_e = 10^{12}$ to $10^{13}\,\mathrm{cm}^{-3}$.

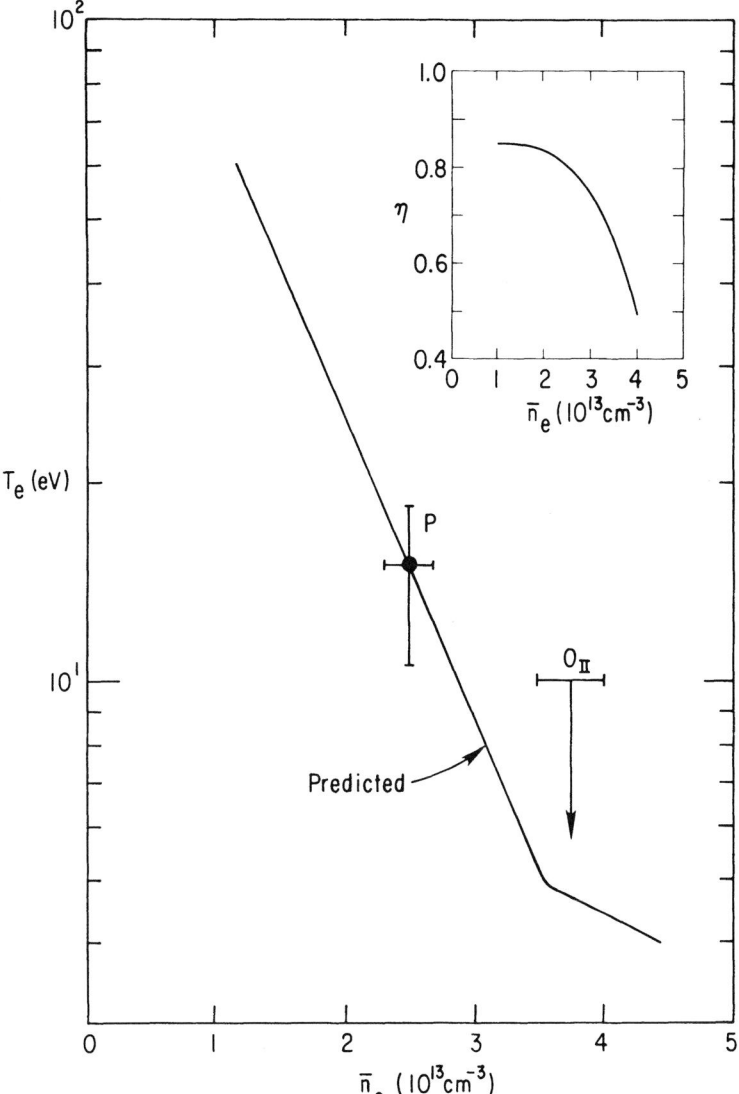

Fig. 26. Divertor electron temperature, T_e, as a function of the line-averaged plasma density in the main chamber. Point P is a Langmuir probe measurement, and OII light disappeared above $\overline{n}_e = 3.5 \times 10^{13} \, cm^{-3}$, indicating that $T_e < 10 \, \text{eV}$.

where m_{D^0} is the D^0 mass, and the factor of two on the right hand side comes from the conversion of atoms to molecules. The resulting contribution, $p'_{D^0_2}$, to the pressure at the end of the tube is then

$$p'_{D^0_2} = n'_{D^0_2} T_{wall} = n_{D^0} \sqrt{\overline{E}_{D^0}/3}.$$

In the PDX divertor, $P'_{D^0_2}$ was about 10% of the total pressure.

The calculated total dome neutral pressures are shown in Fig. 24. The computed and measured neutral pressures agreed quantitively when a flow velocity Mach number of 0.11 was used. The calculated H_α emission rates agreed qualitatively with those measured, though no attempt was made for a quantitative comparison of absolute H_α emission intensities.

We thus felt that we had a self-consistent picture of the neutral-plasma behavior in the divertor, as well as some insight into low temperature plasma recycling behaviour. An indirect measurement of T_e comes from noting that OII (4415Å) emission cuts off as T_e drops below 10 eV, and one experimental measurement did see a cutoff of OII emission at $\overline{n}_e(m) \approx 4 \times 10^{13}$ cm^{-3}. Now, the cooling of T_e to under 10 eV reduces the ionization probability of atoms, making the plasma more transparent to the neutrals. This reduces the ability of the divertor to maintain neutral pressure, and explains the saturation of the pressure as shown in Fig. 24. An extreme example of this may have occurred in the collapse of the pressure seen during high flow gas fueling in the dome, as the plasma loses its ability to ionize, and the divertor channel becomes completely transparent.

Encouraged by our model's achievements, we applied the same methodology to the three phases of PDX discharges exhibiting H-mode behaviour: the ohmic (OH), the pretransition (PT), and the H-mode phases (H) [Heifetz et al (1984)]. We now had extensive data on the plasma parameters [Owens et al]. Using this information we computed two-dimensional distributions of neutral temperatures, densities, and pressures such as the one shown in Fig. 27.

The one variable we focused on was again the measured dome pressure (Table 4), which rose monotonically during the discharge. Our initial calculations resulted in pressures 1.5 - 2 times those measured (Table 4), and even worse, showed a peak pressure during the PT phase. We then questioned the assumption that all device walls were saturated, and did no net pumping during the discharge; that is, that all absorbed particles desorb immediately as molecules. We introduced a pumping fraction, γ, defined as that fraction of sticking atoms which do not desorb during the neutral profile. In particular, the higher the saturation of the wall, the smaller the value of γ.

Results for $\gamma = 0.1$ are also listed in Table 4. The calculated pressures decreased from the $\gamma = 0$ cases, agreeing with the measured values during

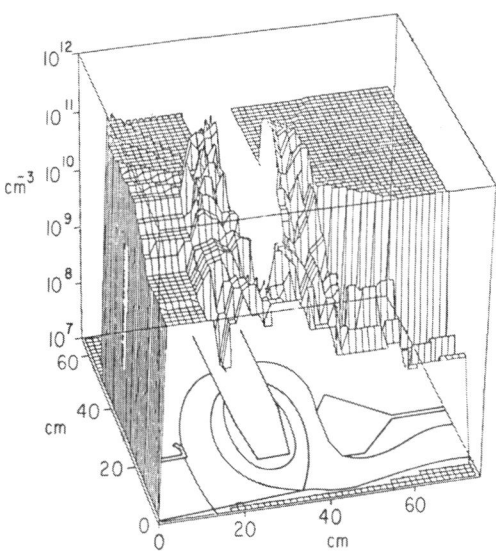

Fig. 27. Two-dimensional variation of the D^0 density during the ohmic phase of a typical H-mode PDX discharge.

Table 4
PDX Dome Pressure During an H-Mode Discharge
(10^{-4} Torr)

Phase	Measured	$\gamma = 0.0^*$	$\gamma=0.1$
OH	1.8	2.4	1.8
PT	2.3	4.9	2.2
H	2.9	4.3	2.1

* γ is the fraction of non-reflected atoms assumed to be retained in the wall during the discharge. Agreement with measured pressures was obtained when $\gamma = 0.09$ in the PT phase and $\gamma = 0.04$ in the H phase.

both the OH and PT phases. However they still peaked in the PT phase. Readjusting γ further, we achieved agreement when $\gamma = 0.09$ in the PT phase, and $\gamma = 0.04$ in the H phase, and convinced ourselves that γ should decrease during the discharge as the walls became more saturated. This wall pumping model resulted in a pumping rate by the walls roughly equal to the rate of dome fueling, 20 Torr-liters /sec, applied during a typical discharge. We also note, however, that we cannot explain the good agreement our modeling of the ohmic discharges achieved described above, where $\gamma = 0$ was used.

Table 5 gives the overall divertor-main discharge refueling balance. Particle plugging, the fraction of neutrals reionized in the divertor, increased from 0.60 to 0.73 from the OH to the PT phases, but then dropped to 0.70 during the H phase. In all cases the fraction of particles returned to the main discharge was between 15-18%. These results are significant in that many scenarios of H-mode functioning are based on the change in the nature of particle refueling during the discharge, and it is not clear from our results how much of a change occurs. Perhaps the energy of the neutrals transmitted through the divertor throat also effects the recycling.

Recent two-dimensional measurements of the distribution of H_α light in the main chamber and divertor dome [Grek et al] showed intensity peaks at a point in the main chamber near the x-point. Our results predicted that the peaks occurred in front of the plates. This is a troublesome qualitative discrepancy since much of the analysis above was based on analyzing the H_α light. One source of H_α light missing from our model above is from Frank-Condon atoms excited during the process of molecular dissociation. Molecules can flow from the dome to the main chamber through openings between the divertor plates, hence can contribute to the H_α emission when they dissociate.

Table 5

Particle Recycling During a PDX H-Mode Discharge

Phase	Fraction Reionized in Divertor	Fraction Absorbed by Walls	Fraction Returned to Main Discharge
OH	0.60	0.22	0.18
PT	0.73	0.12	0.15
H	0.70	0.14	0.16

6.5. Recycling in the the DITE Bundle Divertor

The only bundle divertor on any sizeable machine today is the MkII divertor on the DITE IA tokamak [Axon et al, Johnson et al.] (Fig. 28). As in the case of the PDX divertor (cf. Sec. 6.4), understanding the role of the divertor in the particle and energy balance requires a picture of the recycling within the divertor.

Following [Harbour et al], the rate of change of the molecular hydrogen in the divertor chamber is given by

$$pV \left(\frac{1}{p} \frac{\partial p}{\partial t} - \frac{1}{T_0} \frac{\partial T_0}{\partial t} \right) = \Phi_{div} + P\Phi_{plate} - p\left(S_{pump} + \alpha_D S_{duct} + S_{plasma} \right),$$

where p is the pressure of the gas in the divertor, V is the volume of the plenum, T_0 is the temperature of the gas, Φ_{div} is the rate of gas feed into the divertor, Φ_{plate} is the ion flow rate onto the divertor plate, P is the fraction of Φ_{plate} which escapes from the plasma into the divertor gas, S_{pump} is the pumping speed of the attached pump, S_{duct} is the conductance of the ducts, α_D accounts for the blocking of the ducts by the plasma, and S_{plasma} is the pumping speed of the plasma.

DEGAS was used to estimate P, α_D, and S_{plasma}. Since the shape of the divertor was thought to influence neutral recycling as much as the wall reflection and atomic physics models used, a faithful three-dimensional representation of the divertor was written and used in the calculation. The cryopump was simulated using a partially reflecting surface with an albedo.

The plasma parameters in the calculation were based on experimental measurements, and the two neutral sources included were recycling off the limiter plate and gas puffing through a valve at the center top of the divertor chamber.

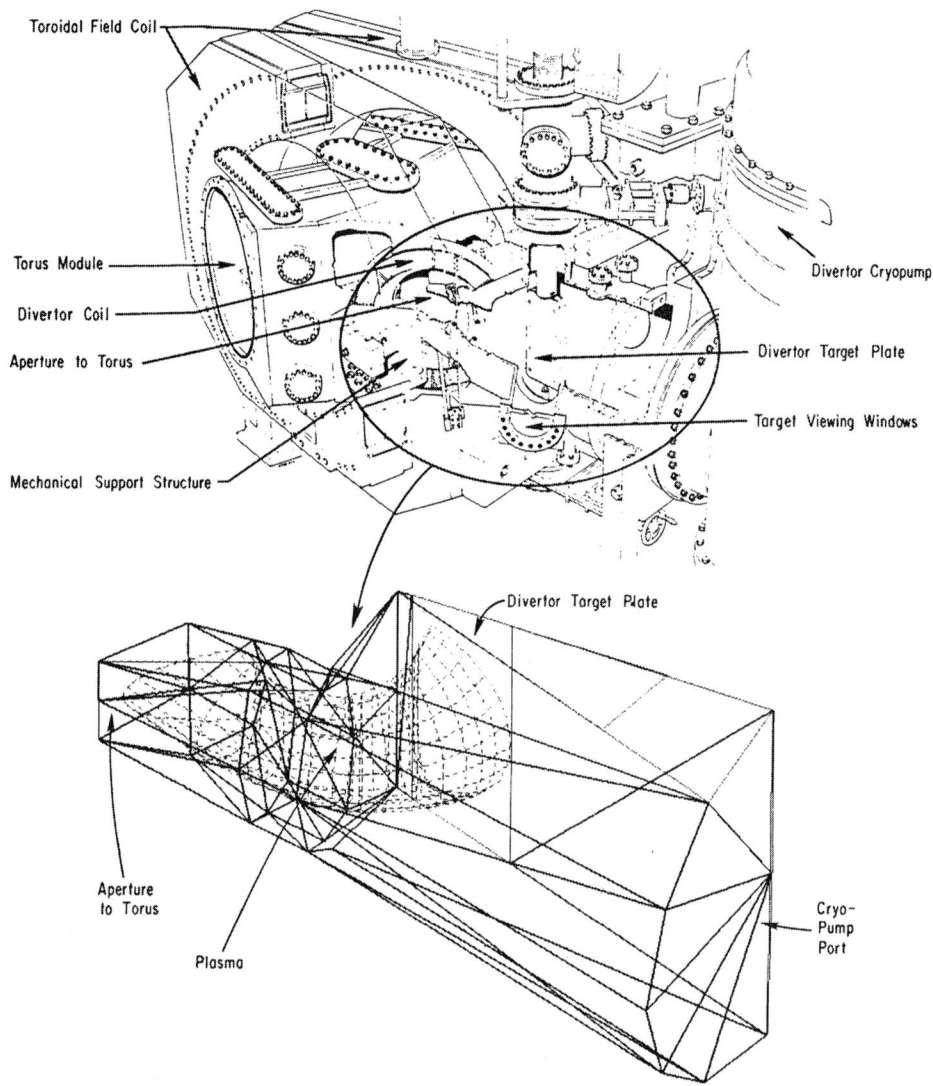

Fig. 28. The DITE MKII bundle divertor design (top), and the model geometry used by the DEGAS code (below). Only half of the divertor is shown in the model geometry, the other half being represented in the calculation by a mirror image. The plasma shape was determined using a field-line tracing program together with photography from the actual experiment.

In the case of both sources the particle flow through the duct, $p\alpha S_{duct}$, was less than 10% of the input source rate, Φ_{div} or Φ_{plate}. The low calculated atomic density within the divertor ducts was qualitatively consistent with H_α measurements from the experiment.

A value of P was also estimated. About half of the neutrals leaving the plate reionize without ever leaving the plasma. However the total actual ionization in the divertor appeared to be much less than calculated by DE-GAS. One possible explanation for this is that a collisional neutral transport model is necessary in order to simulate the high neutral pressures of a few 10^{-2} Torr which have been measured in DITE.

When neutral-neutral interactions are included the calculation becomes nonlinear and hence more difficult. Monte Carlo test flights would be influenced by the ambient bath of neutrals, and a type of iterative scheme for calculating the neutral density would be necessary.

6.6. The ALT-II Experiment on TEXTOR

The Advanced Limiter Test-II (ALT-II) will be a toroidal belt pump limiter experiment expected to be installed in the TEXTOR tokamak in 1986 [Conn et al (1984B)] (Fig. 29). It has been designed to be prototypical of limiters considered for use in the next generation of major tokamak fusion experiments [Boley].

During the design phase of ALT-II a number of limiter/pumping system configurations were modeled using DEGAS. We describe here three of the these calculations: an estimation of the effect on pumping of ballistic particles entering the pump ducts, determination of whether a system with an axisymmetric neutralizer plate would pump more particles than one with descrete plates, and the extent to which the recycling at a neutralizer plate can be perturbed by gas puffing.

Results from the modeling of particle reflection show that the reflection coefficient and polar angle of scattering off a smooth wall is dependent on the incident polar angle [Eckstein and Verbeek]. Thus the transmission probability, K, for particles in a duct with sufficiently smooth walls would be energy dependent. We tested this in the case of the ALT-II pump ducts of dimension $23\,cm \times 15\,cm \times 132\,cm$. Two reflection models were used. The first assumed that the direction of the polar angle for both sticking and reflecting particles leaving the wall obeyed a cosine distribution. In this case the assumptions are the same as for molecular flow in a duct. The second model included incident angle dependence by taking the reflection coefficient to be 1 when the incident angle was greater than 40°, and sampling the reflected polar and azimuthal angles out of a distribution that varied smoothly from a cosine distribution at normal incidence to a purely specular reflection as the incident angle approaches 90°.

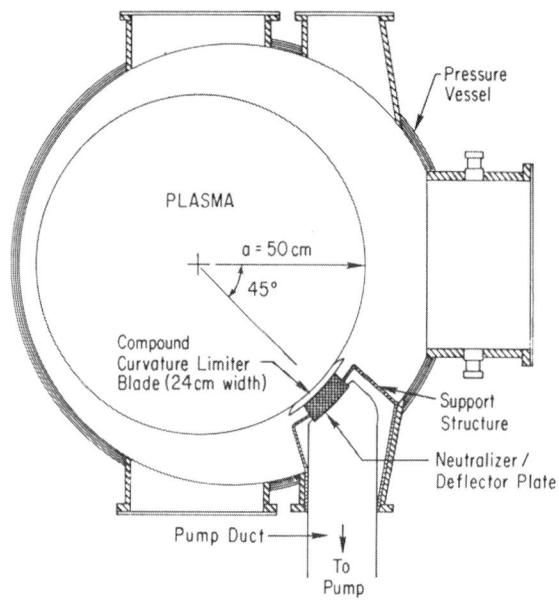

Fig. 29. A poloidal cross section of the TEXTOR vacuum vessel and the axisymmetric ALT-II pump limiter. Eight pump ducts are planned.

The value of $K = 0.144$ computed for the first model agreed well with standard molecular flow theory [Dushman and Lafferty]. Two incident energies, 6 and 37.5 eV, were used in the case of the second reflection model, where it was further assumed that the particles entered the duct with a cosine distribution in direction. At 6 eV, K increased to 0.53, and at 37.5 eV to 0.65, and thus the effect was significant. Of course we do not know how smooth real duct walls would have to be for anything like this effect to occur. The value of this calculation was to give limits for K for use in subsequent pumping calculations.

Originally two designs were proposed for the neutralizer plate system (Fig. 30). One used eight two-sided plates each positioned over a pump duct opening. The other was for an axisymmetric plate above a continuous plenum box which was connected to the eight pump ducts.

In order to compare the pumping performance of these two designs both were modeled using DEGAS. One modification to the stand-alone DEGAS calculation was to follow ionized neutrals along field lines and to reneutralize them if the restrike the limiter. This was a crude attempt to simulate nontrivial recycling without using a full fluid model.

The model geometries were fully three-dimensional (Fig. 30). Transmission of neutrals through the pumping ducts was modeled by an albedo. Initially the albedo was set to $K = 0.144$ for both atoms and molecules. Thus an infinite pumping speed was assumed. The plasma edge density was taken to be 10^{13} cm^{-3}, the edge temperature to be 50 eV, and the ion current leaving the plasma to be 5×10^{21} sec^{-1}.

The discrete limiters were very efficient in sending particles to the duct openings, with over 70% of them striking the duct opening immediately after being neutralized at the plate. A total of 13.6% of the plasma efflux was pumped. In the case of the axisymmetric limiter, neutrals were born uniformly along the plate. Thus on the average they were initially far from the duct openings, which subtended a total of only about 10% of the circumference of the machine. The neutrals in the plenum must therefore have had to be kept there by a roof long enough to find the duct openings. The plenum roof should be as large as possible to do this, without cutting off the initial flow of particles from the plate. The optimal design occurred when the roof closed off approximately three-quarters of the channel opening. However even then only 2.9% of the plasma efflux was pumped.

Two refinements in the calculation were to use a finite pumping speed and to include the energy dependent values for K described above. The average energy of atoms entering the pump duct was 37.5 eV in the discrete limiter design, so K for atoms equaled 0.648, and 6 eV in the axisymmetric design, so $K = 0.53$. The results are shown in Fig. 31. The enhanced

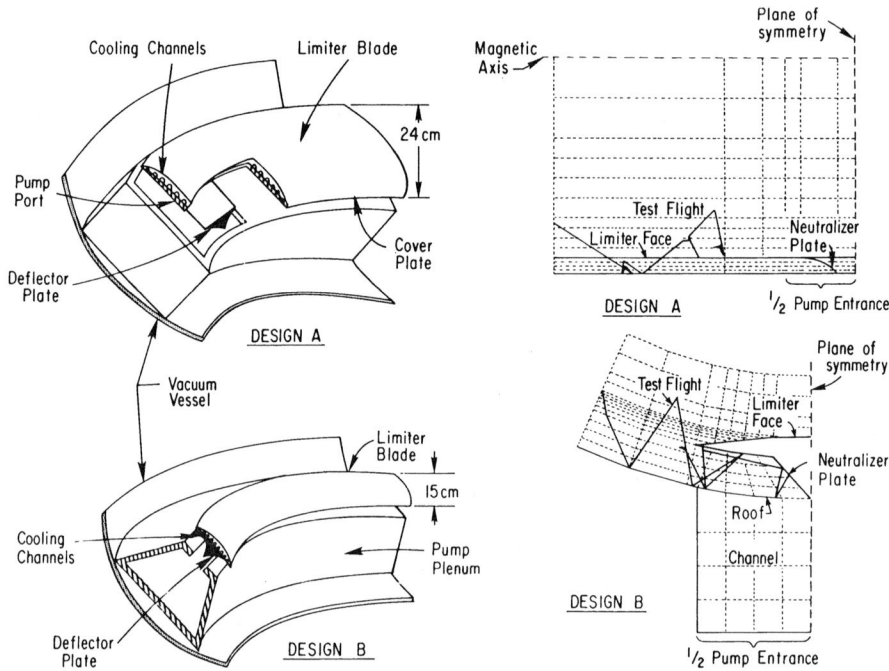

Fig. 30. Two ALT-II neutralizer plate designs (left) and model geometries used in DEGAS (right). In design A plasma strikes eight two-sided plates located directly over pump duct openings. Plasma striking the axisymmetric plate in design B is scattered into a toroidal collection plenum connected to the pump ducts. The design A model cross section was rotated around the magnetic center in the calculation, creating a cylinder around that axis. The limiter face and neutralizer plate however had realistic dimensions. The limiter was 24 cm wide over the discrete 14 cm wide limiter plates. The one dimensional fluid code DCAL [Post] was combined with DEGAS for a self-consistent calculation of the plasma/neutral transport in design A. Plasma transport was calculated independently in the four strips of plasma under the limiter blade which flow into the neutralizer plate.

762

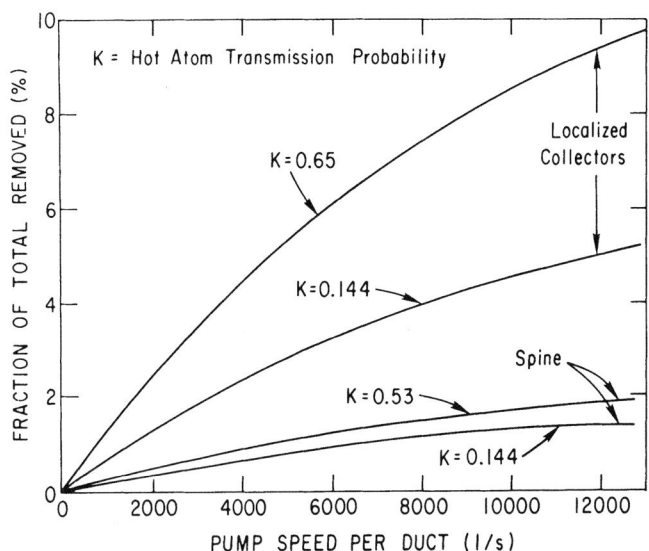

Fig. 31. ALT-II particle removal percentages for the two design options for the cases of ballistic scattering (K=0.65 and 0.53) and standard molecular flow theory (K−0.144).

transmission probability was a smaller effect in the axisymmetric design, because a smaller fraction of particles hitting the duct openings were atoms. The conclusion remained that the discrete limiter plates provided the most pumping.

The pumping performance of ALT-II ultimately depends on the neutral pressure at the pump duct openings. Predicting the ion flux at each of the discrete plates requires a model of both the plasma flow near the plate and the transport of neutral particles produced when the plasma strikes the plate. Such a model would also predict whether a regieme of high particle recycling, such as had been shown for divertors [Petravic et al(1982)], can be maintained in front of the limiter plates.

In order to do this the DEGAS code was coupled to the one-dimensional fluid model DCAL [Post and Lackner]. We briefly review here the assumptions in this fluid model. These are that (1) $T_i = T_e = T$ and $n_i = n_e = n$ (one fluid), (2) the pressure is constant along the field lines, (3) the perpendicular particle and energy sources are uniformly distributed along a field line away from the plate region (at a distance greater than 20 cm), and (4) the particle and power sources due to neutral/plasma interaction are computed using DEGAS, with the fluid and neutral transport codes run iteratively up to convergence.

The three one-dimensional conservation equations for the plasma are

$$\frac{\partial(nv_\parallel)}{\partial s} = S_n \qquad \text{(mass)}$$

$$\frac{\partial}{\partial s}\left(2nT + mnv_\parallel^2\right) = 0 \qquad \text{(pressure)}$$

$$\frac{\partial}{\partial s}\left(5nv_\parallel T + K_e\frac{\partial T}{\partial s} + \frac{1}{2}nv_\parallel^2\right) = S_E \qquad \text{(energy)},$$

where s is the distance along the field line ($s = 0$ is at the point of symmetry, where the net plasma velocity is 0), $v_\parallel$ is the parallel flow velocity, S_n is the mass source due to reionizing neutral particles and ions leaving the core discharge, m is the ion mass, K_e is the parallel heat conductivity, and S_E is the ion and electron energy source described in Sec. 6.1.

The boundary conditions are that the parallel heat, $Q_\parallel$, and particle, $\Gamma_\parallel$, flows are zero at $s = 0$, and at the plate, at $s = L$ the connection length, that $v_\parallel^2 = 2T/m$, $\Gamma_\parallel = nv_\parallel$, and $Q_\parallel = 2(\gamma + 1)Tnv_\parallel$, where γT is the sheath potential.

Neutral transport for the discrete plate design was modeled using the same three-dimensional geometry as in the stand alone DEGAS calculations (Fig. 30). The transmission coefficient was taken to be $K = 0.144$ for all particles.

The ALT-II limiter face was assumed to be at a minor radius of 48 cm and the edge of the blade to be at 51.5 cm. The plasma flow is calculated independently in the four 1 cm strips of plasma which intersect the plate (Fig. 30) at minor radii of 51.75, 52.75, 63.75, and 54.75 cm. The plasma densities and temperatures at $s = 0$ at these radii were taken to be 5.0, 3.6, 2.5, and 1.6×10^{12} cm^{-3}, and 34, 30, 23, and 16 eV respectively.

Results for the electron/ion density and temperature are shown in Fig. 32. The densities near the plate are two to three times those at $s = 0$, and the temperatures drop by more than a factor of two. This suggests that a set of probes set along the field lines would be useful, because measurement with probes of density to within a factor of two is possible.

The particle flux at the plate was also enhanced. Define R as the ratio of particle flux at the plate to the flux 10 cm downstream from the plate. Then $R = 4, 9$, 4.3, 3.1, and 2.0 at minor radii of 51.75, 52.75, 53.75, and 54.75 cm respectively. We note that these values are much higher than observed on the PDX scoop experiment [Budny et al(1984A and B)].

It may be possible to create stronger gradients in density and temperature near the plate by gas puffing beneath the blade directly in front of the plate. To see the effect of such fueling, a source of 10 Torr-liters/sec of 300° K molecules was simulated at the underside of the limiter face. The resulting plasma densities and temperatures are shown in Fig. 32. The densities were more highly peaked at the plate than without puffing. The temperature dropped within the first 20 cm away from the plate by a factor of more than two, to under 7 eV across the entire plate. The value of R increased to 6.4 at the top of the plate. Finally the fraction of the plate flux pumped increased 50% over the non-puffed case, because the plasma temperature dropped well below the ionization potential of hydrogen, making it more transparent to the neutrals in spite of its higher density.

7. Conclusion: Future Development of the Models

The most important need at present is to validate the models by modeling active experiments. Care must be taken to choose data which is a relatively direct measure of neutral transport processes, such as, for example, plenum pressure . Even so, the modeler will find that the experimental data is typically incomplete or inconsistent. For example, even the best diagnosed machines from an edge physics point of view have Langmuir probes at only a handful of sites, giving us only one or two data points to work with. Better coordination is necessary between the experimentalists and the modeler in order to design diagnostics such as probes, pressure gauges, bolometers, spectrometers, and so on, to the best advantage. As a corollary, the experiments themselves should be designed geometrically as simply as

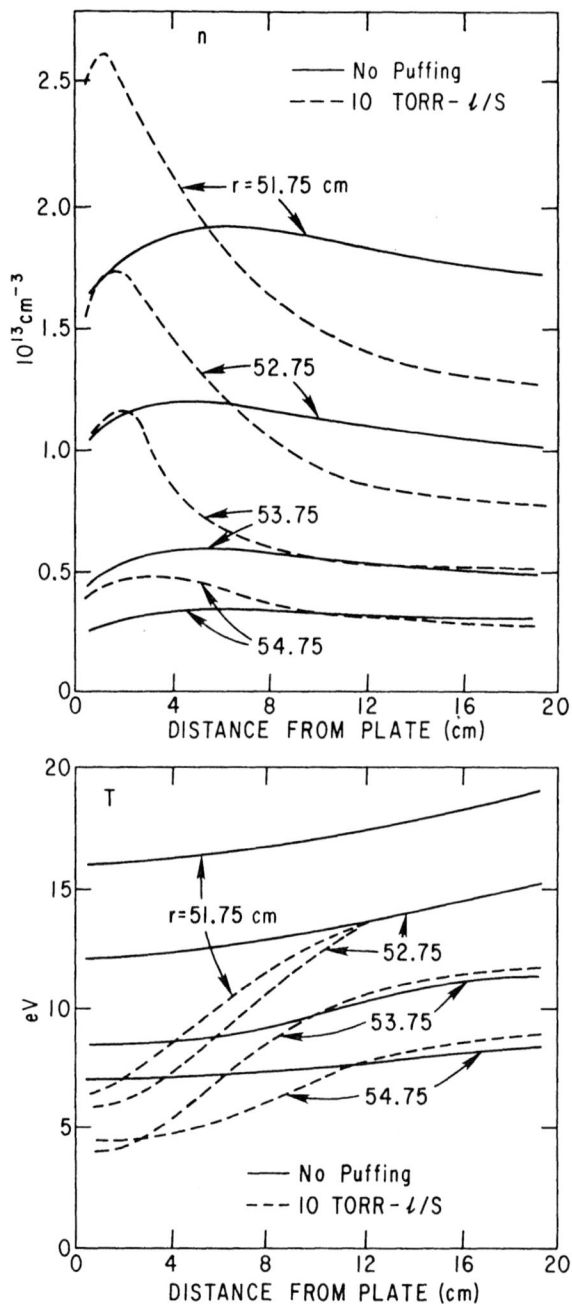

Fig. 32. Scrapeoff layer electron/ion density and temperature in the vicinity of the ALT-II neutralizer plate computed by the combined DCAL + DEGAS codes.

possibly, so that the simpler models may be adquate, and that we may have the best chance to understand what is going on.

In any event, given the limits on the accuracy of available atomic and wall physics data we really have no right to expect good agreement without juggling physical models or including anomalous "effects." However this experience is vital in the limitations of any model.

The best situation is to have data on atomic and wall physics which is as complete and accurate as possible. However, as mentioned in Sec. 2, the data needs often are far greater than what can be supplied by present theory and experiment. Wall reflection, for example, in the PDX divertor was modeled in DEGAS using data from MARLOWE, which assumes smooth, clean walls. However the walls in PDX are neither smooth nor clean. Further, hypotheses in MARLOWE are not valid at the low atomic energies (5-25 eV) found in the divertor. As mentioned, work has begun on modeling such low energy reflection. There has also recently been some work done in describing the process of reflection off rough walls [Sotnikiv, Koborov et al], where roughness was simulated using a "sawtoothed" wall characterized by tooth depth and the distance between peaks. A next step might be to simulate roughness in three-dimensions using the notion of fractals [Mandelbrot, Ruzic], for example,computing the reflection coefficent as a function of fractal dimension. Thus we must proceed with our modeling using incomplete experimental data, and can only eagerly await improvements in the data.

The numerical methods are divided into two groups, the analytic and the Monte Carlo. Analytic methods usually make for the fastest computer codes. For example, the anlytic code of [Tamor] runs over 30 times faster than the comparable Monte Carlo code of [Hughes and Post] (0.3 CRAY-I CPU sec and $\approx$10 CRAY-1 CPU sec respectively). Monte Carlo programs are all the more time-consuming when distributions need to be known in detail to high accuracy. Monte Carlo algorithms, however, can easily contain detailed physical models. The gaps in applicability and computational speed are narrowing. Now the generation of vectorizing computers benefit Monte Carlo codes relatively little because of the many "IF" tests in the tracking algorithms. However the coming generation of multi-processor machines will be ideally suited to linear Monte Carlo schemes, as each processor can follows a test flight.

In conclusion, the numerical methods for modeling neutral transport are maturing to the point where computations including detailed physical models can be performed in a reasonable, though sometimes expensive, fashion. However the experimental data to support the underlying models are sparse, and significant improvements in the validity of these calculations will be possible only when the we have more experimental measurements.

ACKNOWLEDGMENT

The author wishes to thank Dr. D. Reiter for his many useful and interesting suggestions for improvements in this paper. This work was supported by the U. S. Department of Energy Contract No. DE-AC02-76-CHO-3073.

REFERENCES

(Note: The superscript after the list of authors of a work refers to the sections which make references to that work)

Audenaerde, K. et al[4.2], J. Comp. Phys. **34** (1980) 268-284.

Axon, K.B. et al.[6.5], in *Plasma Physics and Controlled Nuclear Fusion Research* (IAEA, Vienna, 1983) Vol. 3, 201-208.

Baskes, W.[6], J. Nucl. Mater. **128-129** (1984) 676-680.

Baskes, W. et al.[6], J. Nucl. Mater. **128-129** (1984) 629-635.

Behrisch, R. and Eckstein, W.[2.1,5.5] "Ion Backscattering from Solid Surfaces," Nato ASI, Val Morin, Canada (1984).

Biersack, J.P. and Haggmark, L.G.[5.5,6], Nucl. Inst. and Methods **174** (1980) 257-269.

Boley, C.[6], J. Nucl. Mater. **128-129** (1984) 127-130.

Boley, C. et al.[6], J. Nucl. Mater. **121** (1984) 316-321.

Budny, R. et al.(1982)[6.2], J. Vac. Sci. Tech. **20(4)** (1982) 1238-1241.

Budny, R. et al.(1984A)[6.1], J. Nucl. Mater. **121** (1984) 294-303.

Budny, R. et al.(1984B)[6.1], J. Nucl. Mater. **128-129** (1984) 425-429.

Budny, R.[6.3], private communication.

Burrell, K.[3], J. Comp. Phys. **27** (1978) 88-102.

Carter, L.L. and Cashwell, C.D.[5], *Particle Transport Simulation with the Monte Carlo Method*, ERDA Critical Review Series, 1975.

Cashwell, E.D. and Everett, C.J.[5], *Monet-Carlo Method for Random Walk Problems*, Pergamon Press, Oxford (1959).

Cecchi et al.[6.2], J. Nucl. Mater. **111-112** (1982) 305-310.

Chodura, R.[2.2], "Plasma Flow in the Sheath and Presheath," Nato ASI, Val Morin, Canada (1984).

Conn R.W. et al.(1984A)[2.2], "Technical Assessment of the Critical Issues and Problem Areas in the Plasma Materials Interaction Field," Vol. 1, U.S. Dept. of Energy. Office. of Fusion Energy Report PPG-765 (January, 1984).

Conn, R.W. et al.(1984B)[6.6], J. Nucl. Mater **121** (1984) 350-362.

Connor, J.W.[4.1], Plasma Phys. **19** (1977) 853-873.

Cupini, E. et al.(1983A)[5], "NIMBUS - Monte Carlo Simulation of Neutral Particle Transport in Fusion Devices," Comm. of the European Communities, Directorate General XII - Fusion programme Report 324/9, Brussels (1983).

Cupini, E. et al.(1983B)[5.2], J. Comp. Phys. **52** (1983) 122-129.

Dnestrovskii, Y.N. et al.[4.1], Atomic Energy **32** (1972) 301.

Doll, D.W. et al[6.3], in *Proceedings of the Ninth Symposium on the Engineering Problems of Fusion Research* (IEEE, New York 1981) Vol. 2, 1654-1657.

Dushman, S. and Lafferty, J.M.[6.6], *Scientific Foundations of Vacuum Technique*, John Wiley and Sons, New York (1962).

Dylla, H.F. et al.[6.4], J. Nucl. Mater. **121** (1984) 144-150.

Eckstein, W. and Verbeek, H.[6], "Data on Light Ion Reflection," Max-Planck-Institute für Plasmaphysik, Garching-bei-München, Report IPP 9/32 (1979).

El-Derini, Z. and Gelbard, E.M.[3], Trans. Amer. Nucl. Soc. **23** (1976) 45.

Evans, K. et al.[6], J. Nucl. Mater. **128-129** (1984) 452-457.

Fielding, S.J.. et al.[2.1,6,6.5], J. Nucl. Mater. **128-129** (1984) 390-394.

Freeman, R.L. and Jones, E.M.[5.2,6.1], "Atomic Collision Processes in Plasma Physics Experiments," Culham Laboratory Report CLM-R137 (1974).

Galambos, J.D. and Peng, Y-K.M.[6.4], J. Nucl. Mater. **121** (1984) 205-209.

Gilligan , J. et al.[3], Nucl. Fusion **18** (1978) 63-85.

Greenspan, E.[3], Nucl. Fusion **14** (1974) 771-778.

Grek, B.[6.4], private communication.

Hammersley, H.H. and Handscomb, D.C.[5,5.4], *Monte Carlo Methods*, Methuen, London, 1964.

Harbour, P.J. et al.[6.5], J. Nucl Mater. **128-129** (1984) 359-367.

Harrison, M.[2.1], "Atomic and Molecular Collisions in the Plasma Boundary," Nato ASI, Val Morin, Canada (1984).

Heifetz, D.B. and Budny, R.[6.1], Bull. Am. Phys. Soc. **29** (1984) 1219.

Heifetz, D.B. and Post, D.[5,6.1], Comp. Phys. Comm. **29** (1983) 287-299.

Heifetz, D.B. and Cecchi, J.[6.2], Bull. Am. Phys. Soc. **27** (1982) 1144.

Heifetz, D.B. et al.(1982)[5,5.5,6], J. Comp Phys. **46** (1982) 309-327.

Heifetz, D.B. et al.(1984)[6.4], J. Nucl. Mater. **121** (1984) 189-193.

Heifetz, D.B. et al.[6.3], "Three-Dimensional Calculations of the Transport of Neutral Hydrogen and Molecular Impuirities in TFTR," presented at the Twelth European Conf. on Controlled Fusion and Plasma Physics, Budapest, Hungary, 1985, to be published.

Hogan, J.T.[5], J. Nucl. Mater. **111-112** (1982) 413-419.

Hsu, W.[2.1], *The Gasesous Divertor Experiment*, Ph.D. Thesis, Princeton University, Princeton (1984).

Hughes, M.H. and Post, D.E.[5,6.1,7], J. Comp. Phys. **28** (1978) 43-55.

James, E.[5], Rep. Prog. Phys. **43** (1980).

Janev, R.K. et al.(1984)[6,6.1,6.3], J. Nucl. Mater. **121** (1984) 10-18.

Janev, R.K. et al.(1985)[6], 'Atomic and Molecular Processes in Hydrogen-Helium Plasmas," preprint (1985).

Johnson, P.C. et al., J. Nucl. Mater. **121** (1984) 210-221.

Jones, E.M.[6.1], "Atomic Collision Processes in Plasma Physics Experiments:

II," Culham Laboratory Report No. CLM-R175 (1977).

Kalos M.H. et al.[5], "Monte-Carlo Methods in Reactor Computations," in *Computing Methods in Reactor Physics*, Gordon and Breach, Boston, 1968.

Kato, T. et al.[2.2], J. Nucl. Mater. **128-129** (1984) 1006-1010.

Kaye, S.M. et al.[6.4], J. Nucl. Mater. **121** (1984) 115-125.

Koborov, N.N. et al.[7], J. Nucl. Mater. **128-129** (1984) 691-693.

Langer, W.E.[2.2], Nucl. Fusion **22** (1982) 751-761.

Langley, R.A. et al.[2.2], "Data Compendium for Plasma-Surface Interactions," Nucl. Fusion Supp. (1984).

Lehnert, B.[4.1], Physica Scripta **12** (1975) 327.

Lewis, E.E. and Miller, W.F.[4.1], *Computational Methods of Neutron Transport*, Wiley, New York, 1984.

Mandelrot, B.B.[7], *The Fractal Geometry of Nature*, Freeman, San Francisco, 1982.

Marable, J.H. and Oblow, E.M.[3], Nucl. Sci. Eng. **61** (1976) 90-97.

McGrath, E.J. and Irving, D.C.[5], "Techniques for Efficient Monte Carlo Simulation," ORNL-RSIC-38 (April, 1975).

Morgan, J.G. and Harbour, P.J.[6.4], in *Fusion Technology 1980*, Pergamon Press, Oxford (1981) Vol. 2, 1187.

Ohyabu, N. et al.[6.4], in *Proceedings of the IEEE International Conf. on Plasma Science* (IEEE San Diego, CA 1983) 52-53.

Oshiyama, T. et al.[2.2], J. Nucl. Mater. **128-129** (1984) 996-998.

Owens, D. et al.[6.4], J. Nucl. Mater. **121** (1984) 29-35.

Ozawa, K.[2.2], J. Nucl. Mater. **128-129** (1984) 999-1000.

Parsons, C. and Medley, S.[1,5], Plasma Phys. **16** (1974) 267-273.

Petravic, M. et al.(1982)[2.1,6], Phys. Rev. Letts. **48** (1982) 326-329.

Petravic, M. et al.(1984A)[6], J. Nucl. Mater. **128-129** (1984) 91-99.

Petravic, M. et al.(1984B)[6], J. Nucl. Mater. **128-129** (1984) 111-113.

Pfeiffer, W.[3], "Calculation of Neutral Transport in a Plasma Using a Neutron Transport Method," General Atomic Co. Rep. GA-A13995 (1976).

Post, D.E. and Lackner, K.[6,6.6], "Plasma Models for Impurity Control Experiments," Nato ASI, Val Morin, Canada (1984).

Post, D.E.(1982)[6.1], J. Nucl. Mater. **111-112** (1982) 383-395.

Post, D.E. et al.(1984)[6.4], J. Nucl. Mater. **121** (1984) 171-178.

Post, D.E. and Singer, C.[2.2], J. Nucl. Mater. **128-129** (1984) 78-90.

Reiter, D. and Nicolai, A.[5], J. Nucl. Mater. **111-112** (1982) 434-439.

Robinson, M. et al.[5.5,6], Phys. Rev. **B9** (1974) 5008.

Roth, J.[2.1], "Chemical Sputtering and Radiation Enhanced Simulation of Carbon," NATO ASI, Val Morin, Canada (1984).

Russell, R.M.[5.6], "The CRAY-1 Computer System," Communications of the ACM (January, 1978) 63-72.

770

Ruzic, D. (1984)[2.2], *Total Scattering Cross-Sections and Interatomic Potentials for Neutral Hydrogen and Helium on Some Noble Gases*, Ph.D. Thesis, Princeton University, Princeton (1984).

Ruzic, D. et al.[6.3], Bull. Am. Phys. Soc. **29** (1984) 1334.

Ruzic, D.(1985)[7], private communication.

Sagara, A. and Kamada, K.[2.2], J. Nucl. Mater. **111-112** (1982) 812-815.

Saito, S. et al.[6.4], J. Nucl. Mater. **121** (1984). 199-204.

Sakharov, A.P.[2], in *Plasma Physics and the Problem of Controlled Thermonuclear Reactions*, Pergamon Press, Oxford (1961) Vol. 1, 21.

Schneider, W. et al.[6], in *Eleventh Conf. on Controlled Fusion and Plasma Physics*, European Physical Soc. **7d** (September, 1983).

Schneider, W. and Lackner, K.[6,6.4], in *Int. Conf. on Plasma Physics*, (Goetborg, Sweden, 1982).

Seki, Y. et al.[5,5.5], Nucl. Fusion **20** (1980) 1213-1226.

Shimada, M. et al.[6,6.4], Nucl. Fusion **22** (1982) 643-655.

Singer, C. et al.(1985A)[6], J. Vac. Sci. and Tech. bf A 3(3) (1985) 1183-1187.

Singer, C. et al.(1985B)[6], "BALDUR: A one-Dimensional Plasma Transport Code," to appear in Comp. Phys. Comms.

Smith, D. et al.[6], in *Ninth Symp. of Eng. Problems of Fusion Research* (IEEE, New York, 1981) Vol. 1, 719-722.

Sobol, I.M.[5], *The Monte Carlo Method*, Mir, Moscow, 1975.

Sotnikiv, V.M.[7], Sov. J. Plasma Phys. **7(2)** (1981) 236-239.

Spanier, J. and Gelbard, E.M.[5], *Monte-Carlo Principles and Neutron Transport Problems*, Addison Wesley Pub. Co., Reading, Mass., 1969.

Stangeby, P.[2.2], "The Plasma Sheath," Nato ASI, Val Morin, Canada (1984).

Tabata, T. et al.[6], "Data on the Backscattering Coefficients of Light Ions from Solids," Inst. of Plasma Physics, Nagoya University Report No. IPP-J-AM-18 (1981).

Tamor, S.[4.2,6.1,7], J. Comp. Phys. **40** (1981) 104-119.

Tendler, M.[2.2], Plasma Physics **25** (1983) 767-779.

Terasawa, M. ct al.[2.2], J. Nucl. Mater. **128-129** (1984) 1001-1005.

Thomas, E.W. et al.[2.2], J. Nucl. Mater. **111-112** (1982) 809-811.

Vernickel, H. et al.[6], J. Nucl. Mater. **128-129** (1984) 71-76.

Wagner, F. et al.[6.4], Phys. Rev. Lett. **49** (1982) 1408.

Weisheit, J.[6], J. Phys. B: Atom Molec. Phys. **8** (1975) 2556-2564.

Williams, M.M.R.[4.1], in *Advances in Nuclear Science and Technology, Vol. 7*, Academic Press, New York (1973) 283.

PARTICLE CONFINEMENT AND CONTROL IN EXISTING TOKAMAKS

S.A. Cohen

Plasma Physics Laboratory, Princeton University
Princeton, New Jersey 08544

ABSTRACT

Measurements of particle losses relevant to plasma-wall interactions in tokamaks are reviewed. The methods commonly used to measure the confinement and loss of thermal plasma, impurity ions, suprathermal electrons and ions, and photons are compared and evaluated. Refueling and particle control techniques, including pulsed gas and pellet injection and gettering and discharge cleaning, are described.

I.   INTRODUCTION

A major objective of tokamak research is the confinement of a high temperature plasma for sufficient time that fusion reactions liberate more energy in the plasma than is lost from it by conduction, convection, and radiation. The particle confinement aspect of the density/temperature/energy-confinement-time triad is pivotal for a number of reasons. Most importantly, the correct density profile must exist in the plasma to maximize the fusion rate. The density must be high where the temperature is high, optimally both occurring as far from material surfaces as possible. The proper ratio of reacting species must be present. For a conventional, thermal, D-T burning reactor, equal deuteron and triton densities are necessary. Contamination by α-particle ash or other impurities must be minimal. This purity requirement is stringent because the fusion rate is proportional to the product of the reacting species' densities, $n_D n_T$. A fully ionized iron atom, Fe^{+26}, at constant plasma electron density, displaces

773

26 hydrogen ions. Thus a reactor requires methods to reduce impurity influx, to flush out impurities and helium ash, and to refuel the reacting species which are lost by burn-up and plasma transport. The goals of improved D-T ion confinement and enhanced impurity ion and α-particle loss may be difficult to achieve simultaneously.

For research tokamaks, where a high fusion rate is not the main aim, species control may mean the correct He^3/D ratio for ICRF experiments[1] or the correct nuclear polarization for quantifying effects on neutron emission.[2] The plasma density profile and species mix are functions of plasma fueling techniques, plasma transport properties, and impurity generation processes. Species mix and nuclear polarization also depend strongly on wall materials[3] and recycling events.

Different fueling methods will result in different plasma behavior such as energy confinement time.[4] Fueling on axis, as is possible by using high velocity pellet or energetic neutral-beam injection, will cause the residence time, $\tau^\dagger$, of the injected particles to be longer than of gas-fueled particles. $\tau^\dagger$ is defined as the time interval in which the number of newly-fueled particles remaining in the plasma drops to 1/e of its initial value. Recycling, R, is assumed to be zero. In Fig. 1, $\tau^\dagger$ is shown[5] for particles initially located on annuli concentric to r = 0. The limiter is located at r = a. The diffusion coefficient, D, was assumed to be independent of r and t, and the boundary (wall) to be perfectly absorbing. As shown in Fig. 1, $\tau^\dagger$ of particles originally on the axis is much longer than of those originally near the edge. Note that the time for the density to drop to e^{-m} of its initial value is <u>not</u> $m\tau^\dagger$, but approximately $m\tau_p$ for large m.

The global particle confinement time, τ_p, defined by

$$\tau_p \equiv \left(\frac{d}{dt}(\ln N)\right)^{-1} = \int_{\substack{torus \\ volume}} n dV / \int_{\substack{torus \\ surface}} \Gamma dA , \qquad (1)$$

is another measure of particle confinement. As the right hand side of Eq. (1) indicates, τ_p is determined by particles lost at the edge (Γ is the outflux from the plasma), and particles confined in the interior (n is the number density in the plasma and N is the total number of particles in the plasma). τ_p is quite different from $\tau^\dagger$. For D(r) = constant and R = 0, particles initially concentrated in a narrow annulus will eventually be so spread out as to form a density profile described by $J_0(P_{01}r/a)$, where $P_{01} = 2.405$. J_0 is shown in Fig. 1 and labeled as $n_e(r)$. For an electron density profile proportional to J_0, the global particle confinement time is

$$\tau_p = \frac{a^2}{(P_{01})^2 D} \,.$$ (2)

Again, cylindrical geometry, no recycling or refueling, and a constant D have been assumed. For $D = 10^4$ cm^2/s and a = 40 cm, τ_p is 27.7 ms, as indicated in Fig. 1.

The global energy confinement time, $\tau_E \equiv [d(\ell nE)/dt]^{-1}$, is affected by the global particle confinement time and other factors. That is, there are several energy loss channels including particle heat conduction and convection, radiation, and charge exchange. Measurements of the local contribution of each to energy transport are extremely difficult in hot plasmas. As an alternative, simulations of transport are performed. Constraints

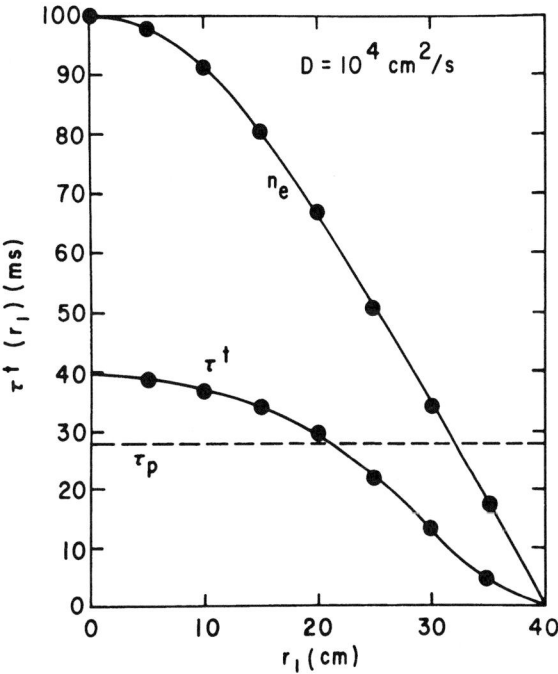

Fig. 1. Characteristic residence time, $\tau^\dagger$, versus initial position, r_1, for diffusing particles in cyclindrical geometry. The wall is at r = 40 cm and is assumed to be perfectly adsorbing. The diffusion coefficient is set to 10^4 cm^{2}s^{-1}. Also shown in the global particle confinement time, τ_p, for the case where the density profile, n_e, is proportional to J_o. (Ref. 5.)

on the transport codes include comparisons with experimentally
determined electron temperature and density profiles, ion
temperatures, and bolometrically measured radiation. Particle
transport at the edge is particularly difficult to measure, in
part, because of severe poloidal and toroidal asymmetries. Thus a
full set of constraints does not exist. Different codes use
different particle transport rates which cause the energy loss
attributable to particle convection to vary. Results of a TRANSP
simulation[6] are shown in Fig. 2 for a beam-heated PDX plasma.
Convective losses are seen to be about 5% of the input power.
BALDUR simulations[7] of the same discharge show convective losses
to be about 15% of the input power because a 3-fold shorter
particle confinement time is assumed in that code. Neither choice
may be correct.

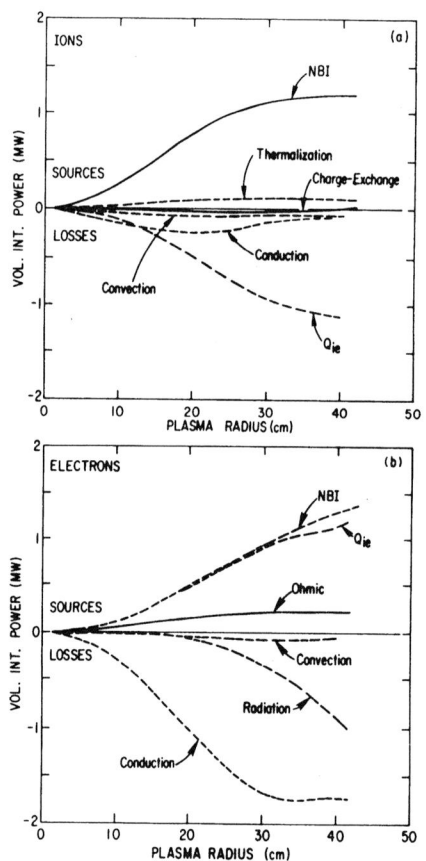

Fig. 2. Calculated volume integrated power flow for (a) ions
and (b) electrons in a beam-heated PDX discharge.
(Ref. 6.)

Detailed observations have been recently made[8] showing a strong positive correlation between particle and energy confinement. Several tokamaks[9-13] have seen simultaneous increases in particle and energy confinement times in beam-heated diverted plasmas. Both τ_p and τ_E increase a factor of 2 ÷ 3 at the onset of improved energy confinement, called the H-mode. But earlier work, for example, on Alcator A in 1978, produced cases where $\tau_E \propto \tau_p^{-1}$ for ohmic discharges. Thus no generalization can yet be made on the link between τ_E and τ_p. However, the correct τ_p is very important because it determines the average energy of the lost particles, to which impurity generation is sensitive.

The transport of particles from the core of the plasma to the edge and ultimately to the tokamak walls may result in impurity generation. The contributing processes, which include thermal, plasma, and binary collision phenomena, depend on the details of particle confinement. Important questions are: What are the fluxes and fluences of the lost particles? What are their energies? Is the plasma potential high? The resulting impurity concentrations are governed by the impurity confinement time, which may differ from both that of the thermal and the suprathermal particles.

This lecture will address several aspects of particle confinement and control in existing tokamaks. The emphasis is placed on particle confinement as it affects plasma-wall interactions. There will be no discussion of the microscopic mechanisms responsible for or the theoretical models describing plasma transport. Instead results of measurements of the confinement of different species are presented. Speculations on new effects caused by photon loss are given. The experimental techniques for determining particle confinement times are reviewed and the strengths of each evaluated. In this way the student may judge the suitability of each and determine the possible need for new techniques. Details of experimental equipment, however, are not presented. These can be found in the Varenna Summer Schools on Plasma Diagnostics. The global equation of particle confinement is discussed to provide a common point for comparing confinement times of different species. The observed effects of density, auxiliary heating, and rf current drive on particle confinement are briefly described. Refueling the plasma by means of gas puffing and pellet injection is also briefly addressed.

Additional methods exist, beyond refueling, for effecting particle control. Among the earliest schemes proposed were limiters and divertors. The physics of these is discussed fully in other lectures in this volume. In this lecture only the behavior of gas fueling efficiency for these two configurations will be discussed. Other means of density control include gettering and limiter and wall conditioning. The latter has been predominantly carried out by discharge cleaning. The different types of cleaning discharges will be reviewed. Both evaporable and bulk getter systems will be described.

II. PARTICLE CONFINEMENT
A. The Global Model

A multispecies plasma cannot be perfectly confined by static electric and magnetic fields. The entropy of the plasma must increase. For a tokamak plasma, relaxation of the profile occurs by Coulomb collisions between the different species, by plasma instabilities driven by free energy sources, and by field "errors," that is, the wandering of field lines from closed magnetic surfaces. The resulting plasma transport is predicted to depend strongly on parameters such as the pressure gradients, field ripple and rotational transform. Detailed theories exist for plasma transport under most conditions relevant for fusion reactors. Experimental verification of any of these theories is lacking, though general trends have been successfully modelled by drift wave theories.[14] For the purposes of predicting plasma-materials interactions, a starting point is to consider only the global confinement time. We first write for each species a continuity equation which lumps all loss processes, $\partial n/\partial t$, occurring in an infinitesimal volume into a single expression related to the flux, Γ, into the volume and the particle creation rate, S. In cylindrical coordinates, the equation for the ith species is

$$\frac{\partial n_i}{\partial t} = - \frac{1}{r} \frac{\partial}{\partial r} (r\, \Gamma_i) + S_i \; , \qquad (3)$$

where

$$\Gamma_i = -D_i \nabla n_i + n_i v_i \; . \qquad (4)$$

The loss is ambipolar, so the equations describing electron and ion losses must be coupled. In Eq. (4), the Fick's law term is a summation of all diffusive processes into a single diffusion coefficient, D_i. This may well vary in time and position. Furthermore, it may be incorrect since it asserts that diffusion only results from the self-gradient. The second term similarly may vary in $\bar{r}$ and t and is a sum of all convective processes into a single convective velocity, v_i. Toroidal and poloidal symmetry

778

are also assumed. Equation (3) is applicable in volumes linked by closed field lines, i.e., field lines which do not terminate on material surfaces. Open field lines allow another loss process, flow along $\bar{B}$. So a term $-n/\tau_\parallel$ is added to Eq. (3) for volumes outside the limiter radius. $\tau_\parallel$ is the average time for flow parallel to $\bar{B}$ out of the annulus.

In most tokamaks it is observed that if external gas sources are shut off the plasma density will decrease. This means two things: firstly, the particle confinement time is shorter than the discharge duration; and secondly, not all particles lost from the plasma return to it. It is possible to maintain a constant plasma density by supplying neutrals to replace those particles lost from the plasma. This is a global description of particle confinement, as compared to the microscopic description embodied in Eq. (3). To turn Eq. (3) into a global model, integrate over the tokamak volume out to the limiter radius, a:

$$\int_0^a dV \, \frac{\partial n_i}{\partial t} = \int_0^a dV \left[-\frac{1}{r} \frac{\partial}{\partial r} \left(r\Gamma_i \right) + S_i \right] , \text{ or}$$

$$\frac{dN_i}{dt} = -\Gamma_i \, A\big|_a + \gamma_E \phi_E + \gamma_R \phi_R \tag{5}$$

where N_i is the total number of i-type particles in the tokamak, A is the surface area of the plasma, ϕ_E and ϕ_R are the external and recycling gas sources respectively, and γ_E and γ_R are the fueling efficiencies. The source term, S_i, for hydrogen ions is given by the sum of additional atoms introduced into the plasma by fueling $\gamma_E \phi_E$ and recycling $\gamma_R \phi_R$ minus the number of ions lost by recombination. This latter term is small and generally neglected. Its quantitative value is discussed in Section IIB-3. The term $\Gamma_i A\big|_a$ represents the loss of particles from the plasma and gives the definition of the global particle confinement time which we saw in Eq. (1),

$$\tau_{ip} \equiv N_i / \Gamma_i A\big|_a . \tag{6a}$$

The recycling source, ϕ_R, is related to the particle outflux $\Gamma_i A\big|_a$ by the recycling coefficient defined as

$$R \equiv \phi_R / \Gamma_i A\big|_a . \tag{6b}$$

779

As described in the lectures of Harrison and Post, the atomic and molecular phenomena which occur during plasma bombardment of neutral species include numerous processes other than ionization. Some of these processes, e.g., charge exchange and Franck-Condon dissociation, will result in a fraction of the incident neutrals (ϕ_E and ϕ_R) being directed back to the wall with several eV of energy. For molecular hydrogen, the highest fraction of ionization is 50%, as shown in Fig. 3. The fueling efficiency is thus defined as the fraction of neutrals which turn into ions and enter the plasma at $r < a$. One way to estimate $\gamma_R \phi_R$ and $\gamma_E \phi_E$ is by a computer simulation which follows the trajectories of molecules from the gas source into the plasma, evaluating the probabilities for each molecule/plasma and molecule/wall interaction along the trajectory.

Recycling refers to the return to the plasma of hydrogen lost to the walls and limiters. The recycling source, ϕ_R, discussed in detail at the microscopic level in the lectures of Behrisch and Roth, represents the sum of several types of events, of which the three main ones are:

1. Direct reflection of hydrogen ions by a solid surface. The ions are generally (95%) neutralized in the process.

2. Reemission as molecules of hydrogen implanted in a solid. This occurs in times ranging from 10^{-4} to 10 seconds.

3. Desorption of adsorbed hydrogen atoms or molecules by impacting particles.[15]

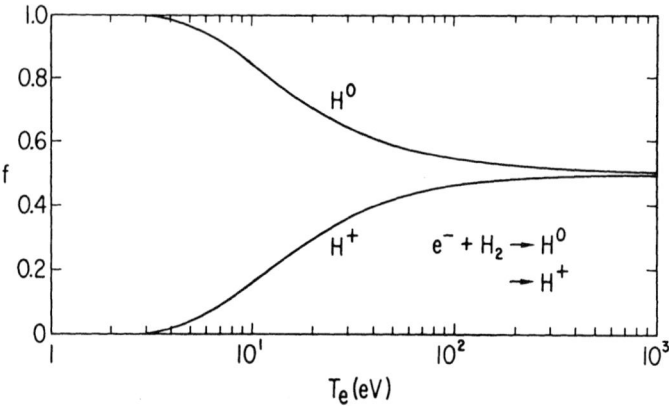

Fig. 3. H^0 and H^+ yield fractions from H_2 dissociation. [D.E. Post, J. Nucl. Mater. 111 & 112 (1982) 383.]

Again, the global model, (Eq. 3), is nothing more than particle conservation. Also, it is not possible to generate a self-consistent model of particle confinement considering only one species. Indeed first simulations even showed two component models to be inadequate. Four component models were then developed.[16] Energy conservation is also not included in the global model. Continuity equations for momentum and heat flow must be added to make a time-dependent self-consistent solution. A related point is that charge-exchange events, which do not alter N, do change the energy distribution and thus affect the plasma ions. It is possible to model this with the Boltzmann equation.[17]

The next five subsections will discuss measurements of the loss from tokamaks of bulk plasma (thermal electrons and hydrogen ions), impurity ions, suprathermal ions, suprathermal electrons, and photons. In each case only the global model will be used since it describes losses from the entire plasma to the walls or limiters. However toroidal and poloidal nonuniformity are also addressed because they impact experiment design and data interpretation. The complexity of tokamak phenomena has not allowed thorough studies across the breadth of tokamak operating regimes. We present several examples which serve to illustrate the techniques, methodologies and results. These examples come from ohmic, beam-heated and ICRF-heated discharges.

B. Thermal Plasma

The discussion of thermal plasma confinement will be divided into sections based on the measurement technique. The four types of measurements considered are the ones most commonly used: electron density behavior, H_α emission, charge-exchange emission, and probe characteristics in the edge plasma. Before proceeding with discussions of these four techniques a point of similarity between three of the techniques is presented to gain insight into a possible pitfall in the data interpretation. To view H_α light, detect charge-exchange particles, or measure a rise in electron density, neutral atoms must be introduced into the plasma. These neutrals may come from an external gas source, ϕ_E, or from recycling, $RN_e/\tau_{ep} \propto RD$. In contrast, the global confinement time is not an implicit function of ϕ_E or R, only of D, and is proportional to D^{-1}. It is clear, then, that R and ϕ_E must be known for τ_{ep} to be evaluated. Recycling and gas feeds are localized phenomena, in part due to nonuniformity in heating or machine configuration, in part due to the scale length for collisions. These must be carefully evaluated when planning and interpreting experiments to measure particle confinement time.

B-1 Electron Density Behavior

Systematic studies of electron density behavior have been performed on many tokamaks. On TFR the experiments[18,19] consist of monitoring by microwave interferometry the line-averaged electron density. The gas feed rate and pressure, P, are also measured. They assume a simplified form of Eq. (5) where $\gamma_E = \gamma_R = 1$. (This may not be valid, particularly for short time scales, $t \lesssim 0.1$ sec. The reason is that the Franck-Condon effect or charge exchange can result in atoms implanted in the walls. In TFR these atoms diffuse out in a time of ~0.1 sec.) So the final equation used for TFR is

$$\frac{dN_e}{dt} = (R-1) \frac{N_e}{\tau_{ep}} + \phi_E - \frac{dP}{dt} (V_T - V_p) \tag{7}$$

where P is the gas pressure in the torus, V_T is the torus volume, and V_p the plasma volume. The gas pressure in the torus is a "mixture" of the external gas feed and recycling. Equation (7) can be rearranged to yield an effective electron confinement time,

$$\tau^*_{ep} \equiv \frac{\tau_{ep}}{1-R} = \frac{N_e}{\phi_E - (dN_e/dt) - (dP/dt)(V_T - V_p)} . \tag{8}$$

So by this technique, τ_{ep} is not determined. If some other means exists for measuring R, then τ_{ep} can be found. Conversely, if τ_{ep} can be found by a different technique, then R is determined.

$$R = 1 - \frac{\tau_{ep}}{\tau^*_{ep}} .$$

The four measured parameters, N_e, dN_e/dt, ϕ_E and P, are shown in Fig. 4 for an ICRF-heated discharge. For these discharges, gas fueling causes a density increase at $0 < t < 0.35$ sec. The effective particle confinement time rises steadily from 1 ms at $t = 0$ to ~20 ms at $t = 0.2$ sec. From probe measurements, to be described later, τ_{ep} is estimated to be ~10 ms. Based on this estimate, R is found to increase from 0.8 at 100 ms to 1.0 at 600 ms. These data are examples of the case mentioned earlier, that in most tokamaks the particle confinement is much shorter than the discharge duration and that some particles lost to the wall do not recycle. It is also clear from Fig. 4 that, for this discharge, the initiation is rather different from the quasi-steady-state portion, 100 to 300 ms.

In Fig. 5, case (c), the recycling coefficient is seen to exceed unity at $t = 0.18$ sec. Using Eq. (7), τ_{ep}^* calculated for this time is negative! $R > 1$ occurs when the tokamak walls are dirty, the discharge duration is long, the density is high, or

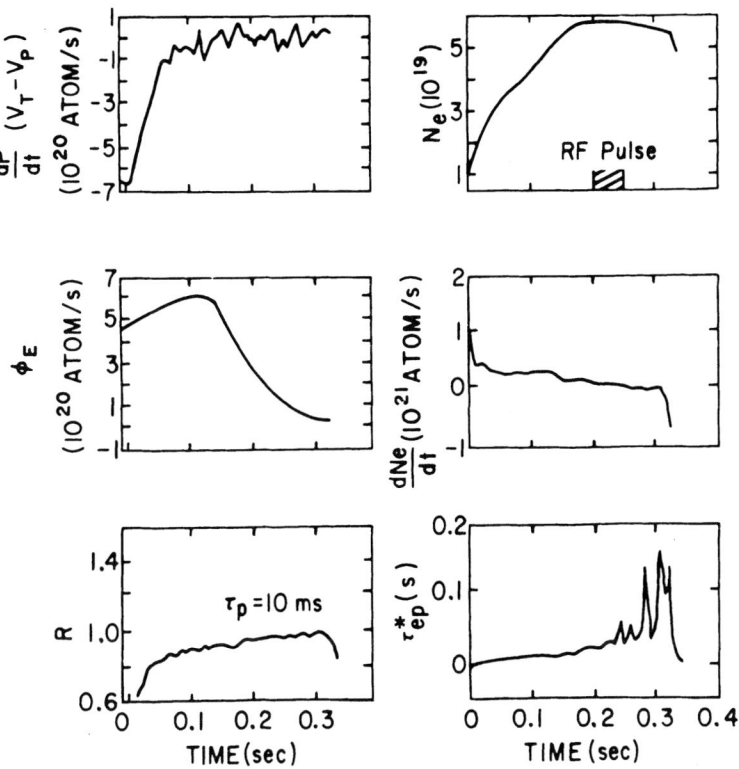

Fig. 4.   Time dependence of N_e, dN_e/dt, ϕ_e, dP/dt and R for an ICRF-heated discharge in TFR-600. The value of τ_p was determined from Langmuir probe data; R and τ_{ep}^* were determined from Eq. (8). (Ref. 18.)

plasma-wall interactions are enhanced, as by neutral-beam or rf heating. For long discharges, wall and limiter saturation with hydrogen may be important. Carbon limiters were used in case (c) while inconel ones were used during the case (a) and (b) discharges. Further studies[19] have been made at TFR of gas recycling variation with carbon and inconel limiters and hot and

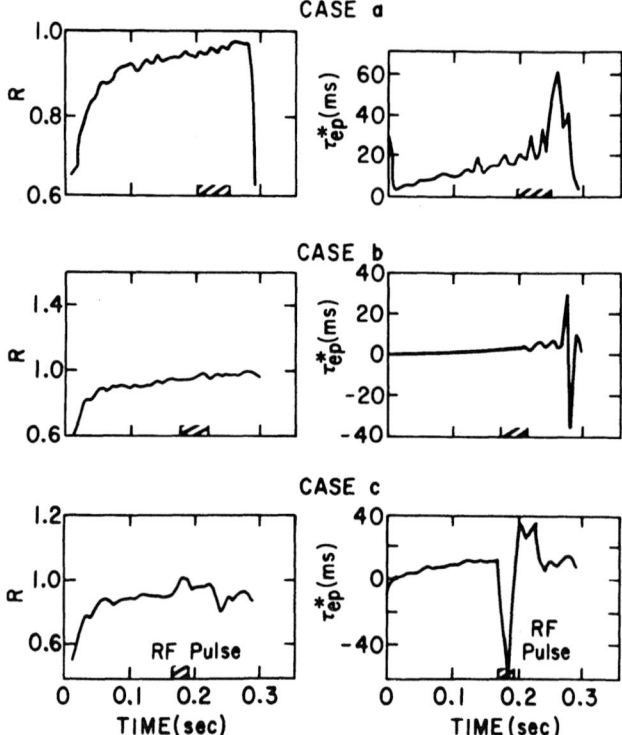

Fig. 5. Recycling and τ_{ep}^* versus time for three different discharges in TFR-600. (Ref. 18.)

cold walls. The main finding is that τ_{ep}^* increases about a factor of two (from 150 to 300 ms) when the wall temperature is raised from 20 °C to 250 °C, regardless of rail limiter material (Fig. 6). This corresponds to an increase of R from 0.9 to 0.95.

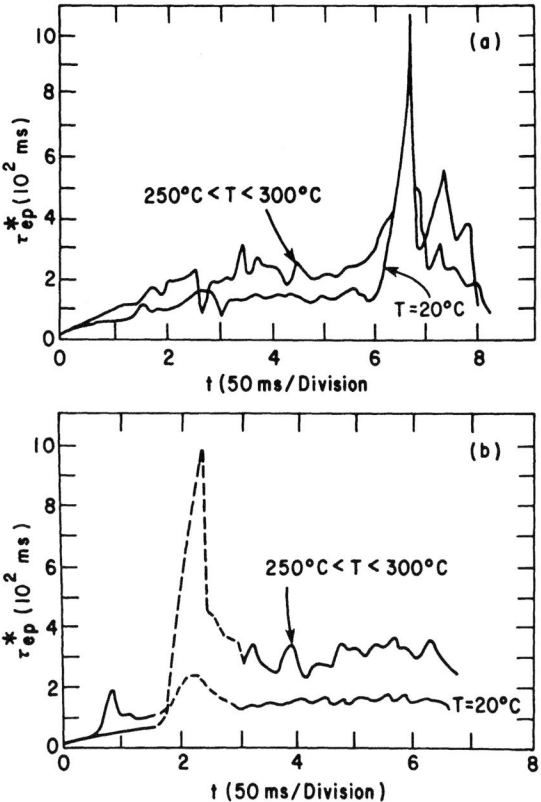

Fig. 6. Time evolution of τ^*_{ep} for TFR discharges with (a) graphite limiters at two different temperatures and (b) inconel limiters at two different temperatures. (Ref. 19.)

Studies of recycling were carried out by measuring n_e behavior in the beam-heated tokamaks PDX,[20,21] ASDEX,[22,23] DITE,[16] and D-III.[24] On PDX and ASDEX, τ^*_{ep} was determined from Eq. (8) at times when there were no external gas feeds, i.e., $\phi_E = 0$. So an important assumption is that τ^*_{ep}, τ_{ep} and R do not change when ϕ_E changes. The solution to Eq. (8) with $\phi_E = 0$, $\gamma_E = \gamma_R = 1$, and $dP/dt \ (V_T - V_p) \ll dN_e/dt$ is

$$N_e(t) = N_e(0) \exp\left(-t/\tau^*_{ep}\right) . \tag{9}$$

The decay of $\bar{n}_e$, the line average electron density, is shown in Fig. 7 for two different types of diverted discharges in ASDEX, one with (DP) and one without (D) gettering in the divertor chamber. ϕ_E is shut off at $t = t_o$. The ensuing density behavior is clearly not exponential as predicted by Eq. (9). To remedy this, τ_{ep}^* is defined by Eq. (1) evaluated at $t = t_o + 0.01$ sec. Gettering in ASDEX reduced τ_{ep}^* by a factor of 7. Both of these discharges had similar energy confinement times. Another point concerning the constancy of τ_{ep}^* and R when $\phi_E = 0$ was also noted by the ASDEX group. Evaluating τ_{ep}^* from the initial slope of dN_e/dt at $t = t_o$ is actually a determination of τ_{ep}^* for the edge plasma. Confinement in the core is much longer than in the edge, as indicated in Fig. 1. Note, in Fig. 7, that the decay time constant of dN_e/dt increases as t increases beyond t_o. How the electron density profile initially changes is shown in Fig. 8. The edge plasma disappears rapidly leaving behind a more peaked profile.

During intense neutral-beam heating the effective confinement time at first was observed to decrease in tokamaks with rail limiters. This was ascertained by examining the behavior of $d\bar{n}_e/dt$ at constant ϕ_E. However, it was later noted in divertor tokamaks that a transition to a different type discharge, the H-mode, could occur and this resulted in a return of τ_{ep}^* to its value during ohmic heating. Figure 9 shows the behavior of the line average electron density measured in a diverted ASDEX plasma along a chord about 20 cm above the midplane.[23] An ohmic discharge ($I_p \sim 320$ kA) was irradiated with 2.5 MW of neutral beams from $\tau = 1.42$ to 1.62 sec. The electron density is observed to first decrease (L-mode). Then, when the H-mode sets in at $\tau \sim 1.48$ sec, the density returns to, and then exceeds its value during the pure ohmic phase. Throughout this discharge ϕ_E was kept constant.

Because the reflection coefficient of H on surfaces is always nonzero, it is possible using pressure gauges to indicate where plasma bombardment of surfaces is most intense. Charge-exchange H^o which leave the plasma may enter the pressure gauge. There they bounce around until accommodated with the surfaces, perhaps forming H_2 in the process. The gauge fills up with gas until the gas outflow balances the charge-exchange influx. From the kinetic theory of gases we derive the relation between the gas pressure and the charge-exchange flux ϕ_{cx}.

$$\phi_{cx} \ (cm^{-2}s^{-1}) = 3.5 \times 10^{22} \ P \ (mm)/[m \ (amu) \ T \ (^0K)]^{1/2} \ . \quad (10)$$

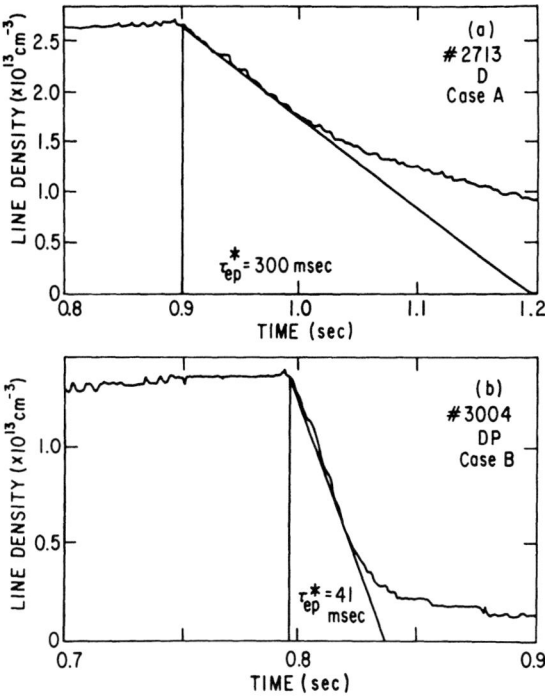

Fig. 7. Decaying line-average electron density after interruption of gas feed. Case A: Ungettered discharge with high recycling; Case B: Gettered discharge with low recycling. (Ref. 22.)

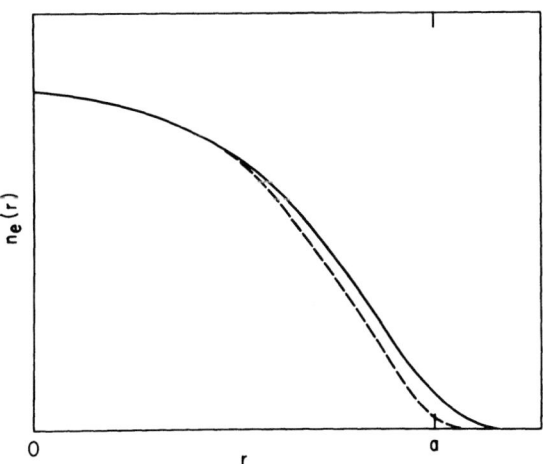

Fig. 8. Electron density profile prior to (solid line) and shortly after (dashed line) external gas sources are shut off.

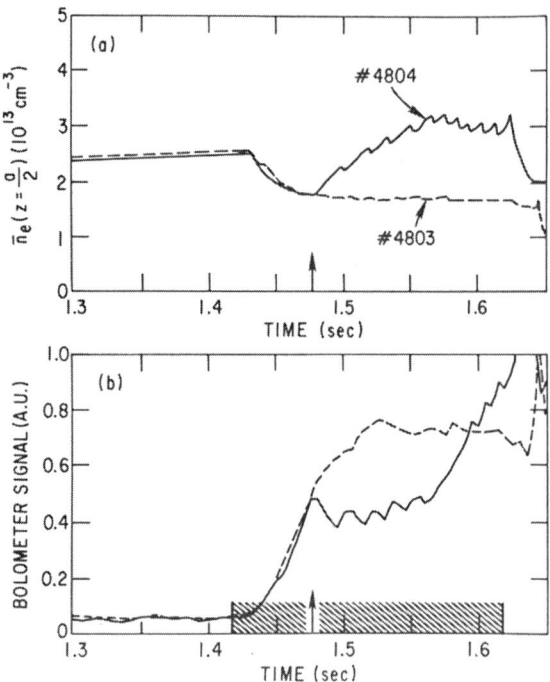

Fig. 9. Time dependence of (a) the line-average electron density and (b) the radiated power for two beam-heated diverted discharges in ASDEX. The dashed line is for an L-mode discharge; the solid line is for an H-mode discharge. The arrow marks the time of transition from L- to H-mode. (Ref. 23.)

Figure 10 shows the variation of gas pressure with time for two different PDX discharges, one with a toroidal bumper limiter, the other with a rail limiter.[20] The low pressure, $\sim 10^{-7}$ T, measured in the divertor chamber for the rail limiter case shows that little particle loss takes place away from the rail limiter, i.e., $\phi_{cx} \lesssim 10^{14}$ cm^{-2}s^{-1}. A similarly low pressure near the wall is found for PDX H-mode discharges and for PLT limiter discharges.[25] Improvements on this technique are made with energy resolved charge-exchange detectors, discussed in Section IIB-3.

Three of the major weaknesses of this simplistic use of the global model to determine τ_{ep} have been indicated above. First, there is no fixed relation between τ_{ep}^*, what is measured, and τ_{ep}, which is the desired quantity. In the ASDEX results, a factor of 7 change in τ_{ep}^* is seen for diverted discharges when recycling in

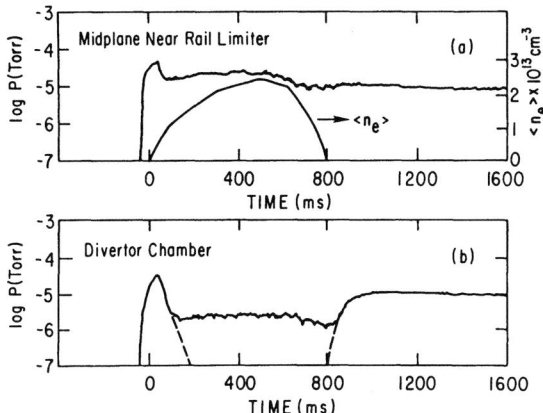

Fig. 10. Time variation of gas pressure (a) at the rail limiter in the PDX midplane and (b) in the PDX divertor chamber. The dashed line in (b) is for rail limiter discharges; the solid line is for inner bumper limiter discharges. (Ref. 20.)

the divertor chamber is reduced by gettering, yet τ_E, and possibly τ_p, did not change. Even larger discrepancies between τ_{ep} and τ_{ep}^* are possible, depending, for example, on wall conditions. In the ALT-1 experiment at TEXTOR,[26] a τ_{ep}^* in excess of 2 sec has been measured while τ_{eE} is less than 50 ms. In TFTR[27] no gas feed is required to sustain n_e in 1 sec duration discharges, thus giving $\tau_{ep}^* \to \infty$. Impurities further complicate the interpretation.

The second problem arises from the assumption that transport will not change as ϕ_E changes. Gas puffing experiments on PLT were performed where the evolution of the entire electron density profile was documented.[28] Detailed studies of the local transport showed a rapid increase of electron density on axis. This could be modeled by $D \simeq 10^4$ cm^2/s and $v \simeq -10^3$ cm/s (inward). (See Fig. 11). The inward velocity is about ×10 higher than required to explain the density profile for $dN_e/dt = 0$.

The third weakness of this measurement is that no estimate of the energy of the particles is possible. The energy of the lost particles is important for assessing the contribution of convective energy loss, and for assessing plasma-material interactions such as impurity generation and reflection.

In summary, measurements of particle confinement by studying the electron density behavior and gas fueling rates yield an effective global electron confinement time, τ_{ep}^*, not the true global confinement time, τ_{ep}, or the local confinement time, $\tau^\dagger$.

Though τ_{ep}^* may be proportional to τ_{eE} and τ_{ep}, as implied by the L- and H-mode ASDEX data, enough counter-examples exist to make this connection suspect. Even for well-conditioned tokamaks, the experimental methodology for determining τ_{ep}^* may change the plasma

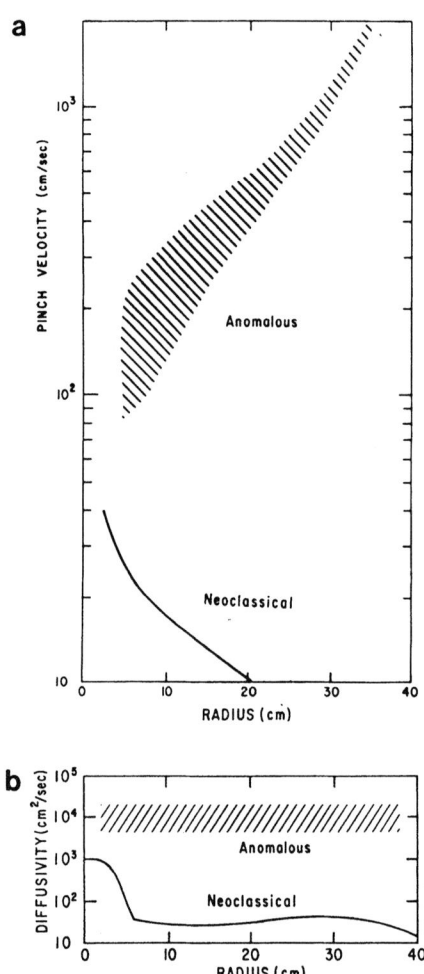

Fig. 11.   (a) Inward pinch velocity (diagonal lines) and (b) diffusion coefficient (diagonal lines) necessary for a code to match the experimental results of a gas-puffing experiment on PLT. Also shown are the predicted velocity and diffusion coefficient based on neoclassical theory. (Ref. 28.)

behavior, e.g., variations in ϕ_E have been observed to cause changes in particle transport. We conclude that τ_{ep}^* is best suited as an indicator of recycling and wall conditions not of true particle confinement.

B-2 Hydrogen Light Emission

A second method to find the particle confinement time is to measure hydrogen photon emission from the tokamak plasma. This is similar to measuring electron density behavior because in both the detected signal is proportional to the product of the electron and neutral densities. From a theoretical point-of-view, Lyman α (L_α) photons would be the best spectra to detect because only the ground and first excited states are involved in computing the neutral hydrogen density from the photon emission rate. But L_α photons are in the near UV and are thus difficult to detect and measure quantitatively. Instead H_α photons are most commonly used. For a pure hydrogen plasma, we start with Eqs. (5) and (6a)

$$\frac{dN_e}{dt} = -\frac{N_e}{\tau_{ep}} + \phi_E + \phi_R \ . \tag{11}$$

ϕ_E and ϕ_R comprise the source term, S_e, which is given by

$$S_e = \int \left(-\alpha(T_e) n_e\, n_+ + \langle \sigma v \rangle_{ion}\, n_e\, n_o \right) dV \ ,$$

where $\alpha(T_e)$ is the recombination rate coefficient, $\langle \sigma v \rangle_{ion}$ is the electron impact ionization rate coefficient, n_o is the neutral density, and n_+ is the hydrogen ion density. The neutral atoms are predominantly in the ground state, $n = 1$. The ionization rate from $n=1$ has been related to the excitation rate[29] to the $n = 3$ level and radiative decay to the $n = 2$ level. The recombination rate coefficient and the inverse ratio of the number of H_α photons to each ionization event are plotted in Fig. 12 as functions of n_e and T_e. Below $n_e \sim 10^{12}$ cm^{-3} this ratio, B_α/S_e, is nearly independent of n_e. Above $n_e \simeq 10^{14}$ cm^{-3} the ratio is nearly proportional to n_e.

Probe measurements in the edge plasma show $T_e(r > a)$ to range from 4 to 40 eV and n_e to reach $\sim 10^{13}$ cm^{-3}. Over this parameter range the H_α yield per ionization event, B_α/S_e, varies a factor of 5.

Thus the equation used to determine τ_{ep} is

$$\tau_{ep} = \frac{N_e}{dN_e/dt + B_\alpha (B_\alpha/S_e)^{-1}} \; , \tag{12}$$

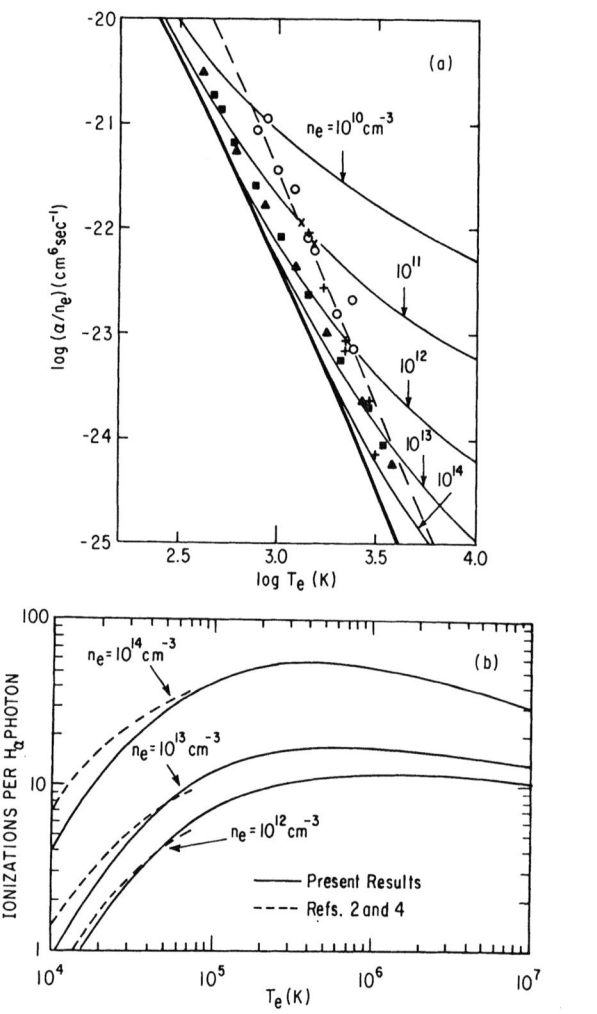

Fig. 12. (a) Radiative recombination rate versus electron temperature for $H^+ + e^- \to H^0$ at several different electron densities from 10^{10} to 10^{14} cm^{-3}. (b) Ratio of ionization events/H_α photons emitted versus electron temperature for electron densities of 10^{12}, 10^{13} and 10^{14} cm^{-3}. (Ref. 29.)

where B_α is the total number of H_α photons emitted per second. Note how this equation differs from Eq. (7) or Eq. (8). In addition to measuring N_e and dN_e/dt (and perhaps ϕ_E and P), the photon emission B_α is recorded. This allows one to determine the recycling, R. Consider a case similar to experience in TFTR: $dN_e/dt = 0$, $dP/dt = 0$, and $\phi_e = 0$. Equations (7) and (12) would respectively be

$$(R-1)\ \frac{N_e}{\tau_{ep}} = 0\ , \tag{7a}$$

$$\tau_{ep} = N_e/B_\alpha (B_\alpha/S_e)^{-1}. \tag{12a}$$

The solutions to Eq. (7a) are either $R = 1$ or $\tau_{ep} = \infty$. Thus τ_{ep} cannot be determined by n_e behavior alone. In contrast, if all the emitted H_α photons were measured, then, by using Fig. 12 and Eq. (12a), τ_{ep} could be determined.

The question of toroidal and poloidal uniformity must be carefully considered. The ionization rate of hydrogen atoms is about 10^{-8} cm^3 s^{-1} for $T_e \sim 10$ eV. For Franck-Condon H^0 with 3 eV average energy, the mean free path for ionization in a plasma of $n_e \sim 10^{13}$ cm^{-3} is about 20 cm. Room temperature molecules have a much shorter mean free path. Tokamaks operating in the late 1960's and early 1970's, such as TM-3, ST, ATC, and ORMAK, did not suffer from problems of nonuniformity since their minor diameters were small and their operating densities low. More recent tokamak plasmas are larger and denser and have marked nonuniformity in H_α emission. Measurements on Alcator-A showed nearly a hundred-fold variation in H_α brightness depending on whether the plasma was viewed near (H_α-high) or away (H_α-low) from the limiter.[30] This is shown in Fig. 13. Even at the gas feed port, the H_α brightness was only 1/6 of that near the limiter. The conclusion is that for that discharge, $\tau_{ep} \ll N_e/\phi_E$. As stated earlier, on Alcator A it was discovered that τ_p decreased as τ_E increased. The behavior of τ_p versus density is shown in Fig. 14. τ_E in Alcator A rises with n_e up to densities of $\sim 6 \times 10^{14} cm^{-3}$. At low n_e, $\tau_{ep} \simeq 2\tau_E$. At high n_e, $\tau_{ep} \simeq 0.1\tau_E$.

Studies of H_α emission were made in PDX during limiter and divertor discharges.[20] Highly localized recycling is evident at the rail limiter (Fig. 15a-I). The recycling spreads out considerably as the plasma is moved to the inner bumper limiter (Fig. 15a-II). Diverted discharges show enhanced H_α emission at the gas injection ports because of the higher ϕ_E required to maintain diverted discharges (Fig. 15a-III). During neutral-beam injection into limiter or L-mode diverted discharges, the H_α light

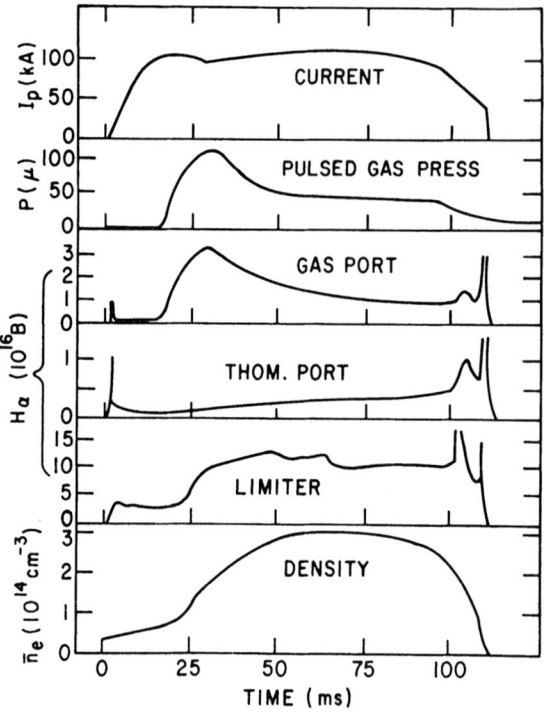

Fig. 13. Plasma current, electron density, and H_α brightness
measured on the Alcator-A tokamak. H_α brightness at
the limiter is even more intense than at the gas
injection port. (Ref. 30.)

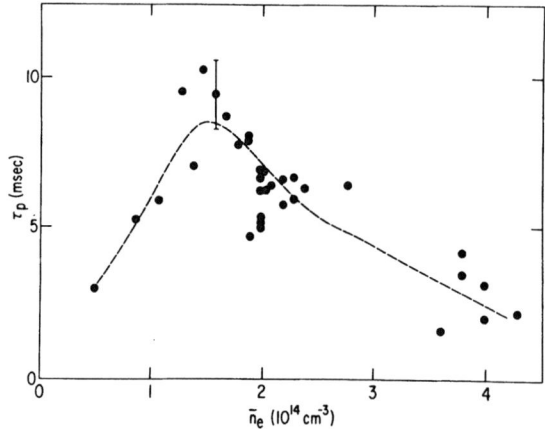

Fig. 14. Particle confinement time, τ_p, versus line-averaged
electron density on Alcator-A. (Ref. 30.)

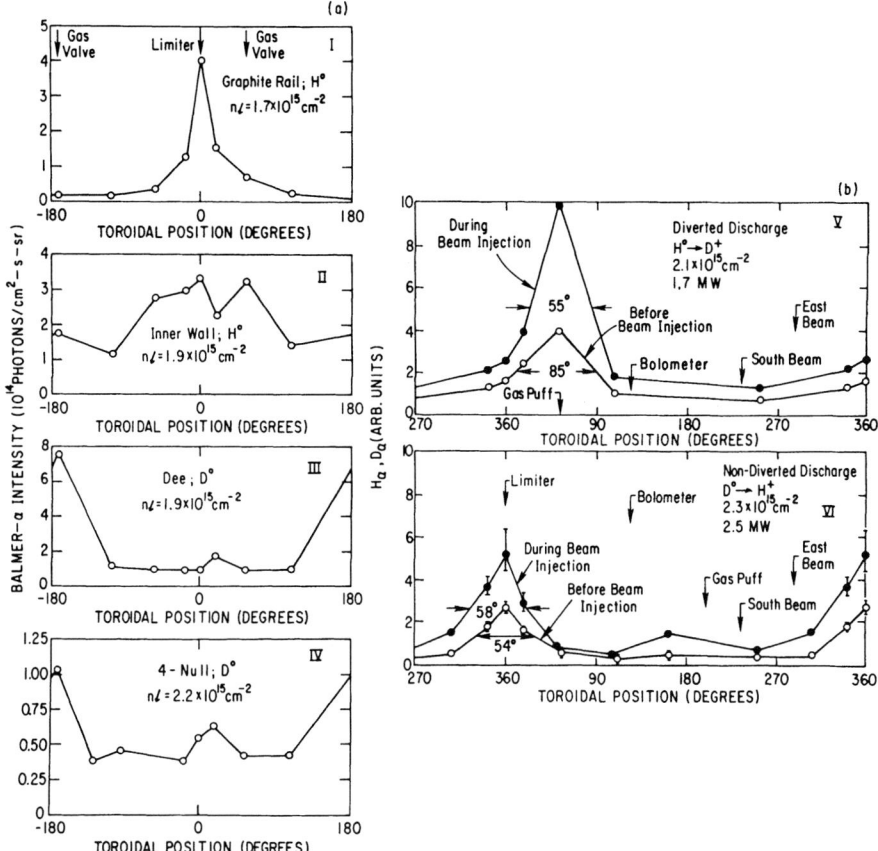

Fig. 15. (a) Toroidal variation of the intensity of H_α/D_α radiation measured at the vessel midplane for various discharge configurations (Ref. 20). (b) Toroidal distribution of H_α emission for divertor and titanium limiter discharges with and without beam injection. (Ref. 31.)

is seen to increase (Fig 15b-I,II). As shown in Fig. 16, from ASDEX, the onset of the H-mode results in a reduction of H_α emission back to pre-injection brightness. Because of this single clear use, i.e., defining the onset of the H-mode, H_α radiation should continue as a favored diagnostic.

The spatial nonuniformity of H_α emission may even be more severe than these line integral measurements indicate. Indeed, cinematic[32] and video[33] recordings show remarkable variations in

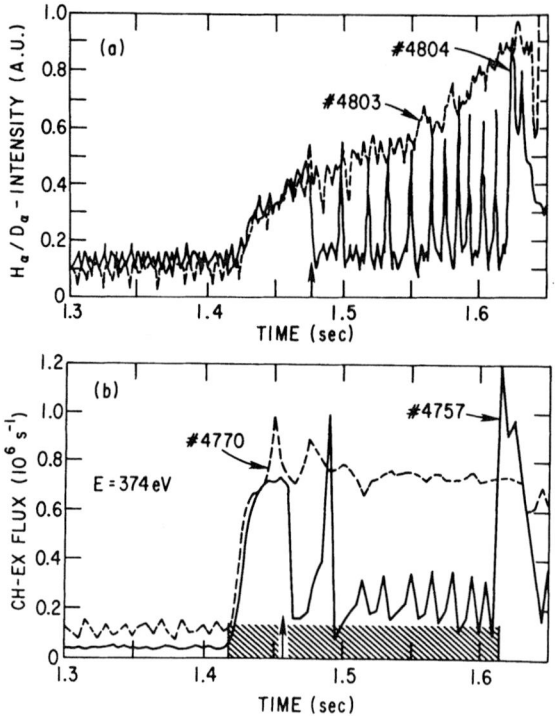

Fig. 16. (a) H_α brightness and (b) charge-exchange flux at 374 eV observed in ASDEX during L- and H-mode discharges (see Fig. 9). (Ref. 23.)

the location and brightness of H_α emission. For example, helical bands of light have been observed around the plasma surface.[32,34] One of the shortcomings of using the H_α line is its variation with T_e and n_e. This could be overcome by observing L_α radiation. Here the ratio of photons per ionization event is approximately 1 and is less sensitive than H_α brightness to n_e and T_e in the relevant parameter range. However, the difficulty in detecting UV radiation has, in the past, made H_α the overwhelming choice. But, because of the aforementioned difficulties, few quantitative studies of τ_{ep} have been made with H_α light.

B-3 Charge-Exchange Efflux

The emission of charge-exchange particles from a plasma results in no change in the ion density. But, as noted earlier, it does result in altered energy distributions, energy loss, particle convection, and wall loading. Because the charge-exchange rate coefficient is about 2-3 times that of the ionization rate coefficient, a neutral hydrogen atom that enters the plasma will probably result in a charge-exchange hydrogen atom hitting the wall. Thus the recycling of ions off the limiter will also cause neutrals to impact the nearby walls.

Neutral hydrogen is an obvious necessity for generating charge-exchange neutrals from the plasma ions. For a plasma with no refueling or recycling, the neutral density would quickly drop to $\sim 10^7$ cm^{-3}, which is the equilibrium value obtained by balancing the recombination rate of $H^+ + e^-$ with the ionization rate at $T_e \simeq$ 1000 eV. For $n_0 = 10^7$ cm^{-3}, charge-exchange losses would be negligible for power loss. However, the confinement is not perfect, R is between 0 and 1, and refueling is necessary to maintain the electron density. Thus the charge-exchange emission is abundant.

The charge-exchange flux observed along a line-of-sight is given by

$$\phi_{cx} = \int_{-a}^{+a} dz \; [n_o n_+ \; <\sigma v>_{cx} - L(z)] \tag{13}$$

where the first term represents the source rate and the second, $L(z)$, the attenuation by such processes as ionization and re-charge exchange.

We have previously identified $\gamma_E \phi_E$ and $\gamma_R \phi_R$ as the number of fueled hydrogen atoms and recycling hydrogen atoms that are ionized in the plasma. The others are lost from the plasma as neutrals and represent the so-called charge-exchange flux. ϕ_{cx} is comprised of Franck-Condon neutrals and charge-exchange neutrals. The total charge-exchange loss, $\Phi_{cx} = \int dA \; \phi_{cx}$, is approximately given by

$$\Phi_{cx} = (1 - \gamma_E) \phi_E + (1 - \gamma_R) \phi_R \; . \tag{14}$$

As shown in Fig. 3, $\gamma_E \simeq \gamma_R \simeq 0.5$. Combining this with Eq. (5) yields

$$\Phi_{cx} \simeq dN_e/dt + \frac{N_e}{\tau_{ep}} \tag{15}$$

or

$$\tau_{ep} = \frac{N_e}{\Phi_{cx} - dN_e/dt} \; . \tag{16}$$

As discussed in Section IIB-1, this flux can be measured by a
simple pressure sensor, such as an ionization gauge. A
quantitative measure of the total charge-exchange flux would
require many ionization gauges placed around the machine. This is
as practical as H_α detectors, which are simply filters and
photomultipliers. What then are the reasons for using more
sophisticated charge-exchange diagnostics, such as stripping cells
with $\vec{E} \times \vec{B}$ analysis or time-of-flight systems? The main reason is
because they are energy dispersive. The energy spectrum of the
efflux is determined, and indirectly the temperature profile of
the plasma. Energy dependent flux spectra, $d\phi_{cx}/dEd\Omega$ are shown in
Fig. 17 for three times during a ohmically-heated PLT discharge:
40, 100, and 200 ms. At discharge initiation, the average energy
may sometimes exceed 300 eV, both because of plasma transparency
at low n_e and because of the formation of a narrow high-
temperature plasma channel. Figure 18 shows the time behavior for
the charge-exchange flux (28 < E < 1000 eV), the average energy,
the reduced power and the calculated sputtered iron influx for the
same PLT discharge.[25,35] The observations were made at a location
100° removed toroidally from the limiter. The high charge-
exchange efflux at $\tau \simeq 50$ ms is due to the high neutral density
during discharge initiation. During the steady-state portion of
the discharge the charge-exchange flux is measured to be about
10^{14} $cm^{-2}s^{-1}$. However ϕ_{cx} is predicted to about 10^{17} $cm^{-2}s^{-1}$ at
the limiter. Discharges with several MW of auxiliary heating have
a several-fold increase in flux but about the same average energy,
~300 eV, that is, when the density is in the range $1.5 < \bar{n}_e < 5 \times$
10^{13} cm^{-3}.

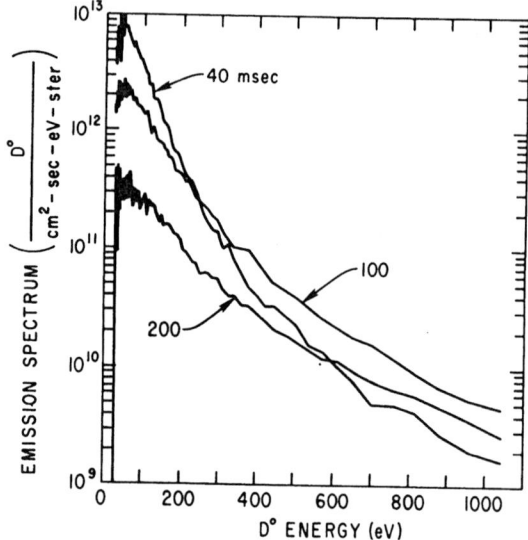

Fig. 17. The differential flux emission function $d\phi_{cx}/dEd\Omega$
versus energy at three different times during an
ohmically-heated PLT discharge. (Ref. 35.)

798

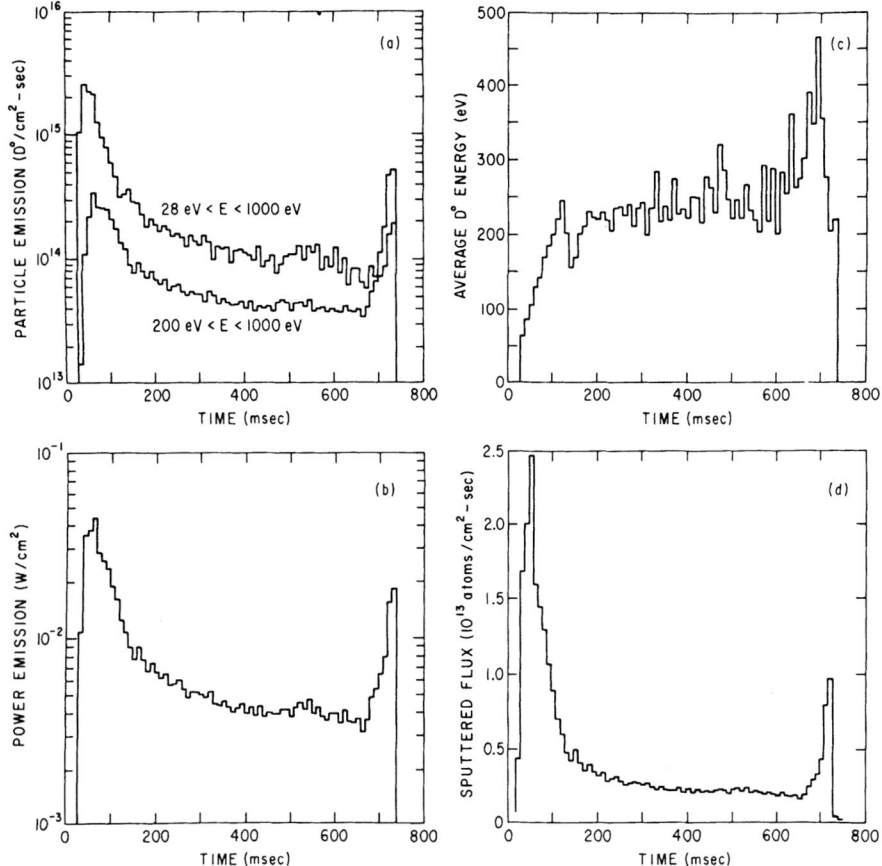

Fig. 18.   (a)   Time evolution of the reduced flux, ϕ_{cx}, is plotted with 10 ms resolution for two energy windows. $T_{i\,axis}$, $\langle n_e \rangle$, and I_p for this discharge were 800 eV, 1.2×10^{13} cm^{-3}, and 450 kA, respectively. (b) Reduced power emission P for the energy window 28-1000 eV. (c) Average D^O energy, $E \equiv P/\phi_{cx}$. (d) Time dependence of the calculated flux of iron atoms sputtered from the stainless steel vacuum vessel wall by D^O neutrals in the energy window 28-1000 eV. (Ref. 35.)

    During ICRF and neutral-beam heating experiments ϕ_{cx} is observed to increase by 10^{15} $cm^{-2}s^{-1}MW^{-1}$, presumably due to increased recycling (decreased τ_{ep}). The time dependences are markedly different for ICRF and neutral beams. For beams the increase in ϕ_{cx} is slow, on the order of τ_E. For ICRF ϕ_{cx}

increases rapidly, in about 1 ms. Figure 19 shows the time and position variation of ϕ_{cx} for observations made near and away from the limiter during an ICRF-heated discharge.[36] The 5-fold increase in ϕ_{cx} during the ohmic portion of the discharge is only a fraction of the poloidal asymmetry of ϕ_{cx}. Measurements taken in ASDEX,[37] Fig. 20, show a hundred-fold increase in ϕ_{cx} at the limiter versus away from it. Results such as these on PLT and ASDEX lead to speculation[25,37a] that the "missing" energy in many tokamaks' energy balance studies could be found in the charge-exchange flux at high recycling locations.

In the preceeding data only the low energy portion of $d\phi_{cx}/dE$ was presented. The charge-exchange spectra at higher energies for ICRF-heated plasmas is shown in Fig. 21. Particles with energies in excess of 100 keV are seen. These are the results of ICRF resonances. In beam-heated plasmas the highest energy of ions need not be the beam energy. Instabilities may accelerate ions above their injection energy. Further discussion of suprathermal particles is later in Section II-E.

A second reason for using sophisticated charge-exchange detectors is that they can be mass dispersive. The ratio of hydrogen isotope abundances can then be readily determined. The

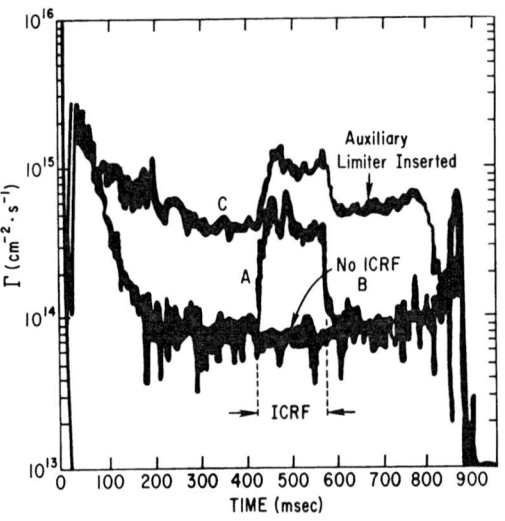

Fig. 19.   Time dependence of the charge-exchange flux ($\phi_{cx} = \Gamma$) during ICRF-heating on PLT. (A) ϕ_{cx} with ICRF at a position away from the limiter. (B) ϕ_{cx} without ICRF [same position as (A)] and (C) ϕ_{cx} with ICRF at a position near the limiter. (Ref. 36.)

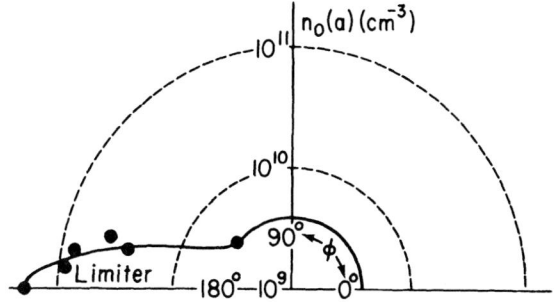

Fig. 20. Toroidal variation of the edge neutral density on ASDEX
unfolded from the toroidal variation of the charge-
exchange efflux. (Ref. 37.)

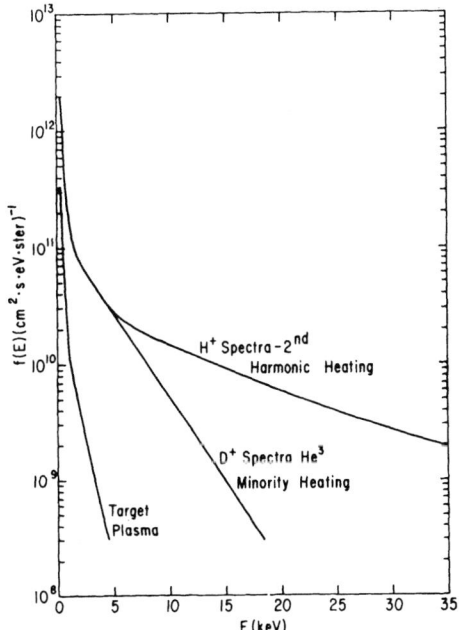

Fig. 21. Differential charge-exchange flux $d\phi_{cx}/dEd\Omega$ for three
different types of plasmas on PLT. The ICRF-heated
cases have a rf power of about 2 MW and an electron
density of about 3×10^{13} cm^{-3}.

sensitivity is orders of magnitude greater than spectroscopic separation of D_α from H_α.

A third reason for using charge-exchange detectors is that plasma-materials interactions can be studied. Neutrals formed by the reflection of ions from limiter or neutralizer plate surfaces can be seen directly,[37] without the need for electron stimulated excitation. This removes the variability from which H_α brightness suffers as a function of n_e and T_e. From measurements such as those on ASDEX, the flux of ions to the neutralizer plate can be estimated.

A question that frequently arises during the use of charge-exchange detectors is calibration. Mass analyzed calibrated beams are one approach. A simpler method[35] has been described wherein the charge-exchange flux is related to the change in electron density by the equations

$$\phi_{cx} \simeq \int_{-a}^{+a} d\ell \; n_o n_+ \langle \sigma v \rangle_{cx} \; , \tag{17}$$

and

$$\frac{dn_e}{dt} = n_e n_o \langle \sigma v \rangle_{ion} \; . \tag{18}$$

Thus by measuring n_e and dn_e/dt and estimating the rate coefficients from T_e and T_i, an effective calibration can be carried out during the start-up of a tokamak.

Other in-situ methods to calibrate the charge exchange include the use of solid-state probes such as silicon collectors and carbon resistance probes (see the lecture of Manos and McCracken). Yet another way is to measure the pressure in a box attached to the tokamak as described in Section IIB-1.

B-4 Probe Measurements

The theories of electric and surface probes have been discussed in the lectures of Stangeby and Manos and McCracken. In principal, the particle confinement time can be determined by measuring the particle flows on the open field lines, those that intersect material surfaces. In practice, accuracy in this type of measurement is difficult to achieve since complete poloidal (or toroidal, depending on the limiter configuration) and radial (r > a) probe arrays must be built. The probe arrays must be positioned at the appropriate location with respect to the limiter or neutralizer plate, because plasma parameters change dramatically along the field lines. As noted by Stangeby and Chodura, plasma flow is not at the sound speed all along open field lines.

Several devices have implemented complete arrays of probes. Among them are D-III[38] and Alcator-C.[39] One set of measurements[18] of $T_e(r)$ and $n_e(r)$ for $r > a$ were performed on TFR in conjunction with τ_{ep}^* determination from gas feed and dN_e/dt measurements. Poloidal symmetry was assumed. The electron parameter profiles are shown in Fig. 22. That the scrape-off distance for thermal plasma, λ_{Th}, is about 1-2 cm has been seen on many tokamaks. For these experiments it is estimated that little ionization took place in the scrape-off plasma because of its low density and narrow width. Thus the flux of charged particles obeys a continuity equation without a source term,

$$\nabla \cdot (F_\perp + F_\parallel) = 0 \ . \tag{19}$$

The particle confinement time is then given by

$$\tau_{pe} = \frac{N_e}{\int F_\parallel ds_1 + \int F_\perp ds_2} \ , \tag{20}$$

where the s_1 integral is over the limiter surface and the s_2 integral is over the wall surface.

The $F_\parallel$ term was found in TFR to be about 3-times as important as the $F_\perp$ term. A plot of τ_{pe} vs time is given in Fig. 23. τ_{pe} is seen to increase with n_e, as the Alcator-A data behaved at $\bar{n}_e \lesssim$ 1.5 × 10^{14}cm^{-3}. Using the probe theory of Sengoku and Ohtsuke,[40] Equip TFR predicted the heat flux to the limiter by subtracting radiation and charge-exchange losses. The value of τ_p agrees well with its value predicted on this basis. Similar agreement between power to the limiter and that predicted by probe theory has been found for DIVA.[40]

A summary of all τ_{ep} measurements using probes is beyond the scope of this lecture. But because of the important implications of the probe data, in closing we note the following two points. As stated in the lectures of Manos and McCracken, the energy of the thermal ions that hit the limiter range from about 20 to 200 eV. And as noted in Section IIB-3, the reflection of these ions from the limiter causes neutrals of 200 eV average energy to hit the nearby walls.

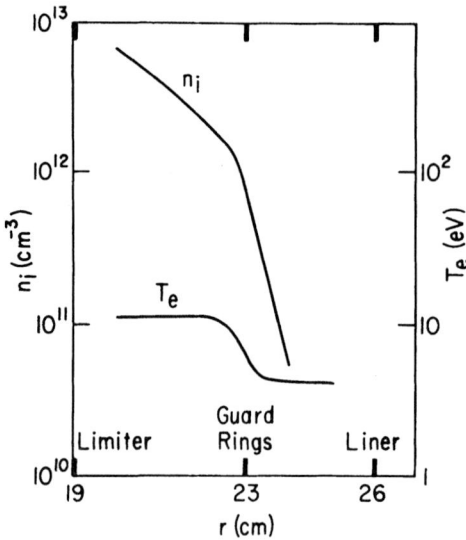

Fig. 22.   Electron temperature and ion density in the boundary
plasma of TFR-600.   (Ref. 18.)

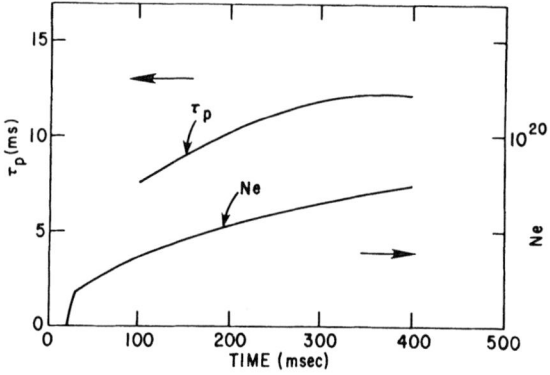

Fig. 23.   Global particle confinement time and total number of
electrons in a TFR-600 discharge versus time.   (Ref.
18.)

B-5   Summary of Thermal Plasma Particle Confinement

The common techniques for measuring thermal plasma particle confinement were reviewed. Of the three techniques which depend on neutral atom densities for their signals, $\bar{n}_e$ measurements alone were seen to yield only the effective particle confinement time τ_{ep}^*, not the true τ_{ep}. The others do give τ_{ep}, but each, i.e., H_α detectors, CX systems and electrostatic probes, requires extensive arrays because of toroidal, poloidal and radial variations. Measures from electron density behavior gave τ_{ep}^* ranging from 40 ms to over 2 sec, the latter resulting from high recycling and dirty walls. H_α and charge-exchange measurements have shown thousand-fold variations in signal. Computer modeling of the "meager" data has taken into account the toroidal, poloidal and radial asymmetries. τ_{ep} from these measurements and models ranges from ~5 ms on Alcator-A to ~70 ms on PLT. The average energy of the charge-exchange particles is about 300 eV. In limiter discharges τ_p measured by probes agrees with that inferred from power balance. The electron temperature on the open field lines is rather constant (~10 eV). Ion temperatures are higher and show more variation, 20-200 eV. The scrape-off distance for thermal plasma is 1 to 2 cm.

C.   Impurity Ion Confinement

The confinement and loss of impurity ions from tokamak plasmas is important because of power balance, depletion of hydrogen ions, and impurity generation processes. Collector probe techniques have had some successes in determining impurity ion fluxes in the plasma edge. However, they have not been able to measure charge states, an important parameter both for determining transport coefficients and impurity ion sputtering. Mass spectrometric measurements are infrequently, if ever, performed. Standard UV spectroscopy techniques have been the most extensively used in studies of transport rates. The main experimental methodology is to inject impurities, usually by the laser blow-off method,[41] and make chordal scans of the resonance lines.[42] In general, nonrecycling impurities are used. Abel inversion of the chordal data can yield radial profiles of the impurity concentrations, but only if the impurity densities have a known (usually cylindrical) symmetry (see Fig. 24). From the time evolution of the low ionization states D_I and v_I [see Eq. (4)] are determined, and from the time evolution of the terminal (high) charge states the global confinement time is usually found. Variations of this theme rely on computer modeling of each charge state's time behavior. In this way it is even possible to determine τ_{IP} from the behavior of low charge states. For a machine such as PLT, the global confinement time of impurities is usually in the range 20 $\rightarrow$ 80 ms for ohmic discharges.

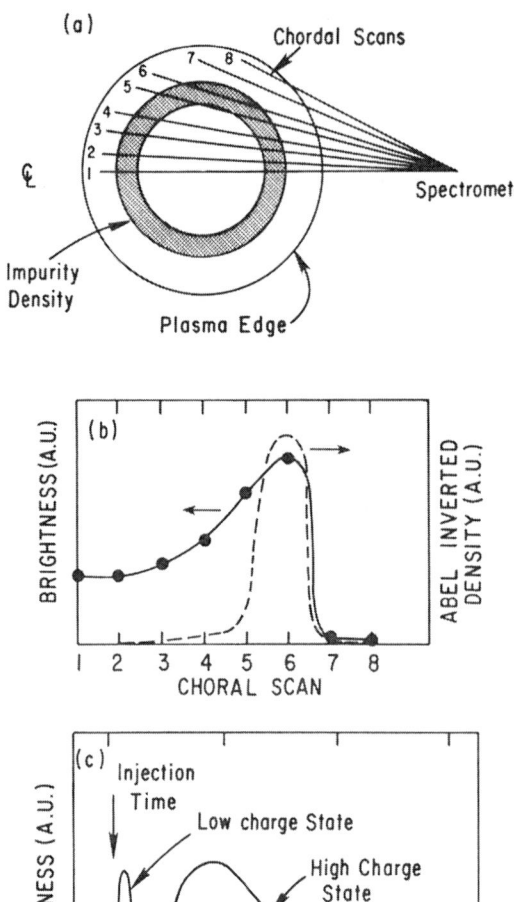

Fig. 24.   Schematic of the experimental arrangement for studying
            impurity transport. (a) A scanning spectrometer makes
            several (8) chordal measurements of the radiation from
            a particular impurity. (b) The brightness and Abel
            inverted density profiles for an impurity concentrated
            near a particular annulus, as in (a). (c) The time
            evolution of radiation from a low and a high charge
            state of an impurity injected into the plasma. The low
            charge state burns through quickly. The decrease in
            brightness from the high charge state is usually
            determined by impurity loss (confinement time) not by
            ionization.

As a point of contrast, consider the time to recombine to lower charge states. This can be found from the recombination rate, which is given by the following approximate expression,[45]

$$\alpha(T_e) \simeq 5 \times 10^{-14} Z(\chi/T_e)^{1/2} [0.4 + \frac{1}{2} \ell n (\chi/T_e)$$

$$+ 0.4(\chi/T_e)^{-1/3}] (cm^3 s^{-1}) . \tag{21}$$

For Al XII recombining to Al XI at $T_e = 50$ eV and $n_e = 10^{13} cm^{-3}$, the recombination time is 20 ms, which is comparable to the confinement time. Hence, an appreciable fraction of impurities, aluminum in this example, ionized to high charge states in the core, are expected to leave the tokamak highly ionized. Results[43] of a detailed calculation for iron are shown in Fig. 25. In case (a) the iron ion distributions plotted were calculated under the assumptions of D = 0 and coronal equilibrium. The temperature and density profiles used in the calculation are shown in Fig. 26. The limiter radius was 40 cm. For case (b), Fig. 25, a diffusion coefficient of $D = 1 \times 10^4$ $cm^2 s^{-1}$ was assumed. For (a), that is, coronal equilibrium and D = 0, ions of charge Z ≤ 6 are present in abundance at the edge. For (b), that is, $D = 1 \times 10^4 cm^2 s^{-1}$, ions with Z up to ~22 are at the edge. These have ionization potentials in excess of 1800 eV.

Experimental verification of these predictions have come slowly. Definitive work has been performed on PDX.[46] The brightness of the UV emission from the high ionization states in the edge plasma was enhanced by populating selected levels through charge transfer using a neutral hydrogen beam. The beam could be scanned across the entire plasma cross section (Fig. 27). The profiles of 0 IX and C VII, determined in this fashion, are shown in Fig. 28. The limiter radius was 30 cm and the peak electron temperature about 800 eV. An appreciable density of these fully ionized ions (I.P. $\simeq$ 500 → 800 eV) is evident at r = a. A diffusion coefficient of $D \sim 1 \times 10^4$ $cm^2 s^{-1}$ fits the data.

Based on this data we can estimate the outflux of fully ionized C and 0 to be

$$\int dA \ \Gamma = \int dA \ D\nabla n$$

$$\simeq 10^{19} \ s^{-1} . \tag{22}$$

807

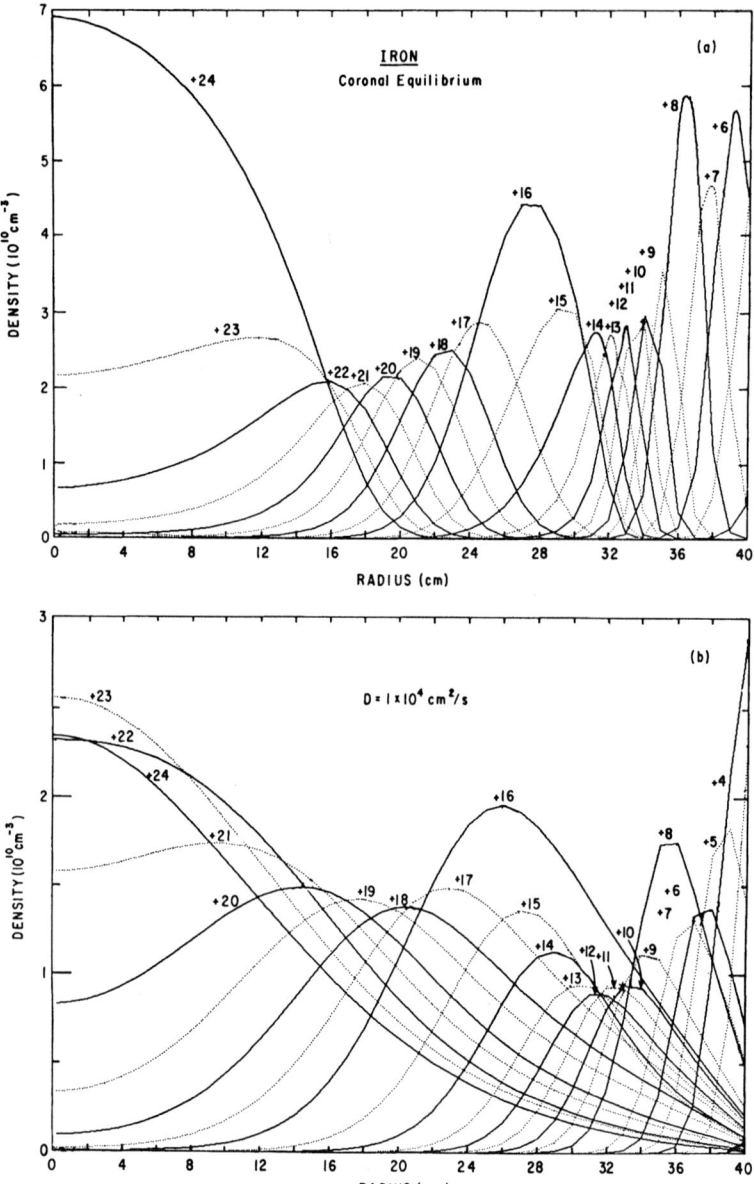

Fig. 25.   One-dimensional   transport   code   steady-state   radial
charge-state density profiles for iron impurity in the
plasma of Fig. 26 for (a) CE and (b) $D = 1 \times 10^{14}$
cm^2/s. The total iron density in both cases is $n_{Fe} = 1$
$\times 10^{11}$ cm^{-3}. Note that the density scales for the two
cases are different. (Ref. 43.)

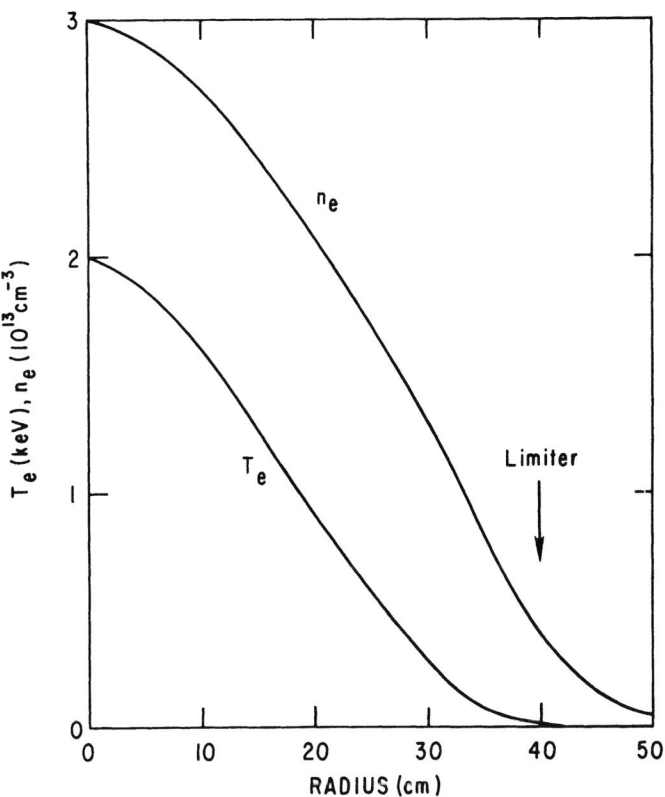

Fig. 26.   Typical tokamak electron temperature, T_e, and density, n_e, radial profiles.   (Ref. 43.)

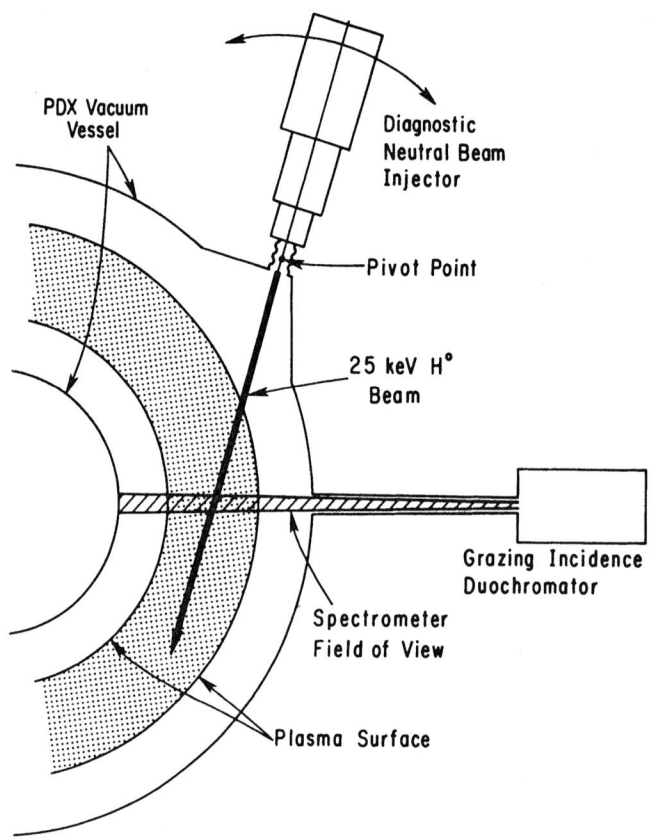

PDX Vacuum
Vessel

Diagnostic
Neutral Beam
Injector

Pivot Point

25 keV H°
Beam

Grazing Incidence
Duochromator

Spectrometer
Field of View

Plasma Surface

Fig. 27.   Experimental arrangement for charge-exchange enhanced
spectroscopy.   A neutral hydrogen beam is scanned
across the plasma cross section.   Photons from the
selectively populated levels of a particular impurity
are detected with a spectrometer. (Ref. 46.)

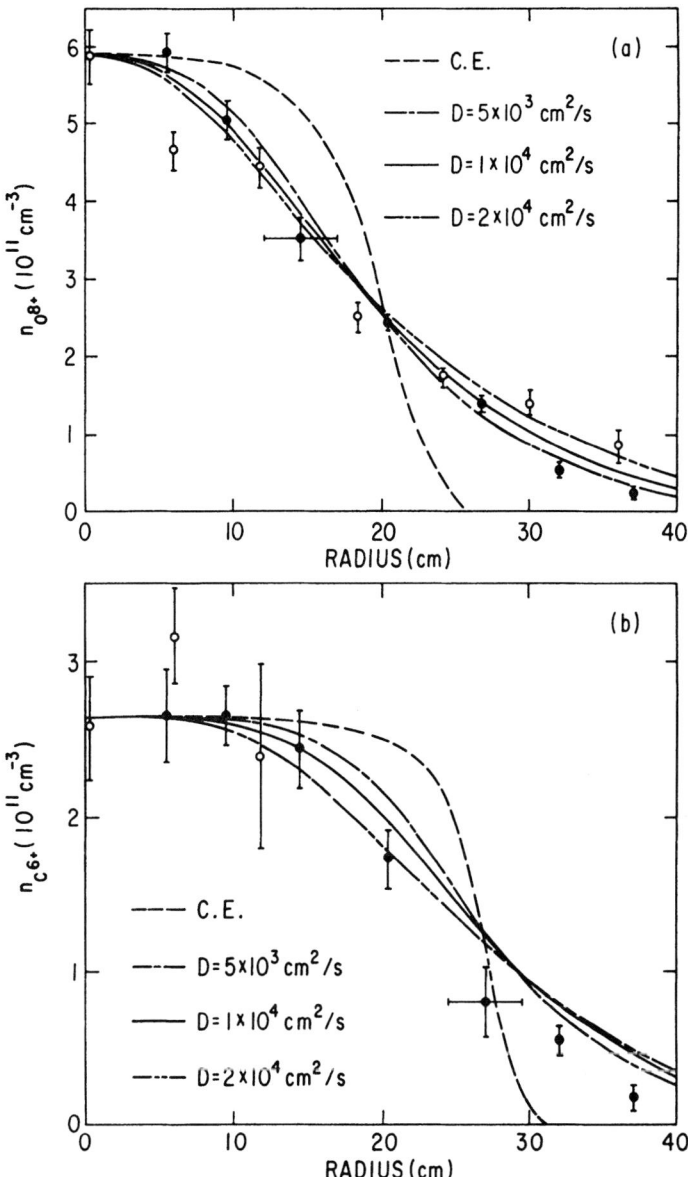

Fig. 28.   Radial profiles of (a) O^{+8} and (b) C^{+6} determined by
charge-exchange enhanced spectroscopy. The data points
are the experimental values; the solid curves are
results of computer calculation with various assumed
impurity diffusion coefficients. The predictions based
on coronal equilibrium are shown as dashed curves.
(Ref. 46.)

The lower charge states should have similar loss rates. From our earlier discussion recall that the loss rate for H^+ was about 5×10^{21} s^{-1}. Because of the higher sputtering yield of C and O, these ions, despite their lower fluxes, could cause more limiter erosion than H ions.

The behavior of impurities can be drastically modified by auxiliary heating. Recent measurements, on ASDEX in particular, show that during H-mode discharges impurity confinement in the plasma core is much longer than in L-mode discharges. Measurements on ISX-B showed a similar improvement in impurity confinement during counter-beam injection.[47] Ohmic discharges, while usually characterized by impurity confinement times comparable to τ_E, have also been observed to make transitions to modes where τ_{Ip} is large, in some cases greater than 300 ms.[48] Studies on Alcator-C[49] have found a positive correlation between τ_{Ip} and $\tilde{B}^{-1}$.

D.   Suprathermal Electron Losses[50]

Electrons with energies 100 times greater than T_e can be generated in tokamaks by dc or rf electric fields. In most situations these energetic electrons have $v_\parallel \gg v_\perp$. (One case where the opposite is true is during electron cyclotron resonance heating.) Runaway electrons are those which gain more energy per circuit around the torus than they lose via collisions. The critical dc electric field above which electrons run away is given by

$$E_{crit} = \frac{4\pi n_e e^3 \ln \Lambda \ (Z_{eff} + 2)}{m v_{th}^2} \tag{23}$$

where Λ is the Coulomb logarithm and $\ln \Lambda \simeq 10$.

For a toroidal electric field, $E = V_L/2\pi R$, and Maxwellian distribution, the fraction of particles with energy such that $E > E_{crit}$ is given by $\exp(-1/\varepsilon)$ where

$$\varepsilon = \frac{E}{E_{crit}} = 6.2 \times 10^{11} \ \frac{V_L \ (\text{volts}) \ T_e \ (\text{eV})}{R \ (\text{cm}) \ n \ (\text{cm}^{-3}) \ \ln \Lambda} . \tag{24}$$

The runaway electron creation rate depends sensitively on T_e and n_e. Results are shown in Figs. 29 and 30 for two different PLT discharges. The creation rate S_R varies 6 orders of magnitude though T_e varies only 10% and n_e a factor of 3.

812

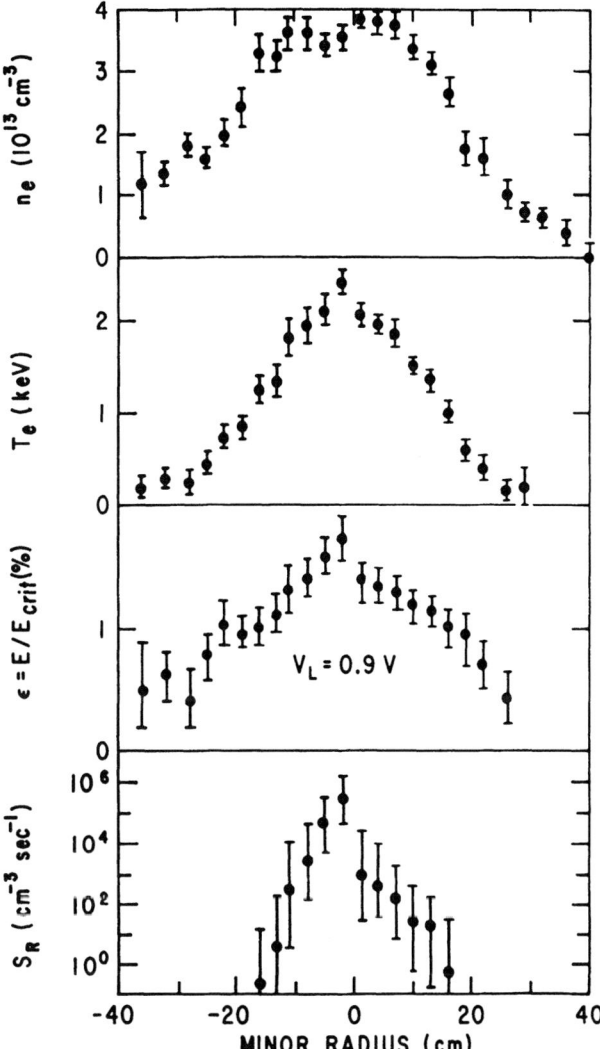

Fig. 29.   Calculated runaway electron creation rate, S_R, for the
measured electron temperature and density profiles and
loop voltage shown.   (Ref. 50.)

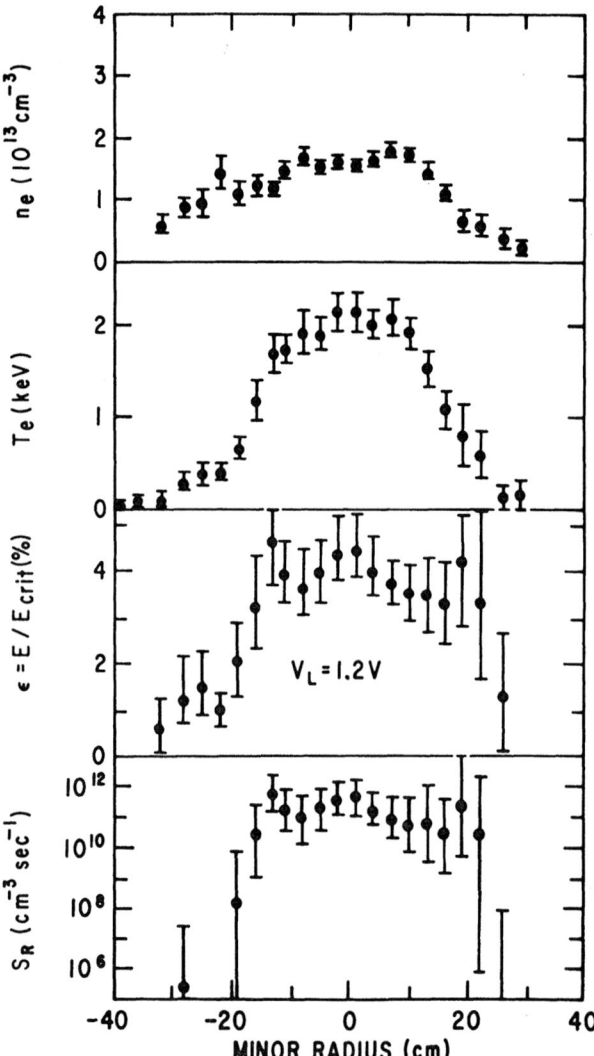

Fig. 30. Calculated runaway electron creation rate, S_R, for the measured loop voltage and electron temperature and density profiles shown. (Ref. 50.)

Because their velocity is nearly the speed-of-light, relativistic electrons gain energy nearly linearly with time. For such a "freely falling" electron (no collisions) in PLT, the rate of energy gain is about 50 MeV/s. Electrons with high energy move off the magnetic surfaces due to various drifts. For passing particles, the distance away from a flux surface is found by solving the equation for conservation of toroidal momentum

$$\frac{d}{dt}(P_\phi) = \frac{d}{dt}(mRv_\phi + \frac{q}{c}RA_\phi) \tag{25}$$

where

$$RA_\phi = -\int_o^r RB_\theta dr' . \tag{26}$$

The result is a shift away from the magnetic axis by an amount

$$\Delta r \equiv d_\gamma = \frac{v_\phi}{\Omega_p} R \ (1 + \frac{1}{2} \frac{v_\perp}{v_\parallel^2}) \ \Delta R , \tag{27}$$

where Ω_p is the cyclotron frequency in the poloidal field. For fast electrons moving anti-parallel to the plasma current, the shift of the orbit center is outwards, as shown in Fig. 31. The result is that runaway electrons most frequently impact the outer limiter. For a 10 MeV electron in PLT and $I_p \sim 400$ KA, the orbit shift is about 5 cm.

Suprathermal electrons may be trapped in ripples in the toroidal field if

$$\frac{v_\perp}{v_\parallel} > (\frac{2(B_{max} - B_{min})}{B_{max} + B_{min}})^{1/2} . \tag{28}$$

They will then ∇B drift vertically with a velocity

$$v_{\nabla B} \ (cm/sec) = \frac{10^8 E \ (eV)}{R \ (cm) \ B_T \ (gauss)} . \tag{29}$$

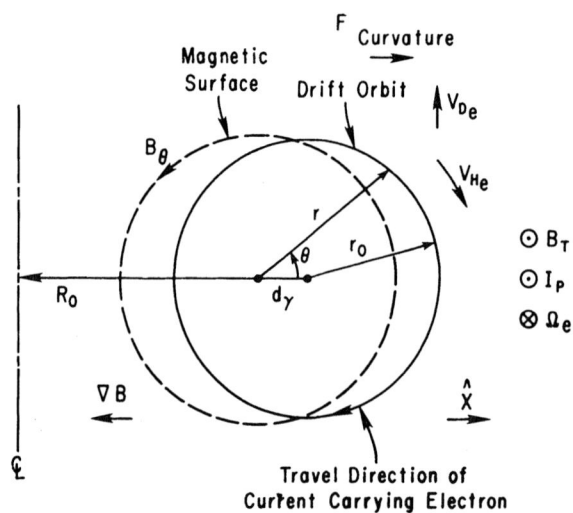

Fig. 31. Schematic of the displacement, d_γ, of a charged particles guiding-center orbit due to the various drifts in a tokamak. (Ref. 50.)

Studies in TFR[51] have shown the loss of electrons trapped in the toroidal field ripple. The observations were carried out with infrared cameras and pressure gauges. About 200 W/cm^2 was deposited on the vessel walls. Since the walls were thin bellows, holes were burned through.

The loss of these fast electrons is not accompanied by recycling, i.e., fast electrons do not reflect from surfaces and secondary electron yields are small. Hot spots, of course, could result in thermionic emission.

Loss of electrons with $v_\parallel \gg v_\perp$ has been studied by techniques[50] based on the interactions of fast electrons with matter. The bremsstrahlung caused by electron impacts on solid target yields much information. Figures 32 and 33 show the dependence of bremsstrahlung on electron energy and target materials. Thus from the x-ray spectra one can deconvolve the energy and flux.

Another method to ascertain the loss of fast electrons is to study the changes caused by these electrons as they impact the limiters. The changes can be simple melting[51] or nuclear transmutations.[50] Nuclear transmutation can result when the bremsstrahlung causes photoneutron emission. The yields for several elements are shown in Fig. 34. The products themselves may be radioactive with half-lives appropriate for detection. Table I and Fig. 35 show data on the activation of selected components of stainless steel. As can be seen, a threshold energy is necessary for these reactions to take place. From the abundance of each radioactive species (Fig. 36) the average runaway electron density can be computed. And then from knowing the loop voltage, τ_p can be determined. A collection of the data on runaway confinement time determined in this manner are shown in Fig. 37. τ_{ep} ranges from ~5 ms at 1 MeV to 1 sec at 10 MeV. The value of τ_{ep} at a particular energy may vary several-fold because of variations in plasma behavior such as MHD activity.

The transport and loss of runaways has been determined by the damage pattern in limiters.[52] The scrape-off distance, found from the melting, is inversely proportional to the parallel velocity and is about 1 mm. Similar results for D and τ as shown in Fig. 36 are obtained. Because of the small scrape-off distance, power loading can be intense even if S_R is small.

The final situation we consider is the confinement of 50–500 keV electrons generated during current drive experiments. Hot spots are seen to develop on the limiter and are attributed to the loss of the suprathermal electrons. From x-ray measurements of the plasma bremsstrahlung the electron distribution in the plasma can be determined. The total number of energetic electrons is readily related to the total plasma current. The loss rate can then be measured from the power flux to the limiter.[53] A suprathermal electron confinement time of 20-50 ms is found, comparable to τ_E.

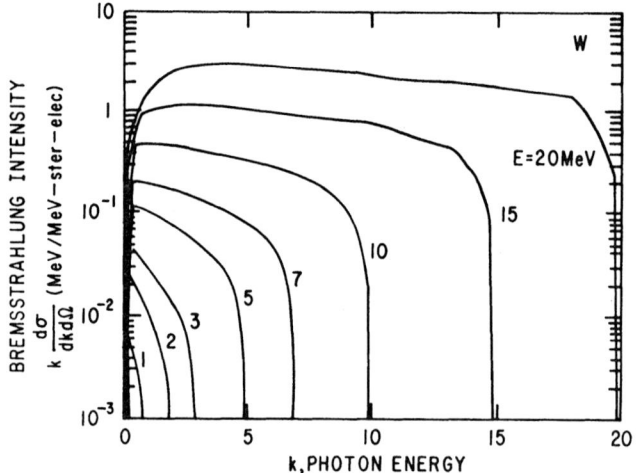

Fig. 32.   Thick target bremsstrahlung intensity for a tungsten target versus photon energy, parametrized by the incident electron energy. (Ref. 50.)

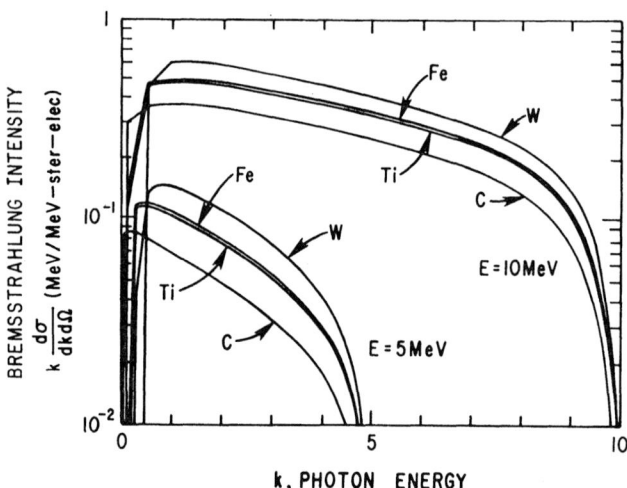

Fig. 33.   Comparison of thick target bremsstrahlung intensity for different materials and different electron energies. (Ref. 50.)

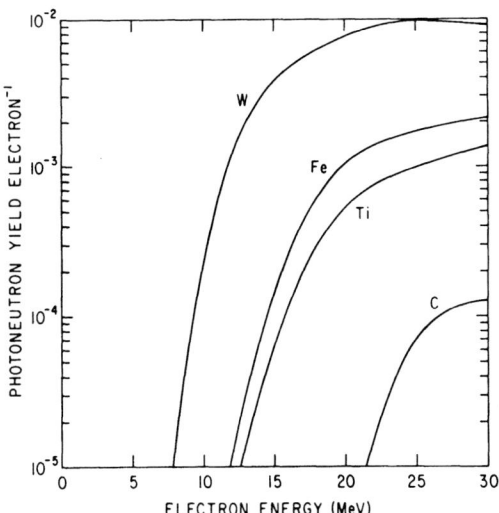

Fig. 34.   Photoneutron yield per electron for various electron
            energies incident on several limiter materials.   (Ref.
            50.)

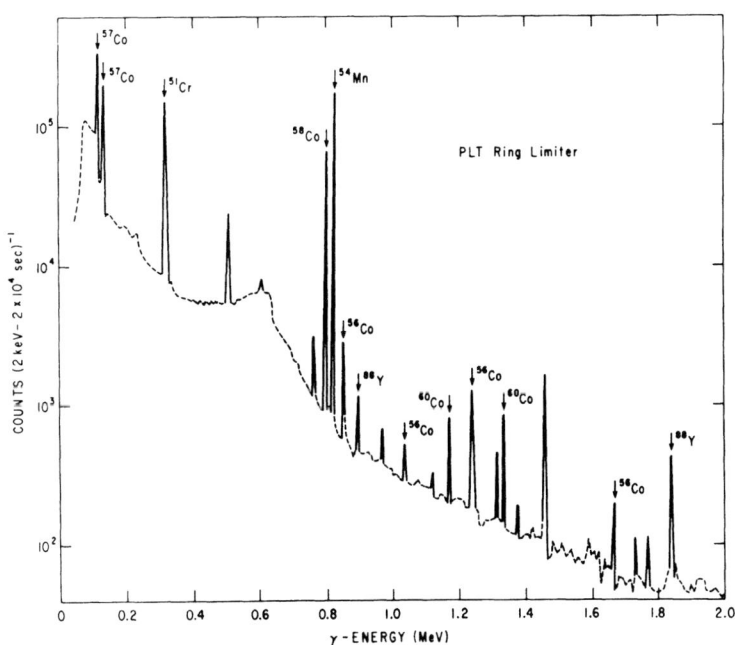

Fig. 35.   Example of the gamma-ray spectrum used to identify
            activation products. The ^{57}Co peaks have been reduced
            by a factor of 10 for this graph. The ^{56}Ni and ^{57}Ni
            lines have relatively short half-lives and have decayed
            away by the time of this spectrum. (Ref. 50.)

Table I. Photonuclear Reactions Produced in PLT Stainless Steel Limiter (Ref.50).

Reaction	Threshold Energy (-Q) MeV	Half-Life Secs	Half-Life Days	Target Abundance in Limiter (Isotopic × Elemental)	Cross Section Weighted by Electron Distribution (mbarns)	Activation Inferred from Midplane Decay Rate
$^{58}Ni(\gamma,p)^{57}Co$	8.17	2.37(7)	(272d)	0.678 × 0.10 = 6.78%	3.51	2.69(12)
$^{61}Ni(\gamma,p)^{60}Co$	9.86	1.66(8)	(1924d)	0.0125 × 0.10 = 0.125%	1.20	8.43(11)
$^{55}Mn(\gamma,n)^{54}Mn$	10.23	2.70(7)	(312.6d)	1.0 × 0.01 = 1.0%	2.46	2.94(12)
$^{52}Cr(\gamma,n)^{51}Cr$	12.04	2.40(6)	(27.72d)	0.838 × 0.19 = 15.9%	4.11	1.97(11)
$^{58}Ni(\gamma,n)^{57}Ni$	12.2	1.30(5)	(1.5d)	0.678 × 0.10 = 6.78%	1.38	1.33(11)
$^{58}Ni(\gamma,pn)^{56}Co$	19.6	6.68(6)	(77.3d)	0.678 × 0.10 = 6.78%	0.41	2.82(9)
$^{58}Ni(\gamma,2n)^{56}Ni$	22.5	5.27(5)	(6.10d)	0.678 × 0.10 = 6.78%	0.063	4.78(7)

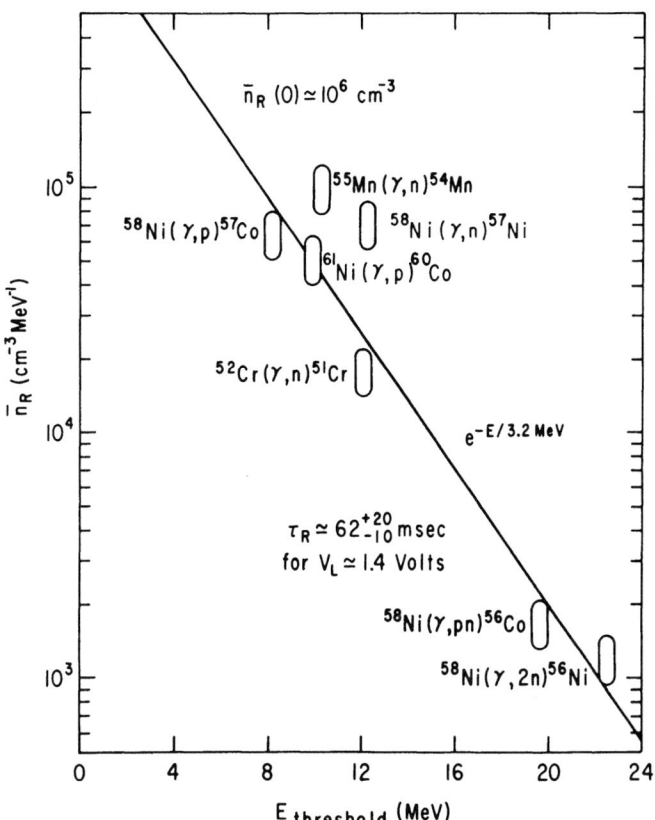

Fig. 36.   Average runaway electron density as a function of
energy.   An exponential fit of the data yield a slope
of $(3.2 \text{ MeV})^{-1}$ which corresponds to a confinement time
$\tau_R \simeq 62$ msec for a loop voltage $V_L \simeq 1.4$ Volts.
Extrapolation of the spectrum back to zero energy gives
a density of $\bar{n}_R(0) \simeq 10^6$ cm^{-3} MeV^{-1}. (Ref. 50.)

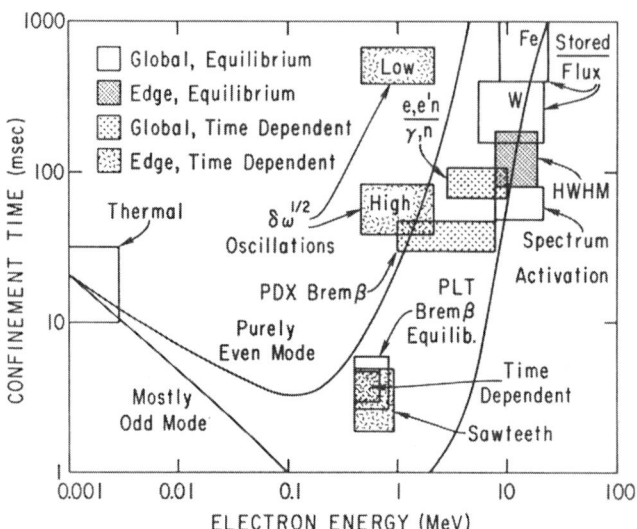

Fig. 37. Measured electron confinement time versus electron energy. "Low" and "High" refer to the range of frequencies. (Ref. 50.)

E. Suprathermal Ion Losses

Suprathermal ions can be generated in tokamaks by auxiliary heating, such as neutral beam injection or ICRF resonances, by fusion reactions, or by instabilities. The energy and velocity distributions of these particles depend on the details of the heating process. The loss of these "fast" ions can come about from the same processes that affect electrons, such as bad orbits[53a] (orbit shifts away from magnetic surfaces), field ripple and diffusion. The mechanisms by which fast ions may leave the plasma include charge exchange and plasma instabilities. Fishbone oscillations in PDX,[54] for example, are seen to cause the ejection of 30 → 50 keV beam ions as well as ~100 keV "pumped" ions. These ions leave the plasma both by charge exchange and intersection of their orbits with the limiter. Thus, both stripping cell analyzers (see Sec. IIB-3), and probes are used to determine the loss rates. A thorough discussion of probe measurements of fast ion losses during both beam and ICRF heating can be found in the lectures of Manos and McCracken. We simply note here that the scrape-off distance for these particles is about 3 mm. Counter-going particles are lost preferentially to the outer limiter. As for suprathermal electrons, the power loading on limiters can be intense because of the small scrape-off distance. No "hard" number has yet been given for the confinement time of fast ions lost by instabilities. This is because both the production and loss rates are not known absolutely.

Fast ions produced by fusion events are now more readily studied because of the higher temperatures and densities attained. Results from PLT[55] have shown a strong dependence of confinement on plasma current. For plasma currents sufficiently high that bad orbits do not cause immediate loss, fast ions (~1 MeV) have τ_{ip}, in general, about twice greater than τ_E.

F. Photon Loss

Photons with a wide spectrum of energies are emitted from tokamaks. At the low energy end of the spectrum are radio-frequency waves generated by plasma instabilities. Emission at these wavelengths is difficult since in the plasma the waves are generally electrostatic, while in the vacuum they are electromagentic. At a thousand-fold higher frequency, $10-10^3$ GHz, are microwave emissions caused by cyclotron orbits. Over a small part of the lower end of this frequency range the emission is blackbody. Eventually this type of emission will be an important energy loss channel. A proposal has been made to drive plasma current through the emission and readsorption of this radiation.[56] Photons emitted in the infrared portion of the spectrum may be generated by molecules entering the plasma.

Visible and UV photons come from electronic transitions between discrete energy levels of atoms and ions. Visible bremsstrahlung is also emitted and has become a useful diagnostic[57] for measuring Z_{eff}. The power radiated in this spectral region ($100 \rightarrow 8000$ Å) has, in some instances, reached ~50% of the input power, i.e., several MW. At this level photon impacts on the wall can have numerous important effects.

One effect is impurity generation by photon desorption.[58] Assuming an average photon energy of 30 eV, the total photon flux is about about 5×10^{23} s^{-1} for 3 MW of radiation. With a photon desorption yield of ~10^{-4}, typical of CO, H_2O or CH_4 on relatively clean surfaces, the impurity influx rate is still seen to be potentially important.

Another effect of photon impacts is the emission of secondary electrons. As shown in Fig. 38, the photoelectron yield is ~10^{-1} between 300 and 1200 Å. Photons in this wavelength range are generated by H and O excitation. Since about one L_α photon is emitted per H^O ionization event, we expect a copious source of cold electrons to be generated from H recycling alone. These cold electrons could effect the plasma potential and plasma transport. Experiments with electron emitting probes[59] already indicate the importance of much processes.

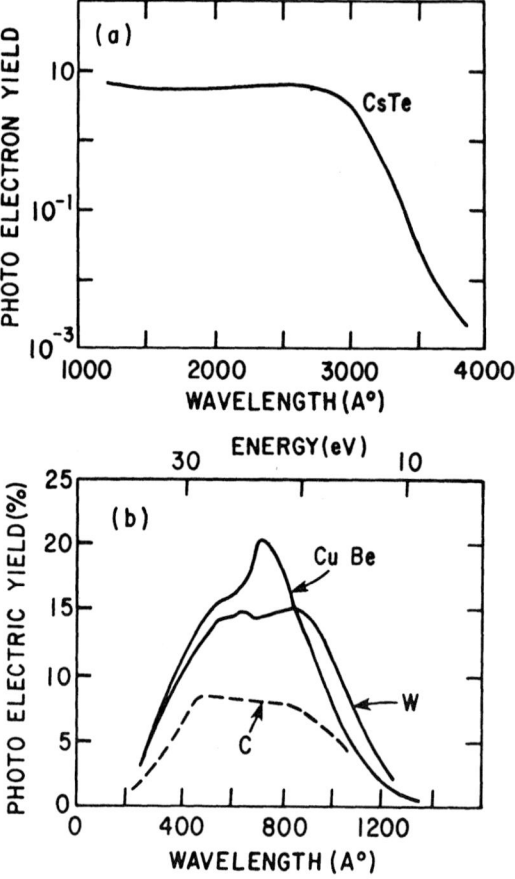

Fig. 38.   Photoelectron yields for (a) CsTe and (b) C, W and BeCu
in the wavelength ranges of 1000-4000 Å and 200-1200 Å,
respectively. [From Sampson and Cairns, J. Opt. Soc.
Am. 56 (1966) 1568.]

Soft and hard x-rays are also emitted in copious numbers,
particularly from impure plasmas. These photons may represent up
to ~10% of the plasma energy loss. Their interaction with matter
is not as important as visible and UV photons because of their
greater penetrating power.

Most photons emitted from the plasma, except for those at
cyclotron resonances, are not   reabsorbed as they pass through
the plasma. The optical depth for reabsorption[45] is given by

$$d_o \ (cm) = \frac{2 \times 10^{16}}{\lambda_o \ (\text{Å}) \ n \ (cm^{-3})} \left(\frac{T \ (eV)}{\mu \ (amu)}\right)^{1/2} \ , \tag{30}$$

and is thus only appreciable for long wavelength radiation or for a high density of absorbers.

G. Summary of Particle Confinement

The understanding of the conceptually simple topics of particle and energy balance in a tokamak requires a dazzling array of complicated diagnostics, each tuned to a different energy and species. The techniques used to determine the confinement time of impurity ions, suprathermal electrons and ions and photons are seen to be similar to those used for thermal plasma particles. These techniques involve the use of spectrometers for particles and photons and include gratings and solid state detectors for photons, and stripping cells with electrostatic, magnetostatic analyzers or bolometer probes for particles. The particle confinement time can vary markedly from species to species. It is also seen to depend strongly on plasma parameters, including instabilities and plasma current.

III. FUELING TOKAMAKS
A. Motivation

Early in fusion research it was recognized[60] that a method must exist to replenish the fuel lost by burn-up and transport. Even then it was clear that simply puffing in gas at the plasma edge would not suffice if the plasma transport was slow or predominantly outward. And, as indicated in Fig. 1, particles at large minor radii have a relatively short confinement time compared with those near $r = 0$, and their loss would be rapid. One proposed alternative was the injection of supersonic pellets of frozen hydrogen. The smaller surface-to-volume ratio of the injected pellets compared to an equivalent amount of gas atoms would enable deeper fueling than gas puffing. Other methods to fuel deeper are to accelerate neutral atoms to high energy,[61] about 100 keV, or conglomerate hydrogen molecules into clusters[62] and accelerate them to a somewhat lower energy, about 10 keV/nucleus. Variations for these approaches, for example to cause ripple in the toroidal field to assist particle penetration,[63] have also appeared.

Detailed experiments[28,64] were performed in the 1970's which showed, however, that the inward plasma transport during gas puffing, especially when it was intense, was adequate to fuel the plasma core and avoid possibly dangerous--though possibly beneficial[65]--inverted density profiles. The question then naturally arises, why use or develop a more complex or expensive fueling technique. There are at least two responses. Firstly, the transport rates and recycling coefficients in future larger tokamaks may not be adequate for gas puffing to be a fueling technique. For example, tritium implantation in the walls might be excessive. Secondly, we are still ignorant of many aspects of the plasma behavior. It is imprudent to summarily dismiss alternative approaches to operating techniques because of model calculations based on incomplete knowledge. As will be seen later, pellet fueling has, at least in one case, resulted in a major surprise, the dramatic improvment in certain plasma parameters.

In the following sections the two main fueling techniques, gas puffing and pellet injection, will be reviewed. The experimental techniques employed in each are presented, and results of tokamak studies of fueling efficiency and other salient features are noted.

B. Gas Puffing

In Section II of this paper, some experimentally required fueling rates were presented. For dirty machines or those with high recycling, virtually no additional fuel was required. But for well-conditioned tokamaks the required gas feed to sustain the density ranged up to 300 Torr liters/s (2×10^{22} atoms s^{-1}). A low rate is for limiter discharges and a high rate for divertor or expanded boundary discharges.[66] For this high throughput, special fast-acting valves had to be employed. In general these valves use piezoelectric crystals to open and close an elastomer seal (Fig. 39).[67] To achieve the highest throughput the pressure required behind the seal may be several atmospheres.

With these types of valves, submillisecond response times are possible. To maintain this, a short conductance length to the tokamaks is necessary, placing the valve in the toroidal magnetic field. Sturdy mechanical supports are needed to protect certain valve components from $\mathbf{J} \times \mathbf{B}$ forces.

As shown in Figs. 13 and 15, the gas puffed into circular limiter discharges quickly flows to the limiter where it continues to recycle. The brightest H_α emission is at the limiter, not near the gas injection port. The toroidal and poloidal variations shown in these figures is probably not a full measure of the severe asymmetry.[68]

As the density is increased, the gas required to maintain that density also increases[20] (Fig. 40), thus showing that the effective particle confinement time has dropped. This occurs in spite of τ_E rising. A similar result is seen in expanded boundary plasmas.[66] These two observations--H_α emission at the limiter being dominant and the decrease of τ_p^* with n_e--are both signs of the same set of phenomena: poor penetration of cold molecules, slow inward transport, and rapid parallel transport to the limiter. In large tokamaks with a single rail limiter or a cold edge, the parallel loss process may be sufficiently slow to allow better fueling efficiency.

The actual fueling efficiency, γ, is difficult to determine because of the earlier-mentioned complications from recycling affecting the measurement of τ_p. However, it is clear that γ is near unity for low density discharges (Fig. 41). Simulations[22] of ASDEX data using the BALDUR code show γ near unity even at high density, undoubtedly due to multiple chances of recycling particles to penetrate the scrape-off. Limiters with low-Z and high hydrogen capacity, e.g., clean graphite, do show lower γ, as low as 0.5. Figure 42 shows γ as a function of plasma radius as a steel limiter is withdrawn from the separatrix. The fueling efficiency rapidly drops from 1 to 0.25 as the plasma becomes diverted instead of limited. Particles that impinge on the neutralizer plate may be reflected, but not back into the core plasma. A thorough comparison between limited and diverted discharges can be found in the lecture of Wagner.

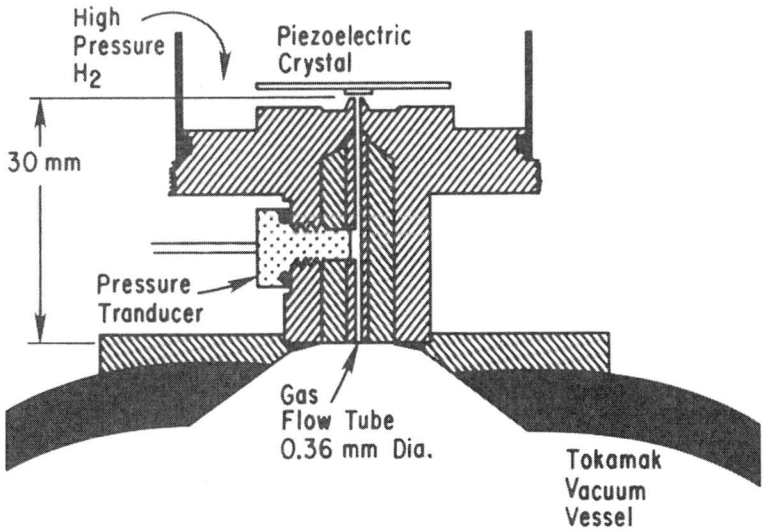

Fig. 39.   Schematic of a high-throughput, fast, pulsed-gas valve. (Ref. 67.)

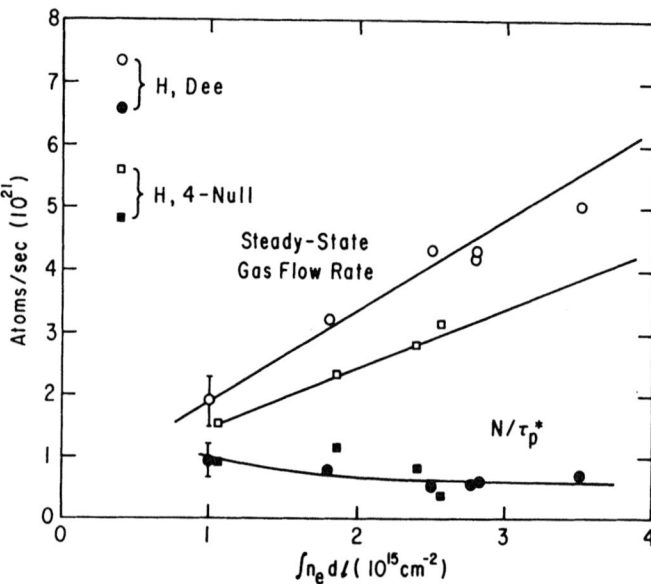

Fig. 40.   Gas fueling curves for diverted PDX plasmas.   (Refs.
20, 21, and 31.)

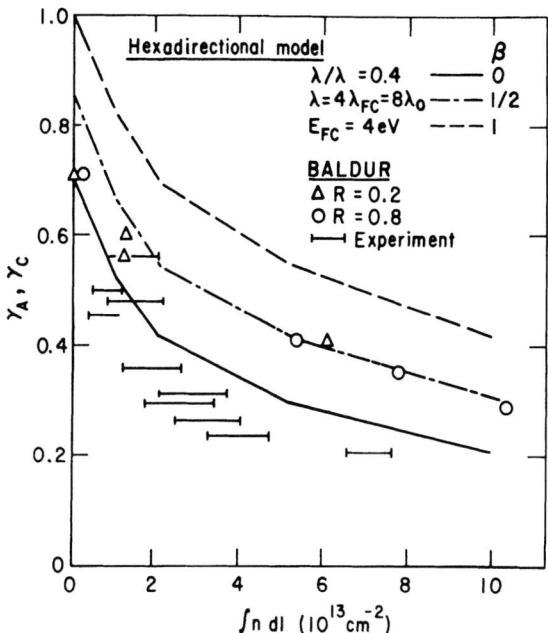

Fig. 41.　Fueling efficiency for Franck-Condon atoms, γ_a, versus
column density for various models including the BALDUR
simulation $(\Delta, 0)$.　The experimental values are γ_c.
(See Ref. 22 for details.)

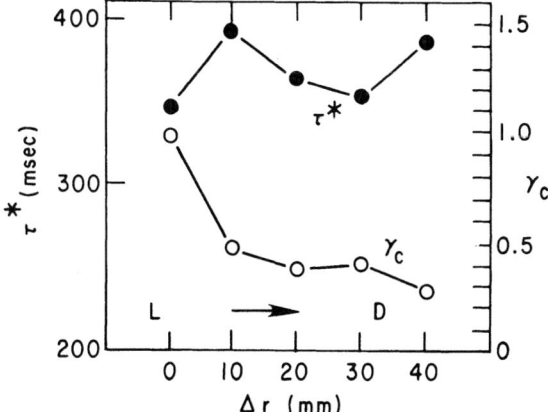

Fig. 42.　Transition from a limiter, L, to a divertor, D,
discharge in ASDEX.　(Ref. 22.)

C.   Pellet Fueling

Discharges sustained by gas-puffing require typical fueling rates of 2×10^{22} atoms/sec, equivalent to the solid hydrogen in a sphere 1.5 mm in radius. (Solid hydrogen has a density of 0.086 gm cm^{-3}.) The speed required for a pellet of this size to reach the plasma center depends on the rate of pellet ablation. Many models of pellet ablation have been proposed (see Ref. 69 for a review), but the one that seems to fit the data best is the so-called neutral-gas shielding model of Parks and Vaslow. In this model the heat flow carried by the thermal ions to the pellet surface is reduced by a cold gas cloud surrounding the pellet. The detailed agreement with the model is only moderate for ohmic plasmas and rather poor for beam-heated or rf current-driven plasmas because of suprathermal particles. An estimate of the velocity required to fuel INTOR was made[69] based on the neutral-gas shielding model and found to be about 10^6 cm/s.

The means to produce mm-sized pellet with this high a velocity does not yet exist. A complete description of all methods used can be found in Milora's article.[69] Therein are also listed the factors which make pellet acceleration to 10^6 cm/s so difficult. For brevity, the only device noted here is the pneumatic variety which has been used successfully on ISX-B,[70] PDX[71] and Alcator-C.[4] In this scheme, see Fig. 43, a pellet of frozen hydrogen is placed into a gun barrel and propelled by a burst of high pressure helium gas. Pressures of 30-bar have accelerated 1 mm diameter pellets to velocities slightly above 10^5 cm/s.

The penetration of such a pellet is monitored by viewing the H_α light emitted. The time evolution is shown in Fig. 44 along with the brightness predicted on the basis of two neutral gas shielding models. Though the time of the peak brightness is accurately predicted, the detailed behavior is rather different. This may be due to a different ablation or atom excitation rate. More thorough studies of the pellet motion were made using cinematic recordings. These showed the pellet trajectories to deviate from straight lines, even in the case of ohmic discharges. For the ohmic discharge an asymmetric ablation was inferred from the deflection of the pellets in the direction antiparallel to the electron drift.

Pellet fueling efficiency is found to be unity. The density rise is as predicted based on the number of atoms in the pellet. Charge-exchange losses, once thought to be a potential problem, are, in fact, negligible. On PDX the particle confinement time was found to increase 30%.

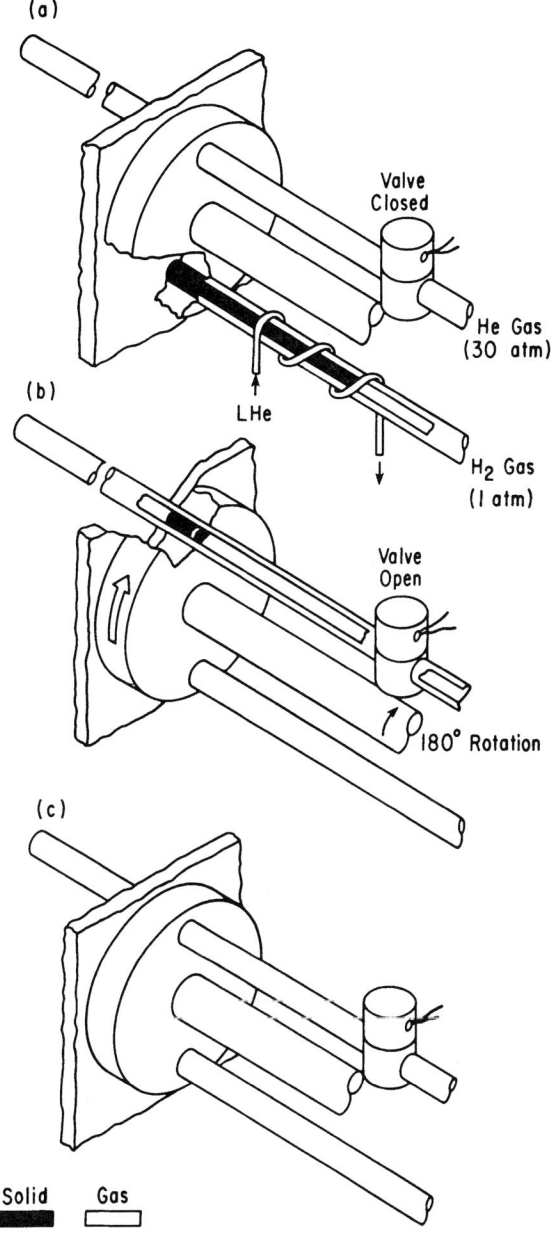

(a)

Valve
Closed

He Gas
(30 atm)

LHe

(b)

H₂ Gas
(1 atm)

Valve
Open

180° Rotation

(c)

Solid   Gas

Fig. 43.   Operating principle of the ORNL pneumatic hydrogen
pellet injector.   (Ref. 69.)

The first pellet fueling experiments were of an exploratory nature. The plasmas produced[70,71] had rather similar energy confinement properties to those fueled by gas puffing. The injection process was adiabatic. However, an unexpected benefit resulted from the injection of pellets into high density Alcator-C discharges. The changes in main plasma parameters are shown in Fig. 45. A sharp rise in n_e and drop in T_e were observed in all

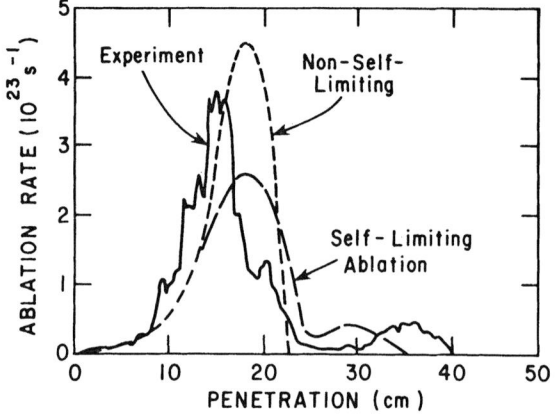

Fig. 44. Experimental and predicted H_α emission (labelled ablation rate) of a pellet into ISX-B. (Ref. 70.)

previous pellet injector experiments. The same is true for these radial profiles of n_e and T_e (Fig. 46). What was novel was the behavior of τ_E. Though the density limit was about the same for gas-fueled discharges, $\bar{n}_e(max) \sim 8 \times 10^{14}$ cm^{-3}, the energy confinement time achieved was about 70% higher. For the first time ever, $n_e(0)\tau_E$ exceeded the Lawson criteria for breakeven (Fig. 47). The cause for this improvement is not known. Speculations are that the reduced edge recycling and the modified density profile are the main contributors to the improved confinement. Regardless of the exact cause, these results clearly illustrate the importance of density control in achieving controlled thermonuclear fusion.

832

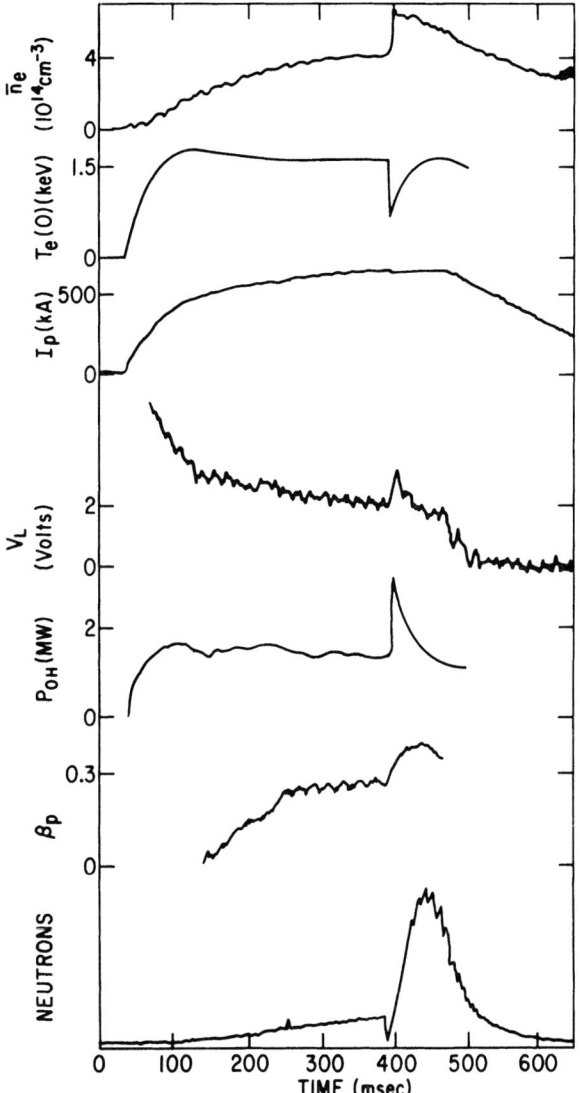

Fig. 45.   Plasma parameters of a typical pellet-fueled Alcator-C
discharge.   (Ref. 4.)

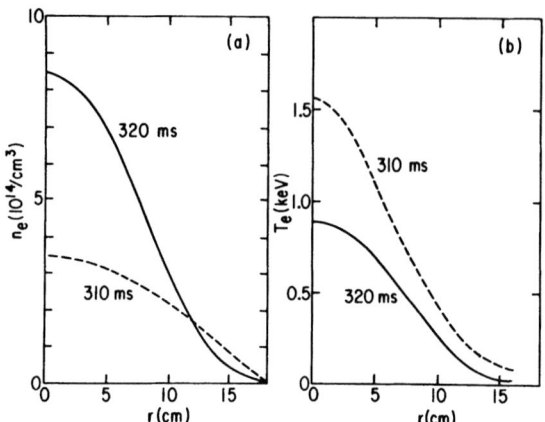

Fig. 46.   Radial profiles of T_e and n_e just prior to and subsequent to pellet injection into Alcator-C.   The pellet was injected at 315 ms.   (Ref. 4.)

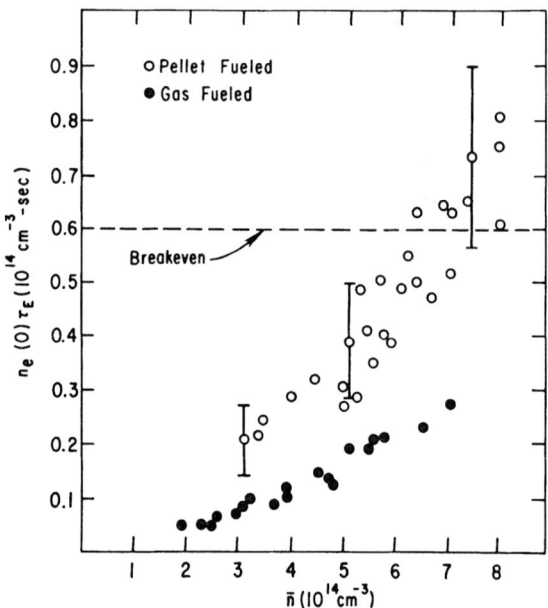

Fig. 47.   Value of the Lawson product, $n_e(0)\tau_E$ as a function of density for gas-fueled and pellet-fueled discharges. (Ref. 4.)

IV. WALL AND LIMITER CONDITIONING

A. Overview

The control of plasma density and purity relys on more than plasma transport properties and fueling techniques. Wall conditioning is also an important aspect of density control because of the large numbers of loosely bound atoms that can reside on a wall surface. The areal density of adsorbed atoms in a surface monolayer is about 10^{15} cm^{-2}, which is 10 times the volume density of ions in a tokamak plasma. Thus, for a tokamak which has a volume to surface ratio of a/2 $\lesssim$ 100, desorption of a single monolayer of atoms from the wall would alter the electron density at least 10% and the ion density Z-times more. For this reason, wall and limiter conditioning have played central roles in the achievement of the best tokamak plasma parameters.

The effects of wall conditioning are more pronounced than numbers in the last paragraph indicate. The release of adsorbed gas is uncontrolled, with the possible result that desorption may be extremely rapid and cause edge cooling, contraction of the current channel and even a major disruption. Furthermore, it is an experimental fact[72,73] that the cleaner a tokamak discharge the lower q values attainable and the wider the operating range. It is clear that loosely bound layers must be completely removed for the tokamak to function well. And, because the cost of running tokamaks is so large, about $1000 per discharge on PLT and $25,000 per discharge on TFTR, any improvement in wall conditioning can be very cost effective.

The main aim of wall conditioning is to produce surfaces which do not emit particles into the plasma, even when bombarded by the plasma. The particles include high-Z impurities, such a Fe, Mo and W, low-Z impurities, such as N, C and O, and plasma species, such as H, D and He. Any of these, particularly the low-Z elements, may be loosely bound to the wall surface. The techniques which have been used to achieve this aim can be categorized under the following headings:

- Thermal,

- Chemical (wet and dry),

- Mechanical,

- Atomic,

- Coating.

Discharge cleaning techniques could fit under the "dry chemical" or "atomic" (sputter cleaning) headings. Evaporable getters belong under the "coating" heading. Bulk getters are also placed there for consistency.

Before embarking on a discussion of the details of discharge cleaning and gettering, it is appropriate to present a typical conditioning sequence, starting with the initial commissioning of the vacuum vessel. This follows a review[74] given by Dylla of the processes used prior to 1979. After the vacuum vessel has been manufactured, the interior is usually mechanically polished with abrasives such as alumina, cleaned with detergents in water and then rinsed with deionized water, trichloroethylene, acetone and finally alcohol. The vacuum vessel is then installed in the toroidal field coils and another solvent rinse is applied. The vessel is pumped down and leak checked. If there are no leaks greater than 10^{-3} scc/s, a mild bakeout is usually performed to outgas the walls of CO, H_2O and oily residues not removed by the rinses. The bakeout temperatures have ranged up to ~400°C, with the highest used on TEXTOR and TFR. At the highest bakeout temperatures some hydrocarbon contaminants dehydrogenate leaving a carboneous residue. Discharge cleaning is then applied. In most recent tokamaks, glow discharge cleaning (GDC) has preceeded pulsed discharge cleaning (PDC). The time to terminate discharge cleaning and attempt high power discharges is determined from the residual gas spectra. The usual criterion is when the H_2O, CO, and CH_4 peaks during GDC have decreased about 100 times from their initial values to a partial pressure of ~10^{-6} T. The time behavior of several residual gases during GDC in ASDEX[75] is shown in Fig. 48 and for PDX[76] in Fig. 49. After any vacuum opening a lesser amount of discharge cleaning than originally used is necessary. After months (or years) of high power discharges, most tokamaks have tried various getters, both to bury impurities on the wall and reduce hydrogen recycling.

B.   Discharge Cleaning

It is possible to thoroughly clean tokamak walls by using plasmas. The cleaning mechanisms are thermal, atomic, and/or dry chemical. Thermal outgassing and sputtering have been addressed in other lectures and won't be discussed here. Instead, our present understanding of dry chemical cleaning will be reviewed. About six methods using plasmas have been developed and adopted in tokamaks. These six methods, high power pulsed discharge cleaning, (PDC), Taylor discharge cleaning (TDC), glow discharge cleaning (GDC), radio frequency assisted discharge cleaning (RFADC), electron cyclotron discharge cleaning (ECDC) and 60 Hz discharge cleaning (ACDC), have exceedingly different parameters, but chemical effects are active in all. Their major parameters are listed in Table II.

836

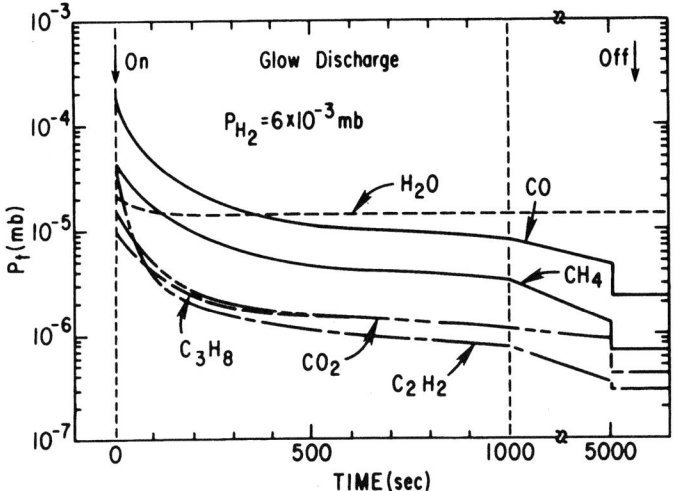

Fig. 48. Partial pressure versus time during a GDC run in ASDEX. The vessel was pumped at a speed of 5000 liters/sec. About 0.1 gm of carbon was removed during this GDC run. (Ref. 75.)

Table II. Types of Cleaning Plasmas used in Tokamaks.

Acronym	n_e (cm^{-3})	T_e (eV)	Power (watts)	Duration	Duty Cycle
PDC	10^{13}	30–100	5×10^5	50 msec	10^{-4}
TDC	5×10^{10}	3–10	5×10^4	10 msec	10^{-2}
GDC	10^{10}	5*	2000	CW	1
RFADC	10^{10}	5	2000	CW	1
ECDC	5×10^{10}	7	2000	CW	1
ACDC	5×10^{11}	8	2000	CW	1

*The sheath causes ions to impact the cathode with several hundred eV of energy.

That the cleaning rates are similar, modulo a factor of about 3, at first is very surprizing considering the wide variations in plasma and other parameters: a factor of 10^4 in duty cycle, 10^3 in electron density, 10^2 in ion and electron temperatures and 10^2 in

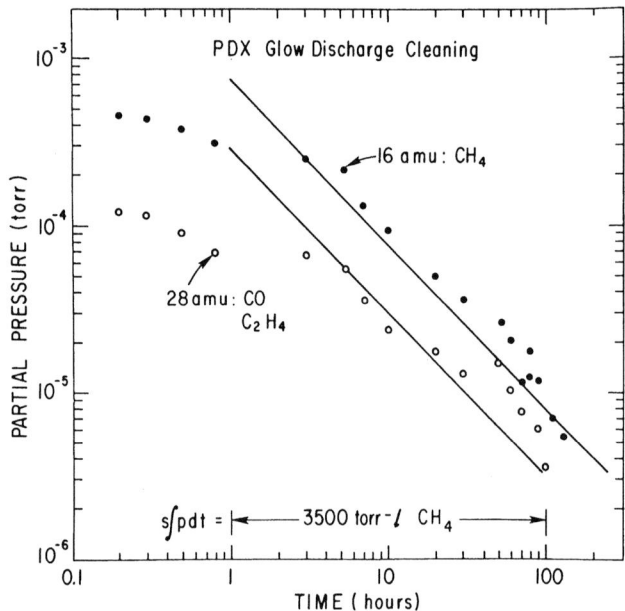

Fig. 49.   Partial pressures of the primary residual gases (CH_4 at 16 amu, CO and C_2H_4 at 28 amu) produced within the PDX torus during H_2 glow discharge conditioning. The H_2 pressure was 3×10^{-2} Torr, and the net pumping speed for the heavy residual gases at this H_2 pressure was $\simeq 280$ liters/sec. (Ref. 76.)

power. Part of the reason for this is that too high plasma energies and densities can mediate against efficient cleaning by causing ionization of the desorbed impurities and implantion back on the walls rather than evacuation out a pump.

That chemical reactions and not thermal or atomic effects are responsible from most of the cleaning can be appreciated by comparing PDC with TDC. In neither case do the tokamak walls get appreciably warmed. Also the charge-exchange fluence of D^O energetic enough to sputter, that is E > 20 eV, has been shown[73] to be a factor of 10^3 higher during each PDC pulse than during each TDC pulse.[77] This is depicted in Fig. 50. The charge-exchange fluence during a PDC pulse is about 10^{14} cm^{-2} with an $\bar{E}$ of about 30 eV, but only 10^{11} cm^{-2} with the $\bar{E}$ of detected particles of 20 eV during TDC.

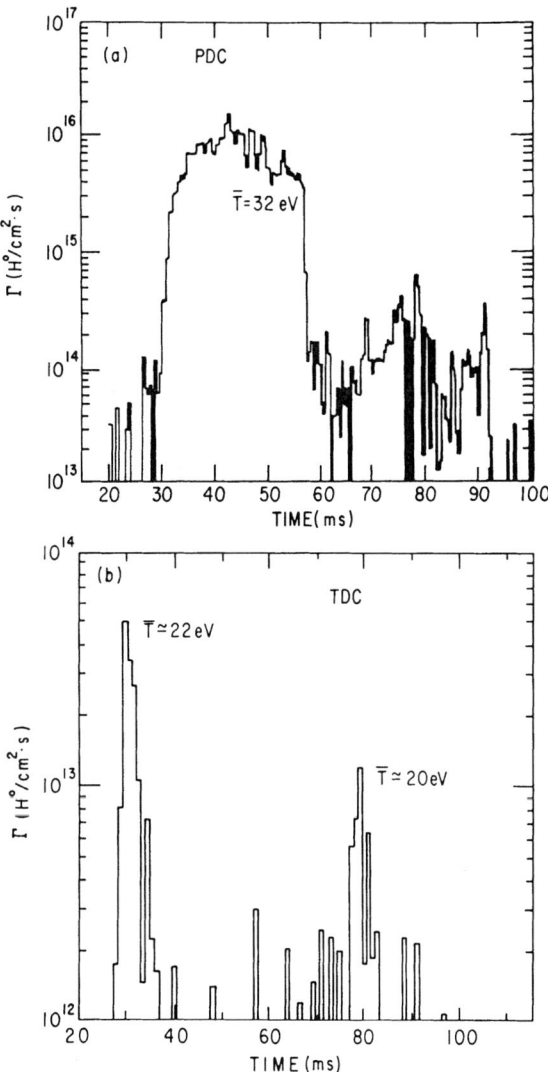

Fig. 50.   (a) Measured neutral hydrogen efflux from PLT during PDC.   (b) Measured neutral hydrogen efflux from PLT during TDC.   In both (a) and (b) the energy range of the detector was 7 to 405 eV.

The mechanism for chemical "scrubbing" during discharge cleaning is the formation and desorption of volatile hydrocarbons and oxides from the wall surfaces. These volatile products are finally removed by the pumping system. Carbon removal from the wall usually proceeds at a steadily decreasing rate, the surface coverage changing with time as shown in Fig. 51 from a PDX GDC run. Oxygen removal is more complicated. For stainless steel a rather refractory oxide, Cr_2O_3, remains after the loosely bound oxides and hydroxides are removed. This remaining oxygen can be at the 20 atomic % level and is sufficiently stable that it is not readily desorbed into high power discharges. Titanium is another material that behaves similarly to Cr, i.e., during GDC it forms a stable oxide.

There are also cases where carbon is also not completely removed. An obvious example of this is for bulk graphitic or carbidic structures used as limiters. These types of carbon structures are not major contaminantors of the plasma because the chemical bond strength of the carbon to the surface is high. Conversely, certain cleaning procedures are able to remove even chromium oxide. At Julich[78,79] thermal H^o bombardment of warm (T $\simeq$ 200 °C) inconel surfaces resulted in oxygen coverages less than 2 atomic %.

Of the six standard plasma cleaning techniques which is the best? Why this question has no simple answer should be clear after the following delineation of the benefits and problems associated with each.

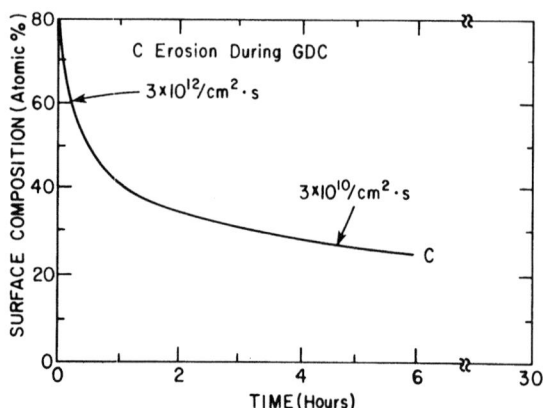

Fig. 51. Carbon removal during H_2 GDC in PDX. The current density to the wall was about 10 µA/cm^2. The initial carbon removal rate was 3 $\times$ 10^{12} cm^{-2}s^{-1}. After 4 hours this dropped to 3 $\times$ 10^{10} cm^{-2}s^{-1}.

PDC[80] requires all the fields and currents used for ordinary tokamak discharges. For this reason it is manpower and cost intensive. Also, discharges of this type show reasonably good confinement, thus remote sections of the vacuum vessel are not rapidly cleaned. In spite of these drawbacks PDC is an essential technique because it is so easy to go from PDC to high power discharges.

TDC[81] and ACDC[82] have rather similar cleaning rates. Both rely on much less energetic plasmas than PDC. Their cleaning efficiency is somewhat faster than PDC, perhaps a factor of two. Again, there are the drawbacks that the tokamak is inaccessible during cleaning because the OH and TF fields are used.

GDC[75,76,77] employs no OH or TF fields. The required power supply, a high voltage (1 kV) high current (5 A) dc supply, is comparatively inexpensive. GDC does reach into remote parts of the vacuum vessel such as the divertor chamber. A main drawback of GDC is that the ions impact the walls with several hundred eV of energy. This causes sputtering which coats windows and insulaters. Also, the high plasma potential is known to cause arcs. Though this may be a benefit by reducing the number of arc sites available to high power discharges, it is also a potential threat to the integrity of insulators, particularly polyimide. The final drawback to GDC is that it loads all structures with hydrogen. This complicates the use of getters and makes density control initially difficult.

Some of the drawbacks of GDC can be overcome by the use of rf-assisted[83,84] GDC. The plasma potential is greatly reduced causing virtually no sputtering or arcing. Hydrogen loading still occurs.

ECDC[85] is an rf discharge cleaning but at a much higher frequency than standardly used for RFADC. The TF is once again used, which again is a drawback. However, control of n_e and T_e is more complete than with simple RFADC. An electron density of above 10^{10} cm^{-3} is possible. The cleaning efficiency for any plasma is predicted to saturate above $n_e \sim 10^{11}$ cm^{-3} because of the effects of "sticky" walls and ionization.

Discharge cleaning has been successful in permitting ohmically-heated tokamaks to operate at low Z and high density and temperature. The effects of auxiliary heating become rather important when the auxiliary power exceeds the ohmic. Under these conditions, sputtering and evaporation of even the most refractory wall material may occur and severe contamination of the plasma may result. Discharge cleaning cannot alter this situation.

C.   Gettering

Gettering refers to the active chemisorption of gases by reactive materials. The gases may dissociate upon adsorption and go into a solid solution within the metal. Such is the case, for example, of H_2 gettered by Ti or Zr/Al. These types of getters reversibly absorb hydrogen. A plot of the equilibrium hydrogen overpressure versus temperature[86,87] is shown in Fig. 52 for Zr/Al, ZrH_2 and TiH_2. Though less hydrogen is reemitted from these getters at low temperature, these getters are usually operated at a moderate temperature to ensure that the absorbed H_2 rapidly diffuses into the bulk leaving the surface free to chemisorb more H_2. The value of the operating temperature is selected such that the equilibrium hydrogen overpressure is less than 10^{-8} Torr. Heating these getters to higher temperature allows for reactivation and other benefits.[88]

For other getter/gas systems the chemisorbed gases may not go into solution, but remain tightly bound on the surface. This is the case for O_2 pumped by Cr. The sticking coefficient[89] for some gases on nonhydride-forming getters is shown in Fig. 53a. As shown in Fig. 53b hydride-forming getters have similar sticking coefficients for O_2 and N_2 but maintain the same sticking coefficients for H_2 and D_2 at high coverages. Both may be selective about which gases are sorbed. Ti, for example, getters most gases commonly found in tokamaks including H_2, D_2, CO, CO_2 and H_2O. Cr, in contrast, does not getter significant amounts of H_2 or D_2. Neither Ti or Cr pumps the noble gases or methane appreciably at or above room temperature. From the above it is clear that the aims of gettering are to control hydrogen density and low Z (O, C, N) impurities.

Getters have one of two configurations, evaporable or bulk. The evaporable type, which include Ti and Cr, are actually deposited by sublimation as a thin film of the getter element onto specially prepared surfaces, such as the tokamak walls. In all but one of the experiments performed to date (ATC,[90] DITE,[91] PLT,[92] ISX-A,[93] JFT-2,[94] PDX,[95] ISX-B,[73,96] Microtorr/Macrotorr,[97] and ASDEX[22]) the wall area coated with getter material was an appreciable fraction (>25%) of the whole wall area. The sole exception was PDX where getters used in the divertor chamber coated about 5% of the total wall area. About 0.2 gm of getter material are required each day. This corresponds to about one Ti atom for each H atom. The sticking coefficient of H_2 on Ti is only a few percent, but for O_2 is near unity. This leads to a pumping speed for the getter of about 0.6 liters $cm^{-2}s^{-1}$.

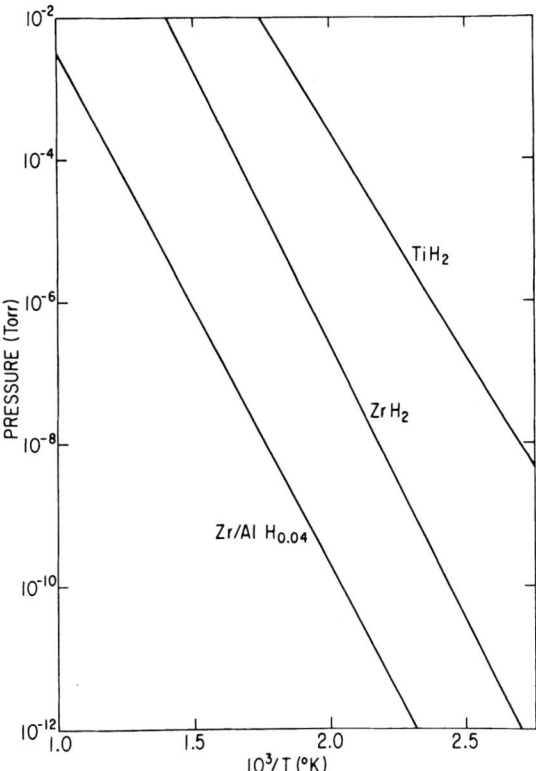

Fig. 52. Equilibrium over-pressure for TiH_2, ZrH_2 and $Zr/Al\ H_{0.04}$. (Refs. 86 and 87.)

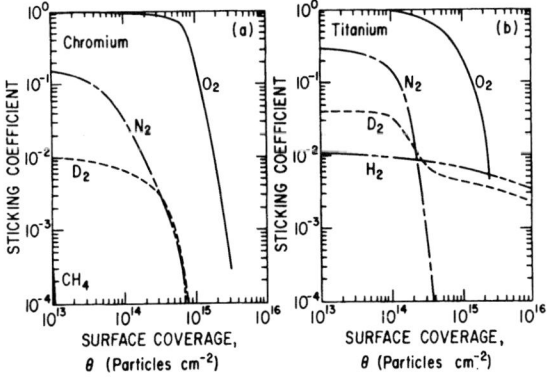

Fig. 53. Sticking coefficients of O_2, N_2, H_2 and D_2 on (a) chromium and (b) titanium as a function of surface coverage by the sorbing gas. (Ref. 89.)

Results of Ti and Cr gettering have been uniformly good, though Cr gettering is not as effective as Ti in reducing H recycling. The effective particle confinement time is seen to decrease in ASDEX[22] from 400 ms to 40 ms, as Ti gettering was applied. This indicates a dramatic increase in the control of impurities and recycling. Another clear indicator that recycling is reduced is that the gas feed required to sustain a discharge is reduced. Figure 54 shows that the difference in the required gas feed rate for two modes of gettering, once-a-day and between shots.

In all gettered tokamaks Z_{eff} is observed to decrease to near unity. The main cause is a reduction in C and O in the plasma. Also, the sublimated Ti covers the stainless steel walls and limiters, resulting in decrease in Fe content, but an increase in Ti. Spectroscopic data from ISX-B[96] (Fig. 55) clearly show this.

ISX-B also documented the enlargement of the operating regime and looked for possible effects on τ_E and the β-limit.[73] Some indication of increased τ_E was evident, but only if small amounts of low-Z impurities, e.g., Ne, were added to the discharge. No improvements in the β-limit was noted.

The bulk getter configuration uses a thin coating of a getter, usually Zr/Al, clad onto a metal substrate. The substrate is bent into a convolved shape[98] to increase the available surface area. A group of substrates are packaged into a module of size 5 to 30 cm in both length and height and 2 to 5 cm in thickness. The modules can be passivated by exposure to air and reactivated

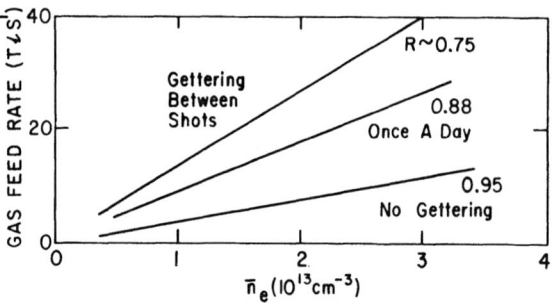

Fig. 54.   Time-averaged gas-feed rate versus $\bar{n}_e$ at 60 ms into ohmically-heated plasmas. The straight lines are fits to data points where τ_p = 75 ms and recycling coefficients, R as shown, were assumed. (Ref. 73.)

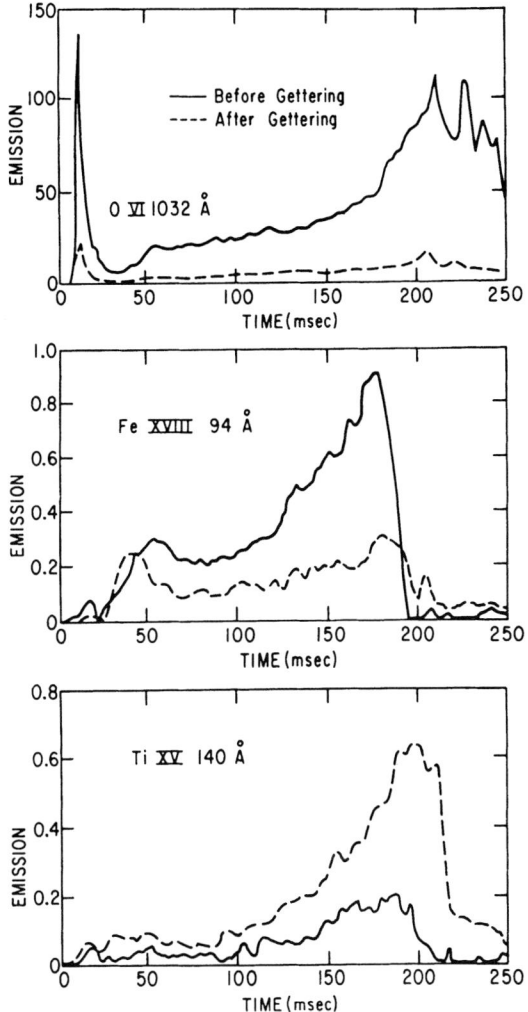

Fig. 55.   Emission rates of impurities in ohmically-heated ISX-B
plasmas. (Ref. 73.)

by heating to about 700 °C which allows the sorbed O_2 and N_2 to
diffuse into the bulk. Because of these properties, Zr/Al getters
have found use behind pumped limiters,[26,99] and near standard
limiters.[100]   For the pumped limiter application, getter modules
offer much a higher pumping speed than possible with more remote
vacuum pumps.   In the first tests of Zr/Al pumps in the main
plasma chamber a reduction of 10% in recycling from 0.9 to 0.81
was achieved.   More extensive use of these modules in TFTR[101]
should soon elucidate further their advantages and limitations.

Getters are not without problems. Evaporable getters may coat sensitive windows or insulators. The films may eventually flake. Certainly they complicate the changeover of hydrogen isotopes and the tritium inventory in reactors. Bulk getters do not have the same problems, although some embrittlement has been observed in at least one situation.[99] Bulk getters are considerably more expensive and more difficult to implement. They can take up much of the interior real estate in tokamaks. But despite these drawbacks, gettering continues to be a favored technique, because it is fast and because of its uniformly high success in enabling the achievement of low Z_{eff} and high density.

V. POSTSCRIPT

The aim of this chapter was to provide students coming into the field of plasma physics with an introduction to the methods used to control and measure particle confinement. The material presented here was drawn from the literature of tokamak experiments, but selected results could just as well apply to other plasma devices. Historically, new methods to measure plasma confinement have, for good reason, always preceeded new ways to control the plasma density. Volume averaged measurements gave way to radial profiles; radial profiles were refined by species discrimination both in type and in energy. At first particle losses were considered at a global level; then local transport and asymmetries gained prominence. Closely following the improved diagnostics were improved ways of controlling the density. Gas pre-fill was changed to pulsed gas fill; and pulsed gas fill was eventually joined by pellet fueling. Bakeout was supplemented with discharge cleaning; gettering and coating provided additional species control. Each step has addressed a broad issue, such as plasma purity or plasma stability. These innovations have gone far in improving the performance of tokamaks. But before tokamaks can become a viable source of power, methods must be developed to further control the plasma density and temperature profiles. Particle loss rates, helium ash removal and impurity influx must be controlled under reactor-like conditions which include hot walls, long pulses, high temperatures and densities, and intense particle and photon fluxes. There are the problems that must be addressed by the next generation of scientists.

REFERENCES

1. J.C. Hosea, N. Bretz, A. Cavallo et al., "High Power ICRF

Heating on PLT and Extrapolation to Future Devices," in 3rd Joint Varenna-Grenoble International Symposium on Heating in Toroidal Plasma (Commission of the European Communities, Brussels, 1982) Vol. 1, 213.

2. R.M. Kulsrud, H.P. Furth, E.J. Valeo et al., "Fusion Reactor Plasmas with Polarized Nuclei," Phys. Rev. Lett. 49 (1982) 1248.

3. H.S. Greenside, D.E. Post, and R.A. Budny, "Depolarization of D-T Plasmas by Recycling in Material Walls," J. Vac. Sci. and Technol. 2 (1984) 97.

4. M. Greenwald, D. Gwinn, S. Milora et al., "Energy Confinement in High-Density Pellet-Fueled Plasmas in the Alcator-C Tokamak," Phys. Rev. Lett. 53 (1984) 352.

5. E.S. Marmar and R. Hulse, private communications.

6. S.M. Kay, R.J. Goldston, M. Bell et al., "Studies of Thermal Energy Confinement Scaling in PDX Plasmas: $D^0 \rightarrow H^+$ Limiter Discharges," Princeton University Plasma Physics Laboratory Report PPPL-2109 (1984) 100 pp.

7. C.E. Singer, M. Redi, D. Boyd et al., "Semiempirical Modelling of H-Mode Discharges," Princeton University Plasma Physics Laboratory Report PPPL-2103 (1984) 55 pp; submitted to Nuclear Fusion.

8. F. Wagner, M. Keilhacker and the ASDEX and NI Teams, "Importance of the Divertor Configuration for Attaining the H-Regime in ASDEX," J. Nucl. Mater. 121 (1984) 103.

9. F. Wagner, G. Becker, K. Behringer et al., "Regime of Improved Confinement and High-β in Neutral-Beam-Heated Divertor Discharges of the ASDEX Tokamak," Phys. Rev. Lett. 49 (1982) 1408.

10. M. Shimada, M. Nagami, K. Ioki et al., "High Density, Single-Null Poloidal Divertor Results in Doublet III," J. Nucl. Mater. 111 & 112 (1982) 362.

11. M.A. Mahdavi, C.J. Armentrout, F.P. Blau et al., "A Review of the Recent Expanded Boundary Divertor Experiments in the Doublet III Device," J. Nucl. Mater. 111 & 112 (1982) 355.

12. M. Shimada, M. Washizu, S. Sengoku et al., "Divertor Studies in High-Power Beam-Heated Discharges in D-III," J. Nucl. Mater. 128 & 129 (1984) 340.

13. S.M. Kaye, M.G. Bell, K. Bol et al., "Attainment of High Confinement in Neutral-Beam-Heated Divertor Discharges in the PDX Tokamak," J. Nucl. Mater. 121 (1984) 115.

14. W.M. Tang, C.Z. Cheng, J.A. Krommes et al., "Anomalous Transport and Scaling Studies in Tokamaks," Princeton University Plasma Physics Laboratory Report PPPL-2142 (1984) 42 pp.

15. R. Bastasz and L.G. Haggmark, "Hydrogen Ion Impact Desorption of Adsorbed Deuterium from Stainless Steel," J. Nucl. Mater. 111 & 112 (1982) 805.

16. S.J. Fielding, G.M. McCracken, and P.E. Stott, "Recycling in Gettered and Diverted Discharges in DITE Tokamak," J. Nucl. Mater. 76 & 77 (1978) 273.

17. H.C. Howe, "Hydrogen Recycle Modeling and Measurements in Tokamaks," J. Nucl. Mater. 93 & 94 (1980) 17.

18. TFR Group, "Recycling and Particle Confinement Characteristics in TFR 600," J. Nucl. Mater. 93 & 94 (1980) 272.

19. TFR Group, "Gas Recycling Studies on the TFR Tokamak," J. Nucl. Mater. 111 & 112 (1982) 199.

20. H.F. Dylla, W.R. Blanchard, R. Budny et al., "Gas-Fueling Studies in the PDX Tokamak," J. Nucl. Mater. 111 & 112 (1982) 211.

21. H.F. Dylla, M.G. Bell, R.J. Fonck et. al., "Gas-Fueling Studies in the PDX Tokamak: II," J. Nucl. Mater. 121 (1984) 144.

22. H.M. Mayer, F. Wagner, G. Becker et al., "Fueling Efficiency of Gas Puffing in ASDEX," J. Nucl. Mater. 111 & 112 (1982) 204.

23. F. Wagner, K. Behringer, D. Campbell et al., "Variation of the Particle Confinement During Neutral Injection into ASDEX Divertor Plasmas," Max-Planck Institut fur Plasmaphysik Report IPP III/78 (June 1982) 35 pp.

24. K.H. Burrell, R. Pratter, S. Ejima et al., "Comparison of Energy Confinement in D-III Limiter and Divertor Discharges with Ohmic, Neutral Beam and Electron Cyclotron Heating," in 10th International Conference on Plasma Physics and Controlled Nuclear Fusion Research 1984 (London, England, September 1984) to be published.

25. D.E. Voss, "Low Energy Neutral Atom Emission and Plasma Recycling in the PLT Tokamak," Ph.D. Thesis, Princeton University (1980).

26. A.E. Pontau, S.E. Guthrie, M.E. Malinowski et al., "Initial ALT-I Pump Limiter Studies on TEXTOR," J. Nucl. Mater. 128 & 129 (1984) 434.

27. H.F. Dylla, private communication.

28. J.D. Strachan, N. Bretz, E. Mazzucato et al., "A Density Rise Experiment on PLT," Nucl. Fusion 22 (1982) 1145.

29. L.C. Johnson and E. Hinnov, "Ionization, Recombination, and Population of Excited Levels in Hydrogen Plasmas," J. Quant. Spectr. & Rad. Trans. 13 (1973) 333.

30. E.S. Marmar, "Recycling Processes in Tokamaks," J. Nucl. Mater. 76 & 77 (1978) 59.

31. M. Bell, V. Arunasalam, M. Bitter et al., "Recent PDX Results" in 10th European Conference on Controlled Fusion and Plasma Physics (Moscow, USSR, 1981) Vol. 2, 21.

32. D.H.J. Goodall, "High Speed Cine Film Studies of Plasma Behavior and Plasma Surface Interactions in Tokamaks," J. Nucl. Mater. 111 & 112 (1982) 11.

33. R. Fonck, private communication.

34. R. Jacobsen, "High Speed Photographic Studies of the Equilibrium and Stability of the ATC Tokamak," Plasma Physics 17 (1975) 547.

35. D.E. Voss and S.A. Cohen, "Low Energy Neutral Outflux from the PLT Tokamak," J. Nucl. Mater. 93 & 94 (1980) 405.

36. S.A. Cohen, D. Ruzic, D.E. Voss et al., "Measurements of Low Energy Neutral Hydrogen Efflux during ICRF Heating," Nucl. Fusion 24 (1984) 1490.

37. F. Wagner, "Neutral Particle Diagnostics for Ohmically and Auxiliary Heated Tokamaks," J. Vac. Sci. Technol. 20 (1982) 1211.

37a C. Muller and F. Wagner, private communiation.

38. C. Kahn, "Probe Studies of Scaling with Plasma Parameters in D-III," GA Technologies, Inc. Report GA-A17576 (1984) 45 pp.

39. B. LaBombard, B. Lipschultz, and P. Pribyl "Poloidal Pressure Asymmetries in the Limiter Shadow Region of Alcator-C," Bull. Am. Phys. Soc. 29 (1984) 1320.

40. S. Sengoku and H. Ohtsuka, "Experimental Results on Boundary Plasmas, Resulting Surface Interactions and Extrapolation to Large Fusion Devices," J. Nucl. Mater. 93 & 94 (1980) 75.

41. E.S. Marmar, J.L. Cecchi, and S.A. Cohen, "System for Rapid Injection of Metal Atoms in Plasmas," Rev. Sci. Instrum. 46 (1975) 1149.

42. S.A. Cohen, J.L. Cecchi, and E.S. Marmar, "Impurity Transport in a Quiescent Tokamak Plasma," Phys. Rev. Lett. 35 (1975) 1507.

43. R.A. Hulse, "Numerical Studies of Impurities in Fusion Plasmas," Nucl. Technol. 3 (1983) 259.

44. G. Fussmann, W. Poschenrieder, K. Bernhardi et al., "Studies on Impurity Retainment in the ASDEX Divertor," J. Nucl. Mater. 121 (1984) 164.

45. D.L. Book, "NRL Plasma Formulary," Naval Research Laboratory (1980).

46. R.J. Fonck, M. Finkenthal, R.J. Goldston et al., "Spatially Resolved Measurements of Fully Ionized Low-Z Impurities in the PDX Tokamak," Phys. Rev. Lett. 49 (1982) 737.

47. R.C. Isler, L.E. Murray, E.C. Crume et al., "Impurity Transport and Plasma Rotation in the ISX-B Tokamak," Nucl. Fusion 23 (1983) 1017.

48. S.A. Cohen, R. Bell, A Cavallo et. al., "Impurity long-τ Experiments on PLT," Bull. Am. Phys. Soc. 28 (1983) 1127.

49. E.S. Marmar, J.E. Rice, J.L. Terry, and F.H. Seguin, "Impurity Injection Experiments in the Alcator Tokamak," Nucl. Fusion 22 (1982) 1567.

50. C.W. Barnes, "Studies of Runaway Electron Transport in PLT and PDX," Ph.D. Thesis, Princeton University (1981).

51. Equipe TFR, in 5th International Conference on Plasma Physics and Controlled Nuclear Fusion Research (IAEA, Vienna, 1975) Vol. I, 135.

52. S.A. Cohen, R. Budny, G.M. McCracken, and M. Ulrickson, "Mechanisms Responsible for Topographical Changes in PLT Stainless-Steel and Graphite Limiters," Nucl. Fusion 21 (1981) 233.

53. S.A. Cohen, R.V. Budny, V. Corso et al., "The PLT Rotating Pumped Limiter," J. Nucl. Mater. 128 & 129 (1984) 430.

54a. W. Bauer, K.L. Wilson, C.L. Bisson et al., "Alpha Transport and Blistering in Tokamaks," Nucl. Fusion 19 (1979) 93.

54. P. Beiersdorfer, R. Kaita, and R.J. Goldston, "Fast-Time-Resolution Charge-Exchange Measurements During the 'Fishbone' Instability in the Poloidal Divertor Experiment," Nucl. Fusion 24 (1984) 487.

55. W.W. Heidbrink, R. Hay, and J.D. Strachan, "Confinement of Fusion Reaction Products During the Fishbone Instability," Princeton University Plasma Physics Laboratory Report PPPL-2135 (1984) 27 pp.

56. P. Kaw and J.M. Dawson, "Current Maintenance in Tokamaks by Use of Synchrotron Radiation," Phys. Rev. Lett. 48 (1981) 1730.

57. K. Kadota, M. Otsuka, and J. Fujita, "Space- and Time-Resolved Study of Impurities by Visible Spectroscopy in the High-Density Regime of JIPP T-II Tokamak Plasma," Nucl. Fusion 20 (1980) 209.

58. D. Lichtman and Y. Shapira, "The Role of Carbon in Photodesorption," J. Nucl. Mater. 63 (1976) 184.

59. L. Oren, L. Keller, F. Schwirzke et al., "Influence Exerted by the Plasma Edge Potential on Recycling, Sputtering and Impurity Accumulation," J. Nucl. Mater. 111 & 112 (1982) 34.

60. L. Spitzer, D.J. Grove, W.E. Johnson et al., "Problems of the Stellarator as a Useful Power Source," USAEC Report NYP-6047 (1954).

61. M.M. Menon, "Neutral Beam Heating Applications and Development," Proc. IEEE 69 (1981) 1012.

62. J. Tachon, private communication (1975).

63. D.L. Jassby and R.J. Goldston, "Enhanced Penetration of Neutral-Beam Injected Ions by Vertically Asymmetric Toroidal Field Ripple," Nucl. Fusion 16 (1976) 613.

64. M. Gaudreau, A. Gondhalekar, M.H. Hughes et al., "High Density Discharges in the Alcator Tokamak," Phys. Rev. Lett. 39 (1977) 1267.

65. W.M. Tang, P.H. Rutherford, H.P. Furth, and J.C. Adam, "Stabilization of Trapped Particles Modes by Reversed-Gradient Profiles," Phys. Rev. Lett. 35 (1975) 660.

850

66. N. Ohyabu, N.H. Brooks, K.H. Burrell et al., "Role of Recycling in Beam-Heated Expanded Boundary Divertor Discharges in D-III," J. Nucl. Mater. 121 (1984) 157.

67. S.C. Bates and K.H. Burrell, "Fast Gas Injection System for Plasma Physics Experiments," Rev. Sci. Instrum. 55 (1984) 934.

68. D. Ruzic, D. Heifetz, S. Cohen, and D. Voss, private communication.

69. S.L. Milora, "Review of Pellet Fueling," J. Fus. Energy 1 (1981) 15.

70. S.L. Milora, C.A. Foster, C.E. Thomas et al., "Results of Hydrogen Pellet Injection into ISX-B," Nucl. Fusion 20 (1980) 1491.

71. S.L. Milora, G.L. Schmidt, W.A. Houlberg et al., "Pellet Injection into Diverted PDX Plasmas," Nucl. Fusion 22 (1982) 1263.

72. S.J. Fielding, J. Hugill, G. McCracken, J.W.H. Paul, R. Prentice, and P.E. Stott, "High Density Discharges with Gettered Torus Walls in DITE," Nucl. Fusion 17 (1977) 1382.

73. A.J. Wooton, P.H. Edwards, R.C. Isler, and P. Mioduszewski, "Gettering in ISX-B," J. Nucl. Mater. 111 & 112 (1982) 479.

74. H.F. Dylla, "A Review of the Wall Problem and Conditioning Techniques for Tokamaks," J. Nucl. Mater. 93 & 94 (1980) 61.

75. W. Poschenrieder, G. Staudenmaier, and P. Staib, "Conditioning of ASDEX by Glow Discharge," J. Nucl. Mater. 93 & 94 (1980) 322.

76. H.F. Dylla, S.A. Cohen, S.M. Rossanagel et al., "Glow Discharge Conditioning of the PDX Vacuum Vessel," J. Vac. Sci. and Tech. 17 (1980) 286.

77. D. Ruzic, S. Cohen, B. Denne, and J. Schivell "Low Energy Neutral Spectroscopy During Pulsed Discharge Cleaning in PLT," J. Vac. Sci. Technol. A1 (1983) 818.

78. K.G. Tsersich and J. Von Seggern "Light Impurity Removal from Stainless Steel by Atomic Oxygen," J. Nucl. Mater. 111 & 112 (1982) 487.

79. F. Waelbroeck, P. Wienbold, and J. Winter, "Thermally Activated Processes in Hydrogen Recycling," J. Nucl. Mater. 111 & 112 (1982) 185.

80. H.F. Dylla, K. Bol, S.A. Cohen et al., "Observation of Changes in Residual Gas and Surface Composition with Discharge Cleaning in PLT," J. Vac. Sci. Technol. 16 (1979) 752.

81. L. Oren and R.J. Taylor, "Trapping and Removal of Oxygen in Tokamaks," Nucl. Fusion 17 (1976) 1143.

82. N. Noda, S. Tanahashi, K. Kawahata et al., "Wall Conditioning of the JIPP T-II Torus by AC Discharge Cleaning," J. Nucl. Mater. 111 & 112 (1982) 498.

83. K.J. Dietz, K. Sonenberg, F. Waelbroeck, et al., "Wall Conditioning in JET," in 9th International Vacuum Congress ASEVA (Imprinta Moderna, Madrid, Spain, 1983) 706.

84. J. Burt, S.J. Fielding, G.M. McCracken, and D.D.R. Summers, "RF Assisted Glow Discharge Cleaning in the DITE Tokamak," Fusion Techn. 6 (1984) 399.

85. Y. Sakamoto, Y. Ishibe, K. Yano et al., "Electron Cyclotron Resonance Discharge Cleaning of JTF-2 Tokamak (JAERI)," J. Nucl. Mater. 94 & 94 (1980) 333.

86. R.J. Knize, J.L. Cecchi, and H.F. Dylla, "Measurement of H_2, D_2 Solubilities in Zr-Al," J. Vac. Sci. Technol. 20 (1982) 1135.

87. J.H. Singleton, "Hydrogen Pumping by Sputter-Ion Pumps and Getter Pumps," J. Vac. Sci. Technol. 8 (1971) 275.

88. J.L. Cecchi, S.A. Cohen, and J.J. Sredniawski, "Transient Getter Scheme of the Tokamak Fusion Test Reactor," J. Vac. Sci. Technol. 17 (1980) 294.

89. J.E. Simpkins, D. Mioduszewski, and L.W. Stratton, "Studies of Chromium Gettering," J. Nucl. Mater. 111 & 112 (1982) 827.

90. P.E. Stott, C.C. Daughney, and R.A. Ellis, Jr. "Control of Recycling and Impurities in the ATC Tokamak by Means of Gettered Surfaces" Nucl. Fusion 15 (1975) 431.

91. P.E. Stott, J. Hugill, S.J. Fielding et al., "High Density Discharges with Gettered Walls in DITE," in 8th European Conference on Controlled Fusion and Plasma Physics (Prague, Czechoslovakia, 1977) 1, 37.

92. K. Bol, V. Arunasalam, C. Barnes et al., "Recent Results from the PLT Tokamak" in 7th International Conference on Plasma Physics and Controlled Nuclear Fusion Research (IAEA, Vienna, 1979) Vol. I, 11.

93. R.J. Colchin, C.E. Bush, and P.H. Edmonds et al., "Plasma-Wall Impurity Experiments in ISX-A," J. Nucl. Mater. 76 & 77 (1978) 405.

94. S. Konoshima, N. Fujisawa, M. Maeno et al., "Improvement of Plasma Parameters by Titanium Gettering in the JFT-2 Tokamak," J. Nucl. Mater. 76 & 77 (1978) 581.

95. D. Meade, V. Arunasalam, C. Barnes et al., "Initial Operation of PDX," in 9th European Conference on Controlled Fusion and Plasma Physics (Oxford, England, 1979) Vol. I, 91.

96. P.H. Edmonds, P. Mioduszewski, J.E. Simpkins, and A.J. Wooton, "Gettering Experience in ISX or Titanium is Good For Your Health," J. Vac. Sci. and Technol. 20 (1982) 1317.

97. R.J. Taylor, R.F. Bunshah, and F. Schwirzke, "Impurity Control in Tokamaks with in-situ Metal Deposition," J. Nucl. Mater. 93 & 94 (1980) 338.

98. B. Ferrario and L. Rosai, "New Types of Volume Gettering Panels for Vacuum Problems in Plasma Machines," in 7th

International Vacuum Congress and 3rd International Conference Solid Surfaces (Vienna, Austria, 1977) 359.

99. P. Mioduszewski, "Experimental Studies on Pump Limiters," J. Nucl. Mater. 111 & 112 (1982) 253.

100. J.L. Cecchi, R. Knize, H.F. Dylla et al., "Reduction of Recycling by Pumping at the PDX Limiter," J. Nucl. Mater. 111 & 112 (1982) 305.

101. J.L. Cecchi, M.G. Bell, M. Bitter et al., "Initial Limiter and Getter Operation in TFTR," J. Nucl. Mater. 128 & 129 (1984) 1.

LIMITERS AND DIVERTOR PLATES

M. Ulrickson
Plasma Physics Laboratory, Princeton University
Princeton, New Jersey 08544

INTRODUCTION

As fusion devices move toward scientific breakeven and
beyond that to ignition, the demands placed upon the limiter and
divertor plates become more severe. The design of these devices
depends upon a careful balance between plasma edge processes and
material thermal, mechanical, and electromagnetic properties.
The insertion of a limiter or divertor plate in the plasma edge
causes a large perturbation of the edge. The physics of the
plasma limiter interaction is discussed in great detail in other
chapters and will not be discussed here so that emphasis can be
placed on the techniques used to design limiters and divertor
plates. The difficulty in the design of a limiter or divertor
plate lies in the simultaneous satisfaction of all the
constraints imposed by real materials and geometries within the
allowed boundaries of the physics of the plasma edge. The
uncertainties in the plasma edge properties makes optimization
of any design more difficult because some constraints are poorly
defined. The purpose of this chapter is to present the basic
physical principles that are important for the design of
limiters and divertor plates. The methods used to find a
solution within the constraints will be described. The
equations that result from the physical principle will, in
general, not be derived since the majority of them are discussed
in detail in other chapters. The main thrust of this chapter is
to show how all of the individual effects interact to lead to a
design. Finally, some remarks concerning how the designed
components are conditioned for optimum performance will be
presented.

The first section describes the various generic types of limiters and divertor plates that have been considered. The normal and abnormal heat and particle fluxes to these generic designs will be examined. The second section covers electromagnetic interactions between the plasma and limiters. The material choices that follow from the thermal and electromagnetic loads are examined in the third section. Conditioning of the device is the subject of the last section.

PLASMA ENERGY AND PARTICLE DEPOSITION

The flux of energy or particles to a limiter or divertor is determined by two things: profiles of density and temperature in the edge of the plasma and the shape of the limiter or divertor plates. The plasma edge profiles are discussed in great detail in other chapters. The shapes that have been proposed for various types of limiters and divertor plates must be considered before the heat and particle flux to a plate can be discussed. The heat and particle loads also depend on the presence of plasma instabilities such as disruptions. This section will consider first limiter and divertor plate shapes then the normal and abnormal heat and particle fluxes.

Limiter types

Limiters can be divided into three generic classes: poloidal limiters, toroidal limiters, and multiple limiters. Examples of all these types can be found on existing fusion devices. The purpose of all these types of limiters is to absorb the energy of the plasma before it can reach the walls of the vacuum vessel. While in principle this is not necessary, it allows for considerable flexibility in the design of both limiters and vacuum vessels making it important from a practical standpoint.

A poloidal limiter is one that has a limited toroidal extent. It may be a complete poloidal ring or a structure having limited poloidal extent. An example of a poloidal limiter is shown in Fig. 1a. The term as it is used in this chapter is restricted to a single toroidal location. However, it may include several blades. The poloidal limiter is the first type of limiter that was used on tokamaks. Poloidal limiters are presently in use on Alcator-C,[1] Doublet-III,[2] ISX-B,[3] PDX,[4] and PLT[5] in the U.S. fusion program. They are also being used during the initial phases of large tokamaks such as JET[6] and TFTR.[7] The main advantage of poloidal limiters are simplicity, low cost, and flexibility. The major disadvantage of a poloidal limiter is that the area wetted by the plasma is limited which results in large heat loads unless the input power is small.

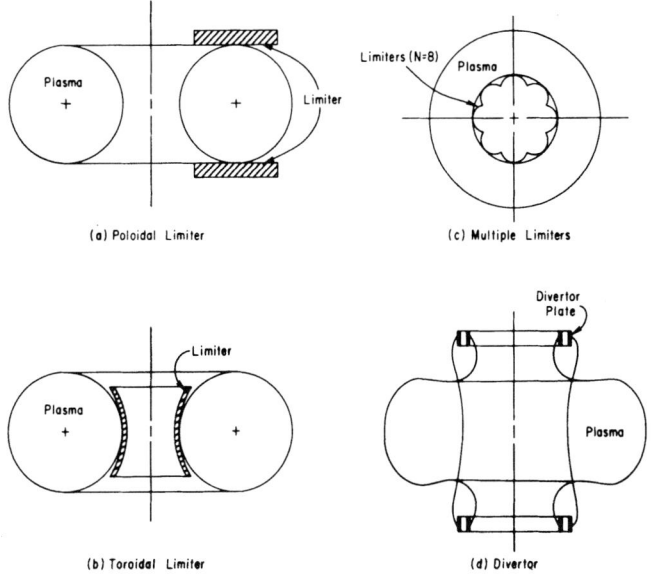

Fig. 1. Schematic diagrams of the general types of limiters and
         divertors used in tokamaks.

A toroidal limiter is one that has a limited poloidal
extent but extends completely around the torus in the toroidal
direction. Toroidal limiters are also referred to as belt
limiters, bumper limiters, and axisymmetric limiters. They have
been proposed both curving toward the plasma and curving away
from the plasma, and all possibilities in between. The
definition is limited to a single belt in this chapter. An
example of a toroidal limiter is shown in Fig. 1b. Both PDX[8]
and ASDEX[9] have used a toroidal limiter. One is planned for
TFTR.[10,11] The major advantage of toroidal limiters are the
large wetted area that can be achieved. This is a particularly
important point as we move toward ignition devices and beyond.
The major disadvantages of toroidal limiters are their
complexity, and their inability to accommodate large variations
in plasma shape and/or size.

Multiple limiters will be used to refer to sets of two or
more of the above two types; e.g., a limiter system consisting
of 20 equispaced poloidal limiters of the type shown in Fig.
1a. An example of a multiple limiter is shown in Fig. 1c.
Multiple limiters are called out as a separate type because
there are unique problems to be faced with them. The connection
length for multiple limiters lies between that for poloidal and
toroidal limiters implying different scrape-off lengths (all
other things being equal). Multiple limiters cast shadows on
each other. This effects the wetted area and, in general, the
heat flux does not decrease as rapidly as expected; i.e., ten

poloidal limiters have more than one tenth of the peak heat flux of a single poloidal limiter because of shadowing. Multiple limiters are being used or proposed by both TFTR[12,13] and JET.[6] In fact, JET is planning the first multiple toroidal limiter.[14] The major advantage of multiple limiters is that some of the flexibility of single poloidal limiters is retained while attaining a larger wetted area. The major disadvantage is higher peak heat fluxes because of shadowing and the gaps between segments (this will be discussed more later).

Divertor plates

Divertor plates serve much the same function as limiters but they are farther away from the main plasma volume. Divertors fall into two general categories: poloidal and bundle. The divertor plates for a poloidal divertor are axisymmetric in the toroidal direction. Poloidal divertors are being used or proposed for PDX,[15] Doublet-III,[16,17], ASDEX,[18] and JT-60.[19] The bundle divertor plates are localized at one toroidal location. An example of a poloidal divertor plate is shown in Fig. 1d. The designs are not as varied because the magnetic topology places more restrictions on allowed shapes.

Energy and particle fluxes

Any discussion of the energy and particle fluxes to a limiter or divertor plate must include a plasma edge model. Since the plasma edge characteristics are discussed in great detail in other chapters, a very simple model for the plasma edge profile will be used here. Any design of a limiter would be proceeded by plasma edge modeling that would predict the expected edge conditions. The simple exponential profiles used here are close to the results of the models in many cases.[20] They are chosen because the important effects can be illustrated without getting into the complications of the physics of the edge plasma. It will be assumed that the plasma edge density and temperature profiles are characterized by simple exponential scrape-off lengths λ_n and λ_T, i.e.,

$$n(r) = n_o \exp\left[-(r-a)/\lambda_n\right] ,$$
(1)

$$T_e(r) = T_i(r) = T_o \exp\left[-(r-a)/\lambda_T\right] .$$
(2)

These apply only for r ≥ a where a is the plasma minor radius. The above expressions are only valid for circular plasma cross sections but they are easily generalized to non-circular cross sections by replacing (r-a) by the distance outside the last closed flux surface (separatrix surface) and adjusting λ to account for the flux surface expansion (or contraction). If the profiles of density and temperature are characterized by exponentials then the flux of energy is also characterized by an exponential. The energy flux is given by

$$\Gamma_Q(r) = \Gamma_o \exp\left[-(r-a)/\lambda_Q\right] ,$$ (3)

where

$$\lambda_Q = \frac{\lambda_n \lambda_T}{3/2\ \lambda_n + \lambda_T} .$$ (4)

These assumed profiles are supported by the results of some experiments, but there are also examples of other types of profile shapes which are discussed in other chapters. These profiles will be used to illustrate how the plasma edge properties are translated into heat and particle flux onto the limiter of divertor plate surfaces.

The heat and particle fluxes that result from a given edge profile depend on the exact shape of the surface of the limiter or divertor plate and the geometry of the plasma surface. If we assume that the plasma fluxes are traveling along field lines then the flux on the surface of a plate is given by

$$\Gamma_w = \Gamma_o\ \hat{n} \cdot \hat{t},$$ (4)

where

$\hat{n}$ = the unit surface normal of the limiter plate,
$\hat{t}$ = the unit tangent vector of the field line,

and

Γ_o = the flux (either particles or energy) normal to a field line [cf. Eq. (3)].

The dot product is the angle between the field line and the plate. If there is a component of plasma transport perpendicular to field lines, then t should be replaced with a unit vector in the direction of the vector sum of the perpendicular and parallel components. Since the plasma fluxes decrease as one moves out through the scrape-off layer (exponentially in the example chosen here) and one can increase the flux on the plate by increasing the angle of incidence ($\hat{n} \cdot \hat{t}$ above) it is possible to design a limiter or divertor plate that has a constant flux of either particles or heat.[21] Both cannot, in general, be made constant because $\lambda_n \neq \lambda_Q$ unless $\lambda_T \gg \lambda_n$. The constant flux can only be maintained out to the point where the flux normal to a field line is equal to the design flux. The uncertainties in the plasma edge properties and fabrication difficulties make constant flux limiters or divertor plates impossible to achieve at this time, but such designs do serve as an important point to compare with a given design. The specific examples below will shed more light on this. The calculation of the heat flux profile on several types of limiter plates will be demonstrated below.

Let us first consider a toroidal limiter having a constant curvature in the poloidal direction (see Fig. 2). In this case, there is only a radial component of the surface normal which allows the toroidal component of the tangent vector to be ignored. In an x-y coordinate system whose origin is at the center of curvature of the limiter the surface normal is given by

$$\hat{n} = \cos\theta\,\hat{i} - \sin\theta\,\hat{j} \; . \tag{5}$$

The tangent vector is given by

$$\hat{t} = -\frac{\ell\sin\theta}{r}\,\hat{i} + \left(1 - \frac{\ell^2\sin^2\theta}{r^2}\right)^{1/2}\hat{j} \; , \tag{6}$$

where

$$r^2 = \ell^2 + (\ell - a)^2 - 2\ell(\ell - a)\cos\theta. \tag{7}$$

The dot product is then

$$\hat{n}\cdot\hat{t} = \frac{\ell\cos\theta\sin\theta}{r} - \left(1 - \frac{\ell^2\sin^2\theta}{r^2}\right)^{1/2}\sin\theta. \tag{8}$$

The normalization for the heat flux is then found by integrating over the surface of the limiter; i.e.,

$$P_{Total} = \int_{-\theta_\ell}^{\theta_\ell} \Gamma_o \; \hat{n}\cdot\hat{t} \; \exp\left[-(r-a)/\lambda_Q\right] 2\pi\,R\,\ell\,d\theta \; , \tag{9}$$

where

$$R = R_o - \ell\cos\theta \; . \tag{10}$$

860

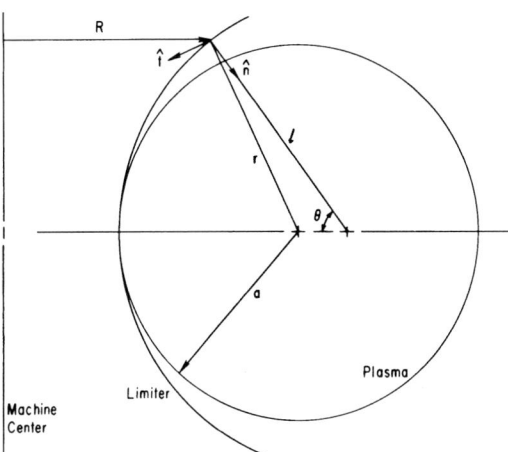

Fig. 2. The geometry of an axisymmetric bumper limiter showing the meaning of the variables used to describe it.

If the limiter is large enough that the integrand vanishes at $\theta = \theta_\ell$ then P_{Total} is simply the total power lost by conduction and convection, otherwise the value of P_{Total} will depend on the nature (shape, size, etc.) of the other objects contacted by the plasma. Typically, P_{Total} is well defined. Taking an example where $a = 85$ cm, $R_o = 265$ cm, $\ell = 103$ cm, $\theta_\ell = \pm 60°$, $\lambda_Q = 1$ cm, and $P_{Total} = 32$ MW, the integration gives $\Gamma_o = 1.528 \times 10^4$ W/cm^2. The heat flux on the upper half of the limiter is shown in Fig. 3. The peak heat flux is 420 W/cm^2. The effect of perpendicular transport on this heat flux profile is shown in Fig. 4 for various values of $v_\perp/v_\parallel$. The value of $v_\perp/v_\parallel$ is typically 10^{-4} to 10^{-3} is most calculations of the plasma edge under normal conditions. Under conditions of strong instabilities such as disruptions there is evidence[22] for profiles such as that for $v_\perp/v_\parallel = 1$. The particle flux has the same shape as the heat flux, but the peak is shifted toward larger θ because $\lambda_n > \lambda_Q$. The symmetry in this particular case results in particularly simple equations, and it has become a classic design to which many other designs are compared.

Before moving on to other more complicated limiter types, let us examine what happens to this case if one actually tries to make such an ideal symmetrical limiter. The above analysis assumes the limiter is perfectly aligned to the toroidal field. If the center of toroidal curvature of the limiter is displaced from the true center of curvature of the toroidal field (or plasma) by an amount $\lambda_Q/2$ then the peak heat flux is doubled. For this effect to be negligible the limiter must be aligned to within about $\lambda_Q/10$ which implies an accuracy

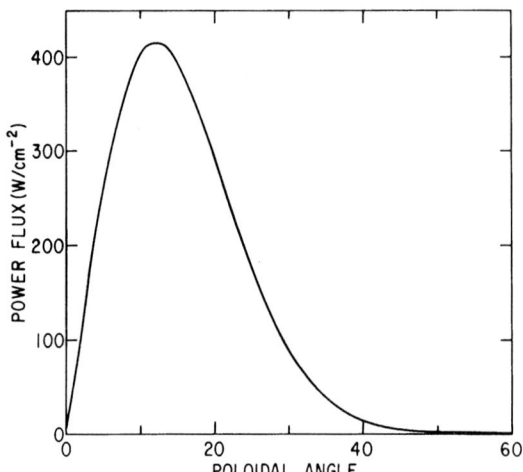

Fig. 3. Heat flux on the upper half of a bumper limiter for a scrape-off length of 1.0 cm. The poloidal angle is measured from the horizontal midplane.

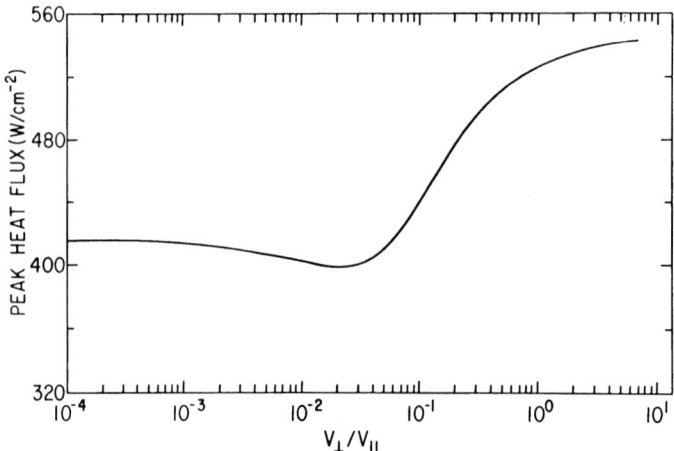

Fig. 4. Variation of the peak heat flux versus the ratio of the perpendicular to the parallel transport in the plasma edge.

of order 1 mm if $\lambda_Q = 1$ cm. This requires a relative field
strength measurement of about 6 parts in 10^4 in a location where
the gradient is about 6×10^{-3}/cm. This is possible but only
with great care. It does imply a segmented design that allows
for adjustment. Segments are also required to get the limiter
into the vacuum vessel. If the edge of a segment were to
protrude beyond the surface of the limiter then the heat flux
would become Γ_o because it would be normal to the flux. From
the above example it can be seen that this value is
intolerable. Hence, such a situation must be avoided. This can
be done by pulling back the leading edges, but this reduces the
effective area which increases the peak heat flux. Other such
depressions in the surface may be required for diagnostics.
Figure 5 shows what happens when such a depression in the
axisymmetric surface is made. The energy that would have hit
the original surface is redistributed on the edges of the
depression. In the example shown in Fig. 5, the energy to be
redistributed is

$$E = (W + 2\ell)\, P_o \, , \tag{11}$$

where P_o is the initial heat flux. If $d \gg \lambda$, then the
effective area of the edge is

$$A = \frac{2\lambda}{d} \left(d^2 + \ell^2 \right)^{1/2} . \tag{12}$$

which gives a new heat flux of

$$P' = \frac{E}{A} = \frac{(W + 2\ell)\, d}{2\lambda(d^2 + \ell^2)^{1/2}} \; P_o \, . \tag{13}$$

For $\ell \gg d$ and $\ell = 2w$ this gives

$$P' = 2.5\, P_o \, .$$

Again showing that substantial increases in the heat flux result
from relatively small perturbations to the simple symmetric
surface.

Considerable variations in the heat flux also result from
the uncertainties in the scrape-off thickness; e.g., if $\lambda_Q = 3$
cm in the above example, the peak heat flux is 200 W/cm^2, and it
occurs at $\theta = 30°$. The uncertainties in the edge properties are

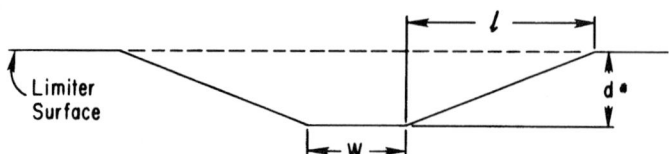

PLASMA

Fig. 5. Geometry used to calculate the heat flux on the edge of
a gap in a bumper limiter. The edges are tapered to
reduce the heat load.

one of the major limitations in using the constant flux shape.
It has been shown[23] that the constant flux shape is more
sensitive to edge variations for $\lambda < \lambda_{design}$ than other simple
designs which have a higher peak heat flux. The trade-off is
between a higher peak heat flux and greater sensitivity to
changes in plasma edge properties. The constant flux shape is
also more sensitive alignment errors since its low heat flux
relies on a very specific angle between the surface normal and
the tangent vector at every position. From these few examples
it is evident that design of even a "simple" axisymmetric
limiter is very complicated when real pieces of hardware are
considered.

While the expressions for $\hat{n}$ and $\hat{t}$ are simple for the case
just considered, in general, the expression for n is either very
complicated or must be found numerically because of the
complexity of the limiter surface. In a toroidal coordinate
system centered on the plasma the expression for $\hat{t}$ is simple as
long as we assume flow along field lines. Then t is given by

$$\hat{t} = \frac{a}{(a^2 + R^2 q^2)^{1/2}} \hat{\theta} + \frac{qR}{(a^2 + R^2 q^2)^{1/2}} \hat{\phi} . \qquad (14)$$

This is often not the best choice of coordinate system in which
to compute the surface normal which also complicates the
expression for $\hat{t}$. Even though the expressions for $\hat{n}$ and $\hat{t}$ do
not give much insight into the important aspects of the heat
flux distribution on a limiter, the results from integrating $\hat{n} \cdot \hat{t}$
over the limiter as in (9) do provide some insight. We will now
look at the results for examples of both poloidal and multiple
limiters.

864

An example of the heat flux to a poloidal limiter which has
a constant curvature both toroidally and poloidally is shown in
Fig. 6. The heat flux is zero at the contact point because the
plasma is tangent there; i.e., n·t = 0. If there is any
perpendicular diffusion then the zero will not be present. The
two peaks arise from the increase in angle of incidence together
with the exponential decrease of power in the edge. The
separation between the peaks is linearly proportional to λ_0.
The displacement of the peaks above and below the horizontal is
due to the rotational transform and is greatest for low q. Heat
flux profiles of this nature have been observed on D-III and
TFTR. The high peak heat flux shown in Fig. 6 illustrates the
problem of using poloidal limiters in high power devices. They
simply do not have sufficient area to absorb large amounts of
energy.

A possible solution to the small area of a single poloidal
limiter is to use an array of such limiters around the torus.
Figure 7 shows an example of an array of limiters of the same
geometry as in the previous paragraph. They touch edges only at
the midplane ($\theta=0$) and have constant width. It might be
expected that such an array of N limiters would reduce the peak
heat flux by 1/N, but the shadows cast by one element on the
adjacent element make this an overly optimistic assumption. The
mapping of the leading edge of one limiter onto the adjacent
limiter along field lines is shown in Fig. 8. The size of the
wetted area depends on the toroidal separation between the
plates. This implies that shadowing is least important for
small N (large $\Delta\phi$). For very large N, the axisymmetric case is
approached where again there is no shadowing. The toroidal

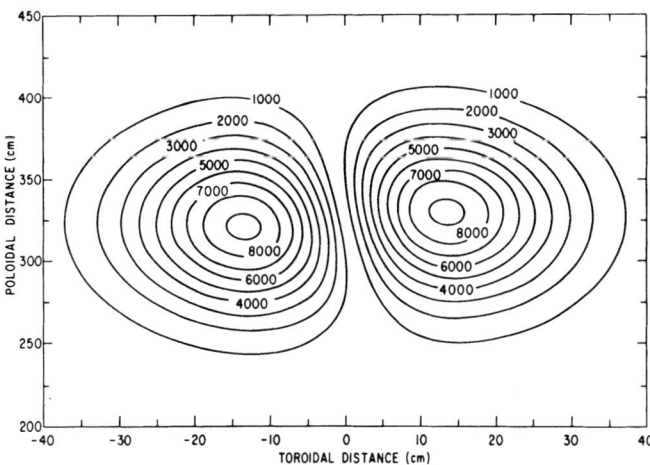

Fig. 6. Contours of constant heat flux on a single poloidal
limiter as described in the text.

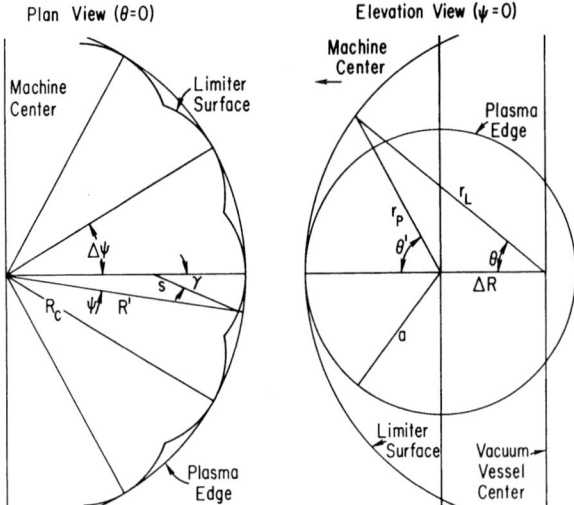

Fig. 7.   The geometry of a set of multiple limiters arrayed in the toroidal direction showing the variables used to describe it.

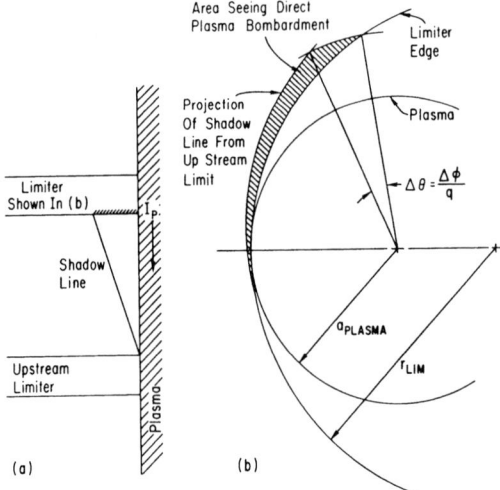

Fig. 8.   Schematic diagram of the shadowing that exists in a set of multiple limiters.

limiter case is obviously the lower limit on the heat flux for a given poloidal extent of the limiter. If there is perpendicular diffusion then there will be deposition in the shadowed region but the scrape-off length will be different than the bulk plasma because the connection length is shorter; i.e., $\lambda_{shadow} < \lambda_Q$. The variation in peak heat flux with the number of limiters is shown in Fig. 9. The asymptotic value is higher than the axisymmetric case because the rings have constant width which results in substantial gaps toward the top and bottom of the machine. These gaps are similar to the cut outs discussed for the axisymmetric limiter above.

Computation of the power density on a divertor plate is done in a very similar fashion to that for a limiter. The distance into the scrapeoff is measured from the separatrix. The tangent vector is more complicated because of the shape of the flux surfaces in the divertor, but the plate geometry is axisymmetric and somewhat simpler. Overall the computation is no more difficult than for a limiter. The power and particle density on the INTOR divertor is shown in Fig. 10. In general, the peak heat flux for both limiters and divertors are similar.

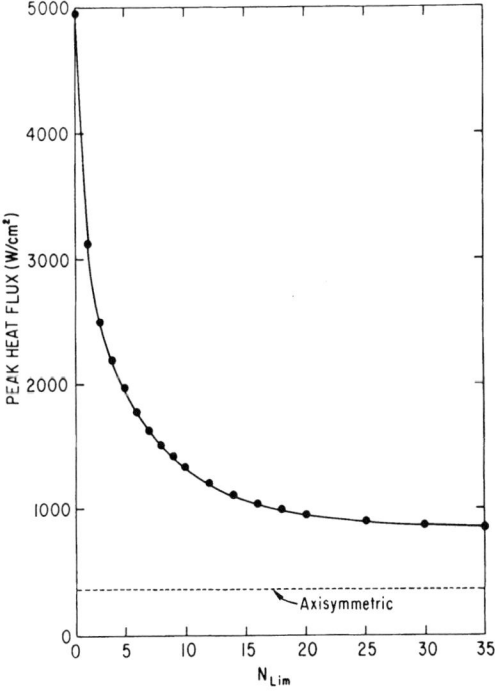

Fig. 9.   Variation of the peak heat flux on a limiter as a function of the number of limiters for constant scrape-off conditions.

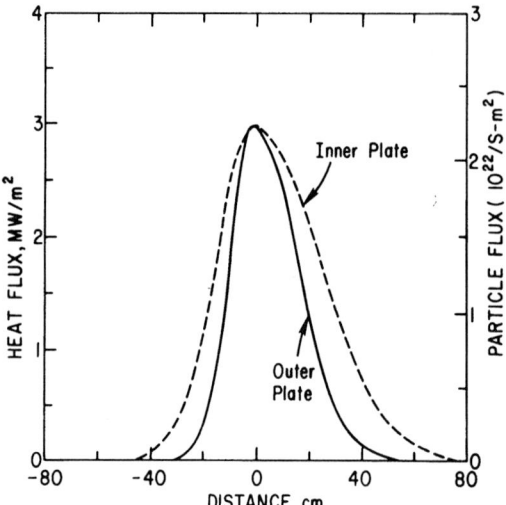

Fig. 10. Heat and particle flux distributions on the inner and
outer divertor plates predicted for the INTOR machine.

THERMAL RESPONSE

Once the power density on the material surface has been
defined using the above techniques, then the thermal response of
the plate can be computed. The temperature of the material is
found by solving

$$\rho c_p \frac{dT}{dt} = \vec{\nabla} \cdot (k \vec{\nabla} T), \tag{15}$$

where

ρ = the material density,
c_p = the material specific heat,
k = the material thermal conductivity, and
T = the temperature.

In general, ρ, c_p, and k are function of the temperature. They
will also vary with position if the plate is made of more than
one material; e.g., a coated or cladded tile on a backing
plate. The thermal diffusivity (D_{th}) can be defined as $k/\rho c_p$.
The boundary conditions include the power density on the plate
as well as radiation and/or conductive or convective
interfaces. There exist many finite element or finite
difference computer programs that can solve Eq. (15) in one,
two, or three dimensions with complicated plate geometries and
boundary conditions (see Ref. 24,26). Most of these codes are

well documented and readily accessible. The sophistication of these codes is required for most practical applications, but considerable insight into important time scales and general ranking of materials can be gained from examining analytic solutions that can be found under certain assumptions.

If the material thermal properties are independent of temperature, and a plate of thickness 2h is heated by a heat flux F_o on the face $z = h$ with no heat flow across $z = -h$ or the edges, and the starting temperature is 0, then the temperature is given[26] by

$$T(z,t) = \frac{F_o h}{12k} \left\{ \frac{24}{\pi^2} \frac{t}{\tau_{th}} + 3\left(\frac{z}{h}\right)^2 + 6\left(\frac{z}{h}\right) - 1 \right.$$

$$\left. - \frac{48}{\pi^2} \sum_{n=1}^{\infty} \frac{(-1)^n}{n^2} \exp\left(-n^2 t/\tau_{th}\right) \cos\left[\frac{n\pi(z + h)}{2h}\right] \right\}, \quad (16)$$

where $\tau_{th} = 4h^2/\pi^2 D_{th}$ and D_{th}, k, ρ, and c_p as above. This case corresponds to an uncooled plate or one that is only cooled between heat pulses. A second case where the face $z = -h$ is held at $T = 0$ has been previously evaluated[27] with the results

$$T(z,t) = \frac{F_o h}{k} \left\{ \frac{z}{h} + 1 - \frac{16}{\pi^2} \sum_{n=0}^{\infty} \frac{(-1)^n}{(2n + 1)^2} \right.$$

$$\left. \times \exp\left[-(2n + 1)^2 t/\tau_{th}\right] \sin\left[\frac{(2n + 1)\pi(z + h)}{4h}\right] \right\}, \quad (17)$$

where $\tau_{th} = 16h^2/\pi^2 D_{th}$. This case corresponds to an actively cooled plate. Examination of Eqs. (16) and (17) reveals that the temperature distributions are of the form

$$T(z,t) = \frac{F_o h}{k} G\left(t/\tau_{th}, z\right). \quad (18)$$

The form of G is such that the maximum temperature always occurs at $z = h$, i.e., the heated face. The value of $G\left(t/\tau_{th}, z\right)$ at $z = h$ is plotted versus t/τ_{th} for Eqs. (16) and (17) in Fig. 11 (a and b, respectively). Figure 11a shows that the temperature rises linearly with time after about two time constants. This is due to the deposited energy being absorbed by the mass of the plate since the plate is uncooled. The plate in Fig. 11b reaches steady state because heat is assumed to be removed from

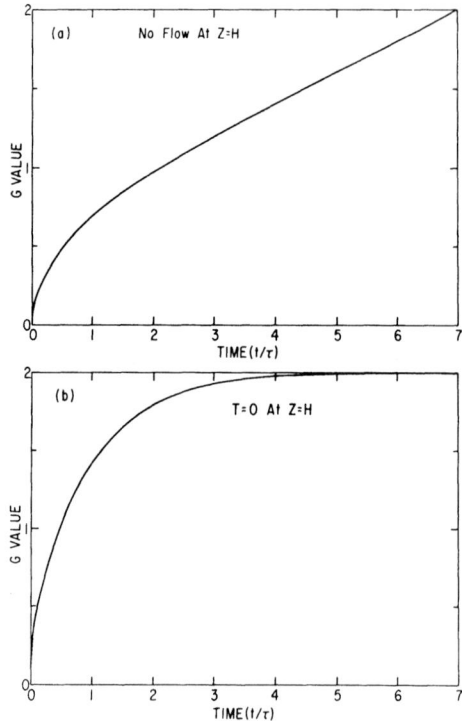

Fig. 11. The value of the G (temperature) function versus
normalized time for (a) the case with no heat transfer
across the face z = h, and (b) the case with T = 0 at z
= h.

the back face as rapidly as it is deposited on the front face.
In both cases for $t/\tau_{th} < 1$, the behavior is characterized by
$t^{1/2}$ (the semi-infinite solution) since the heat wave has not
had time to "discover" the back face of the plate. These two
cases are very simplified but provide a rule of thumb for
deciding whether active cooling is required if average thermal
properties are used and the resultant temperature is compared to
a limiting temperature such as the melting point. However,
detailed computations which include the variation of thermal
properties must be done in the vast majority of all designs of
limiters or divertor plates.

The thermal stresses that arise from a given temperature
distribution are found by dividing the object into finite
elements and specifying the temperature at each nodal point and
the mechanical properties of the material as a function of
temperature then using a program such as NASTRAN.[28] This
procedure involves large amounts of data and computation time.
If some assumptions are made then some analytic results can be

870

obtained which again give some insight into important processes. If the length and width of the plate are much larger than its thickness, the heat flux is uniform over the surface, the plate is unrestrained, and the material mechanical properties are independent of temperature, then the stress away from the edge is given[29] by

$$\sigma_{xx} = \sigma_{yy} = \sigma_{th}(z), \tag{19}$$

$$\sigma_{zz} = \sigma_{xz} = \sigma_{xy} = \sigma_{yz} = 0, \tag{20}$$

with

$$\sigma_{th}(z) = -\frac{\alpha_1 E_y}{1 - \nu_p}\left[T - \frac{1}{2h}\int_{-h}^{h} T\,dz - \frac{3z}{2h^3}\int_{-h}^{h} Tz\,dz\right], \tag{21}$$

where α_1, E_y, and ν_p are the thermal expansion coefficient, elastic (Young's) modulus, and Poisson's ratio. Evaluating the integrals using the temperature distribution in Eq. (16) above results in the following:

$$\sigma_{th}(z,t) = \frac{\alpha_1 E_y}{1 - \nu_p}\frac{F_o h}{k}\left\{-\frac{1}{4}\left(\frac{z}{h}\right)^2 + \frac{1}{12} + \frac{4}{\pi^2}\right.$$

$$\times \sum_{n=1}^{\infty}\frac{(-1)^n}{n^2}\exp(-n^2 t/\tau_{th})\left(\cos\frac{n\pi(z + h)}{2h}\right.$$

$$\left.\left.- \frac{6}{n^2\pi^2}\frac{z}{h}\left[(-1)^n - 1\right]\right)\right\}. \tag{22}$$

For Eq. (17), the result is

$$\sigma_{th}(z,t) = \frac{\alpha_1 E_y}{1 - \nu_p}\frac{F_o h}{k}\left(\frac{16}{\pi^2}\sum_{n=0}^{\infty}\frac{(-1)^n}{(2n + 1)^2}\right.$$

$$\times \exp\left[-(2n + 1)^2 t/\tau_{th}\right]\left\{\sin\left[\frac{(2n + 1)\pi(z + h)}{4h}\right]\right.$$

$$\left.\left.- \frac{6z}{h}\left[\frac{[4(-1)^n - (2n + 1)\pi]}{(2n + 1)^2\pi^2}\right] - \frac{2}{(2n + 1)\pi}\right\}\right). \tag{23}$$

871

In both cases the stress is of the form

$$\sigma_{th}(z,t) = \frac{\alpha_1 E_y}{1 - \nu_p} \frac{F_o h}{k} f(t/\tau_{th}, z).$$ (24)

Figure 12 is a plot of f for the two cases as a function of t/τ_{th} for three different values of z (front face z = h, back face z = -h, and center z = 0). Examination of the plots reveals that a maximum compressive stress occurs on the z = h face if $F_o > 0$ for both cases. This result of compressive stresses on the front and back faces and tensile stresses inside the plate is also seen in most finite element calculations that have similar boundary conditions. The presence of notches or holes can result in stress concentrations that produce stresses much higher than the formulas predict. However, Eqs. (22) and (23) do permit comparison of the relative performance of different materials if the variation of the mechanical properties with temperature is ignored.

The maximum in the compressive stress in Fig. 12 implies that a stress limited heat flux can be defined as

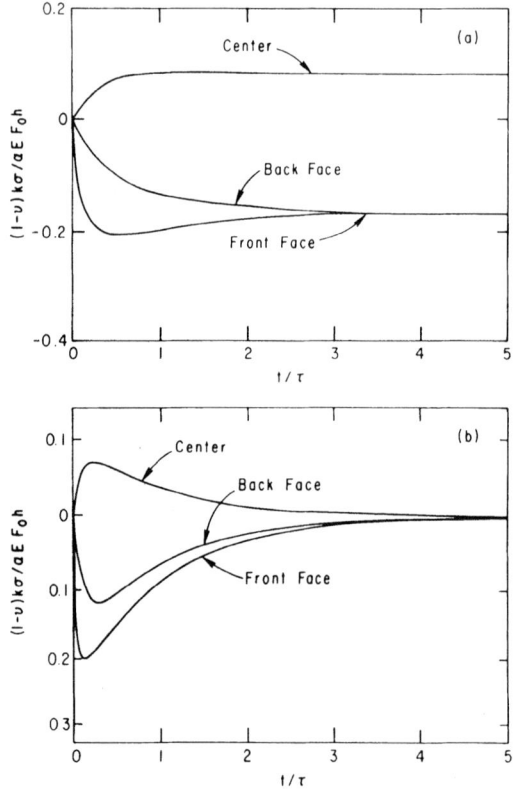

Fig. 12. The value of the f (stress) function versus normalized time for (a) the case with no heat transfer across the face z = h, and (b) the case with T = 0 at z = h.

$$F_o^S = \frac{(1 - \nu_p) k \sigma_Y}{\alpha_1 E_y h f_{max}(t/\tau_{th}, h)} , \qquad (25)$$

where σ_y is the compressive stress limit for the material. Similarly a heat flux limit for melting can be defined for $z = h$ and the same value of t/τ_{th} as the stress maximum. This melting flux limit is given by

$$F_o^M = \frac{k T_{melt}}{h G[(t/\tau_{th})_o, h]} , \qquad (26)$$

where $(t/\tau_{th})_o$ is the value of t/τ_{th} for which the maximum compressive stress is achieved. The ratio of the stress flux limit to the melting flux limit is then

$$R = G\left[(t/\tau_{th})_o, h\right](1 - \nu_p) \ \sigma_y/f_{max} \alpha_1 E_y T_{melt}. \qquad (27)$$

If $R > 1$ the material should melt before cracking whereas if $R < 1$ the material will crack before melting. For $R \simeq 1$ a more detailed analysis is required. The relative rankings obtained by this technique are approximate but do serve as a guide to material selection. The relative ranking of several interesting materials is shown in Table 1. Materials having coatings or

Table 1. Calculated stress and melting flux limits.

Material	F_o^S (kW/cm^2)	F_o^M (kW/cm^2)	R
Tungsten	7.83	18.20	0.43
Molybdenum	4.83	11.78	0.41
Copper	3.90	13.00	0.30
UCAR Graphite Grade ATJS	66.88	20.64	3.24
UCAR Graphite Grade ATJ	22.00	13.72	1.60
POCO Graphite Grade AXF5Q	11.64	13.28	0.88
Niobium	1.60	4.00	0.40
Tantalum	1.65	5.16	0.32
Beryllium	1.93	8.04	0.24
Aluminum	1.13	4.72	0.24
Titanium	0.46	1.10	0.42
Stainless steel type 304	0.17	0.64	0.27
Inconel X-750	0.17	0.51	0.34
Silicon Carbide	5.22	10.54	0.50
Titanium Carbide	0.88	3.40	0.26
Titanium Diboride	0.74	2.32	0.32

cladding must be analyzed using different techniques (most involve experimental tests of the limits).

While thermal stresses are significant, they are not the only stresses which must be considered. The stresses due to mechanical loads such as those due to electromagnetic forces must be coupled with thermal stresses to verify a given design. These mechanical loads and their sources are the subject of a subsequent section.

ABNORMAL CONDITIONS

The disruptive instability in tokamaks is the single most severe heat load that limiters and divertor plates must withstand. Disruptions are the subject of intense investigation but there has yet to emerge a complete understanding. Space is not sufficient here to discuss this phenomenon in any detail. The interested reader is referred to Refs. 30,31. A disruption is characterized by a sudden loss of plasma thermal energy followed by a somewhat slower decay of the plasma current. The thermal decay has been observed to take place in a few hundred microseconds with a deposition profile similar to the normal profile,[32] and with a profile indicative of strong radial transport.[22] In general larger plasmas disrupt more slowly; e.g., D-III observes 100-200 μs while TFTR observes 1-2 ms. As an example of the type of heat load that results from a disruption consider a plasma having a stored energy of 30 MJ and disrupting in 1 ms. Assuming the deposition profile is the same as that shown in Fig. 3, the peak heat flux is 420 kW/cm^2. The temperature rise during the 1 ms of the disruption is 1.2×10^4 °C for graphite if sublimation is ignored. This type of power density is sufficient to melt and/or vaporize any known material. The thermal stresses which result are also enormous. Since there are very few materials that can withstand this type of temperature and stress, the disruption severely limits materials choices. The current decay phase of a disruption also results in design problems as is discussed below.

Abnormal conditions of heat and particle flux can also occur due to the interaction of a disruption with neutral beam heating. If a disruption occurs when the beams are on then the beam would no longer be attenuated before striking the far wall. This can result in heat fluxes of several kilowatts per square centimeter. The duration will be as long as it takes to detect the disruption and turn off the beam. Improved diagnostics are reducing the incidence of this fault, but hardware failures still make it necessary to plan for its occurrence.

ELECTROMAGNETIC INTERACTIONS

In addition to the problems caused by the heat and particle
flux to limiters and divertor plates, there are also strong
electromagnetic interactions taking place. These arise because
of both normal and disruptive conditions. The currents arising
during normal conditions will all be put under the heading of
"resistive" currents. Those currents arising from disruptions
will be put under the heading of eddy currents. Once the
currents have been derived the forces and stresses in a plate
can be calculated. This section will conclude with a
consideration of the coupling between motion of the plate, the
eddy currents, and the magnetic fields that I chose to call
magnetic damping.

Resistive currents

Consider a poloidal limiter like that in Fig. 1a (top and
bottom rail limiters). A voltage will be induced between the
top and bottom of the plasma if there is motion in the
horizontal direction due to the Hall effect. The voltage
induced by the motion is given by

$$V = |\vec{v} \times \vec{B}_T| \ 2a_o \ \text{(volts)}, \tag{28}$$

where B_T is the toroidal field (Tesla), a_o is the plasma minor
radius (m), and v is the velocity of the plasma (m/s).

We must now ask how much current flows because of this
voltage. Experiments on PDX have shown that a limiter of the
type in Fig. 1a behaves like a double probe.[33] If the
resistance connecting the top to the bottom limiter is small
(small is defined later) then the current is given by double
probe theory.[34] In this case, the area of the two probes is
equal if the plasma is top-bottom symmetric. The result is

$$J = \frac{e n_e A}{2} \left(\frac{T_e + T_i}{m_i}\right)^{1/2} \tanh\left(\frac{eV}{2T_e}\right) , \tag{29}$$

where A is the area of one side of the limiter (probe) and V is
the voltage in (28) above. If $V \gtrsim 4 \ T_e$, the current saturates
to a value of

$$J_{sat} = \frac{e n_e A}{2} \left(\frac{T_e + T_i}{m_i}\right)^{1/2} . \tag{30}$$

As an example let us consider a TFTR compression case where v = 40 m/s, a_o = 0.85 m, B_T = 5 T, and take T_i = T_e = 50 eV, and n_e = 10^{13}/cm^3 with A = 3.6 × 10^3cm^2 (the wetted area for λ = 2 cm). The result is then V = 340 V, J_{sat} = 28 kA. This implies the resistance between the two limiters must be less than 12 mΩ. If R is greater than this the usual J = V/R law applies. In the TFTR case the plasma looses contact with the limiter. If we assume an exponential scrape-off thickness then the wetted area decreases exponentially. This implies the time history of the current is

$$J = J_{sat} \exp\left(-t/\tau\right), \tag{31}$$

where $\tau = \lambda/v$. If the plasma velocity is less than about 23.5 m/s then the current is reduced by the hyperbolic tangent term. Plasma velocities of 330 m/s can be inferred from from disruption data. The voltage is then 2.7 kV which implies substantial currents even if the connecting resistor is 10 Ω.

We will now examine what currents will flow if the limiter is one tip of a double probe and the torus wall is the other tip.[35] The electron current is given by

$$J_e = A \frac{en_e}{4} v_{th} \exp\left[\frac{e(V - V_s)}{k_B T_e}\right] , \tag{32}$$

where

$$v_{th} = \left(\frac{8 \, T_e}{\pi \, m_e}\right)^{1/2} , \tag{33}$$

and V_s = space (floating) potential. The ion current is

$$J_i = A \frac{en_e}{2} C, \tag{34}$$

where

$$C = \left(\frac{T_e + T_i}{m_i}\right)^{1/2} . \tag{35}$$

In equilibrium the net currents must be equal which implies

$$J_e^\ell - J_i^\ell = J_i^w - J_e^w = J, \tag{36}$$

where ℓ and w refer to the limiter and wall. The voltage drop across the external resistance connecting the limiter to the wall determines the potential between the two. Thus

$$V_\ell - V_w = JR. \tag{37}$$

We then have

$$\frac{kT_e^\ell}{e} \ell n\left[\frac{J + J_i^\ell}{A_\ell (en_\ell/r) \, v_{th}^\ell}\right] = JR + V - V_s^\ell \, , \tag{38}$$

$$\frac{kT_e^w}{e} \ell n\left[\frac{J_i^w}{A_w (en_w/4) \, v_{th}^w}\right] = V_2 - V_s^w \, . \tag{39}$$

We can take $V_2 = 0$ and then simultaneously solve the two equations with the correct values of n and T_e at the limiter and wall (V_s is typically $3kT_e$). For

$$T_e^\ell = 400 \text{ keV} \, ,$$

$$n_e^\ell = 2 \times 10^{13}/\text{cm}^3 \, ,$$

$$A_\ell = 7 \times 10^3/\text{cm}^2 \, ,$$

$$T_e^w = 15 \text{ eV} \, ,$$

$$n_e^w = 1.4 \times 10^{11}/\text{cm}^3 \, ,$$

$$A_w = 1 \times 10^6/\text{cm}^2 \, ,$$

we have

$$J = 0.7 \text{ A} \quad \text{for } R = 100 \text{ } \Omega \, ,$$

and

$$J = 7.2 \text{ A} \quad \text{for } R = 10 \text{ } \Omega \, ,$$

which is typical of the current observed flowing through the
limiter grounding resistors during the steady-state portion of
the discharge. These currents are modest but if an accidental
short occurs then the current could be 10 to 20 kA. Also during
start-up or disruptions large voltages are observed on the
resistors indicating large currents (10 to 50 times the steady-
state value).

Eddy Currents

As was discussed above, a disruption is characterized by a
rapid loss of plasma energy followed by a somewhat slower loss
of plasma current. For structure like limiter or divertor
plates that are immersed in the plasma magnetic fields, this
loss of current results in eddy currents being induced in the
structure. The eddy currents are found by solving Maxwell's
equations using the Coulomb gauge.[36] This is done by breaking
the structure into finite meshes and solving for the mutual
inductances and the mesh and branch currents.[37] There are many
codes[37,38] for doing this. The important time scales are
determined by the disruption time τ_p and the structure L_s/R time
(τ_L) where L_s is the plate inductance and R its resistance. If
$\tau_L \gg \tau_p$ the flux is frozen in (i.e., a superconducting
solution) and $J_{sc} = \phi_m/L_s$. If $\tau_L \ll \tau_p$ then the solution is
resistive and J is given by $\dot{\phi}_m/R$. For intermediate values, the
result[39] is

$$J = J_{sc} \frac{\tau_L}{\tau_p - \tau_L} \left[\exp\left(-t/\tau_p\right)\right] - \left[\exp\left(-t/\tau_L\right)\right], \qquad (40)$$

$$\text{if } \left(\tau_p \neq \tau_L\right),$$

and

$$J = J_{sc} \frac{t}{\tau_p} \exp\left(-t/\tau_p\right) , \qquad (41)$$

$$\text{if } \left(\tau_p = \tau_L\right).$$

These are for an exponential current decay. Other decay rates
such as linear give slightly different results. This also
assumes the plasma to be stationery during the disruption. If
the plasma moves then very different behavior results. Typical
eddy currents versus time for the TFTR bumper limiter are shown
in Fig. 13 for several conditions. For machines which have been
studied thus far the current decay rates are < 1 MA/ms. This
value was used for TFTR designs.

878

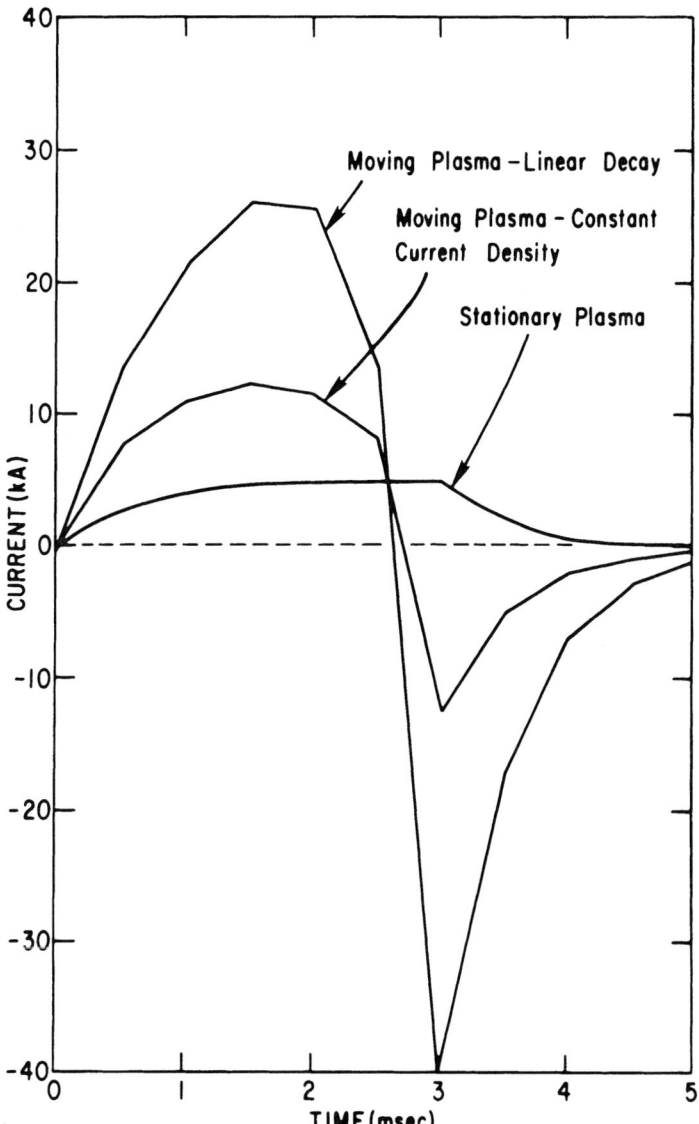

Fig. 13. The time history of the eddy currents in a TFTR bumper limiter plate for several plasma current decay modes.

Forces and Stresses

Once we have determined the currents and current distributions that are important then the resulting forces on and stresses in the structure can be determined. The force is given by

$$\vec{F} = J \, \vec{\ell} \times \vec{B} \, , \qquad\qquad (42)$$

where $\vec{\ell}$ is a unit vector along the current path and $\vec{B}$ is the magnetic field. The force is time varying in most cases. The eddy currents result in forces that have a strong impulse characteristic. In the static case the stress is determined by balancing the force against the strain times the elastic modulus. The time varying force case requires the normal modes of the structure to be determined. The eddy current forces are then applied as a time varying forcing function. The stress codes used for determining the thermal stresses can also be used to find the response in this case. Typically, the stresses due to currents are those which most strongly influence the design. It is very difficult to make general guidelines for designs that will work because of the complex nature of the problem. The only solution is to do the calculation and compare the results to allowable stresses.

Magnetic Damping

Up to this point the calculation of the currents and forces has been done independent of the stresses and displacements. However, the eddy currents induced by the motion of a conducting plate in a steady magnetic field have been known for a long time. This means that the motion of the plate in response to the eddy currents due to the changing field may alter the current distribution and/or time history. That is one may have to solve Maxwell's equations and the stress/displacement equations simultaneously! This is a very formidable task and it has never been done. The important question to answer is when is this necessary.

We will begin by considering a loop of wire in crossed B fields (see Ref. 41 for more details). One component of B will vary with time and the other will be constant. Figure 14 shows the geometry of the loop and the fields. The loop is assumed to

be rigid and rotates only about the x-axis. The equations of
motion are then

$$I_m \frac{d^2\theta}{dt^2} + K_s \theta = \tau_x \ ,$$ (43)

$$L_s \frac{dJ}{dt} + R \ J = \ - \frac{d\phi_m}{dt} \ ,$$ (44)

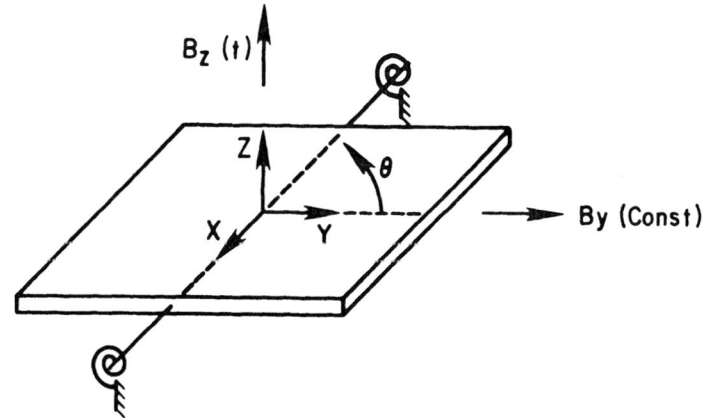

Fig. 14. Geometry of the loop used for the magnetic damping
calculation showing the important parameters.

where

I_m = mass moment of inertia,

K_s = rotational spring constant,

τ_x = torque about the x-axis,

L_s = self-inductance of the loop,

R = resistance of the loop,

ϕ_m = magnetic flux in the loop.

J = current in the loop.

If the loop has an area A then

$$\tau_x = \left(-B_Z \sin\theta - B_y \cos\theta\right) JA ,\tag{45}$$

and

$$\frac{d\phi_m}{dt} = A\left(\frac{dB_Z}{dt} \cos\theta - \frac{d\theta}{dt} B_Z \sin\theta\right.$$

$$\left. - \frac{d\theta}{dt} B_y \cos\theta\right).\tag{46}$$

Mechanical damping would result in a $d\theta/dt$ term in Eq. (43), but this is ignored because it is negligible in most cases. The energy in the system is also of interest, and is given by

$$1/2\ I_m \frac{d\theta^2}{dt} + 1/2\ K_s\theta^2 + 1/2\ L_s J^2 + \int_o^t J^2\ Rdt$$

$$= -\int_o^t \frac{d\phi_m}{dt} J\ dt,\tag{47}$$

where the integrals are along the trajectory given by the equations of motion. Equations (43) and (44) must be solved numerically. If we assume small θ and $B_y \gg B_Z$ we obtain

$$I_m \frac{d\theta^2}{dt} + K\theta = -J A B_y ,\tag{48}$$

$$L_s \frac{dJ}{dt} + R J = -A \frac{dB_Z}{dt} + A B_y \frac{d\theta}{dt} .\tag{49}$$

882

Using Laplace transform methods we obtain

$$\theta(s) = \frac{A^2 B_y [dB_z(s)/dt]}{(I_m s^2 + K_s)(L_s + R) + A^2 B_y^2 s},$$ (50)

$$J(s) = -\frac{A [dB_z(s)/dt](I_m s^2 + K_s)}{(I_m s^2 + K_s)(L_s + R) + A^2 B_y^2 s}.$$ (51)

The roots of the denominator identify the behavior of the system. We can identify a "magnetic" damping coefficient of $A^2 B_y^2/R$. This would be the coefficient of the $d\theta/dt$ term in Eq. (43) above. The dominant term in the frequency is

$$\omega^2 = \frac{K_s}{I_m} + \frac{A^2 B_y^2}{I_m L}.$$ (52)

Since K_s/I_m is the usual natural mechanical frequency, we can identify the second term as a "magnetic spring" term. For large area and large values of B_y the magnetic spring dominates and the damping is approximately critical. If the τ_x term is omitted from Eq. (43) then we have uncoupled equations which have been used in the past. Figure 15 shows the ratio of the maximum deflection for the uncoupled to the coupled cases. It is easy to see that the effect of magnetic damping must be considered at natural frequencies below about 400 Hz for this particular case.

An experiment has been carried out to verify these results (see Refs. [42] and [43]). A copper loop was placed in the FELIX device at Argonne. A vertical field of about 300 Gauss was allowed to decay exponentially with a time constant of about 10 ms. The characteristics of the plate were A = 0.2 m^2, I_m = 1.65 kq·m^2, K_s = 2.8 × 10^3 nt·m/radian, L = 1.2 μH, and R = 56 μΩ. The B_y field was 0.44 Tesla. The response of the plate is shown in Fig. 16. The solid curves are the prediction of Eqs. (43) and (44). The agreement is excellent.

For TFTR we had to take advantage of magnetic damping to find an acceptable design. Since there are no coupled codes that can do the problem, we approximated the damping as a mechanical damping having one-fourth of the damping coefficient found from the simple model above for each of the natural frequencies of the limiter plate. We do not recommend this for future devices and are working on better models. One word of caution is that magnetic damping only works for plates. Cylinders, tubes, and spheres, etc. are not effected by magnetic damping, since their area in the B_y field does not change.

883

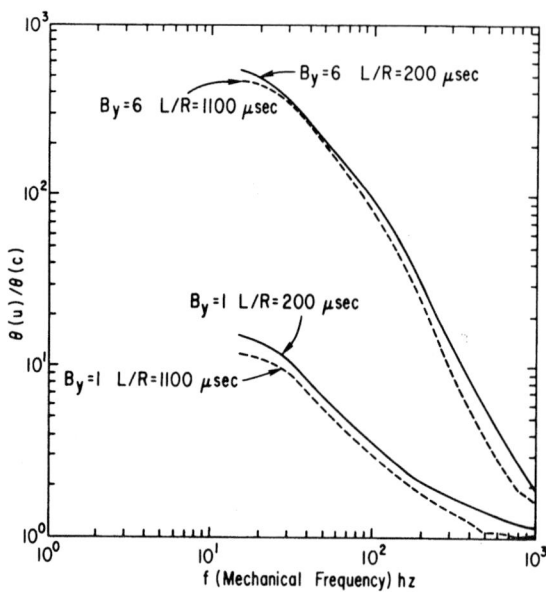

Fig. 15. A comparison of the maximum deflection of the loop with
(coupled) and without (uncoupled) the effect of
magnetic damping.

MATERIAL SELECTION

   The selection of materials for use as the plasma surface or
support plate for limiters or divertor plates depends strongly
on the specific details of a given application (see Ref. 26, for
example). However, some general guidelines can be used to
narrow the range of materials to be considered. The guidelines
are somewhat different for limiters and divertor plates. The
guidelines are dictated by the thermal and electromagnetic
loads, and the physics of impurity generation and transport.

   Plasma modeling studies[45] have shown that there are two
important limits to consider in regard to impurity content. For
medium to high atomic number (Z) the limiting values are given
by radiation by the partially stripped ions for which the limit
is proportional to Z^{-3}. For low Z atoms the limit is due to
depletion of the hydrogen isotopes at the β limit and is
proportional to Z^{-1}. The two limits cross around Z = 13. The
nature of the limits is such that they favor low Z elements for
limiters. For divertor plates the amount of shielding expected
from a given design will determine the highest acceptable Z. In
many cases the shielding is adequate to permit use of high Z
elements for the divertor plate. However, selection of
materials for regions near the divertor throat may not permit
selection of high Z materials depending on the details of a
particular design.

884

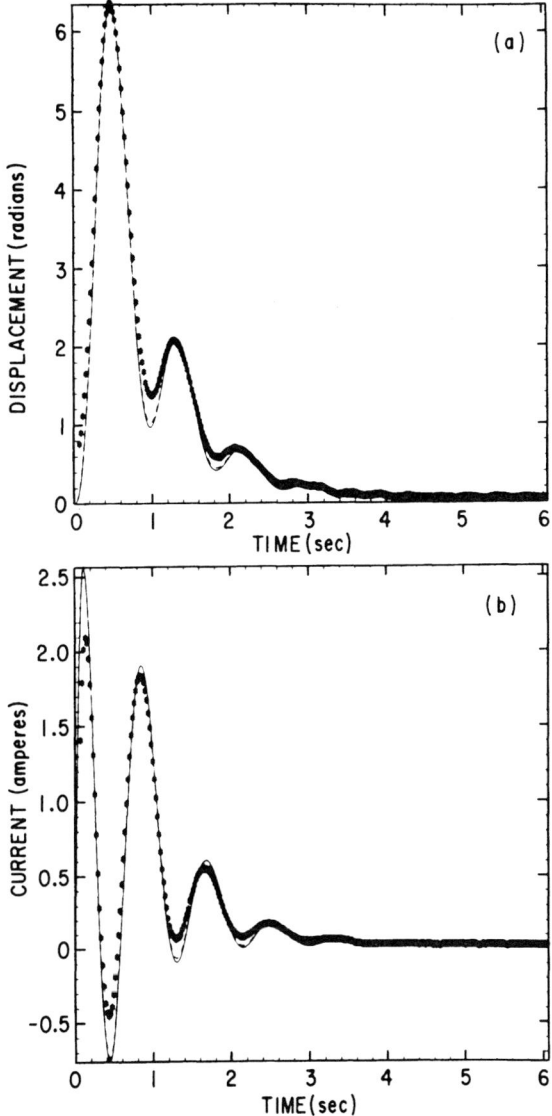

Fig. 16. Results of the FELIX test of magnetic damping showing
the displacement and current waveforms for the
parameters given in the text. The solid curve is the
result of the equations derived in the text. The
agreement is seen to be excellent.

The heat loads that result from normal operation require the use of materials having high thermal conductivity. This is especially important as pulse lengths are extended to where steady-state thermal designs are necessary. The heat loads during disruptions dictate that the material must have good thermal shock resistance. High thermal conductivity mitigates thermal shock to some extent by reducing thermal gradients, but mechanical strength is also important. These considerations are the same for both limiters and divertors.

The presence of disruptions has a strong influence on the desirable electrical properties of both the plasma surface and support plate materials. The disruption induced eddy currents are reduced in materials having large electrical resistance. Magnetic damping, on the other hand, is reduced as the resistance is increased. The proper choice of electrical resistance will thus depend on the exact details of the electromagnetic and mechanical response of the system.

Erosion of the plasma surface material and transport of that material into the plasma have strong influence on both component lifetime and plasma cleanliness. Large erosion implies a short surface material lifetime. Since limiters are in general in more intimate contact with the plasma, limiter materials must have a lower erosion than divertor plates. The important erosion mechanisms are arcing, melting, evaporation, and sputtering (physical and chemical).

The allowed hydrogen isotope retention in a plasma surface material is determined by the allowed inventory or by the desired burn-up percentage. Hydrogen retention may also alter the mechanical behavior of the material, e.g., hydrogen embrittlement. Retention becomes a more important problem as we move toward reactors because of tritium inventory considerations.

There is no one material which ideally satisfies all of the above requirements. Any design is a compromise between conflicting requirements. One additional consideration is design integration. The compatibility of the various materials used both in the limiter/divertor system and the remainder of the vacuum vessel must be considered. Some examples of compatibility are alloy formation due to material transport and expansion coefficient differences. Since complex designs are required to satisfy the requirements, material fabricability is very important. Table 2 shows those materials which have been used recently for limiters and divertor plates. They represent the result of the balancing of all the requirements for the present generation of machines, and are indicative of the "best" choices as we now understand the physics of plasma

Table 2. Materials Being Used for Limiters and Divertor Plates.

I. Materials used on the plasma side

Material	Coating	Limiters	Divertors
Graphite	No	X	
Graphite	TiC	X	
Graphite	SiC	X	
Graphite	C/SiC	X	
Beryllium	No	X	
Titanium	No	X	X
Stainless Steel	No	X	X
Inconel	No	X	
Inconel	TiC	X	
Molybdenum	No	X	X
Molybdenum	TiC	X	X
Copper	Ni	X	

II. Materials used as support plates for the plasma side materials

Stainless Steel
Inconel
Copper

materials interactions and the physics of the plasma edge.

CONDITIONING

The conditioning of limiters and divertor plates is discussed in detail in S. Cohen's chapter. A brief summary is included here for completeness. The most recent data shows that "well conditioned" graphite limiter surface consists of a thin layer of the metals present in the machine together with carbon and oxygen.[45] This modification of the near surface region is accomplished by the use of one or more of the following techniques:

(1) glow discharge cleaning,
(2) pulse discharge cleaning (high power and low power),
(3) high power discharges.

In-situ baking has been found to enhance the cleanup rate for both pulse and glow discharge cleaning. In addition to the above, graphite surface materials are "precleaned" in a vacuum furnace to outgas volatile impurities. Care must be taken to avoid high Z contamination from the furnace heating elements. It has been found that a hydrogen firing cycle ($\sim$ 500 mTorr H_2) together with a careful venting of the furnace; i.e., cooling in vacuum to less than 300°C, will eliminate the problem of high Z contamination.

CONCLUSION

In this chapter we have tried to outline the steps and methods used to design limiters and divertor plates within the constraints imposed by physics and engineering considerations. Several significant advances in our understanding of the important phenomena and techniques for calculating important effects have been made in the last five years. There is room for further significant advances. Reactor designs will require a better understanding of the physics and engineering constraints so that less conservatism can be included as a "factor of ignorance."

ACKNOWLEDGMENT

This work was supported by the U.S. Department of Energy Contract No. DE-AC-02-76-CHO3073.

REFERENCES

1. D. O. Overskei, "Efficiency of Passive-Limiter Pumping of Neutral Particles," Phys. Rev. Lett. 46 (1981) 177.
2. T. Taylor, N. Brooks, K. Ioki, "Changes in Limiter Surface Temperature Profiles for Small Radius Plasmas in D-III," J. Nucl. Mater. 111&112 (1982) 569.
3. P. Mioduszewski, "Experimental Studies on Pump Limiters," J. Nucl. Mater. 111&112 (1982) 253.
4. R. Fonck, M. Bell, K. Bol et al., "Impurity Levels and Power Loading in the PDX Tokamak with High Power Neutral Beam Injection," J. Nucl. Mater. 111&112 (1982) 343.
5. S. A. Cohen, R. Budny, G. M. McCracken, M. Ulrickson, "Mechanisms Responsible for Topographical Changes in PLT Stainless Steel and Graphite Limiters," Nucl. Fusion 21 (1981) 233.
6. K. J. Dietz, "Design Considerations and Layout of the JET Limiters," in Eleventh Symposium on Fusion Technology (Pergamon, NY, 1980) Vol. II p. 1053.
7. D. W. Doll, M. A. Ulrickson, J. L. Cecchi et al., "Design of TFTR Moveable Limiter Blades for Ohmic and Neutral beam Heated Plasmas," in Ninth Symposium on the Engineering Problems of Fusion Research (IEEE, NY, 1981), p. 1654.
8. H. W. Kugel and M. Ulrickson, "The Design of the Poloidal Divertor Experiment Tokamak Wall Armor and Inner Limiter System," Nucl. Technol./Fus. 2 (1982) 712.

9. H. Vernickel, K. Behringer, D. Campbell et al., "Test of A Toroidal Large Area Limiter in the ASDEX Tokamak," J. Nucl. Mater. 111&112 (1982) 317.

10. L. Sevier, G. O'Connor, "Design of the TFTR Phase II Bumper Limiter," in Ninth Symposium on the Engineering Problems of Fusion Research (IEEE, NY, 1981) p. 1589.

11. L. Sevier, M. F. Ho, J. Citrolo et al., "TFTR Bumper Limiter Design," in Tenth Symposium on Fusion Engineering (IEEE, NY, 1982) p. 1072.

12. P. Winkler, S. Fixler, W. V. Timlen, "Design and Analysis of the TFTR Fixed Limiters," in Ninth Symposium on the Engineering Problems of Fusion Research (IEEE, NY, 1981) p. 1383.

13. J. L. Cecchi, "Tokamak Limiter Design," Ibid, p. 1378.

14. D. H. Rebut, K. J. Dietz, "The First Wall in JET-Status and Perspectives," in Twelfth Symposium on Fusion Technology (Pergamon, NY, 1982) Vol. I, p. 85.

15. D. K. Owens, W. Arunasalam, C. Barnes et al., "PDX Divertor Operation," J. Nucl. Mater. 93&94 (1980) 213.

16. M. A. Mahdavi, C. J. Armentrout, F. P. Blau et al., "A Review of the Recent Expanded Boundary Divertor Experiments in the Doublet-III Device," J. Nucl. Mater. 111&112 (1982) 355.

17. M. Shimada, M. Nagami, K. Ioki, "High Density Single Null Poloidal Divertor Results in Doublet-III," Ibid, p. 362.

18. W. Engelhardt, G. Becker, K. Behringer et al., "Divertor Efficiency in ASDEX," J. Nucl. Mater. 111&112 (1982) 337.

19. Y. Shimomura, K. Shimizu, T. Hirayama et al., "JT-60 Program," J. Nucl. Mater. 128&129 (1984) 19.

20. M. Ulrickson, D. E. Post, "Particle and Energy Transport in the Plasma Scrape-off Zone and Its Impact in Limiter Design," J. Vac. Sci. Technol. A1 (1983) 907.

21. P. Mioduszewski, Institut for Plasmaphysik, Kernforschungsanslage, Julich, Report No. JAl-1681 (1980).

22. M. Ulrickson, H. W. Kugel, "Initial Temperature Profiles on the PDX Inner Toroidal Limiter," Nucl. Technol./Fus. 4 (1983) 141.

23. M. Ulrickson, "Optimum Shapes for Pump Limiters," Princeton Plasma Physics Laboratory Report No. PPPL-1901 (1982).

24. J. C. Carslaw and J. C. Jaeger, Conduction of Heat in Solids (Clarendon Press, Oxford, 1959) pp. 466-478.

25. P. J. Burns, "TAC02D-A Finite Element Heat Transfer Code," Lawrence Livermore National Laboratory Report No. UCID-17980, Rev. 2 (1980).

26. M. Ulrickson, "Material Selection for TFTR Limiters," J. Vac. Sci. Technol., 18 (1981) 1037.

27. J. F. Schivell, D. J. Grove, J. Nucl. Mater. 53 (1974) 107.

28. J. F. Gloudeman, "The Modularity of MSC/NASTRAN," in Proceedings of Computer Software and Applications Conference (IEEE, NY, 1978) p. 444.

29. S. P. Timoshenko, J. N. Goodier, Theory of Elasticity, (McGraw-Hill, NY, 1970) p. 433.

30. D. J. Strickler, Y-K. M. Peng, J. B. Miller et al., "Numerical Simulation of the Plasma Current Quench Following a Disruptive Energy Loss," Fusion Technol. 6 (1984) 44.

31. B. Carreras, H. R. Hicks, J. A. Holmes, B. V. Waddel, "Nonlinear Coupling of Tearing Modes with Self-Consistent Resistivity Evolution in a Tokamak," Phys. Fluids 23 (1980) 1811.

32. T. Hino, J. DeGrassie, T. S. Taylor, "Heat Flux to the Limiter During Disruptions and Neutral Beam Injection in Doublet-III," J. Nucl. Mater. 121 (1984) 337.

33. R. Hawryluk, unpublished report.

34. See D. Manos chapter.

35. D. K. Owens, unpublished report.

36. D. W. Weissenberger, "Transient Eddy Currents on Finite Plane and Toroidal Conducting Surfaces," Princeton Plasma Physics Laboratory Report No. PPPL-1517 (1979).

37. D. Weissenberger, "SPARK Verson 1 Reference Manual," Princeton Plasma Physics Laboratory Report No. PPPL-2040 (1983).

38. R. J. Lari, L. R. Turner, "Survey of Eddy Current Programs," Magnetics MAG19 (IEEE Trans., 1983) 2474.

39. J. Bialek, unpublished report.

40. D. W. Weissenberger, U. R. Christensen, "A Network Mesh Method to Calculate Eddy Currents on Conducting Surfaces," Magnetics MAG 18 (IEEE Trans., 1982) 422.

41. J. Bialek, D. Weissenberger, M. Ulrickson, J. Cecchi, "Modeling the Coupling of Magnetodynamics and Elastomechanics in Structural Analysis," in Proceedings of the Tenth Symposium on Fusion Engineering (IEEE, NY, 1983) p. 51.

42. D. W. Weissenberger, J. M. Bialek, G. J. Cargulia et al., "Experimental Observations of the Coupling Between Induced Currents and Mechanical Motion in Torosionally Supported Square Loops and Plates/Part 1: Experimental Analysis" Princeton Plasma Physics Laboratory Report No. PPPL-2158 (1984).

43. D. W. Weissenberger, J. M. Bialek, G. J. Cargulia et al., "Part 2: Data Inventory," Princeton Plasma Physics Laboratory Report No. PPPL-2159 (1984).

44. R. F. Mattas, D. L. Smith, M. A. Abdou, "Materials for Impurity Control," J. Nucl. Mater. 122&123 (1984) 66.

45. J. L. Cecchi, "Impurity Control in TFTR," J. Nucl. Mater. 93&94 (1980) 28.

46. B. L. Doyle, W. R. Wampler, H. F. Dylla, et al., "Characterization of Impurities Deposited on the PDX Graphite Rail Limiters," J. Nucl. Mater. 128&129 (1984) 955.

ADVANCED LIMITERS

P. Mioduszewski

Oak Ridge National Laboratory
Oak Ridge, Tennessee 37831

1. INTRODUCTION

In toroidal fusion devices the plasma is confined by magnetic
fields forming closed magnetic surfaces. Energy and particles
are lost from this confined region by perpendicular diffusion,
eventually across the last closed magnetic surface. The space
between the last closed flux surface and the vessel wall is characterized by large fluxes of particles and energy parallel to the
magnetic field. The heat can be removed in a controlled fashion
either by installing mechanical limiters in the main vacuum vessel
or by diverting the magnetic field lines such that the particle
and heat fluxes are directed into a separate chamber where the
heat is removed with a specially designed divertor target.
Ideally, divertors fulfill three tasks: energy removal, particle
unloading, and impurity control. For impurity control it is
obvious that divertors have an advantage over limiters since the
main plasma-wall interaction takes place in a remote chamber. In
the short-pulse machines of the past, the main task of limiters,
as well as divertors, was to remove the heat in a controlled way.
In future long-pulse devices and especially in D-T-burning devices,
it will be important to remove particles and, in particular,
helium ash from the main plasma chamber. Divertors seem to be
well suited for this task, but their disadvantage is that they
require additional magnetic fields (i.e., additional coils), which
adds to the complexity of a fusion reactor. Therefore, it seems
worthwhile to explore mechanical means of particle removal:
mechanical divertors or pump limiters.

2. SYNOPSIS OF THE MAJOR CONCEPTS OF MECHANICAL PUMP LIMITERS AND MAGNETIC ERGODIC LIMITERS

2.1 Pump Limiters

Pump limiters are mechanical devices for helium ash removal, fuel particle control, and possibly impurity control.

The principal idea of a mechanical divertor/pump limiter was published almost a decade ago by Vershkov and Mirnov in 1973 [1]. They suggested, as an intermediate step, to control the impurity content in tokamaks by improving the limiter. To minimize the impurity evolution at the limiter, they proposed to design it in the form of a rotating ring which would spread the deposited energy over a large area. The authors proposed a "mushroom" cross section of the limiter ring with a titanium getter on the lateral walls to perform some plasma and impurity unloading. The pump limiter concepts that have emerged since are basically variations of this original idea. In 1974 Kelley suggested a combined "Toroidal Band Limiter and Pump for a Tokamak" [2]. This limiter consisted of vertical bars around the outer vacuum chamber wall, and the surface behind it was freshly evaporated titanium. A more detailed proposal of a "Scrapeoff Limiter" for alpha-particle exhaust and wall-originated impurity screening was given by Bieger et al. [3]. They based their discussion on reactor applications and estimated an unloading efficiency of 0.1. A similar design of a scrapeoff limiter was proposed by Schivell [4]. The emphasis of this proposal was on impurity control in a contemporary device like the Princeton Large Torus (PLT), with an extrapolation to the Tokamak Fusion Test Reactor (TFTR).

First pump limiter designs that combine alpha-particle exhaust with adequate heat removal capabilities, for steady-state operation, have been published by Conn et al. [5] and by Brooks et al. [6]. In these designs the limiter surface facing the plasma has a convex curvature. In this way the leading edges are removed a few centimeters behind the plasma edge, and the power loads are reduced correspondingly. Fairly extensive conceptual design studies for pump limiter systems were conducted for the Fusion Engineering Device (FED) [7] and for the International Tokamak Reactor (INTOR) [8]. The FED pump limiter has one leading edge and is based on the idea that the heat load there can be controlled by changing the plasma major radius. The performance of the FED pump limiter has been studied in terms of plasma behavior [7,9] as well as material aspects [10,11]. While the analysis of alpha-particle exhaust and plasma behavior indicates satisfactory performance, a major problem results from the surface erosion due to sputtering and vaporization following plasma

892

disruptions. Depending on the plasma model, limiter lifetimes between a few weeks and a few months have been estimated [7,10,11]. The INTOR work on pump limiters [8] includes studies on the following issues: heat load, erosion, impurity control, and particle removal. The first experimental studies on pump limiters were focused on measurements of gas pressure buildup behind limiter blades. These early experiments without active gas pumping were carried out on Alcator-A [12], Macrotor [13], the Poloidal Divertor Experiment (PDX) [14], and the Impurity Study Experiment (ISX-B) [15]. More recently, modular pump limiters have been installed in ISX-B [16], PDX [17], PLT [18], and TEXTOR [19]. These pump limiters are designed to take the full heat load and are equipped with pumping sufficient to influence the plasma density. Results are discussed in Section 4 of this chapter.

2.2 Ergodic Magnetic Limiters/Divertors

To achieve an even distribution of heat and particle fluxes to limiters and other first-wall components, the characteristic decay lengths of density and temperature between the confined plasma and the wall should be as large as possible. In principle, this can be verified by enhancing transport in the scrapeoff layer with magnetic field line ergodization, which can be achieved by superposition of helical magnetic fields [20]. If the poloidal and toroidal mode numbers are high and in resonance with a rational q value near the edge, where q is the safety factor, only the outer flux surfaces are destroyed without affecting the inner ones. In this configuration the confinement region, given by the closed magnetic surfaces, is separated from the wall by an ergodized boundary layer, which is characterized by convection cooling due to thermal flow along field lines that form oblique angles with the wall. If the vacuum vessel is equipped with helical divertor plates and gas pumping, the configuration is referred to as a helical magnetic divertor [20]. Alternatively, if there are no divertor plates in the vacuum vessel and the particle convection flux recycles fully back into the plasma, forming a recycling loop in the boundary, the case is referred to as an ergodic magnetic limiter [21]. Modeling of the ergodic boundary layer predicts densities which are high enough to provide effective screening of impurities produced at the wall [22,23].

Experimental results on ergodic limiters are scarce. Studies carried out on the Texas Experimental Tokamak (TEXT) use modular helical coils that allow mode numbers of q = 7/2 and q = 7/3. Initial observations indicate changes in the heat load pattern on the limiter, in the confinement in the boundary layer, and shielding of impurities [24]. An ergodic limiter experiment with localized helical perturbation coils is planned for the TEXTOR device [25].

3. PRINCIPLES OF PUMP LIMITER OPERATION

3.1 Pump Limiter Action as a Function of Recycling

Pump limiters combine particle exhaust with heat removal.
The particle exhaust is based on the following processes: The
plasma particles in the scrapeoff layer flow along the magnetic
field lines to the limiter or neutralizer-plate surface, where a
large fraction of the impinging ions is neutralized and reflected.
For typical tokamak conditions (T_i = 10–100 eV at the edge), the
reflection coefficient is approximately 0.5, the neutral fraction
of the reflected particles is close to 1, and their energy peaks
near the incident energy [26,27]. A fraction of the reflected
neutrals, which are no longer confined by the magnetic field, is
then scattered into a pumping channel. For a crude estimate,
neglecting plasma effects, we assume that the neutral pressure in
the pumping channel increases until the influx of fast particles
is balanced by the backflow of room-temperature neutrals. The
achieved density compression in the pumping channel is then
proportional to the ratio of the velocities of incident to back-
flowing particles, which is typically in the range of 10 to 100.
This compression can be expected to be even higher if the limiter
configuration is such that the backflow of neutrals is impeded by
the incident plasma flow ("plasma plugging").

Two possible configurations of pump limiters are shown
schematically in Fig. 1. Characteristic of the configuration in
Fig. 1(a) is the channel behind the limiter. Due to this channel,
the backflow of neutralized particles into the plasma chamber can
be strongly impeded by the incident plasma flow. A consequence of
this design is a high heat load at the edges of the limiter at
which the projected heat flux is perpendicular to the limiter
surface (leading edges). This problem can be avoided with a
configuration as shown in Fig. 1(b), in which no leading edges are
exposed to the plasma flow. However, in this case the collection
efficiencies are low. Only a fairly small fraction of the incident
particle flux is scattered into the pumping duct, and the backflow
into the plasma chamber is assumed to depend largely on the
pumping speed of the applied vacuum pumps. Pump limiters of the
type represented by the configuration in Fig. 1(a) are referred to
as "closed pump limiters," and those of the type shown in Fig. 1(b)
are referred to as "open pump limiters." Although the experimentum
crucis is yet to be performed, it is generally believed that the
heat load problem at the leading edges of closed pump limiters can
be solved. Accordingly, open pump limiters are not of much
interest at the present time due to their low particle collection
efficiency, and we refer to the literature [15] for further
considerations. In this paper we concentrate our discussion on
closed pump limiter configurations.

894

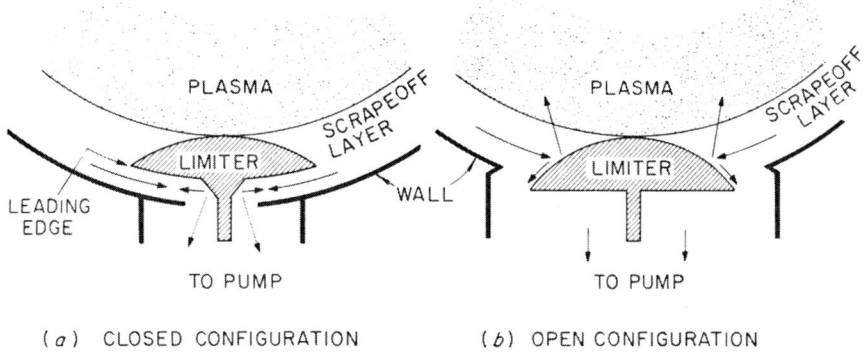

(a) CLOSED CONFIGURATION   (b) OPEN CONFIGURATION

Fig. 1.   Configurations of "closed" and "open" pump limiters.

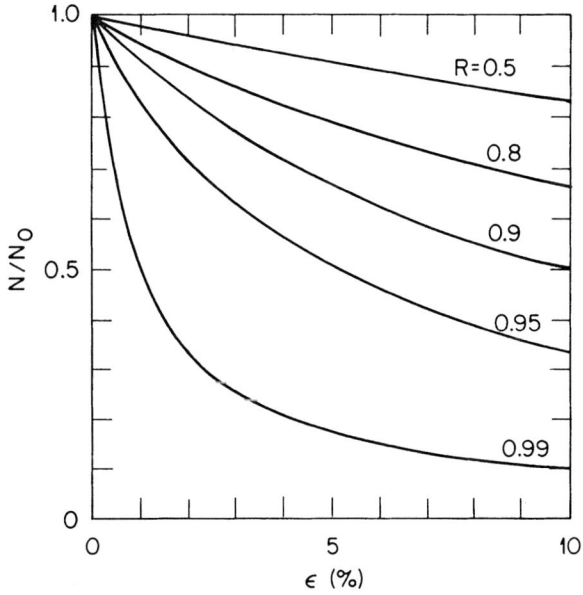

Fig. 2.   Density reduction $N/N_o = (1 - R)/(1 - R + \varepsilon)$ with R as
recycling coefficient and ε as change in recycling due
to the pump limiter exhaust.

Pump limiters are capable of removing approximately 1 to 10% of the total ion flux incident on the limiter. This is sufficient to control the particle density, provided that the global recycling coefficient of the respective species is close enough to unity (0.9-1). The recycling coefficient R is defined as the ratio of the total particle flux coming back from the limiter and wall (direct reflection plus desorption) to the total incident flux. The global particle balance is given by the following equation:

$$\frac{dN}{dt} = \phi + R \frac{N}{\tau_p} - \frac{N}{\tau_p} , \tag{1}$$

where N, ϕ, and τ_p are the total number of particles in the confined plasma, the fueling rate, and the particle confinement time, respectively. Equation (1) states that the rate of change in the number of plasma particles is equal to the balance of sources (external fueling rate ϕ plus recycling rate RN/τ_p) and losses (the total particle flux leaving the confined plasma N/τ_p). In Eq. (1), ϕ is the external plasma fueling rate. If ionization of gas in the scrapeoff layer is negligible, the fueling rate is equal to the external gas flow rate. For further discussion, we assume that sources in the scrapeoff layer are negligible and that the fueling rate ϕ is equal to the gas flow rate. In steady state, dN/dt = 0 and the number of plasma particles is given by

$$N = \frac{\phi \tau_p}{1 - R} . \tag{2}$$

This equation demonstrates that a small change in R entails a large change in N, provided R is close enough to unity. Pump limiter operation is based on this principle: for given source rates and particle confinement, the pump limiter reduces the effective recycling by a small fraction to yield substantial reductions in the particle density. The source rate ϕ can correspond to either external fueling (i.e., gas puffing, pellet injection, or beam injection) or helium generation rates and, possibly, to impurity production rates. Figure 2 illustrates the effect of changing the effective recycling. The graph shows the ratio of densities with and without a pump limiter according to Eq. (2) with constant ϕ and τ_p: $N/N_0 = (1 - R)/(1 - R + \epsilon)$,

where R is the recycling coefficient without a pump limiter and $(R - \epsilon)$ is the reduced global recycling due to the exhaust efficiency ϵ of the pump limiter. The strong amplification resulting from high recycling coefficients can yield exhaust rates equal to fueling rates or helium production rates. The global model discussed here is very useful to demonstrate the principle of pump limiter operation. More detailed discussions of plasma and scrapeoff layer modeling are given elsewhere in this course.

896

3.2 Pump Limiter Design

The design of the pump limiter blade is a trade-off between
high particle exhaust and sufficiently low power fluxes at the
leading edge. For high particle exhaust, the leading edge should
be as close as possible to the plasma boundary to get a high flux
of particles behind the limiter. For power handling, on the other
hand, the leading edge (which is by definition perpendicular to
the projected power flux) should be recessed as far as possible
from the plasma edge. Fortunately, the decay length of the power
flux is shorter than that of the particle flux, which makes it
possible to deposit most of the power in a shallow angle on the
front face of the limiter and still collect sufficient particle
flux behind the leading edge.

The decay of plasma density and temperature outside the last
closed flux surface is approximately exponential:

$$n_e(x) = n_o e^{-x/\lambda_n}$$

and

$$T_e(x) = T_o e^{-x/\lambda_T} \, ,$$

where λ_n and λ_T are the characteristic decay lengths of density
and temperature in the scrapeoff layer. We assume that $T_e = T_i$
and

$$\Gamma_\| = n v_\| \propto n T_e^{1/2}$$

for the parallel particle flux and

$$Q = \delta \Gamma_\| k T_e \propto n T_e^{3/2}$$

for the power flux, where δ is the heat transfer coefficient. The
composite e-folding lengths for the particle flux λ_Γ and power
flux λ_Q are

$$\lambda_\Gamma = \frac{2\lambda_n \lambda_T}{2\lambda_T + \lambda_n} \, ,$$

$$\lambda_Q = \frac{2\lambda_T \lambda_n}{2\lambda_T + 3\lambda_n} \, . \tag{3}$$

For $\lambda_n = \lambda_T$, these equations yield $\lambda_Q/\lambda_\Gamma = 3/5$. Experimentally, it is found that λ_T is usually larger than λ_n. Typical examples for rail limiters in ISX-B are $\lambda_n = 2$ cm and $\lambda_T = 5$ cm.

The decay lengths λ_n and λ_T can be derived theoretically by solving the transport equations in the scrapeoff layer [28]. However, since the transport coefficients are not known well enough, the merits of these calculations are not so much in the absolute results but rather in the insight into the dependence of the scrapeoff layer on plasma parameters and limiter design.

To discuss the correlation between power dissipation and particle removal, we take, as a specific example, a limiter designed for uniform power deposition [29]. Assuming that the power flux in the scrapeoff layer decays exponentially and the characteristic length λ_Q is known, the limiter blade can be shaped for uniform heat load [30,9]. Uniform heat flux over the limiter requires the projection of the heat flux $Q(x)$, where x is the distance into the scrapeoff layer, as a constant:

$$Q(x) = Q_o e^{-x/\lambda_Q} \cos \psi = Q_D \, , \tag{4}$$

where Q_D is referred to as the design heat load and ψ is the angle between the incident flux and the limiter surface normal. From Eq. (4) the equation for the surface contour of a toroidal limiter with uniform heat flux can be derived:

$$y(x) = -\lambda_Q (\sqrt{c^2 e^{-2x/\lambda_Q} - 1} - \arctan \sqrt{c^2 e^{-2x/\lambda_Q} - 1})$$

$$+ \lambda_Q (\sqrt{c^2 - 1} - \arctan \sqrt{c^2 - 1}) \, , \tag{5}$$

where the y coordinate is the tangent to the poloidal cross section of the limiter and the constant c is the ratio of the peak heat flux Q_o perpendicular to the flow direction to the design heat flux Q_D. The power to the toroidal limiter with major radius R_o is $P_L = 4\pi R_o Q_o \lambda_Q$. The shaping constant $c = Q_o/Q_D$ can be computed if the limiter power P_L and the e-folding λ_Q are known: $c = P_L/4\pi R_o Q_D \lambda_Q$. Equation (5) is valid for $0 \leqslant x \leqslant \lambda_Q \ln c$. For $x_m = \lambda_Q \ln c$ the heat flux is perpendicular to the surface and $Q(x_m) = Q_D$. For $x > x_m$ an unshaped surface can be tolerated because the flux has dropped below the design flux: $Q(x) < Q_D$. Putting x_m into Eq. (5) yields the maximum y value, that is, the width of the limiter blade:

$$y_m = y(x_m) = \lambda_Q(\sqrt{c^2 - 1} - \arctan\sqrt{c^2 - 1}) . \tag{6}$$

The power deposited on the shaped part of the limiter is $P_S = P_L(1 - e^{-x_m/\lambda_Q}) = P_L(1 - c^{-1})$.

The fraction of the particle flux that flows behind the limiter is the "exhaust fraction," $f = e^{-x_m/\lambda_\Gamma} = c^{-\lambda_Q/\lambda_\Gamma}$. These considerations are best illustrated by a numerical example. The Fusion Engineering Device [7] has the following parameters: $R_0 = 500$ cm, $P_L = 50$ MW, $Q_D = 250$ W/cm^2, $\lambda_Q = 1.5$ cm, and $\lambda_\Gamma = 2.5$ cm. With these parameters we obtain $c = 21.2$, $x_m = 4.58$ cm, $y_m = 30$ cm, $P_S = 0.95\ P_L$, and $f = 0.16$. This means that 95% of the power flux is deposited on the front surface of the limiter while 16% of the total particle flux still flows behind the blade.

3.3 Particle Exhaust

In the previous section we have defined the exhaust fraction $f = c^{-\lambda_Q/\lambda_\Gamma}$, which represents the fraction of particle flux available for removal. This fraction of the particle flux enters the pump limiter channel and impinges on the neutralizer plate, where it is scattered into the pump limiter vacuum chamber or back into the plasma vessel. The captured particles build up a pressure in the pump limiter vacuum chamber that rises until the back flow into the plasma chamber plus the fraction removed by pumping equals the incident flux. We define a trapping coefficient ε_t which is that part of the exhaust fraction actually removed. In a simple case, neglecting plasma effects, the trapping coefficient is equal to the ratio of applied pumping speed to back-conductance into the plasma chamber. With this concept, trapping coefficients of the order $\varepsilon_t = 0.5$ can be realized. In practice, it is hoped to achieve higher trapping coefficients with lower pumping speeds due to plasma plugging. This effect causes the neutral backflow in the pump limiter throat to be strongly impeded by the incoming plasma flow if the mean free path for ionization is short compared to the throat length.

As a quantitative measure for the particle control capability of a pump limiter, we define the "exhaust efficiency" ε as

$$\varepsilon = f\varepsilon_t , \tag{7}$$

which is the product of the exhaust fraction f and the trapping efficiency ε_t. For our numerical example of the previous chapter, with $f = 0.16$ and $\varepsilon_t = 0.5$, an estimate of the exhaust efficiency would be $\varepsilon = 0.08$.

The upper bound for the exhaust efficiency is determined by compatibility with heat removal at the leading edge. For the uniform heat load limiter, the exhaust fraction is $f = c^{-\lambda_Q/\lambda_\Gamma}$, with $c = P_L/4\pi R_o Q_D \lambda_Q$. Assuming a trapping coefficient of $\varepsilon_t = 0.5$, which is close to experimental findings, the exhaust efficiency becomes

$$\varepsilon = 0.5(P_L/4\pi R_o Q_D \lambda_Q)^{-\lambda_Q/\lambda_\Gamma} . \qquad (8)$$

This equation gives a useful estimate of the exhaust efficiency if the other parameters are known.

In a more practical design approach, one would start with the desired particle exhaust rate ϕ_{EX}, that is, the exhausted flow rate, make an assumption about the achievable pressure p in the pump limiter, and determine the necessary pumping speed $S = \phi_{EX}/p$. A numerical example may be given here for the toroidal pump limiter ALT-II (Advanced Limiter Test-II) which is presently under design for the TEXTOR device [31]. An exhaust rate of 10 to 20 Torr·ℓ/s is desired. According to available experimental experience with pump limiters, one can assume that pressures of the order 1 mTorr can be achieved. This yields necessary effective pumping speeds of $1-2 \times 10^4$ ℓ/s.

3.4 Pumping Considerations

To remove particles permanently from the torus, pump limiters have to be equipped with appropriate vacuum pumps. Type and size of the pumps depend on the specific task. If density control (i.e., hydrogen pumping) is the only objective, turbomolecular pumps, cryogenic pumps, and bulk getter pumps can be employed. For helium ash exhaust, bulk getter pumps are ruled out, and cryogenic pumps can only be used in complicated schemes if large amounts of hydrogen species are present. This leaves turbomolecular pumps, which can handle helium and hydrogen as well but have the disadvantage that they must be removed from strong magnetic fields so that pumping usually becomes conductance-limited.

The size of pumps depends on the exhaust requirements and on the design of the pump limiter. If a long-pulse device with beam injection of 5 to 50 MW is to operate in quasi equilibrium with recycling near unity, an exhaust rate of 10 to 100 Torr·ℓ/s is necessary to remove the beam particles. Assuming a 1-mTorr pressure buildup, pumping speeds of 10^4 to 10^5 ℓ/s are required.

For a D-T-burning device producing fusion power of 500 MW_{th}, approximately 5 Torr·ℓ/s of helium ash has to be removed. Since the helium flux into the pump limiter is at least an order of magnitude lower than the hydrogen flux, pumping speeds have to be correspondingly higher to remove the helium.

The specific design of the pump limiter is an important factor in the pumping considerations. A design that favors plasma plugging is likely to build up pressures on the order of 10 mTorr. This would strongly reduce the pumping requirements. However, if turbomolecular pumps are to be employed, excessively high pressures could lead to reduced pumping speeds. In any case, the choice of the pumping system has to be made in accordance with the specific goals of the pump limiter application.

Finally, when we design pumping systems, we usually base our assumptions on the removal of thermal gases. Future experiments might reveal that the conductance of pump limiter ducts is substantially enhanced due to ballistic particle collection (i.e., the energies of particles scattered down the duct are higher than those corresponding to room temperature). This would be very valuable for the reactor design, because for a given throughput the duct size can be reduced correspondingly.

3.5 Impurity Control with Pump Limiters

For impurity control, pump limiters seem to be inferior to magnetic divertors. Provided there are no external sources (beam injectors, gas injectors), impurities can originate either at the wall or at the limiter (divertor target).

We discuss first the impurities originated at the limiter itself. The dominant part of the limiter (i.e., where most of the power and particle flux goes) faces the plasma. Impurities that are born here have a high probability of being ionized and captured by the plasma. The pump limiter has no immediate effect on these impurities. The pump limiter has the first chance to remove the impurities when they leave the plasma and recycle back into the scrapeoff layer. However, most limiter materials are more likely to plate out instead of recycling. In a magnetic divertor, the impurities produced at the plate are generated in a separate chamber and are shielded from the main plasma by the scrapeoff layer plasma. Only a small fraction of impurities produced at the plate can reach the main plasma. It is obvious that the plasma contamination due to impurity release at the limiter or divertor plate can be expected to be smaller in divertor devices.

Next, we consider impurities released at the vessel wall. These impurities can consist of adsorbed gases, oxide or carbide layers, and the wall material proper. The wall metals either will be ionized in the scrapeoff layer and plate out at the limiter or will penetrate into the main plasma, depending on their energies and the plasma parameters in the scrapeoff layer. The pump limiter is not likely to have an effect on metal impurities: removal by plating out would be the same on a conventional limiter and much more efficient than any pump limiter action. Certain light impurities desorbed from the wall, however, might recycle and accumulate in the main plasma. In this case the pump limiter might interrupt the recycle and prevent accumulation in the plasma.

In the divertor configuration, the wall impurities are, in an ideal case, ionized in the scrapeoff layer and swept into the divertor chamber. Although transport along field lines into the divertor chamber is much faster than perpendicular to the field into the plasma, the characteristic lengths are also very different, that is, several centimeters scrapeoff length versus several tens of meters connection lengths into the divertor. As a result, the net transport across field lines becomes comparable to the parallel transport [32], thus reducing the effective shielding.

We conclude that control of wall impurities is likely to be comparable in pump limiter and divertor machines. The source rate might be higher in the limiter case, however, because recycling at the limiter entails higher charge exchange fluxes to the wall. The major difference between limiter and divertor devices is the location of the main plasma wall interaction: in divertor tokamaks it is removed into a separate chamber with some shielding to the main chamber, while in limiter tokamaks the interaction is in the main chamber facing the bulk plasma.

4. EXPERIMENTAL RESULTS

4.1 Pump Limiter Probes and Pressure Rise Studies

Early experiments were performed to study pressure rise behind limiters or in especially designed "pump limiter probes." These pump limiter probes were basically gas collection chambers without active pumping and were not designed to take the full heat load of the respective device.

The first demonstration of pressure buildup behind limiters was performed on Alcator-A by measuring the neutral density rise near the regular molybdenum limiter [12]. A schematic view of the experimental apparatus is shown in Fig. 3. In the poloidal plane of the limiter, the toroidal vacuum vessel had three rectangular slots, at the end of which the pressure buildup is measured with

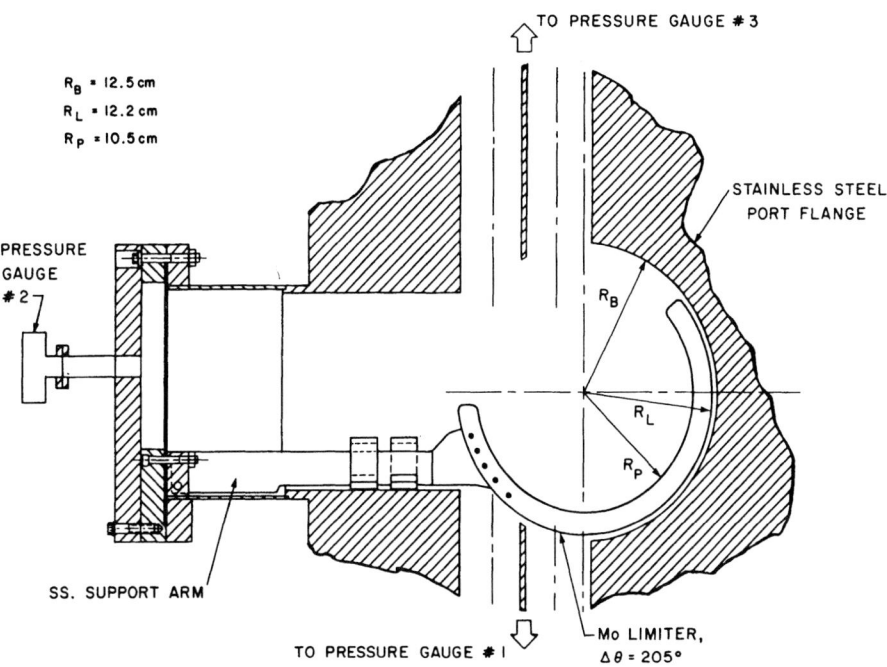

Fig. 3.   Cutaway view of the Alcator-A pump limiter experiment
          (courtesy of D. Overskei).

fast response ($\leqslant$5 ms), absolute pressure gauges. The vacuum
response time for room temperature deuterium was 10 to 15 ms at
the horizontal duct and about 100 ms at the top and bottom ducts.
Typical plasma parameters during these experiments were B_T = 6 T,
plasma current 130 $\leqslant$ I_p $\leqslant$ 200 kA, plasma density 1×10^{14} $\leqslant$
$\bar{n}_e$ $\leqslant$ 5.5×10^{14} cm^{-3}, and the discharge duration between 115 and
140 ms. The filling pressure before plasma initiation was 4 to
8×10^{-5} Torr, and the working gases included H_2, D_2, or He.
The evolution of plasma density $\bar{n}_e$ and limiter port pressures is
shown in Fig. 4, along with plasma current, loop voltage, and gas
puff. The pressure buildup in the three limiter ports is represented by the bottom three traces. The top port, which was not
intersected by any portion of the limiter, did not exhibit any
pressure rise during the discharge, while the side and bottom
ports showed large pressure rises. The different time histories
are explained by the respective vacuum response times. This
indicates that the pressure rise was primarily due to particles
directly reflected from the limiter, and the charge-exchange flux
did not contribute significantly. The maximum pressure rise,
observed at the bottom port, was 29 mTorr of D_2 when $\bar{n}_e$ = 5.1 $\times$
10^{14} cm^{-3}. This was a factor of 5 to 10 higher than the pressure
rise due to puffing cold gas into the torus without plasma.
Pressure measurements in other ports where there were no limiters
showed a behavior similar to the top port in the limiter plane.

The first experiment specifically designed to study particle
collection was performed on Macrotor [13] and was also based on
the collection of reflected particles. Typical plasma parameters
for these experiments were B_T = 0.3 T, I = 80 kA, $\bar{n}_e$ = 10^{12}–
10^{13} cm^{-3}, T_e = 120 eV, and T_i = 50 eV. Due to the low magnetic
field, the Macrotor plasma had a large scrapeoff thickness of
about 10 cm. Pump limiter experiments were carried out with a
small probe (2.3-cm diam) and a large area limiter (20 cm $\times$
60 cm $\times$ 6 cm), the experimental arrangements of which are shown in
Fig. 5. The probe was capable of radial movement and the collection throat had either microchannels or an open end, as shown in
Fig. 5. Approximately 60 cm from the collection throat, the
pressure was measured with a cold cathode pressure gauge. The
vaccum time constant for room temperature molecular hydrogen was
50 ms.

A typical set of data signals is shown in Fig. 6. The
neutral pressure in the collection probe reached about 5 mTorr
with a time delay relative to the density corresponding to the
response time of the tube. The pressure rise as a function of
probe position is shown in Fig. 7. Also plotted is the incident
particle flux as inferred from Langmuir probe measurements. The
maximum observed pressure was 8 mTorr, corresponding to molecular

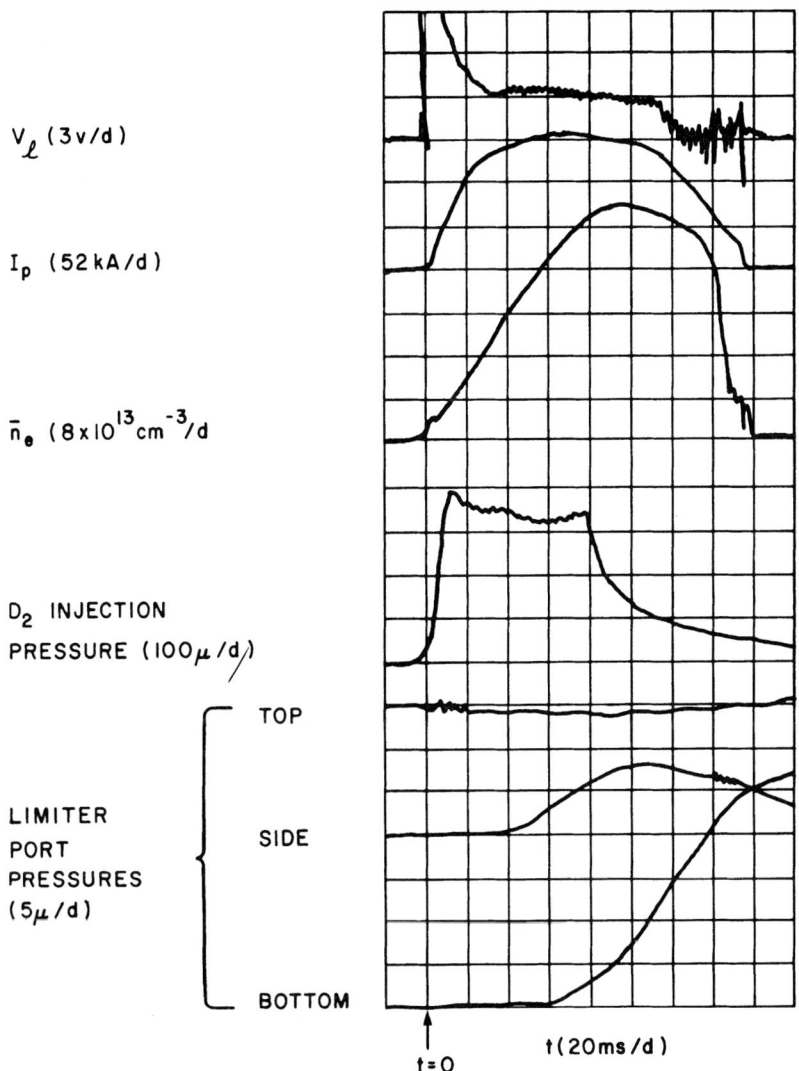

V_ℓ (3v/d)

I_p (52 kA/d)

$\bar{n}_e$ $(8 \times 10^{13} cm^{-3}$/d

D_2 INJECTION
PRESSURE (100μ/d)

LIMITER
PORT
PRESSURES
(5μ/d)

TOP

SIDE

BOTTOM

t=0

t(20ms/d)

Fig. 4. Temporal evolution of a typical deuterium discharge. The
bottom three traces are the signals from the pressure
gauges in the limiter parts (courtesy of D. Overskei).

905

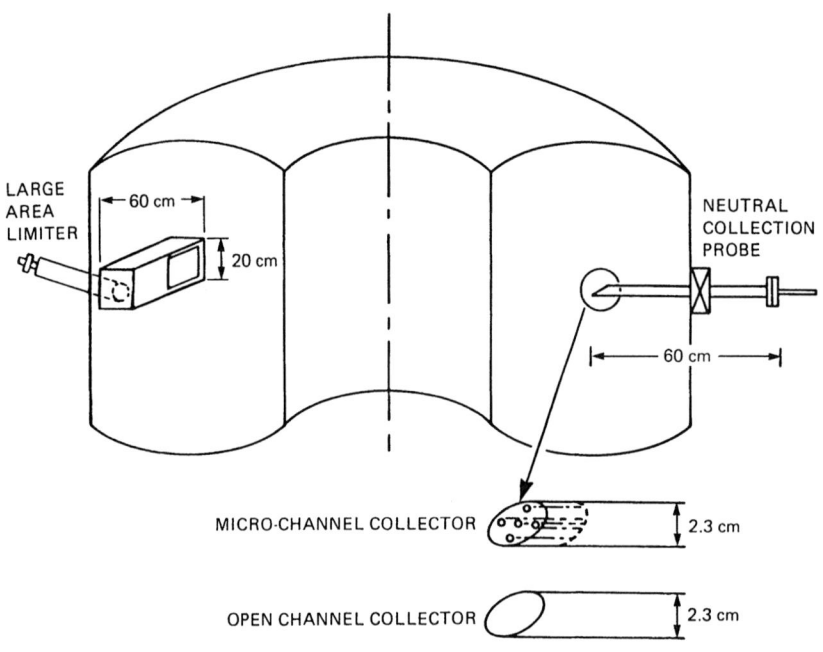

Fig. 5. Experimental arrangement of pump limiters in Macrotor
(courtesy of S. Talmadge).

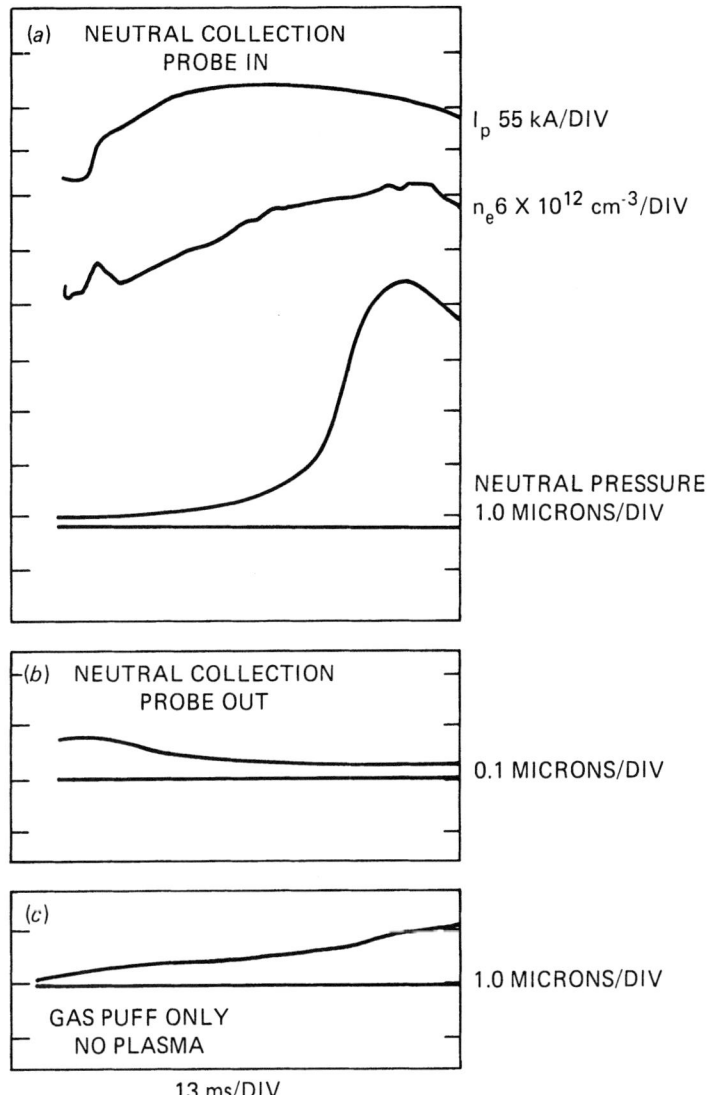

(a) NEUTRAL COLLECTION
PROBE IN

I_p 55 kA/DIV

n_e 6 X 10^{12} cm^{-3}/DIV

NEUTRAL PRESSURE
1.0 MICRONS/DIV

(b) NEUTRAL COLLECTION
PROBE OUT

0.1 MICRONS/DIV

(c)

1.0 MICRONS/DIV

GAS PUFF ONLY
NO PLASMA

13 ms/DIV

Fig. 6. Typical set of raw data of the Macrotor pump limiter
experiment (courtesy of S. Talmadge).

907

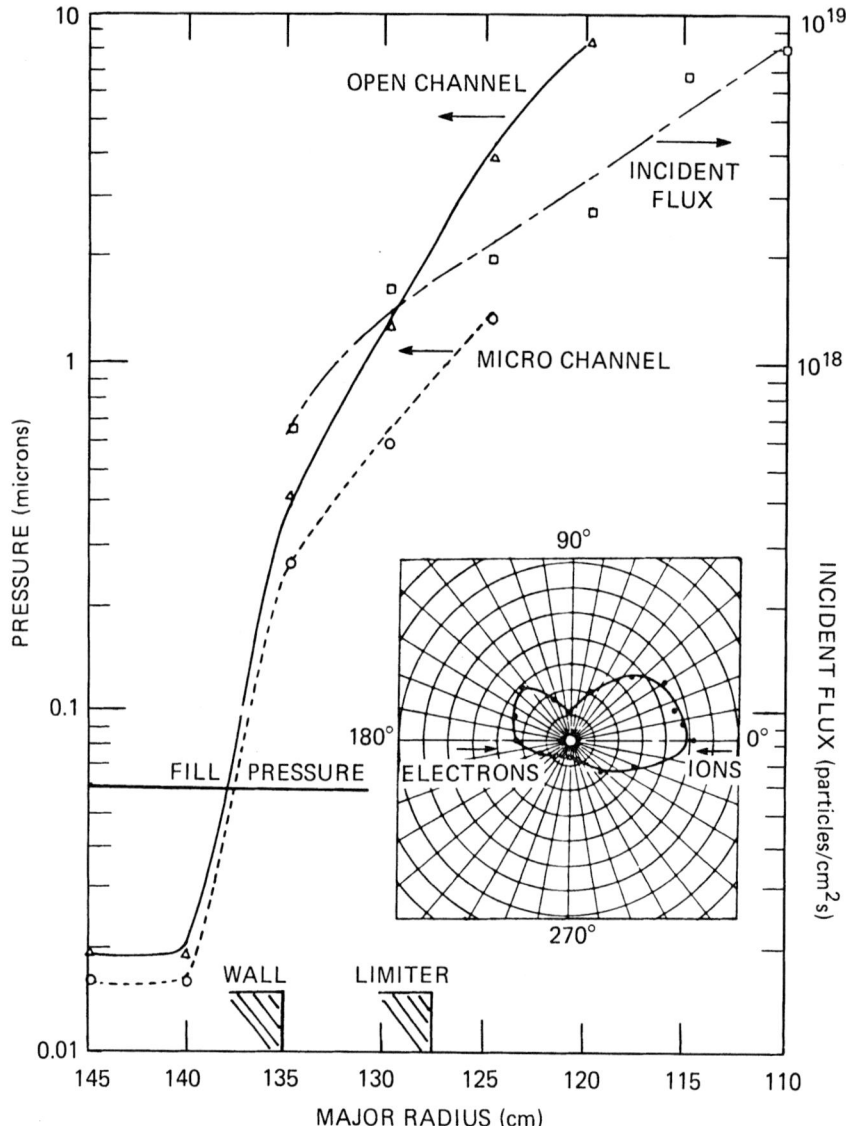

Fig. 7. Peak pressure and incident ion flux as a function of
position in Macrotor. The insert shows a polar plot of
the pressure (courtesy of S. Talmadge).

densities of $4-6 \times 10^{14}$ cm^{-3}, which is more than 100 times the local plasma density. The graph also reveals that the ratio of the pressure increase to the incident flux was a function of position. The insert of Fig. 7 shows a polar plot of the maximum pressure at a given radial position. The pressure obtained on the ion drift side was approximately twice that on the electron drift side. Empirical scaling of the pressure buildup as a function of density showed a stronger-than-linear dependence, that is, $p \propto n^2$ for the open probe and $p \propto n^{1.5}$ for the channeled probe. The authors attribute this behavior to plasma effects, rather than to ballistic collection only.

In ISX-B, studies were performed to explore possibilities of designing pump limiters with sufficient particle removal but without leading edges and the related power removal problems [15]. For this purpose a probe-type pump limiter was built with a geometry simple enough to be modeled [15]. Figure 8 shows the experimental arrangement. The probe consisted of two separate electrically isolated blades intercepting the scrapeoff plasma perpendicularly. A certain fraction of the plasma particles reflected from one of the blades was collected by a tube (3.5-cm diam, 80-cm length), at the outer end of which the pressure rise was measured with a response time of 50 ms for thermal hydrogen. By simultaneously measuring the ion saturation current to the limiter blade, a correlation between incident flux and pressure buildup was found which allows one to deduce a particle collection efficiency. Both limiter blades were equipped with thermocouples to measure the total energy deposition during the discharge.

The plasma parameters during these experiments were $B_T = 1.2$ T, $I_p = 170$ kA, and $\bar{n}_e = 2-4 \times 10^{13}$ cm^{-3}. The pressure rise, ion saturation current, and average energy flux density to the limiter were measured as a function of radial distance from the plasma edge, defined by the main limiter. Figure 9 shows the measured profiles. All data were taken on the electron-drift side of the limiter. The total pressure rise was modest, compared to those seen in other machines, but, referring to the Macrotor results (see insert of Fig. 7), one might expect it to be a factor of 2 higher on the ion-drift side. Pressure, as well as power and particle flux profiles, reveal an approximately exponential decay up to 3 cm behind the plasma edge, with e-folding lengths of $\lambda_Q = 1.5$ cm and $\lambda_\Gamma = 2$ cm. Plasma density and temperature at $r = a$ were measured by Thomson scattering to be $n_e = 2.5 \times 10^{12}$ cm^{-3} and $T_e = 30-40$ eV, respectively. The corresponding profiles in the scrapeoff were calculated from the Thomson scattering values at the edge, the measured profiles of ion saturation current j_{sat}^+ and power flux Q, and the equations $j_{sat}^+ = n_e e (2kT_e/\pi m)^{1/2}$ and $Q = \delta(j_{sat}^+/e)kT_e$. The corresponding e-folding

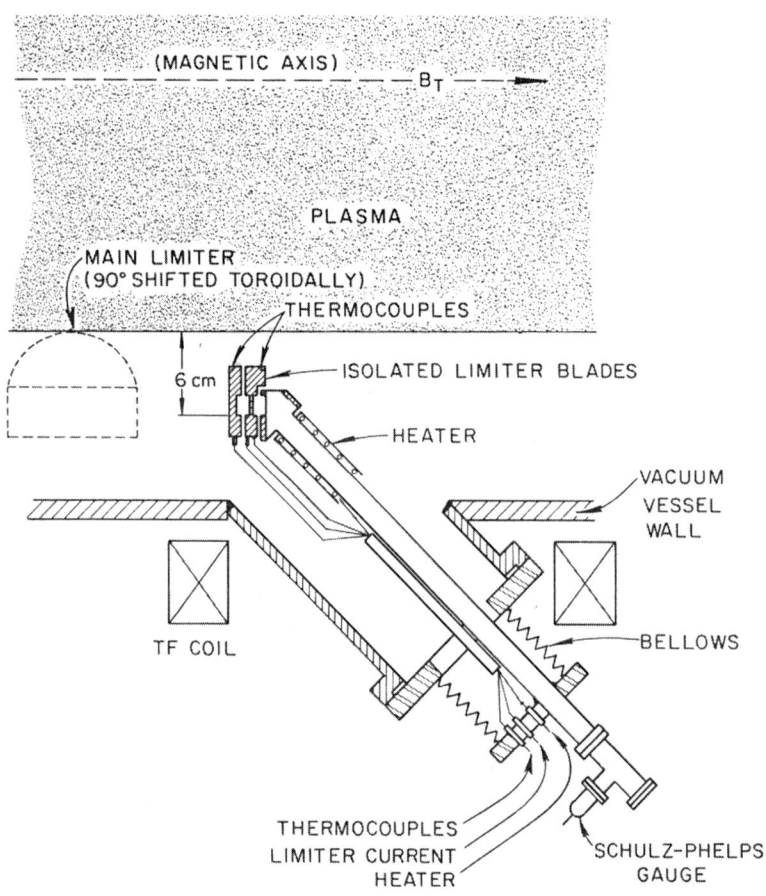

Fig. 8. Schmatic of pump limiter scoping studies in ISX-B [15].

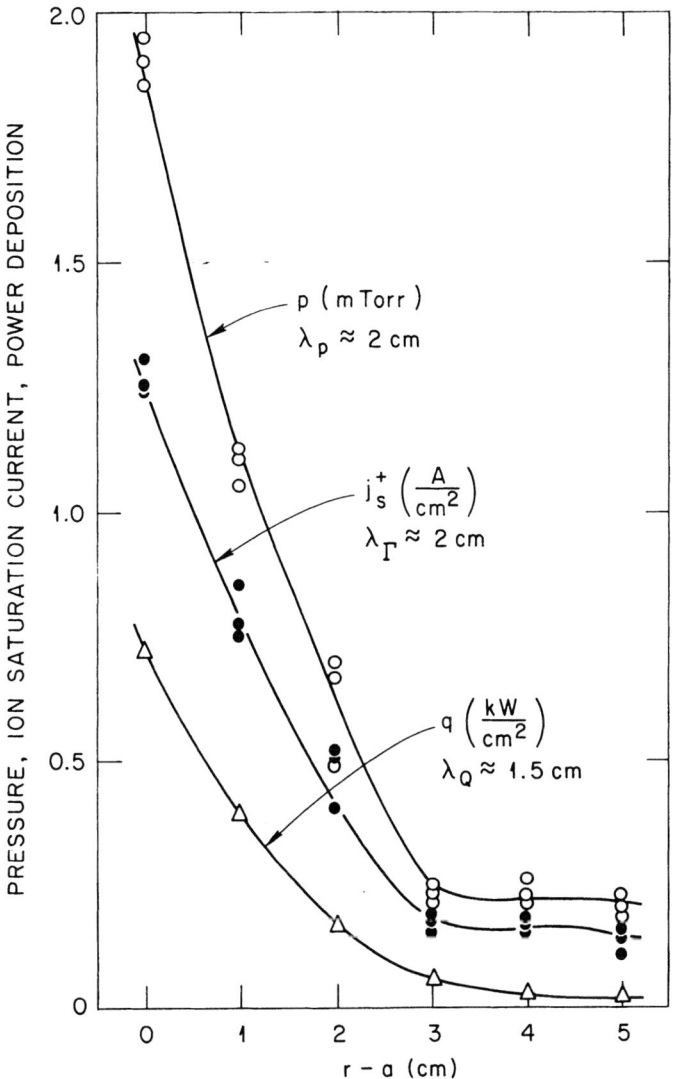

Fig. 9. Pressure rise, ion saturation current density, and average power deposition as a function of radial position of the limiter probe in ISX-B [15].

distances for density and temperatures were found to be λ_n = 2.5 cm and λ_T = 5.0 cm, respectively.

Although the experiments discussed so far had basically open pump limiters, the "scoop limiter" installed on PDX [14] was of the closed pump limiter type and exhibited quite different behavior. The PDX experiments were conducted in ohmically heated deuterium discharges with typically B_T = 1.7 T, I_p = 300 kA, $\bar{n}_e$ = 3 × 10^{13} cm^{-3}, and a standard D configuration. It was found that in this configuration most of the energy was flowing within about 1 cm of the separatrix, while for distances of the order of 5 cm outside the separatrix, the plasma density and temperature were essentially constant. The location of the scoop and the corresponding pressure gauges on PDX are shown in Fig. 10(a); details of the scoop can be seen in Fig. 10(b). The plenum of the scoop was connected through a duct of 130-cm length and 2.5-cm diameter to the pressure gauges. The scoop was moved through the scrapeoff layer and the pressure rise measured as a function of position. The peak pressures for various line-integrated densities are plotted in Fig. 11 as a function of major radius. Although the plasma densities were modest during these experiments, the pressure buildup in the scoop was impressive. The peak pressure was found to increase exponentially with the line-integrated density $\int n \, dl$ of the main plasma, consistent with other measurements on PDX [33], showing that $\int n \, dl$ in the divertor throat was rising exponentially with $\int n \, dl$ measured on the plasma equator. Plotting the maximum achieved pressures as a function of the electron density in the scrapeoff plasma revealed a much stronger than linear dependence, indicating strong plasma plugging effects. This behavior has been modeled successfully with Monte Carlo techniques to yield a trapping efficiency of about 90% and a neutral pressure scaling roughly with the square of the plasma density [34].

4.2 Full-Size Pump Limiters

Modular pump limiters which can take the full heat load of the device have been installed in ISX-B at Oak Ridge [16], in PDX [17] and PLT [18] at Princeton, and in TEXTOR at Jülich [19]. Although all of these pump limiters are different in design and pumping schemes, they have some common features: they represent relatively small, localized modules with correspondingly high power and particle fluxes. A recent review article gives a good summary of results obtained on these devices [35].

The basic features observed on pump limiter experiments, as performed currently, can be illustrated with data obtained on the ISX-B tokamak. The first demonstration of continuous particle

912

SCOOP LIMITER IN PDX

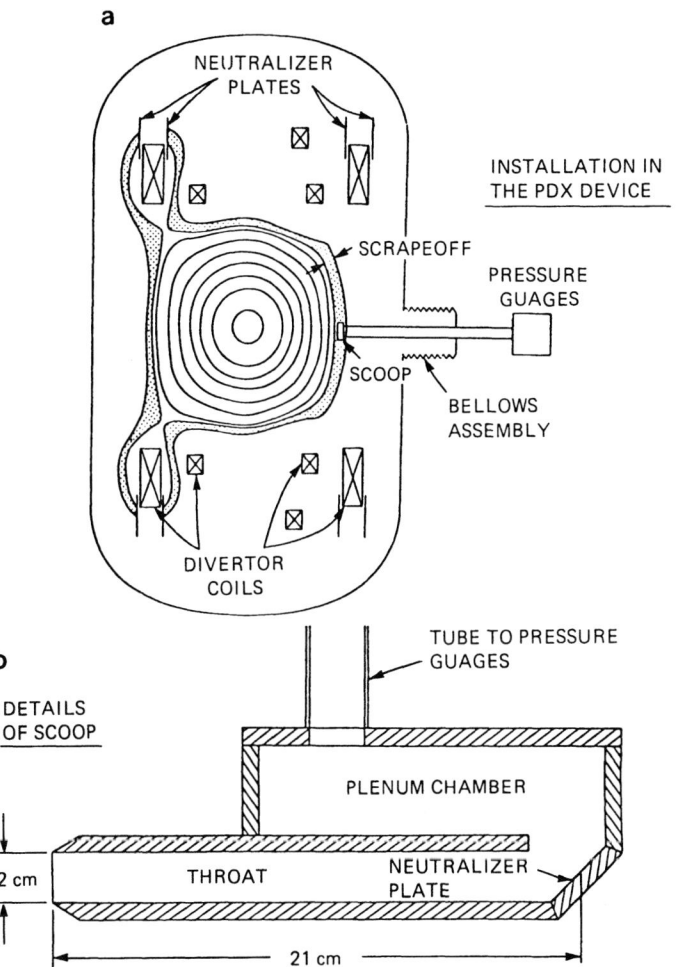

Fig. 10.   (a) Schematic of the small scoop experiment on PDX;
            (b) details of the scoop (courtesy of R. Jacobsen).

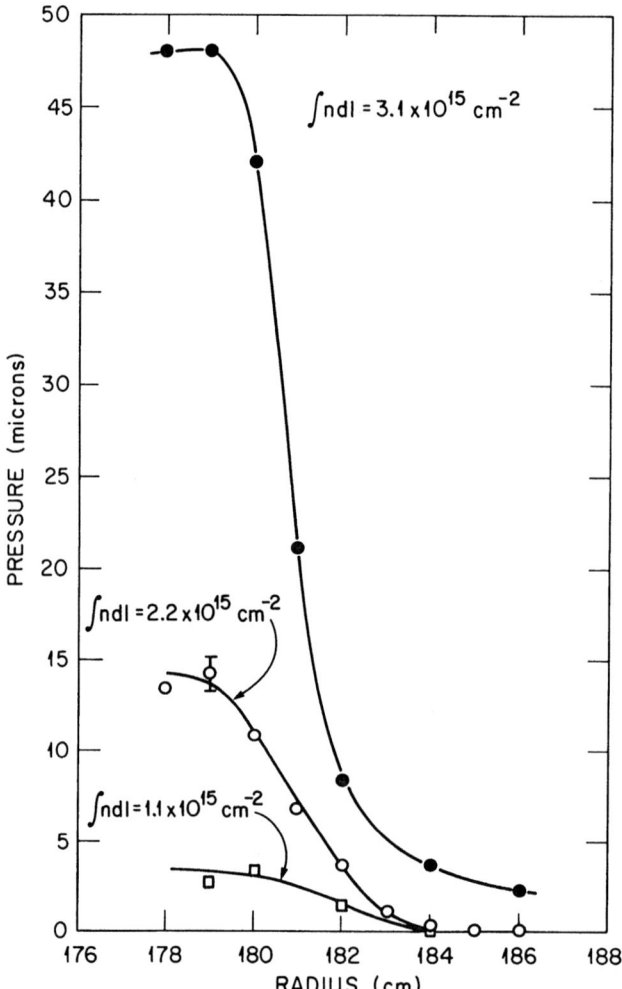

Fig. 11. Peak pressure in the PDX small scoop as a function of
major radius for different line densities (courtesy of
R. Jacobsen).

914

exhaust and density control with pump limiters was performed with the arrangement shown in Fig. 12. Two limiters were installed, one in the top and one in the bottom of one toroidal sector of the tokamak. Each unit had a radial stroke of 25 cm. The limiter heads consisted of TiC-coated graphite, while the rest of the structure was made of stainless steel. One of the limiter modules was equipped with a gas puff nozzle in the center of the limiter head which was intended for impurity injection and fueling studies. Each of the two pump limiter modules was equipped with a Zr-Al getter cartridge rated at 1000 to 2000 ℓ/s and located in the cavity below the limiter. The two pump limiter heads had different designs, as shown in Fig. 13. Type 1 had two slots for particle collection and an average throat length of 4 cm, while type 2 had only one slot and an average throat length of 10 cm. The purpose was to compare particle collection as a function of throat length and orientation of the collection channel (ion-drift side versus electron-drift side).

Experiments were performed with ohmically heated discharges and with neutral beam injection. A typical result obtained in an ohmic discharge is shown in Fig. 14, which shows very clearly the density change caused by the pump limiters. From top to bottom, the data represent the gas flow rate, the measured pressure rise in the limiters, and the line-averaged density in the plasma. The labels "ON" and "OFF" refer to whether the getters were activated or not. For identical gas flow rates, we observed a density decrease by approximately a factor of 2. After activating the pumps, the pressure in the pump limiter modules, as measured with the pressure gauges, dropped on the average by a factor of 6. This measured value was a factor of 2 lower than the average pressure in the cavity due to a strong pressure gradient along the getter cartridge.

Neutral beam injection usually lead to a deterioration in confinement and to a decrease in global recycling. Both of these effects had some impact on pump limiter performance. A decrease in particle confinement caused higher densities and longer decay lengths in the scrapeoff layer at fixed densities in the bulk plasma. As a result, the pump limiter pressure was higher at a given line-averaged density. The exhaust fraction, however, did not change since the total flux to the limiter increased accordingly. If, on the other hand, neutral beam injection had changed the scrapeoff layer plasma such that plasma plugging was induced, the trapping efficiency could have been improved and, correspondingly, the exhaust efficiency. Figure 15 demonstrates the effect of neutral beam injection on the scrapeoff layer particle flux, measured with a Langmuir probe. The corresponding change in pump limiter pressure is shown in Fig. 16. These data were taken with the short-throat limiter.

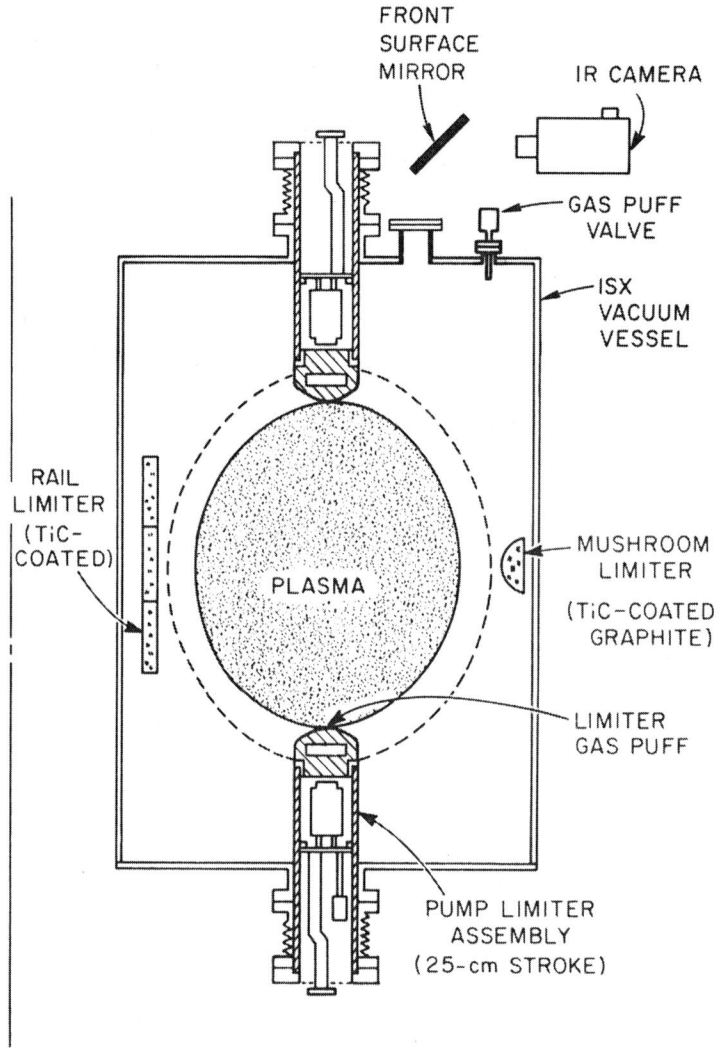

Fig. 12. Experimental arrangement of two pump limiter modules in
ISX-B [16].

916

(a)

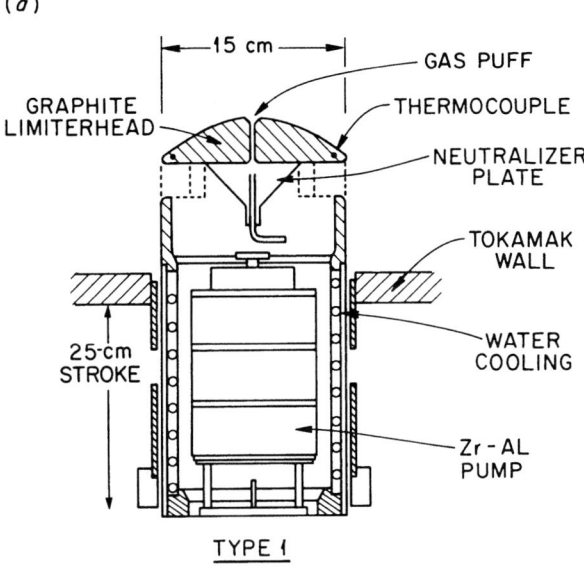

TYPE 1

(b)

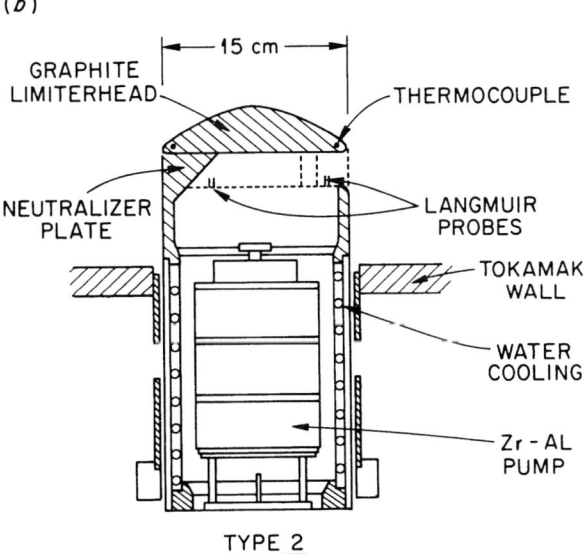

TYPE 2

Fig. 13. The two different designs of the ISX-B pump limiter modules.

917

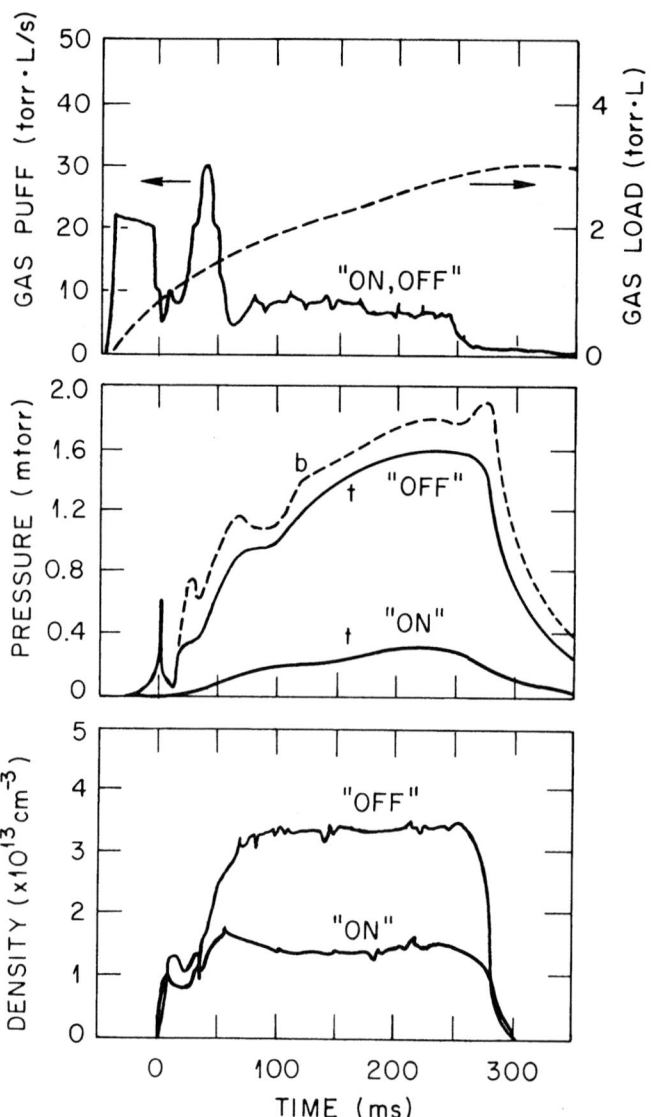

Fig. 14. Gas puff, pressure, and density data of two OH discharges: before activation of the limiter pumps ("OFF")
and after activation of the limiter pumps ("ON");
pressure data for top limiter (t) and bottom limiter
(b).

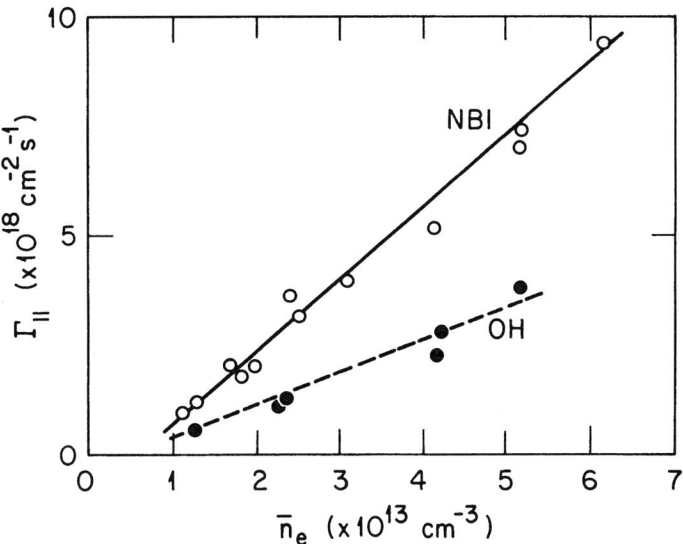

Fig. 15. Scrapeoff layer particle flux for OH and NBI discharges
measured with the Langmuir probe.

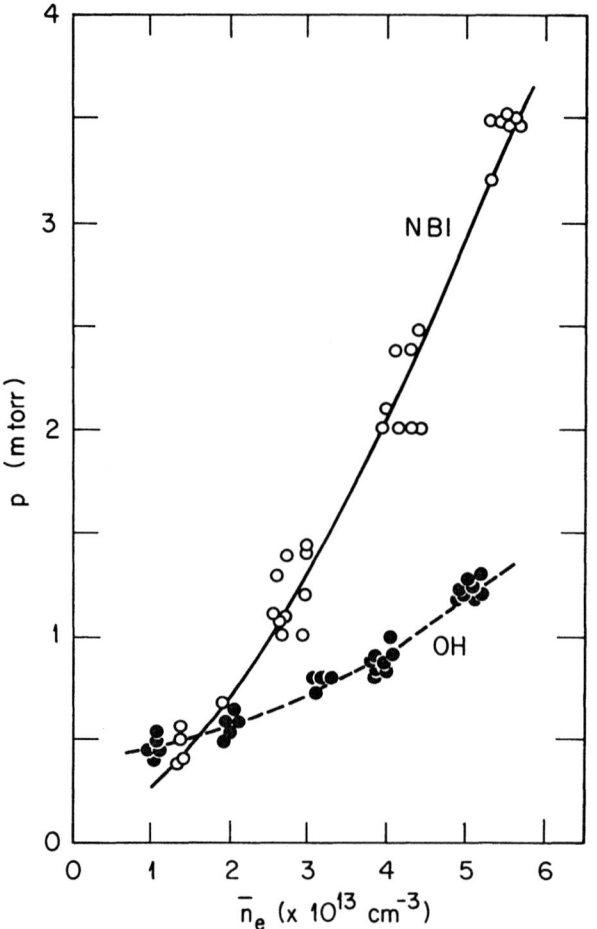

Fig. 16. Pump limiter pressure as a function of line-averaged
density for OH and NBI discharges.

The reduction in recycling due to beam heating diminished the effect of the pump limiter on the overall particle balance. For a given exhaust efficiency, the effect of the pump limiter on particle control became smaller with decreasing recycling, as demonstrated in Fig. 2.

It is interesting to compare the pressure response of the two different limiters (see Fig. 13) with one another. Figure 17 shows some raw data on plasma density and limiter pressures. Beam injection started at 100 ms and ended at 300 ms. While the short throat limiter came to an equilibrium very rapidly, the pressure in the other limiter kept going up until the end of the shot. Part of the higher pressure in the long-throat design was attributed to the difference in solid angle and orientation, but the continuous rise indicated plasma plugging effects. The maximum pressures achieved with neutral beam injection are plotted as a function of scrapeoff layer line density in Fig. 18. The getters were not activated in this case. The figure shows clearly that the pressure in the long-throat limiter rose faster than linear with density, indicating plasma effects. The mean free path for ionization in the limiter throat was 2 to 4 cm ($T_e \cong 10$ eV, $n_e \cong$ $2-4 \times 10^{12}$ cm^{-3}). A neutral born at the neutralizer plate moving back into the plasma chamber had an average path length of 10 cm and, hence, a high probability of being ionized and captured again. This process can lead to high local recycling and buildup of neutral pressure [34].

When the getter pumps were activated, the pressure in the cavity dropped by approximately a factor of 3. This depended on the ratio of the applied pumping speed to the conductance between limiter cavity and plasma vessel. With active getters, the limiter removed particles continuously from the plasma, and the external gas feed rate had to be increased to maintain a given density. This behavior is shown for ohmic discharges in Fig. 19, which shows the line-averaged plasma density as a function of the gas puff rate. The labels "ON" and "OFF" refer to whether the getters were active or not. For ohmic discharges the increased gas puff rate was consistent with the measured pressure. This was not the case for beam-heated discharges. Figure 20 shows the effect of the pump limiters on plasma density and gas feed rate for beam-heated plasmas. The increase in gas flow rate due to the pump limiters was typically a factor of 2 higher than can be accounted for by the measured pressure. The reason for this discrepancy is not quite clear: it could be enhanced pumping due to fast particles or a systematic problem with the pressure measurements. At any rate, the data show clearly that particle exhaust rates of the same order as the gas puffing rates can be achieved.

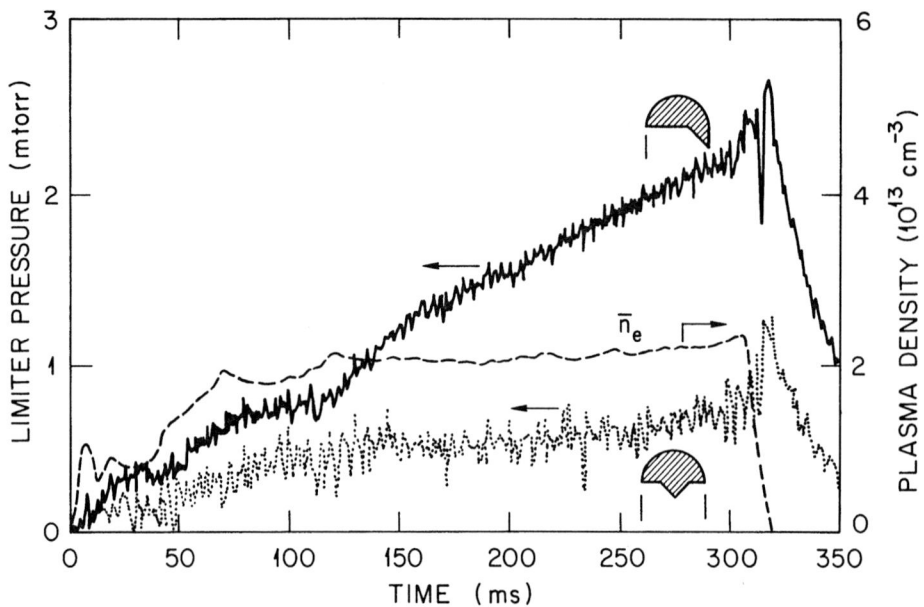

Fig. 17. Pressure buildup during the discharge for two different
designs of ISX-B pump limiters.

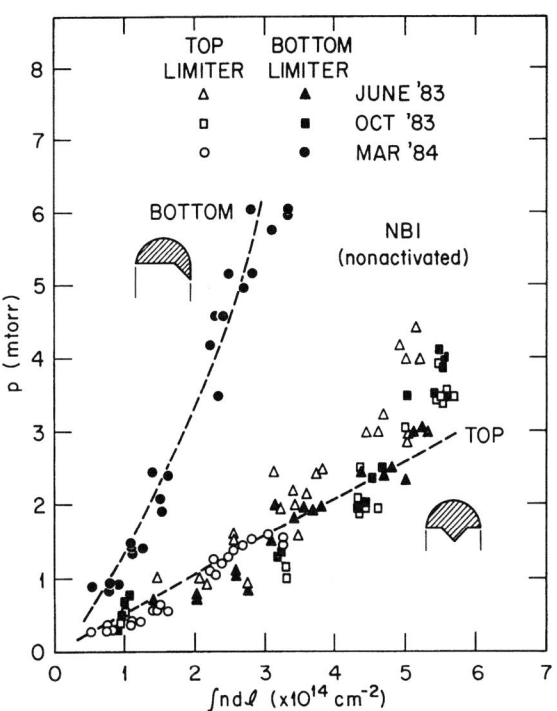

Fig. 18. Maximum pressures as a function of line-averaged
density (getters are not activated).

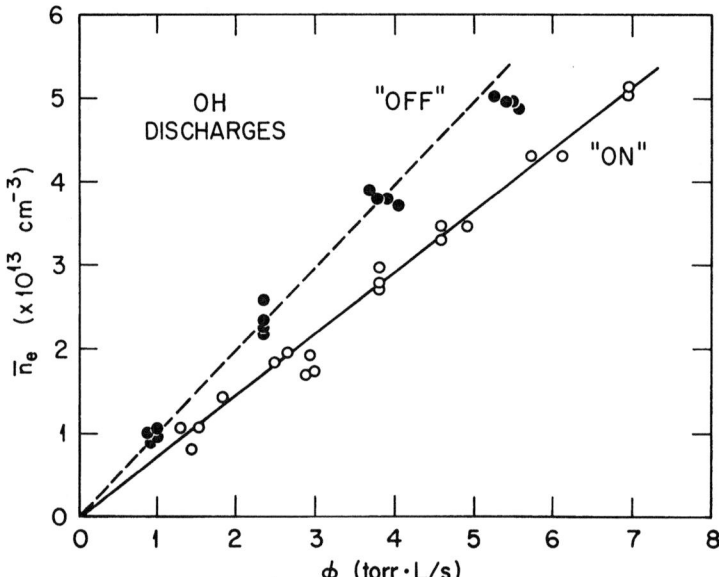

Fig. 19.   Line-averaged plasma density as a function of gas flow
rate at 250 ms into the OH discharge before and after
activation of the pumps.

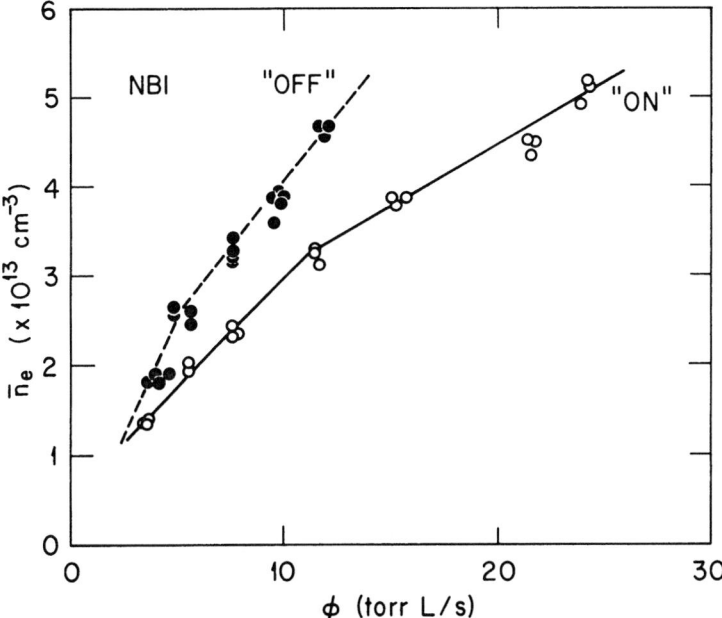

Fig. 20. Line-averaged plasma density as a function of gas flow
rate at 280 ms into the NBI discharge before and after
activation of the pumps.

We have focused our discussion of experimental pump limiter results on particle removal and density control. A problem of similar importance is that of impurity control. Up to now, an experimental data base is nonexistent in this important area. The only impurity experiment with pump limiters reported so far has been performed on ISX. Nitrogen was puffed into the plasma at 150 ms into the discharge, and the total radiation was monitored with the pump limiter getters "ON" or "OFF," respectively. Figure 21 shows the total radiation of two superimposed discharges: one with the pump limiter activated ("ON") and the other one with the pump limiter nonactivated ("OFF"). The first rise of the radiated power occurring at 100 ms was due to the neutral beam, and the sharp increase at approximately 150 ms was due to the nitrogen puff, which lasted about 30 ms. It is obvious from Fig. 21 that the pump limiter did not affect the impurity content immediately, since the initial radiation level was the same with or without pumping. This is to be expected. Even if all the nitrogen were ionized in the scrapeoff layer, only 5 to 10% could be removed in one cycle. We observed that the impurity content decayed faster due to the pump limiter action, indicating reduced recycling. These observations are just a beginning of impurity studies with pump limiters, and systematic investigations have to follow now in order to evaluate the impurity control capabilities of mechanical devices.

5. PROSPECTS FOR FUSION REACTOR APPLICATIONS

In fusion reactors, pump limiters or divertors have to fulfill the following tasks: heat removal, density control, helium ash removal, and impurity control. If mechanical pump limiters could meet these requirements, it would make complex fusion reactor designs simpler. With the present experience of pump limiter experiments, it appears that pump limiters can handle all of these tasks except for one: impurity control. Recycling impurities like light gases will probably be reduced by pump limiters, whereas impurities from the limiter itself are likely to penetrate into the plasma. If impurity control turns out not to be a problem in the large limiter tokamaks like TFTR and JET, pump limiters might be a viable solution. If, on the other hand, impurities present a problem for plasma operation, then — with our present knowledge — pump limiters might not be a sufficient solution to control plasma contamination. Before we are able to answer any of these questions definitively, we have to establish a sufficient data base on pump limiters.

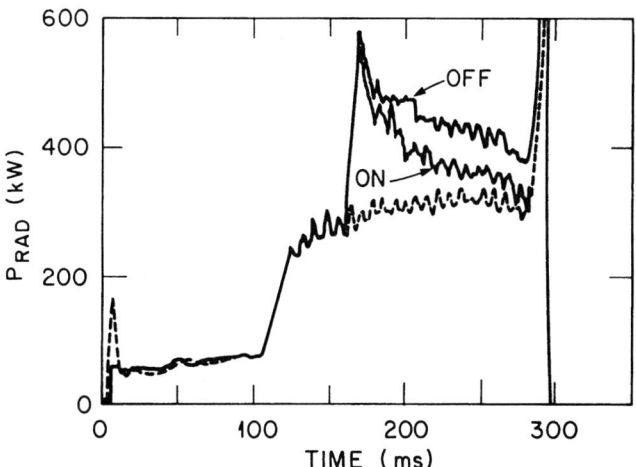

Fig. 21. Total radiation with nitrogen puffing at 150 ms into the
discharge: before ("OFF") and after ("ON") activation
of the pump limiter getters.

REFERENCES

[1] V. A. Vershkov and S. V. Mirnov, "The Role of Impurities in the Present Tokamak Experiments," PPPL Matt-Trans 113, 1973; Nucl. Fusion 14 (1974) 383.

[2] G. G. Kelley, "Modification of the Belt Limiter Concept to Provide for Purging of Impurities," ORMAK II/TTF Design Memo No. 17, Nov. 22, 1974.

[3] W. Bieger, K. H. Dippel, G. Fuchs, and G. Wolf, presented at the Symposium on Plasma-Wall Interaction, Jülich, 1976.

[4] J. F. Schivell, "Method of Plasma Impurity Control Without Magnetic Divertor," PPPL-1342, June 1977.

[5] R. W. Conn, I. N. Sviatoslavsky, and D. K. Sze, in Proc. 8th Symposium on Engineering Problems in Fusion Research, San Francisco, 1979.

[6] J. N. Brooks, C. C. Baker, H. C. Stevens, and C. A. Trachsel, in Proc. 8th Symposium on Engineering Problems in Fusion Research, San Francisco, 1979.

[7] C. A. Flanagan, D. Steiner, G. E. Smith, eds., "Fusion Engineering Device Design Description," ORNL/TM-7948, December 1981.

[8] U.S. Contribution to the Impurity Control Physics, INTOR Workshop, Phase 2A, Session IV, March 1982.

[9] H. Howe, "Physics Considerations for the FED Limiter," ORNL/TM-7803, June 1981.

[10] G. M. Fuller, coordinator, presented at the FED First Wall/Limiter Workshop, Oak Ridge, April 30, 1981.

[11] J. R. Haines, B. A. Cramer, J. P. Davisson, and H. C. Mantz, J. Nucl. Mater. 103 (1981) 22.

[12] D. O. Overskei, Phys. Rev. Lett. 46 (1981) 177.

[13] S. Talmadge et al., Nucl. Fusion 22 (1982) 1369.

[14] R. Jacobsen, Nucl. Fusion 22 (1982) 277.

[15] P. Mioduszewski et al., J. Vac. Sci. Technol. 20 (1982) 1284.

[16] P. Mioduszewski et al., J. Nucl. Mater. 121 (1984) 285.

[17] R. Budny et al., J. Nucl. Mater. 121 (1984) 285.

[18] S. A. Cohen et al., presented at the 6th International Conference on Plasma Surface Interactions in Controlled Fusion Devices, Nagoya, 1984 (to be published in J. Nucl. Mater.).

[19] R. W. Conn et al., in Proc. 10th Symposium on Fusion Technology (SOFT), Vol. 1, Pergamon Press, 1983, 497.

[20] F. Karger, K. Lackner, Phys. Lett. 61A (1977) 385.

[21] W. Feneberg, G. Wolf, Nucl. Fusion 21 (1981) 669.

[22] W. Engelhardt, W. Feneberg, J. Nucl. Mater. 76 & 77 (1978) 518.

[23] A. V. Nedospasov, M. Z. Tokar, J. Nucl. Mater. 93 & 94 (1980) 248.

[24] J. S. DeGrassie et al., presented at the 6th International Conference on Plasma Surface Interactions in Controlled Fusion Devices, Nagoya, 1984 (to be published in J. Nucl. Mater).

[25] G. H. Wolf, presented at the 3d Topical Meeting on Fusion Reactor Materials, Albuquerque, 1983 (to be published in J. Nucl. Mater.).

[26] W. Eckstein and H. Verbeek, J. Nucl. Mater. 76 & 77 (1978) 365.

[27] O. S. Oen and M. T. Robinson, J. Nucl. Mater. 76 & 77 (1978) 370.

[28] R. W. Conn, UCLA ENG 8410, University of California, Los Angeles, 1984.

[29] J. A. Schmidt, TFTR Physics Report No. 12, Princeton Plasma Physics Laboratory, 1979.

[30] P. Mioduszewski, Jül-1681, KFA Jülich, 1980.

[31] R. W. Conn et al., J. Nucl. Mater. 121 (in press).

[32] W. Engelhardt et al., J. Nucl. Mater. 111 & 112 (1982) 337.

[33] D. K. Owens, Bull. Am. Phys. Soc. 25(8) (1980) 941.

[34] M. Petravic, D. Post, D. Heifetz, and J. Schmidt, Phys. Rev. Lett. 48 (1982) 326.

[35] R. W. Conn, presented at the 6th International Conference on Plasma Surface Interactions in Controlled Fusion Devices, Nagoya, 1984 (to be published in J. Nucl. Mater.).

DIVERTOR TOKAMAK EXPERIMENTS

F.Wagner and K.Lackner

Max-Planck-Institut für Plasmaphysik
EURATOM Association, D-8046 Garching

INTRODUCTION

A future tokamak reactor has to fulfill several requirements: good confinement properties to achieve ignition at low external heating power; high stability at minimal toroidal magnetic field strength; and efficient helium removal from the system for continuous burn. These conditions have to be met under circumstances, where the material, surrounding the burning plasma, is loaded to its limits. During steady state burn the first wall will be exposed to neutron fluxes of about 3 MW/m^2. Additionally 600 MW is deposited into the plasma by the fusion α-particles, conducted to the plasma boundary and has to be removed from there at a tolerable power density. The wall is further exposed to a He-flux of $1.2 \cdot 10^{21}$ s^{-1} and a deuterium and tritium flux which is about a factor of 20 larger. An implication of these figures is that several tens of tons of wall material are eroded and re-deposited during one year of reactor operation /1/.

A large fraction of present day tokamak research concentrates on the question how to specifically tailor the boundary layer of the plasma to successfully transfer these fluxes to the wall and to shield the plasma core from impurities entering it from the outside. The most successful scheme devised up to now has been the divertor concept. The original idea was to surround the main plasma by a scrape-off layer which is interspersed by magentic field lines which do not map out closed surfaces like those within the plasma. The open field lines of the scrape-off layer are guided into a chamber separated from the main plasma chamber and intercepted there by target plates. A schematic of the divertor topology is shown in Fig. 1.

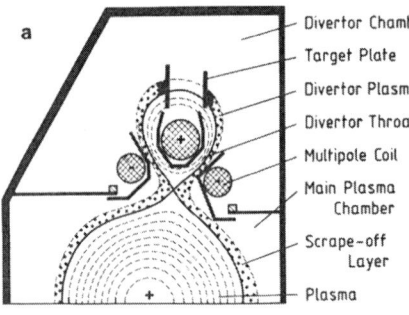

Fig. 1a.

Cross-section of a divertor experiment introducing divertor related installations and terms.

Labels: Divertor Chamber, Target Plate, Divertor Plasma, Divertor Throat, Multipole Coil, Main Plasma Chamber, Scrape-off Layer, Plasma

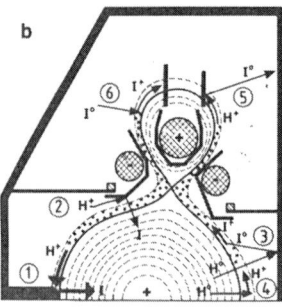

Fig. 1b.
Divertor configuration with an additional limiter to exemplify the location of enhanced impurity erosion and possible divertor effects.

1. Intense plasma wall interaction at the limiter tip. Impurities are eroded by sputtering, arcing, melting etc. and can directly penetrate the main plasma.

2. The corresponding plasma interaction with wall parts surrounding the main plasma occurs far from the plasma surface with cold and low-density plasma.

3. Impurity release by charge exchange sputtering. The impurity atom is ionized in the scrape-off layer and swept into the divertor chamber. This process gives rise to the shielding characteristic of the scrape-off layer.

4. A charge exchange atom is ionized in the scrape-off layer. This process may occur in high density discharges not yet realized at present.

5. An impurity atom released from the target plate sticks to the divertor wall and does not penetrate the main plasma.

6. An impurity atom is ionized in the divertor plasma, swept back to the target plate and effectively retained within the divertor chamber.

The magnetic field line which just separates the closed field lines within the plasma from those within the scrape-off layer - the so-called separatrix - maps out a surface which defines the plasma

932

cross-section. Such a magnetic field configuration is established by the superposition of the magnetic field of the plasma current with that of additional skillfully placed external coils. Particle and energy fluxes from the main plasma are channelled along the scrape-off layer into the divertor chamber. Nearly 100 % of the particles flow into the divertor chamber yielding the high exhaust efficiency of this configuration. Unlike the conventional limiter configuration, the plasma wall interaction does not occur at the immediate plasma surface but at a remote location - at the target plate. Out of geometrical reasons the impurities released from there have a lower probability to penetrate the main plasma (see Fig. 1). Because of the separation of main plasma periphery and divertor plasma in front of the target plate, an effective decoupling of the respective plasma parameters is intended with the aim to cool down the scrape-off plasma below the relevant threshold energies for impurity release.

Those impurities which are nevertheless produced at the wall, encompassing the main plasma (either by insufficient exhaust or by wall bombardement by charge exchange atoms) are ionized within the scrape-off layer and swept into the divertor chamber along with the background plasma. At the time, the present large divertor machines were designed, the divertor plasma was expected to have a high screening efficiency against impurities.

Along with the hydrogen plasma constituents, also α-particles which have slowed down within the plasma and transferred their fusion energy to it are transported into the divertor chamber and accumulate there to be easily pumped and removed out of the system. The divertor concept was expected to offer additionally the technical means for successful He-ash removal and fusion species control.

The objectives of this contribution is to give a survey of the experimental results of present-day divertor tokamaks. The results, presented here, are mostly based on the investigations carried out on the large divertor tokamaks ASDEX, DIII and PDX. These machines started to operate at the end of the seventies so that we can survey a material accumulated over the last five years. In this time span, the divertor tokamaks have been the largest operational tokamaks (minor radius ~0.5 m, major radius ~1.5 m) clearly pointing towards the urgency of the above summarized problems. In agreement with the research programme on these tokamaks, a large fraction of this paper will deal with the impurity aspects. But we will also summarize the magnetic design and stability problems of a plasma, bounded by a magnetic surface, deal with the transport phenomena in the scrape-off layer and discuss differences in the confinement properties of limiter and divertor tokamaks. The last topic will mostly be based on material from neutral beam heated experiments to which divertors had important contributions. Finally, we will shortly present two future divertor experiments under design or construction now.

The manuscript is written in the spirit to search for the best experimental evidence for documenting the various divertor aspects. In those cases where we have (innocently) violated this principle in favour of results from the ASDEX divertor tokamak it happened because both authors belong to its team and are most familiar with this device and its results.

1. MAGNETIC DIVERTOR CONFIGURATION

Displacement of the interaction between the energy carrying part of the charged particle flux and material walls to a certain distance from the bulk plasma requires the diversion of magnetic field lines by externally applied fields. For a topological separation between field lines on closed surfaces and diverted ones, additionally applied divertor fields have to form a stagnation axis, along which the two perpendicular field components vanish. In the only divertor configuration possessing a true symmetry – the axisymmetric divertor – this is done by externally applied poloidal fields, which cancel the fields produced by the plasma currents in one point of the intersection between plasma surface and a meridional plane /2/. At this location, the divertor fields have to be approximately of the magnitude $\mu_o I_p/(2\pi\, a_s)$, with a_s the distance plasma center-stagnation point (Fig. 2, 2a).

An even more effective use of symmetry properties would be possible in an infinite aspect-ratio tokamak with (necessarily concentric) circular flux surfaces. On rational surfaces both field components perpendicular to a closed helical curve vanish, so that arbitrarily small additional fields of proper symmetry suffice to produce a helical stagnation line /3/. To divert however field lines over a radial distance Δ (i.e. to produce magnetic islands with this half width) helical divertor fields with an amplitude

$$b_{r,o} = \frac{\mu_o\, I_p}{2\pi\, a_s} \cdot \frac{\Delta^2}{a_s^2} \cdot \frac{m}{2}$$

of the radial component are needed (m is the poloidal mode number, a multiple of q at the boundary). Thus a significant advantage compared to the poloidal field divertor would be achieved only for Δ substantially smaller than a_s, and probably partly offset by the more complex topology of the field coils. In particular however, the nicely symmetric picture of Fig. 2b is destroyed by finite toroidal curvature and noncircular cross-sections, which will cause the pitch of the resonant field line to become nonuniform, the q-value at a given radius to vary (for constant I_p) with plasma pressure, and will finally break up the structure of helical flux

934

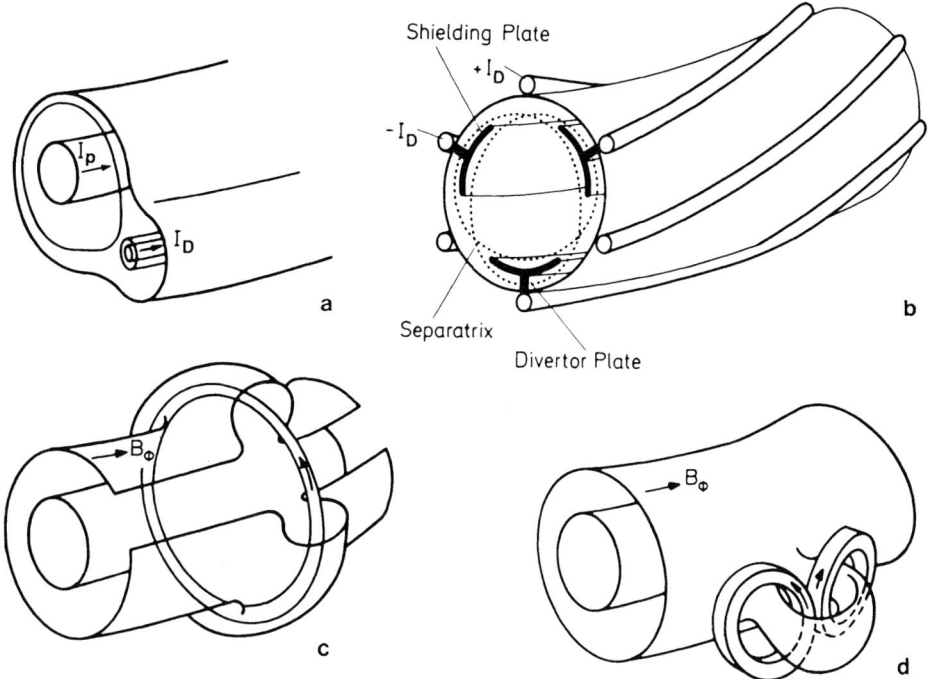

Fig. 2.   Schematic view of magnetic divertor configurations.

    (a)   Poloidal (axisymmetric) divertor
    (b)   Helical divertor
    (c)   Toroidal divertor
    (d)   Bundle divertor

Figures 2a, 2c, 2d are from ref. /7/, 2b from Ref. /3/.

surfaces by ergodization (a kind of random walk of field lines in radial direction, crossing the "unperturbed" flux surfaces). In a modification of this concept, the resulting ergodization has indeed been proposed /4/ (and is currently tested on TEXT /5/) as an alternative solution to the boundary engineering problem, by leading to enhanced diffusion or radial outward drift of particles in an edge layer, to result in reduced hydrogen and impurity confinement times. This concept, however, is not anymore a divertor - the particle and energy fluxes are not channelled away to regions distant from the bulk plasma - and therefore not subject to this lecture.

Other considered divertor schemes do not make use - and thereby conserve - symmetry properties of the tokamak. In their case, the stagnation axis runs approximately perpendicular to the toroidal direction, and divertor fields have to be of the order $\sim B_t$, i.e.

A · q_a times larger than in the poloidal divertor case. For this reason, and in order to keep the magnetic field perturbation localized, coils have to be placed close to the plasma, which in a reactor precludes the use of superconducting coils. The first divertor of this type - and in fact the first of any kind - was the toroidal divertor (schematically shown in Fig. 2c) realized on the C stellarator /6/. To apply it to a tokamak, a substantial change in its basic geometry - e.g. introduction of straight segments into the doughnut - would have to be made, and the proposal, with all physics ramifications, has never been considered very seriously.

A more localized version of such a divertor, and in fact the only alternative to the poloidal divertor in tokamaks so far experimentally realized, is the bundle divertor /7/ (Fig. 2d). It consists of a set of ring-shaped coils, situated close to each other and to the plasma surface, extracting a flux bundle and guiding it into a target chamber. Field lines of the extracted flux bundle, projected over several toroidal transits map out an approximately annular region, whose inner boundary is the last closed flux surface, the separatrix. Principle advantage is its localized geometry, which could allow relative simple insertion between two toroidal field coils. This is however counterbalanced by the need to bring these coils very close to the plasma, and by the large forces exerted on the coils, which interact with the strong main field. On the physics side, relinquishing axisymmetry introduces additional particle drift motions, which particularly at high temperatures or for suprathermal particles could lead to intolerable losses. For both of these reasons, the bundle divertor is currently considered only a rather low-probability reactor option. Irrespective of this, however, the bundle divertor as an experimental device differs so significantly from the poloidal divertor, that its results should correspond to valuable data points for our modelling studies. These differences are essentially twofold:
(1) the total magnetic field strength along the extracted flux bundle varies strongly, leading to a true expansion of the divertor fan towards the target plates, and the possibility of the acceleration of the plasma to Mach numbers $M > 1$ (in contrast the apparent expansion of the divertor fan in poloidal divertors is only a projection effect: the field strength in a flux bundle, and therefore the variation in its cross-section area under flux conservation is overwhelmingly determined by the toroidal field, which varies only $\sim 1/R$). (2) The total cross-sectional area of the plasma fan - which under DITE neutral injection transports an energy flux only a factor ≈ 2 smaller than in e.g. ASDEX - is nearly two orders of magnitude smaller than in poloidal divertor experiments, with expected (and demonstrated) strong consequences for the tolerable hydrogen pressure and the kinetic regimes in the divertor chamber /8/.

In the following more detailed discussion of magnetic problems we will, however, restrict ourselves to the poloidal divertor, for which a range of configurations have been realized already in

936

experiments, a set of next-generation devices is in construction, and rather detailed reactor design studies have been carried out.

2. POLOIDAL DIVERTORS

2.1 Present Experiments

Placing poloidal field (PF) coils inside the toroidal ones and very close to the plasma boundary allows to produce, with restricted effort, a wide variety of field configurations. It has become customary to distinguish between closed and open poloidal divertors, where a working definition is that the target plates are hidden behind the main divertor coil in the former case, but in direct line of sight of the main plasma in the latter one.

Configurations of mid-size divertor tokamaks of the present generation (including, for comparison, also the bundle divertor of DITE) are shown in Fig. 3. For the poloidal divertor, the region comprising the magnetic flux passing at the torus outside in the midplane through a ring of 1.5 cm thickness is shaded, to indicate the shape of the scrape-off plasma fan. ASDEX /9/ and PDX /10/ fall into the category of magnetically closed divertors. To raise experimental flexibility, PDX was originally designed with four divertors, but in later stages operated exclusively with only the inner coil sets activated. ASDEX was designed so as to minimize configurational differences to ordinary circular tokamaks: the effect of the two divertor coil sets is therefore highly localized, and main plasma flux surfaces remain circular up to close to the boundary. A closely spaced coil triplet moreover allows to constrict the plasma fan in the divertor throat, allowing to narrow it and reduce its neutral conductance. The external fields producing a localized divertor are rapidly increasing away from the plasma boundary, so that the stagnation point moves very little when the plasma column is displaced. This feature was extensively exploited in PDX and ASDEX for comparative double- and single-null configuration studies, whereby small radial fields, displacing the plasma column closer to one of the divertors, were used to bring the two stagnation points onto different flux surfaces (see PDX configuration shown in Fig. 3a).

Not designed a priori for divertor operation, but equipped with an extensive set of poloidal field coils allowing flexible feeding, Doublet III was used to realize two open divertor configurations shown in Figs. 3c and 3d, termed SN /11/ and XB /12/, respectively. Distinguishing feature of both is the strong expansion of the scrape-off fan in the vicinity of the wall segments acting as target plates (as mentioned before this expansion in the poloidal cross-section does not imply a net increase of the cross-section area of the single flux tubes but only a deformation, increasing their diameter in the direction perpendicular to but decreasing that parallel to the flux

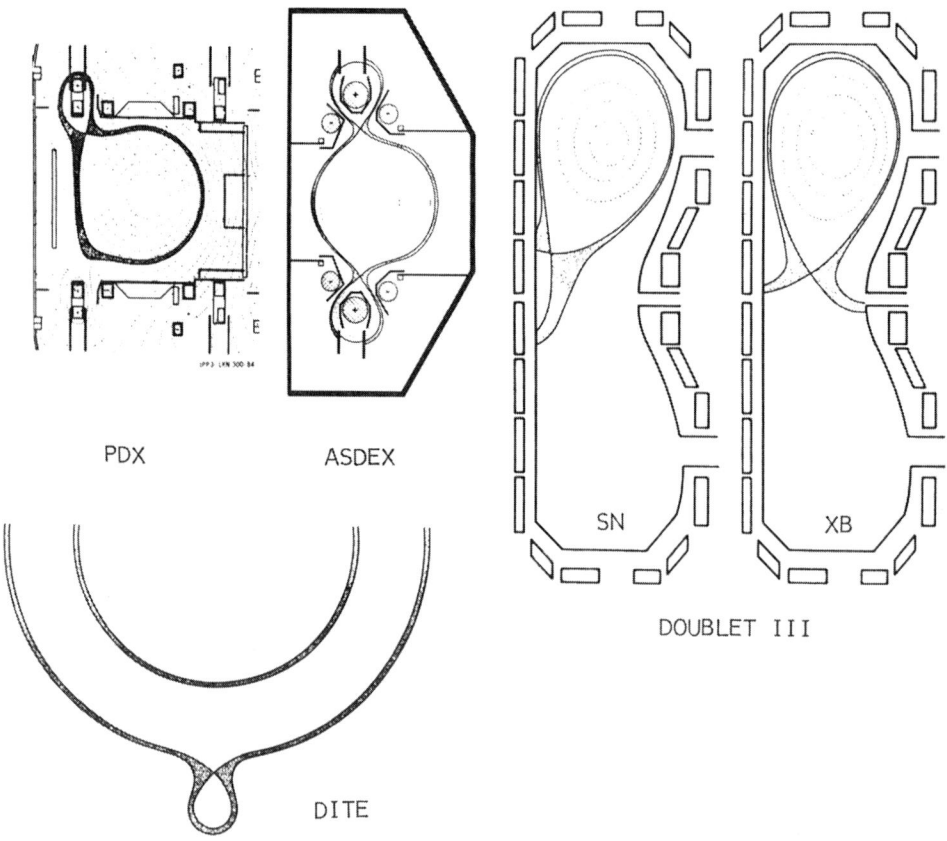

PDX   ASDEX

SN   XB

DOUBLET III

DITE

Fig. 3. Configurations of present-generation, mid-size divertor
tokamaks. Axisymmetric configurations are shown in a
poloidal cross-section, the bundle divertor in the intersection with the toroidal plane. The shaded area gives an
approximate picture of the shape and relative dimensions
of the scrape-off layer.

surfaces). This feature is deemed responsible for their success in
reaching the high recycling regime (see Sec. 4.3) even in the absence of a materially separated divertor chamber.

2.2 Poloidal Divertors in a Reactor

Because of the necessity to shield superconducting coils from
radiation and for assembly/repair reasons it is now clear that PF
coils have to be placed fairly distant from the main plasma and most
probably outside the toroidal field coils of a reactor. This evi-

dently precludes the use of a closed divertor and places also narrow limits on the possible open configurations.

For a principle discussion it is useful to think of the externally applied fields in multipole components. In the stagnation point, the divertor fields have to cancel those of the plasma currents. Omitting for the moment toroidal effects and assuming external conductors arranged at a circular surface with radius a_w, this requires divertor currents whose sum of moduli varies like

$$\Sigma_M |I_M| \sim I_p (\frac{a_w}{a_s})^m$$

with plasma cross-section, conductor distance and multipole order ($m = 2$ for a quadru-, $= 3$ for hexapole etc.). Excluding dipole fields ($m = 1$), whose magnitude is fixed by the equilibrium constraints (and which only for $\beta_p \approx$ aspect ratio become of the right order of magnitude to produce a stagnation point at the plasma boundary) the most economic divertor can be produced by quadrupole fields. Moreover this scaling shows that the penalty for using higher order multipole components rises with a_w/a_s, so that it has become fully appreciated only during the design of INTOR and of other tokamaks intentionally designed under reactor-like topological constraints (ASDEX-Upgrade) /13/.

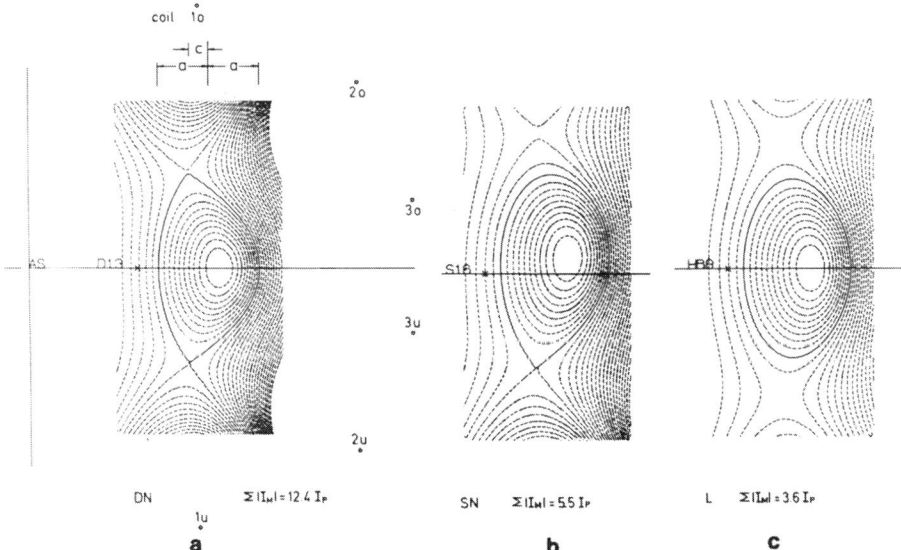

Fig. 4. Plasma cross-sections and required shaping currents (measured by the sum of their absolute values $\Sigma/I_M/$ as multiples of the plasma current I_p) of possible boundary configurations in ASDEX Upgrade: 4a double null divertor, 4b single null divertor, 4c limiter case. Also shown schematically in Fig. 4a are the locations of the active poloidal field shaping coils, which are all placed outside the main field coils.

Magnetic fields of lower multipole order produce at the same time global changes in the plasma configuration, with quadrupole moments (of the orientation considered) leading to vertically elongated and additional hexapole moments to D-shaped plasma columns. Such deformations are however considered essential anyway to realize reactor-relevant ß-values, so that the additional poloidal field effort involved in a divertor is smaller than originally anticipated. A second important development was the experimental proof that the considerable technical advantages of a single null /14/ vs. a double null divertor do not carry penalty in physics performances (see Sec. 9). The more effective vessel-volume utilization of the former is evident from the comparison of Figs. 4a and 4b, showing two possible operating modes of ASDEX Upgrade. At the same time, the values of Σ/I_M - a meaningful figure of merit in the light of future superconducting designs - quoted in the figure caption show a relaxation in the PF effort by more than a factor of 2, which comes from 3 contributions: (1) in order to use the possible midplane diameter - and thereby limit the volume waste - a triangular deformation produced by expensive hexapole fields has to be applied in double null. (2) The top conductors are at a reduced absolute distance from the center of the plasma currents, and (3), even for the bottom conductors for which the absolute magnitude of a_w has grown, the value of a_s to be used in the above expression grows even stronger. Figure 4c shows a pump limiter configuration, also possible in ASDEX Upgrade with the same shape of the inner flux surfaces. The saving in Σ/I_M as compared to the single null divertor configuration is a further 35 %, a result similar to the one obtained in detailed comparative studies for INTOR /15/. These differences are substantial enough to give preference to a pump limiter in a reactor, if it can perform the task, but certainly no knock-out for the poloidal divertor. The unexpected small differences are of course due to the D-shaped plasma cross-sections also deemed necessary in a limiter: to hold in equilibrium a purely circular plasma with coils at similar distance, only about 15 % of the Σ/I_M needed in the D-shaped limiter case would be required. The surprise in the latter number comes as tokamak designs with close poloidal field coils (e.g. JET) give a qualitatively wrong impression about the relative effort involved in producing dipole, quadrupole and hexapole field components under reactor conditions.

2.3 MHD Stability Behaviour of Axisymmetric Divertor Configurations

Plasma with vertically elongated cross-section - both with and without divertor - are subject to an axisymmetric, vertical displacement instability, which if unchecked would grow on an Alfven time scale (typically ~1 µs) /16/. The net displacement of the plasma currents tends however to induce stabilizing currents in the vessel walls or in purposely designed passive conductors (which can be saddle coils with the effective limbs being placed one above and one below the plasma column). If these stabilizing elements are

940

sufficiently close, the motion on the fast time-scale will be suppressed, and only an instability with a time scale of the order of the resistive decay time for currents in the passive stabilizers will persist. This residual instability – whose growth time can typically be brought into the 10 – 100 ms range – has then to be controlled by an active feedback system. This instability is theoretically rather well understood (it is a simple effect, which would be also experienced by a rigid, current-carrying wire substituting the plasma currents) and has been observed and controlled in a number of experiments. Nevertheless, its actual stabilization under reactor conditions is a serious technical problem, as passive conducting elements have to be placed quite close to the plasma boundary to be effective (Fig. 5), and active feedback by poloidal field coils outside the toroidal ones requires substantial powers. However this problem is common to both limiter and divertor configurations.

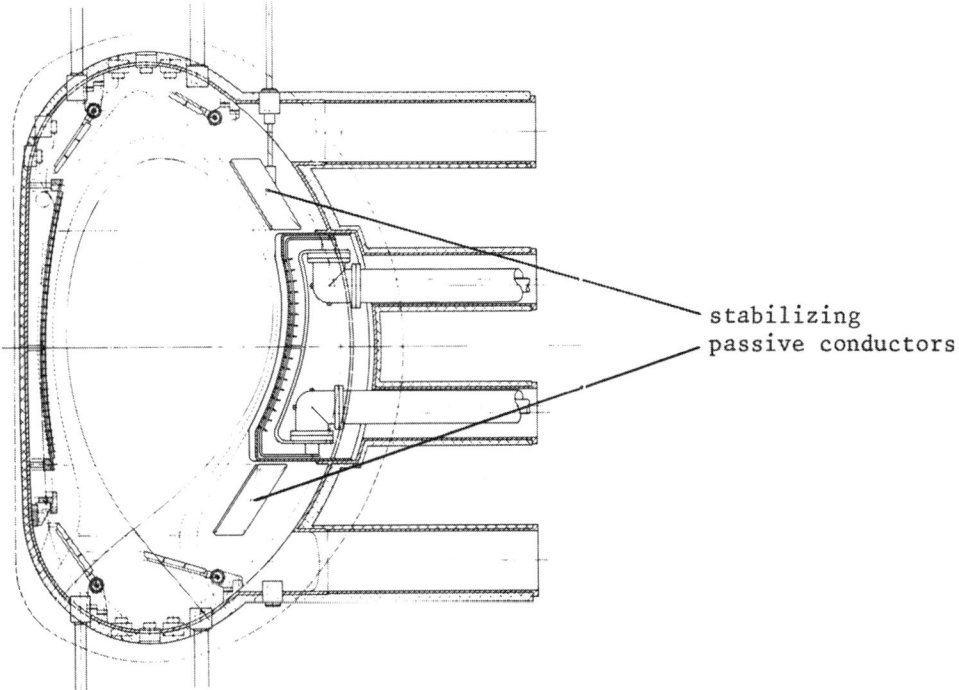

stabilizing
passive conductors

Fig. 5.   Cross-sectional view of ASDEX Upgrade showing the different structures inside the vacuum vessel, in particular the stabilizing passive conductors.

The question has been raised, whether the flux surface shape in the stagnation point vicinity causes additional, divertor-specific axisymmetric instabilities. Experimental evidence is that if the displacement instability is checked by either a field design with circular flux surfaces in the bulk plasma region (ASDEX) or appropriate passive and active stabilization of elongated plasmas (PDX, DOUBLET III), no additional axisymmetric modes are observed. An earlier theoretical analysis using completely flat current distributions had predicted the growth of such higher structured modes /17/; they were later however shown to be strongly stabilized by more realistic peaked current density profiles /18/. Together with the experimental evidence, the conclusion is allowed that a divertor tokamak should behave with respect to axisymmetric modes like a limiter tokamak with similar shape of the interior flux surfaces. This is also intuitively credible, as the axisymmetric modes are driven by forces exerted by the external fields onto the plasma currents. Nearly the whole plasma current flows however in those regions where the flux surface shape and the fields of external conductors are identical in both cases. A possible exception to this might be the case of the regime with good confinement of neutral injection heated plasmas (H-mode, see section 10), where the measured temperature profiles correspond to relatively high current densities in the separatrix vicinity, and where therefore higher order axisymmetric modes - localized near the plasma boundary and specifically in the stagnation point vicinity - can perhaps not be ruled out as explanation of the MHD-activity (so-called ELMs) intermittently observed in such discharges.

No analysis has so far been performed regarding the effect of the specific divertor plasma configuration on non-axisymmetric MHD modes, in particular ideal ballooning and external kink modes. This is simply due to the singularity introduced into the problem by the vanishing of the poloidal field at one point of the plasma boundary. As a consequence the safety factor q, defined for each poloidal flux surface Ψ in an axisymmetric configuration by

$$q(\Psi) = B_t \cdot R \frac{1}{2\pi} \oint \frac{ds}{R^2 B_p}$$

over the intersection with a poloidal plane, diverges at the separatrix. In terms of an effective minor radius $\tilde{r}$ (defined proportional to the square root of the cross-sectional area), this divergence is $\sim \ln(\tilde{r}_s - \tilde{r})$, i.e. significant only in a very localized neighbourhood of the separatrix. The same holds for the shear parameter $\tilde{r}/q \, dq/d\tilde{r}$ which consequently behaves like $\tilde{r}_s/(\ln(\tilde{r}_s - \tilde{r}) \cdot (\tilde{r}_s - \tilde{r}))$. In the usual theory of ballooning modes shear increases the locally possible stable radial pressure gradient according to $(dp/d\tilde{r})_{max} \sim B_t^2 \, \tilde{r}^2/q^2 \cdot \tilde{r}/q \, dq/d\tilde{r}$, but it is not clear to what extent this simple formula can also be applied to a situation where the whole shear is produced in the proximity of a single point in the poloidal cross-section. Large pressure gradients have however been observed experimentally in a narrow zone around the separatrix which strongly exceed those theoretically permitted by ballooning modes in a simple configuration with circular flux surfaces and indeed might suggest a positive effect of the divertor fields on such modes.

942

3. GROSS PLASMA PROPERTIES WITH MAGNETIC LIMITER

Before starting detailed physics experiments with a magnetic limiter configuration more technical aspects related to machine operation such as initiation of a divertor discharge, current built-up and horizontal and vertical stability had to be investigated. It was demonstrated /9,10/ that a divertor discharge can easily be started and built-up in a preprogrammed fashion if sufficient care was taken to cancel all magnetic stray fields (e.g. from the divertor coils) in the plasma center. Under these circumstances, breakdown occurs at low loop voltage of 5V /19/ so that one can dispense with complicated breakdown hardware and additionally save magnetic flux for the subsequent steady-state phase. Figure 6 shows a divertor discharge of 8 sec duration. Together with the plasma current, horizontal and vertical positions are plotted. Apart from initial and final phases, the plasma is well centered.

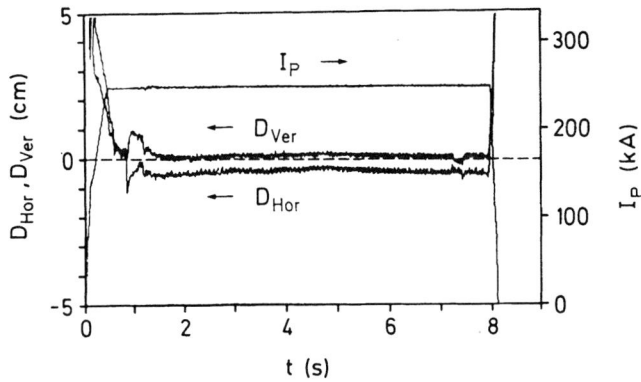

Fig. 6. Plasma current I_p, horizontal (D_{hor}) and vertical (D_{ver}) plasma position during an ASDEX divertor discharge of 8 sec duration.

As described above vertical stability is critical at elongated plasma cross-section. Stable elliptical divertor plasmas with an axis ratio b/a = 1.8 have been produced /20/. Intended vertical displacements of the plasma torus by a radial magnetic field causes the transition from the double-null (in case of a poloidal divertor) to a single-null divertor configuration. These two configurations are compared in Fig. 7. The single null plasma formed in this way is encompassed by two separatrices.

As the divertor plasma is not in touch with a material obstacle in the main plasma chamber, it has the flexibility to move horizontally. This is of advantage in cases of unintended or unavoidable disturbance of the plasma equilibrium. Such a disturbance

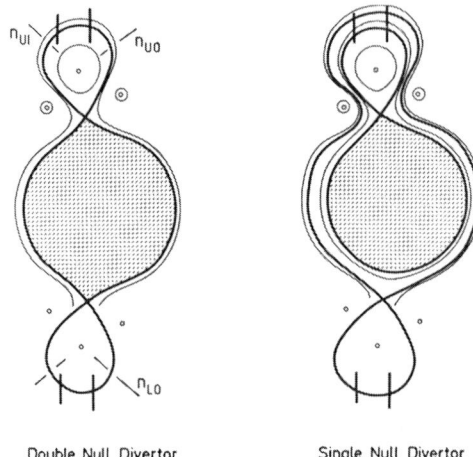

Double Null Divertor   Single Null Divertor

Fig. 7. Comparison of the single-null and double-null configuration
         of ASDEX.

occurs in case of a major disruption when the plasma looses its confinement, is rapidly moved inward and quenches its current. It is
of importance to slow down this process to give the external control
systems time to counteract. It is advantageous when the plasma does
not intercept the limiter right away which might otherwise accelerate
the disruptive process by enhanced radiative cooling.

Another rapid change in plasma equilibrium occurs when neutral
beams are injected into the discharge to auxiliary heat it. The
orbiting beam ions immediately contribute to the plasma pressure so
that the plasma moves outward by a few cm before the control system
can restore the original equilibrium. The plasma movement can induce
enhanced limiter erosion and thereby cause unfavorable initial conditions for a beam heated discharge.

Experiments further show that the separatrix gives rise to a well
defined plasma surface. The extension of the scrape-off layer is
only 1 - 2 cm. Figure 8 shows a photograph viewing the main and
(from a different viewing point) the divertor plasma. The picture
documents a sharp plasma boundary following the envelope of the inner
separatrix. (The outer separatrix is not seen because of low brightness
due to geometry.) The scrape-off layer channelled into the 4 divertor necks is clearly shown. Inside the divertor chamber, the plasma
fan also follows the separatrix and forms a semi-arc. The divertor
plasma is terminated at the target plate.

944

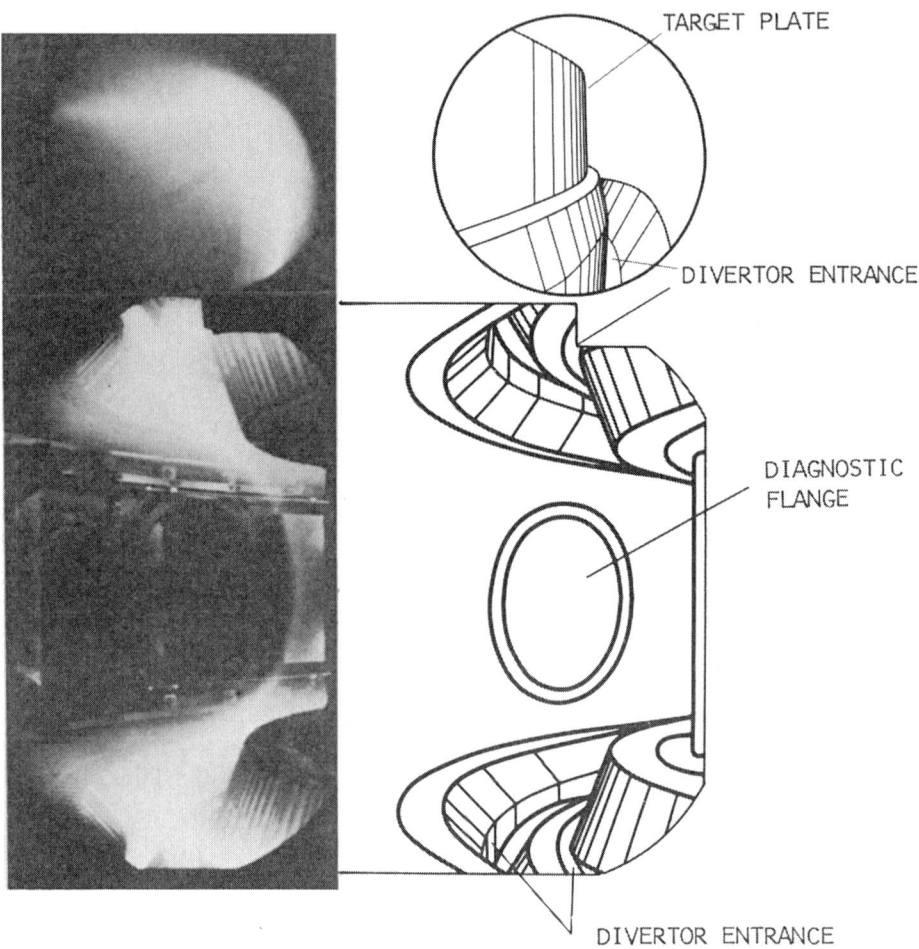

Fig. 8. Photographs taken during ASDEX discharges in the main
plasma and (from a different viewing point) the divertor
chamber (courtesy H. Niedermeyer).

Good tokamak performance is only possible at extremely clean
wall conditions. After a system had been vented (e.g. for maintenance) a sequence of cleaning procedures has to follow before reproducible tokamak discharges are possible again. A divertor

15

machine is quickly brought back to operation giving rise to high
machine availability.

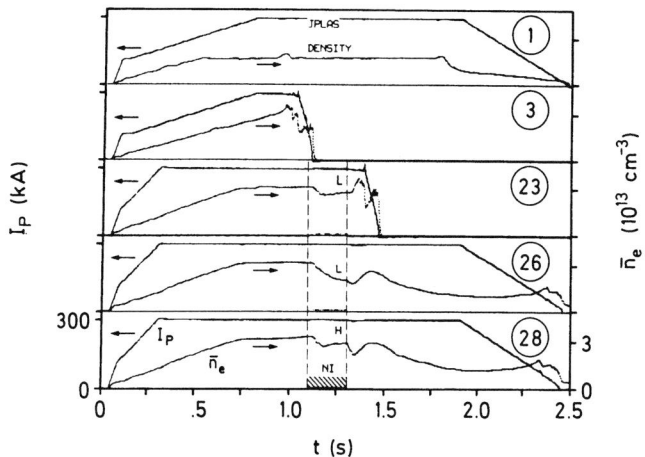

Fig. 9.   Sequence of ASDEX discharges after the vessel had been
          vented for 1 month starting with shot #1.
          Shot #3: Test of density limit; shot #23: with neutral in-
          jection (NI) into hydrogen yielding low confinement
          shot #26: after change of working gas to deuterium with
          NI yielding low confinement; shot #28: The NI-pulse yields
          high confinement (see Sec. 10).

    Figure 9 shows a sequence of discharges after the ASDEX vessel
had been vented and open for 1 month. During this maintenance phase
parts of the inner vessel wall had to be changed and replaced which
required technical personnel to work inside the vessel for several
days. Before the first shot, the vessel was baked (1 week, 120°C)
and shortly cleaned by glow discharge (a few hours). The first shot,
as shown in Fig. 9 lasted already for 2 sec at I_p = 300 kA and
$\bar{n}_e$ = 1.8 · 10^{13} cm^{-3}. In the third shot, the density limit was determined which turned out to be 3.4 · 10^{13} cm^{-3} corresponding to
about 2/3 of the limit in clean discharges. Neutral beam injection
was added to shot 23 which yielded an L-type discharge (see Sec.10).
In shot 26, hydrogen was replaced by deuterium as working gas which
resulted in superior confinement properties (H-regime, see Sec. 10)
from shot 28 on.

4.   PARTICLE TRANSPORT IN THE DIVERTOR CONFIGURATION

    Tokamak operation at clean wall conditions requires external
gas fuelling for maintaining a constant particle content during the

discharge. Under steady state conditions, the refuelling flux just compensates the flux of lost particles which are deposited into wall structures of the surrounding vessel by plasma wall interaction. The global particle blance equation is:

$$\frac{dN}{dt} = - \frac{N}{\tau_p} + \int_V n_n n_e S_i dV_p \qquad (1)$$

(N = plasma particle content, τ_p = particle confinement time of the main plasma called magnetic particle confinement time in the following N/τ_p = flux of charged particles out of the plasma and the integral is the ionization rate within the plasma with n_n = neutral density, n_e = electron density and S_i = ionization rate coefficient). The global ionization rate, as estimated from the toroidally averaged H_α or L_α radiation [21], however, is large in comparison to the external gas flux $\emptyset_G$. On the other hand, the plasma particle content decays when the external gas flux is switched off. The decay time $\tau_{eff} = -N/(dN/dt)$ is large in comparison to τ_p. From these observations we have to conclude that the neutral density within the plasma whose ionization establishes the source of plasma particles, does not predominantly originate from the external gas valve. Part of the loss flux N/τ_p is fuelled back into the discharge after interaction with the wall. The exchange of working gas between plasma and wall is a special aspect of plasma wall interaction and is called recycling. The ratio of global flux from the wall to the one to the wall is defined as recycling coefficient R. Because of recycling, the source term of equ. 1 can be split up into two contributions

$$\int n_n n_e S_i dV_p = \emptyset_G + R \frac{N}{\tau_p} \qquad (2)$$

with RN/τ_p the recycling flux from the wall.

The charged particle fluxes from the plasma are guided along open field lines onto protruding obstacles like limiters or target plates. The fluxes from the wall are predominantly neutral and comprise two components: an energetic component caused by back-scattering and a slow molecular component which originates from the hydrogen, implanted into the limiter and wall material, diffused back to the surface, and subsequently desorbed from there.

The particle content of a discharge is both determined by the magnetic confinement time τ_p and the recycling coefficient R (s. equ. 2). The study of recycling is of importance in order to control the particle content of the discharge and in a future reactor the species mix of the fuel. Such external control is possible only for a global recycling coefficient sufficiently below 1.

As the atomic processes which give rise to backscattering and desorption of the working gas can also produce impurities, the study of this branch of plasma wall interaction is of specific importance for a divertor experiment. In order to keep the impurity erosion processes low, the energy flowing out of the plasma has to be distributed over many particles in order to keep the individual particle energy low and possible below the threshold energy for impurity release. This requirement asks for high recycling conditions with intense particle fluxes across the plasma surface possibly in conflict with the conclusion above. Therefore, an optimal situation has to be found which complies with both requirements.

4.1 Spatial Distribution of the Neutral Density within the Plasma

The atomic hydrogen content of the plasma which constitutes the source for protons and electrons in a clean plasma (see equ. 1) is not affected by the magnetic fields of a tokamak and, therefore, it is not forced to its spatial symmetry. Figure 10a gives a crude notion of the toroidal variation of the edge neutral density $n_n(a)$ at the plasma midplane as inferred from charge exchange measurements /22/. The charge exchange flux $d\emptyset_{cx}/dE$ is proportional to $n_n n_i$ (n_i = proton density); with the help of transport programs for the hydrogen atoms in the plasma /23/, the atom density can be inferred. The toroidal symmetry of $n_n(a)$ is lost at the locations of limiter and gas valve where the distribution bulges out. The variation at the limiter clearly documents the localized ion recycling. With the divertor configuration, the edge neutral denstiy does not change very much apart from limiter and gas valve positions: The local limiter recycling disappears and the external gas flux has to be increased by a factor 3 - 4 to maintain the same plasma density /24/. The global recycling coefficient of a divertor discharge is lower than that of a limiter discharge with the same plasma parameters (if one assumes the same particles confinement times τ_p for the two configurations). There is indeed more toroidal symmetry in the neutral distribution of a divertor discharge but now an additional poloidal asymmetry is observed. Figure 10b compares the poloidal charge exchange flux profiles of limiter and divertor plasmas far from limiter and gas valve. A low particle energy is chosen (E = 230 eV), to achieve a higher sensitivity for the edge neutral density. Starting from the plasma center, the flux increases mostly because the neutral density increases towards the plasma edge (see insert for geometrical reference). After a maximum, the flux decreases again in case of the limiter configuration because the ion temperature decreases towards the edge causing the density of 230 eV particles to decrease. In case of the divertor configuration, this decrease is not observed though the ion temperature profile is roughly independent of configuration (for Ohmic plasmas see Fig. 39). The drop in flux, as caused by the decrease of ion temperature towards the plasma edge, is compensated by an increase in edge neutral density towards the stagnation point of the divertor configuration. We observe high neutral density at the exit of the divertor necks. The poloidal variation of the neutral density and the location of the recycling

948

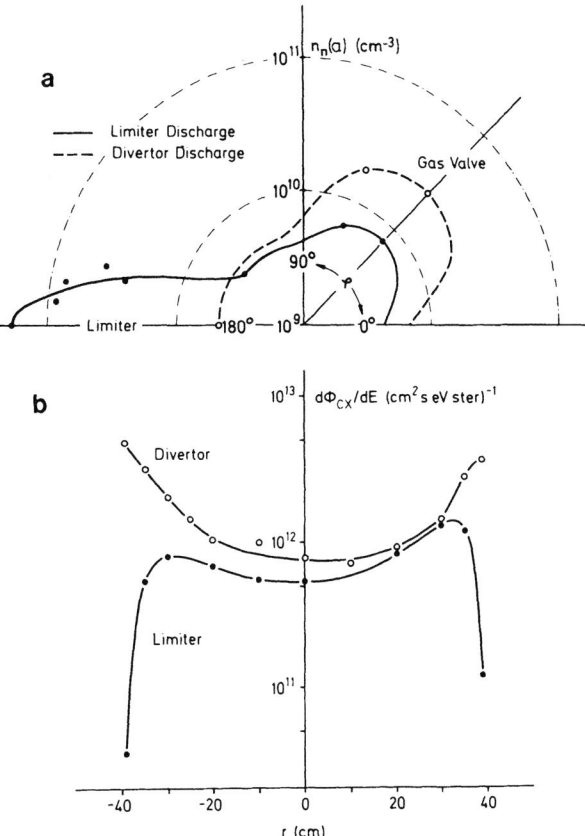

Fig. 10a. Toroidal variation of the edge neutral density of an ASDEX limiter and divertor discharge. The distribution bulges out at limiter and gas-valve location.

10b. Radial variation of the charge exchange flux emitted from a limiter and divertor discharge of ASDEX. The insert shows the experimental geometry. The differences between the limiter and divertor discharge at the plasma edge are mostly due to high neutral density at the stagnation points.

fluxes at the divertor exit is also obvious from the photograph shown in Fig. 8. The intensity of the H_α-radiation which is shown there and which is proportional to the hydrogen ionization rate is bright near the divertor throat.

4.2 Particle Fluxes between Main Plasma and Divertor

Because of the specific magnetic field topology of the divertor configuration, plasma ions and electrons which have diffused across the separatrix are channeled along the open field lines through the divertor throat into the divertor chamber. The cross-field diffusion in the boundary layer is slow enough to give rise to a sharp plasma boundary with a short plasma fall-off length. Therefore, during quiescent phases, the plasma does neither interact with the adjacent vessel walls nor with the channel walls of the divertor throats at an intensity which would contribute to the overall recycling. At a mid-plane fall-off length of 1.5 cm, as it is typical for a poloidal divertor, an annulus width of 3.5 cm is sufficient to accommodate the particles (giving rise to a particle exhaust efficiency >0.9) as the flux surfaces are compressed within the divertor throats by about a factor of 3. Thus ion recycling in the main plasma chamber can largely be neglected. (This statement may be wrong in case of high frequency heating if the antennas are moved close to the plasma surface for good wave coupling). In case of a bundle divertor, the fall-off length is larger than in a poloidal divertor /25/ (because the magnetic field line encircles several times the main plasma till it terminates at the target plate) causing a lower particle exhaust efficiency. For DITE a particle exhaust efficiency of 0.7 is estimated /26/ and the contribution of these ions which hit the edges of the divertor throats to recycling has to be included /27/.

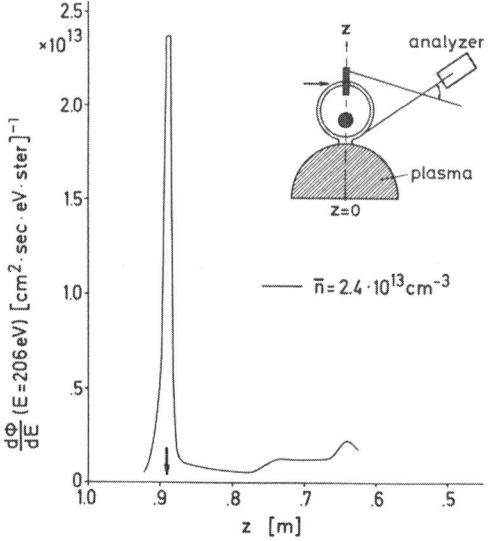

Fig. 11.   Profile of the atomic flux backscattered from the target plate of ASDEX. The insert shows the geometry and denotes the observation range. The arrow points to the intersection of separatrix with target plate (taken from Ref. /22/).

The principal location of the plasma wall interaction is at the intersection of scrape-off layer and target plate. There the ions from the plasma are neutralized and subjected to the two dominant recycling processes: Either they are backreflected as energetic atoms or they are deposited within the target plate material slowly diffusing to the surface from where they finally desorb in molecular form. The energetic part of the backreflected flux with E > 100 eV can be detected by corpuscular techniques /22/. Figure 11 shows the profile of backscattered atoms with E = 273 eV with a sharp peak at the separatrix-target plate intersection. Due to backscattering and desorption, the divertor chamber fills with hydrogen in course of the discharge. Figure 12 plots the preprogrammed density variation with a long plateau phase after the initial built-up phase. Though the density of the main plasma stays constant, the neutral flux (equivalent to neutral density) in the divertor chamber increases continuously /28/. The divertor space plays a similar role as limiter and wall material which, in case of a limiter discharge, store the working gas and exchange it with the plasma.

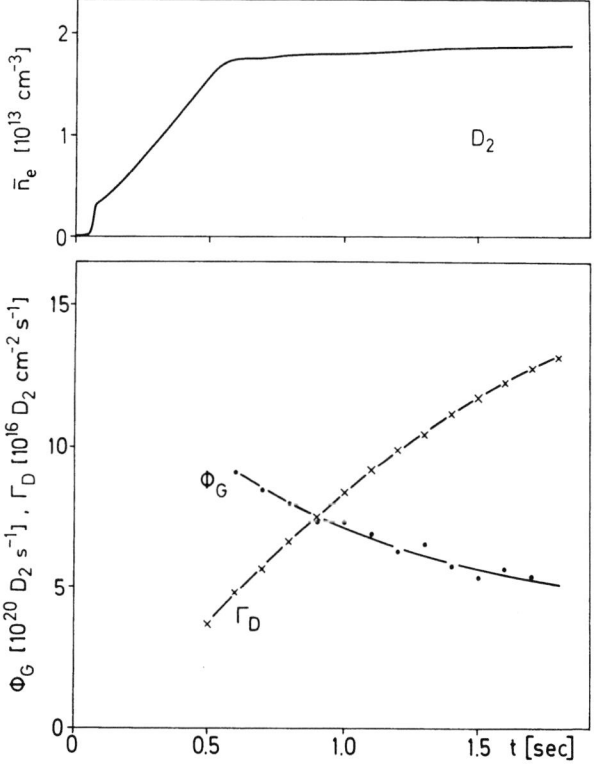

Fig. 12.   Line density $\bar{n}_e$, external gas flux ϕ_G and neutral flux Γ_D in the divertor chamber during an ASDEX divertor discharge. During the density plateau phase, the recycling conditions are non-stationary (taken from Ref. /28/).

Part of the gas, accumulated within the divertor chamber, flows back into the main chamber to refuel the plasma there. As the intensity of the recycling flux increases during the discharge the flux from the external gas valve decreases correspondingly as documented in Fig. 12. In course of a divertor discharge, the ionization rate at the divertor exit increases while its toroidal asymmetry at the gas valve is smoothed out.

After these considerations, we can construct a simplified picture of the recycling pattern during a divertor discharge /29/. The dominant fluxes are the loss flux N/τ_p out of the plasma into the divertor chamber and the backflow of the hydrogen from the divertor into the main plasma chamber. The external gas flux compensates the wall losses and provides the neutral density built-up within the divertor chamber. Figure 13 shows the direction of the different particle fluxes and also introduces the concept of shielding by the boundary layer. The fraction $1 - \gamma_E$ of the external and recycling fluxes is already ionized within the boundary layer and directly swept into the divertor chamber without fuelling the main plasma.

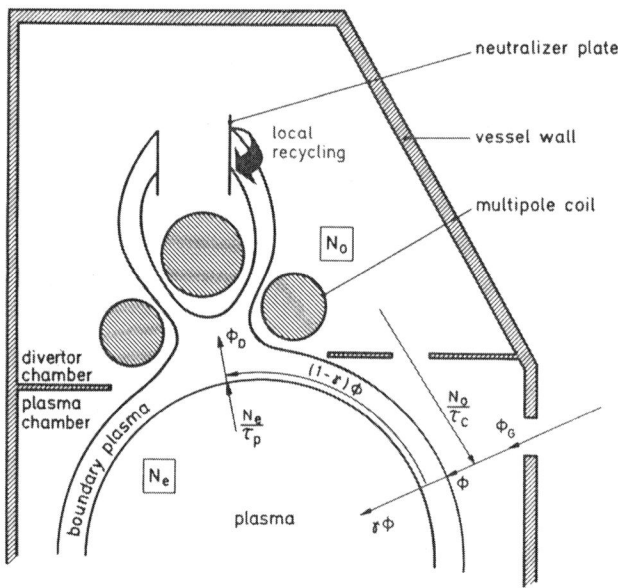

Fig. 13. Recycling model of a divertor discharge indicating the
          different particle fluxes (taken from Ref. /29/).

The recycling fluxes from the divertor can reach the main plasma chamber through additional wall openings (by-pass in Fig.13) or through the divertor necks. Both the open divertor with a by-pass conductance large in comparison to the divertor channel conductance and the opposite limit, the closed divertor configuration, have been studied and will be compared in Sec. 9.

For a divertor discharge, the global particle balance equations are

$$\frac{dN_e}{dt} + \frac{N_e}{\tau_p} = \gamma_E (\emptyset_G + \frac{N_o}{\tau_c}) \tag{3}$$

$$\frac{dN_o}{dt} + \frac{N_o}{\tau_c} = R \left[(1-\gamma) \, (\emptyset_G + \frac{N_o}{\tau_c}) + \frac{N_e}{\tau_p} \right] - \frac{N_o}{\tau_D} \tag{4}$$

The different terms are explained in Fig. 13, R = recycling coefficient at the target plate; N_o/τ_D = flux of atoms into an external pump (e.g. Ti-sublimation). The particle balance of a divertor discharge differs from those of a limiter discharge (equ. 2) by the coupling between plasma and divertor on account of the mutual particle exchange.

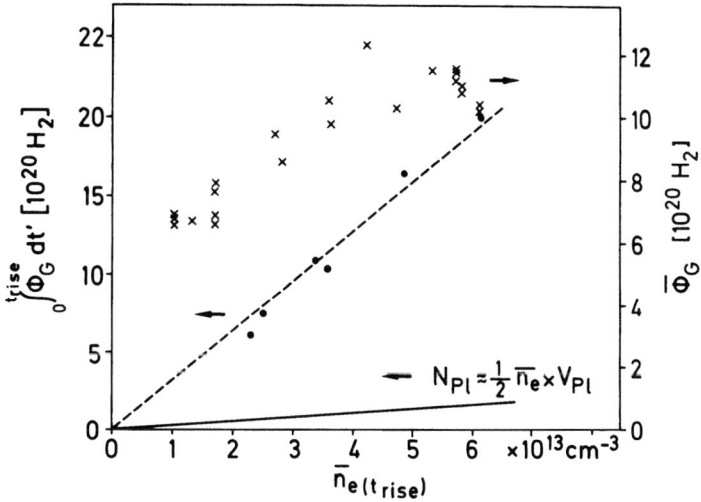

Fig. 14. Density dependence of the gas consumption during plasma built-up in ASDEX (from t = 0 ÷ t = t_{rise}) in comparison to the plasma content when the density plateau $n_e(t_{rise})$ is reached and to the external gas flux $\emptyset_G$ at that instant (taken from Ref. /28/).

During the initial phase of the discharge when the plasma density is ramped up, the divertor chamber has to be filled with neutral

gas in order to establish the recycling flux. Figure 14 compares the
gas consumption during the start-up phase with linear density rise
till the pre-programmed plateau value $\bar{n}_e(t_{rise})$ is reached with the
actual plasma content at this moment /28/. Only a small fraction of
the puffed-in gas is ionized and recovered within the plasma. The
rest is stored mostly within the divertor chamber and its walls.
There is a linear relationship between the total gas consumption and
the desired density plateau value while the averaged gas flux, necessary to keep the density constant, (also shown in Fig. 14) shows
a weaker density dependence. The important stage for the density
built-up and the adjusting of the desired density level is the gas
fluence during the initial phase. The gas flux during the stationary
phase controls the divertor gas density and compensates the wall
losses which seem to be rather density independent.

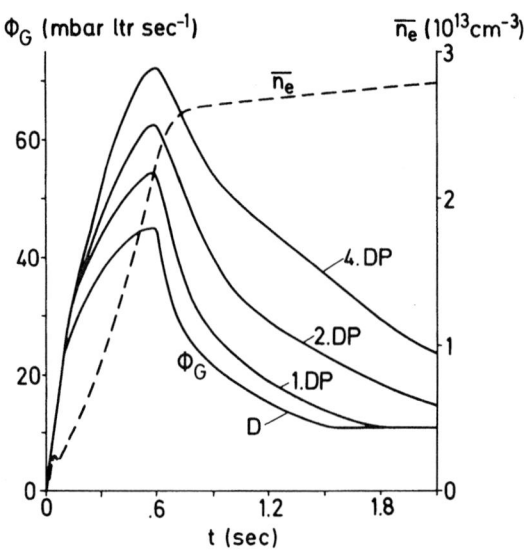

Fig. 15. Increase of the external gas flux Φ_G at the transition
from a divertor (D) to a pumped divertor (DP) discharge
of ASDEX when titanium is evaporated within the divertor
chamber between each discharge. The increased gas flux is
necessary to keep the plasma density variation unchanged.

Despite its rather weak dependence on the momentary gas flux,
the plasma density can easily be controlled. Unlike limiter discharges where the total recycling reservoir is implanted into the walls,
the hydrogen content of the divertor chamber can easily be controlled by external pumps. The action of external pumps gives rise
to an additional loss term in the particle balance for the divertor
content N_0 (see equ. 4) which also affects the main plasma content
because of the coupling terms in equ. 3,4. In present day machines
a simple technical means to achieve high pumping speed within the

divertor is titanium sublimiation /30/. Figure 15 plots the increased external gas flux ϕ_G necessary to maintain constant plasma density when the titanium system of ASDEX has been activated. At invariant plasma density, the neutral divertor density is reduced and the lower recycling fluxes have to be compensated by intensified gas puffing. With titanium evaporation, recycling coefficients as low as 0.2 have been established in divertor discharges /29/.

Apart from the way explicitly shown in equs. 3 and 4, the neutral particle content in the divertor chamber can affect the bulk plasma density also in an indirect form. As we will discuss in the next section (and is described in more detail in the lectures of Ref. /41/), high divertor atom densities give rise to an intensive ionization in the divertor fan, whose influence is felt also by the bulk plasma as an enhancement of its edge density. This change in boundary condition will have an effect also on the magnetic particle confinement time τ_p, which will depend qualitatively on the transport processes in the main plasma. Under extreme conditions – a strong inward drift term in the particle transport equation, and effectively blocked by-passes from the divertor – stationary discharge conditions could then be maintained without any conventional refueling of the main plasma via neutral particles /57/. This so-called ion-refueling model would have interesting and potentially favourable characteristics (e.g. eliminating CX sputtering in the main chamber), but lacks at present clear-cut experimental proof.

In the divertor configuration and with additional pumping in the divertor chamber, the plasma density is well controlled. This low recycling condition gives high experimental flexibility for density variations. An example is shown in Fig. 16. First the density is linearly ramped up to a plateau of $\bar{n}_e = 2.5 \cdot 10^{13}$ cm^{-3}. At 0.8 sec, the external gas valve is switched off so that the density quickly decays to $3 \cdot 10^{12}$ cm^{-3} before it is ramped up again to a second plateau of $\bar{n}_e = 5 \cdot 10^{13}$ cm^{-3}. Recycling control is the premise for reproducible tokamak discharges and it offers the flexibility for investigations at various densities. A rapid and controlled change in density has also some technical advantages. At the end of a tokamak discharge, the density can quickly be reduced in a rate faster than the plasma current. Thereby, dangerous plasma disruptions are avoided. This criterion is of particular importance for large machines with plasma currents in the MA-range. Rapid density changes may also be necessary for successful high frequency current drive in cases with conventional inductive current drive at high plasma density alternating with phases of RF supported recharging of the OH-transformer at low density. At low recycling conditions, the density is quickly reduced below the density threshold for current drive so that neither flux from the Ohmic nor power from the high frequency system is wasted. Furthermore, in a fusion reactor, low recycling conditions ease burn control and allow an exact setting of the fuel ratio.

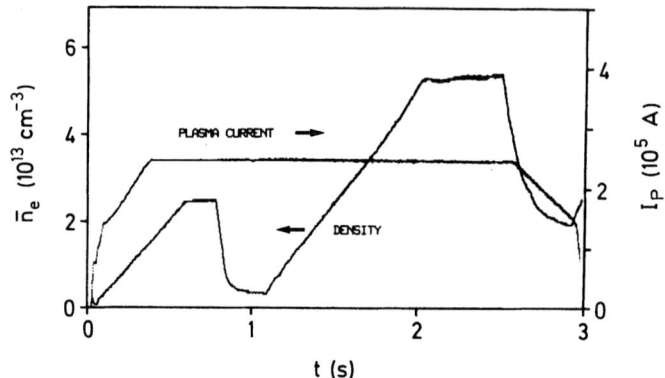

Fig. 16.   Density variation during an ASDEX discharge with well
           controlled recycling. The external gas valve is switched-
           off at 0.8 sec, opened at 1.1 sec from the control system
           to produce the preprogrammed density trace and switched-off
           again at 2.5 sec.

     In this chapter, we have demonstrated that with the divertor
configuration low recycling conditions are obtained which give full
control over the particle content in the discharge. In the next
chapter, we will discuss to what extent this advantage is at the
expense of intense wall erosion concomitant to high edge temperatures.

4.3 The High Recycling Divertor

     Up to now we have discussed the hydrogen flux pattern between
main plasma and divertor chamber with the principal result that
hydrogen gas is preferentially stored within the divertor chamber.
The ratio of hydrogen pressure within the two chambers (= hydrogen
compression) is between 10 and 100. The hydrogen gas flux which
impinges on the plasma within the divertor chamber is re-ionized,
swept onto the target plate and again transformed into neutral
hydrogen. This circular process establishes a separate recycling
vortex within the divertor chamber which is schematically shown in
Fig. 13. At a plasma density of $\bar{n}_e = 5 \cdot 10^{13}$ cm^{-3}, the molecular
pressure within the divertor chamber of ASDEX is about 10^{-3}mbar. The
total flux impinging on the divertor plasma is 1/4 $n_n V_{th} F$ (n_n = molecular or atomic density, V_{th} = average velocity at room temperature and F = surface area of the divertor plasma fan = $2.1 \cdot 10^5$cm^2).
If this flux is fully ionized (in case of no power limitation), the
total ion flux onto the target plate is $8 \cdot 10^{22}$ s^{-1}. The primary

particle flux from the main plasma into the divertor chamber is about $3 \cdot 10^{21}$ s^{-1} (assuming τ_p = 40 ms). Only about 4 % of the particles, hitting the target plate, originate from the main plasma and, therefore, the divertor can be seen as a particle flux amplifier.

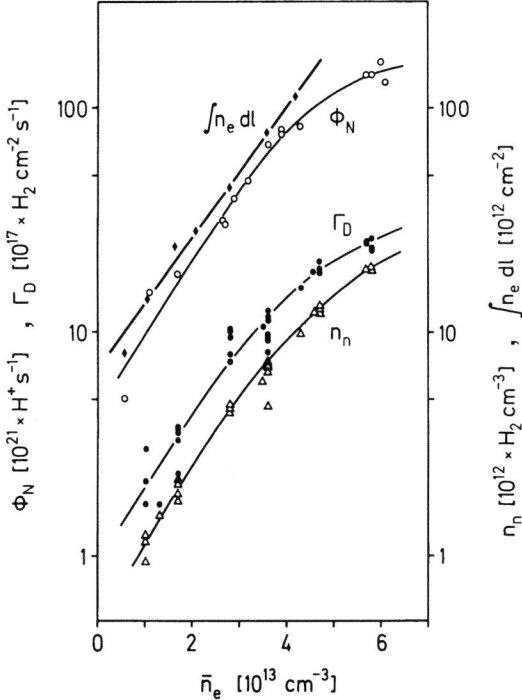

Fig. 17. Dependence of particle flux Φ_N onto the target plate, the hydrogen flux Γ_D and the hydrogen density n_n within the divertor chamber and the line density of the divertor plasma on the line averaged density $\bar{n}_e$ of the main ASDEX plasma. (Taken from Ref. /28/; the divertor line density data are provided by G. Siller).

In Fig. 17 the dependence of the plasma flux onto the target plate, hydrogen flux and density within the divertor chamber together with the line density of the divertor plasma are plotted versus the line averaged density of the main plasma /28/. The divertor parameters rise nearly exponentially with increasing plasma density. Because of the local equilibrium between divertor plasma and surrounding gas, established by neutralization and reionization events and sustained by the power flux into the divertor chamber, densities of the divertor plasma and the divertor gas background are linked.

Approaching the target plate, more and more gas is transferred into plasma causing the divertor density to increase and the temperature to decrease. The temperature drops towards the target plate because part of the available power is dissipated by the different dissociation, ionization and excitation processes occuring at the plasma-gas interface (see Sec. 5.2) and the remaining power is distributed over an increasing number of particles.

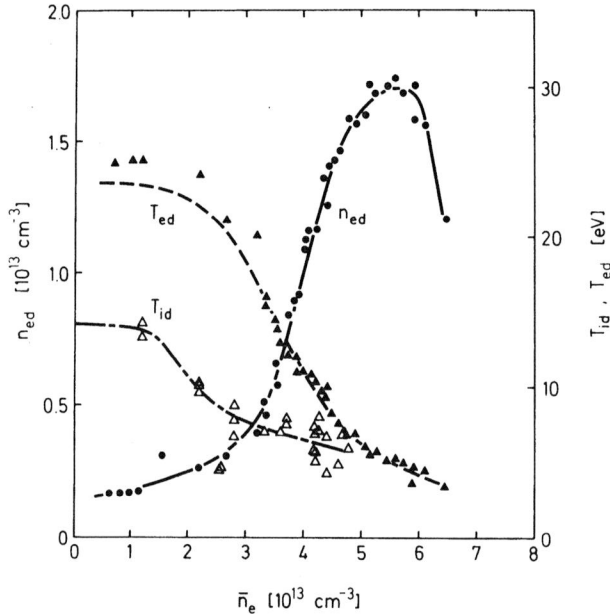

Fig. 18.   Scaling of the electron temperature and density (measured by a Langmuir probe; data taken from Ref. /31/), and the ion temperature deduced from the Doppler-broadened CIII radiation (data points provided by S. Szymanski) of the ASDEX divertor plasma with the main plasma density.

Figure 18 shows the dependence on the main plasma density of the electron temperature T_{ed} and density n_{ed} (measured by a Langmuir probe) /31/ and ion temperature T_{id} (deduced from the Dopplerprofile of CIII radiation) /32/. The rising density causes T_{ed} and T_{id} to drop below 10 eV giving rise to a sheath potential between 10 and 20 eV. Thus, the ion temperature stays well below the threshold energy for impurity generation from the target plate.

The properties as described above - high neutral and plasma density in the divertor chamber which increases non-linearly with the plasma density - characterize the high recycling divertor which

958

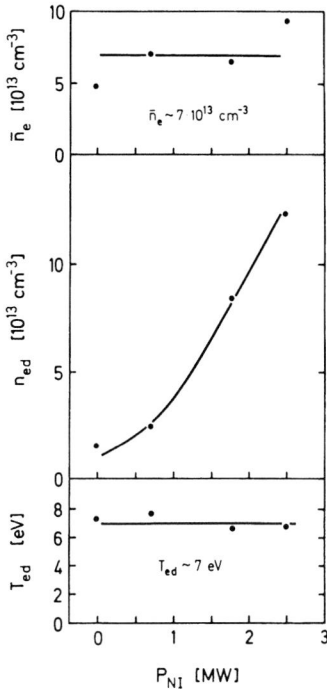

Fig. 19. Variation of divertor temperature and density with neutral
beam power at roughly constant plasma density of ASDEX. The
beam power varies the divertor density such that T_e is
kept constant at a low level (taken from Ref. /31/).

provides the requirements of low temperatures and low sheath potential at the plasma-wall contact zone. Up to now we have described
the divertor recycling conditions and the concomitant plasma parameters in front of the target plate for the low power levels of
Ohmically heated discharges ($\leq$ 0.5 MW). Tt is, however, mandatory
to study the divertor conditions also at higher power levels so as
to be able to scale the behaviour to even higher, reactor like parameters. With neutral injection, plasma behaviour at a power level of
a few MW can be studied. Figure 19 shows the dependence of
electron density and temperature (measured by a Langmuir probe) in
the divertor chamber on the heating power /31/. The investigation
is carried out around a main plasma density $\bar{n}_e = 7 \cdot 10^{13}$ cm^{-3}.
Either allowing the main plasma density to slowly rise, as shown
in Fig. 19, or, alternatively, changing the external gas flux procedure to refueling into the divertor chamber causes the divertor
plasma density to nearly linearly rise with beam power and the temperature to stay below 10 eV. Figure 19 displays a favourable
aspect of the high recycling divertor regime. More power fueled
into the divertor chamber can be used to intensify the local

recycling vortex in such a way that the temperature, which is the important parameter with respect to impurity generation, is kept low. The operational principal of the divertor is the establishment of the high divertor density concomitant to low temperature. The temperature is allowed to drop just to the limit where ionization (= source for plasma) becomes drastically ineffective. This self-regulating mechanism of the divertor is only effective if enough particles are fueled into the divertor and retained there. Therefore, each power level corresponds to a minimal main plasma density above which the divertor temperature is stabilized to ~7 eV (see Fig. 19). We will come back to this dependence when we discuss the aspect of divertor geometry in Sec. 9.

5. ENERGY TRANSPORT IN THE DIVERTOR CONFIGURATION

Under favorable conditions both the Ohmic and auxiliary heating power are deposited in the plasma center where the confinement is best. At steady state conditions, the power flows to the plasma surface. The dominant energy transport mechanism in the bulk of the plasma is anomalous electron heat conduction. Ion heat conduction contributes in Ohmically heated discharges only at high density. At the plasma peripheral zone, a sizable fraction of the power can be radiated away by impurity line radiation or transported to the wall by charge exchange neutrals. These power fluxes are deposited over the whole vessel wall. The power density is low and technically manageable. The rest, which is the dominant fraction in clean discharges, flows along the scrape-off onto limiters or target plates. The radiated power density decreases with the plasma surface $R \cdot a$ while the transport part scales like $1/a \cdot \lambda_p$ in case of a poloidal limiter. $\lambda_p (\sim 1 \text{ cm})$ is the power fall-off length of the plasma scrape-off layer. The reduction in power density in case of a divertor instead of a poloidal limiter is by the factor a/R which could easily be offset by operation with a toroidal limiter /33/. Experiments have shown, however, that volumetric power release mechanisms exist within the divertor chamber, reducing the power density at the target plate, which may not easily be matched by any type of limiter.

5.1 Power Balance with Divertor Operation

The important aspect of the power balance with divertor is that the power flux into the divertor chamber is split up into two branches: Part of the power is released by volumetric processes from the divertor plasma and thereby reduces the power density at the target plate. Figure 20 shows on its right side a schematic of a bolometer camera which views main and divertor plasmas of the SN-configuration of DIII. The left side shows a radiation profile as measured by this array /34/. A sizable fraction of the transported power to the plasma surface is radiated from the divertor plasma

960

and distributed over a large area. The power density radiated from the divertor plasma can be very large and values of 0.3 Wcm^{-3} have been quoted /34/. This characteristic of the divertor has been dubbed "remote radiation cooling".

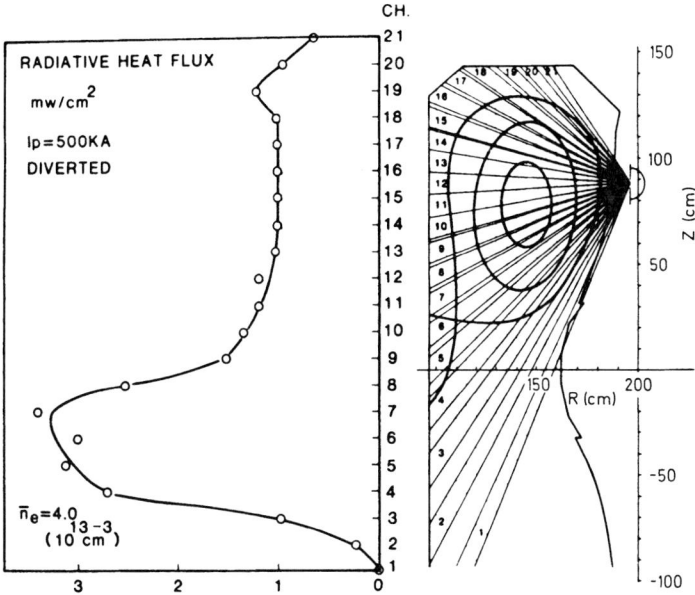

Fig. 20. Line integrated radiation power profile of D III. The figure documents the remote radiation characteristic due to intense divertor radiation. The right side shows the observation geometry (taken from Ref. /34/).

Figure 21 compares the global power balance dependence on main plasma density $\bar{n}_e$ of an Ohmic (P $\sim$ 0.4 MW) with a beam heated diver-tor discharge (P $\sim$ 2 MW) of ASDEX /35/. Both cases document the importance of the divertor radiation contribution which becomes increasingly important at higher auxiliary heating levels. The divertor radiation increases with density and the beam heated case clearly documents that this increase is at the expense of the power deposited onto the target plate. (The sum over the 3 loss channels does not add up to 100 %. This "violation" of the conservation law for energy is a general trait met also in limiter discharges /36/. In the divertor case, it does not necessarily mean that a sizable fraction of the transported power is deposited elsewhere e.g. at the metal plates defining the narrow divertor throat. The discrepancies may be caused by unjustified symmetry assumptions in evaluating the global quantities.)

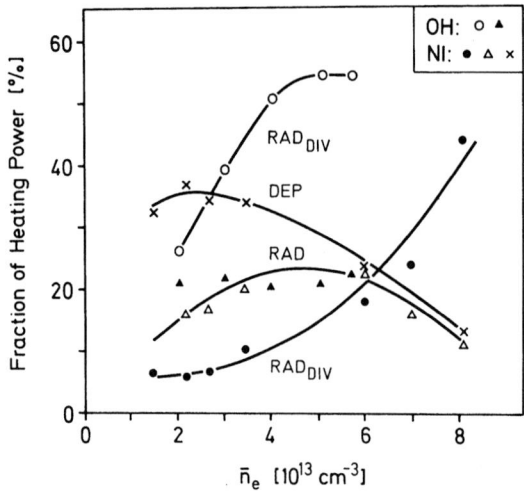

Fig. 21. Fraction of the heating power in the loss channels main
plasma radiation (RAD), divertor radiation (RAD_{DIV}) and
deposition onto the target plates (DEP) for Ohmically
(P_{OH} = 0.4 - 0.5 MW) and beam heated (P_{NI} = 2.1 MW) ASDEX
discharges (taken from Ref. /35/ and Ref. /37/).

In conclusion, the divertor property of remote radiational
cooling contributes two important aspects: It mitigates the otherwise severe technical problem of high power fluxes onto the target
plate by providing intense radiation, which, and this is the second
advantage, is not at the expense of the plasma periphery. At the
surface of the main plasma, the temperatures can stay high which
seems to be a pre-requisite for good confinement properties as we
will discuss in Sec. 10.

5.2 Energy Release Mechanisms within the Divertor

The energy release mechanisms within the divertor chamber are
simple atomic and molecular processes of dissociation, ionization
and excitation of species in low ionization states. Nevertheless,
they are still disputed with sometimes conflicting results.

The important question is whether the favourable feature of
remote radiation cooling is caused by impurities and requires dirty
discharges or whether the inherent hydrogen recycling properties
within the divertor chamber provide it. In case of dirty plasma conditions of DIII, a correlation between bolometrically measured radiation and OI-line radiation in the divertor has been observed
/34/. An analysis shows that high divertor radiation might be possible with low impurity concentration within the main plasma, compatible, for example with the INTOR requirements /34/. In ASDEX,

the question of an oxygen dominated divertor radiation has been investigated by comparing the divertor radiation of normal discharges (D) with those with titanium sublimation within the divertor (DP) /37/. Ti sublimation effectively removes oxygen /38/. From the negligible difference in divertor radiation of D and DP discharge types, it has been concluded that not oxygen but the atomic processes involving the working gas give rise to it. This has also been concluded for clean DIII discharges where the divertor radiation correlates with the local H_α-radiation /34/.

Which energy release mechanism dominates the bolometrically measured radiation when hydrogen molecules are continuously transformed into plasma at a plasma temperature of 10 eV and a density close to 10^{14} cm^{-3} has been investigated on ASDEX in more detail /37/. A 10 mm LiF-filter has been placed in front of the bolometer. This filter allows the transmission of H_α- and L_α-radiation but blocks particles and radiation with $\lambda <$ 1040 Å. With the filter, the signal dropped to 10 %. As the contribution of energetic neutrals (E > 100 eV) could be ruled out from separate observations, it has been concluded, that the divertor radiation is carried by low energy atoms possibly in the energy range of Franck-Condon atoms (3 - 5 eV) /37/.

5.3 Energy Transport in the Scrape-off Layer

The energy transport phenomena in the scrape-off layer are strongly determined by the processes affecting the divertor plasma. The radiation there cools the plasma to T_e-values below 10 eV and causes steep gradients within the scrape-off layer. Figure 22 compares the temperature and density profiles of the scrape-off in the plasma mid-plane and in the divertor in front of the target plate of a beam heated ASDEX discharge /39/. The dashed curves represent the scrape-off and the solid curves the divertor profiles. The scrape-off parameters are measured by laser scattering /40/, those of the divertor plasma are obtained from a Langmuir probe (its essential results are confirmed by laser scattering /32/). The location of the measured profiles are indicated by the insert of Fig. 22. T_e is found to decrease from the mid-plane value of 100 eV down to 10 eV within the divertor. As the boundary plasma behaves like a low Mach number fluid at the prevailing parameters, the pressure is (nearly) constant along the field lines. Therefore, the reduction in T_e is compensated by a corresponding increase in density from about 10^{13} cm^{-3} to 10^{14} cm^{-3}. The steep T_e-gradients in the scrape-off layer cause electron heat conduction parallel to the field lines the dominant energy transport mechanism. Ion heat conductive contributions are low and convective energy transport only plays a role at long mean free paths when simultaneously the fluid-like description of the transport phenomena in the plasma boundary as roughly sketched in the next section is no longer valid.

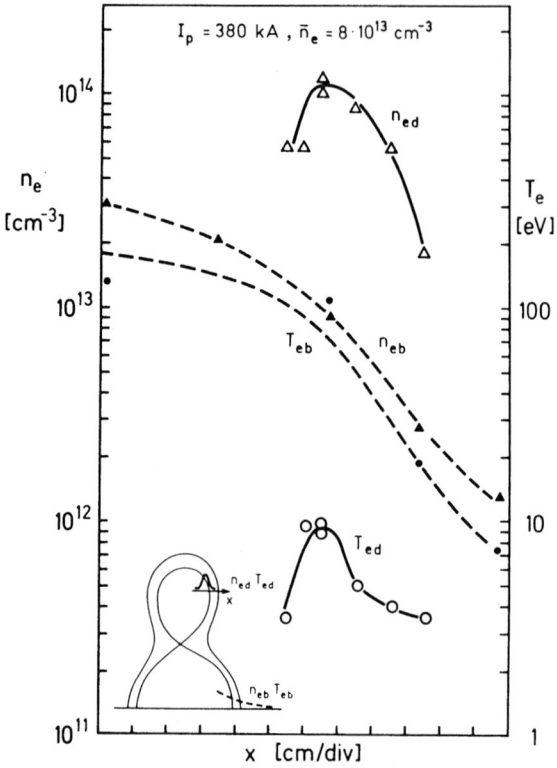

Fig. 22. Radial profiles of temperature T_e and density n_e at two
characteristic positions along the scrape-off layer,
namely in the torus mid-plane (n_{eb}, T_{eb}) and in the divertor chamber (n_{ed}, T_{ed}) for a beam heated ASDEX discharge
(P_{NI} = 2.5 MW). The mid-plane values are measured by laser
scattering, the divertor data are obtained from a Langmuir
probe. The insert shows the location for the measurements.
(Taken from Ref. /39/).

6.   BOUNDARY MODEL

It is not the intention of this contribution to deal with
models describing the transport phenomena within the scrape-off layer
in great detail because other lectures are devoted to this topic
/41/. Nevertheless, a simple mathematical treatment of the transport
within the scrape-off layer may give additional insight in the experimental phenomena and the scaling of edge parameters as discussed
up to now. We will restrict ourselves to a one-dimensional treatment though a more realistic procedure would require two dimensions
because of the rapid parameter changes in radial direction (short
fall-off lengths).

The geometry for the calculations is plotted in Fig. 23. The separatrix is at x=0 with the plasma and the scrape-off layer filling the space of negative and positive x-values, respectively. Because of perpendicular transport (finite confinement), particles and energy flow radially from the plasma across the separatrix. Within the boundary layer on open flux surfaces, the perpendicular transport competes with the parallel transport onto the target plate (see Fig. 23a). As the perpendicular fluxes are determined by density and temperature gradients, the gradients will decrease towards larger x-values (at constant transport coefficients) because the perpendicular fluxes decrease on account of the parallel fluxes into the divertor chamber (see Fig. 23b) which are treated as loss terms.

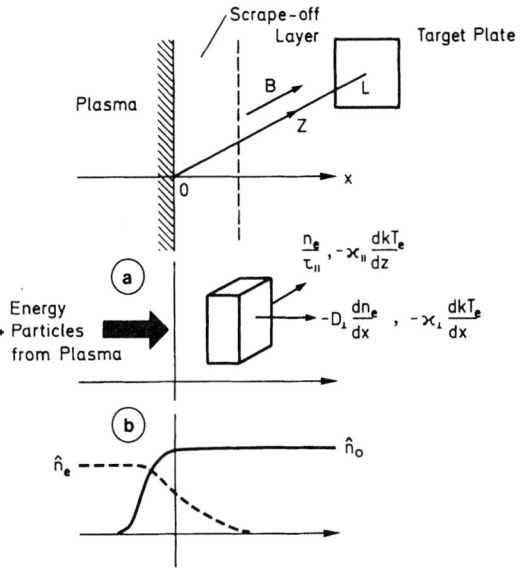

Fig. 23. Geometry for the boundary model calculations denoting
(a) the competing parallel and perpendicular fluxes and
(b) the variation of neutral and plasma density.

6.1 Parallel Parameter Variation

In this section, we discuss the variation of T_e and n_e along the scrape-off layer from the plasma mid-plane ($z = 0$) to the target plate ($z = L$). This problem has been discussed in Ref. /42/ and we will reproduce it with minor modifications. The analysis is based on parallel heat conduction as the dominant energy transfer mechanism. We have seen that the T_e-gradients are caused by remote radiative cooling and recycling within the divertor chamber, so that the one-dimensional heat conductivity equation along the field lines is:

$$\frac{d}{dz} \kappa_{\shortparallel} \frac{dT_e}{dz} = P_{rad} + P_{cx} + \frac{d}{dz} nv \left(\varepsilon_i + \frac{5}{2} k_B (T_e+T_i) + \frac{1}{2} mv^2 \right) \quad (5)$$

$\kappa_{\shortparallel}$ is the parallel electron heat conductivity. Heat conduction with divertor radiation is treated in Ref. /34/. We will simplify the problem by ignoring radiation and other losses and set the right hand side to zero in equ. 5. This can be justified because P_{rad} and P_{cx} play a role close to the target plate after most of the T_e-drop has already occurred and the ionization, convection and kinetic term can be put into the boundary condition.
The boundary conditions at z = 0 and z = L are:

$$z = 0: \quad -\kappa_{\shortparallel} \frac{dT_e}{dz} = \frac{P}{A} \quad (6)$$

We assume 2 divertor openings and that the power flows in an annulus of cross-section $2\pi R\lambda_T$ along the actual field directions with pitch B_p/B_T (~0.1) so that

$$A = 4\pi \, R\lambda_T \, \frac{B_p}{B_T} \quad (7)$$

R is the major radius (= 165 m in case of ASDEX) and λ_T is the radial electron temperature fall-off length in the scrape-off layer. P is the power deposited within the plasma and transported by the electrons across the separatrix and along the scrape-off layer. The boundary condition at

$$z = L: \quad -\kappa_{\shortparallel} \frac{dT_e}{dz} = \delta n_e(L) k_B T_e(L) \left(\frac{k_B T_e(L)}{2\pi m_i} \right)^{1/2} \quad (8)$$

is the sheath condition in front of the target plate. The energy flow onto the target plate is convective within the sheath region given by the particle flux with ion sound velocity. The energy transferred per ion-electron pair is $\delta k T_e(L)$ where δ incorporates the effects of sheath potential, the emission of secondary electrons and the energy backscattering coefficient for ions. We assume $\delta \cong 7-8$ /46/.
The parallel heat conductivity $\kappa_{\shortparallel}$ is given by Spitzer /43/

$$\kappa_{\shortparallel} = \kappa_0 \, T_e^{5/2} \quad (9)$$

with $\kappa_0 = 315/Z_{eff}$ lnΛ which gives $\kappa_0 = 20$ W cm^{-1} eV$^{-7/2}$ for a clean hydrogen plasma. (W, cm, and eV are also the units used in the following.)

Equation 5 (with right hand side = 0) can be integrated to give /42/:

$$T_e(z)^{7/2} = T_e(0)^{7/2} - \frac{7}{2} \frac{P}{A\kappa_o} z \qquad (10)$$

$T_e(0)$ is determined from the boundary condition at $z = L$ if we additionally assume pressure constance along the scrape-off layer, and a drop to

$$n_e(L) \, T_e(L) = \frac{1}{2} \, n_e(0) \, T_e(0) \qquad (11)$$

in the immediate vicinity of the target plate, where the flow is accelerated to sonic speed.

$$T_e(o)^{21/2} = \left(\frac{P}{A} \frac{(2\pi m_i)^{1/2}}{\frac{1}{2}\delta n_e(o) k_B^{3/2}}\right)^7 + \frac{7}{2} \frac{PL}{A\kappa_o} \cdot T_e(o)^7 \qquad (12)$$

For sufficiently high values of $n_e(0)$, the first term in the brackets can be neglected:

$$T_e(o) = \left(\frac{7}{2} \frac{PL}{A\kappa_o}\right)^{2/7} \qquad (13)$$

In order to determine $T_e(L)$, we must rewrite equ. 12:

$$T_e(o)^{7/2} = \frac{7}{2} \frac{PL}{A\kappa_o} \left[1 + \left(\frac{P}{A} \frac{(2\pi m_i)^{1/2}}{\frac{1}{2}\delta n_e(0) k_B^{3/2}}\right)^7 \left(\frac{2}{7} \frac{A\kappa_o}{PL}\right)^3\right]^{1/3} \qquad (14)$$

We can expand the right side of equ. (14), use relation (10) and finally obtain:

$$T_e(L) = 3 \cdot 10^{-4} \frac{2\pi m_i}{\delta^2 k_B^3} \left(\frac{P}{A}\right)^{10/7} \left(\frac{2\kappa_o}{7L}\right)^{4/7} \frac{1}{n_c^2(0)} \qquad (15)$$

The density $n_e(L)$ can be calculated from the pressure balance (equ. 11):

$$n_e(L) = 3.5 \cdot 10^3 \frac{\delta^2 k_B^3}{2\pi m_i} \left(\frac{A}{P}\right)^{8/7} \left(\frac{7L}{2\kappa_o}\right)^{6/7} n_e^3(0) \qquad (16)$$

967

6.2   Radial Parameter Variation

Next we will treat the radial dependence of the plasma density. In a crude model the balance of plasma and neutral particles within the separatrix (x < 0), is described by

$$D_\perp \frac{d^2 n_e}{dx^2} = n_n n_e S_i \qquad \qquad (17)$$

$$V_H \frac{dn_o}{dx} = -n_n n_e S_i \qquad \qquad (18)$$

At steady-state conditions, the radial variation of the particle flux $D_\perp dn_e/dx$ is compensated by the ionization of the neutral background n_n (from gas input and recycling fluxes) with a rate coefficient S_i. Correspondingly, the atom flux $n_n V_H$ into the plasma is absorbed.

In our simplified model, we neglect ionization in the scrape-off layer (this more elaborate case is treated in Ref. /44/). Thus

$$n_n(x) = \hat{n}_n \qquad \qquad x \geq 0 \qquad \qquad (19)$$

and the radial variation of the particle flux is caused only by the parallel loss.

$$D_\perp \frac{d^2 n_e}{dx^2} = \frac{n_e}{\tau_{||}} \qquad \qquad (20)$$

$\tau_{||}$ is a mean particle residence time in the scrape-off layer determined by the particle drift velocity $v_{||}$ onto the target plate and a mean drift length L. A complete description of the scrape-off transport does not require the definition of $\tau_{||}$, but directly gets the parallel fluxes from the balance equations including momentum exchange. For our purposes it is handy to use $\tau_{||}$ which, similar to the global energy confinement time, is a quantity which allows quick comparisons, cross-references and back of the envelope estimates.

The solution of equ. (20), which gives the density profile in the scrape-off layer, is straight forward:

$$n_e = n_e(0) e^{-x/\sqrt{D_\perp \tau_{||}}} \qquad \qquad x \geq 0 \qquad \qquad (21)$$

with $n_e(0)$ being the density value at x = 0 which has to be determined together with the solution for the main plasma.

968

Next, we determine the density variation within the separatrix. Equations (17) and (18) can be combined to

$$D_\perp \frac{d^2 n_e}{dx^2} = -v_H \frac{dn_n}{dx} \tag{22}$$

and integrated to

$$D_\perp \frac{dn_e}{dx} = -v_H n_n$$

The constant of integration is zero because $\frac{dn_e}{dx} = n_n = 0$ for $x = -\infty$. Now we can write down the equation for the density, which is

$$\frac{d^2 n_e}{dx^2} - \frac{S_i}{2v_H} \frac{d(n_e)^2}{dx} = 0 \tag{23}$$

with the solution

$$n_e = C_1 \tanh \left(- \frac{C_1 S_i}{2v_H} x + C_2 \right) \qquad x \leq 0 \tag{24}$$

We get the constants C_1 and C_2 from the edge condition $n_e = \hat{n}_e$ ($x = -\infty$) (whereas we correlate $\hat{n}_e$ with the peak density in the plasma center) and the continuity of the density gradient at $x = 0$:

$$C_1 = \hat{n}_e \tag{25}$$

$$C_2 = \frac{1}{2} \, \text{arcsh} \, \frac{S_i \hat{n}_e \sqrt{D_\perp \tau_{\shortparallel}}}{v_H} \tag{26}$$

The solution for the neutral density is

$$n_n = \hat{n}_n \left[\cosh \left(- \frac{\hat{n}_e S_i}{2V_H} \right) \right]^{-2} \qquad x \leq 0 \tag{27}$$

$\hat{n}_n$ is the outside neutral density (see Fig. 23)

From the solutions, we can deduce a few interesting relations, which we can correlate with our experimental findings:

$$n_e(0) = \hat{n}_e \tanh C_2 \tag{28}$$

$n_e(0)$ varies non-linearly with $\hat{n}_e$ because C_2 itself is a function of $\hat{n}_e$ (equ. 26).

The line integral of the scrape-off density

$$\int_o^\infty n_e \, dx = \hat{n}_n v_H \tau_{"} \tag{29}$$

If we assume that the plasma density is predominantly fueled from recycling fluxes and correlate n_n with the gas density in the divertor chamber, the similar $\bar{n}_e$-dependence of these two quantities, as shown in Fig. 17, reflect the linear relationship between electron line density and atom density of equ. (29).

Next we will discuss the radial T_e-profile which we obtain from the heat conductivity equation. Unlike the parallel density loss, we do not have to condense the parallel transport phenomena into the artificial quantity $\tau_{"}$. In the previous section we have discussed the parallel heat flux due to parallel heat conduction so that we obtain as result /45/.

$$\kappa_\perp \frac{d^2 k_B T_e}{dx^2} = \frac{p_{"}}{1} \frac{2B_p}{B_T} \tag{30}$$

$p_{"}$ is the heat loss along the scrape-off layer (in z-direction), 1 is the extent of the plasma which we identify with $2\pi a$ (a = 40 cm for ASDEX). Unlike $\kappa_{"}$, the perpendicular heat conductivity is a constant. The parallel heat flux is, as we have seen (see equ. (5)), of conductive nature

$$p_{"} = -\kappa_{"} \frac{dT_e}{dz} = -\frac{2}{7} \kappa_o \frac{dT_e^{7/2}}{dz} \tag{31}$$

The temperature derivative is obtained from equs. (10) and (13)

$$\frac{dT_e^{7/2}}{dz} = -\frac{T_e^{7/2}}{L} \tag{32}$$

so that the final heat conduction equation is:

$$\kappa_\perp \frac{d^2 k_B T_e}{dx^2} = \frac{2}{7} \frac{B_p}{B_T} \frac{\kappa_o}{\pi a L} T_e^{7/2} \tag{33}$$

We can solve equ. (33) with the boundary condition

$$-\kappa_\perp \frac{d k_B T_e}{dx} = \frac{P}{4\pi^2 a R} \tag{34}$$

at the plasma surface ($x = 0$); P is the power deposited within the plasma and transported across the surface by the electrons. The solution of equ. (33) is /45/:

$$T_e(x) = 1.9 \ (\frac{\kappa_\perp k_B \pi a B_T L}{B_p \kappa_o})^{2/5} \frac{1}{(\lambda_T + x)^{4/5}} \tag{35}$$

λ_T is a characteristic length for the radial T_e-decay. Unlike the density with $\bar{n}_e$-independent transport parameters, T_e does not follow an exponential profile. The dependence of $\kappa_{\shortparallel}$ on T_e (see equ. (9)), causes T_e to drop quickly away from the separatrix but then gives rise to a long tail at large x-values.

$$\lambda_T = 1.26 \ (\kappa_\perp k_B)^{7/9} (\frac{a\pi B_T L}{B_p \kappa_o})^{2/9} (\frac{4\pi^2 aR}{P})^{5/9} \tag{36}$$

Finally, we calculate the electron temperature at the plasma edge $T_e(0)$:

$$T_e(o) = 1.6 \ (\frac{\pi a B_T L}{B_p \kappa_o \kappa_\perp R})^{2/9} (\frac{P}{4\pi^2 aR})^{4/9} \tag{37}$$

The boundary temperature is weakly dependent on the heating power.

6.3 Comparison with Experimental Results

The comparison of the calculated scrape-off and divertor plasma parameters with those shown in Fig. 22 yields order of magnitude agreement. The mean electron temperature $T_e(z = 0)$ of the mid-plane scrape-off layer (determined from equ. (13)) is 65 eV ($\lambda_T = 1.5$ cm, L = 1500 cm); the peak electron temperature $T_e(x = 0)$ at the separatrix, calculated from equ. (37) is 100 eV ($\kappa_\perp = 4 \cdot 10^{17}$ cm^{-1}s^{-1}). The agreement with the measured values is good. The peak scrape-off density $n_e(x=0)$ can be calculated from equ. (24). The result is $n_e(0) = 4 \cdot 10^{13}$ cm^{-3} ($D_\perp = 6000$ cm^2 s^{-1}, $\tau_{\shortparallel} = 0.5$ ms, $S_i = 1.4 \cdot 10^{-8}$ cm^3 s^{-1}, $v_H = 1.4 \cdot 10^6$ cm s^{-1}) and does not agree very well with the measured value. The mean divertor parameters $n_e(L)$ and $T_e(L)$ are calculated with the assumption, that 50 % of the heating power reaches the divertor plasma: $n_e(L) = 6 \cdot 10^{13}$ cm^{-3} and $T_e(L) = 6$ eV. The agreement with the experimental values is fair.

The model yields a steep dependence of both $T_e(L)$ and $n_e(L)$ on the main plasma boundary density which in itself depends non-linearly on the central density, Thereby, the model reproduces qualitatively the observed sensitive dependence of the divertor parameters on $\bar{n}_e$ as shown in Fig. 17 and Fig. 18. This dependence, as we have seen, is the basis of the advantageous properties of the high recycling divertor.

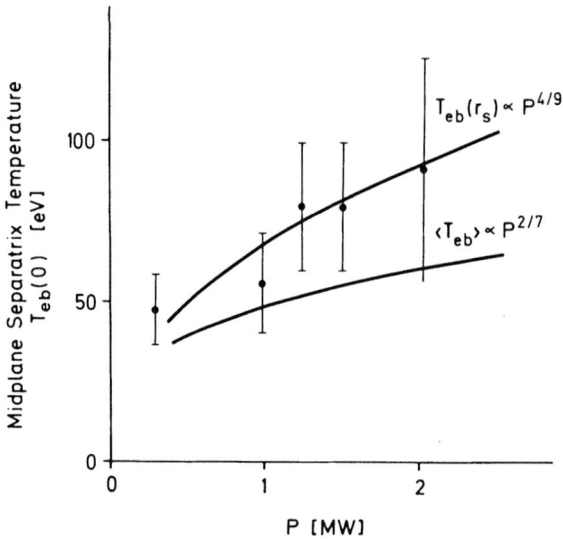

Fig. 24. Variation of the electron temperature at the plasma edge of ASDEX with neutral beam power. The data points are taken from Ref. /45/. The curves are calculated (see text).

The model further reproduces the comparatively weak dependence of the scrape-off values on the heating power P. Figure 24 compares experimental data taken from Ref. /45/ with calculated $T_e(0)$ values taken from equ. (13) and equ. (37) for different heating powers.

A clear discrepancy is that $n_e(L)$ in the model decreases with beam power while experimentally the opposite is observed in case of ASDEX as documented in Fig. 19. But also the experimental situation is not clear because no unique dependence of $n_e(L)$ on P has been observed. In case of DIII, $n_e(L)$ indeed decreases with heating power /47/. The further increase in $n_e(L)$ with beam power achieved in ASDEX may actually be the result of increased external gas puffing, because the main plasma density is allowed to slightly increase with beam power which causes the mid-plane scrape-off layer density and therefore the divertor density also to increase.

The values which are obtained from the model for the divertor parameters are in rough agreement with the experimental findings. Figure 25 compares now radial T_e- and n_e-profiles of the scrape-off layer in the plasma mid-plane measured by different diagnostic techniques on ASDEX. (The data are taken from Ref. /32/). The measurements reproduce the exponential profile in case of the density and the powerlaw-dependence for the electron temperature with the

Te-tail further out. The absolute values are calculated from
$\bar{n}_e = 2 \times 10^{13}$ cm^{-3}, and $P_{OH} = 0.35$ MW. The agreement of the absolute
values is fairly good in particular if one keeps in mind that for
the experiments, the location of the separatrix is hardly known
better than a fall-off length.

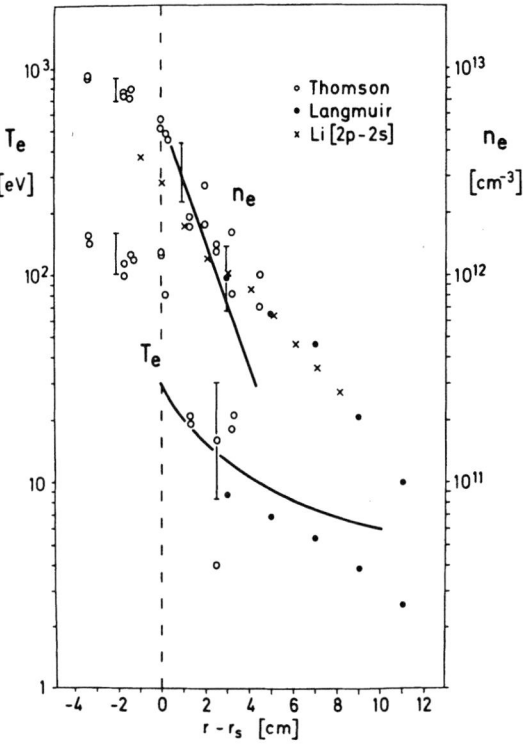

Fig. 25. Plasma edge profiles of n_e and T_c for double null diverted
ASDEX discharges ($\bar{n}_e - 2 \times 10^{13}$ cm^{-3}, $I_p = 316$ kA). The
Li curve is normized at $(r-r_s) = 4$ cm to the Langmuir-
Thomson plots (r_s = separatrix radius). The data points
are taken from Ref. /32/. The solid curves are calculated
(see text).

7. IMPURITY CONTROL

Most important for divertor plasmas is the question whether
impurities can indeed be reduced by this concept. The most instructive technique to demonstrate the potential advantages is the comparison of limiter and divertor discharges of otherwise identical

973

external parameters (e.g. I_p, $\bar{n}_e$, etc.). An important design criterion of the ASDEX magnetic field structure was the achievement of a divertor plasma with circular cross-section which allowed direct comparisons with limiter discharges of the same size. For this purpose, retractable poloidal limiter wings are installed inside the ASDEX vessel which can be moved to the plasma surface between discharges. The gradual transition from a divertor to a limiter discharge can be studied by moving the limiter in steps to the divertor plasma. With a toroidal limiter which had temporarily been mounted into the ASDEX vessel and which bounded the plasma at the bottom side another possibility was given to study gradual divertor to limiter plasma transitions by vertically displacing the plasma column /33/.

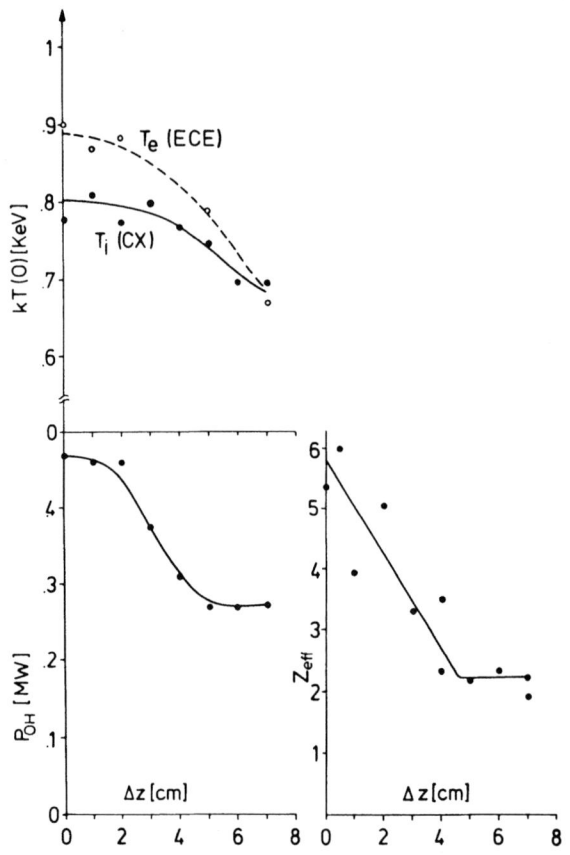

Fig. 26. Variation of the central electron (T_e) and ion (T_i) temperature, the Ohmic power input P_{OH}, and Z_{eff} of ASDEX with the vertical distance Δz between single-null plasma and toroidal limiter (taken from Ref. /33/).

In Fig. 26, results from such a gradual transition are shown
/33/. Plotted are Z_{eff}, the Ohmic heating power P_{OH} and the central
temperatures versus the limiter-separatrix distance Δz. The plasma
density $\bar{n}_e = 2.2 \cdot 10^{13}$ cm^{-3} is kept constant. Z_{eff} increases from 2
of the divertor to 6 of the limiter plasma. As P_{OH} increases with
the plasma resistivity, the central temperatures of a limiter plasma
surpass those of a divertor discharge. Figure 26 clearly demonstrates
that the proximity of a material limiter invariably introduces additional impurities which degradate the plasma performance. Limiter
discharges, however, do not necessarily operate at $Z_{eff} \sim 6$. In particular, high field tokamaks as ALCATOR, FT or PULSATOR, which can
operate at high plasma density and therefore at low edge temperature,
demonstrate Z_{eff}-values comparable to those of divertor plasmas. The
limiter values, shown in Fig. 26, possibly more representative of
the low conditioning state of a large carbon limiter, nevertheless
document the impurity problem of large limiter tokamaks operating
at low toroidal field and consequently at low density.

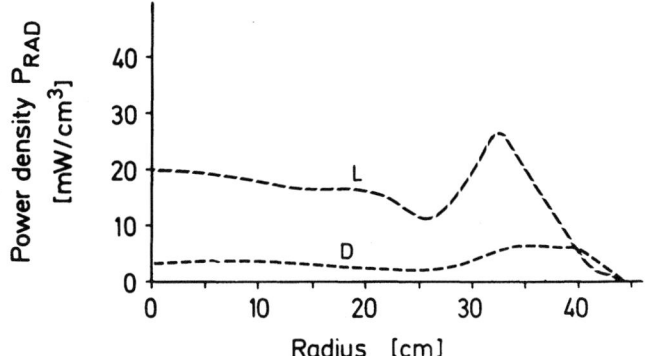

Fig. 27. Comparison of bolometrically measured radiation profiles
          of ASDEX poloidal limiter and divertor discharges. To de-
          monstrate the intrinsic difference in impurity radiation,
          oxygen was removed by titanium evaporation within the
          divertor chamber (taken from Ref. /35/).

The degradation in plasma purity by a limiter is obvious from the
different plasma radiation profiles as measured by a bolometer and
plotted in Fig. 27 /35/. Provision was taken to compare a limiter
with a divertor discharge as clean as possible: The poloidal carbon
limiter was well conditioned and in both cases, titanium was evaporated within the divertor chambers to effectively clean the discharge
from oxygen /38/. Thereby, the unavoidable impurity production
caused by the wall interaction of the hydrogen plasma is documented
for the two configurations unmasked by any non-linear effects of

impurity erosion by light impurities. In both cases, the profiles are rather flat because the oxygen radiation at the plasma edge is missing. The radiation in the limiter discharge is at a 20 mW/cm^3-level while with the divertor, radiation is typically a factor of 5 lower. The lower Z_{eff}-values and the sharply reduced radiation of divertor discharges prove the higher degree of plasma purity of this configuration. Spectroscopic techniques reveal which impurities are particularly affected. The result is astonishing.

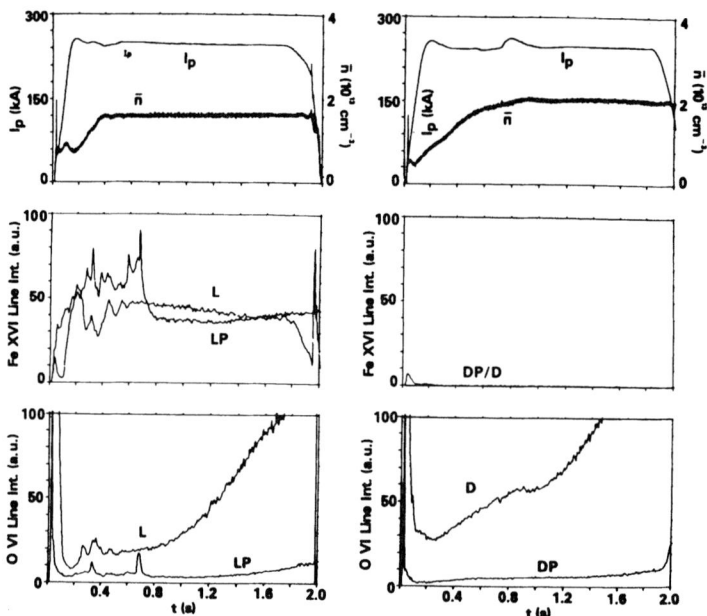

Fig. 28.   Variation of FeXVI and OVI radiation during poloidal li-
           miter (L) and divertor (D) discharges of ASDEX with (P)
           and without titanium sublimation within the divertor cham-
           bers. Plasma current I_p and density $\bar{n}_e$ are additionally
           plotted (taken from Ref. /9/).

Figure 28 compares iron and oxygen radiation of a poloidal limiter and a divertor discharge together with the basic parameters plasma current I_p and line average density $\bar{n}_e$. Iron is chosen, because both limiter and the vacuum vessel of ASDEX are made out of stainless steel. Additionally, for both configurations, the effect of Ti-gettering within the (active or non-active) divertor chambers are documented. The cases with gettering are denoted LP and DP.

The transition from the limiter to the divertor configuration causes the iron radiation to drop drastically already from the beginning of the discharge. Such an improvement has been expected. A surprise, however, was the observation, that oxygen is not reduced by the divertor in a similar way. The removal of oxygen requires

976

additional technical measures as Ti-gettering which can successfully be applied to limiter and divertor discharges.

The pronounced reduction of the iron radiation, as soon as the stainless steel limiter is pulled away, demonstrates the intense plasma interaction with the limiter in comparison to the negligible erosion from the vessel wall during a divertor discharge of the same parameters. The iron content is reduced by a factor of 25 and an iron concentration of 2×10^{-5} is found in Ohmically heated divertor discharges /45/. An increase of the target plate material (Ti in case of ASDEX) has not been noted at the transition to a divertor discharge.

Similar successes in reducing the high-Z impurity component from the limiter have been reported by the other two large poloidal divertor experiments PDX and DIII. In PDX, the limiter material Ti is reduced by a factor 10-30 /48/ and in DIII the inconel component Ni by a factor 10 /11/. The other divertor tokamaks also found that light impurities are not affected by the divertor to the extent of the high-Z limiter component. By using low-Z carbon (or carbon compound) limiters, it became obvious that the different effects on impurities is not a question of the charge state but only of the limiter material. With a graphite limiter, a large drop in carbon radiation is observed as soon as the limiter is withdrawn to give room for a divertor discharge.

The impurity density n_I in a discharge is determined by the impurity influx $\emptyset_i$ which is balanced by the diffusive outflux $D_{\perp}\, dn_I/dx$ at steady state. It has been shown /49/ that the stationary impurity density in the discharge is given by

$$ n_i \simeq \frac{\emptyset_i\, \lambda_i}{D_{\perp}}\, \eta \qquad (38) $$

λ_i is the ionization length for impurities which determines the density gradient and η is the shielding efficiency for impurities. As the scrape-off parameters are sufficient to ionize all sorts of impurities, η is determined by the speed $v_{\shortparallel}$ with which they are swept into the divertor. Thus a crucial parameter which determines η is the residence time $\tau_{\shortparallel}$ of an impurity ion in the boundary layer. Those impurities, which are not swept into the divertor chamber, cross the separatrix and penetrate the main plasma.

From the experimental facts, as presented above, we have to conclude that the shielding properties of limiter and divertor plasmas for impurities entering from the walls are comparable. Those impurities, which do not originate from the limiter are not affected by the configuration. The limiter material on the other hand is strongly affected because the influx $\emptyset_I$ of this specific

977

impurity species is widely different in limiter and divertor discharges. Now, we can draw an important conclusion: The clean divertor discharges are not caused by superb shielding but simply by the fact that there is no obstacle present at the plasma surface which is invariably subjected to high erosion.

The central question of shielding has been investigated at all divertor plasmas in great detail. These studies are generally carried out with artificially introduced impurities in order to decouple the production rate from wall components and the transport effects within the plasma. The shielding studies are favorably carried out with impurity species which do not recycle. After their residence time within the plasma they should stick to the wall. Otherwise, the primary influx again is unknown because it is determined by the impurity recycling processes at the wall. The measurements are carried out with either laser ablated impurities /10/ or gaseous impurities puffed into the discharge /50/. SiH_4 is a good candidate for such studies with Si being studied spectroscopically. The measured radial profiles of impurities of different ionization states can be modelled with one-dimensional multi-species transport codes to yield the impurity distribution. Such an analysis of the experimental profiles yields transport coefficients and the total impurity density profile. A generally observed result is that only a small fraction (~10 %) of the injected amount of impurities actually penetrates the bulk of the plasma irrespective of the plasma configuration /11/. It is noted that also a limiter discharge exhibits good shielding.

Figure 29 shows the result of impurity transport calculations with Si carried out for ASDEX divertor discharges for 2 cases to study the sensitivity of the central impurity density on the shielding efficiency of the scrape-off layer /45/. Together with the total impurity density, the radial profiles of different ionization stages are plotted. The external Si-flux is kept constant for the two cases (note the different ordinates!). $\tau_{\shortparallel}$, the ionized impurity residence time within the scrape-off layer is assumed to be 0.1 msec in case (a) which yields a central density of $1.2 \cdot 10^{10}$ cm^{-3} and 0.5 msec in case (b) giving $n_i = 2.3 \cdot 10^{10}$ cm^{-3}. The comparison of SiH_4-injection experiments into limiter and divertor experiments gave slightly lower values for the limiter case. $\tau_{\shortparallel}$ was found to be 0.5 msec /45/. No improved shielding of the divertor configuration was observed. We have seen that the shielding properties of the scrape-off layer are determined by the competing processes of parallel drift to a sink (limiter, target plate) and the perpendicular diffusion into the main plasma. Experimentally, the loss time $\tau_{\shortparallel} = \pi q \, R/v_{\shortparallel}$ is comparable to the diffusion time $t_D = \lambda^2/D_{\perp}$. λ is the scrape-off layer thickness. As the structure of the scrape-off layer in contact with the main plasma is about the same for limiter and divertor plasmas, no large difference in shielding properties can be expected.

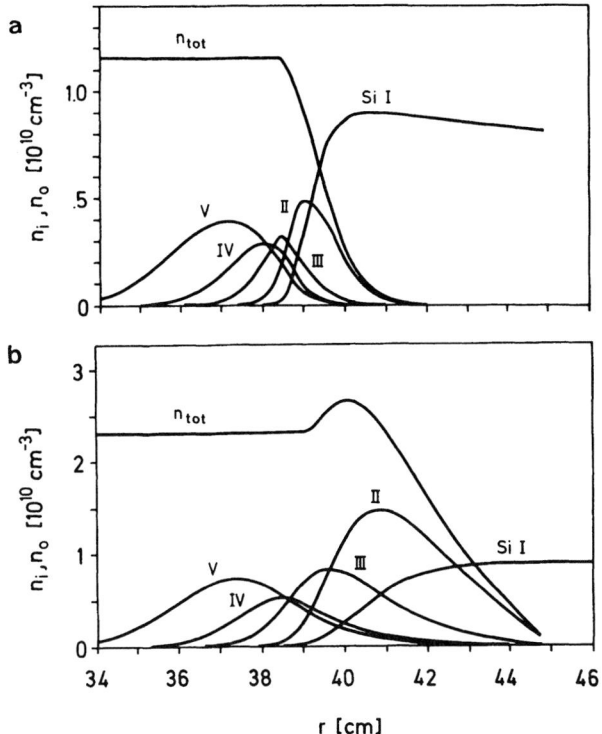

Fig. 29. Calculated radial profiles of the total silicon density
and different ionization stages to demonstrate the effect
of shielding by the scrape-off layer: a) $\tau_{\parallel}$ = 0.1 msec,
(b) $\tau_{\parallel}$ = 0.5 msec. Note the different ordinates! (Taken
from Ref. /45/.

7.1 Impurity Exhaust and Retainment

In discussing recycling and the hydrogen flux pattern of the
divertor configuration, it was shown that the plasma flows along the
scrape-off layer into the divertor chamber establishing the high
exhaust efficiency of the divertor configuration. In this chapter,
we investigate the exhaust efficiency for impurities. Such studies
are carried out with artificial impurities puffed into the discharge. Impurity species are used which recycle so that steady
state conditions are obtained. Favorable gases are Ne and Ar. He
is reserved for specific investigations which we will discuss in
Sec. 8.

Figure 30 compares the NeX-radiation measured after equal
amounts of neon have been puffed into a limiter and divertor discharge of ASDEX of identical parameters /51/. The Ne-valve opened
at 0.8 sec for 6 msec. In the limiter case a high steady state

level of Ne-radiation develops. With the divertor the Ne intensity
rises initially to the limiter level (because of similar shielding)
but then decays to a lower level by a factor 4.5. As the plasma
parameters are similar, this ratio determines the ratio of the Ne-
densities.

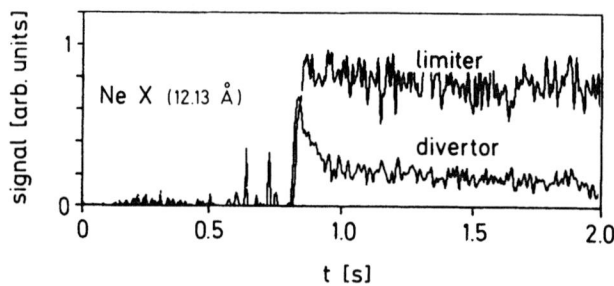

Fig. 30.   Comparison of the NeX-radiation in limiter and divertor
discharges of ASDEX. The Ne gas valve opened at 0.8 sec
for 6 msec (taken from Ref. /51/).

The amount of Ne which is not stored within the plasma is
recovered within the divertor chamber and can be detected by mass-
spectroscopy. The impurity exhaust can be characterized by the ratio
of divertor to limiter concentration which is 22 % for the results
shown in Fig. 30. Similar experiments are carried out on DIII with
Ar puffs /11,42/. The XB configuration of DIII exhibits even better
impurity exhaust than the closed configuration of ASDEX: At high
density an exhaust ratio of 10 % is observed. The open configuration of PDX, on the other hand, does not show any impurity exhaust
efficiency at all /78/. Impurity exhaust efficiency not only requires the transport of impurities in the scrape-off layer along
with the plasma particles but also effective retainment of impurities within the divertor chamber.

The results of the open PDX divertor chamber show, that retainment depends on the flow conductance between divertor and main
plasma chamber. For high retainment and exhaust efficiency, a tight
divertor chamber is necessary. Then the retainment only depends on
the flow conductance through the divertor throat which is blocked
by the divertor fan. The effective conductance of the divertor
throat with plasma has been investigated on ASDEX by puffing in
impurities into the divertor chamber. The time evolution of the
impurity radiation and the steady state distribution in main plasma
and divertor are measured. The puzzling result is that the fluxes
through the divertor throat correspond to the vacuum conductance
despite the presence of the plasma /32/. Additional observations,

however, indicate that this result is fortuitious and does not imply that there is no interaction between the gas flow and the plasma fan. The transport across the throat seems to be the combined effect of ionization processes in the plasma fan and effective transport via wall contact.

Of great importance for future divertor experiments is the demonstration of good impurity exhaust and retainment with the XB configuration D III /11,42/. The impurities are effectively trapped by re-ionization in the plasma fan and subsequently dragged onto the target plate. The requirement for effective trapping is an impurity ionization length smaller than that of hydrogen. This is the case for all impurities apart from He.

7.2 Impurity Production in Divertor Discharges

It was shown that the dominant impurity source term, the erosion at the limiter surface, is avoided in the divertor configuration. In the divertor chamber, the plasma first cools down to low temperatures before it interacts with the target plate. Also the power flow to the target plate is reduced. Nevertheless, erosion of the target plate material is observed and it is shown that the dominant erosion process is sputtering /53/. Figure 31 shows the sputtered Ti-fluxes from the ASDEX target plate as measured by spectroscopic techniques for deuterium and hydrogen plasma versus the line average density /32/. The sputtered fluxes decrease sharply with increasing density. For deuterium, the measured fluxes are compared with those calculated from the plasma parameters in front of the target plate (triangle in Fig. 31). The agreement is fair for deuterium; in case of hydrogen, however, the measured fluxes are an order of magnitude above the calculated ones.

In addition to the cold ions, which establish the high density in front of the target plate, there is a non-thermal, hot ion minority component which can be detected after neutralization at and backscattering from the target plate by copuscular techniques /54/. Such a backscattering profile is shown in Fig. 11. These energetic ions originate from the boundary of the main plasma. Because of their high energy (E > 100 eV) in comparison to the temperatures prevailing in the divertor chamber, they loosely interact with the cold background and propagate with long mean free paths.

From the absolutely measured backscattering spectra, one can estimate the distribution of incident ions and calculate the sputtered Ti-fluxes. The results are also adopted in Figs. 31 and agree with the measured fluxes both for hydrogen and deuterium discharges. One can conclude that besides the cold divertor plasma, energetic ions from the main plasma boundary hit the target plate and cause sputtering. This contribution is also quickly reduced

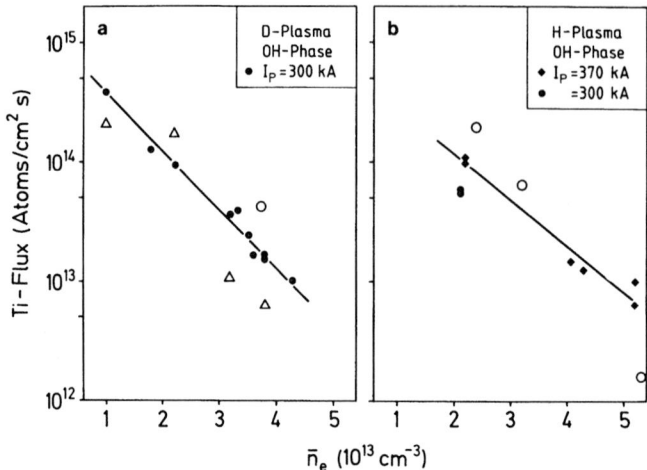

Fig. 31. Titanium flux sputtered off the ASDEX target plate. Solid symbols: spectroscopically measured; open triangles: calculated from the divertor plasma parameters; open circles: sputtering due to the fast ion component of the divertor plasma (taken from Ref. /32/).

at higher density but not because of a temperature reduction. At higher density, these ions suffer predominantly charge exchange processes on their way to the target plate so that their flux is reduced /55/.

The impurities sputtered off the target plate are retained within the divertor chamber like artificially introduced impurities. (In this context, one can attribute the divertor configuration high shielding properties with respect to impurities released from the zone of dominant plasma wall contact.) The retainment together with the low sputtering rates guarantee that target plate impurities can be neglected in the main plasma. The remaining impurities in the main plasma seem to originate from charge exchange sputtering at the main plasma vessel wall. In Sec. 4.1 it is described how the atom density in the main plasma is redistributed at the transition from a limiter to a divertor discharge (see Fig. 10). There is no experimental indication yet (but there is also a lack of precise measurements) that the total amount of hydrogen atoms differs between these two configurations. Therefore, a divertor plasma is subjected to the same charge exchange wall sputtering as limiter discharges. In Ref. /45/ the iron density of a divertor discharge could be explained by the sputtering due to the measured charge exchange fluxes emerging from the plasma. Another indication of charge exchange sputtering is the increase of iron density in the plasma at the transition from hydrogen to deuterium plasmas at otherwise identical plasma parameters /56/.

At high plasma density, when the divertor throat is opaque to the transmission of Franck-Condon atoms "ion refuelling" is expected from the model calculations /57/. At these conditions, the high pressure in the scrape-off layer balances the pressure of the main plasma reducing the net particle outflux across the separatrix. In this case, the particle confinement time is increased giving rise to a corresponding reduction of the ionization source (see equ. 1) and, as a consequence, the charge exchange fluxes to the wall. Up to now, however, there does not exist any convincing experimental evidence for "ion refuelling".

8. He EXHAUST

A special impurity in a fusion reactor is He. It is not an impurity caused primarily by plasma wall interaction and permeated into the discharge from the outside but it is the product of the fusion processes between D and T. Nevertheless, the He concentration in the plasma is of concern and has to be controlled from the out-side. Under steady state conditions, the He-exhaust (in the litera-ture frequently addressed as ash removal) must match the α-particle production rate: 100 MW of fusion power causes a production rate of 2×10^{20} s^{-1}. The α-particle production rate is only determined by the power requirements of the reactor. The steady state He-con-centration in the plasma, however, depends on the intrinsic part-icle confinement properties and on the He-exhaust efficiency. At low efficiency, a fusion reactor operates at unfavourable and un-economic conditions at high plasma pressure but diluted fusion species and therefore reduced fusion reactivity.

The divertor configuration has the potential of a high He exhaust efficiency which relates the He-density at the pumping port (within the divertor chamber) to the one within the plasma. Helium is expected to accumulate within the divertor chamber, similar to hydrogen or impurities, offering the possibility to be pumped at high pressure relieving the requirements for the speed of the ex-ternal pumping system.

Besides the He-exhaust efficiency, the He-enrichment factor is an important system parameter. It is defined as the ratio of He to hydrogen concentration at the pump. A high helium concentration at the pumping port reduces the concomitant pumping of unburnt fuel. Considering the tritium inventory problems of a reactor, the cir-culation of fuel through the system should be kept to a mini-mum.

In present day experiments, these efficiencies cannot yet be studied at the realistic conditions of a burning plasma but have to be simulated by He gas puffing experiments. Under these circumstances, however, the He-source is not in the center but outside of the plasma. But because of the high ionization energy of He (25 eV) giving rise to a low ionization rate coefficient, the scrape-off layer is transparent to He and it is ionized within the main plasma. Like fusion born α-particles, it diffuses to the plasma edge and into the scrape-off layer.

The experimental set-up and instrumentation for He-exhaust measurements is rather simple. Besides a calibrated pulsed gas valve at the main plasma chamber, pressure gauges at the divertor chamber and optionally at the main plasma chamber are necessary. Hydrogen pressure is generally measured by ionization gauges /28/ or by capacitance manometers /58/ while He is measured by a mass-spectrometer or residual gas analyzer. Again the comparison of limiter with divertor behavior is a useful experimental procedure.

8.1 Experimental Results

Figure 32 compares the time dependence of He and the partial pressures both in divertor and main plasma chamber during an ASDEX discharge (I_p = 250 kA, $\bar{n}_e$ = 3 $\cdot$ 10^{13} cm^{-3}) /59/. He was mixed to the H_2 filling gas prior to breakdown in the ratio 1:10 resulting in a He partial pressure of 2 $\cdot$ 10^{-6} mbar. At the beginning of the discharge, H_2 and He are fully ionized and the partial pressures in the plasma vessel initially drop. In course of the discharge, the plasma is refuelled from the outside with a few 10^{21} H_2/sec to compensate the wall and divertor losses and to maintain a density plateau. A He-flux of 6 $\cdot$ 10^{17} He/sec is mixed to the H_2 working gas. Despite the refuelling the pressures in the main plasma chamber stay low. The pressures in the divertor chamber, however, increase in course of the discharge.

Together with the familiar hydrogen accumulation within the divertor chamber, He also accumulates. The compression ratio, defined as pressure ratio of divertor to main chamber is 20 in case of He and 100 for H_2. The divertor is less suitable to compress helium than hydrogen. After the discharge, the pressure differences within the vessel balance. There is an overall H_2-pressure rise after the discharge because of the neutralization of the plasma species. In case of He, the divertor pressure decreases after the discharge because of a lower He content within the plasma than the divertor chamber.

The important aspect of the He exhaust experiments is to relate the He-concentration within the divertor to that within the plasma. At the given α-particle production rate, the He concentration in the plasma should be low while it should be high within the divertor to easily pump it. The most systematic studies have been carried out with D III both in divertor and expanded boundary configuration. Therefore, we will predominantly reproduce these results /60,61/.

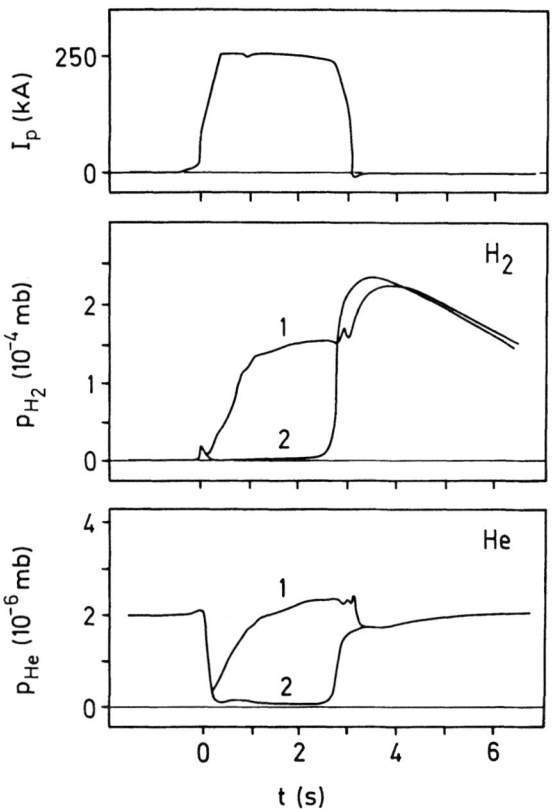

Fig. 32. Hydrogen and helium partial pressures in main plasma (2) and divertor chamber (1) of ASDEX during and after a discharge of 250 kA plasma current (courtesey W. Poschen-rieder).

The experiments are carried out by puffing in a short (5 - 50 msec) He-pulse into the stationary discharge. The He-pressure evolution in the divertor chamber is monitored. No direct measurement of the He-density in the plasma has been tried. The He-content in the plasma is deduced from the increase in plasma density. The amount of gas introduced into the discharge is limited to a density rise typically of $(0.5 - 1) \times 10^{13}$ cm^{-3}.

Figure 33 compares the dependence of H_2 and He partial pressures on main plasma density both for divertor and limiter plasmas. The pressure measurements are carried out after equilibrium is

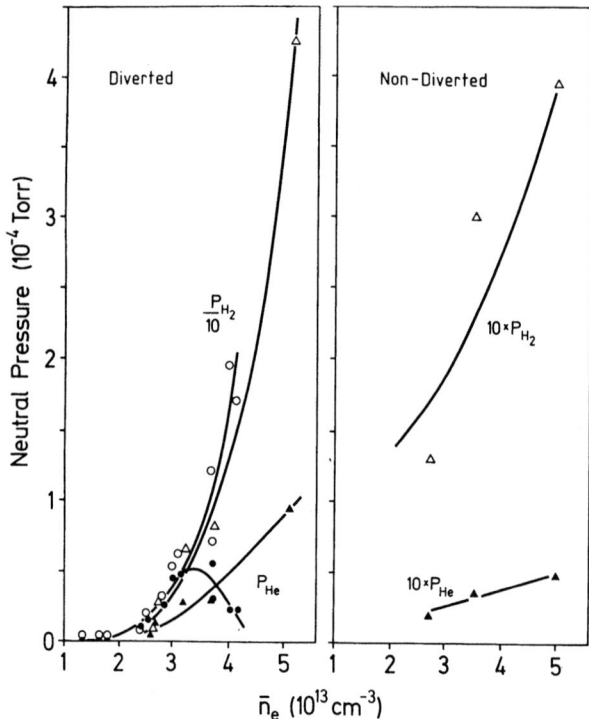

Fig. 33. Partial pressure of hydrogen and helium in the divertor
space of DIII for the SN (triangles) and XB (circles)
configuration. For both configurations, the hydrogen
pressure rises non-linearly with plasma density and
assumes values in the divertor space which are about a
factor 100 above those of non-diverted, limiter discharges. In the XB-configuration, the He-pressure decreases
for $\bar{n}_e > 3 \cdot 10^{13}$ cm^{-3}. (SN-data are taken from Ref. /60/,
XB-data from Ref. /61/.)

established after the He-puff. The He-pressure in the divertor
chamber is more than an order of magnitude higher than in a comparable limiter discharge. This example illustrates the divertor potential to compress He. A critical comparison, however, which allows
a judgement of divertor versus limiter configuration with respect
to He-exhaust has to be based on the He-pressure evolving near an
advanced limiter (e.g. scoop limiter /62/). Such comparisons have
not yet been done experimentally.

The results presented up to now relate to the divertor configuration (SN) of DIII. The XB-configuration shows a somewhat

different behaviour. Though the He-density develops similarly as in the divertor configuration, the He pressure increases at low $\bar{n}_e$ but then decreases towards higher densities after a maximum at $\bar{n}_e = 3 \cdot 10^{13}$ cm^{-3}. The reduction in divertor pressure is accompanied by an increased He concentration within the main plasma. The different behaviour of He-exhaust properties stresses the importance of the actual divertor configuration and the need to optimize it.

The He-enrichment factor η_{He} is determined from

$$\eta_{He} = \frac{n_{He}/n_H(\text{divertor})}{n_{He}/n_H(\text{plasma})} = \frac{P_{He}/2P_{H_2}(\text{divertor})}{\Delta\bar{n}_e/2\bar{n}_e(\text{plasma})}$$

(The 2 in the numerator accounts for the 2 H-atoms per molecule and in the denominator for the 2 electrons per He^{++}). η_{He} is plotted in Fig. 34 for the two configurations. For the divertor configuration η_{He} stays close to one and is nearly independent of density. No He-enrichment but also no pronounced de-enrichment occurs. A different behaviour is found for the XB-configuration. Towards higher densities η_{He} decreases indicating He de-enrichment.

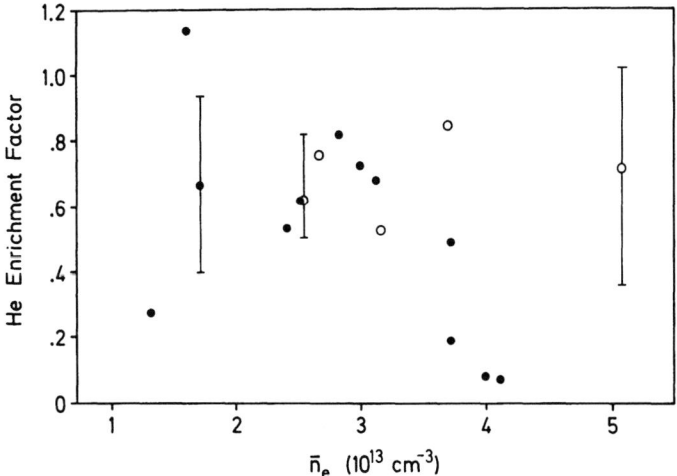

Fig. 34. He enrichment relative to hydrogen for the XB (closed symbols, Ref. /61/) and the SN (open symbols, Ref. /60/) configuration of D III. Both configurations do not show He enrichment, the XB-configuration shows He de-enrichment at higher plasma density.

Now we have to clarify two experimental facts: First, there is no He enrichment in hydrogen and second there is the possibility that He flows out of the divertor contaminating the main plasma and causing the observed He de-enrichment. We follow the argumentation as given in Ref. /61/. For an open divertor configuration the requirements for particles to stay within the divertor chamber is that local reionization occurs after neutralization at the target plate. The ionization length λ_{ion} has to be shorter than the flight path d from the divertor chamber to the main plasma: λ_H, $\lambda_{He} < d$. In order for He enrichment to occur, the requirement $\lambda_{He} < \lambda_H$ has to be fulfilled. Then He is re-ionized within the accretion region in front of the target plate and subjected to the ion-ion frictional forces which sweeps it back to the target plate and effectively traps it within the divertor.

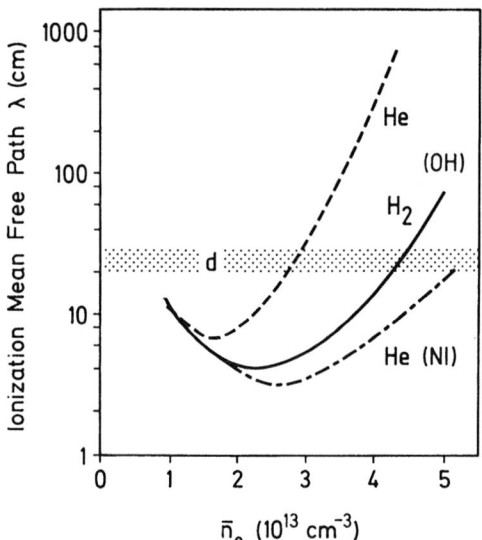

Fig. 35. Mean free path for ionization of helium and hydrogen calculated from Ohmically heated DIII discharges. The results are compared with the characteristic dimension of the divertor d $\sim$ 20 - 30 cm. (Taken from Ref. /61/). Added is the He-curve with 2.0 MW auxiliary heating. The density for de-enrichment shifts from $\bar{n}_e = 2 - 3 \cdot 10^{13}$ cm^{-3} to $5 \cdot 10^{13}$ cm^{-3}.

The density dependence of λ_H and λ_{He} is plotted in Fig. 35 and compared to the characteristic length d of DIII. The variation of

λ_{ion} with $\bar{n}_e$ is determined by the corresponding variation of divertor density and particularly divertor temperature. The difference between λ_H and λ_{He} is caused by the low ionization rate coefficient of He at low temperature /63/ below that of hydrogen, which gives He a chance to leave the divertor plasma and to penetrate the main plasma. Therefore, $\lambda_{He} > \lambda_H$ at all densities contrary to the requirements for He-enrichment. Furthermore, for $\bar{n}_e > 3 \cdot 10^{13}$ cm^{-3}, $\lambda_{He} > d$. The estimated density for He to leave the divertor agrees well with the density where He de-enrichment is observed. The curve, labeled NI, depicts the ionization length for He at an auxiliary heating power of 2 MW and demonstrates that the critical plasma parameters for He de-enrichment to occur are met at higher plasma density. The curve is constructed from the Ohmic curve using the power scaling of equ. 15 and 16.

9.   OPTIMIZATION OF THE DIVERTOR GEOMETRY

We have shown in the previous sections that the requirement for the low plasma temperature in front of the target plate – characteristic for the high recycling and remote radiation mode – is a high gas pressure in the divertor chamber which gives rise to intense energy dissipating and distributing processes at the plasma-gas interface. Because of the non-linear relation to the main plasma density, the high recycling mode is accessible at high plasma densities. In the case of a reactor, however, the plasma density will be fixed by the burn conditions and cannot easily be adjusted to meet the divertor requirements. Nevertheless, in order to operate with a reactor under high recycling conditions, the divertor geometry has to be optimized for this purpose. Therefore, parallel to the basic studies, the divertor operation at different geometries has been investigated. PDX has compared the open and the closed divertor configuration. Both configurations are shown in Fig. 36a,b In case of the open divertor, the divertor and main plasma space are not separated but connected by additional openings which are not filled with plasma. The conductance of these openings is large in comparison to the conductance of the divertor throat to allow easy gas exchange between the two chambers. In case of the closed divertor, the divertor chamber is sealed off from the main chamber by baffels and the only connection is via the divertor throat.

ASDEX has operated in the closed configuration from the beginning. The geometry related studies concentrated on the cross-section of the divertor throat. For a limited sequence of experiments, the divertor duct was narrowed by so-called divertor duct inserts (see Fig. 36c). Thereby, the gap was reduced from 6 cm to 3.5 cm.

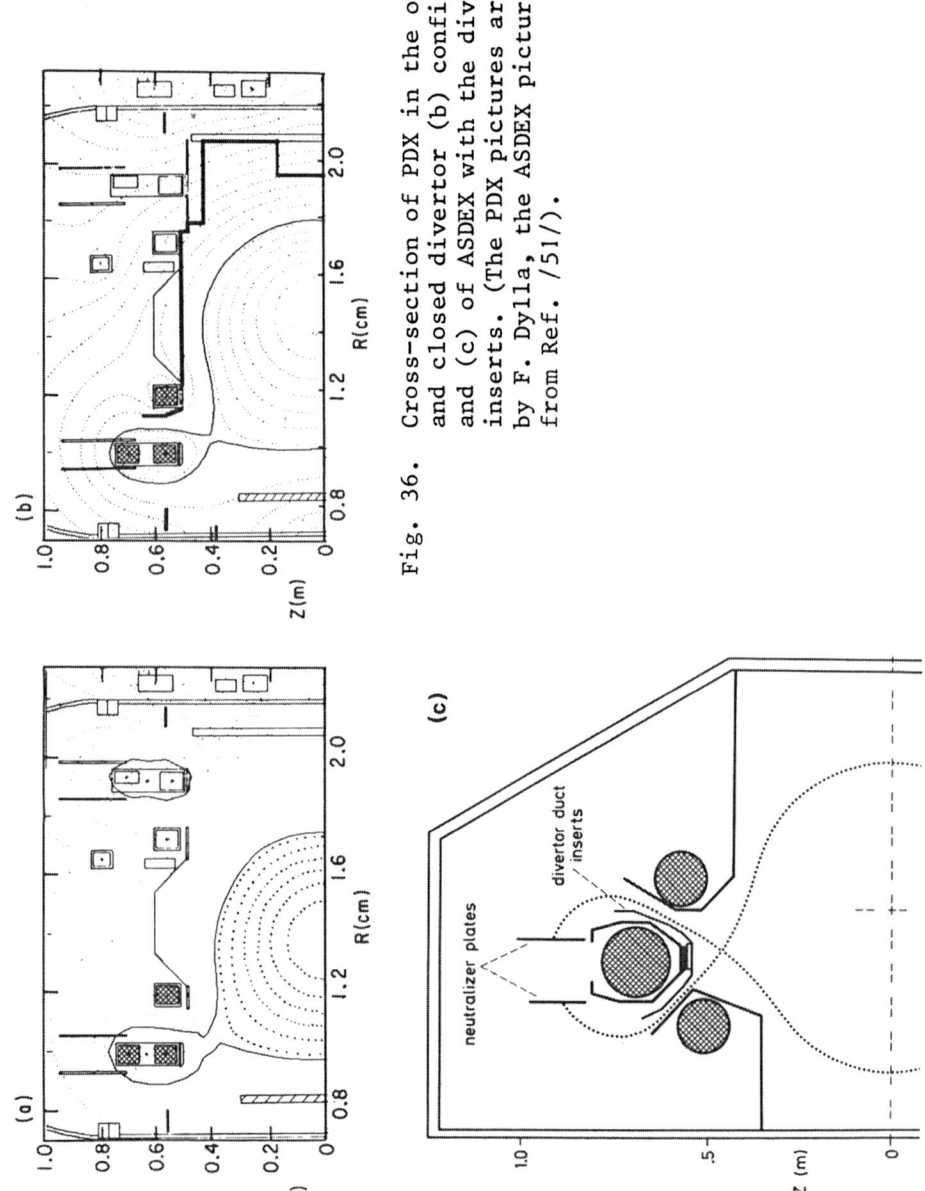

Fig. 36. Cross-section of PDX in the open (a) and closed divertor (b) configuration and (c) of ASDEX with the divertor duct inserts. (The PDX pictures are provided by F. Dylla, the ASDEX picture is taken from Ref. /51/).

The most important contribution to the question of divertor geometry has been made by DIII with its fundamentally different geometry of the expanded boundary. Main plasma and divertor chamber are separated directly by the expanding fan of the scrape-off plasma (see Fig. 3).

With the open divertor of PDX, no hydrogen compression was achieved. The pressures in main plasma and divertor chamber were equal and about 10^{-5} mbar /64/. The power recovered on the target plate was 50 - 60 % of the input power /65/. In case of the closed configuration, the pressure within the main plasma chamber decreased and the pressure in the divertor chamber increased into the 10^{-4} mbar range yielding a compression ratio of 10 - 30 /66/. The power onto the target plate dropped to ~20 % of the input power on account of a sizable fraction radiated within the divertor chamber /65/. But, most spectacular, with the closed configuration, PDX could attain the good confinement regime of neutral injection heated divertor discharges (H-mode, see Sec. 10).

The narrower divertor throats of ASDEX caused a further slight increase in the hydrogen compression. The most spectacular, however puzzling, result has been the observation of a deterioration of the impurity retainment within the divertor. This has been demonstrated by gaseous impurities (Ar, Ne) puffed into the divertor chamber and comparing the steady state level of impurity radiation from the main plasma /51/. A possible explanation, as given in Ref. /51/, is that contrary to expectation the impurity transport along the divertor throat is facilitated by the wall proximity. Impurities, which are carried back onto the target plate by the plasma flow have a chance to hit the throat wall and being neutralized there. After release, they can migrate in the direction of the main plasma with a step size given by the ionization length. These results indicate that there is an optimum width for the divertor throat determined by the requirements of high impurity retainment and high hydrogen compression. A further conclusion is that a diaphragm might be more favorable than an annular geometry for separating divertor and main plasma chambers.

The opposite configuration to a closed divertor with narrow gaps is the expanded boundary configuration of DIII: The divertor fan which widens up away from the stagnation point (see Fig. 3) has sufficient opacity to re-ionize hydrogen atoms with Franck-Condon energies and impurities and effectively localize them within the divertor chamber. No additional baffles are required. We have seen that this configuration leads to high gas compression and low temperatures in front of the target plate and fulfills all criteria of high recycling, remote radiation and, as we will see, good confinement with beam heating.

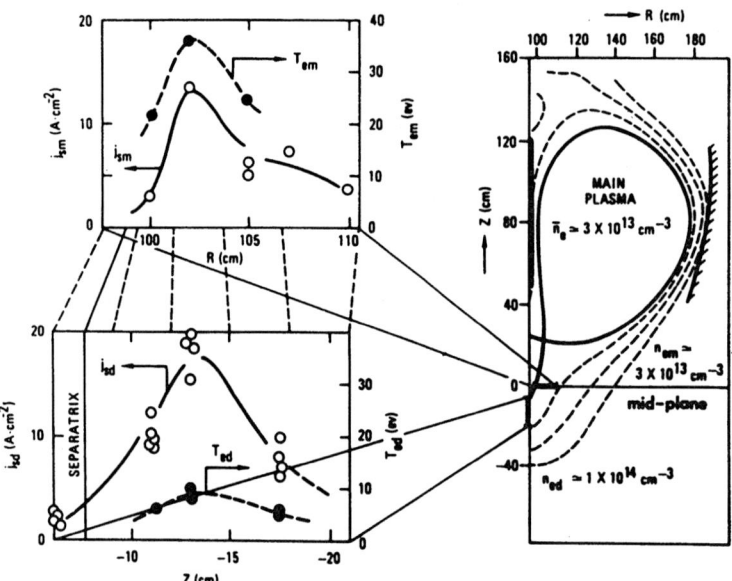

Fig. 37. Horizontal profile of the electron temperature, T_{em}, and ion saturation current j_{sat}, across the lower divertor channel (mid-plane: z = 0 cm) and corresponding vertical profiles of T_{ed} and n_{ed} on the divertor plate at t = 800 ms. The connection of the field lines between both profiles is shown by broken lines. (Taken from Ref. /67/).

This configuration also offers a rather compact divertor design because of the divertor coils which are located outside of the vacuum vessel but also because the high recycling zone is located close to the target plate. This has been documented by Langmuir measurements in the target plate vicinity /67/. Figure 37 compares the results of a horizontal and a vertical scan in front of the target plate. The range probed by the diagnostic device is indicated by heavy lines in the cross-sectional view on the right side. Plotted are profiles of Te and the ion saturation current $j_{sat} \propto n_e \sqrt{T_e}$ (which mostly represents the density distribution). The peak electron temperature drops from 36 eV of the horizontal scan to 8 eV at the target plate. The cooling occurs along the connection length of both points which is 380 cm owing to the glancing incidence of the magnetic field lines on the target plate. The radial separation of the two positions, however, is only about 8 cm and it is this distance which has to be related to the radial extent of the divertor chamber. The divertor chamber of the expanded boundary configuration can be kept small and compact which facilitates its application to the next larger machines and a possible fusion reactor.

The need to use the given space in the vacuum vessel economically asks for a compact divertor structure and a single-null configuration as shown in Fig. 7. Single-null (SN) and double-null (DN) plasmas could be compared on ASDEX on a shot-to-shot basis by vertically displacing the plasma column. Some advantages of the SN configuration have been observed. SN plasmas show a similar degree of plasma purity; at the same plasma density, the gas pressure in the active divertor chamber is higher by a factor of two, reducing the impurity sputtering from the target plate by non-thermal ions as described in Sec. 7.2. The high recycling regime develops at lower plasma density. The good confinement regime of neutral beam heated discharges (H-regime, see Sec. 10) is also observed in the SN-configuration. The accessibility of the H-mode is even improved because its power- and density-thresholds are reduced (from P_{NI} = 2 to 1.2 MW and from $\bar{n}_e$ = 3 · 10^{13} to 2 · 10^{13} cm^{-3}).

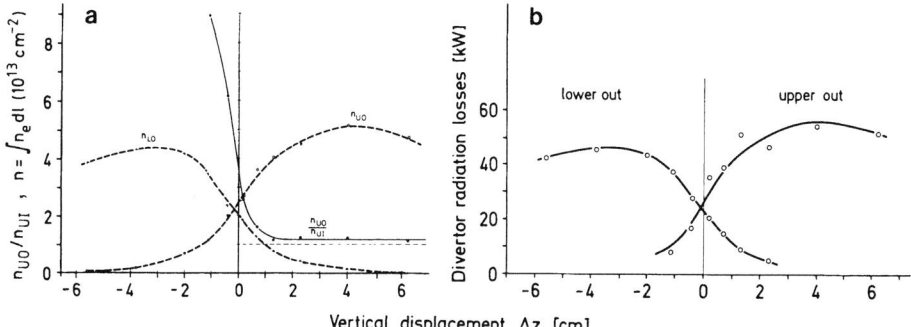

Fig. 38. Effect of the transition from a single-null to a double-null configuration (a) on the line densities of the divertor plasma, the location of the density measurement is shown in Fig. 7a (taken from Ref. /45/) and (b) the bolometrically measured divertor radiation (taken from Ref. /35/).

The power deposited onto 2 instead of 4 target plates as in the DN case could be seen as a principal disadvantage of the SN configuration. Figure 38 shows the transition from the DN to the (up or down) SN configuration examplified by the variation of the scrape-off layer line density of the divertor fans (the location of the measuring chords are shown in Fig. 7) and the bolometrically measured radiation in the two chambers with vertical plasma position z. In the SN configuration (z = 0), the line density in the upper outer (UO) and the lower outer (LO) channels and the upper and lower divertor radiation are the same. Particle and power fluxes along the outer scrape-off layer are symmetric into the two divertor chambers. The upper inner (UI) scrape-off line density, however, is about

1/4 of the outer one. Also the powerflux onto the inner target plate is correspondingly reduced /9/. A vertical upward displacement of the plasma column causes the transition to the SN configuration. UO-line density and upper divertor radiation increases by a factor of 2. Now inner and outer line density and power deposition are symmetric.

We can conclude that also in the DN configuration, most of the particle and energy fluxes are directed only to the two outer target plates. The power and particle fluxes to the inner target plates are reduced typically by a factor 4. The reason for this asymmetry is a radial direction of the fluxes across the separatrix predominantly into the outer scrape-off layer which can only partly be explained by geometrical reasons /79/. The internal flux pattern giving rise to this asymmetry is not well understood. In the SN-configuration, inside and outside scrape-off layers are linked and the discrepancy disappears. Because of the higher gas density in the divertor chamber in this case, outer plasma line density and radiation additionally increase by a factor of 2 making the SN configuration even more attractive than the DN configuration.

10. EFFECT OF THE DIVERTOR ON PROPERTIES OF THE MAIN PLASMA

We have discussed the modified plasma wall interaction due to the divertor configuration and the concomitant reduction of the impurity concentration in the main plasma. In this chapter, we will discuss the extent of other plasma parameter changes as a consequence of the specific divertor properties – high plasma purity and modified field structure at the plasma surface (see Sec. 2.3).

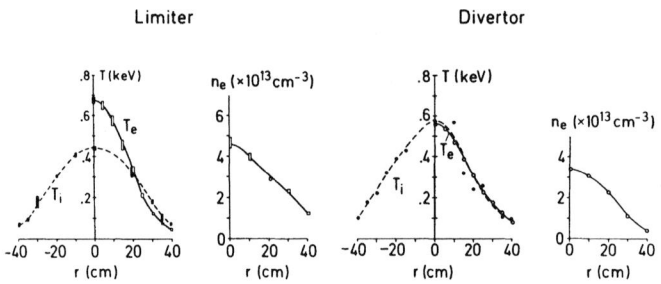

Fig. 39. Electron density and temperature and ion temperature profiles of Ohmically heated ASDEX limiter and divertor discharges.

Figure 39 compares density, electron, and ion temperature profiles of Ohmically heated ASDEX limiter and divertor discharges.

994

In both cases, the profiles are peaked; the central temperatures of
limiter plasmas are above those with divertor because of the higher
Ohmic heating power input. Otherwise, no principal difference is
observed. The volume-integral over the plasma profiles yields the
global energy confinement time τ_E which is plotted in Fig. 40a both
for limiter and divertor discharges /68/. No systematic difference
is observed. We can conclude from these comparisons, that limiter
and divertor discharges are rather similar apart from the different impurity levels. The advantage of the divertor configuration
is not on the expense of any other important plasma aspect. This
similarity only applies to Ohmic discharges. With neutral beam injection, limiter and divertor plasmas have widely differing properties.

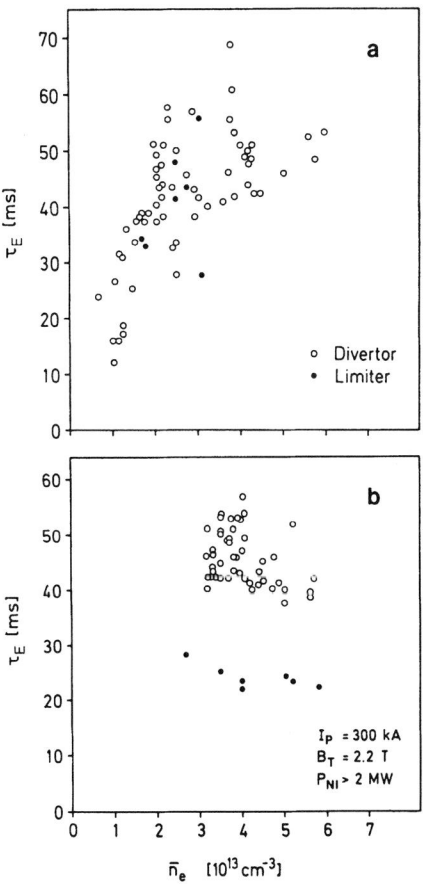

Fig. 40. Density variation of the global energy confinement time
          of ASDEX limiter and divertor discharges with Ohmic and
          beam heating (taken from Ref. /68/ and Ref. /70/).

The attempt to increase the plasma temperatures by auxiliary heating is successful with neutral injection as heating technique. The plasma temperatures can be increased close to the ignition condition /69/. The other important ignition parameter, the energy confinement time, however, is found to decrease so that, in effect, the beam heated plasma does not approach the ignition conditions. This is the situation for limiter discharges. Divertor plasmas, on the other hand, can be operated without degradation in their confinement properties. On the contrary, with ASDEX, confinement times well above those of Ohmic discharges can be achieved.

Figure 40b compares the density dependence of τ_E of beam heated limiter and divertor discharges of ASDEX /70/. τ_E of divertor discharges is about twice that of limiter discharges and comparable with the Ohmic values plotted in Fig. 40a. The two operational regimes are denoted "L" for the low limiter and "H" for the high divertor values. The H-regime was discovered on ASDEX /71/ and subsequently confirmed by the other poloidal divertor tokamaks PDX /72/ and D III /73/. Up to now, no limiter tokamak could be operated in the H-mode and the understanding of the specific divertor characteristic, which allows it, is at present one of the main goals of divertor research.

Also in divertor discharges, the injection pulse first causes deteriorated confinement. Still in the initial phase, when the temperatures are increasing, there is a sudden and distinct transition into the H-phase. It is initiated by the formation of a transport barrier at the separatrix /74/. The barrier is caused by a reduction of cross-field transport coefficients. As a consequence, density and temperature gradients became very steep in this zone. Location and spatial extent of the edge zone with steep gradients coincides with the local shear anomaly of the divertor configuration. As the poloidal field vanishes at the stagnation point, the safety factor q and the shear $S = r/q\ dq/dr$ become infinitely at the separatrix. As large shear can stabilize transport affecting instabilities, it is tentatively concluded in Ref. /74/ that the H-mode is caused by shear stabilization at the edge of the divertor plasma and is therefore inherently a divertor related phenomenon.

High shear is also present in Ohmic divertor discharges, and there, no difference in confinement properties to limiter discharges has been noted (see Fig. 40a). It is found experimentally, that shear stabilization is only effective at increased electron temperature or electrical conductivity /75/. Therefore, the H-phase is preceded by a short L-phase till the edge temperature is sufficiently increased. The threshold temperature at the plasma edge (in the range of 200 eV) cannot be obtained in Ohmic discharges. But also beam heated limiter discharges do not allow large T_e-excursions at the plasma edge. Figure 41 plots T_e at the plasma

periphery versus limiter-separatrix distance ΔR_L. The measurements
are carried out by moving the outside poloidal limiter wing of
ASDEX to the surface of a divertor plasma. The approach of the limiter causes the edge T_e to drop and enforces the transition from
the H to the L mode at a limiter distance of 2.5 cm.

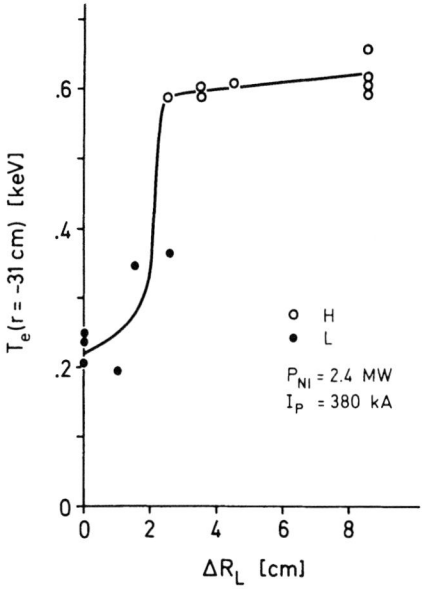

Fig. 41. Variation of the electron temperature T_e in the plasma
periphery of ASDEX when the outside poloidal wing limiter
is moved to the plasma surface. The $\Delta R_L = 0$ data points
are obtained from limiter discharges without energized
divertor coils (taken from Ref. /75/).

The edge conditions of the H-mode give rise to a new class of
plasma profiles. Besides the flat density and temperature profiles,
also the current density profile is found to broaden. This is under-
standable if one visualizes that the electrical conductivity a few
cm inside the separatrix is comparable with the one of an Ohmic
discharge in the plasma center. Figure 42 plots the scaling of T_e
at the plasma periphery and the sawtooth repetition time $\tau_{s.t.}$
(indicative of changes of the current density profile) with the
beam power P_{NI} and compares divertor and limiter results. With increasing beam power, T_e and current density profiles of a divertor
discharge broaden with the consequence that $\tau_{s.t.}$ sharply increases
and for $P_{NI} > 2$ MW, q in the centre increases above 1 so that the

resonance condition for the internal disruption disappears. In the limiter case, the boundary T_e is clamped and the current density profile cannot change. After a slight initial rise in T_e (periphery) and $\tau_{s.t.}$ both quantities saturate pointing towards the existence of a power dependent stabilisation mechanism which could be impurity erosion from the limiter.

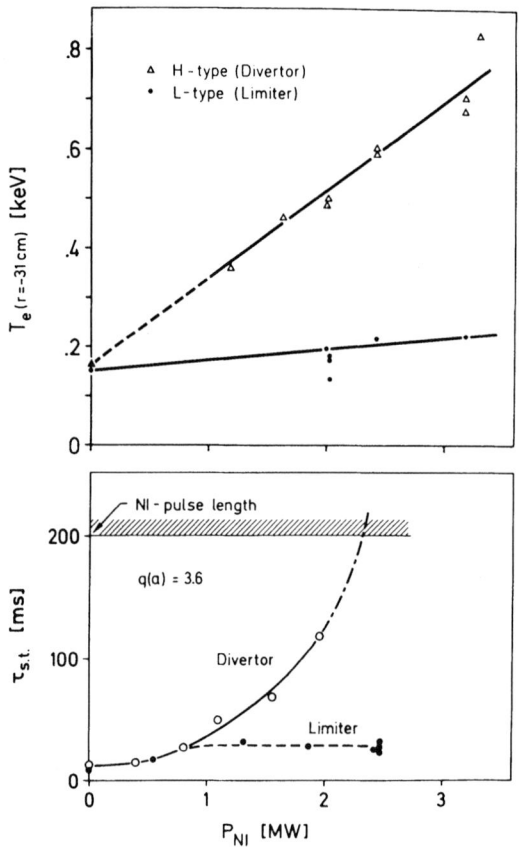

Fig. 42.   Electron temperature T_e at the plasma periphery and the repetition time for the internal disruption $\tau_{s.t.}$ of ASDEX limiter and divertor discharges versus neutral beam heating power P_{NI}. The comparison indicates the large profile change of a divertor discharge. (Taken from Ref. /70/ and Ref. /75/).

In summary, beam heated divertor plasmas are widely different from limiter plasmas with respect to plasma profiles, MHD properties and, most significant, global confinement properties. The H-mode

most clearly demonstrates the advantages of the divertor concept
which decouples the parameters of the scrape-off layer which is in
immediate contact with the wall from those of the periphery of the
main plasma. Because of the low recycling conditions and high purity,
the edge temperature can attain values of a few 100 eV so that the
shear at the plasma edge which is specific to the divertor field
structure can affect transport properties.

11. NEAR-FUTURE DIVERTOR EXPERIMENTS

There are two large divertor experiments under construction:
JT-60 /76/ in Japan which will begin operation 1985 and ASDEX-Upgrade /77/ in Germany scheduled for 1987. The external parameters
of the two machines are given in Table 1, poloidal cross sections
are shown in Fig. 43.

Table 1:

		Asdex-up	JT-60
Major plasma radius	R_0 [m]	1.65	3
Minor plasma radius	a [m]	0.5	0.93
Plasma elongation	b/a	1.6	
Plasma aspect ratio	$A = R_0/a$	3.3	3.2
Plasma current	I_p [MA]	2	2.7
Toroidal magnetic field	B_t [T]	4	4.5
auxiliary heating power	P [MW]	12	20-35

These machines will continue the research programme of present day
experiments. Because of increased auxiliary heating power they
will operate closer to the reactor requirements. Particularly the
power flux density in the scrape-off layer already matches the one
of a reactor. Additionally, confinement and stability aspects of a
near reactor grade plasma will be studied for long auxiliary heating
phases. Because of high toroidal fields, the plasmas can be operated at higher line densities at the edge. The impurity control
studies will be supplemented as even charge exchange atoms from
the hot plasma are absorbed by the boundary layer and do not reach
the wall.

In case of ASDEX-Upgrade, the divertor topology is defined by
external coils which are located outside of the toroidal field
coils. This causes configurational changes (elongated plasma, expanded boundary-like open divertor) which, together with the envisaged single null structure, model in a smaller scale the designs
for reactor-like experiments as INTOR.

999

a

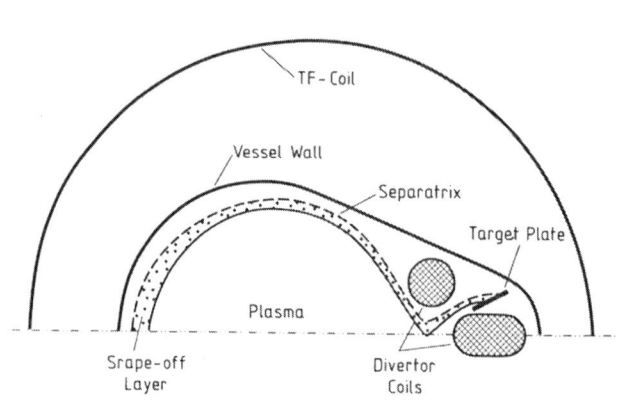

b

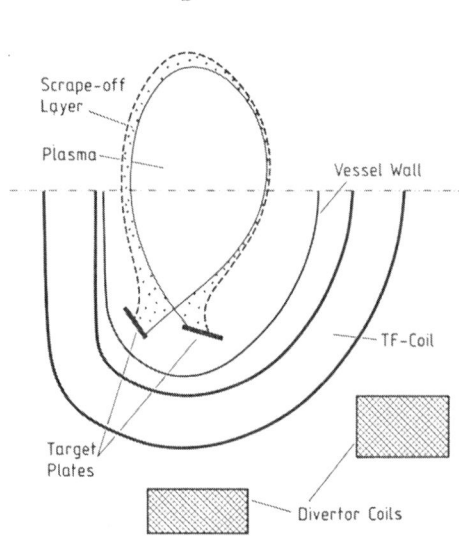

Fig. 43. Plasma cross-section and location of the principal diver-
          tor coils of JT-60 (a) and ASDEX-Upgrade (b).

12. SUMMARY

     Limitation of a tokamak plasma by a magnetic separatrix is a
successful concept. The plasma is stable, its surface is well defined and the radial extent of the scrape-off layer is restricted

1000

to 1 - 2 cm. Particle and energy fluxes from the plasma are guided along the scrape-off layer into the divertor chamber. The exhaust efficiencies both for particles and energy are high. The interaction with the wall structures which surround the main plasma are for present day experiments only via charge exchange atoms.

Though the screening properties of the divertor scrape-off layer for wall produced impurities does not differ from that of a limiter discharge (because of similar scrape-off parameters in the plasma mid-plane), impurity content and radiation are much reduced. The reason is that a limiter which is the dominant impurity source is avoided.

The plasma parameters within the divertor chamber are determined by the local recycling pattern which arises from neutralization of the plasma at the target plate and the subsequent re-ionization of the background gas in the plasma fan. This circular process is fed by the power fueled into the divertor chamber and leads to a sizable reduction of the power flux onto the target plate. The plasma in front of the target plate is cooled down to temperature values below 10 eV. Because of the remote radiation and high recycling properties of the divertor plasma together with the impurity retainment within the divertor chamber, impurities, released from the target plate, do not play any role for the main plasma.

The retainment of impurities within the divertor chamber gives rise to high compression ratios which also applies to He. He, introduced into the discharge to simulate α-particle production, is effectively exhausted from the plasma and compressed in the divertor chamber where it mixes with the hydrogen background so that it can be effectively pumped already at low and technically manageable pumping speeds.

The advantages of decoupling the parameters of the plasma periphery from thoses of the scrape-off layer in immediate contact with the wall are documented by the H-mode which is restricted to neutral beam heated divertor plasmas. The temperature in front of the target plate stays low while that 5 cm inside the separatrix attains values of several 100 eV. Under these circumstances, the specific field topology at the separatrix with high shear effectively stabilizes transport affecting fluctuations giving rise to good confinement at high plasma pressure.

On the basis of the technical means, available at present for studying the most important requirements for the next large machines or the reactor, the divertor concept offers solutions for impurity control, He-ash removal and energy confinement. The magnetic field structure can be produced by external coils, the divertor itself can be compact. The expenses for a divertor solution surpass those of alternative concepts only slightly which, however, have not yet demonstrated comparable efficiency.

ACKNOWLEDGEMENT

The authors have profited from the numerous discussions on various
divertor aspects taking place within the ASDEX-Team and the Tokamak
Physics Group. In particular, thanks are due to G. Fußmann and
W. Poschenrieder for providing experimental results not yet
published, to U. Ditte and S. Cohen for critically reading the
manuscript and to I. Hermann, T. Henningsen and H. Volkenandt
for preparing the manuscript.

REFERENCES

/1/ H. Kotzlowski, et al., Jour. Nucl. Mat. 93/94, 442 (1980).
/2/ W. Feneberg, Phys. Lett. A, 36 (1971) 125.
/3/ F. Karger, K. Lackner, Phys. Lett. A, 61 (1977) 385.
/4/ W. Feneberg, G.H. Wolf, Nucl. Fus. 21 (1981) 669.
/5/ De Grassie, et al., Jour. Nucl. Mat. 128/129 (1984) to be
     published.
/6/ Princeton Plasma Physics Annual Report MATT-Q-21 (1963) p. 7.
/7/ P. E. Stott, C.M. Wilson, A. Gibson, Nucl. Fus. 17 (1977) 481.
/8/ P. J. Harbour, et al., as ref. /5/.
/9/ M. Keilhacker, et al., in Plasma Phys. and Contr. Nucl. Fus.
     Research 1980 (Proc. 8th Intern. Conf. Brussels) Vol. 2,
     IAEA, Vienna (1981), 351.
/10/ D. Meade, et al., ibid, Vol. 1, 665.
/11/ M. Nagami, et al., ibid, Vol. 2, 367.
/12/ N. Ohyabu, et al., Nucl. Fusion 23 (1983), 295.
/13/ O. Gruber, et al., Jour. Nucl. Mat. 121 (1984), 407.
/14/ INTOR, Japanese Contrib. to the 3rd Meeting of the INTOR
     Workshop, 16-28 June 1980, IAEA, Vienna.
/15/ O. Gruber, K. Lackner, Proc. 11th Europ. Conf. on Contr.
     Fusion and Plasma Physics, Aachen 1983, Vol. 2, 443.
/16/ G. Laval, R. Pellat, J.S. Soule, Phys. Fluids 17 (1974)
     835.
/17/ M. D. Rosen, Phys. Fluids 18 (1975) 482.
/18/ G. Becker, K. Lackner, in Plasma Physics and Contr. Nucl.
     Fus. Research 1976 (Proc. 6th Intern. Conf. Berchtesgaden)
     Vol. 2, IAEA, Vienna (1977), 401.
/19/ F. Schneider, Proc. 9th Symp. Eng. Problems of Fusion Research,
     (1981), Vol. I, p. 503.
/20/ H. Yokomizo, et al., "Plasma Physics and Controlled Nuclear
     Fusion Research 1982", Vol. III, IAEA, Vienna (1983) 173.
/21/ D. L. Dimock, et al., Nucl. Fusion 13 (1973) 271.
/22/ F. Wagner, J. Vac. Sci. Technol. 20 (1982) 1211.
/23/ M. H. Hughes, and D.E. Post, Jour. of Comp. Phys. 28 (1978)
     43.
/24/ F. Wagner, "Investigation of Limiter Recycling in the Divertor
     Tokamak ASDEX", Report IPP-III/71 (Garching) 1981.
/25/ G. Proudfoot, and P.J. Harbour, as Ref. 1, p. 413.

/26/ S. J. Fielding, and A.J. Wootton, as Ref. 1, p. 226.
/27/ S. J. Fielding, et al., Jour. Nucl. Mat. 76/77 (1978) 273.
/28/ G. Haas, et al., Jour. Nucl. Mat. 121 (1984) 151.
/29/ ASDEX-Team, presented by F. Wagner, Proceedings of IAEA
     Technical Committee Meeting on Divertors and Impurity
     Control, Garching, July 1981, p. 40.
/30/ G. Haas, et al., Proc. 9th Symp. on Fusion Technology (1976),
     p. 73.
/31/ Y. Shimomura, et al., Nucl. Fusion 23 (1983), 869.
/32/ G. Fußmann, et al., as Ref. 5 and Z. Szymanski, private
     communication.
/33/ H. Vernickel, et al., Jour. Nucl. Mater. 111/112 (1982),
     317.
/34/ M. Shimada, et al., Nucl. Fusion 22 (1982) 643.
/35/ E. R. Müller, et al., Nucl. Fusion 22 (1982) 1651.
/36/ E. B. Meservey, et al., as Ref. 1, p. 267.
/37/ E. R. Müller, et al., as Ref. 28, p. 138.
/38/ P. E. Stott, et al., Nucl. Fusion 15 (1975) 431.
/39/ M. Keilhacker, et al., as Ref. 20, p. 183.
/40/ H. D. Murmann, and M. Huang, "Thomson Scattering Diagnostic
     in the Boundary Layer of ASDEX", Report IPP-III/95
     (Garching) 1983.
/41/ Contributions of D. Post, K. Lackner and of C.E. Singer
/42/ M. A. Mahdavi, et al., Phys. Rev. Lett. 47 (1981) 1602.
/43/ L. Spitzer, "Physics of Fully Ionized Gases", John Wiley and
     Jones Inc., New York (1962).
/44/ W. Engelhardt, et al., as Ref. 33, p. 343.
/45/ M. Keilhacker, et al., Physica Scripta, Vol. T2/2, (1982) 443.
/46/ P. J. Harbour, and M.F. Harrison, Nucl. Fusion 19 (1979) 695.
/47/ M. Shimada, et al., as Ref. 5.
/48/ D. K. Owens, et al., as Ref. 1, p. 213.
/49/ W. Engelhardt, and W. Feneberg, as Ref. 27, p. 518.
/50/ K. Behringer, et al., as Ref. 29, p. 42.
/51/ G. Fußmann, et al., as Ref. 28, p. 164
/52/ J. Neuhauser, et al., Nucl. Fusion 24 (1984) 39.
/53/ B. Schweer, et al., as Ref. 5, p. 71
/54/ F. Wagner, et al., Proc. 3rd All Union Conf. on High Tempera-
     ture Plasma Diagnostics, Dubna 1983.
/55/ F. Wagner, to be published.
/56/ K. Behringer, private communication.
/57/ W. Schneider, et al., as Ref. 28, p. 178.
/58/ H. F. Dylla, Jour. Vac. Sci. Technol. 20 (1982) 119
/59/ W. Poschenrieder, private communication.
/60/ M. Shimada, et al., Phys. Rev. Lett. 47 (1981) 796.
/61/ J. C. DeBoo, et al., Nucl. Fusion 22 (1982) 572.
/62/ P. K. Mioduszewski, this conference.
/63/ W. Lotz, Astrophys. J. Suppl. Ser. 14 (1967) 207.
/64/ H. F. Dylla, et al., as Ref. 33, p. 211.
/65/ M. G. Bell, et al., as Ref. 5.
/66/ H. F. Dylla, et al., as Ref. 5.

/67/ S. Sengoku, et al., Nucl. Fusion 24 (1984) 415.

/68/ O. Klüber, and H. D. Murmann, "Energy Confinement in the Tokamak Devices Pulsator and ASDEX", Report IPP III-72, Garching (1982).

/69/ H. Eubank, et al., Phys. Rev. Lett. 43 (1979) 270.

/70/ F. Wagner, et al., as Ref. 20, p. 43.

/71/ F. Wagner, et al., Phys. Rev. Letters 49 (1983) 1408.

/72/ S. M. Kaye, et al., "Controlled Fusion and Plasma Physics" (Proc. 11th Europ. Conf. 1983), p. 19.

/73/ M. Nagami, et al., Nucl. Fusion 24 (1984) 183.

/74/ F. Wagner, et al., to be published in Phys. Rev. Lett.

/75/ F. Wagner, et al., as Ref. 28, p. 103.

/76/ Y. Shimomura, et al., as Ref. 5

/77/ H. Vernickel, et al., as Ref. 5.

/78/ D. M. Meade, as Ref. 29, p. 1.

/79/ U. Daybelge, Nucl. Fusion 21 (1981) 1589.

PLASMA WALL INTERACTIONS IN HEATED PLASMAS

J. Tachon

Association Euratom - Cea Sur La Fusion
Département de Recherches sur la Fusion Contrôlée
Centre d'Etudes Nucléaires
Boîte Postale n°6. 92260 Fontenay-aux-Roses (France)

INTRODUCTION

With the development of larger, more powerful magnetic
confinement devices, the study of plasma-wall interactions becomes
increasingly relevant to the controlled thermonuclear fusion
program. To increase the plasma temperature in Tokamaks from the
value reached by ohmic heating (1 - 2 keV), toward the value
necessary for self sustaining fusion reactions (5 - 10 keV)
requires a tremendous amount of additional power. This is achieved
either by the injection of high energy neutral particles - called
neutral beam injection (NBI) - or by high frequency electromagnetic
radiation - also called radiofrequency heating (RF). In recent
experiments the amount of power introduced by these "additional
heating" schemes, greatly exceeds the ohmic power (in JET, P_{NBI}
+ P_{RF} = 25 MW, P_{OH} = 2 MW). While substantially increasing the
energy density of the plasma a new set of problems are introduced
which are associated with the increased plasma wall interactions
due to additional heating. In addition to the growth of the energy
density of the confined plasma which obviously increases the energy
flux toward the wall chamber (particles, photons) one has to
consider the impurity influx coming from the heating subsystems
(neutral source, launching structures), the interaction with the
vacuum vessel of fast particles created on the edge and in the core
of the plasma, the power loss on the vacuum chamber surface (for
RF), etc. The understanding of the physical mechanisms which are at
the origin of the impurity production is still incomplete,
particularly in the RF experiments. However, the quantitative
information available in these areas has been growing and the
purpose of these lectures is to give a comprehensive presentation
of the subject. For the sake of clarity the problems associated

with a specific heating method will be presented separately (NBI, RF) even if some common aspects exist.

A – Neutral Beam Injection and Impurities

one of the heating methods used since the earliest investigations of mirror configurations and then successfully applied to tokamaks is the Neutral Beam Injection (NBI). High energy neutral particles are injected into the plasma, where they are ionized, trapped in the magnetic field, and then thermalized by collisions with the main plasma. The most spectacular results to date are the 7 keV ion temperature achieved on the PLT and PDX experiments, the high β (4,7 %) obtained on Doublet III, the high energy confinement regime on ASDEX and the current-free hot plasma in the stellarator W7–A.

A – 1 Conceptual design of an injector

The high energy neutrals can be produced from positive or negative accelerated ions.

A-1-1 With positive ions :

A beam line schema corresponding to ISX–B[1] is shown fig. 1 both with power flow and main operating parameters indicated. The positive ion source delivers a high energy ion beam (here 42 keV) which goes through a gas cell acting as a neutralizer. The fraction of the ion beam which is not transformed into neutrals (here 37 %) is deflected by the bending magnet. A small fraction of the beam power is lost during its propagation through the drift tube (some small losses are not measured) and the overall efficiency is 37 %. As the neutral energy increases this efficiency decreases because of the reduced neutralization cross

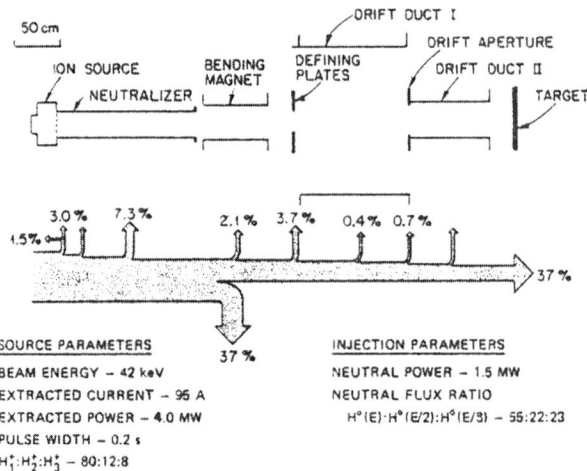

Fig.1. Beam line schematic showing power flow and main operating parameters . (taken from [1])

section. Above 200 keV, which is the energy needed for reactor, this method is probably too inefficient to be applicable.

A-1-2 With negative ions :

   A schematic view of a possible neutral injector based on negative ions is shown fig.2. It consists of three main and well separated sections : i) The negative ion source ; ii) The transport and acceleration section ; iii) The neutralizer.

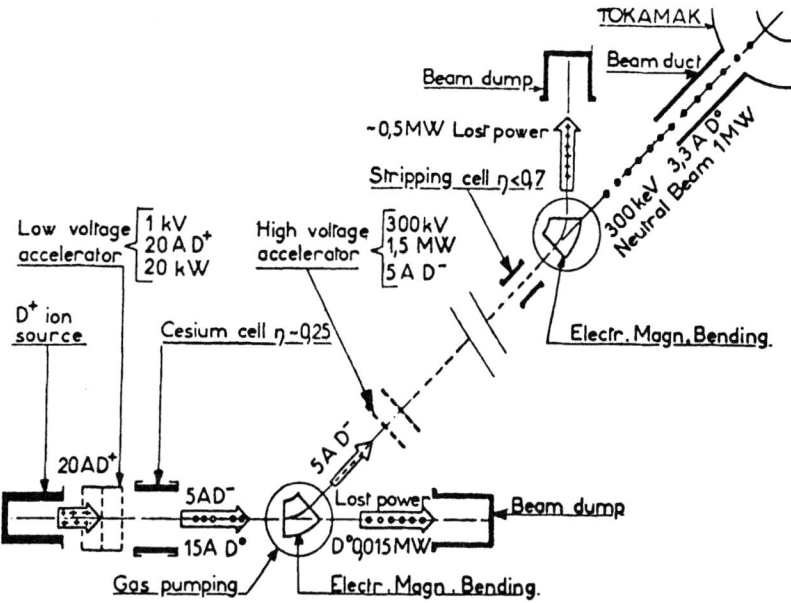

Fig.2.Schematic view of a 300 keV,1 MW Neutral Injector based on negative Deuterons.

i)   The negative ions can be produced by three different ways [2]:
     a) Starting from a positive beam at low energy (1000 eV) one uses the double charge exchange reaction on a cesium vapor.

$$D^+ + Cs \rightarrow D^o + Cs^+$$
$$D^o + Cs \rightarrow D^- + Cs^+$$

After suppressing the electrons, $D^-(H^-)$ ions are accelerated at the full energy.

   b)In a discharge chamber producing a plasma, one places a molybdenum plate, negatively biased relative to the plasma potential. This plate is covered with a cesium film in order to reduce the binding energy of an electron and to convert positive ions striking the plate into negative ions. The negative ions are in turn expulsed outside the source at the collector potential.

1007

c)Hydrogen discharges can be operated in a particular range of parameters to produce dissociative attachment of electrons on molecules highly excited in vibrational or rotational levels. These negative ions can be extracted from the source by an appropriate optics.

ii) The accelerator for negative ions is still its infancy. An electrostatic accelerator, very attractive for fusion reactors uses a transverse field to focus the beam. In this system the beam is transported through an array of suitably biased and curved electrodes ; along the beam path the curvature of the electrodes alternates.

iii) The possible neutralizers are a gas target, a plasma target or a photon target. The third possibility seems to represent the best ultimate choice for an efficient neutralizer. By using an oxygen-iodine chemical laser operated at 1.315μ which is suitable for photo detachment of an electron on H^-, one can avoid the neutralization of the impurity accelerated ions.

A - 2 Impurities coming directly from the beams

Impurities can be generated at the different stages of the neutral beam production : inside the plasma chamber of the ion source, through the metallic grids of the electrostatic optics, in the neutralizer, in the drift tube during the beam transport, and at the port entrance of the confinement device.

If the metallic impurities coming from the filaments (W,Ta...) by evaporation and particle bombardment, and from the grids by sputtering (Cu, Mo,...) are nearly undetectable, the light impurities (such as C,O,H_2O) are frequently observed at a level of several percents. These impurities have two drawbacks : 1) Injected into the confined plasma they radiate with a consequent cooling ;- 2) The dissociation of molecular species (such as C_nH_m, H_xO_y) after their acceleration, produces hydrogen atoms at low energy which are ionized at the plasma edge. These ions have poor confinement properties and have a high probability to produce impurities as they strike the wall.

It is therefore essential to measure the impurity content of the neutral beams and to determine their origin in order to improve the technology associated with their production.

A-2-1 Impurities concentration measurements in the neutral beam line

Three methods have been used:

A carbon collector can be exposed to the beam and the metallic impurities which penetrate into the probe are analysed by Auger spectroscopy and Rutherford back scattering.

.Neutrals are ionized in a gas cell and a magnetic (and elec-trostatic) analyzer receives the charged particles after they pass through a small hole.

.Spectroscopic measurements are made on the optical emission from the beam due to the interaction of the beam neutrals with residual gas particles in the drift tube.

As an example the figure 3 shows some results obtained with this third method [3]. An hydrogen beam accelerated at 25 kV is observed by a spectrometer, 45 degrees with respect to the beam direction. The Doppler shift of the $H\alpha$ and $H\beta$ lines depends on the neutral energy. One can see, besides the neutral at full energy E_B, three peaks corresponding to the dissociation of molecules : H_2 for $E_B/2$, H_3 for $E_B/3$, H_2O^+ for $E_B/18$.

. The impurity content depends largely on the source type. The table I gives for example the impurity concentration (measured by magnetic analyzer) of a duopigatron source and a bucket source [4] measured by the same team.

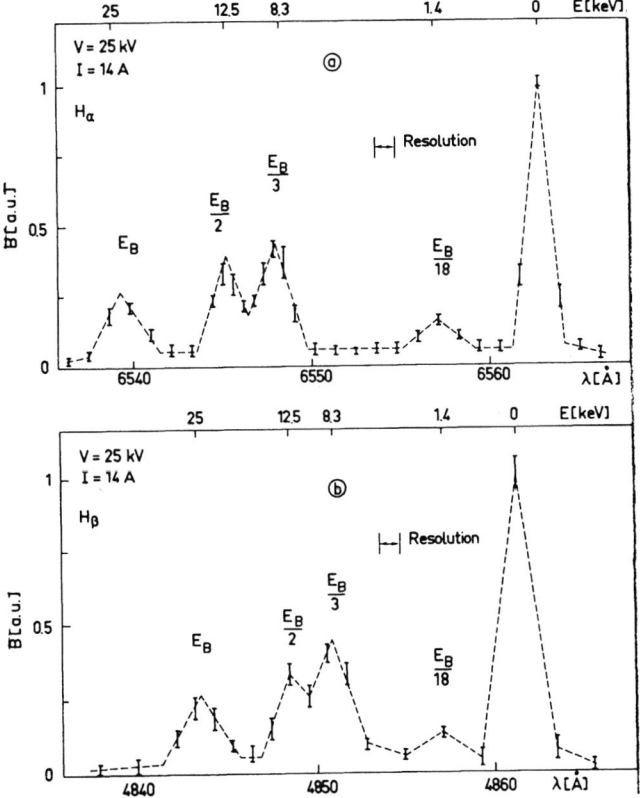

Fig.3. Measured radiances along the central chord, normalized to the unshifted peak, as a function of wavelength. (a) Hα(b)Hβ. The top scale is the energy corresponding to the measured Doppler shift.

TABLE 1

	duo PIGatron	Bucket (Multipole line cusp source)
Light impurities	1.5 %	2.5 %
Cu^+	0.015 %	0.1 %
Zn^+ , Ag^+	0.005 %	0.02 %
W^+	–	0.03 %

A-2-2 Conditioning of the beam line components

The cleanness of the source and the choice of the materials play a major role in determining the light impurity content in the beam. Ultra-high vacuum techniques are now applied in all laboratories : stainless steel for ion source structure, oxygen-free copper and high quality molybdenum for grids, ceramic or glass for insulators, and metallic seals.

In addition, the conditioning of the sources is frequently used. Baking is still limited but conditioning by repeated discharges is efficient. As an example the Doublet III neutral beam preparation is shown[5]

Figure 4 is a spectrum taken immediately after opening the source to the atmosphere. The area under the water peak is about 30 % of the total area of the four peaks. It is estimated that about 25 % of the beam power injected could be in the form of energetic oxygen neutrals.

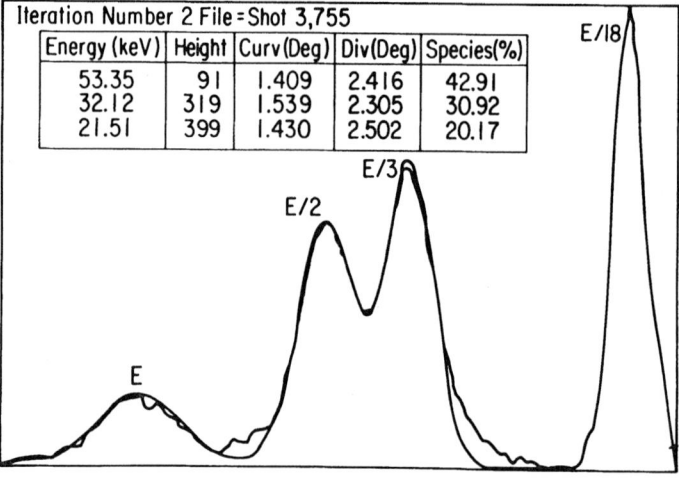

Iteration Number 2 File = Shot 3,755				
Energy (keV)	Height	Curv(Deg)	Div(Deg)	Species(%)
53.35	91	1.409	2.416	42.91
32.12	319	1.539	2.305	30.92
21.51	399	1.430	2.502	20.17

Fig.4. Doppler spectroscopy spectrum taken immediately after air opening. The full, half and third energy peaks are visible as well as the large water peak.(taken from [5])

Figure 5 shows the decay of the water peak after four days of operation. This peak decreases down to the 5 % level.

In order to reduce further the water content it appeared necessary in some cases to use another method which has proved to be successful in fusion machines. As the water coming from a clean metallic wall has its origin in the reduction of the oxides, a fast and effective method of oxygen reduction is to evaporate

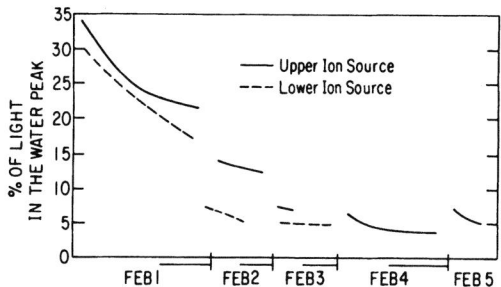

Fig.5. Time behavior of the water peak in the Doppler spectrum as the ion sources are conditioned.(taken from [5])

getter materials as titanium. As for the stellarator W7-A the injected impurities are well confined (see A-2-5), an evaporation of titanium on the inner walls of the ion source has been undertaken. Figure 6 [6] compares the mass spectrum measured by a residual gas analyzer before Ti evaporation, immediately after, and a few shots later. The impurity content was reduced to 0.6 % after Ti evaporation. The additional Titanium contamination of the beam was only 2×10^{-4}. Some modifications appear on the distribution of the hydrogen species, but no deleterious effects were observed either in the ion source or in the ion accelerator after about 1000 shots with titanium evaporation.

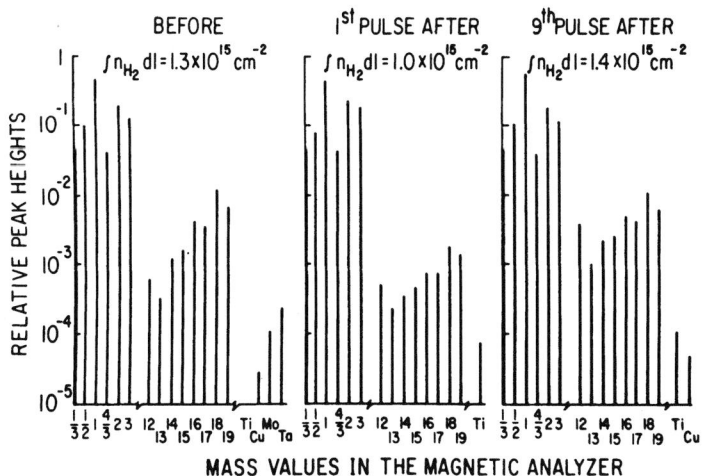

Fig. 6. Influence of Ti evaporation on the mass spectrum. (taken from [6])

1011

A-2-3 Specific problems with negative ion sources :

One specific source of impurities for the first two types of negative ions sources (A-1-2) is the cesium flux which can follow the hydrogen beam. For the double charge exchange source this is minimized by using an ultrasonic cesium jet at high density (10^{15} at/cm^2) ; but, even in this case, a flux of 10^{14} at/cm^2.s along the hydrogen beam has been measured |7|. Also in the surface source a cesium flux of the same magnitude is observed. Nevertheless, this vapor could be cryogenically pumped before the neutralizer.

In addition to the cesium contaminant, light impurities (C^-, O_2^-,O,OH^-) are present. Their level is directly related to the system cleanliness and the time of source operation. For instance, with the surface source, the impurity level reaches about 40 % when the source is first turned on, but drops to 1 % after 10 minutes of operation. In the multi cusp source [2] the total impurity content of negative ions is about 2 % (1.1% OH^- ; 0.6 % O^- ; 0.1 % O_2).

Nevertheless, by using a neutralizer with laser photo detachment, one can avoid the neutralization of those accelerated negative impurity ions. However, these ions will strike the walls of the transport section and therefore must be pumped out before entering to the reactor.

A-2-4 Beam wall interactions

The port entrance of the vacuum chamber is very often a critical passage for high energy beams : it corresponds generally to a reduction of the cross section where the gas pressure can be high (in the 10^{-4} torr range for the Tokamaks) and the magnetic field is fairly large. Consequently, the power density is limited (5 - 8 kW/cm^2) to avoid the beam blocking (as observed in some experiments with an unsufficient pumping speed). Nevertheless the interaction with the port walls is unavoidable and special pieces capable of sustaining high energy fluxes are needed. For example, in JET the port walls are covered with beam scrapers made of hypervapotron water cooled copper plates which can support 1 kW/cm^2

Even with a good choice of the beam parameters (energy, divergence, injection angle, ...) a small fraction of the beam particles "shine through" the plasma without being captured and the walls of the vacuum chamber have to be covered with burial plates made of refractory materials with a low sputtering rate (Mo or C tiles with TiC or SiC). In fact, these plates are generally designed to support the full energy of the beam at least for a limited number of shots, in case of a failure of interlocks between the plasma creation and the beam injection.

A-2-5 Effect of beam-injected impurities in the stellarator W7

In order to illustrate the consequences of the impurities transported by the neutral beams we present data from the W7 experiment [8]. In this device the limitation of the plasma performance by radiation losses was rapidly recognized. Even before it was measured spectroscopically, oxygen, was suspected as the main contaminant. In order to identify the origin of oxygen (beam or walls) several experiments and numerical calculations were made :

a) Sputtered molybdenum in front of the beam dump was measured by laser fluorescence. Its rate is reduced by a factor 10 with the presence of the plasma. As the beam injected oxygen is nearly completely absorbed by the plasma, this reduction can only be understood if accelerated oxygen ions are the main source of sputtering. Indeed, the hydrogen atoms are only 50 % absorbed and would reduce only by a factor 2 the sputtered Mo when the plasma is present.

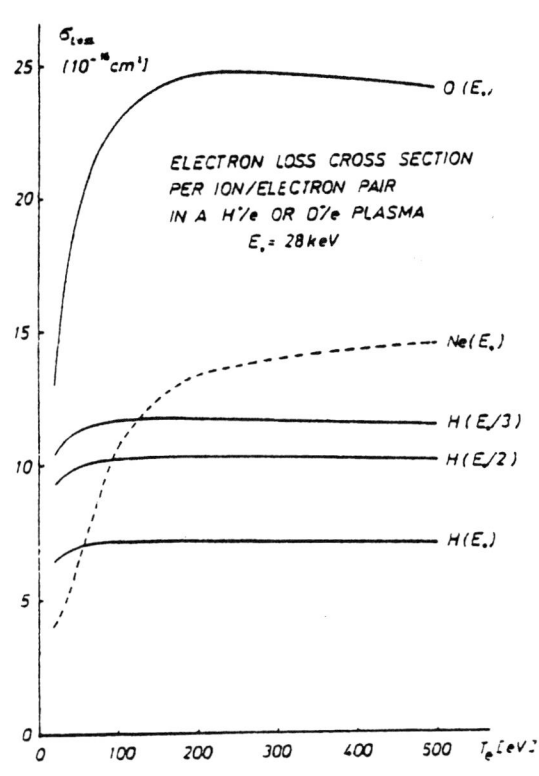

Fig.7. Ionization cross sections of Oxygen, Neon and Hydrogen.

b) Calculations indicate that injected oxygen is absorbed more efficiently by the plasma than is hydrogen [9]. Fig.7 shows that the ionization cross section for oxygen is 2 - 4 times that of hydrogen. Fig. 8 shows the deposition profiles of thermalized oxygen calculated with a Monte Carlo code [10].

c) The reduction of the central radiation losses with titanium gettered sources is substantial as shown on Fig. 9 for the OVIII brightness and soft X-ray radiation (USX) [8]

d) OVIII light increases when a fourth injector with no titanium gettered sources is used in addition to three gettered neutral beams.

e) A contribution of 10 - 20 % of the Fe line radiation is estimated by observing the relative increase of USX radiation during a –laser blow-off injection of Fe.

This effect of impurities brought by high energy neutral beams has therefore been clearly demonstrated in these experiments. The emphasis given by the W7 team to this problem is due to the good particle confinement in the stellarators and to the fairly dirty ion source used for neutral beam injection.

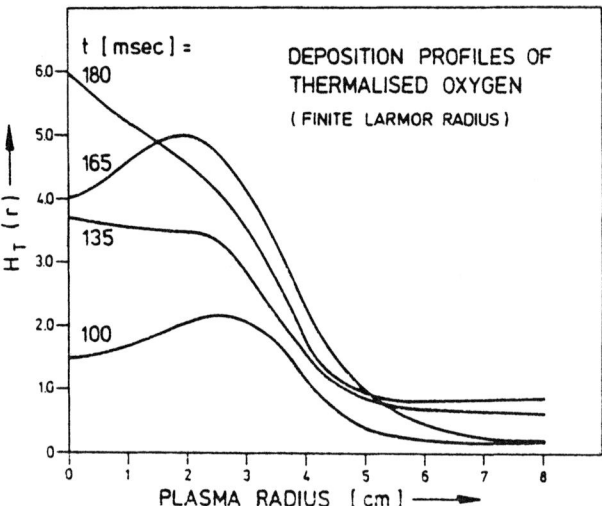

Fig. 8. Calculated deposition profiles of oxygen (25 keV)in W7. The injection pulse starts at 100ms. H_n is proportional to the trapped fraction.(taken from[10])

The same calculations done for the future W7-AS device show that the plasma center will be free of such impurities (n_{imp}/n_e less than 10^{-4}) due to the larger diameter and the higher density 10^{14} cm^{-3} of the plasma. [8]

A - 3 Impurities produced by unconfined ions.

The fast neutrals which are ionized into the confined plasma can be lost before thermalization. Three processes can cause this loss : 1) the initial condidions (place of birth, velocity angle with the magnetic field) put these ions on an unconfined trajectory, 2) some instabilities can expel them outside of the plasma, and 3) a subsequent charge exchange event with the fast ions produces a fast neutral which may escape the plasma.

These ions or neutrals strike the walls at high energy and increase metallic impurities by sputtering. As these effects have clearly been observed in Tokamaks, the results will be presented for this configuration.

A-3-1 Different class of trajectories in Tokamaks

Ignoring colli-sions three types of particle trajectories exist in Tokamaks, de-pending on the velocity angle with the magnetic field : (Fig.10)

   i) Passing par-ticles which follow a magnetic surfa-ce surrounding the ma-gnetic axis.

   ii) Banana-trapped particles which go back and forth between two maxima of magnetic field (due to 1/R variation of the toroidal magnetic field, the projection of the drift surface on a meridian plane looks a banana.)

   iii) Ripple-trap-ped particles (superbanana particles), which are reflected be-tween two maxima of magnetic field associa-ted with the discrete number of toroidal coils. These particles are lost from the plasma by ∇ B-drift on time scales of the order of the vertical drift time.

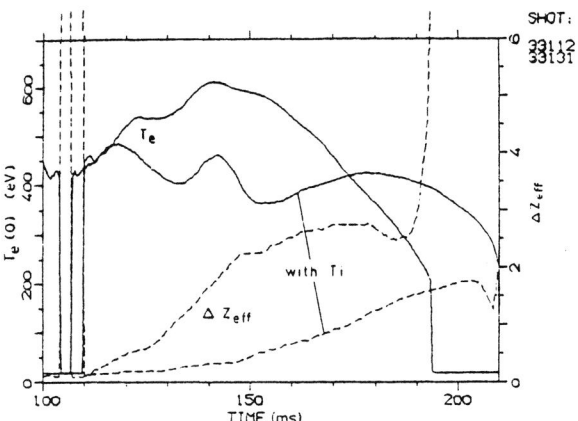

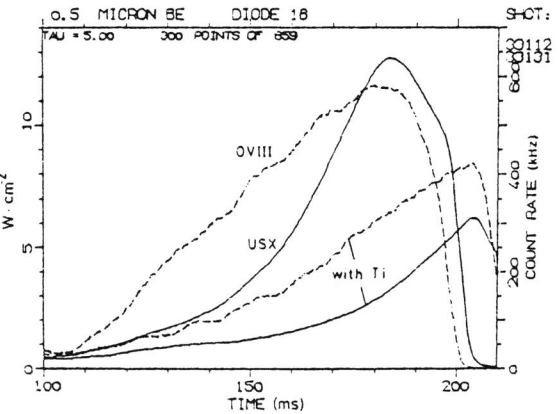

Fig.9. Reduction of Z_{eff} and electron temperature decay rate in the plasma cen-tre (extracted from X-continumm radiation) for "clean" discharge (upper diagram). The lower diagram shows the corresponding effect on the central soft X radiation (> 500 eV) and O VIII line radiation. (taken from [8])

In the velocity space the loss region boundaries depend on each device. Fig. 11 shows the case of DITE [11] in which the bundle divertor increases the peak-to-peak ripple magnitude (5 % on axis up to 100 % at the separatrix radius).

The size of banana-trapped loss region is reduced as the plasma current increases. These ions are lost because the width of the banana (of the order of ion Larmor radius in the poloidal magnetic field) is such that the ion intersects the walls (limiter). The reduction of plasma impurities as the plasma current increases is mainly due to a better confinement of fast ions.

A-3-2 Thermal load on limiter during NBI

The surface temperature of limiters has been determined by infrared measurements. Fig. 12 shows an example, from TFR [12], of the time evolution of the energy flux on the inner limiter with a 700 kW NBI pulse. (This limiter is made of inconel and has a trapezoidal cross section, fig. 13). At the maximum of thermal flux (2600 W/cm^2) corresponds a temperature of 1100°C.

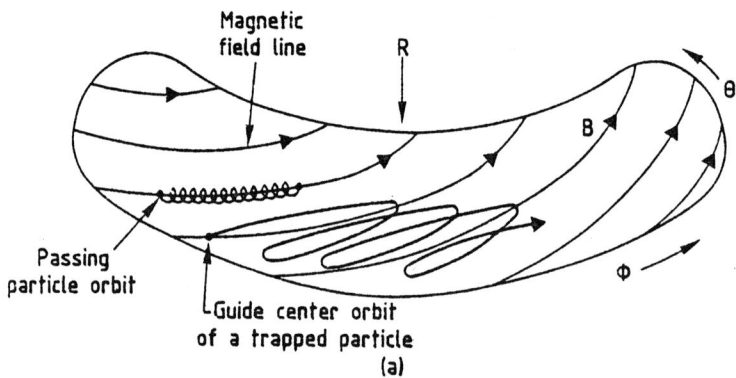

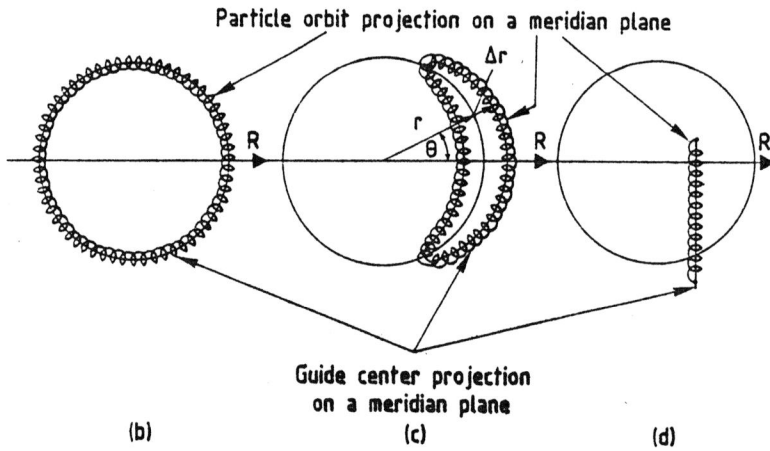

Fig.10. Different class of particle trajectories in a Tokamak:
b) Passing particle which surrounds the magnetic axis ;
c) Trapped particle (banana particle) due to 1/R variation of toroidal field ;
d) Ripple trapped particle which is unconfined.

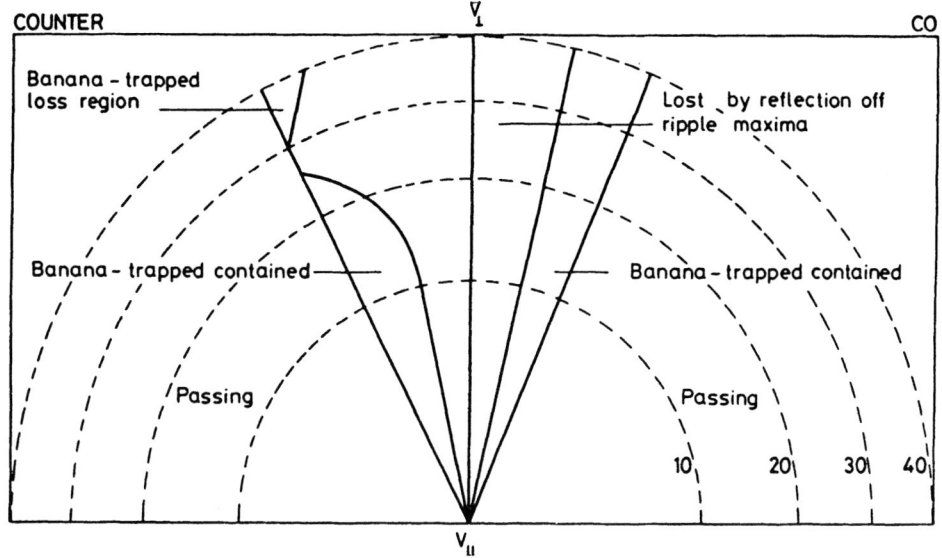

Fig.11. Space velocity domains covered by the different class of trajectories in DITE. The dashed circles correspond to a constant energy in keV.(taken from [11])

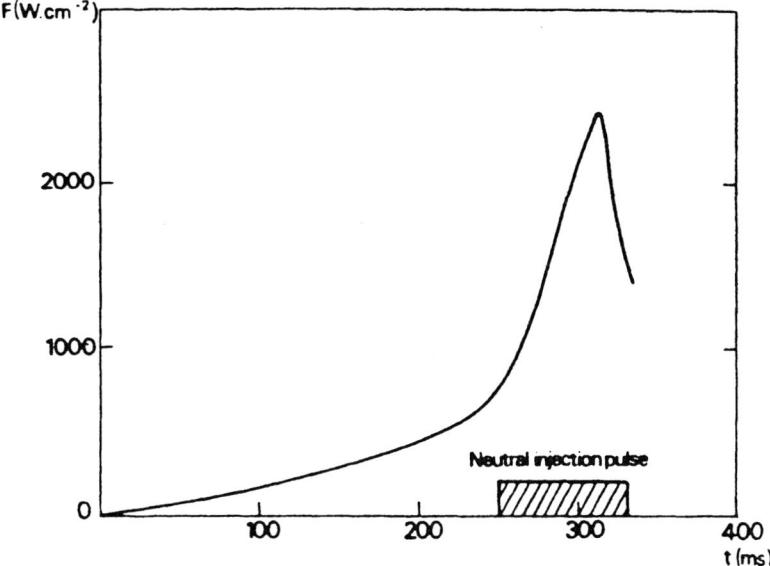

Fig.12. Variation of the thermal flux on the inner limiter during a discharge with neutral injection. (taken from [12])

The surface temperature distribution is shown along the toroidal direction, in the equatorial plane (fig 13). The large temperature gradients are due to the limiter shape. When theses measurements were performed with the same NBI conditions but one month later, the maximum flux was only 1600 W/cm^2. That may be explained by the modifications of the limiter shape (Fig. 14) as a consequence of a superficial melting during NBI pulses and current disruptions (the metal drifted toward the toroidal direction and was deposited at the edge of the sector). The limiter shape tends to adapt itself to the uniform thermal load.

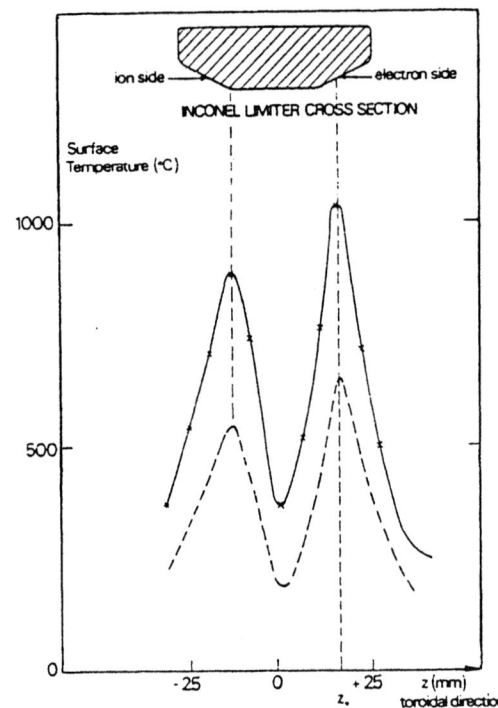

Fig.13. Surface temperature distribution on the inner limiter. The scan is in equatorial plane. The measurement is made before the neutral injection pulse (dashed line) and at the end of the pulse (full line).(taken from [12])

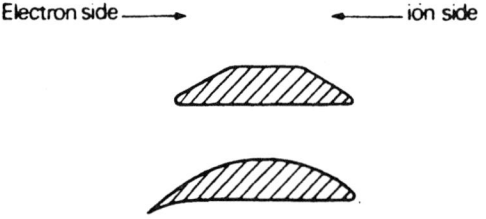

Fig.14. Shape of the inner inconel limiter (cross-section in the equatorial plane) before operation (top) and after a few thousands high power discharges (bottom). (taken from [12])

Fig.15 shows the surface temperature distribution along the poloidal direction. The deposition is more regular than along the toroidal direction and permits to evaluate the fraction of plasma energy deposited on the limiter (∿ 50 %)

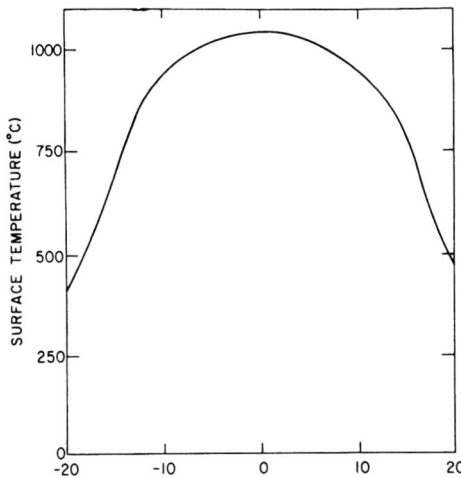

Fig.15. Surface temperature distribution on the inner limiter. The scan in the vertical (poloidal) direction at $Z=Z_o$ (Fig.8) where the heat load is maximum. (taken from [12])

The metallic impurity content in the discharge is clearly correlated with the limiter temperature. This is shown on fig. 16, both for the total radiated power measured by bolometry and for a characteristic line (Ni XVIII) measured spectroscopically. However, the relative part of sputtering and melting for the metallic contamination is unknown. Although it is not possible to rule out that the increase of power deposition

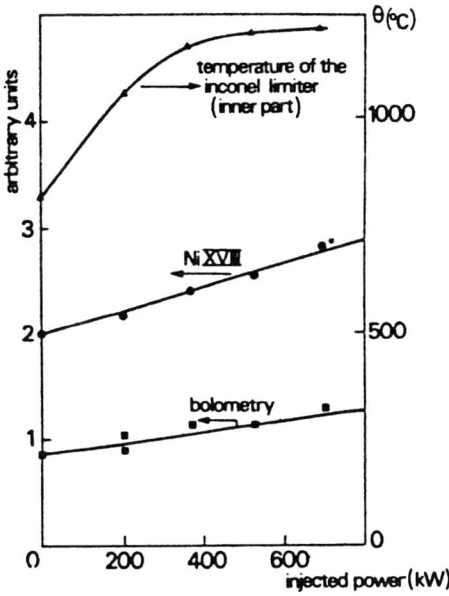

Fig.16. Variation of the total radiated power (bolometry, Ni XVIII line intensity and inner limiter temperature with the neutral injection power. (taken from [12])

on the limiter is associated with the heat flux transport from the hot plasma, the absence of delay between limiter heating and NBI pulse and the linear power dependence are in favour of a power deposition by high energy unconfined ions.

A-3-3 Action of super-banana ions

The neutrals which are ionized at the outer major radius side plasma edge where the ripple is higher may become superbanana ions. In TFR where the injection angle is large (75°) a charge collector placed in the meridian plane between two toroidal coils and near the plasma boundary shows effectively a pulse of positive charges associated with the NBI pulse. This signal disappears as the toroidal magnetic field is inverted. A bolometer looking at the region of the wall which is bombarded by superbanana ions (here the bottom) also detects an increase of signal during NBI pulse (Fig 17). This pulse does not exist on the top side and its amplitude varies proportionally to the NBI power (Fig 18). Spectroscopic measurements show that this radiation comes mainly from the light impurities (C,O). With the increase of NBI power in large Tokamaks it becomes necessary to reduce these superba-

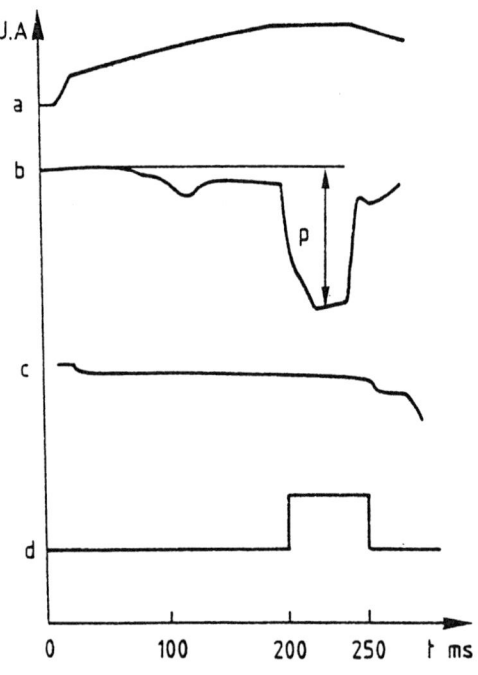

Fig.17. Effect on superbanana particles bombardment on the wall of the vacuum vessel of TFR during NBI pulse.
a) Plasma current
b) Bolometer signal (bottom − 20 cm)
c) Bolometer signal (bottom + 20 cm)
d) NBI pulse.

nana losses by acting on all parameters (low ripple, injection angle, small fraction of neutrals with fractional energy).

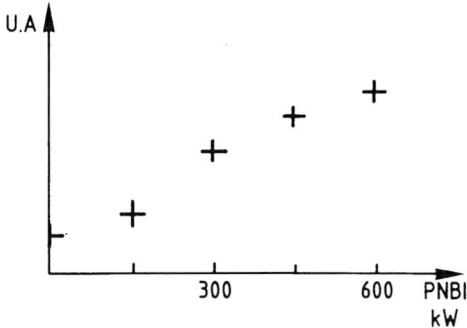

Fig.18. Amplitude of the peak bolometer signal (p on Fig. 17) for different NBI power.

A – 4 Impurity transport and NBI

As soon as the power injected by NBI became a large fraction of the ohmic power, modifications appeared in the particle transport. This subject is an active field of research and still poorly understood.

A-4-1 Experiments on accumulation of impurities in the plasma center

The neoclassical theory of particle transport predicts that in an OH toroidal plasma, impurity ions of charge Z have spatial distributions more peaked than that of the hydrogen :

$$n_z(r) = n_z(o) \left[\frac{n_p(r)}{n_p(o)} \right]^z$$

Very soon, the results for ohmic discharges in Tokamaks showed a discrepancy with this prediction. The impurity radial profile was broad and had no tendency to peak inward during long pulses. Hence, impurity transport was concluded to be non classical or "anomalous". In present simulations of impurity transport the impurity flux is usually assumed to be of the form[13]:

$$\Gamma = - D_A \frac{\partial n_z}{\partial r} - \frac{r}{a} V_A n_z$$

where a is the radius of the cylindrical plasma, and D_A and V_A are arbitrary diffusion coefficient and radial velocity which do not

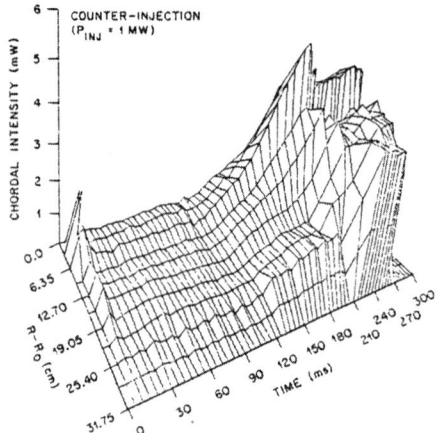

Fig.19. Time evolution of chordal intensity profile for counter-injection with P_{inj}=1MW. Injection of H^o beam into deuterium plasma. Time-resolved profile data using the array of twelve collimated radiometers are available for each discharge; however here the data have been averaged over several discharges. Beam turn-on at t=120ms. (taken from [14])

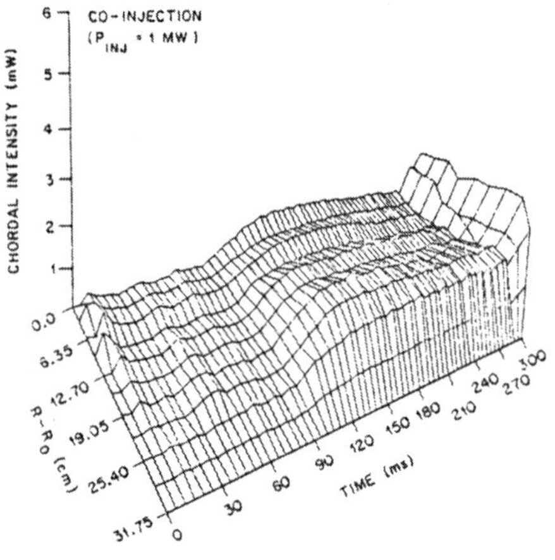

Fig.20. Time evolution of chordal intensity profile of impurity radiation emission for co-injection-heated discharge with P_{inj}=1MW. Injection of H^o beam into a deuterium plasma. Beam turn-on at t = 80 ms. (taken from |14|)

depend on the radius. With large power NBI in stellarators the accumulation of impurities was observed (cf. A-2-5). However, the peaking of the radial profile was not so large as the neoclassical predictions.

In Tokamaks a large difference in the behaviour of impurity exists according to the direction of injection relative to the plasma current (co or counter injection). Compared with purely OH discharges, counter injection enhances the rate of increase of impurity radiation during the discharge, whereas co-injection

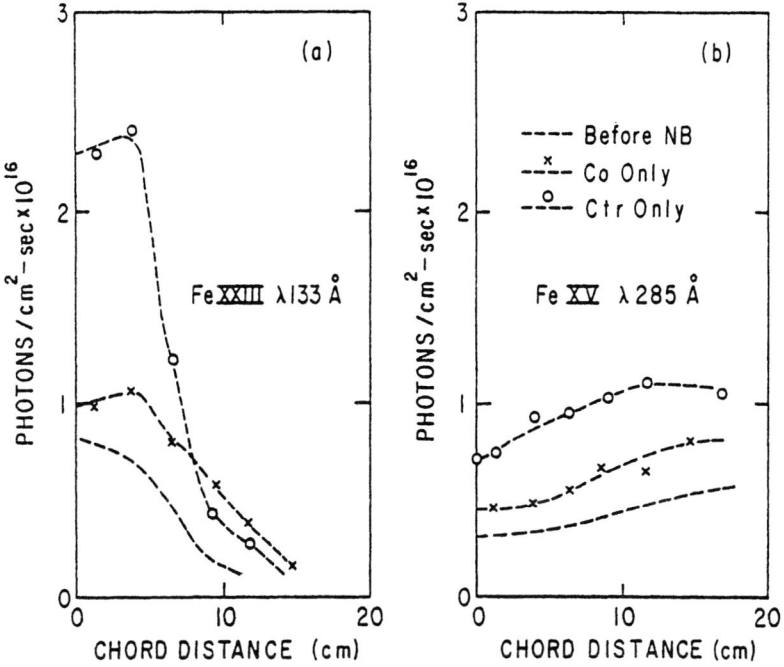

Fig.21. Spatial profiles of iron lines for OH, Co and COUNTER on PLT. (taken from[15])

reduces the impurity radiation from the core plasma. Figures 19 and 20 show the chordal intensity of the radiated power measured by bolometers in ISX [14] with an injected power of 1 MW. In the counter injection case the intensity profile rapidly evolves toward a profile highly peaked at the center and the plasma undergoes a disruption. Spectroscopic measurements of highly stripped impurity ions confirm the large differences between the radial profiles.

Several experiments have shown that the major difference between co and counter injection effects on impurities is a change in impurity transport rather than a change in impurity influx (Although the banana trajectories are less confined in the counter case). Controlled amounts of Ti impurities introduced in ISX [14], by laser blow-off techniques, indicate that confinement is different in the co and counter case. As titanium is injected into a clean OH discharge, the radiative losses increase from 20 % just before impurity injection to 40 % of P_{OH}. As soon as co-injection is turned on, the increase of radiated power is reduced and 50 ms later the profiles become peaked at the edge rather than at the center. On the contrary, with counter injection and the same impurity influx the profile evolves toward a more peaked distribution.

By comparing the time evolution of different ion species of the same impurity one can also rule out an influx increase. Fig. 22 and 23 show in ISX [16], iron and oxygen emissions for co and counter injection. The contrast appears immediately : for co-injection all the line intensities exhibit a relatively low quasi steady state level, for counter injection the radiated levels grow rapidly, first in the inside ions (as Fe XIX) then the lower stages (as Fe IX), that can only be explained by an internal rearrangement which concentrates the higher ionized stages both for the iron and the oxygen in the center of the plasma and by the recombination as a

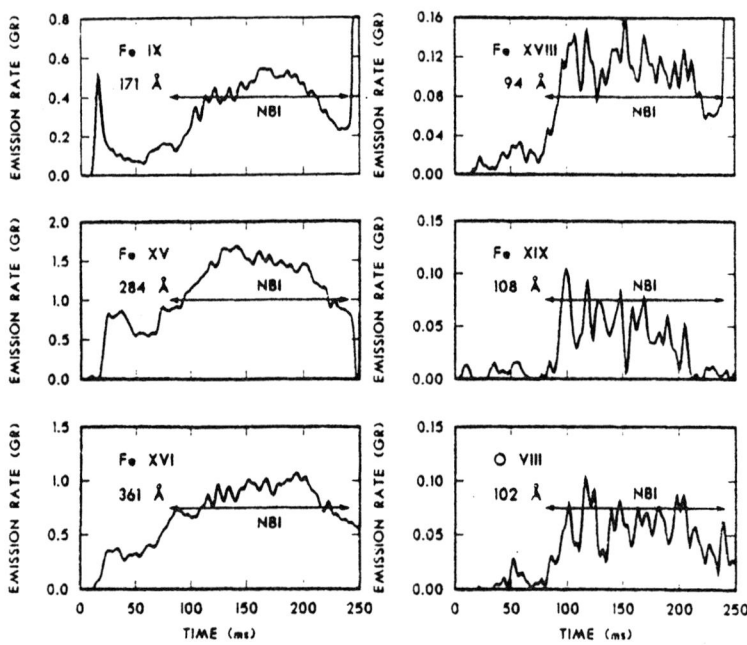

Fig.22. Iron and oxygen emissions during co-injection on ISX. (taken from [16])

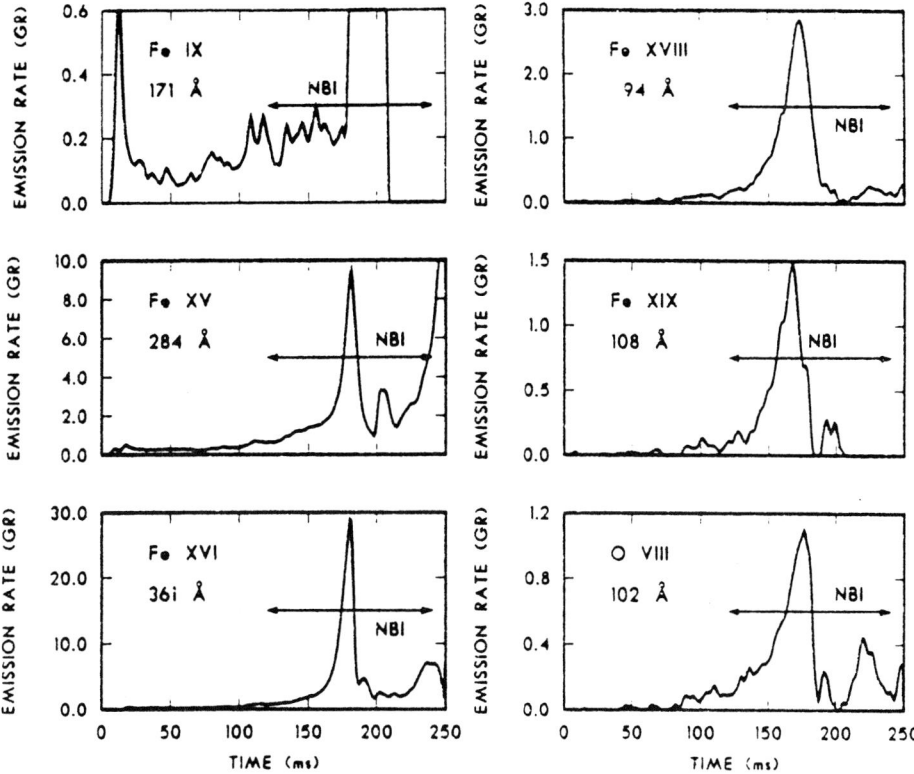

Fig.23. Iron and oxygen emissions during counter-injection on ISX. (taken from [16])

consequence of the plasma cooling for the later increase of the lower ionized stages.

A-4-2   Charge-Exchange between fast neutrals and impurity ions.

A secondary effect of impurities with NBI is the radiation increase as charge exchange recombination of the impurities on injected neutrals.

The process $H^{o} + Z^{n+} \rightarrow H^{+} + Z^{(n-1)+}$ shifts the ionisation equilibrium towards lower-Z. Subsequently two radiation processes contribute to the radiated power : 1) a prompt radiation associated with the recombination itself ; 2) an electron excitation of the recombined ion. The latter is the most important.

This effect has been observed in several devices. For instance in DITE [17] the time variation of some impurity lines is shown on fig. 24. The simultaneous decrease on a short time scale (~ 1 cms) of Ti XVIII and the increase of Ti XIII as the NBI is applied, strongly suggest that neutral injection is reducing the average impurity charge state. A theoretical transport model including these exchange processes is in reasonable agreement with the time dependence of the radiation. Another experimental observation is the toroidal asymmetry of the line emission and total radiation : this can be explained by a highly localized impurity density in lower-charge states near the beam region. Quite spectacular are the computed variation densities as functions of toroidal angle injection from two beam lines and one beam line (fig. 25).

One can make use of these reactions for detecting the presence of the ion Z^{n+} by observing the subsequent prompt variation. In PDX [18] by injection of a highly collimated diagnostic neutral

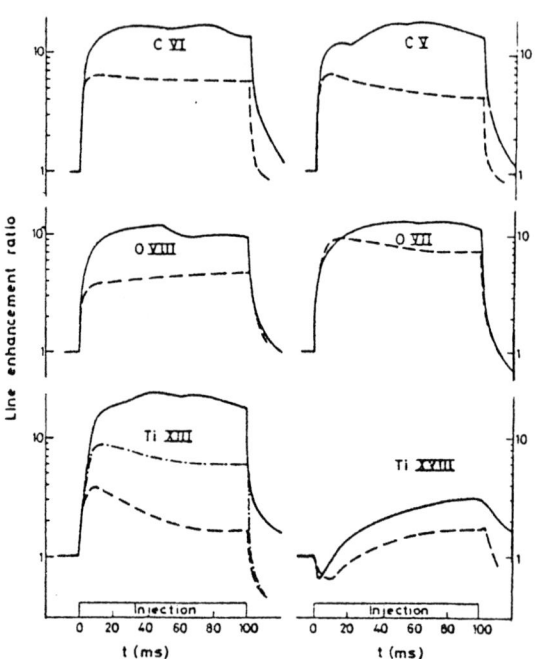

Fig.24. Time variation of line enhancement ratios of carbon, oxygen and titanium impurity ions for low-density discharge. Solid lines are experimental, broken lines are theoretical. Signal-to-background ratios are typically 6 in raw data. (taken from [17])

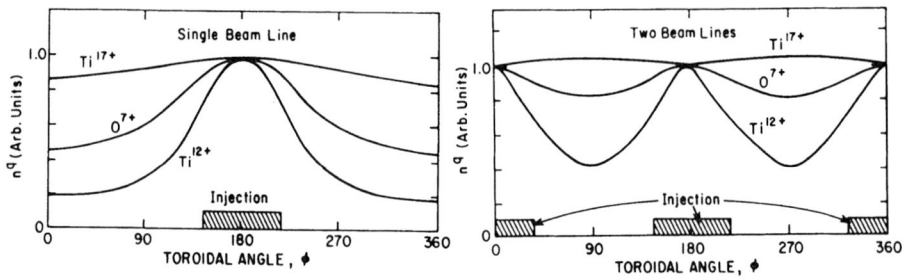

Fig.25. Computed variation in O^{7+}, Ti^{17+} dansities as functions of toroidal angle for neutral injection from two beam lines and one beam line. (taken from [17])

beam, the radial profiles of fully ionized oxygen and carbon have been measured. In order to determine the absolute impurity density one needs to calculate the beam density at the measured radius and the charge-exchange cross sections. On the other hand the Doppler broadening of the radiated line can give the impurity temperature without absolute calibration of the beam or knowledge of cross sections.

B - ICRH and Alfven wave heating

B-1 Launching structures

The excitation of ion cyclotron and Alfven waves in the frequency range $1 \div 150$ MHz, requires a launching structure inside the vacuum chamber as this chamber is generally made of metal. It is indeed a major problem to design an internal launching structure with the conflicting requirements of efficiency, smallness, cleanness and high degree of resistance against mechanical stresses and thermal load.

The antennae are exposed to the outflux plasma (photons, neutrals) but they are also bombarded by the tenuous plasma always existing near the walls. So, the success of RF heating goes necessarily, through an important development of the technology of the launching structure.

The large number of antennae models which have been used excludes an exhaustive presentation of this field of research. We shall limit ourselves to the main lines of the development of the ICRH antennae in TFR and Alfven wave antennae in TCA.

B-1-1 ICRH antennae in TFR [19]

The conceptual design of these antennae is shown on fig. 26. In a meridian plane of the Tokamak a coaxial line feeds a strip line through a coaxial feed through. The extension along the poloidal coordinate depends on different considerations : adaptation of the electrical length with the frequency, desired radiation pattern, room available inside the vacuum chamber.

Two main designs are used :

i) The antenna radiates mainly in the high toroidal field side (HFSA) (type 1 Antenna on fig. 26), and the electromagnetic wave which is excited in the plasma is converted into a electrostatic wave which damps itself both on ions and electrons.

ii) The antenna radiates mainly in the low toroidal field side

(LFSA) (type 2 antenna on fig. 26), and the RF power is deposited on the minority ion species. This scenario uses a two ion species plasma : $\underline{H}$ – D or $\underline{He}^3$ –D. The underlined atom corresponds to the minority ions.

The main elements of the present antenna appear on fig. 27. The central conductor (2) made of inconel 6 cm wide is equipped with strips (3) which act as a capacitor so as to adjust the electrical length of the antenna (generally λ /4). The return conductor (1) constitutes also the mechanical structure of the

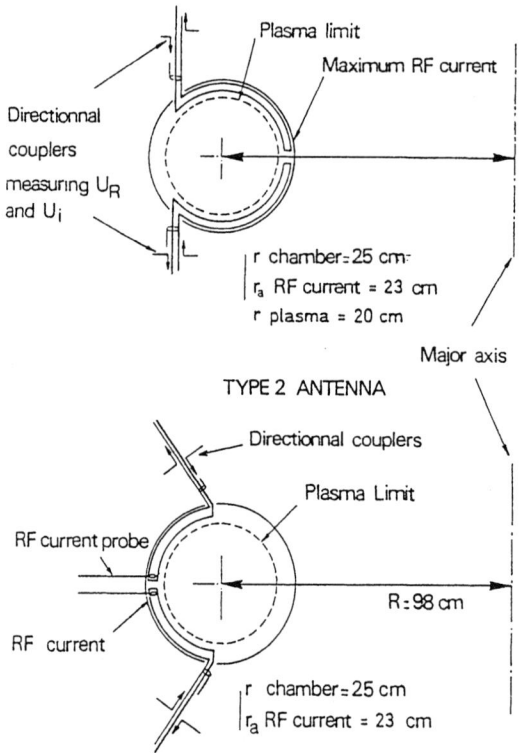

Fig.26. Conceptual design of TFR antennae.

antenna. Between the plasma and the central conductor a double
layer of thin narrow inconel plates which forms a Faraday shield.
Finally a lateral shielding plate (5) protects the active structure
against the plasma flow along the magnetic field lines. All the
antenna components are presently made of conducting materials
(metal or graphite) In the earliest work on TFR an insulating cover
in alumina was placed between the central conductor and the Faraday
shield to prevent the plasma particles from reaching the conductor.
As this cover broke several times due to large electromechanical
forces associated with the plasma disruptions, one tried to operate
without this cover. Since the experimental results were good with
this all-metal antenna this concept has been kept and it is now
largely used in other devices (JET, Textor, Asdex, JFT 2).

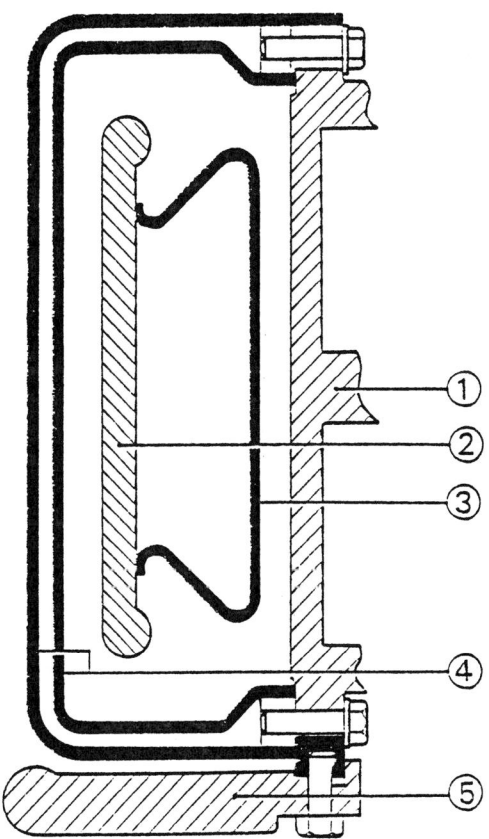

Fig.27. Main elements of a TFR antenna (cross section)
① Return conductor ② Central conductor ③ Distributed capacity
④ Electrostatic screen ⑤ Lateral protections.

The materials chosen to make the antenna are still a matter of development. The central and return conductors having to support stresses are made of inconel (called also incalloy) or stainless steel ; a 50 µ silver coating is plated on the conductor surfaces in order to reduce the RF resistance. The Faraday screen has to sustain the energy flux coming from the plasma and sees all the RF power through it ; for large machine and long pulses it has to be

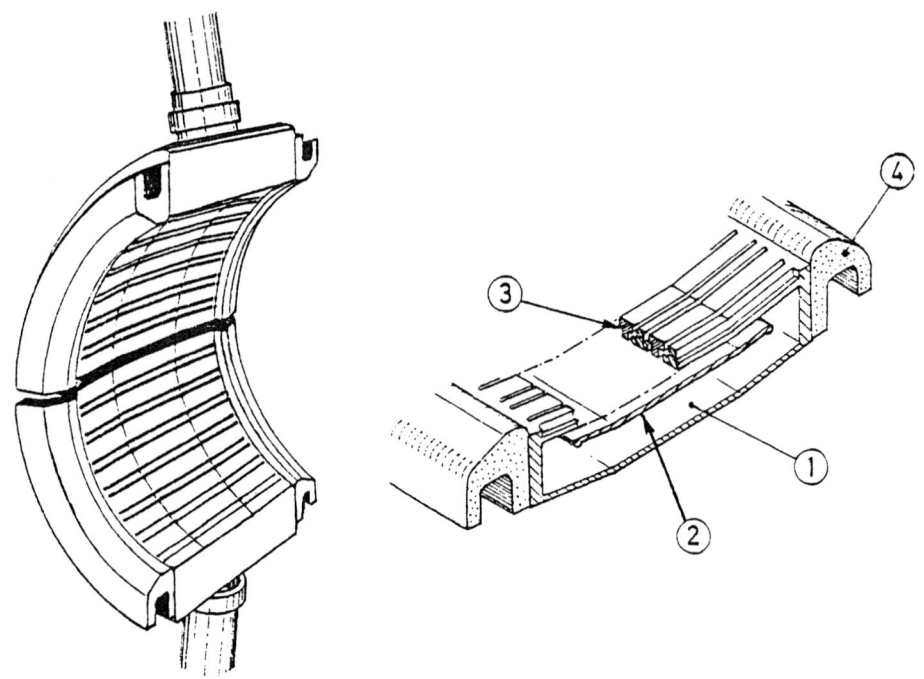

Fig.28. Module of a JET type antenna tested in TFR
① Return conductor ② Central conductor ③ Electrostatic shield ④ Graphite protections.

cooled. Metals (inconel, stainless steel, titanium) and graphite were used. For JET, one foresees to use beryllium. As shown later on (§ B-2-2) the screen does not seem a significant source of impurities (in connection with RF field).

The shielding plates act as a usual limiter and appear always as the most exposed part of the antenna. In the JET antennae design these plates are in fact the main limiters. Fig. 28 shows a module of such an antenna tested in TFR. The lateral parts are made of graphite. The reason for joining the antenna and the limiter is to increase the coupling of the RF launching structure to the plasma by reducing the thickness of the low density non propagating layer

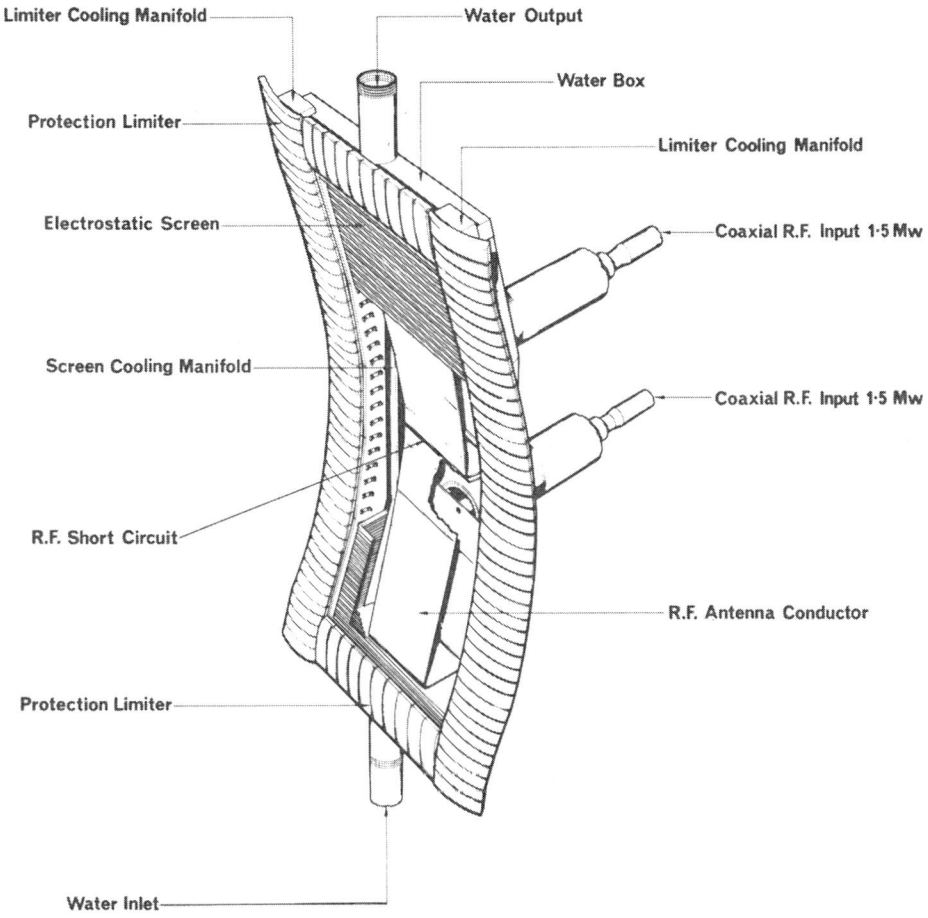

Fig.29. JET ICRF heating antenna

in the vicinity of the antenna. The full extension of this concept to the advanced antennae for JET leads to the belt limiter antenna concept : the limiter consists of two beryllium rings all around the torus and a set of ten antennae (fig. 29) are placed between them.

The coaxial feedthrough needs also to be optimized to reduce impurity production. Fig. 30 shows a 50 ohm (8 cm O.D) alumina vacuum tight feedthrough. It isolates the vacuum chamber from the coaxial line pressurized at 2 bars of nitrogen. The brazed joints are covered with shaping electrodes to prevent corona effects.

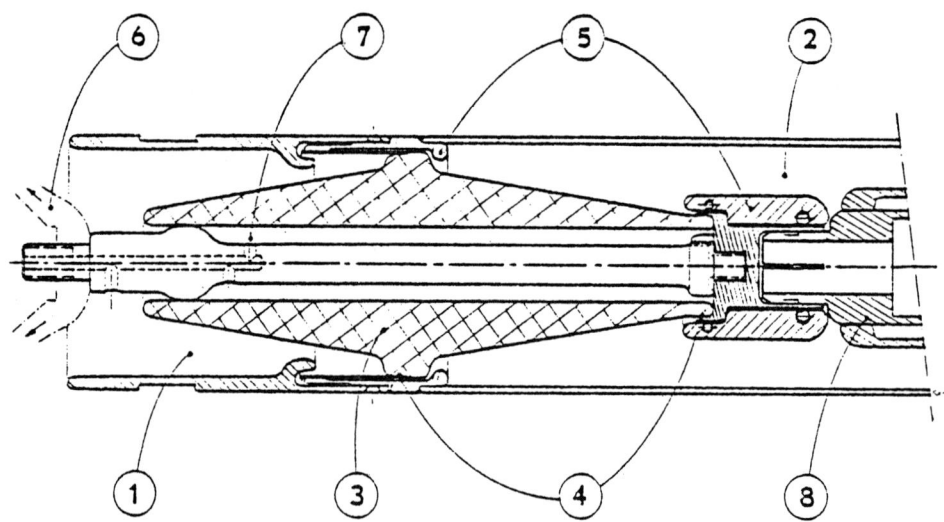

1 . VACUUM SIDE

2 . PRESSURIZED SIDE (+ 2 ATM.)

3 . ALUMINA

4 . Ti-Al$_2$O$_3$ BRAZED JOINTS

5 . CORONNA SUPPRESSORS

6 . PARALLEL INTERNAL CONNECTION OF 2 ANTENNAE

7 . PUMPING HOLES

8 . COAXIAL LINE

Fig.30. Coaxial vacuum -tight feed through.

Conditioning of the antenna :

The antenna has to be conditioned at least as much as any part of the vacuum chamber. In addition to the usual desorption of impurities due to the plasma outflux, the RF power which is dissipated on the structure can produce a lot of light impurities.

In TFR the antenna is baked at 200°C for 2 hours by simultaneously baking the vacuum chamber and joule heating the antenna by the RF current at a level of 1 to 2 kW . The antenna being hot, the gas is admitted into the chamber at about 1 mtorr (hydrogen or deuterium) and RF discharges (100 kW, 20 ms, 1 Hz) are produced with a low toroidal field (4 kG). This procedure is maintained a few hours until the production of water and methane stops decreasing.

The conditioning against RF arcing is made with short pulses (1 to 3 ms) by increasing progressively the voltage. This process takes between one to two hours and its benefit persists during the life of the antenna.

B-1-2   AW antenna in the Tokamak TCA[20]

In this heating method the antennae have to create magnetic field and hence pressure modulations with both a definite frequency and definite wavelength. In TCA the antennae are placed above and below the plasma in order to create a force in the vertical direction which is the most favourable from the heating point of view.

The antennae system comprises 8 groups of 3 plates of stainless steel (Fig. 31) phased in such a way as to excite the mode structure desired. The current feeds are mounted on insulated vacuum feed throughs and the applied voltage is typically 400 V. Due to this low voltage (for ICRH, V = 10 – 20 kV) no electrostatic shield is necessary.

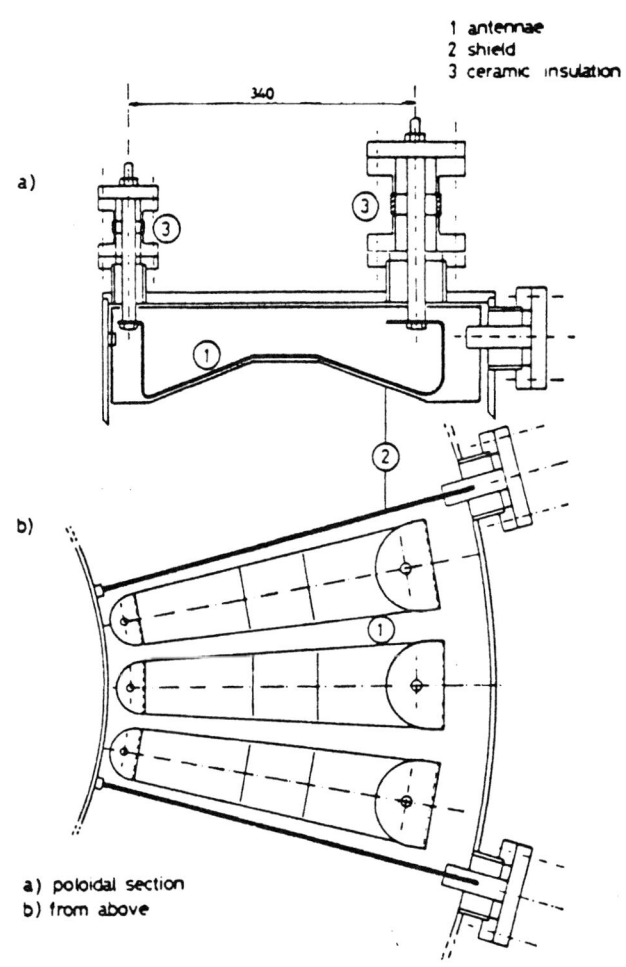

1 antennae
2 shield
3 ceramic insulation

a) poloidal section
b) from above

Fig. 31 Alfven Wave Antennae used on TCA.

1033

A second type of antenna is now installed in TCA. This antenna is manufactu red out of 1 cm diameter stainless steel bars coated with 6 μ thick TiN using the CVD process. (Chemical Vapor Deposition).

The conditioning of both the antennae and the vacuum vessel consists of the usual discharge cleaning in argon, then in the working gas.

B-2 Impurity production during RF heating.

B-2-1 Nature of impurities

A typical example of observations during an ICRH experiment on TFR is shown on fig. 32 [21] (The wave is launched from the high magnetic field side and transfers its energy to the plasma via mode conversion). Electron and deuteron temperatures rise at the beginning of the RF pulse and reach their maxima before the end of the pulse.

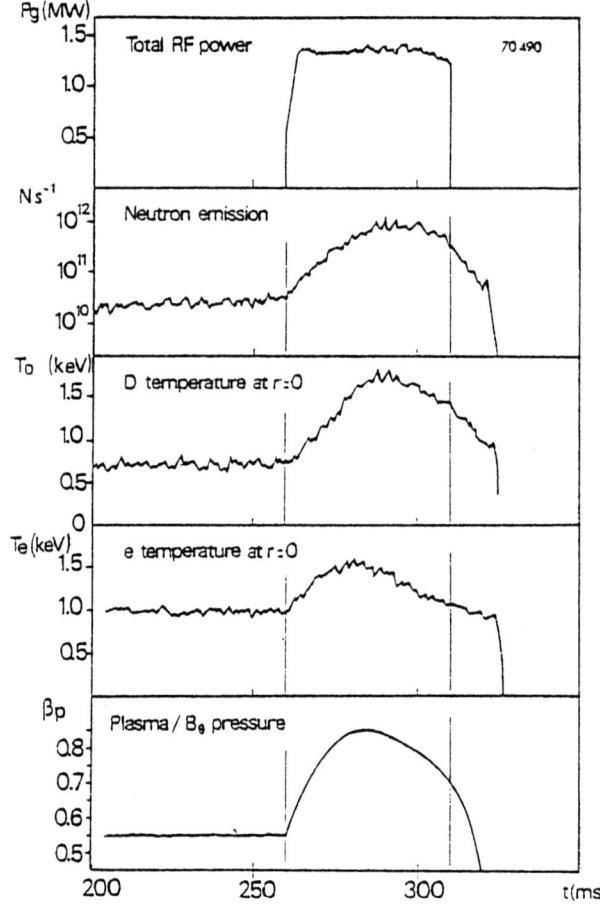

Fig. 32(taken from [21])

Light impurity ions near the plasma periphery (OVI, CIV) present
an abrupt increase, then stay at a constant level indicating a
constant influx of impurities from the outside of the plasma.
Metallic impurity ions (such as Nickel which is the main
constituent of the walls) and of the limiters increase at the very
beginning of the pulse (fig. 34) at the periphery (Ni XII) and more
slowly in the plasma core (Ni XVIII, Ni XXV). The bolometer signal
shows a sharp increase of the radiated power during the RF pulse.
The time evolution of the radial profile of the radiated power
(fig.33) shows the diffusion of the impurities toward the plasma
center. In that particular case the radiated power density on the
axis at t = 50 ms is 2W cm^{-3}, whereas the HF power density is at
most 2.6 W cm^{-3}. The rise of the radiated power during the RF
pulse looks like the cause of the plasma temperature decay after 30
ms. These experiments were performed by using inconel both for
limiters and lateral shieldings of the antennae.

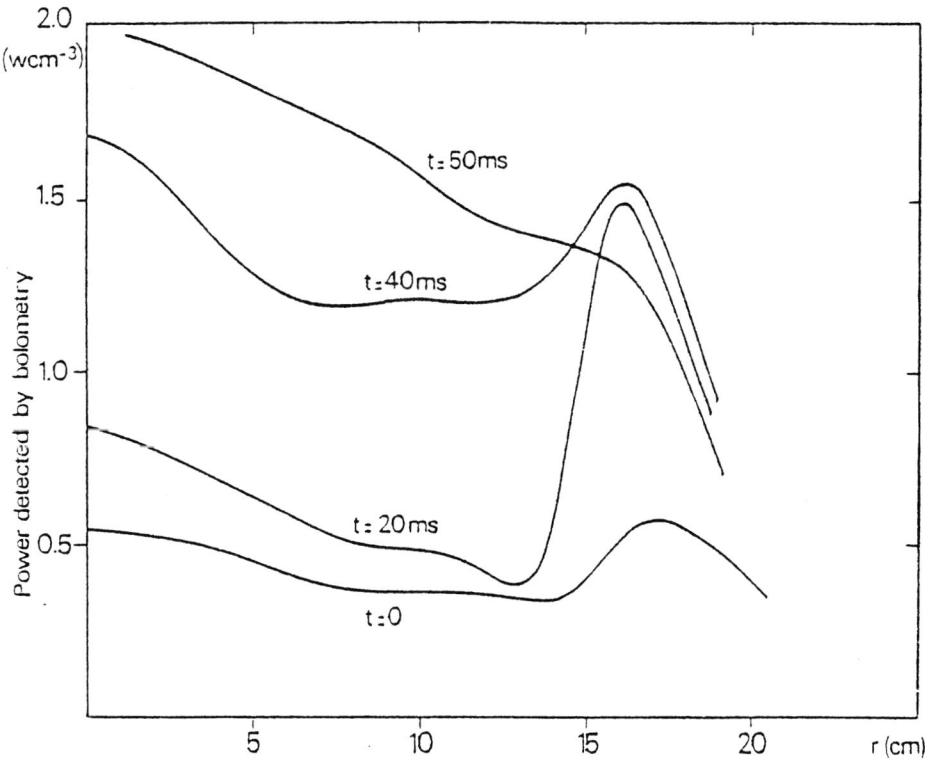

Fig. 33 Radial profiles of the radiated power density as deduced
from the bolometer measurements after Abel inversion.

1035

The influence on the impurity production of a modification of the material for these structures will be presented later on.

[22,23] A similar behaviour is observed in other ICRH experiments [24] but with less amplitude as the size of the device increases. Experiments at low plasma density show always a large increase

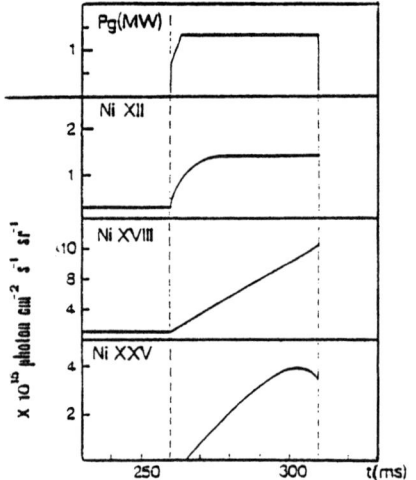

Fig. 34 Impurity line brightness during RF in TFR discharge.

of electron density (Fig. 35[23]). This is probably due to the wall desorption rather than an increase of particle life time as suggested by the increase of low energy charge exchange neutrals during RF pulse. These desorption effects are more important when graphite is used for limiter and in the antenna structures.

B-2-2 Origin of the metallic impurities

. A first way to identify the parts of the structures which play the major role in the impurity influx is to change the structures (form and material). This way has been followed by each experimental team.

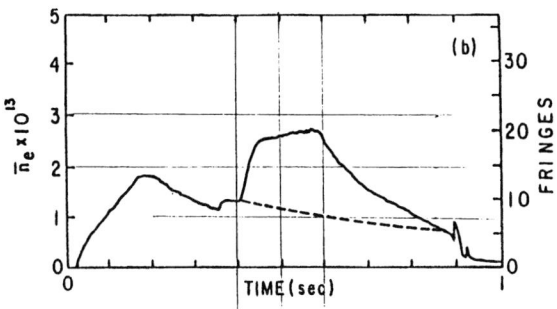

Fig. 35. Line average electron density in PLT, with (-) and without (- - -) RF (taken from [23])

. In TCA, different limiter designs and materials, namely carbon, stainless steel and Ti C coated carbon have been used. Fig. 36 [25] shows the extreme cases for the radiated power density. (Fe - 2) is the case of a large surface area stainless steel limiter where the peaking of the radiated power profile during the RF pulse is dramatic. (C - 2) is the case of a small surface area carbon limiter where the profile is nearly hollow in the center before the RF pulse. The radiated power density has been reduced from 2.6 Wcm^{-3} to 0.6 Wcm^{-3} during the RF pulse. As the antennae are metallic and only 2 cm inside the limiter radius, the heavy metallic impurities are still the most probable cause of the core radiation. A new design of antennae is under development to reduce the contamination.

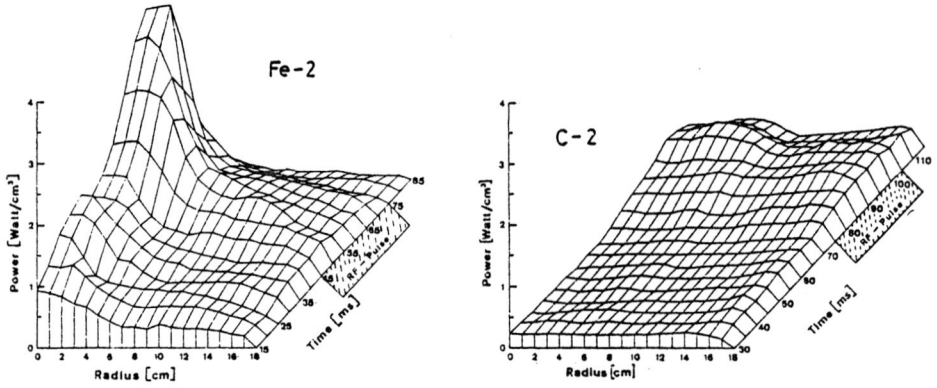

Fig. 36. Radiated power profiles. (taken from [25])

In PLT, a comparison was made between titanium and iron densities in plasmas in which   stainless steel   and   titanium Faraday
shields were used as the antenna protections [26]. The results show
(fig. 37) that the roles of titanium and iron are reversed in the
two cases : the iron density is greater than the titanium density
in the plasmas with stainless steel Faraday screens, both before
and during the RF pulse, and the opposite is true with the titanium
Faraday. Thus the Faraday shields which are not completely shielded
on the lateral parts, are the primary source of metallic impurities
entering the plasma, during ICRF heating and also with ohmic heating.

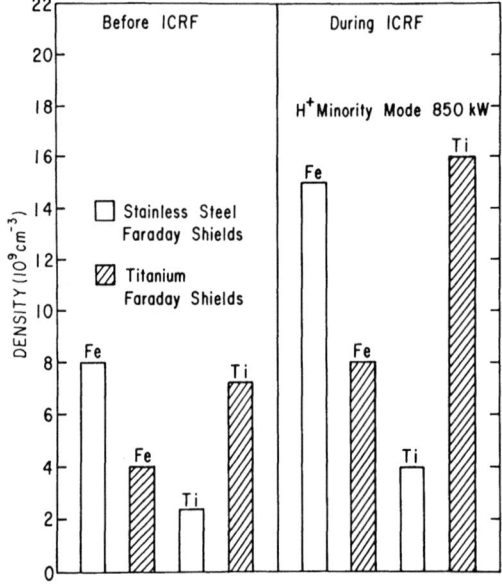

Fig. 37. Effect of the material of the Faraday shields on the metal-
lic impurities concentrations. (taken from [26|)

In TFR a thorough investigation [27] of impurity production has been undertaken by modifying the material of three components : limiter, antenna Faraday shield and lateral antenna protection. Table II summarizes the radiances observed of the resonance lines emitted by the different Na-like ions of Ni, Fe, Cr and Ti in the different experimental conditions. The two numbers given for the radiances are respectively before and at the end of the RF pulse.

TABLE II Impurity radiance associated with the different limiters and antennas

| | YEAR | Antenna | main limiter material | antenna Faraday shield | lateral antenna protection | pulse lenght ms | Impurity radiances $[10^{15}$ ph/cm².s.st] | | | | P_{RF} KW |
							NI XVIII $\lambda=292$ Å	Fe XVI $\lambda=335$ Å	Cr XIV $\lambda=390$ Å	Ti XII $\lambda=480$ Å	
1	1981	HFSA	Incalloy (72%Ni)	Inc.	Inc.	50	>0.5/2.5	0.05/0.16	0.05/0.2		350
2		LFSA	Inc.	Stainless steel	Inc.	50	0.7/5	0.07/0.35	0.06/0.3		100
3	1982	HFSA	C	Ti-V	C	100	0.13/0.52	0.065/0.25	0.065/0.2	~0.05	450
4	1982	HFSA	C	Ti-V	Ti	100	0.06/0.20	0.04/0.16	0.03/0.10	0.08	300
5	1983	HFSA	C	Inc.	Inc.	100	0.2/1.0	0.07/0.3	not measured	<0.04	300
6	1983	LFSA	C	Inc.	C	100	0.18/1.5	0.14/0.5	0.06/0.5	<0.15	300

Comparison of the different situations leads to the following observations :

a) replacing the incalloy by a carbon limiter reduces the Ni XVIII radiance by a factor of at least 2.5. (see cases 1 and 5 where the same incalloy antenna is used).

b) comparing cases 3 (or 4) and 5 indicates that the antennae are responsible for a significant part of the level of metallic impurity release. However, the fact that Ni radiance remains important, even when using a Ti antenna and C limiter, indicates that other parts of the machine contribute to impurity production (in particular the torus wall itself).

c) The contribution of the antennae to the pollution is due for the main part to an interaction with the lateral antenna protection rather than with that part of the shield facing the plasma (this result is indeed not in conflict with the PLT conclusion). That appears from comparison case 2 and 6 where the front part of the Faraday shield of the LFSA is changed (incalloy in place of stainless steel) whereas the Fe XVI radiation does not change very much.

d) The level of metallic impurity radiation increases linearly with the RF power (Fig 38) and with the duration of the pulse at least up to 0.1s. (The same dependence is also observed in PLT for H^+ second harmonic heating Fig 39).

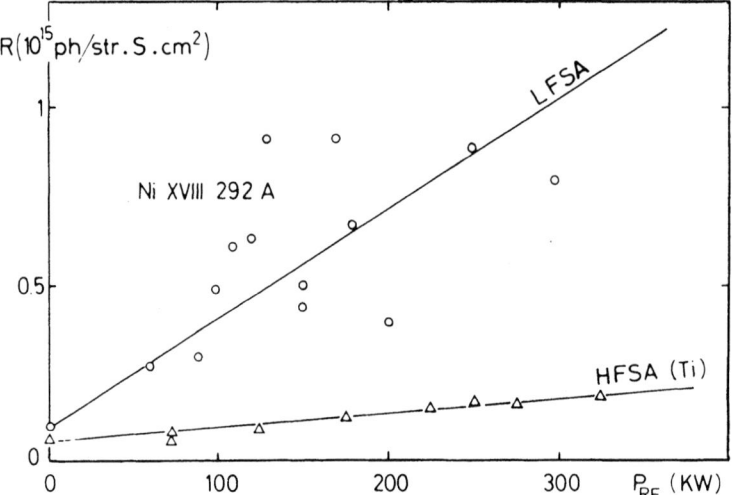

Fig.38. Radiances of Ni XVIII line for RF power emitted by the LFSA and HFSA fit with Ti Faraday shield.(taken from [27])

e) The very low level of metal observed with the HFSA with a Ti shield could not be achieved using the LFSA. As shown in Fig. 38, Ni production differs in those two cases by a factor 5, but a similar difference is observed for every metal.

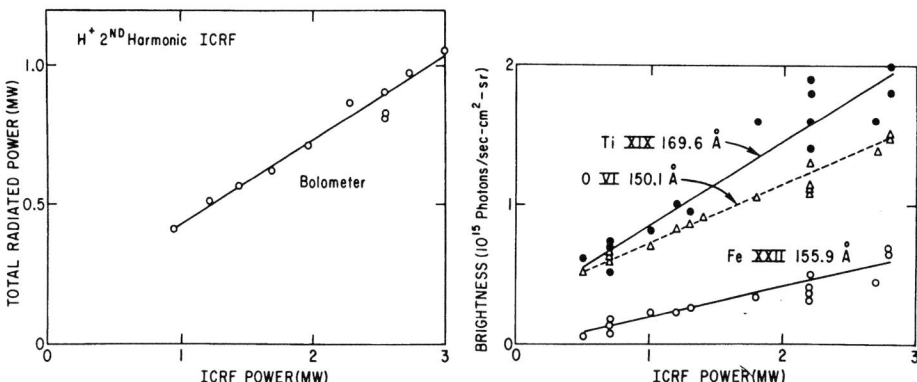

Fig. 39 Radiated power and metallic lines brightness versus ICRF power in PLT.

B-2-3 Experiments to investigate the physical mechanism which generates metallic impurities

Very soon in the RF heating experiments it appeared that impurity release could not only originate from the increased performances of the confined plasma. The investigations were oriented toward the RF energy deposition in the scrape off layer and the existence of fast particles with a poor confinement.

B-2-3-1 Modifications of the scrape off layer (S.O.L)

Fig. 40, 41, 42 show the effect of a RF pulse on the electron density and temperature in the shadow of the limiter for TCA and TFR.

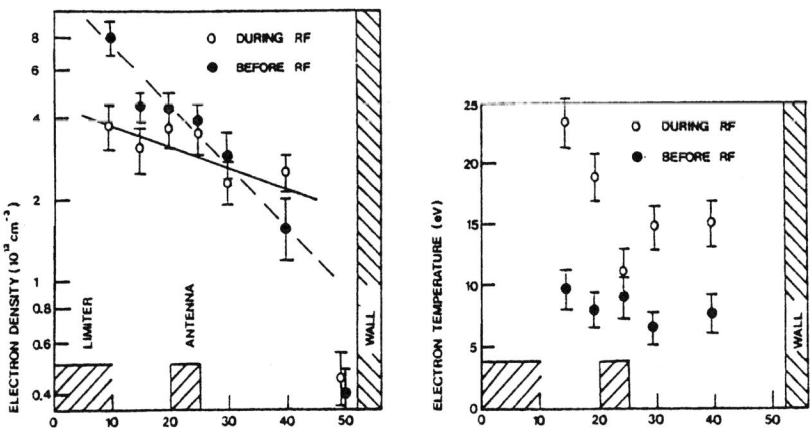

Fig.40. Scrape-off layer measurements (15 kG, D_2, q = 4.3) in TCA.

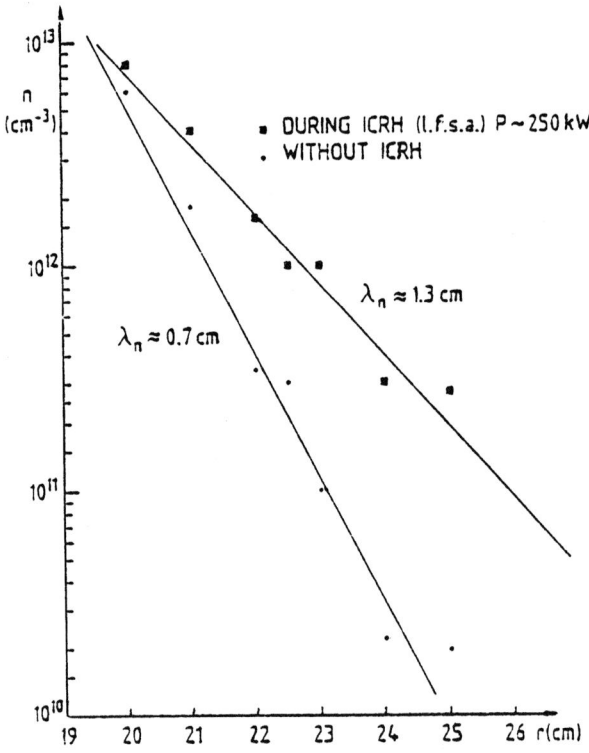

Fig.41. Electron density profiles in the S.O.L. in TFR.

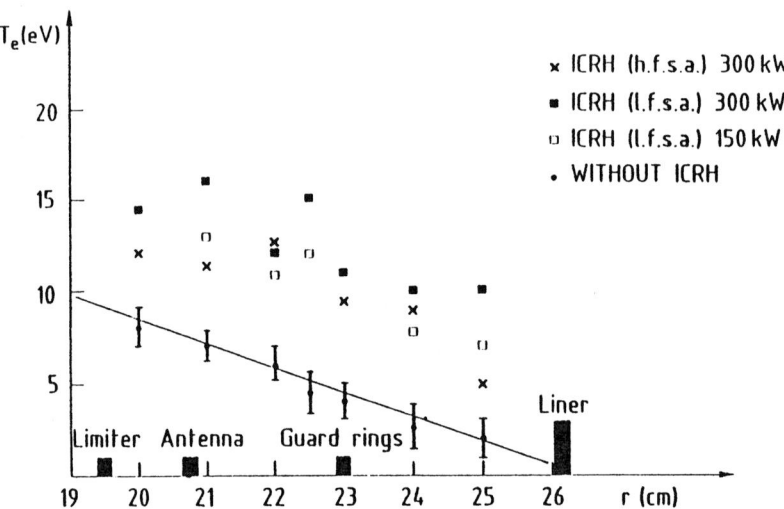

Fig.42. Electron temperature profiles in the S.O.L. in TFR.

On TCA as on TFR the density gradient decreases during RF and the density increase at the wall can be significant. The most relevant fact in relation to the impurity production is the electron temperature increase from typically 8-10 eV to 15-30 eV near the antenna. On TFR the electron temperature increases during the RF pulse at r = 21 cm is shown fig. 43. Near the liner Te reaches and often exceeds 10 eV which is enough through the potential drop of an electrostatic sheath to accelerate the impinging ions above the sputtering threshold. The energy spectra of ions, measured with a E x B analyser on TFR corroborates the Langmuir probe results.

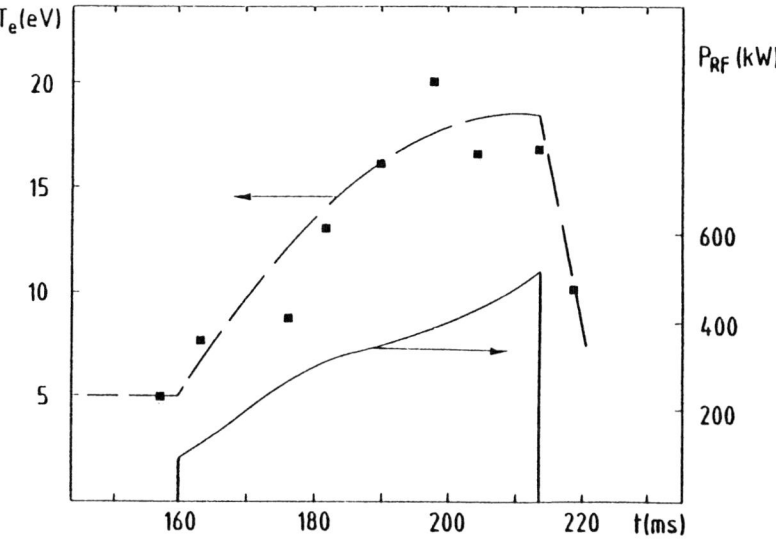

Fig.43. Electron temperature evolution in the scrape-off layer (at r = 21 cm) in TFR, during a RF pulse with a ramp.

Correlated with this energy deposit in the scrape off layer the neutral charge exchange flux grows during RF and on PLT, fast video films show local antenna-related emission in the visible region.

It is worthwhile to notice that same measurements mode with NBI on TFR do not show so large a variation of the scrape off layer characteristics. (First significant increase of electron temperature needs 600 kW of NBI and that corresponds to a low increase of metallic impurities influx).

B-2-3-2 Fast particle generation

The cyclotron damping of the RF wave can create a fast ion tail. This has been particularly observed in the minority and the second harmonic heating scenarios. Fig 44 shows an ion energy distribution observed in PLT[28] with 140 kW of RF power in the second harmonic heating scenario ; Fig. 45 shows the case of the minority ion heating for TFR. In both cases very high energy ions are created and can be lost either the plasma current is too low or as the velocity angle is such that the ion trajectory is unconfined.

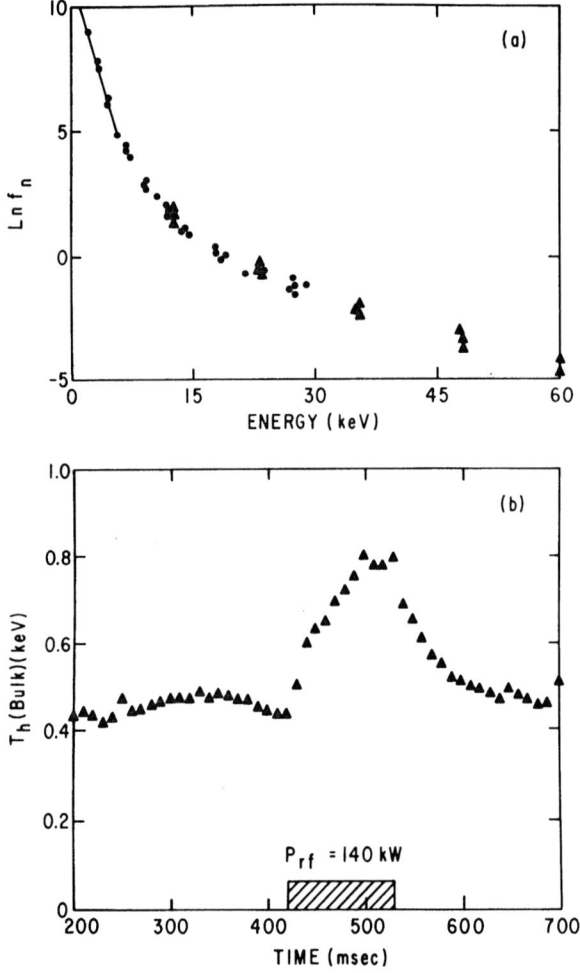

Fig.44. Charge exchange measurement of the proton distribution produced by second harmonic heating of a hydrogen plasma (f = 42 MHz, B_ϕ = 14 kG, $\bar{n}_e$ = 1.7 x 10^{13} cm^{-3}, P_{rf} = 140 kW) and the time evolution of the bulk hydrogen temperature (for the 1.5-5 KeV energy range). Both mass sensitive $\perp$ charge exchange (o) and angle-scanning near$\perp$ charge exchange (Δ) data are shown. (taken from [28])

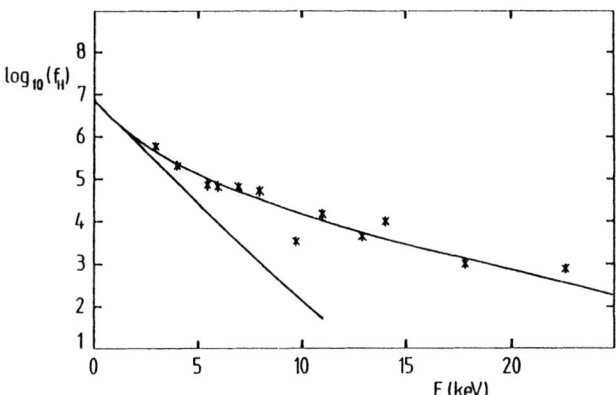

Fig.45. Ion energy distribution obtained during ICRF   for the minority ion heating scenario on TFR.

A calorimeter probe has
been developed in PLT to
measure the losses of the
energetic particles at
the periphery versus the
velocity angle [29]. Re-
sults [24] show (Fig. 46)
in the H minority regime
a peaking near an angle
which is also observed
on a horizontally scan-
ning tangential charge
exchange analyser [30]
(fig.47). That may be
interpreted in terms of
energetic ions with high
perpendicular components
which describe large ba-
nana orbits. These ions,
rapidly lost during RF
pulse could be responsi-
ble for sputtered metal-
lic impurities.

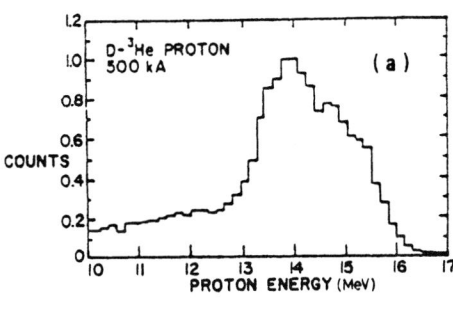

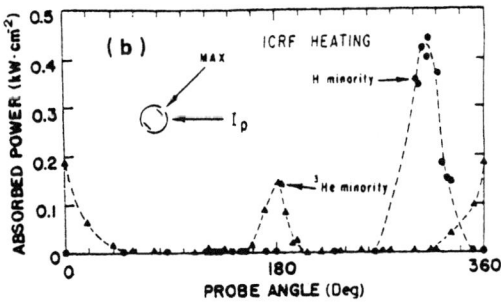

Fig.46. Energetic particle measurements.(a) Proton spec-
trum form D-^{3}He reactions
and (b) probe measurements
for $P_{RF} \sim$ 1MW (for H minority case, the OH level has been sub-
tracted). (taken from [24])

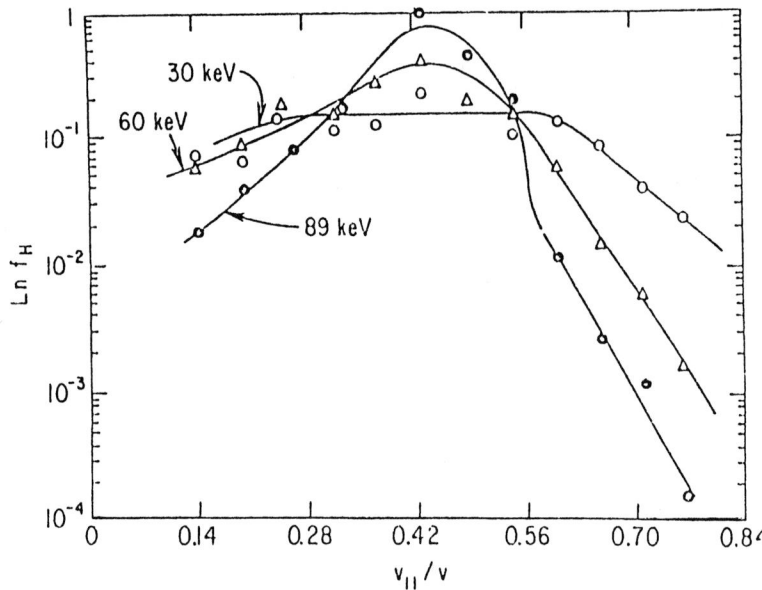

Fig.47.Fast neutral angular distribution for three energies measured with an horizontally scanning charge exchange analyser,
during RF heating on PLT.(taken from [30])

In TFR superbanana losses have been observed with polarized
collector plates in a single ripple located right under the
antennae in the region of maximum RF field. With ICRF, in the
minority heating case the ion current is as much as 60 times
larger than in ohmic case and strongly peaked ; the power lost
through this channel can reach 25 % of the RF power input. In the
mode conversion regime, this loss is reduced by a factor of 15. The
metallic impurity production is also reduced in the mode conversion
regime compared with minority heating.

However, some experimental observations weaken the correlation between an enhanced impurity level and the fast ion tail in
the H minority regime. In fig. 48 the proton velocity distribution
function is shown for two TFR discharges using a module of the JET
antenna ; while varying the ratio n_H/n_D leads to different degrees
of distortion of the proton velocity distribution function this has
no detectable effect on the rate of impurity production.

In TCA the flux and the energy of Fe ions have been
determined in the scrape off layer by analysing their implantation
into gold or carbon foils [32] . The Fe flux increases by a factor

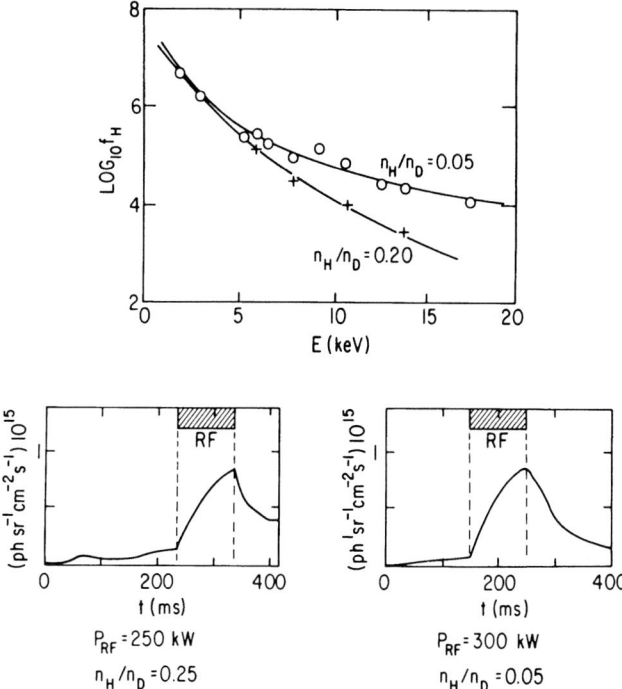

Fig.48. Ion distribution functions for two different scenarios as-
       sociated Ni XVIII brightness n_H/n_D = 0.05. H minority re-
       gime ; n_H/n_D = 0.20 mode conversion regime.

Table III. Ion impact energies of Fe impurities in the scrape-off
           layer of TCA tokamak.

Drift Side	rf Power (kW)	Implantation Depth (Å)	Ion Energy[*] (eV)
electron	0	4.6	130
electron	108	5.1	160
ion	108	3.4	90
ion	108	4.2	120

[*] assuming that projectiles impinge the surface at normal incidence angle.

of ~ 2 during RF heating and the corresponding ion impact energy is given in Table III. These energies are well above the sputtering threshold. One observes that the increase of Fe flux as a function of RF pulse length is non linear (Fig 49) and that may be interpreted in terms of additional sputtering by impurities generated during the RF pulse duration. This flux is reduced considerably (Fig 50) by replacing the steel limiter with a carbon one.

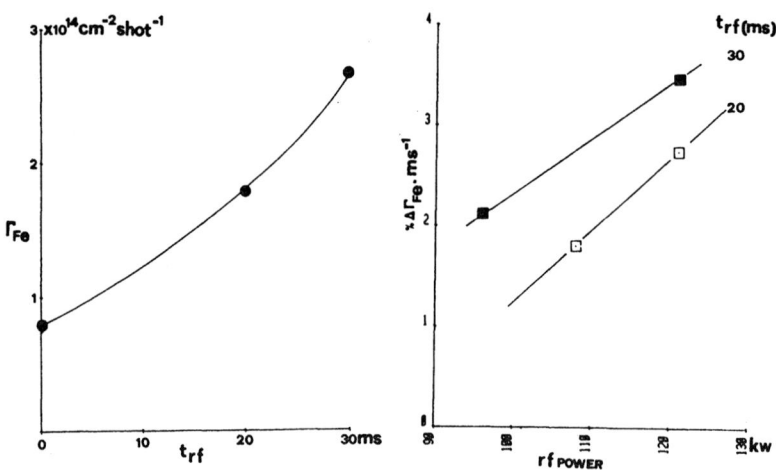

Fig.49.a) Γ_{Fe} as a function of rf pulse time at an rf power of 121 kW and b)%increase in Γ_{Fe} as a function of rf power for t_{rf} = 20 and 30 ms. (taken from [32])

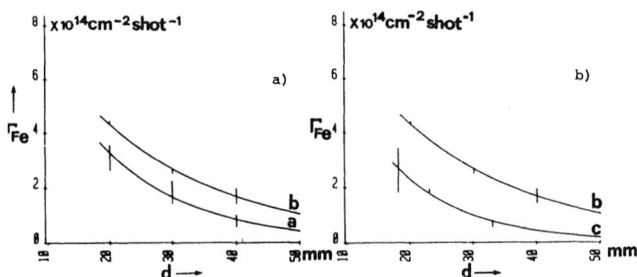

Fig.50. Γ_{Fe} as a function of the radial distance between the probe and the antenna in TCA. (taken from [32])
   a : without RF heating and SS limiter
   b : with RF heating 100 kW SS limiter
   c : with RF heating 100 kW and carbon limiter

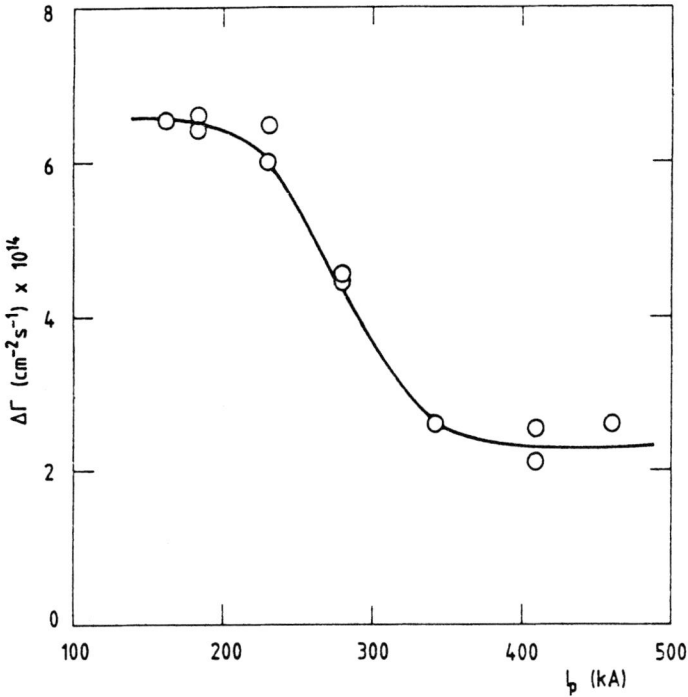

Fig.51.Low energy neutral flux (2 - 2000 eV) during RF heating
at constant power versus the plasma current.(taken from
[23])

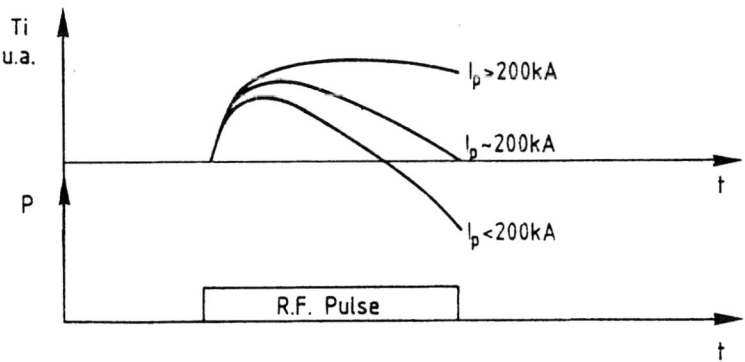

Fig.52. Time evolution of central ion temperature for different
plasma currents.

In order to reduce the fast particle influx to the wall one can increase the bulk plasma density and the plasma   current and also reduce the ripple. The effect of increasing the plasma current has effectively reduced the low energy neutrals increase (in the 2-200 eV range) on PLT during the   RF (fig. 51). In addition Ti saturates during the RF for $I_p >$ 200 kA while it decreases after a maximum for $I_p <$ 200 kA (Fig. 52).

B 2-3-3 Spurious waves excitation

Several authors have suggested that excitation of spurious waves could be responsible for heating of the plasma between the wall and the limiter. We review some theoretical hypothesis to explain the channel by which the impurities are produced :

i) Surface modes propagating at the plasma edge can be excited for particular values of the wave characteristics (frequency and propagation vector). Their energy can be dissipated into a layer where the plasma density is matched to achieve a particular resonance (called lower hybrid resonance). As this resonance depends on the magnetic field such an exploration has been made on TFR. Fig. 53 shows effectively a correlation between the Ni^{17+} brightness and the density of the lower hybrid resonance (LHR).

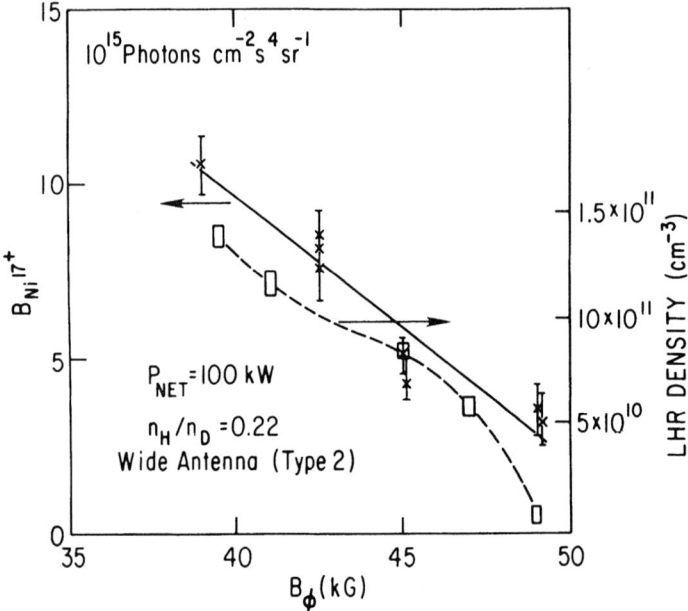

Fig. 53. Radiance of the Ni^{17+} line versus the toroidal magnetic field for the wide antenna. The corresponding density of the lower hybrid resonance is also represented. (taken from [34])

This surface wave needs to be excited, an RF electric field component parallel to the magnetic field. This component does exist if the Faraday shield slits are not strictly parallel to the magnetic field lines. An experiment was made on TFR with a special Faraday shield with inclined slits at the angle (4°) corresponding to the usual angle of the magnetic lines at the edge. The comparison of the impurity production for two directions of the plasma current, the one giving no $E_{\parallel}$ component and the other giving a significant $E_{\parallel}$ component (the angle between the slits and the magnetic lines being 8°) is shown on Fig. 54. No difference appears between these two situations and this mode of excitation of the surface waves cannot be retained.

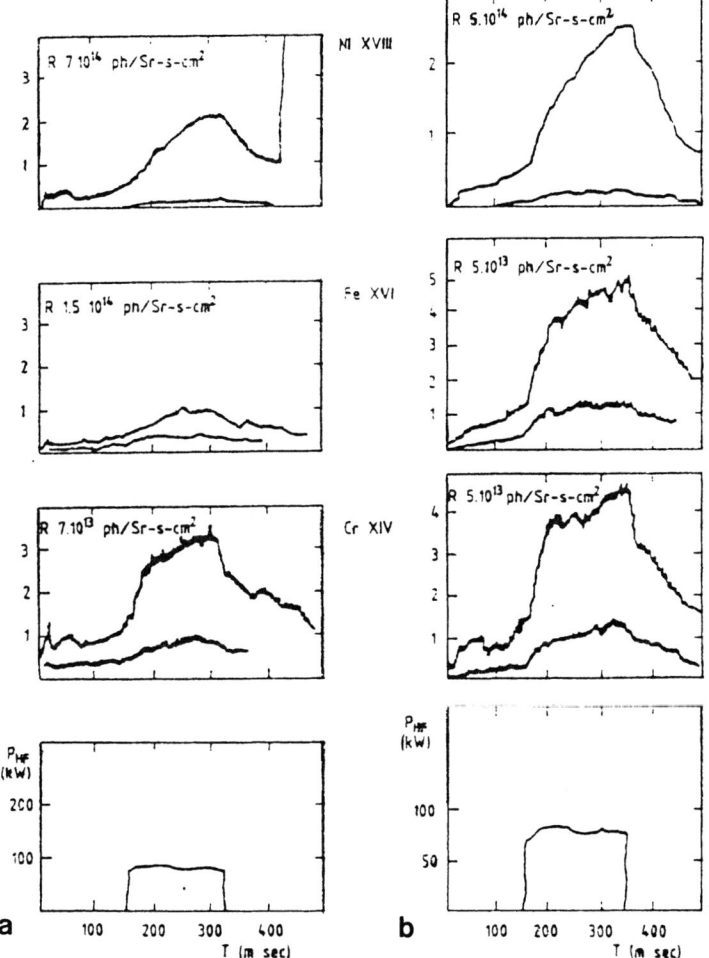

Fig.54. Metallic lines observed on TFR during RF heating
   a) Faraday shield slits makes 8° with regards to the magnetic field lines at the antenna radius.
   b) Faraday shield slits are parallel to these magnetic field lines.
   The lowest traces on each diagram corresponds to the light emitted be sides the lines.

Another way to excite surface waves is by the radial component of the RF current which appears near the connection of the antenna to the coaxial line. Usually this current is zero due to the cancellation from the lower and upper halves of the antenna. By using only one half of the antenna, however, the radial current can be non zero. Nevertheless impurity concentrations in both cases ($j_r = 0$; $j_r \neq 0$) are the same.

ii) <u>Weak absorption</u> of a fraction of the magnetosonic waves excited by the antenna is a possible source of impurity. Indeed the $k_{\parallel}$ spectra of the waves excited by the present antennae presents a maximum near $k_{\parallel} = 0$; in that part of the spectra the waves are weakly absorbed by the plasma. These waves can be reflected several times between the cut off layer and the metallic wall, dissipating a lot of energy at the surface of the wall and in the edge plasma.

New antennae specially designed to reduce the low $k_{\parallel}$ part of the excited wave will be tested on TFR, then on JET.

B - 3 <u>Impurity purification with ICRH</u>

One can take advantage of the linear mode conversion regime appearing in á two-ion-species plasma to effectively pump out a selected impurity. If the two ion hybrid resonance location coincides with an harmonic of the cyclotron frequency of an impurity, a substantial part of the wave energy can be damped on these ions. Monte Carlo calculations have shown that if the RF electric field is strong enough, the resonant particle can gain enough energy to leave the plasma by trajectory effects.

This possibility was explored on TFR [34] by using argon as a test gas. Fig 56 shows the time evolution of two Argon lines Ar VIII and Ar XVI, as RF is applied when the hybrid layer is located at a radius where Ar XVII is the dominant ion and its second harmonic cyclotron frequency corresponds to the frequency of the input wave. By varying the minority ion concentration, the hybrid layer is displaced and the decay of Ar line is smaller coming only from modification of Argon influx as RF is applied.

This method of purification could be applied to the intrinsec impurities by using several frequencies. It needs however to evaluate experimentally the consequences on wall sputtering of the expelled very high energy heavy ions (Monte Carlo calculations indicate that the energy can reach several hundred keV).

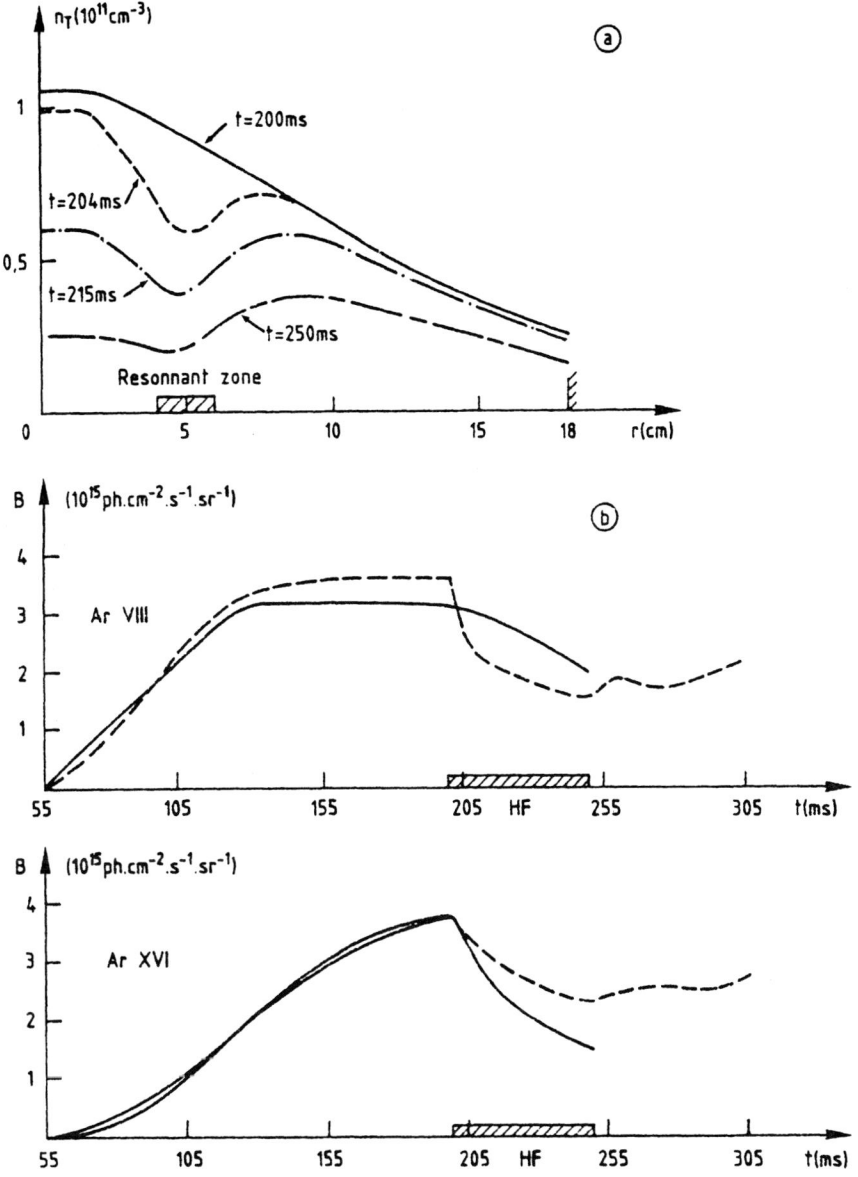

Fig.55. a) Argon density radial profiles versus time RF is applied
between 204 and 250 ms.
(Simulation results)
b) Argon line brightness versus time.
- - - - experimental results.
———— simulation results
(taken from [34])

B - 4 Conclusions :

The mechanisms of impurity production in ICRH are more complex than in NBI. We summarize the major ones below :

- EM wave energy can be directly dissipated on the surface of the metallic walls producing light impurities by thermal effects.

- EM fields parallel to the magnetic field can accelerate charged particles in the shadow of limiter, increasing the sputtered impurities.

- EM surface waves can deposit energy at the plasma edge where the confinement is very poor.

- High energy particles can be created inside the bulk plasma with unconfined trajectories.

- Finally, the increase of energy content of the bulk plasma, which is the final goal of heating methods, does not seem presently to play an important role in impurity production.

Certainly, the next generation of large tokamaks will permit us to rule out some mechanisms by changing the ratio of wavelength to the plasma size. Also, the Tokamaks with divertor will bring new conditions in the vicinity of antenna. However, in order for ICRF to be a viable heating scheme in tokamaks, means of reducing impurity production, -probably the most significant problem to be resolved;must be found.

C - Lower Hybrid Heating and impurities

Among the possible RF wave heating schemes, the waves in the neighbourhood of the lower hybrid resonance frequency (LH) have interesting properties :

i)   due to the range of frequency (~ 1 GHz) the coupling of energy by means of waveguides is possible, thus avoiding mechanical structures inside the vacuum vessel.

ii)   large power is available since klystrons have been developed for communication purposes.

iii)   the region of accessibility inside the plasma coincides with a phase velocity for the wave which can couple the RF energy to the electron tail thus producing a electron current which can take the place of the ohmically induced current, thus saving valuable volt seconds and perhaps offering the possibility of steady state operation.

C-1 Launching structures

All the experiments now working on this subject used to launch
the wave a phase array of waveguides (grill). In the past, due to
the small available port size loops with coaxial feeders have been
used. Fig. 56 shows the WEGA coupling structures [35] used in the 500
MHz range, able to couple 200 kW during 15 ms pulses. However, as
these loops are not extrapolable to multi-MW range which is needed
for large Tokamak they were no longer used.

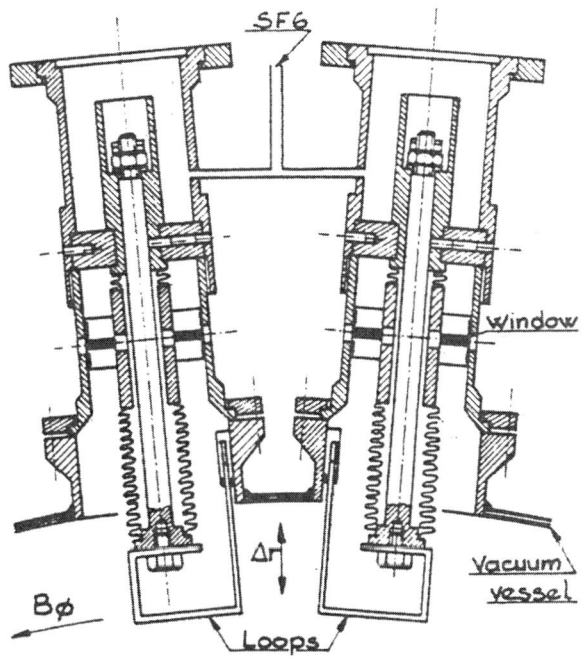

Fig. 56. The 2 loops couplings system used on the tokamak
WEGA. (taken from |35|).

A sketch of a grill used on WEGA is shown on Fig. 57.
This launching structure, made of four waveguides, can
excite into the plasma a wave with a phase velocity

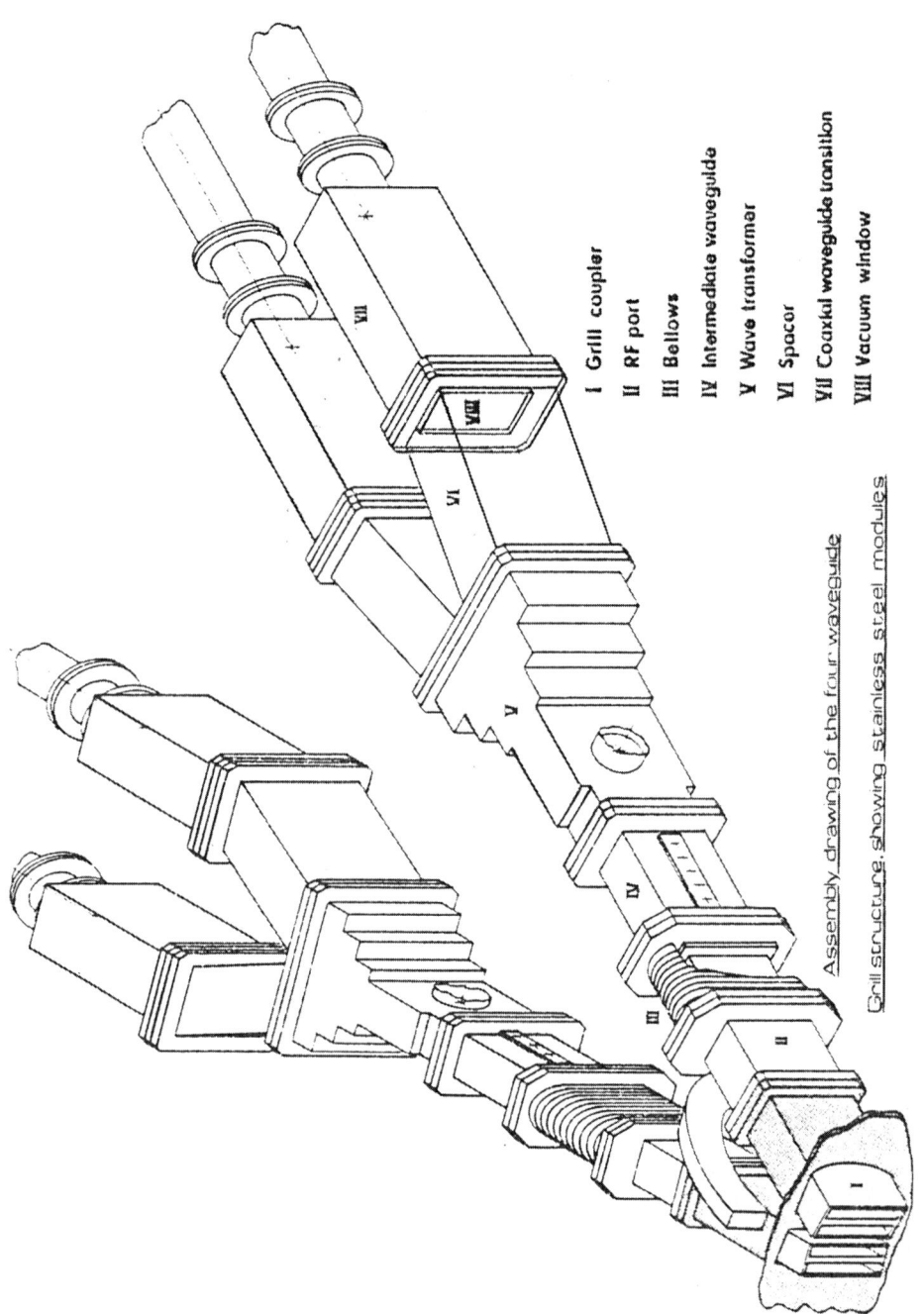

I Grill coupler
II RF port
III Bellows
IV Intermediate waveguide
V Wave transformer
VI Spacer
VII Coaxial waveguide transition
VIII Vacuum window

Assembly drawing of the four waveguide

Grill structure showing stainless steel modules

Fig. 57. The WEGA grill

whose characteristics are determined by the geometry of the waveguides and the difference of phase between each of them. As the room available into the port is fairly narrow, the standard waveguide (VII) has to be progressively reduced by use of a wave transformer. A vacuum window is needed to separate the vacuum vessel from the pressurized waveguide but its position with regard to the plasma depends on each experiment. On the example shown on figure 57 the window is far away the plasma in order to put the electron cyclotron resonance in the part of the waveguide which is at very low and constant pressure during the plasma discharge.

Other people [37,38] have suggested placing the window quite near the plasma to avoid the difficulty of controlling the gas pressure between the window and the plasma. However, this technique will not be applicable under reactor-like conditions due to neutron damage of the window.

A major problem which has limited for a long time the RF power density through the waveguides is the multipactor effect. This appears as electrons in the RF field attain a sufficient energy to create secondary electrons at the wall ; the matching of the distance between the walls of the waveguide, the gas pressure, the frequency and the amplitude of the RF field gives a large multiplication factor and leads to a breakdown. When such breakdowns appear the transmitted RF power is reduced and its phase is significantly changed. Various solutions [39] have been proposed in order to suppress RF breakdown by reducing the secondary electron emission coefficient to values below one :

i)   solid titanium grill + argon glow discharge

ii)   solid titanium grill + Ti pulverisation at a high rate in a relatively high argon pressure + argon glow discharge.

iii)   stainless steel grill + carbon coating

iv)   solid titanium grill baked at high temperature + additional pumping.

v)   solid titanium grill baked at high temperature + additional pumping + argon glow discharge.

vi)   stainless steel grill + cleaning by Taylor discharge cleaning inside the Tokamak vacuum vessel.

These solutions are generally qualified on a test stand by measuring the transmission coefficient, the wave phase and the light emitted during RF breakdown. As an example fig 58 [40] shows for one grill used on WEGA the dependence of the transmission coefficient and emitted light as a function of time during RF conditioning (pulses of 50 kW -10 ms every 10 seconds).

1057

The efficiency of low pressure argon glow discharges [41] is shown in fig 59. The large increase of the maximum effective power being transferred without phase variation, appears since glow discharges are applied.

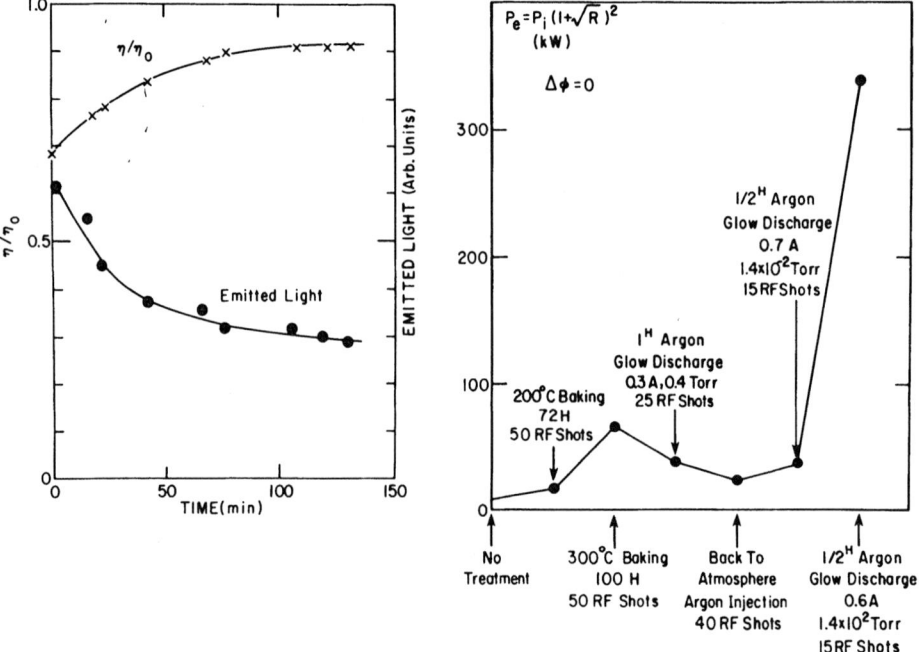

Fig.58. Transmission coefficient and emitted light by gas breakdowns as a function of time during RF conditioning on the WEGA grill. (taken from [40])

Fig.59. Effective transmitted power after different grill conditioning WEGA (taken from [41])

The grill technology is still progressing rapidly and the goal of 5 kW/cm^2 power density at the injection port needed for the multi megawatt projects under construction seems achievable with good reliability. (The highest power density reached on Alcator C is 8 kW/cm^2 for 4.6 GHz and the fig. 60 shows the present results achieved on different experiments).

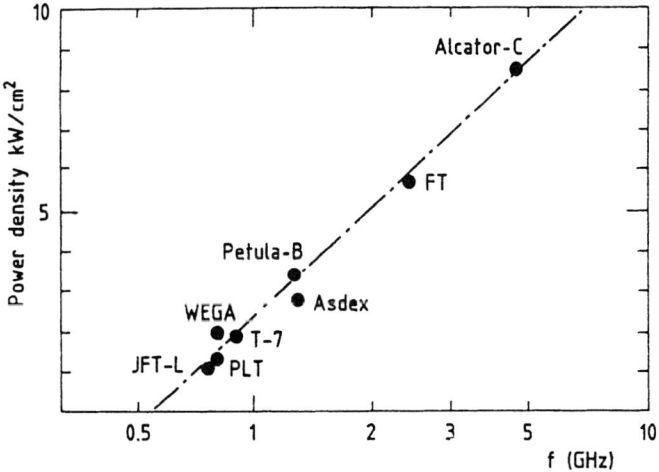

Fig.60. Power density versus frequency (all waveguides being fed)

C - 2 Impurity production during LH heating and current drive

The experimental results are still scarce ; the most extensive studies have been mode on Alcator C [42] and on Petula B [43] .

C-2-1 On Alcator C, it appears that the dominant RF induced impurity influx comes from the limiter. Three types of limiters were used :

- With molybdenum limiters, the major part of the RF power is radiated 850 kW for 950 $_0$ kW RF input. Fig 62 shows the increase of brightness in the 75 A range which includes Mo lines from many ionization states and the rapid Z_{eff} change (deduced from the brightness at 5360 A which is due mainly to free-free bremsstrahlung).

With silicon Carbide coated graphite limiters both Si and C levels are seen to increase substantially . The increase of Z_{eff} larger than with molybdenum limiter indicates also a large amount of low Z impurity influx.

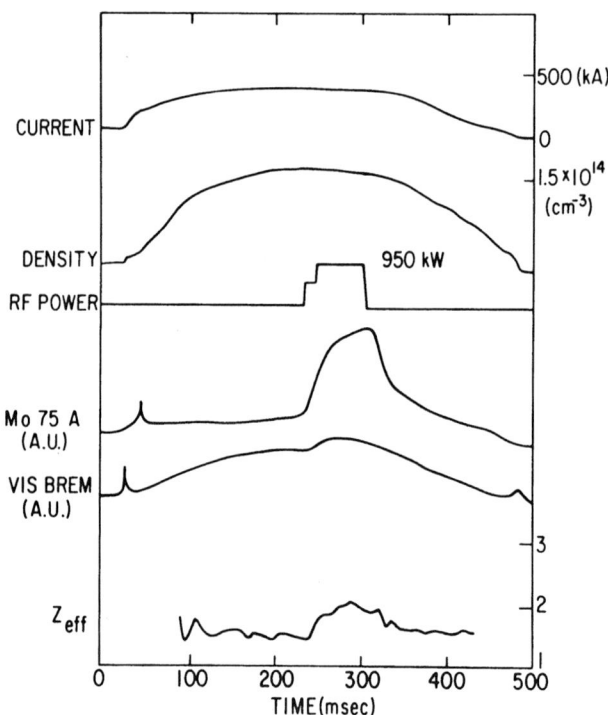

Fig.61. Impurity production in Alcator during LH heating with molybdenum limiters. (taken from [42 |)

The increase of Z_{eff} is a very non linear function of the RF power (Fig 62) ; significant effects appear above 500 kW.

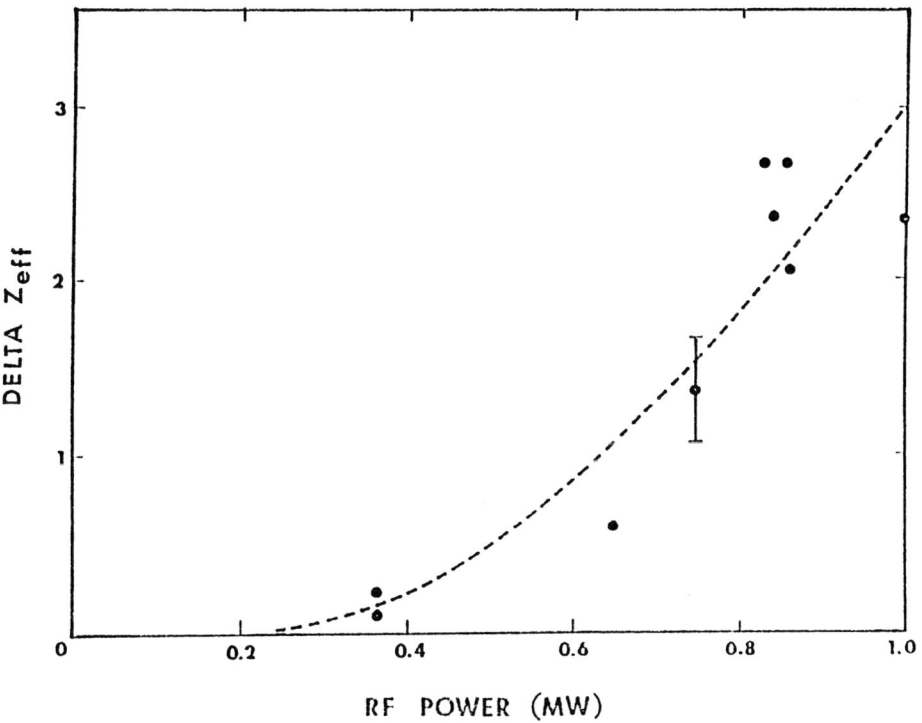

Fig. 62. Z_{eff} increase for LH heating on Alcator with molybdenum limiters. (taken from [42])

The hard X ray signal reveals the existence of non thermal electrons ; their energies extend to over 300 keV. post–mortem examination, and observation of the light emitted at the limiter surface indicate an electron impact in a poloidally localized region of the limiters near the outside mid–plasma. The evaluated power loading on the area affected is of the order of 20 kW/cm^2. This might be the major cause of the impurity generation.

With a shaped graphite–limiter in the outside midplane, designed for a constant power density on its surface the impurity production was considerably reduced for the same RF power (fig 63).

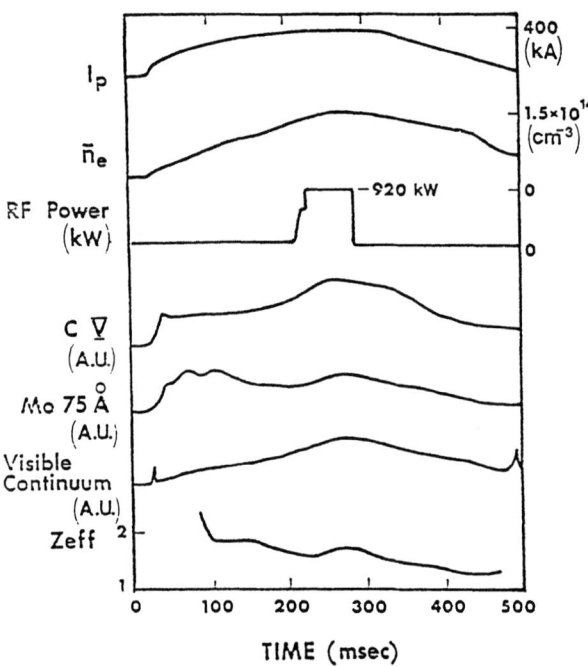

Fig.63. Impurity production in Alcator during LH heating with graphite limiter. (taken from [42])

In order to reduce further the power deposition on the limiter by these fast particles, rotating limiters have been placed on PLT and Alcator and one hopes that this source of impurity influx will be negligible.

Other processes of impurity production similar to those in ICRH, cannot be ruled out. In the shadow of the limiter, probe measurements indicate that the electron temperature increases from about 5 eV before the RF, to about 7.5 eV during. These values are twice as low as in the ICRH case and are probably too low to result in significant sputtering of limiter material due to acceleration of ions across the expected sheath.

C-2-2 On Petula B the impurity behaviour is quite different in the current drive regime and in the ion heating regime.

In the current drive regime observed at low density (n_e <3.10^{13} cm^{-3}) the electron temperature and the plasma density in the scrape off layer decrease during the RF pulse. Consequently, one observes a strong decrease of the impurity content which is maintained even after the RF pulse. The figure 64 shows the time evolution of O VI, Fe XVI and bolometric signals. The radial profiles of O VII (fig. 65) confirms a decrease of impurity in the plasma center. A possible interpretation is a reduction of the transport at the edge as the inductive electric field is short circuited by the RF current. The RF field would act more or less as a RF divertor.

As in Alcator C fast electrons have been observed during the lower hybrid current drive phase. However the lower RF power in Petula B, 140 kW instead of 1 MW in Alcator-, probably indicates that the metallic impurity influx coming from the stainless steel limiter does not dominate the radiated power.

In the ion heating regime the lower hybrid waves create an ion population with high perpendicular energy which is partly unconfined. One observes (fig. 66)) at the beginning of the RF pulse a fast increase of the bolometer signal followed by a slow increase as the impurities penetrate toward the plasma center. The variation of the relative increase of Z_{eff}, Fe XVI and bolometer signals with the RF energy is shown fig. 68. This impurity production caused by the sputtering of fast ions could be avoided with higher plasma currents and would be probably lower in larger tokamaks.

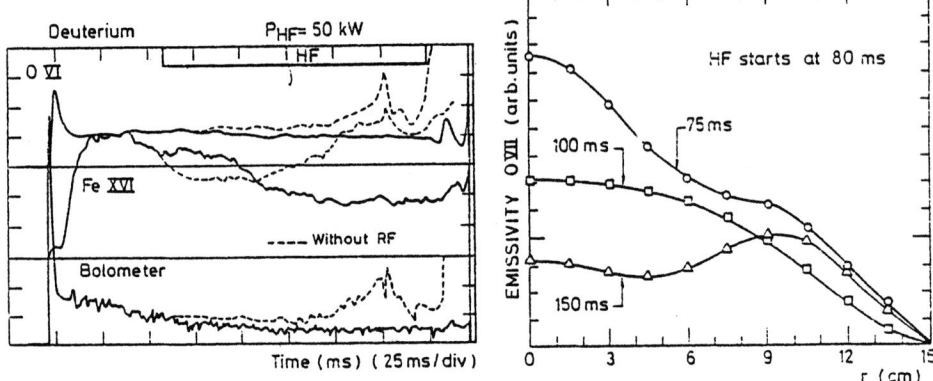

Fig.64. Time evolution of OVI(1032 A), Fe
XVI(335 A), bolometric signal during the current generation. The
dotted lines concern a discharge
without R.F.

Fig.65. Emissivity of O VII(1623 A)
after Abel inversion, as a
function of radius for three
times.

(taken from [43])

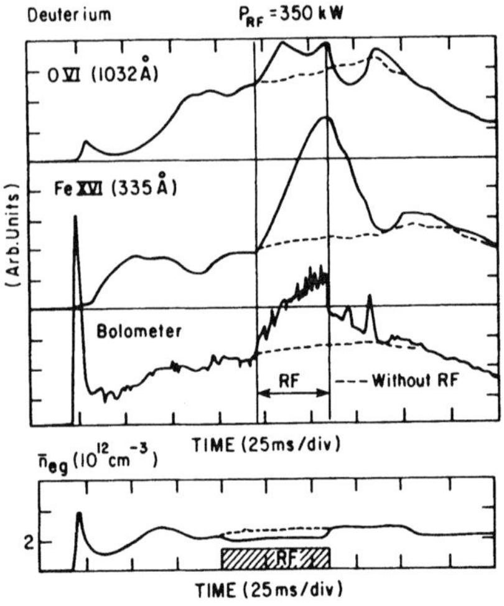

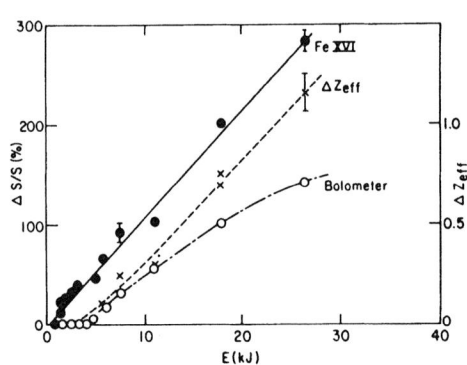

Fig.66. Typical impurity lines obtained from
discharges in the Petula Tokamak with
R.F.. The dotted lines concern discharges without R.F.. The last curve represents the density evolution of the
scrape-off layer plasma.(taken from
[43])

Fig.67. Evolution of Fe XVI and bolometer signals (in percent)
and ΔZ_{eff} as a function of
RF energy.

(taken from [41])

D - CONCLUSIONS

For the past ten years, significant progress has been made in the area of multimagawatt injected power in hot magnetically confined plasmas. These large amounts of power affect the plasma wall interactions and, this subject is now a major issue for the success of Controlled Thermonuclear research. It was certainly expected that the increase of plasma energy would have increased the thermal flux at the plasma frontier. But, other processes, related to an unknown power deposition profile (as in ICRH) or to the unforeseen impurity transport properties (as in NBI) have introduced open and tricky questions. The edge plasma behaviour remains critical to the bulk plasma performances (at least for medium-size tokamaks). But it is yet unknown whether the largest Tokamaks (JET and TFTR) will suffer the same drawbacks. Their first results in ohmic heating show high Z_{eff} which is an indication of large plasma wall interactions.

Thus the need for thorough investigation in this field becomes more and more essential to the progress of magnetic fusion

REFERENCES

1. P.H. Edmonds et al ; Proc. of the 3rd Joint Varenna. Grenoble Int. Symp on Heating in Toroidal plasmas, Grenoble (1982) Vol I, 5

2. K.W. Ehlers Proc. Int. Ion Eng. Congress ISIAT 83, Kyoto (1983) p 59-69

3. J.F. Bonnal et al, Phys. Lett. 75A (1979) 65

4. Y. Okumura et al, Rev. Sci. Instrum. 52 (1), Jan 1981,1

5. A. Colleraine et al, Ibid Ref 1 Vol I, 49

6. W. Ott et al, Rev. Sci. Instrum. 54(1), Jan 1983, 50

7. M. Delaunay et al, Ibid Ref. 1 Vol I, 143

8. W. VII A team, in Plasma Physics and Controlled Nuclear Fusion Research 1982 (Proc. 9th Int Conf Baltimore) Vol II, IAEA, Vienna (1983), 241

8a. F.P. Penningsfeld et al Proc of the 4th Symp. on Heating Toroidal plasmas Rome March 84.

9. W. Ott et al, Ibid Ref 1 Vol III, 813

10. G.G. Lister et al, Ibid Ref 1, Vol I, 103

11. P.C. Johnson et al, in Controlled Fusion and Plasma Physics (Proc 11 th Europ Conf. Aachen 1983) Vol II, Paper C 36.

12. TFR Group, Journal of Nucl. Mat. 105(1) Jan 1982, 62

13. K. Behringer et al, in Proc. IAEA Technical Committee Meeting on Divertors and Impurity Control (Keilhacker, Max Planck-Inst für Plasma Physik Garching (1981) 42).

14. C.E. Bush et al Nucl. Fusion 23 (1) 1983, 67
15. S. Suckewer et al, PPPL 1768 (1981)
16. R.C. Isler et al, Ibid Ref 1, Vol III 1185
17. W.H.M. Clark et al, Nucl. Fusion 22(3) 1982 333
18. R.J. Fonck, Phys. Rev. Letters 49 (10) 1982, 737
19. J. Jacquinot et al, Proc 11th Symp. on Fusion Techn., Oxford Sept 80.
20. A. de Chambrier et al, Ibid Ref 1, Vol I, 161
21. Equipe TFR, Plasma Phys 24 (1982) 615.
22. JFT2 Group, Ibid Ref 1, Vol III, 1191
23. Colestock, P, IAEA Technical Committee, Princeton Oct 81
24. Hwang, D.R. et al. Ibid Ref 8, Vol II, 3
25. R. Behn et al, Plasma Phys and Control Fus 26 (1984) 173.
26. B.C. Stratton et al, to be submitted to Nucl. Fus
27. Equipe TFR, Ibid Ref 25, p 165
28. J. Hosea et al, in Plasma Physics and Controlled Nuclear Fusion Research 1980 (Proc. 8th Int. Conf. Bruxelles) Vol II, IAEA Vienna (1981) 95.
29. Manos. D.M. et al. J. Nucl. Mater 111 and 112 (1982) 130.
30. Kaïta R et al Nucl. Fusion 23 (1983) 1089
31. Equipe TFR, Ibid ref 28, Vol I, 425
32. J.K Gimzewski et al, Ibid ref 11, Vol II, 85, 463
33. Equipe TFR, Ibid ref 1, Vol 1, 225
34. TFR Group, Nucl. Fusion 22 (1982) 956
35. G. Tonon et al, 3rd Int. Congress Waves and Inst. Palaiseau, France 1977 –C6–161
36. G. Tonon et al, Ibid ref 8, Vol I, 179
37. M. Porkolab et al, Ibid ref 8, Vol I, 227
38. F. Alladio et al, Ibid ref 25, 157
39. G. Gormezano, Ibid ref 1, Vol III, 1129
40. G. Gormezano, Ibid ref 1, Vol III, 1141
41. G. Melin, Ibid ref 1, Vol III, 1160
42. E. Marmar et al, PFC/JA–83–87 (Nov 83)
43. M. Clement et al Proc. of the 4th Int. Symp. on Heating in Toroidal plasmas Rome.March 1984.

PLASMA-WALL INTERACTIONS IN TANDEM MIRROR MACHINES

Steven L. Allen and the TMX-U/MFTF-B Experimental Teams

Lawrence Livermore National Laboratory
University of California, Livermore, CA 94550

1. INTRODUCTION

Plasma-surface interactions can play an important role in the present generation of tandem mirror machines, such as the Tandem Mirror Experiment-Upgrade (TMX-U),[1] Phaedrus,[2] Gamma-10,[3] and TARA,[4] which are currently operational; and AMBAL,[5] which will soon be fully operational. The importance of plasma-surface interactions may be even more pronounced in larger, longer-pulse tandem mirrors such as the Mirror Fusion Test Facility (MFTF-B),[6,7] which will be used to investigate plasmas at reactor-like conditions. However, current experiments and models of future experiments indicate that plasma-surface interactions can be minimized in tandem mirror machines. This is in part due to the inherent open field lines of the confinement configuration. In particular, plasma-surface interactions can be controlled at radial surfaces by controlling the axial confinement of the edge plasma.

We present here a description of the plasma-surface interactions in thermal-barrier tandem-mirror machines. The thermal-barrier mode of axial confinement is an integral part of a tandem mirror, and it dictates the required plasma conditions, particularly at the surface of the plasma. For this reason, a qualitative discussion of the thermal barrier is presented first in Section 2. A brief description of the experimental configuration used in tandem mirrors to create the thermal barrier is then examined in detail in Section 3; the TMX-U and MFTF-B machines are used as specific examples. In Section 4, the relevant plasma-surface interaction issues are addressed, and experimental results from currently operating tandem mirror machines are included. Section 5 is both a summary and a discussion of future work concerned with plasma-surface interactions in tandem mirrors.

1067

Several issues are outside the scope of the present discussion and are presented elsewhere. Additional details about experiments concerned with plasma-surface interactions in mirror machines are presented in Refs. 8-10. Reference 11 describes several upgrades to MFTF-B, including steady-state (100-hour) operation with DT fuel and the capability of a very high neutron flux (2 MW/m^2) to test various blanket designs. Issues relevant to tandem mirror reactor designs are discussed in Ref. 12.

2. THE THERMAL-BARRIER PLASMA ENVIRONMENT

To understand the importance of the function of a thermal-barrier mode in a tandem mirror, and in turn to appreciate the requirements it places on the plasma-surface interactions, it is necessary to briefly trace the evolution of mirror experiments. Along each step of mirror development, modifications have been motivated by the need for an efficient reactor concept.

The early single-mirror experiments showed that to control magnetohydrodynamic (MHD) stability, the plasma should be confined in a magnetic well, or "minimum-B" configuration.[13] This requirement motivated the design of highly efficient magnetic geometries, such as the yin-yang quadrupole,[14] which also have good access for neutral beam injection. Similar magnet designs are used in modern day mirror devices.

The next set of experiments was concerned with the effect of high-frequency modes (on the order of the ion gyrofrequency) on a mirror plasma with neutral beam injection. These modes [such as the drift-cyclotron loss cone (DCLC)] are driven by the non-Maxwellian velocity distributions present in mirror plasmas.[15] These modes cause diffusion in velocity space and can thereby increase plasma losses significantly. Several experiments, including 2XIIB, showed that the DCLC mode could be stabilized by introducing a low-energy plasma stream.[16] This was a major breakthough in mirror physics; however, it was also soon realized that a reactor based on a single mirror involved a large amount of circulating power. The expected Q (the ratio of fusion power produced to injected power) from such a reactor was on the order of unity because Coulomb scattering losses imposed a fundamental limit on plasma confinement.[17]

The tandem mirror was invented to overcome the reactor limitations of a single mirror, while incorporating its inherent advantages. The tandem mirror is composed of two mimimum-B mirror cells (end plugs) with high-energy neutral beam injection; these are located at each end of a straight solenoidal section (central cell). The end plugs provide the MHD stability for the system, and the central cell plasma furnishes the required lower energy ions for microstability. This configuration adds another dimension to confinement: if the end cell plasma is more dense than the central cell, an axial ambipolar potential is established that confines the ions in the central cell. Experiments on TMX showed that the confinement time in a tandem mirror machine could be increased at least an order of magnitude compared with a single mirror.[18] The plasma was also nearly microstable, exhibiting only low level fluctuations attributed mostly to the Alfvén ion-cyclotron (AIC) mode.[19]

In their early form, however, tandem mirror reactor designs required high end-plug magnetic fields and a high beam-injection energy. This is because the ion-confining potential is established by the density difference between the central cell and the end plug; it scales logarithmically with density: $\phi = T_e \ln(n_p/n_c)$. On the other hand, the ratio of central-cell fusion power to the injection power required to sustain the end plugs scales as n_c^2/n_p^2. The electron temperature T_e can be increased by auxiliary heating, but it is estimated that to effectively confine a 40-keV central cell plasma with a density of 10^{14} cm^{-3} would require an end plug density of 10^{15} cm^{-3}, which would in turn require neutral beam injection of 600 keV and peak fields of 17 T.[20] These requirements seemed quite restrictive; in addition, the plasma-surface interactions and materials requirements could be quite severe.

The thermal-barrier concept[20] was introduced to ameliorate these requirements. The essence of the idea is to introduce a potential barrier between the end plug and the central cell to more effectively thermally isolate the electrons in these two regions. As a result, each region can be optimized according to its purpose: the central cell plasma can have the high ion energy and density required to produce fusion energy, and the end plug plasma can have a substantially lower ion density but a large ion confining potential to reduce the central cell axial losses. This is achieved by adding rf heating, changing the neutral beam injection angle from perpendicular to the plasma axis to an oblique angle, and adding so-called "ion pumping" of low-energy ions trapped in the thermal barrier.

The thermal-barrier configuration is clarified by comparing the axial plasma potential profiles for three end plug configurations shown in Fig. 1. Note that T_{ec} is end cell temperature, T_e is electron temperature, n_p is plug density, n_p^* is passing density in the barrier region, and n_c is central cell density. The axial magnetic-field profile is shown in Fig. 1a, and a schematic of the standard tandem axial-potential profile is shown in Fig. 1b. When the end-plug beam injection is changed to an oblique angle, the sloshing-ion mode of operation results (Fig. 1c). The value of the ion-confining potential in this case is equal to that of the conventional tandem. However, the beam-injected mirror-trapped ions (so-called sloshing ions) create an ion-density profile that is peaked off the midplane of the plug, thereby causing a potential depression at the midplane; this is advantageous from the standpoint of microstability. It has been shown experimentally that this type of configuration can be created, and that ion fluctuations are low.[21]

The thermal-barrier axial-potential profile is shown in Fig. 1d. The dip in the potential near the midplane is the result of two processes: the formation of a mirror-trapped hot-electron distribution by means of electron-cyclotron resonant heating (ECRH), and the removal of ions from the low-potential region by ion pumping. The potential peak is formed by ECRH heating of the electrons in this region. The ion-confining potential can be made quite large while the central cell density is greater than the end plug density because of the low passing-ion density in the barrier region.

The thermal-barrier configuration has also been recently verified experimentally;[22,23] axial losses of ions can be reduced to nearly zero. Under these conditions, the end plug plasma is quite microstable because of the shape of the axial ion-density profile in the plug (as discussed above) and the central cell ions that pass into the plug region.[24] Control of the neutral pressure in these experiments was shown to be very important, both to minimize charge-exchange losses and to minimize the cold plasma that can fill in the thermal barrier potential region.[25] The edge plasma plays the very important role of shielding the core plasma from neutrals, particularly in the regions where the plasma flux tube is thin. These issues are addressed in detail in Sec. 4.

Several refinements of the thermal barrier concept have recently been made. Again, these were motivated by better understanding of the relevant physics issues and by improved reactor designs. Perhaps one of the most important issues is the control of radial transport. In a tandem mirror with minimum-B end plugs, the ions can be transported radially in the central cell by resonant neoclassical transport. As a plasma ion moves from the central cell to the end plug, it passes through a transition region where the magnetic field has a quadrupole component due to the nonaxisymmetric nature of the end plug magnets. As a result, the ion undergoes a radial displacement ΔR at each end of its axial bounce motion. The major component of this displacement and its direction is proportional to cos 2ψ, where ψ is its azimuthal angle. Therefore, if one end-plug magnet set is rotated 90° with respect to the other end, the radial displacements cancel. However, this cancellation is not complete if an azimuthal drift $\Delta\psi$ is present as a result of radial electric fields in the central cell. Therefore, the causes for resonant neoclassical transport are nonaxisymmetric magnetic fields and azimuthal drifts due to radial electric fields. Detailed explanations of resonant transport are included in Ref. 26.

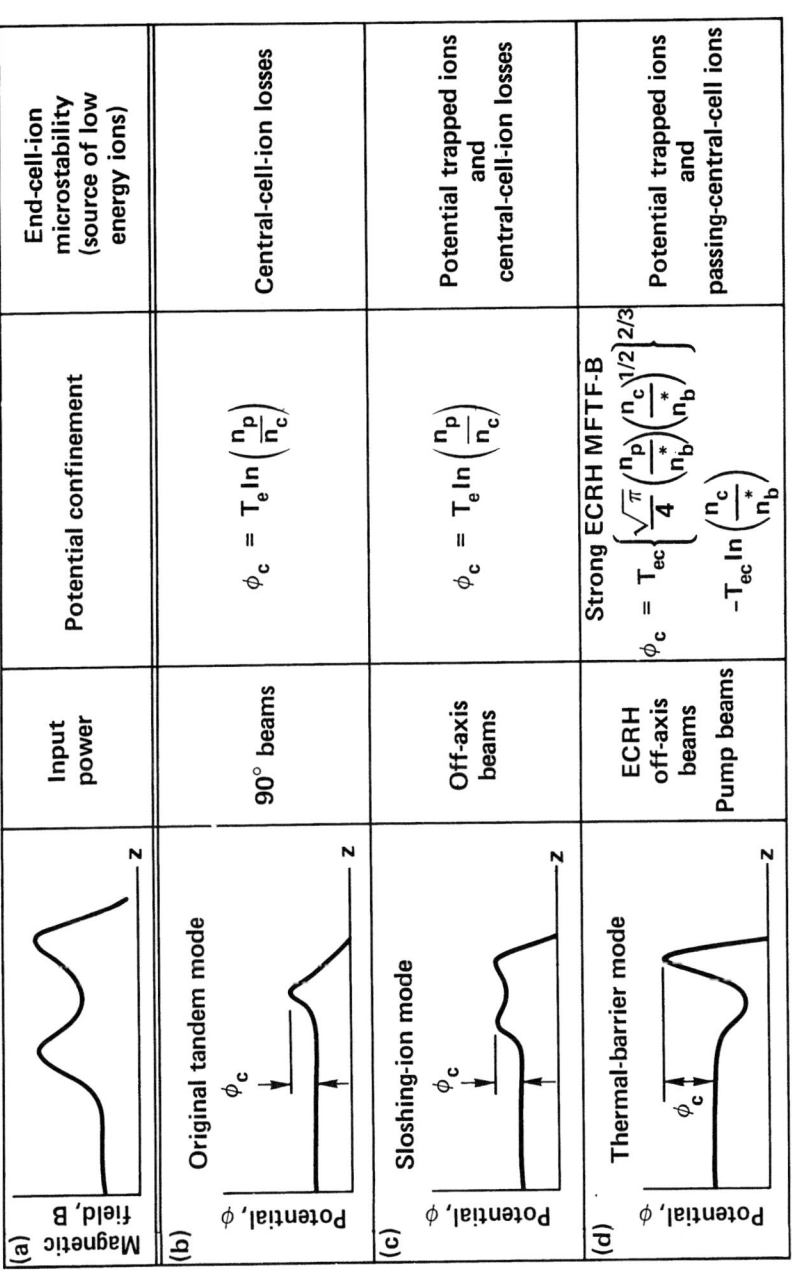

	Input power	Potential confinement	End-cell-ion microstability (source of low energy ions)
(a) Magnetic field, B			
(b) Original tandem mode	90° beams	$\phi_c = T_e \ln\left(\dfrac{n_p}{n_c}\right)$	Central-cell-ion losses
(c) Sloshing-ion mode	Off-axis beams	$\phi_c = T_e \ln\left(\dfrac{n_p}{n_c}\right)$	Potential trapped ions and central-cell-ion losses
(d) Thermal-barrier mode	ECRH off-axis beams / Pump beams	Strong ECRH MFTF-B $\phi_c = T_{ec}\left\{\dfrac{\sqrt{\pi}}{4}\left(\dfrac{n_p}{n_b^*}\right)\left(\dfrac{n_c}{n_b}\right)^{1/2}\right\}^{2/3}$ $-T_{ec} \ln\left(\dfrac{n_c}{n_b^*}\right)$	Potential trapped ions and passing-central-cell ions

Fig. 1. Tandem mirror operating modes illustrate the thermal barrier configuration in the following illustrations: (a) the end-plug axial-magnetic-field profile; (b) the axial potential profile for the conventional tandem mode; (c) changing the end plug neutral beams to off-axis injection results in the sloshing-ion mode profile; (d) addition of ECRH and ion pumping is required for the thermal barrier mode.

Several methods of radial-transport control have been identified. Because one major drive is the shape of the magnetic field, it is desirable to terminate the central cell plasma with an axisymmetric magnet that "throttles" the flow of ions passing into the transition regions. This flow of passing ions is also important for stabilizing trapped particle modes,[27] which are similar to those found in tokamaks. An axisymmetic design is incorporated into most current tandem mirror designs. Errors in the end plug field resulting from misalignment can also cause radial transport. This neccesitates the careful alignment during construction, often with electron beams, and the addition of "trim" coils for fine adjustment after installation.

The drive for radial transport is also reduced by controlling the radial-potential profile, (and therefore the radial-electric field), and hence the azimuthal drift. Reducing the radial fields has the added benefit of reducing the rotational drive for instablilites. This radial-field profile is determined by the electron losses along field lines to the ends of the machine. If a negative bias is applied to the end wall, the potential in the central cell will decrease to offset this change. In practice, this bias must be a function of radius, so segmented plates are used. This technique has been demonstrated qualitatively in current experiments.[28]

Another improvement is the use of drift pumping to remove ions in the thermal barrier regions. Neutral beams--which remove ions by charge exchange--injected nearly parallel to the axis of the machine are the more common method. Drift pumping consists of a field supplied by an external coil; the frequency is resonant with a multiple of the bounce or drift frequency of the ions, resulting in a radial displacement. The cumulative result is that the ion eventually leaves the plasma. This technique is attractive because it is selective: it does not pump the sloshing ions, which are not resonant. It is also effective for removing impurities.

The linear geometry of tandem mirrors also makes it possible to directly generate electrical power from any residual axial plasma losses. The conversion efficiency of these devices is usually about 50%. In the original designs, these converters were very large because they depended on the one-dimensional expansion of the plasma fan and subsequent deceleration through a series of grids. Current designs are much more compact because of the use of recircularizing coils to expand the plasma in two dimensions. Gridless converters can be used because the non-ambipolar radial transport in the plasma separates the charge carriers: electrons are collected in the core, and ions are collected at the edge.

3. THE THERMAL BARRIER EXPERIMENTAL ENVIRONMENT

The previous section dealt with the plasma physics issues of a
tandem mirror machine in a general way. Next we turn to the
experimental environment and how its design addresses these issues.
We use two machines at the Lawrence Livermore National Laboratory
(LLNL) as specific examples, although other tandem mirror machines
are similar in design.

The goal of TMX-U is to demonstrate the thermal barrier
concept. The various regions of the vacuum vessel are shown in
Fig. 2. The first and second injector regions in the machine contain
surfaces that are liquid-nitrogen cooled and titanium gettered to
provide pumping of cold gas. These regions control the influx of
cold gas into the plasma region from sources such as the neutral
beams; they also serve as beam dumps for the energetic neutrals that
are not trapped in the plasma. External beam lines and dumps are
used on other machines such as MFTF-B. The plasma wall is also
gettered.

A schematic of the TMX-U device is shown in Fig. 3a; the axial
profiles of magnetic field $B(z)$, electron density $n(z)$, and
potential $\phi(z)$ are presented in Fig. 3b. Also indicated on these
figures are the heating systems: 20-kV neutral-beam injectors to
establish sloshing-ion distributions in the end plugs and to heat
the central cell, four 28-GHz gyrotrons (200-kW each) to provide
ECRH, and two transmitters (200-kW each) to provide ion-cyclotron
resonant heating (ICRH). The time evolution of the heating systems
and the plasma parameters for a TMX-U plasma shot are shown in
Fig. 4; the east plug, central cell, and west plug parameters as a
function of time are presented in each column. Starting at the top
of the figure, the plotted quantities are: the ECRH power, the
neutral-beam power-supply drain current, the electron density, and
the ion end losses as measured by an array of Faraday cups. The
time history of the gas fueling in the central cell and the ICRH
pulse are also indicated at the top of the figure.

The most notable feature of these data is the dramatic reduction
in the axial ion loss from each end when the neutral beams are
turned on. This occurs after the hot-electron population is
established in the end plugs, and when the sloshing ions are
injected with the neutral beams. This configuration has all of the
qualitative features of a thermal barrier, but another experiment
was performed to verify that the axial potential profile had the
same features as a thermal barrier. This experiment is described in
detail in Ref. 22, and the major result of the measurement is shown
in Fig. 5 from this reference. The experimental setup is shown in
the top of the figure; end-loss energy analyzers were used to measure
the distribution of escaping ions. The model distribution is shown
in the central part of the figure, and the results of the experiment

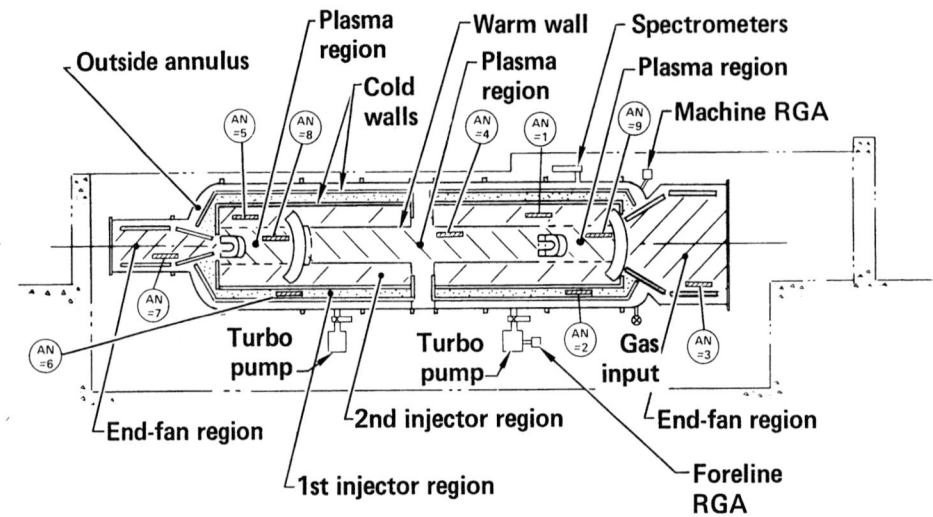

Fig. 2. The TMX-U vacuum vessel. The first and second injector
         regions act as beam lines to reduce the gas from the
         neutral beams. Each region contains titanium getter wires.

are shown at the bottom. Note that the model and the experimental
profiles have the same qualitative features, thereby verifying the
existence of the thermal barrier potential profile. The current
experimental program on TMX-U is focused on achieving these
conditions with a higher central cell electron density. After this
set of experiments, the magnet set will be made more symmetric by the
addition of a throttle coil to examine the issue of radial transport.[29]

An example of the next generation of tandem mirror machines is
MFTF-B, which is now under construction at LLNL. It is a large (64-m
axial length), long-pulse (30-s) machine. The physics goals of
MFTF-B include confinement of reactor-grade central cell plasmas with
ion temperatures up to 15 keV, demonstration of high β (the ratio
of the plasma pressure to the magnetic field pressure) plasmas
approaching 20%, attaining a plasma-density energy-confinement-time
product $n\tau_E$ of approximately 0.6×10^{13} cm^{-3} s, and obtaining
a DT equivalent Q of approximately 0.4. These goals represent a
significant step toward reactor-type parameters.

In addition, just as reactor designs have motivated the evolution
of mirror-machine physics, they have also placed requirements on
engineering technologies. For this reason, another important
function of MFTF-B is to allow us to gain experience with some plasma
technologies that are relevant to reactors. The design and operation
of superconducting magnets with large cryogenic systems will be
demonstrated in MFTF-B. Another goal is to address the issues of

1074

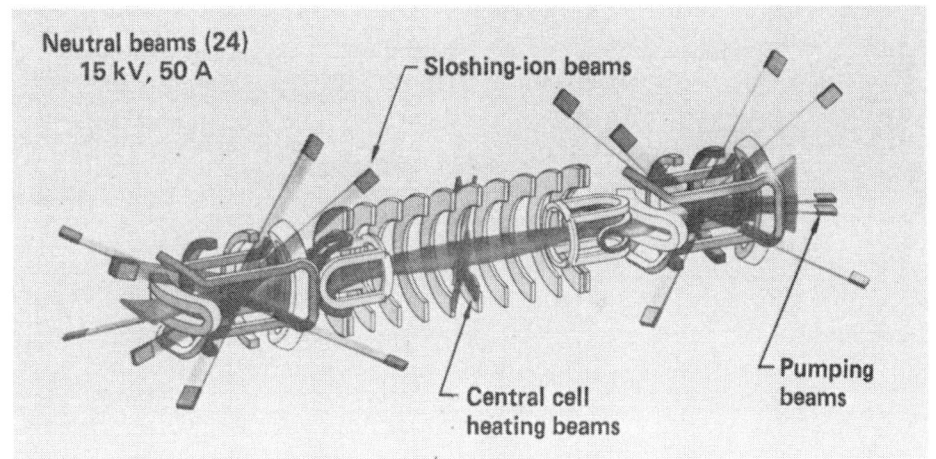

Sloshing-ion beams

Pumping
beams

Central cell
heating beams

a

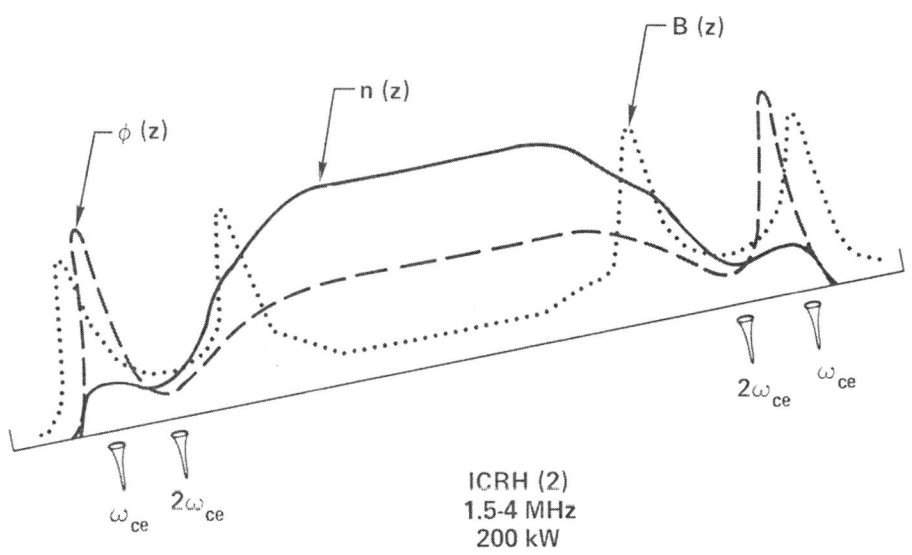

B (z)

n (z)

ϕ (z)

$2\omega_{ce}$

ω_{ce}

ICRH (2)
1.5-4 MHz
200 kW

ω_{ce} $2\omega_{ce}$

ECRH (4)
28 GHz gyrotrons
200 kW (per tube)

b

Fig. 3. (a) A schematic of the TMX-U device, illustrating the
neutral beam locations and the magnet set. (b) The axial
profiles of magnetic field B(z), electron density n(z), the
potential ϕ(z) are shown along with the locations of the
ECRH and ICRH heating systems.

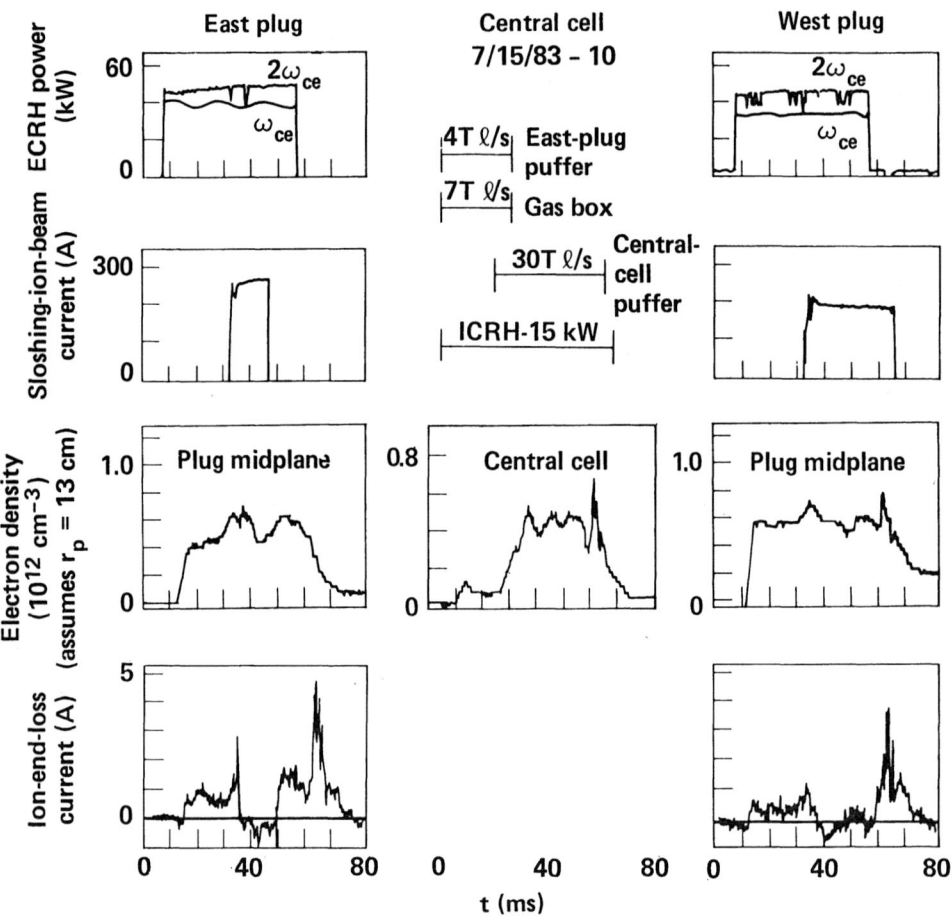

Fig. 4. The time evolution of a plasma shot on TMX-U. Shown from top to bottom are the ECRH power, the neutral-beam power-supply current, the electron density, and the ion end-loss current. The quantities are shown for the east plug, central cell, and west plug. The time history of the gas injection system and the ICRH power is also shown.

long pulse operation. The 30-s plasma duration of MFTF-B puts new technological requirements on all of the heating systems: neutral beam injectors, ECRH, and ICRH. Nearly steady-state plasma-surface interactions can also be investigated.

The thermal barrier mode for MFTF-B is based on the most recent Mirror Advanced Reactor Study (MARS) reactor design is shown in Fig. 6; the steady-state profiles of magnetic field, plasma potential, and density are presented. Such a plasma is created in two steps: a 0.5-s startup sequence followed by 30-s steady-state

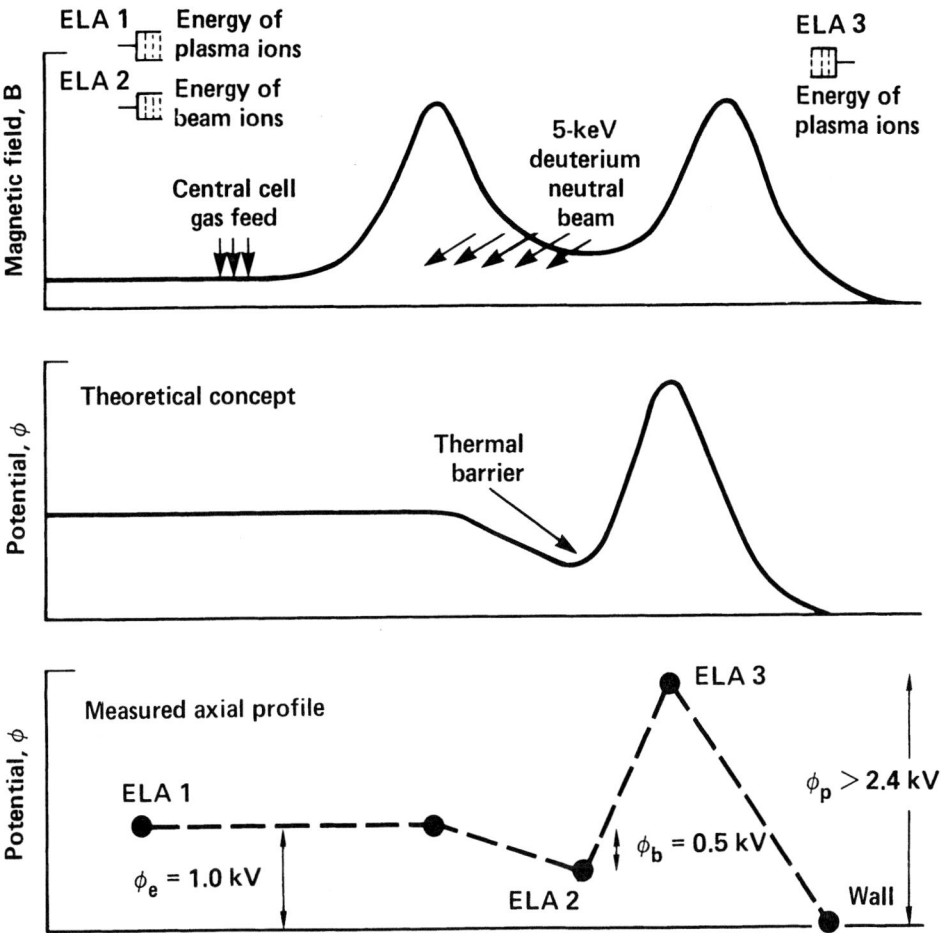

Fig. 5. A schematic representation of the measurement of the
thermal barrier potential profile using three electrostatic
ion analyzers to make simultaneous measurements. The
experimental setup is shown in the top of the figure. The
theoretical and actual axial potential profiles are
compared in the bottom of the figure.

operation. The central cell is designed to have an electron
temperature of 9 keV and an ion temperature of 15 keV. The central-
cell ion-confining potential is about 30 kV. The high magnetic field
region in the axicell decreases resonant radial transport and
"throttles" the central cell passing particles to allow operation in
a regime where trapped particle modes are minimized, as discussed in
the previous section. The yin-yang cell provides MHD stability for
the system. The desired hot electron distribution (with energies up
to 500 keV) is supplied by ECRH injection at three frequencies in the

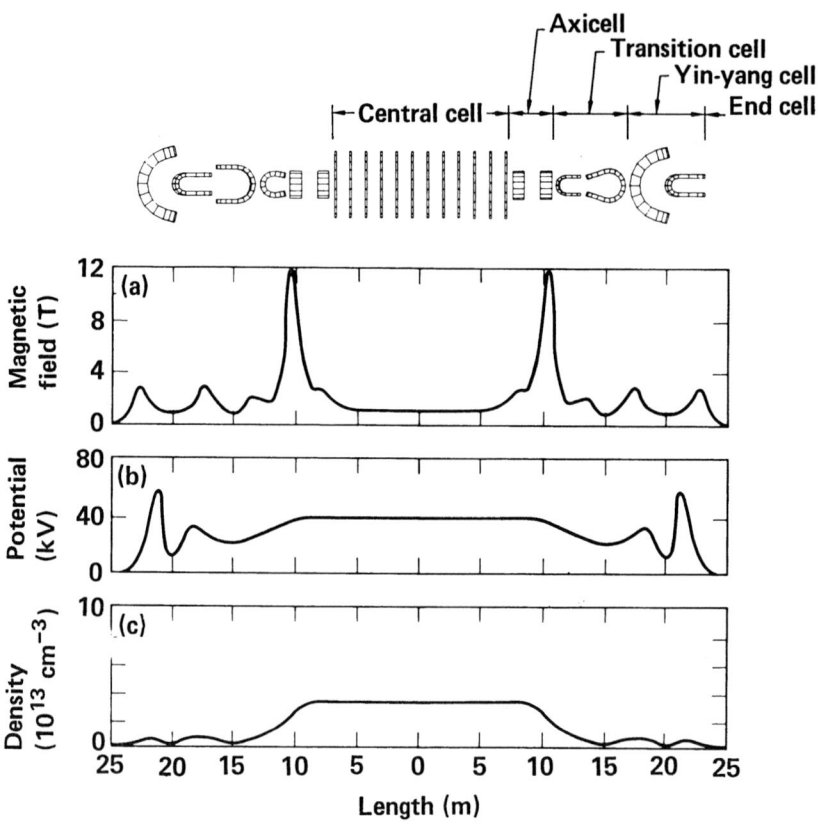

Fig. 6. The axial profiles of magnetic field, plasma potential, and
electron density for MFTF-B. The thermal barrier region is
in the yin-yang cell.

end plug region. Ion pumping is supplied both by neutral beams in
the end region and drift pumping in the transition region.

The vacuum vessel and magnetic field coils for MFTF-B are shown
in Fig. 7. Note that the overall length of the vessel is about 64 m.
The large cylindrical structures on the outside of the machine are
input beam lines and beam dumps for neutral beam injectors. Nearly
all of the pumping both in the beam lines and in the plasma vessel
is by means of liquid-helium-cooled cryopanels. A liquid-nitrogen
guard jacket surrounds each cryopanel and the magnet cases. The
magnet system consists of 24 superconducting Nb-Ti coils and two
high-field Nb_3Sn insert coils to provide the 12-T field for the
axicell region. Also shown in the figure are the trim coils used in
the transition region.

Figure 8 depicts the locations of the neutral beam, ECRH, and
ICRH heating systems. Each end plug has three quasi-optical ECRH

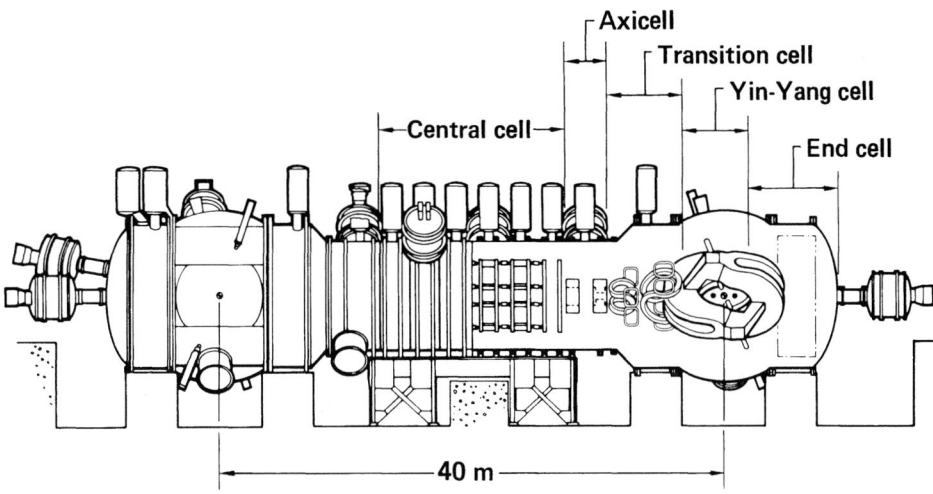

Fig. 7. The vacuum vessel of MFTF-B with its associated neutral beam lines is shown. The right half of the figure is cut away to show the superconducting magnets.

systems. The 56-GHz systems provide 450 kW at a harmonic of the ECRH resonance to provide a mirror-trapped hot-electron distribution. The 28- and 35-GHz systems provide 300 kW near the fundamental ECRH resonance to develop the potential peak. Two frequencies are required because of the β depression by the plasma in this region. These systems are capable of delivering 30-s pulses.

The ICRH transmitters are located off the midplane of the central cell and provide up to 1 MW of power in the frequency range from 6 to 20 MHz in 0.25-MHz steps. A 40% coupling efficiency to the plasma is expected during the steady-state conditions. These transmitters can be operated at this power level for 30 s.

The neutral beams are divided into two classes, 0.5-s beams for startup and 30-s beams for sustaining steady-state operation. The 80-kV startup beams are shown in Fig. 8. There are 10 in the central cell, 1 in each axicell, and 2 in each yin-yang to establish the sloshing ion population. Each beam is designed to provide 50 A of neutral current at the plasma target; the species mix is expected to be 60/20/20% for full/half/third energy components. These beams will have gettered arc chambers to minimize impurity influx. The steady-state 30-s beams are designed to provide 10 A of the full-energy 80-kV component on the plasma target; two each are located in the axicell, the yin-yang or sloshing-ion positions, and in the end fans as high-energy pump beams (HEPB). Magnetic separation is used to decrease the current of lower energy species to very low levels

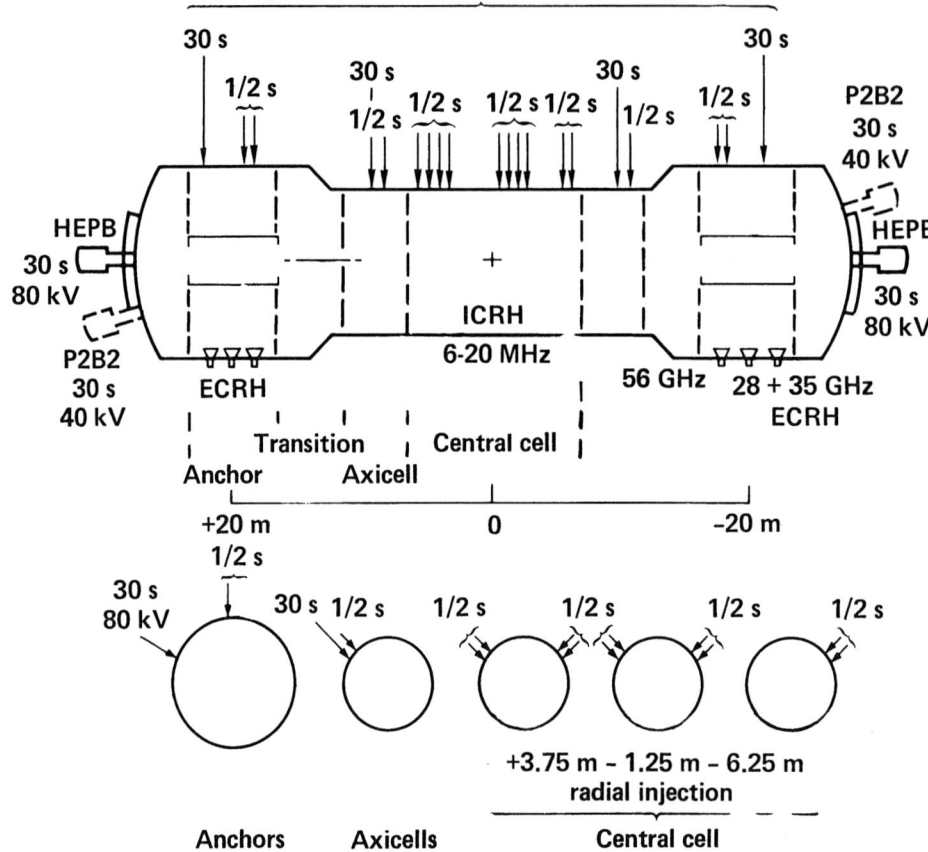

Fig. 8. The neutral beam and rf heating locations for MFTF-B. The
bottom half of the figure indicates the azimuthal angle
where the 0.5-s and 30-s neutral beam sources are located.

to minimize barrier filling. In addition, the impurities are
reduced to very low levels; this is discussed in more detail in a
later section.

Also shown in Fig. 8 are the off-axis 40-kV beams (labeled P2B2)
that had been proposed for ion pumping in the transition region.
Recent analysis has determined that these beams must pass very close
to magnet structures. Beam scrapeoff introduces a source of cold
gas. Baffling the beams is difficult and also reduces the beam
current. For these reasons, drift pumping of ions has replaced
these beams; the physics of the process is described in Ref. 30.
Tests of drift pumping will be carried out on TMX-U. Drift pumping
is also incorporated into recent MARS reactor designs.[12] In
addition, drift pumping has other advantages that are discussed in
Sec. 4.2.

4. PLASMA-SURFACE INTERACTIONS IN TANDEM MIRRORS

In a general sense, the plasma conditions for the central cell of a tandem mirror machine are quite similar to other confinement geometries: a dense plasma is confined and heated by auxiliary methods such as ICRH and neutral beams. We might expect that the similarities in these plasmas would in turn lead to similar plasma-surface interactions. However, this is not the case because even though the core plasmas are similar, the edge plasma in a tandem mirror can be quite different (than toroidal plasmas, for example). In addition, the tandem mirror has an extra region that contains both the thermal barrier region and the end wall. Briefly, the tandem mirror allows the possibility of controlling the edge plasma by means of fueling and transport in this region. This relaxes the severity of the plasma-surface interactions, particularly in the fusion-producing central cell region. The plasma lost to the end regions can be controlled so that it is well isolated from the wall. The control of neutrals and impurities is especially important in the thermal barrier region.

The discussion of the relevant plasma-surface interactions can be divided naturally based on the important regions of the machine: interactions at radial surfaces, processes in the thermal-barrier-forming regions, and interactions at the end walls. Impurity generation can be important in all three regions.

4.1 Plasma-Surface Interactions at Radial Surfaces

The linear geometry of a tandem mirror machine allows another degree of freedom for controlling plasma-surface interactions at radial surfaces: axial confinement. In this context, a mirror machine is a "natural" divertor. That is, a region near the plasma edge is formed that has a much shorter axial confinement time than the core of the plasma. In this way, most of the losses of the particles in this edge region flow to the end wall rather than the radial wall. In addition, cold gas and impurities will not have time to penetrate very far before they arc lost out of the machine axially. This means that most of the surface interactions will be with neutral atoms, photons, and neutrons.

There is an energy cost to support the edge plasma because as the particles are transported axially they can cause power losses by charge exchange, ionization, and other processes such as radiation. In a reactor, part of this energy could be recovered in the direct converter. In current devices, it puts requirements on the cold gas influx into the system from sources such as neutral beams and on the reduction of reflux from the plasma wall. In the central cell the plasma is quite dense and the corresponding neutral attenuation is large. The thermal barrier region is more sensitive and is discussed in a subsequent section.

Control of the plasma radius has been demonstrated experimentally on several mirror machines. The single-cell mirror 2XIIB experiment showed that the radial plasma boundary was determined by the balance of neutral beam fueling and charge-exchange losses on background gas at the plasma edge.[31] In this way, the plasma was not in direct contact with the radial walls. Furthermore, it was shown on 2XIIB that the plasma size could be changed by the aiming of the neutral beams; that is, by varying the distance between the axis of the neutral beam and the axis of the magnetic field. This is illustrated in Fig. 9, which contains data from Refs. 8 and 32. The plasma radius (the radial scale length) is plotted as a function of the position of the guiding center (the center of the orbit) of the average particle deposited by the neutral beams. Note that the plasma radius increases as the position of the guiding center is increased from the magnetic centerline. The reason for this change is that the fueling of the edge plasma is increased. Further evidence of this effect is from EUV measurements,[33] which indicate that the plasma density and temperature drop sharply at the edge of the plasma. These experiments demonstrate that the plasma size is controlled by the neutral beam fueling.

In a conventional tandem mirror such as TMX, each end plug is similar to the single mirror of 2XIIB. In turn, the central-cell plasma radius is controlled by the end-plug radius because the end plug is responsible for establishing the potential and, hence, axial confinement. Specifically, the radial density profile determines the radial dependence of the axial potential profile, which in turn determines the radial profile of axial confinement. In the TMX experiment, the end-plug plasma radius was determined by neutral beam aiming; the radial profile of the line density is shown in Fig. 10a, which contains data from Ref. 8. When the flux tube containing this plasma is mapped into the central cell, it is inside any physical limiters, as shown by the data in Fig. 10b. In this way, the plasma density and temperature are very low near the central-cell radial limiters, thereby minimizing any plasma-wall interactions.

The same type of radius control is possible in a thermal-barrier end plug. The radial extent of the barrier is determined by several processes, including charge exchange of sloshing ions by cold gas, neutralization of the thermal barrier potential by cold ions and electrons, spatially dependent pumping by neutral beams, and spatially dependent ECRH heating. For the most part, these processes favor the formation of a thermal barrier in the core of the plasma, with a lower confinement time at the edge. As shown in Fig. 11, these conditions in the end plug determine the axial confinement of the central cell, and thereby the plasma edge.

Results from TMX-U indicate that this configuration can be established. Under certain conditions, the axial-end-loss current

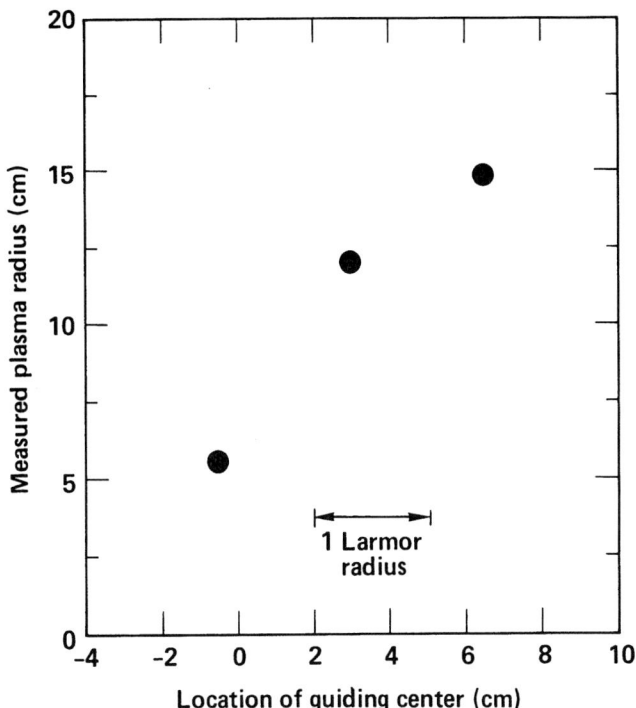

Fig. 9. Data from 2XIIB indicate that the plasma size depends on
the aiming of the neutral beams. The plasma radius (the
radial scale length) is plotted as a function of the
position of the guiding center (the center of the orbit) of
the average particle deposited by the neutral beams.

as measured by an array of Faraday cups on the end wall is small in
the core of the plasma, while the current at the edge is larger. A
second signature of the influence of the edge plasma in these
experiments is the observation of a pressure rise in the end fan
region near the end wall during plasma operation.[25] This
indicates that plasma is being transported to the end wall by the
edge plasma. An example of this data is shown in Fig. 12 from
Ref. 25. Note that the pressure first rises at the beginning of the
plasma shot. When the thermal barrier is formed and the core axial
ion losses decrease at 25 ms, the pressure stabilizes at a fairly
constant level determined by the axial losses of the edge plasma.

Yet another type of radius control is possible based on the
control of radial transport in the plasma by biasing the end wall as
discussed in Sec. 2. If a region near the edge of the plasma is not
biased, then the radial electric field will exist, which can drive
resonant transport. In this way, the particles at a certain radius
experience radial transport into the edge plasma and cannot support

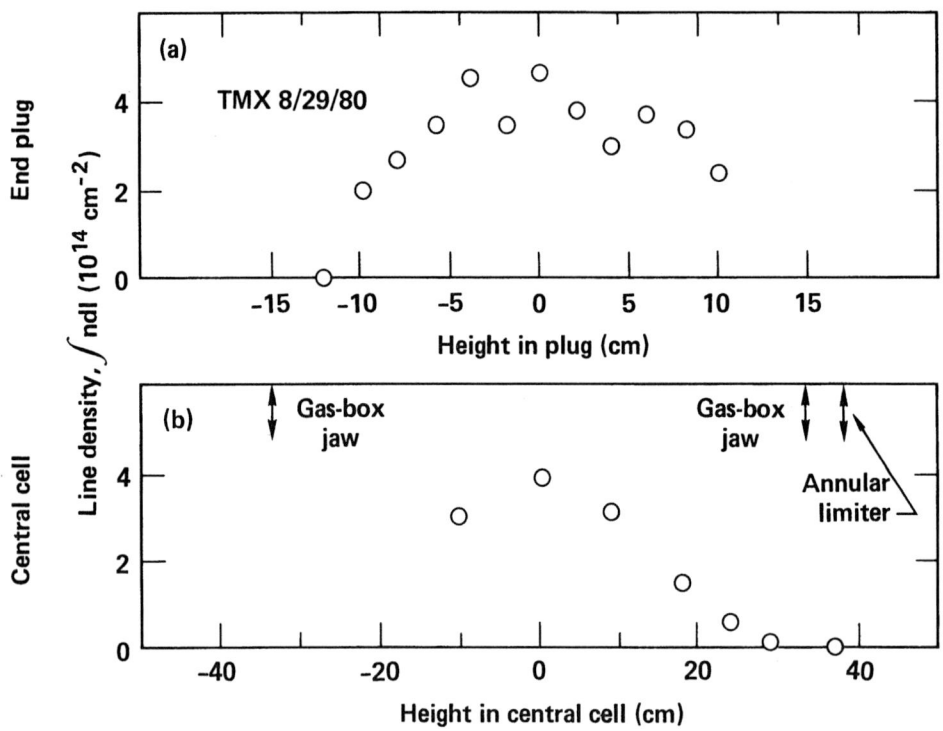

Fig. 10. Data from TMX demonstrate that the central cell plasma can
be isolated from the limiter. The radial profiles of the
end plug (a) and central cell (b) line densities are
shown. Note that the line density is low near the radial
limiters in the central cell. The radius of the end plug
plasma can be used to control the central cell plasma.

the core plasma; this defines the plasma edge. This technique has
the advantage that the plasma radius can be varied without changing
any other parameters. Details of the plasma radius control for
MFTF-B and MARS are presented in Ref. 33.

The edge plasma operation discussed above will only be efficient
if its power requirements are reasonable, that is, if impurity
generation and reflux are minimized. In addition, the machine will
have to be cleaned initially to obtain optimum operation. Plasma
wall conditioning techniques will be required. TMX-U and MFTF-B
contain a plasma wall to shield the plasma from any direct path to a
cold surface such as a cryopanel or magnet case. Several methods
have been used to condition the walls of mirror machines, including:
(1) cleaning by repetitive plasma shots, (2) glow discharge cleaning
(GDC), and (3) titanium gettering.

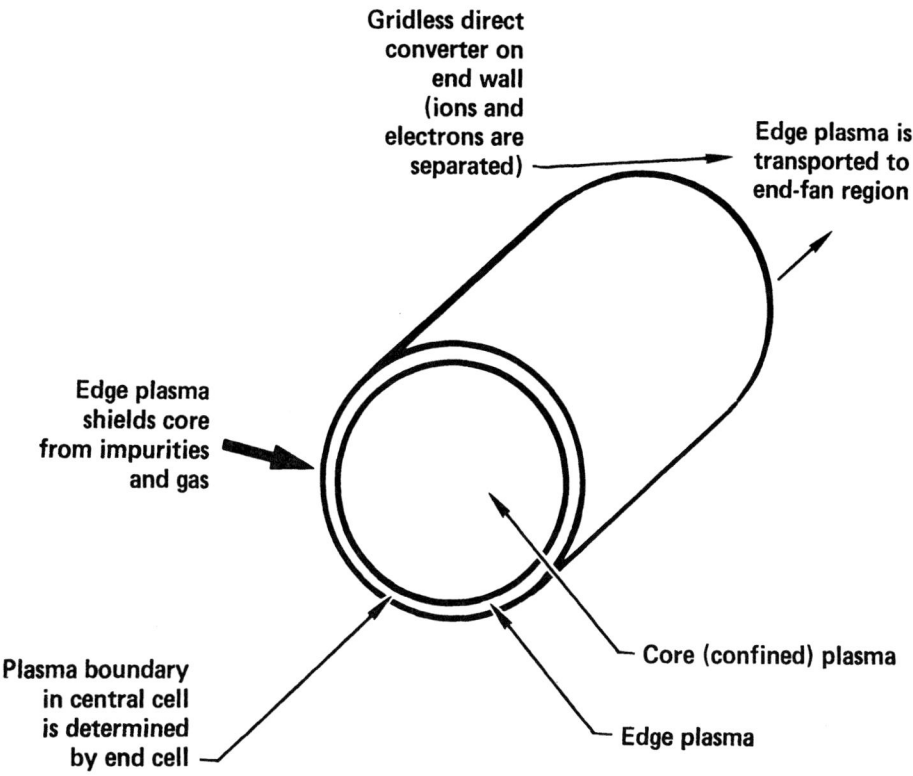

Gridless direct converter on end wall (ions and electrons are separated)

Edge plasma is transported to end-fan region

Edge plasma shields core from impurities and gas

Core (confined) plasma

Plasma boundary in central cell is determined by end cell

Edge plasma

Fig. 11.   A schematic representation of the edge plasma and how it
           can be used to minimize the interactions at radial walls.

The major disadvantage of the first technique in current
operating machines is the low duty cycle. In machines like MFTF-B
where 30-s neutral beam operation is possible, it should be much
more useful. It may also be possible to maintain a cleaning plasma
with other heating methods such as ECRH and ICRH, which could be
designed for even longer pulses.[34,35] The second technique has been
used successfully on TMX-U,[36] essentially replacing method (1) on this
machine. The GDC results have been very similar to those obtained on
tokamaks;[37] a typical curve of the 16 amu peak as a function of time
is shown in Fig. 13.[38] The decay has a t^{-1} dependence, similar to
Ref. 37. The major function of GDC in TMX-U is to remove hydrocarbons.
Some water removal may occur, but the GDC is carried out at room
temperature, so the removal rate is slow.

On MFTF-B, GDC may be somewhat more complicated, as the magnetic
field must be off and the pumping capacity of the cryopumping system may
be exceeded during extended operation at the required pressures. This
requires operation of GDC without the cryopanels and with auxiliary

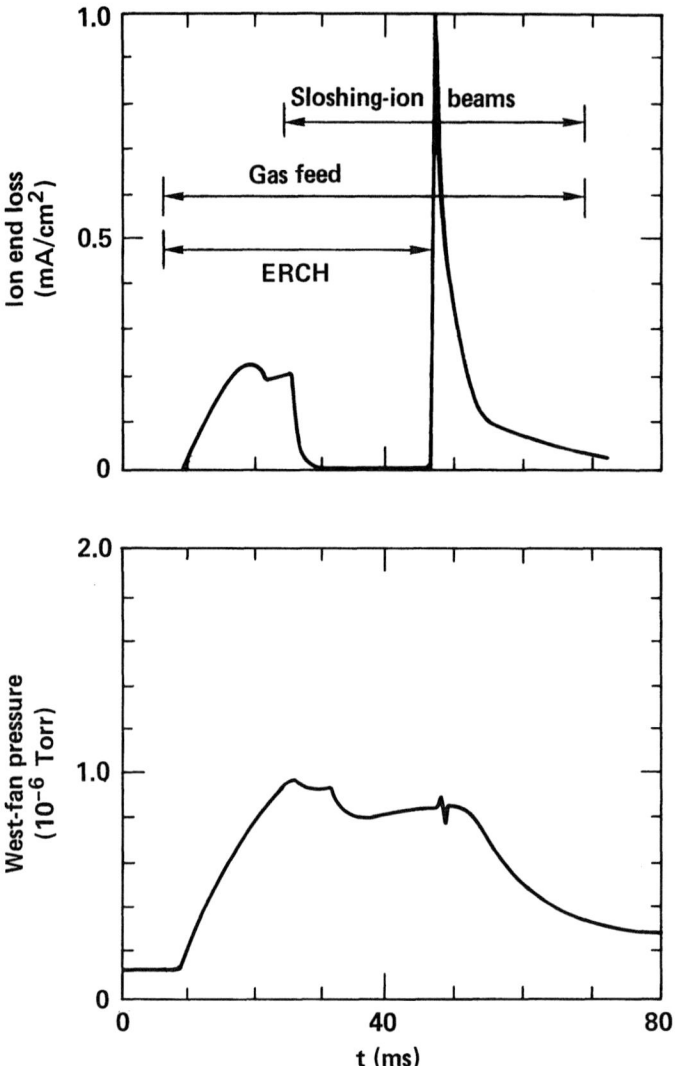

Fig. 12.   The time history of the ion end loss (top) and end fan
pressure (bottom) as a function of time. Note the initial
rise of the end fan pressure, which then stabilizes or
decreases when the end losses decrease.

pumping. In addition, specialized coatings on the cryopanel
surfaces may be degraded by the GDC, although experiments[39] have
shown that electrical isolation of the cryopanels will minimize this
damage.

Another important wall conditioning technique is titanium
gettering. A substantial fraction of the internal TMX-U surface

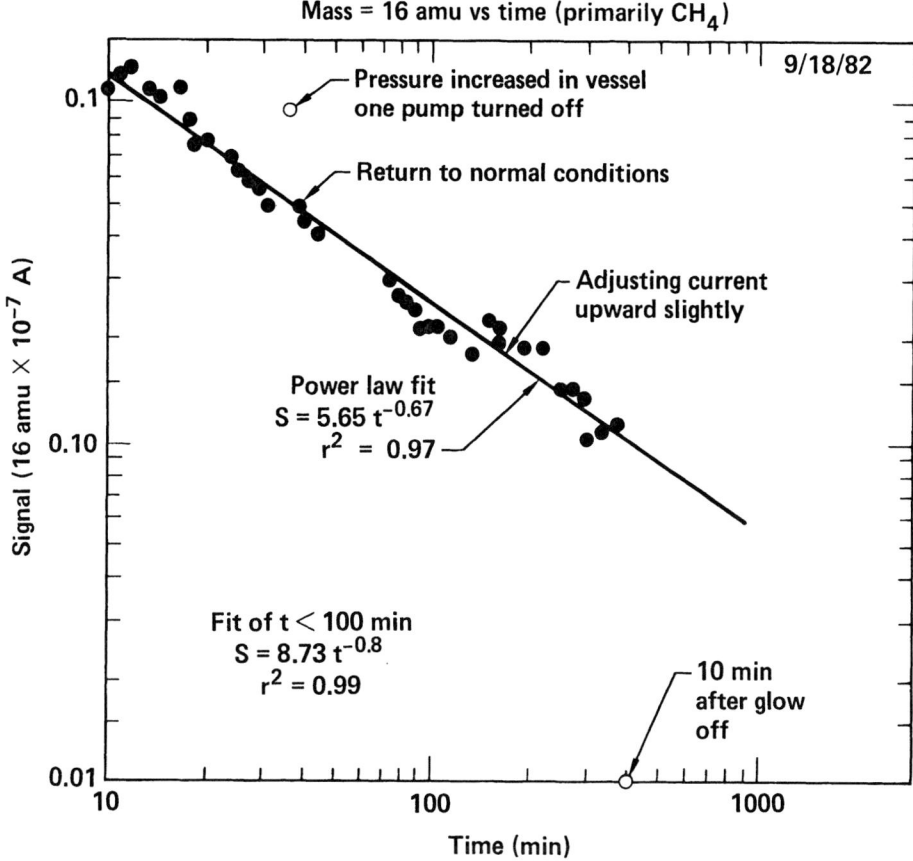

Fig. 13. The RGA signal from 16 amu vs time during GDC is shown. The production rate varies nearly as t^{-1}.

area (>90%) is titanium-gettered. Getter wires are supported between each set of liners and inside the plasma wall so that both sides of these surfaces are gettered. A staged getter cycle, starting with the outside of the machine and working inward, is used to deposit a fresh layer before each plasma shot and pump gas away from the plasma region. Surface analysis of probes inserted near the plasma wall have found that the average film thickness is approximately 6 to 10 monolayers per shot. The resulting pumping speeds exceed 10^6 l/s; details of these measurements are presented in Ref. 40.

Residual gas analyzer (RGA) measurements during gettering indicate that methane is produced presumably by the hot getter wire interacting with hydrogen. A significant fraction of the 10^{-8}-Torr

(10^{-6}-Pa) base pressure before the plasma shot is methane. Cryopumps actively pump the methane, and the liquid-nitrogen-cooled (Ti-gettered) liners pump methane away from the plasma region by physisorption to the surface.

Auger electron spectroscopy (AES) analysis of probes inserted near the wall and exposed to a series of getter and plasma sequences reveals a substantial amount of oxygen ($\sim$30%) present in the gettering. This may be the result of the pumping of water by the titanium films. Further details are presented in Ref. 41; this result is undergoing further study at this time.

In present experiments, impurity concentrations and radiated power are quite low, as indicated by the data in Table 1. This is presumably due at least in part to the extensive titanium gettering. In addition, the edge plasma shields the core from impurities. Shown in Fig. 14 from Ref. 42 is the total oxygen density as a function of radius. These data were obtained during an impurity injection experiment where gas impurities were injected into the edge of the plasma. Several plasma shots were required to measure the radial profiles of all of the ionization states of oxygen. All of these data were then Abel-inverted and divided by the electron density profile to obtain the data in Fig. 14. Note that the total oxygen density is peaked at the edge of the plasma, and does not penetrate efficiently into the core.

4.2 Plasma-Surface Interactions in the Thermal Barrier Region

The thermal barrier region is perhaps the most critical from the standpoint of plasma-surface interactions. To sustain the thermal barrier potential profile, the neutral density and impurity concentrations must be carefully controlled. Specifically, the neutral density must be kept low to minimize the charge-exchange

Table 1. Comparison of impurity concentrations and power loss for the central cell of TMX and of TMX-U.

Species[a]	Impurity Concentration (%)		Power loss (kW)	
	TMX	TMX-U	TMX	TMX-U
Carbon	0.08	0.06	1.5	0.3
Nitrogen	0.10	0.20	1.5	0.7
Oxygen	0.40	0.80	6	2.7
Titanium	0.10	None	10	None
Deuterium			6	0.1

[a]The He-like C V, N VI, and O VII are not included in this table.

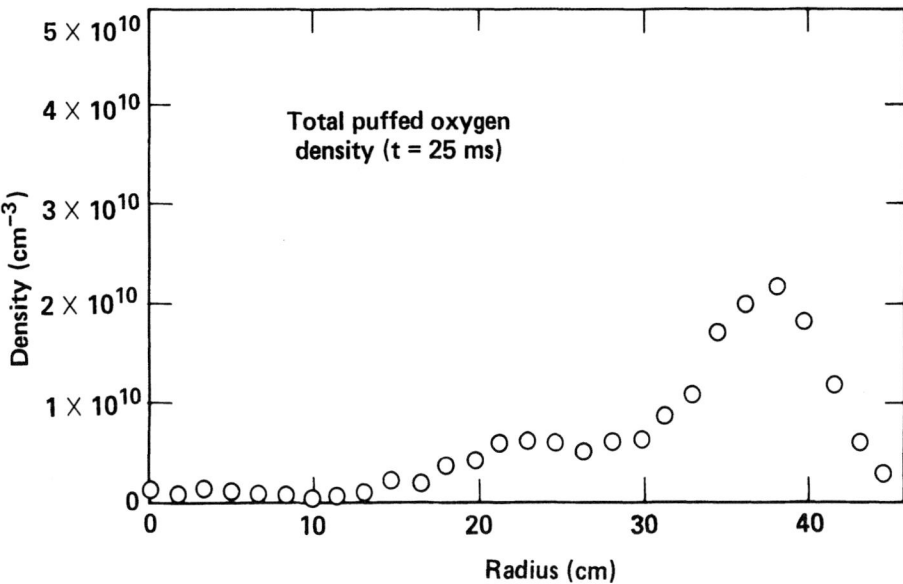

Fig. 14. The radial profile of the total oxygen density as a function of radius is shown. These data were obtained from impurity puffing experiments on TMX. Note that the impurities do not penetrate efficiently into the central cell plasma.

loss of sloshing ions and the trapping of cold electrons and ions in the end plug region. Cold ions neutralize the dip in the thermal barrier potential, and cold electrons decrease the potential peak in the end plug. This means that the combination of shielding and particle removal by the edge plasma, the ion pumping in this region, and the control of cold gas sources from neutral beams and wall reflux must be sufficient to prevent these loss processes from becoming important. In addition, impurity ions can become trapped in the potential dip and neutralize the thermal barrier potential.

The importance of the control of neutral gas is demonstrated by TMX-U results shown in Fig. 15 from Ref. 25; the pressure in the end plug and the ion end losses are shown as a function of time for two cases. In the left half of the figure, the neutral beam gas and wall reflux are not carefully controlled, and the end losses increase when the end plug pressure rises. After these data were taken, extensive baffling of the cold gas from the neutral beams was added to the TMX-U experiment and the wall reflux was reduced by more efficient wall conditioning. (Recall that this experiment does not have separate external beam lines for the neutral beams; they are an integral part of the vacuum vessel.) This change produced the datain the right half of the figure; note that the end plug pressure is lower and the end losses remain at a low value for a longer period of time.

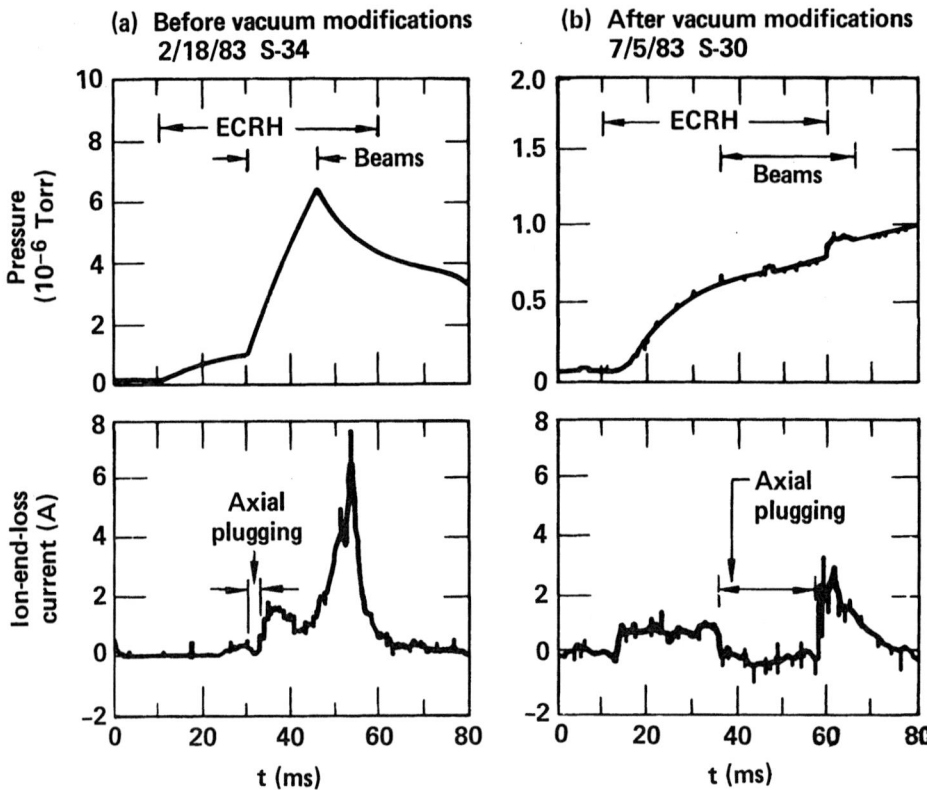

(a) **Before vacuum modifications**
2/18/83 S-34

(b) **After vacuum modifications**
7/5/83 S-30

Fig. 15. Data from TMX-U show that control of the pressure in the end
cells is important for thermal barrier operation. The end
plug pressure and ion end loss is shown for two cases: (a)
before vacuum modifications, and (b) after extensive neutral
beam baffling was added and more effective wall conditioning
was carried out. Note the pressure is lower and the axial
plugging (reduction of end losses) persists for a longer time
after the vacuum improvements.

These same measures have been incorporated into the MFTF-B
vacuum system. Neutral beam injectors, when possible, are connected
to the vessel through a beam line with extensive differential
pumping. When it is necessary to locate the beam closer to the
plasma, the beam line is incorporated into the machine by installing
baffles and pumping. Beam dumps are located in separate tanks
across from the beam, or in large pumping chambers incorporated into
the vessel. A schematic of an external beam line and an internal
beam dump is shown in Fig. 16. Reflux is controlled by wall
conditioning techniques, especially gettering.

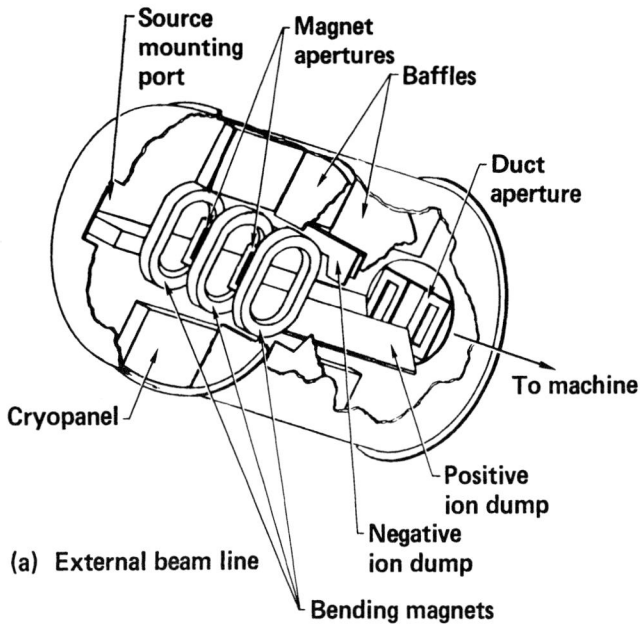

(a) External beam line

Source mounting port — Magnet apertures — Baffles — Duct aperture — To machine — Positive ion dump — Negative ion dump — Bending magnets — Cryopanel

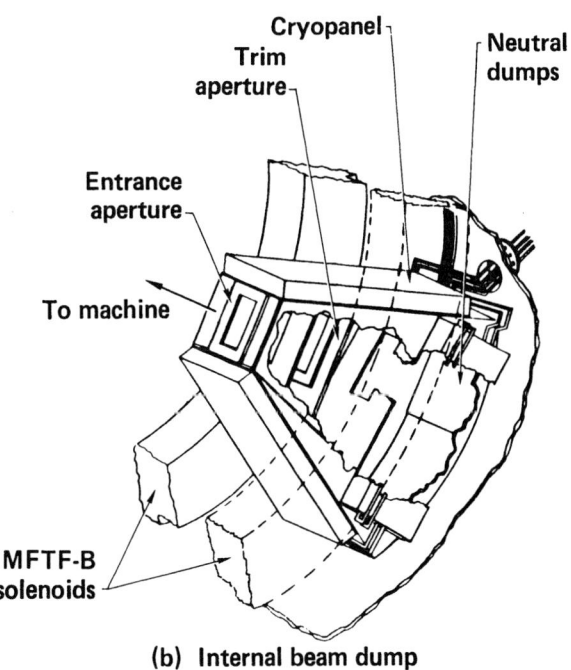

(b) Internal beam dump

Cryopanel — Trim aperture — Neutral dumps — Entrance aperture — To machine — MFTF-B solenoids

Fig. 16.   A schematic representation of the external beam lines and
integral beam dumps for MFTF-B: (a) external beamline,
(b) internal beam dump. These are important for the reduction of gas, particularly in the thermal barrier region.

A model of the gas pressure in the end regions of TMX-U has been developed;[25] gas sources due to beam injection and wall reflux have been included. This model indicates that the pressure outside of the TMX-U plasma should be less than $0.5-1 \times 10^{-6}$ Torr. This model is in qualitative agreement with the experimental results. Similar calculations for MFTF-B have been carried out. Trapping of cold neutrals in the thermal barrier region has been found to be the most important limiting process. This establishes a requirement of maintaining the on-axis neutral density below 10^7 to 10^8 cm^{-3} in this region. A gas penetration code[43] is then used along with the expected plasma parameters to calculate the neutral attenuation by the plasma, which determines the neutral density at the edge plasma. The model indicates that the edge pressure must be in the range from 10^{-7} to 10^{-6} Torr. This pressure must be consistent with the supplied pumping speed of the system and the sources. Calculations indicate that the requirements can be met, as long as wall reflux does not exceed unity by a large amount and gas sources from neutral beams are well controlled.

Impurities can also degrade the thermal barrier because they are efficiently trapped in the potential minimum. Cold impurities are expected to be shielded from the barrier region by the edge plasma. However, impurities present in neutral beams that are used to fuel and pump the thermal barrier can penetrate into the core. Figure 17 shows impurity emissions from several oxygen, nitrogen, and carbon lines during a plasma shot on TMX-U. Note that the nitrogen and oxygen emissions decrease rapidly when the neutral beam is turned off, indicating that the beam is a source of these impurities. Estimates indicate that about 1% of the beam current in normal injectors can be oxygen.[44] Because these impurities can be so detrimental, two steps have been taken in the beam sources. In the TMX-U sources and the 0.5-s beams for MFTF-B, the arc chambers are gettered; tests have shown that this can reduce the impurity current to less than 0.1%. For the 30-s MFTF-B sources, magnetic separation is incorporated into the design, a schematic of such a source is shown in Fig. 18. The expected impurity current from this source is in the range of 0.001 to 0.0001%; these sources are currently being tested. Models have shown than this reduction should be sufficient for thermal barrier operation in MFTF-B.[45]

Finally, it should be noted that other means of pumping the thermal barrier region, such as drift pumping in MFTF-B, should also be advantageous from the standpoint of plasma-surface interactions. In this method, which is not expected to produce any neutral gas or impurity influx, ions are moved radially to the edge plasma, where they are transported to the end fan regions and are pumped. This is in contrast to neutral beam pumping, where ions are charge-exchanged into neutrals that are lost in the pumping region. In addition, it should be possible to selectively pump impurities from the system.

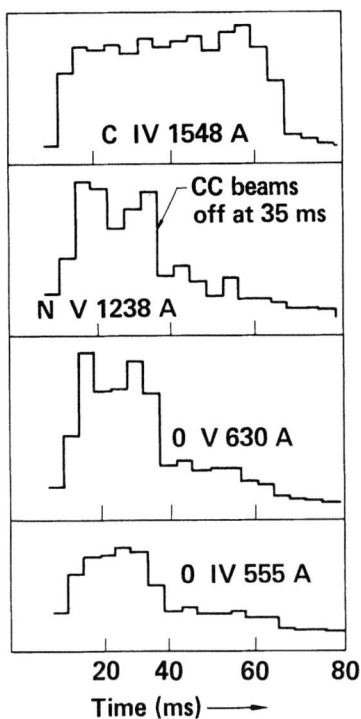

Fig. 17. Data from TMX show the time histories of carbon, nitrogen, and oxygen ionization states. During the plasma shot, the neutral beams were turned off and oxygen and nitrogen emissions decreased dramatically. This indicates that the neutral beam is the source of these impurities.

4.3 Plasma-Surface Interactions at the End Walls

In early mirror machines, a target plasma was injected along field lines by a plasma gun. This created a plasma between the end wall and the magnetic mirror that could cause large power losses.[46] This problem was minimized by the use of a pulsed plasma gun and pulsed magnetic fields, thereby breaking the connection to the end wall. In modern tandem mirrors, ECRH is used to initiate the plasma[47] for thermal barrier operation, and this mode of startup is also planned for MFTF-B. ECRH startup also removes one source of impurities because some plasma guns injected titanium[48] directly into the plasma.

Control of the plasma density in the neighborhood of the end wall is important to minimize power losses. A fraction of hot ions is not confined and is lost to the end wall; the density of these ions can be controlled by expanding the plasma flux tube, which is accomplished by

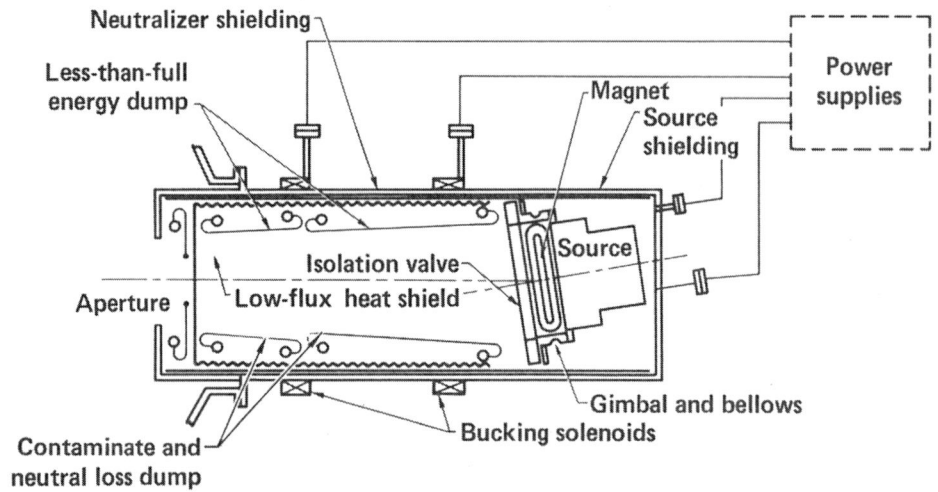

Neutralizer shielding

Less-than-full energy dump

Magnet
Source shielding

Power supplies

Isolation valve
Low-flux heat shield

Source

Aperture

Gimbal and bellows

Contaminate and neutral loss dump

Bucking solenoids

Fig. 18. Schematic of the pure beam injectors for MFTF-B. These beams provide 10 A of full energy, low impurity current.

decreasing the magnetic field from the outside mirror point to the end wall. Explicitly, $n_w = n_m (B_w/B_m)(V_m/V_w)$, where n is the ion density, B is the magnetic field, and V is the velocity parallel to B; the subscripts w and m refer to the wall and the mirror point, respectively. The density reduction was experimentally verified on TMX, where the magnetic field ratio was about 200.[49] On MFTF-B, the magnetic field ratio is equal to 60.

A second source of power loss could be caused by the cold ions produced from recycling at the end wall. The cold neutrals produced by interactions with the end wall must be pumped before they ionize and accumulate in the end region. The plasma density control discussed above plays a role in this process, as the mean time between ionizations will be longer at low density, resulting in a higher pumping probability. A model has been computed for TMX-U and MFTF-B[50] that indicates that gas accumulation should not be a problem with the available pumping.

The potential near the end wall can also be important. It has been shown experimentally that the electron temperature near the end wall can be small, which leads to a small sheath potential, reducing the probability of unipolar arcs.[8] Secondary electron emission at the end walls was once thought to be a major problem in mirror machines, limiting the electron temperature by allowing hot electrons to escape.[46] When space charge limiting effects were considered, as in the model discussed above,[50] it was found that only modest increases in the end loss power would occur. In the case of MFTF-B, an enhancement of only 40% is predicted for a particle reflux coefficient of unity even if the electron temperature near the end wall is equal to the central cell temperature. The actual temperature in the region near the end wall is expected to be much less than this value, so the effect of secondaries is expected to be even smaller.

Impurity generation in this region is not anticipated to be a problem, as ions must penetrate the large axial potential to influence the rest of the plasma region. Radiation losses should be small for impurities in this area because the electron density and temperature are low.

5. SUMMARY

There are indications both from operating experiments and theoretical models that plasma-surface interactions may be minimized in thermal barrier tandem mirror machines. The interactions occur in three primary regions: at radial surfaces, in the thermal barrier, and at the end wall. Interactions at radial surfaces can be minimized by controlling the radial profile of the axial confinement of the plasma. One method of accomplishing this is to control the radial extent of the thermal barrier. In the thermal barrier region, the control of cold gas and impurities is very important, requiring gas control in the neutral beams and also impurity control by gettering and magnetic separation. In the end wall regions, flux tube expansion of the plasma and control of gas recycling can be used to minimize plasma wall interactions.

We have used the results from the TMX-U machine as an example of currently operating devices. The thermal barrier potential profile has been established in this experiment. New experiments are being planned on TMX-U and other experiments to optimize the control of plasma surface interactions, including metal vapor jets to control neutral beam gas, and operating with high temperature plasma walls.[4] Diagnostic techniques are also becoming much more sophisticated, including laser fluorescence measurements of the neutral density.[51]

This work was performed under the auspices of the U.S. Department of Energy by the Lawrence Livermore National Laboratory under contract number W-7405-ENG-48.

REFERENCES

1. T. C. Simonen, D. E. Baldwin, S. L. Allen, "TMX Tandem Mirror
   Experiments and Thermal-Barrier Theoretical Studies," in
   Proc. 9th Intern. Conf. Plasma Physics and Controlled
   Nuclear Fusion Research, held in Baltimore, MD, 1982 (IAEA,
   Vienna, 1983), Vol. 1, p. 519.
2. N. Hershkowitz, R. A. Breun, D. Brouchous, "Recent
   Experiments in the Phaedrus Tandem Mirror," in Proc. 9th
   Intern. Conf. Plasma Physics and Controlled Nuclear Fusion
   Research, Baltimore, MD (1982), IAEA, Vienna, 1983, vol. 1,
   p. 553.
3. M. Inutake, K. Ishii, A. Itakura, "Studies of
   Improvement of Plasma Confinement in Axisymmetrized Tandem
   Mirror," in Proc. 9th Intern. Conf. Plasma Physics and
   Controlled Nuclear Fusion Research, Baltimore, MD, (1982),
   IAEA, Vienna, 1983, vol. 1, p. 545.
4. J. Kesner, R. S. Post, B. D. McVey, and D. K. Smith, "A
   Tandem Mirror with Axisymmetric Central-Cell Ion
   Confinement," Nuc. Fus. 22 549-560 (1982a).
5. G. I. Dimov and G. V. Roslyakov, A Trap With Ambipolar
   Plugs, (Institute of Nuclear Physics, Novosibirsk, USSR,
   Preprint 80-1520), published by Lawrence Livermore National
   Laboratory, Livermore, CA, UCRL-TRANS-11670 (1981).
6. D. E. Baldwin, B. G. Logan, T. C. Simonen, Physics Basis for
   MFTF-B, Lawrence Livermore National Laboratory, Livermore,
   CA, UCID-18496 (1980).
7. K. I. Thomassen, V. N. Karpenko, An Axicell Design for the End
   Plugs of MFTF-B, Lawrence Livermore National Laboratory,
   Livermore, CA, UCID-19318 (1982).
8. R. P. Drake, Nuclear Technology/Fusion 3 405 (1983).
9. W. L. Hsu, "Review of Plasma Wall Interaction Experiments on
   TMX and TMX-U," in Proc. 6th Intl. Conf. on Plasma-Surface
   Interactions in Controlled Fusion Devices, Nagoya, Japan
   (1984). To be published in the J. of Nucl. Instr. and Mater.
10. S. L. Allen and the TMX-U/MFTF-B Experimental Team,
    "Plasma-Surface Interactions in Large Tandem Mirror
    Devices--MFTF-B" in Proc. 6th International Conference on
    Plasma-Surface Interactions in Controlled Fusion Devices,
    Nagoya, Japan (1984). To be published in the J. of Nucl.
    Instr. and Mater.
11. K. I. Thomassen, B. G. Logan, J. N. Doggett, and F. H. Coensgen,
    "A DT-Burning Upgrade to MFTF-B", in Proc. 6th Intl. Conf.
    on Plasma-Surface Interactions in Controlled Fusion Devices,
    Nagoya, Japan (1984).
12. B. G. Logan, MARS Final Report, Lawrence Livermore
    National Laboratory, Livermore, CA, UCRL-53480 (1984).
13. C. C. Damm, J. H. Foote, A. H. Futch, A. L. Gardner, and
    R. F. Post, Phys. Rev. Lett. 13 464 (1964).

14. R. W. Moir and R. F. Post, Nucl. Fusion 9 253 (1969).
15. R. F. Post, in Proc. Intl. Conf. Plasma Confined in Open-Ended Geometry (Gatlinburg, TN, 1967), p. 309; also published as Oak Ridge National Laboratory, Oak Ridge, TN, CONF-671127.
16. F. H. Coengsen, W. F. Cummins, B. G. Logan, A. W. Molvik, W. E. Nexsen, T. C. Simonen, B. W. Stallard, and W. C. Turner, in Proc. 7th Eur. Conf. Cont. Fusion Plasma Phys., Vol. II, p. 167 (1975).
17. R. W. Moir, Standard Mirror Fusion Reactor Design Study, Lawrence Livermore National Laboratory, Livermore, CA, UCID-17644 (1978).
18. D. L. Correll, Nuc. Fus. 22 223 (1982).
19. T. A. Casper and G. R. Smith, Phys. Rev. Lett. 48 1015 (1982).
20. D. E. Baldwin and B. G. Logan, Phys. Rev. Lett. 43 1318 (1979).
21. T. C. Simonen, Phys. Rev. Lett. 50 1668 (1983).
22. D. P. Grubb, S. L. Allen, J. D. Barter, "Thermal Barrier Production and Identification in a Tandem Mirror," submitted to Phys. Rev. Lett.; Lawrence Livermore National Laboratory, Livermore, CA, UCRL-90536.
23. T. C. Simonen, S. L. Allen, T. A. Casper, "TMX-U Experimental Results," in Proc. of Course/Workshop on Mirror-Based Approaches to Magnetic Fusion, Varenna, Italy, Sept. 7-17, 1983; published by Lawrence Livermore National Laboratory, Livermore, CA, UCRL-89286 (1983).
24. T. A. Casper, L. V. Berzins, R. F. Ellis, R. A. James, and C. Lasnier, "Microstability of TMX-U During Initial Thermal Barrier Operation," in Proc. of Intl. Conf. on Plasma Physics, Lausanne, Switzerland (1984); Lawrence Livermore National Laboratory, Livermore, CA, UCRL-90451.
25. W. C. Turner, W. E. Nexsen, S. L. Allen, "Gas Pressure in the End Plug Regions of the TMX-U Thermal Barrier Experiment," submitted to the J. Vac. Sci. Technol. (1983); Lawrence Livermore National Laboratory, Livermore, CA, UCRL-89938.
26. A. A. Marin, S. P. Auerbach, R. H. Cohen, J. M. Gilmore, L. D. Pearlstein, and M. E. Rensink, Nuc. Fus. 23 703 (1983). See also D. E. Baldwin and B. G. Logan (Eds.), Physics Basis for an Axicell Design for the End Plugs of MFTF-B, Lawrence Livermore National Laboratory, Livermore, CA, UCID-19359 (1982).
27. L. D. Pearlstein, D. E. Baldwin, R. H. Cohen, T. K. Fowler, and B. G. Logan, "Stabilization of Tandem-Mirror Trapped-Particle Modes by Incomplete Cancellation of Trapped-Particle Drifts," presented at 1982 Sherwood Meeeting, Annual Controlled Fusion Theory Conference, Santa Fe, NM, April 25-28, 1982.

28. E. B. Hooper, Jr., D. E. Baldwin, T. K. Fowler, R. J. Kane, and W. C. Turner, "Radial Transport Reduction in Tandem Mirrors Using End Wall Boundary Conditions," submitted to Physics of Fluids (1984); Lawrence Livermore National Laboratory, Livermore, CA, UCRL-90639.

29. D. L. Correll, J. A. Byers, T. A. Casper, Throttle Coil Operation of TMX-U, Lawrence Livermore National Laboratory, Livermore, CA, UCID-19650 (1983).

30. Y-J. Chen, D. E. Baldwin, and T. Q. Hua, "Analytic Model and Simulations of RF Drift Pumping of Thermal Barriers," Bull. Amer. Phys. Soc. 28 1195 (1983).

31. B. W. Stallard, F. H. Coensgen, W. F. Cummins, "Plasma Wall Charge Exchange Interactions in the 2XIIB Magnetic Mirror Experiment," in Proc. International Symposium on Plasma Wall Interaction, Jülich, Federal Republic of Germany, Pergamon Press, p. 63. (1976)

32. D. L. Correll, J. F. Clauser, F. H. Coensgen, Nuc. Fus. 20 655 (1980).

33. F. Najmabadi, and R. W. Conn, "Radius Control System for Tandem Mirrors," submitted to Nuclear Technology/Fusion (1984).

34. J. R. Farron, R. A. Breun, S. N. Golovato, "Scaling of RF Sustained Tandem Mirror Parameters With Central Cell Heating Power Including a Central Cell Stand-Alone Mode," Bull. Amer. Phys. Soc. 27 958 (1982).

35. D. Garner, private communications, May 1984.

36. S. L. Allen, C. A. Clower, and W. C. Turner, "The Influence of Vacuum and Wall Conditioning on Plasma Startup in TMX-U," Bull. Am. Phys. Soc. 28 1114 (1983).

37. H. F. Dylla, in Proc. of the 29th National Symposium (American Vacuum Society, Baltimore, MD).

38. S. L. Allen, C. Clower, R. P. Drake, E. B. Hooper, Jr., A. L. Hunt, R. Munger, R. J. Bastasz, W. Bauer, and W. L. Hsu, "Initial Wall Conditioning for the TMX-U Fusion Experiment, J. Vac. Sci. Technol. A 1(2) 916 (1983).

39. J. E. Osher, G. D. Porter, L. E. Valby, Glow Discharge Cleaning Tests of MFTF-B Cryopanel Components, Lawrence Livermore National Laboratory, Livermore, CA, Quarterly Jan.-May, 1981, UCRL-50051-81-1 (1981).

40. W. L. Pickles, in Proc. of the 29th National Vacuum Syposium (American Vacuum Society, Baltimore, MD, 1982).

41. W. Bauer, Bull. Am. Phys. Soc. 27 1137 (1982).

42. O. T. Strand, H. W. Moos, and S. L. Allen, "Experimental Evidence for Outward Radial Transport of Impurities from the Central Cell of TMX," Nuc. Fus. 23 (12) (1983).

43. G. E. Gryczkowski, Neutral Gas Blanket Theory as Applied to the Reference Theta Pinch Reactor, Ph.D. Thesis, Univ. of Michigan, Dept. of Nuclear Engineering (1979).

44. R. P. Drake and H. W. Moos, Nuc. Fus. 19 407 (1979).

45. J. E. Osher, D. P. Grubb, and P. Poulsen, "Impurity Accumulation in a Tandem Mirror," Bull. Am. Phys. Soc. 1119 (1983).

46. L. S. Hall, Nuc. Fus. 17 681 (1977).

47. B. W. Stallard, "ECRH in Tandem Mirror Machines," in Proc. of IEEE Minicourse on RF Heating and Current Drive, San Diego, CA, 1983; Lawrence Livermore National Laboratory, Livermore, CA, UCRL-89276.

48. S. L. Allen, T. L. Yu, and T. J. Nash, "Impurity Characteristics of TMX-U," Lawrence Livermore National Laboratory, Livermore, CA, UCID-20026 (1984).

49. P. Coakley, N. Hershkowitz, and G. D. Porter, "End-Wall Plasma Characteristics in the Tandem Mirror Experiments," Nuc. Fus. 22 1321 (1982).

50. G. D. Porter, "Effect of Gas Recycling and Secondary Emission on the Axial Flow in an Open-Ended Device," Nuc. Fus. 22 1279 (1982).

51. K. Muraoka., M. Maeda, T. Okada, in Proc. of 6th Intl. Conf. on Plasma Surf. Int. in Cont. Fus. Devices 1984.

IMPURITY CONTROL SYSTEMS FOR REACTOR EXPERIMENTS

D. E. Post
Plasma Physics Laboratory, Princeton University
Princeton, N.J. 08544

R. F. Mattas
Argonne National Laboratory
Argonne, IL. 60439

ABSTRACT

Poloidal divertors and pumped limiters are the leading candidates for impurity and particle control systems for reactor tokamak experiments. Such systems must be able to provide heat removal and He pumping while satisfying the requirements for (1) minimum plasma contamination by impurities, (2) reasonable component lifetime (~ 1 year), and (3) minimum size and cost and maximum simplicity. While pumped limiter systems are simpler and cheaper, poloidal divertors offer the possibility of low sputtering rates for the first wall components and modest pumping requirements due to the formation of a cool, dense plasma near the collector plates. Estimates made as part of the INTOR study indicate that the sputtering rates for pumped limiters could be unacceptably large. Both types of systems should be able to provide adequate pumping. Engineering design studies have been carried out for both systems. The study for a poloidal divertor system for INTOR indicates that such a system offers a reasonable solution to the impurity control problem at only a modest increase in total reactor cost (~10%) and complexity compared to a pumped limiter system.

INTRODUCTION

Impurity and particle control is a key issue for the design of any reactor fusion experiment. The impurity and particle control system must be able to absorb the alpha heating power and to exhaust the helium ash without impacting the operational success of the experiment. The two most seriously considered systems are based on poloidal divertors and pumped limiters. A limiter is a piece of material, usually a refractory metal or graphite, at the plasma edge which "limits" the plasma channel. Pumped limiter experiments have been discussed in the chapter by Mioduszewski and divertor experiments have been covered in the chapter by Wagner and Lackner.

Pumped limiters are basically limiters with an opening below the limiter face where neutral gas formed on the limiter backside can be pumped. The "scraped-off" plasma flows underneath the limiter and is neutralized by contact with a surface (Fig. 1). The resulting neutral gas is then pumped. The impurity control aspect of pumped limiters is the same as conventional limiters. Limiters have been used on tokamaks and stellarators since the 1960's. They provide a way to localize the point where the plasma contacts the wall. The limiter has a small enough surface area that it can be made relatively "clean" compared to the vacuum vessel wall so that the plasma can remain relatively clean.

The idea for using a mechanical pumped limiter has been proposed by a number of authors [Vershkov, V., and Mirnov, S., 1973; Kelly, G., 1974; Bieger, W., Dippel, K., Fuchs, G., and Wolf, G., 1976; Schivell, J., 1977]. Pumped limiters were discussed for reactor experiments by Brooks, J., et al., 1979, and Conn, R., et al., 1979, for solving the Helium removal problem.

Poloidal divertors are quite different from limiters. With a poloidal divertor, the edge plasma is deflected magnetically

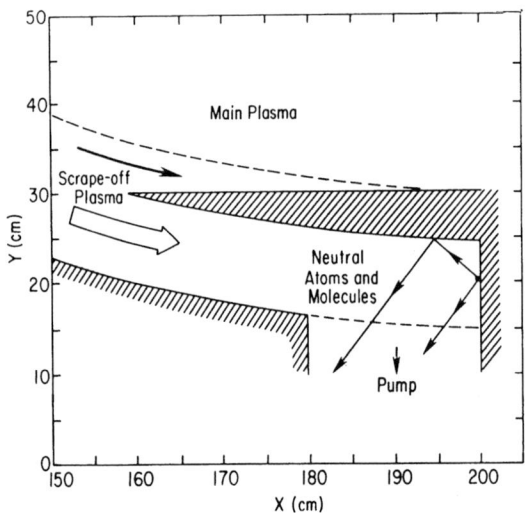

Fig. 1. Schematic illustration of pumped limiter.

POLOIDAL DIVERTOR MAGNETICS

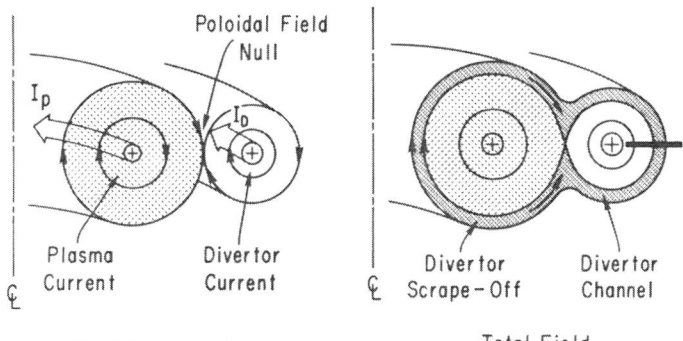

Field Components   Total Field

Fig. 2. Schematic illustration of poloidal divertor magnetics
         [courtesy of D. Meade].

away from the main plasma edge to a region where the interaction
of the plasma and wall can be controlled (Fig. 2). In this way,
the power and particles can hopefully be handled more easily
than at the main plasma edge. Methods for accomplishing this
were not initially well understood. The basic idea was that
impurities born at the wall near the plasma chamber would be
ionized and entrained in the edge plasma, and swept into the
divertor chamber. This concept led to designs with large
divertor chambers and poloidal coils close to the main chamber
(Fig. 3). The basic difficulty still remained that a very hot
plasma hits the divertor plate. One early advance in this
concept was depicted in the Princeton Reactor Design [R. Mills,
et al., 1974] in which the power was radiated by argon
impurities in the divertor chamber plasma before the power
reached the divertor plate. Another feature of this general
approach to divertors is the requirement for large pumping
systems in the divertor chamber to maintain the flux of edge
plasma into the divertor chamber so that the impurities which
would otherwise get to the main plasma from the wall would be
swept into the divertor. The backflow of neutrals and
impurities is prevented by mechanical baffles separating the
divertor chamber from the main plasma chamber. This design
philosophy was embodied in the PDX [Fonck, R., et al., 1982] and
ASDEX [Engelhardt, W., et al., 1982] poloidal divertor
experiments (Fig. 3).

     Poloidal divertor coils cause complications for remote
handling if they are inside the toroidal coils, so designs with
external divertor coils are preferred. With such coils, it is

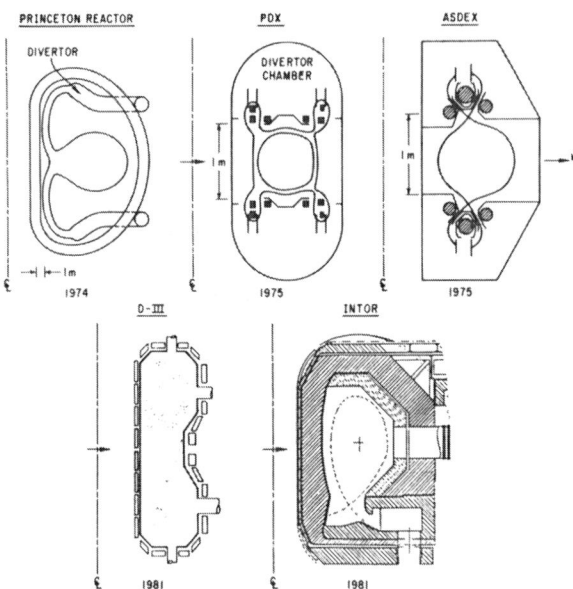

Fig. 3. Evolution of poloidal divertor designs from early
         reactor concepts to the INTOR divertor [Post, D., et
         al., 1984].

difficult to make a closed divertor chamber of the PDX or ASDEX
type shown in Fig. 3. An "open" divertor of the type in the
INTOR design is the most practical design. Such "expanded
boundary" divertors were proposed for the D-III experiment [N.
Ohyabu, 1981] and operated on D-III [Mahdavi, M., 1982; Shimada,
M., 1982]. Current designs for poloidal divertor experiments
such as ASDEX/Upgrade [Vernickel, H., 1984], Big-D [Mahdavi, M.,
et al., 1984], and INTOR [Post, D., et al., 1984; Fujisawa, N.,
et al., 1984] are based on such "open" divertors.

     This paper will first discuss the requirements that an
impurity and particle control system must meet. Then it will
describe the general plasma physics issues relevant to impurity
control and the issues for pumped limiters and poloidal
divertors. In particular for divertors, the physics of the
operation of the "open" divertors described above will be
addressed. The materials issues will be summarized and design
considerations will be described. Examples drawn from the INTOR
design studies will be used as illustrations and to make the
relevant points more specific.

IMPURITY CONTROL REQUIREMENTS FOR IGNITION EXPERIMENTS

     The major requirement of an impurity control system for an
ignited tokamak is that it be able to remove the steady-state
alpha heating power without adversely affecting the performance
of the tokamak. In particular, the impurity contamination of

1104

the main plasma should be keep low so that undesirable radiation losses and fuel dilution are not a problem. Radiation losses reduce the energy confinement making ignition more difficult, and fuel dilution increases the beta requirements [Jensen, R., et al., 1978]. Secondly, erosion of the plasma side components must be kept low to ensure an acceptable lifetime ($\gtrsim$ 1 year).

Another requirement is that the impurity and particle control system will have to exhaust the helium produced from the fusion reactions at the rate the helium is produced. The helium will be accompanied by deuterium and tritium, so the helium exhaust must be accomplished while trying to minimize the tritium handling and inventory requirements. Since space for pumping ports is expensive and neutron shielding is difficult, the pumping system for helium exhaust should be as small as possible.

These two main goals have the constraints that they must be met by a system which: (1) is capable of being maintained by remote handling techniques, (2) is able to survive disruptions, (3) has a minimum cost impact on the device cost, and (4) is as simple as possible. In order to make these concerns and requirements as specific as possible, we will discuss the impurity control problem in the context of the INTOR design effort. INTOR [Post, D., et al., 1984; Fujisawa, N., et al., 1984] is a reactor scale tokamak designed by an international collaborative team from the US, the USSR, Japan, and the Euratom countries and organized by the IAEA. The INTOR design is the basis of the ignition tokamak designs such as NET, FER, and OTR which are being proposed by the Europeans, Japanese, and Soviets. The INTOR impurity and particle control requirements are listed in Table 1. The INTOR pumped limiter design (Fig. 4) is a curved plate which has two leading edges. The limiter scrapes off the edge plasma, and pumps the neutral gas formed below the limiter.

The INTOR divertor design (Fig. 4) has an "open" geometry so that the poloidal field coils can be outside the toroidal field coils to ease the remote maintenance and neutron shielding problems. The divertor is also designed to be as small as possible to minimize the cost.

PLASMA PHYSICS ISSUES

The performance of an impurity and particle control system will depend crucially on the parameters of the plasma that impacts the first wall components. Plasma in contact with a solid material forms a sheath with an electrostatic potential of about three times the electron temperature [see the chapter by P. Stangeby]. The ions are accelerated by the sheath potential

Table 1. INTOR Impurity Control Requirements [INTOR, 1983]

Alpha Heating Power	124 MW
Power Incident on Collector Plates	~ 80 MW
Helium Production Rate	2×10^{20} He/sec
Total Particle Throughput Required	~ 2×10^{21} H2/sec
Maximum Erosion Rate	~ 1 cm/y
Maximum Impurity Levels	
Low-Z (C,O)	~ 1%
Medium-Z (Fe, Ni)	~ 0.1%
High-Z (Mo, W)	~ 0.01%
Maximum Credible Pumping speed	
Available	500,000 liters/sec

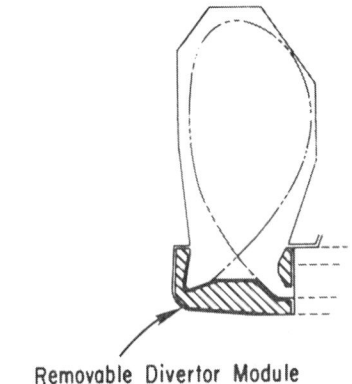

Removable Divertor Module

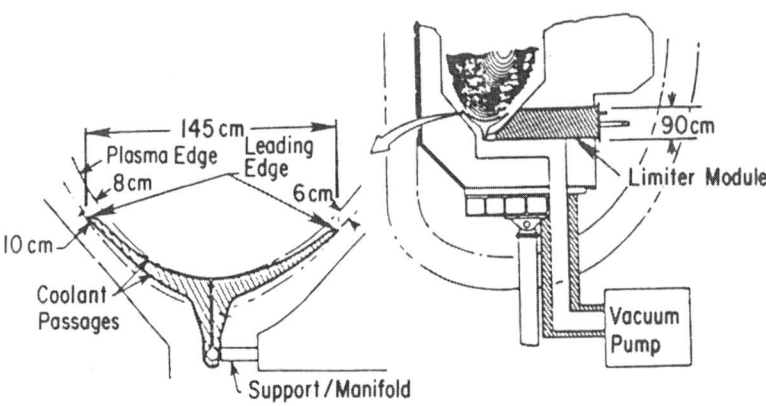

Fig. 4. Schematic outline of the (a) poloidal divertor and (b)
the pumped limiter system for INTOR [Post, D., et al.,
1984].

and the electrons are retarded so that the net current to the
wall is zero. From elementary sheath theory, one can make a
simple estimate of the edge temperature and density. Using the
INTOR heat and particle fluxes [Post, D., et al., 1984;
Fujisawa, N., et al., 1984], one concludes that, at the edge,

$$T \sim 500\text{-}800 \text{ eV},$$

$$n_e \sim 1\text{-}5 \times 10^{12}/\text{cm}^3 \, .$$

More sophisticated estimates using tokamak transport codes
[INTOR, 1983] give (Fig. 5) for INTOR

$$T \sim 200\text{-}300 \text{ eV},$$

$$n_e \sim 10^{13}/\text{cm}^3 \, .$$

Thus the sheath potential is on the order of 600 to 900 eV. The
energy of the ions striking the wall will be larger than the
sputtering threshold of any useful wall materials (~ 100 eV),
with the result that the wall sputtering rate will be large.
Estimates for INTOR indicate that the sputtering rate for a
limiter could be as large as 10 to 50 cm/year. This makes the
design of a long life limiter system difficult. Redeposition of
the sputtered material may reduce the problem, but redeposition
is uncertain and cannot be relied upon at this point.

The low density also implies that the neutral pressure may
be low (~ 0.0001 torr), thus possibly making a large pumping
system necessary to provide adequate helium exhaust ($\sim 500,000$
liters/second).

Thus sputtering of the wall material by ions accelerated by
the sheath is the major problem. The sputtering can be reduced
or eliminated if the kinetic energy of the bombarding ions can
be reduced below the sputtering threshold of the wall material
(~ 100 eV). This requires reducing the sheath potential which
in turn requires reducing the edge temperature. From sheath
theory [see lecture by Stangeby], the temperature of the plasma
hitting the wall is given approximately by

$$T = Q/(8\Gamma) \, ,$$

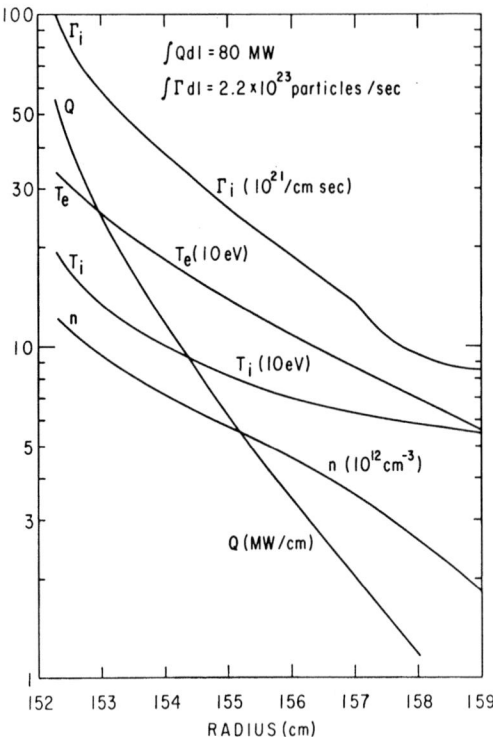

Fig. 5. Calculated edge conditions for INTOR using the BALDUR
tokamak transport code [Post, D., et al., 1984].

where Q is the heat flux and Γ is the particle flux at the
sheath boundary. The temperature can be lowered by either
lowering the heat flux Q or increasing the particle flux Γ.

The heat flux on the limiter or collector plate can be
lowered by impurity radiation from the plasma edge. Such a
mechanism would solve the sputtering problem for a limiter by
reducing the edge temperature. Reducing the sputtering would
require radiating about 80 to 90% of the alpha power from the
edge without radiating much from the plasma center. This
approach actually works in many ohmic heating experiments. As
much as 80% of the ohmic heating energy was lost as radiation
from carbon ions at the plasma edge on PLT [Bol, K., et al.,
1978]. However, a much lower fraction of the heating power is
lost on experiments with high power auxiliary heating. Simple
considerations indicate that the impurity density required to
radiate a given power P increases as P^2 [see Chapter by D. Post
and K. Lackner]. Thus large impurity densities will be required
for the high heating powers for ignited experiments. A second
consideration is that a radiating layer will require a
significant volume of plasma, possibly as much as 20-40% of the

total plasma volume. Thus the total plasma size will have to be increased to obtain a given size of reacting plasma since the radiating layer will be too cold to be part of the reacting plasma. The increase of the cost of the machine due to this will have to be weighed against the costs of competing impurity control techniques. A third consideration is that medium or high-Z impurities will have to be used, since low-Z impurities are too easily ionized. It is difficult to force these impurities to remain at the plasma edge and not reach the plasma center. At the center, they will cause radiation and dilution problems. In summary, a radiating edge is an attractive idea, but does not appear feasible at present since it has problems with contaminating the central plasma, requires an increase in the plasma volume, and does not seem to work well (i.e., does not radiate 90% of the heating power) on current experiments with high power auxiliary heating.

A second way to lower the temperature is to increase the particle flux Γ at the edge since $T = Q/(8\Gamma)$. This can be done by increasing the local recycling. This requires forcing the neutrals formed by the recombination of plasma incident on the wall to be ionized in the plasma flowing to the wall. The continuity equation for particle flow along field lines is

$$\frac{\partial(n v_{\parallel})}{\partial x} = n_o n_e \langle \sigma v \rangle ,$$

where $v_{\parallel}$ is the flow velocity parallel to the field, n_o is the neutral atom density, and $\langle \sigma v \rangle$ is the ionization rate. Integrating this equation from the main plasma to the wall yields

$$\Gamma_{wall} = \Gamma_o + \int_o^{wall} \langle \sigma v \rangle \, n_o n_e \, dx .$$

where Γ_{wall} is the plasma flux at the wall, and Γ_o is the plasma flux from the main plasma. Using this result, we can define a flux amplification factor A as the ratio of the particle flux at the wall to the particle flux from the main plasma

$$A = \frac{\Gamma_{wall}}{\Gamma_o} .$$

Then using $T_{wall} = Q_{wall}/8\Gamma_{wall} \propto (Q/8\Gamma_o) \, 1/A$, we see that the temperature at the wall scales as

$$T_{wall} \propto A^{-1}.$$

The density can be shown to scale as

$$n_e \propto A^{3/2}.$$

Thus increasing the recycling can lower the temperature T and raise the density n. The key to doing this is to increase the local ionization of the recycling neutrals by forcing the neutrals to be ionized many times before they escape back to the main plasma or to a pump. Such operation has been termed "high recycling."

In reality a combination of these two techniques, radiation loss and "high recycling," will be part of any realistic impurity control system.

Limiter Physics

The pumped limiter was examined during the INTOR Phase Two-A, Part One Workshops because it offered the potential for a reduced cost device that might still provide adequate impurity control. The basic configuration is a shaped, double sided pumped limiter located at the bottom of the vacuum vessel (Fig. 4). The potential performance of the pumped limiter was studied by an assessment of pumped limiter experiments and by using large scale computational models to extrapolate from these experiments to INTOR.

The general physics issues for pumped limiters have been discussed in the chapter by P. Mioduszewski. The front surface of the limiter defines the plasma edge. The front face of the limiter absorbs most of the power. The "leading edge" (Fig. 1) of the limiter must be located far enough out in the scrape-off plasma so that the heat flux is reduced to a level that will not result in damage to the limiter. However, the particle flux is thereby reduced, so that a compromise must be made between having the leading edge far out in the scrape-off for an acceptable heat flux and putting the leading edge close to the main plasma for an acceptable particle flux for pumping.

It is to be expected that limiters have less potential for impurity control than divertors, due to higher temperatures (~ 100-150 eV for INTOR conditions) than divertors (~ 25 eV) near the collector plate. However, near term experiments on JET, TFTR, and T-15 with high power auxiliary heating and limiter operation will provide data on impurity control with limiters. At the present time, carbon limiters are able to provide

1110

adequate impurity control on most present tokamaks with high power auxiliary heating. A second issue is the lifetime due to erosion. For a long pulse, high duty factor experiment such as INTOR, erosion will be an issue even if tolerably clean plasmas can be produced with limiters, since the lifetime of the limiter must be of the order of a year or greater. Experimental data will not be soon forthcoming since long pulse, high duty factor machines are not likely to precede INTOR in the immediate future. Predicted sputtering rates for the pumped limiter are in the 5-50 cm/year range. These rates may be reduced by the redeposition of the sputtered material back onto the limiter. Model calculations of this indicate that the net erosion rate may be marginally acceptable in some designs. However, the confidence in our understanding of the physics of the transport of impurities in the plasma edge is not sufficiently high to base the INTOR design on a pumped limiter.

The pumping of helium is a key issue. These issues have been discussed in the chapter by Mioduszewski. Very promising early small scale pumped limiter experiments [Mioduszewski, P., et al., 1982] have been followed by large scale modular pumped limiter experiments with high power auxiliary heating on ISX [Mioduszewski, P., et al., 1984], PDX [Budny, R., et al., 1985], and PLT [Cohen, S., et al., 1985] and with ohmic heating on TEXTOR [Pontau, A., et al, 1984]. With auxiliary heating ($\gtrsim 2$ MW), neutral pressures of $1-5 \times 10^{-3}$ torr in the pumping chamber and particle removal efficiencies of 2-5% were measured. An axisymmetric limiter will be required for an INTOR-sized device to provide steady-state heat exhaust so a key question is how these pressures and particle removal efficiencies will scale when the particle exhaust is spread out on an axisymmetric structure instead of localized on one or two limiters. Experiments on this question are needed.

The performance of these limiter experiments has been modeled with reasonable success [Budny, R., et al., 1984] and [Evans, K., et al., 1985] using Monte Carlo neutral transport codes. These codes have also been used to model the INTOR limiter performance [Heifetz, D., et al., 1982]. These models show that for a large area limiter such as the INTOR limiter, the neutral mean free path is small compared to the limiter. Thus, the neutral density and recycling is localized on the front face of the limiter, and on the "neutralizer plate" underneath the limiter (Fig. 6). The charge-exchange flux falls almost entirely on the limiter, and on the first wall near the limiter tips and near the neutralizer plate. Thus, the erosion due charge-exchange neutrals is localized there. The first wall away from the limiter will have a very low charge exchange flux and therefore will have a very long erosion lifetime for sputtering, perhaps as long as the machine lifetime. This

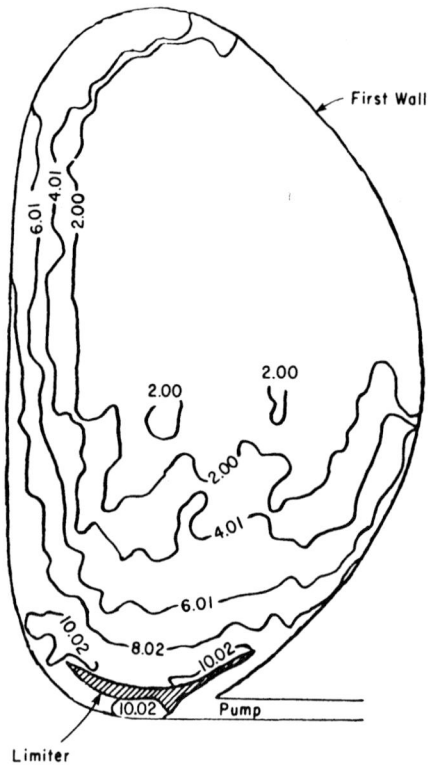

Fig. 6. Equal density contours for the neutral atom density, n_o, $(\log_{10} n_o)$, for the pumped limiter design for INTOR [Heifetz, D., et al., 1982b].

greatly eases the general remote maintenance requirements. There is evidence from PDX and TFTR limiter experiments that this type of localized recycling is a real effect [Budny, R., et al., 1984].

Localized recycling behind the limiter can play a key role in increasing the pumping rate. If the scrape-off plasma is sufficiently dense, a substantial fraction of the neutrals will be ionized locally, and contribute to a locally enhanced level of recycling. This will raise the neutral pressure and reduce the pumping requirement. These effects have been modeled for the TFCX device [Petravic, M., et al., 1985]. The geometry is reasonably realistic, the same as Fig. 1.' The dimensions correspond to the TFCX device; i.e., a major radius of 3.5 m, and minor radius of 1.5 m, and a power flux into the half of the scrape off indicated of 10 MW. In order to keep the number of unknowns to a minimum, no assumptions were made about the main plasmas except that the radial particle and energy outflow from the main plasma is uniform in the poloidal angle. Also, $T_i = T_e$ was assumed to reduce the computing time. To characterize the performance of the pump limiter, the radial outward particle flux was varied between 1.1×10^{22} and 4.4×10^{22}

particles/sec. The perpendicular particle and heat diffusion coefficients were taken to be $D_\perp = 1$ m^2/sec and $\chi_\perp = 4$ m^2/sec. The radial density and temperature profiles along a line passing through the limiter tip are shown in Fig. 7, for a radial particle outflux of 4.4×10^{22} ions/sec.

The pumping performance and plasma parameters depend crucially on the radial transport which determines the number of particles which enter the pumping region. The temperature, density, and parallel particle flux are related by the sheath boundary conditions for a given heat flux. Thus, the limiter performance can be characterized by the density and temperature (Fig. 8) for the edge plasma at the point of contact of the limiter and the main plasma. These calculations indicate that reducing the edge temperature to a value below 20-30 eV would require very high edge temperatures, in the range of 10^{14} cm^{-3}, in qualitative agreement with the results of Singer and Braams, 1985. As described before, helium pumping is an issue, the problem being that keeping the power onto the tip of limiter within reasonable limits leaves little power to drive the

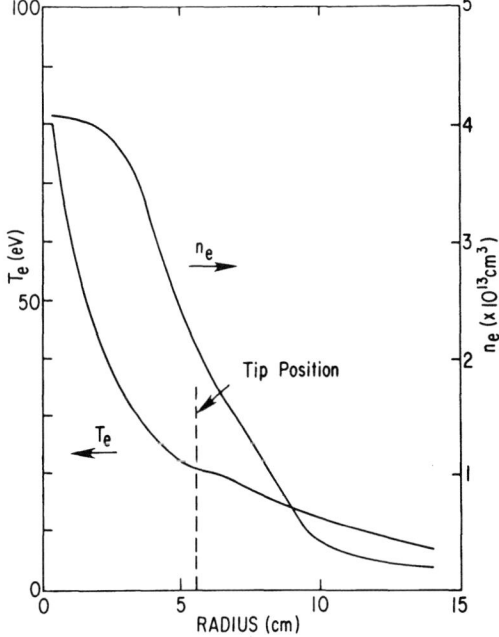

Fig. 7. The radial electron density and temperature profiles taken along a ray going through the limiter tip. The particle flux across the main plasma boundary is 4.4×10^{22} ions/sec, and the power is 10 MW. There is little variation along the field lines in front of the limiter, but the temperature drops further, and the density rises behind the limiter, and towards the stem [Petravic, M., et al., 1985]. 84P0268

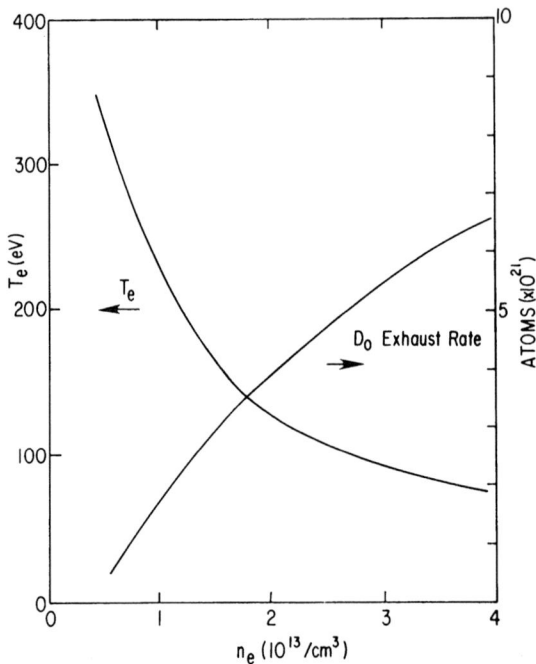

Fig. 8. The maximum electron temperature on the limiter face and the particle removal rate as a function of the electron density on the main plasma boundary. The temperature decreases rapidly with the density, and is estimated to fall below 40 eV at $n_e = 1 \times 10^{14}$ cm^{-3} [Petravic, M., et al., 1985].

pumping behind the limiter. The calculations were done with 200,000 ℓ/sec of pumping beneath the limiter. The hydrogen throughput for the pump is plotted as a function of maximum scrape-off density in Fig. 8. The required throughput for a He concentration of 5% is 1.3×10^{21} atoms/sec. Thus adequate pumping can be achieved for edge densities above 10^{13} cm^{-3}. Although this is dependent on the edge particle transport, there appears to be a significant margin so that adequate helium removal is possible.

DIVERTOR PHYSICS

    The general concept of a poloidal divertor was discussed in the introduction. The major advantage of a poloidal divertor is that it offers the possibility of reducing the temperature of the plasma hitting the collector plate by enhancing the local recycling. The divertor is able to force the neutrals to recycle locally.

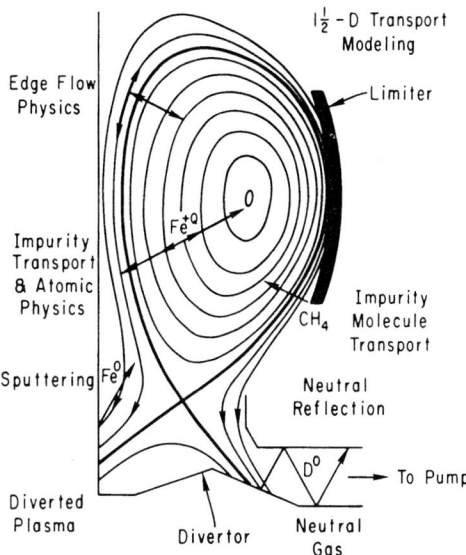

Fig. 9. Schematic outline of the physics in a poloidal divertor for an ignited plasma.

Poloidal divertors have been studied recently both theoretically with analytic and one and two-dimensional computational models and experimentally on D-III, ASDEX, PDX, and PBX. The physics of a poloidal divertor involves a number of complex processes that compete with one another (Fig. 9). Each of these processes must be treated in detail in a serious model of the operation of poloidal divertors. These processes include:

1. Plasma flow both along and across the flux surfaces,
2. Impurity transport and atomic and molecular processes,
3. Sputtering of the wall by plasma ions and neutrals,
4. Neutral gas transport including reflection from the wall and collisions in the plasma,
5. Plasma flow into and through the sheath at the wall,
6. The effects of real geometries of the magnetic surfaces and the first wall collector plates and pumping ports.

The key questions that must be addressed by any examination of the feasibility of divertors for impurity control are:

1. Is the "high recycling" regime with a low temperature and high density credible for divertors?
2. Is the "high recycling" regime credible in an "open" geometry?
3. What is the actual performance of divertors for impurity and particle control on operating experiments?

These questions have been examined for INTOR with large, highly sophisticated, two-dimensional computational models [Petravic M., et al., 1984; Sugihara, M., et al., 1984; Braams, B., et al., 1983; Igitkhanov, Yu., et al., 1984]. The models include much of the physics outline in Fig. 9. They have been used to analyze divertor experiments both to elucidate the physics in the experiment and to calibrate the models. Then they have been used to explore the possible operating parameters of divertor designs for proposed and planned tokamaks. One such code is the PLANET code [Petravic, M., et al., 1984]. The code calculates the two-dimensional transport of plasma at the edge, both along and across the flux surfaces, in a realistic magnetic geometry. It is coupled to the DEGAS code [Heifetz, D., et al., 1982] which calculates the transport of neutral gas in a two or three-dimensional grid with realistic geometries for both the plasma and wall. The DEGAS code includes about twenty atomic and molecular processes for the neutral-plasma collisions and a variety of realistic models for neutral atom reflection from the wall. The PLANET code is thus able to calculate the plasma parameters, wall sputtering, heat fluxes, and neutral fluxes and gas throughputs for pumping systems. This code was used to model the pumped limiter described in the previous section. The details of the codes are described in the chapters by D. Post and K. Lackner, and by D. Heifetz.

The answer to the question as to the credibility of the "high recycling" regime can be rephrased to ask what role does recycling play in determining the plasma conditions in a divertor. This was examined theoretically by modeling the behavior of the original PDX divertor [Fonck, R., et al., 1982]. The divertor chamber was modeled in a rectangular geometry with a pumping chamber (Fig. 10). The pumping speed of the duct was adjusted by varying the size of the pump duct opening. the plasma was 4 centimeters wide and 40 centimeters long. The pump duct opening was varied between 2 centimeters and 12 centimeters. As the opening was made smaller, the pumping speed and thus the fraction of the neutrals (formed by the plasma striking the neutralizer plate) which could escape was reduced. A typical profile of the plasma parameters along the field lines is given in Fig. 11. The ion temperature drops from 170 eV at the divertor entrance to about 15 eV at the plate. The electron temperature was assumed to be constant in this simulation. The particle source due to the ionization of neutrals is localized very close to the plate. The flux of particles to the plate rises from the input value to about four or five times the input value. The density increases near the plate. The effect of varying the recycling can be determined by varying the pump opening (Fig. 12). The recycling is increased as the pump opening is made smaller as shown by the increase in the flux of particles to the plate. The temperature at the

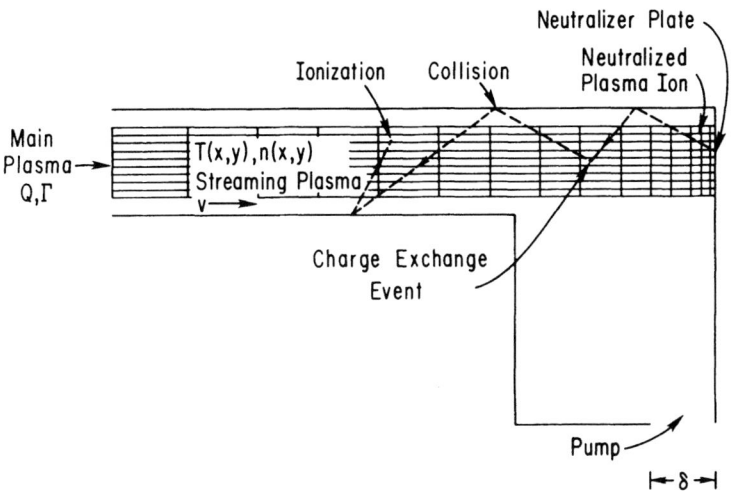

Fig. 10. Model divertor chamber for PDX [Petravic, M., et al., 1982].

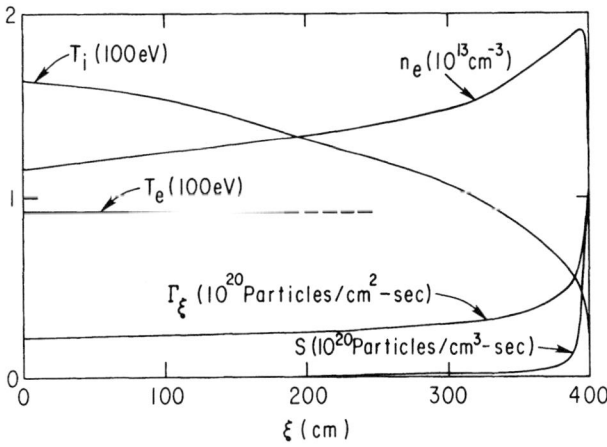

Fig. 11. Calculated plasma parameters along the separatrix in the modified PDX divertor for a pump opening of 4 cm [Petravic, M., et al., 1982].

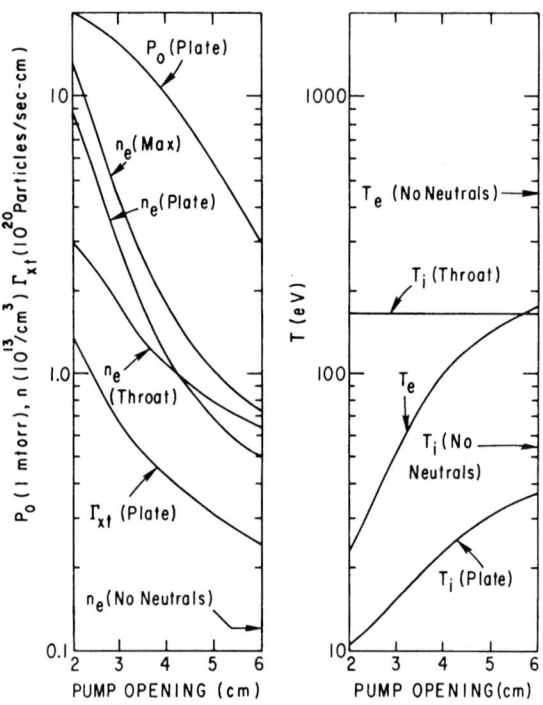

Fig. 12. The neutral pressure P, the plasma density at the
throat and at the plate n, the ion temperature at the
plate T_i, the electron temperature T_e, and the total
particle flux gamma at the plate as a function of the
pump opening for the modified PDX divertor [Petravic,
M., et al., 1982].

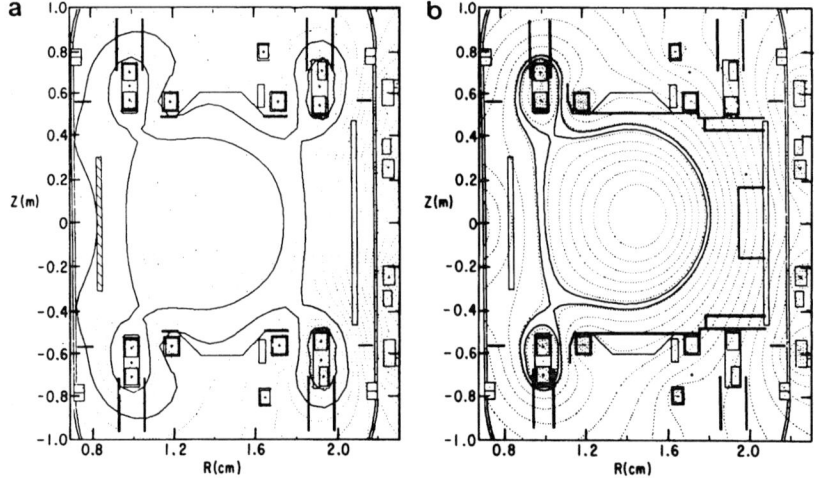

Fig. 13. Flux plots for the unmodified (a) and modified (b) PDX
divertor geometry [Kaye, S., et al., 1983].

plate becomes lower and the density rises. As the density rises, the neutral pressure also rises.

Based on the modeling results, and perhaps, more importantly on the successful operation of a "high recycling" divertor on the ASDEX experiment [Engelhardt, W., et al., 1982], the PDX vacuum vessel was modified to take advantage of increasing the local recycling. The original PDX design had four divertors, two outer and two inner (Fig. 13) [Kaye, S., et al., 1983; Fonck, R., et al., 1984]. Operation with all four divertors was not advantageous. The experiment was generally operated with only the two inner divertors. In this configuration, even though the diverted plasma plugged the inner divertor openings, neutrals formed at the neutralizer plate could easily return to the main plasma chamber through the large openings for the outer divertor. The neutral pressure in the divertor chamber and the plasma chamber was the same. The flow velocity of ions in the scrape-off plasma into the divertor chamber was observed to be sonic. When the openings for the outer divertor were closed, the divertor characteristics changed markedly. The neutral pressure in the divertor chamber was a factor of ten or twenty higher than the pressure in the main chamber. The flow velocity of the scrape-off ions into the divertor chamber was much less than sonic. There were a number of other changes consistent with the behavior predicted by the modeling for a "high recycling" divertor. However, many of the parameters were not changed as much as the modeling suggested they should. This was probably due to the many openings still remaining between the divertor chamber and the main plasma chamber for diagnostics, etc. The ASDEX divertor experiment was able to reach lower temperatures, higher densities and high pressures in the divertor chamber than PDX. This was attributed to the better "sealing" of the divertor chamber in ASDEX.

Thus divertor operation on PDX and ASDEX demonstrates that the "high recycling" regime exists, and that it can produce a cool, dense plasma by proper control of the recycling. However, PDX and ASDEX are able to retard the neutral backflow by mechanical baffles. A practical divertor design will have an "open" geometry of the "expanded boundary" type (Fig. 4) [Ohyabu, N., 1981], since the poloidal coils will have to be outside the toroidal coils. The only feasible way for the recycling to be localized is for the divertor plasma to be several neutral ionization mean free paths wide, so that most of the neutrals formed are ionized again before they can escape from the diverted plasma. This was modeled using a realistic geometry for the INTOR parameters [Petravic, M., et al., 1984b] (Fig. 14). This modeling indicated that a divertor of this size should be able to operate in the "high recycling" regime with the INTOR heat loads and particle confinement times. The

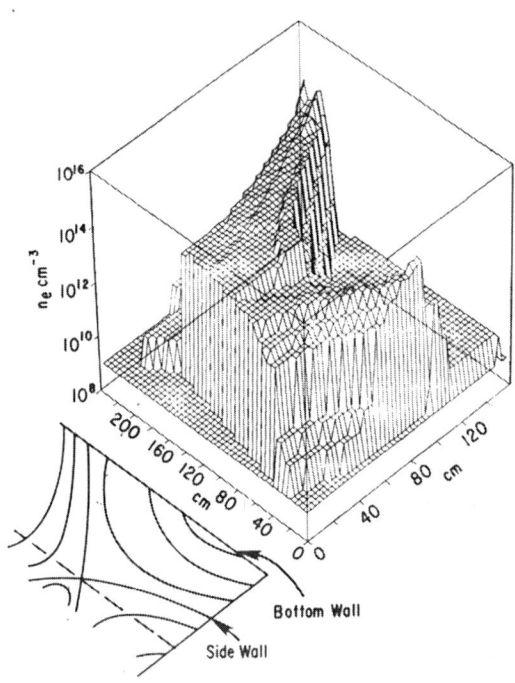

Fig. 14. Two-dimensional calculations of the electron density
for the INTOR divertor using a realistic geometry
[Petravic, M., et al., 1984b].

calculated temperatures at the divertor plate were ~30 eV and
lower, and the densities exceeded $10^{14}/cm^3$.

The question of the viability of the "high recycling"
divertor in an "open" divertor geometry was addressed
experimentally on D-III [Sengoku, S., et al., 1984]. Probe
measurements of the diverted plasma on D-III indicated that the
density at the divertor plate increased (up to $3 \times 10^{14}/cm^3$) and
the temperature decreased (down to 3-4 eV) as the recycling
increased (Fig. 15). The general dependence is the same as the
modeling calculations (Fig. 12). Another key feature of the high
recycling regime is the steep gradients of the density and
temperature along the field lines. The steepness of the
gradient is a measure of the localization of the recycling.
Probe measurements of the D-III divertor indicated that the
temperature dropped from 35 eV to 8 eV in 13 centimeters along a
field line projected onto the poloidal cross section (Fig.
16). These experiments on ASDEX, PDX, and D-III have indicated
that the "high recycling" regime exists, that this regime of
divertor operation produces a cool (~ 5 to 10 eV), dense
(> $10^{14}/cm^3$) plasma near the divertor plate, and that the regime
is possible in the open geometry needed for a divertor for an
ignited tokamak.

1120

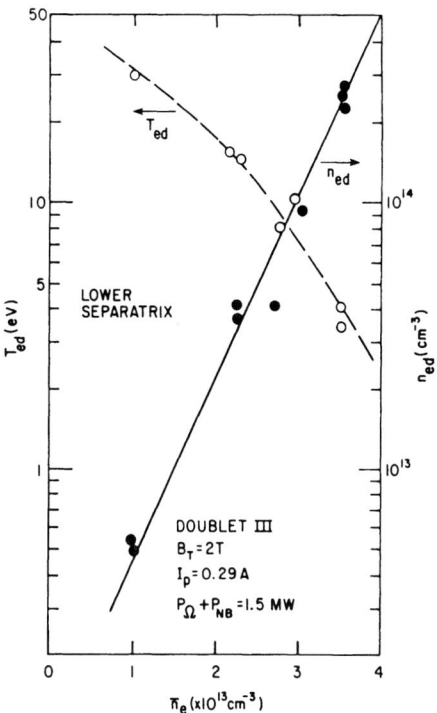

Fig. 15. Electron density n(ed) and temperature T(ed) at the
peaks of the density profile on the divertor plate as a
function of the average electron density of the main
plasma line averaged density at t=500 ms. The
recycling increases as the line average density
increases [Sengoku, S., et al., 1984].

The experiments on D-III, PDX, and ASDEX have been
successfully modeled using the PLANET code and other similar
codes [Petravic, M., et al., 1984a; Schneider, W., et al., 1983;
Shimada, M., et al., 1983; Igitkhanov, Yu. L., et al., 1983].
These same codes predict that a large, high power tokamak should
be able to operate in the "high recycling" regime with a cool,
dense plasma (T ~ 5-20 eV, n_e ~ 1-3 × 10^{14}/cm^3) near the
divertor plate. The low temperature implies that the sheath
potential will be low enough to allow the use of collector plate
materials with sputtering thresholds above the kinetic energy of
the ions. Thus sputtering can be eliminated as a concern for
the impurity control system by the use of properly designed
poloidal divertor.

The high plasma density implies that the neutral density
and consequently the gas pressure will be large also. Thus
helium pumping will be possible with a relatively modest pumping

Table 2. Limiter and Divertor Operating Conditions
[Post, D., et al., 1985]

	DIVERTOR	LIMITER
Total power to collector plates	70 MW	84 MW
Particle	53 MW	80 MW
Radiation	17 MW	4 MW
Presheath ion energy	25 eV	150 eV
Sheath potential	60-80 eV	300-800 eV
Peak power	4.7-7 MW/m^2	2.4 MW/m^2

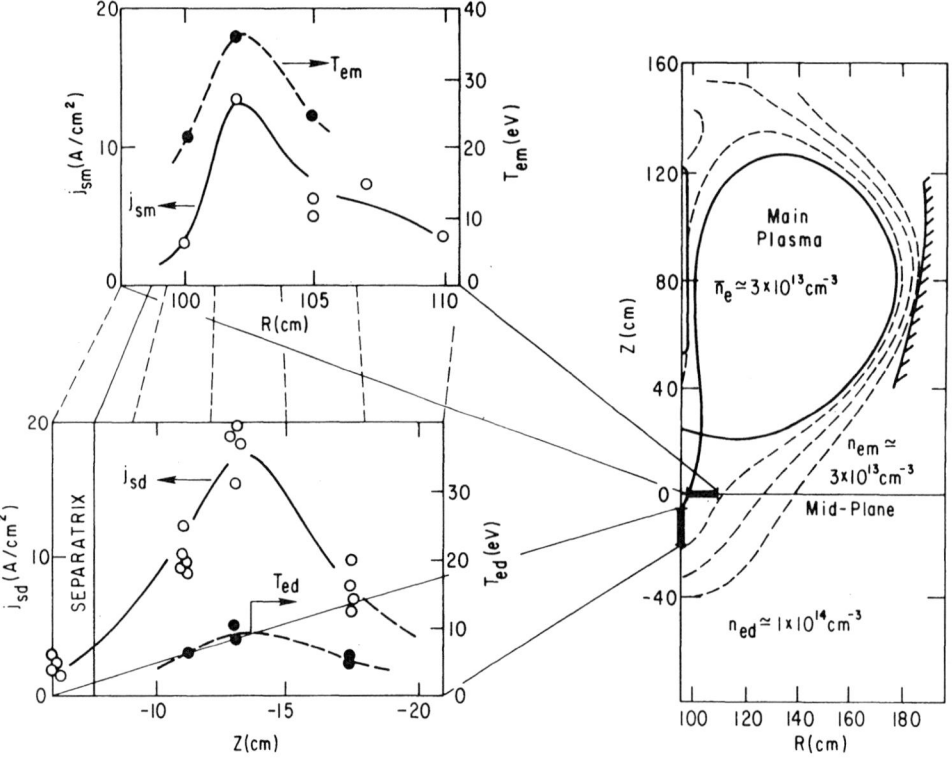

Fig. 16. The horizontal profile of the electron temperature,
T_{em}, and ion saturation current, j_{sm}, across the lower
divertor channel (midplane: Z=0 cm) and corresponding
verticle profiles of T_{ed} and n_{ed} on the divertor plate
at t=800 ms. The connection of the field lines between
both profiles is shown with dotted lines [Sengoku, S.,
et al., 1984].

1122

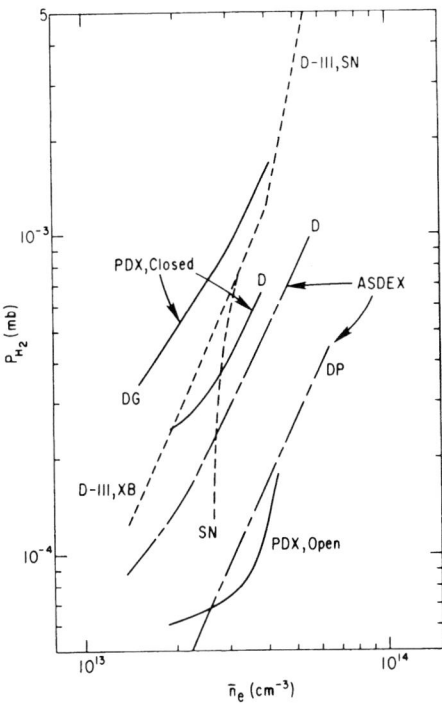

Fig. 17. Comparison of neutral gas pressure in a divertor as a
function of the line-average bulk plasma density for
ohmically heated hydrogen discharges in different
experiments and for different modes of operation (DP,
pumped divertor; D, unpumped divertor; DG, gas feed in
divertor) [Lackner, K., and Keilhacker, M., 1984].

system. The expected pressure is of the order of 0.001 torr, so
the INTOR helium exhaust requirement could be met with only
about 70,000 liters/sec of pumping. Pressure measurements on D-
III, PDX, and ASDEX indicate that divertor chamber pressures in
excess of 0.001 torr can be achieved with high recycling
divertors (Fig. 17). The expected operating parameters for the
INTOR pumped limiter and divertor designs are given in Table 2.

Impurity Behavior

The behavior of impurities is very complicated and evidence
of the effectiveness of impurity control by divertors on tokamak
experiments with high power auxiliary heating is crucial.
Current tokamak experiments rely either on limiters or
divertors. The divertor experiments can also use limiters.
Both approaches have been successful in producing relatively
clean plasmas in the sense that central radiation losses do not
dominate the power balance. However, these experiments almost
all have short pulse lengths. Comparisons of impurity control
with limiters and divertors have been carried out and described

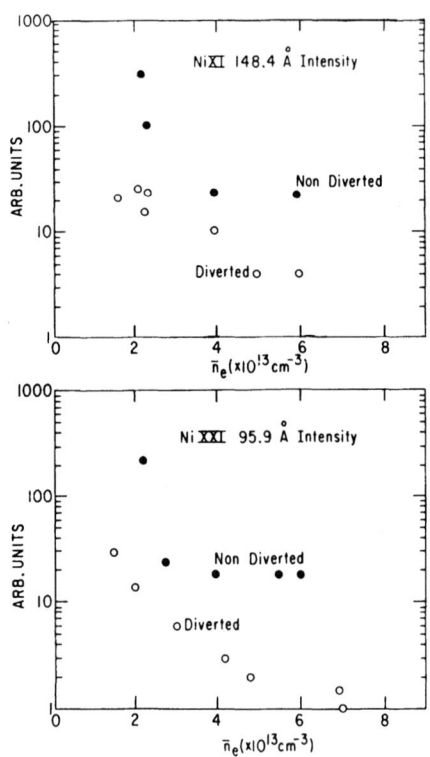

Fig. 18. The divertor reduces the influx and accumulation of
metallic impurities by a factor of 5 to 10 [Nagami, M.,
et al., 1981].

for D-III [Nagami, M., et al., 1981], PDX [Fonck, R., et al.,
1984], and ASDEX [Wagner, F., and Lackner, K., 1985]. The
results indicate that divertors are superior to limiters for
controlling metal impurities, especially for long pulse
operation with high power heating. On D-III, the nickel level
was lower with a divertor than with a limiter (Fig. 18) [Nagami,
M., et al., 1981]. On PDX, the divertor was better than a rail
limiter, particularly for higher heating powers [Fonck, R., et
al., 1984] (Fig. 19). Nevertheless, limiters on current
experiments do achieve adequate impurity control. Upcoming high
power heating experiments on JET, TFTR, and JT-60 will test
limiters at powers and pulse lengths more prototypical of
ignition experiments than are current experiments.

Finally, as an added bonus, it was observed, first on ASDEX
[Wagner, F., et al., 1982], then on PDX [Kaye, S., et al., 1984]
and D-III [Nagami, M., et al., 1984; Burrell, K., et al., 1983],
that the use of "high recycling" poloidal divertors improved the
energy confinement time. With a limiter, the confinement time
usually drops to a value which is one-half and one-third of the
value with ohmic heating when high power auxiliary heating is
used. With a "high recycling" divertor, the confinement time
can be restored to a value close to the ohmic value. This "H-

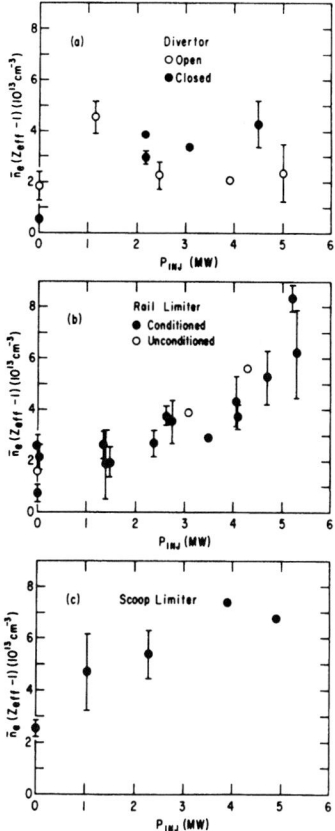

Fig. 19. Dilution-corrected impurity density estimates as a
function of injected power [Fonck, R., et al., 1984].

mode" is accompanied by a change in the edge parameters such as
a steepening of the radial profiles of the edge temperature and
density. The exact cause of this improvement is not well
understood (indeed, the original confinement is not understood
either) but localization of recycling away from the main plasma
seems to be a key ingredient.

There are several other physics issues associated with
divertor magnetic [see chapter by Wagner, F., and Lackner,
K.]. Diverted plasmas are subject to a variety of
instabilities. In particular, axisymmetric stability will
probably require some sort of feedback control. The formation
of a divertor by a coil outside the toroidal field coils
requires a substantial poloidal field coil system. However,
producing a substantial degree of elongation requires almost as
large a coil system and about the same type of coil system, so
the requirement for additional poloidal field coils is less than
might be supposed at first.

MATERIAL ISSUES

The components of any impurity control system will be

exposed to high particle and heat fluxes that can result in high sputtering erosion rates and to high fluxes of 14 MeV neutrons which will degrade the bulk material properties. The materials used for an impurity control system must therefore be resistant to erosion losses and radiation damage, capable of operating at elevated temperatures, and at the same time not be a source of contamination to the plasma.

Among the properties important for material selection [see chapter by Smith and Whitley] are the thermophysical properties, mechanical strength and ductility, fatigue and crack growth behavior, coolant and hydrogen compatibility, radiation swelling and creep, and sputtering erosion behavior [e.g., Mattas, R., et al., 1984]. The desired thermophysical properties are those that minimize the thermal stresses. The mechanical strength and ductility should be adequate to accommodate the weight loads, coolant pressures, thermal stresses, and electromagnetic forces. For high cycle machines such as INTOR, FED, FER, NET, and OTR, the materials should exhibit favorable fatigue, crack growth, and stress corrosion behavior. The materials should also exhibit low radiation swelling and creep rates. The surface sputtering rate should be low enough to provide extended lifetimes and to keep impurities at an acceptable level in the plasma.

A survey of available materials indicates that no one material satisfies both the surface sputtering and structural requirements. Hence, the design of impurity control components incorporates separate plasma side materials which are attached to a structural material selected to meet the strength and radiation damage requirements. The division of plasma side and structural materials allows greater flexibility in the selection of materials but also creates additional difficulties associated with attachment.

The candidate materials considered for impurity control are listed in Table 3. These materials were selected from a larger pool of possible materials based upon the property requirements listed above. The plasma side materials are divided into low-Z, medium-Z, and high-Z materials. At low plasma edge temperatures, (< 50 eV) all materials may be used but high-Z materials are expected to exhibit very low sputtering erosion, and therefore they are predicted to have the greatest lifetimes. At higher edge temperatures, both medium-Z and high-Z materials are unacceptable due to excessive self-sputtering. The permissible plasma side materials are those whose self-sputtering coefficients never exceed unity, which limits the selection to materials whose atomic weights are at or below the atomic weight of SiC. The candidate heat sink materials are copper alloys and transition metal alloys. Several important

Table 3. Candidate Impurity Control Materials
[Post, D., et al., 1985]

PLASMA SIDE MATERIALS		HEAT SINK MATERIALS
Low-Z:	C, Be, B, TiC, SiC, B4C, BeO	Copper Alloys
Medium-Z:	Stainless Steel, Vanadium	Vanadium Alloys
High-Z:	W. Ta, Nb	Niobium Alloys

properties of the candidate alloys are reviewed below.

Besides the thermophysical properties, the effects of interest for plasma side materials are sputtering and their response to neutron irradiation. Excellent compilations of the available physical sputtering data are given in Anderson, H., and Bay, H., (1980) and Roth, J., et al. (1979). For Be, the measured yields compare very well with yields for BeO [Roth, J., et al. (1979)] and are probably more representative of the oxide surface. Physical sputtering of graphite is similar to that of Be. Self-sputtering yields are generally much larger than light ion yields and can lead to catastrophic increases in impurity introduction if they exceed unity. Self-sputtering yields of less than unity are expected for the lighter targets, Be, B, C, and probably SiC at all energies. Self-sputtering yields exceed unity in V, stainless steel, Mo and W for energies above 0.6-1 keV. Normal incidence sputtering yields for tungsten which is representative of high-Z material, are shown in Fig.20 [Abdou, M., et al., (1982)]. The DT sputtering threshold is at 200-300 eV, so that sputtering at low plasma edge temperatures ($\lesssim$ 50 eV) should be negligible. Qualitative estimates of the accuracy of the data range from $\pm$ 30% for light ions in stainless steel to factors of 2-4 in some self-ion cases. The data are more reliable near the peaks in the sputtering curves where the yeilds are larger and less energy dependent. Surface conditions probably represent the largest uncertainty in predicting sputtering yields for fusion devices.

Chemical sputtering of graphite could accelerate erosion in fusion devices. In recent work Roth, J., et al., (1982) studied graphite erosion by H^+, D^+, and He^+ above 1000 K. Figure 21 shows the sputtering yield of carbon as a function of temperature for 1 keV H^+ and D^+ incident energy ions and for 3 keV He^+ ions. The sputtering yield of carbon bombarded by H^+ and D^+ ions increases with increasing temperature reaching a maximum at about 525°C corresponding to the maximum in methane production. As the temperature is raised beyond 525°C, the erosion rate decreases, but above $\sim$ 1000°C, erosion again rapidly increases with no apparent peak. On the other hand, helium bombardment does not produce an erosion peak at 525°C,

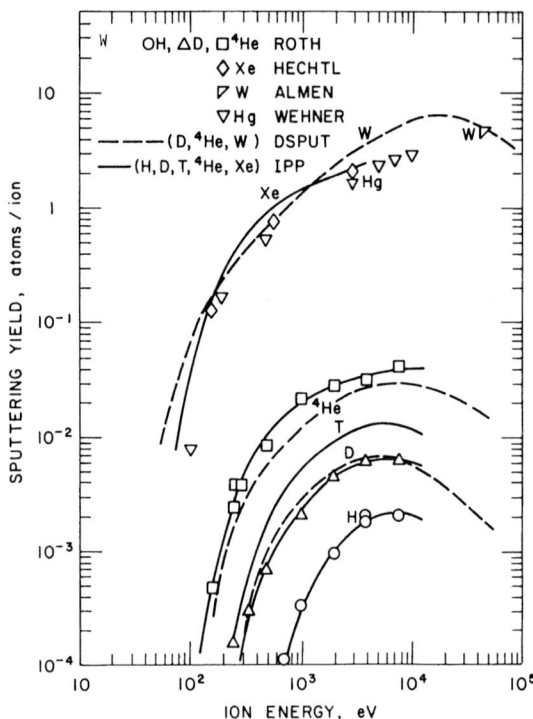

Fig. 20. Normal incidence sputtering yields for W targets.

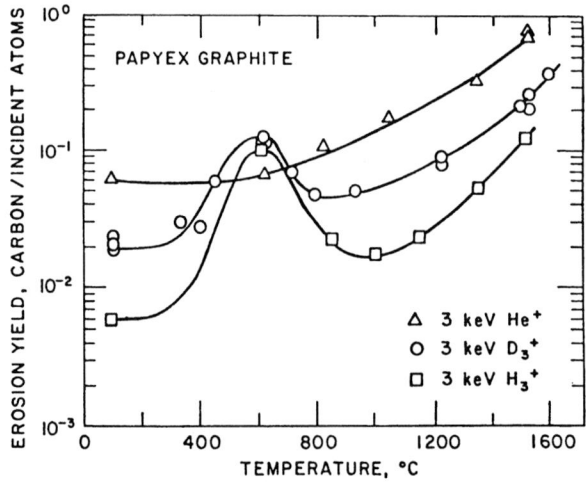

Fig. 21. Temperature dependence of the chemical sputtering yield
of papyex graphite.

but does cause a rapid rise in sputtering above 700°C. High
erosion rates above 1000°C indicate a mechanism other than
physical or chemical sputtering, since no hydrocarbon formation
was detected at the high temperatures.

The data base for neutron irradiation effects in plasma
side materials is generally sparse. Older reviews of
irradiation effects in beryllium [Bush, S., (1965) and
Kangilaski, M., (1971)] based on relatively low fluence
experimental results, conclude that the metal is intrinsically
resistant to purely displacement damage events, and that
observed effects of irradiation at temperatures above the
cryogenic range are due primarily to transmutation helium,
rather than to point defects or defect clusters. Measurements
of resistivity [Blewitt, T., (1958)] and thermal conductivity
[Williams, J., et al., (1972)] in material irradiated at 20 or
77 K showed that recovery was complete, with no residual damage
on annealing to 270 K. More direct evidence was developed by
Carpenter, G., and Fleck, R., (1977) using high-energy electron
bombardment and direct TEM observation of the damage
microstructure. They found that, although visible damage
resulted in bombardments near room temperature, no visible
damage could be developed for bombardments producing 15 dpa at
300°C.

Few data are available on the swelling produced by high
fluence, elevated-temperature irradiation. Figure 22 shows the
swelling that can result when helium-containing material is
annealed at temperatures well above the irradiation
temperatures. These results suggest that high swelling values
can be expected if temperatures much above 700°C are allowed.

The degradation in Be mechanical properties that results
from the elevated-temperature irradiation of beryllium is
produced by the combination of matrix hardening that results
from bubble pinning of dislocations and the grain boundary
weakening that results from helium accumulation at the boundary.
Representative mechanical properties of irradiated material,
taken from the review of Bush S., (1965) are shown in Fig. 23.
The considerable data collected in Fig. 23. The considerable
data collected in Fig. 23 show a decreasing elongation with
increasing fluence as a general theme of the many collected
conditions.

A typical set of swelling design curves for an isotropic
nuclear graphite is given in Fig. 24. The classic definition of
lifetime is that point in time (fluence) when the graphite
distortion returns to its original volume. In actual fact, the
mechanical strength persists for some time after this point.
The generally expected changes in thermophysical and mechanical

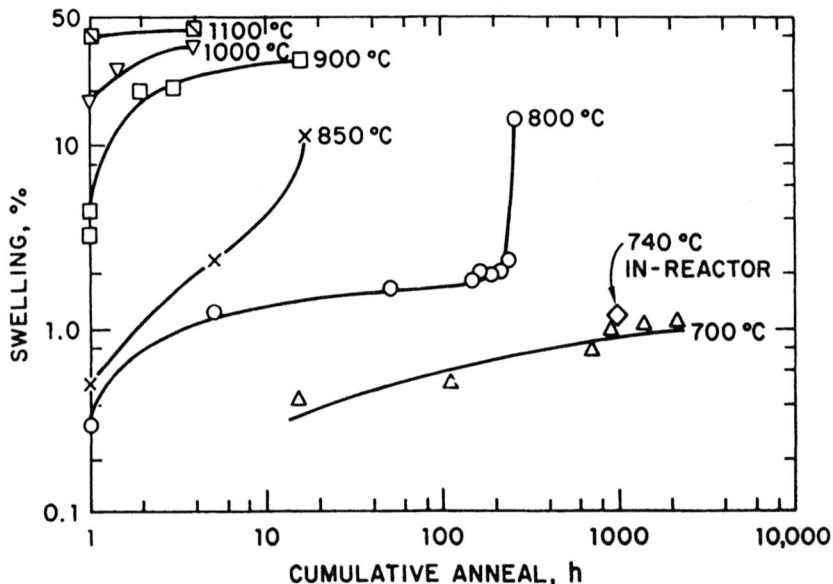

Fig. 22. The effect of time and temperature on the swelling of irradiated graphite.

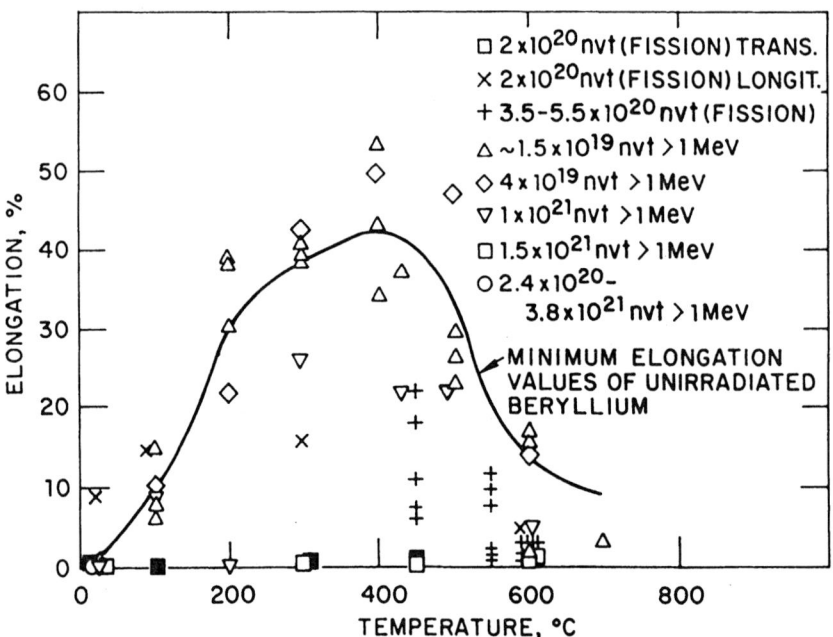

Fig. 23. The influence of irradiation and tensile testing temperature on the elongation of beryllium.

properties that will result from irradiation at ~ 500°C are summarized in Table 4 [Mattas, R., et al., (1984)].

Interest in tungsten for nuclear reactor applications has been very limited; hence, the data base is very sparse. The highest fluence data are those reported by Steichen, J. (1976) for EBR-II irradiations at ~ 385°C (658 K). Tensile specimens of sheet material which had been stress-relieved at 1273 K for half hour to a reported grain size of 5 to 7 μm were irradiated to fluence levels of 0.4 and 0.9 × 10^{26} n/m^2 (E > 0.1 MeV). Post-irradiation tensile tests, conducted in the range 295 to 1200 K, indicated the tensile yield strength was approximately doubled by irradiation over the full temperature range. The ductile-brittle transition temperature, as measured or observed by values of the tensile ductility (i.e., percent elongation or reduction in area), was raised from an initial value of ~ 65°C to ~ 230°C by the irradiation.

Two classes of alloys, copper alloys and refractory metal alloys, have been considered as heat sink materials. Copper alloys were selected because of their high thermal conductivity and their availability. Refractory metal alloys, as typified by V-15Cr-5Ti, have higher temperature capability, acceptable thermal conductivity, and good resistance to radiation damage. They are not, however, readily available. The major properties of interest are the mechanical properties and radiation effects.

The tensile strength of V-15Cr-5Ti has been determined [Gold, R., and Ammon, R., (1981); Mattas, R., et al., (1977); Santhanam, A., et al., (1973)]. Both the ultimate strength and the 0.2% yield strength are approximately independent of temperature in the range from 450-750°C. Above this point, the strength begins to drop off, but reasonable strength remains up to 900°C. The uniform elongation of this alloy is approximately 23% at room temperature, and it is in the range of 11-16% from 450 to 800°C.

In comparison to the vanadium alloy, the yield strength of copper alloys degrades at lower temperatures. The strength begins to decrease at ~ 225°C (498 K) for OFHC copper and at ~ 400°C (673 K) for Cu-Be alloys. The effect of alloying is to increase the strength, but generally at the expense of the thermal conductivity.

The room temperature fatigue behavior of V-15Cr-5Ti and pure copper are compared in Fig. 25 [Gold., R., (1981) and Murphy, M., (1981)]. The V-15Cr-5Ti exhibits superior fatigue behavior with an endurance limit of ~ 0.8%, compared with ~ 0.15% for pure copper. Cu-Be alloys exhibit improved fatigue

Table 4. Extrapolated property values for irradiated GraphNOL N3M.

Property	Generalized Behavior	Numerical Values
Thermal conductivity	Monotonic exponential decrease, saturating at about half-life.	Saturates at about 30 W/(m.K).
Thermal expansion	Rapid rise and fall within first third of life.	Maximum at perhaps 50% over initial value. Saturates at about 75% of initial value.
Tensile strength	Linear falloff with fluence.	About 50% of initial value at end-of-life.
Strain-to-failure	Gradual falloff saturating at about half-life.	Saturates at about 40% of initial value.
Moduli (Young's and shear)	Gradual increase, saturating at about half-life.	Saturates at 2.5.3 times initial value.
Radiation creep	No information; appears to be relatively insensitive to graphite grade.	Assume behaves as other nuclear graphites.

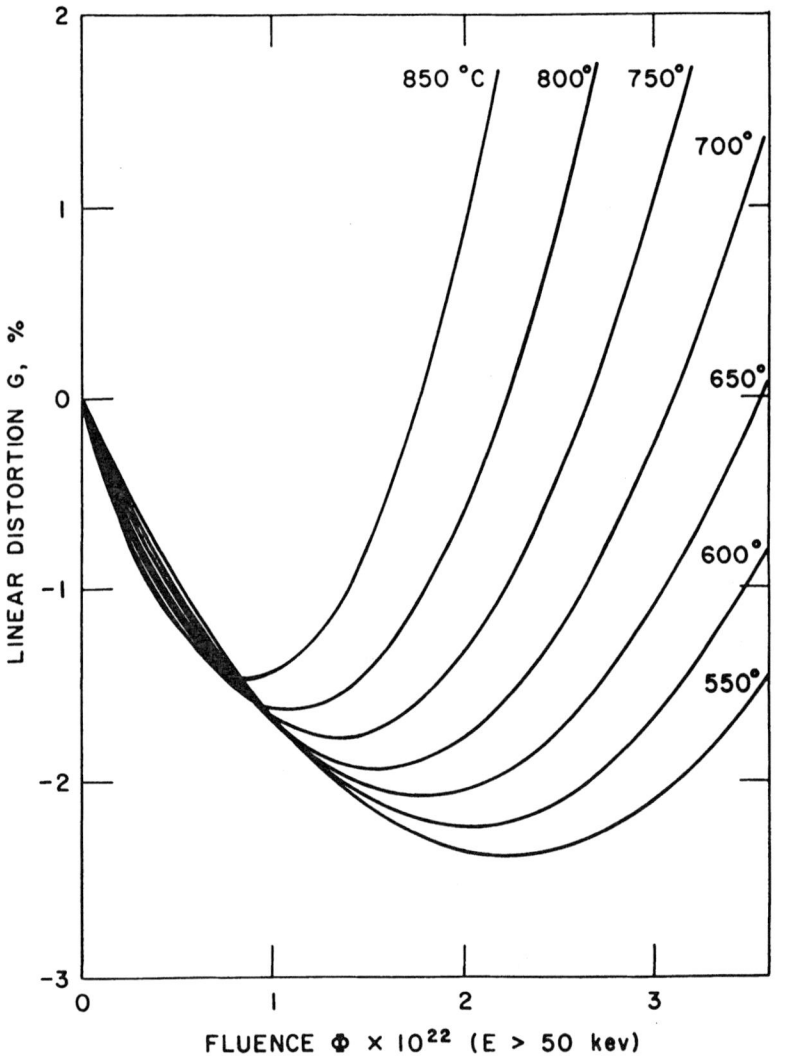

Fig. 24. Typical graphite swelling design curves for a nuclear reactor. Graphite is assumed isotropic.

1132

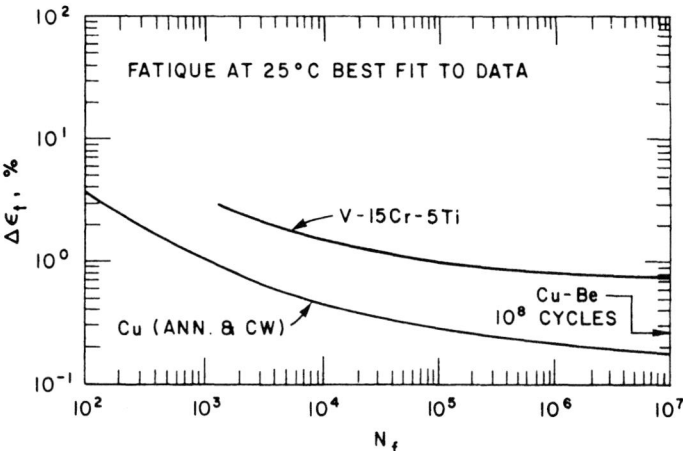

Fig. 25. Fatigue curves for copper and a vanadium alloy at room
        temperature.

behavior over pure copper in the high cycle range, but they are
still considerably inferior to V-15Cr-5Ti.

The temperature dependence for void swelling of the V-15cr-
5Ti alloy, commerical grade vanadium, and 6-pass zone refined
vanadium has been determined for neutron irradiation to 3.6 ×
10^{26} neutrons/m^2 (E > 0.1 MeV). In contrast to the zone refined
and commercial vanadium, the V-15Cr-5Ti alloy shows essentially
no void swelling following irradiation at 450-600°C [Bartlett,
A., et al., (1976); Carlander, R., et al., (1973), and Van
Witzenberg, W., et al., (1981)]. The unalloyed vanadium shows a
peak swelling of ~ 1% at ~ 525°C. In the case of the ternary
alloys, both V-15Cr-5Ti and V-15Ti-7.5 Cr alloys were irradiated
at 425°C to a damage of ~ 14 dpa without the formation of
voids. Irradiation at 600°C resulted in some void formation in
both alloys, but the total swelling was negligible. In another
experiment, V-15Cr-5Ti was irradiated to ~ 20 dpa at 450, 550,
and 600°C [Bentley, J., et al., (1976)]. Except for a few
isolated voids at 550°C, there was no tendency for void
formation or swelling. In a recent experiment, helium was
injected into V-20Ti samples in concentrations if 90 and 900
appm prior to irradiation to 17 dpa at temperatures of 400, 575,
625 and 700°C [Tanaka, M., et al., (1981)]. Numerous cavities
were observed in the samples following irradiation, but the
total swelling in all cases was negligible.

The void swelling of copper is minimal for irradiation
temperatures less than ~ 200°C or greater than ~ 450°C [Adda,
Y., (1972)]. The swelling of copper after neutron irradiation
to a fluence of 5 × 10^{24} neutrons/m^2 (E > 0.1 MeV) has a peak
value of ~ 0.5% at ~ 330°C [Adamson, R., et al., (1980)]. The

effect of substitutional alloying elements on the void swelling
of copper during neutron irradiation have been studied
[Brimhall, J., and Kissinger, H., (1972); Labbe, M., and
Poirier, J., (1973)]. Alloying copper with 1-3 a/o aluminum,
germanium, silicon or nickel can result in a significant
reduction of the void swelling of copper. The swelling of
copper is increased by 1.0 a/o additions of silver or cadmium on
1 MeV electron irradiation at 250°C. However, the alloying of
copper with 1.2 a/o beryllium results in no void formation at
250°C on electron irradiation to 100 dpa. The irradiation of
copper containing 1.35 a/o beryllium at 327-397°C with 300 keV
Cu^+ ions to 8 dpa results in the formation of very fine
precipitates and no voids.

The effects of radiation on tensile properties are less
well studied. The effect of radiation on the tensile properties
of vanadium can be divided into three regimes: low temperatures
(T < 100°C), intermediate temperatures (300° < T < 650°C), and
high temperatures (T > 650°C). Vanadium alloys which are
irradiated and tested at low temperatures generally exhibit
rapid embrittlement due to plastic instability [Wiffen, F.,
1981]. The reduction of uniform elongation to nearly zero by
the early onset of plastic instability in irradiated bcc metals
and alloys has been observed in a number of systems [Wiffen, F.,
1978, and 1973]. Plastic instability has been observed at
fluence levels as low as 0.0095 dpa, where the uniform
elongation was close to zero while the total elongation was ~
8%.

In intermediate temperatures, the effects of radiation on
ductility are minimal, with uniform elongations remaining above
several percent at damage levels up to ~ 18 dpa [Carlander, R.,
et al., 1973]. At high temperatures, the ductility of vanadium
will decrease due to the onset of helium embrittlement. In the
case of V-15Cr-5Ti, helium embrittlement has been studied in
samples where helium was introduced by ion injection and by
tritium decay [Mattas, et al., 1977]. Injected samples which
contained 25 appm He exhibited embrittlement at temperatures >
750°C, whereas the tritium decay samples containing 35 appm He
exhibited embrittlement at temperatures > 700°C.

The data for copper are very sparse. There are indications
that the strength will increase and ductility will decrease as
in other metal systems, but the rates of change are not well
known. Additional work is needed.

For INTOR design study, a specific choice of materials was
made. The material selection focused on the use of low-Z
materials Be and C for plasma edge temperatures greater than 100
eV and the use of the high-Z materials W and Ta for plasma edge

temperatures less than 50 eV. Be is favored over C because C is known to exhibit enhanced chemical sputtering and because C has rather limited irradiation lifetimes. W is favored over Ta because Ta is susceptible to hydrogen embrittlement and thus may not be compatible with the DT environment. Copper alloys have received the most attention as heat sink materials since they are readily available, are easily fabricated, and are capable of operating at the anticipated operating temperature (100 < T < 300 degrees Centigrade).

ENGINEERING DESIGN

Pumped Limiter Design

The INTOR limiter occupies the same location as the divertor, and it takes up somewhat less space than the divertor. The particle and heat flux requirements for the limiter (Table 2) are similar to those for the divertor. The limiter must have adequate heat removal capacity and should have lifetime exceeding ~ 1 y, just like the divertor.

There are some important engineering differences between the limiter and divertor, however. The plasma temperature at the collector plate is 150 eV for the limiter compared with 25 eV for the divertor (Table 2). At 150 eV, low-Z materials must be used to avoid runaway self-sputtering and Be is the favored material. The sputtering erosion, particularly at the leading edge, is predicted to be high. The amount of sputtering and redeposition has been calculated, and the results are shown in Fig. 26 [Abdou, M., et al., 1982]. The gross sputtering rates from D, T, He, and self sputtering are greater than 100 cm/y across the entire limiter. However, the great majority of sputtered particles are predicted to be redeposited back onto the surface such that the net erosion rates are much lower. On the front surface of the limiter, the net erosion rate is calculated to be only 1.3 cm/y, and on the leading edge, the net erosion rate is calculated to be 29 cm/y. There are considerable uncertainties of impurity transport in the scrape-off layer which translate into uncertainties in the redeposition rates. In addition, the properties of redeposited material may be substantially different from the base material. Because of these uncertainties and the lack of experimental data, the INTOR group felt that it is premature to base a design on redeposition. The erosion lifetime will be limited by the maximum allowable thickness of the plasma surface material. The thickness is usually limited by the thermal stresses and fatigue strain that can be tolerated in the structure. For INTOR, the maximum allowable thickness is ~ 2 cm. Its erosion lifetime is approximately 2 y on the top surface but is only ~ 0.15 y at the leading edge. A possible solution to the short lifetime is to

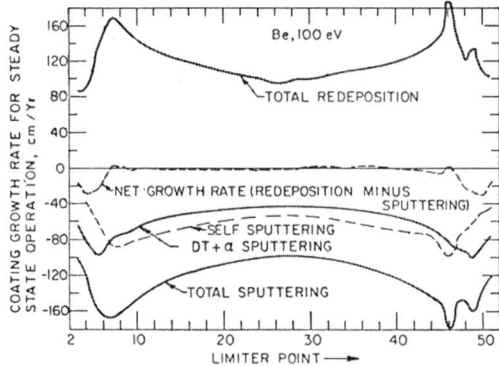

Fig. 26. Erosion (physical sputtering only) and redeposition
rates for berylliumas a function of spatial points at
the limiter surface for a plasma-edge temperature of
100 eV.

LIMITER LEADING EDGE

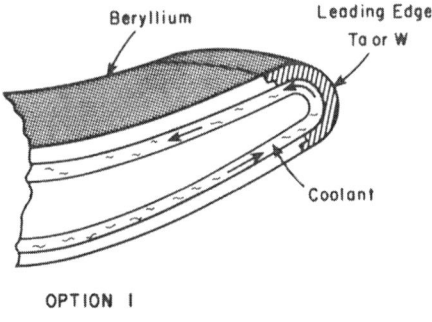

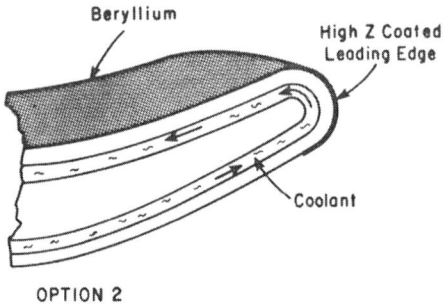

Fig. 27. Schematic of limiter design illustrating two options
for incorporating high-Z materials at the leading edge.

1136

replace the beryllium with tungsten at the leading edge (Fig. 27). The plasma temperature is < 50 eV at this position which is acceptable for the use of sputtering high-Z materials. The use of another material creates additional interface problems, however.

The plasma edge is predicted to have short power and e-folding distances, which would result in high peak heat loads on a flat limiter. In order to reduce peaking, the limiter surface is shaped to spread the power uniformly over the surfaces. Unfortunately, a shaped limiter would be susceptible to nonuniform heating if the plasma shifts position. the leading edges may have lower heat loading limits than the top surface, and therefore the edges have been placed at positions where the peak lead is 1 MW/m^2. At this power level, a double edged limiter is needed to maximize the pumping capability of the system.

Given the likelihood that the erosion rate of the limiter would be large and that the erosion rate of the first wall away from the limiter would be smaller, the limiter and adjacent first wall were made removable in the INTOR design. Thus they could be replaced more often than the rest of the first wall (Fig. 28).

Aside from the sputtering erosion, the lifetime concerns are similar to those of the divertor. Since the data base is sparse it is not possible to adequately characterize the long term response of the limiter.

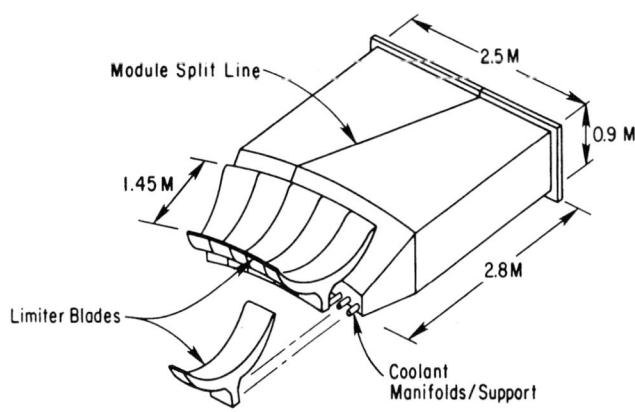

Fig. 28. Shaped, double-edged, bottom limiter.

Poloidal Divertor

As an example of divertor system, we will describe the divertor design for INTOR. The collector plates receive most of the particle flux and power which enters the d vertor, and hence experience the most severe environment of any of the plasma side components. The goals of the design studies such as INTOR have been to develop collector plate designs that can safely and reliably remove the power deposited on the surface, that have extended lifetimes (> 1 year), and that can also satisfy the physics requirements.

The overall configuration of the reference divertor is shown in Fig. 4. The divertor is located at the bottom of the plasma chamber with a continuous toroidal extending around the reactor. The divertor is divided into removable modules with two modules for each TF coil. The modules are required since it is anticipated that replacement of the collector plates will be more frequent than for the first wall. Access limitations with 12 TF coils require the use of two modules per sector. The modules do not incorporate breeding at this time, but it can be included if required. The height of the divertor module is set by the length requirement between the separatrix and the point where the plasma strikes the collector plate. This distance is ~ 80 centimeters for the reference design.

The engineering effort has emphasized design tradeoffs for the divertor collector plates. The basic configuration is an actively cooled flat plate consisting of a high-Z plasma side material bonded to a structural heat sink. The overall thickness of the plates is ~ 3 cm, and the width of the plate is ~ 1 m. Two surface materials, 20% cold worked Typed 316 stainless steel and Cu-0.5Be-2Ni, have been considered. The different design variations are shown in Fig. 29 [Mattas, R., et al, 1985]. The plates are supported by inlet and outlet manifolds at either end of the plate as shown in Fig. 30. The reference design uses W as the plasma side material and Cu-0.5Be-2Ni as the structural material.

The consequence of the low plasma temperature at the divertor plate is that sputtering erosion is almost completely eliminated for high-Z materials. The predicted gross and net sputtering rates for several materials on the collector plates are shown in Table 5 [Brooks, J., et al., 1984]. At 20 eV, the gross erosion rates range from 680 cm/y for Be to 0.07 cm/y for tungsten. When redeposition is included these rates drop to 14 cm/y and ~ 0 for Be and W, respectivley. The gross sputtering rate of tungsten is low enough to result in multiyear sputtering lifetimes, and the net erosion rates, including redeposition, for Ti and Mo, are also predicted to be low. The use of

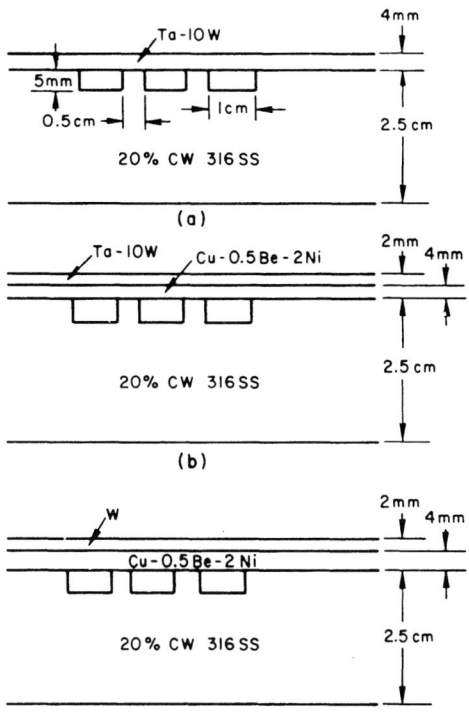

Fig. 29. Divertor collector plate designs [Mattas, R., et al., 1985].

tungsten at the plasma side material provides an additional benefit since no vaporization or melting is predicted to occur for the reference disruption condition. The elimination of erosion on the collector plates means that plasma side material can be a thin layer (~ 1-2 mm). Finite element temperature and stress contributions have been determined for the case of tungsten tiles bonded to a Cu-0.5 Be-2Ni heat sink [Mattas, R., et al., 1985]. Because of the large mismatch in the thermal expansion coefficients of copper and tungsten and the high tungsten elastic modulus, stresses at the tile/substrate interface are expected to be large and may affect the reliability of the INTOR divertor. Large stresses can arise from two sources; residual stresses from fabrication and bonding and thermal stresses during operation. Residual stresses near the bond are induced as the structure is cooled from the attachment temperatures, but as the divertor is heated during plasma operation, these stresses will be reduced.

In order to isolate the effects of the stresses during operation, the stress state due to heat loads was calculated for a structure void of residual stresses. The stresses and temperatures presented are for steady state, which is reached in a few seconds. At 3.3 MW/m^2 the surface temperature of the tungsten is ~ 280°C and the peak temperature in the copper is ~

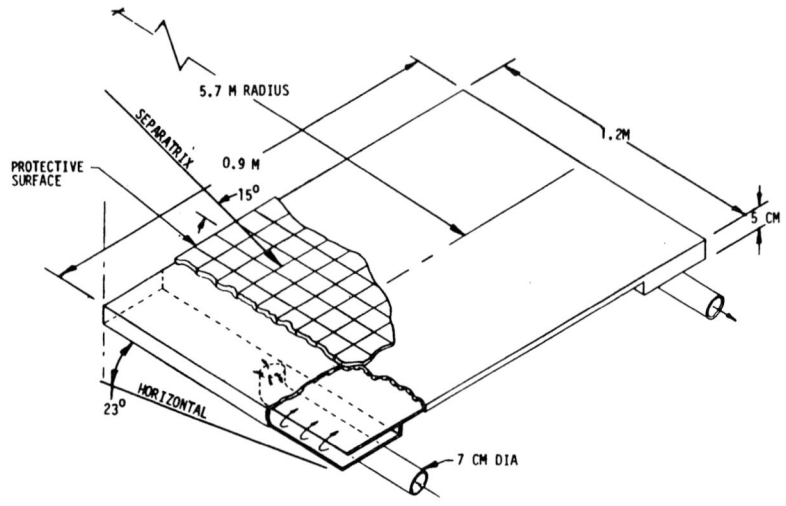

Fig. 30. Outer divertor collector plate configuration [Mattas,
R., et al., 1985].

Table 5. INTOR Divertor Plate Erosion/Redeposition.

Plate Coating Material	Plasma Edge Temperature (At Separatrix) (eV)	Sputtering Coeff. at Separatrix Y_{DT}	Y_α*	Y_{Z_L}	D-T Flux to Divertor Plate Center 10^{23} m^{-2} s^{-1}	Gross Sputtering Rate at Divertor Center Due to DT + He cm/yr	Total Gross Sputtering Rate at Divertor Center cm/yr	Net Erosion Rate at Divertor Center cm/yr
Beryllium	20	.071	.15	.45	1.9	382	680	14.0
	30	.087	.18	.55	1.3	313	688	9.1
	40	.095	.20	.62	.95	256	666	5.4
	50	.098	.21	.66	.76	212	624	4.1
Titanium	20	.011	.038	.49	1.9	134	266	.025
	30	.024	.061	.71	1.3	190	658	.056
	40	.038	.082	.92	.95	225	2750	.42
Molybdenum	20	9.9×10^{-4}	6.3×10^{-3}	.43	1.9	12	21	~0
	30	3.3×10^{-3}	.014	.60	1.3	25	64	~0
	40	5.4×10^{-3}	.023	.79	.95	31	142	~0
Tungsten	20	0	8.4×10^{-5}	.45	1.9	.04	.07	~0
	30	0	.0013	.64	1.3	.40	1.1	~0
	40	0	.0024	.82	.95	.57	3.1	~0
	45	1.5×10^{-3}	.0029	.91	.84	1.3	13.7	~0

* Self-sputtering coefficient for redeposited material.

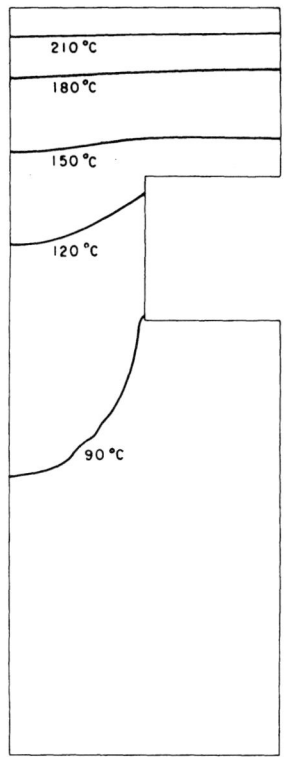

Fig. 31. Temperature contours for W-Cu-0.5 Be-2Ni divertor collector plate for heat load of 3.3 MW/m^2.

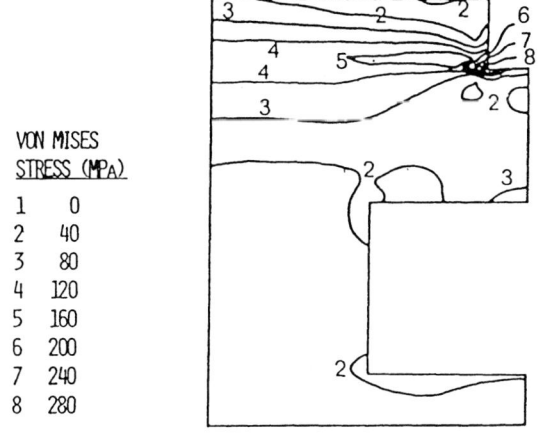

VON MISES
STRESS (MPA)

1	0
2	40
3	80
4	120
5	160
6	200
7	240
8	280

Fig. 32. Von Mises stress contours for W-Cu-0.5Be-2Ni divertor collector plate for heat load of 3.3 MW/m^2.

230°C as shown in Fig. 31. Figure 32 shows the Von Mises stress contours for a total heat flux of 3.3 MW/m^2. The peak stresses are again concentrated at the edge of the tile/substrate interface. The stress in the bulk of the heat sink is below 200 MPa and the peakstress is below 300 MPa, so the response is wholly elastic. The sharp corners of the coolant channels are seen to have some impact on the stress distribution, but the effect is minimal.

The performance of the divertor under heat loads of 7 MW/m^2 has also been examined. The higher flux leads to higher temperatures, causing larger differential expansion of the tile and substrate along with lower yield stresses. This combination leads to slight local plastic deformation at the tiles' edge, which could limit the cycles to failure. The peak Von Mises stresses are 580 MPa and 690 MPa in the copper and tungsten, respectively. The thermomechancial analysis suggests that a heat load of ~ 5 MW/m^2 represents a practical upper limit for impurity control systems. Therefore, the collector plates must be placed at shallow angles with respect to the field lines to reduce the peak heat loads to acceptable levels.

Potential failure modes for the collector plates are erosion, excessive dimensional changes due to radiation swelling or creep, debonding between the plasma side material and heat sink, and severe embrittlement which prevents the system for withstanding off-normal events. The current design eliminates erosion as a life limiting concern, but other concerns such as radiation damage could result in a short lifetime. Unfortunately, the data base for the impurity control materials is sparse, and it is not possible to adequately characterize the long term response of the collector plates. Since the divertor lifetimes could be much shorter than the other nuclear systems, provision is made to replace it independently of the rest of the reactor.

A general comment is that one of the more severe problems with divertors is the problem of high peak heat loads. The best way to reduce peak heat loads is to spread out the heat over a large area. The compactness of the current divertor designs makes that very difficult. Considerable care must be taken to incline the neutralizer plates at as small an angle with respect to the field lines as possible. The peak heat problems associated with limiters are generally less severe since limiters can have larger areas to spread out the heat load.

SUMMARY

Primarily because they offer promising methods for solving the impurity sputtering problem, poloidal divertors offer an attractive solution to the impurity control problem for ignited tokamaks. The INTOR design is based on a poloidal divertor, with a pumped limiter as a back-up. The "high recycling" regime in poloidal divertors has been demonstrated experimentally to produce the low temperature, high density plasma that elimates the sputtering problem and eases the helium removal problem. This type of divertor operation is well understood theoretically, and the theoretical models predict that poloidal divertors in ignited tokamaks should be able to operate in the "high recycling "regime". Table 6 offers a comparison of poloidal divertors and pumped limiters.

In addition to impurity control advantages, divertors offer the promise of improved confinement compared to limiters.

Divertors have evolved considerably since the early 1970's. The old designs tripled the size of the vacuum vessel (Fig. 3). The present designs (Figs. 3 and 4) are very compact. In addition, since most ignited tokamaks require considerable shaping and elongation, much of the poloidal field system required for a divertor is present to produce the elongation [Wagner, F., and Lackner, K., 1985]. Thus the additional cost of divertor compared to a limiter system was judged to be about 10% of the reactor cost.

Divertors are not, however, without problems. They are more complex and costly than limiters. Many of the physics issues are still unresolved. The lifetime issues due to radiation damage are also unresolved, but limiters are no better in this regard. It is difficult to obtain as much collector plate area to dilute the heat load with divertors as with limiters so there may be peak heat load problems. Nonetheless, on the whole, it appears that divertors offer a superior approach to limiters for solving the impurity control problems for ignited tokamaks.

Pumped limiters, however, appear to work well on current experiments, and are being tested on the next generations of tokamaks. There will be a test of an axisymmetric pumped limiter on TEXTOR (ALT-II) in the next few years [R. Conn, 1984]. A high recycling region on the limiter front face and radiative edge cooling may be possible. This would allow the pumped limiter to get around the erosion problem by producing a low plasma temperature at the limiter surface. However, this awaits experimental verification.

Table 6. Comparison of Divertor and pumped Limiter [Post, D., 1985].

ITEM	DIVERTOR	PUMPED LIMITER
1. Plasma parameters in front of collector plate.	high probability of high density ($> 10^{14}$cm^{-3}) and low tmperature ($\leq$ 30 eV)	high probability of medium density (5×10^{12}–5×10^{13}cm^{-3}) and medium temperature (100–200 eV)
2. Impurity control	low sputtering rates and possibility of trapping impurities in divertor chamber	large sputtering rates and easier access to main plasma for impurities
3. Collector plate materials	low or high-Z	low-Z
4. First wall erosion	concentrated near divertor (question about impurity shielding performance of scrape-off plasma)	concentrated near limiter
5. Heat flux limits	2-5 MW/m^2	2-5 MW/m^2 for plate and ~1 MW/m^2 at plate tip
6. Component lifetime	very long (for erosion), redeposition of first wall material potential limitation	short (on the order of 1 year with high degree of redeposition, much less with less lower redeposition)
7. Pumping requirement	(1-10) × 10^4 ℓ/sec	(1-5) × 10^5 ℓ/sec
8. Effects on energy confinement	H-mode	L-mode
9. Torus size	increased torus size due to null points and divertor chamber	lesser increase due to need for pumping chamber
10. Poloidal coil power requirements	increased compared to limiter	
11. Relative cost	~10% more expensive than limiter	

ACKNOWLEDGMENTS

The authors gratefully acknowledge the contributions of the many scientists from the EC, Japan, the US and the USSR who contributed the INTOR critical issue study on impurity control. The authors are particularly grateful to Dr. D. Heifetz, M. Petravic, J. Brooks, M. Abdou, J. Schmidt, and M. Shimada for discussions and contributions. This work was supported by the United States Department of Energy.

REFERENCES

Abdou, M., et al., "Impurity Control and First Wall Engineering," FED-INTOR/ICFW/82-17, U.S. FED-INTOR Activity and U.S. Contribution to the International Tokamak Reactor Phase-2A Workshop (1982).

Adamson, R., Bells, W., and Kelly, P., J. Nucl. Mater. 92, (1980), 149.

Adda, Y., Radiation Induced Voids in Metals, J. W. Corbett and L. C. Ianniello, Eds., CONF-710601 (1972) 31.

Anderson, H., and Bay, H., in Sputtering by Particle Bombardment

<u>I</u>, edited by R. Behrisch, Springer-Verlag (1981); J. Nucl. Mater. <u>93-94</u> (1980) 625.

Bartlett, A., Evans, J., Eyre, B., Terry, E., and Williams, T., in <u>Proceedings of International Conference on Radiation Effects and Tritium Technology for Fusion Reactors 1976</u>, Vol. I, p. 122.

Bentley, J., and Wiffen, F., Nucl. Tech. <u>30</u>, (1976) 376.

Bieger, W., Dippel, K. H., Fuchs, G., and Wolf, G., Sumposium on Plasma-Wall Interactions, Julich (1976).

Blewitt, T., Oak Ridge National Laboratory Report, ORNL-2614 (1958) 64.

Bol., K., Arunaslam, V., Bitter, M., Boyd, D., Brau, K., et al., <u>Plasma Physics and Controlled Nuclear Fusion Research 1978</u>, Vol. 1, 11, IAEA, Vienna, 1979.

Braams, B., Harbour, P., Harrison, M., Hotston, E., and Morgan, J., J. Nucl. Mater. <u>121</u> (1983) 75.

Brimhall, J., and Kissinger, H., Radiation Effects <u>15</u> (1972) 259.

Brooks, J., Mattas, R., Hassanein, A., and Baskes, M., J. Nucl. Mater. <u>128&129</u> (1984) 400.

Brooks, J. N., Baker, C. C., Stevens, H. C., Trachsel, C. A., in Ref. 5

Budny, R., et al., J. Nucl. Mater. <u>121</u> (1984) 294.

Burrell, K., et al., <u>Controlled Fusion and Plasma Physics</u>, Proc. 11th European Conference., Aachen, 1983, Vol. 1, European Phys. Soc. (1983) 11.

Bush, S., <u>Irradiation Effects in Cladding and Structural Materials</u>, American Society for Metals (1965).

Carlander, R., Harkness, S., and Santhanam, A., "Effects of Radiation on Substructure and Mechanical Properties of Metals and Alloys," Americal Society for Testing and Matls. (1973) 399, ASTM-STP 529.

Carpenter, G., and Fleck, R., "Electron Irradiation Damage in Beryllium in a High Voltage Electron Microscope," Paper 26 in Beryllium 1977.

Cohen, S., et al., J. Nucl. Mater. <u>128&129</u> (1984) 430.

Conn, R., J. Nucl. Mater. <u>128 & 129</u> (1984) 407.

Conn, R. W., Sviatoslavsky, I. N., Sze, D. K., Proc. 8th Symposium Engineering Problems in Fusion Research, San Francisco (1979).

Engelhardt, W., Becker, G., Behringer, K., Campbell, D., Eberhagen, A., et al., J. Nucl. Mater. <u>111&112</u> (1982) 337.

Evans, K., et a., J. Nucl. Mater. <u>128&129</u> (1984) 452.

Fonck, R., Bell, M., Bol, K., Brau, K., Budny, R., et al., J. Nucl. Mater. <u>111&112</u> (1982) 343.

Fujisawa, N., et al., J. Nucl. Mater. <u>128&129</u> (1824) 61.

Gold, R., and Ammon, R., ADIP Quarterly Progress Report, DOE/ER-0045/6 (1981) p. 96.

Gold, R., ADIP Semianual Progress Report,DOE/ER/0045/7 (1981) 122.

Heifetz, D., et al., J. Comput. Phys. 42 (1982) 309.

Heifetz, D., et al., J. Nucl. Mater. 111&112 (1982) 2981.

Heifetz, D., Post, D., Ulrickson, M., and Schmidt, J., J. Nucl. Mater. 111&112 (1982b) 298.

Igitkhanov, Yu. L., et al., Plasma Physics and Controlled Fusion Research 1984, IAEA, Vienna.

INTOR, Phase II-A, Part I, IAEA, Vienna, 1983.

Jensen, R., Post, D., and Jassby, D., Nucl. Sci./Eng. 65 (1978) 282.

Kangilaski, Radiation Effects Design Handbook, Structural Alloys, Section 7, NASA CR-1873 (1971)

Kaye, S., Bell, M., Bol, K., Boyd, D., Brau, K., et al., J. Nucl. Mater. 121 (1984) 115.

Kelly, G. G., ORMAK II/TTF Design Memo No. 17 (1974).

Labbe, M., and Poirier, J., J. Nucl. Mater. 46 (1973) 86.

Lackner, K., and Keilhacker, M., J. Nucl. Mater. 128>129 (1984) 368.

Mahdavi, M., Armentrout, C., Blau, F., Bramson, G., Brooks, N., et al., J. Nucl. Mater. 111&112 (1982) 355.

Mahdavi, M., Eames, D., Brown, B., Davis, L., Helton, F., et. al., J. Nucl. Mater. 128&129 (1984) 466.

Mattas, R., Wiedersich, H., Atteridge, D., Johnson, A., and Remark, J., in Proceedings of the 2nd Topical Meeting on th Technology of Controlled Nuclear Fusion 1977, CONF-760935-P1, Vol. 1, p. 199.

Mattas, R. F., Smith, D., and Adbou, M., J. Nucl. Mater. 122&123 (1984) 66.

Mattas, R., et al., U.S. Contribution to INTOR Phase IIA, Part 2, Impurity Control and First Wall Engineering, 1985.

Mills, R. G., et al., "A Fusion Power Plant," MATT-1050, Princeton Plasma Physics Laboratory, Princeton, New Jersey, 1974.

Mioduszewski, P., this volume.

Mioduszewski, P., J. Nucl. Mater. ?? (1982) 253.

Mioduszewski, P., Emerson, L., Simpkins, J., Wootton, A., Bush, C., et al., J. Nucl. Mater. 121 (1984) 285.

Mioduszewski, P., J. Nucl. Mater. 111&112 (1982) 253.

Murphy, M., Fatigue of Engineering Materials and Structures 4 (1981) 199.

Nagami, M., Fujisawa, N., Ioki, K., Kitsunezaki, A., Konoshima, S., et al., Plasma Physics and Controlled Nuclear Fusion Research 1980, Vol. II, IAEA, Vienna (1981) 367.

Nagami, M., et al., Nucl. Fusion 24 (1984) 183.

Ohyabu, N., Nucl. Fusion 21 (1981) 5.

Petravic, M., Post, D., Heifetz, D., and Schmidt, J., Phys. Rev. Lett. 48 (1982) 326.

Petravic, M., Heifetz, D., Heifetz, S., and Post, D., J. Nucl. Mater. 128&129 (1984a) 91.

Petravic, M., Heifetz, D., Kuo-Petravic, G., and Post, D., J. Nucl. Mater. 128&129 (1984b) 111.

Petravic, M., Heifetz, D., and Post, D., Plasma Physics and
    Controlled Nuclear Fusion Research 1985, IAEA, Vienna, Vol. 2.
Pontau, A., et al., J. Nucl. Mater. 121 (1984) 304.
Post, D., for the INTOR Group, "Impurity and Particle Control
    for INTOR," Plasma Physics and Controlled Nuclear Fusion
    Research 1984, IAEA, Vienna, 1985.
Post, D., and Lackner, K., this volume.
Roth, J., Bohdansky, J., and Ottenberger, W., Max Planck
    Institut fur Plasmaphysik, IPP (1979), Report No. IPP-9/26.
Santhanam, A., Taylor, A., and Harkness, S., Nucl. Mater. 18
    (1973) 302.
Schneider, W., Heifetz, D., Lackner, K., Neuhauser, J., Post,
    D., and Rauh, K., J. Nucl. Mater. 121 (1984) 178.
Sengoku, S., et al., Nucl. Fusion 24 (1984) 415.
Shimada, M., and the JAERI Team, J. Nucl. Mater. 121 (1984) 184.
Singer, C., and Braams, B. (private communication, 1985).
Stangeby, P., this volume.
Sugihara, M., Saito, S., Hitoki, S., and Fukisawa, N., J. Nucl.
    Mater. 128&129 (1984) 114.
Tanaka, M., Bloom, E., and Horak, J., J. Nucl. Mater. 103&104
    (1981) 895.
Van Witzenburg, W., Mastenbroek, A., and Elen, J., J. Nucl.
    Mater. 104 (1981) 1187.
Vernicke, H., Blaumoser, M., Ennen, K., Gruber, J., Gruber, O.,
    et al., J. Nucl. Mater. 128&129 (1984) 71.
Vershkov, V. A., and Mirnov, S. V., Princeton Plasma Physics
    Laboratory Report MATT-Trans. 113 (1973); Nucl. Fusion 14
    (1974) 383.
Wagner, F., et al., Phys. Rev. Lett. 49 (1982) 1408.
Wagner, F., and Lackner, K., in Physics of Plasma Wall
    Interactions in Controlled Fusion, eds. D. Psot and R.
    Behrisch, Plenum Press, 1985, NATO ASI Series.
Wiffen, F., ADip Semiannual Progress Report, DOE/ER-0045/7
    (1981) 145.
Wiffen, F., ADIP Quarterly Progress Report, DOE/ET-0058/1 (1978)
    142.
Wiffen, F., in Defects and Defect Clusters in BCC Metals and
    Their Alloys, Nucl. Metallurgy, 18 (1973) 179.

LECTURERS

S.L. Allen   Lawrence Livermore National Laboratory
                           Magnetic Fusion Energy Program
                           P.O. Box 5511 (L-637)
                           Livermore, CA 94550

R. Behrisch   Max-Planck-Institut für Plasmaphysik
                           D-8046 Garching bei München
                           Federal Republic of Germany

R. Chodura   Max-Planck-Institut für Plasmaphysik
                           8046 Garching bei München
                           Federal Republic of Germany

S.A. Cohen   Princeton Plasma Physics Laboratory
                           James Forrestal Campus
                           P.O. Box 451
                           Princeton, NJ 08544

F. Engelmann   FOM-Instituut voor Plasmafysica "Rijnhuizen"
                           Association EURATOM-FOM
                           Nieuwegein
                           The Netherlands

                           and

                           The NET Team
                           Max-Planck-Institut für Plasmaphysik
                           8046 Garching bei München
                           Federal Republic of Germany

M.F.A. Harrison   Culham Laboratory
                           Abingdon, Oxfordshire OX14 3DB
                           United Kingdom

D.B. Heifetz   Princeton Plasma Physics Laboratory
                           James Forrestal Campus
                           P.O. Box 451
                           Princeton, NJ 08544

E. Hintz
Institut für Plasmaphysik
Kernforschungsanlage Julich GmbH
Postfach 1913
D-5170 Julich
Federal Republic of Germany

K. Lackner
Max-Planck-Institut für Plasmaphysik
8046 Garching bei München
Federal Republic of Germany

D.M. Manos
Princeton Plasma Physics Laboratory
James Forrestal Campus
P.O. Box 451
Princeton, NJ 08544

G. McCracken
Culham Laboratory
Abingdon, Oxfordshire OX14 3DB
United Kingdom

P. Mioduszewski
Oak Ridge National Laboratory
P.O. Box X
Oak Ridge, TN 37831

D.E. Post, Jr.
Princeton Plasma Physics Laboratory
James Forrestal Campus
P.O. Box 451
Princeton, NJ 08544

J. Roth
Max-Planck-Institut für Plasmaphysik
8046 Garching bei München
Federal Republic of Germany

C. Singer
Princeton Plasma Physics Laboratory
James Forrestal Campus
P.O. Box 451
Princeton, NJ 08544

M.F. Smith
Sandia National Laboratories
Division 1834
P.O. Box 5800
Albuquerque, NM 87185

P.C. Stangeby
Fusion Plasma-Surface Interaction Group
University of Toronto
4925 Dufferin Street
Downsview, Ontario
Canada M3H 5T6

J. Tachon
Equipe TFR
Association Euratom-CEA sur la Fusion
Centre d'etudes Nucleaires
F92260 Fontenay-aux-Roses
France

M. Ulrickson   Princeton Plasma Physics Laboratory
                        James Forrestal Campus
                        P.O. Box 451
                        Princeton, NJ 08544

F. Wagner   Max-Planck-Institut für Plasmaphysik
                        8046 Garching bei München
                        Federal Republic of Germany

J.B. Whitley   Sandia National Laboratories
                        Division 6248
                        P.O. Box 5800
                        Albuquerque, NM 87185

PARTICIPANTS AND STAFF

M.H. Achard   Department de Recherches
                              sur la Fusion Controlee
                             Centre d'Etudes Nucleaires
                             B.P. No. 6
                             F92260 Fontenay-aux-Roses
                             France

A.B. Antoniazzi   Institute for Aerospace Studies
                             University of Toronto
                             4925 Dufferin Street
                             Downsview, Ontario
                             Canada M3H 5T6

F. Aumayr   Institut für Allgemeine Physik
                             Technische Universität Wien
                             Karlsplatz 13
                             A-1040 Vienna
                             Austria

R. Bastasz   Physical Research Division - 8347
                             Sandia National Laboratories
                             Livermore, CA 94550

W.R. Becraft   Oak Ridge National Laboratory
                             P.O. Box Y
                             Bldg. 9201-2, MS-2
                             Oak Ridge, TN 37831

H. Bergsaker   Forskningsinstitutet for Atomfysik
                             Frescativagen 24
                             S-10405 Stockholm
                             Sweden

S.K. Bilikmen
Physics Department
Middle East Technical University (ODTI)
Ankara, Turkey

C.K. Birdsall
Electrical Engineering & Computer Sciences
Cory Hall
University of California at Berkeley
Berkeley, CA 94720

P. Bogen
Institut für Plasmaphysik
Kernforschungsanlage Jülich GmbH
Postfach 1913
D-5170 Jülich
Federal Republic of Germany

R. Boivin
INRS-Energie
C.P. 1020
Varennes, Quebec
Canada JOL 2PO

J. Borzekowski
Mechanical, Aerospace & Nuclear Engineering
School of Engineering and Applied Science
University of California at Los Angeles
Los Angeles, CA 90024

and

3910 Beethoven St. #8
Mar Vista, CA 90066

C. Boucher
INRS-Energie
1650 Montee Ste-Julie
C.P. 1020
Varennes, Quebec
Canada JOL 2PO

C. Brunet
Industrial Materials Research Institute
75 De Mortagne
Boucheville, Quebec
Canada J4B 6Y4

M. Brunet (Staff)
152 Ste. Anne, Apt. 3
Varennes, Quebec
Canada JOL 2PO

B.L. Cain	2020 N. Mattis Avenue Apt. 207-D Champaign, IL 61821 (University of Illinois)
G. Campbell	Institut für Plasmaphysik Kernforschunganlage Julich GmbH Postfach 1913 D-5170 Julich Federal Republic of Germany (University of California at Los Angeles)
F.F.F. Cap	University of Innsbruck Karl Innerebnerstr 40 A 6020 Innsbruck Austria
E. Carey (Staff)	Princeton Plasma Physics Laboratory P.O. Box 451 Princeton, NJ 08544
M. Carroll	1206 West Green Street, Rm. 144 Urbana, IL 61801 (University of Illinois)
T.E. Cayton	Los Alamos National Laboratory P.O. Box 1663, MS-F642 Los Alamos, NM 87545
C. Copenhaver	Los Alamos National Laboratory P.O. Box 1663 Los Alamos, NM 87545
P. Couture	Institut de Recherche d'Hydro-Quebec (IREQ) 1800 Montee Ste-Julie Varennes, Quebec Canada JOL 2PO
C.D. Croessmann	University of Wisconsin-Madison Engineering Research Building 330 1500 Johnson Drive Madison, WI 53706

E.C. Crume, Jr.

Fusion Energy Division
Oak Ridge National Laboratory
P.O. Box Y
Oak Ridge, TN 37831

J.W. Davis

Institute for Aerospace Studies
University of Toronto
4925 Dufferin Street
Downsview, Ontario
Canada M3H 5T6

K. Dimoff

INRS-Energie
C.P. 1020
Varennes, Quebec
Canada JOL 2PO

U. Ditte

Max-Planck-Institut für Plasmaphysik
D-8046 Garching bei München
Federal Republic of Germany

D. Driemeyer

McDonnell Douglas Astronautics Company
Bldg. 278, Level 1
P.O. Box 516
St. Louis, MO 63166

J.M. Dupouy

NET Team
Max-Planck-Institut für Plasmaphysik
Boltzmanstrasse 2
D-8046 Garching bei Muchen
Federal Republic of Germany

M.J. Embrechts

Rensselaer Polytechnic Institute
Department of Nuclear Engineering
NES Building
Troy, NY 12181

and

20 Parker Avenue
Troy, NY 12180

K. Ertl   Max-Planck-Institut für Plasmaphysik
                        D-8046 Garching bei München
                        Federal Republic of Germany

K.H. Finken   Institut für Plasmaphysik
                        Kernforschungsanlage Julich GmbH
                        Postfach Box 1913
                        D-5170 Julich
                        Federal Republic of Germany

                            and

                        Ludwid Str 24
                        516 Duren 8
                        Federal Republic of Germany

D. Fournier   INRS-Energie
                        C.P. 1020
                        Varennes, Quebec
                        Canada JOL 2PO

J. Galambos   Fusion Engineering Design Center
                        Oak Ridge National Laboratory
                        P.O. Box Y
                        Oak Ridge, TN 37831

E. Gultekin   Cekmece Nuclear Research & Training Center
                        Plasma Physics Laboratory
                        P.K. 1 Havaalani
                        Istanbul, Turkey

J.R. Haines   Fusion Engineering Design Center
                        Oak Ridge National Laboratory
                        P.O. Box Y
                        Oak Ridge, TN 37831

                            and

                        2408 Bishops Bridge Road
                        Knoxville, TN 37922

Y.-K. Ho

Physics Department
Zhengzhou University
Henan Province
People's Republic of China

J.R. Howell

College of Engineering
Department of Mechanical Engineering
Department Office 512
University of Texas at Austin
Austin, TX 78712

C. Kahn

GA Technologies, Inc. TO-415
P.O. Box 85605
San Diego, CA 92139

J.A. Koski

Fusion Technology
Division 6248
Sandia National Laboratories
Albuquerque, NM 87185

B. LaBombard

Plasma Fusion Center
Massachusetts Institute of Technology
167 Albany Street
Cambridge, MA 02139

P.H. LaMarche

Princeton Plasma Physics Laboratory
P.O. Box 451
Princeton, NJ 08544

W.K. Leung

Physics Department
University of Texas at Austin
Austin, TX 78712

B. Lipschultz

Plasma Fusion Center, NW16-284
Massachusetts Institute of Technology
167 Albany Street
Cambridge, MA 02139

G. Maddaluno   Associazione EURATOM - ENEA
Centro Ricerche Energia - Frascati
C.P. 65
00044 Frascati (Rome)
Italy

B.K. Malaviya   Rensselaer Polytechnic Institute (RPI)
JEC 5048
Troy, NY 12180-3590

E.H. Marlinghaus   Ruhr Universitat Bochum
Institut fur Experimentalphysik
Universitatsstr. 150 Postf. 102148
D-4630 Bochum 1
Federal Republic of Germany

G.F. Matthews   Culham Laboratory
Abingdon, Oxfordshire OX14 3DB
United Kingdom

G. Mazzitelli   Associazione Euratom ENEA-CRE
Centro Ricerche Energia - Frascati
C.P. 65
00044 Frascati (Rome)
Italy

B.J. Merrill   Idaho National Engineering Laboratory
EG&G Idaho, Inc.
P.O. Box 1625
Idaho Falls, ID 83415

A.A. Mondelli   Science Applications, Inc.
1710 Goodridge Drive
P.O. Box 1303
McLean, VA 22102

T.J. Morgan   Physics Department
Wesleyan University
Middletown, CT 06457

D. Mueller   Princeton Plasma Physics Laboratory
                            P.O. Box 451
                            Princeton, NJ 08544

R. Neufield   Institut de Recherche d'Hydro-Quebec (IREQ)
                            1800 Montee Ste-Julie
                            Varennes, Quebec
                            Canada JOL 2PO

A.A. Newton   Culham Laboratory
                            Abingdon, Oxfordshire OX14 3DB
                            United Kingdom

E. Oktay   U.S. Department of Energy
                            Office of Fusion Energy
                            ER-55, Germantown
                            Washington, DC 20545

R.R. Parker   Plasma Fusion Center, NW16-288
                            Massachusetts Institute of Technology
                            167 Albany Street
                            Cambridge, MA 02139

V. Philipps   Institut für Nuhlearchemie
                            Kernforschungsanlage Jülich GmbH
                            Postfach 1913
                            517 Jülich
                            Federal Republic of Germany

C.A. Phillips   Princeton Plasma Physics Laboratory
  (Staff)   P.O. Box 451
                            Princeton, NJ 08544

C.S. Pitcher   Institute for Aerospace Studies
                        University of Toronto
                        4925 Dufferin Street
                        Downsview, Ontario
                        Canada M3H 5T6

                                and

                        59 Kingslake Road
                        Willowdale, Ontario
                        Canada M2J 3E4

D. Platts   Los Alamos National Laboratory
                        P.O. Box 1663, MS-K302
                        Los Alamos, NM 87545

A.E. Pontau   Institut für Plasmaphysik
                        Kernforschungsanlage Jülich GmbH
                        Postfach 1913
                        D-5170 Jülich
                        Federal Republic of Germany

                                and

                        Division 8347
                        Sandia National Laboratories
                        Livermore, CA 94550

A.S. Rao   Department of Chemical Engineering
                        Carnegie-Mellon University
                        Schenley Park
                        Pittsburgh, PA 15213

K.-G. Rauh   Max Planck Institut für Plasmaphysik
                        D-8046 Garching bei München
                        Federal Republic of Germany

                                and

                        Lambsheimer Str 43A
                        D-6710 Frankenthal
                        Federal Republic of Germany

D. Reiter   Institut fur Plasmaphysik
                                 Kernforschungsanlage Julich GmbH
                                 Postfach 1913
                                 D-5170 Julich
                                 Federal Republic of Germany

                                 and

                                 Turmstrasse 12
                                 51 Aachen
                                 Federal Repbulic of Germany

G. Ross   INRS-Energie
                                 1650 Montee Ste-Julie
                                 C.P. 1020
                                 Varennes, Quebec
                                 Canada JOL 2PO

D.N. Ruzic   University of Illinois
                                 214 Nuclear Engineering Laboratory
                                 103 South Goodwin Avenue
                                 Urbana, IL   61801

U. Samm   Institut für Plasmaphysik
                                 Kernforschungsanlage Julich GmbH
                                 Postfach 1913
                                 D-517 Julich
                                 Federal Republic of Germany

R. Simonini   JET Joint Undertaking
                                 Abingdon, Oxfordshire OX14 3EA
                                 United Kingdom

B. Spagnolo   Istituto di Fisica
                                 Facolta di Ingegneria
                                 Parco D'Orlean
                                 90128 Palermo
                                 Italy

B.L. Stansfield   INRS-Energie
                                 C.P. 1020
                                 Varennes, Quebec
                                 Canada JOL 2PO

D.D.R. Summers   JET Joint Undertaking
                        Abingdon, Oxfordshire OX14 3EA
                        United Kingdom

A. Tanga   JET Joint European Torus
                        Abingdon, Oxfordshire OX14 3EA
                        United Kingdom

B. Terreault   INRS-Energie
                        C.P. 1020
                        Varennes, Quebec
                        Canada JOL 2PO

G.F. Thomas   Ontario Hydro Research Division
                        800 Kipling Avenue
                        Toronto, Ontario
                        Canada M8Z 5S4

C.C. Tsai   Oak Ridge National Laboratory
                        Building 9201-2, MS-002
                        P.O. Box Y
                        Oak Ridge, TN 37831

G. Van Wassenhove   Ecole Royale Militaire
                        Laboratoire de Physique des Plasmas
                        Av de la Renaissance, 30
                        B1040 Bruxelles
                        Belgium

J.C. Aurele Vitali   INRS-Energie
                        C.P. 1020
                        Varennes, Quebec
                        Canada JOL 2PO

                                and

                        10579 Parthenais
                        Montreal, Quebec
                        Canada H2B-2L9

P. Weber   Los Alamos National Laboratory
                        P.O. Box 1663, MS-F639
                        Los Alamos, NM 87545

S.K. Wong   GA Technologies, Inc.
                             P.O. Box 85608
                             San Diego, CA 92138

                                  and

                             5989 Cozzens Street
                             San Diego, CA 92122

I.S. Youle   Institute for Aerospace Studies
                             University of Toronto
                             4925 Dufferin Street
                             Downsview, Ontario
                             Canada M3H 5T6

R. Zuhr   Solid State Division
                             Oak Ridge National Laboratory
                             Building 3003
                             P.O. Box X
                             Oak Ridge, TN 37831